S0-BLM-214

COMPACT STELLAR X-RAY SOURCES

X-ray astronomy provides the main window onto astrophysical compact objects such as black holes, neutron stars and white dwarfs. In the past ten years new observational opportunities have led to an explosion of knowledge in this field. In sixteen chapters, written by leading experts, this book provides a comprehensive overview of the observations and astrophysics of X-ray emitting stellar-mass compact objects.

Topics discussed in depth include the various phenomena exhibited by compact objects in binary systems such as X-ray bursts, relativistic jets and quasi-periodic oscillations, as well as gamma-ray burst sources, super-soft and ultra-luminous sources, isolated neutron stars, magnetars and the enigmatic fast transients. The populations of X-ray sources in globular clusters and in external galaxies are discussed in detail. This is an invaluable reference for both graduate students and active researchers.

WALTER LEWIN is Professor of Physics at MIT. A native of The Netherlands, Professor Lewin received his Ph.D. in Physics from the University of Delft (1965). In 1966, he went to MIT as a postdoctoral associate in the Department of Physics and was invited to join the faculty as Assistant Professor later that same year. He was promoted to Associate Professor of Physics in 1968 and to full Professor in 1974. Professor Lewin's honors and awards include the NASA Award for Exceptional Scientific Achievement (1978), twice recipient of the Alexander von Humboldt Award (1984 and 1991), a Guggenheim Fellowship (1984), MIT's Science Council Prize for Excellence in Undergraduate Teaching (1984), the W. Buechner Teaching Prize of the MIT Department of Physics (1988) and the Everett Moore Baker Memorial award for excellence in undergraduate teaching (2003). In 1997, he was the recipient of a NASA Group Achievement Award for the Discovery of the Bursting Pulsar. He is a corresponding member of the Royal Netherlands Academy of Arts and Sciences (elected 1993) and Fellow of the American Physical Society.

MICHIEL VAN DER KLIS is Professor of Astronomy at the Astronomical Institute Anton Pannekoek of the University of Amsterdam and winner of the NWO Spinoza Prize (2004) for his pioneering research into X-ray radiation from binary stars. He gained his Ph.D. in 1983 from the University of Amsterdam for his observations of X-ray stars. Following this, he held various positions, including a period at the European Space Research and Technology Centre in Noordwijk. In 1989, he returned to the University of Amsterdam as a senior lecturer and he became a professor there in 1993. In 1987, he received the Bruno Rossi Prize, the highest international award in high-energy astrophysics and in 1990 the Zeldovitch Award for Astrophysics from Space of the International Committee of Space Research (COSPAR). Since 2002 he has been a member of the Royal Academy of Arts and Sciences.

Cambridge Astrophysics Series

Series editors

Andrew King, Douglas Lin, Stephen Maran, Jim Pringle and Martin Ward

Titles available in this series

7. Spectroscopy of Astrophysical Plasmas
   edited by A. Dalgarno and D. Layzer
10. Quasar Astronomy
   by D. W. Weedman
17. Molecular Collisions in the Interstellar Medium
   by D. Flower
18. Plasma Loops in the Solar Corona
   by R. J. Bray, L. E. Cram, C. J. Durrant and R. E. Loughhead
19. Beams and Jets in Astrophysics
   edited by P. A. Hughes
22. Gamma-ray Astronomy 2nd Edition
   by P. V. Ramana Murthy and A. W. Wolfendale
23. The Solar Transition Region
   by J. T. Mariska
24. Solar and Stellar Activity Cycles
   by Peter R. Wilson
25. 3K: The Cosmic Microwave Background Radiation
   by R. B. Partridge
26. X-ray Binaries
   by Walter H. G. Lewin, Jan van Paradijs and Edward P. J. van den Heuvel
27. RR Lyrae Stars
   by Horace A. Smith
28. Cataclysmic Variable Stars
   by Brian Warner
29. The Magellanic Clouds
   by Bengt E. Westerlund
30. Globular Cluster Systems
   by Keith M. Ashman and Stephen E. Zepf
32. Accretion Processes in Star Formation
   by Lee W. Hartmann
33. The Origin and Evolution of Planetary Nebulae
   by Sun Kwok
34. Solar and Stellar Magnetic Activity
   by Carolus J. Schrijver and Cornelis Zwaan
35. The Galaxies of the Local Group
   by Sidney van den Bergh
36. Stellar Rotation
   by Jean-Louis Tassoul
37. Extreme Ultraviolet Astronomy
   by Martin A. Barstow and Jay B. Holberg
38. Pulsar Astronomy 3rd Edition
   by Andrew G. Lyne and F. Graham-Smith
39. Compact Stellar X-ray Sources
   edited by Walter Lewin and Michiel van der Klis

COMPACT STELLAR X-RAY SOURCES

Edited by
Walter Lewin & Michiel van der Klis

CAMBRIDGE UNIVERSITY PRESS
Cambridge, New York, Melbourne, Madrid, Cape Town, Singapore, São Paulo

Cambridge University Press
The Edinburgh Building, Cambridge CB2 2RU, UK
Published in the United States of America by Cambridge University Press, New York

www.cambridge.org
Information on this title: www.cambridge.org/9780521826594

First published 2006

Printed in the United Kingdom at the University Press, Cambridge

A catalog record for this publication is available from the British Library

ISBN-13 978-0-521-82659-4 hardback
ISBN-10 0-521-82659-4 hardback

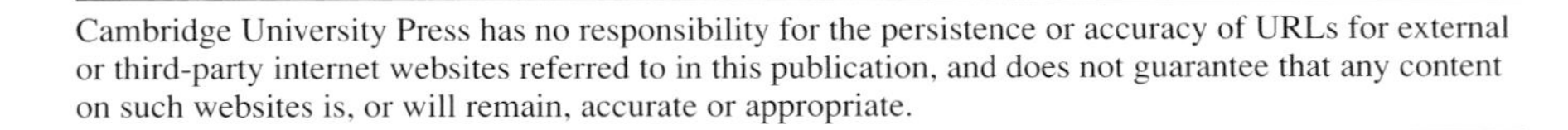

Contents

Contributors

Lars Bildsten
Kavli Institute for Theoretical Physics
University of California, Santa Barbara
Santa Barbara, CA 93106
USA

P. A. Charles
School of Physics & Astronomy
University of Southampton
Highfield
Southampton SO17 1BJ
UK

M. J. Coe
School of Physics & Astronomy
University of Southampton
Highfield
Southampton SO17 1BJ
UK

S. G. Djorgovski
Palomar Observatory
California Institute of Technology
Pasadena, CA
USA

G. Fabbiano
Harvard-Smithsonian Center for Astrophysics
60 Garden St.
Cambridge, MA 02138
USA

Rob Fender
Astronomical Institute Anton Pannekoek
University of Amsterdam
Kruislaan 403
1098 SJ Amsterdam
Netherlands

Alice K. Harding
Sciences and Exploration Directorate
NASA Goddard Space Flight Center
Greenbelt, MD 20771
USA

John Heise
Space Research Organization Netherlands &
Astronomical Institute, Universiteit Utrecht
Sorbonnelaan 2
3584 CA Utrecht
Netherlands

E. P. J. van den Heuvel
Astronomical Institute Anton Pannekoek
University of Amsterdam
Kruislaan 403
1098 SJ Amsterdam
Netherlands

K. Hurley
Space Sciences Laboratory
7 Gauss Way
University of California, Berkeley
Berkeley, CA 94720-7450
USA

Jean in 't Zand
Space Research Organization Netherlands &
Astronomical Institute, Universiteit Utrecht
Sorbonnelaan 2
3584 CA Utrecht
Netherlands

P. Kahabka
Max-Planck-Institut für extraterrestrische Physik
Giessenbachstrasse
85741 Garching
Germany

Victoria M. Kaspi
Physics Department, McGill University
Rutherford Physics Building
3600 University Street
Montreal, QC H3A 2T8
Canada

A. R. King
Theoretical Astrophysics Group
University of Leicester
Leicester LE1 7RH
UK

M. van der Klis
Astronomical Institute Anton Pannekoek
University of Amsterdam
Kruislaan 403
1098 SJ Amsterdam
Netherlands

Erik Kuulkers
ISOC, ESAC/ESA
Apartado 50727
28080 Madrid
Spain

Walter H. G. Lewin
Physics Department
Kavli Institute for Astrophysics and Space Research
Massachusetts Institute of Technology
Cambridge, MA 02139
USA

Jeffrey E. McClintock
Harvard-Smithsonian Center for Astrophysics
60 Garden St.
Cambridge, MA 02138
USA

Andrew Norton
Department of Physics & Astronomy
The Open University
Walton Hall
Milton Keynes MK7 6AA
UK

Dimitrios Psaltis
Department of Physics
University of Arizona
Tucson, AZ 85721
USA

Ronald A. Remillard
Center for Space Research
Massachusetts Institute of Technology
Cambridge, MA 02139
USA

Mallory S. E. Roberts
Physics Department, McGill University
Rutherford Physics Building
3600 University Street
Montreal, QC H3A 2T8
Canada

R. Sari
Department of Theoretical Astrophysics
California Institute of Technology
Pasadena, CA
USA

Axel Schwope
Astrophysikalisches Institut Potsdam
An der Sternwarte 16
11482 Potsdam
Germany

Tod Strohmayer
Laboratory for High Energy Astrophysics
NASA Goddard Space Flight Center
Greenbelt, MD 20771
USA

T. M. Tauris
Astronomical Observatory
Niels Bohr Institute
Copenhagen University
Denmark

C. Thompson
Canadian Institute for Theoretical Astrophysics
60 George St.
Toronto, ON
Canada

Frank Verbunt
Astronomical Institute
Universiteit Utrecht
Postbus 80125
3508 TA Utrecht
Netherlands

Brian Warner
Department of Astronomy
University of Cape Town
Rondebosch 7700
South Africa

N. E. White
NASA Goddard Space Flight Center
Code 660
Greenbelt, MD 20771
USA

P. M. Woods
Universities Space Research Association
National Space Science and Technology Center
Huntsville, AL
USA

Preface

Extra-solar X-ray astronomy began with the historical paper in *Physical Review Letters* by Giacconi, Gursky, Paolini, and Rossi (1962). Now, more than four decades later, X-ray astronomy is central to many aspects of astronomy. In 2002, Riccardo Giacconi was awarded the Nobel Prize in Physics "for pioneering contributions to astrophysics, which have led to the discovery of cosmic X-ray sources". In the decade since the publication of *X-ray Binaries* – the predecessor of the present book – the study of compact stellar X-ray sources has received enormous impetus from observations with the BeppoSAX, Rossi X-ray Timing Explorer (RXTE), Chandra, and XMM-Newton X-ray observatories. In addition, many exciting new results on these X-ray sources have also been produced in the radio, infrared, optical and ultraviolet bands. Highlights include the discovery in low-mass X-ray binaries of millisecond X-ray pulsations, confirming the connection with the millisecond radio pulsars. Millisecond and sub-millisecond quasi-periodic oscillations (QPO) were discovered that are thought to provide a direct view of regions of strong-field gravity near neutron stars and black holes. The discovery of X-ray, optical and radio afterglows of gamma-ray bursts (GRB) firmly established their long-suspected cosmological distances. Super-luminal motion of radio jets was discovered in accreting black-hole binaries. Dozens of ultra-luminous X-ray sources (ULX) have been detected in many galaxies. Their origin is still not clear; some may be accreting intermediate-mass (i.e., of order 10^3 $M_\odot$) black holes (IMBH). Great progress was also made in our understanding of the soft gamma-ray repeaters (SGR) and anomalous X-ray pulsars (AXP). We now know that they are "magnetars", neutron stars with magnetic dipole fields of enormous strength ($10^{14}-10^{15}$ G). This book is a comprehensive and up-to-date survey on compact stellar X-ray sources written by leading experts in the field. It covers in detail the recent developments in X-ray and multi-wavelength observations, and the theory behind them.

1

Accreting neutron stars and black holes: a decade of discoveries

Dimitrios Psaltis
University of Arizona

1.1 Introduction

Since their discovery in 1962 (Giacconi *et al.* 1962), accreting compact objects in the Galaxy have offered unique insights into the astrophysics of the end stages of stellar evolution and the physics of matter at extreme physical conditions. During the first three decades of exploration, new phenomena were discovered and understood, such as the periodic pulsations in the X-ray lightcurve of spinning neutron stars (Giacconi *et al.* 1971) and the thermonuclear flashes on neutron-star surfaces that are detected as powerful X-ray bursts (see, e.g., Grindlay *et al.* 1976; Chapter 3). Moreover, the masses of the compact objects were measured in a number of systems, providing the strongest evidence for the existence of black holes in the Universe (McClintock & Remillard 1986; Chapter 4).

During the past ten years, the launch of X-ray telescopes with unprecedented capabilities, such as RXTE, BeppoSAX, the Chandra X-ray Observatory, and XMM-Newton opened new windows onto the properties of accreting compact objects. Examples include the rapid variability phenomena that occur at the dynamical timescales just outside the neutron-star surfaces and the black-hole horizons (van der Klis *et al.* 1996; Strohmayer *et al.* 1996; Chapters 2 and 4) as well as atomic lines that have been red- and blue-shifted by general relativistic effects in the vicinities of compact objects (Cottam *et al.* 2001; Miller *et al.* 2002b). Accreting neutron stars and black holes have been monitored in broad spectral bands, from the radio to gamma rays, leading to the discovery of highly relativistic jets (Mirabel & Rodriguez 1994; Chapter 9), to the indirect imaging of accretion flows (Horne 1985; Chapter 5), and to the possible identification of neutron stars with masses close to the maximum value allowed by general relativity (Barziv *et al.* 2001). Finally, the theoretical modeling of accretion flows also experienced significant advances, such as the identification of a whole suite of stable solutions for accretion flows beyond the standard model of geometrically thin accretion disks (e.g., Narayan & Yi 1994) and of the most promising avenue towards explaining the very efficient transport of angular momentum in accretion flows (e.g., Balbus & Hawley 1991).

The aim of this chapter is to provide a general overview of these recent advances in the astrophysics of X-ray binaries in our Galaxy. The basic concepts have been reviewed in a number of textbooks (e.g., Shapiro & Teukolsky 1983; Glendenning 2003) and review articles (e.g., White *et al.* 1995) and will only be briefly mentioned here. Several other classes of compact stellar X-ray sources that do not involve accretion onto a neutron star or black hole will also not be discussed in this chapter but are reviewed elsewhere in this volume. These systems include: isolated neutron stars (Chapter 7); cataclysmic variables (CVs; Chapter 10); super-soft sources (SSS; Chapter 11); soft gamma-ray repeaters and

Compact Stellar X-Ray Sources, eds. Walter Lewin and Michiel van der Klis.

anomalous X-ray pulsars (SGRs and AXPs; Chapter 14); and Gamma-ray bursts (GRBs; Chapter 15). Finally, accreting compact objects in other galaxies will be reviewed in Chapter 16.

1.1.1 X-ray binary systems

Whether a compact object in a binary system is accreting mass in a stable long-lived phase or not depends mostly on the mode of mass transfer, the ratio of the mass of the compact object to that of the companion star, and their orbital separation. For example, in the case of a neutron star (with a mass $\sim 1.4-2.0\,M_\odot$), stable mass transfer through the inner Lagrangian point occurs only when the companion fills its Roche lobe and has a mass smaller than that of the neutron star. In such systems, mass is driven by angular momentum losses due to gravitational radiation (for very small masses and orbital separations) and magnetic braking (for orbital periods ≤ 2 day) or by the evolution of the companion star (for orbital periods ≥ 2 day). These sources are significantly brighter in the X-ray than in the optical wavelengths, with the flux at the latter spectral band being mostly due to reprocessing of the X-ray flux from the outer accretion flows. Binary systems with low-mass companions to the neutron stars or black holes are called low-mass X-ray binaries (LMXBs).

A compact object can also accrete matter from a companion star that does not fill its Roche lobe, if the latter star is losing mass in the form of a stellar wind. For this process to result in a compact star that is a bright X-ray source, the companion star has to be massive ($\geq 10\,M_\odot$) in order to drive a strong wind. In this configuration, the optical luminosity of the companion star dominates the total emission from the system and the rate of mass transfer is determined by the strength and speed of the wind and the orbital separation. Such systems are called high-mass X-ray binaries (HMXBs).

The large difference in the companion masses between low- and high-mass X-ray binaries leads to a number of additional differences between these two classes of systems. The lifetimes of HMXBs are determined by the evolution of the high-mass companions and are short ($\sim 10^5-10^7$ yr), whereas the lifetimes of the LMXBs are determined by the mass-transfer process and are longer ($\sim 10^7-10^9$ yr). For this reason, HMXBs are distributed along the galactic plane, as young stellar populations are, whereas LMXBs are found mostly towards the galactic center and in globular clusters (Fig. 1.1). Moreover, because neutron stars in HMXBs accrete for a relatively short period of time, their magnetic fields do not evolve away from their high birth values, and hence these neutron stars appear mostly as accretion-powered pulsars. On the other hand, the prolonged phase of accretion onto neutron stars in LMXBs is believed to be responsible for the suppression of the stellar fields and the absence of periodic pulsations in all but a handful of them.

Finally, in LMXBs, the very small sizes of the companion stars can lead to a number of interesting configurations in systems that are viewed nearly edge on. For example, in the accretion-disk corona (ADC) sources, the X-rays from the central object are scattered towards the observer by electrons in a hot corona that has a size larger than that of the companion, e.g., smoothing out the lightcurves of the X-ray eclipses (White & Holt 1982). On the other hand, in the so-called dippers, the shallow X-ray eclipses may not be caused by the companion star but rather by the stream of mass transfer from the companion star to the accretion disk (see, e.g., White & Swank 1982).

Overall, there are believed to be only a few hundred accreting high-mass and low-mass X-ray binaries in the whole Galaxy. Consequently, these binaries are extremely rare among

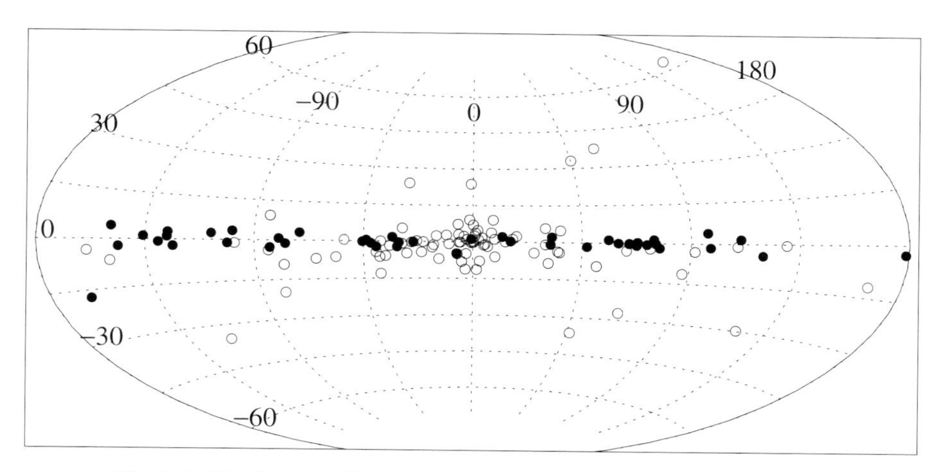

Fig. 1.1. Distribution of low-mass X-ray binaries (open symbols) and high-mass X-ray binaries (filled symbols) in galactic coordinates (Grimm *et al.* 2002).

stellar systems. This is in accord with the large number of improbable evolutionary steps a primordial binary needs to follow in order to become an X-ray source with an accreting compact object. Indeed, the progenitors of the compact objects are believed to be too large to fit in the tight orbits of most X-ray binaries. Moreover, the supernova explosions that precede the formation of the compact objects may disrupt most systems at the phase prior to the formation of the X-ray binary. The resolutions to these and other puzzles on the formation and evolution of X-ray binaries involve exotic and poorly understood binary-evolution processes such as common-envelope evolution of binary stars (Taam & Sandquist 2000), asymmetric supernova explosions that impart recoil velocities to the newborn compact objects, and two- and three-star interactions in the dense stellar fields of globular clusters (see Chapter 8) . The processes that lead to the formation and evolution of X-ray binaries are reviewed in detail in Chapter 16.

1.1.2 Accretion onto compact objects

An X-ray binary is formed when either the companion star transfers matter onto the compact object through the inner Lagrangian point or the compact object captures mass from the wind of the companion star. In both cases, the fate of the transferred mass depends on the amount of angular momentum it possesses, on the physical processes by which it loses angular momentum, and, most importantly, on the radiation processes by which it cools (see Frank *et al.* 2002 for a comprehensive review of this subject).

Beginning in the early 1970s and for the next two decades, most of the modeling effort of accretion flows onto neutron stars and black holes was based on two restrictive assumptions. First, accretion flows were assumed to be losing angular momentum at high rates because of an unspecified process, typically taken to be proportional to the pressure (see, e.g., Shakura & Sunyaev 1973; these solutions are often called α-disks, named after the constant of proportionality). Second, radiation processes were assumed to be very efficient, so that the resulting accretion flows were relatively cool, in the form of geometrically thin accretion disks. The first of these assumptions stemmed from calculations that showed the inefficiency of microscopic viscosity to account for the high inferred rates of mass accretion in the observed sources (see Pringle 1981 for a review). The second assumption, on the other hand, was relaxed in a number of studies (e.g., Shapiro *et al.* 1976) but the resulting solutions were shown to be unstable (e.g., Piran 1978).

During the past decade, theoretical models of accretion flows onto compact objects became increasingly more sophisticated and diverse because of two major developments. First was the identification of a magnetohydrodynamic instability in differentially rotating flows (the magneto-rotational instability, or MRI; Balbus & Hawley 1991, 1998), which allows seed magnetic fields of infinitesimal strength in the flow to get enhanced and tangled. This was shown to lead to a fully developed magnetohydrodynamic turbulence and provide an efficient mechanism of angular momentum transport, as envisioned in the earlier empirical models (Balbus & Papaloizou 1999). Studies of the non-linear development of the instability, its level of saturation (e.g., Sano *et al.* 2004), as well as of its effect on the overall properties of accretion disks (e.g., Armitage *et al.* 2001; Hawley & Krolik 2001; Krolik & Hawley 2002; McKinney & Gammie 2002) require large-scale numerical simulations and are all subjects of intense research efforts.

The second development is related to the discovery of new stable inefficient accretion flows (e.g., Narayan & Yi 1994). In these solutions, the electrons and ions have different and high temperatures, the accretion flows are geometrically thick, and most of their potential energy is not radiated away but is rather advected towards the compact objects; these are the so-called advection-dominated accretion flows (ADAFs). Besides being interesting new theoretical solutions to the hydrodynamics equations, advection-dominated flows provide a framework within which the anomalously low efficiency of accretion onto several black holes in the centers of galaxies (Narayan *et al.* 1995) and in the quiescent states of X-ray transients (Narayan *et al.* 1996) can be understood.

Recently, a number of basic properties of these advection-dominated solutions have been scrutinized. The basic assumption that electrons and ions are coupled inefficiently, mostly due to Coulomb scattering, has been revised, taking into account the effects of magnetic fields (e.g., Quataert & Gruzinov 1999). Advection-dominated flows were shown to be capable of launching strong outflows (advection-dominated inflow/outflow solutions or ADIOS; Blandford & Begelman 1999). Moreover, for a wide area of the parameter space, numerical (Stone *et al.* 1999; Igumenshchev *et al.* 2000) and analytical studies (Narayan *et al.* 2000) showed that the solutions are convectively unstable (convection-dominated accretion flows or CDAFs). Finally, the effects of the magneto-rotational instability on the properties of radiatively inefficient flows have also been investigated recently both in the Newtonian regime (see Fig. 1.2; Stone & Pringle 2001; Igumenshchev *et al.* 2003) and in the general relativistic regime (DeVilliers *et al.* 2003; Gammie *et al.* 2003).

The final ingredient in the models of accretion flows onto compact objects is the interaction of the flows with the objects themselves. This is the region in the accretion flows where most of the high-energy radiation is produced and hence is the one that is probed by observations with X-ray and gamma-ray telescopes. Clearly, the interaction depends on whether the compact object is a black hole or a neutron star, and in the latter case, on whether the neutron star is strongly or weakly magnetic. The main observational manifestation of these differences is the presence or absence of pulsations in the X-ray lightcurves of the systems, which reflects the strength of the magnetic fields of the central objects. In the rest of this chapter the observational properties of the pulsating and non-pulsating X-ray binaries, as well as the current efforts for their theoretical modeling, will be reviewed.

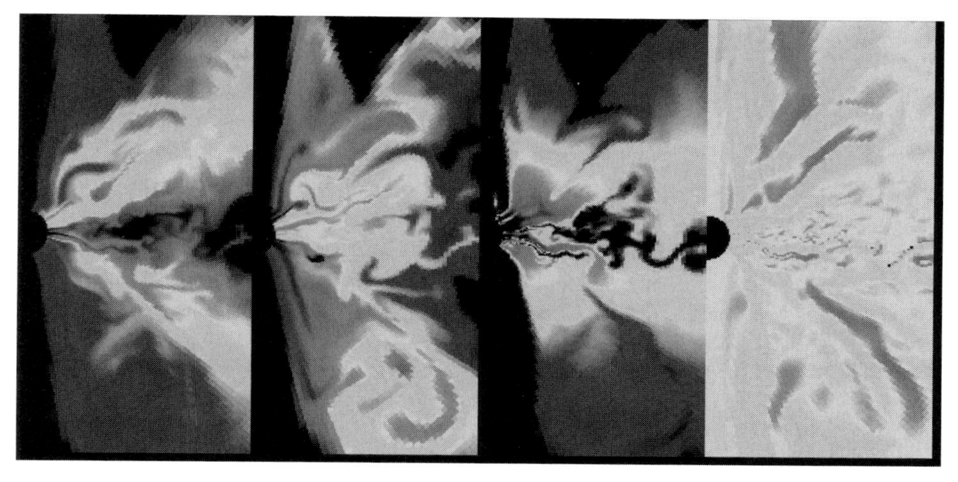

Fig. 1.2. Results of a numerical simulation of a magnetohydrodynamic accretion flow onto a black hole. The panels show the logarithm of the density, specific entropy, square of the toroidal magnetic field, and the $r-\phi$ component of the Maxwell stress tensor (Stone & Pringle 2001).

1.2 Pulsating neutron stars

Neutron stars possess some of the strongest magnetic fields observed in nature. The origin of these magnetic fields is only poorly understood, mostly due to our inability to observe directly the magnetic fields of the cores of pre-supernova stars, which collapse to form the neutron stars, and model their amplification. However, observations of isolated radio pulsars (Chapter 7) and magnetars (Chapter 14) provide strong evidence that neutron-star magnetic fields range between $\sim 10^8$ G and $\sim 10^{15}$ G.

When a strongly magnetic neutron star accretes plasma from a companion star or the interstellar medium, its magnetic field becomes dynamically important close to the stellar surface and determines the properties of the accretion flow. The radius at which the effects of the magnetic field dominate all others is called the Alfvén radius and its precise definition depends on the mode of accretion (i.e., thin-disk vs. quasi-radial), the topology of the magnetic field (i.e., dipolar vs. multipolar), etc. For thin-disk accretion onto a neutron star, the Alfvén radius is defined as the radius at which magnetic stresses efficiently remove the angular momentum of the accreting material (see Ghosh & Lamb 1991 and references therein). For a surface magnetic field strength of 10^{12} G and a mass-accretion rate comparable to the Eddington critical rate, the Alfvén radius is of order 100 neutron-star radii.

The fate of the accreting material after it interacts with the stellar magnetic field near the Alfvén radius depends on the spin frequency of the neutron star. If the stellar spin frequency is smaller than the orbital frequency of matter at the interaction radius, then the accreting material is forced into corotation with the star and is channeled along field lines onto the magnetic poles. As the neutron star spins and the observer sees a different aspect of the hotter magnetic poles, the X-ray flux received is modulated at the stellar spin frequency and an accretion-powered pulsar is produced. On the other hand, if the stellar spin frequency is larger than the orbital frequency of matter at the interaction radius, then the material cannot overcome the centrifugal barrier in order to accrete onto the star. The fate of matter in this case is presently unknown, but it is often assumed that matter eventually escapes the neutron star in the form of a wind. Magnetic neutron stars in this configuration are often said to be in the propeller regime (after Illarionov & Sunyaev 1975).

The neutron star itself also reacts differently to the accretion of matter depending on its magnetic field strength, its spin frequency, and the mass accretion rate. Magnetic field lines rotate at the spin frequency of the star and couple the stellar surface to the accreting material. As a result, they transfer angular momentum from the accreting material to the neutron star, if the former is spinning faster than the latter, or from the neutron star to the accreting material, in the opposite situation. Both situations occur simultaneously in an accreting system, since the orbital frequency of matter decreases with increasing radius. The overall effect is a net torque on the neutron star, which can be either positive (spin-up) or negative (spin-down). The magnitude of the torque on the star is expected to increase with increasing mass accretion rate and with increasing magnetic field strength (see Ghosh & Lamb 1979, 1991). Clearly, for every magnetic field strength and mass accretion rate, there is a critical spin frequency at which the net torque on the star is zero. This frequency corresponds to an equilibrium, towards which the neutron star evolves in its lifetime. For a surface dipolar magnetic field with a strength of 10^{12} G and a mass accretion rate comparable to the Eddington critical rate, the equilibrium spin frequency is of the order of a few tenths of a hertz (Ghosh & Lamb 1992).

Accretion-powered pulsars currently provide the best systems in which the spin frequencies and magnetic field strengths of accreting neutron stars can be studied. Two distinct classes of such pulsars are known: pulsars with periods of order a second, which are found mostly in high-mass X-ray binaries, and pulsars with millisecond spin periods, which are found in binary systems with very short orbital periods (see Section 1.2.2).

1.2.1 Classical (slow) accretion-powered pulsars

The detection of coherent pulsations from an accreting X-ray source, in 1971 (Giacconi *et al.* 1971), provided the strongest evidence, at the time, that the compact objects in many of these sources were neutron stars. Since then, accretion-powered pulsars with periods of the order of one second or more have been studied extensively with every X-ray satellite. In recent years, the long-term monitoring of such pulsars with BATSE as well as the detailed spectral studies with RXTE and BeppoSAX provided a remarkable look into the properties of these systems, resolving some long-standing questions and posing a number of new ones (see Bildsten *et al.* 1997 and Heindl *et al.* 2004 for comprehensive reviews of the accretion-powered pulsars discussed in this section).

The vast majority of slow accretion-powered pulsars are found in high-mass X-ray binaries; only five of them (Her X-1, 4U 1626−67, GX 1+4, GRO J1744−28, and 2A 1822−371) have low-mass companions. Indeed, most low-mass X-ray binaries are old systems and their prolonged phase of accretion is thought to have suppressed the magnetic fields of the neutron stars and to have spun them up to millisecond periods (see Section 1.2.2). On the other hand, high-mass X-ray binaries are younger systems and the neutron stars in them are expected to have magnetic fields that are dynamically important.

The properties of the high-mass binary systems in which slow pulsars reside can be described more easily on the diagram that correlates their spin to their orbital periods (the Corbet diagram; Fig. 1.3). About half of the slow pulsars are orbiting main-sequence Be stars, whereas the remaining pulsars are orbiting evolved OB supergiants. The systems with Be companions are generally eccentric transient systems, in which the companion stars are not filling their Roche lobes and the pulsars become detectable during periastron passages

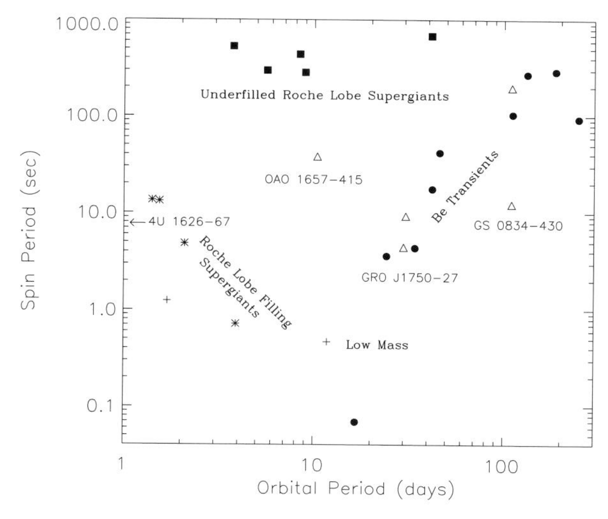

Fig. 1.3. The spin and orbital periods of classical accretion-powered pulsars (the Corbet diagram; after Bildsten *et al.* 1997).

(Chapter 5). The properties of the systems with supergiant companions depend on whether these stars fill their Roche lobes or not. If they do, matter is transfered onto the neutron stars via the inner Lagrangian point of their binary potential possessing significant angular momentum and forming a geometrically thin accretion disk. On the other hand, for companion stars that do not fill their Roche lobes, mass lost via a radiation-driven wind is captured by the neutron star at rates that are typically lower than the disk-fed systems.

1.2.1.1 Spin-period evolution

The mode of mass transfer onto the neutron stars, which depends on the properties of the binary systems, also determines the spin evolution of the neutron stars. The long-term aspect of this dependence is clearly visible in Fig. 1.3. The systems with Roche-lobe filling supergiants have short spin periods that are anticorrelated with the orbital periods; the systems with underfilling supergiants have long spin periods that do not show any correlation with orbital periods; and the Be transient systems have long orbital periods that are positively correlated to the orbital periods. These correlations are believed to depend strongly on the mode and efficiency of mass transfer from the companion stars to the neutron stars but are only poorly understood (see, e.g., Waters & van Kerkwijk 1989).

A clear look into the short-term dependence of the spin periods of neutron stars on the properties of the accretion flows was made possible because of the intense monitoring of several accretion-powered pulsars with the BATSE experiment onboard CGRO. Contrary to earlier results, the measurements with BATSE revealed that transient and persistent sources show two different types of spin-period evolution (Bildsten *et al.* 1997).

Transient accretion-powered pulsars in outburst show a positive dependence of the accretion torque (as measured by the spin-up rate) on the inferred accretion luminosity (Fig. 1.4; left panel). This is consistent with the simple model of disk–magnetosphere interaction (e.g., Ghosh & Lamb 1992), in which, as the accretion rate increases, the rate of angular momentum transfer from the accretion flow to the neutron star increases. At the limit of very low mass accretion rate, the neutron stars are expected to spin down, because the magnetic

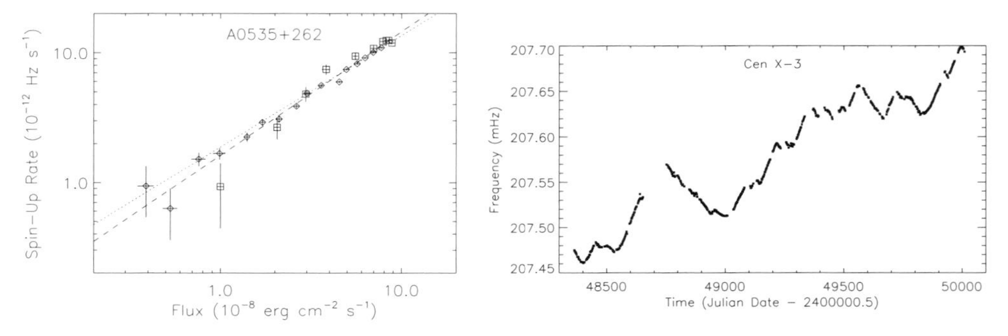

Fig. 1.4. (*Left*) The dependence of the spin-up rate on the pulsed flux for the source A 0535+262, as observed by BATSE; the dashed and dotted curves show the best-fit line and the theoretical prediction, respectively (see text). (*Right*) The evolution of the spin frequency of Cen X-3 as observed by BATSE (Bildsten *et al.* 1997).

field lines that couple to the outer, slower accretion flow remove spin angular momentum from the neutron star. Such spin-down episodes have not been detected by BATSE, although evidence for spin-down in the pulsar EXO 2030+375 has been previously reported based on EXOSAT data (Parmar *et al.* 1989). The very low fluxes down to which the transient sources continue to spin up place strong constraints on the relative importance of angular momentum transfer between the accretion disk and the neutron star via anchored magnetic field lines.

In sharp contrast to the transient sources, persistent disk-fed pulsars show a bimodal behavior in their accretion torques (Bildsten *et al.* 1997; see also Fig. 1.4; right panel). Episodes of spin-up and spin-down of approximately equal accretion torques alternate at timescales that vary from $\sim$10 days (e.g., in Cen X-3) to $\geq$10 yr (e.g., in GX 1+4). The transition between spin-up and spin-down is rapid ($\leq$ a few days) and cannot be resolved with BATSE measurements. Current models of the disk–magnetosphere interaction in accretion-powered pulsars can account for the observed bimodal torques only if one of the physical properties of the accretion flow is also assumed to show a bimodal behavior. Such assumptions include a bimodal distribution of the mass transfer rate onto the pulsar, or a bimodal dependence on accretion rate of the orientation of the disk (Nelson *et al.* 1998; van Kerkwijk *et al.* 1998), of the orbital angular velocity of the accreting gas (Yi & Wheeler 1998), or of the strength and orientation of any magnetic field produced in the disk (Torkelsson 1998). Alternatively, for any given mass accretion rate onto the neutron star, two equilibrium solutions may be possible, one in which the star is spinning up and one in which it is spinning down (Lovelace *et al.* 1999). It is not clear at this point which, if any, of these alternatives is responsible for the observed torque reversals in accreting neutron stars and this remains one of the puzzles of the BATSE monitoring of slow accretion-powered pulsars.

1.2.1.2 Quasi-periodic oscillations

In several accretion-powered pulsars, the power-density spectra in X-rays or in longer wavelengths show a number of quasi-periodic oscillations, in addition to the period of the pulsars (Table 1.1; Fig. 1.5). The frequencies of these oscillations range from $\sim$1 mHz

Table 1.1. *Quasi-periodic oscillations in accretion-powered pulsars*[a]

Source	Spin frequency (mHz)	QPO frequency (mHz)
4U 1907+09	2.27	55
XTE J1858+034	4.5	111
A 0535+26	9.71	27–72
EXO 2030+375	24	187–213
LMC X-4	74	0.65–1.35, 2–20
4U 1626−67	130	1, 48
Cen X-3	207	35
V 0332+53	229	51
4U 0115+63	277	2, 62
Her X-1	807.9	8, 12, 43
SMC X-1	1410	60?
GRO 1744−28[b]	2140	40000

[a] Compilation after Shirakawa & Lai 2002; [b] Zhang *et al.* 1996.

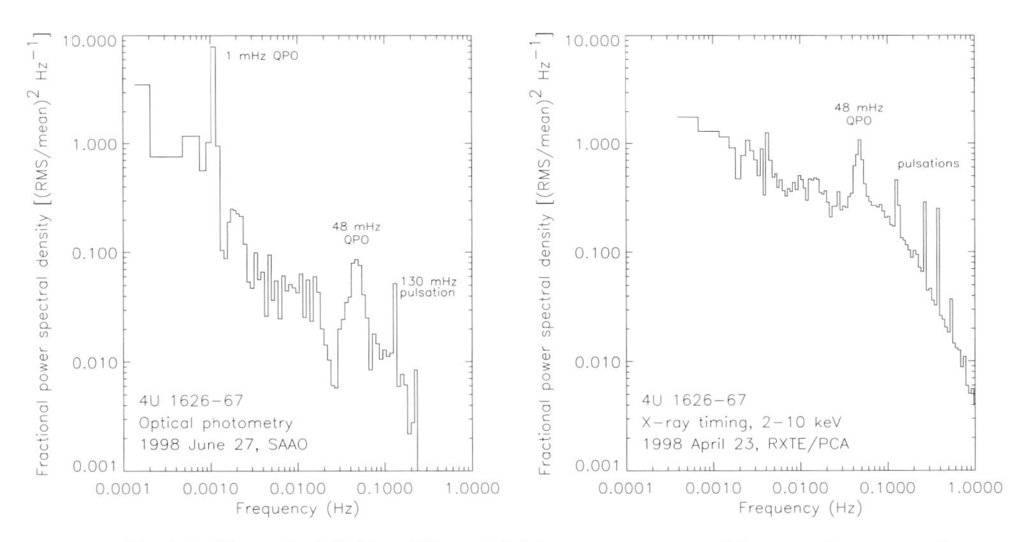

Fig. 1.5. The optical (*left*) and X-ray (*right*) power spectrum of the accretion-powered pulsar 4U 1626−67 showing mHz quasi-periodic oscillations (Chakrabarty *et al.* 2001).

to ∼40 Hz and they can be from ∼100 times smaller to ∼100 larger than the pulsar spin frequencies.

The frequency of the fast oscillations in the transient pulsar EXO 2030+375 was found to be in good agreement with beat-frequency models (Finger *et al.* 1996), in which oscillations occur at the beat frequency between the orbital frequency of matter at the Alfvén radius and the stellar spin frequency (Alpar & Shaham 1985). On the other hand, the low-frequency oscillations observed in 4U 1626−67 appear also as asymmetric sidebands to the pulse period (Kommers *et al.* 1998) and are probably related to a low-frequency modulation of

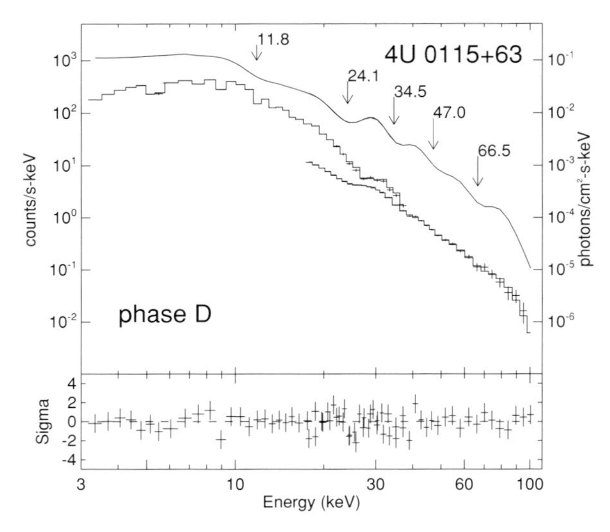

Fig. 1.6. The observed (histogram) and model spectrum (solid line) of the accretion-powered pulsar 4U 0115+63 showing evidence for cyclotron lines with as many as four overtones (Heindl *et al.* 1999).

the accretion flow, possibly due to the presence of a precessing disk warp (Shirakawa & Lai 2002). It is unclear at this point whether all of these quasi-periodic oscillations are related to the same phenomenon or not and what is their physical origin.

1.2.1.3 Cyclotron lines

The X-ray spectra of accretion-powered pulsars are typically described in terms of relatively flat power laws with exponential cutoffs at energies $\geq$ 10 keV. These continuum spectra are believed to be the result of upscattering of soft photons by the hot electrons in the accretion columns above the magnetic polar caps (Meszaros 1992). For neutron-star magnetic field strengths of $\sim 10^{12}$ G, the cyclotron energy on the stellar surface is $\sim$11.6 keV and the continuum spectra are expected to show evidence for harmonically related "cyclotron resonance scattering features" (or simply cyclotron lines) in the X-rays. Observation of such features was anticipated from the early days of X-ray astronomy and expected to lead to direct measurements of the magnetic field strengths of accreting neutron stars (e.g., Trumper *et al.* 1978).

The broadband spectral capabilities of RXTE and BeppoSAX made possible the unequivocal detection of harmonically related features in four accretion-powered pulsars, as well as the detection of single features in ten more sources (Fig. 1.6 and Table 1.2). The widths of the resonance features appear to be correlated to the energies of the continuum cutoffs and to be proportional to their central energies and to the inferred scattering depths (Coburn *et al.* 2002). The central energies of these features provided direct measurements of the surface magnetic fields of accreting pulsars, which can be used in constraining the models of disk–magnetosphere interaction. Future observations of the pulse-phase dependence of the scattering features is also expected to provide more detailed constraints on the geometries of the accretion columns in slow accretion-powered pulsars.

Table 1.2. *Energies of cyclotron lines in accretion-powered pulsars*

Source	Spin period (s)	Energy (keV)	Harmonics
4U 0115+63	3.61	12	Yes
4U 1907+09	438	18	Yes
4U 1538−52	530	20	Yes
Vela X-1	283	25	Yes
V 0332+53	4.37	27	
Cep X-4	66.2	28	
Cen X-3	4.82	28.5	
4U 0352+309	835	29	
XTE J1946+274	15.8	36	
MX 0656−072	160.7	36	
4U 1626−67	7.66	37	
GX 301−2	681	37	
Her X-1	1.24	41	
A 0535+26	105	50 or 110	

After Heindl *et al.* 2004

1.2.2 Millisecond accretion-powered pulsars

The presence of accretion-powered millisecond pulsars in low-mass X-ray binaries had been predicted 17 years before such sources were eventually discovered. Millisecond rotation-powered pulsars were thought to acquire their low ($\sim 10^8$ G) magnetic fields and fast spin frequencies while accreting mass at high rates in low-mass X-ray binaries (Alpar *et al.* 1982; Radhakrishnan & Shrinivasan 1982). These millisecond radio pulsars were most often found in binaries with evolved, low-mass white dwarf companions (Bhattacharya & van den Heuvel 1991), which were thought to be the descendents of LMXBs, and their birthrates were similar (albeit with systematic differences) to those of LMXBs (see, e.g., Kulkarni & Narayan 1988; Lorimer 1995 and references therein). Moreover, circumstantial evidence based on the bursting behavior (Lewin *et al.* 1996), rapid variability (Alpar & Shaham 1985; Ghosh & Lamb 1991), and X-ray spectra (Psaltis & Lamb 1998) of the compact objects in bright LMXBs strongly supported their identification with weakly magnetic, rapidly spinning neutron stars. However, despite intense searches (e.g., Vaughan *et al.* 1994), periodic pulsations could not be detected from any LMXB, making this the holy grail of X-ray binary astrophysics in the pre- RXTE era.

The discovery, with RXTE, of highly coherent pulsations in the X-ray fluxes of LMXBs during thermonuclear X-ray bursts (Strohmayer *et al.* 1996; see also Chapter 3 in this volume) provided the then strongest evidence for the presence of neutron stars with millisecond spin periods in LMXBs. However, the first bona fide millisecond accretion-powered pulsar was discovered only in 1998, in a transient ultracompact binary (see Fig. 1.7). Since then, four additional millisecond pulsars have been discovered in very similar transient binaries (see Table 1.3).

The spin periods and inferred magnetic fields of these five pulsars are indeed consistent with the prediction that such systems are the progenitors of the rotation-powered millisecond pulsars observed in radio wavelengths (see, e.g., Fig. 1.8; Psaltis & Chakrabarty

Table 1.3. *Observed properties of millisecond accreting pulsars*[a]

Source	f_s (Hz)	P_{orb} (m)	a (lt-ms)	$f(M_\odot)$
SAX J1808.4−3658[b]	401.0	120.9	62.809	3.78×10^{-5}
XTE J0929−3314[c]	185.1	43.6	6.290	2.7×10^{-7}
XTE J1751−305[d]	435.3	42.4	10.1134	1.278×10^{-6}
XTE J1807−294[e]	190.6	40.1	4.80	1.54×10^{-7}
XTE J1814−338[f]	314.3	256.5	390.3	2.016×10^{-3}

[a] Spin frequency, orbital period, projected semi-major axis, and mass function. [b] Wijnands & van der Klis 1998 and Chakrabarty & Morgan 1998; [c] Galloway *et al.* 2002; [d] Markwardt *et al.* 2002; [e] Markwardt *et al.* 2003; Markwardt, priv. comm.; [f] Markwardt & Swank 2003; Markwardt, priv. comm.

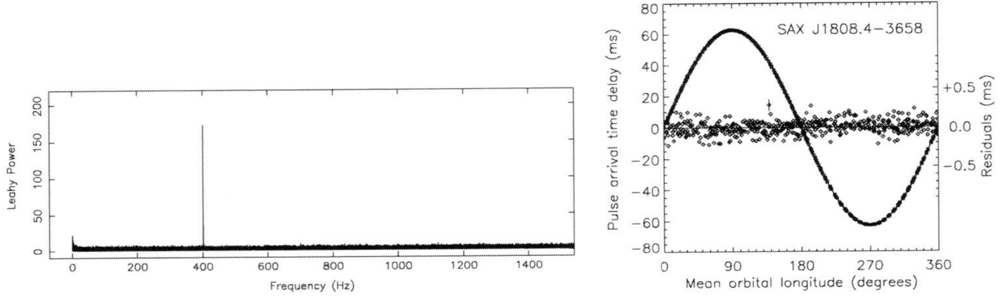

Fig. 1.7. The power-density spectrum of the X-ray flux from SAX J1808.4−3658 showing clearly the ~401 Hz pulsation (*left* Wijnands & van der Klis 1998) and the pulse-time residuals caused by its 2 hr orbit (*right* Chakrabarty & Morgan 1998).

1998). However, the binary systems that all five pulsars belong to have a number of unusual characteristics. First, they are all ultracompact as inferred from their small orbital periods ($\lesssim$2 hr) and projected semi-major axes ($\lesssim$62 light-ms). Moreover, the companions to the neutron stars have masses of only a few hundredths of the solar mass and are, most probably, the remnants of white dwarfs that have lost most of their mass. Finally, the inferred long-term averages of the mass accretion rates onto these neutron stars, as well as their maximum luminosities during outbursts are among the lowest in the known LMXB population (see, e.g., Chakrabarty & Morgan 1998).

Although they resolved one of the long-standing puzzles in X-ray binary astrophysics, these five millisecond pulsars have posed two very important questions: First, why do these particular neutron stars in these special binaries appear as X-ray pulsars, whereas the compact objects in the other LMXBs do not; and second, why does the spin-up of these systems stop short of the sub-millisecond periods that most neutron-star models allow (see, e.g., Glendenning 2003 and references therein). No satisfactory answer to the first question has been found to date, whereas three equally exciting answers to the second question appear plausible.

Whether an accreting neutron star appears as an X-ray pulsar depends mostly on three factors: its magnetic field, the mass accretion rate, and the relative orientation of the binary, spin, and magnetic axes to the direction of the observer. For example, unless the binary axis

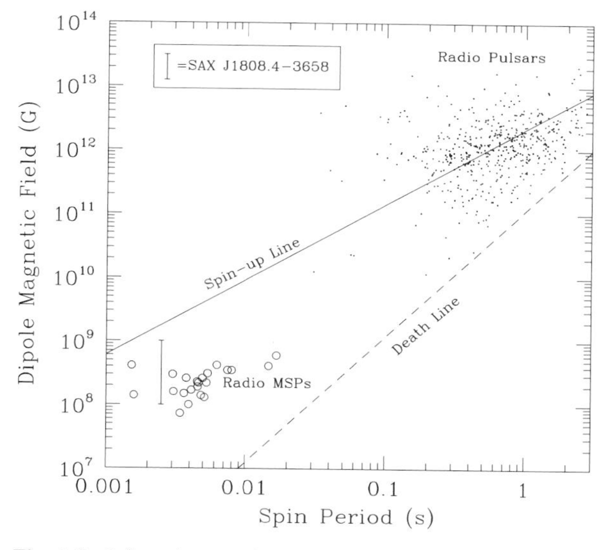

Fig. 1.8. Inferred strengths of the dipole magnetic fields of slow radio pulsars (dots), millisecond radio pulsars (open circles), and of the accretion-powered millisecond pulsar SAX J1808.4−3658 (Psaltis & Chakrabarty 1999).

makes a relatively small angle with the direction of the observer, the accretion flow may block the direct viewing of the polar caps and hence the modulation of the X-ray flux at the stellar spin frequency. Motivated by the very small measured values of the mass function of SAX J1808.4−3658 (see Table 1.3), which suggest a-priori small inclination angles, Psaltis and Chakrabarty (1998) suggested that a favorable viewing geometry is responsible for the fact that this source appears as an X-ray pulsar. However, the subsequent discovery of four additional pulsars in similar binary systems (albeit with equally small mass functions), the discovery of a small modulation of the X-ray flux at the orbital period of SAX J1808.4−3658 (Chakrabarty & Morgan 1998), as well as the optical properties of the reprocessed radiation from the accretion disk (Wang *et al.* 2000) all argue against such a favorable geometry.

Alternatively, if the Alfvén radius is smaller than the size of the neutron star then the stellar magnetic field is nowhere dynamically important and cannot channel the accretion flow preferentially onto the magnetic poles. Cumming *et al.* (2001) argued that in the accreting pulsars the very low rate of mass transfer onto the neutron stars is not sufficient to "bury" their magnetic fields and make them dynamically unimportant, in contrast to the case of the other LMXBs that are accreting at much larger rates on average. This is consistent with the peculiar properties of the binaries in which all five millisecond accreting pulsars reside. It is, however, hard to reconcile with the fact that these five sources, besides being pulsars, have very similar X-ray spectra (Gilfanov *et al.* 1998) and aperiodic variability properties (Wijnands & van der Klis 1998) with many non-pulsing sources.

It appears that, in the five millisecond pulsars, the stellar magnetic fields only introduce a modulation of the X-ray flux at the stellar spin frequency without altering significantly the other timing and spectral properties of the systems, and hence their accretion flows. Consistent with this fact, these five pulsars have spin frequencies in the same range as the ones inferred in most other, non-pulsing LMXBs from the properties of burst oscillations (Chakrabarty *et al.* 2003). Indeed, the spin frequencies of all these sources are limited to be less than ~700 Hz, which is also similar to the spin frequency of the fastest known millisecond, rotation-powered pulsar. The physics that sets this limit is unclear at this point.

As the simplest explanation, general relativity and the equation of state of neutron-star matter may not permit stable neutron stars with spin frequencies faster than ~700 Hz. Albeit reasonable, this limit on the spin frequency would require neutron stars to be about twice as large as predicted by most models and hence, if correct, would point to a more exotic equation of state than any currently discussed (Cook *et al.* 1994). Alternatively, if the neutron stars in LMXBs spin near the magnetic spin equilibrium, there may be a natural upper limit on spin frequency that magnetic accretion can achieve. This would require a lower bound on the stellar magnetic field, since a non-magnetic neutron star will always spin up to the maximum frequency allowed by its equation of state, and hence would provide significant clues for the efficiency of magnetic field decay in accreting neutron stars.

The most exciting alternative, however, is the possibility that neutron stars spinning at rates faster than about 700 Hz rapidly loose their spin angular momenta via emission of gravitational radiation (Bildsten 1998). For this to happen, the distribution of matter inside the neutron star must be non-axisymmetric since the emission of gravitational radiation depends on the time derivative of the mass quadrupole of the star. Temperature anisotropies in the surface layers of accreting neutron stars may result in anisotropies in the crystallization of the material underneath that are sufficient to account for the rapid loss of spin angular momentum (Bildsten 1998). Alternatively, excitation of non-radial modes in the neutron star may provide such a time-dependent mass quadrupole (Andersson *et al.* 1999). If this is the reason why the spin-up of accreting neutron stars stalls at a frequency well below the maximum frequency allowed by their equation of state, then LMXBs may become the first sources detected in the very near future by gravitational wave observatories.

1.3 Non-pulsing neutron stars and black holes

The large majority of accreting compact objects show no evidence for periodic pulsations in their persistent emission. In the case of accreting black holes, this is a direct consequence of the presence of the event horizon, which does not allow for any stable feature to be anchored to the rotation of the compact object. In the case of neutron stars, however, the absence of pulsations requires a rather weak magnetic field ($\lesssim 10^8$ G) so that the accretion flow is not disrupted and channeled onto the magnetic poles.

1.3.1 Transient and persistent sources

Non-pulsing accreting compact objects appear both as persistent and as transient sources. Members of the first class are observed at X-ray fluxes that can be variable by up to several factors of 2 over timescales ranging from milliseconds to months. Their distinguishing characteristic, however, is the fact that they have been at detectable flux levels for most of the history of X-ray astronomy.

Transient sources, on the other hand, are characterized by long periods of inactivity, lasting months to decades, that are interrupted by short outbursts, during which their X-ray brightness increases by several orders of magnitude (Bradt *et al.* 2000). Figure 1.9 shows typical lightcurves of a neutron-star (Aql X-1) and a black-hole transient (GRO J1655−40) as observed by the All Sky Monitor on RXTE (see Chapter 4).

The nature of the compact object in a transient system appears to affect its properties in four ways: the fraction of transients among the black-hole systems is larger than the fraction of transients among neutron-star systems and their outbursts are typically longer and rarer (see,

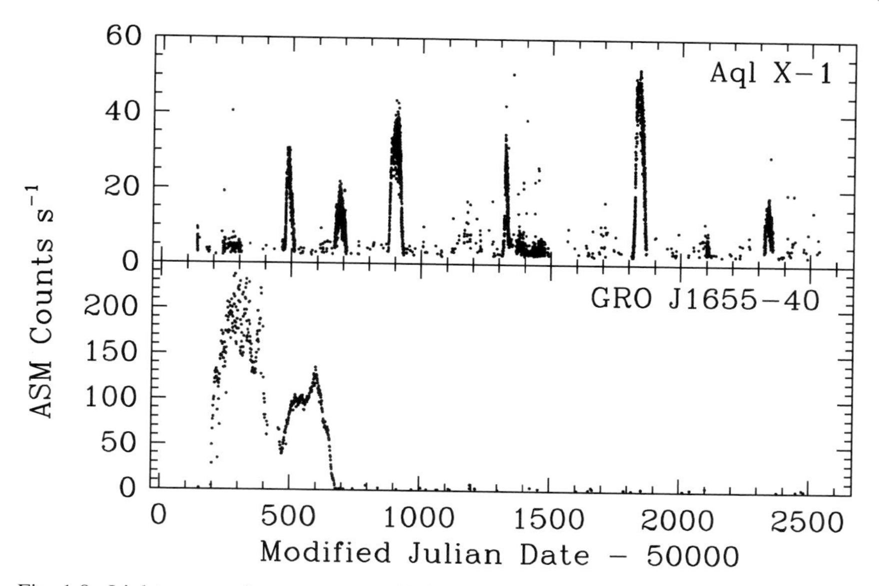

Fig. 1.9. Lightcurves of a neutron-star (Aql X-1) and a black-hole (GRO J1655−40) transient source, as observed by the All Sky Monitor on RXTE.

e.g., Fig. 1.9); moreover, black-hole transients in quiescence are significantly fainter than their neutron-star counterparts. According to our current understanding, the above differences are caused by the different mass ratios of the members of the binary systems between the two populations as well as by the presence of an event horizon in the black-hole systems.

The prevailing model of transient sources is based on the disk instability model of illuminated accretion disks (King *et al.* 1996; van Paradijs 1996): accretion flows that extend to large radii (typically $> 10^9 - 10^{10}$ cm) from the compact object have characteristic temperatures less than $\sim 10^4$ K, at which the anomalous opacity related to the ionization of hydrogen renders them susceptible to a thermal disk instability (see Chapter 13). At the off-cycle of the instability, material piles up at the outer edges of the accretion disk with very little, if any, mass accreted by the central object; this is the quiescent phase of the transient. When the disk becomes unstable, the accretion flow evolves towards the central object at the viscous timescale and the system becomes a bright X-ray source in outburst.

The disk-instability mechanism depends crucially on the temperature of the accretion disk, and hence heating of the disk by illumination can alter the above picture. The details of illumination by the central object or the disk itself are hard to compute, as they depend strongly on how concave the accretion disk is, on the presence of warps that can expose different parts of the disk to radiation, and on the effects of disk winds that can backscatter radiation towards the disk plane (see, e.g., Dubus *et al.* 1999). It is generally expected, though, that neutron stars illuminate their accretion disks more efficiently than black holes, since most of the accretion energy has to be released near their surfaces. The stabilizing effect of illumination can then account for the difference in the outburst properties between neutron-star and black-hole systems.

Studies of the quiescent emission of X-ray transients became possible only after the launch of X-ray telescopes with good resolution and low background, such as ASCA and the Chandra X-ray Observatory (see, e.g., Rutledge *et al.* 2001). It is now well established that the quiescent

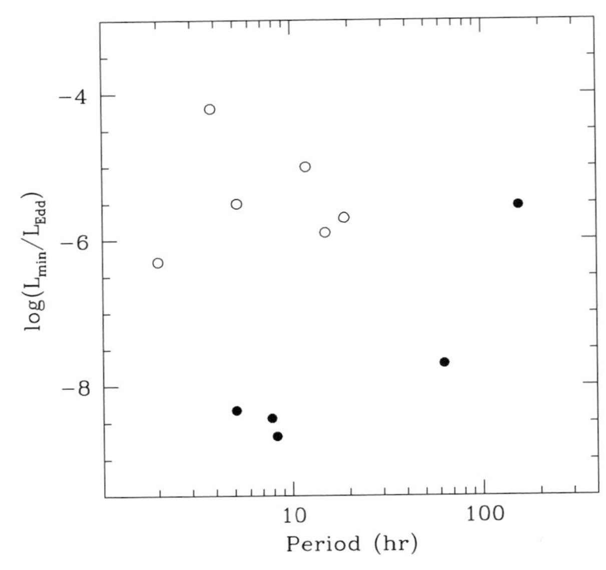

Fig. 1.10. The quiescent X-ray luminosity of neutron-star (open circles) and black-hole transients (filled circles) as a function of their orbital period (after Garcia *et al.* 2001).

emission of black-hole transients is fainter by more than an order of magnitude compared to the quiescent emission of neutron-star sources in similar binary systems (Fig. 1.10). The time variability and non-thermal character of the X-ray brightness in both cases strongly suggests that at least a large fraction of the emission arises from accretion (see, e.g., Narayan *et al.* 2001). However, release of heat buried in the deep layers of a neutron star (Brown *et al.* 1998) and coronal emission from the companion star (Bildsten & Rutledge 2000; Campana & Stella 2000; but see Lasota 2000) are also expected to contribute to the total X-ray brightness. In most cases, the existence of an event horizon in the black-hole sources, which traps a large fraction of the accretion luminosity and does not allow the storage of latent heat, is believed to be responsible for their significantly less luminous quiescent emission (Narayan *et al.* 2001).

The transient nature of these systems had hampered their systematic study until recently. Since the mid-1990s, however, the Wide-Field Cameras on BeppoSAX and the All Sky Monitor on RXTE have revealed and monitored a large number of transient sources. The impact of such monitoring programs has been enormous. The statistics of the transient sources and their recurrence times shed light on the total number of X-ray binaries in the Galaxy, their birthrates, and formation mechanisms. The rise and decay times of their outbursts helped constrain the efficiency of angular momentum transport in accretion disks (e.g., Hameury *et al.* 1998). More significantly, however, optical observations of X-ray transients in quiescence allowed for the measurement of the masses of many compact objects and have thus provided the best evidence for the existence of stellar-mass black holes in the Galaxy (Chapter 4).

1.3.2 Long-wavelength counterparts

X-ray binaries are initially identified from their intense X-ray brightness and hard spectra. However, the properties of the binary systems, such as their orbital periods, the

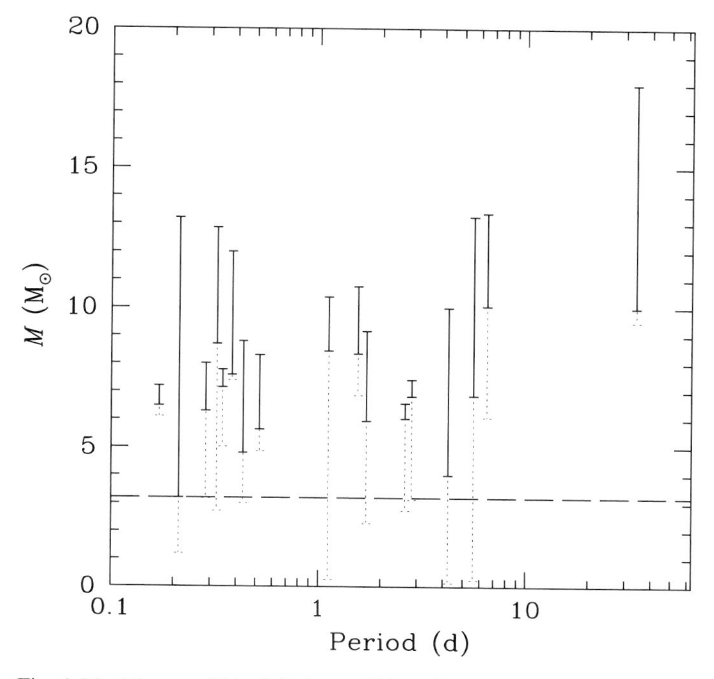

Fig. 1.11. Masses of black-hole candidates in binary systems and their orbital periods; the horizontal dashed line corresponds to the 3.2 $M_\odot$ separatrix (see Section 1.3.8) between neutron stars and black-hole candidates (after Orosz 2002).

masses of the companion stars, and the masses of the compact objects can be determined only if a counterpart of the X-ray source in other wavelengths is also observed (see Chapter 5).

The brightness of luminous X-ray sources in optical wavelengths, especially for compact objects with low-mass companions, is typically positively correlated with their X-ray brightness and orbital periods (van Paradijs & McClintock 1995). This is a strong indication that thermal emission from the outer parts of the accretion flow as well as reprocessing of the X-ray emission at the outer accretion disk and the companion star are responsible for at least part of the optical emission.

Detailed studies of X-ray binaries in IR/optical/UV wavelengths are crucial in measuring the temperature profiles, ionization fractions, and abundances of elements in the accretion disk (see, e.g., Kallman *et al.* 1998) or even the inclinations of the binary systems (see, e.g., de Jong *et al.* 1996; Wang *et al.* 2001). Moreover, if the accretion geometry is that of a geometrically thin disk, such studies most probably provide the only handle in measuring the rate at which matter is transferred towards the compact object (see, e.g., Vrtilek *et al.* 1990).

In transient systems during their quiescent periods as well as in systems with short orbital periods, the optical emission is dominated by the companion star, making possible the measurement of their binary orbits and of the masses of the compact objects. Table 1.4 and Fig. 1.11 summarize the current mass measurements of compact objects in accreting binary systems and compare them to the mass measurements of radio pulsars in non-interacting binaries. There is a clear dichotomy between compact objects clustered around $\sim 1.5/M_\odot$

Table 1.4. *Masses of compact objects*

Source	Period (d)	Mass[a] ($M_\odot$)
Black hole candidates		
GRO J0422+32	0.212	3.66–4.97
A 0620−00	0.323	8.70–12.86
GRS 1009−45	0.285	3.64–4.74
XTE J1118+480	0.170	6.48–7.19
GS 1124−683	0.433	6.47–8.18
4U 1535−47	1.116	8.45–10.39
XTE J1550−564	1.543	8.36–10.76
GRO J1655−40	2.622	6.03–6.57
H 1705−250	0.521	5.64–8.30
SAX J1819.3−2525	2.817	6.82–7.42
XTE J1859+226	0.382	7.6–12.0 (?)
GRS 1915+105	34	10.0–18.0 (?)
GS 2000+25	0.344	7.15–7.78
GS 2023+338	6.471	10.06–13.88
LMC X-3	1.705	5.94–9.17
LMC X-1	4.229	4.0–10.0 (?)
Cyg X-1	5.600	6.85–13.25
Accreting neutron stars		
Vela X-1	8.964	1.78 ± 0.15
4U 1538−52	3.728	$1.06^{+0.41}_{-0.34}$
SMC X-1	3.892	$1.17^{+0.36}_{-0.32}$
LMC X-4	1.408	$1.47^{+0.44}_{-0.39}$
Cen X-3	2.087	$1.09^{+0.57}_{-0.52}$
Her X-1	1.700	$1.04^{+0.75}_{-0.58}$ or $1.47^{+0.23}_{-0.37}$
Cyg X-2	9.844	1.78 ± 0.23
Double neutron stars		
J1518+4904	8.634	$1.56^{+0.13}_{-0.44}$
		$1.05^{+0.45}_{-0.11}$
B1534+12	0.421	1.339 ± 0.003
		1.339 ± 0.003
B1913+16	0.323	1.4411 ± 0.00035
		1.3874 ± 0.00035
B2127+11C	0.335	1.349 ± 0.040
		1.363 ± 0.040
B2303+46	12.34	$1.30^{+0.13}_{-0.46}$
		$1.34^{+0.47}_{-0.13}$
Radio pulsars in binaries		
J0437−4715	5.741	<1.51
J1012+537	0.605	1.7 ± 0.5
J1045−4509	4.084	<1.48
J1713+0747	67.825	1.45 ± 0.31
B1802−07	2.617	$1.26^{+0.08}_{-0.17}$
J1804−2718	11.129	<1.73
B1855+09	12.327	1.41 ± 0.20
J2019+2425	76.512	<1.68
J0045−7319	51.169	1.58 ± 0.34

[a] Black hole masses are from Orosz (2002); accreting neutron-star masses are from van Kerkwijk *et al.* (1995), Orosz & Kuulkers (1999), and Barziv *et al.* (2001); radio pulsar masses are from Thorsett & Chakrabarty (1998).

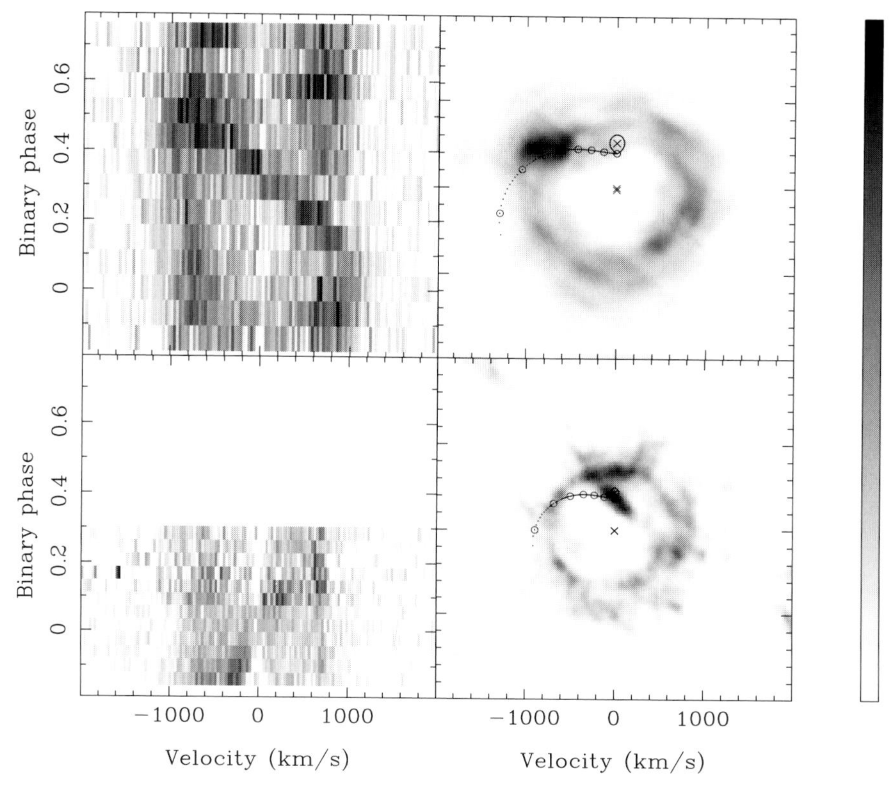

Fig. 1.12. Hα spectra (*left panels*) and reconstructed Doppler images (*right panels*) of the black-hole candidates GS 2000+25 (top) and Nova Oph 1977 (bottom) (Harlaftis 2001).

and objects with significantly larger masses. This is believed to be a direct consequence of the fact that there is an upper limit on the mass of a stable neutron star and any heavier compact object must be a black hole (but see also Section 1.3.8).

Recent developments in observational methods and analysis techniques of long-wavelength data have also led to a number of ways of imaging indirectly the accretion flows around compact objects (see, e.g., Harlaftis 2001; Fig. 1.12). Eclipse mapping is applicable only to high-inclination systems and uses the periodic occultation of different parts of the accretion flow by the companion star during the orbit to infer the relative contribution of different regions to the total emission (see, e.g., Vrielmann 2000 for a review of the method and applications to cataclysmic variables). Doppler tomography uses the dependence on orbital phase of Doppler-shifted atomic lines that originate from different parts of the accretion flow in order to produce a map of the line-emitting regions on the orbital plane of the binary system (see, e.g., Marsh 2000; Harlaftis 2001). Finally, echo tomography uses the time delays between prompt and reprocessed emission at different regions in order to map the geometry of the accretion flows and the binary orbits (see, e.g., Horne 2003).

The above indirect imaging techniques provide the most concrete method of observationally testing accretion flow models. They have been regularly used in mapping the radial temperatures of geometrically thin accretion disks, often showing significantly flatter profiles

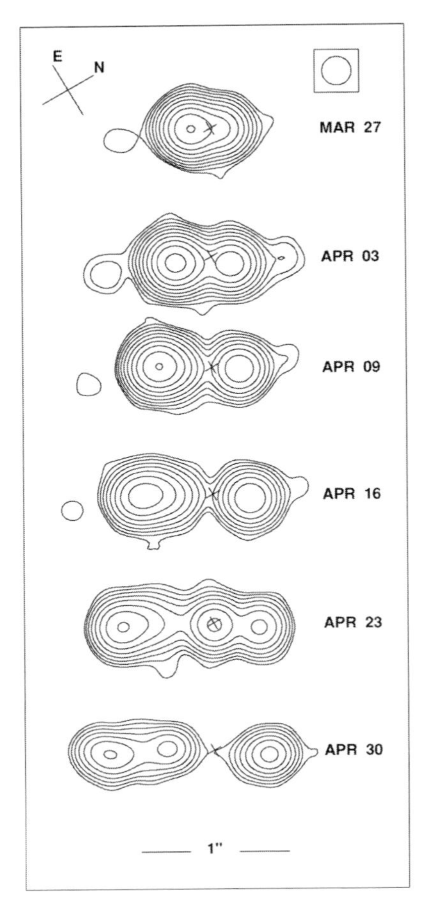

Fig. 1.13. A series of 3.5 cm VLA images of the black-hole candidate source GRS 1915+105, showing the fast-traveling knots of the radio jet (Mirabel & Rodriguez 1994).

compared to accretion theory (see Vrielmann 2000). They can be used to measure the geometrical thickness of the accretion flows, thereby distinguishing between competing models (see, e.g., O'Brien *et al.* 2002). Finally, they provide the best evidence for the presence of spiral structures in accretion disks, thus giving clues to the viscous mechanisms that operate in these flows (see, e.g., Harlaftis 2001).

1.3.3 Jets

Outflows and collimated jets are ubiquitous phenomena in all accreting objects, from young stars to active galactic nuclei. The discovery of mildly relativistic jets from the binary system SS 433 (Spencer 1979) and of superluminal jets from the black-hole candidate GRS 1915+105 (Mirabel & Rodriguez 1994; also Fig. 1.13) revealed that accreting galactic compact objects are no exception to this rule (see Chapter 9).

Jets from accreting stellar-mass black holes in the galaxy share many properties with their counterparts in active galactic nuclei. For example, they have non-thermal, polarized radio spectra indicating the presence of shock-accelerated relativistic electrons that emit synchrotron radiation as they propagate in regions with large-scale magnetic fields (e.g., Eikenberry *et al.* 1998; Mirabel *et al.* 1998). They also show large flux ratios between the approaching and receding sides of the jets, as expected for relativistic flows (Mirabel & Rodriguez 1994).

In addition to revealing such similarities, however, the short dynamical timescales associated with galactic black-hole jets and their proximity made possible an in depth study of their properties, even though only a handful of jet sources is currently known. The orientation of some jets in the sky, with the jet in SS 433 being the main example (Margon 1984), has been observed to precess in real time over large angles with long periods ($\sim$162 days in the case of SS 433). This is believed to be associated with the precession of the underlying accretion disk (see., e.g., Begelman & Rees 1984; Ogilvie & Dubus 2001) and may provide clues towards understanding the launching and collimation of the jets.

The propagation speeds of the jets and their deceleration upon interactions with the interstellar medium can be measured on images by following the kinematics of individual radio knots (see, e.g., Fig. 1.13). Not surprisingly, the inferred speeds appear to depend on the state of the underlying accretion disk from which they are launched, with steady and transient sources showing mildly and highly relativistic jets, respectively (see also Chapter 9).

Coordinated observations of accreting black holes and neutron stars in the radio, infrared, and X-ray wavelengths recently revealed the most striking of jet properties: there is a very clear correlation between the presence of jets and the X-ray spectral state of the accretion flows (see, e.g., Corbel *et al.* 2000). In particular, jets appear when the X-ray spectra of the sources indicate emission from hot electrons ($\sim$100 keV; see next section); on the other hand, when their spectra are typical of cold, geometrically thin accretion disks, the jets are weak or absent. It is unclear at this point what is the causal connection between the radio and X-ray properties of accreting compact objects. The mechanism responsible for the heating of electrons in the accretion flow may be related to the formation of an outflow, as is the case both for magnetically active accretion disks (see, e.g., Blandford & Payne 1982) or for advection-dominated accretion flows (Narayan & Yi 1994). Alternatively, most of the X-ray emission may be produced directly at the base of the jet (Markoff *et al.* 2001).

Finally, recent observations of the X-ray emission from the jet of SS 433 with Chandra confirmed the presence of atomic emission lines, strongly suggesting that heavy ions are also accelerated together with the electrons and positrons that are responsible for most of the jet emission (Marshall *et al.* 2002). The jet in SS 433, however, is only mildly relativistic and it is not clear whether this is related to the presence of ions in the jet or if there is simply an observational selection effect against detecting lines from very relativistic outflows (Mirabel *et al.* 1997).

1.3.4 X-ray and gamma-ray spectroscopy

The design of high-energy missions with broad-band spectral coverage, the numerous campaigns of simultaneous observations with multiple instruments, as well as the advent of high-resolution CCD and grating spectrographs for X-rays have launched a new era of high-energy spectroscopy in astrophysics. Accreting compact objects are often being monitored in all wavelengths, providing a strong handle on their bolometric luminosities as well as placing stringent constraints on accretion models.

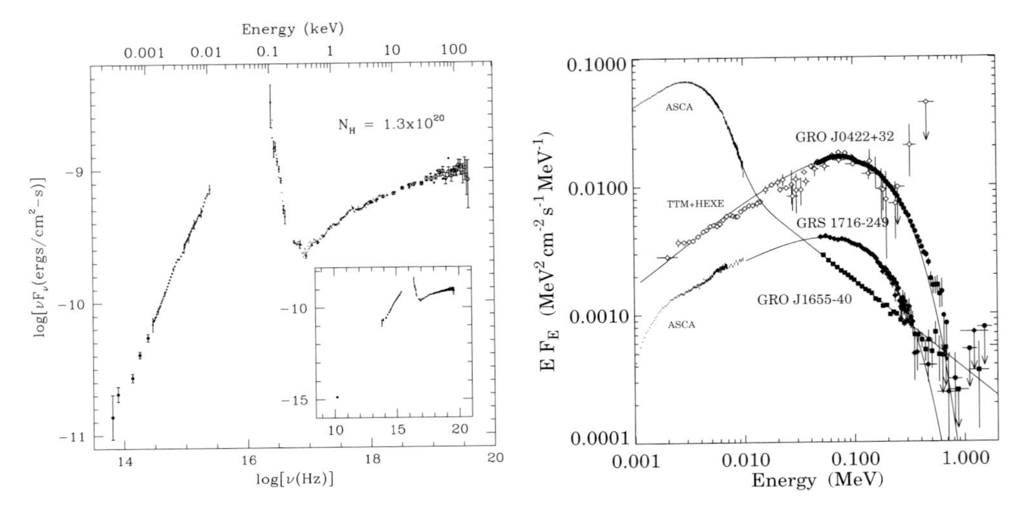

Fig. 1.14. Broad-band spectra of black-hole candidates: (*left*) the infrared to soft X-ray spectrum of XTE J1118+480 (McClintock *et al.* 2001); (*right*) the soft X-ray to gamma-ray spectra of three sources (Grove *et al.* 1998).

Figure 1.14 shows some examples of broad-band spectra of black-hole candidates, from the infrared to the gamma-ray. All such spectra of accreting compact objects show unequivocally that a number of distinct emission mechanisms are responsible for their various features. In fact, simple thermal emission models from geometrically thin accretion disks can produce neither the spectral complexity nor the hard X-ray and gamma-ray fluxes that are observed. Modeling the spectra of accreting compact objects is complicated and appears to be mostly data driven. However, it is also potentially very rewarding as it may lead to the understanding of processes such as the generation of magnetic fields in turbulent flows, viscous heating in magnetic media (see, e.g., Quataert & Gruzinov 1999), and the thermal properties of multi-temperature plasmas (e.g., Coppi 1999).

Most current models of the X-ray emission from accreting neutron stars and black holes (see GRO J0422+32 and GRS 1716−249 in Fig. 1.14) require that a tenuous atmosphere of hot (∼10–100 keV) electrons is present simultaneously with the geometrically thin accretion disks. Such a hot medium may be in the form of a magnetically heated corona (see, e.g., Dove *et al.* 1997), of an advection-dominated accretion flow (ADAF) (Esin *et al.* 1997), or even a jet (Markoff *et al.* 2001). These same electrons are almost certainly responsible also for the longer-wavelength emission (see XTE J1118+480 in Fig. 1.14), which appears to be strongly correlated with the X-ray flux, via radiation processes associated with the magnetic fields generated and sustained by the accretion flow. Finally, the power-law gamma-ray spectral tails of luminous neutron-star and black-hole candidates (see GRO J1655−40 in Fig. 1.14; see also Grove *et al.* 1998; Di Salvo *et al.* 2000) require the existence of a non-thermal population of very energetic electrons, either in the form of a hybrid plasma (Coppi 1999) or of a quasi-radial high-velocity flow (Laurent & Titarchuk 1999).

Perhaps the most eagerly anticipated result of the launch of X-ray telescopes with high spectral resolution, such as ASCA, Chandra, and XMM-Newton, has been the discovery of atomic lines from the spectra of accreting compact objects. The relative strengths and equivalent widths of such lines depend strongly on temperature, density, and ionization flux

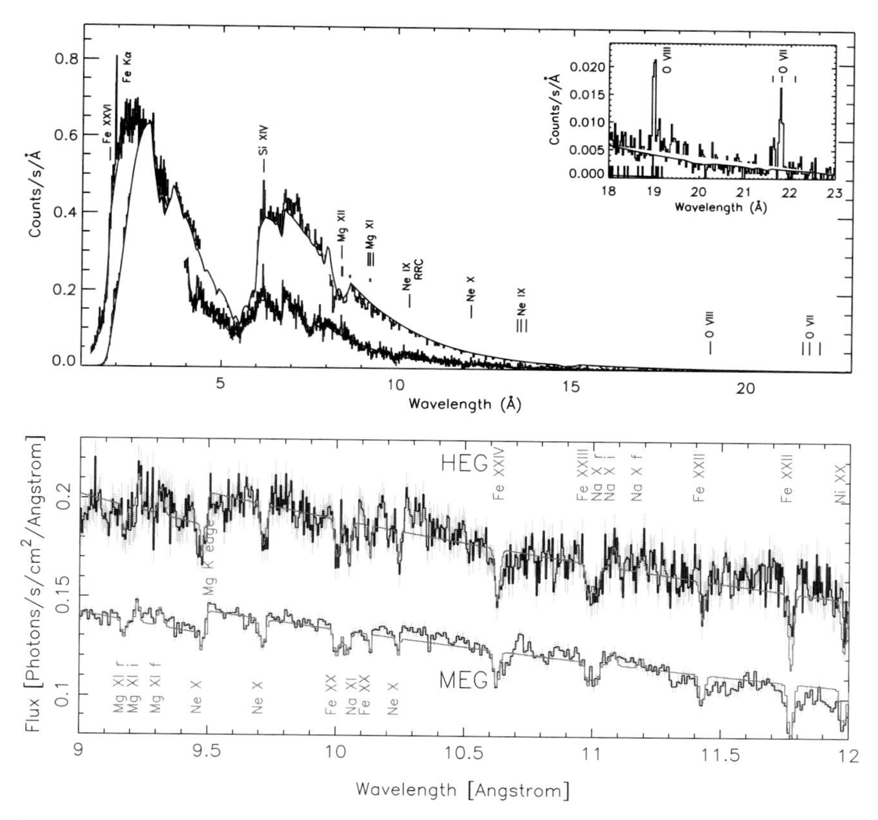

Fig. 1.15. High-resolution X-ray spectra of (*top*) the ADC source 4U 1822−37 (Cottam *et al.* 2001) and (*bottom*) the black-hole source Cyg X-1 as observed by the Chandra X-ray Observatory (Miller *et al.* 2002a).

and hence are valuable probes of the physical conditions in the accretion flows (Liedahl 1999). Moreover, the gravitational redshifts and relativistic broadening of atomic lines generated close to the event horizons of black holes can, in principle, be used to map the spacetimes around the compact objects and measure properties such as their masses and spin angular momenta (e.g., Fabian *et al.* 1989).

The X-ray spectra of many accreting compact objects have, unfortunately, atomic lines that are very weak or even undetectable, largely due to their high temperatures and photonization fluxes. However, studies of atomic lines have proved fruitful in several cases for which the binary configurations are optimal. For example, in accretion-disk corona (ADC) sources, the high inclinations of the binary systems allow for a clear view of the coronal structure away from the central object, where the temperatures are lower and the line emission stronger (see Fig. 1.15; also Cottam *et al.* 2001). In binary systems with companions that exhibit strong winds, the X-ray spectra show a variety of absorption lines and edges that originate at the relatively cooler wind material (Fig. 1.15 and, e.g., Miller *et al.* 2002a). Recent observations of several black-hole candidates have also shown evidence for relativistically broadened iron K lines, similar to those observed in active galactic nuclei (Miller *et al.* 2002b). Finally, a detection of gravitationally redshifted atomic lines during thermonuclear flashes on the surface of an accreting neutron star has also been reported (Cottam *et al.* 2002).

1.3.5 Variability

Accreting compact objects are among the most variable persistent sources in the sky. Even excluding the X-ray transients, the flux from most sources typically varies by factors of 2 over periods ranging from months to fractions of a second (see Fig. 1.16). This is not surprising, given the wide range of characteristic timescales that are involved in the processes that lead to the production of the high-energy emission. For example, the transfer of mass from the binary companion to the compact object is expected to vary at timescales comparable to the orbital period, i.e., hours to days. The inward diffusion of matter in the accretion flow occurs at the viscous timescale, which is slower than the Keplerian orbital frequency at any radius and ranges from days, at the outer edge of the disk, to fractions of a second, close to the compact object. Finally, the interaction of the accretion flow with the central star occurs at the fastest dynamical timescale in the accretion flow, which is of the order of a millisecond.

Over the last two decades, the most unexpected result of timing studies of accreting compact objects has been the discovery of quasi-periodic oscillations (QPO) (see Fig. 1.17) of their X-ray brightness at all these timescales (van der Klis 2000 and Chapter 2). Because of the high degree of variability of the sources as well as observational constraints, the fast (greater than a fraction of a hertz) oscillations of the X-ray brightness are the ones that have been studied more extensively, mostly with the proportional counters on board EXOSAT, Ginga, and RXTE.

The large majority of these oscillations have frequencies that are highly variable, even though they may lose coherence only after tens or hundreds of cycles. When several variable-frequency QPOs are observed simultaneously, their frequencies follow a small number of tight correlations (Psaltis *et al.* 1999). On the other hand, in several luminous black-hole candidates, pairs of high-frequency QPOs have also been detected, with frequencies that are nearly constant and in small integer ratios (i.e., 3:2, 5:3; Strohmayer 2001a,b; Abramowicz & Kluzniak 2001). These distinct and correlated frequencies of the observed QPOs clearly suggest that accretion flows are capable of picking, out of a large pool of alternatives, only a small number of characteristic frequencies at which to vary preferentially.

A number of theoretical interpretations have been put forward so far, in an attempt to account for the mechanism that picks these characteristic frequencies in the accretion flows. Current models of the *variable-frequency* QPOs require the presence of a characteristic radius, across which the properties of the accretion disks change considerably. The observed QPO frequencies are then attributed to the characteristic dynamic or hydrodynamic frequencies of the accretion flow at that radius (e.g., Stella *et al.* 1999; Psaltis & Norman 1999; Titarchuk *et al.* 1999) or, in the case of neutron stars, to the coupling of these frequencies to the stellar spin (Miller *et al.* 1998). On the other hand, models of the *constant-frequency* QPOs in black-hole systems are based on the trapping of oscillatory modes in accretion disks caused by the properties of the relativistic spacetime (Wagoner 1999; Kato 2001) or on the non-linear coupling of different frequencies near the black-hole event horizons (Abramowicz & Kluzniak 2001).

A common feature of all current models of QPOs is the identification of at least some of the observed frequencies with dynamic frequencies in the accretion flows, such as the ones related to the azimuthal orbital motion of plasma, to Lense–Thirring precession, etc. These frequencies correspond to regions very close to the neutron-star surface or the black-hole event horizons and hence offer the possibility of observing, for the first time, effects that

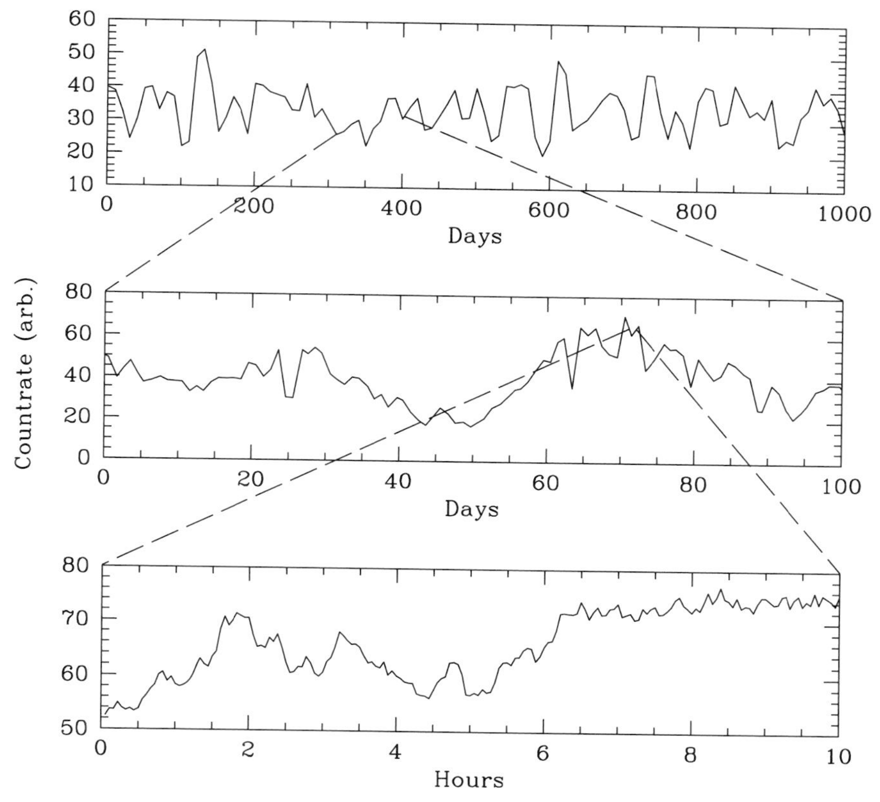

Fig. 1.16. The lightcurve of the neutron-star source Cyg X-2, as observed at different timescales by the All Sky Monitor (upper two panels) and the Proportional Counter Array (bottom panel) on board RXTE.

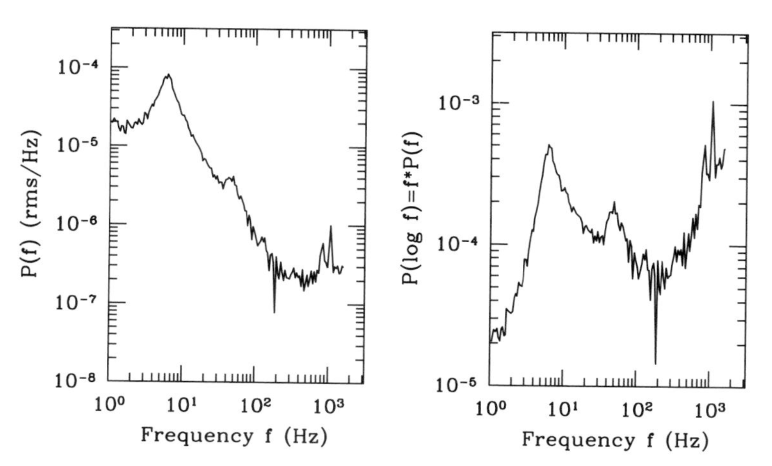

Fig. 1.17. A power-density spectrum of the neutron-star source Sco X-1 (in two commonly used representations) showing multiple, simultaneously detected quasi-periodic oscillations.

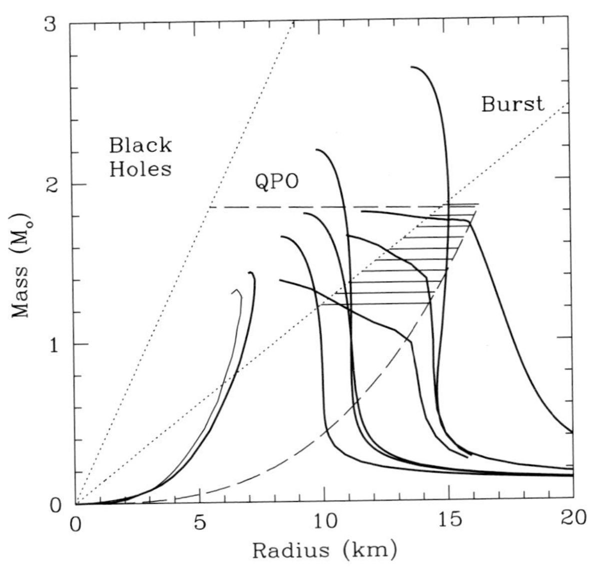

Fig. 1.18. Constraints on the mass and radius of the neutron star in 4U 1636−36 imposed by the observation of a 1193 Hz QPO (dashed line; Jonker *et al.* 2002; Miller *et al.* 1998) and the large amplitudes of burst oscillations (dotted line; Nath *et al.* 2002). The hatched region represents the area of the parameter space that is consistent with all observations and includes stars with baryonic masses larger than $\sim$1.4 $M_\odot$. The solid lines correspond to neutron-star models with different equations of state. Stellar models with condensates correspond to mass–radius relations with a characteristic flattening at small radii; models of strange stars correspond to mass–radius relations that start at the origin.

occur only in strong gravitational fields. In particular, the general relativistic prediction of a radius, inside which no stable circular orbits exist (the so-called innermost stable circular orbit or ISCO), is of fundamental importance for almost all models. This characteristic radius is responsible, in different models, for the saturation of the observed QPO frequencies with accretion rate (e.g., Miller *et al.* 1998; Zhang *et al.* 1998) or the trapping of modes in the inner accretion flows and hence the generation of the QPOs themselves (Wagoner 1999; Kato 2001).

More importantly, the azimuthal orbital frequency at the radius of the innermost circular orbit is thought to provide a natural upper limit on the frequency of any dynamical process that can occur in an accretion flow (see, e.g., Kluzniak *et al.* 1990; Miller *et al.* 1998). Since this azimuthal frequency depends only on the mass and spin of the compact object, it provides, when compared to an observed QPO frequency, an upper limit on the mass (modulo the spin) of the neutron star or black hole. This is illustrated as a horizontal dashed line in Fig. 1.18, for the case of the neutron-star source 4U 1636−36. An additional upper limit on the radius of the neutron star (as a function of its mass) is imposed by the requirement that the maximum observed QPO frequency is less than the azimuthal orbital frequency at the stellar surface (see Fig. 1.18). These two arguments, together with additional bounds imposed by the presence of oscillations during thermonuclear bursts (Nath *et al.* 2002), the detection of redshifted lines from the stellar surfaces (Cottam *et al.* 2002), and the measurement of the mass of the compact objects in the binaries using orbital dynamics will

Table 1.5. *Long-term periods of X-ray binaries*

Source	Orbital period (d)	Long-term period (d)	References
Stable			
LMC X-4	1.4	30.4	
Her X-1	1.7	35	
SS 433	—	164[a]	16
Quasi-periodic			
SMC X-1	3.89	50–60	9
LMC X-3	1.7	∼100–300	11, 12
Cyg X-1	5.6	294	13
Cyg X-2	9.84	∼70–80	1, 2, 3, 14, 15
Cen X-3	2.09	∼120	7
4U 1728−34	—	∼30–70	3, 4
4U 1820−30	0.008	∼170	6, 10
4U 1916−053	0.035	∼80	5
4U 2127+119	0.71	∼37 (?)	8

[a] As inferred from the precession of the jet.
References: 1. Kuulkers *et al.* 1996; 2. Paul *et al.* 2000; 3. Kong *et al.* 1998; 4. Galloway *et al.* 2003; 5. Homer *et al.* 2001; 6. Chou & Grindlay 2001; 7. Priedhorsky & Terrell 1983; 8. Corbet 1997; 9. Wojdowski *et al.* 1998 10. Priedhorsky & Terrell 1984; 11. Cowley *et al.* 1991; 12. Wilms *et al.* 2001; 13. Priedhorsky *et al.* 1983; 14. Smale & Lochner 1992; 15. Kuulkers *et al.* 1999; 16. Margon & Anderson 1999.

provide the most stringent constraints on the properties of neutron-star matter and its equation of state.

As discussed above, the rapid variability properties of accretion flows provide useful probes into the physical conditions close to the compact objects. At the same time, the slow ($\geq$1 day) variability of the same systems can be used in constraining models of the accretion flows at large distances and of the mass transfer process between the members of the binary. Such studies have become possible recently with the systematic observation of the entire X-ray sky using the All Sky Monitor on board RXTE.

The long-term variability of accreting weakly-magnetic neutron stars and black holes is typically aperiodic and reflects the variable nature of the mass transfer and accretion processes. Sources with more systematic slow variability fall in three categories: (a) systems with large amplitude variations of their X-ray flux because of their transient nature (see Fig. 1.9); (b) systems with a periodic modulation at their orbital frequency caused, e.g., by eclipses, by the reflection of the X-ray photons off the binary companion and the accretion stream, or by the variable rate of mass transfer due to a highly elliptical orbit (e.g., Cir X-1; Shirey *et al.* 1996); and (c) systems with quasi-periodic modulations at variable, superorbital periods (see, e.g., Wijnands *et al.* 1996; Heinz & Nowak 2001).

Figure 1.19 shows an example of a variable-period modulation in the source Cyg X-2. This is a neutron-star system with a binary period of 9.84 days, which shows a number of superorbital periods, some of which are nearly integer multiples of 9.84 days (see also Wijnands *et al.* 1996). Such modulation may be related to variable mass transfer of the binary

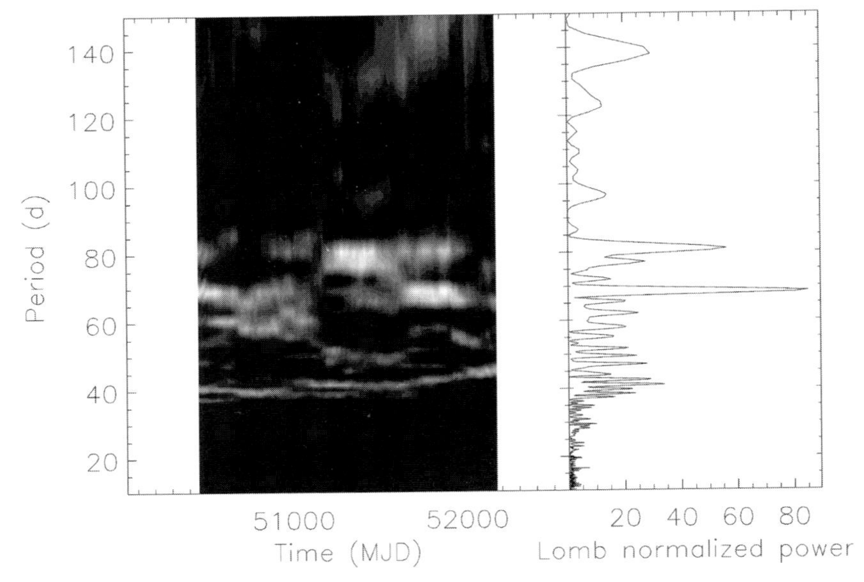

Fig. 1.19. Dynamical periodogram of the neutron-star source Cyg X-2, showing the variable, superorbital oscillations of its X-ray flux, as observed by the All Sky Monitor on RXTE (courtesy D. Galloway).

companion that is driven at the binary orbital period or to reflection off the warped surface of the outer accretion disk. Recent theoretical investigations of the warping of geometrically thin accretion disks caused by the torque of the reflecting X-ray irradiation from the central object or by an asymmetric wind (see, e.g., Pringle 1996; Maloney *et al.* 1998) have indeed shown the possibility of long-lived warping modes that could produce, in principle, the observed modulations.

1.3.6 Thermonuclear bursts

The material that is accreted on the surface of a weakly magnetic neutron star may be compressed to densities and temperatures for which the thermonuclear burning of helium is unstable. The ignition of helium results in a rapid ($\sim$1 s) increase in the X-ray luminosity of the neutron star, followed by a slower (tens of seconds) decay that reflects the cooling of the surface layers that ignited (see Fig. 1.20). The observational manifestations of these thermonuclear flashes are called Type I X-ray bursts (for a review see Lewin *et al.* 1996 and Chapter 3).

For very energetic bursts, the force of the escaping radiation balances gravity, causing the surface layers of the neutron star to expand rapidly. During the expansion phase, the emerging radiation flux remains comparable to the Eddington critical value, at which radiation and gravitational forces are balanced, and the remaining energy of the explosion is given as kinetic and potential energy to the expanding layers (Kato 1983; Nobili *et al.* 1994). These are the so-called Eddington-limited or photospheric radius-expansion bursts. Because the maximum observed flux of an Eddington-limited burst depends, to zeroth order, on the mass and radius of the star (the allowed values of which span a very narrow range) as well as on the distance to the source, these burst can be very useful in constraining all these three

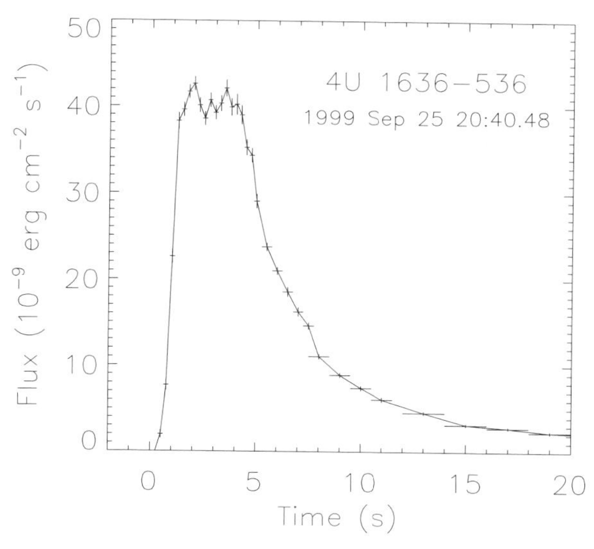

Fig. 1.20. A typical Type I X-ray burst from the neutron-star source 4U 1636−36. The flat top of the burst lightcurve is characteristic of Eddington-limited (or radius-expansion) bursts (courtesy D. Galloway).

parameters and, in particular, in measuring the distances to X-ray bursters (see Kuulkers *et al.* 2003; Galloway *et al.* 2003).

The general properties of Type I X-ray bursts, such as their energetics, peak fluxes and fluences, recurrence times, rise and decay timescales, etc. are qualitatively consistent with the predictions of the helium ignition model. Quantitatively, however, the observations show very little of the regularity that is inherent in the numerical models (Lewin *et al.* 1996; see, however, Galloway *et al.* 2004). This is not surprising given the strong dependence of the burst properties on the time-variable physical conditions in the ignition area. For example, the energetics and timescales of bursts depend on the composition of the accreting material (see, e.g., Cumming & Bildsten 2001), the local rate of accretion, the temperature of the neutron-star core (e.g., Fushiki & Lamb 1987), the presence of ashes from previous bursts, etc. Recently, there has been significant progress towards developing models of thermonuclear flashes that relax many of the limiting assumptions of earlier calculations and incorporate detailed nuclear networks (Schatz *et al.* 2000), the effects of multi-dimensional propagation of burning fronts (Zingale *et al.* 2001; see also Fig. 1.21), and the stellar rotation (Spitkovsky *et al.* 2002).

The long-term monitoring capabilities of BeppoSAX and RXTE brought the discovery of a new type of long (∼hr) bursts with even longer (∼yr) recurrence times, the so-called superbursts (Cornelisse *et al.* 2000; Chapter 3). These are believed to be caused by unstable carbon burning in layers that are deeper than those responsible for the normal Type I bursts, thereby accounting for their longer durations and recurrence timescales (Strohmayer & Brown 2002; Cumming & Bildsten 2002).

During many of the normal Type I X-ray bursts and in one superburst, highly coherent oscillations of the observed X-ray fluxes are often detected (Strohmayer *et al.* 1996; Chapter 3). The frequencies of the oscillations drift by a small percentage during the bursts,

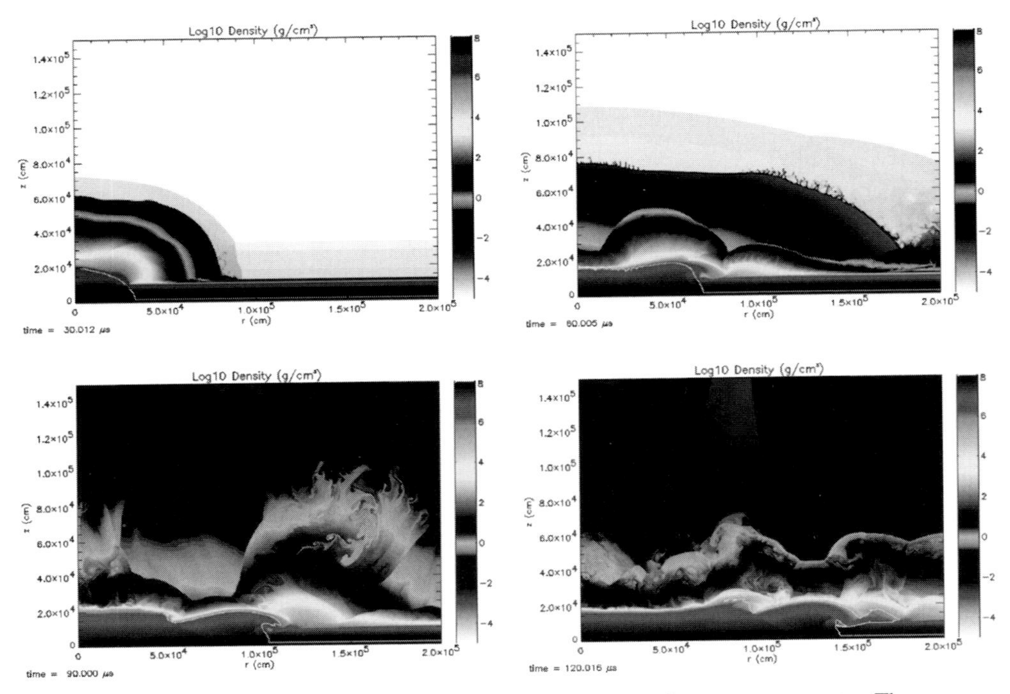

Fig. 1.21. Two-dimensional simulation of helium detonation on a neutron star. The greyscale levels correspond to different densities (Zingale *et al.* 2002)

reaching values that are constant, to within one part in 10^4, between bursts from the same source (Muno *et al.* 2002a). In bursts from two ultracompact millisecond pulsars, in which the spin frequencies of the stars are known, the asymptotic values of the burst oscillation frequencies are nearly equal to the spin frequencies of the stars (see, e.g., Chakrabarty *et al.* 2003 and Chapter 2).

These two properties have led to an interpretation of burst oscillations in which the thermonuclear burning on the neutron-star surface is non-uniform and produces a modulation of the X-ray flux at the stellar spin frequency (Strohmayer *et al.* 1996). The origin of the frequency drift during the rising phase of the bursts, however, is still unresolved. A number of ideas are currently being explored, which involve the decoupling and slowing down of the surface layers from the rest of the neutron star during the rise (Cumming *et al.* 2002), the drift of the burning front with respect to a fiducial azimuth on the stellar surface caused by rotation (Spitkovsky *et al.* 2002), and the excitation of non-radial modes in the burning layers (Heyl 2004).

Despite the lack of a physical model of burst oscillations, their properties have already been used in obtaining stringent constraints on the masses and radii of neutron stars, as well as on the degree of non-uniformity of the thermonuclear burning. For example, during burst rise, the large ($\sim$70%) observed amplitudes of the oscillations require that the neutron stars are not too compact for gravitational self-lensing to wash out the pulsations (Nath *et al.* 2002). Moreover, the lack of detectable harmonics in burst oscillations constrains the emission areas and orientations with respect to the rotation axes of the stars (Muno *et al.* 2002b).

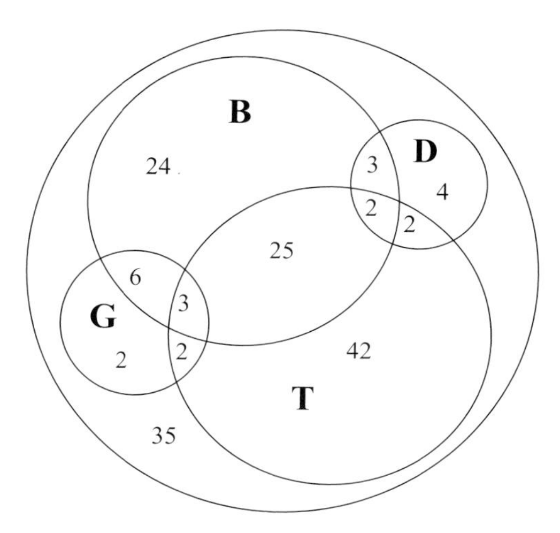

Fig. 1.22. A Venn diagram of the known non-pulsing neutron stars and black holes in binary systems (using the catalog of Liu *et al.* 2001). The initials correspond to B: bursters; G: globular-cluster sources; T: transients; D: dippers. The areas of the circles correspond approximately to the relative number of sources in each category.

1.3.7 A census of non-pulsing neutron stars and black holes

As evident from the discussion in the previous section, non-pulsing neutron stars and black holes are found in a variety of binary systems and configurations. The number of known sources has steadily increased, from 33 in 1983, to 119 in 1995, to 150 in 2000 (Liu *et al.* 2001). This increase is largely due to the discovery of low-luminosity sources with detectors of increasing sensitivity but is also due to the discovery of a large number of transient sources.

Out of 150 sources, 63 exhibit Type I X-ray bursts and can, therefore, be identified as neutron stars. On the other hand, 17 sources have dynamically measured masses in excess of 3.2 $M_\odot$ and can be safely identified as black holes. Half of the known sources (76 out of 150) are transients. Given their relatively low duty cycles, which are of order 10% for neutron-star sources and much lower for black holes, the total number of transient sources in the Galaxy must be significantly larger (King & Kolb 1996).

Finally, about 10% of the objects in each category (5/63 bursters, 4/76 transients, 11/150 total) are dippers. This is consistent with the geometric interpretation of the dips and with an opening angle of the accretion stream (as measured from the central objects) of about 10 degrees.

1.3.8 The nature of the central object in non-pulsing X-ray binaries

The presence of persistent pulsations or of Type I bursts in the X-ray lightcurve of an accreting compact object provide the only unequivocal proof that the central object is a compact *star* and not a black hole. However, the absence of pulsations or bursts does not provide proof that the compact object does not have a surface (see, however, Narayan & Heyl 2002). Indeed, if the magnetic field of the neutron star is dynamically unimportant ($\ll 10^9$ G), it will not appear as an X-ray pulsar. Moreover, Type I X-ray bursts are expected to occur only for a particular range of accretion rates, surface gravity accelerations, compositions of the accreting material, and core temperatures (see, e.g., Fushiki & Lamb 1987). Finally, the presence of pulsations or bursts indicates simply the existence of a stellar surface and does not necessarily prove that the object is a neutron star, as opposed to another type of compact star with yet-to-be-discovered properties. It is in fact extremely difficult to provide a conclusive proof that a non-pulsing object is a black hole or even a neutron star. All current arguments are either empirical or simply attempted proofs by elimination, leaving open the possibility that a viable alternative was simply not considered (see also Abramowicz *et al.* 2002).

Since the discovery of X-ray binaries, a number of empirical tests have been put forward for distinguishing non-pulsing neutron stars from black holes based on their X-ray properties. Recent examples of suggested evidence for black holes are the presence of a hard X-ray spectrum at high luminosities (Barret *et al.* 1996), of a particular type of X-ray spectrum (Done & Gierlinski 2003), or of significant variability power at high Fourier frequencies (Sunyaev & Revnivtsev 2000). Albeit useful as indicators, these empirical tests cannot provide conclusive arguments for the nature of the central objects. Neutron stars and black holes share the general characteristics of a very complex phenomenology of spectral and timing properties and their differences are only in the details (see, e.g., van der Klis 1994 for a discussion).

The measurement of a large gravitational mass for the compact object is currently considered to be the strongest evidence for its identification with a black hole. The reason is that, under three simple assumptions, an optimal upper bound on the mass of any neutron star of $\sim 3.2\ M_\odot$ can be derived (Rhoades & Ruffini 1983). The three assumptions are: (i) the star is non-rotating; (ii) the equation of state at densities below some fiducial value (typically close to the nuclear saturation density) is known; (iii) the speed of sound at larger densities is smaller than the speed of light (the so-called "causality" condition).

Including the effects of rotation introduces only small corrections and affects the limiting mass by $<20\%$ (Friedman & Ipser 1987). The next two assumptions, however, are significantly more constraining. The speed of sound, being a phase velocity, is not bounded by relativity to be less than the speed of light. In fact, the actual condition used is a causality requirement only for cold, non-dispersive material. However, neutron-star matter can be both dispersive and of non-zero temperature. Relaxing this condition and allowing for rotation leads to bounds on the neutron-star mass as large as $\sim 14\ M_\odot$ (Sabbadini & Hartle 1977). Finally, the compact stars under consideration may form a distinct family of objects, which is not a continuation of the normal white-dwarf to neutron-star sequence of equilibrium configurations towards higher central densities. If this is the case, then the second of the above assumptions is irrelevant.

Families of compact objects that are not bound by gravity, such as strange stars (Witten 1984; Alcock *et al.* 1986) and Q stars (Bahcall *et al.* 1990), have been constructed as potential alternatives to neutron stars and black holes. Strange stars can be made practically indistinguishable from neutron stars with respect to masses, radii, and maximum spin frequencies (see, however, Glendenning 1996). On the other hand, Q stars can be constructed to have masses as large as the most massive stellar-mass black-hole candidates, even though this would require extreme changes in our understanding of the properties of matter at densities as low as one-tenth of nuclear saturation (Miller *et al.* 1998). Finally, if gravity is not described by general relativity in the strong-field regime, then the limiting mass of a neutron star may not be $\sim 3.2\ M_\odot$. Metric theories of gravity that are consistent with all Solar System tests but differ from general relativity in the strong-field regime allow for neutron stars with significantly larger mass (see, e.g., DeDeo & Psaltis 2003).

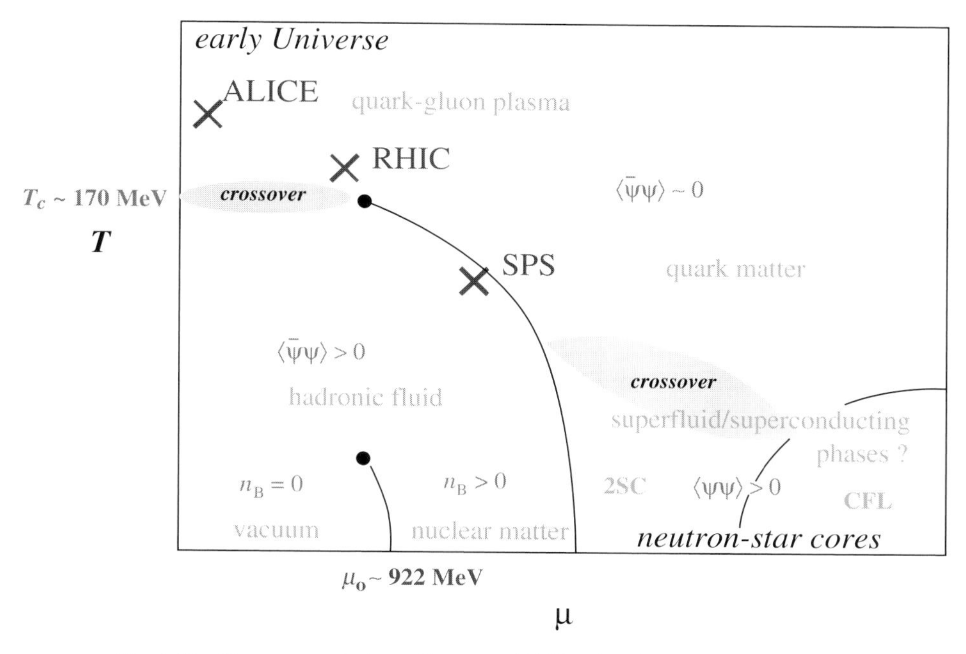

Fig. 1.23. The proposed phase diagram of QCD, showing the position on it of matter in the early Universe, in modern heavy-ion colliders (SPS, RHIC, and ALICE), and in the cores of neutron stars (for details see Hands 2001).

It is important to note here that the existence of black holes is a strong-field prediction of a theory (i.e., general relativity) that has been tested to high accuracy, at least in the weak-field limit (Will 2001). On the other hand, all the other alternatives discussed above are the results of theoretical assumptions that have not been tested (and mostly could not have been tested) with current experiments. Such alternatives provide physically consistent counter-examples to the identification of a compact object as a black hole. However, they will remain simply as thought experiments until experimental evidence shows that our theories of gravity and matter fail to describe extreme physical conditions.

1.4 Accretion-powered X-ray sources in the twenty-first century

Accreting neutron stars and black holes in the galaxy offer the unique opportunity of understanding the properties of matter, electromagnetic, and gravitational fields beyond the conditions found in current terrestrial experiments and other cosmic settings. Indeed, matter in the cores of neutron stars occupies a place in the proposed QCD phase diagram that is distinct from the regions occupied by matter in the early Universe and in modern heavy-ion colliders (Fig. 1.23; Hands 2001). At the same time, the gravitational fields probed by the accretion flows just outside the event horizons of black holes and the surfaces of neutron stars are many orders of magnitude stronger than those probed by other tests of general relativity (Fig. 1.24; Psaltis 2004). X-ray astronomy and the discovery of accretion-powered neutron stars and black holes provide probes with which tests of basic physics theories can be performed in a way that is complementary to other experiments and cosmological studies.

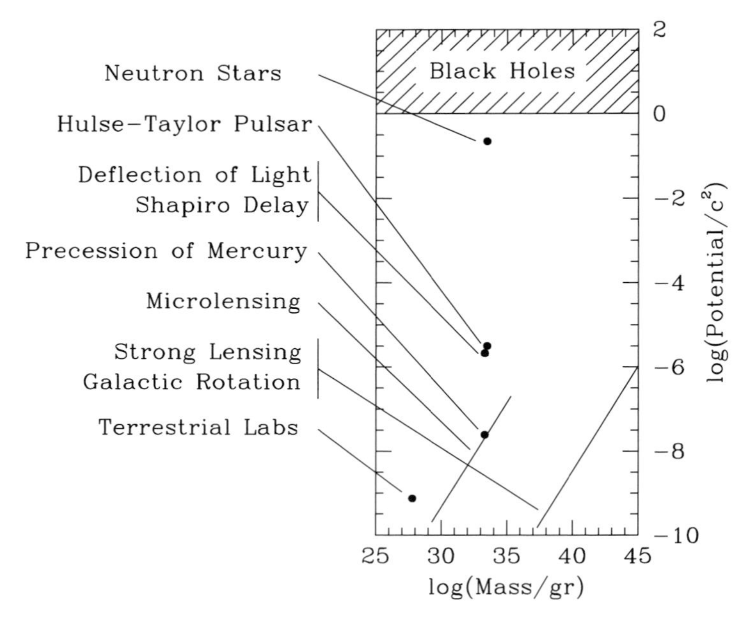

Fig. 1.24. The potential of the gravitational fields probed by different astrophysical observations and tests of general relativity (after Psaltis 2004).

The past four decades have been the period of discovery, in which the astrophysical properties of accreting compact objects were investigated. For his contribution to this effort, Riccardo Giaconni was awarded the 2002 Nobel Prize in physics. In the near future, the observations of neutron stars and black holes with detectors with large surface areas, high spectral resolution, and fast timing capabilities will allow for precise measurements of the physical conditions in the accretion flows, as they vary at the dynamical timescales near the compact objects; moreover, the increase in the computational power and storage capabilities of supercomputers will allow for the development of new tools for modeling radiation–magnetohydrodynamic phenomena in curved spacetimes; and, as has always been the case in compact-object astrophysics, this interplay between theory and observations will offer us a more complete picture of our universe.

Acknowledgements

It is my pleasure to thank a number of people that have helped me in understanding the concepts described in this chapter; I thank especially T. Belloni, D. Chakrabarty, S. DeDeo, D. Galloway, E. Kuulkers, F. Lamb, C. Miller, M. Muno, R. Narayan, F. Özel, and M. van der Klis. I am also grateful to Deepto Chakrabarty for help in planning, writing, and proofreading this chapter, to Duncan Galloway for producing several of the figures, as well as to Martin Pessah and Erik Kuulkers for carefully reading the manuscript.

References

Abramowicz, M. A., and Kluźniak, W. (2001), *A&A* **374**, L19
Abramowicz, M. A., Kluźniak, W., and Lasota, J.-P. (2002), *A&A* **396**, L31
Alcock, C., Farhi, E., and Olinto, A. (1986), *ApJ* **310**, 261

Alpar, M. A., and Shaham, J. (1985), *Nature* **316**, 239
Alpar, M. A., Cheng, A. F., Ruderman, M. A., and Shaham, J. (1982), *Nature* **300**, 728
Andersson, N., Kokkotas, K., and Schutz, B. F. (1999), *ApJ* **510**, 846
Armitage, P. J., Reynolds, C. S., and Chiang, J. (2001), *ApJ* **548**, 868
Bahcall, S., Lynn, B. W., and Selipsky, S. B. (1990), *ApJ* **362**, 251
Balbus, S. A., and Hawley, J. F. (1991), *ApJ* **376**, 214
(1998), *Rev. Mod. Phys.* **70**, 1
Balbus, S. A., and Papaloizou, J. C. B. (1999), *ApJ* **521**, 650
Barret, D., McClintock, J. E., and Grindlay, J. E. (1996), *ApJ* **473**, 963
Barziv, O., Kaper, L., van Kerkwijk, M. H., Telting, J. H., and Van Paradijs, J. (2001), *A&A* **377**, 925
Begelman, M. C., and Rees, M. J. (1984), *MNRAS* **206**, 209
Bhattacharya, D., and van den Heuvel, E. P. J. (1991), *Phys. Rep.* **203**, 1
Bildsten, L. (1998), *ApJ* **501**, L89
Bildsten, L., and Rutledge, R. E. (2000), *ApJ* **541**, 908
Bildsten, L. *et al.* (1997), *ApJS* **113**, 367
Blandford, R. D., and Begelman, M. C. (1999), *MNRAS* **303**, L1
Blandford, R. D., and Payne, D. G. (1982), *MNRAS* **199**, 883
Bradt, H., Levine, A. M., Remillard, R. A., and Smith, D. A. (2000), in *Multifrequency Behaviour of High Energy Cosmic Sources*, eds. F. Giovannelli and L. Sabau-Graziati, astro-ph/0001460
Brown, E. F., Bildsten, L., and Rutledge, R. E. (1998), *ApJ* **504**, L95
Campana, S., and Stella, L. (2000), *ApJ* **541**, 849
Chakrabarty, D., and Morgan, E. H. (1998), *Nature* **394**, 346
Chakrabarty, D., Homer, L., Charles, P. A., and O'Donoghue, D. (2001), *ApJ* **562**, 985
Chakrabarty, D. *et al.* (2003), *Nature* **424**, 42
Chou, Y., and Grindlay, J. E. (2001), *ApJ* **563**, 934
Coburn, W., Heindl, W. A., Rothschild, R. E. *et al.* (2002), *ApJ* **580**, 394
Cook, G., Shapiro, S. L., and Teukolsky, S. A. (1994), *ApJ* **424**, 823
Coppi, P. S. (1999), in *High Energy Processes in Accreting Black Holes*, eds. J. Poutanen and R. Svensson (San Francisco: ASP), 375
Corbel, S., Fender, R. P., Tzioumis, A. K. *et al.* (2000), *A&A* **359**, 251
Corbet, R. (1997), **IAUC** 6632
Cornelisse, R. *et al.* (2000), *A&A* **392**, 885
Cottam, J., Sako, M., Kahn, S. M., Paerels, F., and Liedahl, D. A. (2001), *ApJ* **557**, L101
Cottam, J., Paerels, F., and Mendez, M. (2002), *Nature* **420**, 51
Cowley, A. P. *et al.* (1991), *ApJ* **381**, 526
Cumming, A., and Bildsten, L. (2000), *ApJ* **559**, L127
Cumming, A., Zweibel, E., and Bildsten, L. (2001), *ApJ* **557**, 958
Cumming, A., Morsink, S. M., Bildsten, L., Friedman, J. L., and Holtz, D. E. (2002), *ApJ* **564**, 343, L127
DeDeo, S., and Psaltis, D. (2003), *PRL* **90**, 1101
de Jong, J. A., van Paradijs, J., and Augusteijn, T. (1996), *A&A* **314**, 484
De Villiers, J., Hawley, J. F., and Krolik, J. H. (2003), *ApJ* **599**, 1238
Di Salvo, T. *et al.* (2000), *ApJ* **544**, L119
Done, C., and Gierlinski M. (2003), *MNRAS* **342**, 1041
Dove, J. B., Wilms, J., and Begelman, M. C. (1997), *ApJ* **487**, 747
Dubus, G., Lasota, J., Hameury, J., and Charles, P. (1999), *MNRAS* **303**, 139
Eikenberry, S. S., Matthews, K., Murphy, T. W. *et al.* (1998), *ApJ* **506**, L31
Esin, A. A., McClintock, J. E., and Narayan, R. (1997), *ApJ* **489**, 865
Fabian, A. C., Rees, M. J., Stella, L., and White, N. E. (1989), *MNRAS* **238**, 729
Finger, M. H., Wilson, R. B., and Harmon, B. A. (1996), *ApJ* **459**, 288
Frank, J., King, A. R., and Raine, D. J. (2002), *Accretion Power in Astrophysics* (Cambridge University Press; 3rd edition)
Friedman, J. L., and Ipser, J. R. (1987), *ApJ* **314**, 594
Fushiki, I., and Lamb, D. Q. (1987), *ApJ* **323**, L55
Galloway, D., Chakrabarty, D., Morgan, E. H., and Remillard, R. A. (2002), *ApJ* **576**, L137
Galloway, D., Psaltis, D., Chakrabarty, D., and Muno, M. (2003), *ApJ* **590**, 999
Galloway, D. K., Cumming, A., Kuulkers, E. *et al.* (2004), *ApJ* **601**, 466
Gammie, C. F., McKinney, J. C., and Tóth, G. (2003), *ApJ* **589**, 444

Garcia, M. R., McClintock, J. E., Narayan, R. *et al.* (2001), *ApJ* **553**, L47
Ghosh, P., and Lamb, F. K. (1979), *ApJ* **234**, 296
(1991), in *Neutron Stars: Theory and Observations*, eds. J. Ventura and D. Pines (Dordrecht: Kluwer), 363
(1992), in *X-ray Binaries and Recycled Pulsars*, eds. E. P. J. van den Heuvel and S. A. Rappaport (Dordrecht: Kluwer), 487
Giacconi, R., Gursky, H., Paolini, F. R., and Rossi, B. B. (1962) *PRL* **9**, 439
Giacconi, R., Gursky, H., Kellogg, E., Schreier, E., and Tananbaum, H. (1971), *ApJ* **167**, L67
Gilfanov, M., Revnivtsev, M., Sunyaev, R., and Churazov, E. (1998), *A&A* **338**, L83
Glendenning, N. K. (1996), *Physics of Compact Objects* (Springer-Verlag)
(2003), *Compact Stars* (Springer-Verlag; 2nd edition)
Grimm, H.-J., Gilfanov, M., and Sunyaev, R. (2002), *A&A* **391**, 923
Grindlay, J., Gursky, H., Schnopper, H. *et al.* (1976), *ApJ* **205**, L127
Grove, J. E., Johnson, W. N., Kroeger, R. A. *et al.* (1998), *ApJ* **500**, 899
Hameury, J., Menou, K., Dubus, G., Lasota, J., and Hure, J. (1998), *MNRAS* **298**, 1048
Hands, S. (2001), *Cont. Phys.* **42**, 209
Harlaftis, E. T. (2001), in *Astrotomography, Indirect Imaging Methods in Observational Astronomy* (Springer-Verlag), 359
Hawley, J. F., and Krolik, J. H. (2001), *ApJ* **548**, 348
Heindl, W. A. *et al.* (1999), *ApJ* **521**, L49
Heindl, W. A. *et al.* (2004), in *X-ray Timing 2003: Rossi and Beyond*, eds. P. Kaaret, F. K. Lamb, and J. H. Swank (AIP), astro-ph/0403197
Heinz, S., and Nowak, M. A. (2001), *MNRAS* **320**, 249
Heyl, J. (2004), *ApJ* **600**, 939
Homer, L. *et al.* 2001, *MNRAS* **322**, 827
Horne, K. (1985), *MNRAS* **213**, 129
(2003), in *Astronomical Telescopes and Instrumentation*, astro-ph/0301250
Igumenshchev, I. V., Abramowicz, M. A., and Narayan, R. (2000), *ApJ* **537**, L27
Igumenshchev, I. V., Narayan, R., and Abramowicz, M. A. (2003) *ApJ* **592**, 1042
Illarionov, A. F., and Sunyaev, R. A. (1975), *A&A* **39**, 185
Jonker, P. G., Méndez, M., and van der Klis, M. (2002) *MNRAS* **336**, L1
Kallman, T., Boroson, B., and Vrtilek, S. D. (1998), *ApJ* **502**, 441
Kato, M. (1983), *PASJ* **35**, 33
Kato, S. (2001), *PASJ* **53**, 1
Kerkwijk, M. H. van, van Paradijs, J., and Zuiderwijk, E. J. (1995), *A&A* **303**, 497
Kerkwijk, M. H. van, Chakrabarty, D., Pringle, J. E., and Wijers, R. A. M. J. (1998), *ApJ* **499**, L27
King, A. R., and Kolb, U. (1996), *ApJ* **481**, 918
King, A. R., Kolb, U., and Burderi, L. (1996), *ApJ* **464**, L127
Klis, M. van der *et al.* (1996), *ApJ* **469**, L1
Klis, M. van der (1994), *ApJS* **92**, 511
(2000), *ARA&A* **38**, 717
Kluzniak, W., Michelson, P., and Wagoner, R. V. (1990), *ApJ* **358**, 538
Kommers, J. M., Chakrabarty, D., and Lewin, W. H. G. (1988), *ApJ* **497**, L33
Kong, A. K. H., Charles, P. A., and Kuulkers, E. 1998, *New Astr.* **3**, 301
Krolik, J. H., and Hawley, J. F. (2002), *ApJ* **573**, 574
Kulkarni, S. R., and Narayan, R. (1988), *ApJ* **335**, 755
Kuulkers, E., van der Klis, M., and Vaughan, B. A. (1996), *A&A* **311**, 197
Kuulkers, E., Wijnands, R., and van der Klis, M. (1999), *MNRAS* **308**, 485
Kuulkers, E. *et al.* (2003), *A&A* **399**, 663
Lasota, J.-P. (2000), *A&A* **360**, 575
Laurent, P., and Titarchuk, L. (1999), *ApJ* **511**, 289
Lewin, W., van Paradijs, J., and Taam, R. (1996), in *X-ray Binaries*, eds. W. H. G. Lewin, J. van Paradijs, and E. P. J van den Heuvel (Cambridge University Press)
Liedahl, D. A. (1999), in *X-ray Spectroscopy in Astrophysics*, eds. J. van Paradijs and J. Bleeker, (Springer), 189
Liu, Q. Z., van Paradijs, J., and van den Heuvel, E. P. J. (2001), *A&A* **368**, 1021
Lorimer, D. R. (1995), *MNRAS* **274**, 300
Lovelace, R. V. E., Romanova, M. M., and Bisnovatyi-Kogan, G. S. (1999), *ApJ* **514**, 368

Maloney, P. R., Begelman, M. C., and Pringle, J. E. (1996), *ApJ* **472**, 582
Margon, B. (1984), *ARA&A* **22**, 507
Margon, B., and Anderson, S. F. (1999), *ApJ* **347**, 448
Markoff, S., Falcke, H., and Fender, R. (2001), *A&A* **372**, L25
Markwardt, C. B., and Swank, J. H. (2003), *IAUC* 8144
Markwardt, C. B., Swank, J. H., Strohmayer, T. E., in't Zand, J. J. M., and Marshall, F. E. (2002), *ApJ* **575**, L21
Markwardt, C. B., Smith, E, and Swank, J. H. (2003), *ATEL*, 122
Marsh, T. R. (2000), in *Lecture Notes in Physics: Astrotomography*, eds. H. Boffin and D. Steeghs (Springer-Verlag), astro-ph/0011020
Marshall, H. L., Canizares, C. R., and Schulz, N. S. (2002), *ApJ* **564**, 941
McClintock, J. E., and Remillard, R. A. (1986), *ApJ* **308**, 110
McClintock, J. E. *et al.* (2001), *ApJ* **555**, 477
McKinney, J. C., and Gammie, C. F. (2002), *ApJ* **573**, 728
Meszaros, P. (1992), *High Energy Radiation from Magnetized Neutron Stars* (University of Chicago Press)
Miller, J. C., Shahbaz, T., and Nolan, L. A. (1998), *MNRAS* **294**, L25
Miller, J. M. *et al.* (2002a), *ApJ* **570**, L69
Miller, J. M. *et al.* (2002b), *ApJ* **578**, 348
Miller, M. C., Lamb, F. K., and Psaltis, D. (1998), *ApJ* **508**, 791
Mirabel, I. F., and Rodriguez, L. F. (1994), *Nature* **371**, 46
Mirabel, I. F., Bandyopadhyay, R., Charles, P. A., Shahbaz, T., and Rodriguez, L. F. (1997), *ApJ* **477**, L45
Mirabel, I. F. *et al.* (1998), *A&A* **330**, L9
Muno, M., Chakrabarty, D., Galloway, D., and Psaltis, D. (2002a), *ApJ* **580**, 1048
Muno, M., Özel, F., and Chakrabarty, D. (2002b), *ApJ* **581**, 550
Narayan, R., and Heyl, J. (2002), *ApJ* **575**, L139
Narayan, R., and Yi, I. (1994), *ApJ* **428**, L13
Narayan, R., Yi, I., and Mahadevan, R. (1995), *Nature* **374**, 623
Narayan, R., McClintock, J. E., and Yi, I. (1996), *ApJ* **457**, 821
Narayan, R., Igumenshchev, I. V., and Abramowicz, M. A. (2000), *ApJ* **539**, 798
Narayan, R., Garcia, M. R., and McClintock, J. E. (2001), in *Proc. IX Marcel Grossmann Meeting*, eds. V. Gurzadyan, R. Jantzen, and R. Ruffini (Singapore: World Scientific)
Nath, N. R., Strohmayer, T. E., and Swank, J. H. (2002), *ApJ* **564**, 353
Nelson, R. *et al.* (1998), *ApJ* **488**, L117
Nobili, L., Turolla, R., and Lapidus, I. (1994), *ApJ* **433**, 276
O'Brien, K., Horne, K., Hynes, R. I. *et al.* (2002), *MNRAS* **334**, 426
Ogilvie, G. I., and Dubus, G. (2001), *MNRAS* **320**, 485
Orosz, J. A. (2002) in *A Massive Star Odyssey, from Main Sequence to Supernova*, eds. K. A. van der Hucht, A. Herraro, and C. Esteban (San Francisco: ASP)
Orosz, J. A. and Kuulkers, E. (1999), *MNRAS* **305** 132
Paradijs, J. van (1996), *ApJ* **464**, L139
Paradijs, J. van, and McClintock J. (1995), in *X-ray Binaries*, eds. W. H. G. Lewin, J. van Paradijs, and E. P. J. van den Heuvel (Cambridge University Press)
Parmar, A. N., White, N. E., Stella, L., Izzo, C., and Ferri, P. (1989), *ApJ* **338**, 359
Paul, B., Kitamoto, S., and Makino, F. (2000), *ApJ* **528**, 410
Piran, T. (1978), *ApJ* **221**, 652
Priedhorsky, W. C., and Terrell, J. (1983), *ApJ* **273**, 709
(1984), *ApJ* **284**, L17
Priedhorsky, W. C., Terrell, J., and Holt, S. S. (1983), *ApJ* **270**, 233
Pringle, J. E. (1981), *AR&AA* **19**, 137
(1996), *MNRAS* **281**, 357
Psaltis, D. (2004), in *X-ray Timing 2003: Rossi and Beyond*, eds. P. Kaaret, F. K. Lamb, and J. H. Swank (AIP), astro-ph/0402213
Psaltis, D., and Chakrabarty, D. (1998), *ApJ* **521**, 332
Psaltis, D., and Lamb, F. K. (1998), in *Neutron Stars and Pulsars*, eds. N. Shibazaki, N. Kawai, S. Shibata and T. Kifune (Tokyo: Universal Academy Press), 179
Psaltis, D., and Norman, C. (1999), astro-ph/0001391
Psaltis, D., Belloni, T., and van der Klis, M. (1999), *ApJ* **520**, 262

Quataert, E., and Gruzinov, A. (1999), *ApJ* **520**, 248
Radhakrishnan, V., and Shrinivasan, G. (1982), *Curr. Sci.* **51**, 1096
Rhoades, C. E., and Ruffini, R. (1974), *Phys. Rev. Lett.* **32**, 324
Rutledge, R. E., Bildsten, L., Brown, E. F., Pavlov, G. G., and Zavlin, V. E. (2001), *ApJ* **551**, 921
Sabbadini, A. G., and Hartle, J. B. (1977), *Ann. Phys. (N.Y.)* **104**, 95
Sano, T., Inutsuka, S., Turner, N. J., and Stone, J. M, (2004), *ApJ* **605**, 321
Schatz, H., Bildsten, L., Cumming, A., and Wiescher, M. (2000), *ApJ* **524**, 1014
Schulz, N. S., Chakrabarty, D., Marshall, H. L. *et al.* (2001), *ApJ* **563**, 941
Shakura, N. I., and Sunyaev, R. A. (1973), *A&A* **24**, 337
Shapiro, S. L., and Teukolsky, S. (1983), *Black Holes, White Dwarfs and Neutron Stars* (Wiley-Interscience)
Shapiro, S. L., Lightman, A. P., and Eardley, D. M. (1976), *ApJ* **204**, 187
Shirakawa, A., and Lai, D. 2002, *ApJ* **565**, 1134
Shirey, R. E., Bradt, H. V., Levine, A. M., and Morgan, E. H. (1996), *ApJ* **469**, L21
Smale, A. P., and Lochner, J. C. (1992), *ApJ* **395**, 582
Spencer, R. E. (1979), *Nature* **282**, 483
Spitkovsky, A., Ushomirsky, G., and Levin, Y. (2002), *ApJ* **566**, 1018
Stella, L., Vietri, M., and Morsink, S. M. (1999), *ApJ* **524**, L63
Stone, J. M., and Pringle, J. E. (2001), *MNRAS* **322**, 461
Stone, J. M., Pringle, J. E., and Begelman, M. C. (1999), *MNRAS* **310**, 1002
Strohmayer, T. E. (2001a), *ApJ* **552**, L49
(2001b), *ApJ* **554**, L169
Strohmayer, T. E, and Brown, E. F. (2002), *ApJ* **566**, 1045
Strohmayer, T. E. *et al.* (1996), *ApJ* **469**, L9
Sunyaev, R., and Revnivtsev, M. (2000), *A&A* **358**, 617
Taam, R. E., and Sandquist, E. L. (2000), *ARA&A* **38**, 113
Thorsett, S. E., and Chakrabarty, D. (1999), *ApJ* **512**, 288
Titarchuk, L., Lapidus, I., and Muslimov, A. (1998), *ApJ* **499**, 315
Torkelsson, U. (1998), *MNRAS* **298**, L55
Trumper, J., Pietsch, W., Reppin, C. *et al.* (1978), *ApJ* **219**, L105
Vaughan, B. *et al.* (1994), *ApJ* **435**, 362
Vrielmann, S. (2000), in *Lecture Notes in Physics: Astrotomography*, eds. H. Boffin and D. Steeghs (Springer-Verlag), astro-ph/0012263
Vrtilek, S. D., Raymond, J. C., Garcia, M. R. *et al.* (1990), *A&A* **235**, 162
Wagoner, R. W. (1999), *Phys. Rep.* **311**, 259
Wang, Z. *et al.* (2001), *ApJ* **563**, L61
Waters, L. B. F. M., and van Kerkwijk, M. H. (1989), *A&A* **223**, 196
White, N. E., and Holt, S. S. (1982), *ApJ* **257**, 318
White, N. E., and Swank, J. H. (1982), *ApJ* **253**, L61
White, N. E., Nagase, F., and Parmar, A. N. (1995), in *X-ray Binaries*, eds. W. H. G. Lewin, J. van Paradijs, and E. P. J. van den Heuvel (Cambridge University Press)
Wijnands, R., and van der Klis, M. (1998), *Nature* **394**, 344
Wijnands, R. A. D., Kuulkers, E., and Smale, A. P. (1996), *ApJ* **473**, L45
Will, C. M. (2001), *Living Rev. Relativity* **4**, cited on 15 Aug 2001 (http://www.livingreviews.org/Articles/Volume4/2001-4will/)
Wilms, J. *et al.* (2001), *MNRAS* **320**, 327
Witten, E. (1984), *PRD* **30**, 272.
Wojdowski, P. S. *et al.* (1998), *ApJ* **502**, 253
Yi, I., and Wheeler, J. G. (1998), *ApJ* **498**, 802
Zhang, W., Morgan, E. H., Jahoda, K. *et al.* (1996), *ApJ* **469**, L29
Zhang, W., Smale, A. P., Strohmayer, T. E., and Swank, J. H. (1998), *ApJ* **500**, L171
Zingale, M. *et al.* (2001), *ApJS* **133**, 195

2

Rapid X-ray variability

M. van der Klis
Astronomical Institute Anton Pannekoek, University of Amsterdam

2.1 Introduction

One of the principal motivations for studying X-ray binaries is that accretion onto neutron stars and black holes provides a unique window on the physics of strong gravity and dense matter. Our best theory of gravity, general relativity, while tested, and confirmed, with exquisite precision in weak fields ($GM/R \ll c^2$; e.g., Taylor *et al.* 1992) has not yet been tested by direct observation of the motion of particles in the strong gravitational field near compact objects, where the gravitational binding energy is of order the rest mass. Among the extreme predictions relativity makes for these regions are the existence of event horizons, i.e., black holes (Section 2.4.1), the existence of an inner radius within which no stable orbits exist, strong dragging of inertial frames, and general-relativistic precession at rates similar to the orbital motion itself, $\sim 10^{16}$ times as fast as that of Mercury.

In a neutron star the density exceeds that in an atomic nucleus. Which elementary particles occur there, and what their collective properties are, is not known well enough to predict the equation of state (EOS), or compressibility, of the matter there, and hence the mass–radius (M–R) relation of neutron stars is uncertain. Consequently, by measuring this relation, the EOS of supra-nuclear density matter is constrained. As orbital motion around a neutron star constrains both M and R (Section 2.8.1), measurements of such motion bear on the fundamental properties of matter. Likewise, such motion near black holes constrains the size and spin of black holes of given mass.

For addressing these issues of strong gravity and dense matter, we need to study motion under the influence of gravity within a few Schwarzschild radii[1] of compact objects and map out the strongly curved spacetime there. As the characteristic velocities near the compact object are of order $(GM/R)^{1/2} \sim 0.5c$, the dynamical timescale $(r^3/GM)^{1/2}$ for the motion through this region is short; ~0.1 ms at ~15 km, and ~2 ms at 10^2 km from a 1.4 $M_\odot$ neutron star, and ~1 ms at $3R_{\rm Schw}$ (~10^2 km) from a 10 $M_\odot$ black hole. These millisecond dynamical timescales, the shortest associated with any astrophysical object, form one of the most basic expressions of the compactness of compact objects.

The accretion flow is expected to be turbulent and may show magnetic structures. Its emission will vary in time due to the motions of inhomogeneities through, and with, the flow. This variability can be used to probe the accretion-flow dynamics. For a 10 km object, 90% of the gravitational energy is released in the inner ~10^2 km, hence the bulk of the emission likely comes from within the strong-field region from where we expect the millisecond variability. Temperatures here are $\gtrsim 10^7$ K, so most of this emission is in X-rays.

[1] The radius of a zero-angular-momentum black hole, $R_{\rm Schw} = 2r_{\rm g} = 2GM/c^2 \approx 3\,{\rm km}\,M/M_\odot$.

Compact Stellar X-Ray Sources, eds. Walter Lewin and Michiel van der Klis.
Published by Cambridge University Press.

The transfer of matter towards the compact object usually occurs by way of an accretion disk in which the matter moves in near-Keplerian orbits (Section 13.2). However, the geometry of the innermost part of the flow is uncertain. In many models the Keplerian disk extends down to well into the strong-field region. It is terminated at an inner radius $r_{\rm in}$ of a few $R_{\rm Schw}$ by, for example, relativistic effects, radiation drag, a weak magnetic field (or the neutron-star surface). Advective, or in the case of strongly magnetic neutron stars, magnetically dominated flows feature larger ($\sim 10^2$ $R_{\rm Schw}$) inner disk radii. Within $r_{\rm in}$ the flow is no longer Keplerian, and may or may not be disk-like. Both inside and outside $r_{\rm in}$ matter may leave the disk plane and either flow in more radially, or be expelled. Together these flows constitute what is called the "accretion flow" in this chapter.

The radiation from the Keplerian disk and from a neutron-star surface are expected to be basically thermal, and observations indeed show such thermal X-rays. In addition there is ubiquitous evidence for non-thermal spectral components which may originate in one of the non-disk flows and/or in an energetic "corona" of uncertain geometry (Section 2.5.1) associated with the disk.

Observations of X-ray binaries show considerable variability on a wide range of timescales in all wavelengths, and down to less than a millisecond in X-rays. The study of this variability is called "timing". In this chapter we focus on aperiodic phenomena (QPOs and noise, Section 2.2) that are potential probes of the strong-gravity dominated flow dynamics, i.e., millisecond aperiodic phenomena in weakly magnetic compact objects, with particular attention to the potential for measuring fundamental properties of spacetime and matter. As observationally there are correlated spectral and timing phenomena covering a range of timescales, we look at longer timescale phenomena and relations with X-ray spectral properties as well.

That millisecond variability will naturally occur in the process of accretion of matter onto a stellar-mass compact object is an insight that dates back to at least Shvartsman (1971). Sunyaev (1973) noted that clumps orbiting in an accretion disk closely around a black hole could cause quasi-periodic variability on timescales of about a millisecond. Twenty-five years after these early predictions, millisecond variability was finally discovered, with NASA's breakthrough Rossi X-Ray Timing Explorer (RXTE; Bradt *et al.* 1993, see Sections 2.9.1, 2.10.1, 3.4, and Chapter 1). To this day, RXTE is providing a veritable flood of timing information that is still only partially digested. Hence, contrary to the situation ten years ago it is no longer possible to be exhaustive when reviewing X-ray binary timing (see, e.g., Lewin *et al.* 1988; Stella 1988; Hasinger 1988; Miyamoto 1994; van der Klis 1986, 1989a, 1995a,b, 2000 for preceding timing reviews). In this chapter, Sections 2.2–2.7 provide a broad overview of the phenomenology. We look for common traits in the phenomena of different classes of objects indicating common physics. In Sections 2.8 and 2.12 we examine the ideas that have been put forward to explain the phenomena by, respectively, orbital motions and accretion-flow instabilities. Sections 2.9 and 2.10 give more detail on individual objects and phenomena, with Sections 2.9.1 and 2.10.1 dealing with the most rapid (millisecond) phenomena and the remaining sections summarizing the work on slower variability.

2.2 Timing

The rapid variations diagnosing the inner accretion flow are stochastic, and most effectively dealt with using statistical (random-process) techniques. Fourier analysis is the dominant tool, and the one we focus on here. Some other techniques are mentioned as well.

2.2.1 Fourier analysis

The Fourier power spectrum of the X-ray flux time series provides an estimate of the variance as a function of Fourier frequency ν in terms of the *power density* $P_\nu(\nu)$ (van der Klis 1989b for details). The usual range of ν is mHz to kHz; slower variations are usually studied in the time domain (but see Reig *et al.* 2002, 2003a), as on longer timescales source-state changes and data gaps cause trouble for Fourier techniques. Variations faster than those in the kHz range have not (yet) been detected.

A number of *variability components* or *power-spectral components* together make up the power spectrum (see, e.g., Fig. 2.8). An aperiodic component by definition covers several, usually many, frequency resolution elements. Broad structures are called *noise* and narrow features *quasi-periodic oscillations* (QPOs); "broad-band noise" and "QPO peaks" are common terms. Least-squares fitting techniques are used to measure these components. When a series of power spectra is calculated from consecutive chunks of data, usually the components change: they move through the power spectrum (change frequency from one power spectrum to the next), vary in width and strength, etc., but they remain identifiable as the changes are gradual. This is the empirical basis for the concept of a power-spectral component. The shortest timescale on which changes can typically be followed is seconds to minutes.

The signal in the time series is not completely specified by the power spectrum (and the signals are usually too weak to recover the Fourier phases). The same QPO peak could be due to, e.g., a damped harmonic oscillator, randomly occurring short wave trains, a frequency-modulated oscillation, an autoregressive signal, white noise observed through a narrow pass-band filter or even a closely spaced set of periodic signals. *Time lags* (delays) between signals simultaneously detected in different energy bands can be measured using the cross-correlation function (CCF; Brinkman *et al.* 1974; Weisskopf *et al.* 1975), but if the lags at different timescales differ, the *cross-spectrum* (van der Klis *et al.* 1987c; Miyamoto *et al.* 1988; Vaughan *et al.* 1994; Nowak *et al.* 1999a) performs better. This is the Fourier transform of the CCF and in a sense its frequency-domain equivalent. It measures a *phase lag* (time lag multiplied by frequency) at each frequency. The term *hard lag* means that higher energy photons lag lower energy ones, and vice versa for *soft lag*. *Cross-coherence*[2] is a measure for the correlation between the signals (Vaughan & Nowak 1997).

Power-law noise is noise that (in the frequency range considered) follows a power law $P_\nu \propto \nu^{-\alpha}$. The power-law index (also "slope") α is typically between 0 and 2; for $|\alpha| > 2$ Fourier analysis suffers from power leakage, so measurements of noise steeper than that are suspect (e.g., Bracewell 1986; Deeter 1984). *1/f noise* has $\alpha = 1$, and *white noise* is constant ($\alpha = 0$). *Red noise* is a term variously used for either $\alpha = 2$ power-law noise or any kind of noise whose P_ν decreases with ν.

Band-limited noise (BLN) is defined here as noise that steepens towards higher frequency (i.e., its local power-law slope $-\mathrm{d}\log P_\nu/\mathrm{d}\log\nu$ increases with ν) either abruptly (showing a "break" at *break frequency* ν_{break}) or gradually. BLN whose power density below a certain frequency is approximately constant (white) is called *flat-topped noise*. The term *peaked noise* is used for noise whose P_ν has a local maximum at $\nu > 0$. Various modified power laws (broken, cutoff) as well as broad Lorentzians are used to describe BLN. The precise value of the characteristic frequency associated with the steepening (e.g., ν_{break})

[2] Often just called "coherence", a term that is also used for the sharpness of a QPO peak, below.

differs by factors of order unity depending on the description chosen (e.g., Belloni *et al.* 2002a).

A *quasi-periodic oscillation* (QPO) is a finite-width peak in the power spectrum. It can usually be described with a Lorentzian $P_\nu \propto \lambda/[(\nu - \nu_0)^2 + (\lambda/2)^2]$ with *centroid frequency* ν_0 and full width at half maximum (FWHM) λ. This is the power spectrum of an exponentially damped sinusoid $x(t) \propto e^{-t/\tau} \cos(2\pi\nu_0 t)$, but the underlying signal may well be different from this. λ is related to the *coherence time* $\tau = 1/\pi\lambda$ of the signal, and is often reported in terms of the *quality factor* $Q \equiv \nu_0/\lambda$, a measure for the *coherence* of the QPO. Conventionally, signals with $Q > 2$ are called QPOs and those with $Q < 2$ peaked noise. A *sharp* QPO peak is one with high Q.

The strength (variance) of a signal is proportional to the integrated power $P = \int P_\nu \, d\nu$ of its contribution to the power spectrum, and is usually reported in terms of its *fractional root-mean-squared (rms) amplitude* $r \propto P^{1/2}$, which is a measure for signal amplitude as a fraction of the source flux. It is often expressed in percent, as in "2% (rms)".

The signal-to-noise of a weak QPO or noise component is $n_\sigma = \frac{1}{2} I_x r^2 (T/\lambda)^{1/2}$ (van der Klis 1989b; see van der Klis 1998 for more details), where I_x is the count rate and T the observing time (assumed $\gg 1/\lambda$). As n_σ is proportional to signal amplitude *squared*, if a clear power-spectral feature "suddenly disappears" it may have only decreased in amplitude by a factor of 2 (and gone from, say, 6σ to 1.5σ).

Red and flat-topped BLN may have only one characteristic frequency (e.g., ν_{break}), but peaked noise and a QPO have two (ν_0 and λ). This leads to difficulties in describing the phenomenology when, over time, power-spectral components change in Q between noise and QPO. For this reason, defining the characteristic frequency as $\nu_{\max}$, the frequency at which power-density times frequency (νP_ν, equivalent to νS_ν in spectroscopy) reaches its maximum, has gained some popularity (Belloni *et al.* 2002a). This method is equally applicable to centroids of narrow peaks and breaks in broad-band noise, and smoothly deals with intermediate cases (see Section 2.8.6 for an interpretation). For a Lorentzian, $\nu_{\max} = \sqrt{\nu_0^2 + \Delta^2}$, where $\Delta \equiv \lambda/2$, so a narrow QPO peak has $\nu_{\max} \approx \nu_0$ and a BLN component described by a zero-centered Lorentzian $\nu_{\max} = \Delta$.

Power spectra are in practice presented in a variety of ways, each more suitable to emphasize particular aspects (displaying either P_ν or νP_ν, linearly or logarithmically, subtracting the white Poisson background noise or not; e.g., Figs. 2.6 and 2.8).

2.2.2 *Time-series modeling*

Time-series modeling goes beyond Fourier analysis in an attempt to obtain more detailed information about what is going on in the time domain. One approach is to invent, based on physical hunches, synthetic time series that reproduce observed statistical properties of the variability. Shot noise and chaos (below) are common hunches. Another approach is to refine statistical analysis beyond Fourier analysis. Here, systematic approaches involving higher order statistics, such as skewness and bi-spectra (e.g., Priedhorsky *et al.* 1979; Elsner *et al.* 1988; Maccarone & Coppi 2002b) as well as more heuristic ones (e.g., shot alignment, below) have been tried. "Variation functions" measuring variance as a function of timescale τ (e.g., Ogawara *et al.* 1977; Maejima *et al.* 1984; Li & Muraki 2002) do not contain additional information as is sometimes claimed; they basically provide power integrated over frequencies $<1/\tau$. The calculation of Fourier power spectra on short timescales is a useful technique that has produced interesting results (e.g., Norris *et al.* 1990; Yu *et al.* 2001; Yu & van der Klis 2002; Uttley & McHardy 2001).

Shot noise is a time-series model of randomly occurring identical discrete finite events called shots (Terrell 1972; Weisskopf *et al.* 1975; Sutherland *et al.* 1978; Priedhorsky *et al.* 1979; Lochner *et al.* 1991). The power spectrum is that of an individual shot. Mathematically, the power spectrum of a random process can always be modeled in this way (Doi 1978), so for modeling power spectra the method has little predictive power, but it can be physically motivated in various settings (Section 2.12.3). Heuristic "shot alignment" techniques (Negoro *et al.* 1994, 1995, 2001; Feng *et al.* 1999), like any "fishing in the noise" technique, must be applied with care: they can be misleading if the underlying assumption (i.e., shot noise) is incorrect. Various modifications of pure shot noise have been explored, e.g., involving distributions of different shot profiles. As pointed out by Vikhlinin *et al.* (1994), if the shot occurrence times are correlated (e.g., a shot is less likely to occur within a certain interval of time after the previous shot, certainly plausible in magnetic flare scenarios; Section 2.12.3), then peaked noise or a QPO is produced whose Q increases as the correlation between the shot times increases. Oscillating shots, short wave trains with positive integrated flux, produce a QPO (due to the oscillation) and BLN (due to the shot envelope) in the power spectrum (Lamb *et al.* 1985; Alpar 1986; Shibazaki & Lamb 1987; Elsner *et al.* 1987, 1988; Shibazaki *et al.* 1987, 1988), and fit well within models involving short-lived orbiting clumps (Section 2.8.2).

Given our limited a-priori knowledge of the physical processes producing the rapid variability, mostly mathematically motivated time-series models such as autoregressive, linear phase-state and chaos models (e.g., Lochner *et al.* 1989; Unno *et al.* 1990; Scargle *et al.* 1993; Pottschmidt *et al.* 1998; Timmer *et al.* 2000) usually do not suffiently constrain the physics to conclude much from them (but see Section 2.12.3).

2.3 Spectroscopy

Recent work with Chandra and XMM-Newton suggests that, as previously suspected (e.g., Barr *et al.* 1985; White *et al.* 1985), relativistically broadened Fe lines near 6.5 keV similar to those inferred in AGNs (e.g., Fabian *et al.* 2000) occur in some neutron-star and black-hole binaries as well (e.g., Miller *et al.* 2002; Parmar *et al.* 2002; Section 4.2.3). The gravitational and Doppler distortions of these lines diagnose the dynamics of the same strong-field region as millisecond timing (e.g., Reynolds & Nowak 2003). Combining timing with such spectroscopic diagnostics can enormously improve the grip we have on what is going on in the inner disk, but such work is still in a very early stage, and here we concentrate on the better-explored link between timing and *broad-band spectroscopy*.

Variations in broad-band ($\Delta E/E > 0.1$, usually continuum-dominated) X-ray spectral shape usually just become detectable on the same seconds to minutes timescales on which power-spectral changes are detectable (Section 2.2). Two different techniques of broad-band spectroscopy are used to diagnose these changes: multi-band photometry and spectral fitting. The spectral band used varies somewhat but is usually in the 1–60 keV range and nearly always covers at least the 3–8 keV band.

2.3.1 *Photometric method*

One approach to quantifying broad-band X-ray spectral shape uses *X-ray colors*. An X-ray color is a "hardness" ratio between the photon counts in two broad bands; it is a rough measure for spectral slope. By calculating two X-ray colors (a *hard color* in a higher energy band and a *soft color* in a lower band) as a function of time, a record is obtained of the broad-band X-ray spectral variations that is well matched to the power-spectral variations. Plotted versus one another in a *color–color diagram* (CD), one can observe the source to move through the diagram and, nearly always, create a pattern. It is then possible to study the relation between timing and location in the CD. A *hardness–intensity diagram* (HID) or color–intensity diagram is a similar diagram with a color vs. "intensity"; *X-ray intensity* in this context is nothing but a count rate in some broad X-ray spectral band. Figures 2.4 and 2.5 provide examples of CD and HID patterns. Whether CD or HID presents the "cleanest" pattern depends on source, and on the quality of the data. There are distinct advantages to working with the logarithm of colors and intensities, like magnitudes in the optical, but this is not general practice.

For distinguishing between source states (Section 2.5) the photometric method combined with timing performs well, and the method provides excellent sensitivity to subtle spectral variations. However, this is at the expense of detector dependence. All X-ray detectors are different and change over time, and, contrary to optical photometry, in X-rays there is only one bright standard star (Crab). For the low spectral resolution detectors typically used for timing (proportional counters) it is not possible to completely correct X-ray colors for the detector response. In the absence of a-priori knowledge of the intrinsic spectral shape, all correction methods intended to derive "intrinsic" colors (scaling by Crab colors, unfolding through the detector response matrix either directly or by fitting an arbitrary model) are mathematically imperfect (Kahn & Blisset 1980; Kuulkers *et al.* 1994; Kuulkers 1995; Done & Gierliński 2003).

2.3.2 Spectral fitting method

There are considerable interpretative advantages in describing the X-ray spectral variations instead in terms of physical models fitted to the observed spectra. The drawback is that this involves a description of spectral shape in terms of more numbers (the spectral parameters) than just two X-ray colors, and that the correct models are unknown. This, plus the tendency towards spurious results related to statistical-fit-parameter correlations masking true source variations, usually means that, to obtain sensible results, fitting requires longer integration times than the timescale on which we see the power spectra, and the X-ray colors, change, and that measured parameter-value changes obtained are hard to interpret. Nevertheless, for sufficiently long integration times, and particularly in the case of the black-hole candidates, where the X-ray spectral variations are clearer than in neutron stars, this approach has met with some success. Because both methods have limitations, in practice the interpretation of X-ray spectral variations tends to occur back-and-forth between CD/HIDs and spectral fitting. Techniques such as plotting the predicted colors of model spectra in CDs together with the data, fitting spectra obtained by averaging data collected at different epochs but in the same part of the CD, and approximate intrinsic colors (e.g., Belloni *et al.* 2000; Done & Gierliński 2003), are all being employed to make the link between the two methods.

2.3.3 Presentation and parametrization of CD/HIDs

There is unfortunately no uniformity in either the presentation or the choice of X-ray spectral bands for CD/HIDs. In particular, there is a tendency to present the black-hole diagrams transposed as compared to the neutron-star ones. As illustrated by Jonker *et al.* (2002a) these differences can obscure some of the similarities between neutron-star and black-hole X-ray spectral behavior.

For sources which trace out one-dimensional *tracks* in the CD/HIDs along which a source moves smoothly, position in the track, in a single datum, summarizes its spectral state. Curve length along the track (Hertz *et al.* 1992; Kuulkers *et al.* 1994) is conventionally indicated with a symbol S (S_Z, S_a; see Section 2.5.2). S is defined relative to the track: if the track drifts, the colors of a point with given S_x drift along with it. The patterns observed in a CD are often also recognizable in the corresponding HID obtained by replacing soft color by intensity. Presumably this is the case because intensity is dominated by the more numerous photons in the lower bands, whose dominant thermal component(s) strongly correlate to temperature and hence color.

2.4 Source types

Different sources exhibit similar patterns of timing and spectral properties, allowing to group them into a number of *source types*. Both spectroscopy and timing, preferably in several source states (Section 2.5), are necessary to reliably identify source type. The primary distinction is not between neutron stars and black holes (Section 2.4.1) but between high and low magnetic-field strengths. Black holes and low magnetic-field ($\lesssim 10^{10}$ G) neutron stars potentially have gravity-dominated flows down to the strong-field region (although radiative stresses can also be important, Section 2.8.5). It is these objects, mostly found in low-mass X-ray binaries (LMXBs, Chapter 1), that we shall be mostly concerned with. Strongly magnetic ($\gtrsim 10^{12}$ G) neutron stars are briefly discussed in Section 2.11.

2.4.1 *Neutron stars vs. black holes*

A neutron star is only a few times larger than its Schwarzschild radius, so accretion onto black holes and neutron stars is expected to show similarities. Indeed, based on just the inner-flow diagnostics (timing and spectrum), the distinction between neutron stars and black holes is notoriously difficult to make, with some examples of black-hole candidates turned neutron stars (e.g., Cir X-1, Jones *et al.* 1974, Tennant *et al.* 1986; V0332+53, Tanaka *et al.* 1983, Stella *et al.* 1985; GS 1826–238, Tanaka 1989, Ubertini *et al.* 1999; 4U0142+61, White & Marshall 1984, Israel *et al.* 1994), and in practice it is not easy to prove that a compact object is a black hole. There are two levels of proof: (i) showing that, assuming general relativity, the compact object must be a black hole, and (ii) showing that such an object indeed has the properties general relativity predicts for a black hole. For (i) it is sufficient to prove that a mass is concentrated within its Schwarzschild radius; currently, this mostly relies on dynamical mass estimates combined with theoretical arguments about the maximum mass of a neutron star (e.g., Srinivasan 2002; Section 1.3.8) and not on direct, empirical measurements of radius. For (ii) it is necessary to observe the interaction of the compact object with its surroundings, empirically map out its exterior spacetime, and demonstrate properties such as extreme frame dragging, the existence of an innermost stable orbit, and an event horizon.

In recent work, for well-studied sources, clear differences are discerned between the patterns of correlated spectral and timing behavior (Section 2.5) of neutron stars and black holes which agree with distinctions based on X-ray bursts, pulsations, and dynamical mass estimates. In particular, the most rapid variability (see Section 2.6.1), presumably that produced

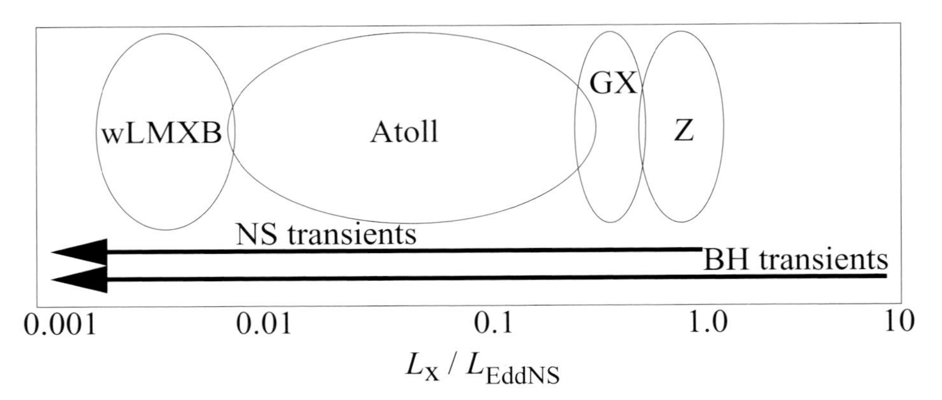

Fig. 2.1. Luminosities attained by Z sources, GX atoll sources, ordinary atoll sources and weak LMXBs, respectively, as well as by neutron-star and black-hole transients. The extent of the L_X overlaps between these source types is undecided in detail, but those shown here are likely.

closest to the compact object and most affected by its properties, is clearly quite different between neutron stars and black-hole candidates. Evidently we are learning to distinguish between black holes and neutron stars based on the properties of the flow in the strong-field gravity region, so there is progress towards the goal of testing general relativity. On the other hand, remarkable spectral and timing similarities exist between certified neutron stars and black-hole candidates, particularly in low luminosity states (Section 2.6.2). To demonstrate the existence of black holes in the sense of general relativity based on an understanding of the accretion phenomena near them is a goal that has not yet been reached. The study of rapid X-ray variability of low-magnetic-field compact objects in X-ray binaries is one of the programs contributing towards this end. Both the actual measurement of compact-object radius, which is part of a level (i) proof, as well as level (ii) proofs are addressed by the research described in this chapter: timing to diagnose motion very near compact objects. I shall use the term *black hole* for objects whose black-hole candidacy is based on a measured mass as well as for those whose patterns of timing and spectral behavior put them into the same phenomenological category as these objects.

2.4.2 Low-magnetic-field object types

The low-magnetic-field neutron-star systems are subdivided into Z sources, atoll sources (Hasinger & van der Klis 1989) and what I shall call the "weak LMXBs". These three main sub-types are closely related (see also Section 2.5.2). *Z sources* are the most luminous (Fig. 2.1), and accrete at an appreciable fraction of the Eddington critical rate (perhaps 0.5–1 L_{Edd}). *Atoll sources*, many of which are X-ray burst sources, cover a much wider range in luminosities, from perhaps 0.001 L_{Edd} (much lower in transients, but below this level timing becomes difficult) all the way up to the range of the Z sources (e.g., Ford *et al.* 2000; L_X overlaps may well occur; distances are uncertain). Ordinary atoll sources are usually in the 0.01–0.2 L_{Edd} range, while the "GX" atoll sources in the galactic bulge (see Section 2.5.2) usually hover at the upper end (perhaps 0.2–0.5 L_{Edd}), and the *weak LMXBs* (see Section 2.5.2) at the lower end ($<$0.01 L_{Edd}) of that range. Weak LMXBs comprise the overlapping groups of faint burst sources, millisecond pulsars and low-luminosity transients

(Chapter 6);[3] many of them appear to be just atoll sources stuck at low L_X. Some faint LMXBs have luminosities that are not well known due to uncertain distances or emission anisotropies (e.g., in dippers, Chapter 1), but many of these are probably weak as well.

While nearly all LMXBs whose X-ray emission is persistent contain a neutron star, nearly all with a black hole (and many with a neutron star as well) are transients, showing intermittent activity intervals called outbursts usually lasting weeks to months and separated by long quiescent intervals. The black holes have no well-defined subdivisions; although one might expect differences in spin (and hence, frame dragging, Section 2.8.1), accretion mode (wind or Roche-lobe overflow, Chapter 1) or between transient and persistent systems to show up, none of these lead to obvious differences in the observable properties attributable to the inner flow. Some black-hole transients remain in the low hard state (Section 2.5) during their entire outburst (e.g., Nowak 1995; Brocksopp 2004), but this can vary from one outburst to another (e.g., Belloni *et al.* 2002b).

Some accreting objects that are *not* the topic of this chapter (or even in some cases this book), notably the ultra-luminous X-ray sources (ULX) observed in some external galaxies, active galactic nuclei (AGN), both of which are thought to contain black holes, and cataclysmic variables (CV), which contain white dwarfs, can have accretion geometries that are similar to those in X-ray binaries (XRB), and it is of interest to compare their variability properties. The ULX are discussed in Sections 9.8.4, 12.3, 12.4.1 and 13.8 (see also Strohmayer & Mushotzky 2003; Cropper *et al.* 2004) and the CV in Chapter 10. In AGN, while QPO detections are still difficult (see Benlloch *et al.* 2001 and references therein, and see also Halpern *et al.* 2003) the best measurements now clearly show band-limited noise with characteristic frequencies consistent with the idea that variability timescales scale with mass (e.g., Edelson & Nandra 1999; Czerny *et al.* 2001; Uttley *et al.* 2002; Markowitz *et al.* 2003; McHardy *et al.* 2004). The comparison of XRBs with AGN is of particular interest as the physics is likely similar but different observational regimes apply. We typically receive more X-ray photons per dynamical timescale from AGN than from XRBs, making it easier, in principle, to study such fast variability in the time domain. X-ray binaries of course have much higher photon fluxes so that background is less of an issue and it is easy to cover very large numbers of dynamical timescales (a hundred million per day) and reliably determine the *parameters* of the stochastic process characterizing the variability rather than observing just one particular *realization* of it. The possibility to compare neutron-star and black-hole systems is unique to X-ray binaries as well.

2.5 Source states

Time variability and spectral properties are found to be correlated, presumably because of their common origin in physical processes in the inner, X-ray emitting part of the accretion flow. *Source states* are qualitatively different, recurring patterns of spectral and timing characteristics. They are thought to arise from qualitatively different, somewhat persistent, inner flow configurations. Both timing and spectroscopy are usually required

[3] Since the discovery in 1998 of millisecond pulsations in the thermonuclear burster and low-magnetic-field neutron star SAX J1808.4–3658 (Chapter 1, Section 2.9.1) the old adage about the mutual exclusion between pulsations and Type I X-ray bursts "pulsars don't burst and bursters don't pulse" (e.g., Lewin *et al.* 1993), which was based on the dichotomy in neutron-star B fields (Section 2.4), no longer holds, because low-magnetic-field neutron stars now can be pulsars, too.

Table 2.1. *Source states of low-magnetic-field compact objects*

HIGH frequencies — disk-dominated; thermal; **SOFT** spectra

↑ Q, and in the short term L_X, generally increase upward

Z	Atoll	Weak LMXB	Black hole
FB	UB		
—	—		
NB	LB		HS
	50		
20	LLB		
HB	*10*		*10*
	IS		
2	*2*		IMS/VHS
	EIS	*1*	
		EIS	*0.5*
	0.2		LS
		0.1	
			0.01

↓ rms and L_{radio} generally increase downward

LOW frequencies — corona-dominated; non-thermal; **HARD** spectra

Q: variability coherences; rms: variability amplitudes; L_{radio}: radio flux (Section 9.5.3). FB: flaring branch; NB: normal branch; HB: horizontal branch; UB: upper banana; LB: lower banana; LLB: lower left banana; IS: island state; EIS: extreme island state; HS: high state; IMS: intermediate state; VHS: very high state; LS low state. Numbers indicate typical BLN frequencies ν_b (in Hz) delineating the states (Section 2.6.2); — indicates the BLN is usually undetected. Reversals in ν_b occur beyond the highest frequencies in Z and atoll sources (Section 2.6.2).

to determine source state. Luminosity and the way in which the power spectrum and X-ray spectrum vary on timescales of minutes and longer (called source *behavior*) can be used as additional state indicators, but note that luminosity does *not* determine source state (below).

The qualitative changes in phenomenology used to define state include the appearance of a spectral or variability component, a sudden step in luminosity, or a clear bend in a CD/HID track, events which often coincide, indicating a qualitative change in the flow. As observations improve, "sudden" transitions become resolved, so eventually somewhat arbitrary boundaries need to be set to make state definitions precise. Depending on which criteria are chosen, authors may differ on what is the "correct" subdivision of the phenomenology into states. Nevertheless, the concept is central in describing the behavior of X-ray binaries, and in this chapter the rapid X-ray variability is discussed as a function of state. How X-ray spectral fit parameters depend on source state and what in detail this might imply is beyond the scope of this chapter (see e.g., di Salvo & Stella 2002; Done & Gierlinsky 2003; Gilfanov *et al.* 2003 for neutron stars and Chapter 4 for black holes).

Table 2.1 lists the source states distinguished in this chapter ordered from spectrally hard, generally characterized by low variability frequencies and luminosities at the bottom, towards soft, with higher frequencies and often higher luminosities at the top. Typical band-limited noise (BLN, Section 2.2) frequencies ν_b corresponding to the state transitions are given

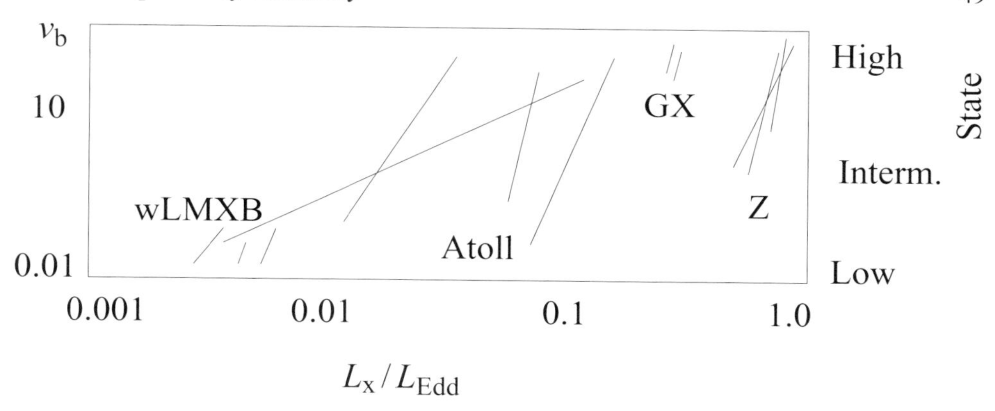

Fig. 2.2. States and luminosities attained by Z sources, GX atoll sources, ordinary atoll sources and weak LMXBs. Similar states occur at very different luminosities; not all states are seen in all object types. The issue of L_X overlaps between these source types is undecided.

(see Sections 2.6.2 and 2.7). These frequencies are indicative, and in practice depend somewhat on source, and on the precise definition of each state. The most striking variability (strong sharp kHz QPOs (Section 2.6.1), black-hole high-frequency QPOs (Section 2.6.1), sharp low-frequency QPOs (Section 2.6.2)) tends to occur in the intermediate states (HB, LLB, IS, IMS, VHS, see Table 2.1) both in neutron stars and in black holes. Note that Z sources so far appear to lack true low-frequency states. That a rough ordering of states such as in Table 2.1 is possible suggests, of course, that there are physical similarities between the states in the different source types (e.g., van der Klis 1994a,b), but a considerable amount of uncertainty still surrounds this issue, as well as the exact nature of the states themselves.

In neutron stars, X-ray luminosity L_X tends to correlate with state only within a source, not across sources, and, in a given source, much better on short (hours to days) than on longer timescales (Sections 2.5.2, 2.9.1.1 and, e.g., van der Klis 2000 and references therein); in black holes, the predictive value L_X has with respect to state is even more tenuous (Section 2.5.1 and, e.g., Miyamoto *et al.* 1995; Nowak *et al.* 2002; Maccarone & Coppi 2003). Note, however, that a really low L_X, $<0.01\ L_{Edd}$ or so, will reliably produce a hard state in most sources. In general, X-ray spectral shape (position in the color diagram) predicts timing characteristics much better than L_X (e.g., van der Klis *et al.* 1990; Hasinger *et al.* 1990; Kuulkers *et al.* 1994, 1996; van der Klis 1994a,b, 1995a; Ford *et al.* 1997b; Kaaret *et al.* 1998; Méndez & van der Klis 1999; Homan *et al.* 2001; Wijnands & Miller 2002; Rossi *et al.* 2004; Belloni *et al.* 2005). The range of correlations in neutron stars between state as defined in Table 2.1 and L_X is schematically illustrated in Fig. 2.2; a full systematic population study has so far only been performed for states with kHz QPOs (Ford *et al.* 2000). Clearly, the overall correlation of state to L_X is not good. What causes neutron stars to exhibit similar states at very different L_X levels, while short-term state changes *correlate* to changes in L_X, and why some sources cover a wider range of states than others, is uncertain. Note that not only the differences between Z and atoll sources need explanation, but also those *among* atoll sources. Differences in magnetic field (e.g., Miller *et al.* 1998a) have been considered. Clearly, this would allow for intermediate cases, but these are rare (Section 2.9.5).

If, as often assumed, accretion rate $\dot{M}$, varying due to processes outside the strong-field region (e.g., instabilities further out in the disk), and increasing from bottom to top in Table 2.1, were to underly the state differences, then $\dot{M}$ would have to govern both timing and spectral properties, but *not* the long-term changes in measured X-ray flux nor the L_X differences between sources (e.g., van der Klis 1995a). This could arise if mass outflows or (in black holes) radiatively inefficient (advective) inflows (Chapter 4) destroy the expected $\dot{M}$–L_X correlation by providing sinks of mass and (kinetic) energy (e.g., Ford *et al.* 2000), or if the flux we measure is not representative for the true luminosity because there are large and variable anisotropies or bolometric corrections in the emission (e.g., van der Klis 1995a). However, in neutron stars, differences in X-ray burst properties as a function of observed L_X (Chapter 3), suggest that the $\dot{M}$–L_X correlation is at least fair, and hence that the $\dot{M}$–state correlation is not so good. Possibly, the $\dot{M}$ that sets source state is not total $\dot{M}$, but only one component of it, e.g., that through the X-ray emitting part of the disk, $\dot{M}_d$, while there is also a radial inflow $\dot{M}_r$ (e.g., Fortner *et al.* 1989; Kuulkers & van der Klis 1995; Kaaret *et al.* 1998; van der Klis 2000, 2001; Smith *et al.* 2002). A more radical solution is that L_X does track $\dot{M}$ but that source state is governed by a physical parameter not correlating well to *any* $\dot{M}$ (perhaps, inner disk radius r_{in}, van der Klis 2000). The question then becomes what, if not a varying accretion rate, *does* cause the changes. A clue is the hysteresis observed in the state transitions of various sources (Sections 2.5.1, 2.5.2), which suggests that the history of a source's behavior affects its current state. S-curve disk-flow solutions (Chapter 13) have this property, but a smoothed response of $\dot{M}_r$ to variations in $\dot{M}_d$ where r_{in} is set by $\dot{M}_d/\dot{M}_r$ may cause hysteresis as well (see Section 2.9.1.1). So, while it seems out of the question that the states *as well* as the differences between sources are all just caused by differences in (instantaneous) $\dot{M}$, differences in $\dot{M}$ *and* in some time average over $\dot{M}$ might still in principle be sufficient. However, the influence of other parameters seems likely. More than one mechanism may be at work. In this chapter, when referring to source states I use the terms *high* and *soft* vs. *low* and *hard* in the general sense implied by Table 2.1, with the understanding that the relation to L_X is complex.

2.5.1 Black-hole states

In black holes the 1–20 keV X-ray spectrum can be decomposed into a *hard*, non-thermal, power-law component with photon index typically 1.5–2 and a *soft* (also: "ultrasoft"), thermal, black-body-like component with $kT < 1$ keV (Section 1.3.4, Chapter 4). The latter is usually attributed to thermal emission from the accretion disk, the former to a *corona* containing energetic electrons. There is no agreement about the nature or energetics of this corona (see e.g., Maccarone & Coppi 2003): it could be located within the inner disk edge or cover part of the disk, it could be quasi-spherical, a thin layer on, or magnetic loops anchored in, the disk, or be the base of the radio jets. It could be quasi-static or part of the accretion flow (azimuthally and/or radially). Its electrons' energy could be thermal or due to bulk motion, and the radiation mechanism could be either Compton or synchrotron.

In the classic black-hole *low state* (LS) in the 2–20 keV band, the hard component dominates (hence it is also called "low hard state"), and strong (up to typically ~50% rms) flat-topped BLN (Section 2.2) is present, which has a low characteristic frequency (down to typically $\nu_b \sim 0.01$ Hz). In the *high state* (HS) the soft component dominates (hence "high soft state") and weak ($\lesssim 3\%$) power-law noise occurs. In the *very high state* (VHS) L_X is high

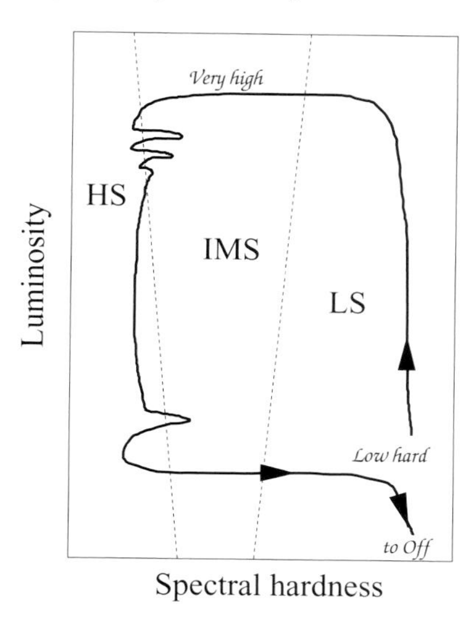

Fig. 2.3. Black-hole states plane. Main states are LS, IMS and HS (low, intermediate and high state) as indicated; locations of classic low/hard, very high and off states are shown as well. Thick curve shows path of typical black-hole transient outburst, inspired by observations of XTE J1550–564 (Homan *et al.* 2001) and GX 339–4 (Belloni *et al.* 2005). Contours of ν_b might be a way to delineate the three states in this plane; as these lines are not accurately known, the dashed lines shown are only indicative. See Fig. 2.8 for typical power spectra observed in these states.

in the range of both spectral components, and strong 3–12 Hz QPOs and BLN that is weaker and higher-frequency (up to ~10 Hz) than in the LS as well as power-law noise stronger than in the HS occur, with rapid transitions between these two noise types (Miyamoto *et al.* 1991). Both LS and HS have turned out to occur over a wide and largely overlapping L_X range in the spectrally hard and soft parts of the HID, respectively (Figs. 2.3, 2.5), and an *intermediate state* (IMS) with timing properties similar to the VHS and at intermediate spectral hardness, but likewise covering a wide L_X range (e.g., Homan *et al.* 2001; see Fig. 2.3) has been identified. To maintain continuity with existing literature (cf. Table 2.7), I continue to refer to these states as LS for the hard low-frequency states and HS for the soft states with weak power-law noise. I take the IMS to include the VHS as the highest-L_X intermediate state in a source (usually attained early in a transient outburst). Somewhat fortuitously, "low", "high" and "intermediate" can also be taken to refer to BLN frequency (although in a full-blown HS the BLN is not detected). Note that in Chapter 4 McClintock and Remillard take another approach: they call HS "thermal dominant state", and depending on the results of X-ray spectral continuum fits classify some instances of the IMS together with the LS as "low hard state" and others as "steep power law state". This subdivision appears to be related, but not identical, to that proposed between hard and soft IMS (Belloni *et al.* 2005; Klein-Wolt 2004b) and the two VHS noise types first noted by Miyamoto *et al.* (1991). Timing in the low-L_X quiescent or "off" state, down to levels where this can be checked, is consistent with the LS but with ν_b down to perhaps 0.0001 Hz (Section 2.10.2). QPOs in both the 1–30 Hz and 100–450 Hz ranges are most prominent in the VHS/IMS, but occasional ~0.01–20 Hz QPOs are also seen in LS and HS.

In a "standard" black-hole transient outburst (Figs. 2.3, 2.5) sources tend to follow a harder trajectory from low to high luminosity than vice versa, and make the main hard–soft (LS→HS) transition at much higher luminosity than the soft–hard (HS→LS) one, behavior that is often called *hysteresis* (Miyamoto *et al.* 1995; Nowak 1995; van der Klis 2001; Nowak *et al.* 2002; Smith *et al.* 2002; Maccarone & Coppi 2003), expressing the idea of a driving force and a memory effect in the response to that force (Section 2.5); whether this is truly what is going on requires further investigation. Transitions to and from the IMS can be rapid, suggesting an obvious delineation between states, but slower transitions occur as well, and in these cases the exact transition point becomes a matter of definition (Section 2.10.2). It might be possible to base such definitions on values of ν_b (cf. Table 2.1), perhaps defining ν_b contours in CD/HIDs as in Fig. 2.3, but it should be noted that the issue is currently in flux and that in particular the best way to deal with the IMS/VHS is surrounded by some controversy (see also Chapter 4). As mentioned above, further subdivisions probably exist in the IMS, but different boundaries are drawn, perhaps depending on whether spectral fit parameters (as in Chapter 4) or rapid variability (Belloni *et al.* 2005; Klein-Wolt *et al.* 2004b) are given precedence in defining the sub-states.

Various spectral characteristics can serve to define the black-hole states plane. Flux, intensity or soft color can track the "luminosity" parameter, which physically might be dominated by the accretion rate through the inner disk $\dot{M}_d$. The "state" parameter, which could be a measure for the strength or size of the corona can be defined as, e.g., hard color or a spectral-component luminosity ratio ("thermal fraction"; L_{soft}/L_{total}, "power-law ratio"; L_{hard}/L_{total}, etc.; e.g., Miyamoto *et al.* 1994; Nowak 1995; Rutledge *et al.* 1999; Remillard *et al.* 2002c).

On physical grounds one might expect spectral and timing properties to be affected not only by state, but also by the large (>factor 10) differences in L_X within each state; but certainly in HS (e.g., Homan *et al.* 2001) and LS (e.g., Belloni *et al.* 2005) the L_X effect appears to be modest. Historically, after LS and HS (Tananbaum *et al.* 1972), the VHS was identified first (Miyamoto *et al.* 1991); the intermediate state was initially noticed as a HS→LS transitional state with 10 Hz BLN similar to that in the VHS in the decay of the transient GS 1124–68 (Belloni *et al.* 1997), and as a 10 Hz BLN state at much lower L_X than the VHS in GX 339–4 (Méndez & van der Klis 1997). Intermediate states with 10 Hz BLN and/or strong power-law noise were then also seen in Cyg X-1 (Belloni *et al.* 1996) and a number of other sources (e.g., Kuulkers *et al.* 1997b), and in XTE J1550–564 during excursions at various L_X levels from the HS to a somewhat harder state characterized by 10 Hz BLN, LF QPOs and occasional rather strong power-law noise (Section 2.10.3, Homan *et al.* 2001). This led to the two-dimensional paradigm of Fig. 2.3. Weak QPOs >100 Hz occur in the VHS (Section 2.6.1) but were also found at least once in the IMS at lower L_X (Homan *et al.* 2001).

That the variability frequencies decrease (Section 2.6) when the spectrum gets harder suggests a relation with inner disk radius r_{in} (stronger corona for larger r_{in}) and some spectroscopic work points into the same direction (Chapter 4), but this entire picture is firmly within the realm of the working hypotheses. Nevertheless, most modeling of black-hole states approximately conforms to this generic framework. The strength of the corona has a relation to that of the radio jets, and as both rely on energetic electrons for their emission, it is natural to suspect a physical relation between these structures (Chapter 9). Even at a purely empirical level, there is still considerable uncertainty associated with the motion of the sources through Fig. 2.3, e.g., why do many sources approximately take the depicted

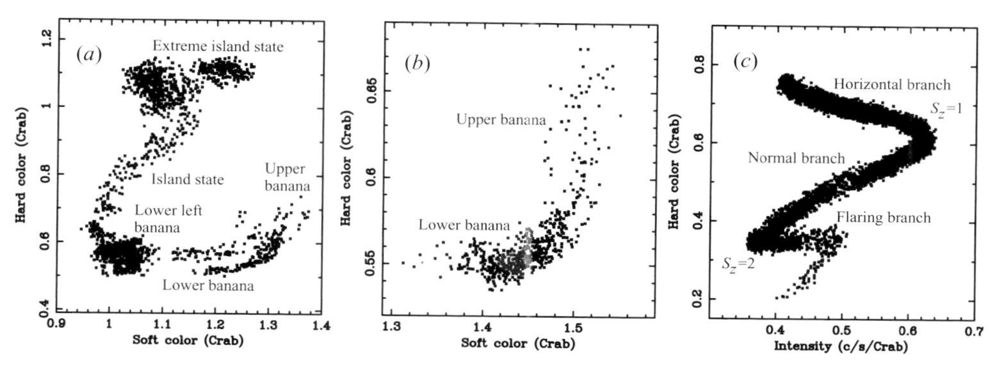

Fig. 2.4. Spectral branches of neutron stars. (a) CD of the atoll source 4U 1608−52, (b) CD of the "GX" atoll source GX 9+1, and (c) HID of the Z source GX 340+0. RXTE/PCA data; soft color 3.5–6/2–3.5 keV, hard color 9.7–16/6–9.7 keV, intensity 2–16 keV, all normalized to Crab. Conventional branch names and S_z values are indicated; for S_a the choice of values differs between authors (though not in sense, see text). Compare van Straaten *et al.* (2003), Reerink *et al.* (2004), Jonker *et al.* (2000a).

"canonical" path while some deviate from this; are all areas in this plane accessible or are some forbidden; what kind of state transitions are possible at each luminosity level? Yet the diagram and its segmentation into three areas representing three main source states provides a useful template embodying the broad correlation between variability frequencies and spectrum occurring largely irrespectively of L_X level, against which black-hole behavior can be matched.

2.5.2 *Low-magnetic-field neutron-star states*

In low-magnetic-field neutron stars spectral decomposition is much less obvious than in black holes, perhaps because of the presence of two thermal emission sites (disk and star) and cooling of the hot electrons in the corona by the stellar flux. The spectra become neither as hard nor as soft as in black holes, but weak hard components are sometimes seen and similarly explained by Comptonization (e.g., Barret & Vedrenne 1994; di Salvo & Stella 2002). Contrary to the case in black holes, CD/HIDs show rather reproducible tracks that embody a one-dimensional states sequence (except at the lowest luminosity levels, see Section 2.5.2.2).

2.5.2.1 Z sources

Z sources on timescales of hours to a day or so trace out roughly Z-shaped tracks (Fig. 2.4*c*) in CD/HIDs consisting of three branches connected end-to-end and called *horizontal branch*, *normal branch* and *flaring branch* (HB, NB, FB). Curve length S_Z (Section 2.3), defined to increase from HB via NB to FB (Fig. 2.4*c*) performs a random walk; it varies stochastically but shows no jumps. Kilohertz QPOs and a 15–60 Hz QPO called HBO occur on the HB and upper NB, a ~6 Hz QPO called NBO on the lower NB, and mostly power-law noise <1 Hz on the FB; see Section 2.6. With increasing S_Z the band-limited noise (LFN, see Section 2.6.2) becomes weaker and flux and frequencies generally increase, although at high S_Z reversals occur (Section 2.6.2). Z tracks differ somewhat between sources (e.g., Hasinger & van der Klis 1989; Kuulkers 1995; Muno *et al.* 2002), and also show slow drifts ("secular motion") that do not much affect the variability and its strong correlation with S_Z (e.g., Hasinger *et al.* 1990; Kuulkers *et al.* 1994; Jonker *et al.* 2002a); some occasionally show shape changes which *do* affect the variability (Kuulkers *et al.* 1996). There is no evidence for hysteresis in either the Z tracks or the rapid variability.

2.5.2.2 *Atoll sources and weak LMXBs*

At high L_X atoll sources trace out a well-defined, curved *banana branch* in the CD/HIDs (Fig. 2.4*a,b*; Hasinger & van der Klis 1989; Reerink *et al.* 2004) along which, like in Z sources, sources move back and forth with no hysteresis on timescales of hours to a day or so, and which sometimes shows secular motion not affecting the variability (e.g., van der Klis *et al.* 1990; di Salvo *et al.* 2003; Schnerr *et al.* 2003). The banana branch is further subdivided into the *upper banana* (UB) where the $<$1 Hz power-law noise (VLFN, Section 2.9.4) dominates, the *lower banana* (LB) where dominant several 10 Hz BLN occurs (Sections 2.6.2, 2.9.3), and the *lower left banana* (LLB) where twin kHz QPOs (Sections 2.6.1, 2.9.1) are observed.

The spectrally harder parts of the CD/HID patterns are traced out at lower L_X. CD motion is often much slower here (days to weeks), and observational windowing can cause isolated patches to form, which is why this state is called *island state* (IS; Hasinger & van der Klis 1989); it is characterized by dominant BLN (Section 2.6.2), becoming stronger and lower-frequency as flux decreases and the $>$6 keV spectrum gets harder. The hardest, lowest luminosity island states (the *extreme island state*, EIS; e.g., Prins & van der Klis 1997; Reig *et al.* 2000a; van Straaten *et al.* 2003 and references therein) are similar to the black-hole LS, with strong, low-frequency flat-topped noise (Section 2.6.2) and a hard power-law X-ray spectrum. Curve length S_a (Section 2.3) increases from EIS to UB, generally anti-clockwise (Fig. 2.4*a*). With increasing S_a the BLN tends to become weaker; frequencies generally increase up to the LB, from where in bright sources ν_b decreases again (Section 2.6.2). Most atoll sources show a banana branch as well as island states but the four GX atoll sources (Sections 2.4.2, 2.5; Reerink *et al.* 2004) are (nearly) always in the banana branch (LB and UB, Fig. 2.4*b*), and the weak LMXBs (nearly) always in the EIS (e.g., Barret *et al.* 2000, Belloni *et al.* 2002a).

In the island states two-dimensional motion (i.e., not following one well-defined track) is often observed, perhaps because secular motion and motion along the branch happen on similar timescales. In transient atoll sources this takes the form of hysteresis (Section 2.5.1) during relatively rapid ($\sim$1 day) transitional IS episodes between EIS and banana states (Fig. 2.5): similarly to black-hole transients, hard to soft transitions occur at higher luminosity than the reverse, but neutron stars lack the hard, high-L_X states (e.g., Barret *et al.* 1996; Olive *et al.* 2003; Maccarone & Coppi 2003). Persistent sources can transit to and from the EIS tracing out a one-dimensional IS branch (Fig. 2.4*a*), suggesting hysteresis is related to the abruptness or speed of the transition. The full picture concerning the EIS, of which only glimpses were seen before (e.g., Mitsuda *et al.* 1989; Langmeier *et al.* 1989; Yoshida *et al.* 1993), is only now becoming clear. In some cases several parallel horizontal EIS branches form, where $>$6 keV spectral hardness, not luminosity or the $<$6 keV spectrum, determines the source state (van Straaten *et al.* 2003). Similarities between this CD behavior in the EIS and the Z-source HB were pointed out by a number of authors (Muno *et al.* 2002, Gierliński & Done 2002b, see also Langmeier *et al.* 1989), but do not extend to the way in which the branches connect, or to the timing behavior (Barret & Olive 2002; van Straaten *et al.* 2003;

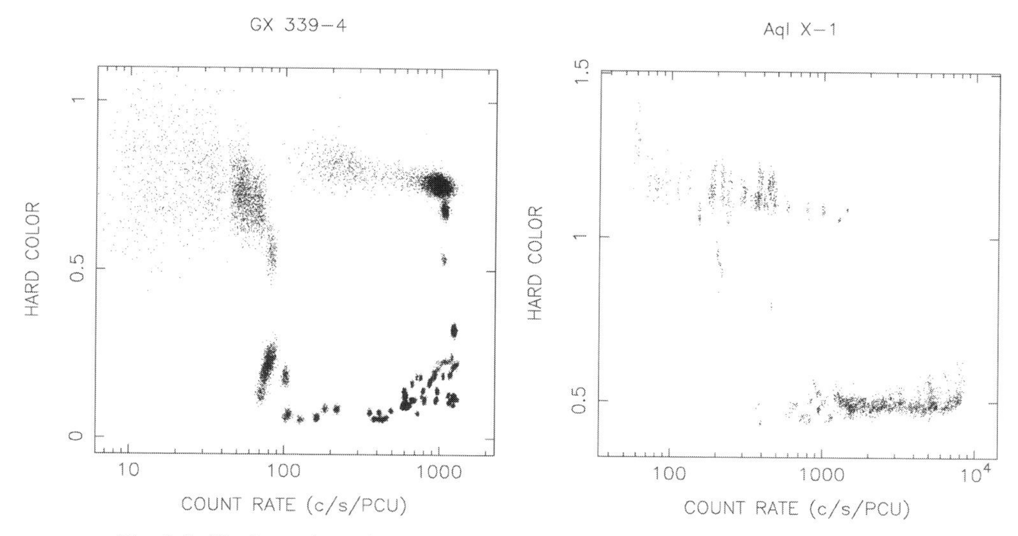

Fig. 2.5. Hardness–intensity diagrams of the black hole GX 339–4 and the neutron-star LMXB transient Aql X-1. Note the orientation of these diagrams, which is similar to that of Fig. 2.4 and transposed compared to Fig. 2.3. Compare Belloni (2004), Reig *et al.* (2004).

Olive *et al.* 2003; Reig *et al.* 2004, Section 2.6.2), where the HB is more like the IS/LLB than like the EIS (Table 2.1).

Similarities of atoll CD/HID behavior with that of Z sources are mostly confined to the banana branch. Similar one-dimensional motion on similar (hours–day) timescales and, in the persistent sources, covering similar L_X ranges takes place there through similarly slowly drifting tracks (which, however, are shaped differently and also intrinsically wider in atoll sources). In island states the CD behavior is quite different from that of Z sources: it shows hysteresis and other forms of two-dimensional motion, and often takes place on longer (days–weeks) timescales over much larger L_X ranges (factors of 5–10, up to 10^3 for transients). This behavior is more similar to that of black holes than that in other neutron-star states, possibly because in low states the inner-disk radius is larger, further away from the compact object.

2.6 Variability components

Table 2.2 summarizes the rapid X-ray variability components seen in low-magnetic-field neutron stars and black holes that we will discuss, grouped into high-frequency phenomena ($\gtrsim$100 Hz), the low-frequency complex (a group of correlated 10^{-2}–10^2 Hz phenomena), power-law components (usually dominating mainly the low frequencies), and other phenomena. Typical characteristic frequency ranges are also given, as well as names used for phenomena in particular source types. Here we introduce the high-frequency and low-frequency-complex components, emphasizing relations across source types. Further details, and discussion of the power-law and other components are provided in Sections 2.9 and 2.10.

2.6.1 High-frequency phenomena

Neutron-star kilohertz QPOs The fastest variability components in X-ray binaries are the *kilohertz quasi-periodic oscillations* (kHz QPOs; van der Klis *et al.* 1996, Strohmayer

Table 2.2. *Variability components*

	Name	Low-magnetic-field neutron stars: Z sources	Atoll sources	Weak LMXBs	Black holes	Typical frequency range[a] (Hz)
High-frequency	kHz QPO	•	•	—		300–1200
	HF QPO				•	100–500
	hHz QPO	○	•	—		100–200
Low-frequency complex	BLN	•[1]	•[b]	•	•[2]	0.01–50
	LF hump/QPO	•[3]	•	•	•	0.1–50
	3rd Lor.	—	•[4]	•[4]	•	1–100
	4th Lor.	—	•[5]	•[5]	○	10–500[c]
Power-law	VLFN	•[d]	•[d]	—	•[6,e]	–[f]
Other	N/FBO	•	○	—	—	4–20
	1 Hz QPO	—	—	•[g]	—	0.5–2
	mHz QPO	—	•	—	•	0.001–0.01

•: observed; ○: some doubt (uncertain, ambiguous, atypical, rare and not clearly seen, etc.); —: variability of similar type as in the other source types not reported. HF: high-frequency; LF: low-frequency; BLN: band-limited noise; Lor.: Lorentzian; VLFN: very-low-frequency noise; N/FBO: normal/flaring branch oscillation. Alternative names: [1] LFN (low-frequency noise), [2] LS (low state) noise, [3] HBO (horizontal branch oscillation), [4] $L_{\ell ow}$, [5] L_u, [6] HS (high state) noise.
[a] ν_{max}, cf. Section 2.2. [b] Sub-components, see Section 2.6.2. [c] See note to Table 2.3. [d] Sub-components, see Section 2.9.4, [e] Stronger power-law noise in IMS, see Section 2.10.3. [f] Often only detected < a few Hz. [g] e.g., dipper QPOs; 1 Hz flaring in SAX J1808.4–3658, Section 2.9.4.

et al. 1996; van der Klis 1998 for a historical account) seen in nearly all Z sources (in HB and upper NB) and atoll sources (in IS and LLB) as well as a few weak LMXBs, including millisecond pulsars (Wijnands *et al.* 2003). Two QPO peaks (the *twin peaks*) occur in the power spectrum (Fig. 2.6*a*) and move up and down in frequency together in the 200–1200 Hz range in correlation with source state (cf. Fig. 2.9). The higher-frequency of these two peaks is called the *upper kHz QPO*, frequency ν_u, the other the *lower kHz QPO* at ν_ℓ; towards the edges of their observed frequency range peaks also occur alone. The several hundred hertz *peak separation* between the peak centroids $\Delta\nu \equiv \nu_u - \nu_\ell$ is typically within 20% of the neutron-star spin frequency, or half that, depending on source, and usually decreases by a few tens of hertz when both peaks move up by hundreds of hertz. Most models involve orbital motion in the disk at one of the kHz QPO frequencies (Section 2.8). Weak sidebands in the kHz domain have been reported in some kHz QPO sources (Section 2.9.1).

Black-hole high-frequency QPOs The fastest black-hole phenomenon is the *high-frequency* (HF) QPOs (Remillard *et al.* 1999c) seen in the IMS (usually VHS). Frequencies ν_{HF} range from 100 to 450 Hz (their relation to QPOs in the 27–67 Hz range, Morgan *et al.* 1997, is unclear), and are reported to usually occur at fixed values different in each source, perhaps inversely proportional to black-hole mass. In a few cases harmonically related (2:3;

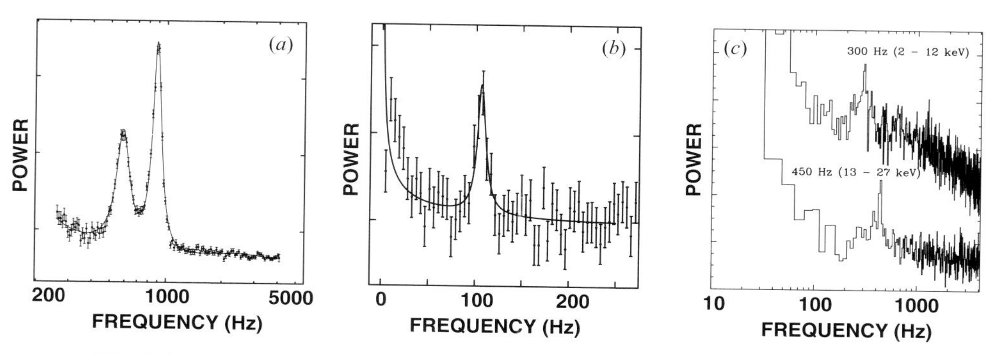

Fig. 2.6. (*a*) Twin kHz QPOs in Sco X-1 (van der Klis *et al.* 1997), (*b*) hectohertz QPO in 4U 0614+09, (*c*) HF QPOs in GRO J1655–40 (Strohmayer 2001a).

Section 2.10.1) frequencies have been seen (Fig. 2.6*c*). The phenomenon is weak and transient so that observations are difficult, and discrepant frequencies occur as well. The constant frequencies might indicate a link with neutron-star hectohertz QPOs (below), but it is not excluded that black-hole HF and neutron-star kHz QPOs can be reconciled within a single explanation (Section 2.10.1).

Neutron-star hectohertz QPOs The *hectohertz* (hHz) QPO (Ford & van der Klis 1998) is a peaked noise phenomenon (sometimes coherent enough to be called a QPO, Section 2.2) with a frequency $\nu_{\rm hHz}$ in the 100–200 Hz range that is seen in atoll sources in most states (Fig. 2.6*b*). It stands out from all other neutron-star components by its approximately constant frequency (Fig. 2.9) which is quite similar across sources, perhaps because $\nu_{\rm hHz}$ derives from compact-object properties and the neutron stars in these systems are all similar (Section 2.8). In addition to a possible link with black-hole HF QPOs, a link with the features between 10 and 100 Hz reported by Nowak (2000) in Cyg X-1 and GX 339–4 has been suggested (van Straaten *et al.* 2002), which may allow a $1/M$ scaling of frequency.

This concludes the summary of the >100 Hz phenomena. Neutron stars have much more broad-band power in the kHz range than black holes (Sunyaev & Revnivtsev 2000; Klein-Wolt *et al.* 2004b), but there is no indication that this is due to components other than those already mentioned here. The "high-frequency noise" reported from neutron stars in earlier work (e.g., Dieters & van der Klis 2000) may variously be due to the (low Q) upper kHz QPO at low frequency, hHz QPO, an HBO harmonic (below) and/or instrumental effects (cf. Berger & van der Klis 1994). Sporadic strong millisecond timescale bursts were reported from Cyg X-1 (Rothschild *et al.* 1974, 1977; Meekins *et al.* 1984; Gierliński & Zdziarski 2003) but detecting such phenomena in a strongly variable source is fraught with statistical difficulties (e.g., Press & Schechter 1974; Weisskopf & Sutherland 1978; Giles 1981); instrumental problems caused some of the reported detections (Chaput *et al.* 2000).

2.6.2 *The low-frequency complex*

In the 0.01–100 Hz range a set of usually two to five band-limited noise, peaked-noise and QPO components is observed whose frequencies all correlate. This *low-frequency complex* is often dominated by strong (up to 60% rms), *flat-topped BLN* with a break at frequency $\nu_{\rm b}$ in the ~0.01–50 Hz range and peaked noise (the *LF hump*) at frequency $\nu_{\rm h}$

Table 2.3. *High-frequency and low-frequency complex component names and symbols*

	Name	Symbol	Frequency (Hz)
High-frequency	Upper kHz	L_u	500–1200
	Lower kHz	L_ℓ	300–1000
	High-frequency	L_{HF}	100–500
	Hectohertz	L_{hHz}	100–200
Low-frequency complex	BLN	L_b, L_{b2}	0.01–50
	LF hump, LF QPO	L_h, L_{LF}	0.1–50
	3rd Lorentzian	$L_{\ell ow}$	1–100
	4th Lorentzian	L_u	10–500[a]

[a] In neutron stars 140–500 Hz, probably the upper kHz QPO at low Q; in black holes 10–100 Hz.

roughly a factor 5 above ν_b (Wijnands & van der Klis 1999a). Both components sometimes feature QPOs located around ν_b and/or around (or instead of) the hump near ν_h. The QPO near ν_h may show several harmonics and is often called the LF QPO (at ν_{LF}). The BLN at ν_b goes under various names (LFN; Section 2.9.3, LS noise; Section 2.10.2, broken power law, flat-topped noise) and is often just called the BLN, or L_b (L for "Lorentzian", Belloni *et al.* 2002a). By analogy, other components go by names such as L_h, L_{LF}, L_ℓ, L_u for the LF hump/QPO, lower/upper kHz QPO, etc.; see Table 2.3.

Low states When L_b and L_h are at low frequency ($\nu_b \lesssim 1$ Hz; black-hole LS and atoll EIS, cf. Table 2.1), a third BLN component is often present above the frequency of the LF hump/QPO. The combination of these three noise components leads to the characteristic power-spectral shapes displayed in Fig. 2.7 that can be remarkably similar between neutron stars and black holes (van der Klis 1994a and references therein; Olive *et al.* 1998; Wijnands & van der Klis 1999a; Belloni *et al.* 2002a), and also show similar time lags (Ford *et al.* 1999). This third component has been suggested to be the lower kHz QPO, L_ℓ, but at frequencies as low as $\sim$10 Hz (Section 2.7), and is designated $L_{\ell ow}$ here. In neutron stars a fourth, $>$100 Hz, BLN component occurs above $L_{\ell ow}$ whose frequency connects smoothly with that of the upper kHz QPO and, like it, is designated L_u.

Low to intermediate states When atoll sources and black holes move out of these low states ($\nu_b \gtrsim 1$ Hz) towards the IMS/VHS or via the island state (IS) to the lower-left end of the banana (LLB), respectively, all components increase in frequency and become weaker and, often, more coherent. By the time that $\nu_b \sim 10$ Hz, L_b and L_h are much weaker and $L_{\ell ow}$ is below detection. Black holes in this state often show narrow 1–30 Hz LF QPO peaks with sometimes several harmonics that differ in strength, phase lags and coherence between odd and even ones, as well as energy dependencies in L_b (and hence ν_b, Section 2.10.2); sometimes no BLN is detected but only power-law noise (Section 2.5.1 and 2.10.3). In the atoll sources L_b attains substructure and often needs to be described by two components, a peaked noise/QPO plus a BLN a factor 2–3 below it, which is sometimes called L_{b2}.

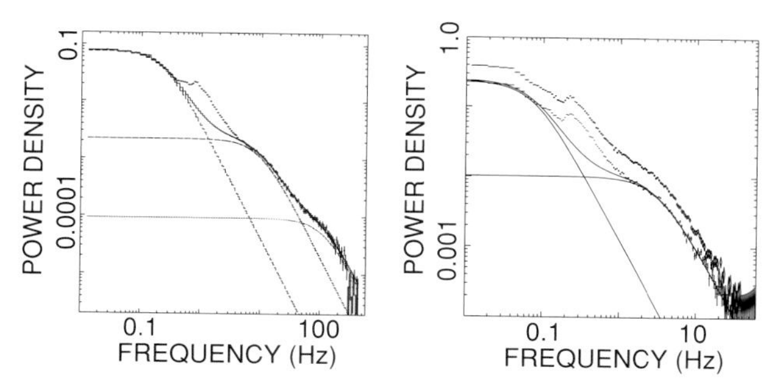

Fig. 2.7. The power spectra of neutron stars (*left*: 1E 1724–3045) and black holes (*right*: GRO J0422+32) in the low state can be strikingly similar. Note, however, the subtle *differences* in characteristic frequencies, power levels and number of detectable components, which may in fact be part of a systematic neutron-star black-hole distinction; cf. Fig. 2.8 LS and EIS. After Olive *et al.* (1998).

Z sources, where ν_b is always >1 Hz, are similar to this in the HB and upper NB. L_b is sometimes called "low-frequency noise" (LFN) and has ν_b varying between 2 and 20 Hz. A 15–60 Hz LF QPO superimposed on this called "horizontal-branch oscillation" (HBO) sometimes has several observable harmonics, with alternating high- and low-Q harmonics somewhat reminiscent of those in black-hole LF QPOs; warped disk geometries have been suggested as a cause of this (e.g., Jonker *et al.* 2000a, 2002a).

Intermediate to high states When atoll sources move further onto the banana, the frequencies of the L_b subcomponents rise (the QPO to 30–80 Hz and the BLN to 20 Hz) and then, in the LB, decrease again (e.g., di Salvo *et al.* 2003; Reerink *et al.* 2004). The low-frequency complex components weaken until in the UB they become undetectable. A similar evolution (including sometimes frequency reversals, e.g., Wijnands *et al.* 1996) is observed in Z sources on the NB, and for black holes moving into the HS, where dominant $\alpha = 0.7$–1.4 power-law noise is also regularly observed at this point (Section 2.10.3), but no clear evidence for frequency reversals has been reported (photon energy dependencies in ν_b in the IMS further complicate the issue). Several more components may occur in the range of the low-frequency complex (Sections 2.9.3 and 2.10.2) and disentangling them is sometimes difficult.

The <100 Hz variabilities of black-hole LS and neutron-star EIS are very similar and there is little doubt that they are physically related. There also obviously are close relations with, and between, variability in black-hole IMS/VHS, atoll source IS/(L)LB, and Z source HB/upper NB, but even empirically the exact nature of these relations is not yet fully established (but see Klein-Wolt *et al.* 2004b for recent progress). An important clue is provided by the correlations between the component frequencies (and strengths), which helps to identify components across sources. These are discussed next.

2.7 Frequency correlations

2.7.1 *Frequency correlations in neutron stars*

Figure 2.9*a* displays the frequency correlations of four well-studied atoll sources and four weak LMXBs (faint burst sources, Section 2.4.2). The frequencies of all components

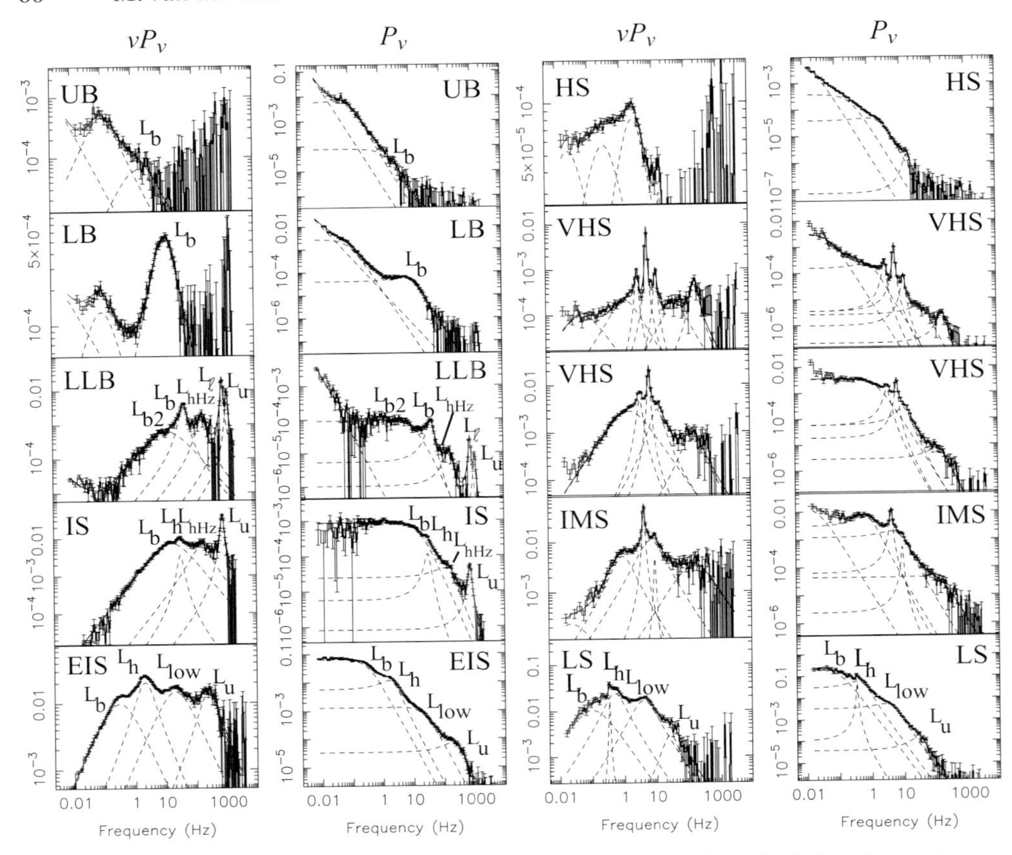

Fig. 2.8. Power spectra in *left*: atoll sources and *right*: black holes in their various states (Section 2.5) in P_ν and νP_ν representations. Two examples of VHS power spectra are shown, with and without strong BLN (high-luminosity hard and soft IMS, respectively, see Section 2.5.1), and one example of a lower-luminosity hard IMS. Atoll source variability component names are indicated (except the lowest-frequency, VLFN, components in UB, LB and LLB; cf. Reerink *et al.* 2004). The analogous broad-band components in the black hole LS are also identified; the narrow component is L_{LF}.

described in Section 2.6 are plotted versus ν_u. Source state designations corresponding to the color–color diagrams in Fig. 2.4 are also indicated. As L_u becomes undetectable $\gtrsim 1200$ Hz, the frequency evolution beyond that point is not covered in Fig. 2.9*a*. The figure includes data from sources covering an order of magnitude in luminosity when in the same state, yet they display very similar power spectra and essentially the same frequency correlations. The tracks of ν_b (and subcomponent ν_{b2}), ν_h, $\nu_{\ell ow}$, ν_{hHz} and ν_ℓ vs. ν_u are clearly recognizable. This plot can usefully serve as a template against which to match the rapid X-ray variability of other objects.

Figure 2.9*b* compares measurements for the Z sources with the atoll source data. The kHz QPOs match, as well as do the HBO and L_h. Differences are that in Z sources no hHz QPO has been reported so far, and that an additional HBO harmonic and "sub-harmonic" occur. How precisely these, and the LFN (the Z source variant of L_b) which varies even among the Z sources themselves, relate to atoll source components is undecided. The dependence

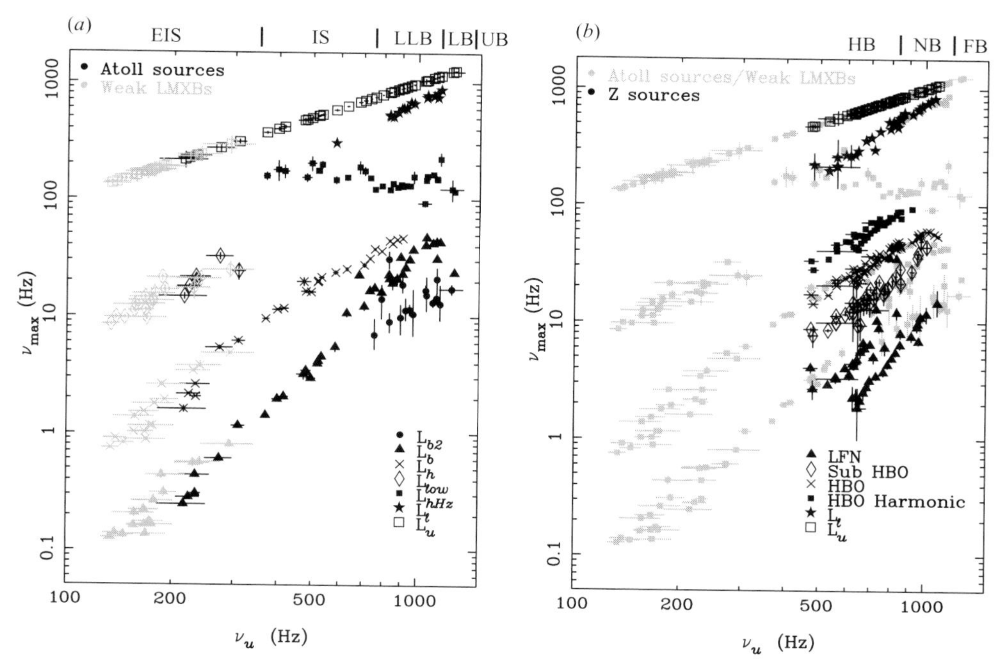

Fig. 2.9. Frequency correlations. (*a*) Atoll sources and weak LMXBs, (*b*) Z sources compared with these objects. The characteristic frequencies (ν_{max}, Section 2.2) of the components are plotted as indicated; approximate source state ranges are indicated at the top.

of ν_h and related QPOs on ν_u is approximately quadratic in both Z and atoll sources (Stella & Vietri 1998 and, e.g., Psaltis *et al.* 1999b; Jonker *et al.* 1998; Homan *et al.* 2002; van Straaten *et al.* 2003; Fig. 2.10) suggesting that Lense–Thirring precession of an orbit with frequency ν_u might be causing the LF QPO (Section 2.8.2), but in GX 17 + 1 at high frequency the HBO frequency (and X-ray flux) start decreasing while ν_u continues increasing (Homan *et al.* 2002).

An interesting discrepancy occurs in the frequencies of some millisecond pulsars: in SAX J1808.4–3658 a pattern of correlated frequencies occurs very similar to Fig. 2.9*a*, but with relations that are offset. At low frequencies, the match can be restored by multiplying the ν_u values and the single measured ν_ℓ with ~1.45, which suggests a link with the 2:3 frequency ratios in black holes (Section 2.6.1); at higher frequencies the identification of the L_b subcomponents is uncertain (van Straaten *et al.* 2005). Of the other millisecond pulsars, XTE J0929–314 and XTE J1807–294 behave in the same way, but XTE J1751–305 and XTE J1814–338 are like ordinary atoll sources in this respect. This suggests ν_u and ν_ℓ form one group of correlated frequencies and the LF complex another, independent one; the case of GX 17+2 above supports this.

2.7.2 *Frequency correlations of black holes compared to neutron stars*

Black holes cannot be directly compared to Fig. 2.9 as L_u is not reliably detected. However, correlations between other frequencies in Fig. 2.9, and one between frequency and power, provide intriguing links between neutron stars and black holes.

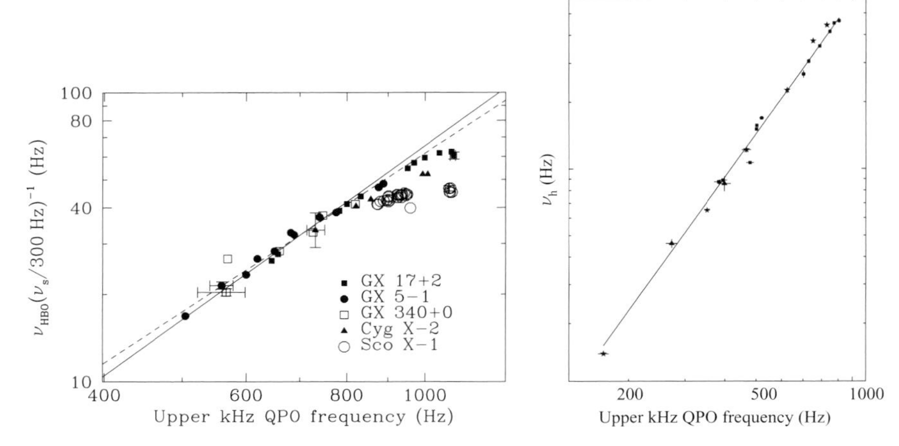

Fig. 2.10. The relation between upper kHz QPO frequency and, *left*: HBO frequency in Z sources (scaled between sources by an inferred spin frequency), drawn line is for a quadratic relationship, dashed line is best power-law fit, *right*: ν_h in atoll sources, line is power law with index 2.01. In a Lense–Thirring interpretation (Section 2.8.1) a value of I_{45}/m of ~4 would be implied in both cases (see also Section 2.8.2). From Psaltis *et al.* (1999b) and van Straaten *et al.* (2003).

BH relation Belloni and Hasinger (1990a) noticed that in Cyg X-1 ν_b and the power-density level $P_{\nu\text{flat}}$ of the BLN flat top anti-correlate (Fig. 2.11*a*), an effect later also seen in other black holes in the LS and neutron stars in the (E)IS (e.g., Méndez & van der Klis 1997; van Straaten *et al.* 2000; Belloni *et al.* 2002a). For $\nu_b < 1$ Hz the relation is consistent with a BLN rms amplitude that remains constant while ν_b varies; this constant amplitude is less for neutron stars than for black holes and might even depend on mass (Belloni *et al.* 2002a), but note that photon-energy dependencies can affect $P_{\nu\text{flat}}$ as well. Above 1 Hz the BLN clearly weakens, but the relation appears to extend to the black-hole I/VHS (van der Klis 1994b) at ν_b up to ~10 Hz, and to $\nu_b > 10$ Hz in neutron stars (van Straaten *et al.* 2000).

WK relation Wijnands and van der Klis (1999a; WK) noted that in atoll sources (including weak LMXBs) and black holes ν_b and ν_h are correlated over 3 orders of magnitude (Fig. 2.11*b*, with the hump at ν_h a factor 8 to 2 above the break, decreasing with ν_b. This relation is that between the two lower traces in Fig. 2.9. Z sources are slightly above the main relation (cf. Section 2.7.1), with ν_h an also decreasing factor 10 to 3 above ν_b. Of course, the QPOs on the break of band-limited noise components mentioned in Section 2.6.2 hug the line $\nu_{BLN} = \nu_{QPO}$; these may be physically the same components and are not shown in Fig. 2.11. Cases of QPOs below the noise break have also been reported (e.g., Brocksopp *et al.* 2001), and while in some instances in the fits they could be reinterpreted as the reverse, this may indicate that the description of the low-frequency complex is not yet complete.

PBK relation Psaltis *et al.* (1999a; PBK) found a rather good correlation between the frequencies of L_ℓ and $L_{\ell ow}$ on the one hand and L_h and L_{LF} on the other spanning nearly three decades in frequency, with the Z and atoll sources populating the ν_ℓ (>100 Hz) range and the weak LMXBs and black holes in the LS the $\nu_{\ell ow}$ (<10 Hz) one, and Cir X-1 (Section 2.9.5) filling in the gap between L_ℓ and $L_{\ell ow}$ (Figs. 2.11*c*, 2.14*b*). The correlation combines

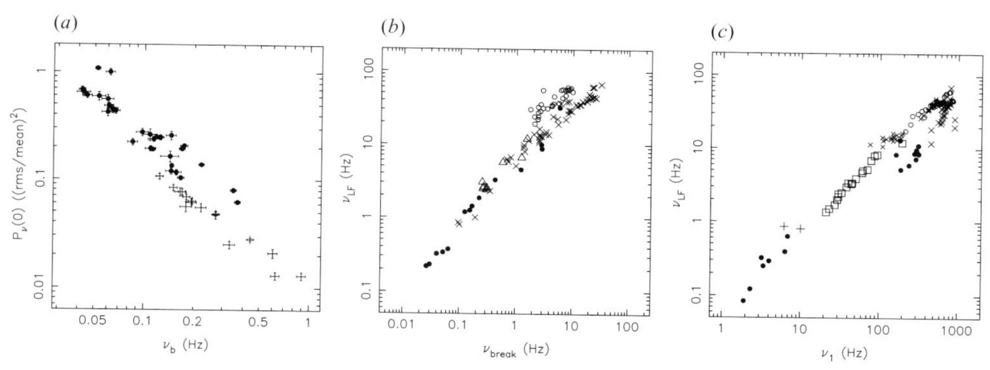

Fig. 2.11. Frequency and power correlations across neutron stars and black holes. (*a*) BLN flat-top power level vs. break frequency (after Belloni *et al.* 2002a), (*b*) low-frequency hump/QPO vs. noise break frequency (after Wijnands & van der Klis 1999a), (*c*) low-frequency hump/QPO vs. ν_ℓ or $\nu_{\ell ow}$ (after Psaltis *et al.* 1999a). Filled circles represent black-hole candidates, open circles Z sources, crosses atoll sources, triangles the millisecond pulsar SAX J1808.4−3658, pluses faint burst sources and squares Cir X-1.

features from different sources with very different Q values with relatively little overlap, and, as Psaltis *et al.* (1999a) note, although the data are suggestive, they are not conclusive. Whether the black-hole HF QPOs contributing to the small branch below the main relation in Fig. 2.11*c* are part of these correlations is questionable in view of their presumed constant frequencies (Section 2.6.1) – on the other hand, in XTE J1550–564 a correlation does occur between *observed* HF and LF QPO frequencies (and spectral hardness; see Section 2.10.1).

Further work (e.g., van der Klis 1994b; Crary *et al.* 1996; van der Hooft *et al.* 1996, 1999a; Belloni *et al.* 1996, 2002a,b; Méndez and van der Klis 1997; Ford & van der Klis 1998; Méndez *et al.* 1998d; Wijnands & van der Klis 1999b; Nowak 2000; Revnivtsev *et al.* 2000c; van Straaten *et al.* 2000, 2002, 2003, 2005; di Salvo *et al.* 2001a; Kalemci *et al.* 2001, 2003; Remillard *et al.* 2002c; Wijnands & Miller 2002; Olive *et al.* 2003; Yu *et al.* 2003; Reig *et al.* 2004; Klein-Wolt *et al.* 2004b) produced many examples of power spectra confirming these correlations, with in particular the weak LMXBs and some ordinary atoll sources, as well as some transient black holes in their decay bridging the gap between neutron stars and black holes in the PBK relation. In black holes a feature perhaps similar to L_u in neutron stars in the 10–100 Hz range was reported (Nowak 2000; Belloni *et al.* 2002a). However, some possible discrepancies also turned up in some of these same works (e.g., Belloni *et al.* 2002a; van Straaten *et al.* 2002, 2005; Pottschmidt *et al.* 2003) and as there is no direct observation of a gradual transition, the identification of the high-Q lower-kHz QPO in Z and atoll sources with the low-Q $L_{\ell ow}$ component in the neutron-star and black-hole low states remains conjectural.

Consequences of the correlations As remarked above, the identification of variability phenomena across source types is difficult, so these results should be interpreted with some caution. Nevertheless, the relations of Fig. 2.11 suggest that physically similar phenomena cause the frequencies plotted there. If so, then these phenomena are extremely tunable, in some cases over nearly three orders of magnitude in frequency, and occur in neutron stars as well as black holes. Warner and Woudt (2002) and Mauche (2002) noted that the PBK relation may even extend to white dwarf systems (Chapter 10), down to frequencies a factor

10^2 below those in X-ray binaries, which would mean strong field gravity is not required for producing the frequencies ν_ℓ and ν_b. However, an alternative is that the systems produce the correlations by a mechanism that complements a given input frequency, which may have a different physical cause in different systems, with an output frequency consistent with the correlation (cf. Abramowicz *et al.* 2004). In either case, the origin of the *correlations* must be a structure common to all these systems, most likely the disk. If one of the kHz QPO frequencies is an orbital motion frequency, orbital motion in the strong-field region of neutron stars and black holes is implicated by the high frequencies observed.

2.8 Orbital and epicyclic frequency models

Because of their direct link with physical timescales, interpretation of observed characteristic frequencies dominates the discussion about the nature of rapid X-ray variability. What modulates the X-ray flux (Section 2.8.7) and why the phenomenon is quasi-periodic rather than periodic (Section 2.8.6) often gets less attention. Many QPO models are essentially models for a periodic phenomenon (e.g., orbital motion) supplemented with a decohering mechanism (e.g., damping). Some broad-band noise models also work in this way, but many (and some QPO models as well) are instead intrinsically aperiodic (e.g., shots), with characteristic frequencies arising through, e.g., correlations in the signal (cf. Section 2.12.3). Another important distinction is that between constant variability frequencies, which may be expressible in parameters of the compact object (M, R, J) alone, and variable ones, which involve flow, or radiation-field parameters.

Phenomena that occur in both neutron stars and black holes cannot rely on physical properties unique to either object, such as a material surface or a horizon, a magnetic field not aligned with the spin or extreme values of J/M. Hence, these essentially require an origin in the accretion flow. Instead, in phenomena unique to either neutron stars or black holes a role for unique compact-object properties is likely.

For accretion-flow phenomena, the most obvious source of variability is the disk, which with Keplerian orbital motion at each radius and various oscillation modes provides a multitude of variability frequencies. Other structures, e.g., the "corona" from X-ray spectral models (Section 2.5.1), a magnetosphere, a neutron-star/disk boundary layer, or a jet (Chapter 9) can contribute as well. Nevertheless, orbital motion (including general-relativistic epicyclic motions, Section 2.8.1) and disk oscillations (Section 2.12.1) are the mechanisms most often considered for QPO phenomena.

In the current section we look at the interpretations of variability frequencies that involve orbital and epicyclic motions in the accretion disk. Some interpretations recur in various models and can be considered model "building blocks". Models based on flow instabilities, and interpretations of other aspects of the variability (amplitude, coherence, phase and their photon-energy dependencies) are discussed in Section 2.12. Some specific models proposed for specific variability components are mentioned in Sections 2.9 and 2.10.

2.8.1 General-relativistic orbital motion

In classical physics, free-particle orbits around a spherically symmetric mass M are closed, and occur with Keplerian frequency

$$\nu_K = \sqrt{GM/r^3}/2\pi \approx 1184\,\mathrm{Hz}\left(\frac{r}{15\,\mathrm{km}}\right)^{-3/2} m_{1.4}^{1/2} \approx 184\,\mathrm{Hz}\left(\frac{r}{100\,\mathrm{km}}\right)^{-3/2} m_{10}^{1/2}$$

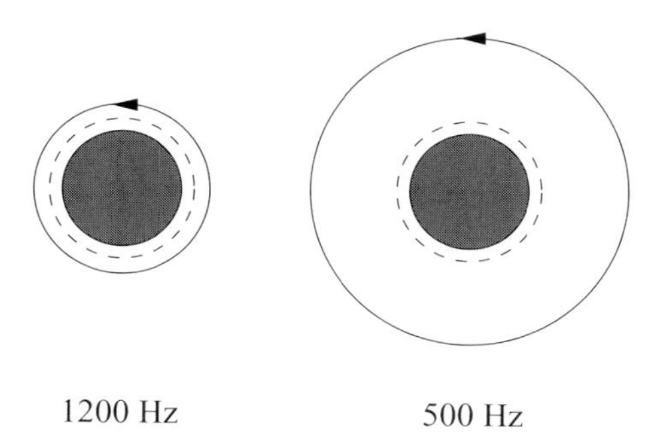

Fig. 2.12. A 10 km radius, 1.4 $M_\odot$ neutron star with the corresponding innermost stable circular orbit (ISCO; dashed circles) and orbits (drawn circles) corresponding to orbital frequencies of 1200 and 500 Hz, drawn to scale.

where $m_{1.4}$ and m_{10} are the compact object's mass in units of 1.4 and 10 $M_\odot$, respectively, and r is the orbital radius. In general relativity, orbits are not closed, as the frequencies of azimuthal, radial and vertical motion differ (Fig. 2.15, e.g., Merloni *et al.* 1999): in addition to the azimuthal motion at the *general-relativistic orbital frequency* ν_ϕ, there are the *radial* and *vertical epicyclic* frequencies ν_r and ν_θ.[4] Due to this, eccentric orbits waltz at the *periastron precession frequency* $\nu_{\mathrm{peri}} = \nu_\phi - \nu_r$ and orbits tilted relative to the equatorial plane of a spinning central mass wobble at the *nodal precession frequency* $\nu_{\mathrm{nodal}} = \nu_\phi - \nu_\theta$.

For equatorial circular orbits in Kerr spacetime (i.e., around a spinning point mass M with angular momentum J) the orbital frequency is given by

$$\nu_\phi = \frac{\sqrt{GM/r^3}/2\pi}{1 + j(r_g/r)^{3/2}} = \nu_K(1 + j(r_g/r)^{3/2})^{-1}$$

where $j \equiv Jc/GM^2$ is the Kerr angular-momentum parameter;[5] $0 < j < 1$ for prograde orbits, $-1 < j < 0$ for retrograde ones, and $r_g \equiv GM/c^2$. For the Schwarzschild geometry, $j = 0$, we have $\nu_\phi = \nu_K$. Infinitesimally tilted and eccentric orbits will have radial and vertical epicyclic frequencies of

$$\nu_r = \nu_\phi \left(1 - 6(r_g/r) + 8j(r_g/r)^{3/2} - 3j^2(r_g/r)^2\right)^{1/2} \quad \text{and}$$

$$\nu_\theta = \nu_\phi \left(1 - 4j(r_g/r)^{3/2} + 3j^2(r_g/r)^2\right)^{1/2}$$

(Fig. 2.15). For arbitrary-inclination orbits, see Sibgatullin (2002). All these frequencies are as observed by a static observer in asymptotically flat spacetime, i.e., at infinity. In a Schwarzschild geometry a local observer, due to gravitational time dilation, sees a frequency higher by $(1 - 2(r_g/r)^{-1/2}$; in the general case, the locally observed frequency also depends on the observer's angular motion.

General relativity predicts (e.g., Bardeen *et al.* 1972) that in a region close to a compact object no stable orbital motion is possible; hence the above expressions are directly useful

[4] Various other terms are used for these motions: e.g., "longitudinal" for ν_ϕ; "latitudinal" or "meridional" for ν_θ.

[5] The quantity j is also sometimes called a_* or, exceptionally, a. Most authors let a denote Jc/GM, some use $a = J/M$, e.g., Stella & Vietri (1998). r_g sometimes denotes $2GM/c^2$.

only outside that region. In a Schwarzschild geometry the *innermost stable circular orbit* (ISCO) or *marginally stable orbit* has a radius

$$r_{\rm ms} = 6r_{\rm g} = 6GM/c^2 \approx 12.5m_{1.4}\,{\rm km} \approx 89m_{10}\,{\rm km}$$

and the corresponding orbital frequency, the highest stable orbital frequency, is

$$\nu_{\rm ms} = c^3/2\pi 6^{3/2}GM \approx (1566/m_{1.4})\,{\rm Hz} \approx (219/m_{10})\,{\rm Hz}$$

For prograde orbital motion in the equatorial plane of a Kerr geometry the ISCO is smaller; as $j \to 1$, $r_{\rm ms} \to r_{\rm g}$ and

$$\nu_{\rm ms} = c^3/4\pi GM \approx (1611/m_{10})\,{\rm Hz}$$

This is still outside the hole, because the horizon radius

$$r_{\rm horizon} = r_{\rm g}(1 + (1 - j^2)^{1/2})$$

shrinks from $2r_{\rm g}$ at $j = 0$ to $r_{\rm g}$ as $j \to 1$ as well. The corresponding frequency increases with j, and hence for a black hole of known mass a high (test-particle) orbital frequency can be used to argue for its spin (Sunyaev 1973; see also Section 2.8.5). The full expression for the ISCO radius (Bardeen *et al.* 1972) follows from the condition $r^2 - 6r_{\rm g}r + 8r_{\rm g}^{3/2}r^{1/2}j - 3r_{\rm g}^2 j^2 \geq 0$ for stable orbits and is somewhat tedious; substitution into the expression for ν_ϕ yields the Kerr geometry ISCO frequency in closed form. To first order in j (Kluźniak *et al.* 1990; Miller *et al.* 1998a,b)

$$r_{\rm ms} \approx (6GM/c^2)(1 - 0.54j) \qquad \text{and} \qquad \nu_{\rm ms} \approx (c^3/2\pi 6^{3/2}GM)(1 + 0.75j)$$

Some disk flows can penetrate down to inside the ISCO before the matter plunges in (e.g., Abramowicz *et al.* 2004; cf. Section 2.8.5), but not beyond the marginally bound orbit at $r_{\rm mb} = r_{\rm g}(2 - j) + 2r_{\rm g}(1 - j)^{1/2}$, which is inside the ISCO (at $4r_{\rm g}$ in the Schwarzschild geometry).

Spacetime outside a spherically symmetric non-rotating star is Schwarzschild. For *spinning* stars the exterior spacetime is Kerr to first order in j; to higher order the metric, and hence the precise frequencies, depend on the mass distribution (Hartle & Thorne 1968; see Miller *et al.* 1998a; Shibata & Sasaki 1998; Morsink & Stella 1999; Marković 2000; Sibgatullin 2002; Abramowicz *et al.* 2003a). Depending on mass and EOS (Section 2.1), spinning neutron stars could have appreciable angular momentum (e.g., $j \sim 0.2$ and $\sim$0.5 for 500 and 1000 Hz spins, respectively, Miller *et al.* 1998a). For small-j holes and slowly spinning neutron stars the first-order expressions above apply. In all cases $\nu_{\rm ms}$ scales as $1/M$.

As ν_r and ν_θ are both $<\nu_\phi$, periastron and nodal precession are both prograde. Periastron precession is a consequence of the non-$1/r^2$ nature of gravity in general relativity; the classic example is Mercury's general-relativistic perihelion precession. Nodal or "Lense–Thirring" precession (Lense & Thirring 1918) is due to the "frame dragging" caused by the central object's spin and does not occur if $j = 0$; this "gravito-magnetic"[6] effect has not yet been detected with certainty in any system. In the weak-field ($r_{\rm g}/r \ll 1$) slow-rotation ($j \ll 1$) limit the prediction is that $\nu_{\rm nodal} = (GM)^2 j/\pi c^3 r^3 = 8\pi^2\nu_\phi^2 I\nu_{\rm spin}/Mc^2$, where I is the neutron-star moment of inertia and $\nu_{\rm spin}$ its spin frequency: by measuring $\nu_{\rm nodal}$, ν_ϕ and

[6] This term does not refer to any magnetic field, but to an analogy with electromagnetism.

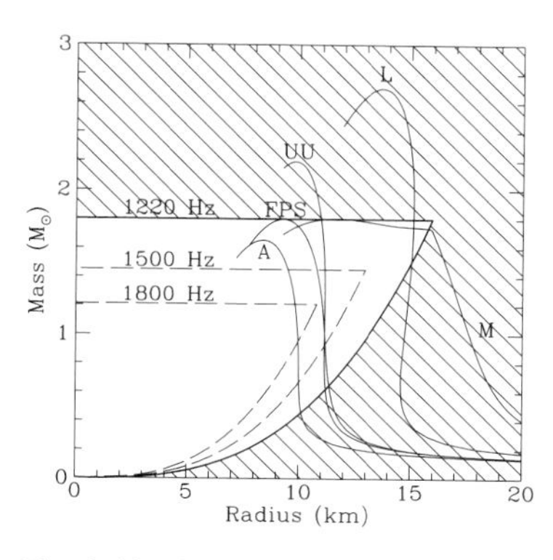

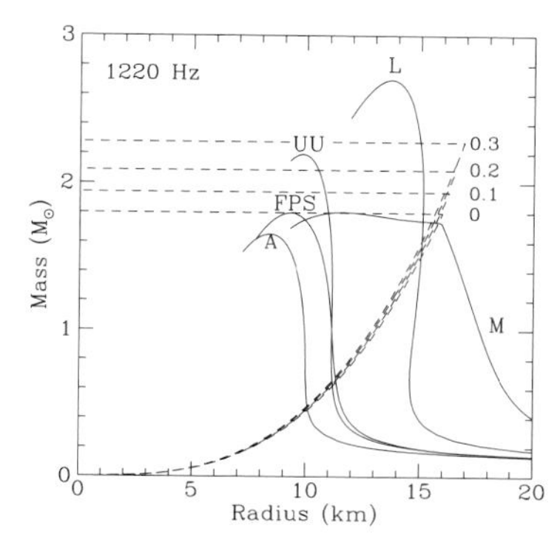

Fig. 2.13. Constraints on neutron-star M and R from orbital motion. *Left*: for $j = 0$, orbital frequencies as indicated; the hatched area is excluded if $\nu_\phi = 1220$ Hz. *Right*: with first-order corrections for frame dragging for the values of j indicated. Mass–radius relations for representative EOS are shown. From Miller *et al.* (1998a).

$\nu_{\rm spin}$ the neutron-star structure-dependent quantity I/M is constrained. In terms of I_{45}/m, where I_{45} and m are I and M in units of 10^{45} g cm^2 and $M_\odot$, respectively,

$$\nu_{\rm nodal} = 13.2\ {\rm Hz}\left(\frac{I_{45}}{m}\right)\left(\frac{\nu_\phi}{1000\ {\rm Hz}}\right)^2\left(\frac{\nu_{\rm spin}}{300\ {\rm Hz}}\right)$$

values of I_{45}/m between 0.5 and 2 are expected (Stella & Vietri 1998).

The detection of stable orbital motion at frequency ν around a neutron star constrains both its mass M and radius R (Miller *et al.* 1998a): (i) R must be smaller than the orbital radius r, and (ii) the ISCO must *also* be smaller than r (assuming the orbit is circular). Condition (i) through the expression for ν_ϕ translates into a mass-dependent upper limit on R, and condition (ii) is an upper limit on M, in Schwarzschild geometry: $6GM/c^2 < (GM/4\pi^2\nu^2)^{1/3} \Rightarrow M < c^3/(2\pi 6^{3/2} G\nu)$. Figure 2.13 shows these limits in the neutron-star mass–radius diagram.

Detection of the ISCO would constitute direct proof of a qualitative strong-field general-relativistic effect and at the same time demonstrate that the neutron star is smaller than the ISCO. This possibility was discussed since early on (e.g., Kluźniak & Wagoner 1985; Paczyński 1987; Biehle & Blandford 1993). Miller *et al.* (1998a) pointed out that when the inner edge of the accretion disk reaches the ISCO, the associated kHz QPO frequency might reach a ceiling while $\dot{M}$ continues rising (see also Section 2.9.1).

2.8.2 Relativistic precession models

The term *relativistic precession model* (Stella & Vietri 1998) is used for a class of models in which observed frequencies are directly identified with orbital, epicyclic, and precession frequencies. These models need additional physics to pick out one or more particular radii in the disk whose frequencies correspond to those observed (Section 2.8.5). Stella and Vietri (1998, 1999) identify the upper kHz QPO frequency $\nu_{\rm u}$ (Section 2.6.1) with the orbital frequency ν_ϕ at the inner edge of the disk (Section 2.8.5), and relate ν_ℓ and $\nu_{\rm h}$ (or related QPOs, Section 2.6.2) with, respectively, periastron precession ($\nu_{\rm peri}$) and nodal precession ($\nu_{\rm nodal}$) of this orbit.

So, $\nu_{\rm h}$ is predicted to be proportional to $\nu_{\rm u}^2$ (Section 2.8.1), which is indeed as observed (Section 2.7.1), and kHz QPO peak separation $\Delta\nu \equiv \nu_r$. Stellar oblateness affects the

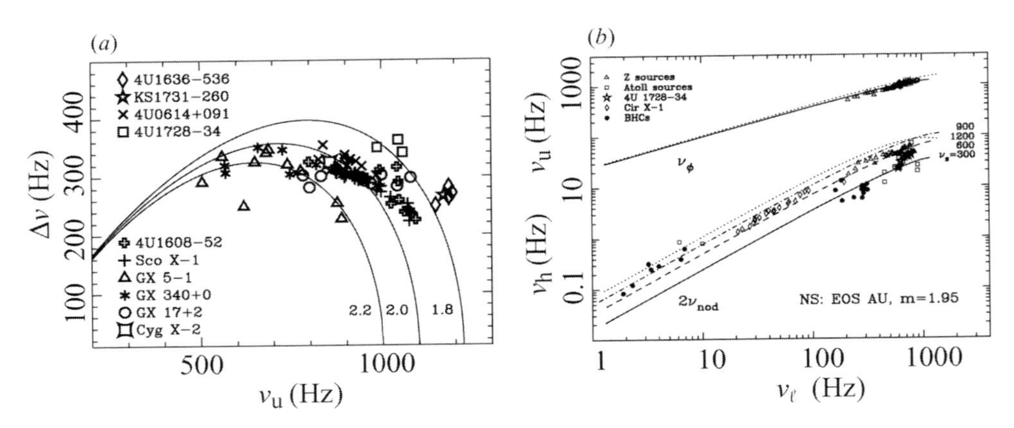

Fig. 2.14. Predicted relations between (*a*) ν_u and $\Delta\nu$ and (*b*) ν_ℓ and ν_h as well as ν_u for different values of the spin frequency (300 Hz, 600 Hz, etc.) in the relativistic precession model, compared with observed values. (Stella & Vietri 1999; Stella *et al.* 1999.) See Section 2.7.2 for a discussion of the data in the right-hand frame.

precession rates and must be corrected for (Morsink & Stella 1999; Stella *et al.* 1999). This model does not explain why $\Delta\nu$ is commensurate with the spin frequency, or how neutron stars with different spins can have the same ν_u–ν_h relation (van Straaten *et al.* 2005). A clear prediction is that $\Delta\nu$ should decrease not only when ν_u increases (as observed) but also when it sufficiently decreases (Fig. 2.14, see also Section 2.9.1.2). For acceptable neutron-star parameters (I/M, see Section 2.8.1), ν_h is several times larger than the ν_{nodal} predicted. In a warped disk geometry ν_h could be $2\nu_{nodal}$ or $4\nu_{nodal}$ (Morsink & Stella 1999). Cui *et al.* (1998b) propose identifications of some black-hole frequencies with ν_{nodal} (see also Merloni *et al.* 1999).

A precise match between model and observations requires additional free parameters. Stella & Vietri (1999) propose that orbital eccentricity systematically varies with orbital frequency. Stella *et al.* (1999), identifying ν_ℓ and $\nu_{\ell ow}$ with ν_{peri}, and ν_h and ν_{LF} with $2\nu_{nodal}$ (cf. Section 2.6), produce an approximate match to the PBK relation (Section 2.7) across neutron stars and black holes (Fig. 2.14*b*) for reasonable black-hole masses and $j = 0.1 - 0.3$, but requiring neutron-star spin rates higher than measured and masses of $\sim 2\,M_\odot$. For critical discussions of these models see Psaltis *et al.* (1999b) and Marković & Lamb (1998, 2000). Vietri & Stella (1998) and Armitage & Natarajan (1999) have performed calculations relevant to the problem of sustaining the tilted orbits required for Lense–Thirring precession in a viscous disk. Miller (1999) calculated the effects of radiation forces on Lense–Thirring precession.

Relativistic precession models are very predictive, as the frequency relations are set by little more than compact-object parameters and general relativity, and in unmodified form most are contradicted by observations (e.g., Homan *et al.* 2002; van Straaten *et al.* 2005). Yet the observed quadratic dependencies between ν_u and ν_h (Section 2.7.2) which suggest Lense–Thirring precession (Section 2.8.1), are striking. Calculations of the theoretically expected QPO sideband patterns produced by luminous clumps in orbits with epicyclic motions (Karas 1999a; Schnittman & Bertschinger 2004), and observations of such patterns, can help to further test models in this class. Certain disk oscillation models are predicted to

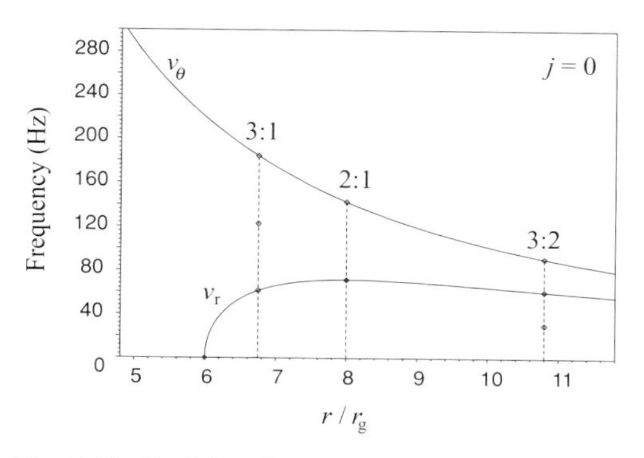

Fig. 2.15. Radial and vertical epicyclic frequencies vs. r/r_g in Schwarzschild geometry. Three resonant radii are indicated. After Abramowicz *et al.* (2004).

produce frequencies close to the orbital and epicyclic ones and may be considered as one way to implement the models described here; hydrodynamic effects are expected to somewhat modify the frequencies and to produce combinations between them, which may allow better fitting of the data. These models are discussed in Section 2.12.1. The general idea that an observed frequency is the orbital frequency ν_ϕ at the appropriate radius in the disk is applied very often, and papers identifying observed frequencies with orbital frequencies of clumps, vortices, etc., are too numerous to cite (see e.g., Chagelishvili *et al.* 1989; Abramowicz *et al.* 1992; Bao & Østgaard 1994, 1995; Karas 1999b; Wang *et al.* 2003 and Section 2.9.1.3).

2.8.3 *Relativistic resonance models*

Relativistic resonance models (Kluźniak & Abramowicz 2001; Abramowicz & Kluźniak 2001) make use of the fact that at particular radii in the disk the orbital and epicyclic frequencies have simple integer ratios or other commensurabilities with each other or with the spin frequency. At these radii resonances may occur which show up in the observations: general relativity itself picks out the frequencies from the disk. Various different physical effects can produce resonance. A type of resonance invoked in some of these models is "parametric resonance": resonance in a system whose eigenfrequency ν_0 is itself perturbed at a frequency ν_1 commensurate with ν_0; resonances occur when $\nu_0/\nu_1 = 2/n$, $n = 1, 2, 3, \ldots$

Various resonant radii have been discussed (Fig. 2.15): the radii where ν_r/ν_ϕ equals 1/2 or 1/3 (Kluźniak & Abramowicz 2001), that where $\nu_r/\nu_\theta = 2/3$ (Kluźniak & Abramowicz 2002; parametric resonance with $n = 3$, the lowest value allowed as $\nu_r < \nu_\theta$), and those where $\nu_\theta = 2\nu_r$ or $\nu_\theta = 3\nu_r$ (Abramowicz & Kluźniak 2004) can all be used to explain the observed approximate 2:3 ratios for HF QPOs in black holes (Section 2.10.1), with other resonant frequencies (1:2:3:5) also being predicted (Kluźniak & Abramowicz 2003). Like the ISCO frequencies, all these resonant frequencies are predicted to scale with $1/M$ (Abramowicz *et al.* 2004), and if the observed resonance is identified they can be used to constrain both M and J.

In the interaction between the neutron-star spin and the disk, resonances could also arise (e.g., Psaltis 2001; van der Klis 2002). Kluźniak *et al.* (2004) suggest several ways in which the epicyclic frequencies can resonate with spin (either $\nu_{\rm spin}$ or $\nu_{\rm spin}/2$; see also Wijnands *et al.* 2003 who suggest $\nu_\phi - \nu_r$ equals $\nu_{\rm spin}$ or $\nu_{\rm spin}/2$ and Lee *et al.* 2004, who consider $\nu_\theta - \nu_r = \nu_{\rm spin}/2$), which is relevant to neutron-star kHz QPOs and their commensurabilities with the spin frequency (Section 2.9.1.2), and also point out that disk *g* modes (Section 2.12.1) may resonate in a frequency ratio tending to $\sqrt{2}$, all of which may be relevant to the twin kHz QPOs observed in SAX J1808.4–3658 (Section 2.9.1.2). Lamb and Miller (2003) propose a resonance at the "spin-resonance" radius where the spin–orbit beat frequency equals the vertical epicyclic one: $\nu_{\rm spin} - \nu_\phi = \nu_\theta$; at this radius the pulsar beam stays pointed at particles in the same *phase* (e.g., an antinode) of their vertical epicyclic motion. This radius is sufficiently far out in the disk that $\nu_\phi \approx \nu_\theta$, i.e., $\nu_\phi \approx \nu_{\rm spin}/2$, which may explain why the observed kHz QPO separation is sometimes the spin frequency, sometimes half that (Section 2.9.1.2).

The radii at which a resonance condition applies are fixed, so these models in principle produce constant frequencies, which is appropriate to black-hole HF QPOs, but analytic (Rebusco 2004) and numerical (Abramowicz *et al.* 2003b) work on "tunable" versions which might be applicable to kHz QPOs has been done. Tunable frequencies can also be produced by beating constant resonant frequencies with a variable one (Lamb & Miller 2003; see Section 2.8.4). Observationally, the 2:3 ratios in black holes (Section 2.10.1), the half-spin separation of some kHz QPOs (Section 2.9.1.2) and the factor ~1.45 offsets of some of the frequency correlations in accreting millisecond pulsars (Section 2.7.1) all suggest that resonances play a role, but it is not yet entirely clear exactly which frequency ratios can occur, when they occur, and what is the incidence of each ratio. Clarifying this will allow testing of the various proposals for frequency commensurabilities that have been put forward.

2.8.4 Beat-frequency models

By "beating" orbital frequencies with the spin frequency of a central neutron star more frequencies can be produced (e.g., Alpar & Shaham 1985; Lamb *et al.* 1985; Miller *et al.* 1998a). This requires that some azimuthally non-uniform structure co-rotating with the spin, e.g., a non-aligned magnetic field or radiation pattern, reaches out to the relevant orbital radius. These models do not work for black holes, for which, in view of the no-hair theorem, no such structures are expected. Additionally there must be some azimuthal structure in orbit to interact with the spin; the mechanism will not work if the flow is completely axi-symmetric.

Spin–orbit interaction then takes place at the difference frequency between the orbital and spin frequencies: the *beat frequency* $\nu_{\rm beat} = \nu_{\rm orb} - \nu_{\rm spin}$. This is the natural disk–star interaction frequency: the frequency at which a particle orbiting in the disk periodically overtakes a given point on the spinning star. An example of an interaction making $\nu_{\rm beat}$ observable would be the periodic illumination of an orbiting blob by a pulsar beam sweeping around at $\nu_{\rm spin}$ (Fig. 2.16). Only *one* beat frequency (a "single sideband") is produced: for prograde orbital motion no signal occurs at $\nu_{\rm orb} + \nu_{\rm spin}$ while for retrograde motion *only* this sum frequency is generated. If the beat interaction occurs at the magnetospheric radius (see Chapter 1) of an accreting magnetic star one expects $\nu_{\rm spin} < \nu_{\rm orb}$, but in other settings, if the relevant orbit is sufficiently far out, $\nu_{\rm spin} > \nu_{\rm orb}$ is possible, and for a variable orbital radius $\nu_{\rm beat}$ could go through zero.

If there is an n-fold symmetric azimuthal pattern associated with spin and/or orbit, frequency multiples occur (e.g., for 2 pulsar beams $\nu_{\rm beat} = 2(\nu_{\rm orb} - \nu_{\rm spin})$). If spin and orbit have an $n_{\rm spin}$ and $n_{\rm orb}$-fold symmetry, respectively, then a degeneracy occurs and $\nu_{\rm beat} = \mu(n_{\rm spin}, n_{\rm orb})(\nu_{\rm orb} - \nu_{\rm spin})$, where $\mu(x, y)$ denotes the least common multiple (Wijnands *et al.*

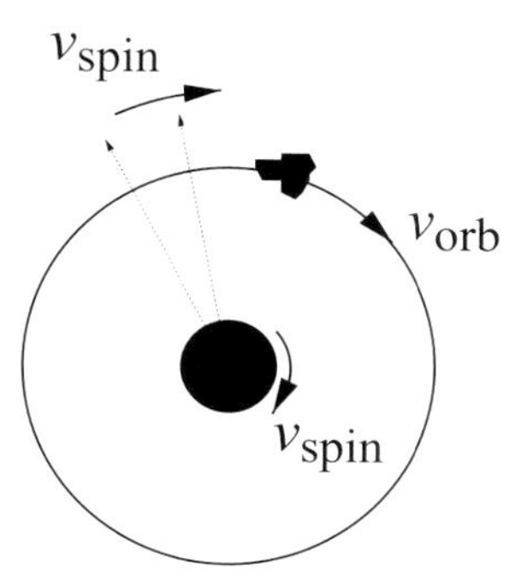

Fig. 2.16. Example of a beat-frequency interaction. The clump orbiting with $\nu_{\rm orb}$ periodically overtakes the magnetic-field lines or the pulsar beam that is sweeping around with $\nu_{\rm spin}$. This happens $\nu_{\rm beat}$ times per unit of time, where $\nu_{\rm beat} = \nu_{\rm orb} - \nu_{\rm spin}$ is the beat frequency.

2003; additionally non-sinusoidal signals include harmonics to the fundamental frequency, and if these also interact a complex spectrum results, Miller *et al.* 1998a). Usually $\nu_{\rm orb}$ is the orbital frequency ν_ϕ, but other azimuthal motions, e.g., periastron and nodal precession, could beat with the spin as well.

In the magnetospheric beat-frequency model, the spin interacts with clumps orbiting just outside the magnetosphere (Section 2.8.5; Alpar & Shaham 1985, see also Alpar & Yılmaz 1997). The model was originally applied to CVs (Patterson 1979) and to the HBO (Section 2.9.3); similar models have been proposed for kHz QPOs (Section 2.9.1) by Cui (2000) and Campana (2000). In the sonic-point beat-frequency model for kHz QPOs (Miller *et al.* 1998a) $\nu_{\rm u}$ and ν_ℓ are identified with respectively ν_ϕ and $\nu_{\rm beat}$ at the sonic radius $r_{\rm sonic}$ (Section 2.8.5). The beat occurs as an X-ray pulsar beam irradiates clumps orbiting at $r_{\rm sonic}$ once per beat period, modulating the rate at which the clumps' material accretes.

In their pure form beat-frequency models predict $\Delta\nu$ to be constant at $\nu_{\rm spin}$, contrary to observations (Section 2.9.1.2). However, in the sonic-point beat-frequency model, if the clumps spiral down while their material accretes, $\Delta\nu$ can be less than this (Lamb & Miller 2001). The observation of twin kHz QPOs separated by half the neutron-star spin frequency is inconsistent with all spin–orbit beat-frequency models of the type discussed above (see Section 2.9.1.2). Lamb and Miller (2003) have proposed that an orbit–orbit beat-frequency interaction between orbital motion at $r_{\rm sonic}$ and an azimuthal structure at the spin-resonance radius (Section 2.8.3) explains ν_ℓ (and maintain that $\nu_{\rm u} = \nu_\phi(r_{\rm sonic})$). In this scenario the beat is between two different radii in the disk. With the spin frequencies of an increasing number of low-magnetic-field neutron stars in LMXBs becoming known (Section 2.9.1.2), further testing of the predictions of beat-frequency models for the QPO frequencies in these systems can be expected to put further rigorous constraints on the theoretical possibilities.

This completes the summary of the "orbital motion" QPO models. How can observations discriminate between these various models? As it turns out, every physical frequency model predicts its own "fingerprint" set of parasitic, weaker frequencies in addition to the strong ones it set out to explain in the first place (e.g., Miller *et al.* 1998a). These predicted patterns of weaker frequencies provide a strong test of each model. Searches for such weaker power-spectral features are very difficult because the limited of the sensitivity of current instrumentation, and the distinct impression of the observers is that there is much still hiding below the formal detection levels. However, a small number of weak sidebands have been detected in neutron stars (Jonker *et al.* 2000b, 2005; Wijnands *et al.* 2003; Section 2.9.1) and the 2:3 ratio HF QPOs in black holes (Strohmayer 2001a; Miller *et al.* 2001; Remillard *et al.* 2003; Section 2.10.1) provide similar constraints on possible models. So far, these detections have led to only a very limited number of proposed precise theoretical explanations in each case (e.g., Psaltis 2000; Kluźniak & Abramowicz 2001; Schnittman & Bertschinger 2004), an unusual situation that testifies to the discriminating power of such additional frequencies. However, too few of these frequencies have been detected yet for the full power of this method to be applied; more sensitivity to weak timing features is required for this. To some extent these fingerprint patterns depend on the way in which the frequencies physically modulate the X-ray flux; models for this are discussed next.

2.8.5 Preferred radii

In models involving frequencies depending strongly on radius (e.g., of orbital and epicyclic motion), *preferred radii* in the disk are required to produce specific frequencies. Constant and variable radii produce corresponding frequencies. Variable radii usually depend on $\dot{M}$ through disk physics (e.g., $r_{\rm mag}$, below), whereas constant radii follow from just the compact object's parameters, perhaps through strong-field gravity effects (e.g., the ISCO radius $r_{\rm ms}$, Section 2.8.1 or resonant radii, Section 2.8.3).

The "inner edge of the disk" at radius $r_{\rm in}$, where the near-Keplerian flow ends and the radial velocity becomes appreciable compared to the azimuthal one, provides a natural preferred radius. The density contrast at $r_{\rm in}$ can be sharp, as the radial velocity of the flow can change suddenly (e.g., Miller *et al.* 1998a); this is a strong-field gravity effect related to the small difference in orbital angular momentum between different radii (see Paczyński 1987). In the absence of magnetic stresses, $r_{\rm ms}$ (for standard thin disks) or $r_{\rm mb}$ (for low radiative efficiency disks such as ADAFs, slim disks and super-Eddington flows; Section 2.8.1) set a lower limit to the flow, but for magnetized flows predictions are uncertain; e.g., Abramowicz *et al.* 1978; Liang & Thompson 1980; Lai 1998; Krolik & Hawley 2002; Watarai & Mineshige 2003a). In neutron stars the surface provides a fixed radius R limiting $r_{\rm in}$; depending on the EOS, R may be smaller or larger than $r_{\rm ms}$, but note that some modulation models (Section 2.8.7) do not work if the disk actually extends down to the star.

Radiation drag by photons emitted from within $r_{\rm in}$ removing angular momentum from the disk flow through the Poynting–Robertson effect can truncate the flow at radius $r_{\rm rad}$ (Miller *et al.* 1998a). As the radiation providing the drag is produced by the same accretion flow that it interacts with on the way out, to first order this radius is fixed, but details of flow and scattering geometries can make it variable. According to Miller *et al.* (1998a) $r_{\rm rad}$ cannot be larger than $\sim 15GM/c^2$, and shrinks when $\dot{M}$ increases. For a sufficiently magnetic neutron star, electromagnetic stresses truncate the disk well outside the ISCO and $r_{\rm rad}$ at the magnetospheric radius (Chapter 1) $r_{\rm mag}$, which shrinks when $\dot{M}$ increases; note that according to some authors orbital motion of part of the accreting matter down to $r_{\rm rad}$ still occurs within $r_{\rm mag}$ (e.g., Miller *et al.* 1998a). Both these disk truncation mechanisms may only work for neutron stars: black holes are not expected to have a sufficiently strong magnetic field (Chapter 1), and may also have difficulty in producing a sufficiently strong, and low specific angular-momentum, radiation field from within $r_{\rm in}$.

Some authors treat the radius of maximum flux from the disk (e.g., Gruzinov 1999) or the radius of maximum pressure in a toroidal flow (Kluźniak *et al.* 2004) as preferred; in such cases one needs to investigate if the function whose extremum is used to define the radius is sufficiently narrow to produce the observed QPO. Laurent and Titarchuk (2001) proposed an interesting model where just the presence of a Comptonizing converging flow makes one particular disk orbital frequency stand out.

2.8.6 Decoherence

The lack of coherence (aperiodicity) of particular observed frequencies attributed to orbital and epicyclic phenomena is usually taken to more or less naturally follow from the sheared and turbulent nature of the flow, where any feature such as a density enhancement ("clump") or a vortex will have a finite lifetime (see Section 2.12.3). Although suggested by the Lorentzians used to describe QPO peaks (Section 2.2), and sometimes implicitly assumed in calculations (e.g., Titarchuk 2002), there is not much evidence that observed signals are produced by (superposition of) exponentially damped harmonic oscillators. Rapid frequency fluctuations, for example related to a varying preferred radius (e.g., a clump spiralling in, Stoeger 1980), or superposition of multiple frequencies, e.g., generated from a finite-width disk annulus, are alternatives (e.g., Abramowicz *et al.* 1992; Nowak 1994; Bao & Østgaard 1994). As noted by Belloni *et al.* (2002a), the distinction (Section 2.2) between a QPO (with two frequencies, ν_0 and λ) and a BLN (with just one break frequency) becomes much less fundamental when interpreted in terms of the width of a disk annulus than when described in terms of damped harmonic oscillators. This may be why the use of $\nu_{\rm max}$ (Section 2.2) to summarize both cases seems to work empirically.

2.8.7 Modulation

Orbital and epicyclic motions must modulate the X-ray flux for them to be observable. A classic mechanism (Cunningham & Bardeen 1972; Sunyaev 1973; see the end of Section 2.8.2 for further references) is that of a self-luminous blob orbiting a black hole in general-relativistic spacetime. Doppler boosting and gravitational light bending produce the modulation. Related models that are considered are centrally illuminated scattering blobs (e.g., Lamb & Miller 2003) and self-luminous turbulent disk flows (e.g., Armitage & Reynolds 2003). Variable obscuration of central emitting regions by matter orbiting further out is often invoked, but is attractive only if the predicted inclination dependencies occur. This is, for example, the case for the ~1 Hz dipper QPOs (Section 2.9.4), where a similar modulation of all emission (persistent and bursts, different photon energies) further strengthens the case for this mechanism (Jonker *et al.* 1999).

A fundamentally different mechanism is modulation of the accretion rate into the inner regions, either taking the form of an actual modulation of total $\dot{M}$ (e.g., Lamb *et al.* 1985) or affecting only the pattern by which the matter accretes (e.g., Miller *et al.* 1998a). Likewise, either the total emitted flux can be modulated (*luminosity modulation*), or only its angular distribution (*beaming modulation*). "Beaming" here is any anisotropy in the emission. Scattering material around the source suppresses a beamed modulation much more effectively than a luminosity modulation (Wang & Schlickeiser 1987; Brainerd & Lamb 1987; Kylafis & Klimis 1987; Bussard *et al.* 1988; Asaoka & Hoshi 1989; Miller 2000). In magnetospheric beat-frequency models the accretion of matter from orbiting clumps (and hence the luminosity) is modulated at $\nu_{\rm beat}$ as accretion is easier near the magnetic poles

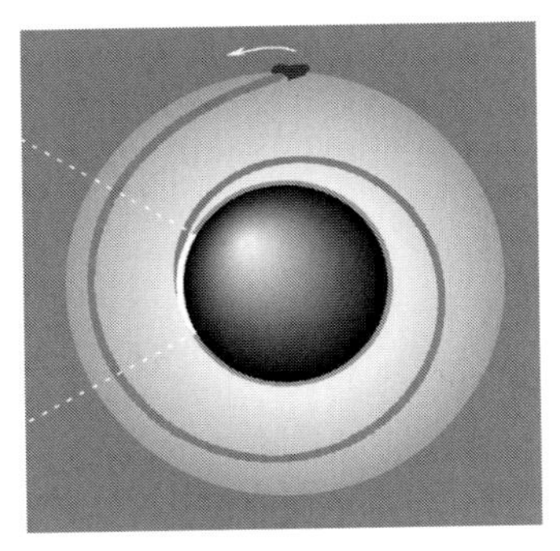

Fig. 2.17. Example of a modulation mechanism. The clump with its spiral flow and the emission from the flow's footpoint (dashed lines) in the Miller *et al.* (1998a) model all rotate at the same angular velocity (that of the clump's orbit), irrespective of the stellar spin, and produce a beaming modulation at that frequency. From Miller *et al.* (1998a).

(Lamb *et al.* 1985), leading to oscillating shots (Section 2.2) in which the luminosity oscillates at the beat frequency and whose lifetime is that of a clump. In the sonic point model of Miller *et al.* (1998a), material from clumps orbiting at $r_{\rm in}$ gradually accretes following a fixed spiral-shaped trajectory in the frame co-rotating with the clumps (Fig. 2.17) at whose "footpoint" accretion, and hence emission, is enhanced. The resulting hot spot moves around the surface at $\nu_\phi(r_{\rm in})$ irrespective of the star's spin, and produces a beaming modulation. Both of these models predict surface hot spots (revolving at $\nu_{\rm spin}$ or $\nu_\phi(r_{\rm in})$, respectively) whose emission is itself modulated at $\nu_{\rm beat}$, leading to the potential for sideband formation.

Fully specified orbital and epicyclic flux-modulation models make specific predictions for fractional amplitude, harmonic content and sideband pattern, and energy dependencies (see also Section 2.12.2) of the variability, so there is no lack of stringent tests to distinguish between such models, and hence the frequency models to which they are appropriate. As the modulation models shape up from the currently still common simple qualitative ideas to quantitatively described mechanisms, it can be expected that many simple models will turn out to need revision.

2.9 Low-magnetic-field neutron stars

The main variability components and the correlations between their frequencies have already been summarized in Sections 2.6 and 2.7. Here (Sections 2.9 and 2.10) a more detailed overview of the observed timing properties of neutron stars and black holes, respectively, is provided. Tables 2.5 and 2.6 summarize the literature on neutron stars and reference codes in the current section are linked to those tables; likewise the codes in Section 2.10 refer to Table 2.7.

2.9.1 Kilohertz quasi-periodic oscillations

Kilohertz QPOs (Section 2.6.1) are seen in a wide range of low-magnetic-field neutron-star sources, including all Z sources, most atoll sources, several transients and two millisecond pulsars (14c, 13b), but not the GX atoll sources (Section 2.4.2), which do not usually reach the relevant states (Section 2.5.2.2). Table 2.5 lists all sources with kHz QPOs, as well as those with burst oscillations (Chapter 3) or millisecond pulsations (Section 2.9.1.2).

Table 2.4. *kHz QPO frequencies (Hz)*

Atoll & weak LMXBs		Z	
ν_ℓ	ν_u	ν_ℓ	ν_u
	(120)		
	(500)	(200)	(500)
300	600	(300)	600
500	800	500	800
800	1000	800	1000
1000	1200		

Typical observed kHz QPO frequency ranges. Exceptions include narrower ranges in GX 5–1 and GX 340+0 (maximum frequencies are 150–200 Hz less; 7b, 16b). Parenthesized values refer to ranges in which usually $Q < 2$.

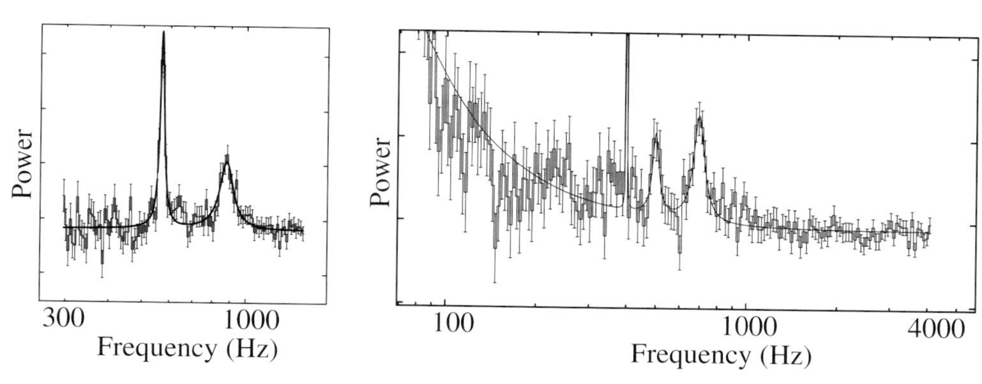

Fig. 2.18. Twin kHz QPOs in *left*: 4U 1608−52 (Méndez *et al.* 1998b) and *right*: the 401 Hz accreting pulsar SAX J1808.4−3658 (Wijnands *et al.* 2003; the 401 Hz pulsar spike is seen as well).

Kilohertz QPO frequencies increase with source state (S_a, S_z, Section 2.5.2) in all cases. Table 2.4 summarizes the typical frequencies seen in well-covered sources; see Section 2.7 for the corresponding states. Below ~500 Hz L_u turns into a BLN component (Section 2.6.2) which is seen (also in weak LMXBs) down to frequencies as low as 120 Hz (14e). A 10–20 Hz BLN ($L_{\ell ow}$, Section 2.6.2) seen in the EIS is sometimes interpreted as a low-frequency version of L_ℓ (4z, but see 14e).

Kilohertz QPO amplitudes increase with photon energy by typically a factor 4 between 3 and >10 keV (8a,b,g,j, 9b,d, 11a, 19a,g, 21a, 27a). In similar bands, the QPOs are weaker in the more luminous sources, with 2–60 keV rms amplitudes ranging from nearly 20% in the weakest atoll sources to typically 2–5% in the Z sources (Jonker *et al.* 2001). They also weaken towards the extremes of their observable range (Fig. 2.19), where often only one peak is detected. At high energy, amplitudes are much higher (e.g., 40% rms >16 keV in 4U 0614+09; 9d). This has been interpreted in different ways, e.g., as an effect of

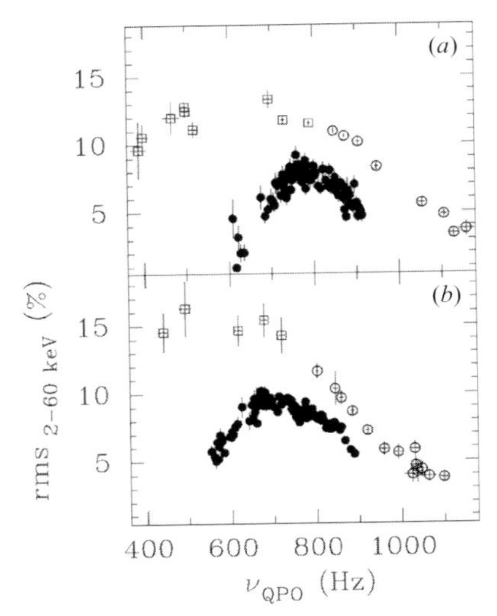

Fig. 2.19. Lower (*filled symbols*) and upper (*open symbols*) kHz QPO amplitudes as a function of their frequency for (*a*) 4U 1728–34 and (*b*) 4U 1608–52 (Méndez *et al.* 2001).

Comptonization in a central corona (Miller *et al.* 1998a) and as due to the combination of variable hard flux from a boundary layer with a constant, softer, disk flux (Gilfanov *et al.* 2003).

Peak widths are affected by variations in centroid frequency during the integration. Typical values are several 10–100 Hz, somewhat broader in Z sources than in atoll sources and usually higher-Q at higher frequency. L_ℓ in atoll sources can be very sharp (Q ~100), then attains excellent signal-to-noise and exhibits soft lags of 10–60 μs (9c, 19c). These lags are opposite to those expected from inverse Compton scattering, so they may originate in the QPO production mechanism rather than in propagation delays.

Weak sidebands 50–64 Hz above ν_ℓ with powers ~0.1 that of L_ℓ have been detected in three objects (Jonker *et al.* 2000b, 2005), as well as an additional high-frequency QPO peak 8–12 Hz above the 401 Hz pulse frequency in SAX J1808.4–3658 (Wijnands *et al.* 2003). In both cases the separation between main peak and sideband increased with QPO frequency, but was different from simultaneous LF QPO frequencies, complicating interpretation. These sidebands are the first examples of the "fingerprints" of weaker parasitic frequencies expected to accompany the kHz QPOs in all models (Section 2.8). Lense–Thirring precession, orbital motion within the inner disk edge (Jonker *et al.* 2000b) and relativistic disk-oscillation modes (Psaltis 2000) have been considered as explanations.

2.9.1.1 *Relation with luminosity*

On timescales of hours and in a given source, kHz QPO frequency ν typically correlates well with L_X, but, remarkably, on longer timescales, and across sources, this ν–L_X correlation is lost, with the result that similar QPO frequencies are observed over two orders of magnitude in L_X (e.g., van der Klis 1997, 8b, 9a; Section 2.5). Due to this, approximately parallel tracks form in ν–L_X diagrams covering either (i) several sources (Fig. 2.20*a*), or (ii) several observations of a single source separated by $\gtrsim$1 day (Fig. 2.20*b*). In a given

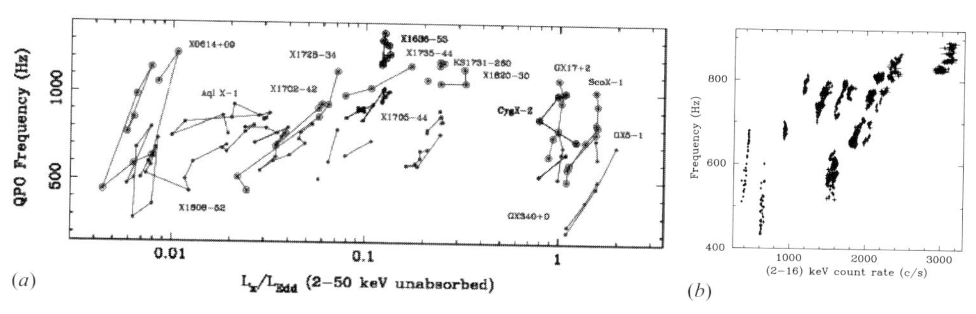

Fig. 2.20. The two parallel-tracks phenomena. (*a*) Across sources (Ford *et al.* 2000; upper and lower kHz peaks are indicated with different symbols). (*b*) In time, in the source 4U 1608–52 (Méndez *et al.* 1999), frequency plotted is ν_ℓ.

source, QPO amplitude is affected much less by the L_X shifts between tracks than predicted if the shifts were caused by an extra source of X-rays unrelated to the QPOs (8g). This is also true across sources.

That different sources have different ν–L_X tracks might be because ν depends not only on L_X, but also (inversely) on a parameter related to average L_X (e.g., van der Klis 1997, 1998; Zhang *et al.* 1997b), such as magnetic-field strength B (White & Zhang 1997) which previously, on other grounds, was hypothesized to correlate to average L_X (e.g., Hasinger & van der Klis 1989). Alternatively, the two parallel-track phenomena might be explained *together* by noting that kHz QPO frequency seems to be governed not by L_X, but by how much L_X deviates from its time average $\langle L_X \rangle$ over a day or so (van der Klis 1999). This could arise if there is an L_X component (e.g., due to nuclear burning or the accretion of a radial inflow) that is proportional to some *time average* over the inner-disk accretion rate $\dot{M}_d$, and whose effect is to make the frequency lower (e.g., via the Miller *et al.* 1998a radiative disk-truncation mechanism; Section 2.8.5). Frequency would then scale with $\dot{M}_d/L_X$ where $L_X = \dot{M}_d + \alpha\langle \dot{M}_d \rangle$, with α the relative efficiency of the additional energy release (cf. Section 2.5). Simulations of this toy model (van der Klis 2001) reproduce the salient features of the observed parallel tracks seen in Fig. 2.20; testing this requires sustained monitoring of the ν–L_X correlation.

Kilohertz QPOs are sensitive to very short-term L_X variations as well. In Sco X-1 amplitudes and frequencies systematically vary on sub-second timescales, in phase with the 6 Hz flux variations of the NBO (Section 2.9.4; 4d). In 4U 1608–52, kHz QPO frequency ν has been observed to increase, as usual, with $\dot{M}$-induced L_X increases, but to decrease in response to probable nuclear-burning ("mHz" QPO, Section 2.9.4) induced L_X increases (8h), confirming a prediction of the Miller *et al.* (1998a) model, where the QPO occurs at ν_ϕ of the inner disk radius r_{in} (Section 2.8.5), which is predicted to increase in response to the extra, nuclear-burning generated luminosity component.

2.9.1.2 *Relation with neutron-star spin*

Peak separation $\Delta\nu$ usually decreases with increasing kHz QPO frequency ν (Fig. 2.21*a*; 4b,c, 7b, 8d, 19f, 22c). In the two cases where a probable positive ν–$\Delta\nu$ correlation ocurred, this was at the lowest detectable frequencies (11b, 22f).

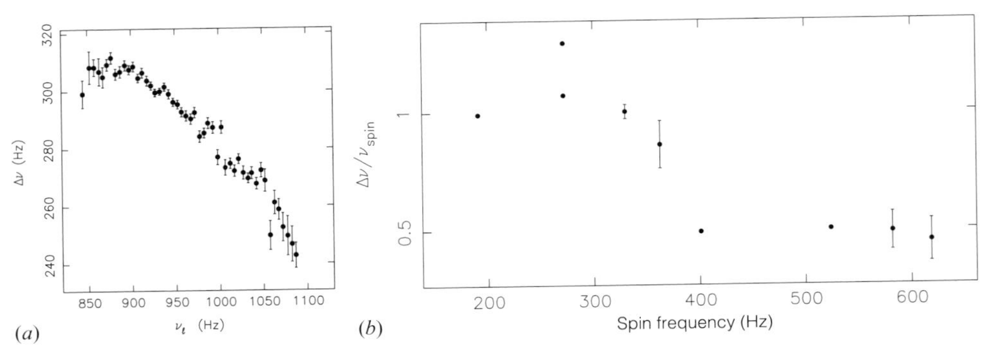

Fig. 2.21. (*a*) The variation in kHz QPO peak separation as a function of the lower kHz QPO frequency in Sco X-1, after Méndez & van der Klis (2000); (*b*) $\Delta\nu/\nu_{\rm spin}$ vs. $\nu_{\rm spin}$, after Table 2.5. Vertical bars indicate the range of variation in $\Delta\nu$.

$\Delta\nu$ is approximately commensurate with $\nu_{\rm spin}$ (Table 2.5), and this is the main motivation for beat-frequency models (Section 2.8.4). Of course, $\Delta\nu$ varies so any commensurability cannot be exact. Spin is measured directly in millisecond pulsars; also, burst oscillations (as verified in two msec pulsars; 14b,d, 15b) likely occur very near $\nu_{\rm spin}$ (Section 3). In the eight sources where both $\Delta\nu$ and spin were measured by one of these two methods (Table 2.5), $\Delta\nu$ was between 0.7 and 1.3 times $\nu_{\rm spin}$ for $\nu_{\rm spin} < 400$ Hz, and 0.36–0.57 times $\nu_{\rm spin}$ for $\nu_{\rm spin} > 400$ Hz (Fig. 2.21*b*); the largest offsets of the ratios from 1.0 and 0.5, respectively, are $\sim 8\sigma$ but usually the discrepancies are much less. In the two pulsars the ratios are 0.49 and 0.99. So, a commensurability with spin does indeed seem to exist where $\Delta\nu \approx \nu_{\rm spin}$ for low, and $\Delta\nu \approx \nu_{\rm spin}/2$ for high $\nu_{\rm spin}$.

That the spin frequency can be twice $\Delta\nu$ was initially suspected based on measurements of burst oscillations, where an alternative explanation, namely that $\nu_{\rm burst} = 2\nu_{\rm spin}$ could not be excluded (see Chapter 3). However, the case of the 401 Hz pulsar SAX J1808.4–3658 now leaves little doubt that $\Delta\nu$ *can* indeed be half the spin frequency (Wijnands *et al.* 2003). This falsifies the direct spin–orbit beat-frequency interpretation (Section 2.8.4) even when allowing for multiple neutron-star hot spots or orbiting clumps, but may be explained by models involving resonances, perhaps combined with a beat (Section 2.8.3; 14c). Seventeen low-magnetic-field neutron-star spins in the millisecond range have now been measured (thirteen burst oscillations and six pulsars, with an overlap of two sources), and ten more spins are known up to a factor of 2. Table 2.5 lists these 27 objects. The spin frequencies are between $\sim$200 and $\sim$700 Hz, suggesting a cutoff well below the limit set by observational constraints and indicating that a braking mechanism limits $\nu_{\rm spin}$ (see Chakrabarty *et al.* 2003). If the stars spin at the magnetospheric equilibrium spin rates (Chapter 1) corresponding to their current $L_{\rm X}$, this predicts a tight correlation between $L_{\rm X}$ and magnetic-field strength B (White & Zhang 1997; a similar possibility came up to explain the similar kHz QPO frequencies at very different $L_{\rm X}$, Section 2.9.1). Another possibility is that gravitational radiation limits $\nu_{\rm spin}$ by transporting angular momentum out as fast as accretion is transporting it in; this predicts these sources to be the brightest gravitational-wave sources, with a known $\nu_{\rm spin}$ facilitating their detection (Bildsten 1998; Andersson *et al.* 1999; 2000; also Levin 1999, and e.g., Bildsten & Ushomirsky 2000; Brown & Ushomirsky 2000; Rezzolla *et al.* 2000; Ushomirsky *et al.* 2000; Yoshida & Lee 2001; Wagoner 2002).

Table 2.5. *Kilohertz QPOs and millisecond neutron-star spins*

Source	kHz QPOs	$\Delta\nu$ (Hz)	$\nu_{\rm burst}$ (Hz)	$\nu_{\rm pulse}$ (Hz)	$\frac{\Delta\nu}{\nu_{\rm spin}}$	Remarks	References
Millisecond pulsars							
XTE J0929–314	—	—	—	185	—	T	10a;14e
XTE J1751–305	—	—	—	435	—	T	12a;14e
XTE J1807–294	2	$\sim$190	—	191	$\sim$0.99	T	13a,b
SAX J1808.4–3658	3[a]	195	401	401	0.49	T	14a-e
XTE J1814–338	—	—	314	314	—	T	15a,b;14e
IGR J00291+5934	—	—	—	599	—	T	80a
Z sources							
Sco X-1(1617–155)	2	245–310	—	—	—		4a-f,z,9g,h
GX 340+0 (1642–455)	2	280–410	—	—	—		16a,b;4f,z;8j;9g,h
GX 349+2 (1702–363)	2	265	—	—	—		17a-c
GX 5–1 (1758–250)	2	240–350	—	—	—		7a,b;4f,z;9g,h
GX 17+2 (1813–140)	2	240–300	—	—	—		11a-d;4f;9g,h
Cyg X-2 (2142+380)	2	345	—	—	—		3a,b;4f,z;9g,h
Atoll sources							
4U 0614+09	2	240–360	—	—	—		9a-h;4f;22h
2S 0918–549	1	—	—	—	—		18a
4U 1608–52	3[b]	225–325	619	—	0.36–0.53	T	8a-j,t;4f;9c,e,g,h
4U 1636–53	4[b]	240–330	582	—	0.41–0.57		19a-g,k;4f;8f,t;9c,g,h
4U 1702–43	2	320–340	330	—	0.97–1.03		20a;9h
4U 1705–44	1	—	—	—	—	T	21a,b;9h
4U 1728–34	3[b]	275–350	363	—	0.76–0.96		22a-h;4f;8f,g,t;9g,h
KS 1731–260	2	260	524	—	0.50	T	23a;4f;9g,h
4U 1735–44	2	295–340	—	—	—		24a,b;4f;9h
SAX J1750.8–2900	2[c]	317[c]	601	—	—	T	32a
4U 1820–30	2	220–350	—	—	—	G	25a-e;4f;9g,h
Aql X-1 (1908+005)	1	—	549	—	—	T	26a-e;8g,t;9g,h
4U 1915–05	2	290, 350[g]	272	—	1.07, 1.29[g]	D	33a,d
XTE J2123–058	2	255–275	—	—	—	T	34a,b
Other sources[d]							
EXO 0748–676	1	—	45	—	—	T, D	27a,c
MXB 1659–298	—	—	567		—	T, D	28a
XTE J1723–376	1[e]	—	—	—	—	T	29a
MXB 1743–29	—	—	589	—	—	[f]	30a;9g
SAX J1748.9–2021	—	—	410	—	—	T, G	31a

References see Table 2.6

kHz QPOs: number of kHz QPO peaks and sidebands with $Q \geq 2$; $\nu_{\rm burst}$: see also Section 3.4. Remarks: T: transient; D: dipper; G: in globular cluster.

Notes: [a] Sideband to pulsation. [b] Sideband(s) to lower kHz QPO. [c] Second QPO tentative. [d] Faint, transient, unidentified, etc., and not unambiguously classified. [e] QPO not confirmed. [f] Source identification uncertain. [g] Two incompatible values of $\Delta\nu$ reported, 33a.

Some caution is advised in interpreting the SAX J1808.4–3658 kHz QPO result, as the twin peaks were observed only once in this pulsar, its $\Delta\nu$ (195 Hz) is lower than in non-pulsars, and other commensurabilities also exist between the observed frequencies (14c). Kilohertz QPOs in XTE J1807–294 have a $\Delta\nu$ close to its 191 Hz spin frequency (13b). The frequency correlations of some pulsars are a factor ~1.45 off the usual ones (Section 2.7). Clearly, further detections of twin kHz QPOs in millisecond pulsars would help to clarify the systematics in this phenomenology. While the kHz QPO centroid frequency ratio ν_ℓ/ν_u is certainly not constant, Abramowicz *et al.* (2003c) propose that ratios near 2:3 may occur more often than others, which might provide a link with black-hole high-frequency QPOs.

2.9.1.3 Interpretations

Kilohertz QPO models include beat-frequency, relativistic precession and relativistic resonance models (Section 2.8) as well as disk oscillations and other flow instabilities (Section 2.12). All proposals involve plasma motion in the strong-field region, and with one exception (photon bubbles, Section 2.12.4) in all models the QPOs originate in the disk. Most identify ν_u with orbital motion at a preferred radius in the disk (in models of Titarchuk and co-workers ν_ℓ has this role), Fig. 2.13 illustrates the constraints the orbital-motion hypothesis provides on neutron-star parameters. The highest measured ν_u of 1329 ± 4 Hz in 4U 0614+09, when interpreted as ν_ϕ for $j = 0$, implies $M_{NS} < 1.65\,M_\odot$ and $R_{NS} < 12.4$ km, imperiling the hardest equations of state; for a 300 Hz spin these numbers become 1.9 $M_\odot$ and 15.2 km (van Straaten *et al.* 2000).

The maximum kHz QPO frequencies in well-studied sources are constrained to a relatively narrow range of $\nu_u = 850$–1330 Hz. If this is the ISCO frequency ν_{ms} (Section 2.8.1), the neutron-star masses are near 2 $M_\odot$ (Zhang *et al.* 1997b; see also Kaaret *et al.* 1997). An apparent leveling off of ν_u as a function of count rate, flux and S_a at ~1060 Hz was found in 4U 1820–30 (25b-d), but further observations cast doubt on this (25e). A tendency to level off (but without a true "ceiling") may be a general feature of the parallel-tracks phenomenon (cf. Fig. 2.20); indeed, the toy model described in Section 2.9.1.1 predicts such a pattern as a consequence of ν_u depending on $\dot{M}_d/\langle\dot{M}_d\rangle$.

Calculations exploring to what extent kHz QPOs constrain the EOS, usually assuming $\nu_u = \nu_\phi$, have further been performed by e.g., Akmal *et al.* (1998), Datta *et al.* (1998, 2000), Kluźniak (1998), Miller *et al.* (1998b), Bulik *et al.* (1999, 2000), Li *et al.* (1999), Schaab & Weigel (1999), Stergioulas *et al.* (1999), Thampan *et al.* (1999), Heiselberg & Hjorth-Jensen (1999), Kalogera & Psaltis (2000), Zdunik *et al.* (2000), Glendenning & Weber (2001), Gondek-Rosinska *et al.* (2001), Ouyed (2002) and Mukhopadhyay *et al.* (2003).

The evidence for orbital motion as the cause of kHz QPOs is not yet iron-clad. Orbital motion could be empirically demonstrated by a number of different possible measurements, such as independent measurements of frequency and orbital radius, where radius could be measured from continuum or line spectroscopy, demonstrating the predicted frequency–radius relation, measurements of orbital and epicyclic frequencies varying together in the way predicted by general relativity, spectroscopic measurements of the Doppler effect in an orbiting hot spot or measurements of the orbital frequency ceiling predicted at the ISCO. Initial explorations of some of these possibilities have already been performed.

2.9.2 *Hectohertz QPOs*

Figures 2.6 and 2.8 show examples of hHz QPOs, and Fig. 2.9 illustrates their approximately constant frequencies (Section 2.6.1) in the 100–200 Hz range. Hectohertz QPOs occur in atoll sources (including the millisecond pulsar SAX J1808.4–3658) when $\nu_b \gtrsim 1$ Hz (8i, 9f, 14a,e, 21b, 22b,e,f,g,h, 26e, see also Fragile *et al.* 2001), but have not been seen in most weak LMXBs (which may not reach the relevant states) or in black holes. They may also occur in Z sources (e.g., 11g). At low ν_b ($\sim$0.2 Hz) confusion occurs with L_u. The phenomenon is usually rather low-Q but occasionally peaks up to $Q > 2$, has an amplitude between 2 and 20% (rms) and becomes weaker and more coherent as ν_b rises. The near-constant frequency allows diskoseismic modes as a model (Section 2.12.1) and suggests a link with black-hole HF QPOs (see Section 2.6.1). Fragile *et al.* (2001) suggest ν_{hHz} could be the orbital frequency at the radius where a warped disk is forced to the equatorial plane by the (strong-field gravity) Bardeen and Petterson (1975) effect. Titarchuk (2003) proposed Rayleigh–Taylor gravity waves at the disk–star boundary layer as the origin of hHz QPOs (cf. Brugmans 1983).

2.9.3 *The low-frequency complex*

Components of the low-frequency complex have been studied in Z and atoll sources since the 1980s (e.g., 7c,3B; van der Klis 1995a for a review). The 15–60 Hz LF QPO in the Z-source HB and upper NB (the "horizontal branch oscillation" or HBO) and the 0.1–30 Hz BLN component (formerly called "low-frequency noise" or LFN in Z sources and confusingly, sometimes "high-frequency noise" in atoll sources, 3B, 8l) were first detected in GX 5–1 (7c); HBO harmonics were sometimes seen (e.g., 11g). Frequencies were found to correlate with spectral state (4i-l,n, 7e,i,j, 38a,b, 3h,q,y,B, 11f-i, 19h). The variations were spectrally hard (e.g., 16d, 7i,k), and the HBO showed hard lags of a few milliseconds (e.g., 7f,k). Similarities with black-hole LS and VHS variability were pointed out (e.g., van der Klis 1994a,b, 8l,m). The BLN was sometimes peaked (in atoll sources, e.g., 38a,b, 25f,g, and in Z sources, e.g., 11b, 17d, 4k,n), but in atoll sources QPOs were only rarely seen (25g, 8l). The classic Z and atoll tracks in CDs and HIDs (Section 2.5.2) were known (3B and e.g., 19h, 7i,j), but the fainter EIS was hardly explored (e.g., 8k, 21c, 19i).

In recent work, with the exploration of the fainter sources and states, similarities in the phenomenology across source types were clarified (Section 2.6), with the frequency correlations between, and among, kHz and LF phenomena (Section 2.7) particularly revealing (e.g., 4a,b,e, 24b, 3a, 20a, 33a, 16b, 7b, 11b, 3b,s, 22g, 21b, 26e), and the EIS–LS match confirmed (Section 2.5). This led to the synthesis summarized in Sections 2.6.2 and 2.7 (4f,y,z, 9c,f, 11c, 22g,h, 8i, 14e). Clear examples of LF QPOs were found in atoll sources (23a, 39a, 24a, 20a, 9f, 33a, 22g,h, 21b, 26e, 11c, 8i) and millisecond pulsars (14e) and these were incorporated into the general picture, although the phenomenology of these QPOs is not yet entirely clear, with apparently both L_b and L_h often producing a narrow ($Q > 2$) peak when they move to high (>30 Hz) frequency, and indications for two harmonically related QPOs in the 0.2–2 Hz range (and once near 80 Hz) very roughly a factor 2 below ν_h (8i,l, 11c, 14e).

Models for LF QPOs (particularly, the Z-source HBO) include the magnetospheric beat-frequency model (Section 2.8.4), Lense–Thirring precession (Section 2.8.2) and various disk oscillations (Section 2.12.1), e.g., the magnetically warped precessing-disk model (Lai 1999; Shirakawa & Lai 2002a). The frequency correlations among low-frequency complex phenomena and with kHz QPOs, and in particular the factor $\sim$1.5 shifts observed in those relations in some millisecond pulsars (Section 2.7.1) are among the main observational features to be accounted for in model scenarios, but presently are too fresh to have affected the models reported here.

2.9.4 *Other phenomena*

A 1–3% rms $Q \sim 2$ QPO near 6 Hz occurs in Z sources in the NB (e.g., 4h,i,j,n, 11b,g,i, 3h,r, 7b,i,j,l, 3o, 16b). $\sim 180^\circ$ energy-dependent phase lags occur in some of these QPOs, which can be interpreted in terms of a quasi-periodically pivoting X-ray spectrum with a pivot point between 3 and 7 keV (3g,r, 4o, 7m). In some sources, this "normal branch oscillation" or NBO seems to jump to 10–20 Hz when the source moves into the FB (e.g., 4j,l,n, 11g,i) while in others it disappears. The 6–14 Hz QPOs rarely seen at the tip of the UB in some atoll sources may be related to this N/FBO (25h, 47a,b, 26e). In Z sources the phenomenon has been modeled as a radiation–force feedback instability in spherical near-Eddington accretion (Fortner *et al.* 1989), but detections in atoll sources occur at luminosities well below Eddington. Disk oscillation models for NBO have also been proposed (e.g., Alpar *et al.* 1992, Wallinder 1995).

An $\alpha = 1.2$–2 power-law component called very-low frequency noise (VLFN) with an rms amplitude of usually a few percent often dominates the variability $\lesssim 1$ Hz. This component is detected in all Z (e.g., 7b,f,i,j, 4j,l,n, 11b,g,i, 3h,o, 17c,e,f) and atoll sources (e.g., 9f, 19d, 22g, 25f, 8i), including some faint and/or transient sources (31a, 47a,b) as well as the Rapid Burster and Cir X-1 (Section 2.9.5). It is rarely detected in the EIS and usually becomes stronger (up to typically 6% rms but sometimes much stronger, 7b) and often also steeper (up to $\alpha = 2$) towards higher states. It has been variously ascribed to accretion-rate variations and unsteady nuclear burning (Section 2.12.4). In recent work, breaks and broad <1 Hz Lorentzians have sometimes been detected in the VLFN range (e.g., 39b, 26e, 38h), which provides some support for models that produce the VLFN from a superposition of finite events such as nuclear "fires" (Bildsten 1995).

In X-ray dip sources (Chapter 1), a 0.6–2.4 Hz QPO occurs (41a, 27b, 40a) which, contrary to nearly all other variability has an amplitude (5–10% rms) that hardly depends on photon energy. Its occurrence in dippers suggests a link with high system inclination; the QPO persists at near-constant fractional amplitude right through X-ray bursts and dips, so quasi-periodic obscuration of central emitting regions by structure above the plane of the disk is an attractive model (cf. 3z). Titarchuk and Osherovich (2000) proposed a specific global normal disk mode that might accomplish this. The 401 Hz pulsar SAX J1808.4–3658 in the late decay of its outbursts sometimes exhibits violent ~ 1 Hz highly non-sinusoidal flaring (14e,j,k) the nature of which is unclear.

Revnivtsev *et al.* (2001a) discovered "mHz" (0.007–0.009 Hz) QPOs that only occurred in a very particular L_X range and at energies <5 keV in 4U 1608–52 and 4U 1636–53 and interpreted them as possible variations in nuclear burning on the neutron-star surface (cf. Chapter 3). These QPOs have been found to affect the kHz QPOs (8h, Section 2.9.1.1).

2.9.5 *Peculiar low-magnetic-field neutron stars*

Cir X-1, apart from behavior attributed to periodic surges of mass transfer at periastron in its 17-d highly eccentric orbit and ill-understood long-term changes (Murdin *et al.* 1980 and references therein), is also peculiar in that it sometimes seems to exhibit the correlated spectral and <100 Hz timing behavior of an ordinary atoll source (37e) and at other

Table 2.6*a*. *Neutron-star variability and states*

Source	HB/NB/FB	LF noise	HB/N-FB/kHz QPOs	References
Z sources				
Sco X−1 (1617–155)	○/•/•	•	•/•/•	4a-A;3A,B;9g,h
GX 340+0 (1642–455)	•/•/•	•	•/•/•	16a-e;3A,B;4y,z;8j,r;9g,h
GX 349+2[a] (1702–363)	○/•/•	•	○/-/•	17a-j;3A,B
GX 5–1 (1758–250)	•/•/•	•	•/•/•	7a-q;3A,B;4f,y,z;9g,h;8r
GX 17+2 (1813–140)	•/•/•	•	•/•/•	11a-l;3A,B;4f,x,y;9g,h;8r
Cyg X−2 (2142+380)	•/•/•	•	•/•/•	3a-B;4f,n,x-z;7d,f;8r;9g,h
Z source candidates[a]				
RX J0042.6+4115	○/○/○	-	-/-/-	35a
LMC X−2 (0521–720)	○/○/○	-	-/-/-	36a-c
Cir X−1 (1516–569)	○/○/○	•	•/-/-	37a-j;4f,x;11c

HB: horizontal branch, NB: normal branch, FB: flaring branch. •: variability of this type observed, ○: some doubt (uncertain, ambiguous, atypical, rare and not clearly seen, etc.), -: not reported.
Note: [a] Z/atoll character somewhat ambiguous, or evidence incomplete.

times, at higher L_X, that of a (somewhat atypical) Z source (37h). However, no kHz QPOs have been reported, which sets it apart from both classes. BLN and 1–30 Hz LF QPOs are seen, as well as broad noise out to several 100 Hz (37a,b,f-h). Both the timing properties and the hard X-ray spectrum make the source resemble a black hole, but the detection of X-ray bursts from the field, almost certainly from the source itself (Tennant *et al.* 1986), strongly argues in favor of a neutron star.

The *Rapid Burster* (MXB 1730–335) is a transient source with thermonuclear bursts (Chapter 3) as well as repetitive "Type II" bursts attributed to a recurrent accretion instability (e.g., Lewin *et al.* 1993) possibly related to similarly interpreted phenomena seen in the high-magnetic-field neutron star GRO J1744–28 (Section 2.11, cf. Lewin *et al.* 1996) and the black hole GRS 1915+105 (Section 2.10.4). In addition to ~0.05 Hz wave trains with gradually increasing frequency apparently excited by the Type II bursts (44i,l), complex 0.5–7 Hz QPOs are observed both inside and outside the Type II bursts (references see Table 2.6) which can be quite strong (up to 35% rms, 44b) with sometimes (usually weak) harmonics. Apart from the similar frequencies and the fact that correlations exist with flux and spectral parameters, no obvious similarities exist between the correlated spectral and timing behavior of the Rapid Burster and other LMXBs. A link with Z-source NBOs was suggested (44d), but the properties of the source do not fit in well with those of either Z or atoll sources, or Cir X-1 (44k). No significant kHz variability has been detected (44m). Models for type II bursts in which the neutron-star magnetic field temporarily interrupts accretion (e.g., Hanawa *et al.* 1989) provide a setting where QPOs can be produced through magnetic interactions with the accretion flow (e.g., Hanami 1988); non-magnetic disk oscillations are a possibility as well (e.g., Cannizzo 1997).

Other sources that do not fit seamlessly within the framework sketched here are GX 13+1 and GX 349+2, which show somewhat ambiguous Z/atoll behavior (cf. 3B and other

Table 2.6*b*. *Neutron-star variability and states*

	EIS/IS/B	Strong BLN	LF/hHz/kHz QPOs	References
Millisecond pulsars				
XTE J0929–314	•/-/-	•	•/-/-	10a;14e
XTE J1751–305	○/-/-	•	-/-/-	12a;14e,j
XTE J1807–294	-/-/○	-	-/-/•	13a,b;14j
SAX J1808.4–3658	•/•/○	•	•/○/•	14a-k;4x,y;8q
XTE J1814–338	○/-/-	•	•/-/-	15a-c;14e,j
Atoll sources				
4U 0614+09	•/•/•	•	○/•/•	9a-k;4f,y;8q;22h
2S 0918–549	○/•/•	•	-/-/•	18a
4U 1608–52	•/•/•	•	•/•/•	8a-t;4x,y;3B;9c,e,g,h
4U 1636–53	○/•/•	•	-/-/•	19a-k;3B;8f,m,p,t;9c,g,h
4U 1702–43	-/-/○	○	•/○/•	20a,b;8p;9h
4U 1705–44	•/•/•	•	-/•/•	21a-e;3B;4x,y;8m,q-s;9h,l
4U 1728–34 (GX354–0)	•/•/•	•	•/•/•	22a-j;3B;4f,y;8f,g,p-r,t;9g,h
GX 9+9 (1758–169)	-/-/•	-	-/-/-	3A,B;38h
KS 1731–260	-/-/•	-	•/-/•	23a,b;4f;8p,q;9g,h
4U 1735–44	-/○/•	-	•/-/•	24a-e;3B;4f,y;9h
GX 3+1 (1744–265)	-/-/•	-	-/-/-	38a-h;3A,B
SAX J1750.8–2900	-/•/•	-	○/-/•	32a
GX 9+1 (1758–205)	-/-/•	-	-/-/-	3A,B;38h
GX 13+1 (1811–171)[a]	-/○/•	-	•/-/-	39a-d;3A,B;8r
4U 1820–30 (NGC6624)	-/•/•	-	-/•/•	25a-i;3B;4f;8r;9g,h
Ser X-1 (1837+049)	-/-/•	-	○/○/-	38c;8r;25i
Aql X-1 (1908+00)	•/•/•	•	○/○/•	26a-f;4x;8g,p,r,s;9g,h
4U 1915–05	○/○/○	○	•/-/•	33a-d
XTE J2123–058	○/•/•	-	-/-/•	34a,b
Weak LMXBs / atoll source candidates[b]				
EXO 0748–676	-/○/○	-	-[c]/-/•	27a,b
4U 1323–62	-/-/-	-	-[c]/-/-	41a
MXB 1659–298	○/-/○	-	-/-/-	28a,b
XTE J1709–267	-/-/-	•	-/-/-	42a
XTE J1723–376	-/-/-	-	-/-/○	29a
1E 1724–3045 (Ter 2)	•/-/-	•	•/-/-	43a;4f,y;8q;11c;14e;23b
MXB 1730–335 (Lil 1; RB)	-/-/-	•	•/-/-	44a-m
SLX 1735–269	•/-/-	•	-/-/-	45a;8q;11c;14e;23b
4U 1746–37 (NGC6441)	-/○/•	-	-[c]/-/-	40a
GRS 1747–312 (Ter 6)	-/○/-	-	-/-/-	46a
SAX J1748.9–2021 (NGC6440)	-/-/○	-	-/-/-	31a,b
XTE J1806–246	○/○/○	•	○/-/-	47a,b
4U 1812–12	-/-/-	•	-/-/-	48a
GS 1826–238	•/-/-	•	•/-/-	49a;4f;8q,r;11c;14e;23b
4U 1850–08 (NGC6712)	-/○/-	-	-/-/-	50a

EIS: extreme island state, IS: island state, B: banana state. •: variability of this type observed, ○: some doubt (uncertain, ambiguous, atypical, rare and not clearly seen, etc.), -: not reported. Strong BLN: > 15% rms. RB: rapid burster.

Notes: [a] Z/atoll character somewhat ambiguous, [b] Evidence incomplete, [c] ~1 Hz "dipper" QPO, Section 2.9.4.

Table 2.6*b* *(Cont.)*

References in Table 2.6*a* and *b*: 3a Wijnands *et al.* 1998a, 3b Kuznetsov 2002a, 3c Branduardi *et al.* 1980, 3d Hasinger *et al.* 1986, 3e Norris & Wood 1987, 3f Hasinger 1987, 3g Mitsuda & Dotani 1989, 3h Hasinger *et al.* 1990, 3i Vrtilek *et al.* 1990, 3j Chiapetti *et al.* 1990, 3k Hjellming *et al.* 1990a, 3l Hirano *et al.* 1995, 3m Kuulkers *et al.* 1995, 3n Focke 1996, 3o Wijnands *et al.* 1997c, 3p Smale 1998, 3q Kuulkers *et al.* 1999, 3r Wijnands & van der Klis 2001, 3s Kuznetsov 2001, 3t Piraino *et al.* 2002, 3u O'Brien *et al.* 2004, 3v di Salvo *et al.* 2002, 3w Done *et al.* 2002, 3x Vrtilek *et al.* 2003, 3y Kuulkers *et al.* 1996, 3z Kuulkers & van der Klis 1995, 3A Schulz *et al.* 1989, 3B Hasinger & van der Klis 1989, 3C Dubus *et al.* 2004; 4a van der Klis *et al.* 1996, 4b van der Klis *et al.* 1997, 4c Méndez & van der Klis 2000, 4d Yu *et al.* 2001, 4e Kuznetsov 2002b, 4f Psaltis *et al.* 1999a, 4g Ilovaisky *et al.* 1980, 4h Middleditch & Priedhorsky 1986, 4i Priedhorsky *et al.* 1986, 4j van der Klis *et al.* 1987b, 4k Hasinger *et al.* 1989, 4l Hertz *et al.* 1992, 4m Kallman *et al.* 1998, 4n Dieters & van der Klis 2000, 4o Dieters *et al.* 2000a, 4p White *et al.* 1985, 4q Hjellming *et al.* 1990b, 4r Vrtilek *et al.* 1991, 4s Augusteijn *et al.* 1992, 4t Barnard *et al.* 2003a, 4u McNamara *et al.* 2003, 4v Santolamazza *et al.* 2003, 4w Bradshaw *et al.* 2003, 4x Done & Gierliński 2003, 4y Wijnands & van der Klis 1999a, 4z Psaltis *et al.* 1999b, 4A Canizares *et al.* 1975, 4B Robinson & Warner 1972; 7a Wijnands *et al.* 1998c, 7b Jonker *et al.* 2002a, 7c van der Klis *et al.* 1985, 7d Elsner *et al.* 1986, 7e van der Klis *et al.* 1987a, 7f van der Klis *et al.* 1987c, 7g Norris *et al.* 1990, 7h Mitsuda *et al.* 1991, 7i Lewin *et al.* 1992, 7j Kuulkers *et al.* 1994, 7k Vaughan *et al.* 1994, 7l Kamado *et al.* 1997, 7m Vaughan *et al.* 1999, 7n van der Klis *et al.* 1991, 7o Tan *et al.* 1992, 7p Blom *et al.* 1993, 7q Asai *et al.* 1994; 8a Berger *et al.* 1996, 8b Yu *et al.* 1997, 8c Méndez *et al.* 1998a, 8d Méndez *et al.* 1998b, 8e Méndez *et al.* 1999, 8f Jonker *et al.* 2000b, 8g Méndez *et al.* 2001, 8h Yu & van der Klis 2002, 8i van Straaten *et al.* 2003, 8j Gilfanov *et al.* 2003, 8k Mitsuda *et al.* 1989, 8l Yoshida *et al.* 1993, 8m Berger & van der Klis 1998, 8n Revnivtsev *et al.* 2001a, 8o Gierliński & Done 2002a, 8p Muno *et al.* 2004, 8q Sunyaev & Revnivtsev 2000, 8r Muno *et al.* 2002, 8s Gierliński & Done 2002b, 8t Méndez 1999; 9a Ford *et al.* 1997a, 9b Ford *et al.* 1997b, 9c Vaughan *et al.* 1997, 1998, 9d Méndez *et al.* 1997, 9e Kaaret *et al.* 1998, 9f van Straaten *et al.* 2000, 9g Psaltis *et al.* 1998, 9h Ford *et al.* 2000, 9i Ford *et al.* 1999, 9j Méndez *et al.* 2002, 9k Singh & Apparao 1994; 10a Galloway *et al.* 2002; 11a Wijnands *et al.* 1997b, 11b Homan *et al.* 2002, 11c Belloni *et al.* 2002a, 11d Kuulkers *et al.* 2002, 11e Stella *et al.* 1987b, 11f Langmeier *et al.* 1990, 11g Penninx *et al.* 1990, 11h Wijnands *et al.* 1996, 11i Kuulkers *et al.* 1997a, 11j Penninx *et al.* 1988, 11k di Salvo *et al.* 2000; 12a Markwardt *et al.* 2002; 13a Markwardt *et al.* 2003, 13b Markwardt 2004 priv. comm; 14a Wijnands & van der Klis 1998b, 14b in 't Zand *et al.* 2001, 14c Wijnands *et al.* 2003, 14d Chakrabarty *et al.* 2003, 14e van Straaten *et al.* 2005, 14f Uttley 2004, 14g Wijnands & van der Klis 1998a, 14h Menna *et al.* 2003, 14i Gierliński *et al.* 2002, 14j Wijnands 2004, 14k van der Klis *et al.* 2000; 15a Markwardt & Swank 2003, 15b Strohmayer *et al.* 2003, 15c Wijnands & Homan 2003; 16a Jonker *et al.* 1998, 16b Jonker *et al.* 2000a, 16c van Paradijs *et al.* 1988a, 16d Penninx *et al.* 1991, 16e Oosterbroek *et al.* 1994; 17a Zhang *et al.* 1998a, 17b Kuulkers & van der Klis 1998, 17c O'Neill *et al.* 2002, 17d Ponman *et al.* 1988, 17e O'Neill *et al.* 2001, 17f Agrawal & Bhattacharyya 2003, 17g di Salvo *et al.* 2001b, 17h Agrawal & Sreekumar 2003, 17i Iaria *et al.* 2004; 18a Jonker *et al.* 2001; 19a Zhang *et al.* 1996a,b; 19b Méndez *et al.* 1998c, 19c Kaaret *et al.* 1999a, 19d di Salvo *et al.* 2003, 19e Méndez 2002a, 19f Jonker *et al.* 2002b, 19g Wijnands *et al.* 1997a, 19h Prins & van der Klis 1997, 19i van der Klis *et al.* 1990, 19j Damen *et al.* 1990, 19k Jonker *et al.* 2005; 20a Markwardt *et al.* 1999a, 20b Oosterbroek *et al.* 1991; 21a Ford *et al.* 1998a, 21b Olive *et al.* 2003, 21c Langmeier *et al.* 1989, 21d Langmeier *et al.* 1987, 21e Barret & Olive 2002; 22a Strohmayer *et al.* 1996, 22b Ford & van der Klis 1998, 22c Méndez & van der Klis 1999, 22d Piraino *et al.* 2000, 22e Migliari *et al.* 2003a, 22f Migliari *et al.* 2003b, 22g di Salvo *et al.* 2001a, 22h van Straaten *et al.* 2002, 22i van Straaten *et al.* 2001, 22j Franco 2001; 23a Wijnands & van der Klis 1997, 23b Barret *et al.* 2000; 24a Wijnands *et al.* 1998b, 24b Ford *et al.* 1998b, 24c Penninx *et al.* 1989, 24d van Paradijs *et al.* 1988b; 24e Corbet *et al.* 1989; 25a Smale *et al.* 1997, 25b Zhang *et al.* 1998b, 25c Kaaret *et al.* 1999b, 25d Bloser *et al.* 2000a, 25e Méndez 2002b, 25f Dotani *et al.* 1989, 25g Stella *et al.* 1987a, 25h Wijnands *et al.* 1999a, 25i Migliari *et al.* 2004; 26a Zhang *et al.* 1998c, 26b Cui *et al.* 1998a, 26c Yu *et al.* 1999, 26d Reig *et al.* 2000a, 26e Reig *et al.* 2004, 26f Yu *et al.* 2003; 27a Homan & van der Klis 2000, 27b Homan *et al.* 1999b, 27c Villareal & Strohmayer 2004; 28a Wijnands *et al.* 2001b, 28b Wijnands *et al.* 2002b; ; 29a Marshall & Markwardt 1999; 30a Strohmayer *et al.* 1997; 31a Kaaret *et al.* 2003, 31b in 't Zand *et al.* 1999a; 32a Kaaret *et al.* 2002; 33a Boirin *et al.* 2000, 33b Bloser *et al.* 2000b, 33c Narita *et al.* 2003, 33d Galloway *et al.* 2001; 34a Homan *et al.* 1999a, 34b Tomsick *et al.* 1999; 35a Barnard *et al.* 2003b; 36a Smale & Kuulkers 2000, 36b Smale *et al.* 2003, 36c McGowan *et al.* 2003; 37a Tennant 1987, 37b Tennant 1988, 37c Ikegami 1986, 37d, Makino 1993, 37e Oosterbroek *et al.* 1995, 37f Shirey *et al.* 1996, 37g Shirey *et al.* 1998, 37h Shirey *et al.* 1999, 37i Qu 2001, 37j Ding *et al.* 2003; 38a Lewin *et al.* 1987, 38b Makishima *et al.* 1989, 38c Oosterbroek *et al.* 2001, 38d Makishima *et al.* 1983, 38e dal Fiume *et al.* 1990, 38f Asai *et al.* 1993, 38g den Hartog *et al.* 2003, 38h Reerink *et al.* 2004; 39a Homan *et al.* 1998, 39b Schnerr *et al.* 2003, 39c Matsuba *et al.* 1995, 39d Homan *et al.* 2003c; 40a Jonker *et al.* 2000c; 41a Jonker *et al.* 1999; 42a Jonker *et al.* 2003; 43a Olive *et al.* 1998; 44a Tawara *et al.* 1982, 44b Stella *et al.* 1988a, 44c Stella *et al.* 1988b, 44d Dotani *et al.* 1990, 44e Kawai *et al.* 1990, 44f Lubin *et al.* 1991, 44g Tan *et al.* 1991, 44h Lubin *et al.* 1992a, 44i Lubin *et al.* 1992b, 44j Lubin *et al.* 1993, 44k Rutledge *et al.* 1995, 44l Kommers *et al.* 1997, 44m Fox *et al.* 2001; 45a Wijnands & van der Klis 1999b; 46a in 't Zand *et al.* 2000; 47a Wijnands & van der Klis 1999c, 47b Revnivtsev *et al.* 1999a; 48a Barret *et al.* 2003; 49a in 't Zand *et al.* 1999b; 50a Kitamoto *et al.* 1992; 80a Galloway *et al.* 2005.

Table 2.7. *Black-hole rapid X-ray variability*

Source	LS/IMS/HS	Strong BLN	LF QPO	HF QPO	References
GRO J0422+32	•/–/–	•	○	—	51a-d;1v;5s;64d
LMC X-1 (0538–641)	–/•/•	—	•	—	52a-d
LMC X-3 (0540–679)	•/–/•	•	•	—	53a-e;52c,d
A0620–00	○/○/○	○	—	—	54a-b,56d,f
XTE J1118+480	•/–/–	•	•	—	55a-j;64d
GS 1124–68	•/•/•	•	•	—	56a-f;1t
GS 1354–64	•/–/–	•	•	—	57a,b;1y;64d
4U 1543–47	•/•/•	•	•	—	58a
XTE J1550–564	•/•/•	•	•	•	6a-r;1A,H;5s,x;61g
4U 1630–47	•/•/–	•	•	•	59a-g;1y;5x,6l;61g
XTE J1650–500	•/•/•	•	•	•	60a-d;5u;61g
GRO J1655–40	•/•/•	•	•	•	61a-h;1v,y,A;5s,x;6e,i,l
GX 339–4 (1659–487)	•/•/•	•	•	—	5a-w;1l,p,t-v,y,z,A,H,K;56b,e
IGR J17091–3624	•/–/–	•	—	—	62a
SAX J1711.6–3808[a]	•/•/–	•	•	—	63a,b
GRO J1719–24 (1716–249)	○/–/–	—	•	—	64a-c
GRS 1737–31	•/–/–	•	—	—	65a
GRS 1739–278	–/•/•	—	•	—	66a,b
1E 1740.7–2942	•/–/–	•	•	—	67a,b;1v,y,K;5v
H 1743–322	–/–/○	—	•	•	68a
XTE J1748–288	•/•/•	•	•	—	69a;1v,y;5x;6l;61g
XTE J1755–324	•/–/○	•	—	—	70a;1v,y
GRS 1758–258	•/–/•	•	•	—	71a,b;1v,A,K;5v;67a,b
XTE J1819–254	○/–/–	•	—	—	72a-d
EXO 1846–031	–/–/•	—	—	—	73a,b
IGR J18539+0727	•/–/–	•	—	—	62a
XTE J1859+226	–/•/–	—	•	•	74a-c;1H
XTE J1908+094	•/–/–	—	○	—	75a
GRS 1915+105	○/○/○	•	•	•	2a-C;1y,A;5s
4U 1957+11	–/–/•	—	—	—	76a-c
Cyg X-1 (1956+35)	•/•/○	•	•	—	1a-L;5p,t,v
GS 2000+25	•/•/•	•	○	—	77a;1t
XTE J2012+381	–/–/•	—	—	—	78a-c
GS 2023+338	•/○/○	•	—	—	79a-c;1u

This is not a list of certified black holes; relevant low-magnetic-field compact-object timing behavior plus an absence of bursts and pulsations suffice for inclusion. LS: low state, IMS: intermediate state (includes VHS), HS: high state. •: variability of this type observed; ○: some doubt (uncertain, ambiguous, atypical, rare and not clearly seen, etc.); —: not reported. Strong BLN: >15% rms.

Note: [a] Contamination by SAX J1712.6–3739, 63b.

References in Table 2.7: 1a Oda *et al.* 1971, 1b Terrell 1972, 1c Nolan *et al.* 1981, 1d Miyamoto & Kitamoto 1989, 1e Belloni & Hasinger 1990a, 1f Negoro *et al.* 1994, 1g Vikhlinin *et al.* 1994, 1h Crary *et al.* 1996, 1i Belloni *et al.* 1996, 1j Cui *et al.* 1997a,b, 1k Revnivtsev *et al.* 1999b, 1l Ford *et al.* 1999, 1m Nowak *et al.* 1999a,b, 1n Gilfanov *et al.* 1999, 1o Gilfanov *et al.* 2000, 1p Nowak 2000, 1q Uttley & McHardy 2001, 1r Churazov *et al.* 2001, 1s Pottschmidt *et al.* 2003, 1t Rutledge *et al.* 1999, 1u Miyamoto *et al.* 1992, 1v Wijnands & van der Klis 1999a, 1w Reig *et al.* 2002, 1x Berger & van der Klis 1998, 1y Sunyaev & Revnivtsev 2000, 1z Miyamoto *et al.* 1988, 1A Li & Muraki 2002, 1B Zdziarski *et al.* 2002, 1C Wen *et al.* 2001, 1D Frontera *et al.* 2001, 1E Gierliński *et al.* 1999, 1F Zhang *et al.* 1997a, 1G Ling *et al.* 1983, 1H Done & Gierliński 2003, 1I Kotov *et al* 2001, 1J Maccarone *et al.*

references in Table 2.6*a*), and the millisecond pulsar XTE J1751–305 (14e). The other millisecond pulsars have properties similar to those of other weak LMXBs (14e,j).

2.10 Black holes

The reference codes in this section refer to those listed with Table 2.7.

2.10.1 *High-frequency QPOs*

The seven black holes from which HF QPOs (Section 2.6.1) >100 Hz have been reported (2x,z,A,C, 6b,f,h,i,m, 59c,g, 60b, 61c,e, 68a, 74a,b) are listed in Table 2.7 (some possibly related <100 Hz oscillations are discussed at the end of this section). HF QPOs are weak, transient, and energy dependent, and are detected only at high count rate. In many cases they hover around formal detection levels, so there is considerable uncertainty about their exact properties, but typically Q ranges between less than 2 and 10, and amplitudes are between 0.5 and 2% rms (2–60 keV). HF QPOs usually occur in the VHS, but in XTE J1550–564 a 250 Hz QPO was seen in an IMS at a count rate 70–85% below the VHS (6f).

←

Table 2.7. *(Cont.)*

2000, 1K Lin *et al.* 2000a, 1L Maccarone & Coppi 2002a; 2a Chen *et al.* 1997, 2b Morgan *et al.* 1997, 2c Paul *et al.* 1997, 2d Belloni 1998, 2e Swank *et al.* 1998, 2f Reig *et al.* 2003b, 2g Trudolyubov *et al.* 1999a,b, 2h Markwardt *et al.* 1999b, 2i Feroci *et al.* 1999, 2j Muno *et al.* 1999, 2k Naik *et al.* 2000, 2l Rao *et al.* 2000a,b, 2m Chakrabarti *et al.* 2000, 2n Reig *et al.* 2000b, 2o Lin *et al.* 2000c, 2p Belloni *et al.* 2001, 2q Strohmayer 2001b, 2r Nandi *et al.* 2001, 2s Tomsick & Kaaret 2001, 2t Muno *et al.* 2001, 2u Trudolyubov 2001, 2v Rodriguez *et al.* 2002a, 2w Ji *et al.* 2003, 2x Remillard & Morgan 1998, 2y Cui 1999, 2z Remillard *et al.* 2002b, 2A Remillard *et al.* 1999b, 2B Belloni *et al.* 2000, 2C Remillard *et al.* 2003; 5a Samimi *et al.* 1979, 5b Motch *et al.* 1982, 5c Motch *et al.* 1983, 5d Maejima *et al.* 1984, 5e Makishima *et al.* 1986, 5f Imamura *et al.* 1990, 5g Grebenev *et al.* 1991, 5h Miyamoto *et al.* 1991, 5i Méndez & van der Klis 1997, 5j Steiman-Cameron *et al.* 1997, 5k Kong *et al.* 2002, 5l Belloni *et al.* 1999a, 5m Smith & Liang 1999, 5n Nowak *et al.* 1999c, 5o Revnivtsev *et al.* 2001b, 5p van Straaten *et al.* 2003, 5q Nowak *et al.* 2002, 5r Nespoli *et al.* 2003, 5s Psaltis *et al.* 1999a, 5t Belloni & Hasinger 1990b, 5u Belloni 2004, 5v Lin *et al.* 2000a, 5w Vaughan & Nowak 1997; 6a Cui *et al.* 1999, 6b Remillard *et al.* 1999a, 6c Wijnands *et al.* 1999b, 6d Cui *et al.* 2000a, 6e Sobczak *et al.* 2000a, 6f Homan *et al.* 2001, 6g Kalemci *et al.* 2001, 6h Miller *et al.* 2001, 6i Remillard *et al.* 2002c, 6j Rodriguez *et al.* 2002b, 6k Belloni *et al.* 2002b, 6l Vignarca *et al.* 2003, 6m Remillard *et al.* 2002a, 6n Kubota & Done 2004, 6o Kubota & Makishima 2004, 6p Gierliński & Done 2003, 6q Rodriguez *et al.* 2003, 6r Sobczak *et al.* 2000b, 6s Reilly *et al.* 2001; 51a Denis *et al.* 1994, 51b Vikhlinin *et al.* 1995, 51c Grove *et al.* 1998, 51d van der Hooft *et al.* 1999a; 52a Ebisawa *et al.* 1989, 52b Schmidtke *et al.* 1999, 52c Haardt *et al.* 2001, 52d Nowak *et al.* 2001; 53a Treves *et al.* 1990, 53b Boyd *et al.* 2000, 53c Boyd *et al.* 2001, 53d Wilms *et al.* 2001, 53e Cowley *et al.* 1991; 54a Carpenter *et al.* 1976, 54b Kuulkers 1998; 55a Revnivtsev *et al.* 2000b, 55b Wood *et al.* 2000, 55c Malzac *et al.* 2003, 55d Frontera *et al.* 2003, 55e Hynes *et al.* 2003b, 55f Belloni *et al.* 2002a, 55g Kanbach *et al.* 2001, 55h Spruit & Kanbach 2002, 55i McClintock *et al.* 2001, 55j Esin *et al.* 2001; 56a Miyamoto *et al.* 1994, 56b Belloni *et al.* 1997, 56c Takizawa *et al.* 1997, 56d Hynes *et al.* 2003a, 56e Miyamoto *et al.* 1993, 56f Esin *et al.* 2000, e7 Esin *et al.* 1997, e8 Ebisawa *et al.* 1994; 57a Revnivtsev *et al.* 2000a, 57b Brocksopp *et al.* 2001; 58a Park *et al.* 2004; 59a Parmar *et al.* 1986, 59b Kuulkers *et al.* 1997b, 59c Remillard & Morgan 1999, 59d Dieters *et al.* 2000b, 59e Tomsick & Kaaret 2000, 59f Trudolyubov *et al.* 2001, 59g Klein-Wolt *et al.* 2004a; 60a Kalemci *et al.* 2003, 60b Homan *et al.* 2003b, 60c Tomsick *et al.* 2004, 60d Rossi *et al.* 2004; 61a Zhang *et al.* 1997c, 61b Méndez *et al.* 1998d, 61c Remillard *et al.* 1999c, 61d Yamaoka *et al.* 2001, 61e Strohmayer 2001a, 61f Sobczak *et al.* 1999, 61g Kalemci *et al.* 2004, 61h Hynes *et al.* 1998; 62a Lutovinov & Revnivtsev 2003; 63a Wijnands & Miller 2002, 63b in 't Zand *et al.* 2002; 64a van der Hooft *et al.* 1996, 64b van der Hooft *et al.* 1999b, 64c Revnivtsev *et al.* 1998a, 64d Brocksopp *et al.* 2004; 65a Cui *et al.* 1997c; 66a Borozdin & Trudolyubov 2000, 66b Wijnands *et al.* 2001a; 67a Smith *et al.* 1997, 67b Main *et al.* 1999; 68a Homan *et al.* 2003a; 69a Revnivtsev *et al.* 2000c; 70a Revnivtsev *et al.* 1998b; 71a Lin *et al.* 2000b, 71b Smith *et al.* 2001; 72a Wijnands & van der Klis 2000, 72b Revnivtsev *et al.* 2002, 72c Uemura *et al.* 2002, 72d Uemura *et al.* 2004; 73a Parmar *et al.* 1993, 73b Sellmeijer *et al.* 1999; 74a Cui *et al.* 2000b, 74b Markwardt 2001, 74c Brocksopp *et al.* 2002; 75a Woods *et al.* 2002; 76a Ricci *et al.* 1995, 76b Nowak & Wilms 1999, 76c Wijnands *et al.* 2002a; 77a Terada *et al.* 2002; 78a Vasiliev *et al.* 2000, 78b Naik *et al.* 2000, 78c Campana *et al.* 2002; 79a Oosterbroek *et al.* 1996, 79b Oosterbroek *et al.* 1997, 79c Życki *et al.* 1999a,b.

In two objects, two peaks at an approximate 2:3 frequency ratio were detected together (GRO J1655–40: ~300 and ~450 Hz, Fig. 2.6c; 61e, see also 6m, 61c; XTE J1550–564: 188 ± 3 and 268 ± 3 Hz, 6h; see also 6m and 6b,f,i,). A similar case (but not simultaneous) has been mentioned but not yet fully reported for GRS 1915+105 (113 ± 3 Hz and 165 ± 3 Hz, 2C; Section 4.4.3). The frequencies of the 2:3 pairs may scale with inverse black-hole mass (6m), but there is some choice in frequencies and the systematic uncertainties in the masses are considerable. Marginal cases involving 1:2 frequency ratios have also been discussed (92 Hz in XTE J1550–564, 6m; 328 Hz in GRS 1915+105, 2z). Usually HF QPO peaks are seen alone, but with a tendency to occur near fixed frequencies. In addition to sources mentioned above, HF QPOs have been reported at 184 ± 5 Hz in 4U 1630–47 (59c,g), 250 ± 5 Hz in XTE J1650–500 (60b), 240 Hz in H1743–322 (68a) and, marginally, at 150^{+17}_{-28} and 187^{+14}_{-11} Hz in XTE J1859+226 (74a,b). However, values well off the nominal values 184 and 276 Hz occur in XTE J1550–564 (e.g., 123 ± 2 Hz, 6f, but see 6m; 141 ± 3 Hz, 6i).

Below 6 keV the HF QPOs are usually not detected, while at higher energies amplitudes up to 5% rms have been measured. The higher-frequency peak is more evident at energies >13 keV in GRO J1655–40 (61e) but not in XTE J1550–564 (6h). In XTE J1550–564 and GRO J1655–40 the higher harmonic becomes stronger when the spectrum becomes softer (6f,m), i.e., while all other variability frequencies increase (Section 2.10.2). In XTE J1550–564 this, together with the HF QPOs observed off the fixed frequencies, produces a correlation between the frequencies of actually observed HF and LF QPOs (6f). An observed correlation with LF QPO type and coherence (6i,m) is related to this.

In GRS 1915+105 a 67-Hz QPO occurs (2b,x) whose frequency varies by only a small percentage in no apparent correlation to factor-several X-ray flux changes. Q is usually around 20 (but sometimes drops to 6; 2A), the amplitude is ~1% (rms), and hard lags up to 130° occur (2y). The spectral characteristics of this QPO depend strongly on the peculiar rapid state changes of this source (Section 2.10.4, 2p). QPOs at 40 and possibly 56 Hz sometimes occur with the 67 Hz one (2q), and a 27 Hz QPO, again strongly dependent on the rapid state changes, occurs on other occasions (2p). Note that 27:40:56:67 is close to 2:3:4:5. The relation of these QPOs to the $\gtrsim$ 100 Hz HF QPOs is unclear, but of course the reported 113/165 Hz pair in GRS 1915+105 (above) complicates their interpretation. In XTE J1550–564, a possibly related Q ~ 4.4 QPO near 65 Hz has been reported (6g).

The high frequencies of the QPOs discussed in this section, similar to ISCO frequencies in stellar mass black holes, strongly suggest an origin in the strong-field region. In view of the reported constant frequencies, diskoseismic modes (Section 2.12.1) are a candidate explanation, but these models do not predict harmonic sets of frequencies such as seen in HF QPOs (Nowak *et al.* 1997; Pérez *et al.* 1997; Ortega-Rodriguez & Wagoner 2000; Silbergleit *et al.* 2001). Rezzolla *et al.* (2003) calculated p modes in a toroidal accretion geometry which can explain some of the key HF QPO properties such as the integer frequency ratios. Cui *et al.* (1998b) proposed that HF QPOs occur at $\nu_{\rm nodal}$ (Section 2.8.2). Relativistic resonance models (Section 2.8.3) predicted constant frequencies in small integer ratios which were then found in GRO J1655–40 (Strohmayer 2001a; Abramowicz & Kluźniak 2001). The previously mentioned models of Schnittman & Bertschinger (2004, see Section 2.8.1) and Laurent & Titarchuk (2001, see Section 2.8.5) both aim at explaining HF QPOs.

HF QPO properties, to the extent that it has been possible to establish them for this somewhat elusive phenomenon, seem quite different from those of neutron-star kHz QPOs: the neutron-star QPOs come in pairs with a separation related to the star's spin, have strongly tunable frequencies and low harmonic content, while the black-hole ones are single, but with high harmonic content, and have a much more stable frequency. Yet, the two phenomena might still be reconciled within a single theoretical description. For example, the variable frequencies in neutron stars might occur because the phenomenon occurs at a variable (e.g., inner disk) radius, set by interaction of the disk flow with either a magnetic field or radiation from the stellar surface, while in black holes, in the absence of these influences, the same phenomenon occurs at a more constant radius (perhaps, close to the ISCO); the second QPO may occur only in neutron stars because it is due to an interaction with the spin; the high harmonic content in black-hole QPOs may be due to relativistic effects on the flow and its emission that become important only near the ISCO (e.g, as a simple example, extreme Doppler boosting near the ISCO of a spinning black hole). Future observations at higher sensitivity will help enormously in understanding this phenomenon and elucidating the relation with the neutron-star kHz QPOs.

2.10.2 The low-frequency complex

The strong BLN in the LS was first noticed in Cyg X-1 (Oda *et al.* 1971) and initially interpreted in terms of a 73 msec periodicity. Later shot-noise models (Section 2.2) were applied to this noise (see also Section 2.12.3). Nolan *et al.* (1981) presented the first power spectrum unambiguously showing the characteristic $\nu_b = 0.01$–1 Hz flat-topped noise shape (Fig. 2.8, LS). The energy dependence of the noise amplitude is generally small in the 2–40 keV range, above that the rms may decrease (e.g., 1m,I, 5n,v, 51a,b,c, 57a,), although background subtraction issues may have affected some of these results. While the noise is well correlated between energy bands (0.1–10 Hz cross-coherence, Section 2.2, is $\sim$1), hard $\sim$0.1 rad lags are seen between <6 and >15 keV, increasing logarithmically with energy. Contrary to predictions from basic Comptonization light travel-time models (Section 2.12.2), time lag is not constant but decreases with Fourier frequency as $\nu^{-0.7}$, with some structure possibly related to the characteristic noise frequencies (ν_b, ν_h, Section 2.6.2; 1d,l,m,u,z, 5n,w, 64b, 67a, 71a). While in some scattering geometries this could be understood (Kazanas *et al.* 1997, Hua *et al.* 1997), lack of the expected smearing of the variability towards higher photon energies by light travel-time effects makes models of this type unlikely (1J,K) and favours propagation models for the lags instead (Section 2.12.2).

In the IMS/VHS, ν_b is higher, from 0.1–1 Hz to >10 Hz (2–60 keV). Energy dependencies in noise amplitude, time lags and coherencies tend to be larger and more complex and variable here than in the LS (1s,t, 5h, 6a,c,d, 52d, 56c,e, 60a, 72b, 79c), possibly due to competition between the contribution of hard and soft spectral components to these quantities, and characteristic frequencies depend on energy as well (e.g., 1i,j,t, 6a,f, 55d, 56b, 60a, 61c). GRS 1915+105 in its hard state ("C state"; 2B) has noise similar to this (i.e., this state probably corresponds to a VHS/IMS, 2f, see also, e.g., 2n,o,t).

Noise amplitude is less in the 6–7-keV (Fe-line) region (1k,o,L, 72b, 79a,). This effect is seen in the IMS, but is more pronounced in the LS, particularly >5 Hz. Perhaps this is due to a larger line-forming ("reflection") region in that state (1k,o, see also 79a).

QPOs ($Q > 2$) with amplitudes of a few to more than 10% rms are sometimes observed superimposed on the noise at frequencies (Section 2.7) a factor 2–8 above, or around, ν_b. This is seen in the LS (0.01–2 Hz, sometimes including a harmonic: 1s, 5c,g,m,o,q,s, 6i, 53b, 55a, 61b, 67a, 71a) where frequency often gradually increases through the initial LS rise of a transient outburst (in the 0.015–0.3 Hz range: 55b,d, 57b, 64a,d). It is also often observed in the IMS (1–33 Hz, often with rich harmonic structure: 1t, 5h,r, 6c,d,f, 56a-c, 58a, 59d-f, 60a,b,d, 61b,c,g, 63a,b, 66a,b, 69a). Transitions in frequency occur in the 0.08–13 Hz range between LS and IMS both in the rise (0.08–13 Hz range; 6a,e,i,s) and in the decay (6–0.2 Hz range; 6g,s, 59e) of some transient outbursts. Transitions that are smooth in frequency may nevertheless be rapid in time and, depending on coverage, seem abrupt. IMS/VHS LF-QPO harmonic structure is intricate, with subsets of 1:2:3:4:6:8 frequency ratios all observed (references above). Time lags of opposite sign and different Q values are sometimes seen between alternating harmonics (56b,c, 6a,c, 74a) reminiscent of what is observed in Z-source HBOs (Section 2.6.2). Source-state dependent QPO sub-types are distinguished based on differences in peak width, time lags and accompanying noise (6c,f,i,m, Chapter 4). In GX 339–4 a 6-Hz LF QPO fluctuated by about 1 Hz on timescales down to 5 s in correlation with X-ray flux variations (5r). QPOs with similar harmonic structure and alternating time lags are also observed in the 0.5–12 Hz range in GRS 1915+105 in its hard state (e.g., 2a,b,g,h,j,l,n,o,s,v). In the HS, weak ($\lesssim 1\%$ rms) 0.02–0.08 and 14–27 Hz QPOs are occasionally seen whose relation to the other LF QPOs is unclear (6e,f, 52a, 61c, 69a).

Correlations between the LF complex and X-ray spectral (usually, Comptonization and reflection) model parameters are mostly an expression of the source-state dependencies already discussed above and in Section 2.6.2: as the spectrum hardens (power-law fraction increases, blackbody flux decreases, coronal compactness increases, photon index decreases, etc.), amplitudes increase and frequencies decrease (e.g., 1n,s, 5o,q, 6g,l, 56a, 59e, 60c, 61g, 77a, also di Matteo and Psaltis 1999). Gilfanov *et al.* (2004) present a tentative correlation between QPO frequency and Fe line width (wider line for higher frequency), which is roughly in accordance with expectations if the QPO is an orbital frequency and the line width is determined by Doppler shifts due to motion in that orbit. Again, in the IMS correlations are more complex, perhaps due to competition between spectral components. In both XTE J1550–564 and GRO J1655–40 the QPO frequency increases with blackbody flux, but in XTE J1550–564 the photon index and total flux then both rise, whereas in GRO J1655–40 both fall (6e,i, who conclude that QPO frequency correlates with $\dot{M}_{\rm d}$, Section 2.5, and that the power-law fraction should be $>$20% for QPOs to be observed).

In XTE J1650–500, during the decay towards the off state, at a luminosity a factor $\sim$500 below that early in the outburst, a $\nu_{\rm b}$ of 0.0035 Hz was measured (60c), so on the way to the off state $\nu_{\rm b}$ apparently continues to decrease. Optical and UV power spectra in the low and off states are similar to those in X-rays, with $\nu_{\rm b}$ as low as 0.0001–0.001 Hz in the off state (55e, 56d). 0.02 Hz BLN and 0.1 Hz QPO were observed simultaneously in UV and X-rays in the LS in XTE J1118+480, with 1–2 s UV lags possibly due to light travel-time delays in reprocessing, but the detailed correlation is complex and not well understood (55c,e,g,h). In GRO J1655–40 in the VHS, 10–20 s UV lags were observed (61h), while in XTE J1819–254 7 min *X-ray* lags were seen, which were attributed to disk propagation effects (72c,d). In GX 339–4 optical BLN and QPOs occur in the 0.005–3 Hz range (e.g., 5b,c,f,j).

The BLN in black holes has been extensively modeled in terms of shot noise, often used as a model for magnetic flares, and chaos- and certain MHD-simulations of disk flows also produce power spectra reminiscent of LS noise (see Section 2.12.3). For the LF QPOs a variety of, nearly exclusively disk-based, models has been considered (orbiting hot spots, Section 2.8.1; Lense–Thirring precession, Section 2.8.2; disk oscillations, Section 2.12.1). Superposition of various oscillation frequencies can also produce BLN (Nowak 1994). Churazov *et al.* (2001) have proposed that ν_b is the orbital frequency of the inner edge of the optically thick disk, and explain the LS and IMS power spectra of Cyg X-1 by supposing that the variability is generated at the local orbital timescale in the hot optically thin regions of the flow located both within the inner edge, and sandwiching, the disk.

2.10.3 Power-law noise

Power-law noise (in this section defined as noise with no clear flat top in P_ν down to several 0.01 Hz) is not normally seen in the LS (strong BLN could mask it). The HS is characterized by weak (0.5–3%) power-law noise with index typically 0.7–1.2 (e.g., 5l, 6f, 56c, 59d, 61b,d, 69a); sometimes a break to $\alpha = 1.5-2$ occurs around 3 Hz (Fig. 2.8, HS; 6f, 61b). The most varied power-law noise is observed in the IMS.

Rapid transitions (minutes to days apart) between $\sim$1 Hz BLN-dominated and few-percent rms $\alpha \sim 1$ power-law dominated power spectra in GS 1124–68 and GX 339–4 were an early defining characteristic of the VHS (5h, 56c). In more recent work (2f, 6f, 52c,d, 59b, 61a-c), a 4–8% rms $\alpha = 0.8$–1.5 power-law component is commonly found, sometimes together with BLN, usually at the "soft side" of the VHS/IMS, where the BLN weakens. While the noise in these various examples is clearly not flat-topped, it sometimes has a discernable curvature or break in the 1–10 Hz range, and steepens towards higher frequencies. When XTE J1550–564 moves from HS to gradually harder intermediate states, $\alpha = 0.7$–1.2 power-law noise, which can be as weak as 0.5% in the HS proper, becomes stronger, up to 5.6% in the VHS, 2.8% in lower-luminosity IMS and up to 8.9% on the way to the LS, with no BLN detected (6f). When the BLN appears, the power-law noise can still be detectable at low frequencies at levels of 1.5–4% rms, but as the BLN gets stronger and ν_b drops, it disappears below the detection level. In what may be a similar phenomenon, Cyg X-1 displays a strong (up to 33% rms, 1j,r) $\alpha = 1$ power law, usually with a clear break to $\alpha = 2$ around 10 Hz at the top of soft flares in what is usually called the "high state" in this source, but is actually more similar to an IMS (1s, 61g); when the spectrum gets slightly harder in this IMS, BLN with $\nu_b = 1$–3.5 Hz is seen together with a flat ($\alpha = 0.3$–0.8) power law at low frequency (1i,j). Because it dominates in the disk-dominated "soft" state, Cui *et al.* (1997b) interpret this noise as originating in the disk, but Churazov *et al.* (2001) argue instead that it originates in the corona.

Like VLFN in neutron stars (Section 2.9.4), power-law noise in black holes may be due to $\dot{M}$ variations, perhaps arising in the outer disk. Numerical simulations can produce power spectra that are red over several decades in frequency (Mineshige *et al.* 1994b; Kawaguchi *et al.* 2000; Hawley & Krolik 2001) but as noted in Section 2.10.2, flat tops are usually discernible in these cases. Clearly, models involving surface phenomena such as nuclear burning do not apply in the black-hole case, so such models for neutron-star power-law noise (Section 2.9.4) predict differences between neutron stars and black holes which should show up in comparative studies.

2.10.4 Other phenomena and peculiar objects

The number of black-hole X-ray binaries that is peculiar with respect to the timing characteristics and source states discussed here is small; mostly what makes systems peculiar is unusual variations on longer timescales, likely originating outside the strong-field region, or *lack* of variability (in possible black holes such as SS433, Cyg X-3 and CI Cam; Grindlay *et al.* 1984; Berger & van der Klis 1994; Belloni *et al.* 1999b; but see van der Klis &

Jansen 1985), usually plausibly attributable to external circumstances. This suggests that the innermost accretion flow, where the rapid variability arises, is dominated by the black-hole properties. XTE J1819–254 (V4641 Sgr) in its brief outburst showed unusually strong (46% rms) power-law noise with rapidly varying α (72a) and rapid optical fluctuations (72c,d). It may be an SS433-like super-Eddington accretor (72b). Objects showing properties not clearly characteristic for either low-magnetic-field neutron stars or black holes include 4U 1957+11 (76a-c), possibly a black hole always in the HS, and 4U 1543–62 and 4U 1556–60, which might be atoll sources (Table 2.6, Farinelli *et al.* 2003).

A small number of rapid variability phenomena observed in black-hole systems do not fit the above categories. Transient oscillations in the 1–1000 mHz range are regularly observed in the very high state ("dips and flip-flops", 5h; e.g., 56c, 59d,f) which may just be relatively rapid source-state transitions (e.g., Belloni *et al.* 2000) due to disk instabilities; the peculiar nature of GRS 1915+105 may be due to the source showing such behavior most of the time. Of course, faster and lower-luminosity phenomena could similarly be due to rapid state transitions (cf. Yu *et al.* 2001), but this is harder to check.

2.11 High-magnetic-field neutron stars

In the high-magnetic-field neutron stars the typically 10^{12} G B field disrupts the disk flow at the magnetospheric radius $r_{\rm mag}\sim 10^8$ cm and channels the plasma to the magnetic poles. The resulting strong periodic pulsations complicate the analysis of the stochastic variability. The dynamical timescale at $r_{\rm mag}$ is of order seconds, hence variability much faster than this is not a-priori expected from the accretion disk (but see e.g., Orlandini & Morfill 1992; Orlandini & Boldt 1993; Klein *et al.* 1996a,b for possible rapid magnetospheric accretion-flow phenomena). Indeed, aperiodic variability tends to be dominated by strong (several 10% rms) flat-topped BLN with ν_b often (but not always) $\sim\nu_{\rm spin}$ (e.g., Frontera *et al.* 1987; Belloni & Hasinger 1990b; Angelini *et al.* 1989; Takeshima *et al.* 1991; Takeshima 1992), suggesting that it arises in the disk outside $r_{\rm mag}$. Possibly, such a BLN component is a characteristic of any disk flow truncated at 10^8 cm, and similar physics underlies this component in high-magnetic-field neutron stars and in low-magnetic-field compact objects in low states (e.g., Hoshino & Takeshima 1993; van der Klis 1994b). However, as noted by Lazzati and Stella (1997), if the pulse amplitude is itself modulated, the strong periodic component can give rise to these noise components, complicating interpretation.

A broad QPO often seen at factors up to $\sim 10^2$ below $\nu_{\rm spin}$ (see Table 1.1 in Section 1.2.1.2; references above and in Shirakawa & Lai 2002b; see also Li *et al.* 1980) is sometimes interpreted as a magnetospheric beat frequency (Section 2.8.4), implying that orbital motion at $r_{\rm mag}$ occurs at approximately the spin frequency. On occasion, when a QPO occurred above $\nu_{\rm spin}$ (e.g., Angelini *et al.* 1989; Finger *et al.* 1996) its luminosity or spin-up-rate related changes in frequency were consistent with either the beat frequency or the orbital frequency at $r_{\rm mag}$, allowing neutron-star parameters to be constrained. Upper and lower sidebands were seen separated from $\nu_{\rm spin}$ by the QPO frequency (Kommers *et al.* 1998; Moon & Eikenberry 2001a,b), implying the pulse amplitude is modulated by the QPO; although this is what a beat-frequency model predicts, asymmetry between the two sidebands suggests that an orbital frequency in the disk contributes to the QPO signal. Disk oscillations (Titarchuk & Osherovich 2000) and warped disk modes (Shirakawa & Lai 2002b) have also been proposed for these QPOs. The peculiar 0.5 s pulsar GRO J1744–28, which showed Type II bursts, (cf. Section 2.9.5) also produced 20, 40 and 60 Hz QPOs (Zhang *et al.* 1996c).

Jernigan *et al.* (2000) reported QPO features near 330 and 760 Hz in the 4.8 s accreting pulsar Cen X-3. They interpreted this in terms of the photon bubble model (Section 2.12.4). This is the only report of millisecond oscillations from a high-magnetic-field neutron star, but note that the QPO features are quite weak, and instrumental effects are a concern at these low power levels.

2.12 Flow-instability and non-flow models

In Section 2.8 we looked at models for observed variability frequencies based on orbital and epicyclic motions supplemented with decoherence and modulation mechanisms in order to fit the properties of observed QPO and noise components. In the current section, we discuss rapid X-ray variability models in which the variability arises in instabilities in the accretion flow. Models of this type often include the modulation mechanism as part of their physics, although the consequences of this have not always been fully explored yet in terms of predictions of observable phenomena. At the end of this section we look at neutron-star boundary-layer and surface phenomena.

2.12.1 Disk-oscillation models

The expressions in Section 2.8.1 are for free-particle orbits. An accretion disk is a hydrodynamic flow, so the particles are not free but exert forces upon each other. This leads to the potential for oscillations in the flow. The resulting vibration modes of disk, corona, and disk–star and disk–magnetosphere boundary layers can lead to QPOs or noise. In a standard disk model several physical timescales are of the order of the local orbital timescale $1/\nu_\phi$ (Shakura & Sunyaev 1973, 1976; Thorne & Price 1975; Pringle 1981; Section 13.2.3). Many analytic as well as numerical explorations have been made of disk oscillation modes. Often the free-particle orbital and epicyclic frequencies can still be recognized (e.g., Wagoner 1999), but combination frequencies with (magneto-)hydrodynamic frequencies (e.g., of sound waves) occur as well (e.g., Psaltis 2000). In Sections 2.9–2.11 some examples of models in this class proposed for specific phenomena have already been mentioned. Here we emphasize models relevant to kHz QPOs and black-hole HF QPOs.

In the strong-field gravity region global oscillation modes with frequencies similar to the free-particle frequencies can occur which are "trapped" in (cannot propagate outside) particular disk annuli (see Kato *et al.* 1998; Nowak and Lehr 1998; Wagoner 1999; Kato 2001; Lamb 2003 for useful summaries). This trapping is a predicted strong-field gravity effect; it does not occur in Newtonian gravity. Of these "diskoseismic" modes, the g, c and p modes occur at approximately constant frequencies, which like the orbital frequency at the ISCO scale as $1/M$. The w modes depend on the inner disk radius and hence can vary in frequency. Diskoseismic modes were not found in simulations including MHD turbulent effects (Hawley & Krolik 2001).

If there is a sharp transition in the disk, according to Psaltis & Norman (2000) the transition radius can act as a filter with a response that has strong resonances near the epicyclic (and orbital) frequencies of that radius; predicted frequencies are similar to those of the relativistic precession model (Section 2.8.2), with hydrodynamic corrections and additional frequencies that might allow a better match to the LF QPO, kHz QPO and sideband data (but see also Marković & Lamb 2000). Oscillations associated with a viscous transition layer between Keplerian disk and neutron star, some involving the star's magnetic field but *not* requiring strong-field gravity, and hence not involving the epicyclic frequencies, have been proposed to apply to high-frequency and other QPO and noise phenomena in neutron stars, black holes and white dwarfs, with at least six different observed frequencies and their correlations being predicted (Titarchuk *et al.* 1998, 1999; Osherovich & Titarchuk 1999; Titarchuk & Osherovich 1999; Titarchuk 2002; Titarchuk & Wood 2002; but see also Miller 2003 for a theoretical discussion of these ideas, and Jonker *et al.* 2000a, 2002a and van Straaten *et al.* 2000, 2003 for discrepancies with the observations). Oscillations associated with instabilities in a disk–magnetosphere boundary have been explored as a model for high-frequency QPOs (Li & Narayan 2004; Kato *et al.* 2001). Hydrodynamic disk modes, some involving the effects of radiation and magnetic fields have further been explored by various authors (e.g., Hanami 1988; Fukue & Okada 1990; Spruit & Taam 1990; Okuda & Mineshige 1991; Alpar *et al.* 1992; Chen & Taam 1994, 1995; Luo & Liang 1994; Abramowicz *et al.* 1995; Wallinder 1995; Ipser 1996; Milsom & Taam 1996, 1997; Marković & Lamb 1998; Lai 1999; Gnedin & Kiikov 2001; Kato *et al* 2001; Shirakawa & Lai 2002a; Varniere *et al.* 2002; Das *et al.* 2003; Mukhopadhyay *et al.* 2003; Watarai & Mineshige 2003b). Damping, or the superposition of many local frequencies (e.g., Nowak 1994) can turn intrinsically periodic disk oscillations into QPO or broad-band noise.

Fluctuations in the density or temperature of a Comptonizing medium can produce a modulation in spectra formed by Comptonization (e.g., Boyle *et al.* 1986; Stollman *et al.* 1987; Alpar *et al.* 1992; Lee & Miller 1998), as can modulation of the rate at which photons are injected into the medium (e.g., Kazanas & Hua 1999; Laurent & Titarchuk 2001; Titarchuk & Shrader 2002). Various types of intrinsic flow instabilities can modulate the accretion rate onto the central object (e.g. hydrodynamic, Taam & Lin 1984; general-relativistic, Paczyński 1987, Kato 1990; and radiation-feedback instabilities, Fortner *et al.* 1989, Miller & Lamb 1992, Miller & Park 1995) and in this way produce relatively large-amplitude modulations in the X-ray flux. All disk oscillations to some extent directly affect the disk luminosity and the photon-energy dependence of the disk emissivity (references above and, e.g., Nowak & Wagoner 1993), yet many of the models that have been proposed essentially are variability-frequency models, predicting the frequency, or the power spectrum, of the fluctuations only in some physical flow parameter rather than in any observable quantity. Clearly, in order to allow further tests of disk-oscillation models and to discriminate between them, such predictions of observables of the oscillations are essential.

2.12.2 Energy dependencies: amplitude, phase, cross-coherence

To the extent that the intrinsically periodic models discussed above and in Section 2.8 include explicit X-ray flux modulation mechanisms they are in principle severely constrained by the observed photon-energy dependencies of amplitude and phase (Section 2.2) of the modulation. To explain why the variability often has a harder spectrum than the average flux (i.e., larger fractional rms, Section 2.2, towards higher photon energies, see e.g., Sections 2.9.1 and 2.9.3, but see also Section 2.10.2) in models where the frequencies arise in the disk, which usually emits a *softer* than average flux, two main types of mechanism have been discussed: (i) modulation of the rate at which the disk is providing soft seed photons to the Comptonizing corona (references below), and (ii) modulation of the accretion rate into an inner region, such as a neutron-star boundary layer or black-hole inner corona whose X-ray spectrum is harder that that of the disk (e.g., Churazov *et al.* 2001, Gilfanov *et al.* 2003). Of course, another possible explanation for this is that the frequencies do not arise in the disk in the first place. QPOs and noise in black holes in the IMS tend to be more strongly energy dependent than in other states or in neutron stars; this may be related to the large differences between the X-ray spectral components that are simultaneously prominent in black holes in this state.

Propagation of an X-ray signal through a hot cloud of Comptonizing electrons produces hard lags: on average, photons emerging later have undergone more scatterings and hence gained more energy; the time lag between bands E_1 and E_2 scales roughly as $\log(E_2/E_1)$ (Payne 1980; Miller 1995; Nowak & Vaughan 1996). This mechanism has been applied many times (see references in Sections 2.2, 2.8.7 and 2.12.1 and, e.g., Wijers *et al.* 1987; Miyamoto *et al.* 1988; Schulz & Wijers 1993; Nowak 1994; Nowak *et al.* 1999b; Hua & Titarchuk 1996; Hua *et al.* 1997, 1999; Böttcher & Liang 1998; Cullen 2000; Nobili *et al.* 2000; Laurent & Titarchuk 2001; Titarchuk & Shrader 2002), with detailed calculations of the effects of the cloud oscillating (e.g., Stollman *et al.* 1987; Fortner *et al.* 1989; Miller & Lamb 1992; Miller 1995; Lee & Miller 1998; Lee *et al.* 2001), and estimates of the effects of other Comptonizing geometries (e.g., Miller *et al.* 1998a; Shibazaki *et al.* 1988; Nobili 2003; Reig *et al.* 2003c). The applicability of these models to black-hole BLN lags has been addressed in Section 2.10.2; lags naturally arising from shot-noise models involving propagating clumps or waves (Section 2.12.3) may be more appropriate to explain these. Reprocessing by the disk of X-rays emitted above the disk plane ("reflection"; Poutanen 2002; see also Kaaret 2000), or by the companion star (Vikhlinin 1999) is another way to affect the energy dependencies in the variability. Körding and Falcke (2004) provide a detailed analysis of time lags produced by a pivoting power-law spectrum.

The observed energy dependencies in variability amplitude and phase are very constraining, and none of the models cited are able to explain everything observed. The unknown geometry of the emitting and scattering regions contributes to the considerable uncertainty with respect to the correct model. Clearly, this is an area where the observations are needed to guide the modeling, yet observational capabilities are not yet at the required level. If, for example, luminous blobs were orbiting the compact object in the region of the disk where the relativistically broadened Fe line is generated, with sufficient sensitivity this would produce a very obvious signal due to Doppler-shifted features moving up and down through the line profile, but present instrumentation is not able to reliably detect signals of this kind. With the larger instruments currently under consideration, such work may become possible.

2.12.3 Intrinsically aperiodic models

The intrinsically aperiodic (Section 2.8) models most often used are in the class of shot-noise models (see also Section 2.2). Shots are sensible in various physical settings, e.g., magnetic flares (usually, storage of magnetic energy followed by release in a reconnection event, e.g., Wheeler 1977; Galeev *et al.* 1979; Pudritz & Fahlman 1982; Aly & Kuijpers 1990; Haardt *et al.* 1994; Mineshige *et al.* 1995; di Matteo *et al.* 1999; Poutanen & Fabian 1999; Merloni *et al.* 2000; Kato *et al.* 2001; Rodriguez *et al.* 2002c; Varniere *et al.* 2002; Krishan *et al.* 2003), infalling blobs or waves producing a flare ("propagation models", Nowak *et al.* 1999b: e.g., Miyamoto *et al.* 1988; Kato 1989; Miyamoto & Kitamoto 1989; Manmoto *et al.* 1996; Böttcher & Liang 1999; Misra 2000; Böttcher 2001) and finite-lifetime orbiting blobs (references see Section 2.2 and Section 2.8.2). In particular the propagation models may be able to produce the frequency-dependent lags and relatively energy-independent high-frequency amplitudes (Lin *et al.* 2000a; Maccarone *et al.* 2000) appropriate to black-hole BLN (Section 2.10.2).

The observation that variability amplitude is linearly related to flux over a wide range of timescales, *including* those of the variability itself in SAX J1808.4−3658 and Cyg X-1 (Uttley & McHardy 2001; Gleissner *et al.* 2004; Uttley 2004; see also Bałucińska-Church *et al.* 1997) is incompatible with straightforward shot-noise models and instead suggests that slower variations are produced first, and are then operated on by faster processes to produce the higher frequency variability. In this context models seem appropriate where slow accretion rate variations originate far out in the disk (where correspondingly long dynamical timescales apply) and become observable after disturbances have propagated inward to the emitting regions, undergoing further modulation producing the faster variations on the way (Vikhlinin *et al.* 1994; Lyubarskii 1997; see also Życki 2002, 2003; King *et al.* 2004).

Among chaos models (see also Section 2.2) for which some physical justification was put forward are the "dripping handrail" model of Scargle *et al.* (1993), and models involving self-organized criticality of a disk as described in a cellular automaton model (Abramowicz & Bao 1994; Mineshige *et al.* 1994a,b; Takeuchi *et al.* 1995; Xiong *et al.* 2000). Autonomous flow instabilities are certainly suggested by direct and to some extent "first principle" magnetohydrodynamic simulation of disks, which tend to produce broad noise spectra (Kawaguchi *et al.* 2000; Hawley & Krolik 2001; 2002; Armitage & Reynolds 2003).

2.12.4 *Non-flow models: boundary layer, stellar surface*

In addition to models associated with various aspects of the accretion flow, neutron-star oscillations or processes occurring on the neutron-star surface or in a boundary layer could produce variability. Clearly, such models exclude any possibility that the same phenomenon also occurs in a black hole. Processes like these are thought to cause the burst oscillations (Section 3.4), but they could also underlie some of the observed variability in the persistent emission. Radial (e.g., Shibazaki & Ebisuzaki 1989) and non-radial oscillations of various kinds (e.g., McDermott & Taam 1987; Epstein 1988; McDermott *et al.* 1988; Bildsten & Cutler 1995; Bildsten *et al.* 1996; Strohmayer & Lee 1996; Bildsten & Cumming 1998), can cause a variation in the surface emission properties; emission would be further modulated as the spin periodically modulates aspect and visibility of the "spots" thus formed. Temporary surface features due to nuclear-burning processes, which were specifically proposed as a model for neutron-star VLFN (Section 2.9.4; Bildsten 1993, 1995, see Chapter 3) or magnetohydrodynamic effects (Hameury *et al.* 1985) have also been considered (see also Inogamov & Sunyaev 1999; Popham & Sunyaev 2001) and a ~0.01 Hz quasi-periodicity in nuclear burning rate has been suggested by Revnivtsev *et al.* (2001a; Chapter 2.9.4). Klein *et al.* (1996a,b) proposed a model where photon bubbles rise up by buoyancy through the accreted material and produce a flash of radiation when they burst at the top as a model for kHz QPOs and for QPOs in high-magnetic-field neutron stars (Section 2.11).

2.13 Final remarks

Since the previous edition of this book (Lewin *et al.* 1995) rapid X-ray variability studies have come closer towards the goals of strong-field gravity and dense-matter physics. For the first time we are seeing, and are able to study in some detail, variability at the dynamical timescale of the strong-field region in accreting low-magnetic-field neutron stars as well as stellar-mass black holes. Strong-field gravity is an integral part of many of the models proposed. The aim of future timing work will be to turn the *diagnostics* of strong-field gravity and dense matter we now have into true *tests* of GR and *determinations* of the EOS. A theoretical framework for interpreting the observables of motion in the strong-field region in detail and for testing strong-field gravity theories is currently emerging (e.g., Weinberg *et al.* 2001; DeDeo & Psaltis 2004). What is needed on the observational side is a considerable increase in timing sensitivity coupled with good spectral capabilities. As timing sensitivity in the relevant regime is proportional to collecting area (not the square root of it; Section 2.2), this is an attainable goal. With future large-area timing instrumentation ($\sim 10\,\mathrm{m}^2$, e.g., XEUS, see Barret 2004, or a dedicated timing array) the predicted patterns of weak sidebands and harmonics ("fingerprints") of the variability phenomena will be mapped out as a spectrum of interrelated frequencies. This should make models that successfully predicted them unassailable, as of course they should be in order to be accepted as true tests of general relativity in the strong-field regime. Current observations are still clearly limited by the drop in QPO amplitude towards the extreme frequencies (Fig. 2.19). With more sensitivity the frequency range over which QPOs are detected would be considerably widened, likely making it possible to follow kHz QPOs up to the ISCO and check for the predicted frequency saturation there. A considerable synergy in terms of testing gravitation theories may occur with the results of gravitational wave instruments.

Combining timing and spectroscopy is another way to clinch the models. For example, in an orbital motion model for a QPO phenomenon in the Fe line region, to zeroeth order the frequency provides the orbital period, and the line profile the orbital velocity, so that we can solve for orbital radius r and central mass M. So, combining spectral and timing measurements will provide strong tests of the models and allow us to start using them to learn more about the curved spacetime near compact objects. By measuring the line-profile changes on short timescales, or equivalently the amplitude and phase differences between QPOs in several spectral bands within the line profile, exciting tests are possible. The line widths are $\sim$keV, so moderate-resolution millisecond spectroscopy is sufficient to do this. Clearly, entirely different signals are expected from QPOs caused by luminous blobs orbiting in the strong-field region and from, e.g., spiral-flow modulated-accretion QPOs: hence, such measurements will decide the emission geometry and constrain the modulation model.

Finally, depending on the precise phenomenon, large-area detectors will make it possible to detect the QPOs either within one cycle ν^{-1}, or one coherence time $(\pi\Delta\nu)^{-1}$, allowing them to be studied in the time domain. Wave-form studies will quantitatively constrain compact object mass, radius and angular momentum, orbital velocity and gravitational light bending by modeling approaches such as described by Weinberg *et al.* (2001) for neutron-star surface hot spots. The opportunities provided by large-area detectors for doing strong-gravity and dense-matter physics by timing X-ray binaries are clearly excellent.

Acknowledgements

It is a pleasure to acknowledge the help of the many colleagues who either made data available before publication, provided figures or produced new ones, read versions of the manuscript or provided insightful discussion at various stages during the writing of this chapter: Marek Abramowicz, Didier Barret, Tomaso Belloni, Deepto Chakrabarty, Eric Ford, Jeroen Homan, Peter Jonker, Fred Lamb, Marc Klein-Wolt, Wlodek Kluźniak, Erik Kuulkers, Tom Maccarone, Craig Markwardt, Mariano Méndez, Simone Migliari, Dimitrios Psaltis, Thomas Reerink, Pablo Reig, Roald Schnerr, Luigi Stella, Steve van Straaten, Phil Uttley, Rudy Wijnands, Wenfei Yu. I am particularly indebted to Cole Miller and Diego Altamirano. All errors are mine. This work was supported in part by the Netherlands Organization for Scientific Research (NWO) and the Netherlands Research School for Astronomy (NOVA).

References

Abramowicz, M. & Bao, G. 1994, *PASJ*, **46**, 523
Abramowicz, M. A. & Kluźniak, W. 2001, *A&A*, **374**, L19
2004, in *Rossi and Beyond*, AIP Conf. Proc., **714** , 21; astro-ph/0312396
Abramowicz, M., Jaroszynski, M., & Sikora, M. 1978, *A&A*, **63**, 221
Abramowicz, M. A., Lanza, A., Spiegel, E. A., & Szuszkiewicz, E. 1992, *Nature*, **356**, 41
Abramowicz, M. A., Chen, X., & Taam, R. E. 1995, *ApJ*, **452**, 379
Abramowicz, M. A., Almergren, G. J. E., Kluźniak, W., & Thampan, A. V. 2003a, gr-gc/0312070
Abramowicz, M. A., Karas, V., Kluźniak, W., Lee, W. H., & Rebusco, P. 2003b, *PASJ*, **55**, 467
Abramowicz, M. A., Bulik, T., Bursa, M., & Kluźniak, W. 2003c, *A&A*, **404**, L21
Abramowicz, M. A., Kluźniak, W., Stuchlik, Z., & Torok, G. 2004, astro-ph/0401464
Agrawal, V. K. & Bhattacharyya, S. 2003, *A&A*, **398**, 223
Agrawal, V. K. & Sreekumar, P. 2003, *MNRAS*, **346**, 933
Akmal, A., Pandharipande, V. R., & Ravenhall, D. G. 1998, *Phys. Rev. C*, **58**, 1804
Alpar, M. A. 1986, *MNRAS*, **223**, 469
Alpar, M. A. & Shaham, J. 1985, *Nature*, **316**, 239
Alpar, M. A. & Yılmaz, A. 1997, *New Astronomy*, **2**, 225
Alpar, M. A., Hasinger, G., Shaham, J., & Yancopoulos, S. 1992, *A&A*, **257**, 627
Aly, J. J. & Kuijpers, J. 1990, *A&A*, **227**, 473
Andersson, N., Kokkotas ,K. D., & Stergioulas, N. 1999, *ApJ*, **516**, 307
Andersson, N., Jones, D. I., Kokkotas, K. D., & Stergioulas, N. 2000, *ApJ*, **534**, L75
Angelini, L., Stella, L., & Parmar, A. N. 1989, *ApJ*, **346**, 906
Armitage, P. J. & Natarajan, P. 1999, *ApJ*, **525**, 909
Armitage, P. J. & Reynolds, C. S. 2003, *MNRAS*, **341**, 1041
Asai, K., Dotani, T., Nagase, F., *et al.* 1993, *PASJ*, **45**, 801
Asai, K., Dotani, T., Mitsuda, K., *et al.* 1994, *PASJ*, **46**, 479
Asaoka, I. & Hoshi, R. 1989, *PASJ*, **41**, 1049
Augusteijn, T., Karatasos, K., Papadakis, M., *et al.* 1992, *A&A*, **265**, 177
Bałucińska-Church, M., Takahashi, T., Ueda, Y., *et al.* 1997, *ApJ*, **480**, L115
Bao, G. & Østgaard, E. 1994, *ApJ*, **422**, L51
1995, *ApJ*, **443**, 54
Bardeen, J. M. & Petterson, J. A. 1975, *ApJ*, **195**, L65
Bardeen, J. M., Press, W. H., & Teukolsky, S. A. 1972, *ApJ*, **178**, 347
Barnard, R., Church, M. J., & Bałucińska-Church, M. 2003a, *A&A*, **405**, 237
Barnard, R., Kolb, U., & Osborne, J. P. 2003b, *A&A*, **411**, 553
Barr, P., White, N. E., & Page, C. G. 1985, *MNRAS*, **216**, 65P
Barret, D. 2004, in *Rossi and Beyond*, AIP Conf. Proc., **714**, 405; astro-ph/0401099
Barret, D. & Olive, J. 2002, *ApJ*, **576**, 391
Barret, D. & Vedrenne, G. 1994, *ApJS*, **92**, 505
Barret, D., McClintock, J. E., & Grindlay, J. E. 1996, *ApJ*, **473**, 963
Barret, D., Olive, J. F., Boirin, L., *et al.* 2000, *ApJ*, **533**, 329
Barret, D., Olive, J. F., & Oosterbroek, T. 2003, *A&A*, **400**, 643
Belloni, T. 1998, *New Astron. Rev.*, **42**, 585
2004, in *The Restless High-Energy Universe*, *Nucl. Phys. B*, **132**, 337; astro-ph/0309028
Belloni, T. & Hasinger, G. 1990a, *A&A*, **227**, L33
1990b, *A&A*, **230**, 103
Belloni, T., Méndez, M., van der Klis, M., *et al.* 1996, *ApJ*, **472**, L107
Belloni, T., van der Klis, M., Lewin, W. H. G., *et al.* 1997, *A&A*, **322**, 857
Belloni, T., Méndez, M., van der Klis, M., Lewin, W. H. G., & Dieters, S. 1999a, *ApJ*, **519**, L159
Belloni, T., Dieters, S., van den Ancker, M. E., *et al.* 1999b, *ApJ*, **527**, 345
Belloni, T., Klein-Wolt, M., Méndez, M., van der Klis, M., & van Paradijs, J. 2000, *A&A*, **355**, 271
Belloni, T., Méndez, M., & Sánchez-Fernández, C. 2001, *A&A*, **372**, 551
Belloni, T., Psaltis, D., & van der Klis, M. 2002a, *ApJ*, **572**, 392

Belloni, T., Colombo, A. P., Homan, J., Campana, S., & van der Klis, M. 2002b, *A&A*, **390**, 199
Belloni, T., Homan, J., Casella, P., van der Klis, M., *et al.* 2005, *A&A*, **440**, 207
Benlloch, S., Wilms, J., Edelson, R., Yaqoob, T., & Staubert, R. 2001, *ApJ*, **562**, L121
Berger, M. & van der Klis, M. 1994, *A&A*, **292**, 175
1998, *A&A*, **340**, 143
Berger, M., van der Klis, M., van Paradijs, J., *et al.* 1996, *ApJ*, **469**, L13
Biehle, G. T. & Blandford, R. D. 1993, *ApJ*, **411**, 302
Bildsten, L. 1993, *ApJ*, **418**, L21
1995, *ApJ*, **438**, 852
1998, *ApJ*, **501**, L89
Bildsten, L. & Cumming, A. 1998, *ApJ*, **506**, 842
Bildsten, L. & Cutler, C. 1995, *ApJ*, **449**, 800
Bildsten, L. & Ushomirsky, G. 2000, *ApJ*, **529**, L33
Bildsten, L., Ushomirsky, G., & Cutler, C. 1996, *ApJ*, **460**, 827
Blom, J. J., in't Zand, J. J. M., Heise, J., *et al.* 1993, *A&A*, **277**, 77
Bloser, P. F., Grindlay, J. E., Kaaret, P., *et al.* 2000a, *ApJ*, **542**, 1000
Bloser, P. F., Grindlay, J. E., Barret, D., & Boirin, L. 2000b, *ApJ*, **542**, 989
Boirin, L., Barret, D., Olive, J. F., Bloser, P. F., & Grindlay, J. E. 2000, *A&A*, **361**, 121
Borozdin, K. N. & Trudolyubov, S. P. 2000, *ApJ*, **533**, L131
Böttcher, M. 2001, *ApJ*, **553**, 960
Böttcher, M. & Liang, E. P. 1998, *ApJ*, **506**, 281
1999, *ApJ*, **511**, L37
Boyd, P. T., Smale, A. P., Homan, J., *et al.* 2000, *ApJ*, **542**, L127
Boyd, P. T., Smale, A. P., & Dolan, J. F. 2001, *ApJ*, **555**, 822
Boyle, C. B., Fabian, A. C., & Guilbert, P. W. 1986, *Nature*, **319**, 648
Bracewell, R. N. 1986, *The Fourier Transform and its Applications*, 2nd edn., McGraw-Hill
Bradshaw, C. F., Geldzahler, B. J., & Fomalont, E. B. 2003, *ApJ*, **592**, 486
Bradt, H. V., Rothschild, R. E., & Swank, J. H. 1993, *A&AS*, **97**, 355
Brainerd, J. & Lamb, F. K. 1987, *ApJ*, **317**, L33
Branduardi, G., Kylafis, N. D., Lamb, D. Q., & Mason, K. O. 1980, *ApJ*, **235**, L153
Brinkman, A. C., Parsignault, D. R., Schreier, E., *et al.* 1974, *ApJ*, **188**, 603
Brocksopp, C., Jonker, P. G., Fender, R. P., *et al.* 2001, *MNRAS*, **323**, 517
Brocksopp, C., Fender, R. P., McCollough, M., *et al.* 2002, *MNRAS*, **331**, 765
Brocksopp, C., Bandyopadhyay, R. M., & Fender, R. P. 2004, *New Astronomy*, **9**, 249
Brown, E. F. & Ushomirsky, G. 2000, *ApJ*, **536**, 915
Brugmans, F. 1983, *Earth Surface Processes and Landforms*, **8**, 527
Bulik, T., Gondek-Rosińska, D., & Kluźniak, W. L. 1999, *A&A*, **344**, L71
Bulik, T., Kluźniak, W. & Zhang, W. 2000, *A&A*, **361**, 153
Bussard, R. W., Weisskopf, M. C., Elsner, R. F., & Shibazaki, N. 1988, *ApJ*, **327**, 284
Campana, S. 2000, *ApJ*, **534**, L79
Campana, S., Stella, L., Belloni, T., *et al.* 2002, *A&A*, **384**, 163
Canizares, C. R., Clark, G. W., Li, F. K., *et al.* 1975, *ApJ*, **197**, 457
Cannizzo, J. K. 1997, *ApJ*, **482**, 178
Carpenter, G. F., Eyles, C. J., Skinner, G. K., *et al.* 1976, *MNRAS*, **176**, 397
Chagelishvili, G. D., Lominadze, J. G., & Rogava, A. D. 1989, *ApJ*, **347**, 1100
Chakrabarti, S. K. & Manickam, S. G. 2000, *ApJ*, **531**, L41
Chakrabarty, D., Morgan, E. H., Muno, M. P., *et al.* 2003, *Nature*, **424**, 42
Chaput, C., Bloom, E., Cominsky, L., *et al.* 2000, *ApJ*, **541**, 1026
Chen, X. & Taam, R. E. 1994, *ApJ*, **431**, 732
1995, *ApJ*, **441**, 354
Chen, X., Swank, J. H., & Taam, R. E. 1997, *ApJ*, **477**, L41
Chiappetti, L., *et al.* 1990, *ApJ*, **361**, 596
Churazov, E., Gilfanov, M., & Revnivtsev, M. 2001, *MNRAS*, **321**, 759
Corbet, R. H. D., Smale, A. P., Charles, P. A., *et al.* 1989, *MNRAS*, **239**, 533
Cowley, A. P., *et al.* 1991, *ApJ*, **381**, 526
Crary, D. J., Kouveliotou, C., van Paradijs, J., *et al.* 1996, *ApJ*, **462**, L71
Cropper, M., Soria, R., Mushotzky, R. F., *et al.* 2004, *MNRAS*, **349**, 39

Cui, W. 1999, *ApJ*, **524**, L59
2000, *ApJ*, **534**, L31
Cui, W., Heindl, W. A., Rothschild, R. E., *et al.* 1997a, *ApJ*, **474**, L57
Cui, W., Zhang, S. N., Focke, W., & Swank, J. H. 1997b, *ApJ*, **484**, 383
Cui, W., Heindl, W. A., Swank, J. H., *et al.* 1997c, *ApJ*, **487**, L73
Cui, W., Barret, D., Zhang, S. N., *et al.* 1998a, *ApJ*, **502**, L49
Cui, W., Zhang, S. N., & Chen, W. 1998b, *ApJ*, **492**, L53
Cui, W., Zhang, S. N., Chen, W., & Morgan, E. H. 1999, *ApJ*, **512**, L43
Cui, W., Zhang, S. N., & Chen, W. 2000a, *ApJ*, **531**, L45
Cui, W., Shrader, C. R., Haswell, C. A., & Hynes, R. I. 2000b, *ApJ*, **535**, L123
Cullen, J. 2000, *Publ. Astron. Soc. Aust.*, **17**, 48
Cunningham, C. T. & Bardeen, J. M. 1972, *ApJ*, **173**, L137
Czerny, B., Nikołajuk, M., Piasecki, M., & Kuraszkiewicz, J. 2001, *MNRAS*, **325**, 865
dal Fiume, D., Robba, N. R., Frontera, F., & Stella, L. 1990, *Nuovo Cimento C*, **13**, 463
Damen, E., Wijers, R. A. M. J., van Paradijs, J., *et al.* 1990, *A&A*, **233**, 121
Das, T. K., Rao, A. R., & Vadawale, S. V. 2003, *MNRAS*, **343**, 443
Datta, B., Thampan, A. V., & Bombaci, I. 1998, *A&A*, **334**, 943
2000, *A&A*, **355**, L19
DeDeo, S. & Psaltis, D. 2004, astro-ph/0405067
Deeter, J. E. 1984, *ApJ*, **281**, 482
den Hartog, P. R., in't Zand, J. J. M., Kuulkers, E., *et al.* 2003, *A&A*, **400**, 633
Denis, M., Olive, J.-F., Mandrou, P., *et al.* 1994, *ApJS*, **92**, 459
Dieters, S. W. & van der Klis, M. 2000, *MNRAS*, **311**, 201
Dieters, S. W., Vaughan, B. A., Kuulkers, E., Lamb, F. K., & van der Klis, M. 2000a, *A&A*, **353**, 203
Dieters, S. W., Belloni, T., Kuulkers, E., *et al.* 2000b, *ApJ*, **538**, 307
di Matteo, T. & Psaltis, D. 1999, *ApJ*, **526**, L101
di Matteo, T., Celotti, A., & Fabian, A. C. 1999, *MNRAS*, **304**, 809
Ding, G. Q., Qu, J. L., & Li, T. P. 2003, *ApJ*, **596**, L219
Di Salvo, T. & Stella, L. 2002, astro-ph/0207219
Di Salvo, T., Stella, L., Robba, N. R., *et al.* 2000, *ApJ*, **544**, L119
Di Salvo, T., Méndez, M., van der Klis, M., Ford, E., & Robba, N. R. 2001a, *ApJ*, **546**, 1107
Di Salvo, T., Robba, N. R., Iaria, R., *et al.* 2001b, *ApJ*, **554**, 49
Di Salvo, T., Farinelli, R., Burderi, L., *et al.* 2002, *A&A*, **386**, 535
Di Salvo, T., Méndez, M., & van der Klis, M. 2003, *A&A*, **406**, 177
Doi, K. 1978, *Nature*, **275**, 197
Done, C. & Gierliński, M. 2003, *MNRAS*, **342**, 1041
Done, C., Życki, P. T., & Smith, D. A. 2002, *MNRAS*, **331**, 453
Dotani, T., Mitsuda, K., Makishima, K., & Jones, M. H. 1989, *PASJ*, **41**, 577
Dotani, T., Mitsuda, K., Inoue, H., *et al.* 1990, *ApJ*, **350**, 395
Dubus, G., Kern, B., Esin, A. A., Rutledge, R. E., & Martin, C. 2004, *MNRAS*, **347**, 1271
Ebisawa, K., Mitsuda, K., & Inoue, H. 1989, *PASJ*, **41**, 519
Ebisawa, K., Ogawa, M., Aoki, T., *et al.* 1994, *PASJ*, **46**, 375
Edelson, R. & Nandra, K. 1999, *ApJ*, **514**, 682
Elsner, R. F., Weisskopf, M. C., Darbro, W., *et al.* 1986, *ApJ*, **308**, 655
Elsner, R. F., Shibazaki, N., & Weisskopf, M. C. 1987, *ApJ*, **320**, 527
1988, *ApJ*, **327**, 742
Epstein, R. I. 1988, *ApJ*, **333**, 880
Esin, A. A., McClintock, J. E., & Narayan, R. 1997, *ApJ*, **489**, 865
Esin, A. A., Kuulkers, E., McClintock, J. E., & Narayan, R. 2000, *ApJ*, **532**, 1069
Esin, A. A., McClintock, J. E., Drake, *et al.* 2001, *ApJ*, **555**, 483
Fabian, A. C., Iwasawa, K., Reynolds, C. S., & Young, A. J. 2000, *PASP*, **112**, 1145
Farinelli, R., Frontera, F., Masetti, N., *et al.* 2003, *A&A*, **402**, 1021
Feng, Y. X., Li, T. P., & Chen, L. 1999, *ApJ*, **514**, 373
Feroci, M., Matt, G., Pooley, G., *et al.* 1999, *A&A*, **351**, 985
Finger, M. H., Wilson, R. B., & Harmon, B. A. 1996, *ApJ*, **459**, 288
Focke, W. B. 1996, *ApJ*, **470**, L127
Ford, E. C. & van der Klis, M. 1998, *ApJ*, **506**, L39

Ford, E., Kaaret, P., Tavani, M., *et al.* 1997a, *ApJ*, **475**, L123
Ford, E. C., Kaaret, P., Chen, K., *et al.* 1997b, *ApJ*, **486**, L47
Ford, E. C., van der Klis, M., & Kaaret, P. 1998a, *ApJ*, **498**, L41
Ford, E. C., van der Klis, M., van Paradijs, J., *et al.* 1998b, *ApJ*, **508**, L155
Ford, E. C., van der Klis, M., Méndez, M., van Paradijs, J., & Kaaret, P. 1999, *ApJ*, **512**, L31
Ford, E. C., van der Klis, M., Méndez, M., *et al.* 2000, *ApJ*, **537**, 368
Fortner, B., Lamb, F. K., & Miller, G. S. 1989, *Nature*, **342**, 775
Fox, D. W., Lewin, W. H. G., Rutledge, R. E., *et al.* 2001, *MNRAS*, **321**, 776
Fragile, P. C., Mathews, G. J., & Wilson, J. R. 2001, *ApJ*, **553**, 955
Franco, L. M. 2001, *ApJ*, **554**, 340
Frontera, F., dal Fiume, D., Robba, N. R., *et al.* 1987, *ApJ*, **320**, L127
Frontera, F., Palazzi, E., Zdziarski, A. A., *et al.* 2001, *ApJ*, **546**, 1027
Frontera, F., Amati, L., Zdziarski, A. A., *et al.* 2003, *ApJ*, **592**, 1110
Fukue, J. & Okada, R. 1990, *PASJ*, **42**, 533
Galeev, A. A., Rosner, R., & Vaiana, G. S. 1979, *ApJ*, **229**, 318
Galloway, D. K., Chakrabarty, D., Muno, M. P., & Savov, P. 2001, *ApJ*, **549**, L85
Galloway, D. K., Chakrabarty, D., Morgan, E. H., & Remillard, R. A. 2002, *ApJ*, **576**, L137
Galloway, D. K., Markwardt, C. B., Chakrabarty, D., Morgan, E. H., Chakrabarty, D., & Strohmayer, T. E., 2005, *ApJ*, **622**, L45
Gierliński, M. & Done, C. 2002a, *MNRAS*, **337**, 1373
2002b, *MNRAS*, **331**, L47
2003, *MNRAS*, **342**, 1083
Gierliński, M. & Zdziarski, A. A. 2003, *MNRAS*, **343**, L84
Gierliński, M., Zdziarski, A. A., Poutanen, J., *et al.* 1999, *MNRAS*, **309**, 496
Gierliński, M., Done, C., & Barret, D. 2002, *MNRAS*, **331**, 141
Giles, A. B. 1981, *MNRAS*, **195**, 721
Gilfanov, M., Churazov, E., & Revnivtsev, M. 1999, *A&A*, **352**, 182
2000, *MNRAS*, **316**, 923
Gilfanov, M., Revnivtsev, M., & Molkov, S. 2003, *A&A*, **410**, 217
Gilfanov, M., Churazov, E., & Revnivtsev, M. 2004, in Rossi and beyond, AIP Conf. Proc. **714**, 97; astro-ph/0312445
Gleissner, T., Wilms, J., Pottschmidt, K., *et al.* 2004, *A&A*, **414**, 1091
Glendenning, N. K. & Weber, F. 2001, *ApJ*, **559**, L119
Gnedin, Y. N. & Kiikov, S. O. 2001, *Astron. Lett.*, **27**, 507
Gondek-Rosińska, D., Stergioulas, N., Bulik, T., Kluźniak, W., & Gourgoulhon, E. 2001, *A&A*, **380**, 190
Grebenev, S. A., Syunyaev, R. A., Pavlinskii, M. N., *et al.* 1991, *Sov. Astron. Lett.*, **17**, 413
Grindlay, J. E., Band, D., Seward, F., *et al.* 1984, *ApJ*, **277**, 286
Grove, J. E., Strickman, M. S., Matz, *et al.* 1998, *ApJ*, **502**, L45
Gruzinov, A. 1999, astro-ph/9910335
Haardt, F., Maraschi, L., & Ghisellini, G. 1994, *ApJ*, **432**, L95
Haardt, F., Galli, M. R., Treves, A., *et al.* 2001, *ApJS*, **133**, 187
Halpern, J. P., Leighly, K. M., & Marshall, H. L. 2003, *ApJ*, **585**, 665
Hameury, J.-M., King, A. R., & Lasota, J.-P. 1985, *Nature*, **317**, 597
Hanami, H. 1988, *MNRAS*, **233**, 423
Hanawa, T., Hirotani, K., & Kawai, N. 1989, *ApJ*, **336**, 920
Hartle, J. B. & Thorne, K. S. 1968, *ApJ*, **153**, 807
Hasinger, G. 1987, in *The Origin and Evolution of Neutron Stars*, IAU Symp. **125**, 333
1988, in *Physics of Neutron Stars and Black Holes*, Tokyo, Y. Tanaka (ed.), p. 97
Hasinger, G. & van der Klis, M. 1989, *A&A*, **225**, 79
Hasinger, G., Langmeier, A., Sztajno, M., Truemper, J., & Lewin, W. H. G. 1986, *Nature*, **319**, 469
Hasinger, G., Priedhorsky, W. C., & Middleditch, J. 1989, *ApJ*, **337**, 843
Hasinger, G., van der Klis, M., Ebisawa, K., Dotani, T., & Mitsuda, K. 1990, *A&A*, **235**, 131
Hawley, J. F. & Krolik, J. H. 2001, *ApJ*, **548**, 348
2002, *ApJ*, **566**, 164
Heiselberg, H. & Hjorth-Jensen, M. 1999, *ApJ*, **525**, L45
Hertz, P., Vaughan, B., Wood, K. S., *et al.* 1992, *ApJ*, **396**, 201
Hirano, A., Kitamoto, S., Yamada, T. T., Mineshige, S., & Fukue, J. 1995, *ApJ*, **446**, 350

Hjellming, R. M., Han, X. H., Cordova, F. A., & Hasinger, G. 1990a, *A&A*, **235**, 147
Hjellming, R. M., Stewart, R. T., White, G. L., *et al.* 1990b, *ApJ*, **365**, 681
Homan, J. & van der Klis, M. 2000, *ApJ*, **539**, 847
Homan, J., van der Klis, M., Wijnands, R., Vaughan, B., & Kuulkers, E. 1998, *ApJ*, **499**, L41
Homan, J., Méndez, M., Wijnands, R., van der Klis, M., & van Paradijs, J. 1999a, *ApJ*, **513**, L119
Homan, J., Jonker, P. G., Wijnands, R., van der Klis, M., & van Paradijs, J. 1999b, *ApJ*, **516**, L91
Homan, J., Wijnands, R., van der Klis, M., *et al.* 2001, *ApJS*, **132**, 377
Homan, J., van der Klis, M., Jonker, P. G., *et al.* 2002, *ApJ*, **568**, 878
Homan, J., Miller, J. M., Wijnands, R., *et al.* 2003a, *ATel*, **162**, 1
Homan, J., Klein-Wolt, M., Rossi, S., *et al.* 2003b, *ApJ*, **586**, 1262
Homan, J., Wijnands, R., Rupen, M. P., *et al.* 2004, *A&A*, **418**, 255
Hooft, F. van der, *et al.* 1996, *ApJ*, **458**, L75
1999a, *ApJ*, **513**, 477
1999b, *ApJ*, **519**, 332
Hoshino, M. & Takeshima, T. 1993, *ApJ*, **411**, L79
Hua, X. & Titarchuk, L. 1996, *ApJ*, **469**, 280
Hua, X., Kazanas, D., & Titarchuk, L. 1997, *ApJ*, **482**, L57
Hua, X., Kazanas, D., & Cui, W. 1999, *ApJ*, **512**, 793
Hynes, R. I., O'Brien, K., Horne, K., Chen, W., & Haswell, C. A. 1998, *MNRAS*, **299**, L37
Hynes, R. I., Charles, P. A., Casares, J., *et al.* 2003a, *MNRAS*, **340**, 447
Hynes, R. I., Haswell, C. A., Cui, W., *et al.* 2003b, *MNRAS*, **345**, 292
Iaria, R., Di Salvo, T., Robba, N. R., *et al.* 2004, *ApJ*, **600**, 358
Ikegami, T. 1986, Ph.D. thesis, University of Tokyo
Ilovaisky, S. A., Chevalier, C., White, N. E., *et al.* 1980, *MNRAS*, **191**, 81
Imamura, J. N., Kristian, J., Middleditch, J., & Steiman-Cameron, T. Y. 1990, *ApJ*, **365**, 312
Inogamov, N. A. & Sunyaev, R. A. 1999, *Astron. Lett.*, **25**, 269
in 't Zand, J. J. M., Verbunt, F., Strohmayer, T. E., *et al.* 1999a, *A&A*, **345**, 100
in 't Zand, J. J. M., Heise, J., Kuulkers, E., *et al.* 1999b, *A&A*, **347**, 891
in 't Zand, J. J. M., Bazzano, A., Cocchi, M., *et al.* 2000, *A&A*, **355**, 145
in 't Zand, J. J. M., Cornelisse, R., Kuulkers, E., *et al.* 2001, *A&A*, **372**, 916
in 't Zand, J. J. M., Markwardt, C. B., Bazzano, A., *et al.* 2002, *A&A*, **390**, 597
Ipser, J. R. 1996, *ApJ*, **458**, 508
Israel, G. L., Mereghetti, S., & Stella, L. 1994, *ApJ*, **433**, L25
Jernigan, J. G., Klein, R. I., & Arons, J. 2000, *ApJ*, **530**, 875
Ji, J. F., Zhang, S. N., Qu, J. L., & Li, T. 2003, *ApJ*, **584**, L23
Jones, C., Giacconi, R., Forman, W., & Tananbaum, H. 1974, *ApJ*, **191**, L71
Jonker, P. G., Wijnands, R., van der Klis, M., *et al.* 1998, *ApJ*, **499**, L191
Jonker, P. G., van der Klis, M., & Wijnands, R. 1999, *ApJ*, **511**, L41
Jonker, P. G., van der Klis, M., Wijnands, R., *et al.* 2000a, *ApJ*, **537**, 374
Jonker, P. G., Méndez, M., & van der Klis, M. 2000b, *ApJ*, **540**, L29
Jonker, P. G., van der Klis, M., Homan, J., *et al.* 2000c, *ApJ*, **531**, 453
2001, *ApJ*, **553**, 335
2002a, *MNRAS*, **333**, 665
Jonker, P. G., Méndez, M., & van der Klis, M. 2002b, *MNRAS*, **336**, L1
Jonker, P. G., Méndez, M., Nelemans, G., Wijnands, R., & van der Klis, M. 2003, *MNRAS*, **341**, 823
Jonker, P. G., Méndez, M., & van der Klis, M. 2005, *MNRAS*, **360**, 921
Kaaret, P. 2000, in *Stellar Endpoints*, AIP-Conf. Proc., **599**, 406; astro-ph/0008424
Kaaret, P., Ford, E., & Chen, K. 1997, *ApJ*, **480**, L27
Kaaret, P., Yu, W., Ford, E. C., & Zhang, S. N. 1998, *ApJ*, **497**, L93
Kaaret, P., Piraino, S., Ford, E. C., & Santangelo, A. 1999a, *ApJ*, **514**, L31
Kaaret, P., Piraino, S., Bloser, P. F., *et al.* 1999b, *ApJ*, **520**, L37
Kaaret, P., Zand, J. J. M. i., Heise, J., & Tomsick, J. A. 2002, *ApJ*, **575**, 1018
2003, *ApJ*, **598**, 481
Kahn, S. M. & Blissett, R. J. 1980, *ApJ*, **238**, 417
Kalemci, E., Tomsick, J. A., Rothschild, R. E., Pottschmidt, K., & Kaaret, P. 2001, *ApJ*, **563**, 239
Kalemci, E., Tomsick, J. A., Rothschild, R. E., *et al.* 2003, *ApJ*, **586**, 419
Kalemci, E., Tomsick, J. A., Rothschild, R. E., Pottschmidt, K., & Kaaret, P. 2004, *ApJ*, **603**, 231

Kallman, T., Boroson, B., & Vrtilek, S. D. 1998, *ApJ*, **502**, 441
Kalogera, V. & Psaltis, D. 2000, *Phys. Rev. D*, **61**, 024009
Kamado, Y., Kitamoto, S., & Miyamoto, S. 1997, *PASJ*, **49**, 589
Kanbach, G., Straubmeier, C., Spruit, H. C., & Belloni, T. 2001, *Nature*, **414**, 180
Karas, V. 1999a, *ApJ*, 526, 953
1999b, *PASJ*, **51**, 317
Kato, S. 1989, *PASJ*, **41**, 745
1990, *PASJ*, **42**, 99
2001, *PASJ*, **53**, L37
Kato, S., Fukue, J., & Mineshige, S. 1998, in *Black-hole Accretion Disks*, Kyoto University Press
Kato, Y., Hayashi, M. R., Miyaji, S., & Matsumoto, R. 2001, *Adv. Space Res.*, **28**, 505
Kawaguchi, T., Mineshige, S., Machida, M., Matsumoto, R., & Shibata, K. 2000, *PASJ*, **52**, L1
Kawai, N., Matsuoka, M., Inoue, H., *et al.* 1990, *PASJ*, **42**, 115
Kazanas, D. & Hua, X. 1999, *ApJ*, **519**, 750
Kazanas, D., Hua, X., & Titarchuk, L. 1997, *ApJ*, **480**, 735
King, A. R., Pringle, J. E., West, R. G., & Livio, M. 2004, *MNRAS*, **348**, 111
Kitamoto, S., Tsunemi, H., & Roussel-Dupre, D. 1992, *ApJ*, **391**, 220
Klein, R. I., Arons, J., Jernigan, G., & Hsu, J. J.-L. 1996a, *ApJ*, **457**, L85
Klein, R. I., Jernigan, J. G., Arons, J., Morgan, E. H., & Zhang, W. 1996b, *ApJ*, **469**, L119
Klein-Wolt, M., Homan, J. & van der Klis, M. 2004a, Ph.D. thesis, University of Amsterdam, p. 113
Klein-Wolt, M., van Straaten, S. & van der Klis, M. 2004b, Ph.D. thesis, University of Amsterdam, p. 165
Klis, M. van der, 1986, in *The Physics of Accretion Onto Compact Objects*, Lect. Notes Phys., **266**, 157
1989a, *ARA&A*, **27**, 517
1989b, in *Timing Neutron Stars*, NATO ASI C262, p. 27
1994a, *ApJS*, **92**, 511
1994b, *A&A*, **283**, 469
1995a, in *X-ray Binaries*, W. H. G. Lewin *et al.* (eds.), Cambridge University Press, p. 252
1995b, in *The Lives of the Neutron Stars*, NATO ASI C450, p. 301
1997, in *Astronomical Time Series*, ASSL **218**, 121
1998, in *The Many Faces of Neutron Stars*, NATO ASI C Proc. 515, p. 337
1999, in *Stellar Endpoints*, AIP Conf. Proc., **599**, 406
2000, *ARA&A*, **38**, 717
2001, *ApJ*, **561**, 943
2002, in *XEUS – Studying the Evolution of the Hot Universe*, MPE rep. 281, 354
Klis, M. van der & Jansen, F. A. 1985, *Nature*, **313**, 768
Klis, M. van der, Jansen, F., van Paradijs, J., *et al.* 1985, *Nature*, **316**, 225
1987a, *ApJ*, **313**, L19
Klis, M. van der, Stella, L., White, N., Jansen, F., & Parmar, A. N. 1987b, *ApJ*, **316**, 411
Klis, M. van der, Hasinger, G., Stella, L., *et al.* 1987c, *ApJ*, **319**, L13
Klis, M. van der, Hasinger, G., Damen, E., *et al.* 1990, *ApJ*, **360**, L19
Klis, M. van der, Kitamoto, S., Tsunemi, H., & Miyamoto, S. 1991, *MNRAS*, **248**, 751
Klis, M. van der, Swank, J. H., Zhang, W., *et al.* 1996, *ApJ*, **469**, L1
Klis, M. van der, Wijnands, R. A. D., Horne, K., & Chen, W. 1997, *ApJ*, **481**, L97
Klis, M. van der, Chakrabarty, D., Lee, J. C., *et al.* 2000, *IAU Circ.*7358, 3
Kluźniak, W. 1998, *ApJ*, **509**, L37
Kluźniak, W. & Abramowicz, M. A. 2001, astro-ph/0105057
2002, astro-ph/0203314
2003, astro-ph/0304345
Kluźniak, W. & Wagoner, R. V. 1985, *ApJ*, **297**, 548
Kluźniak, W., Michelson, P., & Wagoner, R. V. 1990, *ApJ*, **358**, 538
Kluźniak, W., Abramowicz, M. A., Kato, S., Lee, W. H., & Stergioulas, N. 2004, *ApJ*, **603**, L89
Kommers, J. M., Fox, D. W., Lewin, W. H. G., Rutledge, R. E., *et al.* 1997, *ApJ*, **482**, L53
Kommers, J. M., Chakrabarty, D., & Lewin, W. H. G. 1998, *ApJ*, **497**, L33
Kong, A. K. H., Charles, P. A., Kuulkers, E., & Kitamoto, S. 2002, *MNRAS*, **329**, 588
Körding, E. & Falcke, H. 2004, *A&A*, **414**, 795
Kotov, O., Churazov, E., & Gilfanov, M. 2001, *MNRAS*, **327**, 799
Krishan, V., Ramadurai, S., & Wiita, P. J. 2003, *A&A*, **398**, 819

Krolik, J. H. & Hawley, J. F. 2002, *ApJ*, **573**, 754
Kubota, A. & Done, C. 2004, *MNRAS*, **353**, 980
Kubota, A. & Makishima, K. 2004, *ApJ*, **601**, 428
Kuulkers, E. 1995, Ph.D. thesis, University of Amsterdam
1998, *New Astron. Rev.*, **42**, 1
Kuulkers, E. & van der Klis, M. 1995, *A&A*, **303**, 801
1998, *A&A*, **332**, 845
Kuulkers, E., van der Klis, M., Oosterbroek, T., *et al.* 1994, *A&A*, **289**, 795
Kuulkers, E., van der Klis, M., & van Paradijs, J. 1995, *ApJ*, **450**, 748
Kuulkers, E., van der Klis, M., & Vaughan, B. A. 1996, *A&A*, **311**, 197
Kuulkers, E., van der Klis, M., Oosterbroek, T., *et al.* 1997a, *MNRAS*, **287**, 495
Kuulkers, E., van der Klis, M., & Parmar, A. N. 1997b, *ApJ*, **474**, L47
Kuulkers, E., Wijnands, R., & van der Klis, M. 1999, *MNRAS*, **308**, 485
Kuulkers, E., Homan, J., van der Klis, M., Lewin, W. H. G., & Méndez, M. 2002, *A&A*, **382**, 947
Kuznetsov, S. I. 2001, *Astron. Lett.*, **27**, 790
2002a, *Astron. Lett.*, **28**, 73
2002b, *Astron. Lett.*, **28**, 811
Kylafis, N. D. & Klimis, G. S. 1987, *ApJ*, **323**, 678
Lai, D. 1998, *ApJ*, **502**, 721
1999, *ApJ*, **524**, 1030
Lamb, F. K. 2003, in *Jan van Paradijs Memorial Meeting*, ASP Conf. Series, **308**, 221
Lamb, F. K. & Miller, M. C. 2001, *ApJ*, **554**, 1210
2003, astro-ph/0308179
Lamb, F. K., Shibazaki, N., Alpar, M. A., & Shaham, J. 1985, *Nature*, **317**, 681
Langmeier, A., Sztajno, M., Hasinger, G., Truemper, J., & Gottwald, M. 1987, *ApJ*, **323**, 288
Langmeier, A., Hasinger, G., & Truemper, J. 1989, *ApJ*, **340**, L21
1990, *A&A*, **228**, 89
Laurent, P. & Titarchuk, L. 2001, *ApJ*, **562**, L67
Lazzati, D. & Stella, L. 1997, *ApJ*, **476**, 267
Lee, H. C. & Miller, G. S. 1998, *MNRAS*, **299**, 479
Lee, H. C., Misra, R., & Taam, R. E. 2001, *ApJ*, **549**, L229
Lee, W. H., Abramowicz, M. A., & Kluźniak, W. 2004, *ApJ*, **603**, L93
Lense, J., & Thirring, H. 1918, *Phys. Z.*, **19**, 156
Levin, Y. 1999, *ApJ*, **517**, 328
Lewin, W. H. G., *et al.* 1987, *MNRAS*, **226**, 383
Lewin, W. H. G., van Paradijs, J., & van der Klis, M. 1988, *Space Sci. Rev.*, **46**, 273
Lewin, W. H. G., Lubin, L. M., Tan, J., *et al.* 1992, *MNRAS*, **256**, 545
Lewin, W. H. G., van Paradijs, J., & Taam, R. E. 1993, *Space Sci. Rev.*, **62**, 223
Lewin, W. H. G., van Paradijs, J., & van den Heuvel, E. P. J. 1995, *X-ray Binaries*, Cambridge University Press
Lewin, W. H. G., Rutledge, R. E., Kommers, J. M., *et al.* 1996, *ApJ*, **462**, L39
Li, T. P. & Muraki, Y. 2002, *ApJ*, **578**, 374
Li, L. & Narayan, R. 2004, *ApJ*, **601**, 414
Li, F. K., McClintock, J. E., Rappaport, S., Wright, E. L., & Joss, P. C. 1980, *ApJ*, **240**, 628
Li, X., Ray, S., Dey, J., Dey, M., & Bombaci, I. 1999, *ApJ*, **527**, L51
Liang, E. P. T. & Thompson, K. A. 1980, *ApJ*, **240**, 271
Lin, D., Smith, I. A., Böttcher, M., & Liang, E. P. 2000a, *ApJ*, **531**, 963
Lin, D., Smith, I. A., Liang, E. P., *et al.* 2000b, *ApJ*, **532**, 548
Lin, D., Smith, I. A., Liang, E. P., & Böttcher, M. 2000c, *ApJ*, **543**, L141
Ling, J. C., Mahoney, W. A., Wheaton, W. A., Jacobson, A. S., & Kaluzienski, L. 1983, *ApJ*, **275**, 307
Lochner, J. C., Swank, J. H., & Szymkowiak, A. E. 1989, *ApJ*, **337**, 823
1991, *ApJ*, **376**, 295
Lubin, L. M., Lewin, W. H. G., Tan, J., Stella, L., & van Paradijs, J. 1991, *MNRAS*, **249**, 300
Lubin, L. M., Lewin, W. H. G., Dotani, T., *et al.* 1992a, *MNRAS*, **256**, 624
Lubin, L. M., Lewin, W. H. G., Rutledge, R. E., *et al.* 1992b, *MNRAS*, **258**, 759
Lubin, L. M., Lewin, W. H. G., van Paradijs, J., & van der Klis, M. 1993, *MNRAS*, **261**, 149
Luo, C. & Liang, E. P. 1994, *MNRAS*, **266**, 386

Lutovinov, A. A. & Revnivtsev, M. G. 2003, *Astron. Lett.*, **29**, 719
Lyubarskii, Y. E. 1997, *MNRAS*, **292**, 679
Maccarone, T. J. & Coppi, P. S. 2002a, *MNRAS*, **335**, 465
2002b, *MNRAS*, **336**, 817
2003, *MNRAS*, **338**, 189
Maccarone, T. J., Coppi, P. S., & Poutanen, J. 2000, *ApJ*, **537**, L107
Maejima, Y., Makishima, K., Matsuoka, M., *et al.* 1984, *ApJ*, **285**, 712
Main, D. S., Smith, D. M., Heindl, W. A., *et al.* 1999, *ApJ*, **525**, 901
Makino, Y. 1993, Ph.D. thesis, University of Tokyo
Makishima, K., Mitsuda, K., Inoue, H., *et al.* 1983, *ApJ*, **267**, 310
Makishima, K., Maejima, Y., Mitsuda, K., *et al.* 1986, *ApJ*, **308**, 635
Makishima, K., Ishida, M., Ohashi, T., *et al.* 1989, *PASJ*, **41**, 531
Malzac, J., Belloni, T., Spruit, H. C., & Kanbach, G. 2003, *A&A*, **407**, 335
Manmoto, T., Takeuchi, M., Mineshige, S., Matsumoto, R., & Negoro, H. 1996, *ApJ*, **464**, L135
Marković, D., 2000, astro-ph/0009450
Marković, D. & Lamb, F. K. 1998, *ApJ*, **507**, 316
2000, astro-ph/0009169
Markowitz, A., Edelson, R., Vaughan, S., *et al.* 2003, *ApJ*, **593**, 96
Markwardt, C. 2001, *Astrophys. Space Sci. Suppl.*, **276**, 209
Markwardt, C. B. & Swank, J. H. 2003, *IAU Circ.*, 8144
Markwardt, C. B., Strohmayer, T. E., & Swank, J. H. 1999a, *ApJ*, **512**, L125
Markwardt, C. B., Swank, J. H., & Taam, R. E. 1999b, *ApJ*, **513**, L37
Markwardt, C. B., Swank, J. H., Strohmayer, T. E., *et al.* 2002, *ApJ*, **575**, L21
Markwardt, C. B., Smith, E., & Swank, J. H. 2003, *IAU Circ.*, 8080, 2
Marshall, F. E., & Markwardt, C. B. 1999, *IAU Circ.*, 7103
Matsuba, E., Dotani, T., Mitsuda, K., *et al.* 1995, *PASJ*, **47**, 575
Mauche, C. W. 2002, *ApJ*, **580**, 423
McClintock, J. E., Haswell, C. A., Garcia, M. R., *et al.* 2001, *ApJ*, **555**, 477
McDermott, P. N. & Taam, R. E. 1987, *ApJ*, **318**, 278
McDermott, P. N., van Horn, H. M., & Hansen, C. J. 1988, *ApJ*, **325**, 725
McGowan, K. E., Charles, P. A., O'Donoghue, D., & Smale, A. P. 2003, *MNRAS*, **345**, 1039
McHardy, I. M., Papadakis, I. E., Uttley, P., Page, M. J., & Mason, K. O. 2004, *MNRAS*, **348**, 783
McNamara, B. J., Harrison, T. E., Zavala, R. T., *et al.* 2003, *AJ*, **125**, 1437
Meekins, J. F., Wood, K. S., Hedler, R. L., *et al.* 1984, *ApJ*, **278**, 288
Méndez M. 1999, in *19th Texas Symp.*, Paris; astro-ph/9903469
2002a, in *Jan van Paradijs Memorial Meeting*, ASP Conf. Proc., **308**, 289; astro-ph/0207279
2002b, in *The Ninth Marcel Grossmann Meeting*, 2319; astro-ph/0207278
Méndez, M. & van der Klis, M. 1997, *ApJ*, **479**, 926
1999, *ApJ*, **517**, L51
2000, *MNRAS*, **318**, 938
Méndez, M., van der Klis, M., van Paradijs, J., *et al.* 1997, *ApJ*, **485**, L37
1998a, *ApJ*, **494**, L65
Méndez, M., van der Klis, M., Wijnands, R., *et al.* 1998b, *ApJ*, **505**, L23
Méndez, M., van der Klis, M., & van Paradijs, J. 1998c, *ApJ*, **506**, L117
Méndez, M., Belloni, T., & van der Klis, M. 1998d, *ApJ*, **499**, L187
Méndez, M., van der Klis, M., Ford, E. C., Wijnands, R., & van Paradijs, J. 1999, *ApJ*, **511**, L49
Méndez, M., van der Klis, M., & Ford, E. C. 2001, *ApJ*, **561**, 1016
Méndez, M., Cottam, J., & Paerels, F. 2002, astro-ph/0207277
Menna, M. T., Burderi, L., Stella, L., Robba, N., & van der Klis, M. 2003, *ApJ*, **589**, 503
Merloni, A., Vietri, M., Stella, L., & Bini, D. 1999, *MNRAS*, **304**, 155
Merloni, A., Di Matteo, T., & Fabian, A. C. 2000, *MNRAS*, **318**, L15
Middleditch, J. & Priedhorsky, W. C. 1986, *ApJ*, **306**, 230
Migliari, S., Fender, R. P., Rupen, M., *et al.* 2003a, *MNRAS*, **342**, L67
Migliari, S., van der Klis, M., & Fender, R. P. 2003b, *MNRAS*, **345**, L35
Migliari, S., Fender, R. P., Rupen, M., *et al.* 2004, *MNRAS*, **351**, 186
Miller, M. C. 1995, *ApJ*, **441**, 770
1999, *ApJ*, **520**, 256

2000, *ApJ*, **537**, 342
2003, in *Rossi and Beyond*, AIP Conf. Proc., **714**, 365; astro-ph/0312449
Miller, G. S. & Lamb, F. K. 1992, *ApJ*, **388**, 541
Miller, G. S. & Park, M. 1995, *ApJ*, **440**, 771
Miller, M. C., Lamb, F. K., & Psaltis, D. 1998a, *ApJ*, **508**, 791
Miller, M. C., Lamb, F. K., & Cook, G. B. 1998b, *ApJ*, **509**, 793
Miller, J. M., Wijnands, R., Homan, J., *et al*. 2001, *ApJ*, **563**, 928
Miller, J. M., Fabian, A. C., Wijnands, R., *et al*. 2002, *ApJ*, **570**, L69
Milsom, J. A. & Taam, R. E. 1996, *MNRAS*, **283**, 919
1997, *MNRAS*, **286**, 358
Mineshige, S., Takeuchi, M., & Nishimori, H. 1994a, *ApJ*, **435**, L125
Mineshige, S., Ouchi, N. B., & Nishimori, H. 1994b, *PASJ*, **46**, 97
Mineshige, S., Kusnose, M., & Matsumoto, R. 1995, *ApJ*, **445**, L43
Misra, R. 2000, *ApJ*, **529**, L95
Mitsuda, K. & Dotani, T. 1989, *PASJ*, **41**, 557
Mitsuda, K., Inoue, H., Nakamura, N., & Tanaka, Y. 1989, *PASJ*, **41**, 97
Mitsuda, K., Dotani, T., Yoshida, A., Vaughan, B., & Norris, J. P. 1991, *PASJ*, **43**, 113
Miyamoto, S., 1994, *ISAS Research Note* 548
Miyamoto, S. & Kitamoto, S. 1989, *Nature*, **342**, 773
Miyamoto, S., Kitamoto, S., Mitsuda, K., & Dotani, T. 1988, *Nature*, **336**, 450
Miyamoto, S., Kimura, K., Kitamoto, S., Dotani, T., & Ebisawa, K. 1991, *ApJ*, **383**, 784
Miyamoto, S., Kitamoto, S., Iga, S., Negoro, H., & Terada, K. 1992, *ApJ*, **391**, L21
Miyamoto, S., Iga, S., Kitamoto, S., & Kamado, Y. 1993, *ApJ*, **403**, L39
Miyamoto, S., Kitamoto, S., Iga, S., Hayashida, K., & Terada, K. 1994, *ApJ*, **435**, 398
Miyamoto, S., Kitamoto, S., Hayashida, K., & Egoshi, W. 1995, *ApJ*, **442**, L13
Moon, D. & Eikenberry, S. S. 2001a, *ApJ*, **549**, L225
2001b, *ApJ*, **552**, L135
Morgan, E. H., Remillard, R. A., & Greiner, J. 1997, *ApJ*, **482**, 993
Morsink, S. M. & Stella, L. 1999, *ApJ*, **513**, 827
Motch, C., Ilovaisky, S. A., & Chevalier, C. 1982, *A&A*, **109**, L1
Motch, C., Ricketts, M. J., Page, C. G., Ilovaisky, S. A., & Chevalier, C. 1983, *A&A*, **119**, 171
Mukhopadhyay, B., Ray, S., Dey, J., & Dey, M. 2003, *ApJ*, **584**, L83
Muno, M. P., Morgan, E. H., & Remillard, R. A. 1999, *ApJ*, **527**, 321
Muno, M. P., Remillard, R. A., Morgan, E. H., *et al*. 2001, *ApJ*, **556**, 515
Muno, M. P., Remillard, R. A., & Chakrabarty, D. 2002, *ApJ*, **568**, L35
Muno, M. P., Galloway, D. K., & Chakrabarty, D. 2004, *ApJ*, **608**, 930
Murdin, P., Jauncey, D. L., Lerche, I., *et al*. 1980, *A&A*, **87**, 292
Naik, S. & Rao, A. R. 2000, *A&A*, **362**, 691
Naik, S., Agrawal, P. C., Paul, B., *et al*. 2000, *J. Astrophys. Astron*., **21**, 29
Nandi, A., Manickam, S. G., Rao, A. R., & Chakrabarti, S. K. 2001, *MNRAS*, **324**, 267
Narita, T., Grindlay, J. E., Bloser, P. F., & Chou, Y. 2003, *ApJ*, **593**, 1007
Negoro, H., Miyamoto, S., & Kitamoto, S. 1994, *ApJ*, **423**, L127
Negoro, H., Kitamoto, S., Takeuchi, M., & Mineshige, S. 1995, *ApJ*, **452**, L49
Negoro, H., Kitamoto, S., & Mineshige, S. 2001, *ApJ*, **554**, 528
Nespoli, E., Belloni, T., Homan, J., *et al*. 2003, *A&A*, **412**, 235
Nobili, L. 2003, *ApJ*, **582**, 954
Nobili, L., Turolla, R., Zampieri, L., & Belloni, T. 2000, *ApJ*, **538**, L137
Nolan, P. L., Gruber, D. E., Matteson, J. L., *et al*. 1981, *ApJ*, **246**, 494
Norris, J. P. & Wood, K. S. 1987, *ApJ*, **312**, 732
Norris, J. P., Hertz, P., Wood, K. S., *et al*. 1990, *ApJ*, **361**, 514
Nowak, M. A. 1994, *ApJ*, **422**, 688
1995, *PASP*, **107**, 1207
2000, *MNRAS*, **318**, 361
Nowak, M. & Lehr, D. 1998, in *Theory of Black Hole Accretion Disks*, M. A. Abramowicz *et al*. (eds.), Cambridge University Press, 233
Nowak, M. A. & Vaughan, B. A. 1996, *MNRAS*, **280**, 227
Nowak, M. A. & Wagoner, R. V. 1993, *ApJ*, **418**, 187

Nowak, M. A. & Wilms, J. 1999, *ApJ*, **522**, 476
Nowak, M. A., Wagoner, R. V., Begelman, M. C., & Lehr, D. E. 1997, *ApJ*, **477**, L91
Nowak, M. A., Vaughan, B. A., Wilms, J., Dove, J. B., & Begelman, M. C. 1999a, *ApJ*, **510**, 874
Nowak, M. A., Wilms, J., Vaughan, B. A., Dove, J. B., & Begelman, M. C. 1999b, *ApJ*, **515**, 726
Nowak, M. A., Wilms, J., & Dove, J. B. 1999c, *ApJ*, **517**, 355
Nowak, M. A., Wilms, J., Heindl, W. A., *et al.* 2001, *MNRAS*, **320**, 316
Nowak, M. A., Wilms, J., & Dove, J. B. 2002, *MNRAS*, **332**, 856
O'Brien, K., Horne, K., Gomer, R. H., Oke, J. B., & van der Klis, M. 2004, *MNRAS*, **350**, 587
Oda, M., Gorenstein, P., Gursky, H., *et al.* 1971, *ApJ*, **166**, L1
Ogawara, Y., Doi, K., Matsuoka, M., Miyamoto, S., & Oda, M. 1977, *Nature*, **270**, 154
Okuda, T. & Mineshige, S. 1991, *MNRAS*, **249**, 684
Olive, J. F., Barret, D., Boirin, L., *et al.* 1998, *A&A*, **333**, 942
Olive, J., Barret, D., & Gierliński, M. 2003, *ApJ*, **583**, 416
O'Neill, P. M., Kuulkers, E., Sood, R. K., & Dotani, T. 2001, *A&A*, **370**, 479
O'Neill, P. M., Kuulkers, E., Sood, R. K., & van der Klis, M. 2002, *MNRAS*, **336**, 217
Oosterbroek, T., Penninx, W., van der Klis, M., *et al.* 1991, *A&A*, **250**, 389
Oosterbroek, T., Lewin, W. H. G., van Paradijs, J., *et al.* 1994, *A&A*, **281**, 803
Oosterbroek, T., van der Klis, M., Kuulkers, E., *et al.* 1995, *A&A*, **297**, 141
Oosterbroek, T., van der Klis, M., Vaughan, B., *et al.* 1996, *A&A*, **309**, 781
Oosterbroek, T., van der Klis, M., van Paradijs, J., *et al.* 1997, *A&A*, **321**, 776
Oosterbroek, T., Barret, D., Guainazzi, M., & Ford, E. C. 2001, *A&A*, **366**, 138
Orlandini, M. & Boldt, E. 1993, *ApJ*, **419**, 776
Orlandini, M. & Morfill, G. E. 1992, *ApJ*, **386**, 703
Ortega-Rodríguez, M. & Wagoner, R. V. 2000, *ApJ*, **537**, 922
Osherovich, V. & Titarchuk, L. 1999, *ApJ*, **522**, L113
Ouyed, R. 2002, *A&A*, **382**, 939
Paczyński, B. 1987, *Nature*, **327**, 303
Paradijs, J. van, Hasinger, G., Lewin, W. H. G., *et al.* 1988a, *MNRAS*, **231**, 379
Paradijs, J. van, Penninx, W., Lewin, W. H. G., Sztajno, M., & Truemper, J. 1988b, *A&A*, **192**, 147
Park, S. Q., Miller, J. M., McClintock, J. E., *et al.* 2004, *ApJ*, **610**, 378
Parmar, A. N., Stella, L., & White, N. E. 1986, *ApJ*, **304**, 664
Parmar, A. N., Angelini, L., Roche, P., & White, N. E. 1993, *A&A*, **279**, 179
Parmar, A. N., Oosterbroek, T., Boirin, L., & Lumb, D. 2002, *A&A*, **386**, 910
Patterson, J. 1979, *ApJ*, **234**, 978
Paul, B., Agrawal, P. C., Rao, A. R., *et al.* 1997, *A&A*, **320**, L37
Payne, D. G. 1980, *ApJ*, **237**, 951
Penninx, W., Lewin, W. H. G., Zijlstra, A. A., *et al.* 1988, *Nature*, **336**, 146
Penninx, W., Hasinger, G., Lewin, W. H. G., *et al.* 1989, *MNRAS*, **238**, 851
Penninx, W., Lewin, W. H. G., Mitsuda, K., *et al.* 1990, *MNRAS*, **243**, 114
Penninx, W., Lewin, W. H. G., Tan, J., *et al.* 1991, *MNRAS*, **249**, 113
Pérez, C. A., Silbergleit, A. S., Wagoner, R. V., & Lehr, D. E. 1997, *ApJ*, **476**, 589
Piraino, S., Santangelo, A., & Kaaret, P. 2000, *A&A*, **360**, L35
2002, *ApJ*, **567**, 1091
Ponman, T. J., Cooke, B. A., & Stella, L. 1988, *MNRAS*, **231**, 999
Popham, R. & Sunyaev, R. 2001, *ApJ*, **547**, 355
Pottschmidt, K., Koenig, M., Wilms, J., & Staubert, R. 1998, *A&A*, **334**, 201
Pottschmidt, K., Wilms, J., Nowak, M. A., *et al.* 2003, *A&A*, **407**, 1039
Poutanen, J. 2002, *MNRAS*, **332**, 257
Poutanen, J. & Fabian, A. C. 1999, *MNRAS*, **306**, L31
Press, W. H. & Schechter, P. 1974, *ApJ*, **193**, 437
Priedhorsky, W., Garmire, G. P., Rothschild, R., *et al.* 1979, *ApJ*, **233**, 350
Priedhorsky, W., Hasinger, G., Lewin, W. H. G., *et al.* 1986, *ApJ*, **306**, L91
Pringle, J. E. 1981, *ARA&A*, **19**, 137
Prins, S. & van der Klis, M. 1997, *A&A*, **319**, 498
Psaltis, D. 2000, astro-ph/0010316
2001, *Adv. Space Res.*, **28**, 481
Psaltis, D. & Norman, C. 2000, astro-ph/0001391

Psaltis, D., Méndez, M., Wijnands, R., *et al.* 1998, *ApJ*, **501**, L95
Psaltis, D., Belloni, T., & van der Klis, M. 1999a, *ApJ*, **520**, 262
Psaltis, D., Wijnands, R., Homan, J., *et al.* 1999b, *ApJ*, **520**, 763
Pudritz, R. E. & Fahlman, G. G. 1982, *MNRAS*, **198**, 689
Qu, J. L., Yu, W., & Li, T. P. 2001, *ApJ*, **555**, 7
Rao, A. R., Naik, S., Vadawale, S. V., & Chakrabarti, S. K. 2000a, *A&A*, **360**, L25
Rao, A. R., Yadav, J. S., & Paul, B. 2000b, *ApJ*, **544**, 443
Rebusco, P. 2004, *PASJ*, **56**, 553
Reerink, T., Schnerr, R., van der Klis, M., & van Straaten, S. 2004, *A&A*, in press
Reig, P., Méndez, M., van der Klis, M., & Ford, E. C. 2000a, *ApJ*, **530**, 916
Reig, P., Belloni, T., van der Klis, M., Méndez, M., Kylafis, N. D., & Ford, E. C. 2000b, *ApJ*, **541**, 883
Reig, P., Papadakis, I., & Kylafis, N. D. 2002, *A&A*, **383**, 202
2003a, *A&A*, **398**, 1103
Reig, P., Belloni, T., & van der Klis, M. 2003b, *A&A*, **412**, 229
Reig, P., Kylafis, N. D., & Giannios, D. 2003c, *A&A*, **403**, L15
Reig, P., van Straaten, S., & van der Klis, M. 2004, *ApJ*, **602**, 918
Reilly, K. T., Bloom, E. D., Focke, W., *et al.* 2001, *ApJ*, **561**, L183
Remillard, R. A. & Morgan, E. H. 1998, *Nucl. Phys. B, (Proc. Suppl.)*, **69**, 316
1999, *BAAS*, **31**, 1421
Remillard, R. A., McClintock, J. E., Sobczak, G. J., *et al.* 1999a, *ApJ*, **517**, L127
Remillard, R., Morgan, E., Levine, A., *et al.* 1999b, *BAAS*, **31**, 731
Remillard, R. A., Morgan, E. H., McClintock, J. E., Bailyn, C. D., & Orosz, J. A. 1999c, *ApJ*, **522**, 397
Remillard, R. A., Muno, M. P., McClintock, J. E., & Orosz, J. A. 2002a, *ApJ*, **580**, 1030
Remillard, R., Muno, M., McClintock, J. E., & Orosz, J. 2002b, in *4th Microquasars Workshop*, Cargèse, 2002, Durouchoux *et al.* (eds.), p. 49; astro-ph/0208402
Remillard, R. A., Sobczak, G. J., Muno, M. P., & McClintock, J. E. 2002c, *ApJ*, **564**, 962
Remillard, R. A., Muno, M. P., McClintock, J. E., & Orosz, J. A. 2003, abstract AAS/HEAD 7, 30.03
Revnivtsev, M., Gilfanov, M., Churazov, E., *et al.* 1998a, *A&A*, **331**, 557
Revnivtsev, M., Gilfanov, M., & Churazov, E. 1998b, *A&A*, **339**, 483
Revnivtsev, M., Borozdin, K., & Emelyanov, A. 1999a, *A&A*, **344**, L25
Revnivtsev, M., Gilfanov, M., & Churazov, E. 1999b, *A&A*, **347**, L23
Revnivtsev, M. G., Borozdin, K. N., Priedhorsky, W. C., & Vikhlinin, A. 2000a, *ApJ*, **530**, 955
Revnivtsev, M., Sunyaev, R., & Borozdin, K. 2000b, *A&A*, **361**, L37
Revnivtsev, M. G., Trudolyubov, S. P., & Borozdin, K. N. 2000c, *MNRAS*, **312**, 151 see also **315**, 655
Revnivtsev, M., Churazov, E., Gilfanov, M., & Sunyaev, R. 2001a, *A&A*, **372**, 138
Revnivtsev, M., Gilfanov, M., & Churazov, E. 2001b, *A&A*, **380**, 520
Revnivtsev, M., Gilfanov, M., Churazov, E., & Sunyaev, R. 2002, *A&A*, **391**, 1013
Reynolds, C. S. & Nowak, M. A. 2003, *Phys. Rep.*, **377**, 389
Rezzolla, L., Lamb, F. K., & Shapiro, S. L. 2000, *ApJ*, **531**, L139
Rezzolla, L., Yoshida, S., Maccarone, T. J., & Zanotti, O. 2003, *MNRAS*, **344**, L37
Ricci, D., Israel, G. L., & Stella, L. 1995, *A&A*, **299**, 731
Robinson, E. L. & Warner, B. 1972, *MNRAS*, **157**, 85
Rodriguez, J., Durouchoux, P., Mirabel, I. F., *et al.* 2002a, *A&A*, **386**, 271
Rodriguez, J., Corbel, S., Kalemci, E., & Tomsick, J.A. 2002b, astro-ph/0205341
Rodriguez, J., Varnière, P., Tagger, M., & Durouchoux, P. 2002c, *A&A*, **387**, 487
Rodriguez, J., Corbel, S., & Tomsick, J. A. 2003, *ApJ*, **595**, 1032
Rossi, S., Homan, J., Miller, J. M., & Belloni, T. 2004, in *The Restless High-Energy Universe, Nucl. Phys. B*, **132**, 416; astro-ph/0309129
Rothschild, R. E., Boldt, E. A., Holt, S. S., & Serlemitsos, P. J. 1974, *ApJ*, **189**, L13
1977, *ApJ*, **213**, 818
Rutledge, R. E., Lubin, L. M., Lewin, W. H. G., *et al.* 1995, *MNRAS*, **277**, 523
Rutledge, R. E., Lewin, W. H. G., van der Klis, M., *et al.* 1999, *ApJS*, **124**, 265
Samimi, J., Share, G. H., Wood, K., *et al.* 1979, *Nature*, **278**, 434
Santolamazza, P., Fiore, F., Burderi, L., & Di Salvo, T. 2004, *Nucl. Phys. B*, **132**, 644
Scargle, J. D., Steiman-Cameron, T., Young, K., *et al.* 1993, *ApJ*, **411**, L91
Schaab, C. & Weigel, M. K. 1999, *MNRAS*, **308**, 718
Schmidtke, P. C., Ponder, A. L., & Cowley, A. P. 1999, *AJ*, **117**, 1292

Schnerr, R. S., Reerink, T., van der Klis, M., *et al.* 2003, *A&A*, **406**, 221
Schnittman, J. D. & Bertschinger, E. 2004, *ApJ*, **606**, 1098
Schulz, N. S. & Wijers, R. A. M. J. 1993, *A&A*, **273**, 123
Schulz, N. S., Hasinger, G., & Truemper, J. 1989, *A&A*, **225**, 48
Sellmeijer, H. & van der Klis, M. 1999, unpublished
Shakura, N. I. & Sunyaev, R. A. 1973, *A&A*, **24**, 337
1976, *MNRAS*, **175**, 613
Shibata, M. & Sasaki, M. 1998, *Phys. Rev. D*, **58**, 104011
Shibazaki, N. & Ebisuzaki, T. 1989, *PASJ*, **41**, 641
Shibazaki, N. & Lamb, F. K. 1987, *ApJ*, **318**, 767
Shibazaki, N., Elsner, R. F., & Weisskopf, M. C. 1987, *ApJ*, **322**, 831
Shibazaki, N., Elsner, R. F., Bussard, R. W., Ebisuzaki, T., & Weisskopf, M. C. 1988, *ApJ*, **331**, 247
Shirakawa, A. & Lai, D. 2002a, *ApJ*, **564**, 361
2002b, *ApJ*, **565**, 1134
Shirey, R. E., Bradt, H. V., Levine, A. M., & Morgan, E. H. 1996, *ApJ*, **469**, L21
1998, *ApJ*, **506**, 374
Shirey, R. E., Bradt, H. V., & Levine, A. M. 1999, *ApJ*, **517**, 472
Shvartsman, V. F. 1971, *Sov. Astron.*, **15**, 377
Sibgatullin, N. R. 2002, *Astron. Lett.*, **28**, 83
Silbergleit, A. S., Wagoner, R. V., & Ortega-Rodríguez, M. 2001, *ApJ*, **548**, 335
Singh, K. P. & Apparao, K. M. V. 1994, *ApJ*, **431**, 826
Smale, A. P. 1998, *ApJ*, **498**, L141
Smale, A. P. & Kuulkers, E. 2000, *ApJ*, **528**, 702
Smale, A. P., Zhang, W., & White, N. E. 1997, *ApJ*, **483**, L119
Smale, A. P., Homan, J., & Kuulkers, E. 2003, *ApJ*, **590**, 1035
Smith, I. A. & Liang, E. P. 1999, *ApJ*, **519**, 771
Smith, D. M., Heindl, W. A., Swank, J., *et al.* 1997, *ApJ*, **489**, L51
Smith, D. M., Heindl, W. A., Markwardt, C. B., & Swank, J. H. 2001, *ApJ*, **554**, L41
Smith, D. M., Heindl, W. A., & Swank, J. H. 2002, *ApJ*, **569**, 362
Sobczak, G. J., McClintock, J. E., Remillard, R. A., Bailyn, C. D., & Orosz, J. A. 1999, *ApJ*, **520**, 776
Sobczak, G. J., McClintock, J. E., Remillard, R. A., *et al.* 2000a, *ApJ*, **531**, 537
2000b, *ApJ*, **544**, 993
Spruit, H. C. & Kanbach, G. 2002, *A&A*, **391**, 225
Spruit, H. C. & Taam, R. E. 1990, *A&A*, **229**, 475
Srinivasan, G. 2002, *A&A Rev.*, **11**, 67
Steiman-Cameron, T. Y., Scargle, J. D., Imamura, J. N., & Middleditch, J. 1997, *ApJ*, **487**, 396
Stella, L. 1988, *Mem. Soc. Astron. Ital.*, **59**, 185
Stella, L. & Vietri, M. 1998, *ApJ*, **492**, L59
Stella, L. & Vietri, M. 1999, *Phys. Rev. Lett.*, **82**, 17
Stella, L., White, N. E., Davelaar, J., *et al.* 1985, *ApJ*, **288**, L45
Stella, L., Parmar, A. N., & White, N. E. 1987a, *ApJ*, **321**, 418
Stella, L., White, N. E., & Priedhorsky, W., 1987b, *ApJ*, **315**, L49
Stella, L., Haberl, F., Lewin, W. H. G., *et al.* 1988a, *ApJ*, **327**, L13
1988b, *ApJ*, **324**, 379
Stella, L., Vietri, M., & Morsink, S. M. 1999, *ApJ*, **524**, L63
Stergioulas, N., Kluźniak, W., & Bulik, T. 1999, *A&A*, **352**, L116
Stoeger, W. R. 1980, *MNRAS*, **190**, 715
Stollman, G. M., van Paradijs, J., Hasinger, G., *et al.* 1987, *MNRAS*, **227**, 7P
Straaten, S. van, Ford, E. C., van der Klis, M., Méndez, M., & Kaaret, P. 2000, *ApJ*, **540**, 1049
Straaten, S. van, van der Klis, M., Kuulkers, E., & Méndez, M. 2001, *ApJ*, **551**, 907
Straaten, S. van, van der Klis, M., Di Salvo, T., & Belloni, T. 2002, *ApJ*, **568**, 912
Straaten, S. van, van der Klis, M., & Méndez, M. 2003, *ApJ*, **596**, 1155
Straaten, S. van, van der Klis, M., & Wijnands, R. 2005, *ApJ*, **619**, 455
Strohmayer, T. E. 2001a, *ApJ*, **552**, L49
2001b, *ApJ*, **554**, L169
Strohmayer, T. E. & Lee, U. 1996, *ApJ*, **467**, 773
Strohmayer, T. E. & Mushotzky, R. F. 2003, *ApJ*, **586**, L61

Strohmayer, T. E., Zhang, W., Swank, J. H., *et al.* 1996, *ApJ*, **469**, L9
Strohmayer, T. E., Jahoda, K., Giles, A.B., & Lee, U. 1997, *ApJ*, **486**, 355
Strohmayer, T. E., Markwardt, C. B., Swank, J. H., & in't Zand, J. 2003, *ApJ*, **596**, L67
Sunyaev, R. A. 1973, *Sov. Astron.*, **16**, 941
Sunyaev, R. & Revnivtsev, M. 2000, *A&A*, **358**, 617
Sutherland, P. G., Weisskopf, M. C., & Kahn, S. M. 1978, *ApJ*, **219**, 1029
Swank, J., Chen, X., Markwardt, C., & Taam, R. 1998, in *Some Like It Hot*, AIP Conf. Proc., **431**, 327
Taam, R. E. & Lin, D. N. C. 1984, *ApJ*, **287**, 761
Takeshima, T. 1992, Ph.D. thesis, University of Tokyo
Takeshima, T., Dotani, T., Mitsuda, K., & Nagase, F. 1991, *PASJ*, **43**, L43
Takeuchi, M., Mineshige, S., & Negoro, H. 1995, *PASJ*, **47**, 617
Takizawa, M., Dotani, T., Mitsuda, K., *et al.* 1997, *ApJ*, **489**, 272
Tan, J., Lewin, W. H. G., Lubin, L. M., *et al.* 1991, *MNRAS*, **251**, 1
Tan, J., Lewin, W. H. G., Hjellming, R. M., *et al.* 1992, *ApJ*, **385**, 314
Tanaka, Y. 1989, in *23d ESLAB Symp.*, N. E. White *et al.* (eds.), p. 3
Tanaka, Y. & Tenma team, 1983, *IAU Circ.* 3891
Tananbaum, H., Gursky, H., Kellogg, E., Giacconi, R., & Jones, C. 1972, *ApJ*, **177**, L5
Tawara, Y., Hayakawa, S., Hunieda, H., Makino, F., & Nagase, F. 1982, *Nature*, **299**, 38
Taylor, J. H., Wolszczan, A., Damour, T., & Weisberg, J. M. 1992, *Nature*, **355**, 132
Tennant, A. F. 1987, *MNRAS*, **226**, 971
1988, *MNRAS*, **230**, 403
Tennant, A. F., Fabian, A. C., & Shafer, R. A. 1986, *MNRAS*, **219**, 871
Terada, K., Kitamoto, S., Negoro, H., & Iga, S. 2002, *PASJ*, **54**, 609
Terrell, N. J. J. 1972, *ApJ*, **174**, L35
Thampan, A. V., Bhattacharya, D., & Datta, B. 1999, *MNRAS*, **302**, L69
Thorne, K. S. & Price, R. H. 1975, *ApJ*, **195**, L101
Timmer, J., Schwarz, U., Voss, H. U., *et al.* 2000, *Phys. Rev. E.*, **61**, 1342
Titarchuk, L. 2002, *ApJ*, **578**, L71
2003, *ApJ*, **591**, 354
Titarchuk, L. & Osherovich, V. 1999, *ApJ*, **518**, L95
2000, *ApJ*, **542**, L111
Titarchuk, L. & Shrader, C. R. 2002, *ApJ*, **567**, 1057
Titarchuk, L. & Wood, K. 2002, *ApJ*, **577**, L23
Titarchuk, L., Lapidus, I., & Muslimov, A. 1998, *ApJ*, **499**, 315
Titarchuk, L., Osherovich, V., & Kuznetsov, S. 1999, *ApJ*, **525**, L129
Tomsick, J. A. & Kaaret, P. 2000, *ApJ*, **537**, 448
2001, *ApJ*, **548**, 401
Tomsick, J. A., Halpern, J. P., Kemp, J., & Kaaret, P. 1999, *ApJ*, **521**, 341
Tomsick, J. A., Kalemci, E., & Kaaret, P. 2004, *ApJ*, **601**, 439
Treves, A., Belloni, T., Corbet, R. H. D., *et al.* 1990, *ApJ*, **364**, 266
Trudolyubov, S. P. 2001, *ApJ*, **558**, 276
Trudolyubov, S. P., Churazov, E. M., & Gilfanov, M. R. 1999a, *Astron. Lett.*, **25**, 718
Trudolyubov, S., Churazov, E., & Gilfanov, M. 1999b, *A&A*, **351**, L15
Trudolyubov, S. P., Borozdin, K. N., & Priedhorsky, W. C. 2001, *MNRAS*, **322**, 309
Ubertini, P., Bazzano, A., Cocchi, M., *et al.* 1999, *ApJ*, **514**, L27
Uemura, M., Kato, T., Ishioka, R., *et al.* 2002, *PASJ*, **54**, L79
2004, *PASJ*, **56**, 61
Unno, W., Yoneyama, T., Urata, K., *et al.* 1990, *PASJ*, **42**, 269
Ushomirsky, G., Cutler, C., & Bildsten, L. 2000, *MNRAS*, **319**, 902
Uttley, P. 2004, *MNRAS*, **347**, L61
Uttley, P. & McHardy, I. M. 2001, *MNRAS*, **323**, L26
Uttley, P., McHardy, I. M., & Papadakis, I. E. 2002, *MNRAS*, **332**, 231
Varnière, P., Rodriguez, J., & Tagger, M. 2002, *A&A*, **387**, 497
Vasiliev, L., Trudolyubov, S., & Revnivtsev, M. 2000, *A&A*, **362**, L53
Vaughan, B. A. & Nowak, M. A. 1997, *ApJ*, **474**, L43
Vaughan, B., van der Klis, M., Lewin, W. H. G., *et al.* 1994, *ApJ*, **421**, 738

Vaughan, B. A., van der Klis, M., Méndez, M., *et al.* 1997, *ApJ*, **483**, L115
1998, *ApJ*, **509**, L145
Vaughan, B. A., van der Klis, M., Lewin, W. H. G., *et al.* 1999, *A&A*, **343**, 197
Vietri, M. & Stella, L. 1998, *ApJ*, **503**, 350
Vignarca, F., Migliari, S., Belloni, T., Psaltis, D., & van der Klis, M. 2003, *A&A*, **397**, 729
Vikhlinin, A. 1999, *ApJ*, **521**, L45
Vikhlinin, A., Churazov, E., & Gilfanov, M. 1994, *A&A*, **287**, 73
Vikhlinin, A., Churazov, E., Gilfanov, M., *et al.* 1995, *ApJ*, **441**, 779
Villarreal, A. R. & Strohmayer, T. E. 2004, *ApJ*, **614**, L121
Vrtilek, S. D., Raymond, J. C., Garcia, M. R., *et al.* 1990, *A&A*, **235**, 162
Vrtilek, S. D., Penninx, W., Raymond, J. C., *et al.* 1991, *ApJ*, **376**, 278
Vrtilek, S. D., Raymond, J. C., Boroson, B., *et al.* 2003, *PASP*, **115**, 1124
Wagoner, R. W. 1999, *Phys. Rep.*, **311**, 259
Wagoner, R. V. 2002, *ApJ*, **578**, L63
Wallinder, F. H. 1995, *MNRAS*, **273**, 1133
Wang, Y.-M. & Schlickeiser, R. 1987, *ApJ*, **313**, 200
Wang, D., Ma, R., Lei, W., & Yao, G. 2003, *MNRAS*, **344**, 473
Warner, B. & Woudt, P. A. 2002, *MNRAS*, **335**, 84
Watarai, K. & Mineshige, S. 2003a, *PASJ*, **55**, 959
2003b, *ApJ*, **596**, 421
Weinberg, N., Miller, M. C., & Lamb, D. Q. 2001, *ApJ*, **546**, 1098
Weisskopf, M. C. & Sutherland, P. G. 1978, *ApJ*, **221**, 228
Weisskopf, M. C., Kahn, S. M., & Sutherland, P. G. 1975, *ApJ*, **199**, L147
Wen, L., Cui, W., & Bradt, H. V. 2001, *ApJ*, **546**, L105
Wheeler, J. C. 1977, *ApJ*, **214**, 560
White, N. E. & Marshall, F. E. 1984, *ApJ*, **281**, 354
White, N. E., & Zhang, W. 1997, *ApJ*, **490**, L87
White, N. E., Peacock, A., & Taylor, B. G. 1985, *ApJ*, **296**, 475
Wijers, R. A. M. J., van Paradijs, J., & Lewin, W. H. G. 1987, *MNRAS*, **228**, 17P
Wijnands, R. 2004, in *Rossi and Beyond*, AIP Conf. Proc. **714**, 97; astro-ph/0403409
Wijnands, R. & Homan, J. 2003, *ATel*, **165**, 1
Wijnands, R. & Miller, J. M. 2002, *ApJ*, **564**, 974
Wijnands, R. A. D., & van der Klis, M. 1997, *ApJ*, **482**, L65
1998a, *ApJ*, **507**, L63
1998b, *Nature*, **394**, 344
1999a, *ApJ*, **514**, 939
1999b, *A&A*, **345**, L35
1999c, *ApJ*, **522**, 965
2000, *ApJ*, **528**, L93
2001, *MNRAS*, **321**, 537
Wijnands, R. A. D., van der Klis, M., Psaltis, D., *et al.* 1996, *ApJ*, **469**, L5
Wijnands, R. A. D., van der Klis, M., van Paradijs, *et al.* 1997a, *ApJ*, **479**, L141
Wijnands, R., Homan, J., van der Klis, M., *et al.* 1997b, *ApJ*, **490**, L157
Wijnands, R. A. D., van der Klis, M., Kuulkers, E., Asai, K., & Hasinger, G. 1997c, *A&A*, **323**, 399
Wijnands, R., Homan, J., van der Klis, M., *et al.* 1998a, *ApJ*, **493**, L87
Wijnands, R. A. D., van der Klis, M., Méndez, M., *et al.* 1998b, *ApJ*, **495**, L39
Wijnands, R., Méndez, M., van der Klis, M., *et al.* 1998c, *ApJ*, **504**, 35
Wijnands, R., van der Klis, M., & Rijkhorst, E. 1999a, *ApJ*, **512**, L39
Wijnands, R., Homan, J., & van der Klis, M. 1999b, *ApJ*, **526**, L33
Wijnands, R., Méndez, M., Miller, J. M., & Homan, J. 2001a, *MNRAS*, **328**, 451
Wijnands, R., Strohmayer, T., & Franco, L. M. 2001b, *ApJ*, **549**, L71
Wijnands, R., Miller, J. M., & van der Klis, M. 2002a, *MNRAS*, **331**, 60
Wijnands, R., Muno, M. P., Miller, J. M., *et al.* 2002b, *ApJ*, **566**, 1060
Wijnands, R., van der Klis, M., Homan, J., *et al.* 2003, *Nature*, **424**, 44
Wilms, J., Nowak, M. A., Pottschmidt, K., *et al.* 2001, *MNRAS*, **320**, 327
Wood, K. S., Ray, P. S., Bandyopadhyay, R. M., *et al.* 2000, *ApJ*, **544**, L45

Woods, P. M., Kouveliotou, C., Finger, M. H., *et al.* 2002, *IAU Circ.*, 7856, 1
Xiong, Y., Wiita, P. J., & Bao, G. 2000, *PASJ*, **52**, 1097
Yamaoka, K., Ueda, Y., Inoue, H., *et al.* 2001, *PASJ*, **53**, 179
Yoshida, S. & Lee, U. 2001, *ApJ*, **546**, 1121
Yoshida, K., Mitsuda, K., Ebisawa, K., *et al.* 1993, *PASJ*, **45**, 605
Yu, W. & van der Klis, M. 2002, *ApJ*, **567**, L67
Yu, W., Zhang, S. N., Harmon, B. A., *et al.* 1997, *ApJ*, **490**, L153
Yu, W., Li, T. P., Zhang, W., & Zhang, S. N. 1999, *ApJ*, **512**, L35
Yu, W., van der Klis, M., & Jonker, P. G. 2001, *ApJ*, **559**, L29
Yu, W., Klein-Wolt, M., Fender, R., & van der Klis, M. 2003, *ApJ*, **589**, L33
Zdunik, J. L., Haensel, P., Gondek-Rosińska, D., & Gourgoulhon, E. 2000, *A&A*, **356**, 612
Zdziarski, A. A., Poutanen, J., Paciesas, W. S., & Wen, L. 2002, *ApJ*, **578**, 357
Zhang, W., Lapidus, I., White, N. E., & Titarchuk, L. 1996a, *ApJ*, **469**, L17
1996b, *ApJ*, **473**, L135
Zhang, W., Morgan, E. H., Jahoda, K., *et al.* 1996c, *ApJ*, **469**, L29
Zhang, S. N., Cui, W., Harmon, B. A., *et al.* 1997a, *ApJ*, **477**, L95
Zhang, W., Strohmayer, T. E., & Swank, J. H. 1997b, *ApJ*, **482**, L167
Zhang, S. N., Ebisawa, K., Sunyaev, R., *et al.* 1997c, *ApJ*, **479**, 381
Zhang, W., Strohmayer, T. E., & Swank, J. H. 1998a, *ApJ*, **500**, L167
Zhang, W., Smale, A. P., Strohmayer, T. E., & Swank, J. H. 1998b, *ApJ*, **500**, L171
Zhang, W., Jahoda, K., Kelley, R. L., *et al.* 1998c, *ApJ*, **495**, L9
Życki, P. T. 2002, *MNRAS*, **333**, 800
2003, *MNRAS*, **340**, 639
Życki, P. T., Done, C., & Smith, D. A. 1999a, *MNRAS*, **305**, 231
1999b, *MNRAS*, **309**, 561

3

New views of thermonuclear bursts

Tod Strohmayer
NASA Goddard Space Flight Center

Lars Bildsten
University of California, Santa Barbara

3.1 Introduction

Many accreting neutron stars erupt in spectacular thermonuclear conflagrations every few hours to days. These events, known as Type I X-ray bursts, or simply X-ray bursts, are the subject of our review. Since the last review of X-ray burst phenomenology was written (Lewin, van Paradijs & Taam 1993; hereafter LVT), powerful new X-ray observatories, the Rossi X-ray Timing Explorer (RXTE), the Italian–Dutch BeppoSAX mission, XMM-Newton and Chandra have enabled the discovery of entirely new phenomena associated with thermonuclear burning on neutron stars. Some of these new findings include: (i) the discovery of millisecond (300–600 Hz) oscillations during bursts, so-called "burst oscillations"; (ii) a new regime of nuclear burning on neutron stars which manifests itself through the generation of hours-long flares about once a decade, now referred to as "superbursts"; (iii) discoveries of bursts from low accretion rate neutron stars; and (iv) new evidence for discrete spectral features from bursting neutron stars.

It is perhaps surprising that nuclear physics plays such a prominent role in the phenomenology of an accreting neutron star, as the gravitational energy released per accreted baryon (of mass $m_{\rm p}$), $GMm_{\rm p}/R \approx 200$ MeV,[1] is so much larger than the nuclear energy released by fusion ($\approx$5 MeV when a solar mix goes to heavy elements). Indeed, if the accreted fuel was burned at the rate of accretion, any evidence of nuclear physics would be swamped by the light from released gravitational energy. The only way the nuclear energy can be seen is when the fuel is stored for a long period and then burns rapidly (as in Type I bursts and superbursts).

Advances in millisecond timing were enabled by RXTE's combination of large X-ray collecting area, sub-millisecond time resolution and a state-of-the-art data system and telemetry capacity. RXTE's instruments are also unfettered; they can look where the action is for longer, more often, and more quickly than any previous X-ray observatory. The Wide Field Cameras (WFC) on BeppoSAX (Jager *et al.* 1997), in combination with the All Sky Monitor (ASM) on board RXTE (Levine *et al.* 1996), have provided an unprecedented, long-term view of the X-ray sky that opened up new discovery space for rare events. Their capabilities led to the discovery of superbursts as a new manifestation of thermonuclear burning on neutron stars, as well as the detection of rare bursts from neutron stars accreting at very low rates. Indeed, WFC observations alone have led to the discovery of about 20 new bursters (see in 't Zand 2001).

[1] We use "$\approx$" for expressions known to better than a factor of 2, and "$\sim$" when uncertainties are larger, typically a factor of 5 or so.

Compact Stellar X-Ray Sources, eds. Walter Lewin and Michiel van der Klis.
Published by Cambridge University Press.

In addition to these new results, recent discoveries have provided answers to some long-standing questions concerning the sources of thermonuclear bursts, the neutron star low-mass X-ray binaries (LMXB). These include: (i) the discovery of accreting millisecond pulsars in seven LMXB systems, confirming that neutron stars are spun-up to millisecond periods by accretion; (ii) the discovery of sub-millisecond variability from many neutron star LMXBs (see the review by van der Klis in Chapter 2); and (iii) detections of several bursting neutron star transients (e.g., Aql X-1, Cen X-4) in quiescence with the increased sensitivity afforded by Chandra and XMM-Newton. These data have provided important new information on the thermal state of neutron stars, which strongly influences their bursting behavior (see Psaltis in Chapter 1).

Our review focuses on the new phenomena and insights they provide about neutron stars and thermonuclear burning on them. These discoveries have stimulated much new theoretical thinking. In particular, the discovery of millisecond oscillations during bursts has refocused attention on the need to understand the multi-dimensional nature of thermonuclear flame propagation, that is, how burning is initiated and how it spreads around the neutron star. An understanding of flame propagation, in turn, requires an understanding of how thermonuclear burning influences the dynamics of the accreted layers. To understand superbursts, theorists have been forced to think about what happens at much greater depths in the accreted layers of neutron stars, and about the detailed composition of the "ashes" of H/He burning. Finally, observations of bursts from sources with very low mass accretion rates are forcing theorists to confront some physical processes, for example, element diffusion, which have not been considered in earlier work.

Space constraints do not allow us to present an exhaustive review of all X-ray burst phenomenology and theory. For example, we will not describe in detail the Type II (accretion instability) burst phenomena now observed from two sources; the famous Rapid Burster (see for example, LVT); and the Bursting Pulsar (GRO J1744–28, see Kouveliotou *et al.* 1996; Giles *et al.* 1996). Although new observations of Type II bursters have been obtained since the last reviews were written, including the discovery of GRO J1744–28, only the second known Type II burster, they have generally not provided new qualitative insights on the accretion instability. Some of the more recent observations of these sources, and our current understanding of them, are described elsewhere in this volume (see Chapters 8, 1 and 2 by Verbunt & Lewin, Psaltis, and van der Klis).

Despite its energetic disadvantage, we are now confident that sudden nuclear energy release powers bursts and, as we emphasize here, fleshing out the details of the thermonuclear flash model in the context of the observations is probing both neutron star structure and fundamental physics. We try not to segregate our discussions into separate observational and theoretical parts, rather, we integrate the theory and observations as they relate to particular phenomena as much as possible.

We begin in Section 3.2 with a brief theoretical introduction to the relevant physics of thermonuclear burning on neutron stars and how it accounts for the gross properties of bursts. We refer the reader to Bildsten (1998) for relevant theoretical details. We then briefly review the observational characteristics of bursts and sources (Section 3.3). This lays out the fundamentals and topics important for our discussions of the new findings. Here we also include some discussion on new results from burst spectroscopy (Section 3.3.3).

We then move to the new discoveries. In Section 3.4 we introduce millisecond variability during bursts ("burst oscillations"), and theoretical implications motivated by the observations. In Section 3.5 we describe the new observations of superbursts and their theoretical implications. We close in Section 3.6 with a summary and some comments on the future prospects of burst research.

3.2 The physics of hydrogen/helium burning

At the core of the observed phenomena is the "thin shell" instability discovered theoretically by Schwarzschild and Härm (1965) in the helium shell residing above the carbon/oxygen core during the asymptotic giant branch phase of stellar evolution. The driver of this instability is a nuclear energy generation rate that is more temperature sensitive than radiative cooling and is confined to a thin shell. Hansen and Van Horn (1975) showed that burning of the accumulated hydrogen and helium on a neutron star also occurs in radially thin shells that were susceptible to the same instability. Soon thereafter, Type I X-ray bursts from LMXBs were independently discovered by Grindlay *et al.* (1976) and Belian, Conner and Evans (1976), and were quickly associated (Woosley & Taam 1976; Maraschi & Cavaliere 1977; Joss 1977, 1978; Lamb & Lamb 1978) with Hansen and van Horn's (1975) instability. For a brief historical overview see LVT (1993).

The successful association of thermonuclear instabilities with X-ray bursts made a nice picture of a recurrent cycle that consists of fuel accumulation for several hours to days followed by a thermonuclear runaway that burns the fuel in $\sim 10-100$ seconds. It also secured the identification of the accreting objects as neutron stars (NS). The mass donors – the ultimate source of the thermonuclear fuel – are typically old, Population II objects or, in some cases, degenerate helium or perhaps carbon/oxygen white dwarfs (Rappaport, Joss & Webbink 1982). The accreted composition is important, as the nuclear ashes and burst properties depend on the accreted mix of light elements. Unfortunately, in most cases, we have little information on the composition of the accreted fuel. Of the approximately 160 known LMXBs about 70 are observed to produce bursts (see Liu, van Paradijs & van den Heuvel 2001).

A fundamental physical reason for studying the neutron star example of a thin-shell instability is that the timescales are observationally accessible. Hundreds of bursts have been seen from some neutron stars, and study of the burst dependence on accretion rate can be undertaken. Such an exercise is rare to impossible in the other astrophysical site where the instability is observable: the explosion of hydrogen on an accreting white dwarf as a classical nova. For a more detailed theoretical discussion see Bildsten (1998).

3.2.1 *Nuclear burning during accumulation and ignition*

After the accreted hydrogen and helium has become part of the NS, it undergoes hydrostatic compression as new material is piled on. The fresh fuel reaches ignition densities and temperatures within a few hours to days. The resulting compression rate depends on the accretion rate per unit area, $\dot{m} \equiv \dot{M}/A_{\rm acc}$, where $A_{\rm acc}$ is the area covered by fresh material. We will sometimes quote numbers for both $\dot{m}$ and $\dot{M}$. When we give $\dot{m}$, we have assumed $A_{\rm acc} = 4\pi R^2 \approx 1.2 \times 10^{13}$ cm^2. The time it takes for heat transport to cool the accreted envelope (what we call the thermal time) is only 10 seconds at the ignition location, where the pressure is $P \approx 10^{22}-10^{23}$ erg cm^{-3}. This is so much shorter than the time to accumulate the material (hours to days) that the compression is far from adiabatic.

The temperature exceeds 10^7 K in most of the accumulating atmosphere, so that hydrogen burns via the CNO cycle and the proton–proton (pp) cycle can be neglected. At high temperatures ($T > 8 \times 10^7$ K), the timescale for proton captures becomes shorter than the subsequent β decay lifetimes, even for the slowest ^{14}N(p, γ)^{15}O reaction. The hydrogen then burns in the "hot" CNO cycle of Fowler and Hoyle (1965)

$$^{12}\mathrm{C}(\mathrm{p},\gamma)^{13}\mathrm{N}(\mathrm{p},\gamma)^{14}\mathrm{O}(\beta^+)^{14}\mathrm{N}(\mathrm{p},\gamma)^{15}\mathrm{O}(\beta^+)^{15}\mathrm{N}(\mathrm{p},\alpha)^{12}\mathrm{C} \tag{3.1}$$

and is limited to $5.8 \times 10^{15} Z_{\mathrm{CNO}}$ erg g^{-1} s^{-1}, where Z_{CNO} is the mass fraction of CNO. The hydrogen burns this way in the accumulating phase when

$$\dot{m} > 900 \text{ g cm}^{-2} \text{ s}^{-1} (Z_{\mathrm{CNO}}/0.01)^{1/2} \tag{3.2}$$

and is thermally stable. The amount of time it takes to burn the hydrogen is $\approx (10^3/Z_{\mathrm{CNO}})$ seconds, or about one day for solar metallicities. This time is even longer if the donor star has a low metal content or if there is substantial spallation of the incident CNO elements as discussed by Bildsten, Salpeter and Wasserman (1992). For lower $\dot{m}$, the hydrogen burning is thermally unstable and can trigger Type I bursts.

The slow hydrogen burning during accumulation allows for a unique burning regime at high $\dot{m}$. This simultaneous H/He burning occurs when

$$\dot{m} > 2 \times 10^3 \text{ g cm}^{-2} \text{ s}^{-1} (Z_{\mathrm{CNO}}/0.01)^{13/18} \tag{3.3}$$

(Bildsten 1998; Cumming & Bildsten 2000), as at these high rates the fluid is compressed to helium ignition conditions before the hydrogen is completely burned (Lamb & Lamb 1978; Taam & Picklum 1978). The strong temperature dependence of the helium burning rate (and lack of any weak interactions) leads to a thin-shell instability for temperatures $T < 5 \times 10^8$ K and causes the burst. The critical condition of thin burning shells ($h \ll R$) is true before burning and remains so during the flash. Stable burning sets in at higher $\dot{m}$ (comparable to the Eddington limit; Paczynski 1983; Bildsten 1998; Narayan & Heyl 2002) when the helium burning temperature sensitivity finally becomes weaker than the cooling rate's sensitivity (Ayasli & Joss 1982; Taam, Woosley & Lamb 1996). This is consistent with the absence of bursts from high-field X-ray pulsars, which channel the accretion flow onto a small-area polar cap, and thus achieve a high local (and stabilizing) $\dot{m}$ (Joss & Li 1980; Bildsten & Brown 1997).

For solar metallicities, there is a narrow window of $\dot{m}$ where the hydrogen is completely burned before the helium ignites. In this case, a pure helium shell accumulates underneath the hydrogen-burning shell until conditions are reached for ignition of the pure helium layer. The recurrence times of these bursts must be longer than the time to burn all of the hydrogen, so pure helium flashes should have recurrence times in excess of a day and $\alpha \approx 200$ (α is the ratio of the time-averaged persistent to burst luminosity). To summarize, in order of increasing $\dot{m}$, the regimes of unstable burning we expect from NSs accreting at sub-Eddington rates ($\dot{m} < 10^5$ g cm^{-2} s^{-1}) are (Fujimoto, Hanawa & Miyaji 1981; Fushiki & Lamb 1987; Cumming & Bildsten 2000):

(1) Mixed hydrogen and helium burning triggered by thermally unstable hydrogen ignition for $\dot{m} < 900$ g cm^{-2} s^{-1} ($\dot{M} < 2 \times 10^{-10} \mathrm{M}_\odot$ yr^{-1}).
(2) Pure helium shell ignition for 900 g cm^{-2} s^{-1} $< \dot{m} <$ 2×10^3 g cm^{-2} s^{-1} following completion of hydrogen burning.
(3) Mixed hydrogen and helium burning triggered by thermally unstable helium ignition for $\dot{m} > 2 \times 10^3$ g cm^{-2} s^{-1} ($\dot{M} > 4.4 \times 10^{-10} \mathrm{M}_\odot$ yr^{-1}).

The transition $\dot{m}$ are for $Z_{\rm CNO} \approx 0.01$. Reducing $Z_{\rm CNO}$ lowers the transition accretion rates and substantially narrows the $\dot{m}$ range for pure helium ignition. Another effect critical to the burning is the amount of heat flux coming through the burning layer from deeper parts of the NS (Ayasli & Joss 1982; Fushiki & Lamb 1987). The current theoretical estimates (Brown, Bildsten & Rutledge 1998; Brown 2000; Colpi *et al.* 2001) are that between 10 and 100% of the heat released via pycnonuclear reactions occurring at densities greater than 10^7 g cm^{-3} (Haensel & Zdunik 1990) emerges from the surface, proving most important to the burst properties of pure helium accretors such as 4U 1820–30 (Bildsten 1995; Strohmayer & Brown 2002; Cumming 2003).

3.2.2 *Burning during the bursts: the rp-process*

We now briefly discuss what happens as the thermal instability develops into a burst and what observational differences are to be expected between a pure helium ignition and a mixed hydrogen/helium ignition. The flash occurs at fixed pressure, and the increasing temperature eventually allows the radiation pressure to dominate. For a typical ignition column of 2×10^8 g cm^{-2}, the pressure is $P = gy \approx 4 \times 10^{22}$ erg cm^{-3}, so $aT_{\rm max}^4/3 \approx P$ gives $T_{\rm max} \approx 2 \times 10^9$ K. Here, g, y, and a are the surface gravity, the column depth, and the radiation constant, respectively. For pure helium flashes, the fuel burns rapidly (since there are no slow weak interactions) and the local Eddington limit is often exceeded. These conditions lead to photospheric radius expansion (PRE) bursts with durations, set mostly by the time it takes the heat to escape, of order 5–10 seconds.

When hydrogen and helium are both present, the temperatures reached during the thermal instability can easily produce elements far beyond the iron group (Hanawa *et al.* 1983; Wallace & Woosley 1984; Hanawa & Fujimoto 1984; Koike *et al.* 1999; Schatz *et al.* 1999, 2001) via the rapid-proton (rp) process of Wallace and Woosley (1981). This burning starts after the triggering helium flash heats the gas to high enough temperatures to allow the "breakout" reactions ^{15}O$(\alpha, \gamma)^{19}$Ne and ^{18}Ne$(\alpha, p)^{21}$Na (^{18}Ne is made in the chain ^{14}O$(\alpha, p)^{17}$F(p$, \gamma)^{18}Ne(\beta^+)^{18}$F(p$, \alpha)^{15}$O triggered by the high temperatures) to proceed faster than the β decays. This takes these catalysts out of the CNO cycle loop, where they subsequently burn hydrogen via the rp-process: a series of successive proton captures and β decays.

Figure 3.1 shows the dominant path of the nuclei as they move up the proton-rich side of the valley of stability (much like the r-process, which occurs by neutron captures on the neutron-rich side) more or less limited by the β-decay rates. Theoretical work shows that the end-point of this time-dependent burning is far beyond iron (Hanawa & Fujimoto 1984; Schatz *et al.* 1999, 2001; Koike *et al.* 1999) and is set by either the complete burning of the hydrogen or reaching the closed SnSbTe cycle found by Schatz *et al.* (2001). When hydrogen is exhausted prior to reaching the SnSbTe cycle, a rough estimate of the rp-process end-point is made by merely accounting for the ratios of seed nuclei to hydrogen. This produces a large range of heavy nuclei (Schatz *et al.* 1999, 2001). Schatz *et al.* (1999) also showed an additional important point of nucleosynthesis, which is that the hydrogen burns out before the helium is completely burned. Thus, the carbon made during late helium burning remains as carbon since there are no protons available for it to capture. As we will show in Section 3.5, this remaining carbon is the apparent fuel for the recently discovered superbursts. Observationally, the long series of β decays during the rp-process releases energy for at least 150 seconds after the burst

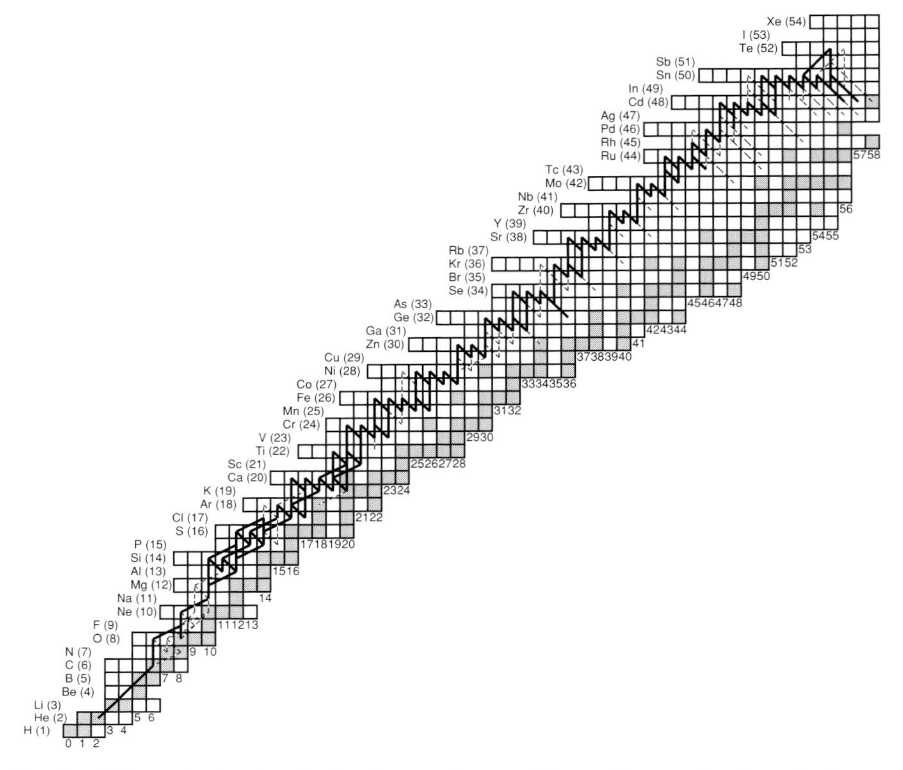

Fig. 3.1. Schematic showing the dominant pathways of the nuclear reaction flows during the rp-process. Elements far beyond ^{56}Fe can easily be reached. Filled squares denote stable nuclides (after Schatz *et al.* 2001).

has started. Thus mixed hydrogen/helium bursts last much longer than pure helium flashes, as nicely exhibited in the work of Woosley *et al.* (2004).

3.2.3 Mixed H/He bursts from GS 1826–238

We discuss shortly that most bursters do not precisely match the theory presented above in a simple way, and initial indications are that this could be due to variations in the area on which new fuel accretes (Bildsten 2000). However, there are times when bursters behave in a limit cycle manner, with bursts occurring periodically (e.g., Robinson & Young 1997) as $\dot{m}$ stays at a fixed value, and we discuss one beautiful example here: the Type I burster GS 1826–238. Ubertini *et al.* (1999) found 70 bursts over a 2.5 year monitoring baseline with the BeppoSAX/WFC. The quasi-periodic recurrence time was 5.76 ± 0.62 hours. Cocchi *et al.*'s (2000) later analysis within observing seasons found that in 1997 and 1998 the recurrence times were even more clock-like, 5.92 ± 0.07 hours and 5.58 ± 0.09 hours. The persistent flux during this bursting period was $F_x \approx 2 \times 10^{-9}$ erg cm^{-2} s^{-1} (Ubertini *et al.* 1999; in 't Zand *et al.* 1999; Kong *et al.* 2000). Since then, the persistent flux has risen, and remarkably, the burst recurrence time has shortened $\propto \dot{M}^{-1}$ (see Fig. 3.2; Galloway *et al.* 2004), implying that the accumulated mass is the same as $\dot{M}$ changes.

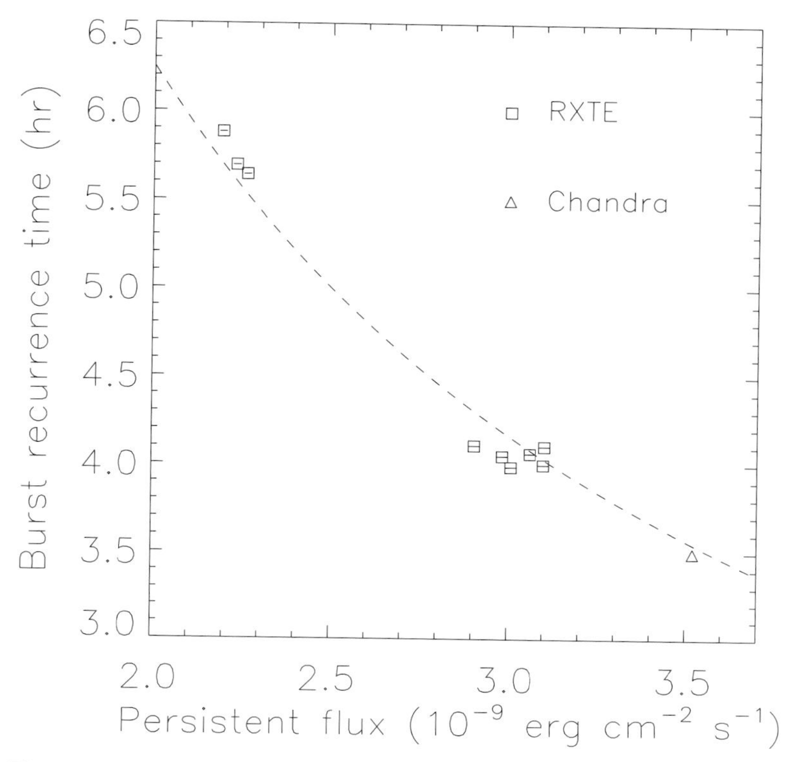

Fig. 3.2. Variation of the burst recurrence time for GS 1826–238 as a function of the persistent flux from RXTE measurements between 1997 and 2000 (squares) and a more recent Chandra/RXTE measurement on 2002 July 29. Horizontal error bars indicate the 1σ errors. The dashed line is the trend expected if the burst recurrence time is $\propto \dot{M}^{-1}$ (after Galloway *et al.* 2004). See also Figure 3 in Cornelisse *et al.* (2003).

The radius of a neutron star emitting a blackbody flux, $F_{\rm bol}^{\infty}$, at temperature $T_{\rm bb}^{\infty}$ (both quantities measured by a distant observer), satisfies the relation $R_{\rm bb}(1 - 2GM/c^2R_{\rm bb})^{-1/2} = d(F_{\rm bol}^{\infty}/\sigma T_{\rm bb}^{\infty\,4})^{1/2}$, where d is the source distance, and the quantity $(1 - 2GM/c^2R_{\rm bb})^{-1/2}$ is the surface redshift correction. From the observed persistent flux and the measured ratio, $R_{\rm bb}/d$ for GS 1826–238 (about 10 km at 10 kpc; Galloway *et al.* 2004), the local accretion rate is $\dot{m} \approx 10^4$ g cm^{-2} s^{-1} when $F_x \approx 2 \times 10^{-9}$ erg cm^{-2} s^{-1}. This is very safely in the mixed H/He burning regime. The estimated $\dot{m}$ gives an accumulated column on the NS prior to the burst of 2×10^8 g cm^{-2}, just what is expected from theory (see Table 2 of Cumming & Bildsten 2000). These periodic bursts allow for a very secure measurement of $\alpha \approx 40$ (Galloway *et al.* 2004), implying a nuclear energy release of 5 MeV per accreted nucleon for a 1.4 $M_{\odot}$, 10 km NS. Energy releases this large can only come about via hydrogen burning and the long ($>$100 s) duration of the bursts is consistent with the expected energy release from the rp-process. Figure 3.3 shows the time profile of a burst seen with the Proportional Counter Array (PCA) on RXTE (see also Kong *et al.* 2000). These data provide an important confirmation of the delayed energy release expected when hydrogen is burning via an rp-process. The resemblance of these profiles to theoretical results of Hanawa & Fujimoto (1984), Schatz *et al.* (2001) and Woosley *et al.* (2004) is striking.

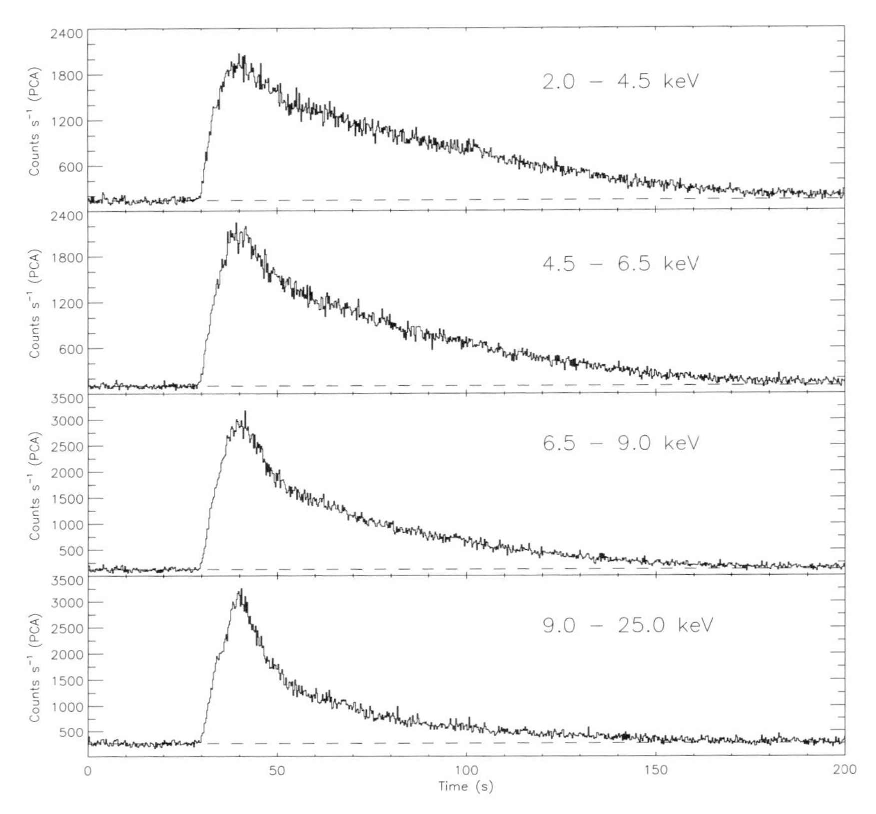

Fig. 3.3. An X-ray burst from GS 1826–238 seen with the RXTE/PCA. The burst is shown in four different energy bands. The long duration is indicative of the delayed energy release from the rapid-proton (rp) process. The dashed line marks the preburst flux level (see also Kong *et al.* 2000).

3.3 Observational overview of bursts

In this section we give a brief overview of important observational characteristics of bursts. We will emphasize recent work and specific topics that are relevant in the context of the new discoveries to be described in later sections. We refer to earlier reviews for some of the details we must omit.

Bursts are most commonly observed from the "atoll" subclass of LMXBs with luminosities above $\approx 10^{36}$ erg s^{-1} (Hasinger & van der Klis 1989; LVT). These systems are accreting at intermediate rates of between 0.01 and 0.3 $\dot{M}_{\rm Edd}$, which is roughly consistent with the accretion rate range of the thermonuclear instabilities described earlier. Interestingly, observations with the WFC on BeppoSAX have recently discovered bursts from 10 LMXBs with persistent X-ray luminosities significantly less than 10^{36} erg s^{-1} (see Cornelisse *et al.* 2002a, b; Cocchi *et al.* 2001; Section 3.3.4). These recently discovered bursts probe the mass accretion rate dependence of thermonuclear burning in a previously unexplored regime. For example, physical processes such as element diffusion, which are not relevant to most bursters, likely become important at such low accretion rates (Wallace, Woosley & Weaver 1982). We will say more about these bursts shortly.

3.3.1 Burst profiles and spectra

Though the time profiles of X-ray bursts are diverse, they do share several characteristic features. (i) Burst rise times are shorter than their decay times. Rise times are typically <2 s, but in some cases can be as long as 10 s. Excluding superbursts, which we discuss later, burst decay times range from about 10 s to several minutes, with most bursts having 10–20 s decays. (ii) Burst profiles are shorter at higher energies. This is a direct result of cooling of the neutron star surface with time. (iii) Burst profiles are generally smooth, showing an exponential or exponential-like intensity decay.

Swank *et al.* (1977) and Hoffman, Lewin and Doty (1977) showed that bursts have thermal (blackbody) spectra. Measurement of the bolometric fluxes and blackbody temperatures can then be used to estimate radii, if the distance is known (see Section 3.2.3). Radii inferred in this manner are typically in the range of ~ 10 km, consistent with cooling of an object having the theoretical size of a neutron star.

Although burst spectra are observationally well described by the Planck function, theoretically they should be harder than a blackbody at the effective temperature of the atmosphere (see London, Howard & Taam 1984, 1986; Ebisuzaki & Nakamura 1988; Madej 1991; Titarchuk 1994). This occurs because electron (Compton) scattering is an important opacity source in a neutron star atmosphere. Direct evidence for this effect comes from the peak blackbody temperatures ($kT_{\rm bb} > 3$ keV) of some bursts, which are significantly higher than the Eddington effective temperature for reasonable neutron star models (see London, Howard & Taam 1984, 1986).

3.3.2 Photospheric radius expansion bursts

In bright bursts the local X-ray luminosity in the atmosphere may reach the Eddington limit

$$L_{\rm Edd} = (4\pi c G M/\kappa)(1 - 2GM/c^2 R)^{-1/2} = 4\pi R^2 \sigma T_{\rm eff}^4 \tag{3.4}$$

where M, R, and κ are the neutron star mass, radius and atmospheric opacity, respectively, and the photospheric layers can be lifted off the neutron star surface by radiation pressure. Note that $L_{\rm Edd}$ depends on the composition of the accreted atmosphere through the opacity κ. In these bursts the blackbody temperature decreases while the inferred blackbody radius simultaneously increases. This all happens while the total X-ray flux stays approximately constant. These bursts are called photospheric radius expansion (PRE) bursts. The moment when the photosphere falls back to the neutron star surface (when the temperature is highest) is called "touchdown." Theoretical work indicates that in such bursts the X-ray flux stays within a small percentage of the Eddington limit, and the excess energy is efficiently transferred into kinetic energy of the outflow (Hanawa & Sugimoto 1982; Paczynski & Anderson 1986; Joss & Melia 1987; Titarchuk 1994; Shaposhnikov & Titarchuk 2002). Ebisuzaki and Nakamura (1988) found evidence for photospheric composition variations when comparing fainter bursts with PRE bursts from the LMXBs 4U 1608–52 and 4U 1636–53. They found differences in the luminosity–color temperature relation between these two classes which they attributed to atmospheric composition changes. They suggested that the luminosity–color temperature relation for faint bursts and PRE bursts could be explained by hydrogen-rich and hydrogen-poor atmospheres, respectively. They further hypothesized that the hydrogen-rich envelope is ejected during bright PRE bursts, and thus the atmospheric composition is pure helium in these cases.

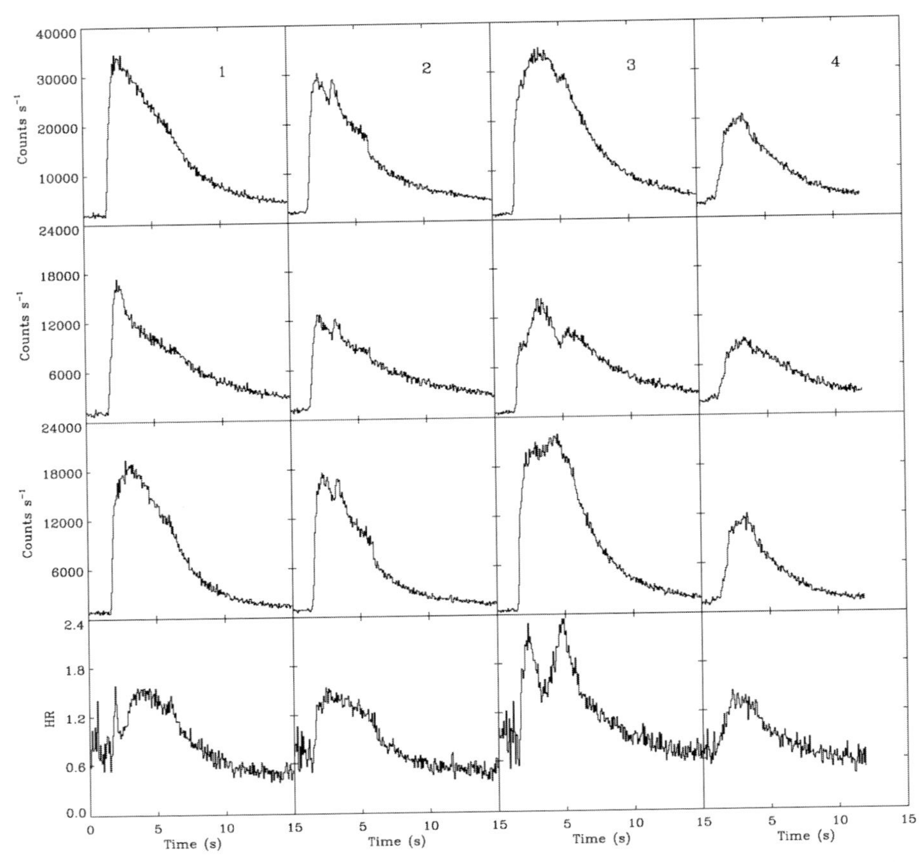

Fig. 3.4. A sample of four X-ray bursts from the LMXB 4U 1728–34 as observed with the RXTE/PCA. Each sequence shows, from top to bottom, the total 2–60 keV countrate, the 2–6 keV countrate, the 6–30 keV countrate, and the hardness ratio (6–30 keV)/(2–6 keV). Bursts 1 and 3 show clear evidence for PRE based on the hardness ratio evolution.

The amount of photospheric uplift can vary dramatically from burst to burst and amongst different sources. In the most powerful bursts the expansion can be large enough to shift the effective temperature of the photosphere entirely below the X-ray band (See Lewin, Vacca & Basinska 1984; Strohmayer & Brown 2002). Such events show "precursors" separated from the main part of the burst by the cooling of the photosphere. Other less powerful bursts may show double peaked profiles as only a portion of the flux is shifted out of the X-ray band by the expansion. Figure 3.4 shows several examples of bursts, both with and without PRE, observed with the RXTE/PCA from the LMXB 4U 1728–34.

Since the Eddington luminosity should impose an upper limit to burst fluxes, it was suggested early on that bursts might provide a "standard candle" (van Paradijs 1978), and that they could be used as distance indicators. Van Paradijs (1981) recognized that globular cluster bursters, with independently known distances, would provide an important test of this idea. Subsequent researchers concluded that the brightest (PRE) bursts likely represent a true limiting luminosity, but some uncertainties remained (Lewin 1982; Basinska *et al.* 1984).

Recently Kuulkers *et al.* (2003) have reexamined this issue using the extensive sets of bursts observed with RXTE and BeppoSAX/WFC, and a uniform set of globular cluster distances. They conclude that the radius expansion bursts can indeed be regarded as standard candles to within about 15%, and derive a critical luminosity of about 3.8×10^{38} erg s^{-1}, consistent with the Eddington limit for hydrogen-poor matter from a neutron star. Since we do not expect that all LMXBs are pure helium accretors, the fact that PRE bursts from the globular cluster sources have approximately the same peak fluxes provides additional support to the idea that in these bursts the hydrogen is blown off in a wind.

Galloway *et al.* (2003) have recently examined the distribution of peak fluxes of bursts from 4U 1728–34. They find that the peak fluxes of PRE bursts are not constant, but show variations of $\approx$44%. However, they find this variation is correlated with the source spectral state just prior to the bursts, and suggest that it may be caused by reprocessing from a precessing, warped disk. Upon removing the correlation they obtain a 3% variation. They use this result to argue that during radius expansion episodes the emission from the photosphere is largely isotropic. Assuming the peak flux is the Eddington limit appropriate for pure helium they derive a source distance of 5.2–5.6 kpc. Smale (1998) and Kuulkers and van der Klis (2000) used radius expansion bursts observed with RXTE from Cyg X-2 and GX 3+1, respectively, to place new constraints on the distances to these sources.

In principle, observations of PRE bursts provide a means to infer the masses and radii of neutron stars, quantities that have profound implications for our understanding of the equation of state of dense matter (Lattimer & Prakash 2001). Here we will only outline the main ideas, and we refer the reader to LVT for a more detailed discussion. As noted above, the Eddington luminosity at the surface of a neutron star depends only on the stellar mass, M, radius, R, and the composition of the atmosphere. The above expression (see Eq. (3.4)) shows that a measurement of the effective temperature, $T_{\rm eff}$, when the luminosity is Eddington limited, gives a constraint on the mass and radius of the neutron star. Indeed, this constraint is independent of the distance to the source. In practice, however, it is model dependent because the observed color temperature (from, say, a blackbody spectral fit), must be corrected to an effective temperature using an atmosphere model. Typically there are no independent constraints on the atmospheric composition, and so the correct model is uncertain. Measurement of the variation of the Eddington luminosity caused by expansion of the photosphere, and hence a change in the gravitational redshift factor, can also, in principle, be used to determine the gravitational redshift from the neutron star surface (see for example, Damen *et al.* 1990; van Paradijs *et al.* 1990; LVT). If the distance to the source is known, then a measure of the burst flux when it is Eddington limited will constrain a slightly different function of M and R. Sources with independent distance constraints, such as the globular cluster bursters, can provide, in principle, the best constraints (see Kuulkers *et al.* 2003).

Many researchers have used these and related methods to try and constrain neutron star masses and radii (see Fujimoto & Taam 1986; Ebisuzaki 1987; Sztajno *et al.* 1987; van Paradijs & Lewin 1987; Chevalier & Ilovaisky 1990; Damen *et al.* 1990; Kaminker *et al.* 1990; van Paradijs *et al.* 1990; Haberl & Titarchuk 1995). In general, the constraints are consistent with a range of neutron star mass–radius relations, but because of systematic uncertainties in the spectra, composition, and burst flux isotropy, they are not generally precise enough to unambiguously constrain the neutron star equation of state. Such efforts have continued using the higher signal-to-noise data from RXTE (see for example, Strohmayer *et al.* 1998;

Smale 2001; Titarchuk & Shaposhnikov 2002; Shaposhnikov, Titarchuk & Haberl 2003), however, systematic uncertainties remain.

3.3.3 Recent progress in burst spectroscopy

The most reliable way to overcome the systematic uncertainties in the interpretation of continuum spectra from bursts is to detect line features from the neutron star surface, and thereby obtain a direct measurement of the gravitational redshift, $1 + z = (1 - 2GM/c^2R)^{-1/2}$. Waki *et al.* (1984), and Nakamura, Inoue and Tanaka (1988) reported absorption lines at 4.1 keV in TENMA data from the LMXBs 4U 1636–53, and 4U 1608–52. Magnier *et al.* (1989) found a similar feature from EXO 1747–214 using EXOSAT data. Waki *et al.* (1984) identified the line with the helium-like iron Lyα transition at 6.7 keV, but gravitationally redshifted from the neutron star surface. The high implied equivalent widths (hundreds of eV), however, led others to suggest an origin outside the neutron star atmosphere, perhaps in the accretion flow (see Day, Fabian & Ross 1992; Foster, Fabian & Ross 1987; Pinto, Taam & Laming 1991). More recent observations, for example, with ASCA, RXTE, and BeppoSAX, have generally not confirmed the presence of such lines in burst spectra. The interpretation, and reality, of these features has therefore remained controversial.

In spite of this controversy, reports of deviations from blackbody spectra during some bursts continue to appear. Kuulkers *et al.* (2002a) found systematic deviations from blackbody spectra in RXTE data of bursts from the high accretion rate Z source GX 17+2. The residuals were most significant during the radius expansion and contraction phases. These variations are quite similar to those described by van Paradijs *et al.* (1990) for a PRE burst from 4U 2129+11, and, more recently, by Franco & Strohmayer (1999) for a burst from 4U 1820–30 (see also Kuulkers *et al.* 2003 for a discussion). Highly significant discrete spectral components, which could be modeled as an $\approx$6.4 keV emission line and $\approx$8 keV absorption edge have recently been found in RXTE/PCA spectra during a superburst from 4U 1820–30 (see Strohmayer & Brown 2002). It seems likely that these features may result from reprocessing (disk reflection) and fluorescence in the accretion disk (Ballantyne & Strohmayer 2004), or perhaps are formed in the burst-driven wind. It is not yet known if the features observed during PRE bursts are directly related to the features observed during the 4U 1820–30 superburst. Until their identifications are more secure it will be difficult to infer neutron star properties with them; however, these recent results have established convincingly that some X-ray burst spectra do have discrete lines. Observations with sufficient collecting area at higher spectral resolution will likely provide the breakthrough needed for reliable interpretation of these features.

The high spectral resolution capabilities of Chandra and XMM-Newton are providing new tools to study neutron star spectra; however, with a few exceptions (see Sanwal *et al.* 2002; Mereghetti *et al.* 2002; van Kerkwijk *et al.* 2004), the high-resolution spectra of isolated neutron stars have been frustratingly devoid of line features (see for example, Burwitz *et al.* 2001; Pavlov *et al.* 2001; Drake *et al.* 2002; Walter & Lattimer 2002). Indeed, bursters may prove to be more promising targets for line searches because of the mix of heavy elements constantly being provided by accretion. For example, Cottam, Paerels and Mendez (2002) have recently reported evidence for redshifted absorption lines in XMM-Newton Reflection Grating Spectrometer (RGS) data of bursts from EXO 0748–676. They co-added data from 28 bursts in order to increase the sensitivity to narrow lines. After modeling the continuum and line features thought to be produced in the highly ionized gas surrounding the neutron star

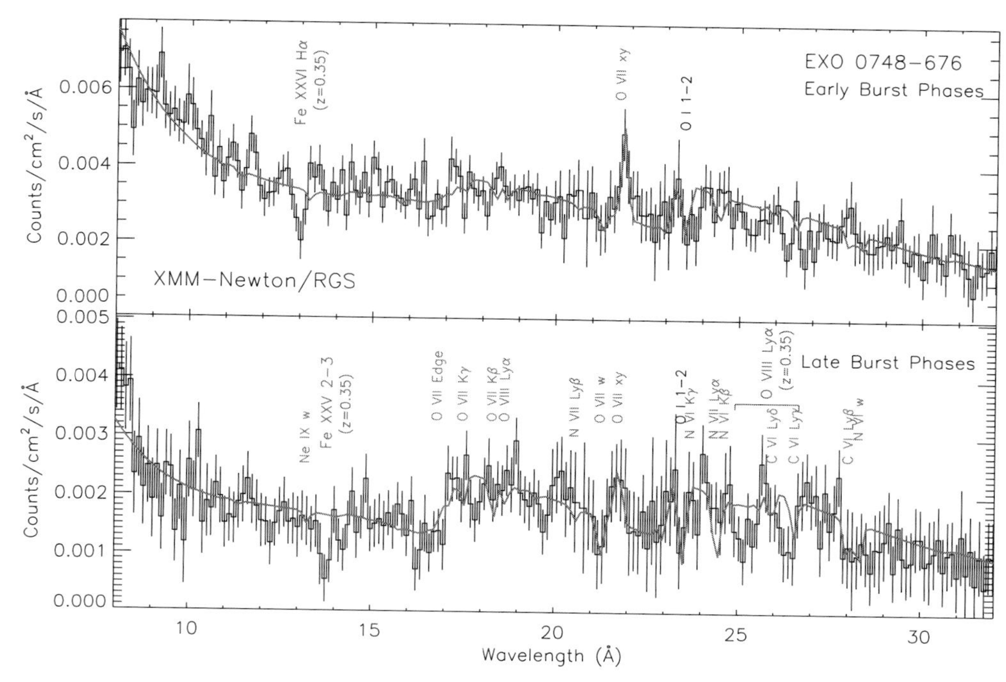

Fig. 3.5. High-resolution X-ray spectra of bursts from the LMXB EXO 0748–676. The XMM-Newton Reflection Grating Spectrometer (RGS) data from the early (*top*) and late (*bottom*) phases of the sum of 28 X-ray bursts are shown. The solid curve shows the best model of continuum and line features due to ionized gas in the vicinity of the neutron star. The unmodeled dips near 13 (top) and 14 (bottom) angstroms are consistent with redshifted lines of Fe XXVI from the neutron star surface. Reprinted by permission from *Nature* (Cottam, Paerels & Mendez 2002) copyright (2002) Macmillan Publishers Ltd.

(see Cottam *et al.* 2001), they found several features that were unaccounted for by their best model. Figure 3.5, from Cottam, Paerels & Mendez (2002), shows their RGS spectra and best model. They interpreted the unmodeled features near 13 and 14 angstroms as redshifted absorption lines of the $n = 2$–3 transitions of hydrogen-like Fe (Fe XXVI) in the early (hotter) part of the bursts, and the same transitions in the helium-like ion in the later (cooler) portions of the bursts. The implied neutron star surface redshift in each case was $z = 0.35$, and is consistent with modern neutron star equations of state with reasonable masses (Lattimer & Prakash 2001). It also implies that the neutron star surface is inside the last stable circular orbit (Kluzniak & Wagoner 1985).

The inferred narrowness of the lines was initially thought to be a problem because rapid neutron star rotation would produce broad, shallow lines (see for example, Özel & Psaltis 2003). However, the recent discovery of a 45 Hz oscillation in bursts from EXO 0748–676 is consistent with this object having a spin frequency small enough to account for the narrow lines (Villarreal & Strohmayer 2004). Another important issue is whether sufficient Fe is present in the atmosphere at the relevant $\dot{M}$ to explain the observed line strengths. Bildsten, Chang and Paerels (2003) have begun to examine this question, exploring how sedimentation, accretion and subsequent spallation (which produces Cr, Mn and V) determines the atmospheric column of heavy elements. As long as $\dot{M} > 4 \times 10^{-13} M_{\odot}\ \mathrm{yr}^{-1}$ during the burst, they find a typical Fe column of $\approx 3 \times 10^{19}$ cm^{-2} with solar abundance accretion. Assuming local thermodynamic equilibrium (LTE), this number is about a factor of 20 lower than the column required to produce the observed line strength. However, non-LTE effects and resonant line transport (Chang, Bildsten & Wasserman 2005) reduce this inconsistency, but do not fully eliminate it, motivating Chang *et al.* (2005) to suggest that the Fe is overabundant due to radiative levitation in the high flux environment of the burst.

3.3.4 Recurrence intervals and the $\dot{M}$ dependences

We previously discussed the one case where the observed Type I bursts matched theory well. However, most bursters are not so easy to understand, and we discuss that more typical case here as it indicates the possibility of non-spherically symmetric behavior. In a particular burning regime, we expect that the time between bursts should decrease as $\dot{M}$ increases since it takes less time to accumulate the critical amount of fuel at a higher $\dot{M}$. Exactly the *opposite* behavior was observed from many low accretion rate ($\dot{M} < 10^{-9} M_\odot$ yr^{-1}) NSs. A particularly good example is 4U 1705–44, where the recurrence time increased by a factor of ≈ 4 when $\dot{M}$ increased by a factor of ≈ 2 (Langmeier *et al.* 1987, Gottwald *et al.* 1989). If the star is accreting matter with $Z_{\rm CNO} = 0.01$ then these accretion rates are at the boundary between unstable helium ignition in a hydrogen-rich environment at high $\dot{M}$ and unstable pure helium ignition at lower $\dot{M}$. The expected change in burst behavior as $\dot{M}$ increases would then be to more energetic and more frequent bursts. This was not observed.

Other NSs showed similar behavior. Van Paradijs, Penninx and Lewin (1988) tabulated this effect for many bursters and concluded: "This suggests that continuous stable burning of a sizeable fraction of the nuclear fuel is a general phenomenon on the surface of accreting neutron stars. This fraction is apparently a gradually increasing function of $\dot{M}$". The following trends were found as $\dot{M}$ increases:

- The recurrence time increases from 2–4 hours to 10 hours.
- The bursts burn less of the accumulated fuel, with α increasing from ≈ 40 to >100.
- The duration of the bursts decreases from ≈ 30 s to ~ 5 s.

More recent RXTE observations found the same trend in KS 1731–260 (Muno *et al.* 2000), but not in 4U 1728–34 (Franco 2001; van Straaten *et al.* 2001).

The low $\dot{M}$ bursts appear to be mixed H/He burning with a large H mass fraction (i.e., energetic and of long duration from the rp-process) whereas the high $\dot{M}$ bursts look more He dominated with less H present (i.e., not so energetic, recurrence times long enough to burn H while accumulating, and short duration due to the lack of much energy release from the rp-process). The simplest explanation is to say that the NS is near the transition from the low $\dot{M}$ mixed burning regime (1 in Section 3.2.1) to the higher $\dot{M}$ pure helium burning (noted as 2 in Section 3.2.1). This would require that the NSs are accreting at $\approx 10^{-10} M_\odot$ yr^{-1} in the lower $\dot{M}$ state and about a factor of 4–5 higher in the high $\dot{M}$ state. However, these estimates are very far away from that observed.

Van Paradijs *et al.* (1988) used the ratio of the persistent flux to the flux during Eddington-limited PRE bursts as a measure of $\dot{M}$ in units of the Eddington accretion rate. They showed that most bursters accrete at rates $\dot{M} \approx (3-30) \times 10^{-10}$ $M_\odot$ yr^{-1}, at least a factor of 3 (and typically more) higher than the calculated rate where such a transition should occur.

Moreover, if the accretion rates were as low as needed, the recurrence times for the mixed hydrogen/helium burning would be about 30 hours, rather than the observed 2–4 hours. Fujimoto *et al.* (1987) discussed in some detail the challenges these observations present to a spherically symmetric model, while Bildsten (1995) attempted to resolve this by having much of the thermally unstable burning occur via slow deflagration fronts that lead to very low frequency noise and not Type I bursts (see also Yu *et al.* 1999). Whether this is possible on rapidly rotating NSs is uncertain (Spitkovsky, Levin & Ushomirsky 2002).

Much of this puzzle might be resolved by allowing the fresh material to cover only a fraction of the star prior to igniting (Bildsten 2000). There are strong observational clues that this may be happening (e.g., Marshall 1982), as the other clear trend found by EXOSAT was an increase in the blackbody radius ($R_{\rm bb}$) as $\dot{M}$ increased. This parameter is found by spectral fitting in the decaying tail of the Type I bursts and, though susceptible to absolute spectral corrections, can likely be trusted for relative sizes (LVT). In a similar vein, van der Klis *et al.* (1990) found that the temperature of the burst at the moment when the flux was one-tenth the Eddington limit decreased as $\dot{M}$ increased (hence a larger area) for the Atoll source 4U 1636–53. In total, these observations suggest the possibility that the covered area increases fast enough with increasing $\dot{M}$ that the accretion rate per unit area actually *decreases*.

By interpreting the measured $R_{\rm bb}$ as an indication of the fraction of the star that is covered by freshly accreted fuel, the quantity $\dot{m} = \dot{M}/4\pi R_{\rm bb}^2$ can be calculated using $F_x = GM\dot{M}/4\pi d^2 R$ and gives $\dot{m} \approx (F_x R/GM)(d/R_{\rm bb})^2$. Bildsten (2000) has argued that the radius increase can offset the $\dot{M}$ increase in the context of data from the burster EXO 0748–676 (Gottwald *et al.* 1986; see Figure 1 in Bildsten 2000).

Where does the material arrive on the NS? We know that these NSs accrete from a disk formed in the Roche lobe overflow of the stellar companion, however, there are still debates about the "final plunge" onto the NS surface. Some advocate that a magnetic field controls the final infall, while others prefer an accretion disk boundary layer. These arguments must also take into account the possible presence of an accretion gap between the inner disk edge and the stellar surface (Kluzniak, Michelson & Wagoner 1990; Kluzniak & Wilson 1991), as the gravitational redshift measurement reported by Cottam *et al.* (2002) implies that the NS surface is inside the last stable orbit. If material is placed in the equatorial belt, it is not clear that it will stay there very long. If angular momentum was not an issue, the lighter accreted fuel (relative to the ashes) would cover the whole star quickly. However, on these rapidly rotating NSs, the fresh matter added at the equator must lose angular momentum to get to the pole. This competition (namely understanding the spreading of a lighter fluid on a rotating star) has been investigated by Inogamov and Sunyaev (1999).

Another remaining conundrum is the burst behavior of the bright Z sources (Sco X-1, Cyg X-2, GX 5-1, GX 17+2, GX 340+0, GX 349+2) which are accreting at $3 \times 10^{-9} - 2 \times 10^{-8}$ $M_{\odot}$ yr^{-1}. At the very upper end of this range, the burning could be thermally stable; however, if unstable, they would clearly be in the regime where the bursts are regularly spaced and of long duration due to mixed hydrogen/helium burning (noted as 3 above and as exhibited by GS 1826–24). However, this is not typically seen; rather, bursts are rare and are clearly not responsible for burning all of the accreted fuel. Though some of the recently studied bursts from GX 17+2 (Kuulkers *et al.* 2002a) do look as expected, not all are easy to understand and we refer the interested reader to Kuulkers *et al.* (2002a) for an excellent summary discussion of the current observational situation at high $\dot{M}$.

Most of the Type I bursts we have discussed occur when the neutron star is accreting at a rate in excess of $10^{-10} M_{\odot}$ yr^{-1}. Recently, however, the BeppoSAX WFCs have discovered a number of new thermonuclear burst sources from which no persistent X-ray flux could be detected down to the 10^{-10} erg cm^{-2} s^{-1} flux limit of the WFCs (see for example, Cocchi *et al.* 2001; Cornelisse *et al.* 2002a; in 't Zand *et al.* 2002). If these objects are closer than 10 kpc then their accretion rate when bursting was less than $10^{-10} M_{\odot}$ yr^{-1} (or $L < 10^{36}$ erg s^{-1}).

At other epochs, some of these sources were observed by ROSAT at luminosities $\approx 10^{35}$ erg s^{-1}, whereas later Chandra observations of several of them revealed persistent luminosities (or upper limits) in the 10^{32-33} erg s^{-1} range (Cornelisse *et al.* 2002b), consistent with flux levels seen from neutron star X-ray transients in quiescence (see, for example, Rutledge *et al.* 2001). Cocchi *et al.* (2001) and Cornelisse *et al.* (2002b) have suggested that these bursters may be either a new class of low persistent emission bursters (in which case there is a huge underlying population) or weak transients with time-averaged accretion rates $< 10^{-11} M_{\odot}$ yr^{-1} (like SAX J1808.4–3658) that just happen to be further away.

Regardless of whether these systems are constantly accreting at such a low rate or are weak transients in outburst, what is important for this discussion is that the bursts occur when the accretion rate is clearly less than $10^{-10} M_{\odot}$ yr^{-1}. Presuming complete covering of the neutron star, the hydrogen burning would be unstable at these accretion rates (regime 1 in Section 3.2.1) and the unstable ignition should lead to long (100–1000 s), infrequent bursts (recurrence times of $\approx$weeks). Some of the recently discovered sources appear to be in this regime (Kaptein *et al.* 2000; in 't Zand *et al.* 2002), however, it is not known if all these "burst-only" sources can be explained this way. In fact, it seems unlikely, since some of the observed bursts are short and therefore inconsistent with the standard low accretion rate theory. It is conceivable that our understanding of these sources may require additional, important physics, such as element diffusion, which is not important for higher accretion rates but becomes more relevant at these rates (Wallace *et al.* 1982, Bildsten, Salpeter & Wasserman 1993).

3.4 Millisecond variability during X-ray bursts

Early theoretical studies noted the likely importance of spreading of the thermonuclear burning front around the neutron star surface (see for example, Joss 1978). Because nuclear fuel is burned in a time much shorter than it takes to accrete a critical pile it is unlikely that ignition conditions will be achieved over the entire surface simultaneously (Shara 1982). It appears more likely that burning is initiated locally and then spreads laterally, eventually engulfing all fuel-loaded parts of the neutron star. For conditions most prevalent in burst sources, the front may spread via convective deflagration, at lateral speeds of up to $\approx 5 \times 10^6$ cm s^{-1} (see Fryxell & Woosley 1982; Hanawa & Fujimoto 1984; Nozakura, Ikeuchi & Fujimoto 1984; Bildsten 1995). Such speeds can account for the sub-second rise times of some bursts, but the time required for burning to engulf the entire star is still long compared to the spin periods of accreting LMXB neutron stars (milliseconds). Moreover, if the burning front is not strongly convective, then a patchy distribution of nuclear fuel is possible (see Bildsten 1995). These considerations suggest that during bursts the rotation of the neutron star can modulate the inhomogeneous or localized burning regions, perhaps allowing for direct observation of the spin of the neutron star.

Mason *et al.* (1980) reported a 36.4 Hz pulsation in an optical burst from 2S 1254–690, and Sadeh (1982) claimed detection of a 12 ms modulation in a burst from 4U 1728–34 with HEAO-1, but these periods were never confirmed by subsequent observations. Murakami *et al.* (1987) reported 1.5 Hz oscillations during the PRE phase of a burst from 4U 1608–52. These oscillations might conceivably be caused by oscillations of the photospheric radius at constant luminosity (Lapidus *et al.* 1994). Schoelkopf and Kelley (1991) reported detection at the $\approx 4\sigma$ level of a 7.6 Hz oscillation in Einstein Monitor Proportional Counter data during a burst from Aql X-1. They suggested that rotation of the neutron star with a non-uniform surface brightness might be responsible, but the signal has not been seen in other bursts. Jongert and van der Klis (1996) searched for high frequency variability in bursts observed with the EXOSAT observatory. They averaged power spectra from multiple bursts from the same source to increase sensitivity, but also found no significant periods. They placed upper limits on average modulation amplitudes in the $\sim$100 Hz range of between 5 and 10%.

3.4.1 *Overview of burst oscillations*

An exciting development in the past decade has been the discovery of high frequency (300–600 Hz) X-ray brightness oscillations during bursts. These modulations are now commonly called "burst oscillations." They were first discovered with the PCA on board RXTE in bursts from the LMXB 4U 1728–34 (Strohmayer *et al.* 1996). As of this writing, burst oscillation detections have been claimed for an additional 13 sources. Two of these objects; SAX J1808.4–3658 (in 't Zand *et al.* 2001; Chakrabarty *et al.* 2003), and XTE J1814–338 (Markwardt, Strohmayer & Swank 2003), are accreting millisecond pulsars whose spin periods are precisely known. Table 3.1 provides a catalog of the known burst oscillation sources, summarizes some of their salient properties, and provides references to the relevant literature.

Burst oscillations were first discovered in Fourier power spectra computed from high time resolution lightcurves of entire bursts. Figure 3.6 shows an example of a Fourier power density spectrum of a burst from 4U 1728–34 with a burst oscillation at 363 Hz (after Strohmayer *et al.* 1996). Although many detailed questions remain, there is now no doubt that spin modulation of the X-ray burst flux is the basic mechanism responsible for these oscillations. The discovery of burst oscillations was closely linked with the discovery of millisecond variability (kHz QPOs) in the persistent X-ray flux from accreting neutron stars. Such oscillations, and some of their inferred connections to burst oscillations, are reviewed in Chapter 2 by van der Klis. Therefore, we will only review the observational properties and current theoretical understanding of burst oscillations and will not discuss in any great detail the possible connections between the two phenomena.

3.4.2 *Oscillations during the burst rise*

In some bursts oscillations are detected during the rising portion of the burst time profile. Indeed, detections can be made within a few tenths of a second after a significant rise in the X-ray flux is seen (see Strohmayer *et al.* 1998a; Miller 1999). Such oscillations can have very large amplitudes. Strohmayer, Zhang and Swank (1997) found that some bursts

Table 3.1. *Burst oscillation sources and properties*

Source	Frequency (Hz)	$\Delta\nu_{qpo}{}^a$	$P_{orb}(hr)^b$	References
4U 1728–34	363	280–363	?	1,2,3,4,5,13,14
4U 1636–53	581	250–320	3.8	6,7,25,26,30
KS 1731–260	524	260	?	10,11,12
Galactic Center	589	?	?	15
Aql X-1	549	?	19.0	16,17
4U 1702–429	330	315–344	?	4,9
MXB 1658–298	567	?	7.1	18,27
4U 1916–053	270	290–348	0.83	19,20
4U 1608–52	619	225–325	?	8,21
SAX J1808.4–3658	401	195	2.0	22,23,28,29
SAX J1750.8–2980	601	?	?	24
XTE J1814–338	314	?	4.3	31,32
SAX J1748.9–2021	410	?	?	33
EXO 0748–676	45	?	3.82	34

(1) Strohmayer *et al.* (1996); (2) Strohmayer, Zhang, & Swank (1997); (3) Mendez & van der Klis (1999); (4) Strohmayer & Markwardt (1999); (5) Strohmayer *et al.* (1998b); (6) Strohmayer *et al.* (1998a); (7) Miller (1999); (8) Mendez *et al.* (1998); (9) Markwardt, Strohmayer & Swank (1999) (10) Smith, Morgan, & Bradt (1997); (11) Wijnands & van der Klis (1997); (12) Muno *et al.* (2000); (13) van Straaten *et al.* (2001); (14) Franco (2001); (15) Strohmayer *et al.* (1997); (16) Zhang *et al.* (1998); (17) Ford (1999); (18) Wijnands, Strohmayer & Franco (2001); (19) Boirin *et al.* (2000); (20) Galloway *et al.* (2001); (21) Chakrabarty (2000); (22) in 't Zand *et al.* (2001); (23) Ford (2000); (24) Kaaret *et al.* (2003); (25) Giles *et al.* (2002); (26) Strohmayer & Markwardt (2002); (27) Wijnands *et al.* (2002); (28) Chakrabarty *et al.* (2003); (29) Wijnands *et al.* (2003); (30) Jonker, Mendez & van der Klis (2002); (31) Markwardt & Swank (2003); (32) Markwardt, Strohmayer & Swank (2003); (33) Kaaret *et al.* (2003); (34) Villareal & Strohmayer (2004)

[a] The frequency separation between pairs of kHz QPO in Hz, if known.

[b] Orbital period of the system, in hours, if known.

from 4U 1728–34 show oscillation amplitudes as large as 43% within 0.1 s of the onset of the burst. Strohmayer *et al.* (1998a) found bursts from 4U 1636–53 with modulations at onset of $\approx$75% (see Fig. 3.7). These early studies also found that the modulation amplitude is anti-correlated with the X-ray intensity during the rise. That is, the amplitude drops as the flux increases toward maximum. This behavior is consistent with simple expectations for spin modulation of an initially localized X-ray "hot spot" which expands in $\sim$1 s to engulf the neutron star. The amplitude is largest when the hot spot is smallest, at onset, and then decreases as the spot engulfs the star and the flux increases (see Nath, Strohmayer & Swank 2002). Time-resolved X-ray spectroscopy during bursts provides additional evidence for localized X-ray emission near burst onset. For example, Strohmayer, Zhang and Swank (1997) tracked the evolution of the bolometric flux and blackbody temperature during bursts from 4U 1728–34. For a spherical blackbody source, the quantity $F_{bol}^{1/4}/kT_{bb}$ is proportional to the square root of the emitting area (see also the discussions in Sections 3.3.1 and 3.3.4). They found that this inferred surface area was smallest at onset and then increased as the

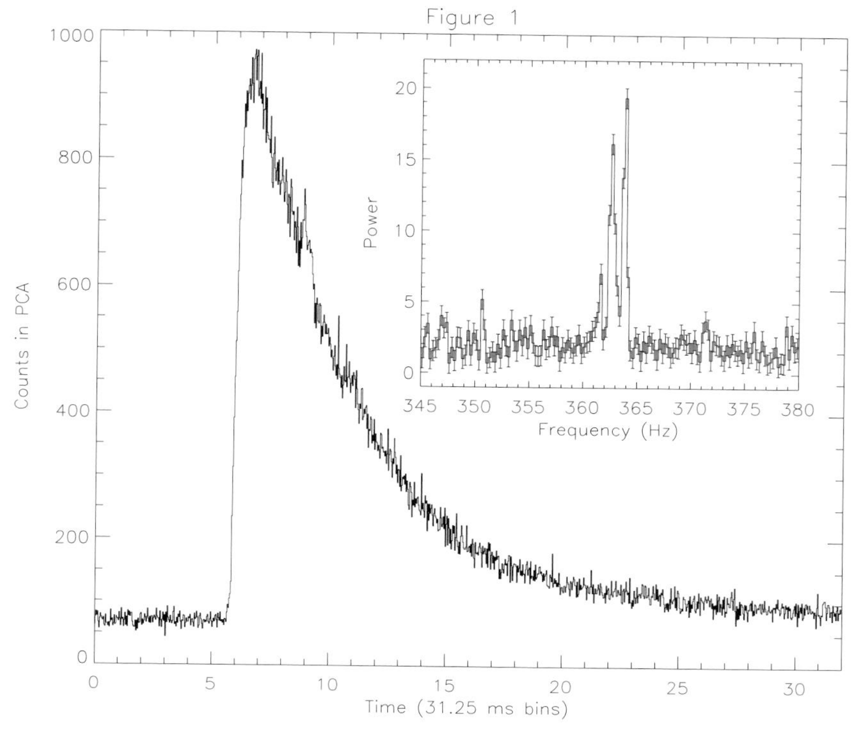

Fig. 3.6. An X-ray burst from 4U 1728–34 observed with the PCA on board RXTE. The main panel shows the X-ray counts observed by the PCA in (1/32) s bins. The inset panel shows the power spectrum in the vicinity of 363 Hz (after Strohmayer *et al.* 1996).

burst flux increased during the rise. The inferred area then stayed approximately constant as the flux declined (see Fig. 3.8).

3.4.3 Oscillations in the decay phase

Oscillations are also observed at late times in the cooling or decay phase of bursts. In many cases oscillations are seen after "touchdown" in PRE bursts (see for example, Smith, Morgan & Bradt 1997; Muno *et al.* 2000), however, bursts without PRE can also show oscillations which persist during the cooling phase (Franco 2001; van Straaten 2001). Typically the amplitude of oscillations seen in the cooling phase is lower than observed during burst rise, but amplitudes as large as 15% (rms) have been observed. Some bursts show oscillations both on the rise and in the cooling phase (see van Straaten 2001; Franco 2001), however, oscillations are not detected during PRE phases, only before and after the PRE episode (see Muno *et al.* 2002). This provides additional evidence that the modulations are associated with processes on the neutron star surface. In bursts where pulse trains are detected during the rise and decay they can sometimes be observed for as long as $\approx$12 seconds. The presence of strong oscillations after the nuclear burning has presumably engulfed the entire neutron star is difficult to account for in the simplest expanding hot spot scenarios. This has motivated a number of recent theoretical ideas to explain the persistence of burst oscillations. We will discuss some of these ideas shortly.

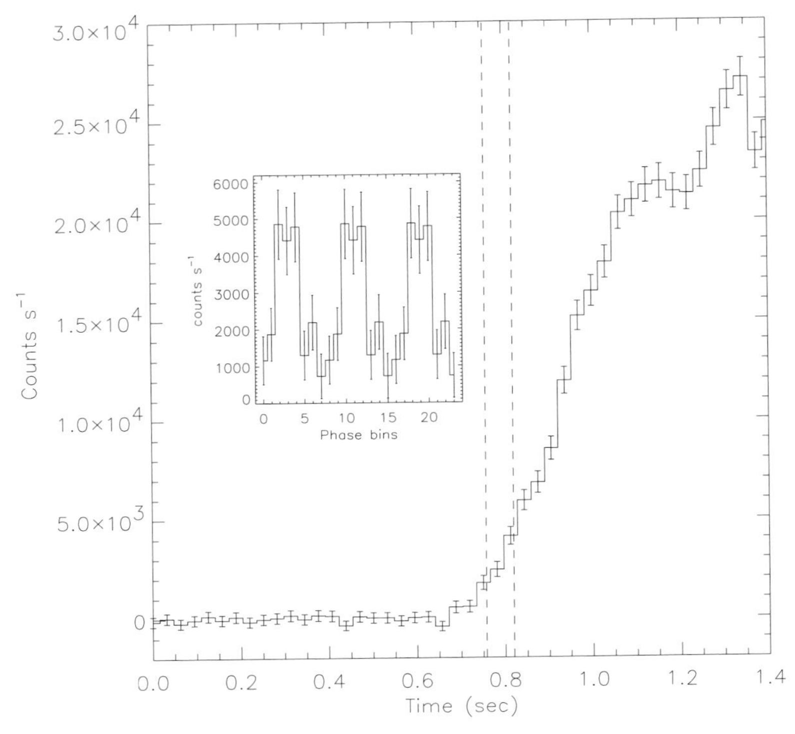

Fig. 3.7. X-ray timing evidence indicating a spreading hot spot at the onset of thermonuclear bursts. The main panel shows a burst from 4U 1636–53 with large amplitude, 581 Hz oscillations on the rising edge of the profile. The inset shows the pulse profile during the interval marked by the vertical dashed lines. The pulse profile is repeated 3× for clarity. Note the large amplitude of the oscillation (after Strohmayer *et al.* 1998a).

3.4.4 Frequency evolution of burst oscillations

The oscillation frequency during a burst is typically not constant. Most commonly the frequency increases as the burst progesses, that is, the evolution can be characterized as a "chirp." Most bursts show spin-up toward some limiting, or asymptotic frequency; however, there are exceptions to this rule of thumb. For example, Strohmayer (1999) and Miller (2000) identified a burst from 4U 1636–53 with spin-down in the cooling tail. This burst also showed an unusually long thermal tail which may have been related to a "reheating" episode having some connection to the spin-down. Muno *et al.* (2000) found spin-down in a burst from KS 1731–260, however, they found no evidence for unusual flux enhancements or spectral variations during the episode. Spin-downs are apparently rare. In a recent study, Muno *et al.* (2002) found some evidence for them in only 3 of 68 bursts examined. The observed frequency drifts are generally $< 1\%$ of the mean frequency, and bursts which have detectable pulsations during the rising phase show the largest frequency shifts. This indicates that the process responsible for the frequency evolution begins with the start of a burst, and not when oscillations are first detected within a burst (Muno *et al.* 2002).

Strohmayer and Markwardt (1999) studied the frequency evolution in bursts from 4U 1702–429 and 4U 1728–34. They found the frequency in these bursts could be modeled

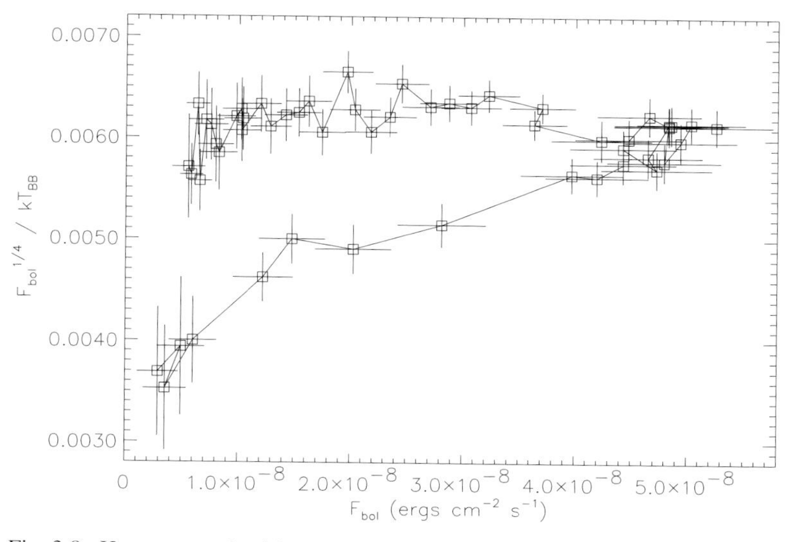

Fig. 3.8. X-ray spectral evidence indicating a spreading hot spot at the onset of thermonuclear bursts. The plot shows the time evolution of the inferred surface area in a burst from 4U 1728–34. The evolution is from lower left to upper right, and then horizontally to the left (after Strohmayer, Zhang & Swank 1997).

as a smooth exponential recovery of the form, $\nu(t) = \nu_0(1 - \delta_\nu e^{-t/\tau})$, where ν_0, δ_ν, and τ are the asymptotic frequency, the fractional frequency drift, and the recovery timescale, respectively. With this form they were able to recover coherent signals, with coherence values, $Q \equiv \nu_0/\Delta\nu_0 > 4500$ in some bursts. Figure 3.9 shows an example of a burst from 4U 1702–429 with exponential frequency evolution. These results support the existence of a reference frame on the neutron star, perhaps the nuclear burning layer, in which the oscillations are coherent or nearly so. This frame, however, cannot be rigidly connected to the bulk of the neutron star, because the torque required to change the spin frequency of the star by $\approx 1\%$ in only 10 seconds is unphysically large. This implies the existence of shearing in the surface layers of the neutron star. In the exponential model, the total amount of phase shearing is simply $\phi_{\rm shear} = \nu_0\delta_\nu(1 - e^{-T/\tau})$, where T is the length of the pulse train. For typical bursts this value ranges from about 4 to 8, suggesting that the burning layer "slips" this many revolutions over the underlying neutron star during the pulsations. The amount of phase shearing has implications for the surface magnetic field strength, as a sufficiently strong field will enforce corotation (Cumming & Bildsten 2000).

Muno *et al.* (2000) carried out a detailed study of the frequency evolution in burst oscillations from KS 1731–260. They used pulsar timing techniques to compute phase-connected timing solutions in order to study the functional form of the frequency evolution. They found that the phase evolution can usually be modeled as a polynomial, and that the exponential relaxation model was adequate for many bursts. More recently, Muno and colleagues (2002) have explored the frequency evolution in a larger sample of bursts (68) from several sources. They confirm many of their earlier findings, however, they do find a subset of bursts for which simple second and third order polynomials are insufficient to explain the phase evolution. In these bursts there is evidence for phase jitter on timescales of seconds.

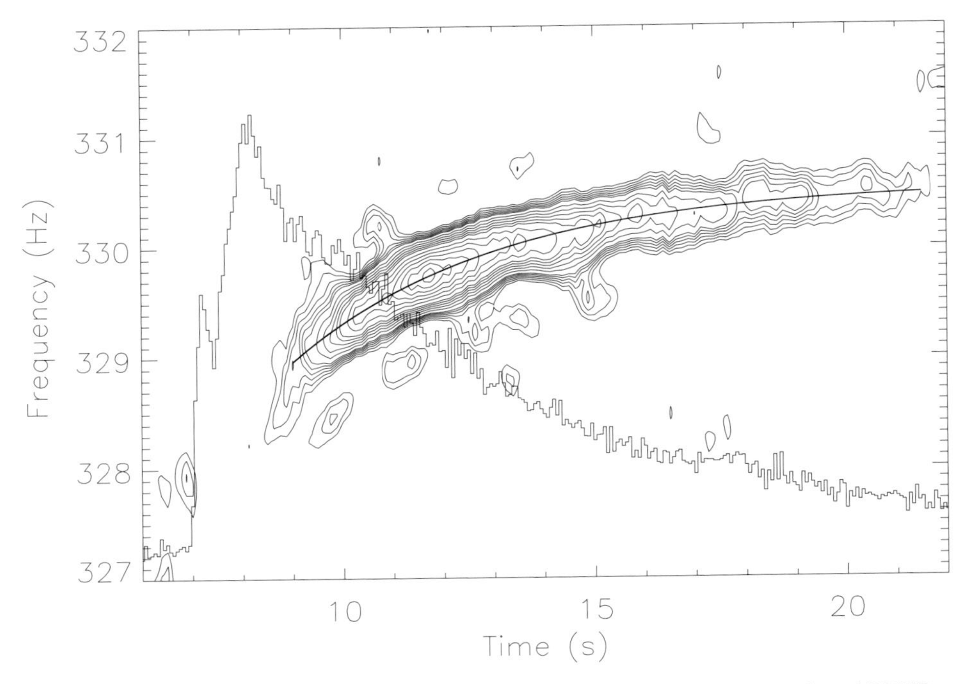

Fig. 3.9. An X-ray burst from 4U 1702–429 observed with the PCA on board RXTE. Shown are contours of constant power spectral density as a function of frequency and time. The solid curve shows the best fitting exponential model. The burst time profile is also shown (after Strohmayer & Markwardt 1999).

Figure 3.10 shows several examples of burst oscillations with complex phase evolution. They suggest this may indicate the presence of some instability in the mechanism that generates the oscillation. One idea is that two signals, or modes, with nearly equal frequency may be simultaneously present (see also Miller 1999). Another possibility is that phase jumps occur on relatively short timescales (see Strohmayer 2001). These behaviors, though infrequent, may provide important clues to the physical mechanism that produces the observed modulations, particularly in the cooling phase.

Strohmayer *et. al* (1997) argued that the observed frequency evolution results from angular momentum conservation of the thermonuclear shell. Burst-induced heating expands the shell, increasing its rotational moment of inertia and slowing its spin rate. Near burst onset the shell is thickest and thus the observed frequency lowest. The shell spins back up as it cools and recouples to the underlying neutron star. Calculations indicate that the $\sim$10 m thick pre-burst shell can expand to $\sim$30 m during the flash (see Joss 1978; Bildsten 1995; Cumming & Bildsten 2000), which gives a frequency shift of $\approx 2\ \nu_{\rm spin}(20\ {\rm m}/R)$, where $\nu_{\rm spin}$ and R are the stellar spin frequency and radius, respectively. For typical burst oscillation frequencies this gives a shift of $\sim$2 Hz, similar to that observed. However, Galloway *et al.* (2001) reported a 3.5 Hz frequency shift in a burst from 4U 1916–053 with 272 Hz oscillations. They suggested that such a large change, $\sim$1.3% might be inconsistent with expansion of the thermonuclear burning layer because of the magnitude of the implied height change of $\sim$80 m. Wijnands, Strohmayer & Franco (2001) found a $\sim$5 Hz frequency shift in a burst from MXB X1658–298 with a 567 Hz oscillation, which may also be uncomfortably large given current estimates of the expansion of the burning layers (Cumming *et al.* 2002).

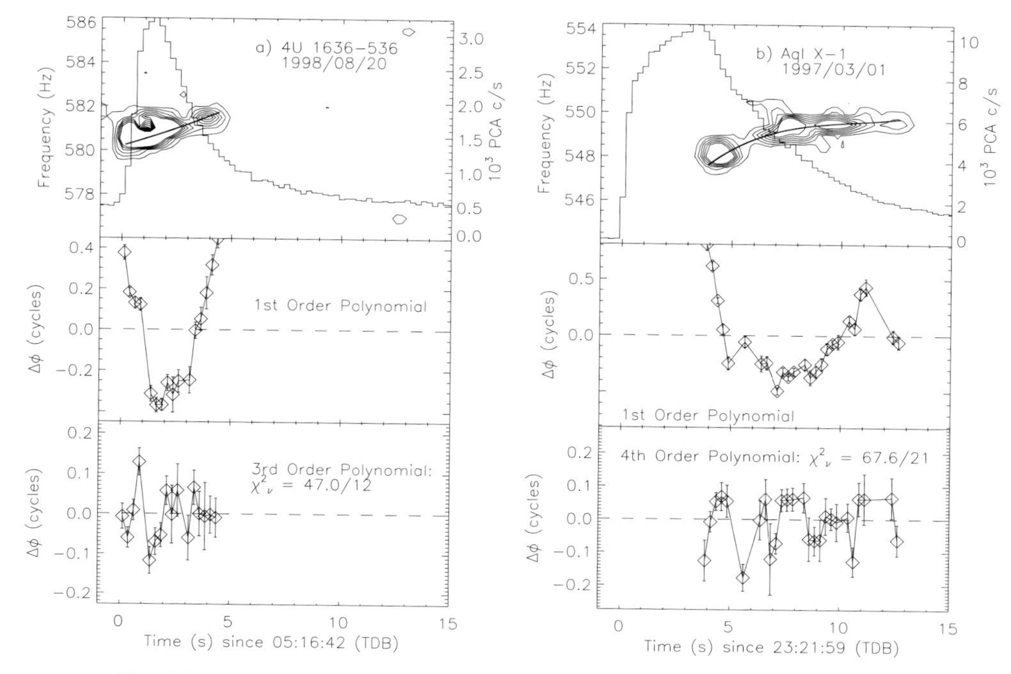

Fig. 3.10. Examples of two burst oscillations that show complex phase evolution. In both cases higher order polynomial phase evolution functions (3rd (*left*) and 4th (*right*) order) still leave significant residuals (after Muno *et al.* 2002).

3.4.5 *Long term stability of burst oscillation frequencies*

Studies of the frequency stability of burst oscillations over years provides constraints on the mechanism that sets the frequency. Strohmayer *et al.* (1998b) carried out one of the first studies of long term stability. They compared the asymptotic burst oscillation frequencies in bursts from 4U 1728–34 and 4U 1636–53. For bursts from 4U 1728–34 spanning a 1.6 year epoch they found that the asymptotic pulsation period was the same to better than 1 μs. This suggests a timescale to change the frequency longer than 23 000 years, and supports a mechanism like rotation with a high degree of intrinsic stability.

The accretion induced rate of change of the neutron star spin frequency in a LMXB has a characteristic value of 1.8×10^{-6} Hz yr^{-1} for a canonical neutron star and typical mass accretion rates for X-ray burst sources. Over a year the accretion induced shift is much smaller than the apparent spin frequency changes that would be caused by the projected orbital motion of the neutron star. This led to the suggestion that, if burst oscillation frequencies were intrinsically stable enough, one might be able to extract the projected orbital velocity of the neutron star from a sample of bursts observed at different orbital phases. Recent studies, however, have confirmed that it will be much more difficult to extract neutron star velocity information from burst oscillation frequencies than initially hoped. For example, Giles *et al.* (2002) studied the burst oscillation frequencies of 26 bursts from the LMXB 4U 1636–53. The highest observed oscillation frequencies of all bursts in their study are stable at the level of 2×10^{-3}, but are not correlated with orbital phase as expected for binary modulation.

In a related study, Muno *et al.* (2002) examined the asymptotic burst oscillation frequencies of bursts from eight different sources, including 4U 1636–53. They quantify the dispersion in

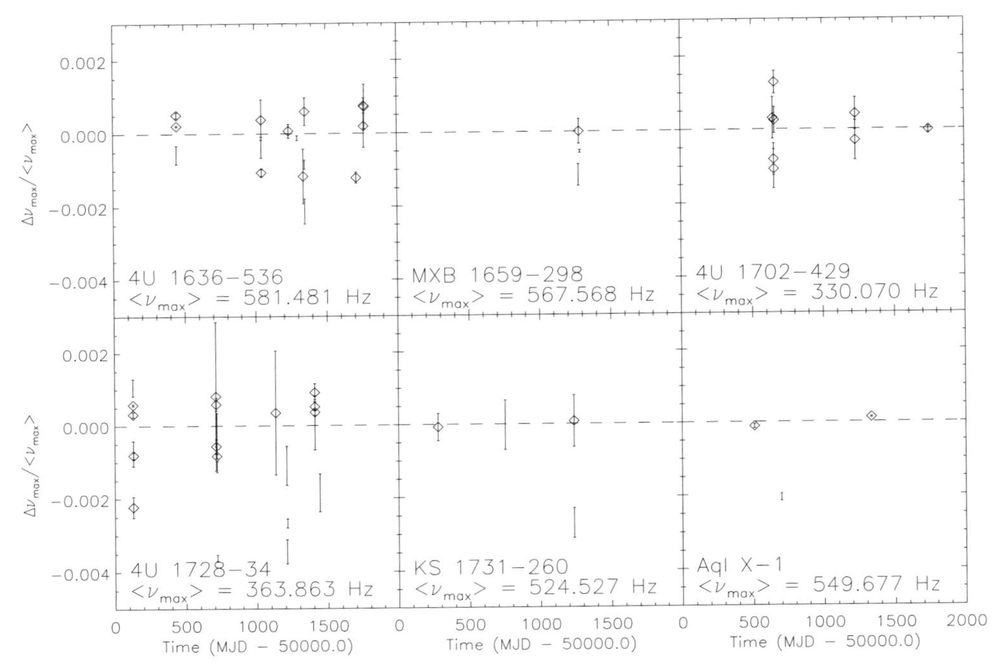

Fig. 3.11. The fractional deviations from the mean of maximum burst oscillation frequencies from six different burst sources. The data are shown as a function of time. The standard deviation, $\sigma_\nu/\langle\nu_{\rm max}\rangle$, is typically $7-10 \times 10^{-4}$ (after Muno *et al.* 2002).

asymptotic frequencies using the standard deviation, σ_ν, of the observed frequencies and find that $\sigma_\nu/\langle\nu_{\rm max}\rangle$ is typically $< 1 \times 10^{-3}$. Figure 3.11 summarizes the distributions of asymptotic burst oscillation frequencies found in the Muno *et al.* (2002) study. These results indicate that the asymptotic burst oscillation frequencies are quite stable, but that there is more variation than can easily be accounted for by binary Doppler modulations alone. This requires that models for the burst oscillations must be able to account for an intrinsic fractional frequency variation perhaps as large as $\approx 5 \times 10^{-4}$.

One physical effect that approaches this magnitude is the change in the thickness of the burning layer from before the burst to after the burst due to the change in mean molecular weight (Cumming & Bildsten 2000, see their Figure 12). A complete burn of hydrogen via the rp-process can cause a fractional change in the rotation rate of $\approx 5 \times 10^{-4}$ (Cumming *et al.* 2002), so that varying levels of burning can lead to differing asymptotic frequencies at this level.

3.4.6 *Burst oscillations and the mass accretion rate*

Not all bursts from a given source have detectable oscillations. Recent work indicates that the mass accretion rate, $\dot{M}$, onto the neutron star has a strong influence on the strength, and therefore the detectability of burst oscillations. This is perhaps not too surprising, since $\dot{M}$ is known to influence other burst properties as well, and, as outlined earlier, is an important ingredient in any description of the nuclear burning physics. It is not unreasonable then to expect that the properties of burst oscillations might also depend importantly on $\dot{M}$.

Muno *et al.* (2000) carried out the first systematic study of burst oscillation properties and source spectral state. They studied bursts from the LMXB and atoll source KS 1731–260. The 524 Hz burst oscillations in this source were discovered by Smith, Morgan and Bradt (1997). Muno *et al.* (2000) studied both the spectral and timing properties of bursts as a function of the position of the source in an X-ray color–color diagram (CD), and found that burst properties, including the presence or absence of burst oscillations, were strongly segregated in the CD. In particular, they found that only bursts which occurred when the source was located on the "banana branch" (see Section 2.5.2) of the atoll pattern (i.e. at high inferred $\dot{M}$) produced detectable oscillations. These bursts also showed PRE, had the highest peak fluxes, and had characteristically short durations (so called "fast" bursts). As discussed earlier, these latter characteristics are an indication that helium is the primary fuel in such bursts (see Section 3.2.2).

Subsequent studies of bursts from 4U 1728–34 by van Straaten *et al.* (2001) and Franco (2001) confirmed some of the main results found by Muno *et al.* (2000), but new findings also complicated the picture. For example, both of these studies found a similar dependence of burst oscillations on source position in the CD. That is, bursts with oscillations are restricted to the banana branches at higher inferred $\dot{M}$. However, the properties of so-called "fast" and "slow" bursts were not as nicely segregated in 4U 1728–34 as for KS 1731–260. For example, the bursts at low inferred $\dot{M}$ (in the so-called island state of the CD) from 4U 1728–34 all showed PRE whereas the corresponding bursts from KS 1731–260 do not. Moreover, some of the "fast" bursts from 4U 1728–34 which show strong burst oscillations show no evidence for PRE. Franco (2001) further characterized the bursts by computing a measure of integrated oscillation strength through the bursts. This integrated oscillation strength increased with inferred $\dot{M}$ on the CD (see Fig. 3.12). A relationship between position in the CD and where during a burst oscillations are detected was also suggested. Bursts with oscillations detected only during the rising phase were found at the highest inferred $\dot{M}$, while bursts with oscillations only in the cooling (decay) phase were found at lower $\dot{M}$ (but still on the banana branch).

3.4.7 *Burst oscillations and photospheric radius expansion*

Soon after the discovery of the first burst oscillation sources a connection between the appearance of oscillations in bursts and PRE was apparent. Smith, Morgan and Bradt (1997) found that oscillations in a PRE burst from KS 1731–260 were first detected after photospheric touchdown. Strohmayer *et al.* (1997) found similar results for bursts with 589 Hz oscillations from a burster near the galactic center. Muno *et al.* (2000) have explored the connection between PRE and oscillations in a large sample of bursts from several sources. They segregated burst oscillation sources into two classes, those with burst oscillation frequencies closer to 300 Hz (so-called "slow" oscillators), and those with frequencies close to 600 Hz (the "fast" group), and found that those bursts with fast oscillations were almost always observed in bursts with PRE episodes, while bursts with slow oscillations only showed PRE about half the time. Muno *et al.* (2000) suggest that this distinction between the fast and slow sources cannot be an observational selection effect. They argue that the distinction could result if burst properties vary differently with $\dot{M}$ in the fast and slow sources. Exactly how this difference comes about and how it is related to the oscillation (or spin) is not yet understood.

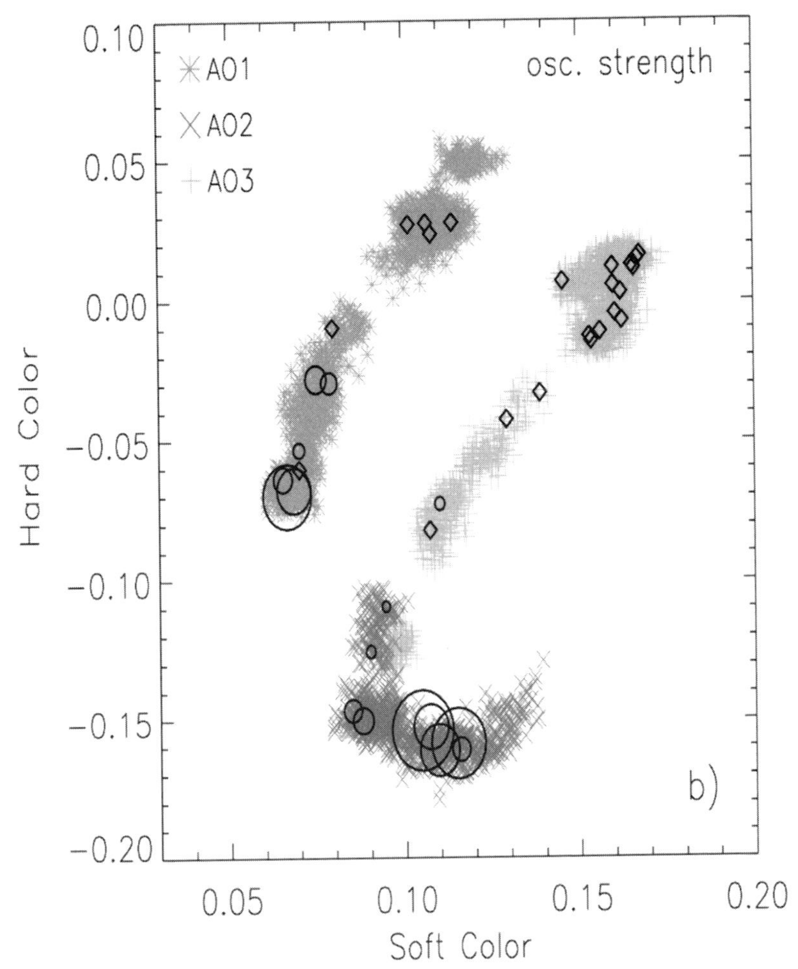

Fig. 3.12. X-ray color–color diagram for 4U 1728–34 showing a typical atoll track identifying, counterclockwise from upper right, the extreme island state (EIS), the island state (IS), the lower banana branch (LB), and the upper banana branch (UB) (see Section 2.5.2). The locations of bursts both with (circles) and without (diamonds) oscillations are shown. For bursts with oscillations, the size of the circle is proportional to the integrated strength of the oscillations (after Franco 2001).

3.4.8 *Harmonics, subharmonics and sidebands*

The pulse shapes of burst oscillations are highly sinusoidal. Strohmayer and Markwardt (1999) found no evidence for harmonics after coherently summing oscillation signals from 4U 1728–34 and 4U 1702–429. They placed limits on the ratio of pulsed amplitude at the fundamental to that at the first harmonic of ≈ 24 and 15 in these sources, respectively. Muno *et al.* (2000) found only weak (2σ) evidence of a signal at the first harmonic of the 524 Hz frequency in bursts from KS 1731–260. Recently, Muno, Özel and Chakrabarty (2002) have explored the amplitude evolution and harmonic content of cooling phase burst oscillations in eight different sources. They find mean amplitudes (rms) of about 5%, and during a typical burst the amplitude can vary in a manner that is uncorrelated with the burst flux. They did not detect any harmonic, or subharmonic signals, and placed upper limits on the fractional amplitudes at integer and half-integer harmonics of less than 5 % and 10 % of the amplitude of the strongest observed signal, respectively. Comparison of these results to theoretical models with one or two circular, antipodal hot spots suggests that if a single spot is present it must lie near the rotational pole or cover a substantial fraction ($\approx 1/2$) of the neutron star in order to be consistent with the limits on harmonic signals. If antipodal spots are present, then the implications are that the spots must lie close to the rotational equator.

Chakrabarty (2002) has found evidence for sidebands separated by 30–50 Hz from the burst oscillation frequency in a few sources. These sidebands do not yet fit cleanly within the context of current models. However, Spitkovsky *et al.* (2002) have argued that zonal flows associated with burst heating and rapid spin of the neutron star may produce a modulation pattern as well as sidebands (see Section 3.4.11).

In many LMXBs a pair of kHz QPOs are observed (see the review by van der Klis in Chapter 2). In six sources the frequency difference between the kHz QPO is close to the observed burst oscillation frequency or one half of the frequency. This closeness of the burst oscillation frequency, and the frequency difference of the kHz QPOs has motivated the "beat frequency" models of kHz QPOs (see Strohmayer *et al.* 1996; Miller, Lamb & Psaltis 1998; Lamb & Miller 2001). In three sources the QPO difference frequency is close to half the burst oscillation frequency. This might be possible if a pair of antipodal hot spots on the neutron star produces the burst oscillation. Based on these considerations, Miller (1999) searched for a subharmonic of the 581 Hz burst oscillation in 4U 1636–53, and claimed detection of a 290 Hz signal at a 4×10^{-5} significance level. Extensive efforts to detect such a signal again have, however, been unsuccessful (see Strohmayer 2001 for a discussion). Based on the recent detections of burst oscillations in the two accreting millisecond pulsars, SAX J1808.4–3658 and XTE J1814–338, it appears certain that the burst oscillation frequency directly reflects the spin and not a harmonic of the spin frequency.

3.4.9 Burst oscillations as probes of neutron stars

Modeling of burst oscillations holds great promise as a new tool for probing the structure of neutron stars and their environs. The emission and propagation of photons from the surfaces of rapidly rotating neutron stars are strongly influenced by general relativistic effects. Gravitational light deflection suppresses the modulation amplitude and reduces the harmonic content of pulses produced by rotational modulation of a hot spot on a rotating neutron star (see Pechenik, Ftaclas & Cohen 1983). The strength of light deflection is a function of the compactness, $m/r \equiv GM/c^2R$. More compact stars produce greater deflections and therefore weaker spin modulations (see Strohmayer *et al.* 1998a; Miller & Lamb 1998). Relativistic motion of the hot spot creates asymmetry and sharpening of the pulse profile, increasing the harmonic content (Chen & Shaham 1989; Miller & Lamb 1998; Braje, Romani & Rauch 2000; Weinberg, Miller & Lamb 2001). Such motion also introduces a pulse-phase dependent Doppler shift in the X-ray spectrum. The magnitudes of these effects are directly proportional to the surface velocity, which is a function of the unknown stellar radius and the known spin frequency. Ford (1999) has analysed data during a burst from Aql X-1 and finds that the softer photons lag higher energy photons in a manner which is qualitatively similar to that expected from a rotating hot spot. Fox *et al.* (2001), however, found that the sense of the phase lags in this burst switched, with hard lags preceding the soft lags found by

Ford (1999). This suggests that photon scattering may play an important role in addition to Doppler beaming.

Several studies have been undertaken to constrain neutron star properties based on modeling of burst oscillations. Miller and Lamb (1998) have investigated the amplitude of rotational modulation pulsations as well as harmonic content assuming emission from a point-like hot spot. They also showed that knowledge of the angular and spectral dependence of the emissivity from the neutron star surface can have important consequences for the derived constraints. Nath, Strohmayer and Swank (2001) have modeled bolometric pulse profiles observed during the rising phase of bursts from 4U 1636–53. They fit the pulse profiles with a rotating, expanding hot spot model that includes light deflection in the Schwarzschild spacetime. They find that the inferred constraints depend very sensitively on whether or not two spots are present. Much more restrictive compactness constraints can be achieved if two spots are present, the main reason being that large amplitudes are much more difficult to achieve with two spots than one. Assuming two hot spots they find a lower limit to the compactness of $m/r < 0.163$ at 90% confidence. This requires a relatively stiff equation of state for the neutron star interior and disagrees with the recently measured value of $m/r = 0.23$ from Cottam *et al.* (2002). If one hot spot is assumed, then the constraint is consistent with the Cottam *et al.* (2002) measurement. Weinberg, Miller and Lamb (2001) have explored the oscillation waveforms and amplitudes produced by rotating neutron stars with single and antipodal hot spots of varying size. They include photon propagation in the Schwarzschild spacetime and consider the effects of relativistic aberration and Doppler shifts induced by the rotational motion of the neutron star surface. They conclude that pulse profile fitting could be a powerful tool in constraining neutron star properties.

3.4.10 Spin modulation: implications for neutron star spin in LMXBs

The discovery of burst oscillations provided the first strong observational evidence for millisecond rotation periods in accreting neutron star LMXBs. Since then seven bona fide accreting millisecond pulsars have been discovered; SAX J1808.4–3658 (Wijnands & van der Klis 1998; Chakrabarty & Morgan 1998), XTE J1751–305 (Markwardt *et al.* 2002), XTE J0929–314 (Galloway *et al.* 2002), XTE J1807–294 (Markwardt, Smith & Swank 2003), XTE J1814–338 (Markwardt & Swank 2003), IGR J00291+5934 (Markwardt, Swank & Strohmayer 2004; Galloway *et al.* 2005), and HETE J1900.1–2455 (Morgan, Kaaret & Vanderspek 2005) with spin frequencies of 401 Hz, 435 Hz, 185 Hz, 191 Hz, 314 Hz, 598, and 377 Hz respectively. Four of these pulsars have extremely low mass companions, with $M_c \approx 0.01\ M_\odot$ (for XTE J1751–305 and XTE J0929–314, Bildsten 2002; Deloye & Bildsten 2003), and they have firmly established the link between millisecond radio pulsars and accreting LMXBs. However, until very recently no source had shown both persistent pulsations and burst oscillations, with the exception of a 3σ detection with the WFC of an oscillation at 401 Hz during a burst from SAX J1808.4–3658 (in 't Zand *et al.* 2001).

This situation changed dramatically in the few months prior to finalizing our review. A new outburst of the accreting millisecond pulsar SAX J1808.4–3658 was discovered with the RXTE/ASM (Markwardt, Miller & Wijnands 2002) on October 13, 2002. Extensive RXTE follow-up observations of the outburst detected four thermonuclear bursts, all of which are PRE bursts. Each burst also shows oscillations, with pulsations detected during the rise and after "touchdown" of the photosphere, as is seen in many other burst oscillation sources that do not show pulsations in their persistent emission (Chakrabarty *et al.* 2003). The amplitude and frequency of the oscillations in the bursts from SAX J1808.4–3658 confirms that the burst flux is modulated at the 401 Hz spin frequency. Oscillations seen in the decaying tails were at a constant frequency very nearly equal to the spin frequency, but exceeding it by an average of about 6 mHz, or about one part in 10^5 (Chakrabarty *et al.* 2003). The pulsations in the tails of the bursts were also found to be phase aligned, to within about 6%, with the persistent pulsations.

In one burst a clear frequency drift of $\approx 4-5$ Hz is detected during the burst rise. This behavior is similar to the frequency increases seen during the rise of bursts from several other sources, although the magnitude and speed of the drift are larger in SAX J1808.4–3658. Chakrabarty *et al.* (2003) argue that the rapid drift in SAX J1808.4–3658 is evidence for a stronger magnetic field than in the non-pulsing LMXBs.

Even more recently, on June 5, 2003, a new accreting millisecond pulsar, XTE J1814–338, was discovered (Markwardt & Swank 2003). Extensive monitoring of the outburst with RXTE resulted in the detection of more than a dozen bursts, all of which also show oscillations at the 314 Hz spin frequency (Markwardt, Strohmayer & Swank 2003). These recent findings, as well as the discovery of coherent pulsations during a superburst from 4U 1636–53 (see Strohmayer & Markwardt 2002; Section 3.5.4 below) conclusively establish that burst oscillations result from spin modulation of the X-ray burst flux. In addition, the bursts from XTE J1814–338 were the first bursts from any source to show significant harmonic structure in their oscillation profiles (Strohmayer *et al.* 2003). Bhattacharyya *et al.* (2005) have carried out detailed modeling of the observed burst oscillation profiles from XTE J1814–338 and use these models to derive constraints on the neutron star compactness.

With confidence that burst oscillations reflect the spins of neutron stars, it is possible to ask what the observed distribution of burst oscillation frequencies implies for the maximum spin rate of accreting neutron stars. Chakrabarty *et al.* (2003) argue that the observed distribution, which spans the range from 45 to 619 Hz more or less uniformly, supports an upper limit to the spin rate of neutron stars of about 760 Hz. This is interesting because the maximum spin rate of a neutron star depends directly on the nuclear equation of state. Only the very stiffest equations of state have a break-up frequency in this range (see Cook, Shapiro & Teukolsky 1994), suggesting the possibility that some other process may limit the spin frequencies of neutron stars. A particularly intriguing possibility is that gravitational radiation losses might be important in removing angular momentum and limiting the spins (see Wagoner 1984, Bildsten 1998).

3.4.11 Theoretical implications of burst oscillations

The spin modulation mechanism requires a slow-moving, non-uniform brightness pattern on the neutron star surface. At burst onset the pattern is most likely a localized "hot spot", whereas we are still mostly in the dark as to the origin of oscillations long after the burning has started. What all of the observations have made clear is that we can no longer persist with spherically symmetric modeling. In many ways, the question we now need to answer is: "What breaks the symmetry?" as prior to and after the bursts, there is no indication of azimuthal variations on the stellar surface.

The initial work on the spreading of a locally ignited (i.e., symmetry broken by hand) "hot spot" away from the ignition site focused on the laminar or convective combustion viewpoint (e.g., Fryxell & Woosley 1982; Bildsten 1995) and neglected the physics of the atmospheric response to transverse pressure gradients. Such an approximation is only appropriate if the matter is very degenerate. In this limit, the observed short rise times can only be explained with convective velocities. However, for most observed bursts the degeneracy is only partial during the flash. In this case, an additional velocity becomes important, which is the shallow water wave speed, $V_{sw} \approx (gh)^{1/2}$, where g is the surface gravity and $h \approx 10$ m is the thickness of the burning layer. This speed is $\approx$4000 km/s, so that, in the absence of rotation, any transverse pressure disturbances create wave-like disturbances with periods as short as 5–10 milliseconds (e.g., Livio & Bath 1982; McDermott & Taam 1987; Bildsten & Cutler 1995; Bildsten & Cumming 1998; Piro & Bildsten 2004, 2005a, 2005b; Lee 2004). However, these periods are longer than the neutron star rotation period, in which case the Coriolis force must be taken into account (see Bildsten *et al.* 1996; Strohmayer & Lee 1996 for the rotational modification of modes) when considering the nature of the modes.

Spitkovsky *et al.* (2002) showed that this same interplay between the shallow-wave speed and the Coriolis force is relevant to the propagation speed of localized burning on a rotating star. In their groundbreaking calculation, they found that the width of the burning front at the leading edge of the hot spot was the Rossby adjustment radius $\sim V_{sw}/\Omega$, and the resulting speed of the front is $V_f \sim V_{sw}/(\Omega t_n) \sim 2-10$ km s^{-1}, where $t_n \sim 0.1-1$ s is the time to burn the fuel (for example, in a helium-rich flash). The spreading time for the whole star is then 0.1 to a few seconds, in the observed range. The dependence on the rotation rate, Ω, would be nice to test, but the current range of measured rotation rates in bursters is probably too small to allow for it.

Spitkovsky *et al.* (2002) also investigated the dynamics of the nuclear ignition on a rotating star and argued that ignition will tend to occur in the equatorial region and propagate to the poles (see Fig. 3.13). The combination of radial uplift and horizontal flows they found may also be able to explain the observed frequency drifts. This initial progress is remarkable, but much remains to be done, including an implementation of more realistic burning during the propagation and an improved understanding of where the accreted fuel resides and where/how ignition really starts in a three-dimensional star. The standard ignition condition is for a spherically symmetric perturbation of a spherically symmetric model and most likely both of those approximations need to be dropped.

In the burst tails the oscillation might conceivably be produced by a mode (e.g.,McDermott & Taam 1987; Heyl 2004; Lee & Strohmayer 2005; Heyl 2005) or perhaps generated dynamically by the interaction of burst heating and cooling with the rapid spin of the star (see Spitkovsky *et al.* 2002; Cumming 2005). Indeed, the pure harmonic content points to an azimuthal perturbation $\propto \exp(im\phi)$. However, any such model must explain both the $\approx 1\%$ frequency drift and the long-term stability. In the context of non-radial oscillations, Piro and Bildsten (2005a) have identified modes capable of satisfying these constraints, a surface wave in the burning layer and a crystal interface mode.

Without specifically addressing the mode question, Cumming and Bildsten (2000) explored in detail the radial uplift mechanism outlined by Strohmayer *et al.* (1997) to explain the frequency drifts. In the context of presuming that the burning layer became disconnected from the underlying material but itself rigidly rotated, they concluded that this process could explain most of the drifts. Heyl (2000) claimed that properly including general relativistic effects would allow for strong constraints on the NS radius, but soon thereafter Abramowicz, Kluzniak and Lasota (2001) and Cumming *et al.* (2002) found an error in Heyl's formulation that strongly reduced the impact of general relativity. However, during this re-evaluation, Cumming *et al.* (2002) found that they had overestimated the radial uplift drift by about a factor of 2. This result, combined with the observations of much larger drifts in several sources (Galloway *et al.* 2001, Wijnands *et al.* 2002) now makes it appear that radial uplift is insufficient to account for all the observed drift if the whole burning layer is rigidly rotating.

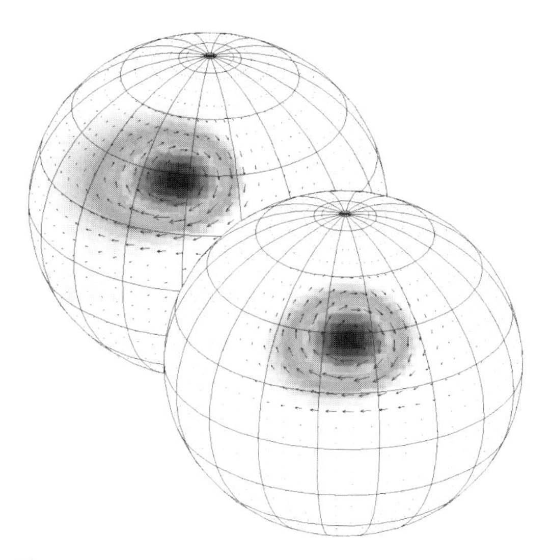

Fig. 3.13. Initial evolution of a burning hot spot ignited off the equator as seen in a frame rotating with the neutron star. Velocity vectors show the circulation of the fluid induced by the Coriolis forces. The hot spot expands due to burning and drifts west-southwest because of the latitude dependence of the Coriolis force (after Spitkovsky, Levin & Ushomirsky 2002).

One possible way out from this conundrum is to allow some fluid elements in the burning layer to keep their initial angular momentum (i.e., not demand rigid rotation of the whole burning layer). This easily produces much larger frequency shifts for the outermost layers (Cumming & Bildsten 2000), but the origin of a coherent signal in the context of the rapid internal differential rotation is then a new puzzle to solve. These uncertainties will likely remain with us until the origin of the asymmetry is resolved.

The strength of the large-scale dipole field that threads the burning layers on the neutron star is unknown. The lack of persistent pulsations during accretion leads to a limit of $B < 10^{8-9}$ G by presuming that the spherical magnetospheric radius is inside the neutron star. However, B might need to be even lower than this to ensure azimuthal symmetry. For example, it is easy to imagine that a permanent asymmetry could result even if the material arrives on the equator, but encounters an ordered field as it tries to spread away as the photospheric pressure is only $\approx 10^{15}$erg cm^{-3}. In addition, during the bursts, if the frequency drift that is observed is due to vertical shear from the radially expanded burning layers, an initially poloidal field as weak as 10^6 G could become dynamically important as it is wound up by differential rotation (Cumming & Bildsten 2000).

3.5 Superbursts: a new burning regime

Since the advent of BeppoSAX and RXTE in 1996, the X-ray sky has been monitored with unprecedented sensitivity and frequency. This capability has opened up the discovery space for burst events with long (years) recurrence times, which were apparently missed by previous missions. Cornelisse *et al.* (2000) reported the first superburst discovery (from BeppoSAX) in the familiar Type I burster 4U 1735–44. This was rapidly followed by reports of more long X-ray flares lasting 3–5 hours from previously known X-ray bursters; 4U 1820–30 (Strohmayer 2000, Strohmayer & Brown 2002); KS 1731–260 (Kuulkers *et al.* 2002c); 4U

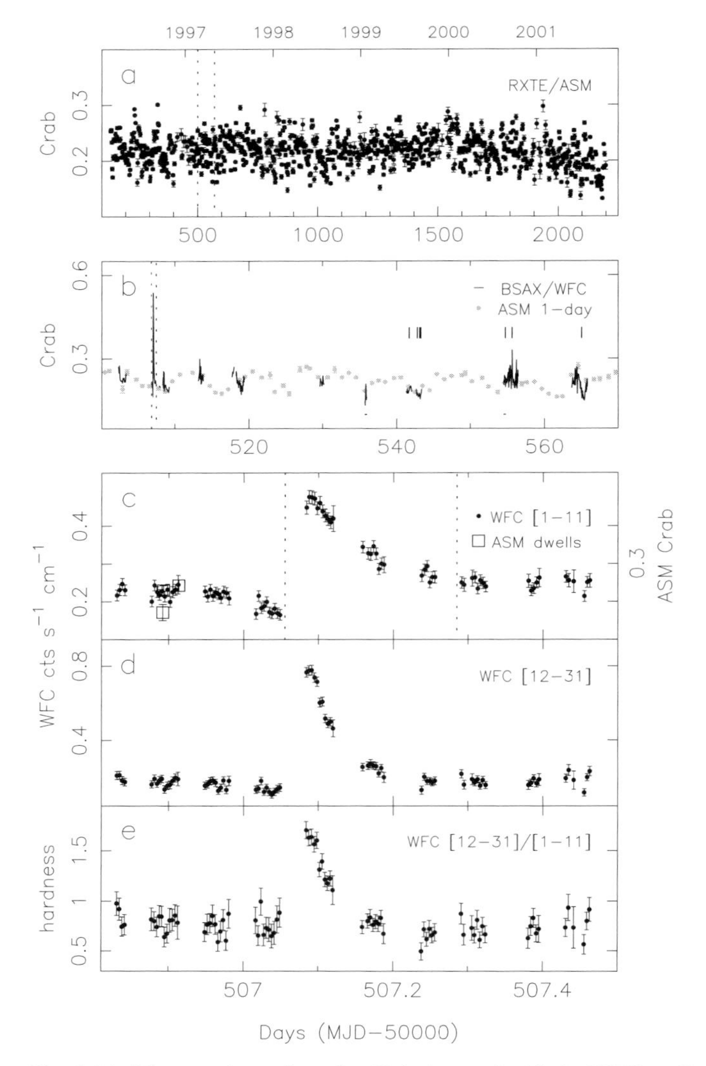

Fig. 3.14. The superburst from Ser X-1 observed with the WFC on BeppoSAX. Note the persistent offset in the flux after the superburst in the 2–5 keV band (c). The hardness ratio (e) drops through the burst, indicative of cooling of the neutron star surface (after Cornelisse *et al.* 2002c).

1636–53 (Wijnands 2001; Strohmayer & Markwardt 2002); Ser X-1 (Cornelisse *et al.* 2002c); GX 3+1 (Kuulkers 2002); and 4U 1254–69 (in't Zand *et al.* 2003). One of the sources, 4U 1636–53, produced two bursts separated by 4.7 years (Wijnands 2001). Two of the events, one from 4U 1636–53, and the other from 4U 1820–30 were observed with the large area PCA on board RXTE (Figs. 3.14 and 3.15 show superbursts seen with the BeppoSAX/WFC and RXTE/PCA, respectively).

Table 3.2 (after Kuulkers *et al.* 2002c) summarizes the important properties of the superbursts observed to date. These flares show all the hallmarks of thermonuclear bursts; they

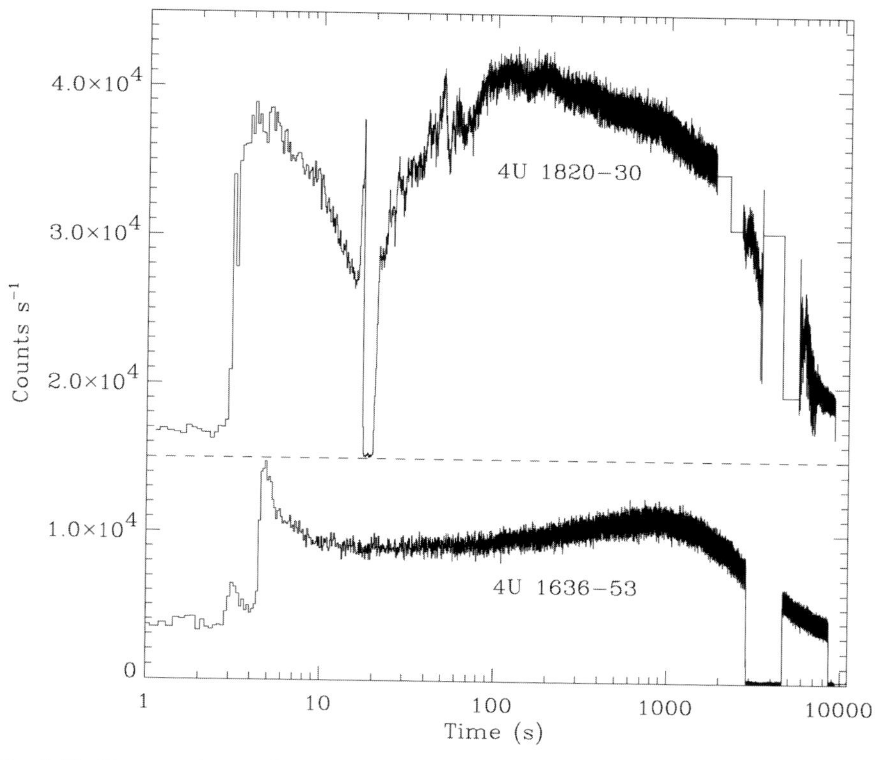

Fig. 3.15. Two superbursts observed with the RXTE/PCA. Shown are the 2–30 keV countrate histories observed in the PCA. Note the shorter precursor events prior to the superbursts. The event from 4U 1820–30 has been displaced vertically for clarity. The horizontal dashed line shows the zero level for this event. The time axis is logarithmic (after Strohmayer & Brown 2002; Strohmayer & Markwardt 2002).

have thermal spectra which soften with time, they show more or less smooth exponential-like decays, and some show PRE episodes. The fundamental distinctions between these events and standard Type I X-ray bursts are their long durations, larger fluences and long recurrence times (i.e., they are rare). Indeed, these events are typically 1000 times longer and more energetic than standard X-ray bursts; because of this, they have come to be referred to as "superbursts," and we will continue to use this appellation. Although the sample size is still relatively small, the systems which have produced superbursts have typical accretion rates in the range $\approx 0.1 - 0.3$ of $\dot{M}_{\rm Edd}$ (Wijnands 2001; Kuulkers *et al.* 2002c). As we describe in more detail shortly, this is an important constraint on theoretical models.

The durations and energetics of superbursts suggest that they result from thermonuclear flashes occurring in fuel layers at much greater depth than for typical X-ray bursts. For example, if the nuclear energy release is 0.3 MeV per accreted nucleon, the accumulated mass required to power a 10^{42} erg superburst is 3.5×10^{24} g, giving a recurrence time of ≈ 2 years for an accretion rate of $\dot{M} \approx 10^{-9}\,{\rm M}_\odot\,{\rm yr}^{-1}$. In the remainder of this section we will summarize the current state of knowledge of superbursts and explore what new insights they are giving us about nuclear burning on neutron stars.

Table 3.2. *Superburst sources and properties (after Kuulkers et al. 2002b)*

Source	4U 1820–30	4U 1735–44	KS 1731–260	4U 1636–53	Ser X-1	GX 3+1	4U 1254–69
Detector[a]	PCA	WFC	WFC, ASM	PCA, ASM	WFC	ASM	WFC
Duration[b]	3	7	12	$> 2-3$	≈ 4	> 3.3	14
Precursor?	yes	?	yes	yes	?	?	yes
$\tau_{\rm exp}$[b]	≈ 1	1.4 ± 0.1	2.7 ± 0.1	1.5 ± 0.1 3.1 ± 0.5	1.2 ± 0.1	1.6 ± 0.2	6.0 ± 0.3
$L_{\rm pers}$[c]	≈ 0.1	≈ 0.25	≈ 0.1	≈ 0.1	≈ 0.2	≈ 0.2	0.13 ± 0.03
$kT_{\rm max}$[d]	≈ 3	≈ 2.6	≈ 2.4	?	≈ 2.6	~ 2	1.8 ± 0.1
$L_{\rm peak}$[e]	3.4	1.5	1.4	1.2	1.6	0.8	0.44 ± 0.2
$E_{\rm b}$[f]	> 1.4	> 0.5	≈ 1	$0.5 - 1$	≈ 0.8	> 0.6	0.8 ± 0.2
$t_{\rm quench}$[g]	?	> 7.5	> 35	?	~ 34	?	< 125
References[h]	S00, SB02	C00	K02c	W01, SM02	C02	K02	I03

[a] Instruments that observed the superburst.
[b] In hours.
[c] Persistent luminosity prior to superburst, in terms of the Eddington luminosity.
[d] In keV.
[e] In units of 10^{38} erg s^{-1}.
[f] in units of 10^{42} erg.
[g] Time following superburst with no normal burst activity, in days.
[h] S00 (Strohmayer 2000); SB02 (Strohmayer & Brown 2002); C00 (Cornelisse *et al.* 2000); K02c (Kuulkers *et al.* 2002c); W01 (Wijnands 2001); SM02 (Strohmayer & Markwardt 2002); C02 (Cornelisse *et al.* 2002c); K02 (Kuulkers 2002); I03 (in't Zand *et al.* 2003).

3.5.1 Time profiles and spectra

The most distinctive characteristic of superbursts is their long durations. The time profiles of most superbursts have been characterized with an exponential decay time, and these values range from about 1 to 3 hours (see Kuulkers *et al.* 2002b). The longest event seen to date was the superburst from 4U 1254–69, which lasted for more than half a day (in't Zand *et al.* 2003). Their time profiles are generally smooth, with a fast rise and exponential decay, however, not all superbursts have been observed with comparable sensitivity and temporal resolution, so comparisons must be made with some caution. For example, the superburst from 4U 1820–30 showed substantial variability during portions of the decay phase (Strohmayer & Brown 2002). Both superbursts observed with the large area PCA show "precursor" events just prior to the start of the superburst (see Strohmayer & Brown 2002; Strohmayer & Markwardt 2002; Fig. 3.15). These precursors look more or less like standard Type I bursts from the respective objects. There is also some evidence for a precursor in the burst from KS 1731–260 observed with the BeppoSAX/WFC (Kuulkers *et al.* 2002c).

Based on the available data, it appears that such precursors are a common feature of superbursts. If the superbursts result from energy release at great depths, then it seems plausible that the superburst flux could trigger a flash in the H/He layers above it as it diffuses outward. There has been some speculation that the precursor may act as a trigger for the superburst, but given the much longer radiative diffusion time at the depth where the superbursts are probably triggered, this seems unlikely (see Strohmayer & Brown 2002).

The X-ray spectra of superbursts are well described by thermal emission, with peak blackbody temperatures in the range from 2 to 3 keV, quite typical for thermonuclear bursts (Cornelisse *et al.* 2000). Interestingly, most of the observed superbursts have peak fluxes that are sub-Eddington, only the superburst from the pure helium accretor 4U 1820–30 had a peak flux consistent with the Eddington limit.

Strohmayer and Brown (2002) found discrete components in the spectra of the superburst from 4U 1820–30; a broad emission line centered near 6 keV and an accompanying absorption edge between 8 and 9 keV (see their Figure 5). They suggested that these features can be explained by reflection of the burst flux from the inner accretion disk around the neutron star. Day and Done (1991) had predicted that such features might be detectable in spectra of bursts. The energy and width of the dominant Fe Kα fluorescence line as well as the energy of the edge can be used as important diagnostics of the ionization state of the disk (see for example, Ross, Fabian & Young 1999; Nayakshin & Kallman 2001). An origin of the features in a burst-driven wind is also possible. Ballantyne and Strohmayer (2004) used theoretical disk reflection models to fit the spectra observed with the RXTE/PCA from the 4U 1820–30 superburst. They concluded that the reflection spectra provided an adequate fit for most of the superburst duration. Variations in the disk ionization state, covering factor and inner radius are suggested by the time-resolved spectral fitting. Their modeling suggests that the inner region of the disk was disrupted by the burst, possibly being heated into a thicker, more tenuous flow, before recovering later in the burst.

Because superbursts last about a 1000 times longer than normal bursts and can give very high signal-to-noise spectra, they would allow much more sensitive searches for discrete spectral lines from neutron star surfaces. Superbursts would therefore make tempting targets for rapidly triggered observations with sensitive, high spectral resolution observatories. The recently launched Swift mission may be able to provide the necessary triggering capability.

3.5.2 *Superburst energetics*

The available X-ray spectroscopy on superbursts provides estimates of the total X-ray energy liberated by the events. For the superburst from 4U 1820–30 a lower limit to the energy fluence was 1.5×10^{42} erg, assuming a distance of 6.6 kpc (Strohmayer & Brown 2002). It is extremely unlikely that unstable helium burning could provide such a large fluence. This could only occur at very low atmospheric temperatures and thus at low mass accretion rates inconsistent with the persistent X-ray flux observed from 4U 1820–30 (Fryxell & Woosley 1982; Zingale *et al.* 2001). Similar fluence limits have been derived for the other superbursts (see Kuulkers *et al.* 2002c). The total energy observed in X-rays, combined with the decay timescales of superbursts strongly argues for a fuel source located at depths below the column density where helium flashes are triggered.

3.5.3 *Quenching of normal burst activity*

There are strong indications from several of the superburst sources that the occurrence of a superburst has a profound influence on the thermal state of the accreted "ocean" on the neutron star. In particular the occurrence of normal (short duration) X-ray bursts appears to be suppressed for some time following superbursts. For example, Kuulkers *et al.* (2002c) and Cornelisse *et al.* (2002c) found that normal bursting ceased for about 35 days following the superbursts from KS 1731–260 and Ser X-1, respectively (see Fig. 3.14, where the tick marks in panel b denote the positions of normal bursts). There are also indications that normal bursting was suppressed for at least a week following the superburst from 4U 1735–44 (Cornelisse *et al.* 2000). A likely explanation of this suppression is that flux from the deep parts of the neutron star ocean remains high enough to quench the Type I bursting activity (Cumming & Bildsten 2001; Cumming & Macbeth 2004).

3.5.4 *Millisecond pulsations during a superburst: 4U 1636–53*

Burst oscillations at ≈582 Hz have been observed in many bursts from the LMXB 4U 1636–53 (see Table 3.1; Zhang *et al.* 1998: Giles *et al.* 2002). Strohmayer and Markwardt (2002) recently discovered coherent pulsations at this frequency during a portion of the February 22, 2001 (UT) superburst from this source. They were detected during an ≈800 s interval spanning the flux maximum of the superburst. The average pulsation amplitude was 1%, which is smaller than the amplitudes of oscillations observed from standard bursts. The pulse trains observed during the superburst are much longer than the typical, 10 s long pulse trains observed in normal bursts. The pulsation frequency was found to increase in a monotonic fashion by less than a part in 10^4 (see Fig. 3.16). This is a much smaller frequency drift than commonly seen in burst oscillations from standard bursts (see Section 3.4.4). The form of the frequency evolution appears consistent with Doppler modulation caused by the known orbital motion of the neutron star around the center of mass of the binary. Strohmayer and Markwardt (2002) showed that a circular orbit model fits the observed frequency evolution well. The best phase evolution model is consistent with a coherent pulsation during the observation interval, and gives a limit on the coherence $Q \equiv \nu_0/\Delta_{\nu_0} > 4.5 \times 10^5$. The orbital fits indicate that the projected neutron star velocity lies between 90 and 175 km s^{-1}. The brevity of the observed pulse train with respect to the orbital period of 3.8 hr does not allow for more precise constraints. The fact that the coherent pulsation frequency during the superburst is within ≈1 Hz of all the measured asymptotic burst oscillation frequencies for 4U 1636–53 indicates that the frequencies are set by the spin of the neutron star.

3.5.5 *Theory of superbursts: ashes to ashes*

Unstable burning of a pure carbon layer (Woosley & Taam 1976; Taam & Picklum 1978; Brown & Bildsten 1998) has been proposed as an explanation for the superburst from the pure helium accretor 4U 1820–30 (Strohmayer & Brown 2002). In this scenario, the accumulated mass of carbon at ignition is $10^{26}-10^{27}$ g, giving decades-long recurrence times and most of the energy released (10^{43}–10^{44} erg) escapes as neutrinos or is conducted into the star, leaving ≈ 10^{42} erg to emerge from the surface within a few hours (Strohmayer & Brown 2002). Shorter recurrence times are possible if a smaller mass of carbon can be triggered somehow. Large carbon fractions in the ocean are expected for 4U 1820–30, since stable helium burning at the higher inferred mass accretion rates, when normal (10–20 s duration) bursts are not observed, will produce lots of carbon.

Pure carbon is unlikely to apply to the superbursts from H/He accretors as Schatz *et al.* (1999, 2001) have shown that only a small amount of carbon remains after the burning of H and He via the rp-process. However, Cumming and Bildsten (2001) (hereafter CB01) showed that even small amounts of carbon can be a promising energy source for the superbursts. They found that burning of this small mass fraction of carbon is thermally unstable at low accumulated masses when the ocean contains heavy ashes from the rp-process.

CB01 proved the important role played by the rp-process ashes. Their low thermal conductivity gives a large temperature gradient in the ocean, so that the trace carbon ignites at accumulated masses comparable to that observed (see recent work by Brown 2004; Cooper and Narayan 2005). The resulting energies, recurrence times, and conductive cooling times can easily accommodate the observed properties of superbursts (see Fig. 3.17), especially now that an extra energy source has been found by Schatz, Bildsten and Cumming (2003). They showed that the conversion of rp-process nuclei back to the iron group elements during

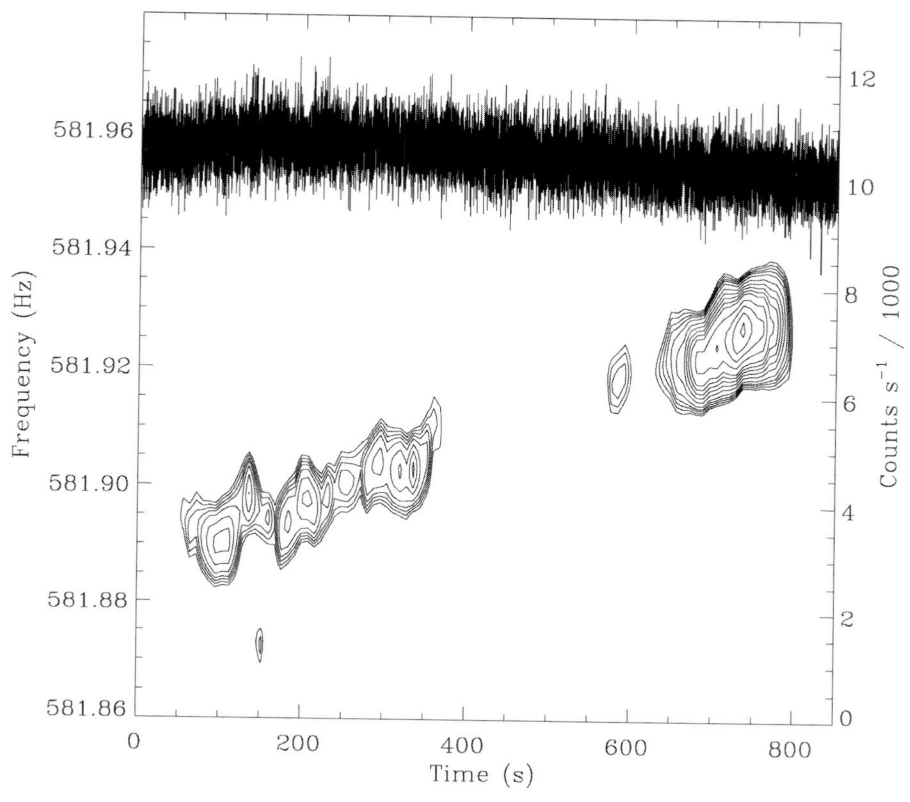

Fig. 3.16. Dynamic power spectrum of a portion of the February 22, 2001 superburst from 4U 1636–53 showing 582 Hz pulsations. Shown are contours of constant Fourier power versus frequency (left axis), and the RXTE/PCA countrate (right axis), both as a function of time. The monotonic increase in the pulsation frequency is consistent with binary orbital modulation (after Strohmayer & Markwardt 2002).

the carbon-triggered flash can enhance the energy release from that given in CB01 by factors of 4.

The instability requires that $\dot{M} > 0.1\dot{M}_{\rm Edd}$ when the carbon mass fraction is less than 10%. Lower $\dot{M}$'s stably burn the carbon. Though the instability is present at accretion rates $\approx \dot{M}_{\rm Edd}$, those flashes provide less of a contrast with the accretion luminosity, thus explaining why detection is easier when $\dot{M} \approx (0.1-0.3)\dot{M}_{\rm Edd}$. Detecting one of these flares from a rapidly accreting Z source requires flux sensitivity at the 10% level on a timescale of a few hours and spectral sensitivity to distinguish that the flux rise is from extra thermal emission (in't Zand, Cornelisse & Cumming 2004). This should be carried out and would confirm the notion of trace carbon ignition in the heavy rp-process ashes.

The energy from these mixed flashes takes a long time to escape the star, possibly explaining the persistent "offset" in the flux nearly a day after the superburst in Ser X-1 (see Fig. 3.14; Cumming & Macbeth 2004). CB01 also attributed the halting of regular Type I bursts after the superbursts to thermal stabilization of the H/He burning layers by the large heat flux from the cooling ashes of the carbon burning. Paczynski (1983) and Bildsten (1995) have shown that luminosities in excess of the helium burning flux (or $L > L_{\rm accr}/100$) will stabilize the

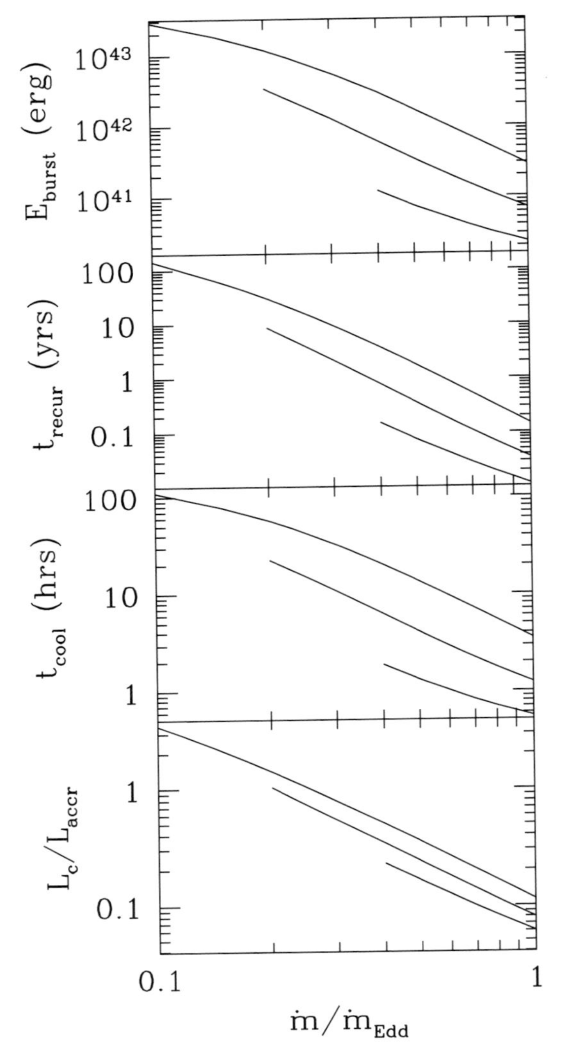

Fig. 3.17. Results of theoretical calculations of carbon flashes in the heavy element ocean on a neutron star. Shown from top to bottom are the burst energies, recurrence times, luminosity decay times, and the ratio of cooling luminosity to accretion luminosity, as a function of $\dot{m}$ for a carbon fraction, $X_{12} = 0.1$, in a ^{104}Ru ocean. In each panel three curves are shown for different values of the flux, $Q_{17} = 0.5, 0.1, 0.2$ (from top to bottom), due to electron captures and pycnonuclear reactions in the deep ocean (Haensel & Zdunik 1990; Brown 2000). The curves begin at the lowest accretion rate at which the thermal instability occurs in the heavy element (^{104}Ru) ocean (after Cumming & Bildsten 2001).

H/He burning. This thus provides additional evidence that the superbursts are from burning deep within the neutron star.

An alternative to carbon scenarios has been discussed by Kuulkers *et al.* (2002c). They suggest that electron captures on protons might be able to supply sufficient energy to power superbursts. The Fermi energy at column depths required to account for superburst energetics is close to the proton–neutron mass difference, so protons can easily capture electrons. The resulting neutrons can then be captured on heavy nuclei, supplying about 7 MeV per nucleon (Bildsten & Cumming 1998). It is not yet clear whether this process is thermally unstable under the relevant conditions, and it is not certain that sufficient hydrogen can survive the initial H/He burning. Further theoretical work should be able to answer these questions.

The initial theoretical understanding of superbursts is an exciting development as it further motivates the nucleosynthesis studies during Type I bursts and connects the ashes from the rp-process burning to the superburst explosions. We also hope that future observations will find the equivalent of superbursts from the higher accretion rate Z sources, or even from accreting X-ray pulsars.

3.6 Summary and future prospects

There has clearly been tremendous growth in our observational understanding of thermonuclear bursts in the past decade. Indeed, we think it is fair to say that the observations have gotten significantly ahead of the theory at the time of this writing. Yet, the promise of probing neutron star structure, and the fundamental physics needed to describe it, with burst observations has never been greater. Our understanding, though still incomplete, of the new phenomena, as well as insights drawn from previous observations, are sufficient to show that researchers must really begin to explore fully three-dimensional and time-dependent calculations of nuclear energy production and radiation transport in the neutron star surface layers in order to fully exploit the new phenomena as probes of neutron star structure and fundamental physics. The days of spherically symmetric, static models being adequate are long gone in our opinion.

In the context of burst oscillations it seems clear that the spreading of the nuclear burning front has important observational consequences, particularly for oscillations seen near the onset of bursts. A thorough theoretical understanding will only come with multi-dimensional dynamical studies of nuclear ignition and propagation. Indeed, higher signal-to-noise observations of the oscillations during burst rise could in principle provide us with a snapshot view of how burning spreads. Such studies will require even larger X-ray collecting areas than RXTE. A future goal would be to have detailed theoretical models of front propagation available for comparison with new data by the time such observations are eventually made. Moreover, the persistence of oscillations in the cooling phase suggests that the dynamics of burst heating and rapid rotation combine to produce quasi-stable patterns in the surface layers that can persist for thousands of rotation periods (Spitkovsky *et al.* 2002). It is a triumph of modern X-ray astronomy that we are able to infer changes in the surface layers on neutron stars on the scale of only tens of meters at distances of kiloparsecs! Helioseismology has revolutionized study of the solar interior, perhaps future improvements in sensitivity will allow a similar revolution in neutron star studies by enabling the detection of global oscillation modes of neutron stars. RXTE observations in particular have shown that new insights on burst phenomena can result from precision timing measurements. A prime goal for a future larger area timing mission would be to capitalize on the new probes of neutron stars pioneered by RXTE. For example, high signal-to-noise pulse profiles of burst oscillations could provide a breakthrough in constraining the neutron star mass–radius relation and hence the dense matter equation of state.

As we have shown, an understanding of superbursts probably requires knowledge of the by-products of the nuclear burning. Detailed models of the heat and radiation transport from the deep ocean to the surface will be required to accurately model surface fluxes and time profiles and should be able to inform us about the location and conditions where superbursts are triggered. Better constraints on the recurrence times for superbursts will provide information on the heating and cooling of the deep ocean, perhaps constraining rapid neutron star cooling scenarios. The existing RXTE observations with high signal to noise but low spectral resolution have shown that superbursts are promising targets for higher resolution spectroscopy, one of the goals of which should be to detect line features from neutron star surfaces. If such observations are to be possible, then X-ray monitoring instruments must be in place which can generate the necessary triggers. The discovery of superbursts was made possible with intensive all sky X-ray monitoring. Future efforts to study superbursts, for example, to better constrain the recurrence times, will require continued and improved X-ray monitoring capabilities. Missions that should be able to address these needs include, Integral, Swift, Lobster-ISS, and MAXI.

High spectral resolution studies of bursts have begun with the new capabilities of Chandra and XMM/Newton. Frustratingly, the present data suggest that isolated neutron stars may be poor targets for line searches perhaps because of a dearth of metals in the atmosphere. Accreting neutron stars could be more tempting targets, with a continuous supply of metals furnished by the mass donor. Although the new generation of X-ray observatories have impressive capabilities, they still lack the collecting area necessary to obtain high signal-to-noise spectra from single bursts. To do better will require larger area missions, such as NASA's Constellation-X and ESA's XEUS.

Acknowledgements

We thank Erik Kuulkers, Craig Markwardt, Deepto Chakrabarty, Hendrik Schatz and Jean Swank for comments on the manuscript. We are indebted to Mike Muno, Andrew Cumming, Jean Cottam, Lucia Franco, Hendrik Schatz, Duncan Galloway, Remon Cornelisse, Erik Kuulkers and Anatoly Spitkovsky for either providing graphics or allowing us the use of previously published figures. We sincerely thank them all. We would like to dedicate this work to the memory of John C. L. Wang, a friend and colleague who left us much too soon. This work was supported by the National Science Foundation under grants PHY99-07949 and AST02-05956, NASA through grant NAG 5-8658. LB is a Cottrell Scholar of the Research Corporation.

References

Abramowicz, M. A., Kluzniak, W. & Lasota, J. P. 2001, *A&A*, **374**, 16
Ayasli, S. & Joss, P. C. 1982, *ApJ*, **256**, 637.
Basinska, E. M. *et al*. 1984, *ApJ*, **281**, 337.
Belian, R. D., Conner, J. P. & Evans, W. D. 1976, *ApJ*, **206**, L135.
Bildsten, L. 1995, *ApJ*, **438**, 852.
1998, *ApJ*, **501**, L89.
Bildsten, L. 1998, in *The Many Faces of Neutron Stars*, NATO ASIC Proc. 515: 419.
2000, in *Cosmic Explosions*, eds. S. S. Holt & W. W. Zhang (New York: AIP), p. 359.
2002, *ApJ*, **577**, L27.
Bildsten, L. & Brown, E. F. 1997, *ApJ*, **477**, 897.
Bildsten, L. & Cumming, A. 1998, *ApJ*, **506**, 842.
Bildsten, L. & Cutler, C. 1995, *ApJ*, **449**, 800.
Bildsten, L., Salpeter, E. E., & Wasserman, I. 1992, *ApJ*, **384**, 143.
1993, *ApJ*, **408**, 615.
Bildsten, L., Ushomirsky, G. & Cutler, C. 1996, *ApJ*, **460**, 827.
Bildsten, L., Chang, P. & Paerels, F. 2003, *ApJ*, **591**, L29.
Boirin, L. *et al*. 2000, *A&A*, **361**, 121

Braje, T. M., Romani, R. W. & Rauch, K. P. 2000, *ApJ*, **531**, 447.
Brown, E. F. 2000, *ApJ*, **531**, 988
2004, *ApJ*, **614**, L57
Brown, E. F. & Bildsten, L. 1998, *ApJ*, **496**, 915.
Brown, E. F., Bildsten, L. & Rutledge, R. E. 1998, *ApJ*, **504**, L95.
Burwitz, V., Zavlin, V. E., Neuhauser, R. *et al.* 2001, *A&A*, **379**, L35.
Chakrabarty, D. 2000, Talk presented at AAS HEAD meeting, Honolulu, HI.
2002, American Physical Society, Meeting ID: APR02, abstract #S11.003, 11003.
Chakrabarty, D. & Morgan, E. H. 1998, *Nature*, **394**, 346.
Chakrabarty, D., Morgan, E. H., Muno, M. P. *et al.* 2003, *Nature*, **424**, 42.
Chang, P., Bildsten, L. and Wasserman, I. 2005, *ApJ*, **629**, 998.
Chen, K. & Shaham, J. 1989, *ApJ*, **339**, 279.
Chevalier, C. & Ilovaisky, S. A. 1990, *A&A*, **228**, 119.
Cocchi, M. *et al.* 2000, in *Proceedings of the Fifth Compton Symposium*, eds. M. L. McConnell & J. M. Ryan (New York:AIP), p. 203.
2001, *A&A*, **378**, L37.
Colpi, M., Geppert, U., Page, D. & Possenti, A. 2001, *ApJ*, **548**, L175.
Cook, G. B., Shapiro, S. L. & Teukolsky, S. A. 1994, *ApJ*, **424**, 823.
Cooper, R. L. & Narayan, R. 2005, *ApJ*, **629**, 422
Cornelisse, R., Heise, J., Kuulkers, E., Verbunt, F. & in't Zand, J. J. M. 2000, *A&A*, **357**, L21.
Cornelisse, R. *et al.* 2002a, *A&A*, **392**, 885.
Cornelisse, R., Verbunt, F., in 't Zand, J. J. M., Kuulker, E. & Heise, J. 2002b, *A&A*, **392**, 931.
Cornelisse, R., Kuulkers, E., in't Zand, J. J. M., Verbunt, F. & Heise, J. 2002c, *A&A*, **382**, 174.
Cornelisse, R. *et al.* 2003, *A&A*, **405**, 1033.
Cottam, J., Kahn, S. M., Brinkman, A. C., den Herder, J. W. & Erd, C. 2001, *A&A*, **365**, L277
Cottam, J., Paerels, F. & Mendez, M. 2002, *Nature*, **420**, 51.
Cumming, A. 2003, *ApJ*, **595**, 1077.
2005, *ApJ*, **630**, 441.
Cumming, A. & Bildsten, L. 2000, *ApJ*, **544**, 453.
2001, *ApJ*, **559**, L127.
Cumming, A. & Macbeth, J. 2004, *ApJ*, **603**, L37.
Cumming, A., Zweibel, E. & Bildsten, L. 2001, *ApJ*, **557**, 958.
Cumming, A., Morsink, S. M., Bildsten, L., Friedman, J. L. & Holz, D. E. 2002, *ApJ*, **564**, 343.
Damen, E. *et al.* 1990, *A&A*, **237**, 103.
Day, C. S. R., Fabian, A. C. & Ross, R. R. 1992, *MNRAS*, **257**, 471.
Day, C. S. R. & Done, C. 1991, *MNRAS*, **253**, 35P.
Deloye, C. J. and Bildsten, L. 2003, *ApJ*, **598**, 1217.
Drake, J. J. *et al.* 2002, *ApJ*, **572**, 996.
Ebisuzaki, T. & Nakamura, N. 1988, *ApJ*, **328**, 251.
Ebisuzaki, T. 1987, *Publ. Astron. Soc. Japan*, **39**, 287.
2000, *ApJ*, **535**, L119.
Ford, E. C. 1999, *ApJ*, **519**, L73.
Foster, A. J., Fabian, A. C. & Ross, R. R. 1987, *MNRAS*, **228**, 259.
Fowler, W. A. & Hoyle, F. 1965, *Nucleosynthesis in Massive Stars and Supernovae* (Chicago: University of Chicago Press).
Fox, D. W., Muno, M. P., Lewin, W. H. G., Morgan, E. H. & Bildsten, L. 2001, American Astronomical Society Meeting, 198.
Franco, L. 2001, *ApJ*, **554**, 340.
Franco, L. M. & Strohmayer, T. E. 1999, *BAAS*, **31**, 1556.
Fryxell, B. A. & Woosley, S. E. 1982, *ApJ*, **261**, 332.
Fujimoto, M. Y. & Taam, R. E. 1986, *ApJ*, **305**, 246.
Fujimoto, M. Y., Hanawa, T. & Miyaji, S. 1981, *ApJ*, **247**, 267.
Fujimoto, M. Y., Sztajno, M., Lewin, W. H. G. & van Paradijs, J. 1987, *ApJ*, **319**, 902.
Fushiki, I. & Lamb, D. Q. 1987, *ApJ*, **323**, L55.
Galloway, D. K., Chakrabarty, D., Muno, M. P., & Savov, P. 2001, *ApJ*, **549**, L85.
Galloway, D. K., Chakrabarty, D., Morgan, E. H. & Remillard, R. A. 2002, *ApJ*, **576**, L137.
Galloway, D. K., Psaltis, D., Chakrabarty, D. & Muno, M. P. 2003, *ApJ*, **590**, 999.

Galloway, D. K., Cumming, A., Kuulkers, E. *et al*. 2004, *ApJ*, **601**, 466.
Galloway, D. K., Markwardt, C. B. , Morgan, E. H., Chakrabarty, D. & Strohmayer, T. E. 2005, *Apj*, **622**, L45.
Giles, A. B. *et al*. 1996, *ApJ*, **469**, L25.
Giles, A. B., Hill, K. M., Strohmayer, T. E. & Cummings, N. 2002, *ApJ*, **568**, 279.
Gottwald, M. *et al*. 1989, *ApJ*, **339**, 1044.
Gottwald, M., Haberl, F., Parmar, A. N., & White, N. E. 1986, *Apj*, **308**, 213.
Grindlay, J. E. *et al*. 1976, *ApJ*, **205**, L127.
Haberl, F. & Titarchuk, L. 1995, *A&A*, **299**, 414.
Haensel, P. & Zdunik, J. L. 1990, *A&A*, **227**, 431.
Hanawa, T. & Fujimoto, M. Y. 1984, *Publ. Astron. Soc. Japan*, **36**, 119.
Hanawa, T. & Sugimoto, D. 1988, *Publ. Astron. Soc. Japan*, **34**, 1.
Hanawa, T., Sugimoto, D. & Hashimoto, M. 1983, *Publ. Astron. Soc. Japan*, **35**, 491.
Hansen, C. J. & van Horn, H. M. 1975, *ApJ*, **195**, 735.
Hasinger, G. & van der Klis, M. 1989, *A&A*, **225**, 79.
Heyl, J. S. 2000, *ApJ*, **542**, L45.
2004, *ApJ*, **600**, 939.
2005, *MNRAS*, **361**, 504.
Hoffman, J. A., Lewin, W. H. G. & Doty, J. 1977, *ApJ*, **217**, L23
Inogamov, N. A. & Sunyaev, R. A. 1999, *Astron. Lett.*, **25**, 269.
in 't Zand, J. J. M. 2001, in *Exploring the Gamma-ray Universe*, eds. A. Gimenez, V. Reglero & C. Winkler (ESA Pub. Div), p. 463.
in 't Zand, J. J. M., Heise, J., Kuulkers, E. *et al*. 1999, *A&A*, **347**, 891.
in 't Zand, J. J. M. *et al*. 2001, *A&A*, **372**, 916.
2002, *A&A*, **389**, L43.
in 't Zand, J. J. M., Kuulkers, E., Verbunt, F., Heise, J. & Cornelisse, R. 2003, *A&A*, **411**, L487.
in 't Zand, J. J. M., Cornelisse, R. & Cumming, A. 2004, *A&A*, **426**, 257.
Jager, R., Mels, W. A., Brinkman, A. C. *et al*. 1997, *A&AS*, **125**, 557.
Jongert, H. C. & van der Klis, M. 1996, *A&A*, **310**, 474.
Jonker, P. G., Mendez, M. & van der Klis, M. 2002, *MNRAS*, **336**, L1.
Joss, P. C. 1977, *Nature*, **270**, 310.
1978, *ApJ*, **225**, L123.
Joss, P. C. & Li, F. L. 1980, *ApJ*, **238**, 287
Joss, P. C. & Melia, F. 1987, *ApJ*, **312**, 700.
Kaaret, P, in 't Zand, J. J. M., Heise, J. & Tomsick, J. A. 2003, *ApJ*, 598, 481.
Kaminker, A. D. *et al*. 1990, *Astrophys. Space Sci.*, **173**, 171.
Kaptein, R. G. *et al*. 2000, *A&A*, **358**, L71.
Kerkwijk, M. H. van *et al*. 2004, *ApJ*, **608**, 432.
Klis, M. van der *et al*. 1990, *ApJ*, **360**, L19.
Kluzniak, W. & Wagoner, R. V. 1985, *ApJ*, **297**, 548.
Kluzniak, W. & Wilson, J. R. 1991, *ApJ*, **372**, 87.
Kluzniak, W., Michelson, P. & Wagoner, R. V. 1990, *ApJ*, **358**, 538.
Koike, O., Hashimoto, M., Arai, K. & Wanajo, S. 1999, *A&A*, **342**, 464.
Kong, A. K. H. *et al*. 2000, *MNRAS*, **311**, 405.
Kouveliotou, C. *et al*. 1996, *Nature*, **379**, 799.
Kuulkers, E. 2002, *A&A*, **383**, L5.
Kuulkers, E. & van der Klis, M. 2000, *A&A*, **356**, L45.
Kuulkers, E., den Hartog, P. R., in 't Zand, J. J. M. *et al*. 2003, *A&A*, **399**, 663.
Kuulkers, E., Homan, J., van der Klis, M., Lewin, W. H. G. & Mendez, M. 2002a, *A&A*, **382**, 947.
Kuulkers, E. *et al*. 2002b, *A&A*, **382**, 503.
Lamb, D. Q. & Lamb, F. K. 1978, *ApJ*, **220**, 291L.
Lamb, F. K. & Miller, M. C. 2001, *ApJ*, **554**, 1210.
Langmeier, A. *et al*. 1987, *ApJ*, **323**, 288.
Lapidus, I., Nobili, L. & Turolla, R. 1994, *ApJ*, **431**, L103.
Lattimer, J. M. & Prakash, M. 2001, *ApJ*, **550**, 426.
Lee, U. 2004, *ApJ*, **600**, 914.
Lee, U. & Strohmayer, T. E., 2005, *MNRAS*, **361**, 659.
Levine, A. M. *et al*. 1996, *ApJ*, **469**, L33.

Lewin, W. H. G. 1982, in *Accreting Neutron Stars*, eds. W. Brinkman & J. Trümper, MPE Report 177, ISSN 0340-8922, 176.
Lewin, W. H. G., Vacca, W. D. & Basinska, E. M. 1984, *ApJ*, **277**, L57.
Lewin, W. H. G., van Paradijs, J. & Taam, R. E. 1993, *Space Sci. Rev.*, **62**, 223.
Liu, Q. Z., van Paradijs, J. & van den Heuvel, E. P. J. 2001, *A&A*, **368**, 1021.
Livio, M. & Bath, G. T. 1982, *A&A*, **116**, 286.
London, R. A., Howard, W. M. & Taam, R. E. 1984, *ApJ*, **287**, L27.
1986, *ApJ*, **306**, 170.
Madej, J. 1991, *ApJ*, **376**, 161
Magnier, E. *et al.* 1989, *MNRAS*, **237**, 729.
Maraschi, L. & Cavaliere, A. 1977, in *Highlights in Astronomy*, ed. E. A Müller: (Dordrecht: Reidel), Vol. 4, Part I, 127.
Markwardt, C. B. & Swank, J. H. 2003, *IAUC*, 8144.
Markwardt, C. B., Strohmayer, T. E., & Swank, J. H. 1999, *ApJ*, **512**, L125.
Markwardt, C. B., Miller, J. M. & Wijnands, R. 2002, *IAUC*, 7993.
Markwardt, C. B., Strohmayer, T. E. & Swank, J. H. 2003, *The Astronomer's Telegram*, 164.
Markwardt, C. B., Smith, E. & Swank, J. H. 2003, *IAUC*, **8080**, 2.
Markwardt, C. B., Galloway, D. K., Chakrabarty, D., Morgan, E. H. & Strohmayer, T. E. 2004, *The Astronomer's Telegram*, **360**, 1.
Marshall, H. L. 1982, *ApJ*, **260**, 815.
Mason, K. O., Middleditch, J., Nelson, J. E. & White, N. E. 1980, *Nature*, **287**, 516.
McDermott, P. N. and Taam, R. E. 1987, *ApJ*, **318**, 278.
Mendez, M. & van der Klis, M. 1999, *ApJ*, **517**, L51.
Mendez, M., van der Klis, M. & van Paradijs, J. 1998, *ApJ*, **506**, L117.
Mereghetti, S., De Luca, A., Caraveo, P. A., *et al.* 2002, *ApJ*, **581**, 1280.
Miller, M. C. 1999, *ApJ*, **515**, L77.
2000, *ApJ*, **531**, 458.
Miller, M. C. & Lamb, F. K. 1998, *ApJ*, **499**, L37.
Miller, M. C., Lamb, F. K. & Psaltis, D. 1998, *ApJ*, **508**, 791.
Morgan, E. H., Kaaret, P. & Vanderspek, R. 2005, *The Astronomer's Telegram*, **523**, 1.
Muno, M. P., Fox, D. W., Morgan, E. H. & Bildsten, L. 2000, *ApJ*, **542**, 1016.
Muno, M. P., Chakrabarty, D., Galloway, D. K. & Savov, P. 2001, *ApJ*, **553**, L157.
Muno, M. P., Özel, F. & Chakrabarty, D. 2002, *ApJ*, **581**, 550.
Muno, M. P., Chakrabarty, D., Galloway, D. K. & Psaltis, D. 2002, *ApJ*, **580**, 1048.
Murakami, T., Inoue, H., Makishima, K. & Hoshi, R. 1987, *Publ. Astron. Soc. Japan*, **39**, 879.
Nakamura, N., Inoue, H. & Tanaka, Y. 1988, *Publ. Astron. Soc. Japan*, **40**, 209.
Narayan, R. & Heyl, J. S. 2002, *ApJ*, **574**, L139.
Nath, N. R., Strohmayer, T. E. & Swank, J. H. 2002, *ApJ*, **564**, 353.
Nayakshin, S. & Kallman, T. R. 2001, *ApJ*, **546**, 406.
Nozakura, T., Ikeuchi, S. & Fujimoto, M. Y. 1984, *ApJ*, **286**, 221.
Özel, F. & Psaltis, D. 2003, *ApJ*, **582**, L31.
Paczynski, B. 1983, *ApJ*, **264**, 282.
Paczynski, B. & Anderson, N. 1986, *ApJ*, **302**, 1.
Paradijs, J. van 1978, *Nature*, **274**, 650.
1981, *A&A*, **101**, 174.
Paradijs, J. van & Lewin, W. H. G. 1987, *A&A*, **172**, L20.
Paradijs, J. van, Penninx, W. & Lewin, W. H. G. 1988, *MNRAS*, **233**, 437.
Paradijs, J. van *et al.* 1990, *Publ. Astron. Soc. Japan*, **42**, 633.
Pavlov, G. G. *et al.* 2001, *ApJ*, **552**, L129.
Pechenick, K. R., Ftaclas, C. & Cohen, J. M. 1983, *ApJ*, **274**, 846.
Pinto, P. A., Taam, R. E. & Laming, J. M. 1991, *BAAS*, **23**, 1321.
Piro, A. L. & Bildsten, L. 2004, *ApJ*, **603**, 252.
2005a, *ApJ*, **619**, 1054.
2005b, *ApJ*, **629**, 438.
Rappaport, S., Joss, P. C. & Webbink, R. F. 1982, *ApJ*, **254**, 616.
Robinson, E. L. & Young, P. 1997, *ApJ*, **491**, L89.
Ross, R. R., Fabian, A. C. & Young, A. J. 1999, *MNRAS*, **306**, 461.
Rutledge, R. E., Bildsten, L., Brown, E. F., Pavlov, G. G. & Zavlin, V. E. 2001, *ApJ*, **577**, 405.

Sadeh, D. *et al.* 1982, *ApJ*, **257**, 214.
Sanwal, D., Pavlov, G. G., Zavlin, V. E. & Teter, M. A. 2002, *ApJ*, **574**, L61.
Schatz, H. *et al.* 1998, *Phys. Reports*, **294**, 167.
Schatz, H., Bildsten, L., Cumming, A. and Wiescher, M. 1999, *ApJ*, **524**, 1014.
Schatz, H. *et al.* 2001, *Phys. Rev. Lett.*, **86**, Number 16, 3471.
Schatz, H., Bildsten, L. & Cumming, A. 2003, *ApJ*, **583**, L87.
Schoelkopf, R. J. & Kelley, R. L. 1991, *ApJ*, **375**, 696.
Schwarzschild, M. & Härm, R. 1965, *ApJ*, **142**, 855.
Shaposhnikov, N. & Titarchuk, L. 2002, *ApJ*, **567**, 1077.
Shaposhnikov, N., Titarchuk, L. & Haberl, F. 2003, *ApJ*, **593**, L35.
Shara, M. M. 1982, *ApJ*, **261**, 649.
Smale, A. P. 1998, *ApJ*, **498**, L141.
2001, *ApJ*, **562**, 957.
Smith, D., Morgan, E. H. & Bradt, H. V. 1997, *ApJ*, **479**, L137.
Spitkovsky, A., Levin, Y. & Ushomirsky, G. 2002, *ApJ*, **566**, 1018.
Straaten, S. van *et al.* 2001, *ApJ*, **551**, 907.
Strohmayer, T. E. 1999, *ApJ*, **523**, L51.
2000, AAS/High Energy Astrophysics Division, 32.
2001, *Adv. Space Res.*, **28**, 511.
Strohmayer, T. E. & Brown, E. F. 2002, *ApJ*, **566**, 1045.
Strohmayer, T. E. & Lee, U. 1996, *ApJ*, **467**, 773.
Strohmayer, T. E. & Markwardt, C. B. 1999, *ApJ*, **516**, L81.
2002, *ApJ*, **577**, 337.
Strohmayer, T. E. *et al.* 1996, *ApJ*, **469**, L9.
Strohmayer, T. E., Jahoda, K., Giles, A. B. & Lee, U. 1997, *ApJ*, **486**, 355.
Strohmayer, T. E., Zhang, W. & Swank, J. H. 1997, *ApJ*, **487**, L77.
Strohmayer, T. E., Swank, J. H. & Zhang, W. 1998, *Nuclear Phys B* (Proc. Suppl.), **69/1-3**, 129.
Strohmayer, T. E. *et al.* 1998a, *ApJ*, **498**, L135.
1998b, *ApJ*, **503**, L147.
Strohmayer, T. E., Markwardt, C. B., Swank, J. H., & in't Zand, J. 2003, *ApJ*, **596**, L67.
Swank, J. H. *et al.* 1977, *ApJ*, **212**, L73.
Sztajno, M. *et al.* 1987, *MNRAS*, **226**, 39.
Taam, R. E. & Picklum, R. E. 1978, *ApJ*, **224**, 210.
Taam, R. E., Woosley, S. E. & Lamb, D. Q. 1996, *ApJ*, **459**, 271.
Titarchuk, L. 1994, *ApJ*, **429**, 340.
Titarchuk, L. & Shaposhnikov, N. 2002, *ApJ*, **570**, L25.
Ubertini, P. *et al.* 1999, *ApJ*, **514**, L27.
Villarreal, A. R. & Strohmayer, T. E. 2004, *ApJ*, **614**, L121.
Wagoner, R. V. 1984, *ApJ*, **278**, 345.
Waki, I. *et al.* 1984, *Publ. Astron. Soc. Japan*, **36**, 819.
Wallace, R. K. & Woosley, S. E. 1981, *ApJS* **43**, 389.
1984, in *High Energy Transients in Astrophysics*, ed. S. E. Woosley (New York: AIP), p. 273.
Wallace, R. K. Woosley, S. E. & Weaver, T. A. 1982, *ApJ* **258**, 696.
Walter, F. M. & Lattimer, J. M. 2002, *ApJ*, **576**, L145.
Weinberg, N., Miller, M. C. & Lamb, D. Q. 2001, *ApJ*, **546**, 1098.
Wijnands, R. 2001, *ApJ*, **554**, L59.
Wijnands, R. & van der Klis, M. 1997, *ApJ*, **482**, L65.
1998, *Nature*, **394**, 344.
Wijnands, R., Strohmayer, T. E. & Franco, L. M. 2001, *ApJ*, **549**, L71.
Wijnands, R. *et al.* 2002, private communication.
2002, *ApJ*, **566**, 1060.
2003, *Nature*, **424**, 44.
Woosley, S. E. & Taam, R. E. 1976, *Nature*, **263**, 101.
Woosley, S. E. *et al.* 2004, *ApJS*, **151**, 75.
Yu, W., Li, T. P., Zhang, W. & Zhang, S. N. 1999, *ApJ*, **512**, L35.
Zhang, W. *et al.* 1998, *ApJ*, **495**, L9.
Zingale, M. *et al.* 2001, *ApJS*, **133**, 195.

4

Black hole binaries

Jeffrey E. McClintock
Harvard–Smithsonian Center for Astrophysics

Ronald A. Remillard
Center for Space Research, MIT

4.1 Introduction

4.1.1 Scope of this review

We focus on 18 black holes with measured masses that are located in X-ray binary systems. These black holes are the most visible representatives of an estimated ~300 million stellar-mass black holes that are believed to exist in the Galaxy (van den Heuvel 1992; Brown & Bethe 1994; Timmes *et al.* 1996; Agol *et al.* 2002). Thus the mass of this particular form of dark matter, assuming $\sim 10\,M_\odot$ per black hole, is ~4% of the total baryonic mass (i.e., stars plus gas) of the Galaxy (Bahcall 1986; Bronfman *et al.* 1988). Collectively this vast population of black holes outweighs the galactic-center black hole, SgrA*, by a factor of ~1000. These stellar-mass black holes are important to astronomy in numerous ways. For example, they are one endpoint of stellar evolution for massive stars, and the collapse of their progenitor stars enriches the Universe with heavy elements (Woosley *et al.* 2002). Also, the measured mass distribution for even the small sample of 18 black holes featured here is used to constrain models of black hole formation and binary evolution (Brown *et al.* 2000a; Fryer & Kalogera 2001; Nelemans & van den Heuvel 2001). Lastly, some black hole binaries appear to be linked to the hypernovae believed to power gamma-ray bursts (Israelian *et al.* 1999; Brown *et al.* 2000b; Orosz *et al.* 2001).

This review concentrates on the X-ray timing and spectral properties of these 18 black holes, plus a number of black hole candidates, with an eye to their importance to physics as potential sites for tests of general relativity (GR) in the strongest possible gravitational fields. There are several current areas of research that probe phenomena in these systems that are believed to occur very near the event horizon. These X-ray phenomena include quasi-periodic oscillations (QPOs) at high frequency (40–450 Hz) observed from seven systems, relativistically broadened iron lines from the inner accretion disk, and thermal disk emission from near the innermost stable circular orbit allowed by GR. We also comment on evidence for the existence of the event horizon, which is based on a comparison of black-hole and neutron-star binaries and on models for advective accretion flows.

The black hole binaries featured here are mass-exchange binaries that contain an accreting black hole primary and a non-degenerate secondary star. They comprise about 10% of all bright X-ray binaries. For background on X-ray binaries, see Chapter 1, and references therein. For comprehensive reviews on black hole binaries, see Tanaka & Lewin (1995; hereafter TL95) and Tanaka & Shibazaki (1996; hereafter TS96). In this review, we emphasize the results of the past decade. Throughout we make extensive use of the extraordinary database amassed since January 1996 by NASA's Rossi X-ray Timing Explorer (Swank 1998). Due

Compact Stellar X-Ray Sources, eds. Walter Lewin and Michiel van der Klis.
Published by Cambridge University Press.

to space constraints several important topics are omitted such as X-ray reflection studies (Done & Nayakshin 2001), the estimated number of black-hole X-ray novae in the Galaxy (which is ~1000; van den Heuvel 1992; TS96; Romani 1998), and the present and projected rates of discovery of black hole binaries.

The main elements of the review are organized as follows. In the remainder of this section we catalog a total of 40 black hole binaries and candidate black hole binaries and discuss several introductory subjects, notably the physics of accretion onto black holes. In Section 4.2 we present X-ray light curves for 20 systems. In Section 4.3 we provide modified definitions of the canonical states of black hole binaries. Both low- and high-frequency quasi-periodic oscillations are discussed in Section 4.4. A recurring theme of the review is the importance of these black holes as potential sites for tests of general relativity.

4.1.2 The 18 black hole binaries

The first black hole binary (BHB), Cygnus X-1, was established by Webster and Murdin (1972) and Bolton (1972); the second, LMC X-3, was identified by Cowley *et al.* (1983). Both of these systems are persistently bright in X-rays; furthermore, they are classified as high-mass X-ray binaries (HMXBs) because their secondaries are massive O/B stars (White *et al.* 1995). The third BHB to be established, A0620–00, is markedly different (McClintock & Remillard 1986). A0620–00 was discovered as an X-ray nova in 1975 when it suddenly brightened to an intensity of 50 Crab[1] to become the brightest non-solar X-ray source ever observed (Elvis *et al.* 1975). Then, over the course of a year, the source decayed back into quiescence to become a feeble (1 μCrab) source (McClintock *et al.* 1995). Similarly, the optical counterpart faded from outburst maximum by $\Delta V \approx 7.1$ mag to $V \approx 18.3$ in quiescence, thereby revealing the optical spectrum of a K-dwarf secondary.

During the past 20 years, black holes (BHs) have been established in 15 additional X-ray binaries. Remarkably, nearly all these systems are X-ray novae like A0620-00. Thus, in all there are now 18 confirmed BHBs. They are listed in Table 4.1 in order of right ascension. The coordinate name of each source is followed in the second column by its variable star name (or other name, such as Cyg X-1), which is useful for web-based literature searches. For X-ray novae, the third column gives the year of discovery and the number of outbursts that have been observed. As indicated in the table, among the confirmed BHBs, there are six recurrent X-ray novae and three persistent sources, Cyg X-1, LMC X-3 and LMC X-1. As indicated in the fourth column, these latter three sources are high-mass X-ray binaries and are the only truly persistent sources among the BHBs.

Two of the X-ray novae are peculiar: GRS 1915+105 has remained bright for more than a decade since its eruption in August 1992; GX339–4 (1659–487), which was discovered by Markert *et al.* in 1973, undergoes frequent outbursts followed by very faint states, but it has never been observed to reach the *quiescent* state (Hynes *et al.* 2003). Columns 5 and 6 give the peak X-ray flux and a distance estimate for each source. The orbital period is given in the seventh column (a colon denotes an uncertain value) and the spectral type in the eighth. The first one or two references listed for each binary contain information on the orbital period and the spectral type. The remaining references support the distance estimates.

[1] 1 Crab = 2.6×10^{-9} erg cm^{-2} s^{-1} keV^{-1} (averaged over 2–11 keV) = 1.06 mJy at 5.2 keV for a Crab-like spectrum with photon index $\Gamma = 2.1$.

Table 4.1. *Confirmed black hole binaries: primary properties*

Source	Alternative name[a]	Year[b]	Type[c]	FX,max (μJy[d])	D (kpc)	P_{orb} (hr)	Spec.	References
0422+32	V518 Per	1992/1	L,T	3000	2.6 ± 0.7	5.1	M2V	1,2
0538–641	LMC X-3	–	H,P	60	50 ± 2.3	40.9	B3V	3,4
0540–697	LMC X-1	–	H,P	30	50 ± 2.3	101.5	O7III	3,5,6
0620–003	V616 Mon	1975/2	L,T	50000	1.2 ± 0.1	7.8	K4V	7,8,9,10
1009–45	MM Vel	1993/1	L,T	800	5.0 ± 1.3	6.8	K7/M0V	11,12
1118+480	KV UMa	2000/1	L,T	40	1.8 ± 0.5	4.1	K5/M0V	13,14
1124–684[e]	GU Mus	1991/1	L,T	3000	5 ± 1.3	10.4	K3/K5V	15,15a,16
1543–475	IL Lupi	1971/4	L,T	15000	7.5 ± 0.5	26.8	A2V	17,18
1550–564	V381 Nor	1998/5	L,T	7000	5.3 ± 2.3	37.0	G8/K8IV	19
1655–40	V1033 Sco	1994/2	L,T	3900	3.2 ± 0.2	62.9	F3/F5IV	20,21,22
1659–487[f]	V821 Ara	1972/[f]	L,T	1100	4	42.1:	–	23,24
1705–250	V2107 Oph	1977/1	L,T	3600	8 ± 2	12.5	K3/7V	7,25,26
1819.3–2525	V4641 Sgr	1999/1	L,T	13000	7.4–12.3	67.6	B9III	27
1859+226	V406 Vul	1999/1	L,T	1500	11	9.2:	–	28,29
1915+105	V1487 Aql	1992/1	L,T	3700	11–12	804.0	K/MIII	30,31,32,33
1956+350	Cyg X-1	–	H,P	2300	2.0 ± 0.1	134.4	O9.7Iab	34,35
2000+251	QZ Vul	1988/1	L,T	11000	2.7 ± 0.7	8.3	K3/K7V	35a,7,36,37
2023+338	V404 Cyg	1989/3	L,T	20000	2.2–3.7	155.3	K0III	38,39,40

[a] Name recognized by the SIMBAD Database and the Astrophysics Data System (ADS).
[b] Year of discovery/number of outbursts observed (Chen *et al.* 1997; this work).
[c] 'H' – HMXB, 'L' – LMXB, 'T' – transient, 'P' – persistent; Liu *et al.* 2000, 2001; this work.
[d] $1\ \mu\text{Jy} = 10^{-29}$ erg cm^{-2} s^{-1} Hz^{-1} $= 2.42 \times 10^{-12}$erg cm^{-2} s^{-1} keV^{-1}.
[e] Commonly known as Nova Muscae 1991.
[f] Commonly known as GX339–4; number of outbursts ~ 10 (Kong *et al.* 2002). References: [1]Esin *et al.* 1997; [2]Filippenko *et al.* 1995a; [3]Freedman *et al.* 2001; [4]Cowley *et al.* 1983; [5]Hutchings *et al.* 1987; [6]Cowley *et al.* 1995; [7]Barret *et al* . 1996b; [8]Gelino *et al.* 2001b; [9]Marsh *et al.* 1994; [10]McClintock & Remillard 2000; [11]Barret *et al.* 2000; [12]Filippenko *et al.* 1999; [13]McClintock *et al.* 2001a; [14]Wagner *et al.* 2001; [15]Orosz *et al.* 1996; [15a]Gelino *et al.* 2001a; [16]Shahbaz *et al.* 1997; [17]Orosz *et al.* 2002b; [18]Orosz *et al.* 1998; [19]Orosz *et al.* 2002a; [20]Hjellming & Rupen 1995; [21]Orosz & Bailyn 1997; [22]Shahbaz *et al.* 1999; [23]Cowley *et al.* 1987; [24]Hynes *et al.* 2003; [25]Remillard *et al.* 1996; [26]Filippenko *et al.* 1997; [27]Orosz *et al.* 2001; [28]Zurita *et al.* 2002; [29]Filippenko & Chornock 2001; [30]Mirabel & Rodriguez 1994; [31]Fender *et al.* 1999b; [32]Greiner *et al.* 2001a; [33]Greiner *et al.* 2001b; [34]Mirabel & Rodrigues 2003; [35]Gies & Bolton 1982; [35a]Harlaftis *et al.* 1996; [36]Filippenko *et al.* 1995b; [37]Casares *et al.* 1995; [38]Shahbaz *et al.* 1994; [39]Casares & Charles 1994; [40]Casares *et al.* 1993.

The data in Table 4.1 reveal considerable diversity among the BHBs. For example, these 18 binaries range in size from tiny XTE J1118+480 with $P_{orb} = 0.17$ days and a separation between the BH and its companion of $a \approx 2.8\ R_\odot$ to GRS 1915+105 with $P_{orb} = 33.5$ days and $a \approx 95\ R_\odot$. Only 6 of these 18 systems were established as BHBs a decade ago (van Paradijs & McClintock 1995). As indicated in Table 4.1, all of the 15 X-ray novae are low-mass X-ray binaries (LMXBs), which typically contain a secondary with a mass of roughly $1\ M_\odot$ or less (White *et al.* 1995). The BHBs 4U1543–47 and SAX J1819.3–2525 have relatively massive secondaries: $2.7 \pm 1.0\ M_\odot$ and $2.9 \pm 0.2\ M_\odot$, respectively (Orosz *et al.*

2002b). We classify them as LMXBs because their secondary masses are comparable to the mass of the secondary of Her X-1 (2.3 $\pm$ 0.3 $M_\odot$; Reynolds *et al.* 1997), which is a well-known LMXB (Liu *et al.* 2001). Furthermore, a 2–3 $M_\odot$ secondary is much less massive than the O/B secondaries ($\gtrsim$10 $M_\odot$) found in HMXB systems.

BHBs manifest themselves in five rather distinct spectral/temporal states defined in the 1–10 keV band (e.g., van der Klis 1994; TL95; TS96). The three most familiar are (1) the *high/soft* (HS) state, a high-intensity state dominated by thermal emission from an accretion disk; (2) the *low/hard* (LH) state, a low-intensity state dominated by power-law emission and rapid variability; and (3) the *quiescent* state, an extraordinarily faint state also dominated by power-law emission. The remaining two states, (4) the *very high* (VH) state and (5) the *intermediate* state, are more complex; recently they have come to the fore, as they have now been observed in an appreciable number of sources. These five BH states and the transitions between them are the focus of Section 4.3, where they are redefined and illustrated in detail.

Additional data specific to the BH primaries are contained in Table 4.2. Of special importance is the mass function, $f(M) \equiv P_{\rm orb} K_2^3/2\pi G = M_1 \sin^3 i/(1+q)^2$. The observables on the left side of the equation are the orbital period, $P_{\rm orb}$, and the half-amplitude of the velocity curve of the secondary, K_2. On the right, the quantity of most interest is the BH mass, M_1; the other parameters are the orbital inclination angle, i, and the mass ratio, $q \equiv M_2/M_1$, where M_2 is the mass of the secondary. Thus a secure value of the mass function can be determined for a quiescent X-ray nova or an HMXB by simply measuring the radial velocity curve of the secondary star. The mass function values are given in the second column of Table 4.2 An inspection of the equation for $f(M)$ shows that the value of the mass function is the absolute minimum mass of the compact primary. Thus, for 12 of the 18 BHBs, the very secure value of $f(M)$ alone is sufficient to show that the mass of the compact X-ray source is at least 3 $M_\odot$ (Table 4.2), which is widely agreed to exceed the maximum stable mass of a neutron star (NS) in GR (Rhoades & Ruffini 1974; Kalogera & Baym 1996). For the remaining half-dozen systems, some additional data are required to make the case for a BH. The evidence for BHs in these 18 systems is generally very strong (see Chapter 5). Thus, assuming that GR is valid in the strong-field limit, we choose to refer to these compact primaries as BHs, rather than as BH candidates. We note, however, our reservations about three of the systems listed in Table 4.2: (1) The mass function of LMC X–1 (0540–697) is quite uncertain; (2) the orbital period of XTE J1859+226 is not firmly established (Filippenko & Chornock 2001; Zurita *et al.* 2002); and (3) the dynamical data for GX 339–4 (1659–487) were determined in outburst by a novel technique, and neither the orbital period nor the velocity amplitude are securely determined (Hynes *et al.* 2003).

The mass of a BH can be derived from the measurement of its mass function in combination with estimates for M_2 and $\sin i$. The ratio of the mass of the BH to the mass of the secondary star is usually deduced by measuring the rotational velocity of the latter. The binary inclination angle can be constrained in several ways; commonly, one models the photometric variations associated with the gravitationally distorted secondary star that is seen to rotate once per binary orbit in the plane of the sky (e.g., Greene *et al.* 2001). Mass estimates for the known BHBs are given in the third column of Table 4.2. In an astrophysical environment, a BH is completely specified in general relativity by two numbers, its mass and its specific angular

Table 4.2. *Confirmed black hole binaries: X-ray and optical data*

Source	$f(M)^a$ ($M_\odot$)	M_1^a ($M_\odot$)	f(HFQPO) (Hz)	f(LFQPO) (Hz)	Radio[b]	E^c_{max} (MeV)	References
0422+32	1.19±0.02	3.2–13.2	–	0.035–32	P	0.8, 1–2:	1,2,3,4,5
0538–641	2.3±0.3	5.9–9.2	–	0.46	–	0.05	6,7
0540–697	0.14±0.05	4.0–10.0:	–	0.075	–	0.02	8,7
0620–003	2.72±0.06	3.3–12.9	–	–	P,J?	0.03:	9,10,11,11a
1009–45	3.17±0.12	6.3–8.0	–	0.04–0.3	–[d]	0.40, 1:	12,4,13
1118+480	6.1±0.3	6.5–7.2	–	0.07–0.15	P	0.15	14,15,16,17
1124–684	3.01±0.15	6.5–8.2	–	3.0–8.4	P	0.50	18,19,20,21
1543–475	0.25±0.01	7.4–11.4[e]	–	7	–[f]	0.20	22,4
1550–564	6.86±0.71	8.4–10.8	92,184,276	0.1–10	P,J	0.20	23,24,25,26,27
1655–40	2.73±0.09	6.0–6.6	300,450	0.1–28	P,J	0.80	28,29,30,31,54
1659–487	> 2.0[g]	–	–	0.09–7.4	P	0.45, 1:	32,33,4,13
1705–250	4.86±0.13	5.6–8.3	–	–	–[d]	0.1	34,35
1819.3–2525	3.13±0.13	6.8–7.4	–	–	P,J	0.02	36,37
1859+226	7.4±1.1	7.6–12:	190	0.5–10	P,J?	0.2	38,39,40,41
1915+105	9.5±3.0	10.0–18.0:	41,67,113,168	0.001–10	P,J	0.5, 1:	42,43,44,4,13
1956+350	0.244±0.005	6.9–13.2	–	0.035–12	P,J	2–5	45,46,47,48,49
2000+251	5.01±0.12	7.1–7.8	–	2.4–2.6	P	0.3	18,50,51
2023+338	6.08±0.06	10.1–13.4	–	–	P	0.4	52,53

[a] Orosz *et al.* 2002b, except for 1659–487; colon denotes uncertain value.
[b] Radio properties: 'P' – persistent over 10 or more days and/or inverted spectrum; 'J' – relativistic jet detected.
[c] Maximum energy reported; colon denotes uncertain value.
[d] No observations made.
[e] Orosz, private communication.
[f] Very faint (e.g., see IAUC 7925).
[g] For preferred period, $P = 1.76$ days, $f(M) = 5.8 \pm 0.5\ M_\odot$; Hynes *et al.* 2003.

References: [1]van der Hooft *et al.* 1999; [2]Vikhlinin *et al.* 1992; [3]Shrader *et al.* 1994; [4]Grove *et al.* 1998; [5]van Dijk *et al.* 1995; [6]Boyd *et al.* 2000; [7]Nowak *et al.* 2001; [8]Ebisawa *et al.* 1989; [9]Owen *et al.* 1976; [10]Kuulkers *et al.* 1999; [11]Coe *et al.* 1976; [11a] Marsh *et al.* 1994; [12]van der Hooft *et al.* 1996; [13]Ling *et al.* 2000; [14]Wood *et al.* 2000; [15]Revnivtsev *et al.* 2000b; [16]Fender *et al.* 2001; [17]McClintock *et al.* 2001a; [18]Rutledge *et al.* 1999; [19]Belloni *et al.* 1997; [20]Ball *et al.* 1995; [21]Sunyaev *et al.* 1992; [22]This work; [23]Remillard *et al.* 2002b; [24]Corbel *et al.* 2001; [25]Wu *et al.* 2002; [26]Corbel *et al.* 2003; [27]Sobczak *et al.* 2000b; [28]Remillard *et al.* 1999; [29]Strohmayer 2001a; [30]Hjellming & Rupen 1995; [31]Hannikainen *et al.* 2000; [32]Revnivtsev *et al.* 2001; [33]Corbel *et al.* 2000; [34]Wilson & Rothschild 1983; [35]Cooke *et al.* 1984; [36]Hjellming *et al.* 2000; [37]Wijnands & van der Klis 2000; [38]Cui *et al.* 2000a; [39]Markwardt 2001; [40]Brocksopp *et al.* 2002; [41]Dal Fiume *et al.* 1999; [42]Morgan *et al.* 1997; [43]Strohmayer 2001b; [44]Mirabel & Rodriguez 1994; [45]Vikhlinin *et al.* 1994; [46]Cui *et al.* 1997b; [47]Stirling *et al.* 2001; [48]Ling *et al.* 1987; [49]McConnell *et al.* 2002; [50]Hjellming *et al.* 1988; [51]Sunyaev *et al.* 1988; [52]Han & Hjellming 1992; [53]Sunyaev *et al.* 1991b; [54]Tomsick *et al.* 1999.

momentum or spin, $a = J/cM_1$, where J is the BH angular momentum and c is the speed of light (e.g., Kato *et al.* 1998). The spin value is conveniently expressed in terms of a dimensionless spin parameter, $a_* = a/r_g$, where the gravitational radius is $r_g \equiv GM/c^2$. The value of a_* lies between 0 for a Schwarzschild hole and 1 for a maximally rotating Kerr hole. The radius of the BH event horizon depends on both M_1 and a_*; this topic is discussed further in Section 4.1.5.

High-frequency QPOs (HFQPOs) in the range 40–450 Hz are observed for the four systems with data given in the fourth column. Low-frequency QPOs (LFQPOs) are also observed from these systems and ten others, as indicated in column 5. Most of the systems were at one time or another detected as radio sources and at least five have exhibited resolved radio jets (column 6; see Mirabel & Rodriguez 1999). An X-ray jet (two-sided) has been observed from XTE J1550–564 (Corbel *et al.* 2002). As shown in column 7, power-law emission extending to $\sim$1 MeV has been observed for six sources. The references listed in the last column support the X-ray QPO, radio, and $E_{\max}$ data given in columns 4–7.

The dynamical evidence for massive ($M > 3$ $M_{\odot}$) collapsed stellar remnants is indisputable. However, this evidence alone can never establish the existence of BHs. At present, the argument for BHs depends on the assumption that GR is the correct theory of strong gravity. To make an airtight case that a compact object is a BH with an event horizon, we must make clean quantitative measurements of relativistic effects that occur near the collapsed object. That is, we must measure phenomena that are predicted by GR to be unique to BHs. A possible route to this summit is to measure and interpret the effects of strong-field gravity that are believed to be imprinted on both HFQPOs (Section 4.4.3) and X-ray emission lines (Section 4.2.3). Another approach, which is discussed in Section 4.3.4, is based on sensing the qualitative differences between an event horizon and the material surface of a NS.

4.1.3 Black hole candidates

Certain characteristic X-ray properties of the 18 established BHs are often used to identify a candidate BH when the radial velocities of the secondary cannot be measured (e.g., because the secondary is too faint). These frequently observed characteristics of the established BHBs, which are discussed in detail by TL95, include an ultrasoft X-ray spectrum (1–10 keV), a power-law spectral tail that extends beyond 20 keV, characteristic state transitions (Section 4.1.2; Section 4.3), and rapid temporal variability. However, none of these putative BH signatures has proved to be entirely reliable; each of them has been forged by one or more systems known to contain a NS primary (TL95). This is not surprising since the X-ray spectral/temporal properties originate in an accretion flow that is expected to be fairly similar whether the primary is a BH or a weakly magnetized NS. Another characteristic of all BHBs is a complete absence of either periodic pulsations or Type I X-ray bursts, which are the very common signatures of NS systems.

Nevertheless, despite the limitations of these spectral/temporal identifiers, they have served as useful guides and have allowed the identification of a number of probable BH primaries, which we refer to as BH candidates or BHCs. In order to economize on acronyms, we also use BHC to refer to the binary system that hosts the BH candidate. A list of 22 BHCs is given in Table 4.3 These systems are less well known than the BHBs listed in Table 4.1. Therefore, to aid in their identification we have included in most cases the prefix to the coordinate source name that identifies the discovery mission (e.g., EXO = EXOSAT). For the four sources with variable star names, the prefix is omitted. In columns 2–4, we give the best available coordinates and their uncertainties (if they exceed 10″). These coordinates are also an excellent resource for interrogating ADS and SIMBAD.

In column 5 we list the characteristics that indicate the BHC classification, e.g., an observation of one or more of the canonical states of a BHB. A source classified as "ultrasoft" in the older literature (e.g., White & Marshall 1984) is assumed here to have been observed in the HS state. Three BHCs with resolved radio jets are also noted in column 5. Variability

Table 4.3. *Candidate black hole binaries*[a]

Source	RA(2000)	DEC(2000)	r_X[b]	BH trait[c]	Grade[d]	References
1354–645 (BW Cir)	13 58 09.74	−64 44 05.2		LH,HS	A	1,2,3,4
1524–617 (KY TrA)	15 28 16.7	−61 52 58		LH,HS	A	5,6,7
4U 1630–47	16 34 01.61	−47 23 34.8		LH,HS	A	8,9,10,11
XTE J1650–500	16 50 01.0	−49 57 45		LH,HS,VH	A	12,13,14,15,16
SAX J1711.6–3808	17 11 37.1	−38 07 06		LH,HS	B	17,18
GRS 1716–249[e]	17 19 36.93	−25 01 03.4		LH	B	19,20,21
XTE J1720–318	17 19 59.06	−31 44 59.7		LH:,HS	C	22,23,24
KS 1730–312	17 33 37.6	−31 13 12	30″	LH,HS	C	25,26
GRS 1737–31	17 40 09	−31 02.4	30″	LH	B	27,28,29
GRS 1739–278	17 42 40.03	−27 44 52.7		LH,HS,VH	A	30,31,32,33,34
1E 1740.7–2942	17 43 54.88	−29 44 42.5		LH,HS,J	A	35,36,37,38,39
A 1742–289	17 45 37.3	−29 01 05		HS:	C	40,41,42,43
H 1743–322	17 46 15.61	−32 14 00.6		HS,VH	A	44,45,46,80,81,82
XTE J1748–288	17 48 05.06	−28 28 25.8		LH,HS,VH,J	A	47,48,49,50,51
SLX 1746–331	17 49 50.6	−33 11 55	35″	HS:	C	52,53,54
XTE J1755–324	17 55 28.6	−32 28 39	1′	LH,HS	B	55,56,57,58
1755–338 (V4134 Sgr)	17 58 40.0	−33 48 27		HS	B	59,42,60,61,62
GRS 1758–258	18 01 12.67	−25 44 26.7		LH,HS,J	A	63,38,64,65,66
EXO 1846–031	18 49 16.9	−03 03 53	11″[f]	HS	C	67
XTE J1908+094	19 08 53.08	+09 23 04.9		LH,HS	B	68,69,70,71
1957+115 (V1408 Aql)	19 59 24.0	+11 42 30		HS	C	72,42,73,74,75
XTE J2012+381	20 12 37.70	+38 11 01.2		LH,HS	B	76,77,78,79

[a] For additional references and information, see TL95 and Liu *et al.* 2001.
[b] Positional uncertainty given if $r_X > 10''$.
[c] LH – low/hard state, HS – high/soft state, VH – very high state, J – radio jet.
[d] Qualitative grade indicating the likelihood that the candidate is in fact a BH.
[e] GRS 1716–249 = GRO J1719–24 = X-ray Nova Oph 1993.
[f] Alternative position also possible; see Parmar *et al.* 1993. References: [1]Kitamoto *et al.* 1990; [2]Brocksopp *et al.* 2001; [3]Revnivtsev *et al.* 2000a; [4]Fender 2001; [5]Murdin *et al.* 1977; [6]Kaluzienski *et al.* 1975; [7]Barret *et al.* 1992; [8]Hjellming *et al.* 1999; [9]Tomsick & Kaaret 2000; [10]Dieters *et al.* 2000; [11]Augusteijn *et al.* 2001; [12]Groot *et al.* 2001; [13]Kalemci *et al.* 2002; [14]Homan *et al.* 2003a; [15]Miller *et al.* 2002b; [16]Sanchez–Fernandez *et al.* 2002; [17]in 't Zand *et al.* 2002a; [18]Wijnands & Miller 2002; [19]Mirabel *et al.* 1993; [20]van der Hooft *et al.* 1996; [21]Revnivtsev *et al.* 1998b; [22]Rupen *et al.* 2003; [23]Remillard *et al.* 2003a; [24]Markwardt & Swank 2003; [25]Borozdin *et al.* 1995; [26]Vargas *et al.* 1996; [27]Ueda *et al.* 1997; [28]Cui *et al.* 1997a; [29]Trudolyubov *et al.* 1999; [30]Marti *et al.* 1997; [31]Vargas *et al.* 1997; [32]Borozdin *et al.* 1998; [33]Wijnands *et al.* 2001; [34]Greiner *et al.* 1996; [35]Cui *et al.* 2001; [36]Churazov *et al.* 1993; [37]Smith *et al.* 1997; [38]Smith *et al.* 2002; [39]Marti *et al.* 2000; [40]Davies *et al.* 1976; [41]Wilson *et al.* 1977; [42]Branduardi *et al.* 1976; [43]Kennea & Skinner 1996; [44]Gursky *et al.* 1978; [45]Cooke *et al.* 1984; [46]White & Marshall 1984; [47]Hjellming *et al.* 1998b; [48]Revnivtsev *et al.* 2000c; [49]Kotani *et al.* 2000; [50]Miller *et al.* 2001; [51]Rupen *et al.*1998; [52]Skinner *et al.* 1990; [53]White & van Paradijs 1996; [54]Motch *et al.* 1998; [55]Remillard *et al.* 1997; [56]Ogley *et al.* 1997; [57]Revnivtsev *et al.* 1998a; [58]Goldoni *et al.* 1999; [59]Bradt & McClintock 1983; [60]White *et al.* 1988; [61]Pan *et al.* 1995; [62]Seon *et al.* 1995; [63]Rodriguez *et al.* 1992; [64]Sunyaev *et al.* 1991a; [65]Smith *et al.* 2001; [66]Rothstein *et al.* 2002; [67]Parmar *et al.* 1993; [68]Rupen *et al.* 2002; [69]in't Zand *et al.* 2002b; [70]Woods *et al.* 2002; [71]Chaty *et al.* 2002; [72]Margon *et al.* 1978; [73]Wijnands *et al.* 2002; [74]Yaqoob *et al.* 1993; [75]Nowak & Wilms 1999; [76]Hjellimg *et al.* 1998a; [77]Campana *et al.* 2002; [78]Vasiliev *et al.* 2000; [79]Hynes *et al.* 1999; [80]Revnivtsev *et al.* 2003; [81]Steeghs *et al.* 2003; [82]Homan *et al.* 2003b.

characteristics (Section 4.4) and broad Fe Kα emission lines (Section 4.2.3) can also be important BH indicators, but these are not explicitly noted in Table 4.3. Selected references are cited in the last column. The coordinate data can be found in the first reference. The following few references support the BH characteristics given in column 5. For several sources, a few additional references provide limited information on optical/IR/radio counterparts, etc.

We have assigned a grade to each BHC that is based both on how thoroughly the candidate has been observed and on the characteristics it displayed. In fact, this grade is qualitative and subjective and meant only as a guide. Our sense of the grade is as follows: We would be surprised if even one of the nine A-grade BHCs contains a NS, but not surprised if one of the six C-grade BHCs contains a NS. In compiling Table 4.3, we have been selective. For example, we did not include SAX J1805.5–2031 (Lowes *et al.* 2002) and XTE J1856+053 (Barret *et al.* 1996a) primarily because we judged that the available information was too scanty. Thus the total number of BH or BHC cases considered here is 40. For narrative discussions about many of the systems in Tables 4.1 and 4.3, see TL95. For additional data and as a supplement to the list of references given in Table 4.3, see Liu *et al.* (2001).

4.1.4 X-ray novae

If we include the X-ray novae observed during the past decade, then about 300 bright binary X-ray sources are known (van Paradijs 1995; Liu *et al.* 2000, 2001). More than half of them are LMXBs, and roughly half of each type (i.e., LMXB and HMXB) are classified as transient sources. The HMXB transients are neutron-star/Be-star binaries, which are not relevant to this review. The well-studied LMXB transients, on the other hand, include all of the X-ray novae that are listed in Tables 4.1–4.3. The principal hallmarks of an X-ray nova include both the discovery of the source during a violent outburst and a very large ratio of maximum to minimum X-ray intensity. The behavior of A0620–00, described in Section 4.1.2 provides a classic example of an X-ray nova. Indeed, it is the extreme faintness of the quiescent accretion disk in these systems that allows one to view the companion star, leading to secure dynamical measurements of the BH mass (Sections 4.1.2, and 4.3.4; Chapter 5).

Recurrent eruptions have been observed for several of the X-ray novae listed in Table 4.1: e.g., A0620–00 in 1917 and 1975; H1743–322 in 1977 and 2003; GS 2023+338 in 1938, 1956 and 1987; 4U 1543–47 in 1971, 1983, 1992 and 2002; and 4U 1630–472 and GX 339–4 at ~1–2 year intervals. (Outbursts prior to 1970 are inferred from studies of optical plates *ex post facto*.) Most X-ray novae, however, have been observed to erupt only once. Nevertheless, all X-ray novae are thought to be recurrent, with cycle times for some possibly as long as several centuries or more. TS96 suggest an average cycle time of 10–50 years. The infrequent outbursts are due to a sudden surge in the mass accretion rate onto the BH. The generally accepted mechanism driving the outburst cycle is that described by the disk instability model, which was developed initially for dwarf novae (Smak 1971; Osaki 1974; Cannizzo 1993; Lasota *et al.* 2001) and extended to X-ray novae (e.g., Dubus *et al.* 2001).

4.1.5 Accretion onto black holes

The desire to understand observations of BHBs compels us to model the hydrodynamics and radiation processes of gas orbiting in the gravitational potential of a compact object (see Chapter 13 for a detailed review). The best-known such model is the thin accretion disk (Pringle & Rees 1972; Novikov & Thorne 1973; Shakura & Sunyaev 1973; Lynden-Bell & Pringle 1974). For nearly all of the systems included in Tables 4.1–4.3, the companion star fills its Roche equipotential lobe and a narrow stream of gas escapes the star through the inner Lagrangian (L_1) point. This gas has high specific angular momentum and cannot accrete directly onto the BH. It feeds into a thin disk of matter around the BH known as an accretion disk. Once entrained in the disk, the gas moves in Keplerian orbits with angular velocity $(GM/R^3)^{1/2}$. However, viscous dissipation slowly taps energy from the bulk orbital motion, and viscosity transports angular momentum outward. As a result, the gas gets hotter as it sinks deeper into the gravitational potential well of the BH. Near the BH the disk terminates because there are no stable particle orbits possible in the extreme gravitational field. The existence of an innermost stable circular orbit (ISCO) and other properties of BHs are discussed in many texts (e.g., Shapiro & Teukolsky 1983; Kato *et al.* 1998). A defining property of a BH is its event horizon, the immaterial surface that bounds the interior region of spacetime that cannot communicate with the external universe. The radius of the event horizon of a Schwarzschild BH ($a_* = 0$) is $R_S \equiv 2R_g \equiv 2(GM/c^2) = 30\,\mathrm{km}(M/10\,\mathrm{M}_\odot)$, the ISCO lies at $R_{\mathrm{ISCO}} = 6R_g$, and the corresponding maximum orbital frequency is $\nu_{\mathrm{ISCO}} = 220\,\mathrm{Hz}(M/10\,\mathrm{M}_\odot)^{-1}$ (see Section 4.1.2 for the definition of a_*). For an extreme Kerr BH ($a_* = 1$), the radii of both the event horizon and the minimum stable (prograde) orbit are identical, $R_K = R_{\mathrm{ISCO}} = R_g$, and the maximum orbital frequency is $\nu_{\mathrm{ISCO}} = 1615\ \mathrm{Hz}(M/10\ \mathrm{M}_\odot)^{-1}$. For the Kerr BH, it is well known that the rotational energy can be tapped electromagnetically (Blandford & Znajek 1977). The gas flows driven by this process are both anisotropic and self-collimating (Blandford 2002, and references therein), and they may be the source of the relativistic jets seen from several BHBs and BHCs (Tables 4.2–4.3).

Even for XTE J1118+480, the smallest system in Table 4.1, the outer radius of the accretion disk is expected to be roughly one solar radius, $\sim 10^5 R_g$, vastly larger than the BH event horizon. Thus a gas element of mass m that is destined to enter the BH via the thin disk starts far out with negligible binding energy. When it reaches the ISCO it will have radiated $0.057mc^2$ for a Schwarzschild BH or a theoretical maximum of $0.42mc^2$ for an extreme Kerr BH. Moreover, 90% of this colossal energy is radiated within about $20R_g$ of the center. At all disk radii, the energy liberated by viscous dissipation is radiated locally and promptly and results in a gas temperature that increases radially inward reaching a maximum of $T \sim 10^7$ K near the BH. This picture is the basis of the standard thin accretion disk model (Shakura & Sunyaev 1973). A non-relativistic approximation to this thin disk spectrum has been formulated conveniently as a multi-temperature blackbody (Mitsuda *et al.* 1984; Makishima *et al.* 1986). The total disk luminosity in a steady state is $L_{\mathrm{disk}} = GM\dot{M}/2R_{\mathrm{in}}$, where $\dot{M}$ is the mass accretion rate and R_{in} is the radius of the inner edge of the disk. This model, often referred to as the multicolor disk (MCD) model, is used to describe the thermal component that is dominant in the HS state and is also present in the VH state (Sections 4.2.2, 4.3.5 and 4.3.7). The MCD model has been available within XSPEC ("diskbb"; Arnaud & Dorman 2002) for many years and has been widely used. Despite the successes of the MCD model, it is important to note a significant limitation, namely, the neglect of a torque-free boundary condition at the ISCO (Gierlinski *et al.* 1999). In the MCD model, the temperature profile is simply $T(R) \propto R^{-3/4}$, which rises to a maximum at the ISCO, whereas a proper inner boundary condition produces null dissipation at the ISCO with the temperature peaking at $R > R_{\mathrm{ISCO}}$ (Pringle 1981; Gierlinski *et al.* 1999). The current MCD model requires attention and improvement.

Efforts have also been made to include relativistic corrections to the MCD model (e.g., Ebisawa *et al.* 1991; Zhang *et al.* 1997a; Gierlinski *et al.* 2001). However, quantitative analyses (e.g., a spectroscopic measurement of the inner disk radius when the source distance and inclination angle are known) will require more sophisticated models. Accretion disk models are being developed to incorporate MHD effects in the context of GR (e.g., McKinney & Gammie 2002), with additional considerations for radiation pressure and radiative transfer (Turner *et al.* 2002; Shimura & Takahara 1995). Early results show that magnetic fields may couple matter in the "plunging region" to matter at radii greater than the ISCO, and thereby extract energy from very near the horizon (Agol & Krolik 2000). In the case of a rapidly rotating Kerr hole, the spin/electromagnetic effects mentioned above may not only drive relativistic jets, but also may modify grossly the spectrum of the inner accretion disk (Wilms *et al.* 2001b; Miller *et al.* 2002b).

At lower mass accretion rates corresponding to several percent of the Eddington luminosity, a BHB usually enters the LH (i.e., *low/hard*) state and at very low accretion rates it reaches the *quiescent* state, which may be just an extreme example of the LH state. In both of these states, the spectrum of a BHB is dominated by a hard, non-thermal power-law component (photon index ~1.7; Sections 4.3.4 and 4.3.6), which cannot be accounted for by a thermal accretion disk model; it is most plausibly explained as due to Comptonization of soft photons by a hot optically thin plasma. Early models for the spectrum of the quiescent state postulated that the disk does not extend all the way down to the ISCO (Narayan 1996; Narayan *et al.* 1996). The disk is truncated at some larger radius (~100 R_S; Esin *et al.* 2001), and the interior volume is filled with a hot ($T_e \sim 100$ keV) advection-dominated accretion flow or ADAF (Narayan & Yi 1994, 1995; Narayan *et al.* 1996; Quataert & Narayan 1999). In an ADAF, most of the energy released via viscous dissipation remains in the accreting gas rather than being radiated away (as in a thin disk). The bulk of the energy is advected with the flow. Only a small fraction of the energy is radiated by the optically thin gas before the gas reaches the center. Consequently, the radiative efficiency of an ADAF (which depends on the uncertain fraction of the viscous energy that is channeled to the electrons) is expected to be only ~0.1–1%, whereas the radiative efficiency discussed above for disk accretion is definitely $\geq 5.7\%$. There is wide agreement that these radiatively inefficient flows have been observed in quiescent BHBs (Section 4.3.4) and from galactic nuclei (Baganoff *et al.* 2001; Loewenstein *et al.* 2001). However, the theoretical picture has become complex with variant models involving winds (ADIOS; Blandford & Begelman 1999) and convection (CDAFs; Igumenshchev & Abramowicz 1999; Stone *et al.* 1999; Narayan *et al.* 2000; Quataert & Gruzinov 2000).

One attempt has been made to unify four of the five states of a BHB using both the MCD and ADAF models (Esin *et al.* 1997). This approach is illustrated in Fig. 4.1, which shows how the geometry of the accretion flow changes as the mass accretion rate $\dot{m}$ varies ($\dot{m}$ is the mass accretion rate expressed in Eddington units). The scenario indicates how a BH system progresses through five distinct states of increasing $\dot{m}$ from the quiescent state to the VH state. In the three states at lower $\dot{m}$, the flow consists of two zones (disk and ADAF), as described above. For the two states of highest $\dot{m}$, the disk extends down to the ISCO. In all five states, the disk is bathed in a corona that is a seamless continuation of the ADAF. Apart from the VH state, the model treats consistently the dynamics of the accreting gas, the thermal balance of the ions and electrons in the ADAF and corona, and the radiation processes. The model

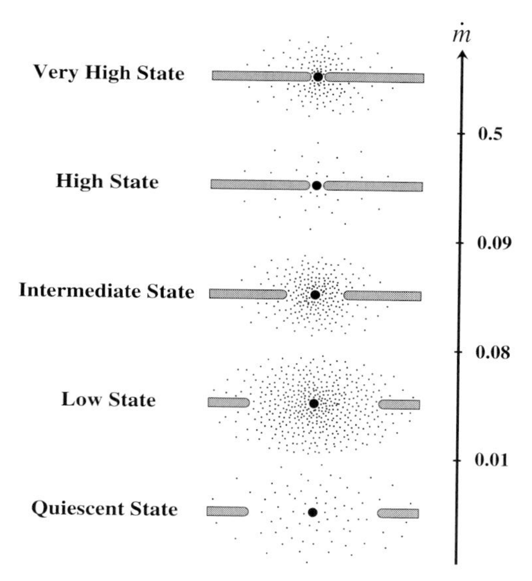

Fig. 4.1. Schematic sketch of the accretion flow in different spectral states as a function of the total Eddington-scaled mass accretion rate $\dot{m}$. The ADAF is represented by dots and the thin disk by the horizontal bars. The very high state is illustrated, but it is not included in the unification scheme (Esin *et al.* 1997).

has had significant successes in describing the spectral evolution of several BHBs (Esin *et al.* 1997, 1998; for further discussion of the ADAF model, see Section 4.3.4.)

However, the multi-state model of Esin *et al.* (1997) has important limitations. For example, it does not unify the most luminous state, the *very high* state, which is characterized by an unbroken power-law spectrum extending out to a few hundred keV or more (Section 4.3.7). Also, this simple ordering of the states by $\dot{m}$ or luminosity is naive (Section 4.3). Moreover, the model does not account for the dynamic behavior of the corona, including strong flares and powerful low-frequency quasi-periodic oscillations (Section 4.4), nor does it account for the radio emission observed from most BHBs (Table 4.2). Finally, the "evaporation" process by which the cold gas in a thin disk feeds into a hot ADAF is at best qualitatively understood, and there is no quantitative model relating the disk truncation radius to the accretion rate $\dot{m}$ (Narayan 2002, and references therein).

There are alternative models of the X-ray states, and many of them invoke a dynamic accretion disk corona that is fed by MHD instabilities in the disk. For example, in the model of Merloni and Fabian (2001a, 2001b) the hot corona that generates the power-law component is intimately connected with the thin accretion disk. Magnetic energy generated (presumably) by the sheared Keplerian disk creates magnetic flares that rise out of the disk because of the Parker instability. Within the framework of this model, Di Matteo *et al.* (1999) present a magnetic flare model for the two common states of GX339–4. In the HS state, the flares occur near the disk and heat it. The disk reradiates the observed soft thermal component, whereas the faint hard component is produced by Comptonization of the soft flux. In the LH state, the flares occur far above the disk and the density of soft seed photons is greatly reduced. Thus, the system is photon-starved, and the resultant Comptonized spectrum is hard.

Merloni and Fabian (2002) also consider coronae as sources of powerful jets/outflows. They find that such outflows can render a source radiatively inefficient even if advection of energy into the BH is unimportant.

It is generally agreed that the temperatures in the corona are in the range 100–300 keV, with optical depths of 0.1–1 (e.g., Merloni & Fabian 2001b). There is little agreement, however, on the geometry and physical properties of the corona. Thus, a wide range of coronal models have been proposed (e.g., Haardt & Maraschi 1991; Dove *et al.* 1997a, 1997b; Kawaguchi *et al.* 2000; Rozanska & Czerny 2000; Meyer *et al.* 2000; Liu *et al.* 2002; Nowak *et al.* 2002). Liu *et al.* conclude that the primary difficulty in modeling the corona is the magnetic field that produces time variations and spatial inhomogeneities; in addition, one must consider complicated radiation/energy interactions between the disk and the corona. Because of the complexity of the problem, some students of the corona choose to apply their models to large quantities of X-ray (and radio) data, an approach that has proved fruitful (Wardzinski *et al.* 2002; Zdziarski *et al.* 2002, 2003).

Timing studies have been used to isolate the fast X-ray variability in BHBs as primarily emanating from the corona rather than from the accretion disk (Churazov *et al.* 2001). Timing studies have also created requirements to explain X-ray QPOs, such as the strong low-frequency QPOs that are prevalent in the VH state. Many coronal models do not deal with fast variability or QPOs explicitly, although there are exceptions, such as the "accretion–ejection instability" model (Tagger & Pellat 1999). In this model, magnetic spiral waves heat and accelerate material at those locations where the spiral waves and the disk are in corotation. During BHB outbursts, it has become exceedingly clear that spectral and timing characteristics at both X-ray and radio frequencies may change dramatically and abruptly. Therefore, no single model can account for all of the complex behavior, and our goal in this review is to summarize what is known about each X-ray state and to then discuss physical models in the context of a given state.

In addition to the models discussed above, we mention briefly the jet model (e.g., Falcke & Biermann 1995). This model is motivated by the observations of resolved radio and X-ray jets (Tables 4.2–4.3), by observations of radio/X-ray correlations (e.g., Hannikainen *et al.* 2001; Corbel *et al.* 2000), and by successes in modeling the broadband spectra of some systems as synchrotron radiation.

4.1.6 Some consequences of an event horizon

The properties of BHs and BH accretion flows are discussed in many texts (e.g., Shapiro & Teukolsky 1983; Abramowicz 1998; Kato *et al.* 1998). As mentioned above, a defining feature of a BH is its event horizon. Since BHs lack a material surface, some effects observed for NSs (e.g., Type I bursts) are absent for a BH. Similarly, a BH cannot sustain a magnetic field anchored within it, and hence cannot generate periodic X-ray pulsations, which are observed for many NSs.

Both Type I bursts and periodic pulsations are considered firm signatures of a NS (TL95). It is interesting to ask what fraction of cataloged, bright sources show either pulsations or Type I bursts. We examined this question using the catalog of van Paradijs (1995), selecting only the brighter sources ($F_X > 30$ μJy) that have been optically identified. We excluded the confirmed and candidate BH systems listed in Tables 4.1–4.3. For the 21 HMXB systems that met our selection criteria, we found that 18 out of 21 (86%) pulse. For 21 LMXB systems, we found that 12 burst and 2 pulse (67% burst or pulse). Thus a very high fraction of these sources manifest behavior that identifies them as NSs. It is also interesting to consider which systems have failed to produce detectable bursts or pulsations because they are either an unusual NS source or they contain a BH. The three non-pulsing HMXB sources are 1700–37, 1947+30 and Cyg X-3; the six corresponding LMXB sources are LMC X-2, 1543–62, Sco X-1, GX349+2, GX9+9 and 1822–00.

4.1.7 The Rossi X-ray Timing Explorer (RXTE): 1996.0–present

RXTE has been in continuous operation since its launch on 1995 December 30. Its prime objective is to investigate the fundamental properties of BHs, NSs and white dwarfs by making high time resolution observations of extremely hot material located near BH event horizons or stellar surfaces. It is the largest X-ray detector array ever flown, and it provides energy coverage from 2 to 200 keV. It features high throughput (up to $\sim$150 000 counts s^{-1}) and $\sim$ 1 μs time resolution. RXTE's year-round, wide-sky coverage and its fast response time have made it an important vehicle for the study of transient phenomena and for the support of multiwavelength science.

The observatory comprises two large-area instruments (PCA and HEXTE) that act in concert, viewing the sky through a common 1° field of view. The third instrument is an All-Sky Monitor (ASM) that surveys about 80% of the sky each orbit. For descriptions of the instruments, see Levine *et al.* (1996), Swank (1998), Rothschild *et al.* (1998), and Bradt *et al.* (2001). The Proportional Counter Array (PCA), which has a total net area of 6250 cm^2, is the chief instrument. The High Energy X-ray Timing Experiment (HEXTE) covers the energy range 20–200 keV and has provided important spectral information beyond the reach of the PCA. However, the modest count rates of the HEXTE (e.g., 290 counts s^{-1} for the Crab) limit its use for timing studies. Indeed, it is the PCA (12,800 counts s^{-1} for the Crab with 5 PCUs) that has achieved groundbreaking results in high-energy astrophysics.

The ASM comprises three wide-field proportional counters that are mounted on a rotating boom. On a daily basis, it surveys about 90% of the sky to a sensitivity of about 10 mCrab for sources with known positions. New sources can be located down to 20 mCrab on a weekly basis with a typical position accuracy of $\sim 5'$. The ASM has proved indispensable in alerting PCA/HEXTE observers to targets of opportunity, and the ASM public archive containing the continuous lightcurves of $\sim$400 X-ray sources (both galactic and extragalactic) has been invaluable in studying the multiyear behavior of X-ray sources and in supplementing X-ray and multiwavelength (e.g., Chandra and HST) studies of individual sources. The data are also important to observers of AGN and gamma-ray bursts.

4.2 X-ray lightcurves, spectra and luminosity data

4.2.1 ASM lightcurves of BH binaries and BH candidates

To illustrate the diversity of behavior among BHBs and BHCs, we show in Figs. 4.2–4.5 the 1.5–12 keV ASM lightcurves and the (5–12 keV)/(3–5 keV) hardness ratio (HR2) for 20 of the systems listed in Tables 4.1–4.3 that were active during the past seven years. These lightcurves should be compared to the heterogeneous collection of $\sim$20 X-ray lightcurves collected by Chen *et al.* (1997). In making such comparisons, note that we use a linear intensity scale, whereas Chen *et al.* and most authors use a log scale. The lightcurves in Figs. 4.2–4.5 are ordered by RA, although we discuss them roughly in order of increasing complexity. An intensity of 1 Crab corresponds to 75.5 ASM c s^{-1}. A hardness ratio (HR2) of 0.5 (1.5) generally corresponds to the HS state (LH state). All good data are included, although the time interval for binning the counts has been tailored to the source intensity. The hardness ratio is not plotted in the absence of a detectable 5–12 keV flux. References are cited sparingly in the narrative descriptions of the lightcurves given below; see Tables 4.1–4.3 for further references. In the following, the term "classic" lightcurve refers to an outburst profile that exhibits a fast rise and an exponential decay, like those observed for several pre-RXTE BH X-ray novae, including A0620–00, GS/GRS 1124–68, GS 2000+25 and GRO J0422+32 (TL95; Chen *et al.* 1997). Such classic lightcurves often show a secondary maximum (roughly a doubling of intensity) that occurs during the decline phase about 40–100 days after the onset of the outburst (TL95; Chen *et al.* 1997).

4.2.1.1 Black hole binaries

Figure 4.2 shows the lightcurves of all six of the X-ray novae with short outburst cycles that were detected by the ASM (Tables 4.1–4.2). *4U 1543–47*: An exceptionally clean example of a classic lightcurve with an e-folding decay time of ≈14 days. It lacks a secondary maximum, presumably because the outburst is so brief. For details and references on three prior outbursts of this source, see Chen *et al.* (1997). *XTE J1859+226*: A second example of a classic lightcurve that does show a secondary maximum (at about 75 days after discovery). Note the intense variability near the primary maximum. *XTE J1118+480*: One of five X-ray novae that remained in a hard state throughout the outburst and failed to reach the HS state. Note the prominent precursor peak.

Also shown in Fig. 4.2 are the complex and similar lightcurves of two X-ray novae with long orbital periods, GRO J1655–40 and XTE J1550–564. *GRO J1655–40*: This source has undergone two outbursts since its discovery in 1994 July. Shown here is the full lightcurve of the second, 16-month outburst. The double-peaked profile is quite unlike the classic profile of 4U 1543–47. During the first maximum in 1996, the source exhibited strong flaring and intense non-thermal emission (VH state). In 1997 the source spectrum was soft and thermal except for a hard episode at the very end of the outburst (Sobczak *et al.* 1999). Note the several absorption dips in the lightcurve of this high inclination system (Kuulkers *et al.* 1998). *XTE J1550–564*: The complex profile includes two dominant peaks during 1998–9, followed several hundred days later by a smaller peak in 2000. Not shown here are three very small outbursts (LH states) that have occurred subsequently. Some unusual characteristics include the slow 10-day rise following the source's discovery on 1998 September 6, followed by the dominant X-ray flare (6.8 Crab; September 19–20), and the abrupt ~10-day decay timescale following each outburst. The source was predominantly in the VH or *intermediate* states during the first peak (1998), softening to the HS state during the second peak (1999). The spectral evolution through all of the X-ray states occurred much more rapidly during the small peak in 2000. *SAX J1819.3–2525 = V4641 Sgr*: As this extraordinary lightcurve shows, the source became active at a low level in the spring of 1999 (dashed line shows intensity × 50). Five months later it underwent a brief, violent flare during which the 1.5–12 keV intensity increased very rapidly (within 7 hours) from 1.6 to 12.2 Crab. Within two hours thereafter, the intensity declined to less than 0.05 Crab (Wijnands & van der Klis 2000, and references therein).

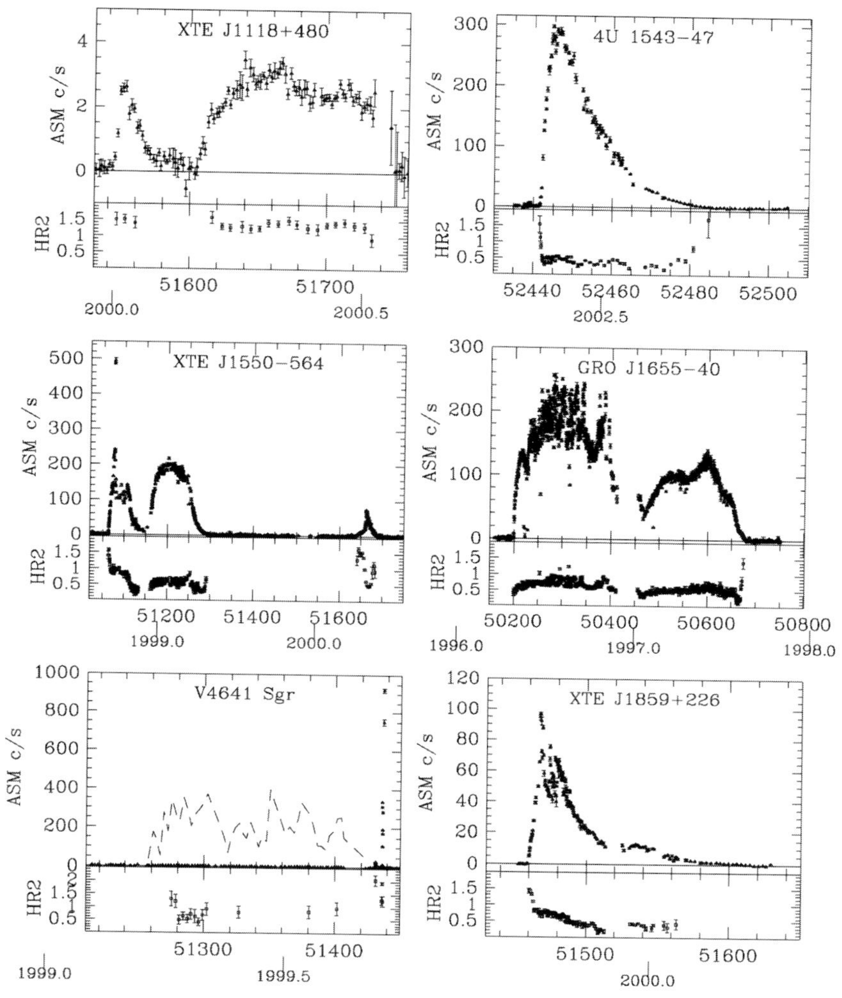

Fig. 4.2. Transient ASM lightcurves of six black hole binaries. For V4641 Sgr, the dashed line shows intensity ×50 in order to highlight the low-level activity that preceded the violent and short-lived flare.

Figure 4.3 shows the lightcurves of the two persistent BHBs (LMC X-3 and Cyg X-1) and two other BHBs that have been active throughout the RXTE era. *LMC X–3*: The lightcurve shows the large-amplitude cyclic modulation in the flux reported by Cowley *et al.* (1991); however, their ∼198 day cycle time is shorter than is indicated by these data. As the hardness ratio plot shows, the source remains in the HS state most of the time. However, at the local intensity minima, where there are gaps in the HR2 plot due to statistical limitations, the PCA observations show transitions to the LH state (Wilms *et al.* 2001a). *GX 339–4*: As shown, the source underwent major eruptions into a soft spectral state (HR ≈ 0.5) in 1998 and 2002. In the time between these two outbursts, the source was very faint (<2 mCrab) compared to its customary hard-state intensity of ∼30 mCrab, which it enjoyed prior to its 1998 outburst.

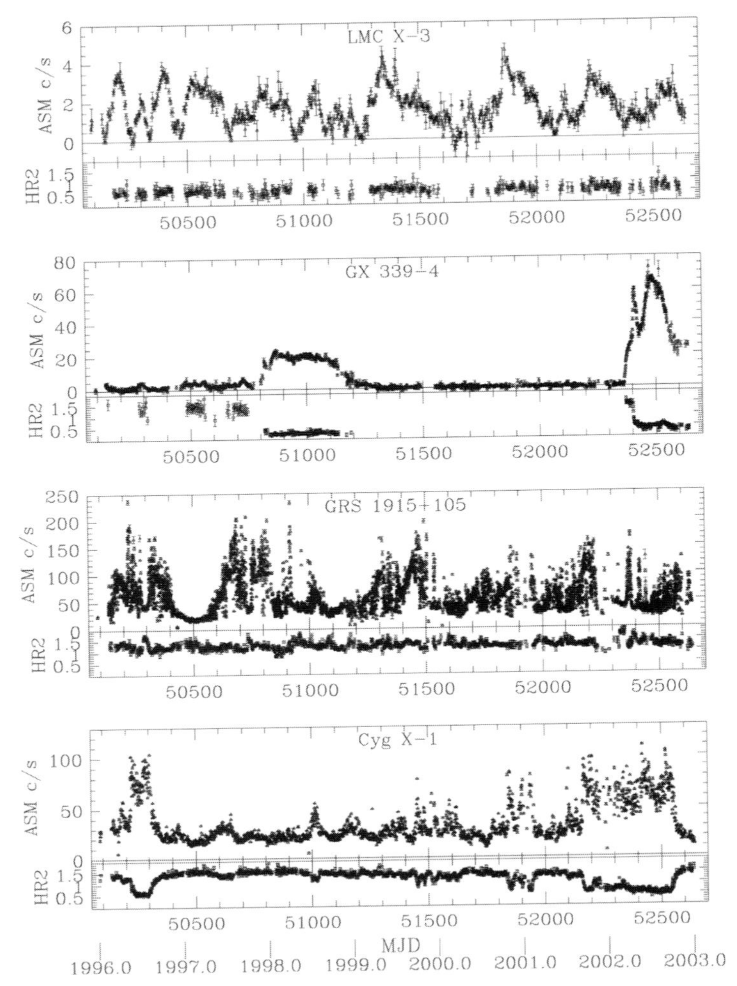

Fig. 4.3. Seven-year ASM light curves of four black hole binaries.

Remarkably, this transient BHB has never been observed in a fully *quiescent* state (Hynes *et al.* 2003). *GRS 1915+105*: This source exhibits extraordinary variations in both the X-ray and radio bands. Astonishing and yet repetitive patterns are sometimes seen in the X-ray lightcurves (Muno *et al.* 1999; Belloni *et al.* 2000; Klein-Wolt *et al.* 2002). For a 1992 X-ray lightcurve showing the birth of this source, see Chen *et al.* (1997). *Cyg X–1*: Transitions between the LH and HS states were first observed in this archetypal BH source (Section 4.3.1). However, as this record shows, there are both gradual and rapid variations in the hardness ratio that suggest both rapid state transitions and intermediate conditions between the HS and hard states (see Section 4.3.9).

4.2.1.2 *Black hole candidates*

Figure 4.4 displays the lightcurves of six BHCs with short outburst cycles. *XTE J748–288*: This short duration outburst that begins and ends in a hard spectral state is similar

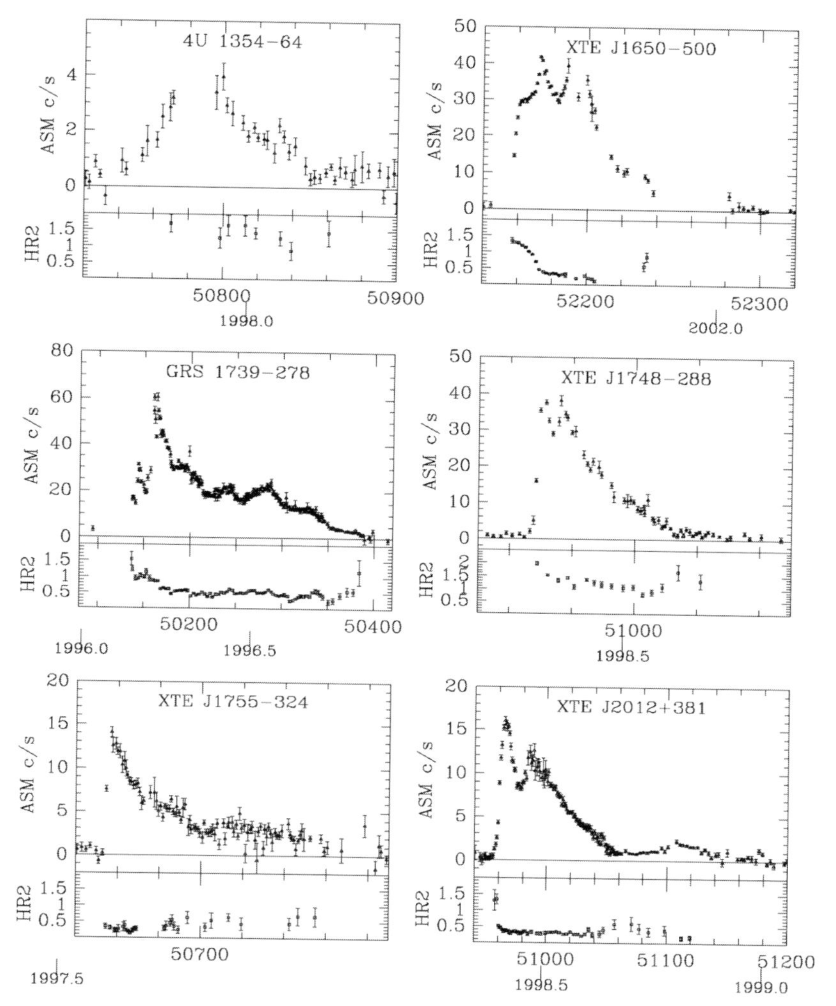

Fig. 4.4. Transient ASM light curves of six black hole candidates.

to the classic lightcurve of 4U 1543–47 (Fig. 4.2). The $1/e$ decay time is $\approx$16 days. This source is heavily absorbed, and the values of the hardness ratio are consequently increased. *GRS 1739–278*: A somewhat longer duration outburst with a nearly classic profile that includes a precursor peak, $\sim$40% variability near maximum, and undulations in intensity during the decay. Again, the outburst begins and ends with a hard spectrum. *XTE J1755–324*: This brief outburst, which follows an abrupt rise, provides yet another example of a classic lightcurve. This outburst is locked in the HS state. *XTE J2012+381*: This unusual lightcurve combines a classic rise to maximum followed directly by a precipitous drop in intensity. Also unusual are a large secondary maximum occurring just 30 days after discovery plus an additional late maximum at $\sim$140 days. *XTE J1650–500*: A complex lightcurve. At the onset of the outburst there is a very rapid rise followed by a slow rise. This unusual behavior is accompanied by a remarkable, slow ($\sim$15 day) transition from a hard spectral state to a soft

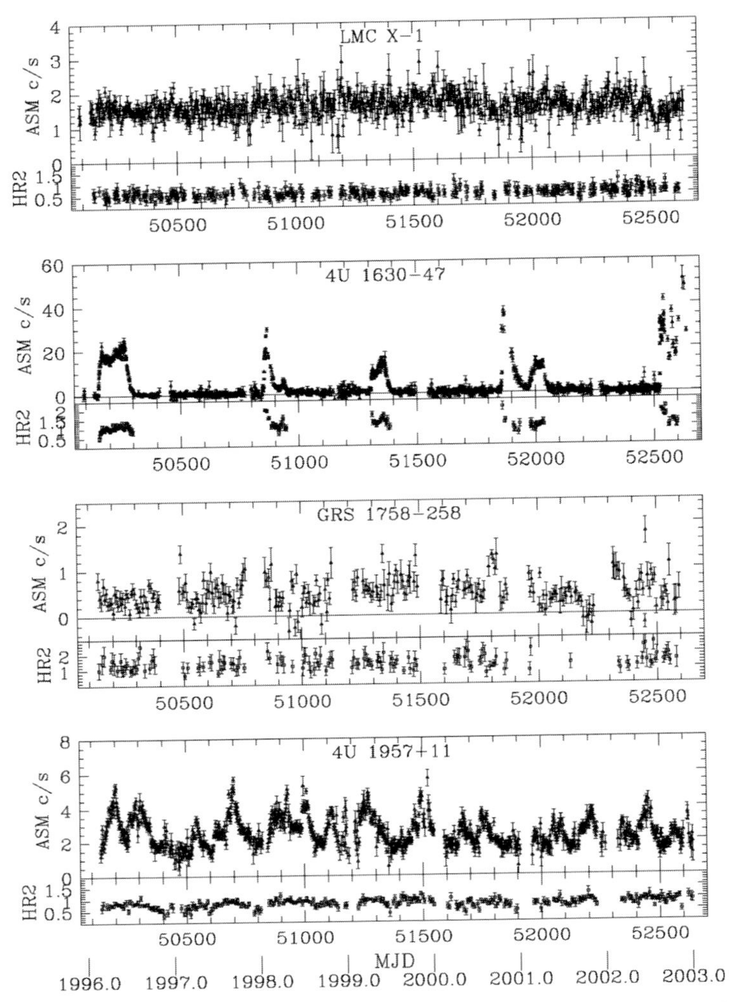

Fig. 4.5. Seven-year ASM light curves of three black hole candidates and the black hole binary LMC X-1.

one. *4U 1354–64*: A slow rise followed by a rather rapid decline. During this outburst, the source remained in the LH state (Brocksopp *et al.* 2001), whereas the HS state was reached during a brighter outburst in 1987 (Kitamoto *et al.* 1990). This source may be identical to Cen X-2, which reached a peak intensity of 13 Crab in 1967 (Brocksopp *et al.* 2001).

In Fig. 4.5 we show the lightcurves of the BHB LMC X-1 and the lightcurves of three BHCs. *LMC X-1*: The source is continually in a soft spectral state and maintains a relatively steady intensity. *4U 1630–47*: The ~600 day recurrence time has been known for 25 years (Jones *et al.* 1976). Here we see five, nearly equally spaced outbursts. Note the very different profiles and fluences of the outbursts. Note also that the 1996 outburst starts with a soft spectrum. *GRS1758–258*: This hard galactic center source was discovered by GRANAT/SIGMA in 1991. Its spectrum extends to at least 300 keV (Sunyaev *et al.* 1991a). During 2001 it underwent a transition to an unusual soft state of very low intensity (Smith *et al.* 2001; Miller *et al.* 2002d); the low flux level during that event is evident in the ASM record shown here. *4U 1957+11*: The source has long been considered a BHC based on its "ultrasoft" spectrum (White & Marshall 1984), although Yaqoob *et al.* (1993) have argued that the primary is a NS. The source displays a consistent flaring behavior and a soft spectrum over the 7 year interval.

It is often said that during its initial rise the spectrum of a BH X-ray nova transitions from a hard spectral state to a soft one. For several sources, the data in Figs. 4.2–4.5 support this view: 4U 1543–47, XTE J1550–564, XTE J1859+226, XTE J1650–500, GRS 1739–278, and XTE J2012+381. However, there are two clear counter-examples, sources whose spectra *hardened* during the initial rise: GRO J1655–40 (Fig. 4.2) and 4U 1630–47 (Fig. 4.5; first of five outbursts).

4.2.2 Synoptic studies of selected black hole binaries

It is important to follow the several-month spectral evolution of individual sources and to construct unified spectral models that can be used to represent the energy spectra of all BHBs. The necessary elements of such models can be deduced from a simple appraisal of the observational data: e.g., Cyg X-1 and other BHBs in the LH state show that the model must contain a non-thermal component, which can be well represented by a power-law function (TL95). On the other hand, the soft spectrum observed for most BH X-ray novae in the HS state is most widely modeled as a multi-temperature blackbody, which approximates the emission from an optically thick (relativistic) accretion disk. Many studies of spectral evolution therefore choose a composite model comprising disk blackbody and power-law components. Although this simple model has significant limitations, nevertheless it has proven to be widely applicable and quite effective in monitoring the spectra of BHBs, as we now discuss briefly by pointing to two examples.

Important studies of Ginga spectra using this model were made by Ebisawa *et al.* (1991, 1993, 1994). The authors added one refinement to the model, namely, a broad absorption feature above 7 keV. This feature is associated with the reflection of X-rays by an optically thick accretion disk (Ebisawa *et al.* 1994). With this model a successful and quantitative comparison was made of the spectra of Cyg X-1, LMC X-3, GS 2000+25, LMC X-1, GX339–4 and GS/GRS 1124–68 (Nova Mus 1991). This latter source, which exhibited a classic lightcurve, was observed 51 times in 1991 over a span of 235 days using the Ginga Large Area Counter (LAC). We direct the reader's attention to Figure 15 in Ebisawa *et al.* (1994), which shows the evolution over a full outburst cycle of the spectral parameters and fluxes for GS/GRS 1124–68. In their figure, it is evident that an important transition occurs 130 days into the outburst. For example, the photon spectral index suddenly decreases from 2.2–2.6 to a value near 1.6, the hard flux increases substantially and the soft disk flux decreases. This characteristic behavior, which has been observed for a number of BH X-ray novae, marks the transition from the HS state to the LH state.

We now compare the results obtained for GS/GRS 1124–68 to the results of an analogous study of the irregular BH X-ray nova XTE J1550–564, which was observed extensively by RXTE during its 1997–8 outburst (Sobczak *et al.* 2000b). A total of 209 pointed observations spanning the entire 255-day outburst were made using the PCA and HEXTE detectors. The 1998–9 lightcurve of the source is complex and includes a slow (10 day) rise to maximum, an intense (6.8 Crab) flare that occurred early in the outburst, and a "double-peaked" profile that roughly separates the outburst into two halves of comparable intensity (Fig. 4.2). Sobczak *et al.* adopted very nearly the same spectral model and methodology as Ebisawa *et al.* (1994). We direct the reader's attention to Figure 4 and the accompanying text in Sobczak *et al.* (2000b), which describes a complex course of evolution relative to that of GS/GRS 1124–68. Very briefly, Sobczak *et al.* show that during the first half of the outburst QPOs are ubiquitous and the spectrum is dominated by the power-law component, which are conditions that mark the very high/intermediate state. In contrast, during the second half of the outburst QPOs are scarce and the spectrum is dominated by emission from the accretion disk, which corresponds to the HS state. During this state, the inner disk radius (Section 4.1.5) remained nearly constant for 4 months (Sobczak *et al.* 2000b). Very similar behavior has been observed for GS/GRS 1124–68 (Ebisawa *et al.* 1994) and for several other sources (TL95). The constancy of the disk inner radius is remarkable, given the accompanying, large variations in luminosity that are usually observed (TL95).

In overview, in both GS/GRS 1124–68 and XTE J1550–564 there are observations where thermal emission from the disk dominates the spectrum; such instances would surely be interpreted as the HS state. At other times, the disk spectrum is substantially modified and a *power-law component may dominate at either high or low luminosity, with a spectral shape that may appear either soft or hard.* In the case of Cyg X-1, the non-thermal LH state may be stable for many months or years. However, its soft-state spectrum was revealed to be a steep power law rather than a thermal spectrum (see Section 4.3.1), adding further ambiguity to any physical interpretation of the HS state. These results highlight the need to restructure our definitions of X-ray states, while avoiding the confusing terminology that has developed during the past 30 years. Accordingly, in Section 4.3 we address this issue and describe definitions of the X-ray states based on quantified properties of the X-ray energy spectra and power density spectra.

4.2.3 Relativistic iron emission lines

Strong evidence for accretion disks in active galactic nuclei has come from X-ray observations of broad iron Kα lines. In particular, in some Seyfert galaxies the very asymmetric profile of the Fe Kα line (e.g., its extended red wing) suggests strongly that the emission arises in the innermost region of a relativistic accretion disk (for reviews see Fabian *et al.* 2000; Reynolds & Nowak 2003). The good energy resolution of ASCA provided the first clear evidence for such a line profile (Tanaka *et al.* 1995). In some cases, the line profile indicates the presence of an accretion disk extending down to the ISCO (Weaver *et al.* 2001). The broad Fe Kα fluorescence line is thought to be generated through the irradiation of the cold (weakly ionized) disk by a source of hard X-rays (likely an optically thin, Comptonizing corona). Relativistic beaming and gravitational redshifts in the inner disk region can serve to create an asymmetric line profile.

In fact, the first broad Fe Kα line observed for either a BHB or an AGN was reported in the spectrum of Cyg X-1 based on EXOSAT data (Barr *et al.* 1985). It was this result that inspired Fabian *et al.* (1989) to investigate the production of such a line in the near vicinity of a Schwarzschild BH, a result that was later generalized by Laor (1991) to include the Kerr metric. Other early studies of relativistically smeared Fe Kα lines from BHBs were conducted by Done *et al.* (1992) and others. Their work was one part of a broader examination of the accretion geometry that is produced as hard X-rays from an overlying corona illuminate an

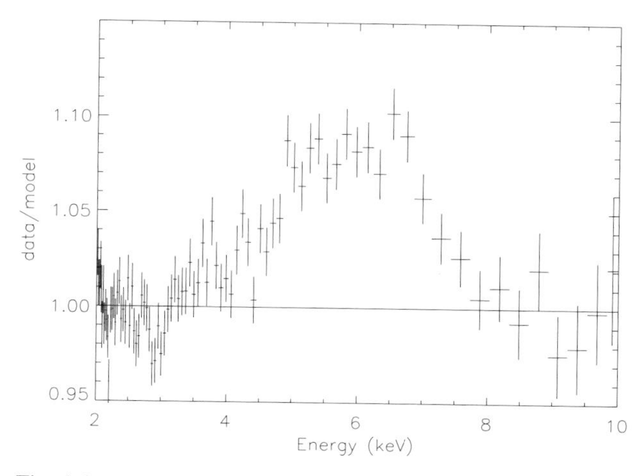

Fig. 4.6. Data/model ratio for XTE J1650–500. The model consists of multicolor disk blackbody and power-law components (Miller *et al.* 2002b). Note the non-Gaussian shape and low-energy extent of the line profile.

optically thick accretion disk. An Fe Kα line and a reflected continuum are always generated in this case (George & Fabian 1991; Matt *et al.* 1991). One example is a Ginga study of the reflected spectrum of V404 Cyg (Zycki *et al.* 1999a, 1999b). A limitation of this study is the use of proportional counter detectors (e.g., the Ginga LAC and RXTE PCA, which have an energy resolution of only FWHM $\approx$ 1.2 keV at Fe Kα). Such Fe Kα studies suffer both because the energy resolution is marginal and because the response matrices of the detectors are uncertain at the 1–2% level, while the Fe line profile in BHBs is typically only 1–5% above the X-ray continuum. Consequently, the results from these instruments must be considered with caution. Some Fe Kα sources that have been studied with RXTE include: GRO J1655–40 (Balucinska–Church & Church 2000); XTE J1748–288 (Miller *et al.* 2001); and GX 339–4 (Feng *et al.* 2001; Nowak *et al.* 2002).

More telling studies of the Fe Kα line have been achieved using the MECS and HP-GSPC detectors aboard BeppoSAX, which have a resolution of $\approx$0.6 keV at Fe Kα. Broad line profiles ($\sim$4–9 keV) have been observed for SAX J1711.6–3808 (in't Zand *et al.* 2002a) and XTE J1908+094 (in't Zand *et al.* 2002b); however, these profiles are rather symmetric and may be more a product of Compton scattering than relativistic broadening. BeppoSAX studies have also revealed other systems with broad, asymmetric Fe K profiles that resemble the predictions of relativistic smearing: GRS 1915+105 (Martocchia *et al.* 2002) and V4641 Sgr (Miller *et al.* 2002a). Interestingly, for both of these systems the inner disk radius deduced from the line profile is consistent with the radius of the ISCO for a Schwarzschild BH, suggesting that rapid spin is not required.

Spectral studies at higher resolution have been made recently with XMM-Newton and Chandra. Using the XMM EPIC-MOS1 detector, Miller *et al.* (2002b) find for XTE J1650–500 a broad, skewed Fe Kα emission line (Fig. 4.6) which suggests the presence of an extreme Kerr BH and indicates a steep radial falloff of disk emissivity with radius. An observation of Cyg X-1 with the HETGS grating and ACIS-S detector aboard Chandra revealed a broad line centered at $\approx$5.82 keV with a FWHM of $\approx$1.9 keV (Miller *et al.* 2002c). Also present was a smeared Fe edge at $\approx$7.3 keV. The authors conclude that the line is predominantely shaped by Doppler/gravitational effects and to a lesser degree by Compton scattering due to reflection.

4.2.4 *Super-Eddington luminosities*

Recently there has been considerable interest in ultraluminous X-ray sources (ULXs) in external galaxies with 0.5–10 keV luminosities in the range $10^{39} - 10^{40.5}$ erg s^{-1} (Makishima *et al.* 2000; Fabbiano *et al.* 2001; Humphrey *et al.* 2003; Miller *et al.* 2003b; Chapter 12). The luminosities of ULXs greatly exceed the Eddington limit of a 1.4 $M_\odot$ NS: $L_{Edd} = 1.3 \times 10^{38}(M_1/M_\odot)$ erg s^{-1}. The most luminous systems also exceed by a factor of $\sim$20 the Eddington luminosity of a typical 10 $M_\odot$ BH. This fact has led to the suggestion that the most luminous ULXs are a new class of accreting BHs with masses $\sim$100 $M_\odot$ (Makishima *et al.* 2000; Fabbiano *et al.* 2001). Alternatively, it has been suggested that the ULXs are powered by conventional stellar-mass BHs that radiate anisotropically (King *et al.* 2001; Chapter 13). We examine this question by comparing as directly as possible the luminosities of the ULXs to the luminosities of the 18 BHBs listed in Tables 4.1–4.2. We also mention briefly apparent differences between ULXs and BHBs in their spectra and duty cycles.

In terms of the maximum flux density of a BHB, $F_{X,max}$ (Table 4.1), the 2–11 keV luminosity is: $L_X/L_{Edd} \approx 2.6 \times 10^{35} \times F_{X,max}(\mu Jy) \times (D/10 \text{ kpc})^2$ (Bradt & McClintock 1983). For a Crab-like spectrum (Section 4.1.2), the flux in the 0.5–10 keV (ULX) band is a factor of 1.9 greater (although this is somewhat of an overestimate compared to the use of a thermal spectrum). Including this factor and using the distances and peak fluxes in Table 4.1, we find that the three most luminous BHBs are V4641 Sgr (6.2×10^{39} erg s^{-1}), 4U 1543–47 (4.2×10^{39} erg s^{-1}) and GRS 1915+105 (2.4×10^{39} erg s^{-1}). Thus, at peak luminosity these three BHBs appear to be in the same league as many of the ULXs observed in external galaxies. Moreover, using the mass measurements from Table 4.2, it appears that all three BHBs were super-Eddington at maximum (0.5–10 keV): L_X/L_{Edd} = 7.0, 3.5 and 1.4 for V4641 Sgr, 4U 1543–47 and GRS 1915+105, respectively.

This comparison of BHBs and ULXs is somewhat problematic: First, the distances of the BHBs are uncertain, and we have no direct measurements of their fluxes in the 0.5–2 keV band. Nevertheless, the results quoted above suggest that the peak luminosities of a few BHBs approach the peak luminosities observed for ULXs to within a factor of $\approx$5 (e.g., Miller *et al.* 2003b). Second, the luminosity shortfall of BHBs may be ascribable to the small sample of 18 BHBs compared to the much larger sample of comparable systems that have likely been detected in surveys of external galaxies. Finally, some ULXs exhibit spectral properties unlike those of BHBs. In particular, the cool disk spectra and the longevity at high luminosity of the most luminous ULXs may distinguish them from BHBs in the Milky Way (Miller *et al.* 2003a, 2003b). The differences among ULX spectra also suggest that the sample may be heterogeneous.

We conclude by noting that super-Eddington luminosities have plainly been observed for a few NS systems. The most clear-cut case is A0535–668, the "LMC transient." This pulsating NS binary with a firm distance of D = 50 kpc (Freedman *et al.* 2001) achieved a peak luminosity of $L_X \approx 1.2 \times 10^{39}$ erg s^{-1} , assuming isotropic emission (Bradt & McClintock

1983, and references therein). This is 6.9 times the Eddington luminosity of a canonical 1.4 $M_\odot$ NS or 3.8 times the Eddington luminosity of a hypothetical 2.5 $M_\odot$ NS.

4.3 Emission states of black hole binaries

As discussed in Sections 4.1 and 4.2, BHBs exhibit thermal and non-thermal components of X-ray emission, both of which can vary widely in intensity. It has long been recognized that BHBs undergo transitions between quasi-stable states in which one or the other of these components may dominate the X-ray luminosity. In the past, the study of BH emission states was based almost exclusively on X-ray spectral and timing studies. More recently, however, the results of X-ray studies have been supplemented with critical contributions by radio, optical and gamma-ray observers to give us a more physical and fruitful framework for regarding the emission states. In the following sections, we review the characteristic behavior that defines each of the principal X-ray states of BHBs. We then describe the current picture of each state in terms of physical structures and the nature of the accretion flow. Finally, we discuss the prospects of using these states to deduce the properties of BHs.

4.3.1 Historical notes on X-ray states

In the spring of 1971, Tananbaum *et al.* (1972) observed a remarkable X-ray state change in Cyg X-1 during which the average soft flux (2–6 keV) decreased by a factor of 4 and the average hard flux increased by a factor of 2. Simultaneously, the radio counterpart of Cyg X-1 brightened. It was later found that luminous X-ray novae such as A0620–00 exhibited similar spectral transitions, suggesting that common emission mechanisms were at work in both persistent and transient BHCs (Coe *et al.* 1976). These early results suggested that such global spectral changes might signify important changes in accretion physics.

As the many lightcurves in Section 4.2 illustrate, the soft X-ray state is generally seen at higher luminosity, motivating frequent references to the *high/soft* (HS) state. In this state, the spectrum may also display a "hard tail" that contributes a small percentage of the total flux. As shown by the synoptic studies discussed in Section 4.2.2, the soft state is best explained as ~1 keV thermal emission from a multi-temperature accretion disk (see Section 4.1.5), as foreseen in the standard theory for accretion in BHBs (Shakura & Sunyaev 1973). However, it has been found that the soft state of Cyg X-1 is not consistent with a thermal interpretation (Zhang *et al.* 1997b), and this has caused considerable confusion as to the proper way to understand Cyg X-1 and/or describe the HS state. In seeking a physical basis for describing X-ray states, it turns out that Cyg X-1 is not a good choice as a prototype, and further remarks about the states in Cyg X-1 are given in a separate section below (Section 4.3.9).

In the hard state the 2–10 keV intensity is comparatively low, prompting the name *low/hard* (LH) state. The spectrum is non-thermal and conforms to a power law with a typical photon index $\Gamma \sim 1.7$ (2–20 keV). In this state, the disk is either not detected at 2–10 keV (e.g., Belloni *et al.* 1999), or it appears much cooler and larger than it does in the soft state (Wilms *et al.* 1999; McClintock *et al.* 2001b).

An additional X-ray state of BHBs was identified in the Ginga era (Miyamoto *et al.* 1993). It is characterized by the appearance of QPOs in the presence of both disk and power-law components, each of which contributes substantial luminosity (e.g., $> 0.1 L_{\rm Edd}$; van der Klis 1995). In this state, which is referred to as the *very high* (VH) state, the power-law component is observed to be steep ($\Gamma \sim 2.5$). Initially, there were only two BHBs (GX339–4 and Nova

Mus 1991) that displayed this behavior, but many additional examples have been seen in the RXTE era.

It was first thought that the two non-thermal states (i.e., LH and VH states) could be distinguished through differences in their photon spectral indices, luminosities, and power density spectra. However, as shown below, the latter two differences have become blurred; nevertheless, the spectral index continues to be a valid discriminator. The importance of distinguishing between the LH and VH states was emphasized in a ~40–500 keV study of seven BHBs with the OSSE instrument aboard the Compton Gamma-Ray Observatory (Grove *et al.* 1998). The gamma-ray spectra of these sources separate naturally into two distinct groups which correspond to the LH state and the VH state, respectively. (1) For the first group it was shown that the X-ray LH state corresponds with a "breaking gamma-ray state" in which the spectrum below ~100 keV is harder than that of the VH state, but then suffers an exponential cutoff near 100 keV. (2) The second group exhibits a power-law gamma-ray spectrum with photon index $2.5 < \Gamma < 3.0$ over the entire range of statistically significant measurements. This gamma-ray photon index is consistent with the X-ray photon index of the VH state. Furthermore, contemporaneous X-ray observations (e.g., with ASCA) confirmed that a luminous thermal component coexists with the power-law component, which is one characteristic of the VH state noted above.

4.3.2 *X-ray states as different physical accretion systems*

Several recent developments in the study of BHBs have taken us beyond a largely phenomenological description of X-ray states to one based on physical elements (e.g., accretion disk, ADAF, jet, and corona). Although this work is still incomplete, the fundamental distinctions between the states are becoming clearer. For example, a key development in this regard is the recognition of a persistent radio jet associated with the LH state that switches off when the source returns to the HS state (Section 4.3.6; Chapter 9). Another example, revealed by gamma-ray observations, is the very different coronal structure that is responsible for the clear-cut distinction between the LH state and the VH state (Section 4.3.1). Arguably, each X-ray state can be regarded as a different accretion system that can be used in unique ways to study accretion physics and the properties of accreting BHs.

In the sections below, we review each of the four canonical X-ray states of BHBs, including the long-lived *quiescent* state. We illustrate the uniform X-ray properties of each of the three active states by showing X-ray spectra and power spectra for several BHBs and BHCs observed by RXTE. We also examine a possible fifth X-ray state, the *intermediate* state, as part of our discussion of the VH state. While presenting this overview, we suggest an alternative set of state names that are motivated by the kinds of emergent physical pictures mentioned above. Although the new state names depend critically on multi-wavelength results (i.e., radio to gamma-ray), we nevertheless attempt to define them on the basis of X-ray data. Furthermore, based on extensive observations of many sources with RXTE (e.g., see Section 4.2.2), *we abandon luminosity as a criterion for defining the states of BHBs* (with the exception of the quiescent state). We do not deny that there are correlations between states and luminosity in many sources, in particular the tendency of the HS to occur at higher luminosity compared with the hard state. However, as shown below, there are clear exceptions to these trends and each X-ray state has now been observed to span a range of two or more decades in X-ray luminosity.

4.3.3 *Notes on X-ray spectral analyses*

Many spectral models have been developed to describe one or more of the spectral states of BHBs. Most models for the non-thermal continuum components invoke inverse Compton or synchrotron emission, but these mechanisms can be applied with many different assumptions and geometric details. Some models closely constrain the relationship between spectral components (e.g., thermal emission providing seed photons for Comptonization), while others allow the spectral components to vary independently. In presenting this review, we have adopted the following pragmatic and generic strategy. As discussed in Section 4.2.2, the spectra of BHBs are well described by a model consisting of a multi-temperature accretion disk component and a power-law component (which may require an exponential cutoff at high energy). This model provides a robust, first-order description of BHB spectra that covers all of the emission states, and we adopt it to help define the states and to compare the luminosities of the thermal and non-thermal components. We also include additional features in the model, such as Fe line emission and a disk reflection component, when the normalization parameter for such a feature is significant at the level of 5σ. In the following discussions of each X-ray state, we comment on some physical interpretations and controversies related to the problem of determining the origin of the power-law component.

4.3.4 *Quiescent state*

A BHB spends most of its life in a quiescent state that can be summarized as *an extraordinarily faint state* ($L_X = 10^{30.5}$–$10^{33.5}$ erg s^{-1}), *with a spectrum that is distinctly non-thermal and hard* ($\Gamma = 1.5 - 2.1$). The first short-period X-ray nova to be detected in quiescence was A0620–00 ($P_{\rm orb} = 7.8$ hr); its X-ray luminosity was several times 10^{30} erg s^{-1}, which is only $\sim 10^{-8}$ of its outburst luminosity (McClintock *et al.* 1995; Narayan *et al.* 1996). The long-period systems, however, are significantly more luminous in quiescence because their mass transfer rates are driven by the nuclear evolution of their secondaries rather than by gravitational radiation (Menou *et al.* 1999). For example, the X-ray luminosity of V404 Cyg ($P_{\rm orb} = 155.3$ hr) is typically $L_X \sim 10^{33}$ erg s^{-1} (Kong *et al.* 2002), but can vary by an order of magnitude in one day (Wagner *et al.* 1994).

It is now possible to make sensitive measurements in the quiescent state using Chandra and XMM-Newton. The minimum quiescent-state luminosities (0.5–10 keV) of five BHs and stringent upper limits on two others have been reported by Garcia *et al.* (2001) and Narayan *et al.* (2002). Three additional BHBs were observed in quiescence more recently (Sutaria *et al.* 2002; Hameury *et al.* 2003; McClintock *et al.* 2003a). Considering only the five short-period systems ($P_{\rm orb} \lesssim 1$ day; see Narayan *et al.* 2002) that have been detected, one finds that four of them (XTE J1118+480, GRO J0422+32, GS 2000+25, and A0620–00) have Eddington-scaled luminosities that are $\approx 10^{-8.5}$. GS 1124–68 is more luminous by about an order of magnitude. As Garcia *et al.* (2001) and Narayan *et al.* (2002) show, the Eddington-scaled luminosities of several ostensibly similar X-ray novae that contain NS primaries are about 100 times higher, a conclusion that is quite robust. In the context of the advection-dominated accretion flow model, these authors argue that the relative faintness of the BH X-ray novae provides strong evidence that they possess event horizons. For a thorough discussion of their model and several alternative models, see Narayan *et al.* (2002). Apart from any specific model, it appears quite reasonable to suppose that the established faintness of quiescent BHs is somehow connected with the existence of the event horizon.

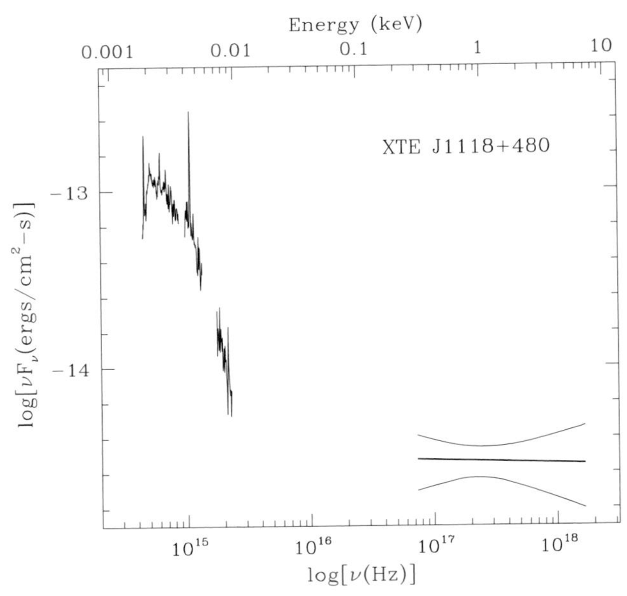

Fig. 4.7. Spectrum of XTE J1118+480 in the quiescent state based on simultaneous, multiwavelength observations. The optical spectrum of the mid-K dwarf secondary has been subtracted. Note the Planckian shape of the optical/UV continuum, which is punctuated by a dominant Mg II 2800 Å line and a strong Hα line on the far left. The best-fit X-ray model is indicated by the heavy, horizontal line; the 90% error box is defined by the flanking curved lines.

The quiescent spectra of BHBs are well fitted by a single power-law plus interstellar absorption. The best-determined photon spectral indices are consistent with the value $\Gamma \approx 2$. Specifically, for A0620–00 and XTE J1118+480, respectively, one has $\Gamma = 2.07(+0.28, -0.19)$ (Kong *et al.* 2002) and $\Gamma = 2.02 \pm 0.16$ (McClintock *et al.* 2003), where the column density is determined from the optical extinction. Only V404 Cyg allows a useful determination of both the column density and the photon index: $N_{\mathrm{H}} = (6.98 \pm 0.76) \times 10^{21}$ cm^{-2} and $\Gamma = 1.81 \pm 0.14$ (Kong *et al.* 2002).

A multiwavelength spectrum of XTE J1118+480 in quiescence is shown in Fig. 4.7 (McClintock *et al.* 2003). This shortest-period BHB ($P_{\mathrm{orb}} = 4.1$ hr) is located at b= 62°, where the transmission of the ISM is very high (e.g., 70% at 0.3 keV). A very similar multiwavelength spectrum was observed earlier for A0620–00 (McClintock & Remillard 2000), which implies that the spectrum shown in Fig. 4.7 represents the canonical spectrum of a BHB radiating at $10^{-8.5} L_{\mathrm{Edd}}$. The spectrum comprises two apparently disjoint components: a hard X-ray spectrum with a photon index $\Gamma = 2.02 \pm 0.16$, and an optical/UV continuum that resembles a 13 000 K disk blackbody spectrum punctuated by several strong emission lines.

The ADAF/disk model (see Section 4.1.5) accounts well for the following properties of BHBs in the quiescent state: (1) The hard power-law spectra (Narayan *et al.* 1996, 1997; Hameury *et al.* 1997; Quataert & Narayan 1999; McClintock *et al.* 2003); (2) the faintness of BHs relative to NSs (Narayan *et al.* 1997, 2002; Garcia *et al.* 2001; (3) the several-day delay in the optical/UV lightcurve when X-ray novae go into outburst (Hameury *et al.* 1997); and

(4) the broadband spectrum shown in Fig. 4.7 (McClintock *et al.* 2003). Especially significant is the prediction, confirmed by observations, that the accretion disk is truncated at a large inner radius in both the quiescent and LH states (Narayan 1996; Esin *et al.* 1997; McClintock et al. 2001b, 2003a).

4.3.5 *Thermal-dominant (TD) state or high/soft (HS) state*

As discussed in Section 4.1.5, considerations of basic principles of physics predict that accreting BHs should radiate thermal emission from the inner accretion disk (Shakura & Sunyaev 1973). It was therefore readily accepted that the soft X-ray state of BHBs represents thermal emission. Confirmations of this picture are largely based on the successful ability to describe the soft X-ray component using the simple multi-temperature accretion disk model (MCD model; Section 4.1.5). In Figs. 4.8 and 4.9 we show the energy spectra and power spectra, respectively, of 10 BHBs in the "thermal-dominant" (TD) or high/soft (HS) X-ray state as observed by RXTE. The thermal component of the model, where it can be distinguished from the data, is shown as a solid line, and the power-law component is shown as a dashed line. Typically, below about 10 keV the thermal component is dominant. With a few exceptions, the temperature of this component is in the range 0.7–1.5 keV (Table 4.4). The power-law component is steep ($\Gamma = 2.1-4.8$) and faint. In GRS 1915+105 the power-law falls off even more steeply with an e-folding cutoff energy of 3.5 keV; similar behavior has been reported for GRO J1655–40 (Sobczak *et al.* 1999).

In Fig. 4.8 we feature data for BHBs obtained during pointed observations with RXTE, while data for BHCs are shown in Fig. 4.9. The spectra of four of the sources in the figures (4U 1543–47, GX 339–4, XTE J1755–324, and XTE J2012+381) correspond to the maximum 2–20 keV luminosities observed during RXTE pointings (i.e., considering all possible states). The power density spectra (PDS) in Figs. 4.8 and 4.9 show that the variability in the TD state is either weak or the power scales roughly as ν^{-1}, which is a characteristic of many physical processes including turbulence (Mandelbrot 1999). The total rms power (r) integrated over 0.1–10 Hz in the PDS (2–30 keV) is in the range $0.01 \lesssim r \lesssim 0.06$, which is significantly below that of the LH state. QPOs in the range 0.1–30 Hz are generally not seen in individual TD observations, but large groups of PDS in the TD state have yielded weak QPOs in two cases. A 0.3% (rms) QPO at 27 Hz was seen in a sum of 27 observations ("1997 soft state") of GRO J1655–40 (Remillard *et al.* 1999), and a similar (0.3%) QPO at 17 Hz was seen in 69 observations of XTE J1550–564 in the soft state during 1998–9 (MJD 51160–51237; Homan *et al.* 2001).

In the following section, we show that a transition to the LH state is followed by the appearance of a hard and dominant power-law spectrum, while the accretion disk, if visible, shows a substantial decrease in temperature. On the other hand, a transition to the VH state (Section 4.3.7) is marked by a steeper power-law spectrum accompanied by either a normal ($\sim$1 keV) disk or one that appears hot with a small inner radius. This latter transition is also accompanied by the presence of QPOs that appear when the disk contribution to the total, unabsorbed flux at 2–20 keV falls below the level of 0.75 (Sobczak *et al.* 2000a). *We thus define the TD state as the set of conditions for which the disk-flux fraction is above 75% (2–20 keV), the PDS (2–30 keV) shows no QPOs or very weak features (rms $\ll$ 1%), and the power continuum is also weak: $r \lesssim 0.06$ integrated over 0.1–10 Hz.*

As further support for a thermal interpretation of the soft X-ray component, many studies have found evidence for disk luminosity variations in which the inner disk radius (which

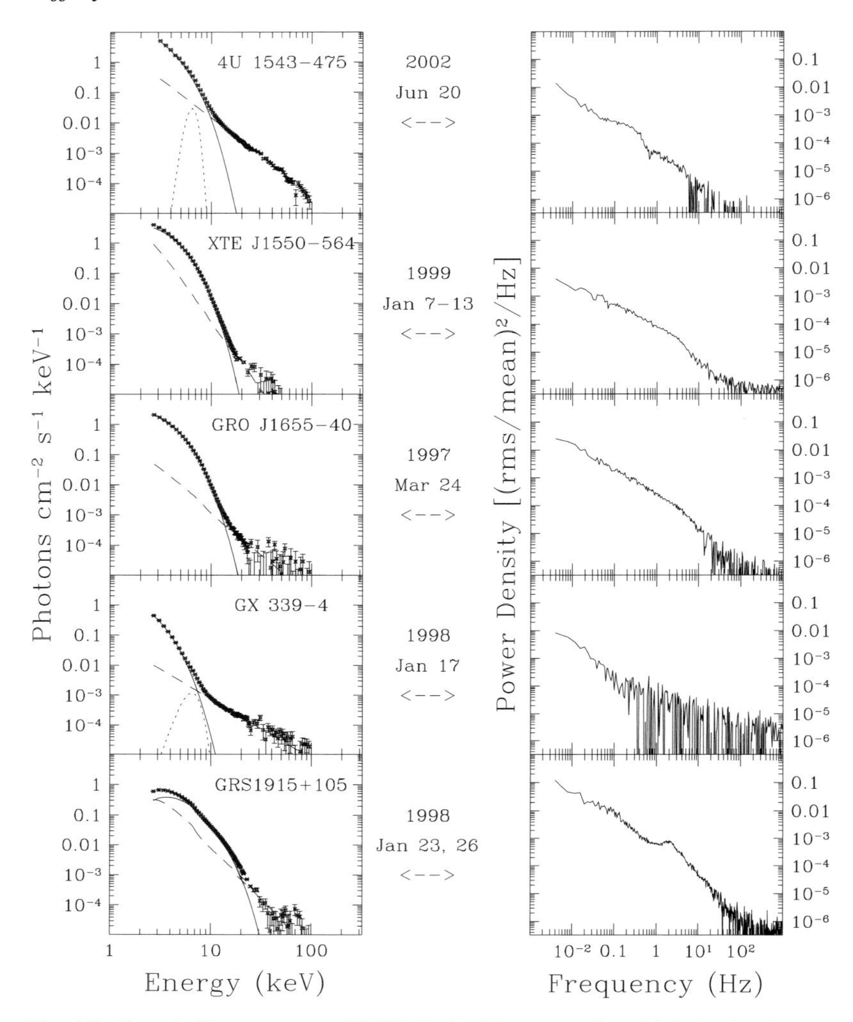

Fig. 4.8. Sample X-ray spectra of BHBs in the X-ray state for which the dominant component is thermal emission from the accretion disk. The energy spectra (*left*) are decomposed into a thermal component, which dominates below ~10 keV (solid line), and a faint power-law component (dashed line); GRS 1915+105 is modeled with a cutoff power-law (see text). For two of the BHBs, an Fe line component is included in the model (dotted line). The corresponding power spectra (2–30 keV) are shown in the panels on the right.

scales as the square root of the MCD normalization parameter) appears constant while the luminosity variations depend only on changes in temperature (Section 4.2.2). This effect is reported in a study of LMC X-3 (Kubota *et al.* 2001) and is illustrated in Fig. 4.10. The measured disk flux and apparent temperature successfully track the relation $L \propto T^4$ (solid line) expected for a constant inner disk radius. The figure also shows gross deviations from this relation associated with the VH state of GRO J1655–40, a topic that is addressed below in Section 4.3.7.

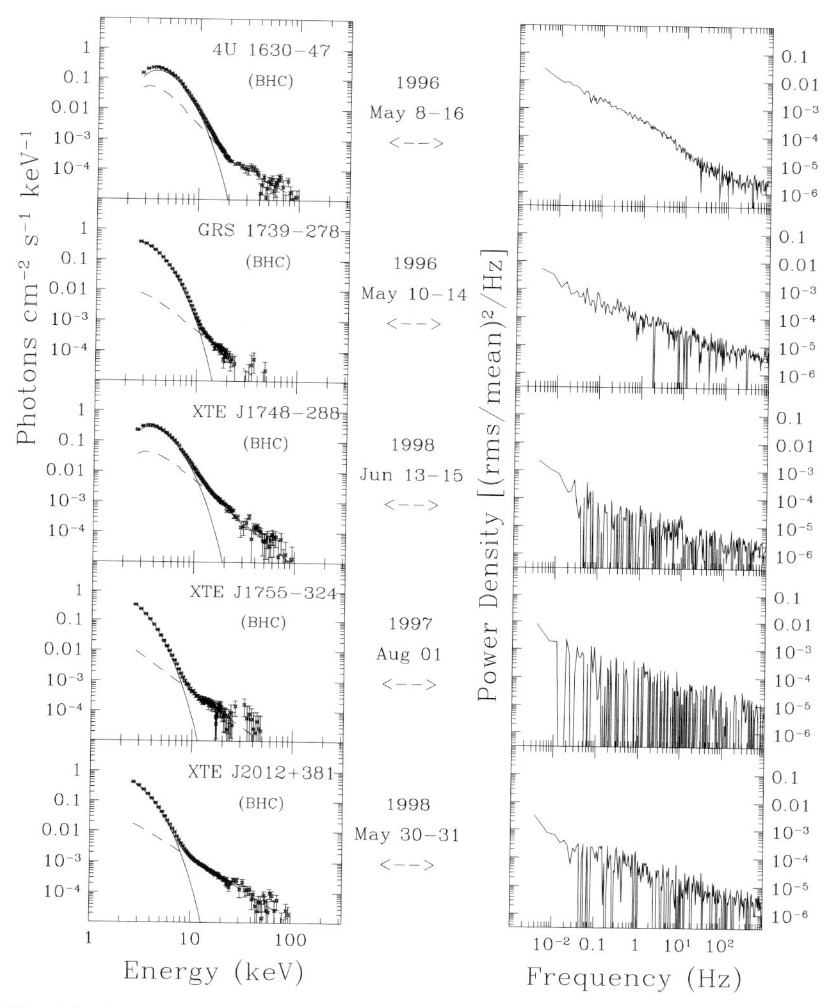

Fig. 4.9. Sample X-ray spectra of BHCs in the TD X-ray state. In the panels on the left, the energy spectra are shown deconvolved into a thermal component due to the accretion disk (solid line) and a power-law component (dashed line). The corresponding power spectra (2–30 keV) are shown in the right half of the figure.

Because of the successes of the simple MCD model, the inner disk radius has been used to provide a type of spectroscopic parallax. In principle, the inner disk radius can be deduced for those sources for which the distance and disk inclination are well constrained from the disk normalization parameter, $(R_{\rm in}/D)^2 \cos\theta$, where $R_{\rm in}$ is the inner disk radius in kilometers, D is the distance to the source in units of 10 kpc, and θ is the inclination angle of the system (e.g., Arnaud & Dorman 2002). However, the MCD model is Newtonian, and the effects of GR and radiative transfer need to be considered. GR predicts a transition from azimuthal to radial accretion flow near $R_{\rm ISCO}$, which depends only on mass and spin and ranges from 1 to 6 $R_{\rm g}$ for a prograde disk (Section 4.1.5). Therefore, for a system with a known distance and inclination, in principle it may be possible to estimate the spin parameter via an

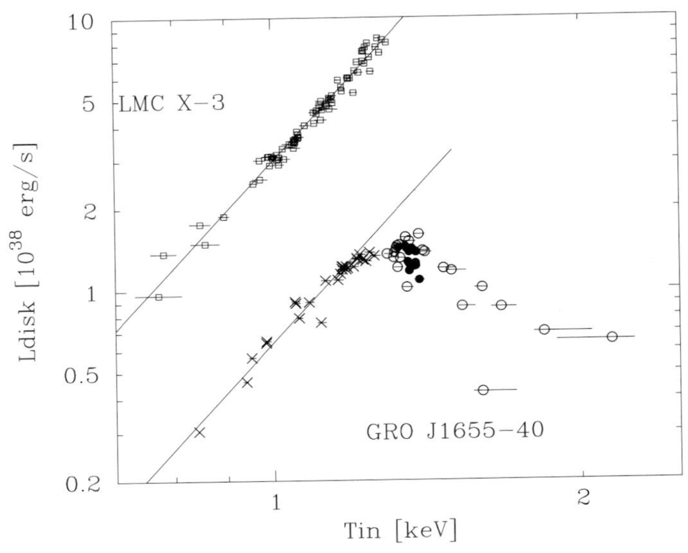

Fig. 4.10. The accretion disk luminosity vs. temperature at the inner accretion disk (Kubota *et al.* 2001). For GRO J1655–40, the symbol type denotes the time periods: early outburst (filled circles), first phase (open circles), and second phase (crosses). The solid lines represent the $L_{\rm disk} \propto T_{\rm in}^4$ relation.

X-ray measurement of the inner radius and an optical determination of the mass. In lieu of a fully relativistic MHD model for the accretion disk, one could attempt to correct the MCD model parameters to account for the effects of radiative transfer through the disk atmosphere (Shimura & Takahara 1995) and for modifications on the structure and emissivity of the inner disk due to GR (Zhang *et al.* 1997a). However, the accuracy of these corrections has been challenged (Merloni *et al.* 2000; Gierlinski *et al.* 1999), and at present the chief value of the MCD model is in monitoring the temperature and the fractional flux contribution from the accretion disk.

4.3.6 *Hard X-ray state associated with a steady radio jet*

The conventional name for this state is the *low/hard* (LH) state; however, henceforth we refer to it simply as the *hard* state for the reasons given in Section 4.3.2. As noted earlier (Section 4.3.1), Cyg X-1 and many transient sources have been observed to undergo transitions to a hard, non-thermal X-ray spectral state. This usually occurs at luminosities below that of the TD state, and the spectrum can be modeled (e.g., in the range 1–20 keV) as a power-law function with a photon index ~1.7. In some sources, such as Cyg X-1 and GS 1354–64, this hard state is accompanied by a broad enhancement at 20–100 keV, which is interpreted as reflection of the power-law component from the surface of the inner accretion disk (Di Salvo *et al.* 2001). This component is discussed further at the end of this section.

In recent years there have been rapid advances in associating the X-ray hard state with the presence of a compact and quasi-steady radio jet (for a thorough review, see Chapter 9). This relationship, which constitutes one of the foundations of the "disk:jet" connection, is based on at least three arguments. First, VLBI radio images have shown a spatially resolved radio jet during episodes of quasi-steady radio and hard X-ray emission from GRS 1915+105 (Dhawan *et al.* 2000) and Cyg X-1 (Stirling *et al.* 2001). In both instances the radio spectrum was flat or inverted. In the case of GRS 1915+105, the radio jet in the hard state appears smooth at many different spatial scales, in contrast to the rapidly separating knots seen for ballistic, superluminal jets from the same source. Second, more generally (i.e., when VLBI images are not available), X-ray sources that remain in the hard state for prolonged periods (weeks to years) are highly likely to show correlated X-ray and radio intensities and a flat radio spectrum (Fender *et al.* 1999a; Fender 2001; Corbel *et al.* 2000, 2003; Klein-Wolt *et al.* 2002; Marti *et al.* 2002). In one of these examples, GX 339–4, the jet interpretation is further supported by the detection of 2% linear radio polarization with a nearly constant position angle (Corbel *et al.* 2000). Finally, it is now routine to witness the quenching of the persistent radio emission whenever an X-ray source exits the hard state and returns to the TD state (Fender *et al.* 1999b; Brocksopp *et al.* 1999; Corbel *et al.* 2000).

In Fig. 4.11, the five BHBs shown in the hard state are the same sources shown earlier in the TD state (Fig. 4.8). In the following cases, the X-ray data have been selected to coincide with specific radio observations: Flat-spectrum radio emission was reported for both XTE J1550–564 on 2000 June 1 during the decay of its second outburst (Corbel *et al.* 2001) and for GX 339–4 on 1999 March 3 (Corbel *et al.* 2000). In addition, both the radio and X-ray emission of GRS 1915+105 are fairly steady on 1997 October 22, which coincides with one of the days on which the core radio image shows the extended structure of a nuclear jet (Dhawan *et al.* 2000).

A second sample of BHBs and BHCs in the hard state is shown in Fig. 4.12. The hard state for XTE J1748–288 occurred during decay from the HS state (Revnivtsev *et al.* 2000c). On the other hand, the data shown for XTE J1118+480 and GS 1354–64 correspond with outburst maxima; these outbursts never reached the HS state (Revnivtsev *et al.* 2000a; Frontera *et al.* 2001b). Associated radio emission for these two sources was reported with jet interpretations by Fender *et al.* (2001) and Brocksopp *et al.* (2001), respectively. Finally, GRS 1758–258 and Cyg X-1 spend most of their time in the hard state, and their energy spectra are shown in the bottom two panels of Fig. 4.12. These RXTE observations happened to coincide with radio observations that confirm a flat radio spectrum. In the case of GRS 1758–258, core radio emission (as distinct from the extended radio lobes observed for this source) was reported on 1998 August 3 and 5 by Marti *et al.* (2002), and a clear detection of Cyg X-1 during 1997 December 12–17 is evident in the public archive of the Greenbank Interferometer available on the NRAO website.

The physical condition of the accretion disk in the hard state is a subject of great significance in the effort to build a detailed physical model for both the jet and the X-ray source. Observations with ASCA, which provided sensitivity in the range 0.5 to 9 keV, showed that the hard states of both GX 339–4 (Wilms *et al.* 1999) and Cyg X-1 (Takahashi *et al.* 2001) exhibit power-law spectra with an additional soft X-ray excess that can be modeled as a large and cool ($\sim$0.1–0.2 keV) accretion disk. The spectral decompositions illustrated in Fig. 4.11 provide some evidence for a soft disk component in both GRO J1655–40 and GRS 1915+105, although RXTE is much less sensitive than ASCA to thermal spectra with temperatures well below 1 keV. By far the best direct measurements of the temperature and inner radius of an accretion disk in the hard state have been made for XTE J1118+480, which has an extraordinarily small interstellar attenuation (e.g., only 30% at 0.3 keV). Based on simultaneous HST, EUVE and Chandra observations made in outburst, it was determined that

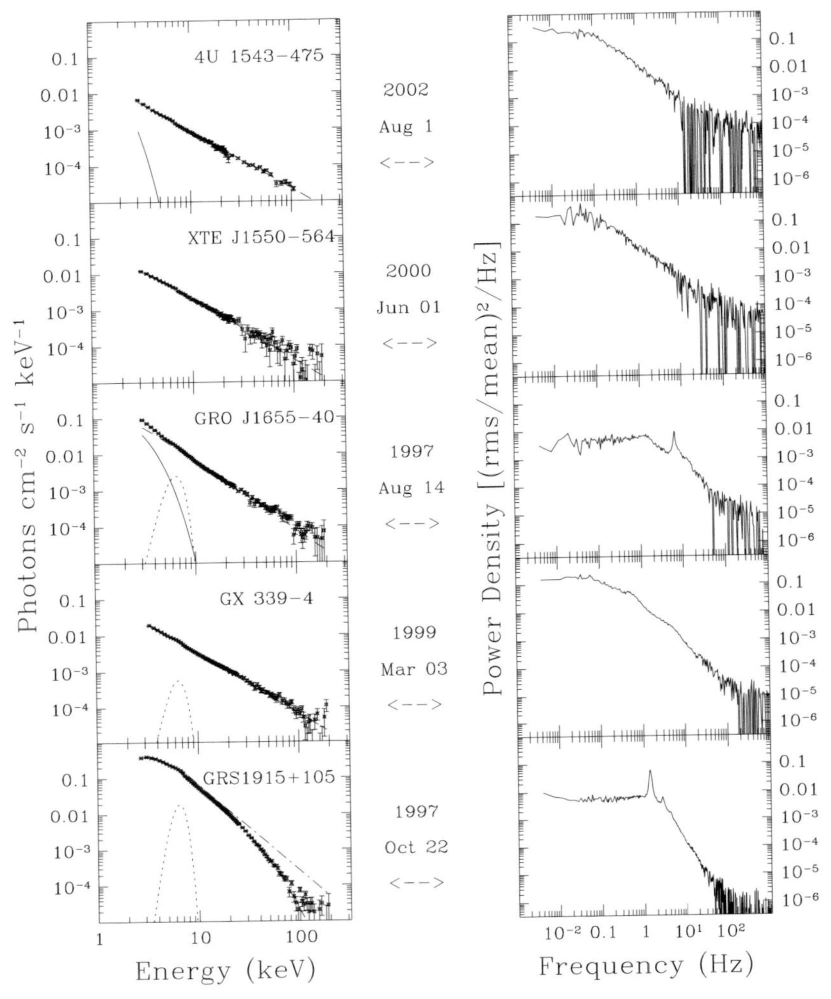

Fig. 4.11. Sample spectra of BHBs in the hard state. The selected X-ray sources are the same BHBs shown in Fig. 4.8. The individual spectral components include the power-law (dashed line) and, if detected, the accretion disk (dotted line) and a reflection component (long dashes). The energy spectra are characterized by a relatively flat power-law component that dominates the spectrum above 1 keV. A second characteristic of the hard state is the elevated continuum power in the PDS, here shown for the range 2–30 keV. This state is associated with the presence of a steady type of radio jet (see text). Radio emission was detected during three of the five observations shown here, but there was no radio coverage for GRO J1655–40 and 4U 1543–47.

the inner disk radius and temperature for the MCD model were $\gtrsim 100\ R_g$ and ≈ 0.024 keV, respectively (McClintock *et al.* 2001b). Somewhat higher temperatures ($\approx$0.035–0.052 keV) have been inferred from observations using BeppoSAX (Frontera *et al.* 2003).

While it seems clear that the blackbody radiation appears truncated at a large radius ($\sim 100 R_g$) in the hard state, the physical state of the hot material within this large radius is still a matter of debate. Is the disk density truncated at this radius, as envisioned in the ADAF

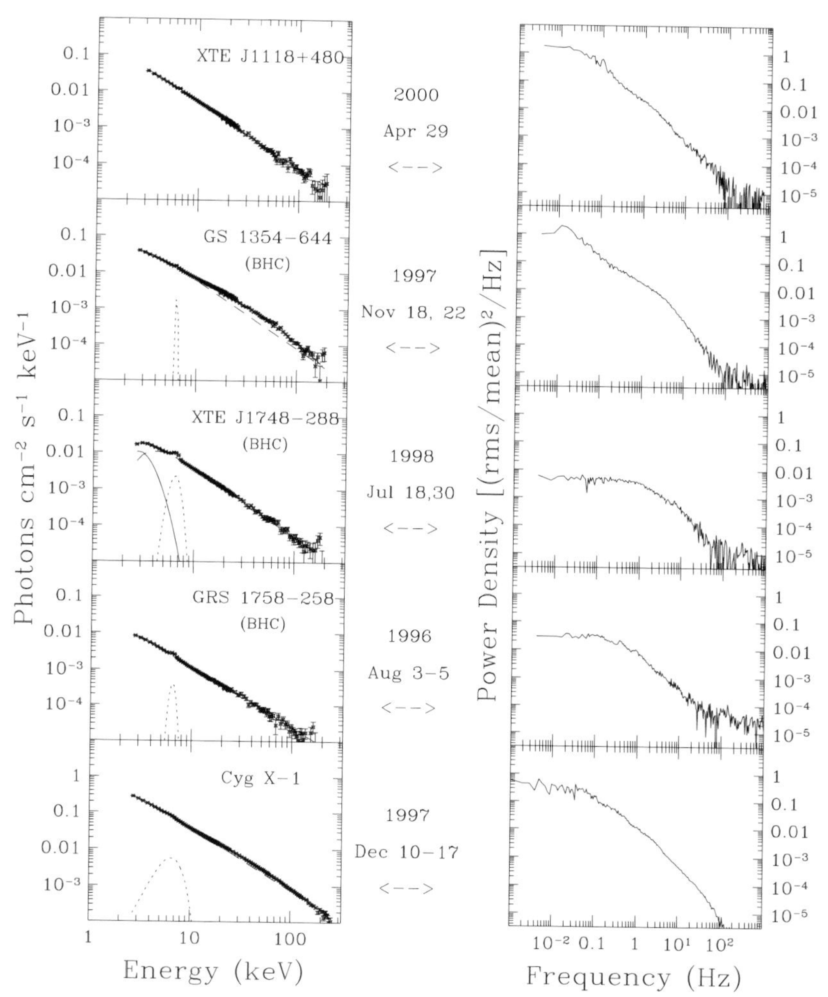

Fig. 4.12. A second sample of BHBs or BHCs seen in the hard state. The observations of XTE J1748–288 occurred during outburst decay, while XTE J1118+480 and GS 1354–64 are seen in the hard state at the peaks of their respective outbursts. GRS 1758–258 and Cyg X-1 spend most of their time in the hard state. There is radio coverage that confirms the presence of a flat radio spectrum for all the sources except XTE J1118+480. The line types denoting the spectral components follow the convention of Fig. 4.11

model, or is a substantial amount of matter present in a relativistic flow that is entrained in a jet (e.g., Markoff *et al.* 2001)? Or is the inner disk basically intact and either depleted of energy or veiled in some type of Compton corona? For the latter possibility, the properties of the corona would be strongly constrained by the absence of any normal ($\sim$1 keV) thermal component in the hard state of XTE J1118+480 (Esin *et al.* 2001; Frontera *et al.* 2003).

Guidance in sorting out these options may eventually come from other types of investigations, such as the study of correlated optical/X-ray variability (Malzac *et al.* 2003). Also promising are spectral analyses that focus on broad Fe emission features (Section 4.2.3) or the X-ray reflection component (Done & Nayakshin 2001). These spectral features depend on substantial density in the inner disk, while the Fe line additionally reveals the pattern of Keplerian flow modified by effects due to GR. Systematic studies of these features during different BHB states and transitions could help to determine the physical changes in the inner disk associated with the hard state. The reflection component is most apparent when the disk is observed nearly face-on. One such system is Cyg X-1, and a reflection analysis has been reported by Done and Zycki (1999). They find that the disk is physically truncated, but the transition radius (tens of R_g) is not as far from the event horizon as suggested by the value of R_{in} inferred from the disk spectrum in the hard state (Takahashi *et al.* 2001).

The origin of the X-ray power law is another aspect of the controversy concerning the appropriate physical model for the hard state. As noted above, we regard the power-law fit as a general signature of non-thermal radiation; however, the observed spectrum can be produced by several different radiation mechanisms. This is well illustrated in the case of XTE J1118+480, where the X-ray spectrum has been fitted by an ADAF model (Esin *et al.* 2001), a synchrotron model (Markoff *et al.* 2001) and a thermal-Comptonization model (Frontera *et al.* 2001b).

Despite the large uncertainties that remain for physical models, it would appear that the association of the hard state with a steady radio jet is an important step forward. And it does remain possible to identify the hard state solely from X-ray spectral and temporal properties, as had been done in the past (TL95). Using Figs. 4.11 and 4.12 and Table 4.4, *we conclude that the hard state is well characterized by three conditions: the spectrum is dominated (>80% at 2–20 keV) by a power-law spectrum, the spectral index is in the range* $1.5 < \Gamma < 2.1$, *and the integrated power continuum (0.1–10 Hz) is strong and typically in the range* $0.1 < r < 0.3$.

4.3.7 *Steep power-law (SPL) state or very high (VH) state*

There are times when BHBs become exceedingly bright ($L_X > 0.2 L_{Edd}$), and the X-ray spectrum again displays substantial non-thermal radiation, which may constitute 40–90% of the total flux. In such cases the photon index is typically $\Gamma \geq 2.4$, which is steeper than the index ($\Gamma \sim 1.7$) seen in the hard state. The strength of this steep power-law component also coincides generally with the onset of X-ray quasi-periodic oscillations (QPOs) in the range 0.1–30 Hz. This suite of characteristics was initially seen in only two instances: during a bright outburst of GX 339–4 (Miyamoto & Kitamoto 1991) and near the time of maximum flux in X-ray Nova Muscae 1991 (= GS/GRS 1124–68; Miyamoto *et al.* 1993). At the time, this very high (VH) state was interpreted as a signature of the highest rate of mass accretion in a BHB system (van der Klis 1995).

As mentioned previously (Section 4.3.1), the high-energy spectra of several BHCs observed with OSSE on CGRO (40–500 keV) reinforced the distinction between the X-ray hard and VH states, showing that the hard-state spectra exhibit a steep cutoff near 100 keV (Grove *et al.* 1998). On the other hand, the VH-state spectra showed no evidence for a high-energy cutoff, while the photon index in the X-ray and gamma-ray bands is the same ($\Gamma \sim 2.5$–3.0). The unbroken power-law spectra observed by OSSE for five sources suggested that these BHBs had been observed in the VH state, although most of the observations were not accompanied by X-ray observations that could assess the presence of QPOs.

Table 4.4. *Spectral fit parameters*

X-ray name	N_H (10^{22})	T_{DBB} (keV)	± (keV)	N_{DBB}	±	Γ_{PL}	±	N_{PL}	±	Fe FWHM	N_{Fe}	±	χ^2_ν	Additional details
TD state: Figs. 4.8 & 4.9														
4U 1543−475	0.3	1.01	0.02	7419	165	2.57	0.02	5.42	0.21	0.61	.0479	.0031	3.62	feature at 4.4 keV
XTE J1550−564	2.0	1.12	0.03	3289	74	4.76	0.04	152	17	–	–	–	0.98	smedge at 9.2 keV
GRO J1655−40	0.9	1.16	0.03	1559	21	2.85	0.23	1.01	0.65	–	–	–	2.00	smedge at 8.0 keV
GX 339−4	0.2	0.71	0.03	2520	62	2.02	0.04	0.08	0.01	1.05	.0032	.0003	1.45	
GRS 1915+105	6.0	2.19	0.04	62	5	3.46	0.02	33.4	1.61	–	–	–	3.13	smedge at 6.7 keV
4U 1630−47	11.0	1.33	0.03	315	7	3.75	0.03	17.4	1.40	–	–	–	1.06	break at 20.8 keV to $\Gamma = 1.9$
GRS 1739−278	3.0	0.95	0.04	972	23	2.65	0.15	0.210	0.008	1.11	.0068	.0008	1.42	
XTE J1748−288	10.4	1.79	0.02	42.4	2.1	2.60	0.02	14.6	0.4	–	–	–	1.18	
XTE J1755−324	0.2	0.75	0.08	1486	133	2.40	0.15	0.11	0.04	–	–	–	1.78	
XTE J2012+381	0.8	0.85	0.05	1176	56	2.06	0.04	0.16	0.015	–	–	–	1.32	
Hard state: Figs. 4.11 & 4.12														
4U 1543−475	0.3	0.38	0.07	645	1338	1.67	0.02	0.041	0.001	–	–		1.57	
XTE J1550−564	2.0	–	–	–	–	1.70	0.10	0.108	0.021	–	–	–	1.13	
GRO J1655−40	0.9	0.77	0.02	228	37	1.93	0.02	0.571	0.021	1.00	.0065	.0006	1.83	
GX 339−4	0.2	–	–	–	–	1.75	0.02	0.168	0.028	0.90	.0013	.0003	0.98	plus reflection
GRS1915+105	6.0	–	–	–	–	2.11	0.02	0.231	0.043	0.91	.0458	.0003	2.39	plus reflection
XTE J1118+480	0.01	–	–	–	–	1.72	0.04	0.267	0.024	–	–	–	1.23	
GS 1354−644	0.7	–	–	–	–	1.48	0.09	0.470	0.032	0.1	.0008	.0002	1.15	plus reflection
XTE J1748−288	10.4	0.48	0.05	5302	479	1.88	0.09	0.293	0.065	0.66	.0045	.0003	1.65	
GRS 1758−258	1.0	–	–	–	–	1.67	0.07	0.053	0.010	0.36	.0004	.0001	1.84	
Cyg X-1	0.5	–	–	–	–	1.68	0.07	0.446	0.025	1.44	.0206	.0018	3.40	plus reflection
SPL state: Figs. 4.13 & 4.14														
4U 1543−47	0.3	0.93	0.07	3137	138	2.47	0.02	6.85	0.21	0.82	.0347	.0034	1.90	
XTE J1550−564	2.0	3.31	0.20	7.76	0.70	2.82	0.05	200	1.5	1.30	.2136	.0314	2.16	smedge at 8.5 keV
GRO J1655−40	0.9	2.22	0.20	9.89	1.6	2.65	0.05	75.3	1.1	1.32	.2321	.0157	4.45	
GX 339−4	0.2	0.89	0.08	1917	109	2.42	0.02	2.34	0.08	0.97	.0178	.0017	1.39	
GRS 1915+105	6.0	1.19	0.07	115	31	2.62	0.08	28.5	0.6	0.90	.0396	.0065	4.13	
4U 1630−47	11.0	1.73	0.02	46.0	2.4	2.65	0.02	17.0	0.4	–	–	–	1.10	
GRS 1739−278	3.0	1.01	0.06	1116	38	2.61	0.03	2.95	0.19	1.53	.0341	.0021	0.94	
XTE J1748−288	10.4	1.36	0.02	210	11	2.92	0.02	26.2	1.2	–	–	–	0.96	
XTE J1859+226	0.5	1.03	0.02	1164	91	2.55	0.08	14.5	0.31	1.33	.0426	.0060	1.36	
Cyg X-1	0.5	0.49	0.03	55708	2962	2.68	0.03	7.65	0.37	0.73	.0270	.0016	1.78	breaks to $\Gamma = 1.83$ above 12 keV
Unusual spectra: Fig. 4.15														
XTE J1550−564	2.0	0.74	0.02	6932	562	2.24	0.02	23.0	0.87	1.03	.121	.008	1.34	p.l. cutoff energy 51.7 keV
GRS 1915+105	6.0	0.88	0.03	775	156	1.91	0.02	4.51	0.18	1.28	.053	.004	1.45	p.l. cutoff energy 50.0 keV
SAX J1819.3−2525	0.3	1.63	0.06	38	6	0.59	0.03	0.25	0.02	0.47	.038	.003	1.46	p.l. cutoff energy 39.3 keV

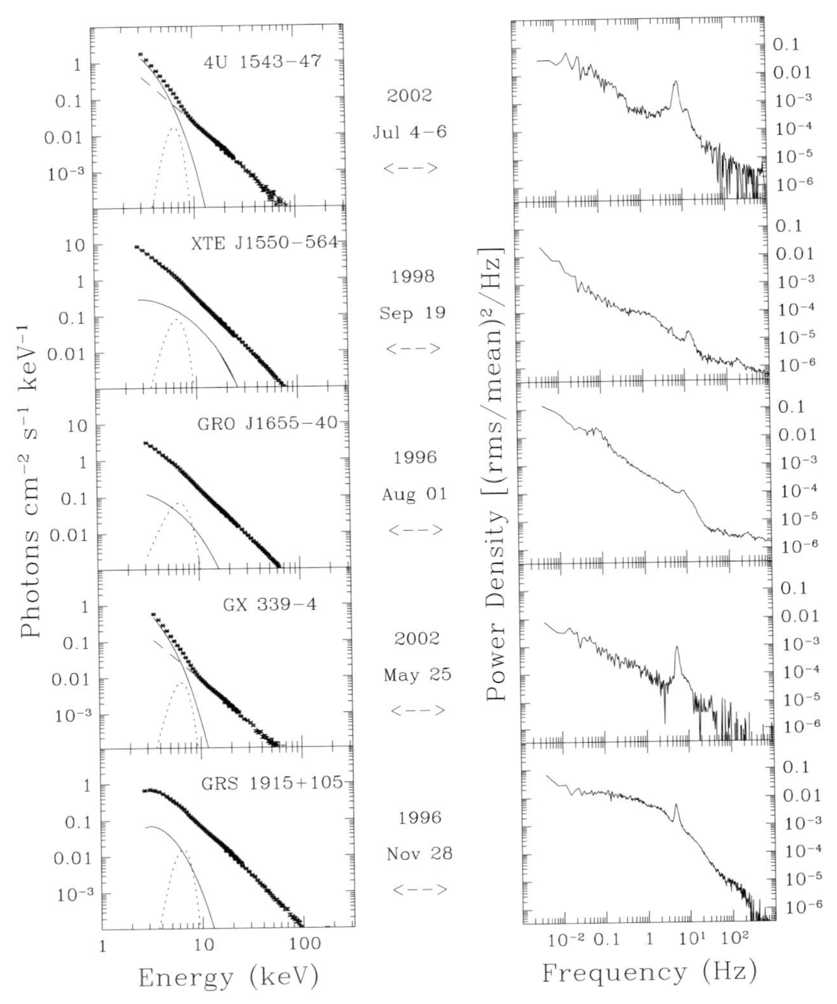

Fig. 4.13. X-ray spectra of BHBs in the SPL state, which is characterized by a strong and steep power-law component in the energy spectrum, along with the presence of X-ray QPOs. The dashed and dotted lines follow the convention of earlier figures.

The monitoring programs of RXTE have shown that the VH state is both more common and more complicated than originally envisioned. Some sources display both X-ray QPOs and a steep power-law component at luminosity levels that are well *below* the maxima seen even in their TD (HS) state (Remillard *et al*. 2002b; Section 4.5). This topic is considered further in Section 4.3.8 below. In response to these developments, we hereafter refer to this state as the "steep power–law" state, or the SPL state, rather than the VH state. We adopt this new name because the steep power-law is a fundamental property of this state, whereas a very high luminosity is not. We view the presence of QPOs as a confirming property of the SPL state.

In Fig. 4.13 we show examples of the SPL state for the same five BHBs considered in Figs. 4.8 and 4.11. The photon index of the steep power-law component covers the range

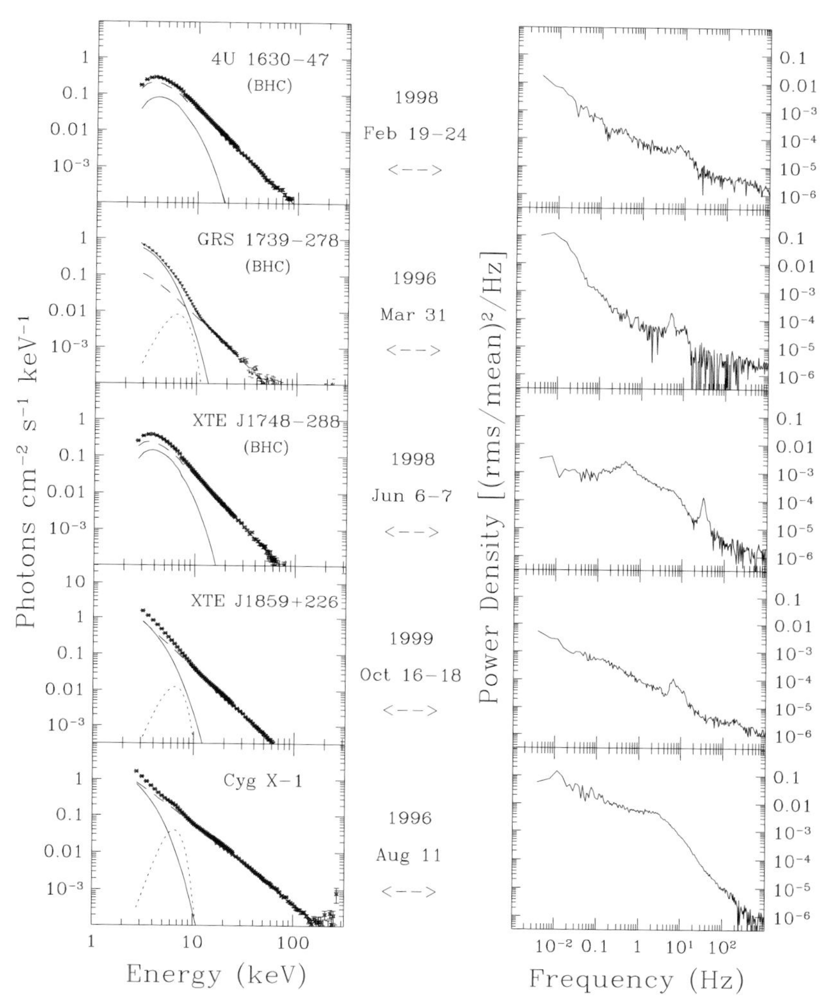

Fig. 4.14. Additional examples of sources in the SPL state. The observations were selected near the times of global or local maxima in the X-ray flux. Cyg X-1 is exceptional for its relatively low luminosity and for the absence of QPOs (see Section 4.3.9). The dashed and dotted lines follow the convention of earlier figures.

$2.4 < \Gamma < 3.0$ as shown in Table 4.4, and QPOs are present with central frequencies over the range 5–13 Hz. The spectra of XTE J1550–564 and GRO J1655–40 (Fig. 4.13) correspond to the highest luminosities observed for these sources during pointed observations with RXTE (Sobczak *et al.* 2000b, 1999). Moreover, both observations revealed the presence of high-frequency QPOs, 186 Hz and 300 Hz, respectively (Remillard *et al.* 2002b). The relationship between the SPL state and HFQPOs will be discussed further in Section 4.4.3.

In Fig. 4.14 we show spectra of the SPL state for three BHCs and two additional BHBs. For three of the sources we have selected observations near the time of maximum flux, as seen in RXTE pointed observations. For the two remaining sources we consider local maxima: the 1996 soft-state episode of Cyg X-1 and the second outburst of 4U 1630–47. X-ray QPOs are seen in the top four panels (5–8 Hz in three cases and 31 Hz in the case of XTE J1748–288). The power spectrum of Cyg X-1 contains broad continuum features but no X-ray QPOs, although the energy spectrum indicates that the source is in an SPL-like state.

We can encompass all of these SPL examples with the following criteria (see also Sobczak *et al.* 2000a). *The SPL state is defined first by the presence of a power-law component in the X-ray spectrum with photon index* $\Gamma > 2.4$. *Second, either there are X-ray QPOs present (0.1–30 Hz) while the power-law contributes more than 20% of the total (unabsorbed) flux at 2–20 keV, or the power-law contributes more than 50% of the total flux without detections of QPOs. Third, we adopt an upper limit for the integrated rms power (0.1–10 Hz) in the PDS (2–30 keV),* $r \lesssim 0.15$. This latter criterion avoids one type of intermediate condition in which a strong, band-limited PDS and a high-energy spectral break suggest evolution away from the SPL state, even while the photon index remains steep (i.e., $\Gamma > 2.4$). Intermediate state conditions are further discussed in Section 4.3.8.

Transitions between the TD and hard states frequently pass through intervals of the SPL state, and this has led some authors to suggest that the SPL (or VH) state is itself an "intermediate state" that lies between the TD and hard states (Rutledge *et al.* 1999; Homan *et al.* 2001). However, this appears to be a radical suggestion that gives inadequate weight to other SPL observations associated with (1) the episodes of highest absolute luminosity in many BHBs, (2) a distinct gamma-ray spectrum, and (3) occurrences of QPOs at both high and low frequency. We therefore conclude that the SPL state is a bona fide state of BHBs, and we note that nothing in our state definitions constrains the order in which state transitions should occur.

The radio properties of the SPL state are an important and complicated topic that calls for the heightened attention of observers. Here we highlight several salient results. The brightest days of GRO J1655–40 during its 1996–7 outburst occurred in the SPL state (1996 August) when the source appeared radio quiet (e.g., Tomsick *et al.* 1999). Similarly, Cyg X-1 becomes radio quiet whenever the spectrum switches from the hard state to the SPL state. On the other hand, the SPL state is also associated with the explosive formation of radio jets. For example, the giant X-ray flare in XTE J1550–564 (shown in Figs. 4.2 and 4.13) has been linked to a relativistic mass ejection seen as a superluminal separation of bipolar radio jets (Hannikainen *et al.* 2001). However, there is a distinct possibility that the X-ray flare in the SPL state may have occurred after the moment of ejection, at a time when the radio properties of the core (i.e., the inner disk) are unknown. This conjecture is supported by radio observations of the same source during the 2000 outburst when a radio flare was observed to decay below detectable levels while the source remained in the SPL (VH) state (Corbel *et al.* 2001). We conclude that the best available evidence suggests that the SPL state is essentially radio quiet, while the instability that causes impulsive jets is somehow associated with the SPL state.

The physical origin of the SPL spectrum remains one of the outstanding problems in high-energy astrophysics. The SPL spectrum extends to ~1 MeV in several sources: e.g., Cyg X-1, GRO J1655–40 and GRO J0422+32 (Table 4.2). The spectrum may extend to even higher energies, but present investigations are limited by photon statistics. At stake is our understanding of accretion physics at the extraordinary times of peak BHB luminosity. Equally important is our need to interpret the high-frequency QPOs associated with the SPL state.

Most models for the SPL state invoke inverse Compton scattering as the operant radiation mechanism (e.g., Zdziarski 2000). The MeV photons suggest that the scattering occurs in a non-thermal corona, which may be a simple slab operating on seed photons from the underlying disk (Gierlinski *et al.* 1999; Zdziarski *et al.* 2001). Efforts to define the origin of the Comptonizing electrons has led to more complicated geometric models with feedback mechanisms, such as flare regions that erupt from magnetic instabilities in the accretion disk (Poutanen & Fabian 1999). One early model, which was applied to AGN, invokes a strongly magnetized disk and predicts power-law spectra extending to tens of MeV (Field & Rogers 1993). An analysis of extensive RXTE observations of GRO J1655–40 and XTE J1550–564 has shown that as the power-law component becomes stronger and steeper, the disk luminosity and radius appear to decrease while maintaining a high temperature (see Fig. 4.10). These results can be interpreted as an observational confirmation of strong Comptonization of disk photons in the SPL state (Kubota *et al.* 2001; Kubota & Makishima 2003).

There are a number of alternative models for the SPL state. For example, bulk motion Comptonization has been proposed in the context of a converging sub-Keplerian flow within 50 R_g of the BH (e.g., Titarchuk & Shrader 2002). Turolla *et al.* (2002) have suggested pair production as a means of extending the photon spectrum beyond the $\sim$350 keV limit initially calculated for this model.

As noted above, Comptonization models are hard-pressed to explain the origin of the energetic electrons. As a further difficulty, a viable model must also account for the QPOs in the SPL state. This is important since strong QPOs (see Figs. 4.13 and 4.14) are common in this state. SPL models and X-ray QPOs are discussed further in Section 4.4.

4.3.8 *Intermediate states*

The four states described above capture the behavior observed for most BHBs on many occasions. However, other forms of complex behavior are often seen, and these dispel the notion that every BHB observation can be classified via these X-ray states. In the following three subsections, we consider observations that challenge or combine elements of the four-state framework described above.

The hard-state energy spectra in Figs. 4.11 and 4.12 show little or no contribution from the accretion disk, while the PDS exhibit a "band-limited" power continuum (i.e., a flat power continuum at low frequencies that breaks to a steeper slope between 0.1 and 10 Hz). However, a band-limited power spectrum is sometimes seen in combination with a stronger contribution from the accretion disk. This condition of a BHB has been interpreted as an *intermediate* state that lies between the hard and TD states (e.g., Mendez & van der Klis 1997). We agree that the spectra in Fig. 4.11 do appear to show characteristics of both the hard and TD states; however, it lies within the boundaries we define for the hard state.

Other cases seem to display a different type of intermediate or hybrid emission state. For example, dozens of observations of XTE J1550–564 yielded energy spectra and QPOs that resemble the SPL state (Sobczak *et al.* 2000b), but the PDS showed band-limited continuum power that is reminiscent of the intermediate state described above (Homan *et al.* 2001). One example is shown in Fig. 4.15, along with a similar observation of GRS 1915+105. These observations can be described as SPL states with band-limited power continua. The significance of this PDS shape is yet to be fully understood. It could suggest an intermediate state linking the hard and SPL states, a detail further supported by the fact that the inner edges of the disks in the two systems appear to have cooler temperatures and larger radii than are observed for the TD state. However, such speculations may be ill-advised without

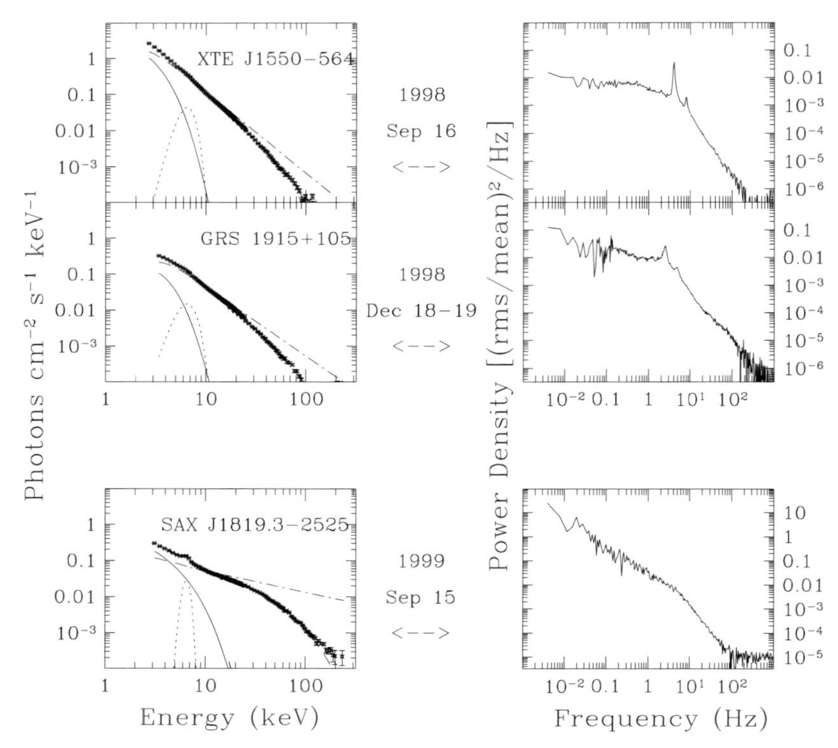

Fig. 4.15. Unusual spectra of three BHBs. The observations of XTE J1550–564 and GRS 1915+105 show spectral properties of the SPL state, but the PDS show a band-limited power continuum that is customarily seen as a characteristic of the hard state, rather than the SPL state. In the bottom panel, the flares and rapid fluctuations seen in SAX J1819.3–2525 (V4641 Sgr) do not coincide with any of the typical X-ray states of BHB systems.

considerations of sensitive radio measurements. We further note that the presence or absence of band-limited power observed in individual SPL-state observations of XTE J1550–564 is closely coupled to the amplitudes and phase lags of the associated QPOs (of types A, B, and C; Remillard *et al.* 2002a). Finally, while the energy spectra of XTE J1550–564 and GRS 1915+105 in Fig. 4.15 are distinctly steep, there is some ambiguity as to whether they are best modeled as a steep power law or as a flatter power law with an unusually low cutoff energy ($\sim$50 keV). In Table 4.4, we show the spectral parameters for the latter model, which is statistically preferred in the case of GRS 1915+105.

We conclude that it is inappropriate to refer to both the observations of XTE J1550–564 (Fig. 4.15) and the very different observations of GRO J1655–40 (Fig. 4.11) as representing a single BHB state, namely, the intermediate state. On the other hand, state transitions and hybrid emission properties are to be expected and *X-ray spectra and PDS should be interpreted as intermediate states when necessary, while specifying which states can be combined to yield the observed X-ray properties.* In summary, we describe the spectra of GRO J1655–40 (Fig. 4.11) as representing primarily a hard state or perhaps an intermediate state between hard and TD. On the other hand, the spectra and PDS of XTE J1550–564 and GRS 1915+105 considered here appear to show an intermediate state related to the SPL and hard states. Since transitions between these latter two states are not generally seen, this hybrid combination merits further scrutiny.

4.3.9 X-ray states of Cygnus X-1 and GRS 1915+105

In this section we briefly summarize the efforts to integrate the behavior of two uncommon BHBs within the framework of the TD, hard/radio jet, and SPL states. We first return to the issue of Cyg X-1 and the nature of its transitions to a soft state of high intensity, which is unlike the canonical TD (HS) state. As noted earlier (Sections 4.2.1.1 and 4.3.1), contrary to expectations the 1996 soft-state spectrum of Cyg X-1 revealed a power-law spectrum, rather than a TD spectrum (Zhang *et al.* 1997a; Frontera *et al.* 2001a; Fig. 4.14). Cyg X-1 has never been seen in the TD state, and the transition from a hard X-ray spectrum to a soft one must be seen as a transition from the hard state to the SPL state (Gierlinski *et al.* 1999; Zdziarski *et al.* 2001). However, this SPL-like state is unusual in two respects: the absence of QPOs and the relatively low temperature of the accretion disk (Table 4.4). Less surprising is the low luminosity of the SPL state (Zhang *et al.* 1997b), since this is also seen in other sources (e.g., XTE J1550–564; Remillard *et al.* 2002a). Whether the SPL state in Cyg X-1 requires a higher mass accretion rate than the hard state is a matter of controversy (Zhang *et al.* 1997b; Frontera *et al.* 2001a).

In the unique case of GRS 1915+105, the wildly varying X-ray lightcurves indicate an imposing number of instability modes (Belloni *et al.* 2000). Nevertheless, within this complexity it is often possible to identify the canonical states of a BHB (Muno *et al.* 1999; Belloni *et al.* 2000). About half of the observations of GRS 1915+105 show fairly steady X-ray flux (rms $< 15\%$ in 1 s bins at 2–30 keV), and most of these intervals yield spectra and PDS that resemble either the TD or hard states (Muno *et al.* 1999; Fender 2001; Klein-Wolt *et al.* 2002). There are, however, some noteworthy anomalies encountered while interpreting the behavior of GRS 1915+105 in terms of the canonical X-ray states. First, the condition of steady radio and X-ray emission extends to $L_X > 10^{38}$ erg s^{-1}, which is a factor ~ 100 higher than is observed for other BHBs in the hard state. Second, the X-ray photon index ($\Gamma \sim 2.2$) is steeper than usual, although it remains flatter than the index seen in in the SPL state ($\Gamma \geq 2.4$) or the hard tail of the TD state ($\Gamma \approx 3$; see Table 4.4). In short, the spectral index for GRS 1915+105 in the hard state appears shifted to a somewhat higher value compared to other BHBs.

Overall, the spectral and temporal properties of Cyg X-1 and GRS 1915+105 are best integrated into the standard description of BH X-ray states if we relax the assumptions regarding the relative or absolute luminosity ranges that are appropriate for the various states. As we will see in Section 4.5.1, the spectral evolution of XTE J1550–564 motivates a similar conclusion, since the luminosity in the SPL state can lie well below that of the TD state. Undoubtedly, there is an overall correlation between spectral states and luminosity intervals in accreting BHB systems. However, *the canonical X-ray states are most usefully defined in terms of the properties of the energy spectrum and PDS, rather than in terms of luminosity.*

4.3.10 Anomalous behavior of SAX J1819.3–2525

Finally, we consider whether BHBs exhibit characteristics that fall entirely outside the X-ray states considered thus far. The most challenging case may be the BHB SAX J1819.3–2525 (V4641 Sgr; Fig. 4.2). The PCA observations of 1999 September 15.9 show the source in a unique flaring state (Wijnands & van der Klis 2000). The spectral hardness ratio remains remarkably constant throughout the brightest ~500 s of this observation, despite the dramatic intensity fluctuations. The spectrum and PDS for this central time interval are shown in Fig. 4.15. The spectral analysis reveals a hot accretion disk and a broad Fe line, while the photon index is extraordinarily hard: $\Gamma = 0.60 \pm 0.03$ with a cutoff energy of 39 keV (Table 4.4). This source is also distinguished as a new prototype for a "fast X-ray nova" (Wijnands & van der Klis 2000) because it exhibited a 12 Crab flare that appeared and decayed in less than 1 day (Fig. 4.2). The multifrequency spectrum near X-ray maximum has been interpreted in terms of super-Eddington accretion with the binary immersed in an extended envelope (Revnivtsev *et al.* 2002).

4.4 Fast temporal variations: QPOs and broad power peaks

As shown in the preceding sections on X-ray states (Section 4.3.6–8), QPOs are prevalent in the SPL state, and they are sometimes seen in the hard state when thermal emission from the disk contributes some flux above 2 keV (e.g., GRO J1655 – 40 in Fig. 4.11). In this section we briefly consider the QPOs of BHBs and BHCs in greater detail. For references on PDS computation, defining QPOs, and QPO characteristics of NS systems, see Chapter 2. We supplement this work by discussing X-ray timing results for BHBs. Following van der Klis, we define QPOs as features (usually modeled as a Lorentzian function) in the PDS that have coherence parameter $Q = \nu/\Delta\nu > 2$, where $\Delta\nu$ is the FWHM of the QPO. Features with significantly lower Q values are regarded as "broad power peaks" and are discussed separately.

4.4.1 Low-frequency QPOs and radiation mechanisms

The X-ray PDS of many BH transients display low-frequency QPOs (LFQPOs) roughly in the range 0.1 to 30 Hz. The significance of these oscillations can be summarized as follows.

(1) LFQPOs are almost always seen during the SPL state, and they are often seen in the SPL-hard intermediate state. They can be exceedingly strong (see Figs. 4.13–4.15) with rms amplitudes (expressed as a fraction of the mean count rate) as high as $r > 0.15$ for sources such as GRS 1915+105 (Morgan *et al.* 1997) and XTE J1550–564 (Sobczak *et al.* 2000a). More generally, they are seen with $0.03 < r < 0.15$ whenever the steep power law contributes more than 20% of the flux at 2–20 keV (Sobczak *et al.* 2000a). LFQPOs have been observed at energies above 60 keV (Tomsick & Kaaret 2001).

(2) In several sources, the LFQPO frequency is correlated with the total disk flux (but not with temperature or inner disk radius; Sobczak *et al.* 2000a; Muno *et al.* 1999; Trudolyubov *et al.* 1999). This behavior, in combination with the role of the steep power law mentioned directly above, suggests that LFQPOs may provide a vital clue to the mechanism that couples the thermal and SPL components.

(3) LFQPOs can be quasi-stable features that persist for days or weeks. For example, in GRS 1915+105 QPOs at 2.0–4.5 Hz persisted for 6 months during late 1996 and early 1997 (Muno *et al.* 2001).

(4) In a general sense, it can be argued that oscillations as distinct and strong as these QPOs (often with $Q > 10$), represent global requirements for an organized emitting region. For example, in the context of models in which thermal radiation originates from MHD instabilities, one cannot accept the common picture of numerous and independent magnetic cells distributed throughout the inner disk.

The effort to tie LFQPOs to the geometry and flow of accreting gas is complicated by the fact that LFQPO frequencies are much lower than the Keplerian frequencies for orbits in the inner accretion disk. For example, for a BH mass of 10 $M_{\odot}$, an orbital frequency near 3 Hz coincides with a disk radius near $100 R_g$, while the expected radius for maximum X-ray emission lies in the range 1–10 R_g (depending on the value of the BH spin parameter).

For the strongest QPOs in GRS 1915+105, the individual oscillations were tracked to determine the origin of frequency drifts and to measure the average "QPO-folded" oscillation profile (Morgan *et al.* 1997). The results show a random walk in QPO phase and a nearly sinusoidal waveform. The ramifications of these results for QPO models remain uncertain.

There are now a large number of proposed LFQPO mechanisms in the literature, and we mention only a few examples here. The models are driven by the need to account for both the QPO frequency and the fact that the oscillations are strongest at photon energies above 6 keV, i.e., where only the power-law component contributes substantially to the X-ray spectrum. The models include global disk oscillations (Titarchuk & Osherovich 2000), radial oscillations of accretion structures such as shock fronts (Chakrabarti & Manickam 2000), and oscillations in a transition layer between the disk and a hotter Comptonizing region (Nobili *et al.* 2000). Another alternative, known as the "accretion–ejection instability model," invokes spiral waves in a magnetized disk (Tagger & Pellat 1999) with a transfer of energy out to the radius where material corotates with the spiral wave. This model thereby combines magnetic instabilities with Keplerian motion to explain the observed QPO amplitudes and stability.

Further analyses have revealed phase lags associated with LFQPOs and their harmonics. The analysis technique uses Fourier cross-spectra to measure both the phase lags and the coherence parameter (versus frequency) between different X-ray energy bands, e.g., 2–6 vs. 13–30 keV. Unexpectedly, both positive and negative phase lags have been found (Wijnands *et al.* 1999; Cui *et al.* 2000b; Reig *et al.* 2000; Muno *et al.* 2001), and efforts have been made to classify LFQPOs by phase lag properties. The expansion of LFQPO subtypes may not be widely viewed as a welcome development. Nevertheless, it has been shown in the case of XTE J1550–564 that the properties of the phase lags clarify how LFQPO parameters correlate with both the accretion-disk and high-frequency QPO parameters (Remillard *et al.* 2002a).

4.4.2 *Broad power peaks and comparisons of BH and NS systems*

The study of broad features in the PDS has led recently to important developments. In many NS systems and in the hard state of BHBs, the PDS can be decomposed into a set of four or five broad power peaks (Nowak 2000; Belloni *et al.* 2002), generally with $0.5 < Q < 1.0$. The evolution of these features has been linked to major behavioral changes in Cyg X-1. For example, the disappearance of the third broad power peak occurred just at the time Cyg X-1 left the hard state and its steady radio jet was quenched (Pottschmidt *et al.* 2003).

Broad PDS features are also involved in renewed efforts to contrast accreting BHB and NS systems via their variability characteristics. It has been proposed that the observed high-frequency limit of the power continuum provides a means to distinguish accreting BH and NS systems (Sunyaev & Revnivtsev 2000), since only the latter exhibit intensity variations above 500 Hz.

Finally, some studies have used both QPOs and broad power peaks to examine the relationship between low- and high-frequency features and to make comparisons between different

NS subclasses and BHBs. There have been claims of a unified variability scheme that encompasses all X-ray binary types (Psaltis *et al.* 1999; Belloni *et al.* 2002). The bottom line of this scheme is that all of the oscillations must originate in the accretion disk. However, important aspects of this work remain controversial, particularly the handling of BH HFQPOs and their association with the lower kHz QPO observed for NSs (Remillard *et al.* 2002a).

4.4.3 High-frequency QPOs and general relativity

The topic of high-frequency QPOs (HFQPOs) in BHBs (40–450 Hz) continues to evolve in the RXTE era. These transient QPOs have been detected in seven sources (four BHBs and three BHCs). HFQPOs have rms amplitudes that are generally ~1–3% of the mean count rate in a given energy band. Remarkably, three sources exhibit pairs of QPOs that have commensurate frequencies in a 3:2 ratio (Remillard *et al.* 2002b, 2003b; Table 4.2). As shown in Fig. 4.16, GRO J1655 − 40 and XTE J1550 − 564 each exhibit a single such pair of frequencies. GRS 1915+105, on the other hand, shows two pairs of HFQPOs; the complete set of four QPO pairs is shown in Fig. 4.16. In addition (see references in Tables 4.2 & 4.3), single-component HFQPOs have been observed in 4U1630–47 (184 Hz), XTE J1859+226 (190 Hz), XTE J1650–500 (250 Hz), and H 1743–322 (240 Hz). Their profiles are similar to the 300 Hz QPO for GRO J1655–40 shown in the top left panel of Fig. 4.16. HFQPOs occur in the SPL state, except for the 67 Hz QPO in GRS 1915+105 (see Fig. 4.16) which appears in the TD state, especially when $L_X > 10^{38}$ erg s^{-1}.

The preponderance of evidence indicates that HFQPOs do not shift freely in frequency in response to luminosity changes, as do the kHz QPOs in NS systems (see Chapter 2). Instead, they appear to exhibit an "X-ray voiceprint". That is, the QPOs occur in harmonics of an unseen fundamental frequency which has a unique value for each BH. For the three cases that show 3:2 frequency pairs, the relationship between the HFQPO frequencies vs. BH mass scales as M_1^{-1}. This relationship is shown in Fig. 4.17, where we have plotted the frequency of the stronger feature (i.e., $2\times\nu_0$), since the fundamental is generally not seen. These results offer strong encouragement for seeking interpretations of BH HFQPOs via GR theory, since each type of GR disk oscillation under strong gravity varies as M_1^{-1}, assuming the sampled BHs have similar values of the dimensionless spin parameter (a_*).

These commensurate frequencies can be seen as strong support for the idea that HFQPOs may represent some type of resonance phenomenon involving oscillations describable by GR, as originally proposed by Abramowicz and Kluzniak (2001). Resonances in some form may be applicable to both BH and NS systems (Abramowicz *et al.* 2003). We note that the 3:2 harmonic pattern cannot be attributed to a distorted sine wave with harmonic content because the individual detections (in a given energy band, on a given day) generally appear as a single peak in the PDS, and the presence of a pair of commensurate frequencies is recognized only when the ensemble of results is examined.

Coordinate frequencies and their differences (i.e., beat frequencies) in GR were proposed earlier to explain some of the X-ray QPOs from both NSs and BHs (Stella *et al.* 1999); however, this work did not treat the commensurate frequencies of interest here, which were discovered subsequently in three BHB systems. In the resonance hypothesis (Abramowicz & Kluzniak 2001), these harmonic frequencies are discussed in terms of accretion blobs following perturbed orbits in the inner accretion disk. Unlike Newtonian gravity, GR predicts independent oscillation frequencies for each spatial coordinate for orbits around a rotating compact object, as seen from the rest frame of a distant observer. Over the range of radii in

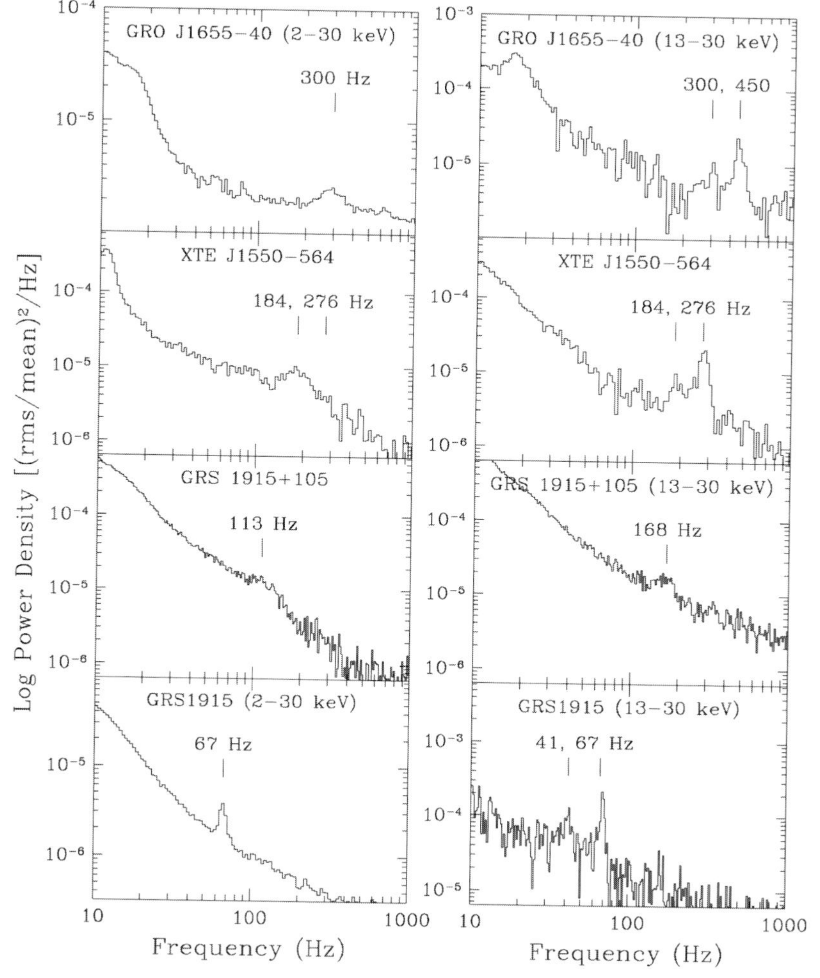

Fig. 4.16. Four pairs of HFQPOs observed in three black hole binary systems. The energy band is 6–30 keV unless otherwise indicated. These usually subtle oscillations are only visible during a fraction of the observations for each source.

the accretion disk where X-rays are expected to originate in the standard disk model, pairs of GR coordinate frequencies have varying, non-integral ratios. Therefore, the discovery of HFQPOs with commensurate frequencies can be seen to suggest enhanced emissivity at a particular radius where a pair of coordinate frequencies are in some type of resonance. Unlike the azimuthal and polar coordinate frequencies, the radial coordinate frequency reaches a maximum value and then falls to zero as the radius decreases toward $R_{\rm ISCO}$ (Kato 2001; Merloni *et al.* 2001). This ensures the possibility of commensurate coordinate frequencies somewhere in the inner disk. For example, there is a wide range in the dimensionless spin parameter, a_*, where one can find a particular radius that corresponds to a 2:1, 3:1, or 3:2 ratio in the orbital and radial coordinate frequencies. A resonance between the polar and radial coordinate frequencies is also possible. In the resonance scenario, linear perturbations may

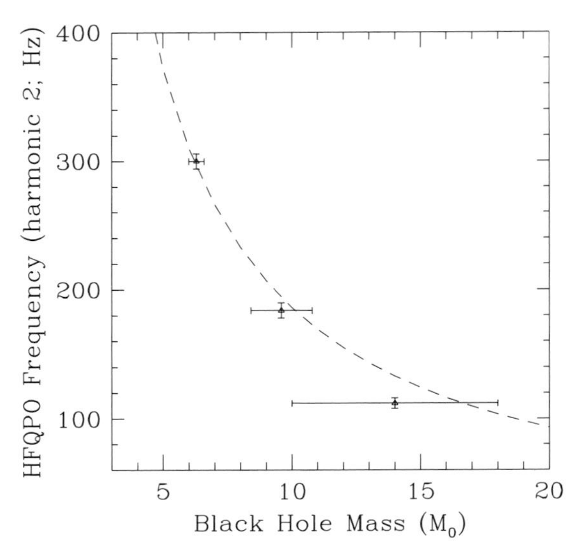

Fig. 4.17. Relationship between HFQPO frequency and BH mass for XTE J1550–564, GRO J1655–40, and GRS 1915+105. These three systems display a pair of HFQPOs with a 3:2 frequency ratio. The frequencies are plotted for the stronger QPO that represents $2 \times \nu_0$. The fundamental is generally not seen in the power spectra. The dashed line shows a relation, $\nu_0(\text{Hz}) = 931(M/\text{M}_\odot)^{-1}$, that fits these data.

grow at these radii, ultimately producing X-ray oscillations that represent some combination of the individual resonance frequencies, their sum, or their difference. However, there remain serious uncertainties as to whether such structures could overcome the severe damping forces and emit X-rays with sufficient amplitude and coherence to produce the observed HFQPOs (Markovic & Lamb 1998).

Models for "diskoseismic" oscillations adopt a more global view of the inner disk as a GR resonance cavity (Kato & Fukue 1980; Wagoner 1999). This paradigm has certain attractions for explaining HFQPOs, but integral harmonics are not predicted for the three types of diskoseismic modes derived for adiabatic perturbations in a thin accretion disk. Clearly, there is a need to investigate further the possibility of resonances within the paradigm of diskoseismology. Other models must be considered as well, e.g., the p-mode oscillations of an accretion torus surrounding a black hole (Rezzolla *et al.* 2003).

It has been argued by Strohmayer (2001a) that HFQPO frequencies are sufficiently high that they require substantial BH spin. For example, in the case of GRO 1655–40 the 450 Hz frequency exceeds the maximum orbital frequency (ν_ϕ) at the ISCO around a Schwarzschild BH (i.e., $a_* = 0$; Section 4.1.5) of mass $M_1 = 6.3 \pm 0.5$ M$_\odot$ (Greene *et al.* 2001). *If* the maximum Keplerian frequency is the highest frequency at which a QPO can be seen, then the results for GRO J1655–40 require a Kerr BH with prograde spin, for example, $a_* > 0.15$. However, the conclusion that spin is required may not be valid if the QPO represents the sum of two beating frequencies. On the other hand, even higher values of the spin parameter may be required if the QPO represents either resonant coordinate frequencies ($a_* > 0.3$; Remillard *et al.* 2002b) or diskoseismic oscillations ($a_* > 0.9$; Wagoner *et al.* 2001).

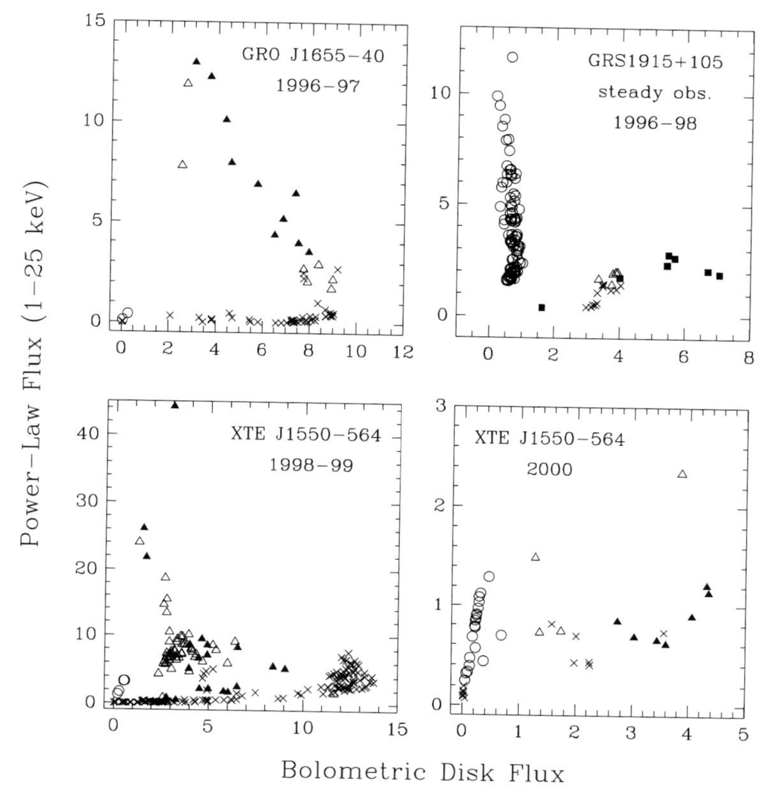

Fig. 4.18. Radiation energy division between the accretion disk and the power-law component. The symbol types denote the X-ray state as TD (×), hard (○), or SPL(Δ). Furthermore, in the SPL state, an open triangle is used when there is only an LFQPO, while a filled triangle denoted the additional presence of an HFQPO. Finally, the TD-like observations that show 67 Hz QPOs in GRS 1915+105 are shown with a filled square. All fluxes are in units of 10^{-8} erg cm^{-2} s^{-1} and corrected for absorption.

Accurate and sensitive measurements of X-ray HFQPO frequencies may lead to a determination of the GR mechanism that is responsible for these oscillations, and ultimately to secure measurements of the spin parameter, a_*, for a number of BHs. These spin measurements would be very valuable in assessing the role of BH rotation in the production of jets (Blandford & Znajek 1977). This is especially true since the three systems with paired frequencies have a history of relativistic mass ejections during some (but not all) of their outbursts.

4.5 Energetics and key variables determining BHB radiation

4.5.1 Division of spectral energy: disk and power-law components

A revealing view of the behavior of a BHB over the course of its entire outburst cycle can be obtained by plotting the flux in the power-law component vs. the flux from the accretion disk. Four such plots are shown in Fig. 4.18, where the various plotting symbols identify the emission states described in Section 4.3. The figure shows the flux diagrams for three BHBs, while considering two of the outbursts of XTE J1550–564 (Muno *et al.* 1999; Remillard *et al.* 2001, 2002b). It has been shown that the emerging patterns in these flux diagrams are largely unaffected by the choice of integration limits, e.g., whether the energy measurements are computed in terms of bolometric flux or the integrated flux within the 2–25 keV PCA bandpass (Remillard *et al.* 2002a).

The TD points ($\times$ symbol; Fig. 4.18) for GRO J1655–40 and XTE J1550–564 (1998–9) appear well organized; they can be described as horizontal tracks in which accretion energy is freely converted to thermal radiation from the accretion disk. These tracks correspond to the standard accretion disk model (Shakura & Sunyaev 1973), and they may also convey moderate Comptonization effects expected from MHD turbulence (e.g., Hawley & Krolik 2001) as these tracks curve upward at high luminosity. The TD points at the highest flux levels correspond to Eddington luminosities of 0.2 for GRO J1655–40, 0.6 for XTE J1550–564 (1998–9), and 0.4 for GRS 1915+105, using the values for mass and distance given in Tables 4.1 and 4.2.

Observations in the hard state (i.e., a dominant power-law component with $1.4 < \Gamma < 2.2$) are plotted as circles in Fig. 4.18. These points form vertical tracks with only minor flux contributions from the disk. The hard-state points coincide with fairly steady radio emission for GRS1915+105 (Muno *et al.* 2001) and XTE J1550–564 in 2000 (Corbel *et al.* 2001), as expected for this state.

Observations in the SPL state (defined by a power-law index $\Gamma > 2.4$ and by QPOs) are plotted as triangles in Fig. 4.18; a solid symbol is used when there is an additional detection of an HFQPO above 100 Hz. The wide diversity in the relative contributions from the disk and power-law components within the SPL state are especially apparent for XTE J1550–564. It is these results, combined with the wide range in the luminosities of the TD and hard tracks shown in Fig. 4.18, that have caused us to abandon the luminosity requirements in defining the X-ray states of BHBs (Section 4.3). Finally, detections of the 67 Hz QPO in GRS 1915+105 are shown as filled squares. They appear to extend the TD branch out to 0.6 L_{Edd}.

4.5.2 Key variables determining BHB radiation

Prior to the RXTE era, it was thought that the X-ray states primarily represent a simple progression in luminosity, and that the emission properties depend only on the accretion rate and the BH mass (e.g., Tanaka & Lewin 1995). The behavioral complexity of sources such as XTE J1550–564 (Fig. 4.18) have challenged this viewpoint (Homan *et al.* 2001; Remillard *et al.* 2002b) because it now appears that any of the three active states of accretion may occur at a given luminosity.

Finally, what other parameters must be considered in the theory of BH accretion? It is clear that BH spin and the angle between the spin and disk axes must also be considered in a complete theory of BH accretion, but these would not help to explain the types of rapid state transitions seen in many systems. As noted in several previous sections, MHD instabilities are widely invoked to explain non-thermal radiation, and magnetic fields are additionally expected to play a leading role in the formation and collimation of jets. It then seems relevant to question whether a parameter of magnetism, such as the ratio of magnetic to gas pressure, should be considered a key variable in accretion physics. The global magnetic field geometry may also play an essential role in the formation of some X-ray states. The continued development of 3-D MHD simulations is expected to be very fruitful in gaining a deeper understanding of the radiation emitted by black hole binaries.

4.6 Concluding remarks

As reviewed in Section 4.4.3, high-frequency QPOs at 40–450 Hz have been observed with RXTE for four BHBs and three BHCs. Furthermore, three of these sources show harmonic (3:2) pairs of frequencies that scale as M_{BH}^{-1}. The models for these QPOs (e.g., orbital resonance and diskoseismic oscillations) invoke strong-field GR effects in the inner accretion disk, and they depend on both the mass and spin of the BH. On the other hand, studies of broadened Fe Kα emission lines (Section 4.2.3), which can also reveal the conditions in the very inner accretion disk, may prove as revealing as timing studies. The line photons that reach a distant observer are gravitationally redshifted and Doppler and transverse-Doppler shifted. Encoded in the line profile are the mass and spin of the BH. The most provocative result has been obtained for the BHC XTE J1650–500 (Section 4.2.3), where the line profile suggests the presence of an extreme Kerr BH.

The behavior of the massive compact objects reviewed herein supports the view that they are bona fide BHs, which are described by GR and were formed by the complete gravitational collapse of matter. Our challenge is to prove that this conclusion is correct by making clean quantitative measurements of relativistic effects in the strong gravitational fields near these objects. At present, no one can say which future mission can best help us meet this challenge: LISA, MAXIM, Constellation-X, or an X-ray Timing Observatory (XTO). Perhaps all of the approaches will be required. In any case, an XTO with effective area and telemetry ten times that of RXTE and with improved energy resolution, would be a powerful probe of physics near the event horizon. Consider that RXTE, with just 1.6 times the effective area of Ginga, broke through to discover kHz QPOs in NSs, a discovery that led quickly to hard constraints on dense matter.

Recent optical studies have promoted two of the black hole candidates listed in Table 4.3 to the status of black hole binaries: (1) GS 1354–64 (Casares *et al.* 2004), and (2) XTE J1650–500 (Orosz *et al.* 2004). Thus the number of confirmed black hole binaries is now 20 (see Tables 4.1 & 4.2).

Acknowledgements

We thank Jon Miller, Mike Muno, David Smith, Jean Swank, John Tomsick, and Rudy Wijnands for their help in assembling the catalogue of candidate BHBs presented in Table 4.3, and Keith Arnaud, George Field, Mike Garcia, Aya Kubota, Kazuo Makishima, and Mike Revnivtsev for valuable comments. We are indebted to Andy Fabian, Jon Miller, Ramesh Narayan, Jerry Orosz and Andrej Zdziarski for a careful reading of portions of the manuscript and their valuable comments. We are especially grateful to Jeroen Homan and John Tomsick for their detailed comments on a near-final draft of the manuscript. We thank Ann Esin for supplying Fig. 4.1, Jon Miller for supplying Fig. 4.6, John Tomsick and Emrah Kalemci for the use of unpublished data on 4U1543–47, John Huchra for advice on the distance to the LMC, and Suresh Kumar for his assistance in typesetting the manuscript. This work was supported in part by NASA under Grants NAG5-10813 and by the NASA contract to MIT for support of RXTE.

References

Abramowicz, M. A. (1998) in *Theory of Black Hole Accretion Discs*, eds. M. A. Abramowicz, G. Bjornsson and J. E. Pringle (Cambridge, Cambridge University Press), 50–60

Åbramowicz, M. A. and Kluzniak, W. (2001), *A&A* **374**, L19–L20

Abramowicz, M. A., Karas, V., Kluzniak, W., *et al.* (2003), *PASJ* **55**, 467–471

Agol, E. and Krolik, J. H. (2000), *ApJ* **528**, 161–170

Agol, E., Kamionkowski, M., Koopmans, L.V. E. and Blandford, R. D. (2002), *ApJ* **576**, L131–L135

Arnaud, K. and Dorman, B. (2002), XSPEC (An X-ray Spectral Fitting Package), version 11.2x (NASA/GSFC/HEASARC, Greenbelt)

Augusteijn, T., Kuulkers, E. and van Kerkwijk, M. H. (2001), *ApJ* **375**, 447–454

Baganoff, F. K., Bautz, M. W., Brandt, W. N., *et al.* (2001), *Nature* **413**, 45–48

Bahcall, J. N. (1986), *ARA&A* **24**, 577–611

Ball, L., Kesteven, M. J., Campbell-Wilson, D., *et al.* (1995), *MNRAS* **273**, 722–730

Balucinska-Church, M. and Church M. J. (2000), *MNRAS* **312**, L55–L59

Barr, P., White, N. E. and Page, C.G. (1985), *MNRAS* **216**, 65P–70P

Barret, D., Roques, J. P., Mandrou, P., *et al.* (1992), *ApJ* **392**, L19–L22

Barret, D., Grindlay, J. E., Bloser, P. F., *et al.* (1996a), *IAU Circ.* 6519

Barret, D., McClintock, J. E. and Grindlay, J. E. (1996b), *ApJ* **473**, 963–973

Barret, D., Olive, J. F. and Boirin, L. (2000), *ApJ* **533**, 329–351

Belloni, T., van der Klis, M., Lewin, W. H. G., *et al.* (1997), *A&A* **322**, 857–867

Belloni, T., Mendez, M., van der Klis, M., *et al.* (1999), *ApJ* **519**, L159–L163

Belloni, T., Klein-Wolt, M., Mendez, M., *et al.* (2000), *A&A* **355**, 271–290

Belloni, T., Psaltis, D. and van der Klis, M. (2002), *ApJ*, **572**, 392–406

Blandford, R. D. (2002), in *Lighthouses of the Universe*, eds. M. Gilfanov, R. Sunyaev, *et al.* (Berlin, Springer) (astro-ph 0202265)

Blandford, R. D. and Begelman, M. C. (1999), *MNRAS* **303**, L1–L5

Blandford, R. D. and Znajek, R. L. (1977), *MNRAS* **179**, 433–456

Bolton, C. T. (1972), *Nature* **240**, 124–126

Borozdin, K. N., Aleksandrovich, N. L., Aref'ev, V. A., *et al.* (1995), *AstL* **21**, 212–216

Borozdin, K. N., Revnivtsev, M. G., Trudolyubov, S. P., *et al.* (1998), *AstL* **24**, 435–444

Boyd, P. T., Smale, A. P., Homan, J., *et al.* (2000), *ApJ* **542**, L127–L130

Bradt, H., Levine, A. M., Remillard, R. A. and Smith, D. A. (2001), *MmSAI* **73**, 256–271

Bradt, H. V. D. and McClintock, J. E. (1983), *ARA&A* **21**, 13–66

Branduardi, G., Ives, J. C., Sanford, P. W., *et al.* (1976), *MNRAS* **175**, 47P–56P

Brocksopp, C., Fender, R. P., Larionov, V., *et al.* (1999), *MNRAS* **309**, 1063–1073

Brocksopp, C., Jonker, P. G., Fender, R. P., *et al.* (2001), *MNRAS* **323**, 517–528

Brocksopp, C., Fender, R. P., McCollough, M., *et al.* (2002), *MNRAS* **331**, 765–775

Bronfman, L., Cohen, R. S., Alvarez, H., *et al.* (1988), *ApJ* **324**, 248–266

Brown, G. E. and Bethe, H. A. (1994), *ApJ* **423**, 659–664

Brown, G. E., Lee, C.-H, Wijers, R. A. M. J. and Bethe, H. A. (2000a), *Phys. Rep.* **333–334**, 471–504

Brown, G. E., Lee, C.-H., Wijers, R. A. M., *et al.* (2000b), *NewA* **5**, 191–210

Campana, S., Stella, L., Belloni, T., *et al.* (2002), *A&A* **384**, 163–170

Cannizzo, J. K. (1993), *ApJ* **419**, 318–336

Casares, J. and Charles, P. A. (1994), *MNRAS* **271**, L5–L9

Casares, J., Charles, P. A., Naylor, T. and Pavlenko, E. P. (1993), *MNRAS* **265**, 834–852

Casares, J., Charles, P. A. and Marsh, T. R. (1995), *MNRAS* **277**, L45–L50

Casares *et al.* 2004, *ApJ* **613**, L133–L136

Chakrabarti, S. K. and Manickam, S. G. (2000), *ApJ* **531**, L41–L44

Chaty, S., Mignani, R. P. and Israel, G. L. (2002), *MNRAS* **337**, L23–L26

Chen, W., Shrader, C. R. and Livio, M. (1997), *ApJ* **491**, 312–338

Churazov, E., Gilfanov, M., Sunyaev, R., *et al.* (1993), *ApJ* **407**, 752–757

Churazov, E., Gilfanov, M. and Revnivtsev, M. (2001), *MNRAS* **321**, 759

Coe, M. J., Engel, A. R. and Quenby, J. J. (1976), *Nature* **259**, 544–545

Cooke, B. A., Levine, A. M., Lang, F. L., *et al.* (1984), *ApJ* **285**, 258–263

Corbel, S., Fender, R. P., Tzioumis, A. K., *et al.* (2000), *A&A* **359**, 251–268

Corbel, S., Kaaret, P., Jain, R. K., *et al.* (2001), *ApJ* **554**, 43–48
Corbel, S., Fender, R. P., Tzioumis, A. K., *et al.* (2002), *Science* **298**, 196–199
Corbel, S., Nowak, M. A., Fender, R. P., *et al.* (2003), *A&A*, **400**, 1007–1012
Cowley, A. P., Crampton, D., Hutchings, J. B., *et al.* (1983), *ApJ* **272**, 118–122
Cowley, A. P., Crampton, D. and Hutchings, J. B. (1987), *AJ* **92**, 195–199
Cowley, A. P., Schmidtke, P. C., Ebisawa, K., *et al.* (1991), *ApJ* **381**, 526–533
Cowley, A. P., Schmidtke, P. C., Anderson, A. L. and McGrath, T.K. (1995), *PASP* **107**, 145–147
Cui, W., Heindl, W. A., Swank, J. H., *et al.* (1997a), *ApJ* **487**, L73–L76
Cui, W., Zhang, S. N., Focke, W. and Swank, J. H. (1997b), *ApJ* **484**, 383–393
Cui, W., Shrader, C. R., Haswell, C. A. and Hynes, R. I. (2000a), *ApJ* **535**, L123–L127
Cui, W., Zhang, S. N. and Chen, W. (2000b), *ApJ* **531**, L45–L48
Cui, W., Schulz, N. S., Baganoff, F. K., *et al.* (2001), *ApJ* **548**, 394–400
Dal Fiume, D., Frontera, F., Orlandini, M., *et al.* (1999), *IAU Circ.* 7291
Davies, R. D., Walsh, D. and Browne, I. W. A., *et al.* (1976), *Nature* **261**, 476–478
Dhawan, V., Mirabel, I. F. and Rodriguez, L. F. (2000), *ApJ* **543**, 373–385
Dieters, S.W., Belloni, T., Kuulkers, E., *et al.* (2000), *ApJ* **538**, 307–314
Dijk, R. van , Bennett, K., Collmar, W., *et al.* (1995), *A&A* 296, L33–L36
di Matteo, T., Celotti, A. and Fabian, A. C. (1999), *MNRAS* **304**, 809–820
Di Salvo, T., Done, C., Zycki, P. T., *et al.* (2001), *ApJ* **547**, 1024–1033
Done, C. and Nayakshin, S. (2001), *MNRAS* **328**, 616–622
Done, C. and Zycki, P. T. (1999), *MNRAS* **305**, 457–468
Done, C., Mulchaey, J. S., Mushotzky, R. F. and Arnaud, K. A. (1992), *ApJ* **395**, 275–288
Dove, J. B., Wilms, J. and Begelman, M. C. (1997a), *ApJ* **487** 747–758
Dove, J. B., Wilms, J., Maisack, M. and Begelman, M. C. (1997b), *ApJ* **487**, 759–768
Dubus, G., Hameury, J.-M and Lasota, J.-P. (2001), *A&A* **373**, 251–271
Ebisawa, K., Mitsuda, K. and Inoue, H. (1989), *PASJ* **41**, 519–530
Ebisawa, K., Mitsuda, K. and Tomoyuki, H. (1991), *ApJ* **367**, 213–220
Ebisawa, K., Makino, F., Mitsuda, K., *et al.* (1993), *ApJ* **403**, 684–689
Ebisawa, K., Ogawa, M., Aoki, T., *et al.* (1994), *PASJ* **46**, 375–394
Elvis, M., Page, C. G., Pounds, K. A., *et al.* (1975), *Nature* **257**, 656–657
Esin, A. A., McClintock, J. E. and Narayan, R. (1997), *ApJ* **489**, 865–889
Esin, A. A., Narayan, R., Cui, W., *et al.* (1998), *ApJ* **505**, 854–868
Esin, A. A., McClintock, J. E., Drake, J. J., *et al.* (2001), *ApJ* **555**, 483–488
Fabbiano, G., Zezas, A. and Murray, S. S. (2001), *ApJ* **554**, 1035–1043
Fabian, A. C., Rees, M. J., Stella, L. and White, N. E. (1989), *MNRAS* **238**, 729–736
Fabian, A. C., Iwasawa, K., Reynolds, C. S. and Young, A. J. (2000), *PASP* **112**, 1145–1161
Falcke, H. and Biermann, P. L. (1995), *A&A* **293**, 665–682
Fender, R. P. (2001), *MNRAS*, **322**, 31–42
Fender, R., Corbel, S., Tzioumis, T., *et al.* (1999a), *ApJ* **519**, L165–L168
Fender, R. P., Garrington, S. T., McKay, D. J., *et al.* (1999b), *MNRAS* **304**, 865–876
Fender, R. P., Hjellming, R. M., Tilanus, R. P. J., *et al.* (2001), *MNRAS* **322**, L23–L27
Feng, Y. X., Zhang, S. N., Sun, X., *et al.* (2001), *ApJ* **553**, 394–398
Field, R. D. and Rogers, R. D. (1993), *ApJ* **403**, 94–109
Filippenko, A. V. and Chornock, R. (2001), *IAU Circ.* 7644
Filippenko, A. V., Matheson, T. and Ho, L. C. (1995a), *ApJ* **455**, 614–622
Filippenko, A. V., Matheson, T. and Barth, A. J. (1995b), *ApJ* **455**, L139–L142
Filippenko, A. V., Matheson, T., Leonard, D. C., *et al.* (1997), *PASP* **109**, 461–467
Filippenko, A. V., Leonard, D. C., Matheson, T., *et al.* (1999), *PASP* **111**, 969–979
Freedman, W. L., Madore, B. F., Gibson, B. K., *et al.* (2001), *ApJ* **553**, 47–72
Frontera, F., Palazzi, E., Zdziarski, A. A., *et al.* (2001a), *ApJ* **546**, 1027–1037
Frontera, F., Zdziarski, A. A., Amati, L., *et al.* (2001b), *ApJ* **561**, 1006–1015
Frontera, F., Amati, L., Zdziarski, A. A., *et al.* (2003), *ApJ* **592**, 1110–1118
Fryer, C. and Kalogera V. (2001), *ApJ* **554**, 548–560
Garcia, M. R., McClintock, J. E., Narayan, R., *et al.* (2001), *ApJ* **553**, L47–L50
Gelino, D. M., Harrison, T. E. and McNamara, B. J. (2001a), *AJ* **122**, 971–978
Gelino, D. M., Harrison, T. E. and Orosz, J. A. (2001b), *AJ* **122**, 2668–2678
George, I. M. and Fabian, A. C. (1991), *MNRAS* **249**, 352–367

Gierlinski, M., Zdziarski, A. A., Poutanen, J., *et al.* (1999), *MNRAS* **309**, 496–512
Gierlinski, M., Maciolek-Niedzwiecki, A. and Ebisawa, K. (2001), *MNRAS* **325**, 1253–1265
Gies, D. R. and Bolton, C. T. (1982), *ApJ* **260**, 240–248
Goldoni, P., Vargas, M., Goldwurm, A., *et al.* (1999), *ApJ* **511**, 847–851
Greene, J., Bailyn, C. D. and Orosz, J. A. (2001), *ApJ* **554**, 1290–1297
Greiner, J., Dennerl, K. and Predehl, P. (1996), *A&A* **314**, L21–L24
Greiner, J., Cuby, J. G. and McCaughrean, M. J (2001a), *Nature* **414**, 522–525
Greiner, J., Cuby, J. G., McCaughrean, M. J., *et al.* (2001b), *A&A* **373**, L37–L40
Groot, P., Tingay, S., Udalski, A. and Miller, J. (2001), *IAU Circ.* 7708
Grove, J. E., Johnson, W. N., Kroeger, R. A., *et al.* (1998), *ApJ* **500**, 899–908
Gursky, H., Bradt, H., Doxsey, R., *et al.* (1978), *ApJ* **223**, 973–978
Haardt, F. and Maraschi, L. (1991), *ApJ* **380**, L51–L54
Hameury, J.-M., Lasota, J.-P., McClintock, J. E. and Narayan, R. (1997), *ApJ* **489**, 234–243
Hameury, J.-M., Barret, D., Lasota, J.-P., *et al.* (2003), *A&A* **399**, 631–637
Han, X. and Hjellming, R. M. (1992), *ApJ* **400**, 304–314
Hannikainen, D. C., Hunstead, R. W., Campbell-Wilson, D., *et al.* (2000), *ApJ* **540**, 521–534
Hannikainen, D., Campbell-Wilson, D., Hunstead, R., *et al.* (2001), *ApSSS* **276**, 45–48
Harlaftis, E. T., Horne, K. and Filippenko, A. V. (1996), *PASP* **108**, 762–771
Hawley, J. F. and Krolik, J. H. (2001), *ApJ* **548** 348–367
Heuvel, E. P. J. van den (1992), in *ESA, Environment Observation and Climate Modelling Through International Space Projects* (SEE N93–23878 08–88)
Hjellming, R. M. and Rupen, M. P. (1995), *Nature* **375**, 464–468
Hjellming, R. M., Calovini, T. A. and Han, X. H. (1988), *ApJ* **335**, L75–L78
Hjellming, R. M., Rupen, M. P. and Mioduszewski, A. J. (1998a), *IAU Circ.* 6924
(1998b), *IAU Circ.* 6934
Hjellming, R. M., Rupen, M. P., Mioduszewski, A. J., *et al.* (1999), *ApJ* **514**, 383–387
Hjellming, R. M., Rupen, M. P., Hunstead, R. W., *et al.* (2000), *ApJ* **544**, 977–992
Homan, J., Wijnands, R., van der Klis, M., *et al.* (2001), *ApJS* **132**, 377–402
Homan, J., Klein-Wolt, M., Rossi, S., *et al.* (2003a), *ApJ* **586**, 1262–1267
Homan, J., Miller J. M., Wijnands, R., *et al.* (2003b), *ATEL* 162
Hooft, F. van der, Kouveliotou, C., van Paradijs, J., *et al.* (1996), *ApJ* **458**, L75–L78
(1999), *ApJ* **513**, 477–490
Humphrey, P. J., Fabbiano, G., Elvis, M., *et al.* (2003), *MNRAS* **344**, 134–148
Hutchings, J. B., Crampton, D., Cowley, A. P., *et al.* (1987), *AJ* **94**, 340–344
Hynes, R. I., Steeghs, D., Casares, J., *et al.* (2003), *ApJ* **583**, L95–L98
Hynes, R. I., Roche, P., Charles, P.A and Coe, M. J. (1999), *MNRAS* **305**, L49–L53
Igumenshchev, I. V. and Abramowicz, M. A. (1999), *MNRAS* **303**, 309–320
in't Zand, J. J. M., Markwardt, C. B., Bazzano, A., *et al.* (2002a), *A&A* **390**, 597–609
in't Zand, J. J. M., Miller, J. M., Oosterbroeck, T. and Parmar, A. N. (2002b), *A&A* **394**, 553–560
Israelian, G., Rebolo, R. and Basri, G. (1999), *Nature* **401**, 142–144
Jones, C., Forman, W., Tananbaum, H. and Turner, M. J. L. (1976), *ApJ* **210**, L9–L11
Kalemci, E., Tomsick, J. A., Rothschild, R. E., *et al.* (2002) *ApJ* **586**, 419–426
Kalogera, V. and Baym, G. (1996), *ApJ* **470**, L61–L64
Kaluzienski, L. J., Holt, S. S., Boldt, E. A., *et al.* (1975), *ApJ* **201**, L121–L124
Kato, S. (2001), *PASJ* **53**, 1–24
Kato, S. and Fukue, J. (1980), *PASJ* **32**, 377–388
Kato, S., Fukue, J. and Mineshige, S. (1998), *Black-Hole Accretion Disks* (Japan, Kyoto University Press)
Kawaguchi, T., Shimura, T. and Mineshige, S. (2000), *NewAR* **44**, 443–445
Kennea, J. A. and Skinner, G. K. (1996), *PASJ* **48**, L117–L117
King, A. R., Davies, M. B., Ward, M. J., *et al.* (2001), *ApJ* **552**, L109–L112
Kitamoto, S., Tsunemi, H., Pedersen, H., *et al.* (1990), *ApJ* **361**, 590–595
Klein-Wolt, M., Fender, R. P., Pooley, G. G., *et al.* (2002), *MNRAS* **331**, 745–764
Klis M. van der (1994), *ApJS* **92**, 511–519
(1995), in *X-ray Binaries*, eds. W. H. G. Lewin, J. van Paradijs and E. P. J. van den Heuvel (Cambridge, Cambridge University Press), 252–307
Kong, A. K. H., Charles, P. A., Kuulkers, E. and Kitamoto, S. (2002), *MNRAS* **329**, 588–596
Kong, A. K. H., McClintock, J. E., Garcia, M. R., *et al.* (2002), *ApJ* **570**, 277–286

Kotani, T., Kawai, N., Nagase, F., *et al.* (2000), *ApJ* **543**, L133–L136
Kubota, A. and Makishima, K. 2003, *ApJ*, **601**, 428–438
Kubota, A., Makishima, K. and Ebisawa, K. (2001), *ApJ* **560**, L147–L150
Kuulkers, E., Wijnands, R., Belloni, T., *et al.* (1998), *ApJ* **494**, 753–758
Kuulkers, E., Fender, R. P., Spencer, R. E., *et al.* (1999), *MNRAS* **306**, 919–925
Laor, A. (1991), *ApJ* **376**, 90–94
Lasota, J.-P. (2001), *NewAR* **45**, 449–508
Levine, A. M., Bradt, H., Cui, W., *et al.* (1996), *ApJ* **469**, 33–36
Ling, J. C., Mahoney, W. A., Wheaton, W. A. and Jacobson, A. S. (1987), *ApJ* **321**, L117–L122
Ling, J. C., Wheaton, W. A., Wallyn, P., *et al.* (2000), *ApJS* **127**, 79–124
Liu, B. F., Mineshige, S. and Shibata, K. (2002), *ApJ* **572**, L173–L176
Liu, Q. Z., van Paradijs, J. and van den Heuvel, E. P. J. (2000), *A&As* **147**, 25–49
Liu, Q.Z., van Paradijs, J. and van den Heuvel, E. P. J. (2001), *A&A* **368**, 1021–1054
Loewenstein, M., Mushotzky, R. F., Angelini, L., *et al.* (2001), *ApJ* **555**, L21–L24
Lowes, P., in 't Zand, J. J. M., Heise, J., *et al.* (2002), *IAU Circ.* 7843
Lynden-Bell, D. and Pringle, J. E. (1974), *MNRAS* **168**, 603–637
Makishima, K., Maejima, Y., Mitsuday, K., *et al.* (1986), *ApJ* **308**, 635–643
Makishima, K., Kubota, A, Mizuno, T., *et al.* (2000), *ApJ* **535**, 632–643
Malzac, J., Belloni, T., Spruit, H. C. and Kanbach, G. (2003), *A&A* **407**, 335–345
Mandelbrot, B. B. (1999), *Multifractals and 1/f Noise: Wild Self-Affinity in Physics* (Heidelberg, Springer-Verlag)
Margon, B., Thorstensen, J. R. and Bowyer, S. (1978), *ApJ* **221**, 907–911
Markert, T. H., Canizares, C. R., Clark, G. W., *et al.* (1973), *ApJ* **184**, L67–L70
Markoff, S., Falcke, H. and Fender, R. (2001), *A&A* **372**, L25–L28
Markovic, D. and Lamb, F. K. (1998), *ApJ* **507**, 316–326
Markwardt, C. (2001), *ApSS* **276**, 209–212
Markwardt, C. B. and Swank, J. H. (2003), *IAU Circ.* 8056
Marsh, T. R., Robinson, E. L. and Wood, J. H. (1994), *MNRAS* **266**, 137–154
Marti, J., Mirabel, I. F., Duc, P.-A. and Rodriguez, L. F. (1997), *A&A* **323**, 158–162
Marti, J., Mirabel, I. F., Chaty, S. and Rodriguez, L. F. (2000), *A&A* **363**, 184–187
Marti, J., Mirabel, I. F., Rodriguez, L. F. and Smith, I. A. (2002), *A&A* **386**, 571–575
Martocchia, A., Matt, G., Karas, G., *et al.* (2002), *A&A* **387**, 215–221
Matt, G., Perola, G.C. and Piro, L. (1991), *A&A* **247**, 25–34
Mendez, M. and van der Klis, M. (1997), *ApJ* **479**, 926–932
McClintock, J. E. and Remillard, R. A. (1986), *ApJ* **308**, 110–122
(2000), *ApJ* **531**, 956–962
McClintock, J. E., Horne, K. and Remillard, R. A. (1995), *ApJ* **442**, 358–365
McClintock, J. E., Garcia, M. R., Caldwell, N., *et al.* (2001a), *ApJ* **551**, L147–L150
McClintock, J. E., Haswell, C.A., Garcia, M. R., *et al.* (2001b), *ApJ* **555**, 477–482
McClintock, J. E., Narayan, R., Garcia, M. R., *et al.* (2003), *ApJ* **539**, 435–451
McConnell, M. L., Zdziarski, A. A., Bennett, K., *et al.* (2002), *ApJ* **572**, 984–995
McKinney, J. C. and Gammie, C. F. (2002), *ApJ* **573**, 728–737
Mendez, M. and van der Klis, M. (1997), *ApJ* **479**, 926–932
Menou, K., Esin, A. A., Narayan, R., *et al.* (1999), *ApJ* **520**, 276–291
Merloni, A. and Fabian, A. C. (2001a), *MNRAS* **321**, 549–552
(2001b), *MNRAS* **328**, 958–968
(2002), *MNRAS* **332**, 165–175
Merloni, A., Fabian, A. C. and Ross, R. R. (2000), *MNRAS* **313**, 193–197
Merloni, A., Vietri, M., Stella, L. and Bini, D. (2001), *MNRAS* **304**, 155–159
Meyer, F., Liu, B. F. and Meyer–Hofmeister, E. (2000), *A&A* **361**, 175–188
Miller, J. M., Fox, D. W., Di Matteo, T., *et al.* (2001), *ApJ* **546**, 1055–1067
Miller, J. M., Fabian, A. C., in't Zand, J. J. M., *et al.* (2002a), *ApJ* **577**, L15–L18
Miller, J. M., Fabian, A. C., Wijnands, R., *et al.* (2002b), *ApJ* **570**, L69–L73
(2002c), *ApJ* **578**, 348–356
Miller, J. M., Wijnands, R., Rodriguez–Pascual, P. M., *et al.* (2002d), *ApJ* **566**, 358–364
Miller, J. M., Fabbiano, G., Miller, M. C., and Fabian, A. C. (2003a), *ApJ* **585**, L37–L40
Miller, J. M., Zezas, A., Fabbiano, G. and Schweizer, F. (2003b), *ApJ* **609**, 728–734 (astro-ph/0302535)

Mirabel, I. F. and Rodriguez, L. F. (1994), *Nature* **371**, 46–48
(1999), *ARA&A* **37**, 409–443
(2003), *Science* **300**, 1119–1120
Mirabel, I. F., Rodriguez, L. F. and Cordier, B. (1993), *IAU Circ*. 5876
Mitsuda, K., Inoue, H., Koyama, K., *et al*. (1984), *PASJ* **36**, 741–759
Miyamoto, S. and Kitamoto, S. (1991), *ApJ* **374**, 741–743
Miyamoto, S., Iga, S., Kitamoto, S. and Kamado, Y. (1993), *ApJ* **403**, L39–L42
Morgan, E. H., Remillard, R. A. and Greiner, J. (1997), *ApJ* **482**, 993–1010
Motch, C., Guillout, P., Haberl, F., *et al*. (1998), *A&AS* **132**, 341–359
Muno, M. P., Morgan, E. H. and Remillard, R. A. (1999), *ApJ* **527**, 321–340
Muno, M. P., Morgan, E. H., Remillard, R. A., *et al*. (2001), *ApJ*, **556**, 515–532
Murdin, P., Griffiths, R. E., Pounds, K. A., *et al*. (1977), *MNRAS* **178**, 27P–32P
Narayan, R. (1996), *ApJ* **462**, 136–141
Narayan, R. (2002), in *Lighthouses of the Universe*, eds. M. Gilfanov, R. Sunyaev, *et al*. (Berlin, Springer)
Narayan, R. and Insu, Y. (1994), *ApJ* **428**, L13–L16
(1995), *ApJ* **444**, 231–243
Narayan, R., McClintock, J. E. and Yi, I. (1996), *ApJ* **457**, 821–833
Narayan, R., Garcia, M. R. and McClintock, J. E. (1997), *ApJ* **478**, L79–L82
Narayan, R., Igumenshchev, I. V. and Abramowicz, M. A. (2000), *ApJ* **539**, 798–808
Narayan, R., Garcia, M. R. and McClintock, J. E. (2002), in *Proc. Ninth Marcel Grossmann Meeting*, eds. V.G. Gurzadyan *et al*. (Singapore, World Scientific), 405–425
Nelemans, G. and van den Heuvel, E. P. J. (2001), *A&A* **376**, 950–954
Nobili, L., Turolla, R., Zampieri, L. and Belloni, T. (2000), *ApJ* **538**, L137–L140
Novikov, I. D. and Thorne, K. S. (1973), in *Black Holes*, eds. C. DeWitt and B. DeWitt (New York, Gordon & Breach)
Nowak, M. A. (2000), *MNRAS* **318**, 361–367
Nowak, M. A. and Wilms, J. (1999), *ApJ* **522**, 476–486
Nowak, M. A., Wilms, J., Heindl, W. A., *et al*. (2001), *MNRAS* **320**, 316–326
Nowak, M. A., Wilms, J. and Dove, J. B. (2002), *MNRAS* **332**, 856–878
Ogley, R. N., Ash, T. D. C. and Fender, R. P. (1997), *IAU Circ*. 6726
Orosz, J. A. and Bailyn, C. D. (1997), *ApJ* **477**, 876–896
Orosz, J. A., Bailyn, C. D., McClintock, J. E. and Remillard, R. A. (1996), *ApJ* **468**, 380–390
Orosz, J. A., Jain, R. K., Bailyn, C. D., *et al*. (1998), *ApJ* **499**, 375–384
Orosz, J. A., Kuulkers, E., van der Klis, M., *et al*. (2001), *ApJ* **555**, 489–503
Orosz, J. A., Groot, P. J., van der Klis, M., *et al*. (2002a), *ApJ* **568**, 845–861
Orosz, J. A., Polisensky, E. J., Bailyn, C. D., *et al*. (2002b), *BAAS* 201, 1511
Orosz *et al*. (2004), *ApJ* **616**, 376–382
Osaki, Y. (1974), *PASJ* **26**, 429–436
Owen, F. N., Balonek, T. J., Dickey, J., *et al*. (1976), *ApJ* **203**, L15–L16
Pan, H. C., Skinner, G. K., Sunyaev, R. A. and Borozdin, K. N. (1995), *MNRAS* **274**, L15–L18
Paradijs, J. van (1995), in *X-ray Binaries*, eds. W. H. G. Lewin, J. van Paradijs and E. P. J. van den Heuvel (Cambridge, Cambridge University Press), 536–577
Paradijs, J. van and McClintock, J. E. (1995), in *X-ray Binaries*, eds. W. H. G. Lewin, J. van Paradijs and E. P. J. van den Heuvel (Cambridge, Cambridge University Press), 58–125
Parmar, A. N., Angelini, L., Roche, P. and White, N. E. (1993), *A&A* **279**, 179–187
Pottschmidt, K., Wilms, J., Nowak, M. A., *et al*. (2003), *A&A* **407**, 1039–1058
Poutanen, J. and Fabian, A. C. (1999), *MNRAS* **306**, L31–L37
Pringle, J. E. (1981), *AR&A* **19**, 137–162
Pringle, J. E. and Rees, M. J. (1972), *A&A* **21**, 1–9
Psaltis, D., Belloni, T. and van der Klis, M. (1999), *ApJ* **520**, 262–270
Quataert, E. and Gruzinov, A. (2000), *ApJ* **539**, 809–814
Quataert, E. and Narayan, R. (1999), *ApJ* **520**, 298–315
Reig, P., Belloni, T., van der Klis, M., *et al*. (2000), *ApJ* **541**, 883–888
Remillard, R. A. and Morgan, E. H., (1999), *Bull. AAS* **31**, 1421
Remillard, R. A., Orosz, J. A., McClintock, J. E. and Bailyn, C. D. (1996), *ApJ* **459**, 226–235
Remillard, R., Levine, A., Swank, J. and Strohmayer, T. (1997), *IAU Circ*. 6710
Remillard, R. A., Morgan, E. H., McClintock, J. E. *et al*. (1999), *ApJ* **522**, 397–412

Remillard, R. A., Morgan, E. H. and Muno, M. (2001), in *Proc. Ninth Marcel Grossmann Meeting*, eds. V. G. Gurzadyan, *et al.* (Singapore, World Scientific), 2220–2223
Remillard, R. A., Muno, M. P., McClintock, J. E. and Orosz, J. A. (2002a), *ApJ* **580**, 1030–1042
Remillard, R. A., Sobczak, G. J., Muno, M. P. and McClintock, J. E. (2002b), *ApJ* **564**, 962–973
Remillard, R. A., Levine, A. M., Morgan, E. H., *et al.* (2003a), *IAU Circ.* 8050
Remillard, R. A., Muno, M. P., McClintock, J. E. and Orosz, J. A. (2003b), *BAAS* **35**, 648
Revnivtsev, M., Gilfanov, M. and Churazov, E. (1998a), *A&A* **339**, 483–488
Revnivtsev, M., Gilfanov, M., Churazov, E., *et al.* (1998b), *A&A* **331**, 557–563
Revnivtsev, M., Borozdin, K. N., Priedhorsky, W. C. and Vikhlinin, A. (2000a), *ApJ* **530**, 955–965
Revnivtsev, M., Sunyaev, R. and Borozdin, K. (2000b), *A&A* **361** L37–L39
Revnivtsev, M., Trudolyubov, S. P. and Borozdin, K. N. (2000c), *MNRAS* **312**, 151–158
Revnivtsev, M., Gilfanov, M. and Churazov, E. (2001), *A&A* **380**, 520–525
Revnivtsev, M., Gilfanov, M., Churazov, E. and Sunyaev, R. (2002), *A&A* **391**, 1013–1022
Revnivtsev, M., Chernvakova, M., Capitanio, F., *et al.* (2003), *ATEL* 132
Reynolds, C. S. and Nowak, M. A. (2003), *Phys. Rep.* **377**, 389–466
Reynolds, A. P., Quaintrell, H., Still, M. D., *et al.* (1997), *MNRAS* **288**, 43–52
Rezzolla, L., Yoshida, S., Maccarone, T. J. and Zanotti, O. (2003), *MNRAS* **344**, L37–L41
Rhoades, C. E. and Ruffini, R. (1974), *Phys. Rev. Lett.* **32**, 324
Rodriguez, L. F., Mirabel, I. F. and Marti, J. (1992), *ApJ* **401**, L15–L18
Romani, R. W. (1998), *A&A* **333**, 583–590
Rothschild, R. E., Blanco, P. R., Gruber, D. E., *et al.* (1998), *ApJ* **496**, 538–549
Rothstein, D. M., Eikenberry, S. S., Chatterjee, S., *et al.* (2002), *ApJ* **580**, L61–L63
Rozanska, A. and Czerny, B. (2000), *A&A* **360**, 1170–1186
Rupen, M. P., Hjellming, R. M. and Mioduszewski, A. J. (1998), *IAU Circ.* 6938
Rupen, M. P., Dhawan, V. and Mioduszewski, A. J. (2002), *IAU Circ.* 7874
Rupen, M. P., Brocksopp, C., Mioduszewski, A. J., *et al.* (2003), *IAU Circ.* 8054
Rutledge, R. E., Lewin, W. H. G., van der Klis, M., *et al.* (1999), *ApJS* **124**, 265–283
Sanchez-Fernandez, C., Zurita, C., Casares, J., *et al.* (2002), *IAU Circ.* 7989
Seon, K., Min, K., Kenji, Y., *et al.* (1995), *ApJ* **454**, 463–471
Shahbaz, T., Ringwald, F. A., Bunn, J. C., *et al.* (1994), *MNRAS* **271**, L10–L14
Shahbaz, T., Naylor, T. and Charles, P. A. (1997), *MNRAS* **285**, 607–612
Shahbaz, T., van der Hooft, F., Casares, J., *et al.* (1999), *MNRAS* **306**, 89–94
Shakura, N. I. and Sunyaev, R. A. (1973), *A&A* **24**, 337–366
Shapiro, S. L. and Teukolsky, S. A. (1983), *Black Holes, White Dwarfs and Neutron Stars: The Physics of Compact Objects* (New York, Wiley)
Shimura, T. and Takahara, F. (1995), *ApJ* **445**, 780–788
Shrader, C. R., Wagner, R. M., Hjellming, R. M., *et al.* (1994), *ApJ* **434**, 698–706
Skinner, G. K., Foster, A. J., Willmore, A. P. and Eyles, C. J. (1990), *MNRAS* **243**, 72–77
Smak, J. (1971), *Acta Astron.* **21**, 15–21
Smith, D. M., Heindl, W. A., Swank, J., *et al.* (1997), *ApJ* **489**, L51–L54
Smith, D. M., Heindl, W. A., Markwardt, C. B. and Swank, J. H. (2001), *ApJ* **554**, L41–L44
Smith, D. M., Heindl, W. A. and Swank, J. (2002), *ApJ* **578**, L129–L132
Sobczak, G. J., McClintock, J. E., Remillard, R. A., *et al.* (1999), *ApJ* **520**, 776–787
(2000a), *ApJ* **531**, 537–545
(2000b), *ApJ* **544**, 993–1015
Steeghs, D., Miller, J. M., Kaplan, D. and Rupen, M. (2003), *ATEL* 146
Stella, L., Vietri, M. and Morsink, S.M. (1999), *ApJ* **524**, L63–L66
Stirling, A. M., Spencer, R. E., de la Force, C. J., *et al.* (2001), *MNRAS* **327**, 1273–1278
Stone, J. M., Pringle, J. E. and Begelman, M. C. (1999), *MNRAS* **310**, 1002–1016
Strohmayer, T. E. (2001a), *ApJ* **552**, L49–L53
(2001b), *ApJ* **554**, L169–L172
Sunyaev, R. and Revnivtsev, M. (2000), *A&A* **358**, 617–623
Sunyaev, R. A., Lapshov, I.Yu., Grebenev, S. A., *et al.* (1988), *SvAL* **14**, 327–333
Sunyaev, R., Gilfanov, M., Churazov, E., *et al.* (1991a), *SvAL* **17**, 50–54
Sunyaev, R. A., Kaniovskii, A. S., Efremov, V. V., *et al.* (1991b), *SvAL* **17**, 123–130
Sunyaev, R., Churazov, E., Gilfanov, M., *et al.* (1992), *ApJ* **389**, L75–L78
Sutaria, F. K., Kolb, U., Charles, P., *et al.* (2002), *A&A* **391**, 993–997

Swank, J. (1998), in *The Active X-ray Sky: Results from BeppoSAX and Rossi–XTE*, eds. L. Scarsi, *et al.*, *Nucl. Phys. B Proc. Suppl.* (astro–ph/9802188)
Tagger, M. and Pellat, R. (1999), *A&A* **349**, 1003–1016
Takahashi, K., Inoue, H. and Dotani, T. (2001), *PASJ* **53**, 1171–1177
Tanaka, Y. and Lewin, W. H. G. (1995), in *X-ray Binaries*, eds. W. H. G. Lewin, J. van Paradijs and E. P. J. van den Heuvel (Cambridge, Cambridge University Press), 126–174, TL95
Tanaka, Y. and Shibazaki, N. (1996), *ARA&A* **34**, 607–644, TS96
Tanaka, Y., Nandra, K. and Fabian, A. C. (1995), *Nature* **375**, 659–661
Tananbaum, H., Gursky, H., Kellogg, E., *et al.* (1972), *ApJ* **177**, L5–L10
Timmes, F. X., Woosley, S. E. and Weaver, T. A. (1996), *ApJ* 457, 834–843
Titarchuk, L. and Osherovich, V. (2000), *ApJ* **542**, L111–L114
Titarchuk, L. and Shrader, C. (2002), *ApJ* **567**, 1057–1066
Tomsick, J. A. and Kaaret, P. (2000), *ApJ* **537**, 448–460
Tomsick J. A. and Kaaret, P. (2001), *ApJ* **548**, 401–409
Tomsick, J. A., Kaaret, P., Kroeger, R. A. and Remillard, R. A. (1999), *ApJ* **512**, 892–900
Trudolyubov, S., Churazov, E., Gilfanov, M., *et al.* (1999), *A&A* **342**, 496–501
Turner, M. J. L., Thomas, H. D., Patchett, B. E., *et al.* (1989), *PASJ* **41**, 345–372
Turner, N. J., Stone, J. M. and Sano, T. (2002), *ApJ* **566**, 148–163
Turolla, R., Zane, S. and Titarchuk, L. (2002), *ApJ* **576**, 349–356
Ueda, Y., Dotani, T., Uno, S., *et al.* (1997), *IAU Circ.* 6627
Vargas, M., Goldwurm, A., Paul, J., *et al.* (1996), *A&A* **313**, 828–832
Vargas, M., Goldwurm, A., Laurent, P., *et al.* (1997), *ApJ* **476**, L23–L26
Vasiliev, L., Trudolyubov, S. and Revnivtsev, M. (2000), *A&A* **362**, L53–L56
Vikhlinin, A., Finoguenov, A., Sitdikov, A., *et al.* (1992), *IAU Circ.* 5608
Vikhlinin, A., Churazov, E., Gilfanov, M., *et al.* (1994), *ApJ* **424**, 395–400
Wagner, R. M., Starrfield, S. G., Hjellming, R. M., *et al.* (1994), *ApJ* **429**, L25–L28
Wagner, R. M., Foltz, C. B., Shahbaz, T., *et al.* (2001), *ApJ* **556**, 42–46
Wagoner, R. V. (1999), *Phys. Rep.* **311**, 259–269
Wagoner, R. V., Silbergleit, A. S. and Ortega-Rodriguez, M. (2001), *ApJ* **559**, L25–L28
Wardzinski, G., Zdziarski, A. A., Gierlinski, M., *et al.* (2002), *MNRAS* **337**, 829–839
Weaver, K. A., Gelbord, J. and Yaqoob, T. (2001), *ApJ* **550**, 261–279
Webster, B. L and Murdin, P. (1972), *Nature* **235**, 37–38
White, N. E. and Marshall, F. E. (1984), *ApJ* **281**, 354–359
White, N. E. and van Paradijs, J. (1996), *ApJ* **473**, L25–L29
White, N. E., Stella, L. and Parmar, A. N. (1988), *ApJ* **324**, 363–378
White, N. E., Nagase, F. and Parmar, A. N. (1995), in *X-ray Binaries*, eds. W. H. G. Lewin, J. van Paradijs and E. P. J. van den Heuvel (Cambridge, Cambridge University Press), 1–57
Wijnands, R. and Miller, J. M. (2002), *ApJ* **564**, 974–980
Wijnands, R. and van der Klis, M. (2000), *ApJ* **528**, L93–L96
Wijnands, R., Homan, J. and van der Klis, M. (1999), *ApJ* **526**, L33–L36
Wijnands, R., Mendez, M., Miller, J. M. and Homan, J. (2001), *MNRAS* **328**, 451–460
Wijnands, R., Miller, J. M. and van der Klis, M. (2002), *MNRAS* **331**, 60–70
Wilms, J., Nowak, M. A., Dove, J. B., *et al.* (1999), *ApJ* **522**, 460–475
Wilms, J., Nowak, M. A., Pottschmidt, K., *et al.* (2001a), *MNRAS* **320**, 327–340
Wilms, J., Reynolds, C. S., Begelman, M. C., *et al.* (2001b), *MNRAS* **328**, L27–L31
Wilson, A. M., Carpenter, G. F., Eyles, C. J., *et al.* (1977), *ApJ* **215**, L111–L115
Wilson, C. K. and Rothschild, R. E. (1983), *ApJ* **274**, 717–722
Wood, K. S., Ray, P. S., Bandyopadhyay, R. M., *et al.* (2000), *ApJ* **544**, L45–L48
Woods, P. M., Kouveliotou, C., Finger, M. H., *et al.* (2002), *IAU Circ.* 7856
Woosley, S. E., Heger, A. and Weaver, T. A. (2002), *RvMP* **74**, 1015–1071
Wu, K., Soria, R., Campbell–Wilson, D., *et al.* (2002), *ApJ* **565**, 1161–1168
Yaqoob, T., Ebisawa, K. and Mitsuda, K. (1993), *MNRAS* **264**, 411–420
Zdziarski, A. A. (2000), in *Highly Energetic Physical Processes*, Procs. IAU Symposium #195, eds. C.H. Martens, S. Tsuruta and M. A. Weber, ASP, 153–170, (astro-ph/0001078)
Zdziarski, A. A., Grove, J. E., Poutanen, J., *et al.* (2001), *ApJ* **554**, L45–L48
Zdziarski, A., Poutanen, J., Paciesas, W.S. and Wen, L. (2002), *ApJ* **578**, 357–373

Zdziarski, A. A., Lubinski, P., Gilfanov, M. and Revnivtsev, M. (2003), *MNRAS* **342**, 355–372
Zhang, S. N., Cui, W. and Chen, W. (1997a), *ApJ* **482**, L155–L158
Zhang, S. N., Cui, W., Harmon, B. A., *et al.* (1997b), *ApJ* **477**, L95–L98
Zurita, C., Sanchez-Fernandez, C., Casares, J., *et al.* (2002), *MNRAS* **334**, 999–1008
Zycki, P. T., Done, C. and Smith, D. A. (1999a), *MNRAS* **305**, 231–240
(1999b), MNRAS **309**, 561–575

5

Optical, ultraviolet and infrared observations of X-ray binaries

P. A. Charles and M. J. Coe
University of Southampton

5.1 Introduction

In the 35 years since the first X-ray binary was optically identified (Sco X-1) the basic division of X-ray binaries into the high-mass (HMXBs) and low-mass (LMXBs) systems has become firmly established. The nomenclature refers to the nature of the mass donor, with HMXBs normally taken to be $\geq 10\ M_{\odot}$, and LMXBs $\leq 1\ M_{\odot}$. However, the past decade has seen the identification and measurement of a significant number of X-ray binaries whose masses are intermediate between these limits. Nevertheless, the nature of the mass-transfer process (stellar wind dominated in HMXBs, Roche lobe overflow in LMXBs) produces quite different properties in the two groups and so this chapter will be divided into two main sections on HMXBs and LMXBs. A more complete introduction can be found in Chapter 1.

While the nature of the compact object and its properties are largely determined from X-ray studies, longer-wavelength observations allow detailed studies of the properties of the mass donor. This is most straightforward for the intrinsically luminous early-type companions of HMXBs, which provide the potential for a full solution of the binary parameters for those systems containing X-ray pulsars. This is particularly important for HMXB evolution in that it allows a comparison of the derived masses with those obtained for neutron stars in the much older binary radio pulsar systems (Thorsett & Chakrabarty 1999).

However, when HMXBs are suspected of harboring black holes (e.g., Cyg X-1), the mass measurement process runs into difficulties. By definition, there will be no kinematic features (such as pulsations) associated with the compact object, and hence the analysis is based entirely on the mass-losing companion. Most importantly, our knowledge of the mass of the compact object depends on the accuracy to which that of the companion is known. The situation for LMXBs, such as Sco X-1, is completely different, in that their short orbital periods require their companion stars to be of much lower mass. In Section 5.3 we collect together and discuss the results of mass determinations in all X-ray binaries. We also emphasise those properties that LMXBs have in common with cataclysmic variables (Chapter 10), most of which are related to the similar structure accretion disks.

Recently,the advent of near-IR spectroscopic instrumentation on large telescopes has allowed the first detailed investigations of X-ray binaries in the Galactic Bulge and other regions of high extinction.

Compact Stellar X-Ray Sources, eds. Walter Lewin and Michiel van der Klis.

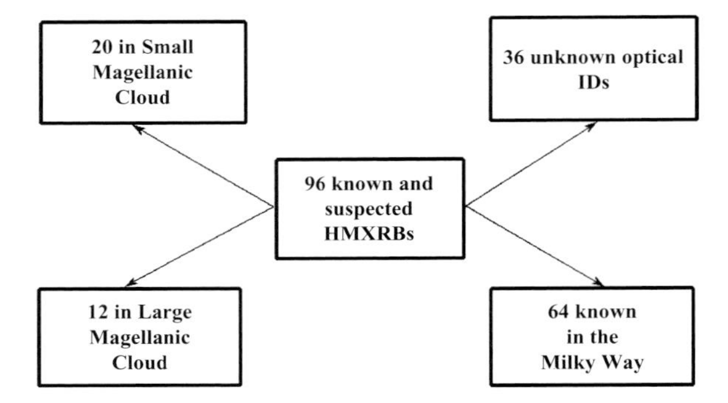

Fig. 5.1. Distribution of different HMXB populations.

5.2 High-mass X-ray binaries

5.2.1 *Supergiant and Be/X-ray binaries*

There are two main sub-groups of HMXBS – the supergiant counterparts (normally of luminosity class I or II), and the Be/X-ray (or BeX) binary systems (normally luminosity class III or V). Both sub-groups involve OB type stars and are commonly found in the galactic plane and the Magellanic Clouds. They differ, however, in accretion modes, with the supergiant systems accreting from a radially outflowing stellar wind, and the BeX binaries accreting directly from the circumstellar disk (possibly with some limited Roche lobe overflow on rare occasions). As a result the supergiants are persistent sources of X-rays, whilst the BeX systems are very variable and frequently much brighter.

Reig and Roche (1999) have suggested a third sub-group of systems; the X Per like systems. Their main characteristics are: long pulse periods (typically 1000 s), persistent low L_X ($\sim 10^{34}$ erg s^{-1}) and low variability, and rare uncorrelated weak X-ray outbursts. The properties of all three groups are summarized in Table 5.1.

About 100 known or suspected HMXBs had been identified by August 2003 (see Fig. 5.1). Surprisingly, nearly one-third of these lie in the Magellanic Clouds. This very large fraction, particularly noticeable in the SMC, will be discussed later. About one-third of all the HMXBs have no known optical counterpart, simply due to the historical inaccuracy of the X-ray location. Current facilities such as the Chandra X-ray Observatory are very effective tools to rectify this situation.

Progress towards a better understanding of the physics of these systems depends on multi-wavelength studies. From observations of the Be star in the optical, UV and IR, the physical conditions under which the neutron star is accreting can be determined. In combination with hard X-ray timing observations, this yields a near complete picture of the accretion process. It is thus vital to identify the optical counterparts and obtain UV to IR measurements for these X-ray systems in order to further our understanding.

5.2.2 *Be/X-ray binaries*

The BeX binary systems represent the largest sub-class of HMXBs. Of 96 proposed HMXB pulsars, 57% are identified as BeX type. The orbit of the Be or supergiant star and the compact object, presumably a neutron star, is generally wide and eccentric. X-ray outbursts

Table 5.1. *General properties of HMXBs*

Type[a]	Percentage of all HMXB	Optical luminosity class	Typical pulse period (s)	Typical binary period (d)	Typical binary eccentricity	Log L_X (erg s^{-1})
Be	57	III–V	0.05–500	2–260	0.3–0.9	36–38
SG	25	I–II	200–700	3–40		34–35
XP	10	III–V	200–1400	250	0.03	34–35
Others	8	—	—	—	—	

[a] Be: Be star binaries, SG: supergiant systems, XP: systems similar to X Per

are normally associated with the passage of the neutron star through the circumstellar disk. The physics of accretion-powered pulsars has been reviewed previously (e.g., White *et al.* 1995; Nagase 1989; Bildsten *et al.* 1997). This section will concentrate on the optical, UV and IR observational properties of BeX systems and address how these have contributed to our understanding of Be star behavior.

X-ray behavioral features of BeX systems include:

- regular periodic outbursts at periastron (*Type I outbursts*);
- giant outbursts at any phase probably arising from a dramatic expansion of the circumstellar disk (*Type II outbursts*);
- "missed" outbursts (i.e., looked for but not seen) frequently related to low Hα emission levels (hence a small disk), or other unknown reasons (e.g., perhaps centrifugal inhibition of accretion, see Stella *et al.* 1986);
- shifting outburst phases due to the rotation of density structures in the circumstellar disk (Wilson *et al.* 2002).

For supergiant systems the X-ray characteristics tend to be much less dramatic. Because of the predominantly circular orbits and the much steadier wind outflow, the X-ray emission tends to be a regular low-level effect. Rarely outbursts may occur, but never Type I.

One particularly interesting piece of astrophysics emerged early on in the study of BeX systems – the Corbet diagram (Corbet 1986). This diagram showed a strong correlation between their orbital and pulse periods. An updated version of this diagram is presented in Fig. 5.2. For a system to be detected as an X-ray source it is necessary that the pressure of infalling material be sufficient to overcome centrifugal inhibition (see Chapter 1). In other words, the Alfvén radius must be inside the corotation radius. For BeX systems the correlation is understood in terms of the wider orbits exposing the neutron stars to a lower wind density on average and hence weaker accretion pressures. Thus it is only those systems that have a correspondingly weaker centrifugal inhibition (i.e., slower spin rates) that permit accretion to occur and hence permit their possible discovery as an X-ray source.

A powerful feature of this diagram is that it allows one to predict either the pulse, or the binary period, if one knows the other parameter. Alternatively, if both periods are determined (from, e.g., X-ray measurements) then the class of object may be defined, even in the absence of an optical counterpart.

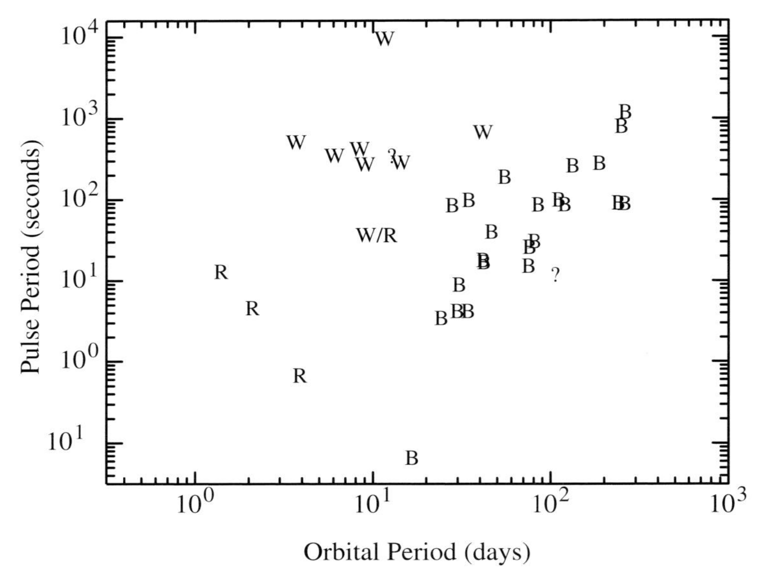

Fig. 5.2. An updated version of the Corbet diagram (Corbet 1986) provided by Corbet (private communication). The three classes of object illustrated are: B, BeX binaries; W, wind-fed; R, Roche-lobe overflow. Two systems are indicated by ? symbols because their optical properties are not yet clear.

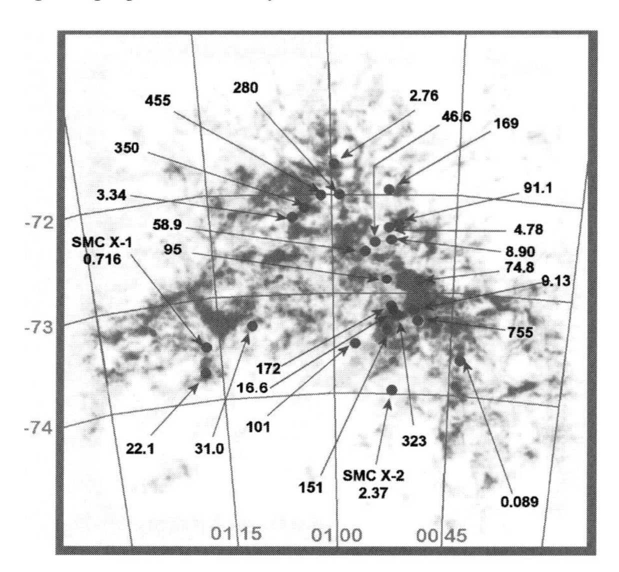

Fig. 5.3. An HI image of the SMC (Stanimirovic *et al.* 1999), on which is superposed the location of 25 known X-ray binary pulsars. The numbers indicate their pulse period in seconds.

5.2.3 Magellanic Cloud HMXBs

It has come as a great surprise to discover that there are a large number of BeX binaries in the SMC. Figure 5.3 shows the distribution of almost all the known HMXB pulsars superimposed upon an HI image of the SMC.

Table 5.2. *Properties of X-ray pulsars in the SMC*

Short name	Full or alternative name(s)	Spectral class	V mag	Period (s)
SXP0.09	AX J0043–737	Be?		0.087
SXP0.72	SMC X-1	B0Ib	13.25	0.716
SXP0.92	PSR0045–7319	B	16.00	0.92
SXP2.16	XTE SMC pulsar			2.16
SXP2.37	SMC X-2	B1.5V	16.00	2.37
SXP2.76	RX J0059.2–7138	B1III	14.10	2.76
SXP3.34	AX J0105–722, RX J0105.3–7210			3.343
SXP4.78	XTE J0052–723, [MA93]537		15.80	4.782
SXP5.44	CXOU J010042.8–721132	AXP?	17.80	5.44
SXP7.70	XTE SMC pulsar			7.7
SXP8.90	RX J0051.8–7231, 1E0050.1–7247	B1III-V		8.9
SXP9.13	AX J0049–732			9.132
SXP15.3	RX J0052.1–7319	B1III-B0V	14.70	15.3
SXP16.6	RX J0051.8–7310			16.6
SXP22.1	RX J0117.6–7330	B0.5III	14.20	22.07
SXP31.0	XTE J0111.2–7317	O8-B0V	15.35	31
SXP46.4	XTE SMC pulsar			46.4
SXP46.6	1WGA 0053.8–7226, XTE J0053–724	B1-2III-V	14.90	46.6
SXP51.0	XTE SMC pulsar			51
SXP58.9	RX J0054.9–7226, XTE J0055–724			58.95
SXP74.8	RX J0049.1–7250, AX J0049–729	Be?	15.90	74.8
SXP82.4	XTE J0052–725			82.4
SXP89.0	XTE SMC pulsar			89
SXP91.1	AX J0051–722, RX J0051.3–7216		15.00	91.1
SXP95.2	XTE SMC pulsar			95.2
SXP101	AX J0057.4–7325, RX J0057.3–7325			101.4
SXP152	CXOU J005750.3–720756, [MA93]1038			152.1
SXP164	XTE SMC pulsar			164.7
SXP169	XTE J0054–720, AX J0052.9–7158		15.50	169.3
SXP172	AX J0051.6–7311, RX J0051.9–7311			172.4
SXP280	AX J0058–72.0			280.4
SXP304	CXOU J010102.7–720658, [MA93]1240			304.5
SXP323	AX J0051–73.3, RXJ0050.7–7316	B0-B1V	15.40	323.2
SXP349	SAX J0103.2–7209, RX J0103–722	O9-B1III-V	14.80	349.9
SXP455	RX J0101.3–7211			455
SXP564	CXOU J005736.2–721934, [MA93]1020			564.8
SXP755	AX J0049.4–7323, RX J0049.7–7323	B1-3V		755.5

It is possible to estimate the number of systems one would expect based upon the relative masses of our Galaxy and the SMC. This ratio is $\sim$50, so with 64 known or suspected systems in our Galaxy we would only expect 1 or 2 in the SMC. However, Maeder *et al.* (1999) have shown that the fraction of Be to B stars is 0.39 in the SMC compared with 0.16 in our Galaxy. So this raises the expected number of BeX systems to $\sim$3 – but we now know of $\geq$35 (Table 5.2)!

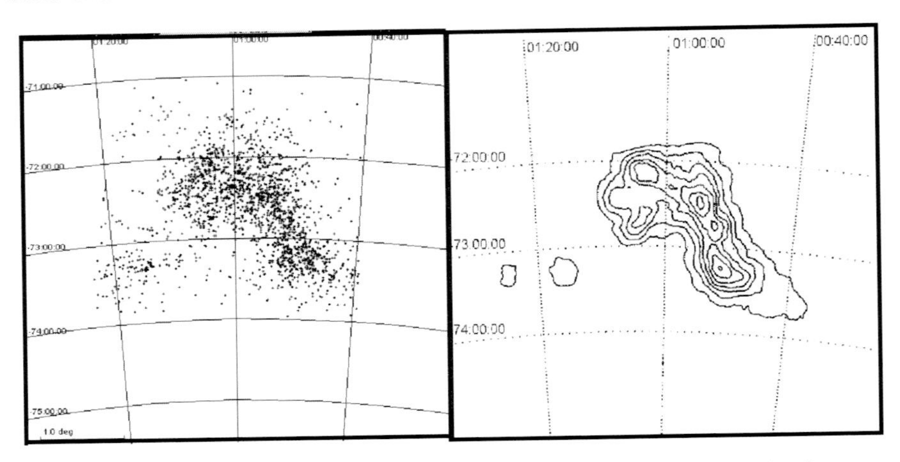

Fig. 5.4. *Left*: spatial distribution of emission line stars in the SMC (Meyssonnier & Azzopardi 1993). *Right*: isodensity contours for SMC stars aged 8–12 Myr (Maragoudaki *et al.* 2001).

The reason for this almost certainly lies in the history of the Magellanic Clouds. Detailed HI mapping by Stavely-Smith *et al.* (1997) and Putman *et al.* (1998) has shown the existence of a significant bridge of material between the Magellanic Clouds and between them and our own Galaxy. Furthermore, Stavely-Smith *et al.* have demonstrated the existence of a large number of supernova remnants of similar age (~5 Myr), strongly suggesting enhanced starbirth has taken place as a result of tidal interactions between these component systems. Consequently it seems very likely that the previous closest approach of the SMC to the LMC ~100 Myr ago may have triggered the birth of many new massive stars, which have given rise to the current population of HMXBs. In fact, other authors (e.g., Popov *et al.* 1998) claim that the presence of large numbers of HMXBs may be the best indication of starburst activity in a system.

Excellent support for this interpretation comes from both the work of Meyssonnier and Azzopardi (1993) in cataloguing emission line stars, and the work of Maragoudaki *et al.* (2001) in identifying stars of age in the range 8–12 Myr. Figure 5.4 shows the results of both of these surveys and demonstrates a strong spatial correlation with the distribution of X-ray pulsars shown in Fig. 5.3.

As a result, the SMC now provides us with an excellent sample of BeX systems in a relatively compact and easily observable region of the sky. In addition, the similarity between the population sizes seen in the Galaxy (~50) and the SMC (~30) permits interesting comparisons to be made. Figure 5.5 from Laycock *et al.* (2004) shows the binned population distributions per decade in period. Applying the K-S test to the two populations gives a probability of 97% that the two samples were drawn from different distributions (although varying extinction conditions may significantly affect the homogeneity of the galactic population).

In complete contrast to the BeX dominated population of the SMC, the HMXB population of the LMC seems very representative of the sample found in the Milky Way. A comprehensive review by Negueruela and Coe (2002) of all identified optical counterparts in the LMC found that the overall spectral distribution of the population looks very similar to that of our Galaxy

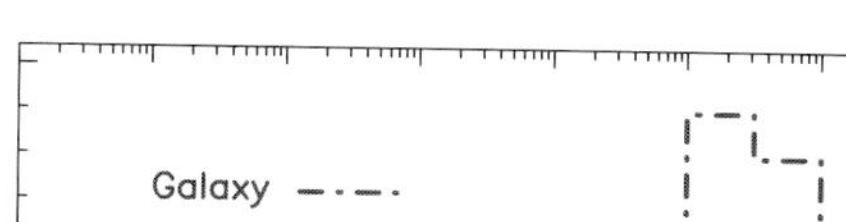

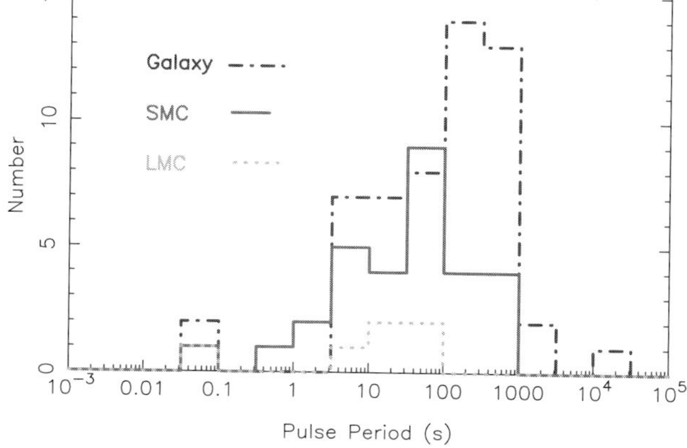

Fig. 5.5. Comparison of the X-ray pulsar populations of the Galaxy, LMC and SMC (Laycock *et al.* 2004).

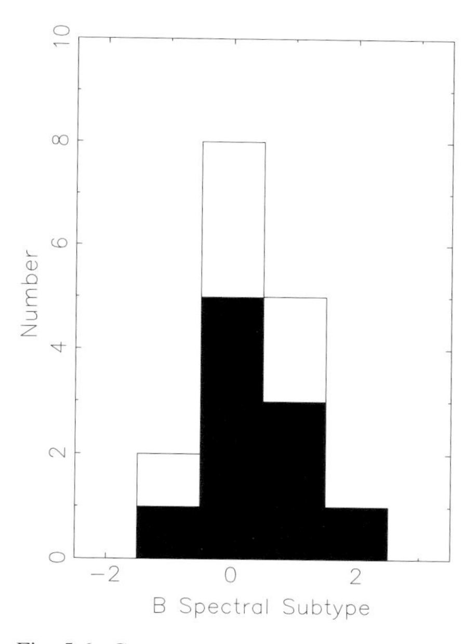

Fig. 5.6. Comparison of the X-ray pulsar optical counterparts in the Galaxy (white) and LMC (black) (Negueruela & Coe 2002). Negative spectral types represent O-type stars.

(Fig. 5.6). They found that the spectral distribution for BeX binaries in both the Milky Way and the LMC is sharply peaked at spectral type B0 (roughly corresponding to $M_* \approx 16\,M_\odot$) and does not extend beyond B2 ($M_* \approx 10\ M_\odot$), which strongly supports the existence of supernova kicks. So the BeX binaries in the LMC must have preferentially formed from moderately massive binaries undergoing semi-conservative evolution in the same manner as most galactic systems.

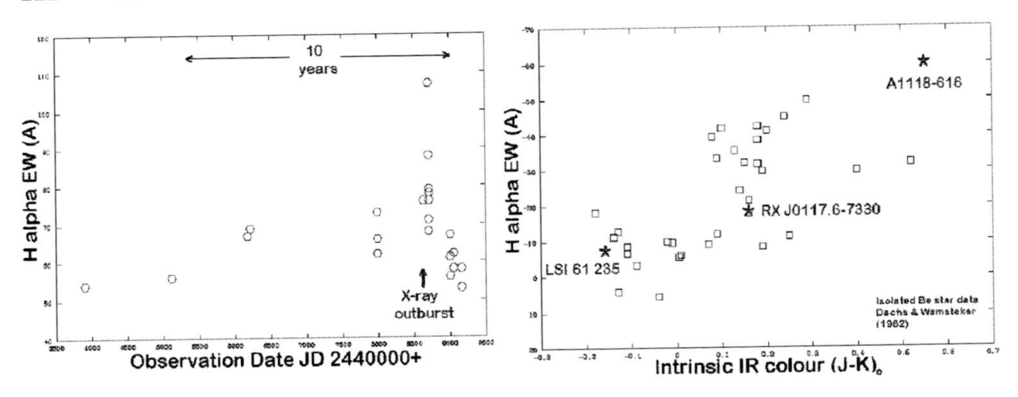

Fig. 5.7. *Left*: the history of the Hα emission from A1118–616 showing the large rise coincident with the X-ray outburst. *Right*: The relationship between the intrinsic IR colors and Hα equivalent width. Three BeX systems are compared to a set of isolated Be stars (Coe *et al.* 1994).

5.2.4 *Properties of BeX circumstellar disks*

Typically the optical star exhibits Hα line emission and continuum free–free emission (revealed as excess IR flux) from a disk of circumstellar gas. Therefore Hα measurements provide very useful information on the disk's physical state. A1118–616 provides an excellent example of a classic BeX system that can be probed effectively in Hα. Discovered by chance in 1974 while Ariel 5 was observing the nearby source Cen X-3 (Eyles *et al.* 1975), the flux increased by more than an order of magnitude during the outburst and became the dominant hard X-ray source in the Centaurus region for 7–10 days. The system then disappeared for 27 years before re-emerging in the same dramatic style in 1991 (Coe *et al.* 1994).

The explanation for these sudden massive X-ray outbursts on very long timescales lies in the Hα observations carried out prior to, during, and after the 1991 outburst. Note, no similar observations exist for the 1974 outburst since the optical counterpart was not identified at that time. The history of the Hα observations over a 10 year period is shown in Fig. 5.7, from which there is a strong clue as to why A1118–616 went into outburst when it did – the Hα equivalent width had reached an exceptionally high value ≥100 Å. Clearly this exceptionally large disk was more than sufficient to trigger accretion on to the neutron star and produce the outburst. Since the pulse period is 404 s the Corbet diagram (Corbet 1986) suggests $P_{\rm orb}$ in the range 200–300 d, therefore it is extremely unlikely that the two recorded outbursts relate to binary motion (Type I outbursts).

It is far more likely that these are Type II outbursts, and the normal Type I outbursts are either very minor or absent. This could simply be due to the orbit not normally taking the neutron star through the circumstellar disk. However, during these abnormal levels of Hα activity, the disk probably expands to include the neutron star's orbit, and hence accretion immediately begins. Note the strikingly rapid decline of Hα immediately after the outburst suggesting that the whole period of activity was probably just due to one major mass ejection event from the Be star.

It is interesting to compare the size of the circumstellar disk seen in A1118–616 with that observed in other systems. Figure 5.7 shows a plot of Hα EW against the intrinsic IR color (also thought to arise from free–free and free–bound emission in the circumstellar disk), and an obvious and unsurprising correlation clearly exists between these two parameters. The quiescent location of A1118–616 is shown together with two other BeX systems, and compared with data from a sample of isolated Be stars (Dachs & Wamsteker 1982). It is interesting to note that even in quiescence A1118–616 is at the extreme edge of the diagram, and in fact its peak Hα EW value of ~110 Å may be one of the very largest recorded values for any Be star.

With one notable exception discussed below, the Hα profiles follow the behavioral patterns seen in isolated Be stars. The profile shape is dominated by the circumstellar disk structure and its inclination to our line of sight. A range of possible profiles are observed and these are well documented and classified elsewhere (e.g., Hanuschik *et al.* 1988; Hanuschik 1996). A good example of a set of varying Hα profiles from X Per is shown in Fig. 5.8 (Clark *et al.* 2001). Not only does this range of profiles indicate significant variations in disk size, there is strong evidence that the blue and red components are varying with respect to each other – known as V/R variations. These V/R changes can only be explained by non-uniformities in the disk structure which are rotating around the system.

The structure of these disks has been explained by Okazaki and Negueruela (2001) and others. They have investigated the tidal torques on these circumstellar (or "decretion" disks and found that a natural truncation occurs at one of certain resonance points (radii at which the local Keplerian period is an integer fraction of the orbital period). Beyond this truncation point matter cannot be transported. Furthermore, HMXBs with very circular orbits are truncated at a fixed size, often smaller than the Roche lobe, making accretion on to the neutron star very unlikely. As a result these systems tend to only exhibit persistent low-level X-ray emission from the stellar wind plus occasional Type II outbursts. Conversely, systems with high eccentricities permit the size of the disk to be orbital-phase dependent and during periastron passage the disk can extend well into the orbital path of the neutron star triggering a Type I X-ray outburst.

Finally, direct Hα imaging has shown itself to be a powerful tool in determining the environment around HMXBs in the case of Vela X-1. The spectacular image reported by Kaper *et al.* (1997; reproduced here as Fig. 5.9) provides direct evidence of a bow shock around this system. This is clear indication that Vela X-1 is traveling at supersonic velocities through the ISM. Kaper *et al.* used the symmetry of the shock to derive the direction of motion and the origin and age of the system. Their results support the Blaauw scenario (Blaauw 1961) whereby a supernova explosion of one binary component can result in a high space velocity of the companion (an "OB runaway"). Surprisingly, this runaway star has a high probability of remaining bound to its companion and thereby producing systems like Vela X-1.

5.2.5 *Long-term optical monitoring*

Of great value in the study of BeX systems has been the availability of long-term observational archives such as OGLE (Udalski *et al.* 1997) and MACHO (Alcock *et al.* 1996). Such databases have permitted the long-term optical variability to be studied, sometimes in conjunction with similar timescale X-ray databases. Two excellent examples of the science achievable from these optical archives are the results on A0538–66 (Alcock *et al.* 2001) and AX J0051–733 (Coe *et al.* 2002). In the case of A0538-66 the MACHO data are shown in Fig. 5.10 and clearly indicate a very regular modulation of ~420 d, but also evidence for activity on the much shorter 16.6 d binary period (already well-known from X-ray data,

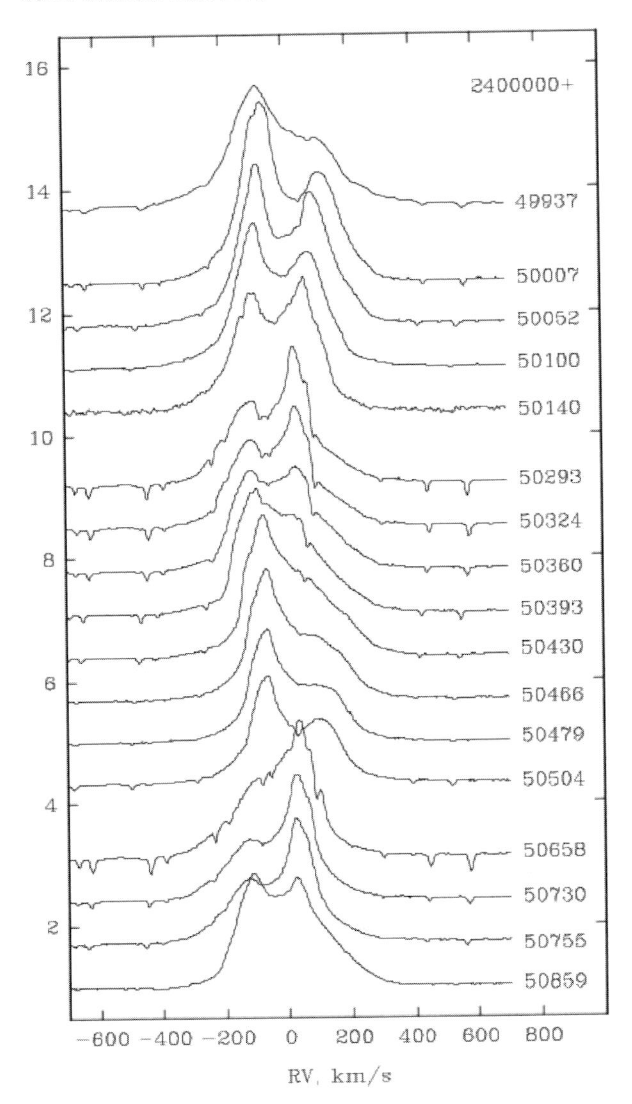

Fig. 5.8. Normalized Hα profiles of X Per as a function of velocity relative to the rest wavelength. Different dates given on the right hand side of the figure (from Clark *et al.* 2001).

Skinner 1980, 1981). Remarkably, the shorter period outbursts *only* occur during the *minima* in the 420 d cycle. McGowan and Charles (2003) interpret this as being due to evolution in the circumstellar disk.

The second example of a source studied using these kinds of data is AX J0051–733. A combination of MACHO and OGLE data (Coe *et al.* 2002) revealed an extremely short modulation period of either 0.7 d or double that value – see Fig. 5.11. The shape of this optical modulation is driven both by tidal distortion of the envelope of the mass donor (see, e.g., Avni & Bahcall 1975 and Avni 1978), and by X-ray heating effects (Orosz & Bailyn

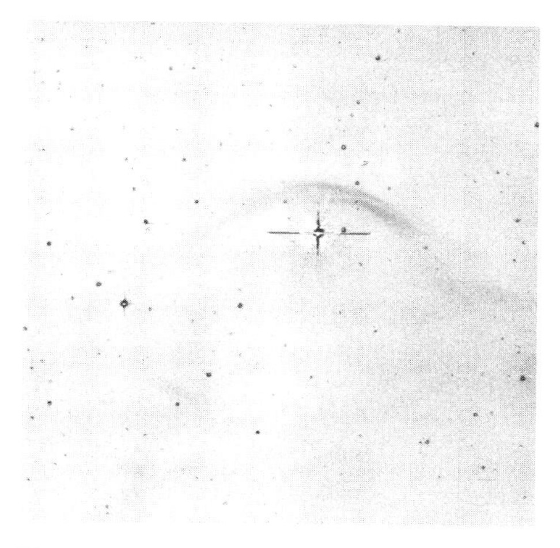

Fig. 5.9. $9' \times 9'$ Hα image of the Vela X-1 bow shock (Kaper *et al.* 1997).

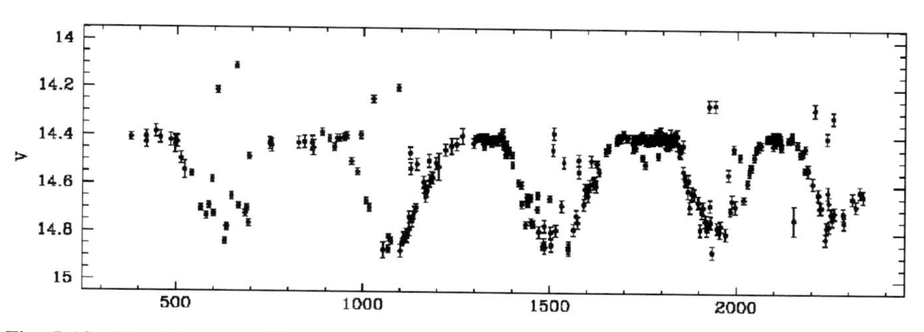

Fig. 5.10. The 5.5 yr MACHO optical lightcurve for A0538–66 showing a regular modulation at $P = 421$ d (from McGowan & Charles 2003).

1997). Orbital parameters, such as inclination and mass ratio, may be determined from these lightcurves.

5.2.6 *Spectral classification of the mass donors*

The evolutionary history of Be stars in HMXBs is somewhat different from that of their isolated colleagues. The widely accepted evolutionary path, based upon conservative mass transfer, was developed by van den Heuvel (1983) and Verbunt and van den Heuvel (1995). The important consequence of this scenario is that wide binary orbits (200–600 d) are produced before the final supernova (SN) explosion. Hence, any small asymmetries in the subsequent SN explosion will then produce the frequently observed wide eccentric orbits.

Of particular interest is the narrow range of spectral class revealed through blue spectroscopy of the identified counterparts (see Negueruela 1998). In contrast to the sample of isolated Be stars found in the Bright Star Catalogue, there are no known BeX objects beyond spectral class B3 – see Fig. 5.12. Most commonly these have counterparts in the B0-B2 group.

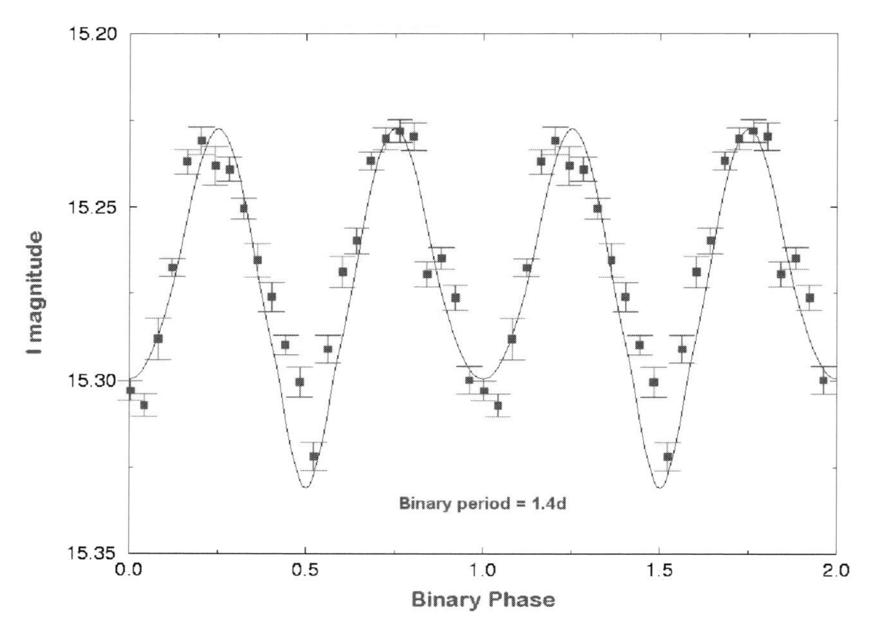

Fig. 5.11. Comparison of a representative ellipsoidal model ($i = 60°$, $Q = 0.20$, $f = 0.99$, and $r_d = 0.33$) and the folded I-band data for RX J0050.7–7316. The amplitude and relative depths of the minima are roughly matched by the model. However, there are relatively large deviations, especially between phases 0.4 and 0.6 (Coe & Orosz 2000).

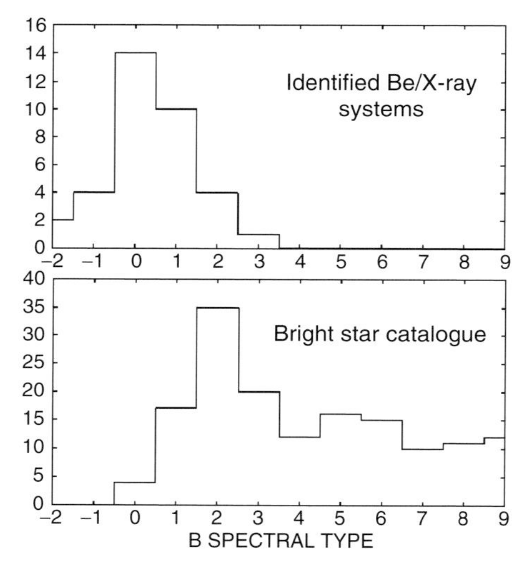

Fig. 5.12. Comparison of the spectral classes of Be stars in BeX binaries (*upper*) with those in isolated systems (*lower*). (Adapted from Negueruela 1998.)

The explanation offered by van Bever and Vanbeveren (1997) for this phenomenon is that the wide orbits produced by the evolutionary models are very vulnerable to disruption during the SN explosion. This will be particularly true for the less massive objects that would make up the later spectral classes. Hence the observed distribution confirms the evolutionary models.

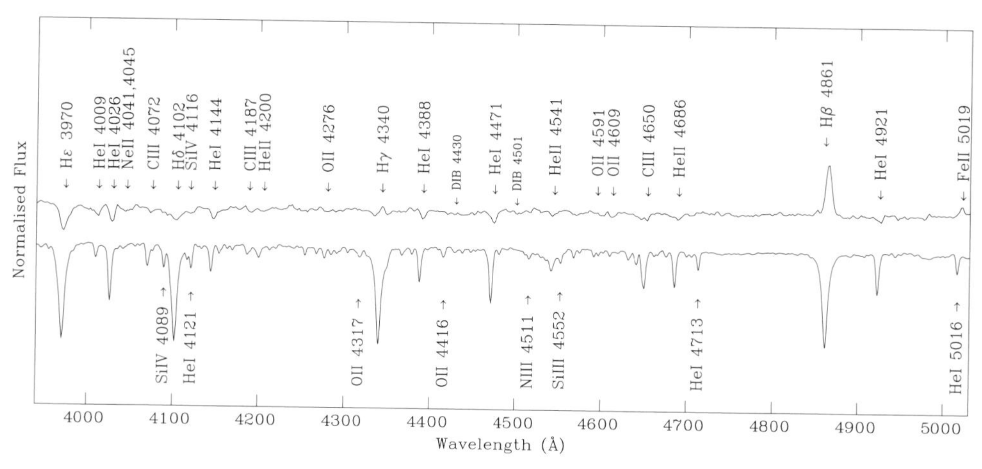

Fig. 5.13. Spectrum of the optical counterpart of RX J0052–7319 (type B0Ve) in the classification region compared to the B0V standard ν Ori (Covino *et al.* 2001).

An excellent example of using optical spectroscopy to investigate an HMXB is the work of Covino *et al.* (2001) on a pair of SMC objects. The blue spectrum of RX J0052–7319 (Fig. 5.13) is compared to a standard star of the same spectral type, where it is clear that some of the weaker He I lines are not visible, presumably filled in by emission. All the spectral features are very shallow and broad, which is typical of a BeX counterpart and the presence of weak He II absorption places it close to B0. Furthermore, if the object is on the main sequence, the presence of weak HeII λ4200 Å and the condition HeII λ4541 $>$ SiIII λ4552 Å give a spectral type of B0Ve.

5.2.7 *IR spectroscopy of HMXBs*

The observations summarized in earlier sections reveal that many of the counterparts to HMXBs exhibit mass outflow to the extent of creating a circumstellar disk of material around the mass donor. Free–free and bound–free IR emission from this disk show themselves as a significant excess over the normal stellar spectrum at all wavelengths greater than the V band. This IR signature, often quantified as a J-K color excess, is important for the following reasons:

- in confirming the identity of a Be star in the absence of optical spectral information;
- in providing an estimate of the size of the circumstellar disk (this is often directly related to the magnitude of the X-ray emission);
- through IR spectroscopy providing a channel for spectral identification in the case of objects suffering heavy optical extinction.

One system that has been the subject of extensive IR observations over the years is X Per. Detailed study by Telting *et al.* (1998) found that the density of the disk varies along with the brightness of X Per, and that in optical high states the disk in X Per is among the densest of all Be stars: $\rho_0 = (1.5 \pm 0.3) \times 10^{-10}$ g cm^{-3}. The disk density at the photosphere varies by a factor of at least 20 from optical high to low states (see Fig. 5.14 for large differences taking place in the IR).

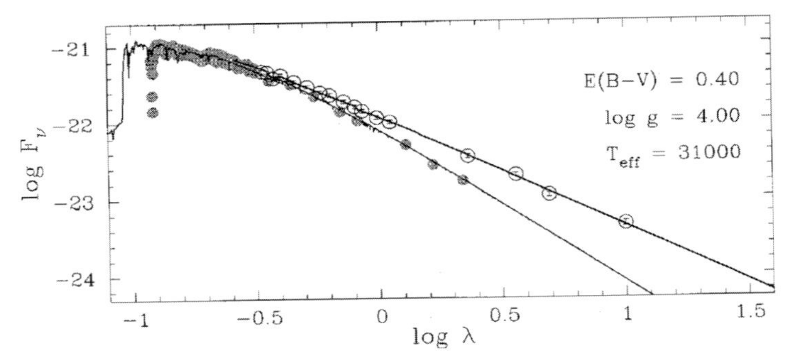

Fig. 5.14. Comparison of the optical/IR emission from X Per in two different states: with (upper curve) and without (lower curve) a substantial circumstellar disk (Telting *et al.* 1998).

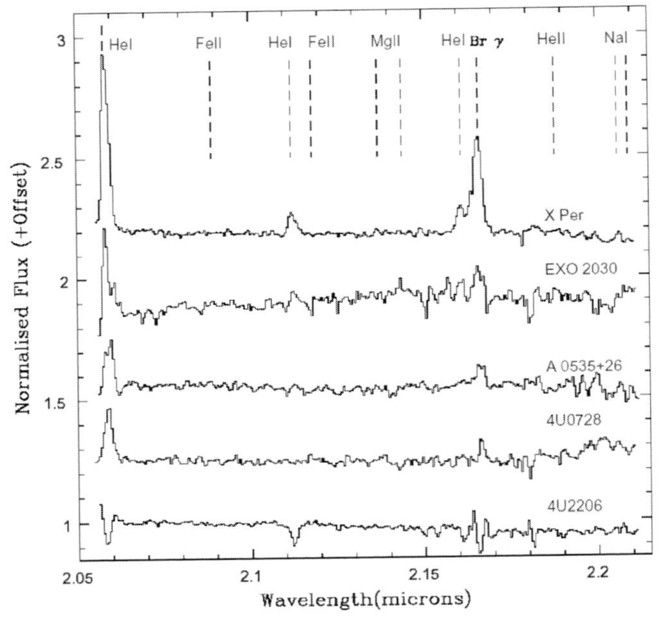

Fig. 5.15. K band spectra of BeX binaries (Clark *et al.* 1999). The positions of prominent H, He and metallic transitions are marked.

With the recent advent of IR spectroscopy on large telescopes, the possibilities of obtaining spectra despite high levels of extinction have become a reality. This is a very powerful tool for exploring previously inaccessible objects, but also directly gleaning information on the circumstellar environment. A good example of the strength of this new tool may be seen in Clark *et al.* (1999), which shows the annotated IR spectra of five HMXBs (Fig. 5.15). Just as the optical spectrum is dominated by Balmer emission, we see the same effect from the Brackett series emission in the IR. In addition, emission lines from He and metallic transitions are prominent. This approach, therefore, offers great potential for identifying and classifying

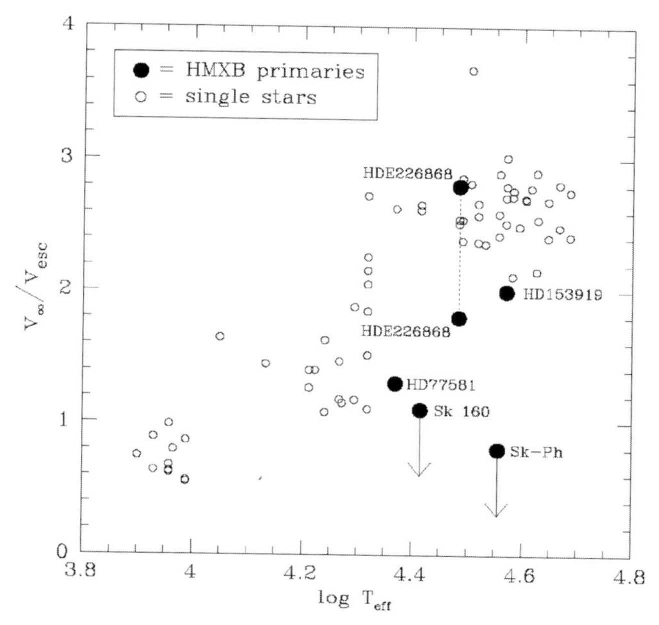

Fig. 5.16. The ratio of terminal over escape velocity for the mass donor stars in five HMXBs compared to isolated stars. The HMXBs have low terminal velocities for their effective temperatures (van Loon *et al.* 2001).

the more optically obscured systems, as well as revealing the state of the circumstellar disk.

5.2.8 UV spectroscopy of HMXBs

UV spectroscopy of the mass donors in HMXBs has proved a powerful tool for understanding these systems. Since these are hot young stars they tend to be extremely UV bright. Important information about the stellar outflow may be determined from P Cygni profiles, as well as direct estimates of interstellar absorption from the 2200 Å feature, which is valuable in decoupling local and ISM extinction.

Van Loon *et al.* (2001) carried out a comprehensive study of five HMXBs using extensive IUE UV data. Using radiation transfer codes they derived the terminal wind velocities and sizes of the Strömgren zones created by the X-ray source from modeling the UV resonance lines. They found that these sizes were in good agreement with that expected from standard Bond–Hoyle accretion and scaled with L_X. However, the determined terminal velocities were lower than expected from single stars with the same T_{eff} – see Fig. 5.16. These lower velocities could be due to the presence of the X-ray Strömgren sphere inhibiting the wind acceleration.

In addition, blue and UV spectra are essential tools for determing the precise spectral class of the mass donor, and hence information about the compact object. Clark *et al.* (2002) used high-resolution IUE UV spectra of 4U1700–37 to carry out detailed non-LTE modeling of the O type primary. Using the modeling code of Hillier and Miller (1998), which utilises radiative transfer processes in a spherical extended atmosphere, Clark *et al.* were able to precisely define the mass donor star to be of spectral type O6.5Iaf. Combining this with binary information from eclipse monitoring led to an accurate determination of an unusually high neutron star mass of $2.44 \pm 0.27\ M_\odot$. This is one of the highest accurate mass determinations for a neutron star, but is not alone in pushing upwards the mass of neutron stars in HMXBs. Barziv *et al.* (2001) have $1.87 \pm 0.20\ M_\odot$ for the compact object mass in Vela X-1, whereas Orosz and Kuulkers (1999) find a value of $1.78 \pm 0.23\ M_\odot$ for Cyg X-2. However, it is still conceivable that there are systematic effects at work here given how the neutron star "mass" correlates with spectral type.

5.3 Low-mass X-ray binaries

The review by van Paradijs and McClintock (1995) remains an excellent introduction to the optical characteristics of both HMXBs and LMXBs, and Liu *et al.* (2000, 2001) provide comprehensive catalogues of their observed properties. However, this is a field that has extended both in wavelength (with high-quality UV spectroscopy available from the HST – Hubble Space Telescope – STIS, and cooled grating spectrographs allowing IR spectroscopy) and sensitivity (the advent of large telescopes such as Keck, VLT, Subaru and VLT) during the last decade. Consequently this section will focus on these later developments, and in particular on the sub-class of *X-ray transients*, which have allowed us to gain a far more detailed knowledge of the properties of the mass donor in LMXBs than was possible a decade ago. Additionally, gains in CCD technology now permit high time resolution photometry and spectroscopy and a consequential improvement in our understanding of luminous accretion disks.

5.3.1 Fundamental properties

The presence of a "steady" X-ray source implies that the companion fills its Roche lobe, and so (Paczynski 1971)

$$R_2/a = 0.46(1+q)^{-1/3} \tag{5.1}$$

where the mass ratio $q = M_X/M_2$. Combined with Kepler's Third Law, then

$$\rho = 110/P_{\rm hr}^2 \tag{5.2}$$

where ρ (in g cm^{-3}) is the mean density of the secondary.[1]

These stars are presumed to be on or close to the lower main sequence, then $M_2 = R_2$ and hence $M_2 = 0.11 P_{\rm hr}$ (a detailed observational analysis of the secondary stars in CVs and LMXBs can be found in Smith & Dhillon 1998). Hence short-period X-ray binaries must be LMXBs, their companion stars will be faint, and the optical light will be dominated by reprocessed X-radiation from the disk (or heated face of the companion star, although this is likely substantially shielded by the disk itself; see van Paradijs & McClintock 1994). This obliteration of the companion in "steady" LMXBs removes one of the key pieces of dynamical information about the binary. The optical spectra of LMXBs consist of hot, blue continua ($U - B$ typically -1) on which are superposed broad H and He emission lines, with velocity widths typical of the inner disk region (~500–1000 km s^{-1}), making accurate

[1] NB the Paczynski relation is only valid for $q > 1$, and there is a more accurate algorithm (Eggleton 1983) which is valid for all q.

velocity information about the motion of the compact object difficult to obtain. Consequently accurate dynamical information about both components in LMXBs has been very difficult to acquire, yet is essential if accurate masses are to be determined (see next section). Hence, evidence on the nature of the compact object in most bright LMXBs is indirect, e.g., X-ray bursting behavior for neutron stars (as few are X-ray pulsars) or fast variability as first seen in Cyg X-1 (and used as a possible black hole diagnostic, see Chapters 2 and 4).

5.3.2 *Dynamical mass measurements in X-ray novae*

Progress in identifying the nature of compact objects in LMXBs requires dynamical mass measurements of the type hitherto employed on X-ray pulsars in HMXBs (see, e.g., White *et al.* 1995). Unfortunately, all that can be measured is the mass function

$$f(M) = \frac{PK^3}{2\pi G} = \frac{M_X^3 \sin^3 i}{(M_X + M_2)^2} \tag{5.3}$$

where K is the radial velocity amplitude. If $M_2 \geq M_X$ (which is the case for Cyg X-1, and the other two HMXBs suspected of harboring black holes, LMC X-1 and LMC X-3), then M_X is poorly constrained due to the wide range of uncertainty ($\sim$12–20 $M_\odot$) in M_2 as a result of the unusual evolutionary history. In LMXBs, however, $M_2 \ll M_X$, which allows for a much more physically useful constraint on M_X to be obtained.

LMXB transients (usually referred to as *soft X-ray transients, SXTs,* or *X-ray novae, XRN*) provide ideal opportunities to study the properties of the secondary stars in great detail. Detected in outburst by X-ray all-sky monitors, they usually fade within timescales of months, after which the companion star becomes visible. The prototype of this sub-class is A0620–00, still the brightest (in X-rays and optical) of all XRN when it erupted in 1975, the data from which were comprehensively re-examined by Kuulkers (1998) producing the revised lightcurve shown in Fig. 5.17. Currently $\sim$25% of XRN are confirmed neutron star (NS) systems (they display Type I X-ray bursts), the remainder are all black hole candidates (BHCs), the highest fraction of any class of X-ray source (see, e.g., Chen *et al.* 1997). Their X-ray properties are discussed in Chapter 4, and their optical/IR properties are summarized in Table 5.3.

5.3.3 *Quiescence studies of X-ray novae*

In quiescence, XRN become a valuable resource for research into the nature of LMXBs. Typically their optical brightness has declined by a factor of $\geq$100, with all known XRN having quiescent V $\sim$ 16–23, which is now dominated by the companion. With large-telescope optical/IR spectroscopy it is possible to determine spectral type, period and radial velocity curve, the latter two giving $f(M)$ (Eq. (5.3)), see Table 5.3. These are separated into three sections: BHCs with cool secondaries, those with earlier spectral types and the neutron star XRN. Table 5.3 shows the significance of such work, since all are LMXBs with $M_X > M_2$. NB $f(M)$ represents the absolute *minimum* values for M_X since we must have $i < 90^\circ$ and $M_2 > 0$.

Some dynamical information can be derived even during outburst, given spectroscopic data of sufficient resolution. Casares *et al.* (1995) observed GRO J0422+32 during an X-ray mini-outburst and found intense Balmer and HeII λ4686 emission modulated on what was subsequently found to be P_{orb}. Furthermore, a sharp component within the (highly complex)

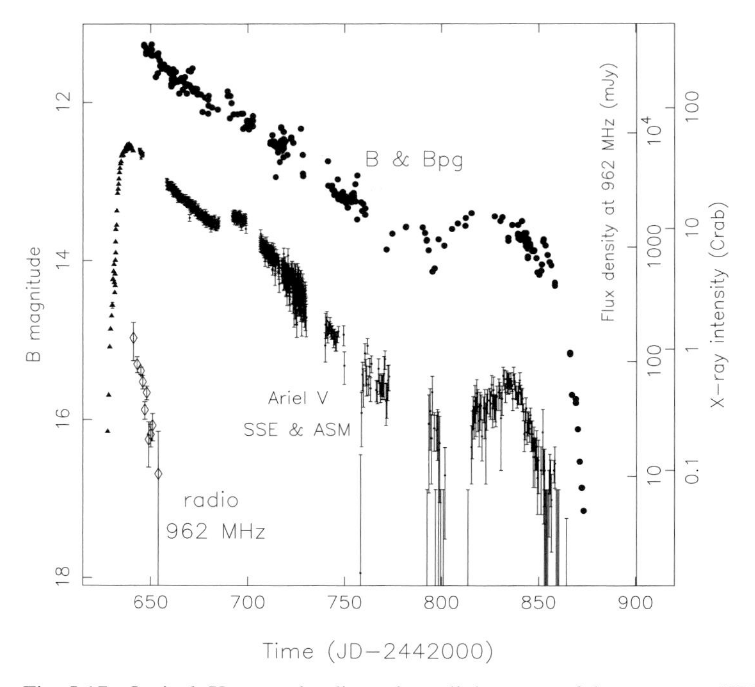

Fig. 5.17. Optical, X-ray and radio outburst lightcurves of the prototype XRN A0620–00 (Nova Mon 1975) showing the typical fast rise, and exponential decay on a timescale of months (adapted from Kuulkers 1998).

HeII emission profile displayed an S-wave that was likely associated with the accretion hotspot.

However, to determine M_X, additional constraints are needed in order to infer values for M_2 and i.

5.3.3.1 *Rotational broadening of companion spectrum*

In short-period interacting binaries the secondary is constrained to corotate with the primary and must also be filling its Roche lobe (in order for mass transfer to occur). Hence R_2 is given by Eq. (5.1), and then (Wade & Horne 1988)

$$v_{\rm rot} \sin i = \frac{2\pi R_2}{P} \sin i = K_2 \times 0.46 \frac{(1+q)^{2/3}}{q} \tag{5.4}$$

from which q can be derived if $v_{\rm rot}$ is measured. Typically $v_{\rm rot} \sim$ 40–100 km s^{-1}, which requires high resolution and high signal-to-noise spectra of the secondary.

The technique is demonstrated in Fig. 5.18 for V404 Cyg (Casares & Charles 1994) and based on Marsh *et al.* (1994). The Doppler-corrected and summed spectrum of V404 Cyg is dominated by broad Hα emission from the disk, but the cool companion absorption features are visible and clearly broader than those of the template K0IV spectrum. The template is then broadened by different velocities (together with the effects of limb darkening), subtracted from that of V404 Cyg and the residuals χ^2 tested, giving $v_{\rm rot} \sin i = 39 \pm 1$ km s^{-1} and hence $q = 16.7 \pm 1.4$. Note that small radial velocity amplitudes have also been seen in the

Table 5.3. *Optical/IR properties of LMXB transients and BH HMXBs*

X-ray source	Optical/IR counterpart	b[I]	Outbursts	P (hrs)	Sp. Type	E_{B-V}	V	K (quiesc)	vsini	K_2 (km s^{-1})	γ	f(M) ($M_\odot$)	q (=M_X/M_2)	i (°)	M_X ($M_\odot$)	M_2 ($M_\odot$)
Black holes																
LMXBs																
GRS1915+105[1]	V1487 Aql	−0.2	1991+"on"	816	K III	~10	–	13	–	140	−3	9.5±3	12	70±2	14±4	1.2±0.2
J1859+226[2]	V406 Vul	+8.6	1999	9.2	–	0.58	23.3	–	–	570	–	7.4±1.1	–	–	>5	–
J1550−564[3]	V381 Nor	−1.8	1998, 00, 03	37.0	G8−K4IV	1.6	22	–	90±10	349	−68	6.9±0.7	–	72±5	9.6±1.2	–
J1118+480[4]	KV UMa	+62.3	2000	4.1	K7−M0	<0.02	19.6		114±4	701	−15	6.1±0.3	~20	81±2	6.8±0.4	0.25±0.15
GS2023+338[5]	V404 Cyg	−2.2	1938, 56, 89	155.3	K0IV	1	18.4	12.5	39±1	208.5	0	6.08±0.06	17±1	55±4	12±2	0.6
GX339−4[6]	V821 Ara	−4.3	frequent	42.1	–	1.1	~21	–	–	317[†]	~ +30	5.8±0.5	–	–	>2.0	–
GS2000+25[7]	QZ Vul	−3.1	1988	8.3	K5V	1.5	21.5	17	86±8	520	+19	4.97±0.10	24±10	56±15	10±4	0.5
H1705−25[8]	V2107 Oph	+9.1	1977	12.5	K	0.5	21.5	–	≤79	441	−54	4.86±0.13	>19	60±10	6±2	0.3
GRS1009−45[9]	MM Vel	+9.3	1993	6.8	K7−M0	0.2	~22			475	+41	3.17±0.12	7±1	67±3	5.2±0.6	0.7
N Mus 91[10]	GU Mus	−7.1	1991	10.4	K0−4V	0.29	20.5	16.9	106±13	421	+20	3.34±0.15	6.8±2	54^{+20}_{-15}	6^{+5}_{-2}	0.8
A0620−00[11]	V616 Mon	−6.5	1917, 75	7.8	K5V	0.35	18.3	14.5	83±5	433	+4	2.91±0.08	15±1	37±5	10±5	0.6
J0422+32[12]	V518 Per	−11.9	1992	5.1	M1V	0.25	22.3	17.4	90±25	372	+11	1.19±0.02	$9.0^{+2.2}_{-2.7}$	45±2	4±1	0.3
IMXBs																
J1819.3−2525[13]	V4641 Sgr	−4.8	1999, 02	67.6	B9III	0.32	13.7		99±2	211	+107	3.13±0.13	2.31±0.08	75±2	7.1±0.3	3.1±0.3
J1655−40[14]	V1033 Sco	+2.5	1994−96	62.9	F6IV	1.2	17.2	–	88±5	228	−148	2.73±0.15	2.39±0.15	70±2	6.6±0.5	2.8±0.3
GS1354−47[15]	BW Cir	−2.8 (+1966, 71 ?)	1987, 97	61.3	G0−5III	~1	~20.5	–	71 ± 4	279	+99	5.75 ± 0.3	7.7 ± 1.3	< 77	>7.8	>1.0
4U1543−47[16]	IL Lup	+5.4	1971, 83, 92, 02	27.0	A2V	0.5	16.6	–	–	124	−87	0.25±0.01	3.6±0.4	21±2	9.4±1	2.5
HMXBs																
LMC X−3[17]	*1	−32.1	"steady"	40.8	B3Ve	0.06	17.2	–	130	235	+310	2.3±0.3	2.2–4.4	67±3	4–7	1–3
Cyg X−1[18]	HDE 226868	+3.1	"steady"	134.4	O9.7Iab	1.06	8.9	–	155	74.9	−1	0.244±0.005	0.4–0.8	27–67	>4.8	>11.7
LMC X−1[19]	*32	−31.5	'steady"	101.3	O8III	0.37	14.5	–	~150	68	+221	0.14±0.04	~0.5	2.5–6	8–20	
Neutron stars																
J2123−058[20]	LZ Aqr	−36.2	1998	6.0	K7V	0.12	21.8	–	121	299	−95	0.68±0.05	2.7±1.0	73±4	1.5±0.3	0.5±0.3
Cen X−4[21]	V822 Cen	+23.9	1969, 79	15.1	K3−5V	0.1	18.4	14.8	43±6	150	+184	0.22±0.01	5.9±1.6	43±11	1.5±1.0	0.3±0.2

References: [a] mostly from Wachter 1998; [1]Greiner *et al.* 2001; [2]Filippenko & Chornock 2001; Hynes *et al.* 2002b; Zurita *et al.* 2002b; [3]Orosz *et al.* 2002a; [4]Wagner *et al.* 2001; Hynes *et al.* 2000; Orosz 2001; Zurita *et al.* 2002a; [5];Casares *et al.* 1992; Casares & Charles 1994; Shahbaz *et al.* 1994b; [6]Hynes *et al.* 2003b;[†]Bowen fluorescence velocity, see section 5.3.9; [7]Filippenko *et al.* 1995a; Beekman *et al.* 1996; Harlaftis *et al.* 1996; [8]Filippenko *et al.* 1997; Remillard *et al.* 1996; Martin *et al.* 1995; Harlaftis *et al.* 1997; [9]della Valle *et al.* 1997; Filippenko *et al.* 1999; Gelino 2003; [10]Orosz *et al.* 1996; Casares *et al.* 1997; Shahbaz *et al.* 1997; Gelino *et al.* 2001a; [11]McClintock & Remillard 1986; Orosz *et al.* 1994; Marsh *et al.* 1994; Shahbaz *et al.* 1994a; Gelino *et al.* 2001b; [12]Filippenko *et al.* 1995b; Beekman *et al.* 1997; Harlaftis *et al.* 1999; Webb *et al.* 2000; Gelino & Harrison 2003; [13]Orosz *et al.* 2001; Orosz 2003; [14]Orosz & Bailyn 1997; van der Hooft *et al.* 1997, 1998; Hynes *et al.* 1998; Shahbaz *et al.* 1999b, 2000; Greene *et al.* 2001; Beer & Podsiadlowski 2002; Shahbaz 2003; [15]Orosz *et al.* 2002b; [16]Kuiper *et al.* 1988; [17]Herrero *et al.* 1995; Brocksopp *et al.* 1999; [18]Hutchings *et al.* 1983, 1987; [19]Tomsick *et al.* 2001, 2002; Hynes *et al.* 2001a; Casares *et al.* 2002; Shahbaz *et al.* 2003; [20]McClintock & Remillard, 1990; Shahbaz *et al.* 1993; Torres *et al.* 2002

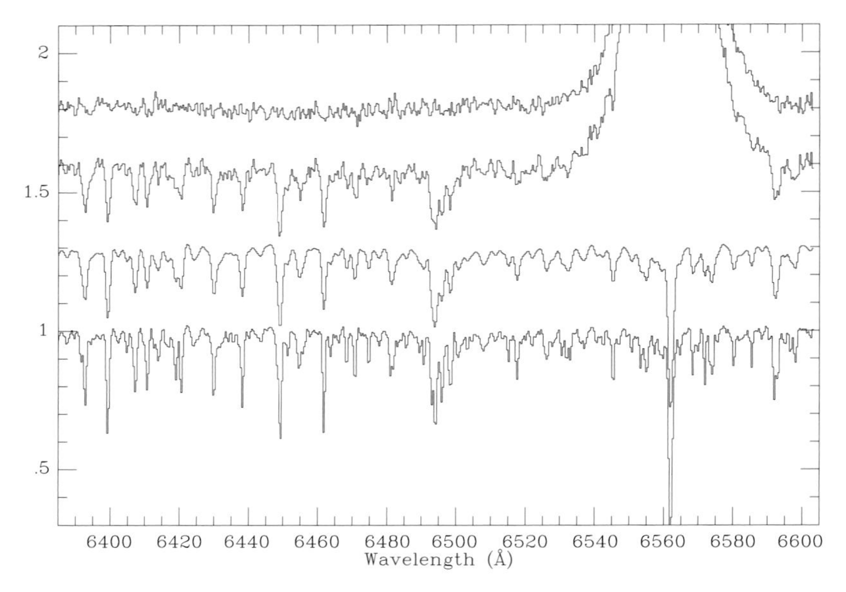

Fig. 5.18. Determining the rotational broadening in V404 Cyg. From bottom to top: the K0IV template (HR8857); the same spectrum broadened by 39 km s^{-1}; Doppler corrected sum of V404 Cyg (dominated by disk Hα emission); residual after subtraction of the broadened template (from Casares & Charles 1994).

Hα emission (e.g., Orosz *et al.* 1994; Soria *et al.* 1998), but their interpretation is complex as there is a small phase offset ($\simeq$0.1) relative to the companion (somewhat larger in GRS1009–45, see Filippenko *et al.* 1999), and so until this is fully understood it cannot be used to determine q.

The combination of q and $f(M)$ then yields the mass constraints (Fig. 5.19), and the only remaining unknown is i. To date, none of the SXTs is eclipsing (although GRO J1655–40 and XTE J2123–058 show evidence for grazing eclipses), and so it is the uncertainty in i that dominates the final mass measurement. Nevertheless there are methods by which i can be estimated.

5.3.3.2 *Ellipsoidal modulation*

The secondary star in interacting binaries has a peculiarly distorted shape which gives rise to the so-called *ellipsoidal modulation* due to its varying projected area as viewed around the orbit. This leads to the classical double-humped lightcurve (Fig. 5.20), which is well defined by theory (i.e., the form of the Roche lobe) and the observed lightcurve depends principally on q and i. For large q values ($\geq$5) the ellipsoidal modulation is largely insensitive to q and hence can provide excellent constraints on i. For full details of the lightcurve modeling see Tjemkes *et al.* (1986), Orosz & Bailyn (1997) and Shahbaz *et al.* (2003), the results from which are in Table 5.3.

5.3.4 ***Mass determinations and limitations***

To derive full XRN orbital solutions requires assuming that the quiescent secondary fills its Roche lobe (reasonable as Doppler tomography of quiescent XRN reveals ongoing

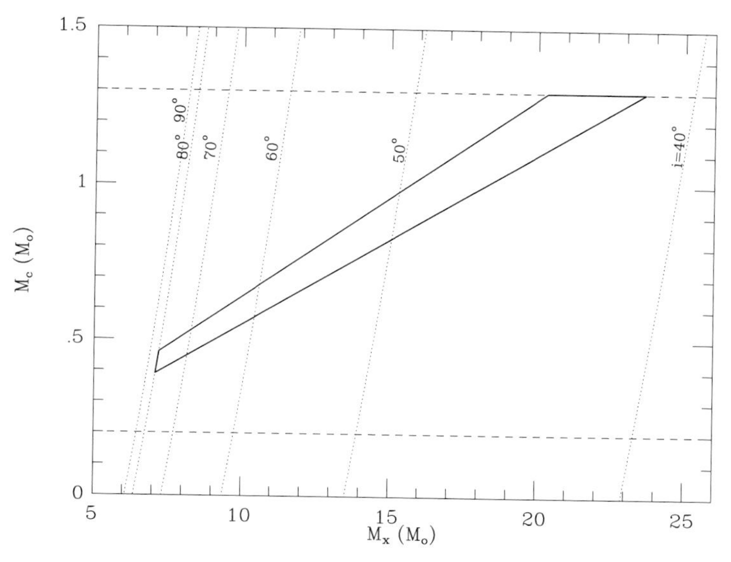

Fig. 5.19. Constraints on M_X and M_2 for a range of values of i in V404 Cyg based on the radial velocity curve ($f(M) = 6.1$ $M_\odot$ and determination of q (= 16.7, from rotational broadening). It is the limited constraint on i (absence of eclipses) that leads to a wide range of M_X (Casares & Charles 1994).

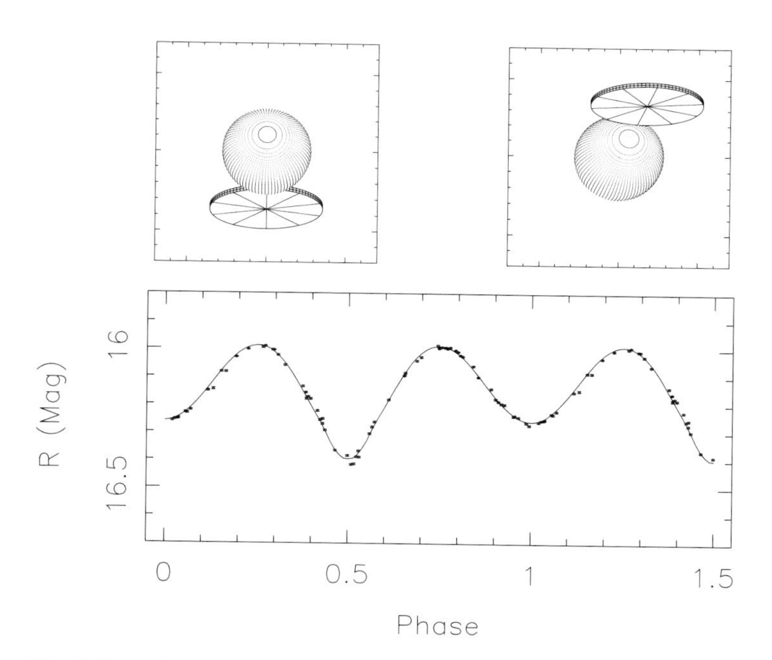

Fig. 5.20. R-band lightcurve of GRO J1655–40 in quiescence (van der Hooft *et al.* 1997) together with schematics of the system orientation (Orosz & Bailyn 1997) which show the grazing eclipses required by the fits to the lightcurve.

mass transfer; Marsh *et al.* 1994), and that the lightcurve is *not* contaminated by any other emitting region. This latter assumption is questionable, since emission lines are definitely attributable to a quiescent disk, but residual X-ray heating and coronal activity on the secondary might also be present (e.g., Bildsten & Rutledge 2000). For this reason, secondary lightcurves are usually derived in the IR whenever possible. The disk contribution in the optical around Hα is obtained as a by-product of the spectral type determination, and is typically $\leq$10%. It should therefore be even less in the IR for a typical disk spectrum. However, the outer disk can be an IR emitter in CVs (Berriman *et al.* 1985) and hence contaminate the lightcurves (Sanwal *et al.* 1996). This is potentially significant, since a contaminating (and presumably steady) contribution will reduce the amplitude of the ellipsoidal modulation, and hence a lower i would be inferred, leading to an erroneously high mass for the compact object.

Nevertheless Shahbaz *et al.* (1996, 1999a) showed via IR K-band spectroscopy of V404 Cyg and A0620–00 that any contamination must be small and hence the masses derived need (at most) to be reduced by only small amounts ($\leq$20%). Furthermore, in studying the non-orbital optical variability in V404 Cyg, Pavlenko *et al.* (1996) found that (as first noted by Wagner *et al.* 1992) the ellipsoidal modulation could be discerned underlying the substantial (short-term) flickering in the lightcurve. Interpreting the flickering as a completely independent component (recently verified by Hynes *et al.* 2002a), Pavlenko *et al.* showed that the *lower envelope* of this lightcurve (rather than the mean) produced an ellipsoidal lightcurve which, when fitted as described above, gave essentially identical results to those obtained from the IR ellipsoidal fitting, thereby providing further weight to the significance of the final mass determinations in Table 5.3. Note that the values for Cen X-4 and XTE J2123–058 (both neutron star XRN, identified on the basis of their Type I X-ray bursts) have been derived exactly by the method outlined here and yield values in excellent accord with those expected for a neutron star. Table 5.3 contains only those systems for which a dynamical study of the companion star has been performed (all via direct detection of the spectral signature of the mass donor, except for GX339–4 – see Section 5.3.9). There are a number of additional BHCs (based on their X-ray and other properties) for which such studies have not yet been possible (secondary too faint or too heavily reddened), and these are listed in Chapter 4.

For completeness, we have also included in Table 5.3 the dynamically determined HMXB compact object mass determinations. In spite of an extremely accurately determined $f(M)$, there are still large and systematic uncertainties in the mass of Cyg X-1's supergiant primary HDE 226868 and the difficulty in constraining the binary inclination (it does not eclipse). This has most recently been addressed by Herrero *et al.* (1995) whose high-resolution optical spectroscopy and detailed atmospheric modeling give a mass range of 12–19 $M_\odot$ for the OBI primary. With i relatively poorly constrained to 28–67 degrees, the compact object mass must be in the range 4–15 $M_\odot$. HDE 226868 has also been studied by LaSala *et al.* (1998) and Brocksopp *et al.* (1999) who have derived a much improved orbital ephemeris, and Balucińska-Church *et al.* (2000), who used this new ephemeris to study the distribution of X-ray dips as a function of orbital phase (there is a strong orbital modulation). In spite of this large uncertainty in compact object mass (a feature shared by the LMC HMXBs LMC X-1 and LMC X-3), Cyg X-1 remains a very strong BHC, and its X-ray properties are still invoked as black hole characteristics (see Chapter 4).

Interestingly most of the compact object masses so far determined lie in the range 6–14 $M_\odot$, so that even the lowest value is well above the theoretical neutron star maximum mass of

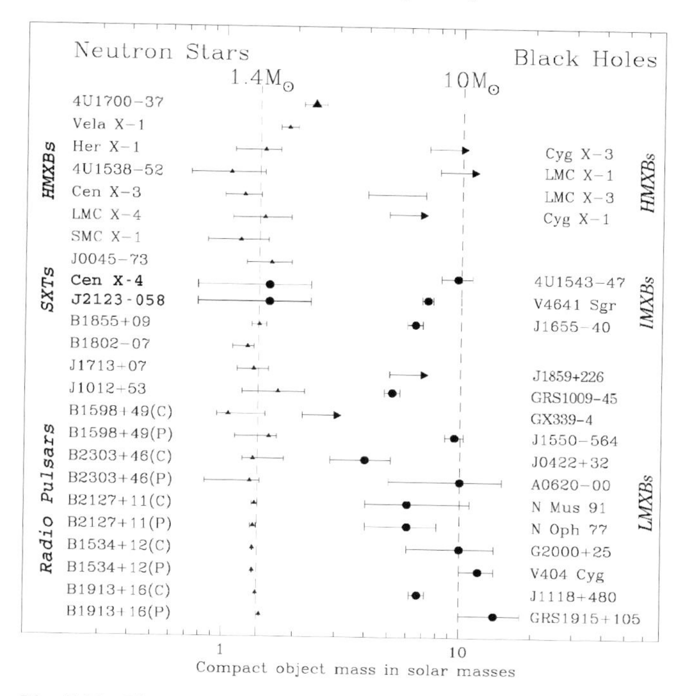

Fig. 5.21. The mass distribution of neutron stars and black holes. Note the remarkably narrow spread of NS masses, and the large factor by which the BH masses exceed the (canonical) maximum NS mass of 3.2 $M_\odot$.

$\sim$3.2 $M_\odot$ (and even further from the directly measured neutron star masses of 1.35 $M_\odot$; see Thorsett & Chakrabarty 1999 and Fig. 5.21). There is only one compact object mass currently in the range 2–5 $M_\odot$ (the recently determined value for GRO J0422+32 by Gelino & Harrison 2003, but GRS1009–45 is close), which is curious and may be a selection effect. Certainly V518 Per and MM Vel are two of the faintest systems in quiescence, but it may also be possible that for low black hole masses transient behavior is suppressed. In such cases mass determination becomes difficult as the disk dominates the optical light, but alternative methods are now possible in certain circumstances (see Section 5.3.9), or if the donor is sufficiently bright to be detectable even against the glare of the disk (GRS 1915+105, Cyg X-2).

5.3.5 *Abundance analyses of the mass donors*

The absence of (significant) X-ray heating allows for detailed chemical analyses of the mass donors in the brighter of the quiescent XRN. An immediate (and unexpected) byproduct of the high-resolution radial velocity study of V404 Cyg was the discovery of strong LiI λ6707 absorption (Martín *et al.* 1992). Li is typically present in young, pre-main sequence and T Tau stars, but is destroyed by subsequent convection in late-type stars, and normal (solar) abundances are a thousand times lower. Similar Li enhancements are found in many (but not all) quiescent XRN (see Martín *et al.* 1996), including the NS XRN Cen X-4, so this is *not* a potential black hole (BH) signature.

As highly evolved objects that are extremely unlikely to have retained such high Li abundances, mechanisms have been sought that create Li within XRN, which could be important for the wider study of the galactic Li enrichment relative to the halo. Li has been seen in XRN with a wide range of P (and hence donor sizes), but their common feature is the recurrent, high L_X X-ray outbursts (CVs do *not* exhibit Li, Martín *et al.* 1995), which led Martín *et al.* (1994) to suggest that spallation could produce Li in large quantities close to the compact object, and might explain the 476 keV gamma-ray feature observed during the N Mus 1991 outburst (Sunyaev *et al.* 1992; Chen *et al.* 1993). Originally interpreted as the gravitationally redshifted $\bar{e}$–e^+ 511 keV line (see also Kaiser & Hannikainen 2002), it might instead be associated with the ^{7}Li 478 keV line. Some of this Li would be transferred to the secondary during subsequent large mass outflows. However, Bildsten and Rutledge (2000) point out that the XRN Li abundances are only slightly higher than those detected in the chromospherically active RS CVn systems and also the pre-CV binary V471 Tau. This suggests a possible link with coronal activity, although it has been argued by Lasota (2000) that this cannot account for the levels of X-ray emission seen in quiescent XRN.

More dramatic processes, and a possible link of XRN with hypernovae, have resulted from the discovery by Israelian *et al.* (1999) of substantial (factors 6 to 10) increases in the α-element abundances (O, Mg, Si, S) in very high resolution spectra of the F sub-giant donor in GRO J1655–40. Other elements (such as Fe) show entirely normal (solar) abundances, and so Israelian *et al.* infer that the companion star was bathed in these elements which were created by explosive nucleosynthesis during the supernova that formed the compact object. Interestingly, Israelian *et al.* point out that the current BH mass estimate (>5.5 M$_\odot$; Table 5.3) requires a hypernova explosion in order to account for the observed abundances (see also Podsiadlowski *et al.* 2002).

5.3.5.1 UV spectroscopy

The evolutionary history of XRN can also be inferred from relative abundances of nuclear processed material, e.g., through UV spectroscopy. A comparison by Haswell *et al.* (2002) of J1118+480 and J1859+226 (both observed with HST/STIS near outburst peak) revealed remarkably similar strengths of NV and SiIV, yet J1118+480 exhibited no CIV or OV (Fig. 5.22). This is very surprising since CIV and NV are both resonance lines of Li-like ions and therefore produced under similar physical conditions. This dramatic difference between these two short-period systems implies that the mass donor in J1118+480 must have lost its outer layers, thereby exposing inner material which has been mixed with CNO-processed matter from the core (the evolution of the C/N ratio as a function of M_2 is shown in Chapter 13 and Ergma & Sarna 2001). Such an interpretation requires that mass transfer began at an initial secondary mass of $\sim$1.5 M$_\odot$ and P of $\geq$12 h, but this raises interesting questions as to how the binary then evolves to its current $P = 4.1$ h (see Chapter 16). However, this does support the results of Smith and Dhillon (1998) whose detailed analysis of the secondary properties in CVs shows that (up to $P \sim$ 7–8 h) they are indistinguishable from main-sequence stars in detached binaries, whereas the XRN secondaries have much larger radii and are therefore evolved.

The HST spectra of J1859+226 (Hynes *et al.* 2002b) also display a strong λ2175 feature that allows for an accurate determination of *E (B–V)*. Combined with optical spectroscopy, Hynes *et al.* find that the UV/optical region can be well fit by a standard, but *irradiated* disk. As the outburst progresses, the irradiated component declines and the implied outer disk

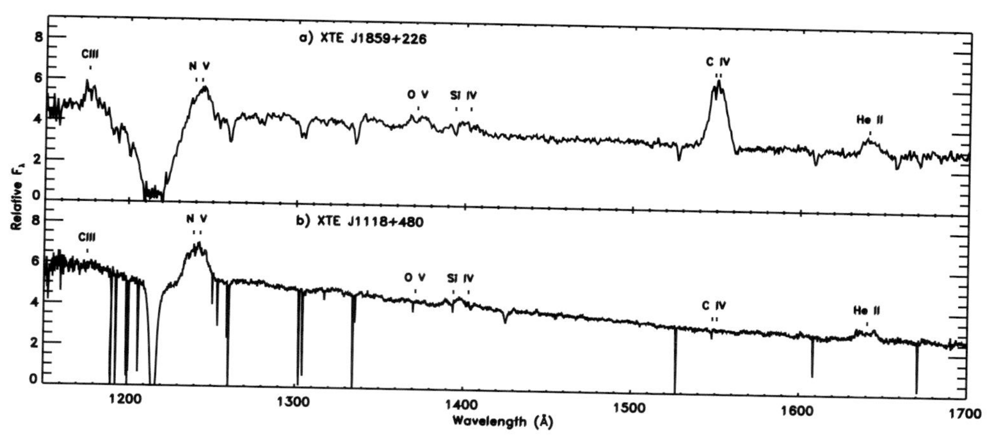

Fig. 5.22. HST/STIS UV spectra of XTE J1859+226 (*upper*), which shows normal line ratios (CIV more prominent than NV), whereas XTE J1118+480 (*lower*) has CIV and OV absent, yet NV is prominent (from Haswell *et al.* 2002).

radius reduces, indicating that the system fading is a result of a *cooling wave* moving inwards (see also Lasota 2001).

5.3.6 Outburst/decline properties

5.3.6.1 Lightcurve shapes and average properties

The wide variety of X-ray lightcurve shapes and outburst behavior has been reviewed by Tanaka & Shibazaki (1996), Chen *et al.* (1997) and updated in Chapter 4. The basic shape of a fast rise and exponential decay (Fig. 5.17) has been interpreted as a disk instability (see Cannizzo 1998, King & Ritter 1998 and Chapter 16) and is a natural consequence of an X-ray irradiated disk being maintained in a hot (viscous) state (and hence producing much longer outbursts than in dwarf novae, see King & Ritter). Furthermore, Shahbaz *et al.* (1998) have shown that SXTs can exhibit both exponential and linear decays, the latter occurring if the SXT outburst L_X is insufficient to ionize the disk outer edge (see Fig. 5.23 for XRN with known $P_{\rm orb}$ and well-defined lightcurve shape).

Additionally, Shahbaz and Kuulkers (1998) have shown (Fig. 5.24) that the optical outburst amplitude ΔV is related to $P_{\rm orb}$ (if $P_{\rm orb} < 1$ day) according to:

$$\Delta V = 14.36 - 7.63 \log P_{\rm orb}(\rm hrs) \tag{5.5}$$

which is essentially due to longer $P_{\rm orb}$ systems having larger and hence brighter mass donors (since they must be Roche-lobe filling). Hence a guide to $P_{\rm orb}$ can be immediately obtained once ΔV is known.

5.3.6.2 Do superhumps occur in LMXBs?

The high q values of XRN are also implied by the detection of *superhumps* in their optical lightcurves (during outburst decay; O'Donoghue & Charles 1996). As discussed in Warner (1995 and references therein), superhumps occur during *superoutbursts* of SU UMa-systems and are attributed to tidal stressing of the accretion disks in high q interacting binaries.

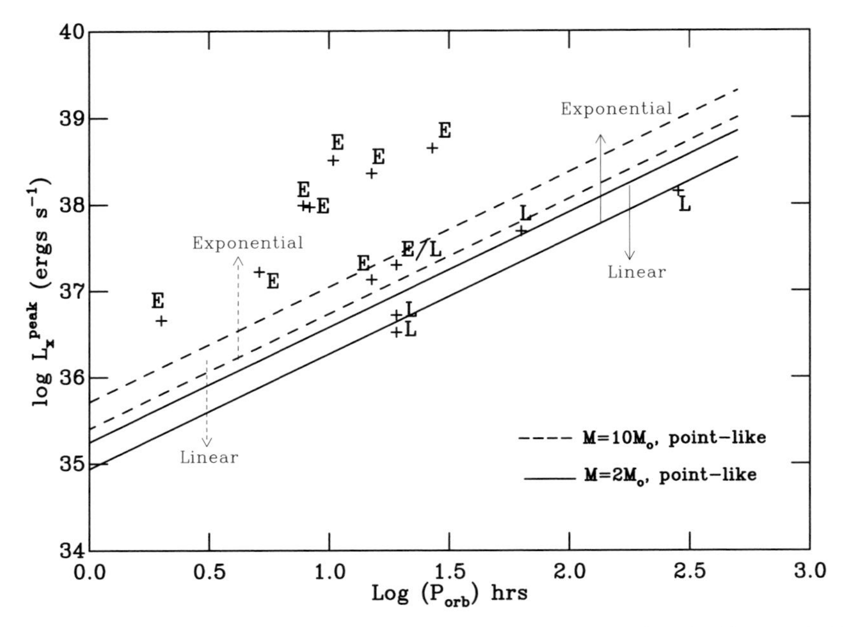

Fig. 5.23. The critical L_X needed to ionize the entire accretion disk (Shahbaz *et al.* 1998), for total masses of 2 and 10 $M_\odot$ corresponding to NS and BH SXTs respectively. These L_X are a factor 2 smaller for exponential (E) decays compared to linear (L) ones, due to the difference in the disk circularization and tidal radii. The SXTs shown are SAX J1808.4–3658, GRO J0422+32, A0620–00, GS2000+25, GS1124–68, Cen X-4, Aql X-1, 4U1543–47, GRO J1655–40 and GRO J1744–28.

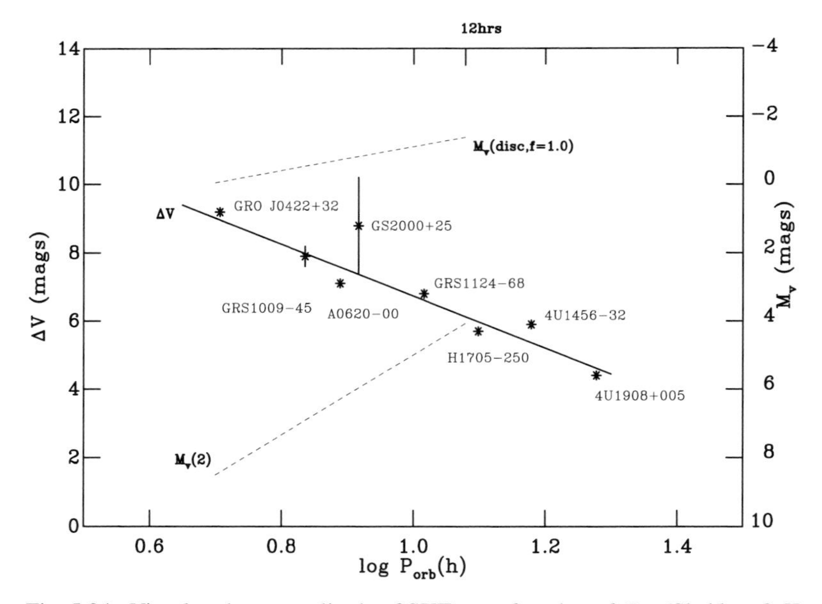

Fig. 5.24. Visual outburst amplitude of SXTs as a function of $P_{\rm orb}$ (Shahbaz & Kuulkers 1998). The dashed lines show the calculated disk M_V in an outbursting SXT and the donor M_V according to Warner (1995).

A wide range of models were proposed to account for this remarkable feature, but it is now widely accepted that SU UMa superhumps are due to tidal energy release in an elliptical, precessing disk, which occurs when the disk expands beyond its stability radius (Whitehurst & King 1991). However, this requires a high q (≥ 3) or else the stability radius is outside the tidal radius, and hence matter is simply lost from the system. With white dwarf masses typically ≤ 1 $M_\odot$ this implies very low mass secondaries and is expected to occur in short P (≤ 2 h) systems (almost all SU UMas are below the period gap; see Chapter 16). Mineshige *et al.* (1992) showed that the period excess $\Delta (= (P_{SH} - P_{orb})/P_{orb})$ is related to q via

$$\Delta = \frac{3}{4} q^{-1/2} (1 + q)^{-1/2} \left(\frac{R_D}{a} \right)^{3/2} \tag{5.6}$$

where R_D is the disk radius and a the orbital separation, and hence there is the potential for deriving some dynamical information from a simple measurement of the superhump Δ.

This basic model for the superhump phenomenon makes no requirement on the nature of the compact object, merely the value of q, implying that high q LMXBs, in which the compact object is now a NS or BH, should also display this effect. However, the much greater depth of the compact object potential well ($\geq 10^3$) in LMXBs implies that their intrinsic L_X exceeds that of CVs by the same factor, and hence that X-ray irradiation of the disk produces optical emission that far exceeds the *intrinsic* light of the disk, i.e., the tidal forces that produce superhumps should be swamped by irradiation. Nevertheless, observations have revealed effects related to the high q in both XRN and double-degenerate LMXBs (such as X1626–671 and X1916–053, see Section 5.3.12). While the latter are confirmed (via X-ray bursts) as NS, their ultra-short P (41 and 50 mins respectively) require a degenerate, and hence very low mass donor, so that q is very high in both cases.

This potentially important effect has been explained by Haswell *et al.* (2001) who utilized the results of disk simulations in extreme q binaries by Murray (2000). These simulations clearly demonstrated the onset of the 3:1 instability that leads to the precessing, elliptical disk (see Whitehurst & King 1991), but they also showed that the effective *area* of the disk is also modulated on the superhump period. And if the disk area is modulated at a given period, then so will be the reprocessed light, which is the dominant factor in LMXBs. This effect is also independent of i, as observed, unlike reprocessing in the heated face of the companion. The significance of this feature in transient LMXBs where the compact object is suspected of being a black hole is that its observation can lead to an estimate of the compact object mass even while the outburst is ongoing. However, there are many "steady" LMXBs (those in bold in Table 5.4) where the periodicity observed has not yet been confirmed as orbital in origin (and requires either eclipses or a radial velocity curve). Such systems are ideal targets for future application of the Bowen fluorescence mechanism (Section 5.3.9).

5.3.7 *Outburst spectroscopy of GRO J1655–40*

Obtaining dynamical information during the outburst/decline phases would be extremely valuable, as the majority of quiescent SXTs are extremely faint. The superluminal transient GRO J1655–40 (see Mirabel & Rodriguez 1999) exhibited extended intervals of activity in 1994 and 1996, during which Soria *et al.* (1998) followed the orbital behavior of

Table 5.4. *LMXB modulation properties*

Source	Period[a] (hrs)	Nature of modulation	X-ray type
X1820–303	**0.19**	X-ray	Burster, globular cluster
4U 1850–087	**0.34**	UV	Burster, globular cluster
X1626–673	**0.7**	opt sideband	Burster, Pulsar
X1832–330	**0.73**	UV	Burster, globular cluster (see also Section 8.2.1)
X1916–053	**0.83**	X-ray, opt	Burster, Dipper
J1808.4–3658	2.0	pulsation RV	Burster, Pulsar, Transient
X1323–619	**2.9**	X-ray dip	Burster, Dipper
X1636–536	**3.8**	opt	Burster
X0748–676	3.8	eclipsing	Burster, Dipper, Transient
X1254–690	**3.9**	X-ray dip	Burster, Dipper
X1728–169	**4.2**	opt	
X1755–338	**4.4**	X-ray dip	Dipper
X1735–444	**4.6**	opt	Burster
J0422+32	5.1	opt RV	BH, Transient
X2129+470	**5.2**	opt	ADC
X1822–371	5.6	eclipsing	ADC
J2123–058	6.0	eclipsing	Burster, Transient
N Vel 93	6.9	opt RV	BH, Transient
X1658–298	**7.2**	X-ray dip	Burster, Dipper
A0620–00	7.8	opt RV	BH, Transient
G2000+25	8.3	opt RV	BH, Transient
A1742–289	8.4	eclipsing	Burster, Transient
X1957+115	**9.3**	opt	
N Mus 91	10.4	opt RV	BH, Transient
N Oph 77	12.5	opt RV	BH, Transient
Cen X-4	15.1	opt RV	Burster, Transient
X2127+119	17.1	eclipsing	Burster, ADC, globular cluster
Aql X-1	**19**	opt	Burster, Transient
Sco X-1	**19.2**	opt	Prototype LMXB
X1624–490	**21**	X-ray dip	Dipper
N Sco 94	62.6	opt RV	BH, Transient
V404 Cyg	155.4	opt RV	BH, Transient
2S0921–630	216	eclipsing	ADC
Cyg X-2	235	opt RV	Burster
J1744–28	283	pulsation RV	Burster (Type II), Pulsar, Transient

Adapted from Charles (2001), van Paradijs & McClintock (1995) and van Paradijs (1998)
[a] Those in boldface have not yet been confirmed as orbital in origin via eclipses or radial velocity curves (see text).

its strong, complex Hα emission, but the asymmetric profile precluded its use as a dynamical tracer of the compact object. However, the broad wings of the double-peaked HeII λ4686 emission were found to be representative of the inner disk regions, giving a radial velocity curve in anti-phase with the companion star to within $9 \pm 20°$. Unfortunately, it is still difficult to use this for dynamical information as the HeII mean velocity is –182 km s^{-1}, whereas the secondary absorption lines give –142 km s^{-1} (both with very small errors). This systematic blue shift in the disk (emission) lines suggests the presence of a substantial disk wind during X-ray high states.

Combined with one of the earliest spectral types (mid-F), the high systemic velocity suggests that GRO J1655–40 might have been formed via accretion-induced collapse of a neutron star (Brandt *et al.* 1995). Furthermore, after its initial 1994 outburst, a subsequent optical rebrightening began ~6 days before the X-rays, indicating an "outside-in" outburst of the accretion disk (Orosz *et al.* 1997). This substantial delay is due to the inner disk needing to be refilled before accretion onto the compact object begins again, it having been evaporated by hard X-rays from the ADAF flow during quiescence (see Hameury *et al.* 1997).

GRO J1655–40 is extremely important amongst the XRN because of its brightness in quiescence, thereby yielding high-quality photometric lightcurves and high-resolution phase-resolved spectroscopy, from which the orbital system parameters can be derived (Orosz & Bailyn 1997; but note the error analysis of van der Hooft *et al.* 1997). The early spectral type also means that GRO J1655–40 has a low mass ratio ($q \sim 3$), and hence the ellipsoidal modulation is sensitive to both q and i (the latter being constrained by the observed grazing eclipse, see Fig. 5.20). The advanced evolutionary state of the secondary (which has been addressed by Kolb *et al.* 1997, Kolb 1998 and Beer & Podsiadlowski 2002) is driving the mass transfer rate, but it returns to the transient domain following temporary drops in $\dot{M}$.

5.3.8 *Effects of irradiation*

5.3.8.1 *Echo mapping*

Many of the X-ray transients are sufficiently bright during outburst that rapid X-ray variability provides a signature that allows "echo mapping" to be employed, which can probe the binary geometry. Although well-developed for mapping AGN structure (see, e.g., Horne 1999a), its application to LMXBs requires high time resolution ($\leq$ seconds) at optical/UV wavelengths, because of the ~seconds light-travel time across short-period LMXBs. This imposes a severe constraint on current technology, and is provided by few observatories due to the limitations of normal CCD cameras. However, HST does provide this facility, and Hynes *et al.* (1998) obtained simultaneous X-ray (Rossi X-ray Timing Explorer; RXTE) high-speed optical/UV FOS spectroscopy of GRO J1655–40 during its extended outburst, which showed the optical lagging the X-ray by ~10–20 s at times of X-ray flaring activity (see Fig. 5.25). With a relatively long $P_{\rm orb}$ (2.6 d) and well-established orbital ephemeris, this optical delay can be explained entirely by reprocessing within the accretion disk and not heating of the secondary.

More recently, Kanbach *et al.* (2001) and Hynes *et al.* (2003a) performed a similar study (using fast ground-based photometry and HST/STIS UV spectroscopy respectively) on XTE J1118+480. While the absolute HST timing is still uncertain, these data do show the longer UV wavelengths (~2700 Å) lagging the shorter (~1400 Å) by about 0.25 s. However, the short timescale of variability suggests that the bulk of this component is due to synchrotron emission, and not the accretion disk.

The long-term goal of these studies is to follow such behavior throughout several orbital cycles, which in principle would allow the system geometry to be fully mapped. Unfortunately, this is a very difficult observational programme, particularly given that at times of X-ray outburst the binary orbital ephemeris is rarely known. But given the typical recurrence

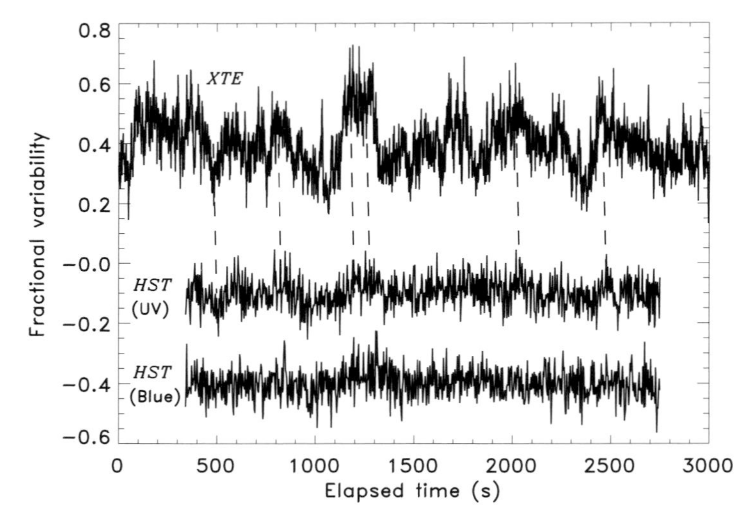

Fig. 5.25. X-ray (RXTE) and optical/UV (HST) lightcurves of GRO J1655–40 obtained in 1996 by Hynes *et al.* (1998), which show the ~10–20 s delay of the optical response to the X-ray flares, and are connected by dashed lines.

timescales of XRN of 10–50 years, such opportunities will grow in future and echo mapping could become an important technique.

5.3.8.2 Distortion of the radial velocity curve

With the X-ray variability signature clearly visible in the disk's optical/UV emission, it is likely that the (very high) outburst L_X has a significant impact on the secondary's atmosphere. The majority of XRN have donors that, in any case, are completely undetectable during outburst, but the small sub-group of intermediate mass donors (see Table 5.3) are sufficiently luminous to be easily spectroscopically visible at all times. This is the case with GRO J1655–40 and the original dynamical study (Orosz & Bailyn 1997) should, ideally, have been undertaken during X-ray quiescence. But in order to obtain full orbital phase coverage they did use some spectra taken during outburst. Under such circumstances, care must be taken to ensure that irradiation of the secondary has not systematically distorted the radial velocity curve.

Phillips *et al.* (1999) tested for the presence of this effect by attempting to fit an elliptical orbit to the Orosz and Bailyn data, and found an inferred, highly significant eccentricity of $e = 0.12 \pm 0.02$, which must be due to irradiation, since short-period, interacting binaries must in fact be circular. With the ratio of incident X-ray to local stellar flux at the donor surface of ~7, it is not surprising that such effects can be important. Fitting an irradiated model of the secondary with free parameters q, i and disk opening angle (which shadows the secondary's equatorial regions), Phillips *et al.* obtained $q = 2.8$ and M_X in the range $4.1 - 6.6\ M_\odot$. A completely quiescent radial velocity curve was subsequently obtained by Shahbaz *et al.* (1999, 2000) which confirmed this analysis by constraining q to the range 2.29–2.97 and M_X to 5.5–7.9 $M_\odot$, and updated further by Shahbaz *et al.* (2003) to the values given in Table 5.3. Hence, it is still possible (Brandt *et al.* 1995) that the high γ-velocity of GRO J1655–40 might indicate the formation of a relatively low mass BH via accretion-induced collapse of a neutron star.

5.3.9 *Spectroscopy of luminous LMXBs*

5.3.9.1 *Sco X-1 and X-ray irradiation of the companion*

The detailed dynamical analysis of the XRN in Section 5.3.3 is only possible because of their very long quiescent intervals, during which the optical emission is dominated by the companion. This accounts for the remarkable fact that XRN account for the majority of detailed LMXB mass constraints (Table 5.3). And since few LMXBs are X-ray pulsars, remarkably little is known of the fundamental parameters of the X-ray luminous ("steady") LMXBs, as their companion stars are perpetually hidden from view in the glare of the X-ray illuminated accretion disk. Indeed, in spite of being the brightest LMXB in the sky (both optically and in X-rays), key parameters of Sco X-1 have only been determined quite recently. VLBA observations of its twin radio lobes give $d = 2.8 \pm 0.3$ kpc (Bradshaw *et al.* 1999) and $i = 44 \pm 6$ degrees (Fomalont *et al.* 2001). However, the donor has not been detected spectroscopically in either the optical (Schachter *et al.* 1989) or IR (Bandyopadhyay *et al.* 1997, 1999), and the optical spectrum is dominated by the X-ray irradiated disk. This situation has been transformed by Steeghs and Casares (2002) who obtained the first high-resolution optical spectroscopy that revealed very sharp emission features from the Bowen $\lambda\lambda$4640–50 emission blend (Fig. 5.26).

These broad emission components produce a radial velocity curve (LaSala & Thorstensen 1985) whose phasing with respect to the lightcurve (Gottlieb *et al.* 1975) confirms their production in the disk. However, the sharp components move in *anti-phase* to the broad emission and are therefore associated with the companion's *heated* face (Fig. 5.27). Note that the Bowen motion is not strictly sinusoidal, as the irradiation dominates on the inner Roche lobe that faces the X-ray source (although there will also be some disk shadowing which will ameliorate this), and hence only the limit $K_2 > 77$ km s^{-1} is obtained. However, Doppler tomography (see Marsh 2001 for a detailed review of this technique) of Sco X-1 with these data (Fig. 5.28) shows strong emission from the secondary and indicates $K_2 > 87$ km s^{-1}. There is a slight offset with the anti-phased HeII emission ($\Delta\phi \simeq 0.1$; likely due to hot spot contamination as in CVs, see Warner 1995), but the amplitude implies $K_1 < 53$ km s^{-1}. The Doppler tomograms show that there is a HeII component on the secondary as well (but not HeI or Balmer emission, these arise completely on the disk), and this distorts the disk radial velocity curve. Searching only for the symmetric component gives $K_1 \simeq 40$ km s^{-1}, and hence $q > 2.2$. However, Steeghs and Casares argue that the likely offset of the heated face combined with the radio-determined i yield $M_2 \sim 0.4\ M_\odot$ if the neutron star has a canonical 1.4 $M_\odot$. This requires an evolved secondary and hence $\dot{M} \sim 8 \times 10^{-10}\ M_\odot$ yr^{-1} (King *et al.* 1996), which gives a disk $T > 6500$ K which is stable.

5.3.9.2 *GX339–4 and XRN in outburst*

This powerful new technique has already been applied to the pseudo-transient LMXB BHC GX339–4 (Hynes *et al.* 2003b). First proposed as a BHC (Samimi *et al.* 1979) based on its Cyg X-1-like X-ray fast variability, GX339–4 moves regularly through high, low and "off" (or quiescent) states (e.g., Kong *et al.* 2000; Zdziarski *et al.* 2004) on timescales of a year or two. Associated with these, the optical counterpart (V821 Ara) varies between

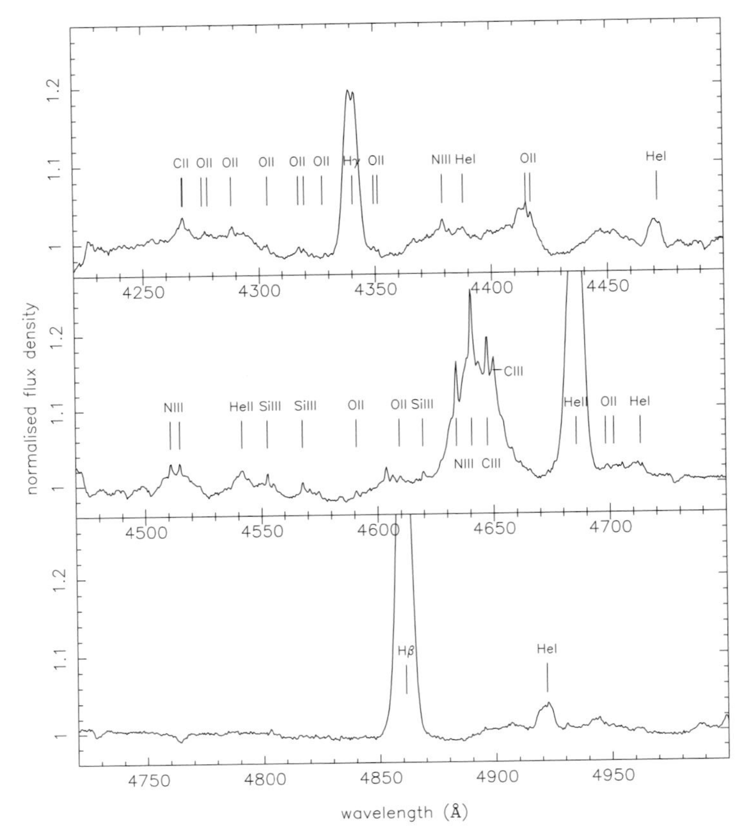

Fig. 5.26. High-resolution blue spectrum of Sco X-1 (Steeghs & Casares 2002) which demonstrates the complex emission profile of the λλ4640–50 Bowen blend. Note particularly the very sharp components of NIII and CIII.

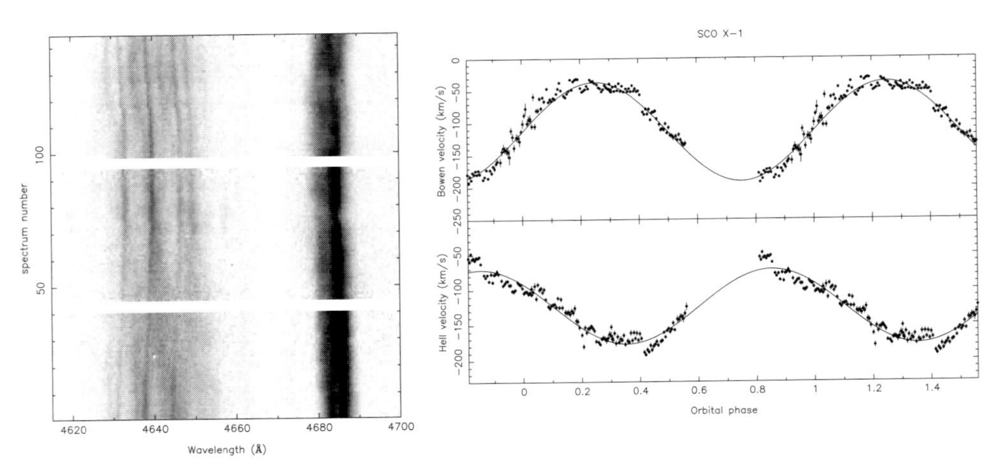

Fig. 5.27. Trailed spectrogram (left) and radial velocity curves (right) of the HeII and Bowen emission from Sco X-1 (Steeghs & Casares 2002).

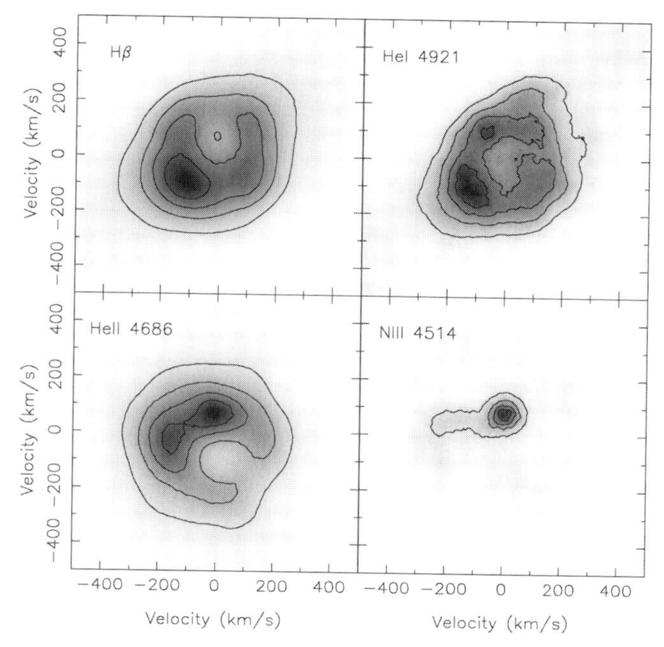

Fig. 5.28. Doppler tomograms of Sco X-1 (Steeghs & Casares 2002) where the origin is the binary center of mass. Note the strong NIII emission located on the companion, whereas the other emission lines' ring-like structure is typical of accretion disks (although HeII also has a component on the companion).

V $<$ 15 at peak to $>$20 at minimum. Yet even at its faintest, there has been no detection of the (presumed) cool mass donor (Shahbaz *et al.* 2001). Consequently, even $P_{\rm orb}$ has been ill-constrained, and so the return to an X-ray bright phase in mid-2002 provided an opportunity to apply the Bowen fluorescence technique.

The Bowen blend is indeed strong when the source is X-ray bright, and smoothly moving sharp features were detected which suggest $P = 1.76$ d, compared to the (non-confirmed) 14.8 h period of a decade earlier (Callanan *et al.* 1992; see also Cowley *et al.* 2002). The resulting K_2 velocity leads to the $f(M)$ given in Table 5.3, finally confirming its BHC status. What is curious in this case is the fact that the sharp features are not always visible (Hynes *et al.* 2003b), suggesting that there are significant intervals during which the face of the donor is effectively shielded from direct illumination. This could arise if the disk either is warped or is elliptical and precesses in such a way that it approaches the donor (Rolfe *et al.* 2001; Foulkes *et al.* 2004). Nevertheless, these results demonstrate the potential power of this technique and we anticipate substantial progress in the coming decade in determining, for the first time, the dynamical properties of the luminous LMXBs.

5.3.10 *Orbital lightcurves and tomography of LMXBs*

5.3.10.1 XTE J2123–058

The 1998 NS XRN J2123–058 is an ideal system with which to demonstrate the range of LMXB optical lightcurves as a function of X-ray luminosity. It has a relatively

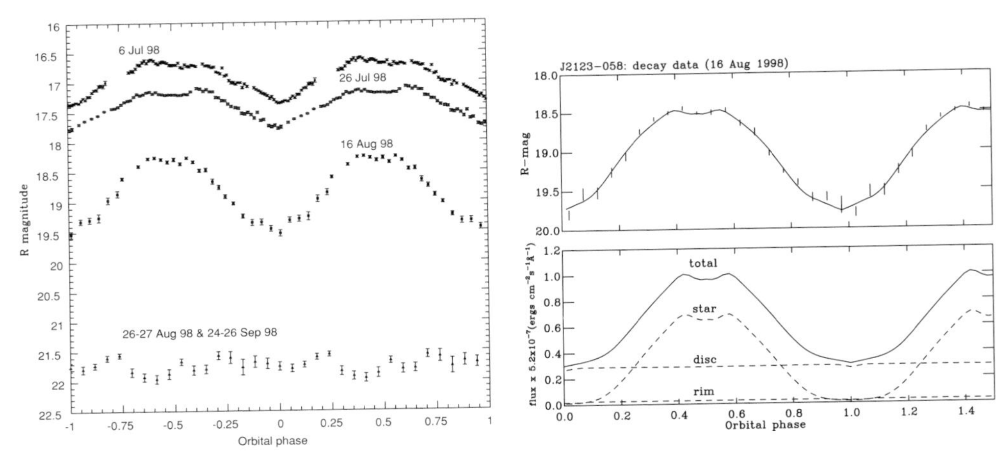

Fig. 5.29. *Left*: orbital lightcurves of J2123–058 (folded on $P = 6$ h) from close to outburst peak (July 1998) through to quiescence several months later, showing the dramatic lightcurve evolution as L_X declines (Zurita *et al.* 2000). *Right*: model fit to the August decay data with the individual model components plotted below (Shahbaz *et al.* 2003).

short (6 h) period and orbital lightcurves were obtained from outburst peak through decline and into quiescence by Zurita *et al.* (2000; see also Soria *et al.* 1999). The high i produces a spectacular modulation that evolves from triangular to double-humped (Fig. 5.29) as the relative contributions of the disk, X-ray heated companion and ellipsoidal components change during the decline. These exquisite lightcurves were modeled by Zurita *et al.* (see also the reanalysis by Shahbaz *et al.* 2003) to constrain i to be $73 \pm 4°$. J2123–058 also produced a value of $\xi = B_0 + 2.5 \log F_X(\mu\text{Jy}) = 21.9$, which is the parameter used by van Paradijs and McClintock (1995) to characterize the ratio of X-ray to optical flux. This value compares remarkably well with the canonical value for LMXB X-ray heating of 21.8 ± 1 from van Paradijs and McClintock (1994). The observation of an X-ray burst also requires the neutron star to be directly visible, and hence provides a tight constraint on the disk flare angle ($\alpha < 90 - i$).

In addition to the outburst lightcurves of J2123–058, time-resolved spectroscopy (Hynes *et al.* 2001a) through two orbital cycles permitted the application of Doppler tomography in order to provide an *image* of the accretion disk, but with surprising results. Multiple S-waves are easily visible in HeII line profiles, but the Doppler maps (Fig. 5.30) show this to be located in the lower left region (on the *opposite* side of the NS from the companion), and hence definitely *not* the donor's heated face.

Combined with transient Hα absorption between orbital phases 0.35–0.55, Hynes *et al.* interpret J2123–058 as the NS analog of the SW Sex phenomenon (a sub-class of CVs, see Hellier 2000). While other LMXBs, both transient (e.g., J0422+32, Casares *et al.* 1995) and steady (e.g., X1822–371, Harlaftis *et al.* 1997b; Casares *et al.* 2003) display their main emission at the stream–disk impact region, there are now several LMXBs with similar properties to J2123–058 (see Fig. 5.31). While this region of the Doppler tomogram can be a result of stream overflow and subsequent re-impact on the disk (Shafter *et al.* 1988), it requires a strongly flared disk in order to prevent the overflowing stream producing absorption at *all* phases, and this is already very tightly constrained by the lightcurve models described above.

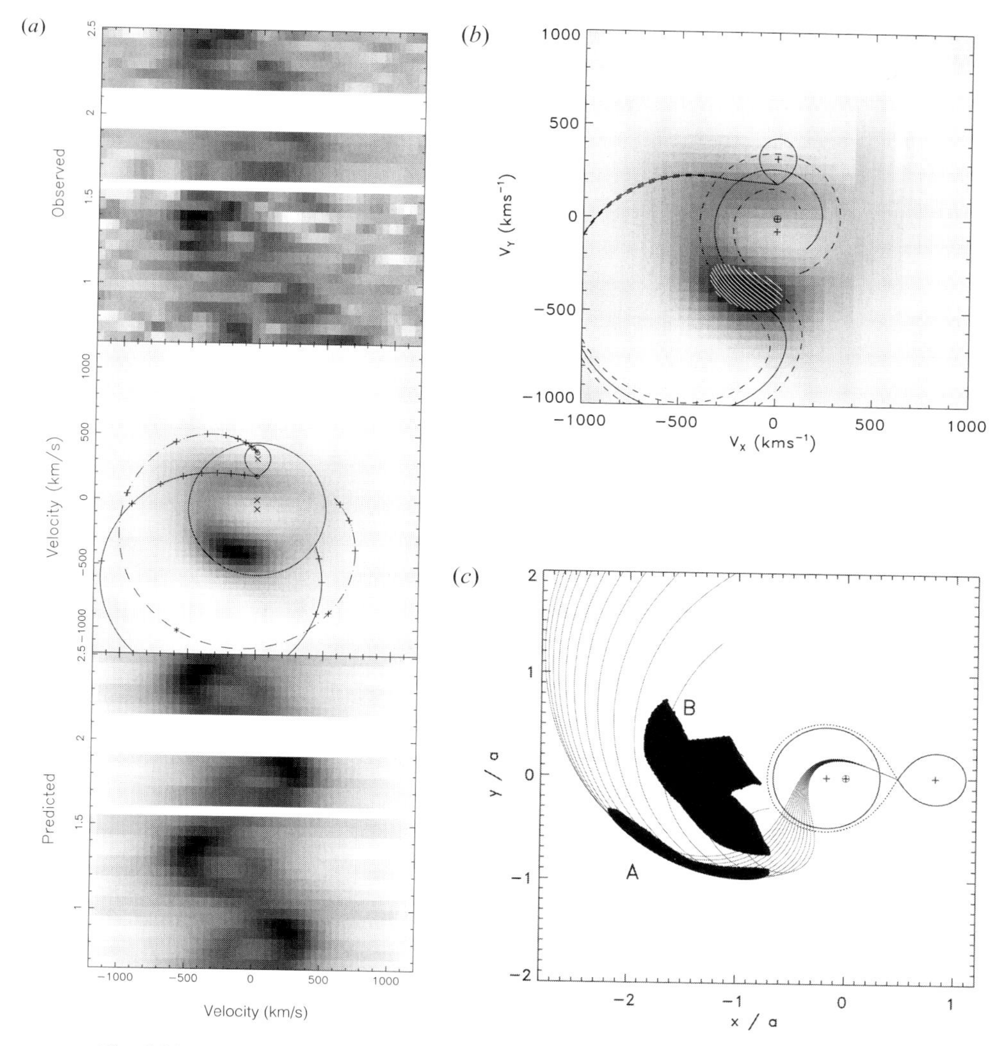

Fig. 5.30. Doppler tomography of J2123–058 during outburst (Hynes *et al.* 2001a). *(a)* (from top) Actual data, tomogram, reconstruction. In the center panel, the solid line is the ballistic stream trajectory, the dashed line the Keplerian velocity at the stream position and the large circle is the Keplerian velocity at the disk edge. *(b)* Comparison between magnetic propeller trajectories and the tomogram. *(c)* Spatial plot corresponding to *(b)*. Points in region A produce emission with kinematics corresponding to the tomogram's white hatched region, whereas those in B produce absorption with the Hα phasing and velocities.

A more likely explanation is to invoke the *magnetic propeller* model of AE Aqr (Eracleous & Horne 1996; Horne 1999b), which involves a strong magnetic field anchored to a rapidly spinning white dwarf that accelerates matter out of the system and into a region beyond the compact object where the streams collide (Fig. 5.30*c*). However, J2123–058 contains a rapidly spinning neutron star ($P_{\rm spin} \sim 4$ ms; Tomsick *et al.* 1999) and would present far too small an area to be effective because of the very small light-cylinder radius, beyond which

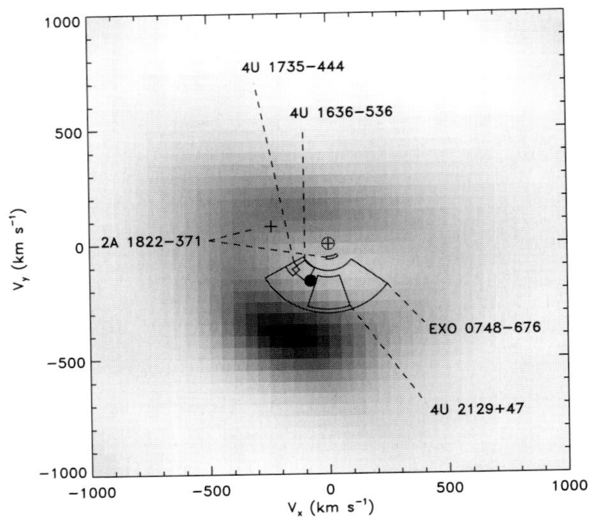

Fig. 5.31. Comparison of J2123–058's tomogram with other systems. Boxes represent uncertainties in velocity semi-amplitude and phasing. The large point is the best fit semi-amplitude and phase for J2123–058 (Hynes *et al.* 2001b).

the field will be wound up. Consequently, Hynes *et al.* suggest that this process could still be occurring but with a magnetic field anchored in the disk.

5.3.10.2 Millisecond pulsar transients

Whilst the discovery of kHz QPOs in LMXBs (see Chapter 2) finally demonstrated that high L_X LMXBs could indeed spin up their neutron stars to very high rates (and hence are the ancestors of radio millisecond pulsars, MSPs), it was not until 1998 that RXTE discovered the first X-ray MSP, J1808.4–3658, with $P_{spin} = 2.49$ ms (Wijnands & van der Klis 1998). X-ray bursts were also seen by SAX (in't Zand *et al.* 1998) and this fast LMXB pulsar is in a 121 minute orbit about a very low mass ($<0.1\ M_\odot$) secondary (Chakrabarty & Morgan 1998). This short-lived transient (only tens of days) peaked at $V \sim 16.6$ (Roche *et al.* 1998), but was below $V \sim 20$ only 6 weeks later (Giles *et al.* 1999). However, Giles *et al.* did detect a $\Delta V \sim 0.1$ modulation on P_{orb} which was anti-phased with the pulsar and hence dominated by the heated face of the cool donor. Even in quiescence at $V \sim 21$, Homer *et al.* (2001a) found a similar, sinusoidal modulation still present, which was due to the heated face, and *not* the double-humped ellipsoidal lightcurve seen in XRN, i.e., the secondary is *not* detectable. This can be explained by the fact that J1808.4–3658 has $F_X/F_{opt} \sim 10$ in quiescence, compared to a more typical $\sim$1000 in outburst, i.e., the quiescent source is still bright optically, which means that the Homer *et al.* lightcurve is due to a combination of the remnant (irradiated) disk and the heated face of the donor.

Furthermore Bildsten and Chakrabarty (2001) have determined that the donor has $M_2 = 0.05\ M_\odot$ and $R_2 = 0.13\ R_\odot$ in order to fill its Roche lobe, and hence it must be heated by the NS (see also Burderi *et al.* 2003), confirming that J1808.4–3658 is the progenitor of an MSP binary. The low average $\dot{M}$ of $\sim 10^{-11}\ M_\odot\ yr^{-1}$ implies that such systems can be extremely long-lived and may still be present as, e.g., low L_X sources in globular clusters (as seen in 47 Tuc, see Chapter 8). This class of object is extremely important as a test-bed for understanding the magnetic properties of neutron stars (Bhattacharya 2002), and will be enhanced with the recent discovery with RXTE of four more LMXB MSPs (Chapters 1 and 2; Markwardt *et al.* 2002; Galloway *et al.* 2002; Bildsten 2002; Markwardt *et al.* 2003; Kirsch & Kendziorra 2003), all with $P_{\rm spin} \sim 2$–5 ms.

5.3.11 *Infrared spectroscopy of LMXBs*

It is an unavoidable consequence of the galactic distribution of LMXBs (White & van Paradijs 1996), in which half of all LMXBs are within 20 degrees of the Galactic Center, that the majority of these are heavily obscured in the optical and hence are accessible only in the IR (since the *V*/*K* extinction ratio is $\sim$10). The most luminous of these form a subgroup known as the "Galactic Bulge Sources" (or GBS) which lie within a strip of $l = \pm 15^\circ$, $b = \pm 2^\circ$ about the Galactic Center. Several of the GBS exhibit QPOs, but rarely X-ray bursts (see Chapter 3), and no clearly defined $P_{\rm orb}$, hence their nature as LMXBs or HMXBs was unclear in many cases. Consequently, the accurate X-ray locations for the GBS were surveyed by Naylor, Charles and Longmore (1991) with the first generation of IR arrays. While the technology has improved over the last decade, IR spectroscopy is still limited to much brighter sources than in the optical, and so much remains to be done. Here we discuss the advances made on particular sources (e.g., Bandyopadhyay *et al.* 1997, 1999), key properties of which are in Table 5.5 together with details of other highly obscured LMXBs (apart from Sco X-1 and Cyg X-2 which are included for reference):

GX1 + 4 (X1728–247) is unusual in being in a region of low reddening, where the X-ray source is coincident with a bright ($K = 8.1$) M6III star at $d \sim 3$–6 kpc and hence $L_X \sim 10^{37}$ erg s^{-1} (Chakrabarty & Roche 1997). A_V is steady (at 5.0 ± 0.6), but the X-ray column is variable, implying the presence of a wind. Its Hα emission is not uncommon for such an object, and the identification was only secure when the Hα flux was found to be pulsing with the same period (114 s) as in X-rays (Jablonski *et al.* 1997). While GX1+4 was, for 25 years, the brightest hard X-ray source in the Galactic Center region (and the pulsar was spinning up), it turned "off" in the 1980s and is now on at a low level (and spinning down). Most HMXBs display an orbital modulation superposed on the pulsar's long-term $\dot{P}$ trend, but not GX1+4. There have been suggestions of a $\sim$304 d variation in the accretion torque history (Cutler *et al.* 1986) and spin-down rate (Pereira *et al.* 1999), but there is no correlation with the X-ray flux and so these claims require confirmation. Assuming a NS mass of 1.4 M$_\odot$ and requiring the M6III to fill its Roche lobe implies that $P \geq 100$ d. GX1+4 is of interest as the *only* NS symbiotic system (but note 4U1700+24, Masetti *et al.* 2002, which may be similar, although much less luminous), and hence a potential progenitor of very wide ($P \gg 10$ d) radio pulsar binaries. These must have been very wide LMXBs and hence required K or M giant donors. Such rare systems (with a high $\dot{M}$) would appear just like a symbiotic.

GX13+1 was identified with a $K \sim 12$ star by Naylor *et al.* (1991) on the basis of its variability and strong Brγ emission (Fig. 5.32). More importantly the spectrum also shows CO absorption bands typical of cool (K–M) stars, and the presence of ^{13}CO indicates that it is evolved. At $d \sim 9$ kpc and $E(B\!-\!V) = 5.7$, then $M_K = -4.6$ and so it must be $\sim$K2–M5III. Possible periodicities in the range 13–20 d have been suggested (Garcia *et al.* 1992; Bandyopadhyay *et al.* 2002) which would be consistent with such a classification, but further

Table 5.5. *LMXB IR properties*

Source	b^{II}	K	P (d)	IR spectrum: Emission	IR spectrum: Absorption	Notes
Cir X-1[1,2]	0	9–12	16.54	Brγ, HeI HeII	–	Super$L_{\rm Edd}$, highly eccentric orbit
X1630–47[3]	+0.3	16.1	–	–	–	~600–690$^{\rm d}$ recurrent BH SXT
GX1+4[4,5]	+4.8	8	~300	Brγ	M6III	Symbiotic, X-ray pulsar (114 s)
Z-sources						
Sco X-1[4,6]	+24	11.9	0.79	Brγ, HeI HeII	–	"Steady" prototype LMXB
GX340+0[7]	−0.1	17.3	–	–	–	Luminous bulge source
Sco X-2[6]	+2.7	14.6	0.93/14.9	Brγ, HeI	–	Luminous bulge source
GX5-1[8,9]	−1.0	13.5	–	Brγ, HeI	–	Luminous bulge source
GX17+2[10–12]	+1.3	14.9–18.5	~20	–	–	**not** NP Ser
Cyg X-2[7]	−11.3	13.3	9.8	–	A9III	IMXB
Atolls						
X1608–522[7,13]	−0.9	16.5	0.54	–	–	NS transient
GX9+9[7]	+9.0	16.0	0.17	–	–	Very short period
GX3+1[14]	+0.8	15.1	–	–	–	Candidate ID
GX9+1[14]	+1.1	16.2	–	–	–	Candidate ID
GX13+1[4,6,12,15]	+0.1	11.9–12.3	~25d	Brγ, HeI	CaI, MgI, ^{12}CO, ^{13}CO	(See Fig. 5.32)

[1] Johnston *et al.* 1999; [2] Clark *et al.* 2003; [3] Augusteijn *et al.* 2001; [4] Bandyopadhyay *et al.* 1997; [5] Chakrabarty & Roche 1997; [6] Bandyopadhyay *et al.* 1999; [7] Wachter 1998; [8] Jonker *et al.* 2000; [9] Bandyopadhyay *et al.* 2003; [10] Deutsch *et al.* 1999; [11] Callanan *et al.* 2002; [12] Bandyopadhyay *et al.* 2002; [13] Wachter *et al.* 2002; [14] Naylor *et al.* 1991; [15] Charles & Naylor 1992

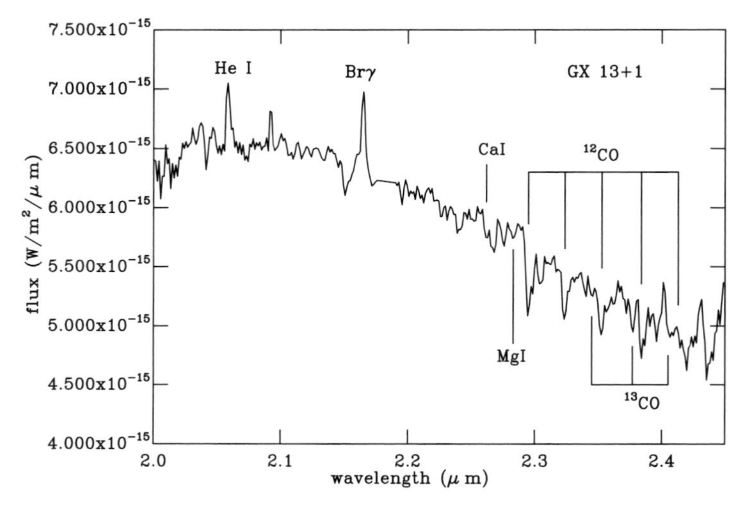

Fig. 5.32. K-band spectrum of GX13+1 (Bandyopadhyay *et al.* 1999) showing strong HeI and Brγ emission, together with CO absorption bands. Note the presence of ^{13}CO, which indicates it is both cool and evolved.

advances in understanding GX13+1 require an IR radial velocity study, which is now feasible, although observationally challenging.

GX17+2 is a bright Z source which also bursts (and is hence a NS). It benefits from an accurate radio location (Hjellming 1978) which aligns with NP Ser, a $V \sim 17$, completely normal G star. This conundrum was settled by an HST/NICMOS image which revealed that NP Ser is actually outside the radio error circle (by 0.7 arcsecs), while two much fainter ($H \sim 20, 21$) stars are consistent with the radio location (Deutsch *et al.* 1999). Keck observations (Callanan *et al.* 2002), however, reveal a $K = 15$ star at that position, ~3.5–4 mags brighter than the NICMOS stars! Furthermore, K-band monitoring by Bandyopadhyay *et al.* (2002) shows that this object displays large amplitude variability once corrections are made for (the unresolved) NP Ser. What is remarkable about GX17+2 is that (according to RXTE/ASM monitoring) the X-ray flux level is extremely high *and* steady, in spite of the huge IR fluctuations.

The two SXTs **X1608–522 and X1630–47**, are unusual in that they both have semi-regular outbursts on timescales of ~600 days (Augusteijn *et al.* 2001; Wachter *et al.* 2002; and references therein). They both suffer from extinction, so their optical counterparts are faint, even in outburst, but both have now been observed in the IR, with X1608–522 recently revealing a 0.54 d (single peak, i.e., X-ray heated) modulation during an extended low X-ray intensity state following an outburst. Such states are compared with Z Cam "standstills", which are seen in CVs and GX339–4 (Kong *et al.* 2002), and have also been seen in Aql X-1, another frequently outbursting NS SXT.

5.3.12 *Long periods and disc structure in high inclination LMXBs*

The first X-ray dipper discovered, X1916–053 (Walter *et al.* 1982), produced one of the first orbital periods of any LMXB. With Type I X-ray bursts indicating the presence of a neutron star, and the ultra-short $P_{\rm orb}$ of 50 min implying that the mass donor was a $\sim 0.1\ M_\odot$ degenerate dwarf, then $q \sim 15$ and hence it should also be susceptible to the SU UMa-type disk precession (Section 5.3.6.2). Indeed an X-ray quasi-periodicity of ~199 d had already been suggested (Smale & Lochner 1992).

In spite of its faintness ($B \sim 21$; due to a combination of reddening and the low-mass donor) and crowding, an optical analogue of the X-ray dip period was found, but remarkably it was 1% longer (3027.6 s versus 3000.6 s)! Both periods are now precisely known and the difference is highly significant (Callanan *et al.* 1995; Chou *et al.* 2001; Homer *et al.* 2001b), but debate continues as to which is orbital. Classical SU UMa-type behavior would imply it is the X-ray, and precession induces the slightly longer optical modulation. However, unlike in SU UMa systems (where the superhump period actually drifts during the ~2 week superoutburst), the optical period in X1916–053 was extremely stable, suggesting that it might be the orbital period. Other models have also been invoked (e.g., a triple system, Grindlay *et al.* 1988, Grindlay 1989).

Furthermore, detailed X-ray and optical lightcurves (Fig. 5.33; see Homer *et al.* 2001b; Chou *et al.* 2001) reveal that the lightcurve morphology evolves dramatically over several days, on a timescale similar (but not necessarily equal) to the X-ray–optical beat period (Fig. 5.33). The X-ray dipping is known to be due to azimuthal structure in the disk edge (e.g., around the stream–disk impact region) viewed at an appropriately high i (Parmar & White 1988). Any variation in this structure, or in the angle it presents relative to our line of

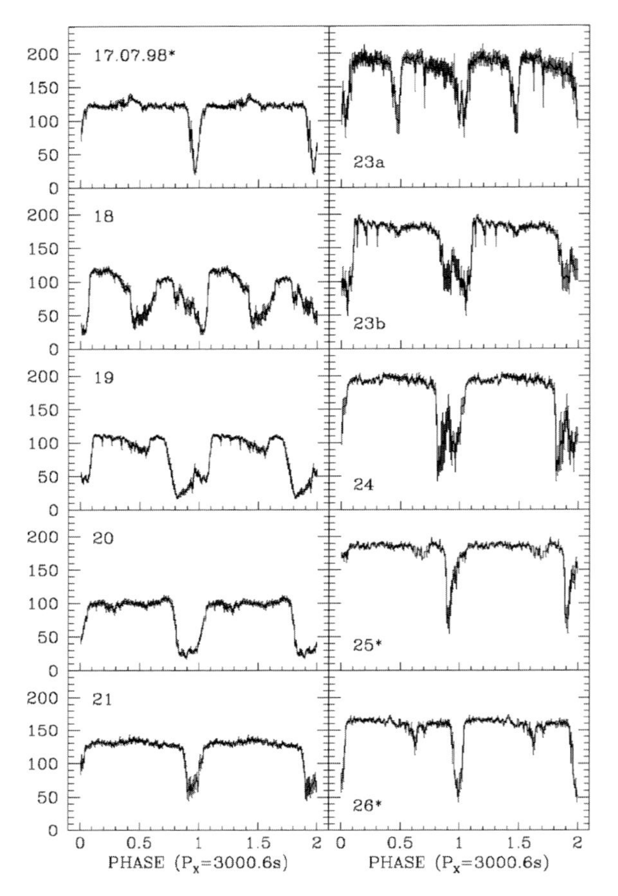

Fig. 5.33. RXTE PCA X-ray lightcurves of X1916–053 folded on the X-ray period of 3000.6 s, in which the dipping structure clearly evolves on a timescale of days, eventually repeating (adapted from Homer *et al.* 2001b).

sight, would cause the observed lightcurve to change. This has led to speculation that the disk in X1916–053 not only precesses but is warped, an effect expected due to X-ray irradiation (Wijers & Pringle 1999; Ogilvie & Dubus 2001). This is presumably similar to the 35 d on–off cycle in Her X-1, where the tilted disk precesses relative to the observer (see, e.g., Schandl & Meyer 1994). This cycle is not precise, it has some "jitter", as is also seen in the precession of SS433 (Margon 1984).

Such a lightcurve modulation can also be explained by invoking a triple system (producing dips predominantly when the mass transfer rate is enhanced), which would have to be formed in a globular cluster core that has subsequently evaporated (Grindlay *et al.* 1988). A further problem is the stability of the triple given the observed periodicities, and this model is now deemed unlikely (Homer *et al.* 2001b; Chou *et al.* 2001). The effects observed are far closer to those expected from the precessing disk model related to the well-established high q. Initially the stability of the optical period was considered a difficulty, but since the mass transfer in this LMXB is basically steady (unlike in SU UMa systems) it is possible that the instability will also stay fixed at a single period, i.e., X1916–053 is a *permanent superhumper*. An alternative is to consider X1916–053 as a *negative superhumper* in which the optical period is indeed orbital, and the X-rays are now the superhump. While this is seen in a number of CVs (see, e.g., Patterson 1999) the effect is not well understood (Retter *et al.* 2002).

With its short P_{orb} and accessible (~days) lightcurve evolution, X1916–053 is an ideal system in which to study accretion disk edge structure and how it might be influenced by irradiation-driven warping. Such results are likely to have application well beyond this immediate field. Possibly similar behavior has been reported in the BHC X1957+11 (Hakala *et al.* 1999) and the system MS1603+2600 (Hakala *et al.* 1998), the latter at high latitude and with a compact object whose nature is still controversial (see, e.g., Mukai *et al.* 2001).

5.3.12.1 The halo transient, XTE J1118+480

This topic has been taken even further with extensive observations of the SXT J1118+480 (Remillard *et al.* 2000). With its low extinction and relative proximity (~ 1 kpc), the late decline phase was monitored in unprecedented detail (Zurita *et al.* 2002a). As expected once the disk contribution faded, the ellipsoidal modulation became visible (Fig. 5.34). However, the accretion disk component was still very significant (~50–70%), and temporal analysis revealed that, in addition to the ellipsoidal variation (which appears to peak at $P_{orb}/2$), there is a further modulation at a slightly longer period (visible as a distorting "wave" in the nightly lightcurves). Subtracting a theoretical ellipsoidal fit from the data leaves this unusually structured modulation, an effect that continues well past the original outburst (as also now seen in CVs, e.g., Patterson 1995). More surprisingly here, the period excess is extremely small, at only $\Delta \sim 1\%$.

However, such a small Δ implies (Eq. (5.6)) that q must be extremely high (~40), and hence this interpretation must be treated with great caution (Fig. 5.34). At such extreme q values the classical 3:1 resonance would be replaced by the 2:1 resonance (Whitehurst & King 1991).

5.3.13 Special cases

In this final section we discuss several well-known sources whose classification as either HMXB, LMXB or even binary is still a subject of controversy, mostly due to the absence of any direct signature of the mass donor. However, in three of these cases there is now circumstantial evidence that the secondaries are quite massive, but the very high implied $\dot{M}$ gives them properties akin to LMXBs, so they transcend a simple classification.

5.3.13.1 SS433: the link with ULXs?

SS433 is the original source for studying the detailed physical processes involving relativistic jets on a galactic scale, and is still the only continuously emitting galactic microquasar (see Chapter 9). The key feature of SS433 is its 160 d precessing jet, observed for several decades (in optical, IR and X-ray) as periodic emission-line shifts which can be fitted to high precision by the *Kinematic Model* (see Margon 1984 for details and a full historical review). However, in spite of more than 20 years of detailed study, remarkably little is known about the fundamental system properties of SS433. This is a consequence of two factors: (i) the enormously powerful and broad, stationary and moving emission lines (which first brought it to attention) obliterate any spectral features of the companion; (ii) the high extinction that has severely biased most optical spectroscopy towards the red (which

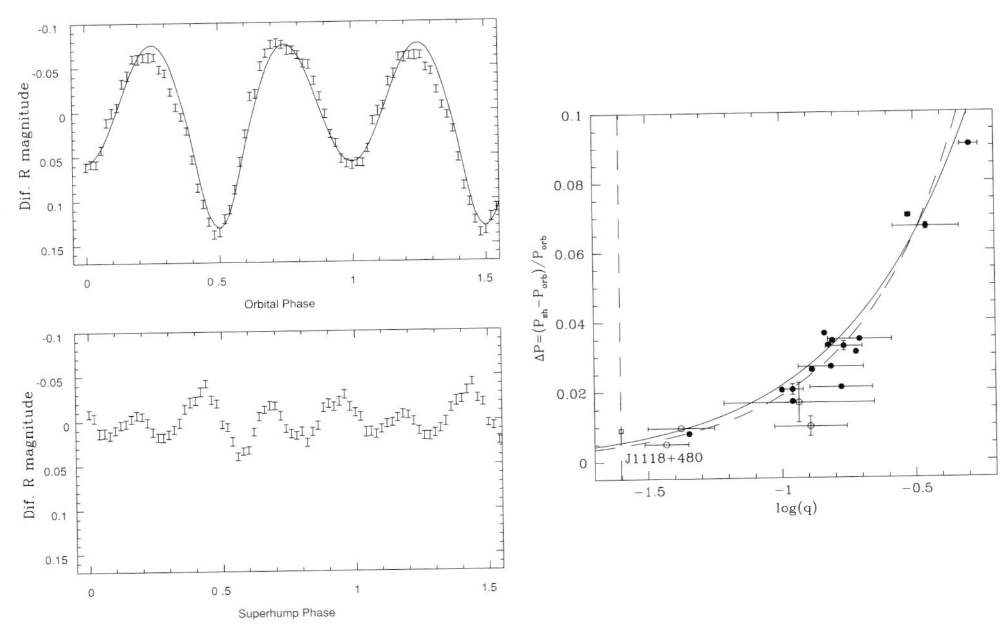

Fig. 5.34. *Left*: Detrended lightcurve (upper) of J1118+480 folded on $P_{\rm orb} = 4.078$ h, together with an ($i = 75°$ and $q = 25$) ellipsoidal model, with residuals plotted below after folding on $P_{\rm sup} = 4.092$ hours. *Right*: Combined results on LMXBs (J1118+480, X1916–053) and CVs (Patterson 2001) by Zurita *et al.* (2002a) that display superhumps, where Δ is plotted against q (NB used by Zurita *et al.* as the *inverse* of q as defined in this chapter). Models are plotted as solid (Patterson) and dashed (Osaki 1985) curves, but are not expected to be valid below $q = 0.025$.

Goranskii *et al.* 1998 show is in any case dominated by an erratically varying component) and hence away from where intrinsic spectral features of a massive donor might be observed. The properties of the micro-quasars, such as SS433 and GRS1915+105, are of great importance in furthering our understanding and interpretation of the ultra-luminous (ULX) sources in nearby galaxies (see Chapters 9, 12 and 13). While SS433 appears to us as an extremely low L_X ($\sim 10^{36}$ erg s^{-1}) source compared to GRS1915+105 (which peaks at $\sim 10^{39}$ erg s^{-1}, and possibly up to a hundred times greater than this, once appropriate correction for extinction has been made, e.g., Greiner *et al.* 1998), it is possible that the high i and extremely high $\dot{M}$ of SS433 are preventing us from viewing its intrinsic L_X. In any case the observed L_X is far exceeded by the jet power, as is true in pulsar wind nebulae, but not in the other micro-quasars (Safi-Harb 2003).

Nevertheless, the orbital period is known (13 d) and it is eclipsing in optical and X-ray lightcurves, but with no signature of the mass donor there exists no constraint on the dynamical masses comparable to those of the XRN described earlier. Indeed, there are published "determinations" of the SS433 compact object mass ranging from 0.8 $M_\odot$ (d'Odorico *et al.* 1991) to 62 $M_\odot$ (Antokhina & Cherepashchuk 1985)! These are all based on radial velocity curves of the HeII emission line, of which the most detailed study is that of Fabrika and Bychkova (1990), who used ~4 yr of 6-m telescope data taken around precessional phase, $\Psi \sim 0$ (which gives the lowest scatter in their radial velocity curve), when the disk normal was closest to our line of sight. Assuming $e = 0$ they obtain $f(M) \simeq 8\ M_\odot$, which implies a 10 $M_\odot$ donor for a NS compact object, or 15 $M_\odot$ for a 6 $M_\odot$ BH (and which would also better explain the extremely high $\dot{M} \sim 10^{-4}\ M_\odot\ yr^{-1}$, see also Fuchs *et al.* 2003).

But perhaps the best attempt so far to directly observe the mass donor has come from Gies *et al.* (2002) who, using the interpretation that SS433 is a binary embedded in an expanding thick disk wind, which explains the equatorial radio disk of Blundell *et al.* (2001) and the stationary lines, also decided to observe SS433 spectroscopically at precessional phase, $\Psi \sim 0$, and in the blue where the expected high-mass donor would be more easily detectable (and has historically been avoided due to the consequent need to use larger telescopes). During eclipse, Gies *et al.* report the possible detection of (weak) spectral features that are typical of an A supergiant, which is consistent with photometry (Antokhina & Cherekpashchuk 1987; Goranskii *et al.* 1997) that shows the donor to be cooler than the disk, which still dominates even in the blue. The velocities of these features and known orbital phase are combined with an assumed $e = 0$ and Fabrika and Bychkova's K-velocity for HeII (assumed to represent the disk) to yield $q = M_2/M_X = 0.57$, $M_X = 11 \pm 5\ M_\odot$, $M_2 = 19 \pm 7\ M_\odot$, $R_2 = 31 \pm 3\ R_\odot$(assuming it is Roche-lobe filling), hence requiring a BH of similar mass to those found in the XRN. However, this result must be treated with caution until a full radial velocity curve has been obtained, which will be extremely difficult given the large contribution from the bright and hot disk.

There also now exists a large database of SS433 V-band photometry (from 1979 to 1996) that Fabrika and Irsmambetova (2002) have reanalysed by using (mostly) radio coverage to separate into active and passive states, which were then examined on both orbital and precessional phases. The most remarkable result is that the largest (optical/radio) flares occur almost entirely at orbital $\phi \sim 0.3$ (with a few at $\sim$0.8), which implies a non-circular orbit! However, the well-known nodding motions in the jet (Katz *et al.* 1982) set an independent upper limit of $e < 0.05$, but Fabrika and Irsmambetova point out that an eccentricity as small as $e \simeq 0.01$ would be enough to modulate the Roche volume and thereby perturb the disk; it being the disk, and not $\dot{M}$, which is already extremely high, that will modulate the flux, following an $\sim$1 d delay in propagating the $\dot{M}$ change through the disk. Such a small value of e would be undetectable in current optical spectroscopic studies.

5.3.13.2 Cyg X-3: a Wolf–Rayet–BH X-ray binary

Cygnus X-3 is a bright HMXB with the exceptionally short P_{orb} of 4.8 h. It is also a strong radio source showing extensive outbursts associated with mass ejection events in a relativistic jet (Geldzahler *et al.* 1983). Because of its high extinction, it took the advent of IR spectroscopy to identify the nature of the mass donor when van Kerkwijk *et al.* (1992) showed the IR emission lines to be arising from the strong stellar wind of a Wolf–Rayet (WN7) star, the only one currently identified with an X-ray binary. Subsequently, van Kerkwijk (1993), van Kerkwijk *et al.* (1996) and Fender *et al.* (1999) carried out detailed studies of the IR line profiles and explained the double and P-Cygni shapes to be associated with the high $\dot{M}$ from the WR star and *not* the jets. Mid-IR (ISOPHOT) 2–12 μm observations (Koch-Miramond *et al.* 2002) infer an expanding wind with $\dot{M} \sim 1.2 \pm 0.5 \times 10^{-4}\ M_\odot\ yr^{-1}$ from the $\sim$WN8 donor.

X-ray spectroscopy by Paerels *et al.* (2000) using the Chandra HETGS revealed a rich spectrum of photoionization-driven excited lines. An overall net redshift of $\sim$750 km s^{-1}, independent of binary phase, was interpreted as an ionized region detached from the main mass outflow. Furthermore, detailed Chandra imaging (Heindl *et al.* 2003) revealed an X-ray structure extending some 16 arcsecs from the nucleus, which could be evidence for the strong stellar wind impacting the ISM, thereby producing bremsstrahlung X-ray emission.

The nature of the compact object within Cyg X-3 remains a puzzle. No direct pulsations have ever been detected that would indicate a NS. The presence of radio relativistic jets encourages many to include it in the galactic micro-quasar group, and hence, by implication, suggest that the compact object is a BH. And while van Kerkwijk (1993) and van Kerkwijk *et al.* (1996) interpreted the IR line profile variations as due to geometrical variations in the X-ray illuminated hemisphere of the WR wind (supported by the correlation of line strength and wavelength shift), Schmutz *et al.* (1996) obtained higher resolution spectra where the profiles did not follow the variation expected of this model. Instead, they interpret them as due to the straightforward orbital motion of the WR star, which then implies $f(M) = 2.3\ M_\odot$. With WN7 star masses in the range 10–50 $M_\odot$(Cherepashchuk & Moffat 1994), or more conservatively 5–20 $M_\odot$ (Massey 1982), and i in the range 30–90° then $M_X > 7\ M_\odot$, which makes Cyg X-3 an important system as potentially the *only* currently identified WR/BH system (although two further candidates have been proposed in the LMC, Wang 1995, and in IC10, Clark & Crowther 2004), suggesting either a rare formation mechanism (e.g., van den Heuvel & de Loore 1973) or a short lifetime, which might be expected given the extremely high inferred $\dot{M}$. A more detailed analysis of the Fender *et al.* data is given in Hanson *et al.* (2000) which reveals, during outburst, an absorption component in HeI that is out of phase (by ~0.25 cycle) with the emission lines. They present a compelling case for interpreting this as the radial velocity curve of the WR wind, yielding similar masses to those above.

5.3.13.3 Cir X-1: HMXB or LMXB?

Since its discovery in the early 1970s (Margon *et al.* 1971) the nature of the X-ray binary Cir X-1 has remained elusive, in spite of the known $P_{\rm orb} = 16.6$ d inferred from the modulation of the X-ray (Kaluzienski *et al.* 1976), near-IR (Glass 1994) and radio fluxes (Haynes *et al.* 1978) and the discovery of Type I X-ray bursts identifying the compact object as a NS (Tennant *et al.* 1986). High extinction has made spectral classification of the donor impossible: the best optical spectra to date are dominated by asymmetric, HI and HeI emission lines (Johnston *et al.* 1999, 2001), with no photospheric features evident.

The lack of spectral type has made determination of the accretion mode difficult. Murdin *et al.* (1980) interpreted the X-ray and radio modulations as evidence for direct accretion from a massive stellar wind. Indeed the X-ray behavior of Cir X-1 is similar to that of the 16.65d BeX A0538–66, in which wind-fed accretion is enhanced by pseudo Roche-lobe overflow at periastron passage (Charles *et al.* 1983).

Conversely, Johnston *et al.* (1999) suggest that a similar scenario is also possible with an intermediate mass (3–5 $M_\odot$) companion. Furthermore, spectral-timing studies of Cir X-1 with RXTE/PCA (Shirey *et al.* 1999) show Z-source X-ray QPO behavior, suggesting an LMXB classification (cf. van der Klis 1995). However, the proposed association with the young SNR G321.9–0.3, which would support a low-mass classification has recently been disproved by a proper motion study (Mignani *et al.* 2002). Detailed IR spectroscopy (Clark *et al.* 2003) reveals a spectrum that superficially resembles that of a high-luminosity mid-B supergiant, however it is concluded that the true spectrum of the donor remains obscured by the emission from the disk or accretion driven outflows.

Clearly a consensus on the nature of Cir X-1 has yet to be reached.

5.3.13.4 *CI Cam: a fast transient B[e] HMXB?*

In 1998 RXTE detected an extremely bright (~2 Crab) new transient, J0421+560 (Smith *et al.* 1998) with a positional error circle that included the known, bright ($V \sim 11$) B[e] supergiant CI Cam. Optical spectroscopy at the time of outburst (Wagner & Starrfield 1998) revealed HeII emission superposed on the already rich emission line spectrum, which strongly suggested CI Cam was responsible for the X-ray source. Contemporaneous photometry (Hynes *et al.* 2002c) revealed CI Cam to be 2–3 magnitudes brighter than usual, and this combined with a radio detection by Hjellming and Mioduszewski (1998) confirmed the identification. However, the extremely rapid decay (~2 d) of the outburst was highly unusual, with only V4641 Sgr of the XRN described earlier exhibiting similarly fast variations.

Hynes *et al.* (2002c) have presented a major review of this system, but there are still some major questions left unanswered. It has been demonstrated that CI Cam is a sgB[e] which underwent a major mass outburst phase, and that this ejected material interacted with a compact object (presumed to be a NS or BH). The ejected mass underwent supercritical accretion on to the compact object resulting in the subsequent ejection of much of the material. This thereby gave rise to the observed radio emssion and the broadened optical emission lines. Hynes *et al.*, however, point out that there is, so far, no evidence of binarity in this system and that, although unlikely, the intriguing possibility exists that the interaction with the compact object was merely a chance encounter between two unbound objects. CI Cam's high luminosity makes it a galactic counterpart to the Magellanic Cloud sgB[e] stars, but it does not fit at all well into the classical Be/OB HMXB classification, and it has even been suggested (Ishida *et al.* 2004) that the variable soft X-ray component might indicate that it is analogous to the supersoft sources (Chapter 11), which would make the compact object a white dwarf.

Acknowledgements

We are grateful to our many colleagues who commented on earlier drafts of the manuscript and sent information in advance of publication. In particular, we would like to thank Reba Bandyopadhyay, Jorge Casares, Mike Garcia, Dawn Gelino, Rob Hynes, Lydie Koch-Miramond, Tariq Shahbaz, and John Tomsick.

References

Alcock, C. *et al.* 1996, *ApJ*, **461**, 84.
2001, *MNRAS*, **321**, 678.
Antokhina, E.A. & Cherepashchuk, A.M. 1985, *Soviet Astron.*, **11**, 4.
1987, *Soviet Astron.*, **31**, 295.
Augusteijn, T. *et al.* 2001, *A&A*, **375**, 447.
Avni Y. 1978, in *Physics and Astrophysics of Neutron Stars and Black Holes*, eds. Giacconi & Ruffini (Amsterdam: North-Holland), 42.
Avni Y. & Bahcall J. N. 1975, *ApJ*, **197**, 675.
Balucińska-Church M. *et al.* 2000 *MNRAS*, **311**, 861.
Bandyopadhyay, R. *et al.* 1997, *MNRAS*, **285**, 718.
1999, *MNRAS*, **306**, 417.
2002, *ApJ*, **570**, 793.
2003, *MNRAS*, **340**, L13.
Barziv, O. *et al.* 2001, *A&A*, **377**, 925.
Beekman, G. *et al.* 1996, *MNRAS*, **281**, L1.
1997, *MNRAS*, **290**, 303.
Beer, M.E. & Podsiadlowski, Ph. 2002, *MNRAS*, **331**, 351.

Berriman, G. *et al.* 1985, *MNRAS*, **217**, 327.
Bever, J. van & Vanbeveren, D. 1997, *A&A*, **322**, 116.
Bhattacharya, D. 2002, *JApA*, **23**, 67.
Bildsten, L. 2002, *ApJ*, **577**, L27.
Bildsten, L. & Chakrabarty, D. 2001, *ApJ*, **557**, 292.
Bildsten, L. & Rutledge, R.E. 2000, *ApJ*, **541**, 908.
Bildsten, L. *et al.* 1997, *ApJS*, **113**, 367.
Blaauw, A. 1961, *Bull. Astron. Inst. Netherlands*, **15**, 265.
Blundell, K.M. *et al.* 2001, *ApJ*, **562**, L79.
Bradshaw, C.F. *et al.* 1999, *ApJ*, **512**, L121.
Brandt, W.N. *et al.* 1995, *MNRAS*, **277**, L35.
Brocksopp, C. *et al.* 1999, *A&A*, **343**, 861.
Burderi, L. *et al.* 2003, *A&A*, **404**, L43.
Callanan, P.J. *et al.* 1992, *MNRAS*, **259**, 395.
2002, *ApJ*, **574**, L143.
1995, *PASJ*, **47**, 153.
Cannizzo, J.K. 1998, *ApJ*, **494**, 366.
Casares, J. & Charles, P.A. 1994, *MNRAS*, **271**, L5.
Casares, J. *et al.* 1992, *Nature*, **355**, 614.
1995, *MNRAS*, **274**, 565.
1997, *New Astron.*, **1**, 299.
2002, *MNRAS*, **329**, 29.
2003, *ApJ*, **590**, 1041.
2004, *ApJ*, **613**, 133.
Chakrabarty, D. & Morgan, E.H. 1998, *Nature*, **394**, 346.
Chakrabarty, D. & Roche, P. 1997, *ApJ*, **489**, 254.
Charles, P.A. 2001, in *Black Holes in Binaries and Galactic Nuclei*, eds. Kaper, van den Heuvel & Woudt (Springer).
Charles, P.A. & Naylor, T. 1992, *MNRAS*, **255**, L6.
Charles, P.A. *et al.* 1983, *MNRAS*, **202**, 657.
Chen, W. *et al.* 1993, *ApJ*, **408**, L5.
1997, *ApJ*, **491**, 312.
Cherepashchuk, A.M. & Moffat, A.F.J. 1994, *ApJ*, **424**, L53.
Chou, Y. *et al.* 2001, *ApJ*, **549**, 1135.
Clark, J.S. *et al.* 1999, *A&A*, **348**, 888.
2001, *A&A*, **380**, 615.
2002, *A&A*, **392**, 909.
Clark, J.S. & Crowther, P.A. 2004, *A&A*, **414**, L45.
Clark, J.S. *et al.* 2003, *A&A*, **400**, 655.
Coe, M.J. & Orosz J.A. 2000, *MNRAS*, **311**, 169.
Coe, M.J. *et al.* 1994, *A&A*, **289**, 784.
2002, *MNRAS*, **332**, 473.
Corbet, R.H.D. 1986, *MNRAS*, **220**, 1047.
Covino, S. *et al.* 2001, *A&A*, **374**, 1009.
Cowley, A.P. *et al.* 2002, *AJ*, **123**, 1741.
Cutler, E.P. *et al.* 1986, *ApJ*, **300**, 551.
Dachs, J. & Wamsteker, W. 1982, *A&A*, **107**, 240.
della Valle, M. *et al.* 1997, *A&A*, **318**, 179.
Deutsch, E.W. *et al.* 1999, *ApJ*, **524**, 406.
d'Odorico, S. *et al.* 1991, *Nature*, **353**, 329.
Eggleton, P.P. 1983, *ApJ*, **268**, 368.
Eracleous, M. & Horne, K. 1996, *ApJ*, **471**, 427.
Ergma, E. & Sarna, M.J. 2001, *A&A*, **374**, 195.
Eyles, C.J. *et al.* 1975, *Nature*, **254**, 577.
Fabrika, S.N. & Bychkova, L.V. 1990, *A&A*, **240**, L5.
Fabrika, S. & Irsmambetova, T. 2002, in *New Views on Microquasars*, eds. Durouchoux, Fuchs & Rodriguez (Kolkata, India: Center for Space Physics), 268.

Fender, R.P. *et al.* 1999, *MNRAS*, **308**, 473.
Filippenko, A.V. & Chornock, R. 2001, *IAUC* 7644.
Filippenko, A.V. *et al.* 1995a, *ApJ*, **455**, L139.
1995b, *ApJ*, **455**, 614.
1997, *PASP*, **109**, 461.
1999, *PASP*, **111**, 969.
Fomalont, E.B. *et al.* 2001, *ApJ*, **558**, 283.
Foulkes, S.B. *et al.* 2004, *MNRAS*, **349**, 1179.
Fuchs, Y. *et al.* 2003, in *New Views on Microquasars*, eds. Durouchoux, Fuchs & Rodriguez (Kolkata, India: Center for Space Physics), 269, astro-ph/0208432.
Galloway, D.K. *et al.* 2002, *ApJ*, **576**, L137.
Garcia, M.R. *et al.* 1992, *AJ*, **103**, 1325.
Geldzahler, B.J. *et al.* 1983, *ApJ*, **273**, L65.
Gelino, D.M. 2004, *RevMexAA*, **20**, 214.
Gelino, D.M. & Harrison, T.E. 2003, *ApJ*, **599**, 1254.
Gelino, D.M. *et al.* 2001a, *AJ*, **122**, 971.
2001b, *AJ*, **122**, 2668.
Gies, D.R. *et al.* 2002, *ApJ*, **578**, L67.
Giles, A.B. *et al.* 1999, *MNRAS*, **304**, 47.
Glass, I.S. 1994, *MNRAS*, **268**, 742.
Goranskii, V.P. *et al.* 1997, *Astron. Rep.*, **41**, 656.
1998, *Astron. Rep.*, **42**, 336.
Gottlieb, E.W. *et al.* 1975, *ApJ*, **195**, L33.
Greene, J. *et al.* 2001, *ApJ*, **554**, 1290.
Greiner, J. *et al.* 1998, *New Astron. Rev.*, **42**, 597.
2001, *Nature*, **414**, 522.
Grindlay, J.E. 1989, *in Proc. 23rd ESLAB Symposium on Two Topics in X-ray Astronomy* (ESA SP), 121.
Grindlay, J.E. *et al.* 1988, *ApJ*, **334**, L25.
Hakala, P.J. *et al.* 1998, *A&A*, **333**, 540.
1999, *MNRAS*, **306**, 701.
Hameury, J.-M. *et al.* 1997, *ApJ*, **489**, 234.
Hanson, M.M. *et al.* 2000, *ApJ*, **541**, 308.
Hanuschik, R.W. 1996, *A&A*, **308**, 170.
Hanuschik, R.W. *et al.* 1988, **189**, 147.
Harlaftis, E.T. *et al.* 1996, *PASP*, **108**, 762.
1997a, *AJ*, **114**, 1170.
1997b, *MNRAS*, **285**, 673.
1999, *A&A*, **341**, 491.
Haswell, C.A. *et al.* 2001, *MNRAS*, **321**, 475.
2002, *MNRAS*, **332**, 928.
Haynes, R.F. *et al.* 1978, *MNRAS*, **185**, 661.
Heindl, W.A. *et al.* 2003, *ApJ*, **588**, 97.
Hellier, C. 2000, *New Astron. Rev.*, **44**, 131.
Herrero, A. *et al.* 1995, *A&A*, **297**, 556.
Heuvel, E.P.J. van den 1983, in *Accretion Driven Stellar X-ray Sources*, eds. Lewin & van den Heuvel (Cambridge University Press).
Heuvel, E.P.J. van den & de Loore, C. 1973, *A&A*, **25**, 387.
Hillier, D.J. & Miller, D.L. 1998, *ApJ*, **496**, 407.
Hjellming, R.M. 1978, *ApJ*, **221**, 225.
Hjellming, R.M. & Mioduszewski, A.J. 1998, *IAUC* 6862.
Homer, L. *et al.* 2001a, *MNRAS*, **325**, 1471.
2001b, *MNRAS*, **322**, 827.
Hooft, F. van der *et al.* 1997, *MNRAS*, **286**, L43.
1998, *A&A*, **329**, 538.
Horne K. 1999a, in *Quasars and Cosmology* ASP Conf. Series, **162**, eds. Ferland & Baldwin (San Francisco: ASP), 189.

1999b, in *Magnetic Cataclysmic Variables* ASP Conf. Series, **157**, eds. Mukai & Hellier (San Francisco: ASP), 349.
Hutchings, J.B. *et al.* 1983, *ApJ*, **275**, L43.
1987, *AJ*, **94**, 340.
Hynes, R.I. *et al.* 1998, *MNRAS*, **299**, L37.
2000, *ApJ*, **539**, L37.
2001a, *MNRAS*, **324**, 180.
2001b, in *Astrotomography, Indirect Imaging Methods in Observational Astronomy*, Lecture Notes in Physics, **573**, eds. Boffin, Steeghs & Cuypers (Springer), 378.
2002a, *MNRAS*, **330**, 1009.
2002b, *MNRAS*, **331**, 169.
2002c, *A&A*, **392**, 991.
2003a, *MNRAS*, **345**, 292.
2003b, *ApJ*, **583**, L95.
in't Zand, J.J.M. *et al.* 1998, *A&A*, **331**, L25.
Ishida, M. *et al.* 2004, *ApJ*, **601**, 1088.
Israelian, G. *et al.* 1999, *Nature*, **401**, 142.
Jablonski, F.J. *et al.* 1997, *ApJ*, **482**, L171.
Johnston, H.M. *et al.* 1999, *MNRAS*, **308**, 415.
2001, *MNRAS*, **328**, 1193.
Jonker, P.G. *et al.* 2000, *MNRAS*, **315**, L57.
Kaiser, C.R. & Hannikainen, D.C. 2002, *MNRAS*, **330**, 225.
Kanbach, G. *et al.* 2001, *Nature*, **414**, 180.
Kaluzienski, L.J. *et al.* 1976, *ApJ*, **208**, L71.
Kaper, L. *et al.* 1997, *A&A*, **327**, 281.
Katz, J.I. *et al.* 1982, *ApJ*, **260**, 780.
Kerkwijk, M.H. van 1993, *A&A*, **276**, L9.
Kerkwijk, M.H. van *et al.* 1992, *Nature*, **355**, 703.
1996, *A&A*, **314**, 521.
King, A.R. & Ritter, H. 1998, *MNRAS*, **293**, L42.
King, A.R. *et al.* 1996, *ApJ*, **464**, L127.
Kirsch, M.G.F. & Kendziorra, E. 2003, *ATel*, 148.
Klis, M. van der 1995, in *X-Ray Binaries*, eds. Lewin, van Paradijs & van den Heuvel (Cambridge University Press), 252.
Koch-Miramond *et al.* 2002, *A&A*, **396**, 877.
Kolb, U. 1998, *MNRAS*, **297**, 419.
Kolb, U. *et al.* 1997, *ApJ*, **485**, L33.
Kong, A.K.H. *et al.* 2000, *MNRAS*, **312**, L49.
2002, *MNRAS*, **329**, 588.
Kuiper, L. *et al.* 1988, *A&A*, **203**, 79.
Kuulkers, E. 1998, *New Astron. Rev.*, **42**, 613.
LaSala, J. & Thorstensen, J.R. 1985, *AJ*, **90**, 2077.
LaSala, J. *et al.* 1998, *MNRAS*, **301**, 285.
Lasota, J.-P. 2000, *A&A*, **360**, 575.
2001, *New Astron. Rev.*, **45**, 449.
Laycock S. *et al.* 2004, *ApJ* (in press), astro-ph/0406420.
Liu, Q.Z. *et al.* 2000, *A&A Suppl.*, **147**, 25.
2001, *A&A*, **368**, 1021.
Loon, J.Th. van *et al.* 2001, *A&A* **375**, 498.
Maeder, A. *et al.* 1999, *A&A*, **346**, 459.
Maragoudaki, F. *et al.* 2001, *A&A*, **379**, 864.
Margon, B. 1984, *ARAA*, **22**, 507.
Margon, B. *et al.* 1971, *ApJ*, **169**, L23.
Markwardt, C.B. *et al.* 2002, *ApJ*, **575**, L21.
2003, *ATel*, 164.
Marsh, T.R. 2001, in *Astrotomography, Indirect Imaging Methods in Observational Astronomy*, Lecture Notes in Physics, **573**, eds. Boffin, Steeghs & Cuypers (Springer), 1.

Marsh, T.R. *et al.* 1994, *MNRAS*, **266**, 137.
Martin, A.C. *et al.* 1995, *MNRAS*, **274**, L46.
Martín, E.L. *et al.* 1992, *Nature*, **358**, 129.
1994, *A&A*, **291**, L43.
1995, *A&A*, **303**, 785.
1996, *New Astron.*, **1**, 197.
Masetti, N. *et al.* 2002, *A&A*, **382**, 104.
Massey, P. 1982, *IAU Symp.*, **99**, 251.
McClintock, J.E. & Remillard, R.A. 1986, *ApJ*, **308**, 110.
1990, *ApJ*, **350**, 386.
McGowan, K.E. & Charles, P.A. 2003, *MNRAS*, **339**, 748.
Meyssonnier, N. & Azzopardi, M. 1993, *A&AS*, **102**, 451.
Mignani, R.P. *et al.* 2002, *A&A*, **386**, 487
Mineshige, S. *et al.* 1992, *PASJ*, **44**, L15.
Mirabel, I.F. & Rodriguez, L.F. 1999, *ARAA*, **37**, 409.
Mukai, K. *et al.* 2001, *ApJ*, **561**, 938.
Murdin, P. *et al.* 1980, *A&A*, **87**, 292.
Murray, J. 2000, *MNRAS*, **314**, L1.
Nagase, F. 1989, *PASJ*, **41**, 1.
Naylor, T. *et al.* 1991, *MNRAS*, **252**, 203.
Negueruela, I. 1998, *A&A*, **338**, 50.
Negueruela, I. & Coe M.J. 2002, *A&A*, **385**, 517.
O'Donoghue, D. & Charles, P.A. 1996, *MNRAS*, **282**, 191.
Ogilvie, G.I. & Dubus, G. 2001, *MNRAS*, **320**, 485.
Okazaki, A.T. & Negueruela, I. 2001, *A&A*, **377**, 161.
Orosz, J.A. 2001, Astron.Tel. No. 67.
Orosz, J.A. 2003, *Proc. IAU Symp.*, **212**, 365, eds. K.A. van der Hucht, A. Herrero & C. Esteban.
Orosz, J.A. & Bailyn C.D. 1997, *ApJ*, **477**, 876 (and *ApJ*, **482**, 1086).
Orosz, J.A. & Kuulkers E. 1999, *MNRAS*, **305**, 132.
Orosz, J.A. *et al.* 1994, *ApJ*, **436**, 848.
1996, *ApJ*, **468**, 380.
1997, *ApJ*, **478**, L83.
2001, *ApJ*, **555**, 489.
2002a, *ApJ*, **568**, 845.
2002b, *BAAS*, **201**, 1511.
Osaki, Y. 1985, *A&A*, **144**, 369.
Paczynski, B. 1971, *ARAA*, **9**, 183.
Paerels, F. *et al.* 2000, *ApJ*, **533**, 135.
Paradijs, J. van 1998, in *The Many Faces of Neutron Stars*, eds. Buccheri, van Paradijs & Alpar (Kluwer).
Paradijs, J. van & McClintock, J.E. 1994, *A&A*, **290**, 133.
1995, in *X-Ray Binaries*, eds. Lewin, van Paradijs & van den Heuvel (Cambridge University Press), 58.
Parmar, A.N. & White, N.E. 1988, *Mem. It. Astr. Soc.*, **59**, 147.
Patterson, J. 1995, *PASP*, **107**, 1193.
1999, in *Close Binary Systems*, Frontiers Science Series, No. 26, 61 eds. S. Mineshige & J.C. Wheeler.
2001, *PASP*, **113**, 736.
Pavlenko, E.P. *et al.* 1996, *MNRAS*, **281**, 1094.
Pereira, M.G. *et al.* 1999, *ApJ*, **526**, L105.
Phillips, S.N. *et al.* 1999, *MNRAS*, **304**, 839.
Podsiadlowski, Ph. *et al.* 2002, *ApJ*, **567**, 491.
Popov, S.B. *et al.* 1998, *Astron. Rep.*, **42**, 29.
Putman, M.E. *et al.* 1998, *Nature*, **394**, 752.
Reig, P. & Roche, P. 1999, *MNRAS*, **306**, 100.
Remillard, R.A. *et al.* 1996, *ApJ*, **459**, 226.
2000, *IAUC* 7389.
Retter, A. *et al.* 2002, *MNRAS*, **330**, L37.
Roche, P. *et al.* 1998, *IAUC* 6885.
Rolfe, D.J. *et al.* 2001, *MNRAS*, **324**, 529.

Safi-Harb, S. 2003, in *New Views on Microquasars*, eds. Durouchoux, Fuchs & Rodriguez (Kolkata, India: Center for Space Physics), 243.
Samimi, J. *et al.* 1979, *Nature*, **278**, 434.
Sanwal, D. *et al.* 1996, *ApJ*, **460**, 437.
Schachter, J. *et al.* 1989, *ApJ*, **340**, 1049.
Schandl, S. & Meyer, F. 1994, *A&A*, **289**, 149.
Schmutz, W. *et al.* 1996, *A&A*, **311**, L25.
Shafter, A.W. *et al.* 1988, *ApJ*, **327**, 248.
Shahbaz, T. & Kuulkers, E. 1998, *MNRAS*, **295**, L1.
Shahbaz, T. *et al.* 1993, *MNRAS*, **265**, 655.
1994a, *MNRAS*, **268**, 756.
1994b, *MNRAS*, **271**, L10.
1996, *MNRAS*, **282**, 977.
1997, *MNRAS*, **285**, 607.
1998, *MNRAS*, **301**, 382.
1999a, *A&A*, **346**, 82.
1999b, *MNRAS*, **306**, 89.
2000, *MNRAS*, **314**, 747.
2001, *A&A*, **376**, L17.
2003, *ApJ*, **585**, 443.
Shirey, R.E. *et al.* 1999, *ApJ*, **517**, 472.
Skinner, G.K. 1980, *Nature*, **288**, 141.
1981, *Space Sci. Rev.*, **30**, 441.
Smale, A.P. & Lochner, J.C. 1992, *ApJ*, **395**, 582.
Smith, D. & Dhillon, V.S. 1998, *MNRAS*, **301**, 767.
Smith, D. *et al.* 1998, *IAUC* 6855.
Soria, R. *et al.* 1998, *ApJ*, **495**, L95.
1999, *MNRAS*, **309**, 528.
Stanimirovic, S. *et al.* 1999, *MNRAS*, **302**, 417.
Stavely-Smith, L. *et al.* 1997, *MNRAS*, **289**, 225.
Steeghs, D. & Casares, J. 2002, *ApJ*, **568**, 273.
Stella, L. *et al.* 1986, *ApJ*, **308**, 669.
Sunyaev, R.A. *et al.* 1992, *ApJ*, **389**, L75.
Tanaka, Y. & Shibazaki, N. 1996, *ARAA*, **34**, 607.
Telting, J.H. *et al.* 1998, *MNRAS*, **296**, 785.
Tennant, A.F. *et al.* 1986, *MNRAS*, **221**, 27.
Thorsett, S.E. & Chakrabarty, D. 1999, *ApJ*, **512**, 288.
Tjemkes, S.A. *et al.* 1986, *A&A*, **154**, 77.
Tomsick, J.A. *et al.* 1999, *ApJ*, **521**, 341.
2001, *ApJ*, **559**, L123.
2002, *ApJ*, **581**, 570.
Torres, M.A.P. *et al.* 2002, *MNRAS*, **334**, 233.
Udalski A. *et al.* 1997, *Acta Astron.*, **47**, 319.
Verbunt, F. & van den Heuvel, E.P.J. 1995, in *X-ray Binaries*, eds. Lewin, van Paradijs & van den Heuvel (Cambridge University Press).
Wachter, S. 1998, Ph.D. thesis, University of Washington, Seattle.
Wachter, S. *et al.* 2002, *ApJ*, **568**, 901.
Wade, R.A. & Horne, K. 1988, *ApJ*, **324**, 411.
Wagner, R.M. & Starrfield, S.G. 1998, *IAUC* 6857.
Wagner, R.M. *et al.* 1992, *ApJ*, **401**, 97.
2001, *ApJ*, **556**, 42.
Walter, F.M. *et al.* 1982, *ApJ*, **253**, L67.
Wang, Q.D. 1995, *ApJ*, **453**, 783.
Warner, B. 1995, *Cataclysmic Variables* (Cambridge University Press), 117.
Webb, N.A. *et al.* 2000, *MNRAS*, **317**, 528.
White, N.E. & van Paradijs, J. 1996, *ApJ*, **473**, L25.

White, N.E. *et al.* 1995, in *X-ray Binaries*, eds. Lewin, van Paradijs & van den Heuvel (Cambridge University Press).
Whitehurst, R. & King, A.R. 1991, *MNRAS*, **249**, 25.
Wijers, R.A.M.J. & Pringle, J.E. 1999, *MNRAS*, **308**, 207.
Wijnands, R. & van der Klis, M. 1998, *ApJ*, **507**, L63.
Wilson, C.A. *et al.* 2002, *ApJ*, **570**, 287.
Zdziarski, A.A. *et al.* 2004, *MNRAS*, **351**, 791.
Zurita, C. *et al.* 2000, *MNRAS*, **316**, 137.
2002a, *MNRAS*, **333**, 791.
2002b, *MNRAS*, **334**, 999.

6

Fast X-ray transients and X-ray flashes

John Heise and Jean in 't Zand
Space Research Organization Netherlands

6.1 Introduction

The X-ray sky is extremely variable. Transient sources occur on probably all timescales. Historically, the easiest timescales to study are seconds to minutes, and days to months. Most of these transients are Galactic in origin and have been shown to be powered by gravitational energy release during accretion of matter onto a compact object or by thermonuclear runaway processes on neutron stars (see Chapter 3 by Strohmayer and Bildsten). Less well understood are the transients with timescales between a minute and a day, the so-called "fast X-ray transients" (FXTs). These often occur off the Galactic plane and are, therefore, sometimes also referred to as "high latitude transients" (see Fig. 6.1). In general an FXT is loosely defined as a new temporary X-ray source that disappears on a timescale of less than a day and is not related to a known persistent X-ray source. The term, however, has been used as a repository for all sorts of ill-understood flares, bursts, flashes and related phenomena.

6.2 The early detections of fast X-ray transients

The very first X-ray satellite, Uhuru, detected fast transient X-ray sources (Forman *et al.* 1978), but it was through studies with Ariel-V (Pye & McHardy 1983) and the High Energy Astronomy Observatory-1 (HEAO-1) (Ambruster & Wood 1986; Connors *et al.* 1986) that FXTs came to be recognized as a separate class of transients. These early missions scanned the sky and the term "fast" covered transients that lasted one to a few satellite orbits (the orbital period of low-Earth satellites being 1.5 hr).

In the Ariel-V sky survey (Pye & McHardy 1983) 27 events were detected in 5.5 years of observations at a time resolution of one 1.5 hr-long satellite orbit (Fig. 6.1). The highest peak flux was 5×10^{-9} erg cm^{-2} s^{-1} (2–10 keV). About 20% of the sources are identified with active coronae in RS CVn systems and these are the longest ones, lasting of the order of an hour. Another 20% could be identified with a variety of of other objects (mostly Be X-ray binaries). One case occurred simultaneously with a gamma-ray burst (GRB; for a recent review see Chapter 15 by Hurley *et al.*). Pye and McHardy conclude that most of the unidentified cases are consistent with as yet unknown coronal sources. The all-sky event rate was estimated at one FXT every $\sim$3 days above 4×10^{-10} erg cm^{-2} s^{-1} (2–10 keV).

Ambruster and Wood (1986) observed 10 FXTs above $\sim 7 \times 10^{-11}$ erg cm^{-2} s^{-1} (0.5–20 keV) with the A-1 instrument on HEAO-1, with durations between 10 s and 1.5 hr and peak fluxes up to 4×10^{-9} erg cm^{-2} s^{-1}. Four FXTs were identified: three flare stars and one RS CVn system; one is proposed to be a gamma-ray weak GRB (see also Ambruster *et al.* 1986). With the A-2 instrument on the same platform, operative between 2 and 60 keV, a further

Compact Stellar X-Ray Sources, eds. Walter Lewin and Michiel van der Klis.

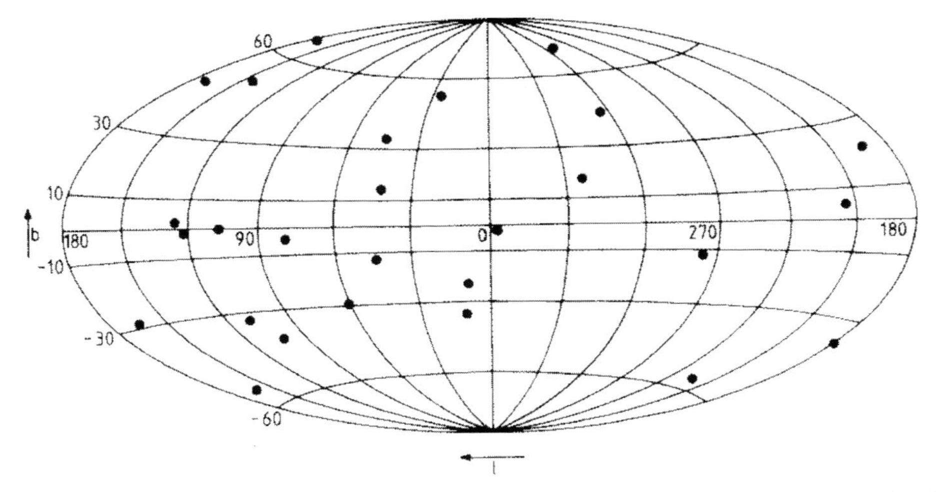

Fig. 6.1. A map of the 27 FXTs detected with Ariel-V, in Galactic coordinates (Pye & McHardy 1983).

eight FXTs were detected (Connors *et al.* 1986) with similar characteristics. Connors *et al.* suggest that six of the events are hard coronal flares from dMe-dKe stars, with a flare rate of 2×10^4 per year above $\sim 10^{-10}$ erg cm^{-2} s^{-1}. They note a possible association with the X-ray counterparts of GRBs in some cases, although there are no proven temporal coincidences.

Imaging missions such as Einstein, ROSAT and ASCA are sensitive to events considerably fainter but suffer from a much smaller field of view and a bias towards softer events. One of the first efforts (Gotthelf *et al.* 1996) finds suggestive evidence for the detection of 42 short duration X-ray events, although they are of a different character than those found in the above-mentioned non-imaging surveys, being shorter (20 s at the longest) and fainter. The implied all-sky event rate is a few million per year. However, another similar search on ROSAT-PSPC data (Vikhlinin 1998) does not confirm such a large population of faint X-ray transients.

Grindlay (1999) suggests that some FXTs may be the early X-ray afterglow of GRBs. Greiner *et al.* (2000) searched the ROSAT-PSPC all-sky survey data for FXTs on this premise. They found 23 FXTs with durations between 10 s and a few hours and 0.1–2.4 keV peak fluxes between 10^{-10} and 10^{-9} erg cm^{-2} s^{-1}. Based on optical follow-up, Greiner *et al.* believe in a flare star origin in many, if not all, cases rather than GRB afterglows. The all-sky event rate is 4×10^3 yr^{-1}.

6.3 Two types of fast X-ray transients

The most recent and perhaps most comprehensive sky survey of FXTs has been performed with the two Wide Field Cameras (WFCs) aboard BeppoSAX (Jager *et al.* 1997). The suitability of this instrument is provided by a large field of view (40×40 square degrees) while maintaining a good angular resolution (5′ full width at half maximum) and a fairly high sensitivity ($\sim 10^{-8}$ erg cm^{-2} s^{-1} on a timescale of 10 s over the energy range 2–25 keV). Throughout the six-year long mission, with a net observing time of 1×10^8 s per camera, the WFCs identified a large number of transient sources with sufficient positional accuracy (2–5′) to undertake meaningful searches for counterparts. A systematic analysis of the

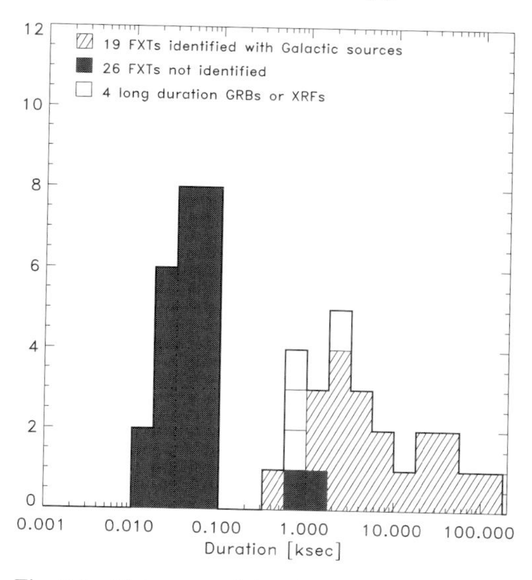

Fig. 6.2. Histogram of durations of FXTs detected by the WFC aboard the BeppoSAX satellite and mostly found at high latitudes. Transients that can be clearly identified with X-ray binaries and other bright Galactic X-ray sources are excluded. The distribution of durations appears to be bimodal. The short duration events are the X-ray flashes (XRFs).

complete database is underway, but the preliminary results are as follows (see also Heise *et al.* 2001a,b).

A total of 104 FXTs (Fig. 6.2) were identified in 90% of the WFC database. This excludes, for the sake of this discussion, five Type-I X-ray bursts for which no persistent emission could be detected (e.g., Cornelisse *et al.* 2002). Of these 104, 55 are identified as X-ray counterparts to GRBs by noting temporal coincidence with events detected in the Gamma-Ray Burst Monitor (GRBM, Frontera *et al.* 1997) or, on an incidental basis, in other gamma-ray burst devices. The remaining set of 49 FXTs appears to exhibit a bimodal distribution of durations (Fig. 6.2).

The long duration events last between 2×10^3 and 2×10^5 s and plausible quiescent counterparts (coronal sources, flare stars, RS CVn stars) can be identified for most of these sources. These identifications are consistent with the results of previous studies and the remaining long duration sources are also likely of coronal origin.

In contrast, the short duration FXTs are not identifiable with any quiescent X-ray source or known coronal source. As dramatically illustrated by the lightcurves of one such short duration event (e.g., Fig. 6.3), neither are these events counterparts of conventional GRBs. Furthermore, as can be seen from Fig. 6.5, much of the energy of this event is radiated in the X-ray band. Following Heise *et al.* (2001a,b) such FXTs are referred to as X-ray flashes (XRFs). They constitute a new class of sources.

A phenomenological definition of XRFs is based on the following criteria:

- Strong non-thermal emission in the 2–10 keV band. This criterion distinguishes XRFs from Type I X-ray bursts (e.g., Chapter 3 by Strohmayer & Bildsten).
- Weak emission in the traditional gamma-ray band, 50–300 keV; this explains why typical XRFs do not trigger gamma-ray burst monitors. Specifically, we require $R > 1$ where

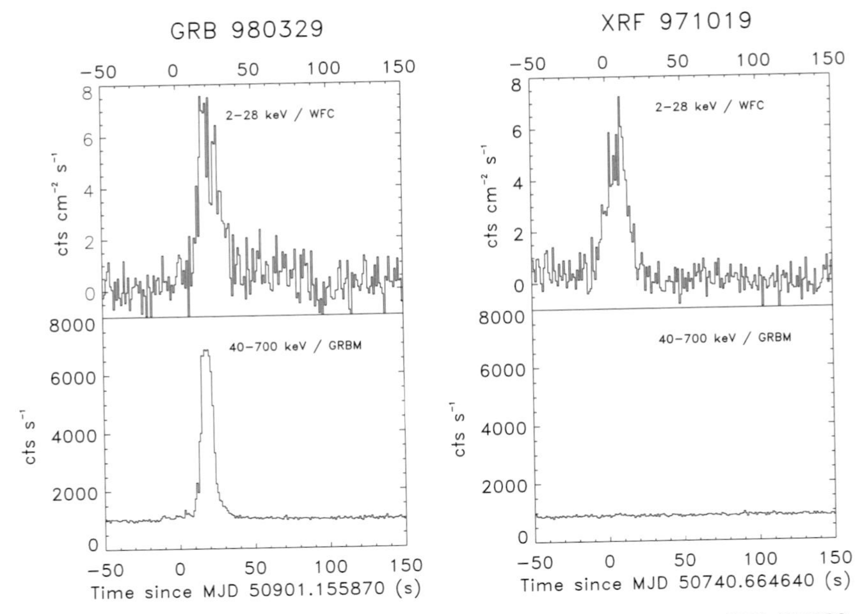

Fig. 6.3. The lightcurves on the same scale of a typical gamma-ray burst GRB 980329 (in 't Zand *et al.* 1998) (*left*) and a prototypical X-ray flash 971019 (*right*) as recorded by the Wide Field Camera (WFC) and the Gamma-Ray Burst Monitor (GRBM), aboard the Italianen–Dutch satellite BeppoSAX. GRB 980329 produces a strong signal in the GRBM (40–700 keV) as is expected of a gamma-ray burst. In contrast, the event of 971019 does not produce any signal in the GRBM.

$R = F_{2-10}/F_{50-300}$ with F_{2-10} the fluence in the 2–10 keV range and F_{50-300} the fluence in 50–300 keV range. In these terms, GRBs in the range $0.5 < R < 1$ may be called X-ray rich.

- Durations less than a few thousand seconds. This differentiates XRFs from the other FXTs, namely stars with flares (as discussed earlier) and also some binaries containing a compact object.
- Absence of a quiescent optical/IR counterpart with optical to X-ray flux ratios in excess of 0.01. This criterion distinguishes XRFs from X-ray binaries that sometimes exhibit short bright flares (such as V4641 Sgr and a number of Be X-ray binaries) and from stars with strong coronal activity (e.g., strong flares from late-type stars, such as dMe stars, RS CVn stars, and pre-main sequence stars).

XRFs observed with the WFC are bright X-ray sources with typical peak fluxes of 1×10^{-8} erg cm^{-2} s^{-1}, in the 2–25 keV range. XRF spectra (Fig. 6.5) can be adequately described by either a power law indicative of Comptonization, or a "Band" spectrum consisting of two smoothly joined power-law components (Band *et al.* 1993). Both are characterized by a peak energy E_p in the νf_ν spectrum, where E_p is of the order of 20 keV, which is much lower than observed typically in GRB spectra.

The mean V/V_{max} of XRFs is 0.27 ± 0.15, which compares to 0.31 ± 0.06 for the X-ray counterparts of GRBs in the BeppoSAX sample. Thus, XRFs share both isotropy and spatial inhomogeneity properties of GRBs. XRFs are not a minor contribution to the cosmic explosion rate. The observed all-sky rate is about one-half of that of GRBs.

6.4 Stellar origin of Galactic fast X-ray transients

X-ray emission from stellar coronae had been anticipated prior to the discovery of the first non-solar X-ray source Sco X-1. However, the Sun scaled to stellar distances implies low X-ray fluxes and it was not until 1975 that the first X-ray source was explicitly identified with a stellar corona (Mewe *et al.* 1975). The source is Capella and was detected with the Astronomical Netherlands Satellite (ANS). Capella had previously been seen in a rocket-flight experiment (Catura *et al.* 1975), but no mention was made of the coronal origin. Around the same time, flaring behavior in stellar coronal sources was discovered on dMe stars (Heise *et al.* 1975). More X-ray flares on dMe stars were found with HEAO-1 (Kahn *et al.* 1979). Decay times in such systems are typically between 5 and 120 min. It was shown with EXOSAT that Algol and RS CVn systems exhibit bright flares ($10^{-10} - 10^{-8}$ erg cm^{-2} s^{-1}) with decay times in the range 30 to 300 min, which is longer than the typical decay time in dMe stars. It took several more years until the Einstein Observatory detected similar X-ray flaring events on young T Tauri stars (Montmerle *et al.* 1983). Nowadays, X-ray flares are known to occur on stars all over the H–R diagram, with a wide distribution of luminosities and durations (for a few examples, see Fig. 6.4). Unfortunately, an up-to-date comprehensive review on this subject does not exist.[1] The cumulative distribution of the total radiated energy in flares is a power law, so observations depend on the detection limit. The class of long duration fast X-ray transients with duration of order an hour or longer (Fig. 6.2) can in general be assumed to originate from stellar coronal activity. In many cases there are variable, late type stars within the error box of the WFC (such as BY Dra, EQ Peg, AT Mic, V857 Cen). Large and rare flares may be expected from previously unknown coronally active sources. We conclude that a significant fraction of the long duration FXTs have a Galactic stellar origin.

6.5 X-ray flashes versus gamma-ray bursts

Observationally, XRFs are a distinct class of transients. However, XRF and GRB properties appear to have overlapping distributions: durations (Fig. 6.2), fluences, peak fluxes and X-ray spectra. Furthermore, the existence of X-ray rich GRBs is suggestive of a continuum between GRBs and XRFs.

A further quantification of the association between the two groups can only be achieved through a combined X/gamma-ray spectral analysis. Because of the spectral softness of the XRFs, this necessitates the most sensitive gamma-ray device, namely the Burst And Transient Source Experiment, BATSE (Fishman *et al.* 1989). In the 3.8 years when BATSE and WFC were operating simultaneously, 13 XRFs were potentially observable by BATSE. Faint signals were detected for 12 of these in 25–100 keV. The remaining one is most probably a Type-I X-ray burst (Cornelisse *et al.* 2002). Apart from studying the broad-band spectrum, the BATSE coverage is valuable in that XRFs can be placed in the context of the vast BATSE database of GRBs.

Kippen *et al.* (2001, 2003) undertook a joint analysis and modeling of the WFC-BATSE data on XRFs and GRBs. Examples of this analysis are shown in Fig. 6.5, as well as a histogram of the peak energies E_p of the νF_ν spectra. XRFs are clearly distinguished from GRBs by their distinctly smaller E_p (Fig. 6.5). Thus, there is no sharp drop in events with E_p below 40 keV as was suggested in earlier studies (e.g., Preece *et al.* 2000).

[1] The reader might find the *Proceedings of IAU Colloquium 151* useful. It contains a short review by Pallavicini (1995) including references to older reviews.

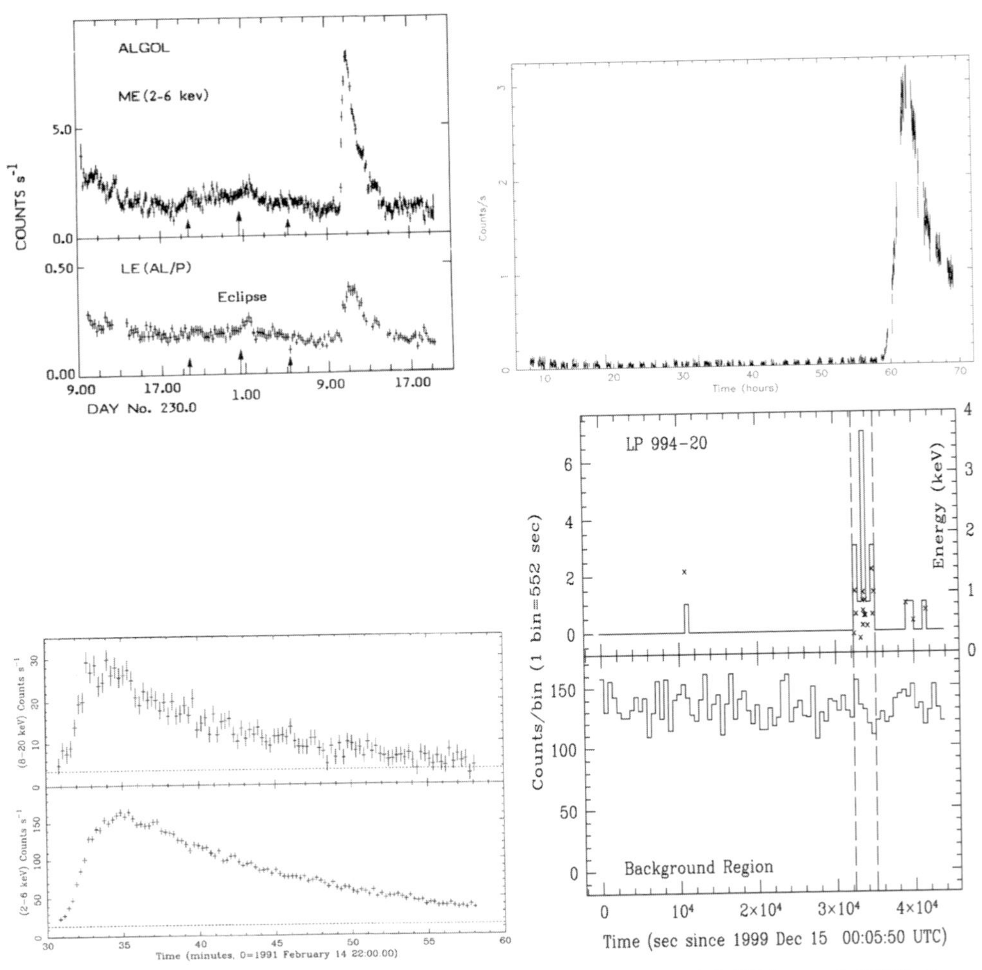

Fig. 6.4. Stellar flares observed in Algol with EXOSAT (*top left*: White *et al.* 1986), the pre-main sequence star ROX-31 in the Rho Oph Cloud with ASCA (*top right*: Skinner 2000), the dMe star EQ 1839.6+8002 with Ginga (*bottom left*; Pan *et al.* 1997), and the brown dwarf LP 994–20 with Chandra (*bottom right*; Rutledge *et al.* 2000).

Prompt X-ray emission from GRBs has been studied in a systematic manner in two other missions. One was carried by Ginga, operational between 1987 and 1991, and consisted of two devices for GRB detection (Murakami *et al.* 1989): a proportional counter (2–25 keV) and a scintillation counter (15–400 keV). Strohmayer *et al.* (1998) carried out a systematic analysis on 22 of the 120 GRBs detected. Forty percent of those 22 cases have E_p below 20 keV. However, much higher (and typical) E_p are recovered if the analysis is restricted to the BATSE range. *More importantly*, the X-ray to gamma-ray fluence ratio, R, of Ginga bursts is less than one and by our definition these events are not XRFs. This is probably related to a trigger range that is insensitive to XRFs (50–400 keV). The second mission is the High Energy Transient Explorer-2 (HETE-2, Ricker *et al.* 2003), which was launched in October 2000 and carries a gamma-ray spectrometer sensitive in the 6–400 keV range, a 2–25 keV

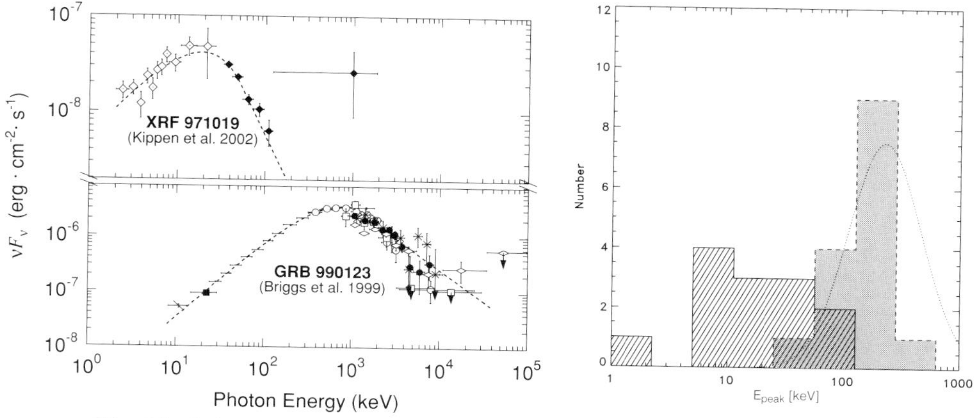

Fig. 6.5. *Left*: Broad-band spectrum of an X-ray flash (*top*) and a bright gamma-ray burst (*bottom*). Both are characterized by a peak energy E_p in the νf_ν spectrum. *Right*: A histogram of E_p for XRFs (shaded) and GRBs (dark) is shown for events seen in both the WFC and BATSE. The dotted line is the distribution of E_p from BATSE alone.

wide-field X-ray camera (similar to the WFC) and a soft X-ray camera which operates at 0.5–10 keV. Sakamoto *et al.* (2004b) report on the global characteristics of X-ray flashes and X-ray rich GRBs (XRRs) observed with HETE. Out of about 45 events, the number of XRFs, XRRs and GRBs are comparable. The HETE instruments confirm the durations and sky distributions of XRFs and XRRs to be similar to GRBs, except that the values of the observed peak energies of the burst spectrum in νF_ν, the peak energy flux and the energy fluence of XRFs are smaller than those of GRBs. The distribution of all three kinds of bursts forms a continuum in three different planes, namely the fluence 2–30 keV versus the fluence 30–400 keV plane; the fluence (2–40 keV) versus the peak energy plane and also in the peak flux (50–300 keV) versus the peak energy plane. The results confirm and provide further evidence that all three kinds of burst arise from a similar phenomenon.

6.6 Extra-galactic origin of X-ray flashes

Four features – isotropy, non-recurrence, non-thermal emission and lack of quiescent emission – distinguish GRBs from other transients. These four key features, within the limitations set by small number statistics, are also shared by XRFs. The overlap in duration and the continuity of E_p between XRFs and GRBs are highly suggestive of a strong linkage between GRBs and XRFs and an extra-galactic origin for the XRFs.

An important clue as to the nature of XRFs came from the discovery of afterglows in XRF011030 in the X-ray (Harrison *et al.* 2001) and radio (Taylor *et al.* 2001a,b). The blue host galaxy found at this position constrains the distance to $z < 3.5$ (Bloom *et al.* 2003). At the time of writing the number of afterglows detected is still very limited. The HETE-2 trigger on XRF 030723 (Prigozhin *et al.* 2003) led to an optical afterglow (Fox 2003). The X-ray flash XRF 020427, observed in the last few days of the operational life of BeppoSAX, led to an X-ray afterglow event observed with BeppoSAX and Chandra (Amati *et al.* 2004).

The wide-field and soft X-ray instruments on board the High Energy Transient Explorer-2 (HETE-2) detected the bright (7.2×10^{-8} erg cm^{-2} s^{-1}) X-ray flash XRF 020903 (Sakamoto *et al.* 2004a). It was the softest event ever detected by HETE-2, with a ratio of X-ray fluence

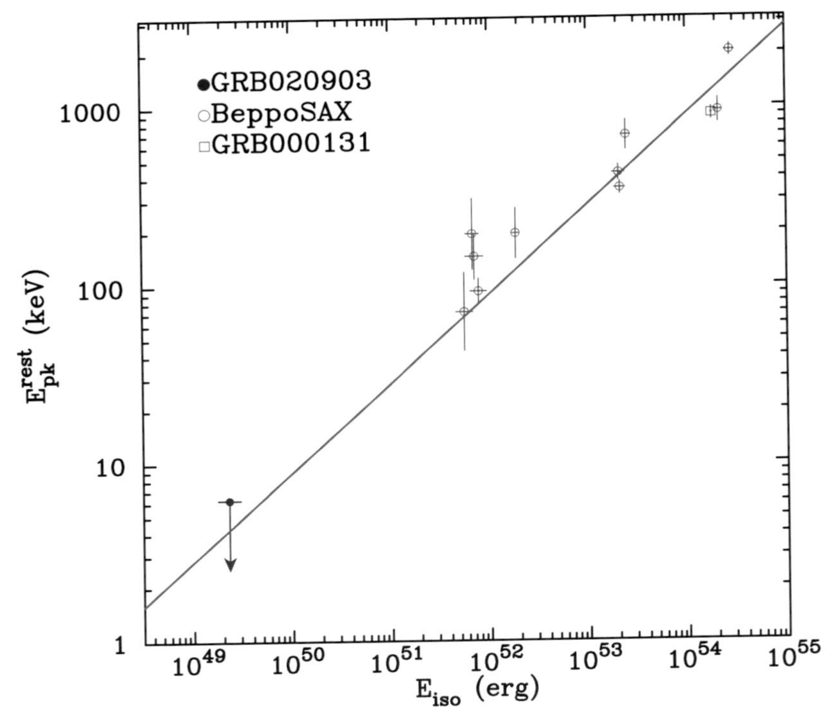

Fig. 6.6. The relation of the peak energy E_{peak} versus the isotropic-equivalent radiated energy E_{iso} (Amati *et al.* 2002), showing the extrapolation by 3 orders of magnitude to XRF020903 (from Sakamoto *et al.* 2004a).

(S_X) to gamma-ray fluence (S_γ) of $\log(S_X/S_\gamma) = 4.3$. Soderberg *et al.* (2004) discovered an optical afterglow very similar to the afterglows of GRBs. Spectroscopic observations show the source to lie in an active star-forming galaxy at $z = 0.251$, very similar to GRBs. This distance, together with the X-ray fluence, implies an isotropic equivalent energy $E_{iso} \approx 1.1 \times 10^{49}$ erg, 3 to 6 orders of magnitude lower than isotropic energies observed in GRBs. Amati *et al.* (2002) observed a relation between the break energy and the isotropic energy release E_{iso} in GRBs. If one extrapolates this relation to an isotropic energy release which is three orders of magnitude smaller, then the low E_{peak} for XRF020903 fits this relation as well (see Fig. 6.6).

A radio afterglow was also found and showed a peak luminosity comparable to that found in GRB afterglows. The radio data indicate a similar total energy release in the explosion of XRF 020903 as in GRBs, which corrected for beaming is about 10^{51} erg for all GRBs.

Kumar (2000) and Freedman and Waxman (2001) have shown that the late time X-ray afterglow is independent of the density of the circumburst medium and can, with some assumptions of the microphysics (equipartition factors, power-law index of the shocked electrons), yield the energy of the explosion. The overlap of the X-ray afterglows of XRF 011030 and XRF 020427 with those of GRBs indicates that the energies involved are similar (provided that the redshift distribution of XRFs is similar to that of GRBs).

To summarize, the redshift determination of one optical transient and constraint on one host galaxy, the evidence in the form V/V_{max}, scintillation of the radio afterglow of XRF 011030 and the similarity between the X-ray afterglows of XRF 011030 and XRF 020427 and those of GRBs lead us to conclude that XRFs are, like GRBs, cosmologically located explosions. The recent XRF-SN association of XRF 031203 (Watson *et al.* 2004) with SN 2003lw (Thomsen *et al.* 2004) links the explosion with the death of massive stars and further strengthens the connection of XRFs with GRBs.

6.7 Oddball FXTs

The sample of FXTs is too small to allow on statistical grounds identification with other types. However, we note that not yet identified FXTs could include for example very fast variability in certain AGNs. Therefore, a possible association of XRF 961229 (see in 't Zand *et al.* 2004) with a double-lobed radio galaxy PKS J0234–0049 with unknown redshift is potentially very interesting and, if real, remarkable. The X-ray event lasted 540 s and is one of the longest XRFs seen in the WFCs. It is difficult to explain it as a Type-I X-ray burst (atypical time profile), stellar flare (too short) or short-duration accretion event on a local compact object (the galactic latitude is -54°). It is possibly a very long GRB (there were no BATSE data at that time).

Three other long events (in 't Zand *et al.* 2004, indicated as white blocks in Fig. 6.2) were covered *and* detected by BATSE, suggesting there may be a substantial number of bursts still hidden in the BATSE database. A spectral analysis of these three events shows that, though soft, they are not typical XRFs.

6.8 Discussion

At least two types of fast (sub-day) X-ray transients seem to exist: a Galactic class associated with active stellar systems and the more puzzling extra-galactic class, the XRFs. Here we focus the discussion on XRFs.

Several features – isotropy, spatial distribution, non-recurrence, non-thermal emission, lack of quiescent emission – are shared between GRBs and XRFs and set them apart from other transients. The characteristics of XRFs point to a close connection with GRBs, the distinguishing property being the softness of the event expressed as the low peak energy in νf_ν. The question has been: are they the same and only appear different for geometrical reasons (redshifted GRBs, different viewing angle in the beamed emission, etc.) or are they intrinsically different?

A model for XRFs in which they are simply highly redshifted GRBs would require a redshift 5 to 10 to explain the observed low peak energies in the spectra of XRFs, a shift from typically $E_{\rm peak} \sim$ 100–200 keV in GRBs to 5–50 keV in XRFs. Such large distances also imply time-stretching of the bursts and one would expect XRFs on average to last longer than GRBs, contrary to the observations. However, the same cosmological signatures – the redshift of the peak energy and the time-stretching of bursts with known redshift – has been difficult to see in localized GRBs with measured redshift, indicating the wide diversity of the GRB sample. The first upper limit $z < 3.5$ already seems to exclude this model. Note, however, that the converse may still be true: distant GRBs may be observed as dim XRFs.

Models to explain XRFs fall into two categories, the off-axis jet scenario or the baryon loaded fireball scenario. They come in different flavors, e.g., off-beam uniform jet model (Ioka & Nakamura, 2001); wide opening angle uniform jet model (Lamb *et al.* 2005); Gaussian jet model (Zhang *et al.* 2004a); power-law jet model (Mészáros *et al.* 1998; Fan *et al.* 2004); two-component jet model (Zhang *et al.* 2004b, Huang *et al.* 2004).

The break in the lightcurve of GRB afterglows indicates that GRBs are produced by collimated jets. The isotropic energy release, corrected for this collimation, is a constant for all GRBs for which such a calculation could be performed (Frail *et al.* 2001). The jet energy structure could be such that Lorentz factors peak on axis and fall off with increasing angle with respect to the central axis. In this view (Yamazaki *et al.* 2002), XRFs are the same as GRBs but viewed from an off-axis position, dominated by lower Lorentz factors resulting in lower peak energy. Granot *et al.* (2002) have shown that in this case the lightcurve, which has a monotonic decay for an on-axis view, should show a maximum after a few days depending on the beam angle. This type of lightcurve has not been observed so far (Amati *et al.* 2004).

In the baryon loaded fireball scenario, the total burst energy is shared with a larger baryon load and, thus, lower Lorentz factor. Indeed, such circumstances were anticipated as soon as the fireball scenario came into consideration (Katz 1994) and the existence of X-Ray flashes (and possible UV-flashes at even lower energy) was in this respect predicted in 1994. In models where the engine of the explosion is derived from a stellar implosion such as a hypernova, a larger baryon load is readily expected.

The first spectroscopic distance scale for XRFs has settled the matter, at least in the case of XRF 020903. The radio observations of XRF 020903 show that the total energy release is similar to that of GRBs. These similar energies strongly suggest a common origin for cosmic explosions that manifest themselves in XRFs and GRBs. In the same way as Berger *et al.* (2003) showed for the abnormal gamma-ray burst GRB 030329, the explosion energy in XRF 020903 is dominated by a mildly relativistic outflow of matter, while significantly less energy (by at least two orders of magnitude) couples to the extreme relativistic motion generating the prompt X-ray and gamma-ray emission. Mildly relativistic motion with the same total energy input is for example, obtained in events where this energy is shared by a larger baryon load.

The significance of the peak energy was shown by Amati *et al.* 2002. $E_{\rm peak}$ correlates with the isotropic energy $E_{\rm iso}$ roughly as $E_{\rm peak} \sim \sqrt{E_{\rm iso}}$. Sakamoto *et al.* (2004) have shown that this relation extends by three more orders of magnitude in the case of XRF 020903, thus emphasizing the intimate relation between GRBs and XRFs. The implication of this relation is discussed by Zhang *et al.* (2004).

6.9 Conclusions

Fast (sub-day) X-ray transients are a mix of X-ray counterparts to gamma-ray bursts, flares from nearby coronal sources, X-ray flashes, which are probably counterparts to a peculiar class of gamma-ray bursts, and possibly other types of objects. It seems that in all cases the ultimate energy reservoir is rotational energy. Stellar rotation creates the magnetic fields which cause coronal activity and subsequently the X-ray flares. The extra-galactic FXTs may also draw on rotational energy. If indeed the X-ray flashes have the same origin as gamma-ray bursts, it is believed that the energy originates in newly formed rotating black holes.

X-ray flashes are the least understood. Enhancing the understanding of these requires successful fast (within a day) multi-wavelength follow ups for a number of flashes, similar to that of gamma-ray bursts. Triggers for this must come from X-ray rather than gamma-ray devices.

References

Amati, L., Frontera, F., Tavani, M., *et al.* 2002, *A&A*, **390**, 81
Amati, L., Frontera, F., in 't Zand, J. J. M. *et al.* 2004, *A&A*, **426**, 415
Ambruster, C. W. & Wood, K. S. 1986, *ApJ*, **311**, 258
Ambruster, C. W., Wood, K. S., Meekins, J. F., *et al.* 1986, *ApJ*, **269**, 779
Band, D., Matteson, J., Ford, L., *et al.* 1993, *ApJ*, **413**, 281
Berger, E., Kulkarni, S. R., Pooley, G., *et al.* 2003, *Nature*, **426**, 154
Bloom, J. S., Fox, D., van Dokkum, P. G., *et al.* 2003, *ApJ*, **599**, 957
Briggs, M. S., *et al.* 1999, *ApJ*, **524**, 82
Catura, R. C., Action, L. W., & Johnson, H. M. 1975, *ApJ*, **196**, L47
Connors, A., Serlemitsos, P. J., & Swank, J.H. 1986, *ApJ*, **303**, 769
Cornelisse, R., in 't Zand, J. J. M., Verbunt, F., *et al.* 2002, *A&A*, **392**, 885
Fan, Y. Z., Wei, D. M., & Wang, C. F. 2004, *A&A*, **424**, 474
Fishman, G. J., Meegan, C. A., Wilson, R. B., *et al.* 1989, *Proc. GRO Science Workshop*, NASA, 2
Forman, W., Jones, C., Cominsky, L., *et al.* 1978, *ApJS*, **38**, 357
Fox, D. W. 2003, *GCN Circ.* 2323
Frail, D. A., Kulkarni, S. R., Sari. R., *et al.* 2001, *ApJ*, **562**, L155
Freedman, D. L. & Waxman, E. 2001, *ApJ*, **547**, 922
Frontera, F., Costa, E., dal Fiume, D., *et al.* 1997, *A&AS*, **122**, 357
Gotthelf, E. V., Hamilton, T. T. & Helfand, D. J. 1996, *ApJ*, **466**, 779
Granot, J., Panaitescu, A., Kuman, P., & Woosley, S.E. 2002, *ApJ*, **570**, L61
Greiner, J., Hartmann, D. H., Voges, W., *et al.* 2000, *A&A*, **353**, 998
Grindlay, J. E. 1999, *ApJ*, **510**, 710
Harrison, F. A., Yost, S., Fox, D., *et al.* 2001, *GCN Circ.* 1143
Heise, J., Brinkman, A. C., Schrijver, J., *et al.* 1975, *ApJ*, **202**, L73
Heise, J., in 't Zand, J. J. M., Kippen, R. M., Woods, P. M., *et al.* 2001a, in *Proc. Gamma-ray Bursts in the Afterglow Era II*, eds. E. Costa, F. Frontera & J. Hjorth (Berlin: Springer), 16
Heise, J., in 't Zand, J. J. M., Kulkarni, S. R., & Costa, E. 2001b, *GCN Circ.* 1138
Huang, Y. F., Wu, X. F., Dai, Z. G., Ma, H. T., & Lu, T., 2004, *ApJ*, **605**, 300
in 't Zand, J. J. M., Amati, L., Antonelli, L. A., *et al.* 1998, *ApJ*, **505**, L119
in 't Zand, J. J. M., *et al.* 2004, *Proc. 3rd Rome Workshop 2002*, eds. L. Piro *et al.*, astro-ph/0305361
Ioka, K. & Nakamura, T., 2001, *ApJ*, **554**, L163
Jager, R., Mels, W. A., Brinkman, A. C., *et al.* 1997, *A&AS*, **125**, 557
Kahn, S. M., Mason, K. O., Bowyer, C. S., *et al.* 1979, *ApJ*, **234**, L107
Katz, J. I. 1994, *ApJ*, **432**, L107
Kippen, R. M., Woods, P. M., Heise, J., *et al.* 2001, in *Proc. Gamma-ray Bursts in the Afterglow Era II*, eds. E. Costa, F. Frontera & J. Hjorth (Berlin: Springer), 22
Kippen, R. M. *et al.* 2003, *GRBs and Afterglow Astronomy 2001*, AIP Conf. Proc. **662**, p. 244
Kumar, P. 2000, *ApJ*, **538**, L125
Lamb, D. Q., Donaghy, T. Q., Graziani, C. 2005, *ApJ*, **620**, 355
Mészáros, P., Rees, M. J., & Wijers, R. 1998, *ApJ*, **499**, 301
Mewe, R., Heise, J., Gronenschild, E. H. B. M., *et al.* 1975, *ApJ*, **202**, L67
Montmerle, T., Koch-Miramond L., Falgarone, E., *et al.* 1983, *ApJ*, **269**, 182
Murakami, T., Fuijii, M., Hayashida, K., *et al.* 1989, *PASJ*, **41**, 405
Pallavicini, R. 1995, *Flares and Flashes* (IAU Coll. 151), eds. J. Greiner *et al.*, 148
Pan, H.-C., Jordan, C., Makishima, K., *et al.* 1997, *MNRAS*, **285**, 735
Preece, R. D., Briggs, M. S., Mallozzi, R. S., *et al.* 2000, *ApJS*, **126**, 19
Prigozhin, G., Butler, N., & Crew, G. 2003, *GCN Circ.* 2313
Pye, J. P. & McHardy, I. M. 1983, *MNRAS*, **205**, 875
Ricker, G., *et al.* 2003, *GRBs and Afterglow Astronomy 2001*, AIP Conf. Proc. 662, p. 3
Rutledge, R. E., Basri, G., Martín, E. L., & Bildsten, L. 2000, *ApJ*, **538**, 141
Sakamoto, T., Lamb, D. Q., Graziani, C., *et al.* 2004a, *ApJ*, **602**, 875
2004b, astro-ph/0409128
Skinner, S. L. 2000, in *Proc. IAU Symposium 200*, eds. B. Reipurth and H. Zinnecker, p. 97
Soderberg, A. M., Kulkarni, S. R., Berger, E., *et al.* 2004, *ApJ*, **606**, 994
Strohmayer, T. E., Fenimore, E. E., Murakami, T., & Yoshida, A. 1998, *ApJ*, **500**, 873

Taylor, G. B., Frail, D. A., & Fox, D. 2001a, *GCN Circ*. 1122
Taylor, G. B., Frail, D. A., & Kulkarni, S. R. 2001b, *GCN Circ*. 1136
Thomsen, B., Hjorth, J., Watson, D., *et al*., 2004, A&A, **419**, L21
Watson, D., Hjorth, J., Levan, A., *et al*., 2004, *ApJ*, 605, L101
Vikhlinin, A. 1998, *ApJ*, **505**, L123
White, N. E., Culhane, J. L., Parmar, A. N., *et al*. 1986, *ApJ*, **301**, 262
Yamazaki, R., Ioka, K., Nakamura, T., *et al*. 2002, *ApJ*, **572**, L31
Zhang, B., Dai, X., Lloyd-Ronning N. M., & Mészaros, P. 2004a, *ApJ*, **601**, L119
Zhang, W. Q., Woosley, S. E., Heger, A, 2004b, *ApJ*, **608**, 365

7

Isolated neutron stars

Victoria M. Kaspi and Mallory S. E. Roberts
McGill University

Alice K. Harding
NASA Goddard Space Flight Center

7.1 Introduction

This chapter deals with X-ray emission from isolated neutron stars for which the energy for the observed X-rays is thought to originate from the rotation of the neutron star, or from an internal heat reservoir following formation. Rotation power can manifest itself as pulsed emission, or as nebular radiation produced by a relativistic wind of particles emitted by the neutron star. Residual heat of formation is observed as soft X-ray emission from young neutron stars. Such thermal radiation, however, can also be produced as a result of reheating from internal or external sources. Rotation-powered pulsed and nebular X-ray emission, as well as thermal emission, can often be observed in a single object simultaneously; this is both fascinating and annoying, as one invariably contaminates the study of the other. There are also a handful of neutron stars for which the origin of the observed X-ray emission is unclear but may be related to the above processes; we will discuss those as well.

Rotation-powered neutron stars are generally referred to as "radio pulsars" since it is at radio wavelengths that the vast majority of the catalogued[1] population (currently numbering $\sim$1400) is observed. However, the radio emission is energetically unimportant, and we now know of several rotation-powered neutron stars that are not detected as radio sources in spite of deep searches (e.g. Crawford *et al.* 1998; McLaughlin *et al.* 2001). We therefore use the more physically motivated term "rotation-powered."

The total available spin luminosity in a rotation-powered pulsar is given by the rate of loss of rotational kinetic energy, $\dot{E} \equiv I\Omega\dot{\Omega} \equiv 4\pi^2 I\dot{P}/P^3$, where I is the stellar moment of inertia and $\Omega \equiv 2\pi/P$ is the angular frequency with P the spin period. Thus a simple measurement of P and $\dot{P}$ for an isolated neutron star determines the available rotational power in a model-independent way, assuming a value for I, typically taken to be 10^{45} g cm^2. Also generally inferred for these sources are their surface dipolar magnetic fields, $B = 3.2 \times 10^{19}(P\dot{P})^{1/2}$ G (for B on the equator), and their characteristic spin-down ages, via $\tau_c = P/2\dot{P}$. Although these latter two inferences are model-dependent (B because it assumes simple magnetic dipole braking in a vacuum – which is almost certainly not the case – and τ_c because it represents the true physical age only for the same assumption and also for a negligible initial spin period), the important point is that P and $\dot{P}$ provide, in addition to $\dot{E}$, at least estimates of other important physical information.

It is for this reason that P–$\dot{P}$ diagrams are so useful. Figure 7.1 shows the P–$\dot{P}$ phase space, with all catalogued rotation-powered neutron stars with measured $\dot{P}$ indicated as dots. Lines of constant $\dot{E}$ and B are indicated. Note the two different populations of rotation-powered

[1] http://www.atnf.csiro.au/research/pulsar/psrcat/

Compact Stellar X-Ray Sources, eds. Walter Lewin and Michiel van der Klis.
Published by Cambridge University Press.

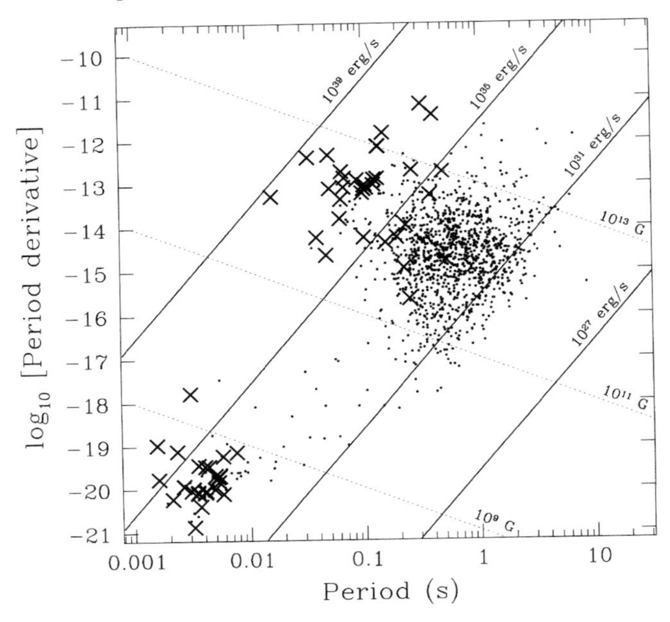

Fig. 7.1. P–$\dot{P}$ diagram for the 1403 currently catalogued rotation-powered pulsars. The 66 X-ray detected sources are indicated with an $\times$. Solid lines show constant $\dot{E}$, and dotted lines show constant inferred surface dipolar magnetic field. "X-ray-detected" means pulsed, unpulsed or nebular emission, be it thermal or non-thermal. X-ray detected sources are generally those with the greatest spin-down luminosity, although the correlation is not perfect because of the wide range of source distances and observational selection effects. The indicated sources are summarized in Tables 7.1, 7.2, and 7.5. Note that the X-ray-detected millisecond pulsars in the globular cluster 47 Tucanae do not have observable intrinsic $\dot{P}$ due to contamination by acceleration in the cluster potential; for those sources (numbering 15), we have used inferred $\dot{P}$ (see Freire *et al.* 2001; Grindlay *et al.* 2002).

neutron stars: those with periods between ~16 ms and ~8 s, having $B \gtrsim 10^{11}$ G, and the lower magnetic field ($B \lesssim 10^{10}$ G), very rapidly rotating class ($P \lesssim 100$ ms), the millisecond pulsars. The division is made particularly obvious when binarity is considered, as the latter sources are mostly in binaries, while the former are only rarely in binaries. The millisecond pulsars are often referred to as "recycled" pulsars as they are thought to be formed in a past episode of mass transfer as part of an accreting X-ray binary phase (see Chapter 1 by Psaltis). Finally, note the absence of sources having $B \gtrsim 10^{14}$ G. Neutron stars having fields higher than this are thought to be "magnetars," discussed in Chapter 14 by Woods and Thompson.

All sources that have been detected in X-rays are identified with a $\times$ in Fig. 7.1. For the purposes of this figure, any X-ray detection, be it pulsed, unpulsed or nebular, thermal or non-thermal, is counted. Note the clear correlation with $\dot{E}$ – most X-ray identified rotation-powered neutron stars have large $\dot{E}$ and vice versa. The correlation is not perfect; the main reason for this is that the distances to these sources vary by over two orders of magnitude. Distant high-$\dot{E}$ sources are harder to detect, and nearby low-$\dot{E}$ sources may be detected through their thermal emission. Note also the large number of millisecond pulsar detections is due to many of them being in X-ray studied globular clusters having larger scale heights, hence smaller line-of-sight absorbing column densities.

Tables 7.1, 7.2 and 7.5 provide a summary list of all X-ray-detected rotation-powered neutron stars, along with their most important properties. Note that the number of X-ray detected rotation-powered pulsars has more than doubled in the past 5 years (from 27 in the 1997 review by Becker and Trümper, to 66 in 2004). This is due to improvements in X-ray sensitivity, in particular from the Rossi X-ray Timing Explorer (RXTE), the Chandra X-ray Observatory and the XMM-Newton Telescope, and also improvements in radio telescope sensitivity, most notably at the Parkes, Arecibo, and Green Bank radio telescopes, which have permitted discoveries of sources that could later be followed up with X-ray observations.

7.2 Magnetospheric emission

7.2.1 *Motivation*

After the discovery (Staelin & Reifenstein 1968; Comella *et al.* 1969) of the Crab pulsar in 1968 and the realization (Ostriker & Gunn 1969; Finzi & Wolf 1969) that the spin-down energy loss of the neutron star, as estimated from its measured P and $\dot{P}$, could power all the visible radiation of the Crab nebula, it was evident that neutron stars are energetically important sources of particle production and acceleration. It is now believed that the radiating particles in pulsar wind nebulae (see Section 7.6) are electron–positron pairs produced by electromagnetic cascades within the pulsar magnetosphere. Since the pulsar magnetosphere is also believed to be the origin of the pulsed non-thermal emission, understanding this emission will lead us to an understanding of the energetics of these sources and ultimately how the pulsar converts its spin-down power into both the pulsed and unpulsed observable radiation. However, the pulsed radio emission, which first enabled the discovery of pulsars, makes up a tiny ($\sim 10^{-6}$) portion of the energy budget. Pulsed emission at high energies, although harder to detect, is energetically much more significant (making up $\sim 10^{-3}$ of the spin-down power in the Crab and up to 0.1 in other pulsars) and thus is a more direct probe of the particle acceleration in the magnetosphere.

Aside from the important issue of energetics, we also want to study the non-thermal emission at high energies in order to understand the physical processes in the pair cascades that generate the electron–positron pairs. It is widely believed that generation of a pair plasma is required for the instabilities that produce coherent radio emission in pulsars (Melrose 2000). Why radio pulsars die after they have spun down to periods of several seconds may possibly be understood as their inability to produce pair plasmas when particle acceleration becomes too feeble (Ruderman & Sutherland 1975; Hibschman & Arons 2001; Harding *et al.* 2002a). Studying the radiation and pair cascades at high energy is also a study of the physical processes, such as one-photon pair creation and photon splitting, that occur only in the strong magnetic fields of neutron stars. Finally, the generation of non-thermal emission is intimately linked to the polar-cap heating which contributes to the thermal emission, particularly in older pulsars.

7.2.2 *Summary of observations*

At X-ray energies, 66 rotation-powered pulsars have been detected, nearly half of which have periods $P < 6$ ms. Many of these detections were first made by ROSAT (Becker & Trümper 1997) and ASCA (Saito 1998), but the number has increased rapidly in recent years with new observations by RXTE, Chandra and XMM-Newton (Becker & Aschenbach 2002). Table 7.1 lists the pulsars thought to show non-thermal emission components in their X-ray

Table 7.1. *Non-thermal X-ray detected rotation-powered pulsars*[a]

PSR	P (ms)	$\dot{P}$ (10^{-15} s s^{-1})	d (kpc)	$\log(B)$ (G)	$\log(\tau_c)$ (yr)	$\log(\dot{E})$ (erg s^{-1})	$\log(L_X)^p$ (erg s^{-1})	$\log(L_X)^t$ (erg s^{-1})	Detections[b] R	O	G	Telescope	Reference
J1846−0258	324.00	7097.1	19	13.99	2.88	36.92	34.61	36.32				ASCA/RXTE	Gotthelf *et al.* 2000
B0531+21	33.40	421	2	12.88	3.12	38.65	35.85	37.02	×	×	×	ASCA	Saito 98
B1509−58	150	1536	4.3	13.49	3.21	37.26	34.60	34.80	×	×	×	ASCA	Saito 98
B0540−69	50	479	49.4	13.00	3.24	38.18	36.10	36.92	×	×		ASCA	Saito 98
J1930+1852	136	750.57	5.00	13.31	3.48	37.07	32.63		×			Chandra	Camilo *et al.* 02a
J0537−6910	16.10	51.2	49.4	12.26	3.72	38.69	35.52					RXTE	Marshall *et al.* 98
J0205+6449	65.70	193	2.6	12.86	3.75	37.43		32.20	×			Chandra	Murray *et al.* 02
J1617−5055	69	140	3.3	12.80	3.91	37.23	33.48	33.76	×			ASCA	Torii *et al.* 98
J2229+6114	52	78.300	3	12.61	4.04	37.34		33.01	×			ASCA	Halpern *et al.* 01
B0833−45	89	124	0.25	12.83	4.08	36.84	31.60	32.70	×	×	×	RXTE	Harding *et al.* 02b
J1420−6048	68	83.167	7.69	12.68	4.13	37.02	34.46		×			ASCA	Roberts *et al.* 01
B1800−21	134	134	3.94	12.93	4.22	36.34		33.06	×			ROSAT	Becker & Trümper 97
B1706−44	102	92.2	1.43	12.79	4.27	36.54	32.16	33.15	×		×	Chandra	Gotthelf *et al.* 02
J1811−1926	64.70	44	5	12.53	4.39	36.81	34.39					ASCA	Torii *et al.* 97
B1951+32	39.53	5.837	2.5	11.99	5.05	36.57		33.79	×		×	ASCA	Saito 98
B0656+14	384	55	0.76	12.97	5.07	34.59	30.26	32.98	×	×	×	ASCA	Greiveldinger *et al.* 96
Geminga	237	11.4	0.16	12.52	5.54	34.53	29.56	29.79	×	×	×	ASCA	Halpern & Wang 97
B1055−52	197	5.8	1.38	12.34	5.75	34.48	29.48	33.42	×	×	×	ASCA	Greiveldinger *et al.* 96
J2042+2740	96	1.20	1.8	11.8	6.12	34.70	31.3	31.5	×			XMM	Becker *et al.* 04
B1929+10	226	1.15	0.17	12.01	6.52	33.60	29.51	30.00	×	×		ASCA	Saito 98
B0823+26	530	1.30	0.34	12.2	6.83	32.50	29.2	29.5	×			XMM	Becker *et al.* 04
B0950+08	253	0.250	0.262	11.7	7.23	32.80	29.0	29.5	×	×		XMM	Becker *et al.* 04
B1821−24	3	1.60E-03	5.5	9.65	7.49	36.37	31.80	33.24	×			ROSAT	Becker & Trümper 97
J0751+1807	3.48	7.27E-04	2.0	9.51	7.90	35.80	30.8	31.3	×			XMM	Webb *et al.* 04
J1012+5307	5.26	7.27E-04	0.52	9.6	8.08	35.30	30.5	30.6	×			XMM	Webb *et al.* 04
B1937+21	1.55	1.00E-04	3.60	8.90	8.41	36.03	31.66	32.8	×			BeppoSax	Nicastro *et al.* 04
J0218+4232	2.32	7.50E-05	5.85	8.93	8.71	35.38	32.11	32.75	×		×	ROSAT	Kuiper *et al.* 98
J0437−4715	5.75	5.70E-05	0.18	9.06	9.23	34.08	30.48	30.86	×			Chandra	Zavlin *et al.* 02
J2124−3358	4.93	1.10E-05	0.25	8.67	9.87	33.56	29.80	30.35	×			ROSAT/ASCA	Becker & Trümper 99
J0030+0451	4.86	1.00E-05	0.23	8.65	9.91	33.54	30.26		×			ROSAT	Becker & Trümper 99

[a] $(L_X)^p$ is pulsed, non-thermal X-ray luminosity and $(L_X)^t$ is total X-ray luminosity.
[b] R, radio; O, optical; G, gamma-ray.

spectra with observed parameters, including detected emission at other wavelengths. The vast majority are also radio pulsars; many of the target sources were chosen from radio pulsar catalogs. However, there have been some detections of X-ray pulsars, such as PSR J0537−6910 (Marshall *et al.* 1998), PSR J1846−0258 (Gotthelf *et al.* 2000), PSR J1811−1926 (Torii *et al.* 1997), and PSR J0205+6449 (Murray *et al.* 2002), without a radio ephemeris. Radio pulses have since been detected from only one of these (PSR J0205+6449; Camilo *et al.* 2002c). Clearly, short-period pulsars seem to have the highest levels of non-thermal X-ray emission. Most of the field millisecond pulsars have purely non-thermal spectra. On the other hand, the millisecond radio pulsars in the globular cluster 47 Tucanae that have been detected as point sources by Chandra (Grindlay *et al.* 2002) have spectra that are quite soft, and most likely thermal. The longer-period pulsars with detected X-ray emission have spectra that are dominated by thermal emission.

7.2.2.1 Spectra

Generally, the X-ray spectra of spin-powered pulsars show a mix of thermal and non-thermal components. Often several thermal components (discussed in Section 7.3) and sometimes even several non-thermal components can be identified. Cleanly separating the non-thermal, or power-law, components from the thermal components can be difficult, especially in sources where the thermal components dominate (e.g. Geminga, PSRs 0656+14 and B1055−52; Halpern & Wang 1997; Greiveldinger *et al.* 1996). Detectors with sensitivity at energies up to at least 10 keV, such as RXTE, Chandra and XMM, have been able to make the best measurements of non-thermal emission components. This non-thermal emission probably originates from the radiation of particles accelerated in the pulsar magnetosphere. Models for this emission will be discussed below.

All of the gamma-ray pulsars that have been detected by the Compton Gamma-Ray Observatory (CGRO) are also X-ray pulsars with measured non-thermal emission components. One interesting question is whether the non-thermal X-ray spectrum is part of the same emission as that detected at gamma-ray energies. For the youngest of these pulsars (Crab, PSR B1509−58), the total power in pulsed emission peaks in the hard X-ray band (Fig. 7.2). There is strong emission through the entire X-ray and gamma-ray bands and a smooth connection between the two. However, in the case of the Crab the smoothness of the spectrum is misleading as the situation may be more complicated. Kuiper *et al.* (2001) have argued for several separate emission components in the Crab optical to gamma-ray spectrum, evidenced by the strong frequency dependence of the interpeak emission and the peak 2 to peak 1 ratio, both of which have a maximum around 1 MeV. The middle-aged pulsars have comparatively weak non-thermal emission in the X-ray band since their power peaks at GeV energies, and there is a gap in the detected spectrum between the X-ray and gamma-ray bands (Fig. 7.3). In several cases (Vela, PSR B1055−52) an extrapolation between the two is plausible, but in others (Geminga, PSR B1706−44; Gotthelf *et al.* 2002) a connection is not clear. In the case of Vela (Fig. 7.3), there is strong evidence for two separate non-thermal X-ray components (Harding *et al.* 2002b), one connecting to the gamma-ray (OSSE) spectrum and the other possibly connected with the optical spectrum.

Millisecond pulsars seem to be bright X-ray emitters, as a relatively large number of the known millisecond radio pulsars are X-ray pulsars. Most have hard power-law spectra with photon indices 1.5–2.0 (Becker & Aschenbach 2002), with the exception of a few (e.g., PSRs J0437−4715, J2124−3358; Zavlin *et al.* 2002; Sakurai *et al.* 2001) that also have

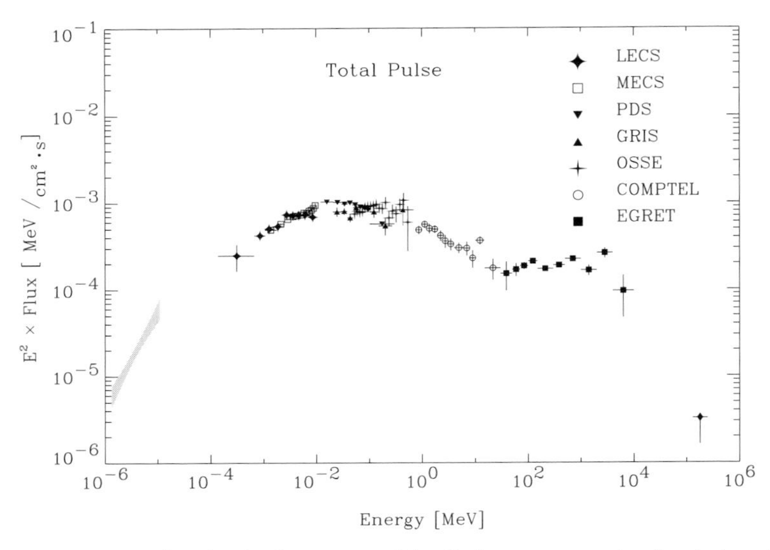

Fig. 7.2. Broadband pulsed spectrum of the Crab, as an example of emission characteristics of a young pulsar (Kuiper *et al.* 2001).

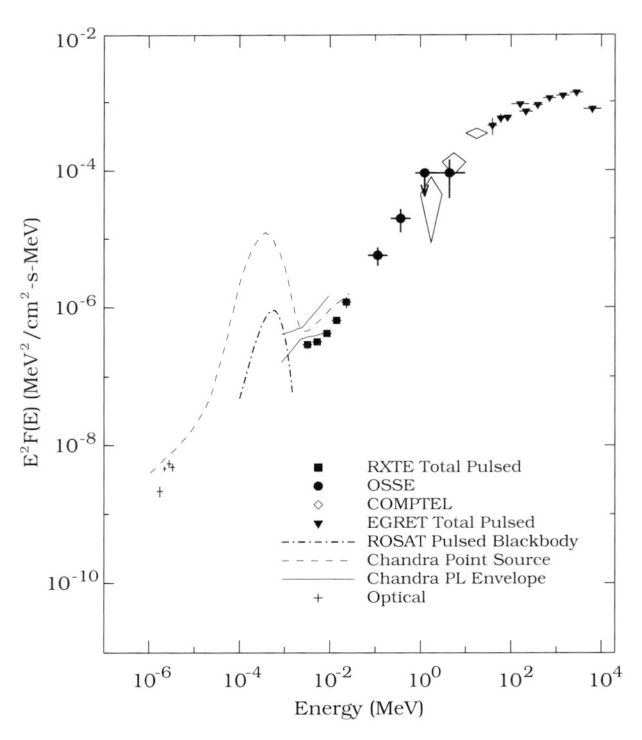

Fig. 7.3. Broadband pulsed spectrum of the Vela pulsar, as an example of emission characteristics of a middle-aged pulsar (Harding *et al.* 2002b).

thermal components. Since they are too old to have detectable cooling emission, the presence of pulsed thermal emission from millisecond pulsars probably requires polar-cap heating.

There are a number of newly discovered X-ray pulsars (PSR J1420−6048; Roberts *et al.* 2001), (PSR J1930+1852; Camilo *et al.* 2002a), (PSR J2229+6114; Halpern *et al.* 2001), (PSR J1105−6107; Kaspi *et al.* 1997) that lie in or near the error circles of EGRET unidentified gamma-ray sources (Hartman *et al.* 1999). Since these pulsars were discovered after the end of the CGRO mission and EGRET detected too few photons to search for pulsations, it is not known whether they are also gamma-ray pulsars. There are also some X-ray pulsars discovered earlier (PSR B1823−13, PSR B1800−21) that are associated with EGRET sources and are candidate gamma-ray pulsars (Kanbach 2002).

7.2.2.2 Pulse profiles

The pulse profiles of the non-thermal X-ray emission usually show much more modulation and narrower pulses than does the thermal emission. The pulsed fraction for many pulsars with mixed emission rises from around 10–20% in the (0.1–1 keV) band where the thermal emission dominates, to 80–90% where the non-thermal emission dominates (Becker & Aschenbach 2002). There is often an accompanying phase shift between the thermal and non-thermal pulses of the middle-aged pulsars.

There seems to be no pattern in the relative phases of the non-thermal X-ray, radio and gamma-ray pulses. The non-thermal X-ray and radio pulses of many pulsars that are young or have very short periods, including the Crab (with the exception of the radio precursor), PSRs J1617−5055, J0205+6114, and J1930+1852 and most of the millisecond pulsars, are in phase and/or the profiles look very similar (see, e.g. Fig. 7.4). Somehow for these pulsars, the radio and high-energy emission must originate from the same region in the magnetosphere. However, the profiles of these pulsars vary from double-peaked to both broad and narrow single-peaked shapes, and a few young X-ray pulsars (PSRs J1811−1926, J0537−6910 and J1846−0258) have no detected radio counterparts. The non-thermal X-ray pulses of middle-aged pulsars, on the other hand, are not in phase with their radio pulses, indicating a different location in the magnetosphere of the two emission components (e.g. Fig. 7.5). The same trend seems to be true for the X-ray and gamma-ray pulses of the gamma-ray pulsars, with the profiles of the fastest, including millisecond (PSR J0218+4232, Kuiper *et al.* 2000), pulsars being alike and in phase while the profiles of the slower pulsars are not in phase (Kanbach 2002). Thus, a fast rotation rate seems to be instrumental in causing alignment of the pulses across the spectrum.

7.2.3 Emission models

Particle acceleration inside the pulsar magnetosphere gives rise to pulsed non-thermal radiation and possibly also thermal emission from backflowing particles that heat the neutron-star surface. Rotating, magnetized neutron stars are natural unipolar inductors, generating huge vacuum electric fields. However, as was first noted by Goldreich and Julian (1969) in their classic paper, a rotating neutron star will not be surrounded by vacuum, since a large surface charge will build up on the star. This surface charge is unstable, because the induced vacuum electric field has a component parallel to the magnetic field at the stellar surface that exceeds the gravitational force by many orders of magnitude and is capable of pulling charges out of the star. If charges are not trapped in the surface by binding forces (i.e., if the crust is solid) a charge density, known as the Goldreich–Julian or corotation charge density,

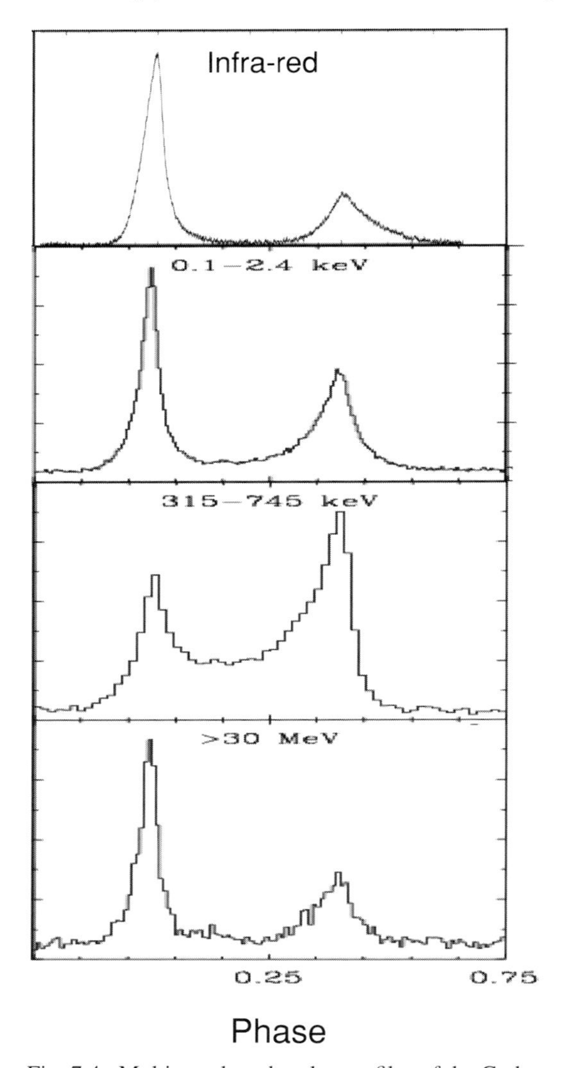

Fig. 7.4. Multiwavelength pulse profiles of the Crab, as an example of emission characteristics of a young pulsar (Kanbach 2002).

$\rho_{\rm GJ} \simeq -\Omega \cdot \mathbf{B}/(2\pi c)$, builds up in the neutron-star magnetosphere. If the magnetospheric charge reaches the Goldreich–Julian value everywhere, it is able to short out the electric field parallel to the magnetic field ($E_{\parallel}$), and the dipole magnetic field will corotate with the star. Corotation of the pulsar magnetosphere must break down at large distances from the neutron star due to particle inertia. Where and how this happens is not yet understood, but it is certain that corotation cannot persist past the speed-of-light cylinder radius, $R_{\rm LC} = c/\Omega$, which is the distance at which the corotation velocity reaches c (see Fig. 7.6). It is believed that in the outer parts of the magnetosphere there must be a transition to the wind zone, where the energy density of the particles is large enough to distort the poloidal dipole field into a toroidal relativistic wind flow that carries the spin-down energy of the pulsar, in the form

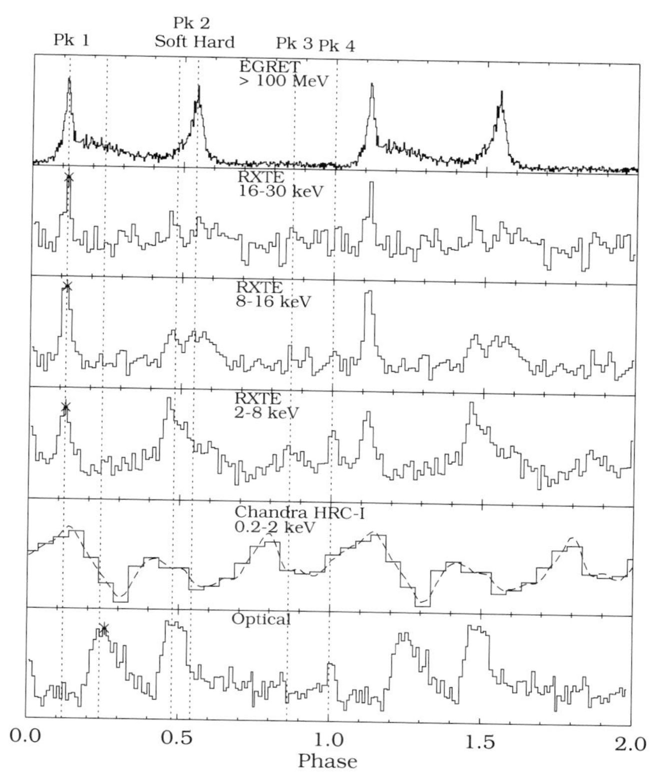

Fig. 7.5. Multiwavelength pulse profiles of the Vela pulsar, as an example of emission characteristics of a middle-aged pulsar (Harding *et al.* 2002b).

of magnetic and particle energy, into a surrounding nebula where (at least in the case of the Crab pulsar) it is dissipated as synchrotron radiation (Rees & Gunn 1974; Kennel & Coroniti 1984b). Although the standard picture of a pulsar magnetosphere described above is usually accepted on faith, global magnetospheric simulations (Krause-Polstorff & Michel 1985; Pétri *et al.* 2002; Spitkovsky & Arons 2002) have not yet been able to show whether and how a pulsar magnetosphere reaches the nearly force-free (ideal MHD) state envisioned by Goldreich and Julian. However, if it is assumed that a force-free, corotating magnetosphere has been achieved, then Contopoulos *et al.* (1999) have found a solution that smoothly connects the corotating magnetosphere to a relativistic wind. Inside the magnetosphere and at the light cylinder, the ratio (known as σ) of the magnetic energy density to the particle energy density is large, primarily because according to the acceleration models to be discussed below, the particles receive only a fraction of the full potential drop across the open field lines. Therefore, σ is large at the start of the wind flow, but must drop to 1 or below to accelerate the particles at the wind-termination shock. The pulsed emission is presumed to originate inside the corotating magnetosphere, and strong $E_{\parallel}$ may develop to accelerate particles at two possible sites where $\mathbf{E} \cdot \mathbf{B} \neq 0$. These sites have given rise to two classes of high-energy emission models: polar-cap models, where the acceleration and radiation occur near the magnetic poles in the inner magnetosphere, and outer-gap models, where these processes occur in the outer magnetosphere.

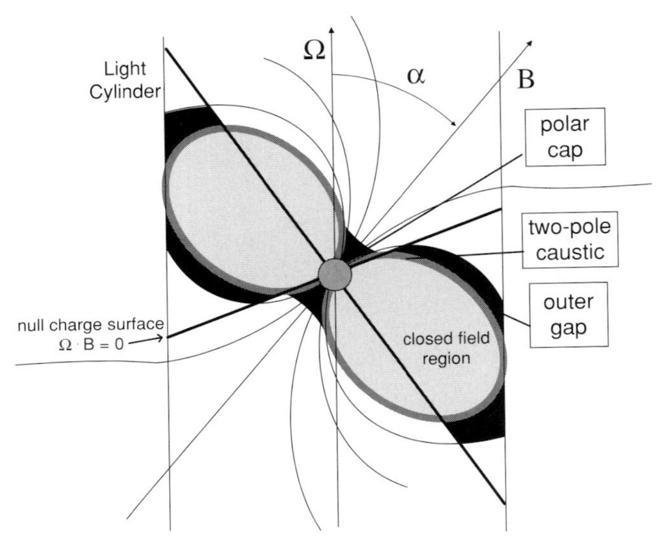

Fig. 7.6. Schematic 2D view of the high-energy emission geometry of several proposed models.

7.2.3.1 Polar-cap model

Polar-cap models (Daugherty & Harding 1982; Usov & Melrose 1995; Daugherty & Harding 1996) advocate that particle acceleration occurs near the neutron-star surface and that high-energy emission results from a curvature radiation or inverse Compton induced pair cascade in a strong magnetic field. There is some variation among polar-cap models, with the primary division being whether or not there is free emission of particles from the neutron-star surface. The subclass of polar-cap models based on free emission of particles of either sign, called space-charge limited flow models, assumes that the surface temperature of the neutron star (many of which have now been measured to be in the range $T \sim 10^5 - 10^6$ K, see Section 7.3) exceeds the ion and electron thermal emission temperatures. Although $E_{\parallel} = 0$ at the neutron-star surface in these models, the space charge along open field lines above the surface falls short of the corotation charge, due to the curvature of the field (Arons & Scharlemann 1979) and to general relativistic inertial frame dragging (Muslimov & Tsygan 1992). The $E_{\parallel}$ generated by the charge deficit accelerates particles, which radiate inverse Compton (IC) photons (at particle Lorentz factors $\gamma \sim 10^2 - 10^6$) and curvature (CR) photons (at Lorentz factors $\gamma \gtrsim 10^6$). Because lower Lorentz factors are required to produce pairs through IC emission, an IC pair formation front (PFF) will form first, close to the surface. However, it is found (Harding & Muslimov 2002) that the IC PFFs do not produce sufficient pairs to screen the $E_{\parallel}$ completely, thus allowing acceleration to the Lorentz factors $\sim 10^7$ sufficient to produce a CR PFF, where there are sufficient pairs to completely screen the $E_{\parallel}$. The CR pair front will therefore limit the particle acceleration voltage and determine the high-energy emission luminosity. The CR photons can produce secondary electrons and positrons through one-photon pair production in the strong magnetic field near the neutron-star surface. These pairs are produced in excited Landau states and radiate synchrotron photons that are energetic enough to produce more pairs. The combined curvature and synchrotron radiation from such pair cascades has been shown to produce high-energy phase-resolved spectra similar to observed spectra of X-ray and gamma-ray Vela-like pulsars (Daugherty & Harding 1996). The gamma-ray spectra cut off very sharply (as a "super-exponential") due to one-photon pair production attenuation, at the pair escape energy, i.e., the highest energy at which photons emitted at a given location can escape the magnetosphere without pair producing. A rough estimate of this cutoff energy, assuming emission along the polar-cap outer rim, $\theta_{\rm PC} \simeq (2\pi r/cP)^{1/2}$, at radius r, is

$$E_{\rm c} \sim 2 \text{ GeV } P^{1/2} \left(\frac{r}{R}\right)^{1/2} \max\left\{0.1,\ B_{0,12}^{-1} \left(\frac{r}{R}\right)^3\right\} \tag{7.1}$$

where $B_{0,12}$ is the surface magnetic field in units of 10^{12} G and R is the stellar radius. Although many high-energy photons are attenuated well below 1 GeV near the surface, in polar-cap cascades much of the radiation occurs at least 1–2 stellar radii above the surface, where the photon escape energy for pair production is several GeV (causing a spectral cutoff there). At all but the highest fields there is a prediction that the spectral cutoff energy should be inversely proportional to surface field strength, or $B_0 = 6.4 \times 10^{19}(P\dot{P})^{1/2}$ G for a dipole field at the pole. In fields above $\sim 2 \times 10^{13}$ G, photon splitting, in which a single photon splits into two lower-energy photons, becomes the dominant attenuation process and lowers the photon escape energy (Baring & Harding 2001). The cascade radiation produces a hollow cone of emission around the magnetic pole, with opening angle determined by the polar-cap half-angle, $\theta_{\rm PC}$, at the radius of emission r. The polar-cap opening angle is very small (a few degrees) unless the emission occurs more than a few stellar radii above the surface. But the pair cascades over most of the polar cap occur within several stellar radii of the stellar surface. Therefore, the wide double-peaked pulses of observed gamma-ray pulsars like the Crab, Vela and Geminga, which require beam opening angles of the order of the magnetic inclination angle in this model, cannot be produced unless the pulsar is nearly aligned.

Recently, this seemingly impossible requirement of polar-cap models has been eased somewhat by the realization that emission at the rim of the polar cap will occur at higher altitudes. The height of the CR pair front is within a stellar radius of the neutron-star surface over most of the polar cap. However, near the edge of the polar cap, at the last open field line which is assumed to be a conducting boundary, the parallel electric field drops to zero. Particles therefore need a longer distance to reach the Lorentz factors necessary to produce pairs, causing the PFF to rise to higher altitude. A slot gap (Arons 1983) is formed as the PFF turns up to asymptotically approach the last open field line. The pair cascades that form on the edge of the slot gap at altitudes of several stellar radii above the neutron-star surface form a wide hollow cone of high-energy emission (Muslimov & Harding 2003). Even so, the inclination angles required in this model to reproduce the Crab and Vela profiles are less than $10-20^\circ$. This may be at odds with the $\sim 60^\circ$ viewing angle inferred from the Chandra image of the Crab nebula (Weisskopf *et al.* 2000). However, acceleration of primary electrons in the slot gap can continue to high altitudes (Muslimov & Harding 2004), where the emission may be viewed at larger angles due to the flaring of field lines.

In polar-cap models, surface heating will occur as a result of the screening of the parallel electric field by pairs. Above the pair front, positrons will decelerate and some fraction will turn around and accelerate downward to the neutron-star surface. Because polar-cap heating emission is intimately tied to the pulsar acceleration mechanism, thermal X-ray emission provides very strong constraints on pulsar models. In a space-charge limited flow model, the trapped positrons needed to screen the $E_\parallel$ are only a small fraction of the number of primary electrons because this field was created by a small charge imbalance in the first place. The polar-cap heating luminosity is predicted by space-charge limited flow models to be only a small percentage of the luminosity of the non-thermal magnetospheric emission, and will fall in the soft X-ray band (Arons 1981). Recent self-consistent models of polar-cap heating in a space-charge limited flow model (Harding & Muslimov 2001, 2002) have predicted heating fluxes from positrons trapped at CR and IC pair fronts. The heating from CR pair fronts predicts surface X-ray luminosity

$$L_{+}^{(\mathrm{CR})} \simeq 10^{31}\ \mathrm{erg\,s^{-1}} \begin{cases} 0.4P^{-6/7}\tau_6^{-1/7} & \text{if } P \lesssim 0.1B_{0,12}^{4/9} \\ 1.0P^{-1/2} & \text{if } P \gtrsim 0.1B_{0,12}^{4/9} \end{cases} \tag{7.2}$$

where τ_6 is the pulsar characteristic age in units of 10^6 yr. Polar-cap heating may make the dominant contribution to thermal emission from old and middle-aged pulsars. The heating from IC pair fronts, with predicted X-ray luminosity

$$L_{+}^{(\mathrm{IC})} \simeq 2.5 \times 10^{27}\ \mathrm{erg\,s^{-1}}\ P^{-3/2} \tag{7.3}$$

can account for thermal emission components detected in the spectrum of some millisecond pulsars.

Polar-cap models predict that emission from the open field line region is visible at all pulse phases. This is because the particles that are accelerated and initiate pair cascades at low altitudes radiate curvature (and possibly inverse Compton) emission at higher altitude on field lines extending to the light cylinder. Such off-pulse emission may have been detected at gamma-ray energies by EGRET (Fierro *et al.* 1998) and at X-ray energies by Chandra (Tennant *et al.* 2001).

7.2.3.2 *Outer-gap model*

Outer-gap models (Cheng *et al.* 1986; Romani 1996) assume that acceleration occurs in vacuum gaps that develop in the outer magnetosphere, along the last open field line above the null charge surfaces, where the Goldreich–Julian charge density changes sign (see Fig. 7.6), and that high-energy emission results from photon–photon pair production-induced cascades. The gaps arise because charges escaping through the light cylinder along open field lines above the null charge surface cannot be replenished from below. Pairs from the polar-cap cascades, which flow out along the open field lines, may pollute the outer gaps to some extent (and vice versa), but this effect has yet to be investigated. The electron–positron pairs needed to provide the current in the outer gaps are produced by photon–photon pair production. In young Crab-like pulsars, the pairs are produced by CR photons from the primary particles interacting with non-thermal synchrotron X-rays from the same pairs. In older Vela-like pulsars, where non-thermal X-ray emission is much lower, the pairs are assumed to come from the interaction of primary particles with thermal X-rays from the neutron-star surface. Some of the accelerated pairs flow downward to heat the surface and maintain the required thermal X-ray emission. The modern outer-gap Vela-type models (Romani 1996; Zhang & Cheng 1997) all adopt this picture.

Although there seems to be agreement on the radiation processes involved in the outer gap, the full geometry of the gap is still not solved. Two approaches to such a solution are underway, but neither is near to defining the complete three-dimensional gap geometry. One group (Zhang & Cheng 1997; Cheng *et al.* 2000) solves the 1D Poisson equation perpendicular to the magnetic field lines, resulting in a gap geometry for young pulsars that is a long, thin sheet bounded below by the last open field line and above by production of pairs. The other group (Hirotani & Shibata 2001; Hirotani *et al.* 2003) obtains solutions to the 1D Poisson equation along the magnetic field (assuming a gap width across field lines) and finds that the gap is limited parallel to the field by pair creation. The actual gap geometry is probably somewhere in between. The long narrow outer-gap geometry of Romani and Yadigaroglu (1995) and Cheng *et al.* (2000), has been successful in reproducing the observed double-peaked pulse profiles for large inclination angles. In these models, the electrons/positrons flowing outward/inward along the last open field line radiate outward/inward from the null surface. The emission forms a wide fan beam which produces in general a wide double-peaked profile, much like the observed Crab profile. The narrow peaks are the result of caustics formed in the observer's frame by the cancelation of the phase shifts from relativistic aberration and light travel time, so that emission from a large range of altitudes is compressed into a narrow phase range. However, due to the fact that no outward emission originates below the null surface, the profile falls off very abruptly at the outer edges (see Fig. 7.7) with no leading and trailing or off-pulse emission. The inability to produce trailing and off-pulse emission, as is seen in high-energy pulsar profiles, is thus a serious problem for current outer-gap models. The models of Hirotani and Shibata (2001) and Hirotani *et al.* (2003) allow external currents to flow through the outer gaps, thus producing a possible extension of the gap below the null surface. In the case of the Crab however, the extent and magnitude of the emission below the null surface is small and would not produce much off-pulse emission. The formation of the leading peak in double-peaked profiles is also problematic since it requires emission very close to the light cylinder, where the structure of the magnetic field has not been determined.

The spectrum of radiation from the outer gap is a combination of synchro-curvature radiation from pairs inside and outside of the gap and inverse Compton scattering of the synchrotron emission by the pairs. Unlike in polar-cap models, pair production in outer-gap models is essential to the production of the high-energy emission: it allows the current to flow and particle acceleration to take place in the gap. Beyond a death line in period–magnetic field space, and well before the traditional radio-pulsar death line, pairs cannot close the outer gap and the pulsar cannot emit high-energy radiation. This outer-gap death line for high-energy pulsars (Chen & Ruderman 1993) falls around $P = 0.3$ s for $B \sim 10^{12}$ G. Geminga is very close to the outer-gap death line and recent self-consistent models (Hirotani *et al.* 2003) have difficulty accounting for GeV gamma rays from pulsars of this age. Polar-cap models, on the other hand, predict that all pulsars are capable of high-energy emission at some level, so that detection as a radio-loud high-energy pulsar is thus a matter of sensitivity.

Half of all the particles produced in the outer gaps are accelerated back toward the neutron-star surface and radiate curvature photons that can produce pairs in the strong magnetic field near the surface, initiating pair cascades (Zhang & Cheng 1997; Wang *et al.* 1998). But the pairs still have enough residual energy to heat the surface at the footpoints of open field lines that thread the outer gaps. The resulting thermal emission has $T \sim 1$ keV, which is much higher than the thermal emission temperatures observed in pulsars. According to the model, this emission is not observed directly (except right along the poles) but only through the blanket of pairs produced by the downward-going particle cascades. The 1 keV photons are reflected back to the surface by the pair blanket through cyclotron resonance scattering (Halpern & Ruderman 1993), and are re-radiated from the entire surface at a temperature around 0.1 keV. Thus, these outer-gap models predict three X-ray emission components: hard thermal emission from direct heating of the polar caps (seen only along the poles), soft thermal emission reflected from the pair blanket, and non-thermal emission from the downward pair cascades (Cheng & Zhang 1999). The components actually observed from a particular pulsar depend on inclination and viewing angle. The predicted X-ray luminosities for pulsars in the ROSAT band can account for the observed $L_X = 10^{-3}\dot{E}$ relation.

7.2.4 *It's the geometry ...*

Theoreticians have spent years building pulsar emission models by starting from the fundamental electrodynamics of the particle acceleration and radiation processes, predicting observable pulsar characteristics from the bottom up. This has had some success, but both polar-cap and outer-gap models as currently formulated also have some basic, unresolved problems. Ultimately, it is the emission geometry required by the observations, which tells us the distribution of radiating particles, that will drive us to the correct understanding of pulsar high-energy emission.

Putting together the results of a number of attempts, in different types of models, to reproduce the narrow and often double-peaked pulse profiles of high-energy pulsars, one common thread seems to emerge. Nearly all successful geometrical models of pulsar high-energy emission have assumed enhanced particle acceleration and radiation along the last open field lines of a magnetic dipole. In their polar-cap emission model, Daugherty and Harding (1996) were led to assume strongly enhanced particle flow on the rim of the polar cap. Acceleration in the slot gap along the polar-cap rim later provided a physical basis for this idea (Muslimov & Harding 2003). The outer-gap models of Romani and Yadigaroglu (1995) and Cheng *et al.* (2000) assumed acceleration and emission essentially only along the last open field line between the null charge surface and the light cylinder in a thin outer gap. But these models cannot produce the non-thermal off-pulse emission seen through the whole pulse phase of the Crab optical and Vela and Geminga gamma-ray profiles. More recently, Dyks and Rudak (2003) have explored a purely geometrical emission model, known as the "two-pole caustic" model in which they assume that particles radiate all along the last open field lines of both poles, from the neutron-star surface to the light cylinder (see Fig. 7.6). Taking into account relativistic effects of aberration and light travel time, the phase shifts of photons emitted at different altitudes along the trailing edge of the polar open field line region cancel, so that the emission in the observer's frame is compressed into a small range of phase, forming a caustic peak in the profile. The caustic peaks in the two-pole caustic model have the same origin as the caustic peaks that form from outward emission above the null charge surface in the outer-gap models. What makes the two-pole caustic model fundamentally different is the addition of outward emission below the null surface, which allows an observer to view caustic emission from both poles as the neutron star rotates past, in contrast to outer-gap models where an observer can view high-energy emission from only one pole. The resulting profiles for viewing angles crossing caustics from both poles are astonishingly similar to the observed profile of the Crab (see Fig. 7.7) and several other high-energy pulsars, and seems to do better than most of the other "bottom-up" models in reproducing the observed geometry. This seems to be telling us that both conventional polar-cap and outer-gap models are missing something critical in their description of the basic electrodynamics. Recently, Muslimov and Harding (2004) have shown that high-altitude radiation from the slot gap could form a physical basis for the two-pole caustic model, producing very similar profiles.

The geometry of emission at different wavelengths is also different in various models. In polar-cap models, the high-energy and radio emission are physically connected, since the electron–positron pairs from the polar-cap cascades are thought to be a necessary ingredient for coherent radio radiation. However, the relative geometry of the two emission regions is not very constrained by observation. Models for radio emission morphology (Rankin 1993; Lyne & Manchester 1988) consist of core and conal components, emitted within a distance of about 10–100 stellar radii from the surface. If the polar-cap high-energy emission cone occurs at a lower altitude than the radio emission, as would be expected, then the leading edge of the radio cone would lead the first high-energy peak. The trailing edge of the radio conal emission would then have to be undetected in the CGRO pulsars. This type of radio geometry for young pulsars was suggested by Manchester (1996) and a class of pulsars with one-sided radio conal emission was proposed by Lyne and Manchester (1988). This picture has received some observational support by Crawford *et al.* (2001) and Crawford and Keim (2003).

High-energy emission in the outer gap is generally radiated in a different direction from the radio emission, which allows these models to account for the observed phase offsets of the radio and high-energy pulses. At the same time, there will be fewer radio–high-energy coincidences and thus a larger number of radio-quiet high-energy pulsars. In the Romani and Yadigaroglu (1995) geometrical outer-gap model, the observed radio emission originates from the magnetic pole opposite to the one connected to the visible outer gap. Many observer lines-of-sight miss the radio beam but intersect the outer-gap high-energy beam, having a much larger emission solid angle. When the line-of-sight does intersect both, the radio pulse leads the high-energy pulse, as is observed in most gamma-ray and some X-ray pulsars.

In the two-pole caustic model of Dyks and Rudak (2003), trailing-edge radio conal emission would arrive in phase with the high-energy emission due to the fact that emission from a wide range of altitudes is compressed into a narrow phase range to form the caustic. This may be a nice explanation for why the radio and high-energy emission tends to be phase-aligned in some of the fastest rotators, such as the Crab and the millisecond pulsars.

7.2.5 Polarization properties

Phase-resolved polarimetry of rotation-powered pulsars has had enormous diagnostic capability at radio and optical wavelengths and could also be a powerful diagnostic in the X-ray range. Several X-ray polarimeters are planned for the near future (AXP, POGO and MEGA) which could measure X-ray polarization characteristics of the brightest sources, such as the Crab pulsar, Cyg X-1 and Her X-1. The pulsed non-thermal radiation from relativistic particles in the magnetosphere is tightly beamed along the neutron-star magnetic field lines and thus the emitted radiation is believed to be highly polarized either parallel or perpendicular to the field lines. Since the field well inside the speed of light cylinder rotates as a solid body with the star, measurement of the polarization properties as a function of pulse phase can provide a multidimensional mapping of the pulsar emission. The expected signature of emission near the poles of a dipole field, an S-shaped swing of the polarization position angle through the pulse profile (Radhakrishnan & Cooke 1969), has been seen from many radio pulsars and has generally been taken as proof that the radio emission originates from the open field lines of a magnetic dipole.

The different high-energy emission models share the common emission mechanisms of curvature, synchrotron and inverse Compton radiation from highly relativistic particles, albeit

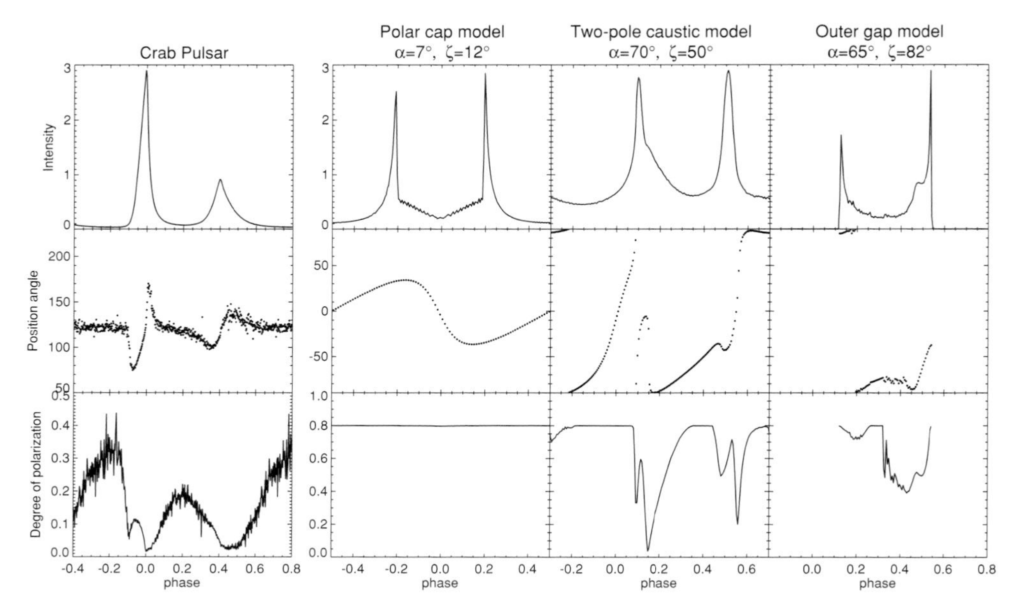

Fig. 7.7. Pulse profiles, polarization position angle and degree of polarization as a function of pulse phase for the Crab pulsar. The left panel is the observed optical pulse profile and polarization characteristics as measured by OPTIMA (Kellner 2002). The right-hand panels are predicted profiles and corresponding polarization characteristics of the polar-cap, two-pole caustic and outer-gap model geometries (Dyks *et al.* 2004).

in very different locations. These mechanisms all intrinsically produce highly polarized radiation (up to 70%, depending on the particle spectrum). Each of the models predicts a distinct variation of position angle and degree of polarization across the pulse profile. Figure 7.7 shows simulations of expected position angle and percentage polarization as a function of pulse phase for outer-gap and polar-cap models of the Crab pulsar. Simulations are also shown for the two-pole caustic model (Dyks & Rudak 2003). The predicted position-angle swing in the polar-cap model shows the classic S-shaped pattern of the rotating vector model since the observer is sweeping near the magnetic pole (phase 0). The most rapid change in position angle occurs near the magnetic axis and there is no predicted decrease in degree of polarization. The two-pole caustic model predicts rapid swings of position angle through the peaks, due to the large change in field orientation, as the viewing angle sweeps past the pole (phase 0) and approaches the trailing edge field line that is the origin of the caustic peaks (Dyks *et al.* 2004). There is also a predicted decrease in percentage polarization on the trailing edges of the peaks, as the observer is viewing emission in the caustics and from rotationally distorted field lines near the light cylinder that fold back and overlap. The outer-gap model (Romani & Yadigaroglu 1995) predicts position angle swings through the peaks, though they are smaller and occur only for carefully selected combinations of inclination and viewing angle (Dyks *et al.* 2004). There would also be a predicted dip in degree of polarization, now primarily at the second peak, since the second peak in the outer-gap model originates from the same caustic that is the origin of the dips at both peaks in the two-pole caustic model. The signature of caustic emission is thus a dip in the percentage polarization and a rapid swing of the position angle at the pulse peak. A recent measurement by OPTIMA (Kellner 2002) of optical polarization position angle and degree of polarization as a function of pulse phase for the Crab pulsar exhibits sharp swings in position angle in the peaks and dips in degree of polarization just following each peak. There appears to be a constant, unpulsed background component (at about 120° position angle and 30%) which, if subtracted from the data, best matches the predictions of the two-pole caustic model.

7.3 Thermal emission from cooling neutron stars

7.3.1 Motivation and background

The existence of magnetospheric emission from neutron stars, as described in Section 7.2, came as an observational surprise to neutron-star theorists. By contrast, not long after the existence of neutron stars was proposed (Baade & Zwicky 1934), it was also pointed out that such stars would form hot, with surface temperatures over 10^6 K, and that surface emission might be detectable (Zwicky 1938). The discovery of cosmic X-ray sources in the 1960s provided hope that neutron stars might one day be observable, and detailed modeling of their potential observational properties began (Morton 1964; Tsuruta 1964). Bahcall and Wolf (1965) pointed out that the equation of state of matter at densities comparable to those in atomic nuclei, which cannot be studied in terrestrial laboratories, might be constrained by observations of neutron-star cooling. This is because the magnitude of the stellar neutrino luminosity, which, in the early history of the neutron star, should far exceed that of the photon luminosity, should depend strongly on what species are present in the core, and on how they interact. Prodigious neutrino emission would result in more rapid cooling than photon cooling alone. Studying neutron-star cooling began in quantatitive earnest with the computation of the first cooling curves (Tsuruta & Cameron 1966).

The first actual detections of thermal emission from the surfaces of neutron stars came much later, following the launch of the Einstein (Helfand *et al.* 1980; Harnden *et al.* 1985; Matsui *et al.* 1988) and EXOSAT (e.g., Brinkmann & Ögelman 1987) X-ray observatories in the 1980s. The situation pre-1990s was reviewed by Tsuruta (1986). The launch of ROSAT in 1990 heralded the next major progress (e.g., Finley *et al.* 1992; Ögelman *et al.* 1993; Halpern & Ruderman 1993), with important contributions made by ASCA (e.g., Halpern & Wang 1997; Zavlin *et al.* 1998) as well. By 1997, four bona fide detections of thermal emission from young or middle-aged pulsars had been reported (from PSRs B0656+14, B0833−45, B1055−52, and the Geminga pulsar) along with interesting upper limits (see Becker & Trümper 1997, and references therein). The recent launches of the Chandra and XMM-Newton observatories have already yielded important results (e.g., Pavlov *et al.* 2001a; Marshall & Schulz 2002; Kargaltsev *et al.* 2002; Sanwal *et al.* 2002b; Bignami *et al.* 2003). Recent reviews of relevant observations include Becker & Pavlov (2002); Pavlov *et al.* (2002b); Pavlov & Zavlin (2003). Reviews of theory include Tsuruta (1998), Page (1998), and Yakovlev *et al.* (1999).

As discussed in Section 7.2, thermal emission can also be produced as a byproduct of emission processes in the magnetosphere. Such emission is believed to be unrelated to that originating from initial cooling. Other heating mechanisms are predicted to matter only in the photon-cooling era, i.e., for $t > 10^6$ yr (e.g., Alpar *et al.* 1984; Umeda *et al.* 1993; van Riper *et al.* 1995; Larson & Link 1999). One notable exception is magnetars, in which the decay of the field is thought to provide a major energy source that dominates over all others (see Chapter 14).

7.3.2 Theory of neutron star cooling

Neutron stars are born in the violent collapse of stars having ZAMS mass ~8–20 $M_\odot$, which have core temperatures prior to collapse of $T_c \sim 10^9$ K. Following collapse, it is expected that the neutron star forms with central temperature $T_c \sim 10^{11}$ K, although prodigious neutrino emission results in $T_c \sim 10^9$–10^{10} K within a day or so. Neutrino emission dominates over photon emission until $T_c \sim 10^7$–10^8 K (depending on leading neutrino emission process), which is reached at least $\sim 10^3$ yr, and more likely $\sim 10^5$–10^6 yr after birth. The thermal behavior before 10^6 yr depends strongly on cooling model. After $t \simeq 10^6$ yr, various internal and external heating mechanisms dominate thermal emission over the initial cooling. Therefore, young, i.e., having age $\lesssim 10^6$ yr, neutron stars for which the surface temperatures T_s are expected to drop either slowly or quickly from $T_s \sim 2 \times 10^6$ to $\sim 2 \times 10^5$ K, with core temperatures dropping from $T_c \sim 10^9$ to 10^7–10^8 K, are interesting targets for studies of neutron-star interior physics. Note that, in spite of these apparently high temperatures, from a statistical mechanics point of view, neutron stars can be considered to be at $T = 0$, since for their average densities, the Fermi energy is ~100–500 MeV, while the internal temperature of a one-year old star is certainly below 0.1 MeV. Next we consider in turn the main elements that determine how a neutron star cools.

7.3.2.1 The stellar core

During the first $\sim 10^5 - 10^6$ yr, a neutron star cools mainly via neutrino emission from its innermost regions (which have densities $\rho \gtrsim 10^{10}$ g cm^{-3}). For ages $t \gtrsim 10-50$ yr, the internal layers are isothermal. They have an enormous density gradient which results in different neutrino emissivities at different radii. As a rule, the most efficient neutrino emission is produced in the stellar core, which extends from $\rho \approx \rho_0/2$ to the stellar center $\rho \sim (3-15)\rho_0$. Here, $\rho_0 \approx 2.8 \times 10^{14}$ g cm^{-3} is the density of nuclear matter in atomic nuclei. The composition and equation of state of supranuclear matter ($\rho \gtrsim 2\,\rho_0$) in neutron-star cores are unknown but are of considerable interest to basic physics.

The main cooling mechanism in the core is neutrino emission via the "Urca" process.[2] In its simplest version adopted for neutron-star cores, the Urca process is a sequence of a β-decay and an electron capture, $\mathrm{n} \longrightarrow \mathrm{p} + \mathrm{e} + \overline{\nu}_\mathrm{e}$, and $\mathrm{p} + \mathrm{e} \longrightarrow \mathrm{n} + \nu_\mathrm{e}$. This is the so-called direct Urca process (Lattimer *et al.* 1991). However, to proceed at interesting rates in the neutron-star interior for a wide range of expected interior properties, the Urca process requires a spectator reactant to simultaneously conserve energy and momentum. The process with the spectator is called the "modified" Urca process (Chiu & Salpeter 1964). Neutrons, protons, and other baryon spectators greatly suppress the direct Urca rate (typically, by 5–7 orders of magnitude) because of the lack of phase space due to degeneracy. The most efficient direct Urca cooling, without any spectator, is possible if the proton fraction is sufficiently high. This may happen in the cores of sufficiently massive stars. If hyperons are present in the core, they initiate additional direct Urca processes (Prakash *et al.* 1992), as efficient (or nearly as efficient) as the basic direct Urca process with nucleons. If the direct Urca process is forbidden (e.g., if the proton fraction is insufficiently high) but the dense matter contains pion or kaon condensates, weaker processes similar to the direct Urca

[2] Urca is the name of a long-since-closed casino in Rio de Janeiro, and was adopted as a name for these reactions by Gamow and Schoenberg (1941) who saw a parallel between how casinos extract money from players and how nature extracts energy in these reactions.

ones (though stronger than the modified Urca ones) can open (see, e.g., Pethick 1992, for a review).

7.3.2.2 *Equation of state*

An equation of state (EOS) is the pressure–density–temperature relationship of matter. In a neutron star, most of the mass is at densities two or three times ρ_0 (the nuclear matter density). At these densities, the EOS is unknown. Nuclear physics can constrain some aspects of the high-density EOS, however certain fundamental parameters, such as the compression modulus, the bulk symmetry energy and the effective nucleon mass remain weakly constrained (Lattimer 1992; Lattimer & Prakash 2001). The EOS is crucial to understanding neutron-star structure; it determines, among other things, the mass-to-radius relationship, the stellar binding energy, the stellar moment of inertia and the relative moments of different components, the minimum and maximum masses, and the maximum angular velocity. From a cooling point of view, the most important parameters determined by the EOS are the density, pressure and temperature profiles of the star, as well as its composition. The profiles are important because, for example, the location, amount and type of superfluidities present depend on the density profile, via a critical temperature/density relation (see Section 7.3.2.3). The composition is important as different species of particles will react differently.

Many EOSs have been proposed, and the range of possible properties of neutron-star matter is large. For example, the range of predicted pressure of matter having density twice ρ_0 is approximately a factor of 5, depending on the choice of plausible EOS, where plausible means consistent with available laboratory data on atomic nuclei and nucleon scattering (Lattimer & Prakash 2001). Phase transitions to kaon-, pion- or hyperon-rich matter are all possible depending on the core density.

Figure 7.8 (left) shows predicted cooling curves for a commonly (e.g., Yakovlev *et al.* 2001) assumed core EOS (originally from Prakash *et al.* 1988) for three different masses. Curves for $M < 1.35\ M_\odot$ are very similar to that for $M = 1.35\ M_\odot$. Above this mass (to be more exact, above $1.358\ M_\odot$), for this EOS, direct Urca cooling sets in; this is clear in the curves. Also shown in the figure are two schematic curves assuming a pion condensate (but no direct Urca if the pion condensate were absent) in the core at $M > 1.35\ M_\odot$. The effect is clearly enhanced cooling, but not as much as for direct Urca.

7.3.2.3 *Superfluidity and superconductivity*

Proton and neutron superfluidity in the neutron-star interior can play a very significant role in cooling. (Note that proton superfluidity means superconductivity.) This is because it suppresses traditional neutrino processes involving nucleons. On the other hand, it produces a specific type of neutrino emission (Flowers *et al.* 1976) associated with Cooper pairing of nucleons. Superfluidity also affects nucleon heat capacity. However, its onset, determined by a critical temperature, is very poorly known at neutron-star densities. Overall, superfluidity in different forms can mimic enhanced cooling, challenging the promise of using cooling to determine the EOS. See Kaminker *et al.* (2002) and Yakovlev *et al.* (2002a, 2004) for recent reviews of the subject.

The relevance of superfluidity can be summarized as follows. In very low-mass stars ($M \lesssim 1.35\ M_\odot$ for the EOS assumed here), the effect of core singlet-state ($^1S_0$) proton superfluidity is well determined and nearly independent of mass or EOS. These stars cool very slowly as even modified Urca processes are suppressed, so that nucleon–nucleon bremsstrahlung, a

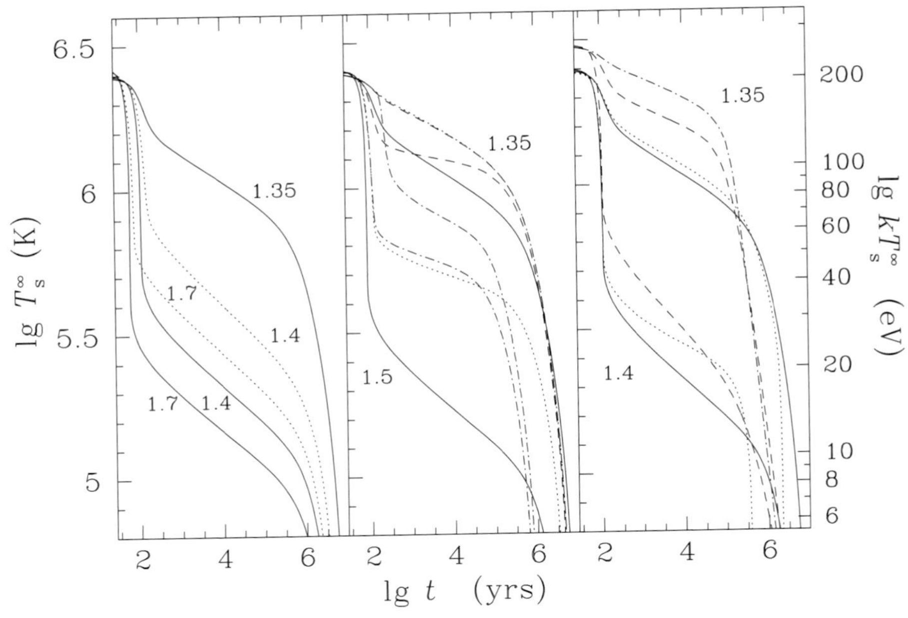

Fig. 7.8. Neutron-star cooling curves for a variety of parameters and models (courtesy D. Yakovlev). All curves are for EOS "A" of Yakovlev *et al.* (2001), or "model I" of Prakash *et al.* (1988) with compression modulus of saturated nuclear matter 240 MeV. *Left*: Curves for non-superfluid stars for three masses (labelled in solar masses) are shown with solid lines. Dotted lines are for pion condensate in the core, for 1.4 and 1.7 $M_\odot$ stars (see Section 7.3.2.2). *Middle*: Curves for superfluid neutron stars of masses 1.35 and 1.5 $M_\odot$. Four curves are shown for each mass: solid curves are for non-superfluid stars (as in the left plot); dotted curves are for proton superfluidity model "1p" of Kaminker *et al.* (2002); dashed curves are for proton superfluidity model "2p" of the same reference; dot-dashed curves are for the combination of proton superfluidity model "p1" and neutron superfluidity model "3nt," also from Kaminker *et al.* (2002). Note that the 1p and 2p curves for the 1.35 $M_\odot$ star coincide as both suppress modified Urca cooling. Also, note that the 2p curve for the 1.5 $M_\odot$ star sits among the 1.35 $M_\odot$ star curves. *Right*: Curves for non-superfluid stars for masses 1.35 and 1.4 $M_\odot$ for different accreted envelope masses and dipolar magnetic fields: solid curves are for zero accreted mass and zero magnetic field B (as in the left plot); dotted curves are for $B = 10^{14}$ G at the pole; dashed curves are for $B = 0$ and accreted envelope mass 10^{-7} $M_\odot$. The dot-dashed curve is for a 1.35 $M_\odot$ solar mass star having $B = 0$ and accreted envelope mass 10^{-7} $M_\odot$, but with proton superfluidity included. It is shown to demonstrate the approximate upper bound for temperatures of neutron stars having ages between ~1 and ~100 kyr.

slower process, dominates. The hottest young neutron stars are therefore plausibly explained as having the lowest mass. On the other hand, for very high-mass neutron stars ($M \gtrsim 1.5\,M_\odot$), proton superfluidity becomes inefficient, and one expects very rapid, model-independent cooling, because direct Urca is too strong to be suppressed. A neutron star that has cooled so rapidly has not yet been observed (see Section 7.3.4).

For intermediate-mass neutron stars, the situation is highly model dependent, and by assuming an EOS and proton superfluidity model, one can, at least in principle, use observed surface temperatures to determine neutron-star masses (e.g., Yakovlev *et al.* 2001). In these stars, direct Urca processes would proceed unimpeded were it not for moderate proton superfluidity suppression. Figure 7.8 (middle) shows two different models for proton superfluidity (see Kaminker *et al.* 2002, and references therein for definitions). Mild $^3P_2$ pairing of neutrons in the stellar core results in very strongly enhanced cooling of stars with $M \lesssim 1.35\,M_\odot$ that is inconsistent with the observations thus far (see Section 7.3.3).

7.3.2.4 *The stellar envelope: composition and magnetic field*

Interpreting observed thermal emission from neutron stars means mapping the observed surface temperature T_s to that of the stellar core T_c. The temperature gradient from the surface inward is mainly in the stellar envelope, the layer beneath the atmosphere down to the isothermal internal region which effectively acts as a thermal insulator for the bulk of the star. The insulating envelope is commonly defined as the region having density $\rho \lesssim 10^{10}$ g cm^{-3}. This has thickness only a few tens of meters, much smaller than the stellar radius. The insulating envelope thickness decreases as the star cools.

For envelopes made of iron, for $T_s \simeq 10^6$ K, $T_c \simeq 10^8$ K with $T_s \overset{\propto}{\sim} T_c^{1/2}$ for a typical surface gravity (Gudmundsson *et al.* 1983). However, the mapping is highly sensitive to envelope composition. Chabrier *et al.* (1997) show that even a tiny ($\sim 10^{-13}$ $M_\odot$) amount of accreted light-element matter, such as from fallback after the supernova, from the interstellar medium or from a binary companion, can have a substantial effect on T_s for a given T_c. Specifically, accreted envelopes have lower thermal conductivity. As a consequence, they lead to higher T_s in the neutrino-cooling era but lower T_s in the photon-cooling era. Figure 7.8 (right) shows the effect of an accreted envelope on predicted cooling curves of two different mass neutron stars.

The magnetic field in the envelope can also play a major, though complicated, role (van Riper & Lamb 1981; Page 1995; Shibanov & Yakovlev 1996; Heyl & Hernquist 1997; Potekhin & Yakovlev 2001; Heyl & Hernquist 2001). A magnetic field affects all plasma components, especially the electrons. In particular, electron motion perpendicular to the field lines becomes suppressed by classical Larmor rotation. In addition, this motion is quantized in Landau levels. Classical and quantum effects greatly modify the electron thermal and electrical conductivities. Thus the effect of a magnetic field on the thermal structure of the insulating envelope depends strongly on the field geometry, with radial fields effectively reducing the insulation by the envelope (quantum effect), and fields tangential to the surface increasing it (classical effect). Thus a dipolar field will have a very different effect than a (often assumed) radial field (see Potekhin & Yakovlev 2001, for a review), with the tangential and radial regions potentially cancelling local deviations when averaging over the surface (Page 1995; Shibanov & Yakovlev 1996; Potekhin & Yakovlev 2001; Heyl & Hernquist 2001; Potekhin *et al.* 2003). Figure 7.8 (right) shows the effect of a dipolar magnetic field of 10^{14} G (at the pole) on cooling curves for two different masses. Curves for 10^{12} G are close to the zero field case. The net effect is clearly much more subtle than for light elements, mass, or EOS. Note that the above consideration assumes no active magnetic field decay of the type invoked in the "magnetar" model (see Chapter 14).

7.3.2.5 *The stellar atmosphere*

The neutron-star atmosphere, defined as the region having density $\rho \lesssim 10^2$ g cm^{-3}, is typically only 0.1–10 cm thick, yet ultimately determines many of the properties of the emerging thermal photon flux. Although in studying thermal radiation the assumption of a blackbody spectrum is often made, every realistic atmosphere model predicts a significant deviation from a Planckian curve. The nature and degree of the deviation, both from the

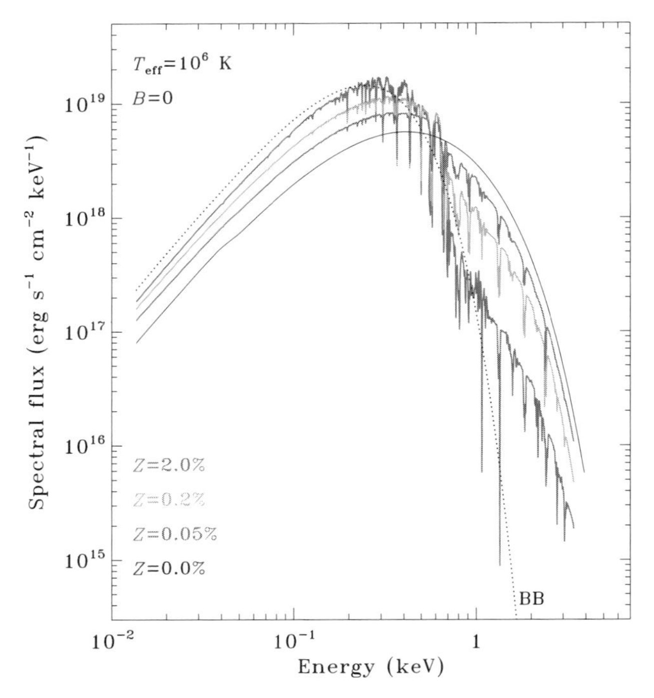

Fig. 7.9. Spectra of emergent radiation in non-magnetic neutron-star atmospheres having $T_{\rm eff} = 10^6$ K for different metallicities Z ($Z = 2.0\%$ corresponds to solar metallicity). The corresponding blackbody is shown with a dotted line. (From Zavlin & Pavlov 2002) For color version see Plate 7.9.

continuum and in the form of lines (see Section 7.3.3.4) depends strongly on atmospheric composition as well as on magnetic field. A nice review of this subject is given by Zavlin and Pavlov (2002). We summarize the basic issues below.

An unadulterated neutron star might be expected to have an iron atmosphere, in keeping with the pre-collapse composition of the progenitor core. Heavy-element spectra, with even tiny (i.e., $\sim$0.05%) metallicities, are expected to contain many spectral features and photoionization edges produced by ions at various stages of ionization (Romani 1987; Rajagopal & Romani 1996; Zavlin *et al.* 1996; Rajagopal *et al.* 1997). However, blurred to lower resolution, a heavy-element atmosphere resembles a blackbody of comparable effective temperature. Figure 7.9 (see also color plate section) shows sample atmospheres containing different metal abundances, for comparison with a blackbody spectrum.

On the other hand, a spectrum emerging from a light-element (H or He) atmosphere is predicted to have few absorption lines at X-ray energies (e.g., Ho *et al.* 2003) – in fact none if the atoms are fully ionized, apart from cyclotron lines. Nevertheless, such an atmosphere has a clear signature: a high-energy tail in the spectrum, resulting in a significant overestimate, by as much as a factor of $\sim$3, of the surface temperature if a blackbody model is fitted instead (see Fig. 7.9; Romani 1987; Rajagopal & Romani 1996; Zavlin *et al.* 1996). For example, fitting a blackbody model to the X-ray spectrum of the Vela pulsar yields $kT_{\rm s}^{\infty} = 127$ eV, versus $kT_{\rm s}^{\infty} = 59$ eV for a hydrogen atmosphere (Pavlov *et al.* 2001a). Also, blackbody fits to hydrogen atmosphere spectra yield emitting radii that are much smaller than those inferred with hydrogen atmosphere models; for example, Marshall and Schulz (2002), as revised by Brisken *et al.* (2003b), find $R^{\infty} = 156$ km and 8.5 km for hydrogen and blackbody fits to the X-ray spectrum of PSR B0656+14, respectively. The high-energy tail is produced because light-element opacities decrease rapidly with increasing energy, so that at higher energies one is observing deeper layers of the star which are at higher temperatures. Although an unadulterated neutron star should have an iron atmosphere, even tiny amounts (e.g. 10^{-20} $M_{\odot}$) of accreted interstellar material should result in a pure hydrogen atmosphere, because of the short timescale for gravitational settling.

As in the envelope, an appreciable magnetic field can also have a major influence on the atmosphere. The most important issue is the huge difference that exists between the opacities for different polarizations. Specifically, the magnetic atmosphere is much more transparent to photons with electric-field vector perpendicular to the $\mathbf{k} \cdot \mathbf{B}$ plane than is a non-magnetic atmosphere. Thus, the emergent radiation is expected to be highly polarized (see Section 7.3.3.3). Furthermore, the emergent flux depends on the direction of the magnetic field; hence for non-radial field geometries, pulsations are expected even for uniform temperature (which is, in any case, implausible, since variations in surface magnetic field strength result in temperature variations).

For magnetic field strengths greater than a quantum critical field $B_{\rm Q} \equiv 2\pi m_{\rm e}^2 c^3/(eh) = 4.414 \times 10^{13}$ G, so-called "vacuum polarization," that is the effective dielectric constant of the vacuum, must be accounted for in the radiative transfer equations, given the plasma densities in the envelope. This is interesting physically as, in spite of being predicted by QED theory, this effect has not been confirmed experimentally. The net effect of vacuum polarization is to soften the high-energy tail of the hydrogen spectrum, though the emission is still harder than the blackbody spectrum of the same effective temperature (Ho & Lai 2001, 2003; Özel 2001, 2003; Zane *et al.* 2001; Lai & Ho 2002, 2003).

7.3.3 *Determinations of neutron-star temperatures and ages*

7.3.3.1 *Measuring neutron-star temperatures*

There are several challenges to measuring neutron-star temperatures. First, the temperature measured by a distant observer $T_{\rm s}^{\infty}$ is not that at the neutron star surface $T_{\rm s}$. This is due to gravitational redshifting; the two are related via $T_{\rm s}^{\infty} = T_{\rm s}[1 - 2GM/(Rc^2)]^{1/2}$. Here, R is a circumferential neutron-star radius (which determines the proper length $2\pi R$ of the stellar equator). Similarly, the stellar luminosity as measured by a distant observer is given by $L_{\gamma}^{\infty} = 4\pi\sigma T_{\rm s}^4 R^2[1 - 2GM/(Rc^2)]$, where the γ specifies that this is the bolometric photon (as opposed to neutrino) luminosity. The apparent radius that would be measured by a distant observer if the latter could resolve the star is $R^{\infty} = R/[1 - 2GM/(Rc^2)]^{1/2}$. Thus the determination of $T_{\rm s}$ from $T_{\rm s}^{\infty}$ requires knowledge of M/R, which is unknown, and which depends on the unknown EOS in the stellar core for any M or R that could be determined independently.

Particularly problematic in the pre-Chandra era was that young neutron stars that are pulsars with high $\dot{E}$ are ubiquitously associated with pulsar wind nebulae (see Section 7.6). These extended sources surround the neutron star and contaminate measurements of the thermal emission from the point source. Also, sources with high $\dot{E}$ produce copious non-thermal X-rays from magnetospheric processes (see Section 7.2). This emission, for example, completely overwhelms any thermal emission from the very young Crab pulsar, hence the availability of only an upper limit on its thermal emission, obtained off-pulse (Table 7.2). Mitigation of both the above problems is best done by avoiding high $\dot{E}$ sources, even though they are often the youngest and hottest (see Section 7.4).

An additional important difficulty is the degeneracy between the equivalent neutral hydrogen absorption toward the source, $N_{\rm h}$, and the measured temperature. For an absorbed blackbody-like spectrum, the low-energy exponential cutoff due to $N_{\rm h}$ results in higher inferred $T_{\rm s}^{\infty}$ for smaller $N_{\rm h}$, and vice versa. Generally it is necessary to constrain $T_{\rm s}^{\infty}$ via a contour plot in the $T_{\rm s}^{\infty}/N_{\rm h}$ plane (e.g., Halpern & Wang 1997; Zavlin *et al.* 1999), although frequently authors quote $T_{\rm s}^{\infty}$ for only a single assumed $N_{\rm h}$ (see Table 7.2).

7.3.3.2 Hot spots and cyclotron resonance scattering

In many instances (see Table 7.2 and Section 7.5), the effective radius of the star as determined by either a blackbody or an atmosphere spectral fit is much smaller than the expected radius of a neutron star, regardless of assumed EOS. In at least one case, RX J1856.5−3754, this has led to claims that the thermal spectrum therefore implies the target is not a neutron star at all, but a quark star (Drake *et al.* 2002).

However, localized backheating of a neutron-star surface by relativistic electron–positron pairs formed from conversion of gamma rays produced in the magnetosphere in the polar-cap or outer-gap accelerators (see Section 7.2) can produce "hot spots" which emit thermal radiation. Such hot spots have areas much smaller than the stellar surface and could in principle outshine surrounding thermal emission from initial cooling, thereby dominating the X-ray spectrum. This is the case for the thermal emission detected from millisecond pulsars (e.g., PSRs J0030+0451, J0437−4715, J2124−3358; see e.g., Becker 2001, for a review), which have ages of $\sim 10^9$ yr hence should have cooled long ago. When hot spots dominate, studying thermal emission is mainly constraining the physics of the magnetosphere, although upper limits on initial cooling emission could still be interesting. For example, the Vela pulsar is a strong gamma-ray source, yet a very faint thermal X-ray source (see Tables 7.1 and 7.2); magnetospheric-acceleration-driven hot spots may well dominate the thermal spectrum, but its low effective temperature is still interesting (see Fig. 7.11). Thus, measured temperatures that are associated with small emitting areas are most conservatively taken to be upper limits on cooling temperatures.

Although one might expect stars with hot spots to have higher pulsed fractions than those without, this is not necessarily the case due to strong gravitational light bending expected in neutron stars (see Section 7.3.3.3). This and a less-than-favourable viewing angle can significantly reduce the pulsed fraction seen from a star with hot spots, as has been argued for RX J1856.5−3754 (e.g., Ransom *et al.* 2002). The detection of a bow-shock Hα nebula around this source supports it being an off-beam rotation-powered pulsar (see Section 7.5; Kerkwijk van & Kulkarni 2001). The same is true of other "isolated neutron stars" (see Section 7.4) as well as of the "central compact objects" (see Section 7.5).

Ruderman (2003) has argued that cyclotron-resonance scattering of thermal photons by electrons and positrons within a few stellar radii of an energetic rotation-powered pulsar results in a Planck-like X-ray spectrum that is quite changed from the seed thermal spectrum. A pair plasma is thought to surround the star, maintained by conversion of gamma rays from the star's polar-cap and/or outer-gap accelerators. This is because a thermal X-ray photon's energy, ~ 1 keV, is less than the cyclotron resonance energy at the stellar surface, $eBh/2\pi mc = 11.6$ keV for $B = 10^{12}$ G. Thus, as the photon passes through the magnetosphere, it goes through a resonance, and the optical depth to cyclotron-resonance scattering becomes large. Ruderman (2003) describes the situation as if the X-rays are reflected back from the surface of a "lightly leaky *Hohlraum* container" and only escape after multiple re-absorptions and re-emissions from the stellar surface. Thus, he argues, measurements of thermal emission from neutron stars teach us only about the stellar magnetosphere, not about initial cooling.

7.3.3.3 Pulsations and polarization from thermally cooling neutron stars

The pulsations from the surface of a cooling neutron star have low pulsed fraction and low harmonic content. This is in strong contrast to the high harmonic content and pulsed fractions seen from the magnetospheric component of the emission, which is presumed to be produced well above the stellar surface (see Section 7.2). Often young neutron stars exhibit emission from both mechanisms, as is clear from the energy dependence of their lightcurves (e.g., Fig. 7.5).

The low harmonic content results primarily from the presumably low variation of temperature over the neutron-star surface. However, pulsations are further reduced by the strong gravitational bending of light near the surface of the neutron star (e.g., Page 1995; Psaltis *et al.* 2000). This becomes important as the general relativistic compactness parameter $p \equiv Rc^2/2GM \rightarrow 1$ (a Newtonian star is described by $p \rightarrow \infty$). Due to bending, although the observer may be facing one side of the neutron star, she/he may still detect significant emission from the opposite side. Simulations of well-defined infinite-contrast spots on neutron stars show that the effect is so strong that even for $p = 4$ the maximum pulsed fraction is less than 0.4 for *any* size emitting spot (Psaltis *et al.* 2000).

Measured temperature is expected to be dependent on pulse phase, as the thermal conductivity of the neutron-star surface depends on the magnetic field strength, which varies over the surface (Page 1995; Psaltis *et al.* 2000). However, generally, only phase-averaged temperatures are measured, both because of faint signals, and also because gravitational light bending makes it difficult to do otherwise. Clearly this complicates the interpretation since the averaged value depends on the distribution of temperature over the surface. However, this effect is likely to be small compared to other systematic effects.

X-ray emission from the surface of a cooling neutron star endowed with a magnetic field at the surface of magnitude $\gtrsim 10^{11}$ G is expected to be significantly polarized. This is because atmospheric opacities depend strongly on polarization when the photon energies are much smaller than the electron cyclotron energy. The opacity to light with its electric-field vector oriented perpendicular to the magnetic field is smaller by a factor of approximately the squared ratio of the photon energy to the cyclotron energy, if the ratio is small, relative to that for light with electric-field vector parallel to B (Lodenqual *et al.* 1974). Thus, radiation with electric-field vector perpendicular to B escapes from greater atmospheric depths where the temperature is higher, hence results in a higher flux. Polarizations from localized regions on the surface should be very high, $\gtrsim 50\%$ (Kanno 1975; Pavlov & Shibanov 1978). However, an observer sees radiation averaged over a hemisphere, and even beyond due to gravitational light bending, which is strongest for large M/R. This will reduce the observed polarization fraction (e.g., Pavlov & Zavlin 2000). Similarly, the different magnitudes and directions of the lines of a dipolar B field at the neutron-star surface might be thought to reduce polarization (Pavlov & Zavlin 2000). However, the QED effect of vacuum polarization, in which the region surrounding the star is effectively birefringent, has recently been shown to counteract the GR and varying B-field effects, because the observed polarization direction of rays from the surface is correlated with the B-field direction far from the surface; a ray bundle that we observe passes through only a small B-field solid angle as it leaves the magnetosphere (Heyl & Shaviv 2002). This results in very large net polarizations as a function of pulse phase, and even phase-averaged polarization fractions of $\sim 10-20\%$ (Heyl *et al.* 2003; Lai & Ho 2003). Measurement of polarization lightcurves should thus allow sensitive constraints on the B field, as well as on M/R and the magnetic geometry of the neutron star. Hence, thermally cooling neutron stars and magnetars (Chapter 14) are expected to be very interesting targets for future X-ray polarimetry experiments.

7.3.3.4 Features in neutron star thermal spectra

Features in the X-ray spectrum of neutron stars have been an astrophysical "Holy Grail" since their X-ray emission was discovered. This is because the presence of lines of identifiable origin, with known inertial reference frame energies, would allow the determination of the stellar redshift, z, and hence, at least in principle, the mass-to-radius relationship via $M/R = (c^2/2G)[1-(1+z)^{-2}]$, which would constrain EOS. (The above is true only under the assumption that general relativity correctly describes the spacetime around the neutron star; Dedeo & Psaltis 2003). Atmosphere models with even as little as 0.05% metal abundance predict prolific line production due to the many transitions available for the various atomic ionization stages (Rajagopal *et al.* 1997). Hydrogen atmosphere models, by contrast, predict few lines. The relevance of spectral features to constraining the neutron-star age is thus that (i) they can help determine the atmosphere composition, which can help establish what continuum model to fit; and (ii) they could, in principle, distort the continuum given poor spectral coverage or resolution.

Sanwal *et al.* (2002b), using Chandra's ACIS CCD detectors, observed two absorption features, at 0.7 and 1.4 keV, in the X-ray spectrum of 1E 1207.4−5209, in the supernova remnant PKS 1209−51/52 (G296.5+10.0). The nature of this object is not understood; see Section 7.5. Mereghetti *et al.* (2002a) confirmed the existence of the lines using XMM-Newton. A subsequent long XMM-Newton exposure revealed an additional absorption feature at 2.1 keV, as well as a possible feature at 2.8 keV (see Fig. 7.10; Bignami *et al.* 2003; De Luca *et al.* 2004). These suggest harmonics of a cyclotron line at 0.7 keV. A cyclotron fundamental at this energy implies a magnetic field of 8×10^{10} G for an electron line, or 2×10^{14} G for a proton line. However, Sanwal *et al.* (2002b) and Hailey and Mori (2002) argue against the cyclotron interpretation, and suggest possible spectral line identification for the most secure 0.7 and 1.4 keV features. In any case, why such lines are not seen, in spite of many comparably sensitive observations of other neutron stars, remains a mystery. This source is discussed further in Section 7.5.

More recently, features in spectra of three "isolated neutron stars" have been reported. These are discussed in Section 7.4.

7.3.3.5 Neutron star ages

The standard age estimator for rotation-powered pulsars assumes frequency evolution of the form $\dot{\Omega} = k\Omega^n$, where n is the "braking index," and k is a constant that depends on the magnetic moment of the neutron star. Assuming k and n to be constant, the braking index can be determined from a measurement of the second time derivative of the frequency,

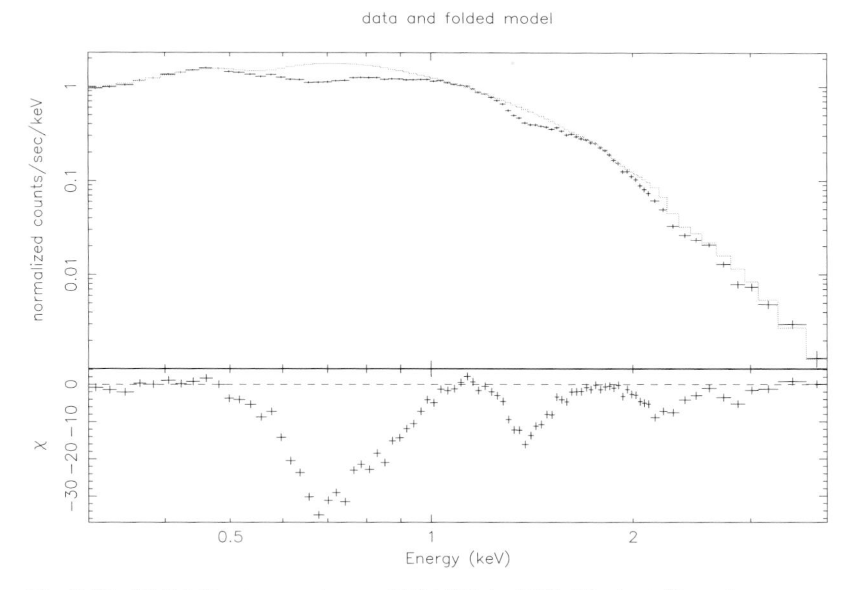

Fig. 7.10. XMM-Newton spectrum of 1E 1207.4−5209. The best-fit continuum curve is represented by the sum of two blackbody functions. Four absorption features appear at the harmonically spaced energies of ~0.7 keV, ~1.4 keV, ~2.1 keV and ~2.8 keV (De Luca *et al.* 2004).

$\ddot{\Omega}$, via $n = \Omega\ddot{\Omega}/\dot{\Omega}^2$. The age derived by integrating the above differential equation is

$$\tau = \frac{P}{(n-1)\dot{P}}\left[1 - \left(\frac{P_0}{P}\right)^{n-1}\right] \tag{7.4}$$

where P_0 is the spin period of the pulsar at the time it became a dipole rotator, generally presumed to coincide with the supernova event. For a simple vacuum dipole spin-down model, $n = 3$. For $P_0 \ll P$ and $n = 3$, Eq. (7.4) reduces to $\tau_c = P/2\dot{P}$, the often-used pulsar characteristic age. For the five pulsars for which a constant value of n has been measured (Lyne *et al.* 1988, 1996; Kaspi *et al.* 1994; Deeter *et al.* 1999; Camilo *et al.* 2000), the observed values are in the range 1.4–2.91. Though pulsars clearly do not rotate like perfect vacuum dipoles, the range of braking indexes is limited and observationally well constrained.

The situation is less clear for the initial spin period, P_0. This can be determined from Eq. (7.4) if the age is known and n measured. This is only the case for the Crab pulsar, whose estimated $P_0 \sim 19$ ms has led to the generally made assumption that $P_0 \ll P$ for all but the very fastest pulsars. However, the initial spin period distribution of neutron stars is not well predicted by theory, since the rotation rates of the cores of the massive progenitors are largely unknown (Endal & Sofia 1978). Also, circumstances at core collapse could significantly affect the neutron-star spin independent of the angular momentum properties of the progenitor (e.g., Spruit & Phinney 1998). As an example, consider PSR J1811−1925 in the supernova remnant G11.2−0.3 (Fig. 7.16; Torii *et al.* 1999; Kaspi *et al.* 2001b). The supernova remnant age is well determined to be ~2 kyr, while $\tau_c = 24$ kyr, implying $P_0 \simeq P = 65$ ms for reasonable n.

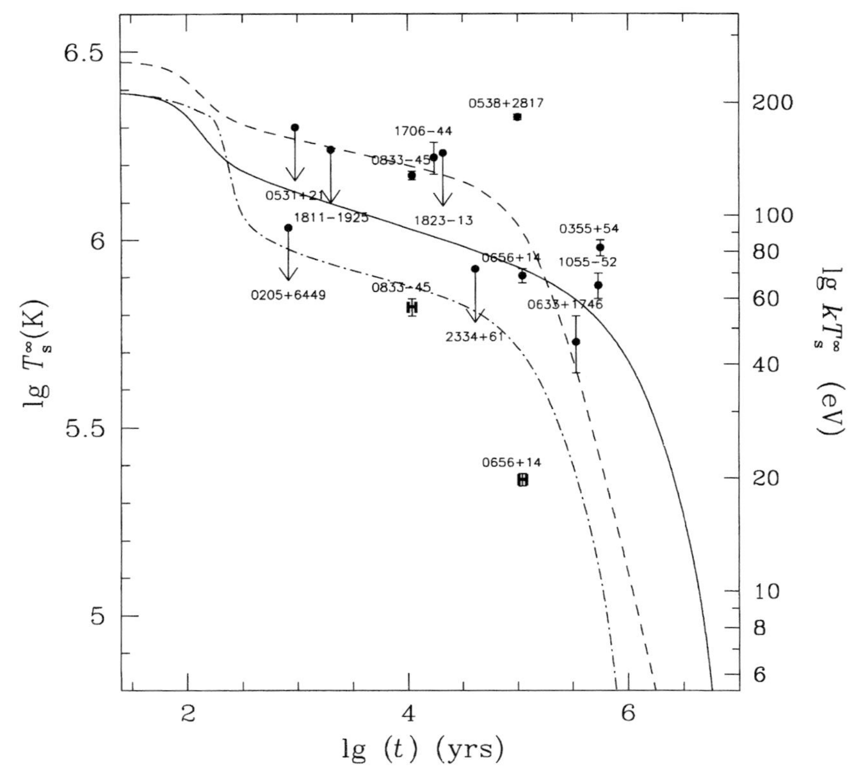

Fig. 7.11. Current best measured rotation-powered pulsar temperature as a function of best age estimate (see Table 7.2). "H" symbols denote that a $B = 0$ hydrogen atmosphere model was used; otherwise, a $B = 0$ blackbody model was assumed. The cooling curves shown are for 1.35 $M_\odot$ stars, as in Fig. 7.8: (solid) non-superfluid star with $B = 0$ and no accreted mass (Fig. 7.8, *left*); (dot-dash) a combination proton and neutron superfluidity model (Fig. 7.8, *middle*); (dash) non-superfluid star with $B = 0$ and accreted envelope mass 10^{-7} $M_\odot$ (Fig. 7.8, *right*). The curves are illustrative only and do not represent fits to the data.

In this case, τ_c is certainly over an order of magnitude different from the true age. Similar arguments hold for PSR J0205+6449 in the historic supernova remnant 3C 58 (Murray *et al.* 2002).

7.3.4 Current status of observations vs. neutron star cooling theory

Chandra and XMM-Newton have made by far the best measurements of neutron-star spectra thus far, and so offer the best opportunity for constraining models of neutron-star cooling. The current status of observations of thermally cooling young neutron stars has recently been reviewed by Kaminker *et al.* (2002), Yakovlev *et al.* (2002b) and Tsuruta *et al.* (2002), and is summarized in Fig. 7.11. There, the data points are actual temperature measurements or upper limits; those marked with an "H" are for hydrogen atmosphere models, while the rest are for simple blackbody models. The uncertainty from possible nebular contamination is not indicated; for sources with nebulae, the observations made with Chandra, given its superb spatial resolution, are superior to those with other telescopes. Age uncertainties are not indicated, but as discussed in Section 7.3.3.5, can be an order of magnitude. For PSRs J0538+2817, B0656+14, and B0833−45 (Vela), we show both the blackbody and hydrogen atmosphere temperatures, to illustrate the substantial difference. The plotted data are summarized in Table 7.2.

A striking property of the data is their scatter. Although the above caveats must be applied, the plot remains log–log, and it is hard to explain all the data with identical neutron stars sharing a common mass and hence EOS and interior structure. Kaminker *et al.* (2002) argue that on the basis of the data together with models that account for most of the phenomena affecting cooling (they do not consider "exotic" particles, but do consider neutron and proton superfluidity in detail), there are effectively three types of cooling neutron stars (see also Tsuruta *et al.* 2002): (i) low-mass, slowly cooling stars where there is no direct Urca process occurring; (ii) medium-mass, moderately cooling stars in which some nucleon direct Urca processes occur, but for which proton superfluidity suppresses the bulk of the possible Urca processes; and (iii) high-mass, rapidly cooling stars for which suppression of direct nucleon Urca processes is weak. Of course the mass cuts corresponding to these different categories depend strongly on the EOS and on the possible types of superfluidity. As an example, for the EOS "A" of Yakovlev *et al.* (2001), for neutron stars having age 4 kyr, for a 2p proton superfluid, the low-mass/medium-mass cut is 1.36 $M_{\odot}$, while the medium/high mass cut is 1.64 $M_{\odot}$.

What is clear is that we have not yet seen any very rapidly cooling objects. However this may be less of a statement about the interior physics as it is about the mass function of neutron stars, i.e., there may be very few high-mass neutron stars produced for purely astrophysical reasons. According to Kaminker *et al.* (2002), the Vela pulsar (PSR B0833−45, fit with a hydrogen atmosphere), and Geminga (PSR J0633+1746, fit with a blackbody), as well as presumably PSR J0538+2817 (with a hydrogen atmosphere), the coolest objects, are prime examples of moderately cooling stars for which some direct Urca reactions are occurring, and their masses could be determined assuming some EOS, a model of proton superfluidity (but see Tsuruta *et al.* 2002) and the dependence of the critical temperature for proton superfluidity on density. Note that the hydrogen-atmosphere model fit to PSR B0656+14 is generally ignored, as it yields a radius much greater than is possible for a neutron star. The apparently higher temperature objects, like PSR B1055−52, must, in this case, be low-mass neutron stars, possibly affected by some internal heating mechanism or by polar-cap heating. The pulsar in 3C 58 recently received fanfare for its relatively low temperature in spite of having likely been born in AD 1181, hence being very young (Slane *et al.* 2002). Though the Vela and Geminga pulsars nominally have lower reported temperatures, hence might be argued as being closest to rapidly cooling objects (Yakovlev *et al.* 2002b), in fact the 3C 58 upper limit was for a blackbody spectrum. A hydrogen-atmosphere fit to the data (as was done for Vela) could possibly bring the upper limit down to a very constraining level. Indeed, the Vela temperature when fit with a blackbody is well above that of the 3C 58 pulsar.

7.4 Isolated neutron stars

A relatively recently discovered class of X-ray sources is believed to be thermally cooling neutron stars. Their nature is still not well understood. These are currently being referred to in the literature as "isolated neutron stars" (INSs). This at first glance may seem like a misnomer since this title does not distinguish them from isolated rotation-powered pulsars. Actually, as we describe below, it is possible that at least a subset of the INSs are

Table 7.2. *Observations of thermally cooling young rotation-powered pulsars*[a]

Pulsar name	SNR	τ_c/SNR age (kyr)	d (kpc)	N_h (10^{21} cm^{-2})	kT^∞ (eV)	R^∞ (km)	$L^{\infty b}$ (erg s^{-1})	Model[c]	Telescope	Reference[d]
J0205+6449	3C 58	5.4/0.82	(3.2)	(3.86)	<93	(12)	$<1.4 \times 10^{33}$	BB	Chandra	Slane *et al.* 2002
B0355+54	–	562/–	(2.1)	(2)	82 ± 4	(10)	$5.8 \pm 0.3 \times 10^{32}$	BB	ROSAT/Einstein	Slane 94
B0531+21	Crab	1.3/0.949	(2)	(3)	<172	(15.6)	$<2.7 \times 10^{34}$	BB	Chandra	Weisskopf *et al.* 03
J0538+2817	S147	620/∼100	(1.2)	(2.51)	183 ± 3	1.68 ± 0.05	4.6×10^{32}	BB	XMM	McGowan *et al.* 03
J0633+1746	–	340/–	0.16	0.08–0.18	38–54	4–8	$1.5-1.7 \times 10^{31}$	BB	ASCA/ROSAT	Halpern & Wang 97
B0656+14[e]	–	110/–	0.288	0.17 ± 0.02	69 ± 3	8.5	2.1×10^{32}	BB	Chandra	Brisken *et al.* 03b
B0833−45	Vela	11/∼10	0.21	0.33 ± 0.03	57 ± 3	(13)	$2.6 \pm 0.2 \times 10^{32}$	*B*-H	Chandra	Pavlov *et al.* 01a
B1055−52[e]	–	540/–	(0.7)	?	65	13	3.8×10^{32}	BB	Chandra	Pavlov & Zavlin 03
B1706−44	G343.1−2.3?	17.4/?	(2.5)	5.5	143^{+14}_{-13}	3.6 ± 0.9	6.8×10^{32}	BB	Chandra	Gotthelf *et al.* 02
J1811−1925	G11.2−0.3	24/2	(5)	(21.3)	<150	(10)	6.5×10^{33}	BB	Chandra	this work
B1823−13	–	21/–	(4)	(20)	<147	(12)	$<8.7 \times 10^{33}$	BB	XMM	Gaensler *et al.* 03b
B2334+61	G114.3+0.3	41/?	(2.5)	(2)	<72	(10.4)	$<6 \times 10^{32}$	BB	ROSAT	Becker *et al.* 96

[a] Numbers in parentheses represent values assumed by the authors.
[b] All luminosities are bolometric.
[c] BB is a blackbody model; *B*-H is a magnetic hydrogen atmosphere model.
[d] Most recent relevant reference; references contained therein may also be relevant.
[e] The spectrum is best fit with two blackbodies. We report here the softer of the two, as that component has inferred radius comparable to the size of the neutron star, as expected for whole-surface thermal emission.

Table 7.3. *Reported isolated neutron stars*

Name	P (s)	ROSAT PSPC rate (cps)	m_V	$\log f_X/f_{\rm opt}$
RX J0420.0−5022	3.45	0.11	–	>3.3
RX J0720.4−3125	8.39	1.69	26.8	4.6
RX J0806.4−4132	11.4	0.38	–	>3.4
RX J1308.6+2127	10.31	0.29	28.7	5.0
RX J1605.3+3249	–	0.88	27.1	4.4
RX J1856.5−3754	–	3.64	25.8	4.4
1RXS J214303.7+065419	–	0.18	–	>3

Data from Treves *et al.* (2000), Kaplan *et al.* (2003a), and Haberl (2003, 2004); see text for more information.

conventional rotation-powered pulsars. However, the sources known as INSs are currently distinguished as showing no evidence for radio pulsations, nebulosity, or accretion, and X-ray emission characterized by very soft spectra that are well described by a blackbody, with no apparent magnetospheric contributions. These very properties, as well as their proximity, make them enticing targets, since many of the observational difficulties inherent in using thermal emission to constrain the nuclear EOS that are described above in Section 7.3 are not present. The name INS therefore does apply to rotation-powered pulsars as well, however the latter are a very special class of the INS population: long after rotation-powered pulsars have spun down past the "death line," beyond which they no longer produce observable magnetospheric emission (this typically takes $\sim$30 Myr), "dead" neutron stars will still populate the Galaxy. INSs, if identified as this "dead" component, may well number as high as 10^9 in the Galaxy, much more than the $\sim 10^5$ active radio pulsars (Lyne *et al.* 1998).

Prior to the launch of the ROSAT satellite, theoretical predictions following an idea first put forth by Ostriker *et al.* (1970) were that several thousand otherwise dead neutron stars would be detectable in soft X-rays by ROSAT, due to their accretion of material from the interstellar medium (Treves & Colpi 1991; Blaes & Madau 1993). In fact, well after the rise and fall of ROSAT, only seven objects (Table 7.3; see Treves *et al.* 2000, for a review) having the predicted properties (such as high X-ray to optical flux ratio) have been found in spite of very careful searches (e.g. Danner 1998a,b; Rutledge *et al.* 2003). This is likely due mainly to the much higher than expected average neutron-star space velocity (e.g. Lyne & Lorimer 1994; Hansen & Phinney 1997; Arzoumanian *et al.* 2002), which hampers standard Bondi–Hoyle accretion, although other explanations have been suggested (e.g. Perna *et al.* 2003).

In fact, of the objects in Table 7.3, it now appears likely that *none* is accreting from the ISM; rather, they are visible because of thermal emission from the neutron-star surface. If so, these are unlikely to be extremely old neutron stars as they are too bright; rather they would have to be either off-beam conventional rotation-powered pulsars having ages $\sim 10^6$ yr and magnetic fields 10^{11}–10^{12} G, or even younger quiescent magnetars having magnetic fields 2–3 orders of magnitude higher.

Although 4/7 of the INSs have now been shown to exhibit pulsations, none has had its spin-down rate measured. Their ages are thus not known, which hinders efforts to use them in cooling studies. However, in these sources, one hope is that spectral modeling, coupled with sensitive, high-resolution X-ray observations, could determine their radii. If spectral lines were detectable as well, the stellar masses could in principle be known independently from the lines' gravitational redshift (Section 7.3.3.4), thus providing a direct constraint on the neutron-star mass/radius relation and testing EOSs. Spectral modeling in these sources is greatly aided by their low equivalent neutral hydrogen column densities, typically 1–2 $\times$ 10^{20} cm^{-2}. Although these ambitious goals have not yet been realized, the recent discoveries of absorption features in the spectra of three INSs are quite encouraging (Haberl *et al.* 2003, 2004; van Kerkwijk *et al.* 2004).

The current situation is nicely illustrated by the example of RX J1856.5$-$3754, which is the best studied of the INSs. This object was discovered in ROSAT data (Walter *et al.* 1996) and optical observations with the Hubble Space Telescope revealed a faint blue counterpart (Walter & Matthews 1997). This suggested emission that is the Rayleigh–Jeans tail of a $kT \simeq 50-60$ eV blackbody. The HST detection enabled an astrometric study of the source that detected both parallax and proper motion (Walter 2001). The source is evidently extremely close to the Sun, having parallax 8.5 ± 0.9 mas (Kaplan *et al.* 2002a; Walter & Lattimer 2002) or distance 117 ± 12 pc. The proper motion of the source, if traced backward, appears to intersect the Upper Sco OB association, if it has a radial velocity of $\sim$30 km s^{-1} (Walter 2001; Kaplan *et al.* 2002a; Walter & Lattimer 2002). Assuming the neutron star was born in the association implies an age of 0.4 Myr.

Because of the proximity of the source and the chance for the discovery of spectral features, a 500 ks Director's Discretionary Time observation with the Chandra X-ray Observatory was performed in 2002. The X-ray emission was found to be unpulsed, with stringent upper limits set on the pulsed fraction ($<$1.3% for frequencies less than 50 Hz; Burwitz *et al.* 2003). The spectral results of the DDT observation were a surprise. Not only were no features detected, but the resulting X-ray spectrum is extremely well modeled by a simple blackbody, with no evidence for the atmospheric distortion so clearly predicted by theory (see Section 7.3.2.5; Drake *et al.* 2002; Braje & Romani 2002; Burwitz *et al.* 2003). In particular, the emitting radius for a single blackbody fit is $4.3(d/117$ pc) km, smaller than expected for any plausible neutron star. This led to speculation that the object is not a neutron star but rather a self-bound quark or strange star (Drake *et al.* 2002). A very recently revised parallax distance of 175 pc would imply a radius of 7.2 km, more in line with expectations for a neutron star (D. Kaplan, personal comm.). In any case, it is possible that the emitting region is simply a hot spot on a larger stellar surface; this is not completely precluded by the low pulsed fraction upper limits because of the strong effects of gravitational light bending (Burwitz *et al.* 2003). Interestingly, the optical emission is too bright to be the low-energy extrapolation of the X-ray spectrum. A two-component blackbody model is necessary: the lower temperature, perhaps from the bulk of the surface and accounting for the optical emission, has $kT^{\infty} < 33$ eV, and the higher temperature, perhaps from a hot spot and accounting for the X-ray emission, has $kT^{\infty} \simeq 63.5$ eV. In this case, a strange star is not required (Drake *et al.* 2002; Braje & Romani 2002; Burwitz *et al.* 2003). Of course, the standard caveats regarding the use of blackbody models apply (see Section 7.3.2.5). Van Kerkwijk and Kulkarni (2001) discovered, using Hα imaging, a cometary shaped nebula having RX J1856.5$-$3754 at its apex, similar to those detected around some conventional radio pulsars (see Section 7.6). This suggests that RX J1856.5$-$3754 is a conventional neutron star whose emission beam does not intersect our line of sight.

Other sources in Table 7.3 are not yet as well characterized as RX J1856$-$3754. We discuss each briefly, in order of the degree to which it has been studied.

RX J0720.4−3125 is an 8.39-s soft-X-ray pulsar (Haberl *et al.* 1997) plausibly identified as an off-beam conventional rotation-powered pulsar and not an accretor (see Kaplan *et al.* 2003b, and references therein). No spin-down rate has yet been determined, although interesting upper limits on $\dot{P}$ have been established, and imply that $B \lesssim 3.5 \times 10^{13}$ G (Cropper *et al.* 2004). Its optical through X-ray spectrum is well modeled by two blackbodies (as in RX J1856−3754; Motch *et al.* 2003) plus a possible power-law component, similar to what is observed in young and middle-aged radio pulsars (e.g., PSR B0656+14). A large inferred velocity seems to preclude it being an accretor (Motch *et al.* 2003). Intriguingly, Haberl *et al.* (2004) recently report the discovery of a phase-dependent absorption line in this pulsar's spectrum. The line has energy ~271 eV and an equivalent width that varies from −31 eV to −58 eV depending on pulse phase. The line appears best interpreted as cyclotron resonance of charged particles, presumably either electrons or protons. For the latter, a $\sim 5 \times 10^{13}$ G magnetic field is implied. For electrons, the field must be 2000 times smaller.

RX J1308+2127 (RBS 1223) was identified by Schwope *et al.* (1999) in ROSAT data as a relatively bright INS candidate, based on its soft spectrum and the high X-ray to optical flux ratio made possible by the good positional localization. Haberl *et al.* (2003) showed the source is pulsed with $P = 10.3$ s. Kaplan *et al.* (2002b) reported a possible optical counterpart having flux in excess of the extrapolation of the X-ray blackbody spectrum, as in RX J1856−3754 and RX J0720.4−3125. A broad absorption line feature has been reported in this source (Haberl *et al.* 2003), and interpreted as a proton cyclotron line in a $> 10^{13}$ G field. Haberl *et al.* (2003) reported the discovery of a broad absorption feature in this object's X-ray spectrum. The line energy is at ~300 eV and has equivalent width ~−150 eV. This line is interpreted by the authors as being due to cyclotron resonance absorption, in the $10^{13} - 10^{14}$ G range if protons, and in the $10^{10} - 10^{11}$ G range if electrons.

RX J1605.3+3249 (RBS 1556) was identified by Motch *et al.* (1999) as being a possible INS, because of the soft source spectrum ($kT \simeq 92$ eV), the lack of X-ray variability, and the absence of any optical counterpart. Kaplan *et al.* (2003a) reported the detection of a faint, blue optical counterpart, which they argue confirms the INS interpretation. The optical counterpart has flux that is over an order of magnitude above the Rayleigh–Jeans extrapolation of the X-ray blackbody. This is like, though somewhat larger than, the excesses seen in the other sources, and argues either for the X-rays being from a hot polar cap or for the optical emission having a non-thermal component. Either is consistent with an off-beam rotation-powered pulsar. Very recently, Kerkwijk van *et al.* (2004) report the detection of a broad absorption feature at 450 eV. As for RX J0720.4−3125 and RX J1308+2127, the origin of the line is as yet unclear, although Ho and Lai (2004) show on the basis of approximate calculations that all can plausibly be explained by proton cyclotron absorption.

RX J0806.4−4132 was reported by Haberl *et al.* (1998) as an interesting INS candidate. It was seen in two ROSAT All Sky Survey observations at the same flux, and having a soft spectrum, with $kT \simeq 78$ eV. Haberl *et al.* (1998) detected no optical counterpart, indicating a high X-ray to optical flux ratio. Haberl and Zavlin (2002) using XMM-Newton observations of this source, report a possible 11.4 s periodicity at the 3.5σ level having a small, ~6%, pulsed fraction, and a blackbody temperature of $kT = 94$ eV. The improved XMM-Newton position should allow deeper searches for a possible optical counterpart.

RX J0420.0−5022 is a ROSAT soft-spectrum source showing evidence for a 22.7 s periodicity, a high X-ray to optical flux ratio, no apparent variability, and a soft spectrum that is well described by a blackbody of temperature ~57 eV (Haberl *et al.* 1999). All argue for it

Table 7.4. *Central compact objects*

Name	SNR	P	$kT^{\infty}_{\rm bb}$ (keV)	$R^{\infty}_{\rm bb}$ (km)	kT^{∞} (keV)	R^{∞} (km)
CXO J082157.5−430017	Pup A	–	0.4	1.4	0.2	10
CXO J085201.4−461753	G266.1−1.2	–	0.4	0.3	0.3	1.5
RX J121000.8−522625	G296.5+10.0	0.424 s	0.25	1.6	0.1	1
CXO J161736.3−510225	RCW 103	6 hr ?	0.4–0.6	0.2–1.6	0.3	1–8
CXO J232327.9+584843	Cas A	–	0.5	0.5	0.3	1

See Pavlov *et al.* (2002a), Pavlov *et al.* (2002b) and references therein.
Note: $kT^{\infty}_{\rm bb}$ and $R^{\infty}_{\rm bb}$ refer to blackbody models, while kT^{∞} and R^{∞} refer to magnetic hydrogen atmosphere models.

being an INS, although the period is much longer than has been seen for any rotation-powered pulsar.

7.5 Central compact objects

There are several putative compact objects that have been studied extensively and have potential to constrain neutron-star cooling theory (Section 7.3), yet are poorly understood. By definition, these sources, known as "central compact objects" (CCOs), are found near the centers of supernova remnants (SNRs), although this could well be a selection effect resulting from SNR X-ray studies. The CCOs share the following distinguishing properties: unusual X-ray spectra that, fit with blackbodies and, in some cases, hydrogen atmospheres, imply very small radii, and very high effective temperatures (see Table 7.4). The small radii are inconsistent with these objects being neutron stars for any EOS, while the temperatures are inconsistent with any neutron-star cooling models. Other distinguishing characteristics are high X-ray to optical luminosity ratios, no evidence for pulsations (except in one case – see below), no evidence for a wind as seen in conventional young rotation-powered pulsars (Section 7.6), and no evidence for any companion star that could be powering the X-ray emission via accretion. We discuss the five best studied CCOs briefly below. For a more detailed review, see Pavlov *et al.* (2002a).

The "first light" of Chandra targeted the 300-year old supernova remnant Cassiopeia A, and revealed, for the first time, the presence of an X-ray point source near the center (Fig. 7.12 (see also color plate section); Tananbaum 1999). This object, CXO J232327.9+584843, seems likely to be the long-sought-after compact object formed in this oxygen-rich remnant, clearly a result of the supernova explosion of a massive star. The spectral properties of this source, as well as the absence of any detectable pulsations, make it very different from canonical young neutron stars like the Crab pulsar. Specifically, Pavlov *et al.* (2000) and Chakrabarty *et al.* (2001) showed that for either a simple blackbody or hydrogen atmosphere model, the temperature of the emission $kT = 0.25-0.35$ keV is very high, and the emitting area ($R = 0.2-0.5$ km) too small to be consistent with surface thermal emission from initial cooling. A power-law fit yields a photon index $\Gamma = 2.6-4.1$, significantly higher than those seen in pulsars. The difference is further evidenced by the absence of any synchrotron nebula around the object (Fig. 7.12), as well as the absence of any radio emission from the source (McLaughlin *et al.* 2001). The very high X-ray to optical flux ratio, however, does implicate a

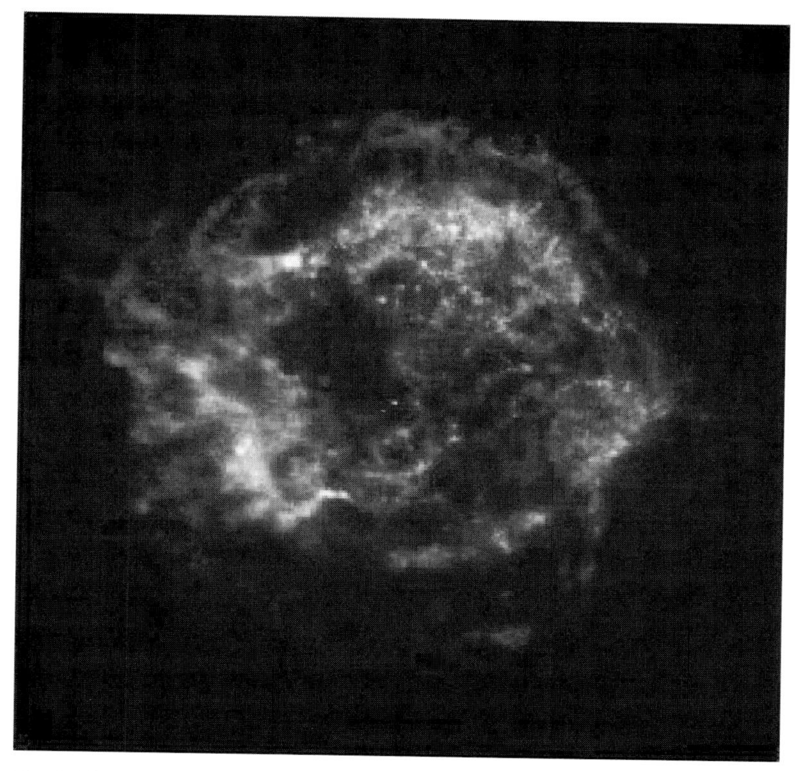

Fig. 7.12. "First light" Chandra X-ray image of the supernova remnant Cas A (after Tananbaum 1999). The image is 6′ on a side. The previously unknown point source at the center of the remnant is obvious. For color version see Plate 7.12.

compact object and render an accreting binary scenario problematic (Kaplan *et al.* 2001; Ryan *et al.* 2001). The source spectrum is consistent with it being an "anomalous X-ray pulsar" (see Chapter 14), that is, endowed with an ultrahigh magnetic field ($10^{14} - 10^{15}$ G), or with it being a conventional $B \sim 10^{12}$ G neutron star having a cooling iron surface and hot hydrogen or helium polar caps (Pavlov *et al.* 2000; Chakrabarty *et al.* 2001). More recent XMM-Newton observations of the source support these conclusions (Mereghetti *et al.* 2002b). However, if it is an AXP, why it should be so faint and show no pulsations, and, if it has hot polar caps, why they should be so hot in the absence of any evidence for the pulsar mechanism, is a mystery.

CXO J082157.5−430017 is located in the Puppis A SNR, though significantly off-center (Petre *et al.* 1982). From ROSAT and ASCA observations (Petre *et al.* 1996; Zavlin *et al.* 1999) the object's X-ray spectrum is well modeled by a blackbody, however the best-fit radius, 1.4 km, is much smaller than that of a neutron star. A magnetized ($B \gtrsim 6 \times 10^{12}$ G) hydrogen atmosphere model yields a more neutron-star-like radius of 10 km, and a temperature more in line with that expected from cooling emission (Zavlin *et al.* 1999). However, neither pulsations nor evidence for a surrounding nebula has been seen (Gaensler *et al.* 2000a; Pavlov *et al.* 2002a), arguing that if this is a rotation-powered pulsar, it must have very low $\dot{E}$. This argues against the heated-polar-cap interpretation, as in Cas A.

SNR G266.1−1.2, also known affectionately as "Vela Jr," is in the direction of the southeast corner of the Vela SNR (Aschenbach 1998). The possible detection of a radioactive ^{44}Ti line suggested an age of ~700 yr, and a distance of ~200 pc. However, subsequent ASCA observations yielded N_h more consistent with a distance of $\gtrsim 1$ kpc, and an age of several thousand years (Slane *et al.* 2001). The point source CXO J085201.4−461753 is near the center of the SNR, and has an X-ray to optical flux ratio only consistent with a compact object identification (Pavlov *et al.* 2001b). This source is not seen to pulsate and shows no evidence of a pulsar wind nebula in Chandra data (Kargaltsev *et al.* 2002). The source spectrum, fit with a blackbody model, yields a radius of only ~0.3 km. Atmosphere models do not yield more reasonable radii (Kargaltsev *et al.* 2002).

RX J121000.8−522625 is a compact source near the center of the large SNR G296.5+10.0 (also known as PKS 1209−51/52). The source was first noticed in Einstein IPC observations made by Helfand and Becker (1984). Mereghetti *et al.* (1996) used ROSAT to show that a blackbody model fit the spectral data well. They further demonstrated the absence of any extended emission, and set an optical limit of $V \gtrsim 25$. They did not detect any radio emission from the source at 4.8 GHz, and Kaspi *et al.* (1996) saw no pulsed radio emission at 436 MHz. Vasisht *et al.* (1997) and Zavlin *et al.* (1998) presented ASCA observations of the source which again supported the neutron-star interpretation. Chandra observations suggested a periodicity in the source, with $P = 424$ ms (Zavlin *et al.* 2000). Pavlov *et al.* (2002c), using a second *Chandra* observation, reported a spin-down age of 200–900 kyr, much larger than the 3–20 kyr age inferred for the SNR. However, subsequent observations have not found simple spin down as in rotation-powered pulsars, unless the source is an active glitcher (Zavlin *et al.* 2004). This is currently not understood. The source spectrum is thermal, and the best-fit continuum blackbody model yields a small 1.6 km emitting radius (Pavlov *et al.* 2002c). This source also shows absorption features in its X-ray spectrum (Fig. 7.10; Section 7.3.3.4).

1E 161348−5055, the central source in the supernova remnant RCW 103 was the first proposed cooling neutron star (Tuohy & Garmire 1980). Gotthelf *et al.* (1999) showed that the source is X-ray variable by as much as an order of magnitude on a timescale of years. A ~6 hr periodicity was tentatively reported in data from Chandra and ASCA observations (Garmire *et al.* 2000), and a 2002 Chandra observation showed a clear 6.4 hr periodicity and apparently phase-correlated X-ray flux variations (Sanwal *et al.* 2002a). This strongly suggests that this object is, in fact, an accreting neutron star having a very low-mass companion.

7.6 Pulsar wind nebulae

7.6.1 *Physical overview*

Only a small fraction ($\lesssim 10\%$) of the spin-down energy of a young pulsar is converted into observable pulsed emission (Section 7.2). It is generally accepted that most of the energy leaves the pulsar's magnetosphere in the form of a magnetized wind (Michel 1969). In the ideal ($\mathbf{E} \cdot \mathbf{B} = 0$) MHD approximation to the aligned rotating magnetic dipole (Goldreich & Julian 1969; Contopoulos *et al.* 1999), it appears that the poloidal field ($B_{\rm p}$, the field perpendicular to the toroidal component B_ϕ) goes from a nearly dipolar structure within the light cylinder to a split monopole structure outside the light cylinder (i.e., $B_{\rm p} = B_{\rm r} \propto r^{-2}$, a purely radial field whose sign abruptly changes at the equator). The toroidal component B_ϕ, which is small out to the light cylinder, grows rapidly outside the light cylinder so that for radii greater than a few times $R_{\rm LC}$, $B_\phi \gg B_{\rm p}$. "Cold" (i.e., non-radiating) charged particles flow outward with this magnetic field forming a magnetized wind which is ultimately accelerated to very high energies ($\gamma \gg 1$). This highly relativistic magnetized wind eventually interacts with the surrounding medium, and emits synchrotron radiation from radio to gamma-ray wavelengths. This synchrotron emission caused by the pulsar wind is what is generally meant by the term pulsar wind nebula (PWN). However, the winds from rapidly moving pulsars may produce a bow shock in the ISM causing Hα emission in what is essentially a thermal process. Optical filaments in the Crab nebula also appear to be thermal emitters (e.g., Fesen & Kirshner 1982), and in principle there could also be thermal X-ray emission at the outer edges of young PWNe, although this is difficult to distinguish from the surrounding supernova remnant (SNR) and has yet to be observed unambiguously.

The details of the structure and luminosity of the PWN should depend on the pulsar's spin-down energy history and space velocity as well as the density profile of the surrounding medium. They may also depend upon the magnetic inclination angle of the pulsar, although how the wind properties depend on this angle is poorly understood and it is possible that the deviation from the aligned rotator model is negligible in terms of overall growth and energetics of the PWN. What we observe is dependent on the observer's viewing angle, with Doppler boosting being a non-negligible factor in the surface brightness distribution of the PWN.

Since a pulsar is born in a supernova, the initial environment that the pulsar wind encounters is the (nearly) freely expanding supernova ejecta. The pulsar may be born with a large space velocity and overtake the expanding ejecta shell within a few tens of thousands of years, after which the wind environment is probably the ISM. There appears to be a minimum spin-down energy necessary to create bright PWNe with a significant drop in PWN emission efficiency from pulsars with $\log \dot{E} \lesssim 36$ (Frail & Scharringhausen 1997; Gaensler *et al.* 2000b; Gotthelf 2003), and generally only the youngest isolated pulsars are observed to have PWNe. However, millisecond pulsars can also have a substantial $\dot{E}$, and both Hα (Bell *et al.* 1995) and X-ray (Stappers *et al.* 2003) nebulae have been observed around them.

The standard theoretical picture for PWNe from young pulsars (Pacini & Salvati 1973; Rees & Gunn 1974; Reynolds & Chevalier 1984; Kennel & Coroniti 1984a) is of a synchrotron bubble being blown at the center of an expanding supernova remnant. Initially, the outer edge expands supersonically into the supernova ejecta, so is presumably bounded by a forward shock ($R_{\rm P}$ of Fig. 7.13). The corresponding reverse shock ($R_{\rm T}$) is near the center of the nebula, where the cold relativistic wind from the pulsar is terminated. (Note that the wind is cold only in the sense that $kT \ll (\gamma - 1)mc^2$ and hence in the comoving frame of the bulk flow the particles have very little time to radiate before encountering the shock.) The bulk flow energy is then converted into random particle motion with a power-law distribution of particle energies. The wind continuously injects high-energy electrons and positrons as well as magnetic field into the bubble. The particles gyrate in the magnetic field, emitting synchrotron radiation from the radio through soft gamma-ray regions of the spectrum.

7.6.2 *Observational properties of PWNe*

Except for a few cases (notably the Crab and Vela nebulae), we have data on PWN emission only at centimeter wavelengths and/or in X-rays below $\sim$10 keV. The radio properties of PWNe have been outlined by Weiler and Shaver (1978) among others. A source can be identified as a radio PWN if it has an amorphous (i.e., non-shell) morphology, flat

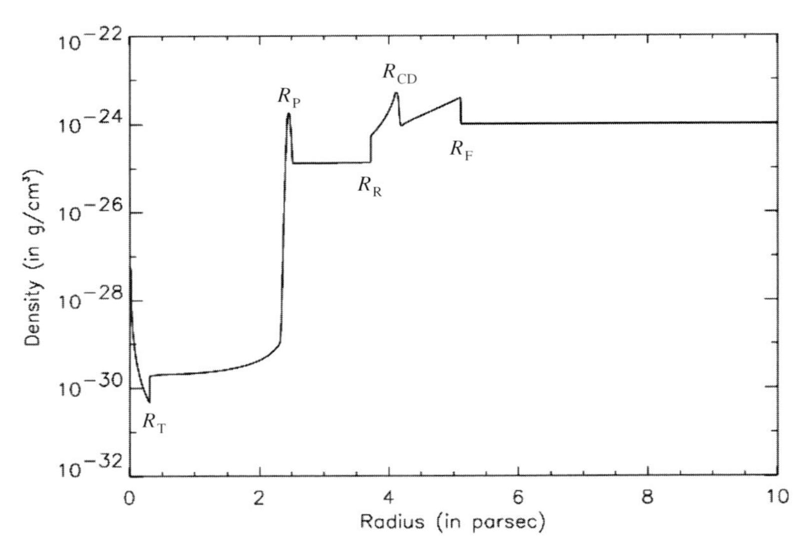

Fig. 7.13. Density profile of a spherically symmetric PWN expanding within a pre-Sedov SNR showing the PWN termination shock R_T, the PWN forward shock R_P, the SNR shell reverse shock R_R, the contact discontinuity between the SNR ejecta and the ISM R_{CD}, and the SNR shell forward shock R_F. (Adapted from van der Swaluw *et al.* 2001)

non-thermal spectrum (energy spectral index $\alpha \simeq 0.0-0.3$, where flux $S = \nu^{-\alpha}$[3]), and high ($\gtrsim$10%) fractional linear polarization. The X-ray emission also has a spectrum described by a power law, except the photon index $\Gamma \equiv 1 + \alpha$ is generally between $\sim$1.5 and 2.2. Hence there is a need for one or more spectral breaks such that the net change in spectral index is $\Delta\alpha = 0.5-0.9$. The X-ray nebula tends to be smaller than the radio nebula, and generally has a more defined morphology, sometimes taking the form of a thick torus with perpendicular jets.

The efficiency with which the spin-down energy is converted into synchrotron radiation can be quite high. However, it is often difficult to determine since the observations tend to be restricted to two narrow bands. In general, one can find four observed quantities, with associated errors, quoted in the literature (or minor variants): radio flux density at $\sim$1 GHz, $S_{1\text{GHz}}$, radio energy spectral index α_r, 1–10 keV X-ray flux F_X (corrected for interstellar absorption), and X-ray photon spectral index $\Gamma = 1 + \alpha_X$. From these, one can infer a break frequency where the spectral slope changes:

$$\nu_b \equiv \left(\frac{S_X \nu_X^{\alpha_X}}{S_r \nu_r^{\alpha_r}} \right)^{1/(\alpha_X - \alpha_r)} \qquad (7.5)$$

where S_X and S_r are the flux densities at ν_X and ν_r, respectively. Possible physical meanings of this break frequency will be discussed in Section 7.6.4. At centimeter wavelengths, the spectral index $\alpha_r < 1$, hence the bulk of the energy emitted in this band is at higher frequencies. The radio luminosity L_r therefore depends on what is chosen as the upper frequency limit ν_{ul} of

[3] In radio work, the spectral index is often defined as $S = \nu^{\alpha}$. However, we use the minus sign here to be consistent with the X-ray definition.

the radio band:

$$L_r = 1.1 \times 10^{30} \frac{S_{1\text{GHz}} d^2 \nu_{ul}^{1-\alpha_r}}{1-\alpha_r} \text{ erg s}^{-1} \tag{7.6}$$

where d is in kpc, $S_{1\text{GHz}}$ is in Jansky, and ν_{ul} is in GHz. A natural upper limit would be the break frequency, but that is usually not well known due to uncertainties in the radio and X-ray spectral indices and the large extrapolation. Often, an upper limit of 100 GHz is randomly chosen in order to compare different nebulae (Gaensler *et al.* 2000b).

The X-ray luminosity L_X can be defined as the total power emitted from the break energy up to a high-energy cutoff E_C:

$$L_X = 1.1 \times 10^{32} \frac{F_{X12} d^2 \left(E_C^{2-\Gamma} - E_B^{2-\Gamma}\right)}{10^{2-\Gamma} - 1} \text{erg s}^{-1} \tag{7.7}$$

where F_{X12} is the 1–10 keV flux in units of 10^{-12} erg cm^{-2} s^{-1} and E_C and $E_B = h\nu_b$ are in keV. If the photon index $\Gamma < 2$, the X-ray luminosity depends critically on the high-energy cutoff, which is only known in a very few cases. For the Crab and Vela PWNe, it is in the few to tens of MeV range. If $\Gamma > 2$, then the bulk of the energy is emitted below the X-ray band, peaking near the break frequency. Note that the X-ray luminosity is usually much higher than the radio luminosity. Since the synchrotron lifetime of the X-ray emitting particles is often much shorter than the age of the nebula, the properties of the X-ray emission can be highly dependent on the current spin-down energy of the pulsar. The radio emitting particles generally have synchrotron lifetimes longer than the age of the nebula, and are therefore an indication of the total number of lower-energy relativistic particles injected into the nebula throughout its lifetime.

All known pulsars with detected X-ray or radio PWNe are listed in Table 7.5. Table 7.6 lists other X-ray or radio PWNe from which pulsations have not been detected, while Table 7.7 lists pulsars with Hα PWNe. Note that the isolated neutron star RX J1856.5$-$3754 also has a Hα nebula associated with it (see Section 7.5). A source is considered a PWN if it has one or more of the following observational signatures: a known pulsar embedded in a radio and/or X-ray nebula that appears to be morphologically related to the pulsar; an isolated, filled-center radio nebula with the properties mentioned above; a composite supernova remnant defined as a SNR shell with either a central radio component whose spectrum is flatter than the shell, a compact but extended central X-ray component with a power-law spectrum, or both; an X-ray point source with a very high ratio of X-ray to optical flux and associated extended X-ray emission with a power-law spectrum. The X-ray morphologies are quite varied, but can be loosely placed within the following categories: torus + weak jet (T), jet/trail dominated (J), and mixed or uncertain (M). In the mixed case, sometimes there appears to be a bow shock or torus as well as a bright trail or jet-like component, but sometimes there is just an indistinct blob. Although a dominant toroid is generally assumed to be the typical structure, only a few of the brightest sources have been clearly determined to have toroidal morphologies. A reference for X-ray morphology is given where available, otherwise for the radio morphology.

There seems to be a change in PWN morphology from toroidal to more of a jet or trail when the pulsar spin-down energy drops below $\log \dot{E} \sim 36.5$ (in erg s^{-1}). This probably simply reflects the on-average greater age of these pulsars so they have moved substantially from their birth sites. The radio and X-ray PWNe largely disappear once the spin-down energy drops below $\log \dot{E} \sim 35.5$. Those few sources listed as being PWNe whose pulsars'

Table 7.5. *Known pulsars with synchrotron wind nebulae*

Name	Pulsar	$\log(\dot{E})$ ($\mathrm{erg\,s^{-1}}$)	d (kpc)	M[a]	S[b]	γ[c]	Reference
N157B	J0537−6910	38.7	50	M	?	Y	Wang *et al.* 01
Crab	B0531+21	38.7	2.0	T	N	Y	Weisskopf *et al.* 00
SNR 0540−69.3	B0540−69	38.2	50	T	Y	Y	Gotthelf & Wang 00
3C 58	J0205+6449	37.4	4.5	M	N	N	Murray *et al.* 02
G106.6+3.1	J2229+6114	37.4	4.0	T	N?	Y	Halpern *et al.* 01
G320.4−1.2	B1509−58	37.2	5.0	M	Y	Y	Gaensler *et al.* 02a
G292.0+1.8	J1124−5916	37.1	6.5	T	Y	N	Hughes *et al.* 01
G54.1+0.3	J1930+1852	37.1	9.1	T	N	N	Lu *et al.* 02
Kookaburra	J1420−6048	37.0	5.6	M	?	Y	Roberts *et al.* 01
Kes 75	J1846−0258	36.9	21	M	Y	N	Helfand *et al.* 03b
Vela X	B0833−45	36.8	0.29	T	Y	Y	Helfand *et al.* 01
G11.2−0.3	J1811−1925	36.8	5	M	Y	N	Roberts *et al.* 03
CTB 80	B1951+32	36.6	3.2	M	Y	Y	Safi-Harb *et al.* 95
G343.1−2.3	B1706−44	36.5	2.3	M	N	Y	Gotthelf *et al.* 02
GeV J2020+3658	J2021+3651	36.5	12	T	?	Y	Hessels *et al.* 03
G18.0−0.7	B1823−13	36.5	3.9	J	N	?	Gaensler *et al.* 03b
Duck	B1757−24	36.4	5.2	J	O	N	Kaspi *et al.* 01a
3EG J1027−5817	J1016−5857	36.4	8.0	M	?	Y	Camilo *et al.* 01
Mouse	J1747−2958	36.4	5.0	J	N	?	Gaensler *et al.* 04
G292.2−0.5	J1119−6127	36.4	6	M	Y	N	Gonzalez & Safi-Harb 03
G308.8−0.1	J1341−6220	36.1	11	?	Y	N	Kaspi *et al.* 92
G270.3−1.0	B0906−49	35.7	2.6	?(J)	N	N	Gaensler *et al.* 98
W44	B1853+01	35.6	3.1	J	Y	Y	Petre *et al.* 02
G341.2+0.9	B1643−43	35.6	5.8	?	Y	N	Giacani *et al.* 01
Black Widow	B1957+20	35.0	2.5	J	N	N	Stappers *et al.* 03
Geminga	J0633+1746	34.5	0.16	J	N	Y	Caraveo *et al.* 03

[a] X-ray morphology. T: torus + weak jet, J: jet or trail dominated, M: mixed or uncertain. A ? implies that no X-ray nebula has been clearly detected yet, but if there is a clear radio or Hα morphology, that is indicated in parentheses.
[b] Indicates if there is an associated SNR shell. O means it is outside the shell, ? means there is a nearby structure that could be interpreted as a related SNR shell.
[c] Indicates coincident gamma-ray source. ? implies outside nominal gamma-ray error box but possibly still related.

spin-down energies are below this tend to have barely detectable X-ray PWN. The fact that they are observable at all, along with the existence of Hα nebulae seen around pulsars whose $\log \dot{E} < 34.0$ attest to the persistence of the relativistic wind generation mechanism long after the effects on the surrounding medium are readily apparent.

7.6.3 *Evolution of a PWN in a SNR*

The standard evolutionary model of PWNe is a spherical bubble being blown at the center of a spherical SNR shell (Fig. 7.13). Observationally, this picture does not accurately represent the true state of affairs. Many PWNe, including the "canonical" example of the Crab, do not have observable SNR shells, and so there is no observational evidence for the expanding blast wave caused by the supernova. However, presumably there is an external

Table 7.6. *PWN with no detected pulsar*

Nebula	Other name	*d* (kpc)	M	S	γ	Reference
G0.13−0.11		?	M	N	?	Wang *et al.* 02
G0.9+0.1		8	T	Y	N	Porquet 03
G7.4−2.0	GeV J1809−2327	1.9	J	?	Y	Braje *et al.* 02
G16.7+0.1		2.2	M	Y	N	Helfand *et al.* 03a
G18.5−0.4	GeV J1825−1310	4.1	J	?	Y	Roberts *et al.* 01
G20.0−0.2		5.4	?	N	N	Becker & Helfand 85
G21.5−0.9		5.5	M	N	N	Slane *et al.* 00
G24.7+0.6			?	N	N	Reich *et al.* 84
G27.8+0.6			?	N	N	Reich *et al.* 84
G39.2−0.3	3C 396	7.7	M	Y	Y	Olbert *et al.* 03
G63.7+1.1		3.8	?	N	N	Wallace *et al.* 97
G74.9+1.2	CTB 87	12	M	N	Y	Mukherjee *et al.* 00
G119.5+10.2	CTA 1	2.1	M	Y	Y	Slane *et al.* 03
G189.1+3.0	IC 443	1.5	J	Y	?	Olbert *et al.* 01
G279.8−35.8	B0453−685	50	M	Y	N	Gaensler *et al.* 03a
G291.0−0.1	MSH 11−62	?	J	Y	Y	Harrus *et al.* 03
G293.8+0.6			?	Y	N	Whiteoak & Green 96
G313.3+0.1	Rabbit	?	J	N	Y	Roberts *et al.* 99
G318.9+0.4			?	Y	N	Whiteoak & Green 96
G322.5−0.1			?	Y	N	Whiteoak & Green 96
G326.3−1.8	MSH 15−56	4.1	?	Y	N	Dickel *et al.* 00
G327.1−1.1		8.8	M	Y	N	Bocchino & Bandiera 03
G328.4+0.2	MSH 15−57	>17	M	N	N	Hughes *et al.* 00
G359.89−0.08		8	J	N	Y	Lu *et al.* 03

See footnotes to Table 7.5.

Table 7.7. *Pulsars with Hα bow-shock nebulae*

Name	Pulsar	$\log(\dot{E})$ ($\mathrm{erg\,s^{-1}}$)	*d* (kpc)	*v* ($\mathrm{km\,s^{-1}}$)	Reference
	J0740−28	35.1	1.9	260	Jones *et al.* 02
Black Widow	B1957+20	35.0	2.5	220	Stappers *et al.* 03
	J0437−4715	33.6	0.14	94	Bell *et al.* 95
	J2124−3358	33.6	0.27	61	Gaensler *et al.* 02b
Guitar	B2224+65	33.1	3.2	1725	Romani *et al.* 97

blast wave although the surface brightness of the shell might be very low. PWNe appear to be highly aspherical, sometimes with thick torii, sometimes dominated by narrow, jet-like features. How this asphericity may affect the evolution of the PWN is not well understood, although it is plausible that the mere addition of a filling factor to the standard equations could adequately account for the morphological variations (Roberts *et al.* 2003). Pulsars are generally high-velocity objects, so they rapidly leave their birth site, eventually overtaking the supernova blast wave. The interaction of the pulsar wind with the swept-up shell material can re-energize the shell, which seems to be occuring in the case of CTB80 (Shull *et al.* 1989; van der Swaluw. *et al.* 2003a). To reach the shell usually takes a few tens of thousand years, and so for the first few thousand years assuming the pulsar is near the center of the SNR is reasonable. Well before the pulsar reaches the SNR shell, it will overtake the outer edge of its original bubble, severely distorting the shape, changing from a spherical/toroidal structure to more of a bow-shock. Since the forward shock of the PWN is generally accelerating within the expanding ejecta (Reynolds & Chevalier 1984), this will usually not happen until after the PWN forward shock encounters the SNR reverse shock. Analytical and numerical models of this transition suggest it occurs at roughly half the pulsar crossing time of the nebula (van der Swaluw *et al.* 2004).

Therefore, for the first few thousand years of a pulsar's life, the assumptions (other than sphericity) of the standard models are probably valid. The evolutionary picture has been described analytically (e.g., Pacini & Salvati 1973; Reynolds & Chevalier 1984) and numerically (e.g., van der Swaluw *et al.* 2001; Blondin *et al.* 2001). Since the content and structure of the pulsar wind are poorly understood, it is generally assumed that a spherical outflow of energy equal to the spin-down energy of the pulsar (the observed pulsed electromagnetic radiation being a negligibly small fraction of the energy) continuously flows outwards from the pulsar magnetosphere. This energy is split between the kinetic energy of the bulk particle flow and the magnetic-field energy. At some radius, which is small compared to the total radius of the nebula, the electron/positron component of the bulk flow is randomized and takes on a power-law distribution of energies. If this transition is caused by a standing shock, then for efficient shock acceleration of the particles to occur, the flow energetics must be dominated by the particle component at the shock. Kennel and Coroniti (1984a), expanding on the work of Rees and Gunn (1974), constructed a steady state MHD model of the Crab nebula, assuming a positronic wind terminated by an MHD shock whose downstream properties are determined by boundary conditions imposed by the size, X-ray luminosity and expansion velocity of the nebula. The upstream flow properties can then be determined from the Rankine–Hugoniot relations. The ratio of magnetic to particle energy flux σ in the unshocked wind of the Crab is inferred to be $\sigma \sim 0.003$ (Kennel & Coroniti 1984b). This leads to the well-known "sigma problem" of pulsar wind theory, since mechanisms for accelerating the particles within the magnetosphere generally require $\sigma > 1$ (see Section 7.2). This suggests that the bulk of the acceleration occurs somewhere in the wind zone outside of the pulsar magnetosphere, for which there is no generally accepted model (e.g., Melatos 1998; Kirk & Skjæraasen 2003).

Although shock theory requires the ratio of magnetic to particle energy just downstream of the wind termination shock to be small, it is expected that the magnetic field will grow rapidly towards equipartition in the downstream flow. This allows the assumption that the magnetic and particle energy densities are roughly equal throughout the bulk of the PWN. For the first few days after the supernova explosion, the magnetic-field density rapidly increases within the nebula, but then the expansion of the nebula causes the magnetic-field strength to decrease with time despite continuous injection of magnetic energy.

The subsequent evolution of the PWNe can then be separated into two eras of the pulsar spin-down $\dot{E}$ and two eras of the SNR expansion (e.g., Reynolds & Chevalier 1984). For times $t < \tau \equiv P_0/2\dot{P}_0$, the initial spin-down time of the pulsar (note that this is probably the case for systems where the true age $t \ll \tau_c \equiv P/2\dot{P}$, the *observed* characteristic age of the pulsar), energy will be injected at a nearly constant rate ($\dot{E} \simeq \dot{E}_0$), and the radius of the PWN expands as $R \propto t^{6/5}$ while the average magnetic-field strength will decrease as $B \propto t^{-1.3}$. For large initial spin periods ($P_0 \gtrsim 50$ ms), the time of nearly constant energy input can last for several thousand years. During this time, the outer edge of the PWN bubble is moving supersonically compared to the surrounding medium, and so a forward shock should form, heating the swept-up ejecta that are in a thin shell around the PWN. The total mass of the swept-up material is initially fairly small, and so the thermal luminosity of this material may not be very large. At later times ($t > \tau$ or $t \sim \tau_c$) when $\dot{E} \ll \dot{E}_0$, the expansion slows to $R \propto t$ if the SNR is still in its free expansion phase.

The blast wave of the surrounding SNR shell initially expands almost freely into the ISM (or into a surrounding bubble blown by the wind of the pre-supernova progenitor star, Chevalier 1982), shock heating the swept-up material. As the swept-up mass builds, a reverse shock forms and begins to propagate towards the center of the SNR, heating the ejecta. This reverse shock will reach the center of the SNR when the swept-up ISM mass is much greater than the ejecta mass, marking the beginning of the so-called Sedov–Taylor phase of the SNR (Sedov 1959). Until this time, there are four shocks in the system: the forward shock of the SNR blast wave propagating into the ISM, the reverse shock of the SNR blast wave propagating back into the freely expanding ejecta interior to the shock, the forward shock of the PWN bubble, and the PWN reverse shock/wind termination shock (Fig. 7.13). When the SNR reverse shock encounters the PWN forward shock, numerical simulations show the PWN is initially compressed, and then its radius oscillates a few times over the next several thousand years before settling down into a more compact and probably distorted shape (van der Swaluw *et al.* 2001). During the compression, the magnetic field is enhanced causing an increase in the synchrotron luminosity and filaments can develop due to Rayleigh–Taylor instabilities (Blondin *et al.* 2001). If the pulsar wind luminosity is still near its initial luminosity ($\dot{E} \simeq \dot{E}_0$) after this compression, the PWN begins to grow again, but now the expansion is subsonic with $R \propto t^{11/15}$. If, on the other hand, $\dot{E} \ll \dot{E}_0$ then the expansion of the PWN goes as $R \propto t^{1/3}$ (Reynolds & Chevalier 1984).

7.6.4 Emission spectra of PWNe

The relativistic particle gas that fills the PWN bubble loses energy through synchrotron radiation and adiabatic expansion. Since the observed radiation spectrum has the form of a broken power law, a power-law electron energy spectrum is suggested, with the particle spectral index s (defined as $N(E) \propto E^{-s}$) related to the radiation spectral index α by $s = 2\alpha + 1$. The simplest model is of a single electron population injected into a uniformly magnetized nebula with a single spectral index S and a high-energy cut-off E_{ec} (Chevalier 2000). Synchrotron radiation is usually the most important cooling process for the X-ray emitting electrons, with the typical timescale t_c for cooling dependent on the electron energy and the magnetic field. At high energies, $t_c \ll t$, the age of the PWN, and an approximate steady state can be assumed between the injection and cooling of the electrons. In this regime, the average particle spectrum in the nebula is $S + 1$, which corresponds to a steepening of the observed spectrum by $\Delta\alpha = 0.5$. A break in the particle spectrum occurs at an energy where $t_c \sim t$, resulting in a break in the observed photon spectrum at

$$\nu_b = \frac{1.68}{B^3 t^2} \text{ GHz}$$

where B is in Gauss and t is in years. Note that since B decreases with time, this break frequency actually increases as the PWN ages. At low energies ($t_c \gg t$), cooling losses are negligible, and the particle spectral index is simply that of the injection spectrum. At late times when $\dot{E} \ll \dot{E}_0$, the initial cooling of the large number of particles injected at early times can leave an imprint on the particle spectrum in the form of a separate break below ν_b whose frequency decreases with time due to losses from the adiabatic expansion of the nebula. This results in an intermediate particle spectral index $S > s > S + 1$ between the two breaks (Reynolds & Chevalier 1984). The passage of the reverse shock causes a brightening of the nebula and an enhancement of the magnetic field. The high-energy electrons all rapidly lose their energy, resulting in a spectral cutoff of the electron population. After the reverse shock passes, the low-energy spectrum is dominated by the old electrons with enhanced emission, while the high-energy spectrum is from newly injected particles, with a middle range below the post-reverse-shock cooling break which could have a fairly steep spectrum, i.e., $s_m > S + 1$.

It is tempting to identify the radio emission with the uncooled $s = S$ spectrum and the X-ray emission with the cooled $s = S + 1$ spectrum. However, this predicts $\Delta\alpha = 0.5$ and in many cases it is significantly greater (Woltjer *et al.* 1997). If the X-ray or radio emitting particles were in the adiabatically cooled regime, $\Delta\alpha < 0.5$. The only time a single injection spectrum would fit this situation is after the reverse shock passage if the cooling break is above the soft X-ray band. In this case, there should be a spectral hardening somewhere in the hard X-ray or soft gamma-ray band. Another possibility is that the injection spectrum has changed over time. However, a correlation between PWNe X-ray spectral indices and $\dot{E}$ showing a softening with increasing $\dot{E}$ has been claimed (Gotthelf 2003), implying the spectral index is decreasing with age. This would again lead to $\Delta\alpha < 0.5$. A third possibility is that more than one electron population is being injected into the nebula, one dominant at low energies and a second at high energies (Bandiera *et al.* 2002). It is now clear that many, if not most, PWNe consist of both equatorial outflows and polar outflows, the latter in the form of collimated jets. It might be expected that the two flows would inject particles into the PWN with different spectral slopes. Such a situation would often require the polar flows to be relatively underluminous in the X-ray region. A more fundamental problem is the flatness of the PWN spectra in radio. Both theoretical models (Kirk *et al.* 2000) and observations of other astrophysical shocks that produce radio emitting electrons (e.g., SNR shells and AGN jets) tend to produce emission spectral indices of $\alpha \sim 0.6$ and then cool to $\alpha \sim 1.1$. The only workable mechanism for producing the radio spectra of PWN currently in the literature is one developed for the Crab where there is a low-energy cutoff to the injected particle spectrum and the radio emission comes from particles which were cooled during an early rapid spin-down era (Atoyan 1999). However, this requires some fine tuning of the magnetic-field evolution of the PWN as well as a non-standard spin-down history for the pulsar, and it is not clear whether this can be consistently applied to other PWN. Alternative models for the acceleration of the radio emitting particles at sites in the nebula other than the termination shock should be explored.

In the above discussion, it has been implicitly assumed that the electrons have been injected evenly throughout the nebula, the so-called one-zone model (Chevalier 2000). In a real PWN, the electrons are injected at the termination shock within a few tenths of a parsec from the pulsar and then propagate outwards, initially at the typical post-shock velocity $v \simeq c/3$ but then with the average bulk velocity typically decreasing as $1/r^2$ (this is true whatever the emission geometry, since the nebula will have an r^2 expansion as long as the external confinement is spherically symmetric; this is probably not the case for some rPWN – see Section 7.6.7). If the PWN bubble is large enough, the X-ray emitting particles will have a chance to cool significantly before reaching the outer edge, and it is expected that the X-ray spectral index will increase as a function of distance from the pulsar. For all energies where the travel time to the outer edge of the PWN is more than the synchrotron cooling time, then the PWN should decrease in size with increasing energy. Therefore, we should expect the radio PWN to be larger than the X-ray PWN. Kennel and Coroniti (1984b) modeled this in detail, and found the integrated X-ray spectral index should be $\Delta\alpha = (4 + \alpha/9)$ larger than that of the uncooled particles emitting just downstream of the termination shock. For reasonable values of $\alpha = 0.0 - 1.0$, $\Delta\alpha = 0.44 - 0.56$, very close to the simple, one-zone model estimate of $\Delta\alpha = 0.5$. The spatial resolution of Chandra and XMM-Newton has allowed an increase in spectral index with radius to be observed in several PWNe (e.g., Slane *et al.* 2000; Weisskopf *et al.* 2000; Porquet *et al.* 2003).

7.6.5 *The Crab and other classic "plerions"*

Until recently, observational and theoretical studies of PWNe have been dominated by the Crab nebula. It contains the most energetic pulsar known in the Galaxy, it is bright at all wavelengths, has a precisely known age due to historical records of the associated supernova, and has been intensely studied for over 100 years. It had been the defining member of the class of "supernova remnants" with no observable shells, sometimes called *plerions*, meaning filled center (Weiler 1978), and provided the conclusive proof that neutron stars are born in Type II supernovae. Ironically, although it has been known as the remnant of SN1054 for many years, it is something of a misnomer to call it a SNR since the nebula is a result of the pulsar wind and not the blast wave of the supernova itself. Why there is no observable radio shell is still uncertain, with explanations including that it is in a low-density region of the ISM and that SN1054 was an anomalously low-energy event (Frail *et al.* 1995). There are several other PWNe in the Galaxy with no observable shell, typified by 3C 58 (Reynolds & Aller 1985). They tend to have much lower X-ray luminosities and harder spectra than the Crab. The inferred break energies between the radio and X-ray spectra of these other PWNe also tend to be much lower than that of the Crab's breaks (Woltjer *et al.* 1997). Whether these differences can be ascribed completely to the difference in spin-down energies is debatable, but since observations of the Crab tend to be of much higher quality than of all other PWNe, attempting to interpret what is viewed by scaling from the Crab is somewhat inevitable. The PWN whose properties are most similar to the Crab's is that around PSR B0540−69, often referred to as the Crab's twin, in the Large Magellenic Cloud (Chanan *et al.* 1984; Manchester *et al.* 1993; Gotthelf & Wang 2000).

At radio energies, the Crab appears as an ovoid blob with a superimposed filamentary structure. The radio spectral index is remarkably spatially uniform (Bietenholz *et al.* 1997), and the emission is mostly constant except for a slow expansion (Velusamy *et al.* 1992) and a long-term overall decline in its flux (Aller & Reynolds 1985b). Recently, variations in the form of small ripples in the nebula have been detected whose general morphology corresponds to the regions of activity seen in the optical and X-ray nebulae (Bietenholz *et al.* 2001). The radio emission is highly linearly polarized and well organized, showing that the emission mechanism is almost certainly synchrotron (Bietenholz & Kronberg 1990).

At optical energies the Crab is somewhat smaller than in the radio, and consists of thermal filaments superimposed on a polarized non-thermal continuum. Near the pulsar, there is an underluminous region probably associated with the unshocked wind. Outside of this zone, there is a series of bright, non-thermal enhancements, referred to as wisps, that have been known for a long time to be variable on timescales of months (Lampland 1921; Oort &

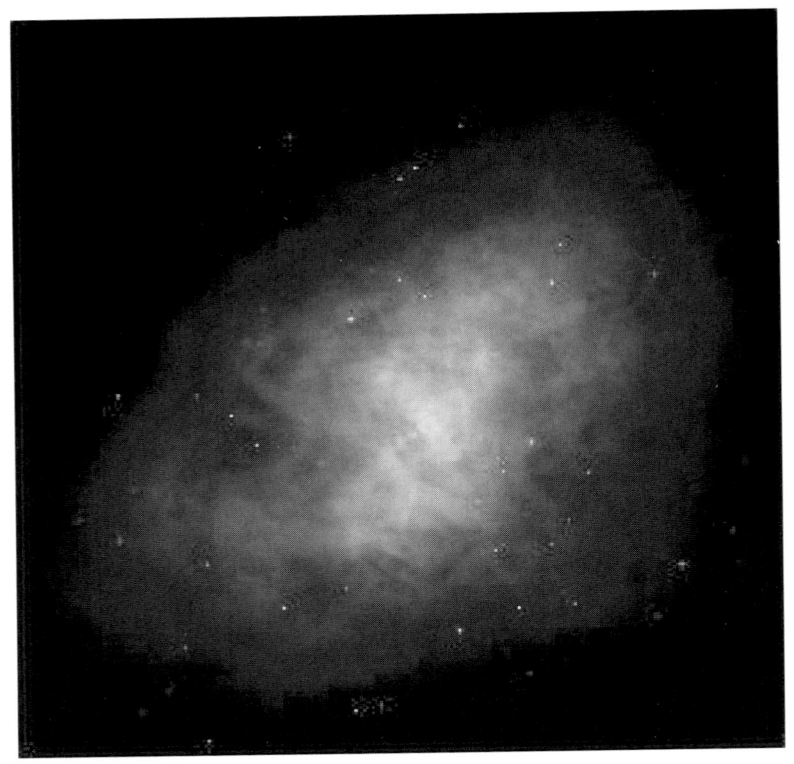

Fig. 7.14. A composite image of the Crab Nebula showing X-ray in blue, optical in green, and radio in red. (Credits: X-ray: NASA/CXC/ASU/J. Hester *et al.*; Optical: NASA/HST/ASU/J. Hester *et al.*; Radio: VLA/NRAO; from Chandra website http://chandra.harvard.edu) For color version see Plate 7.14.

Walraven 1956; Scargle 1969). Monitoring observations with the Hubble Space Telescope have shown the enhancements, the innermost one starting at the edge of the underluminous zone, which is presumably at the wind termination shock, move outwards over several weeks to months (Hester *et al.* 1996). One model of the radial placement of the wisps posits that if the wind contains a significant fraction of ions, the post-shock ions would cause enhancements in the magnetic field at multiples of their cyclotron radius, and hence be sites of enhanced synchrotron emission (Gallant & Arons 1994).

In X-rays, the Crab has a very definite toroidal morphology with polar jets. Again, surrounding the pulsar is an underluminous zone ending in a bright ring with knotty enhancements, which have been observed with Chandra to form and move outwards at $v \sim 0.5c$ and then diffuse, similar to the optical wisps (Hester *et al.* 2002). Outside of this region is a thick torus, which ends rather abruptly well within the boundaries of the optical and radio nebula (Fig. 7.14, see also color plate section). There is clear evidence of spectral steepening with increasing radius in the torus. The jets emit a much smaller fraction of the X-ray luminosity than does the torus, and overall have a harder spectrum. Outward motion of features within the jets has also been observed. The proper motion of the pulsar is along the line of the jets, suggesting the natal kick was along the spin axis of the pulsar (Caraveo & Mignani 1999; Weisskopf *et al.* 2000).

The unpulsed, and hence presumably nebular, emission spectrum is seen to extend smoothly from the X-rays all the way to ~25 MeV. Above this is a sharp, possibly variable cutoff (de Jager *et al.* 1996a) indicating some limiting factor in the particle acceleration mechanism. Above ~100 MeV, a rising unpulsed component is seen which can be observed all the way into the TeV energy range (Weekes 1991). This is most likely due to inverse Compton scattering off the synchrotron emitting electrons in the nebula.

Since the Crab is energetically unique among Galactic pulsars, it is valid to wonder if its properties can be legitimately scaled down over two orders of magnitude to other pulsars with more typical spin-down energies. 3C 58 was the second plerion to be identified and is thought to be a result of SN1181. Its pulsar, which was only recently discovered through X-ray and very deep radio observations (Murray *et al.* 2002; Camilo *et al.* 2002c), has a spin-down energy of 2.7×10^{37} erg s^{-1}, second only to the Crab among Galactic PWNe. Although its spin-down energy is only a factor of ~15 less than the Crab's and the nebular radio luminosity is a factor of ~10 less than the Crab nebula, the nebular X-ray luminosity is a factor of ~1000 less (Becker *et al.* 1982; Murray *et al.* 2002). There appears to be an unusually sharp spectral break (Green & Scheuer 1992) at ~50 GHz. The radio luminosity appears to be increasing, suggesting the unseen SNR reverse shock may already be affecting the PWN (Aller & Reynolds 1985a).

The Vela pulsar is generally considered to be the prototype of pulsars with $\tau_c \sim 10\,000$ yr and spin-down energies in the $10^{36} - 10^{37}$ erg s^{-1} range. Its PWN appears very bright and can be well-resolved at X-ray energies because of its proximity to Earth, being only ~290 pc away (Dodson *et al.* 2003a). There is an associated radio SNR shell, but it is patchy, uneven, and diffuse, as is typical of older remnants (Clark & Caswell 1976), having three sections called Vela X, Y, and Z. Weiler and Panagia (1980) noted that the Vela X region had radio properties more similar to the Crab and other plerions than to ordinary SNR shells. Due to the age of the Vela SNR, it is likely that the reverse shock encountered the PWN long ago, although there may still be some transient effects. The morphology of the Vela PWN in X-rays is a double arc with narrow, jet-like enhancements, again near the line of proper motion. While this has sometimes been interpreted as a bow-shock morphology, it is now usually interpreted as a double torus (Helfand *et al.* 2001). An apparently variable, extended jet feature is sometimes prominent (Pavlov *et al.* 2003) outside of the arcs. There is no clear indication of an underluminous region near the pulsar, although that may be due to an unfortunate viewing angle, and the observed double-arc system may be the location of the termination shock.

The physical size of the bright part of the Vela X-ray PWN is about an order of magnitude smaller than the Crab, and its X-ray efficiency in the 1–10 keV band is more than two orders of magnitude less. The X-ray spectrum is flatter and the high-energy cutoff is probably less than 10 MeV (de Jager *et al.* 1996b). Whether these are indicative of a general trend of lower efficiency, spectral hardening, and lower maximum particle energies with lower $\dot{E}$ is difficult to say, although there is some evidence for a dependency of spectral index on $\dot{E}$ (Gotthelf 2003) and the maximum potential drop across the open field lines $E_{\rm max} \propto \dot{E}^{1/2}$ so there is a theoretical expectation for a lower high-energy cutoff (de Jager *et al.* 1996a). The X-ray PWN is centered between two radio lobes, which are much larger than the X-ray PWN (Fig. 7.15;

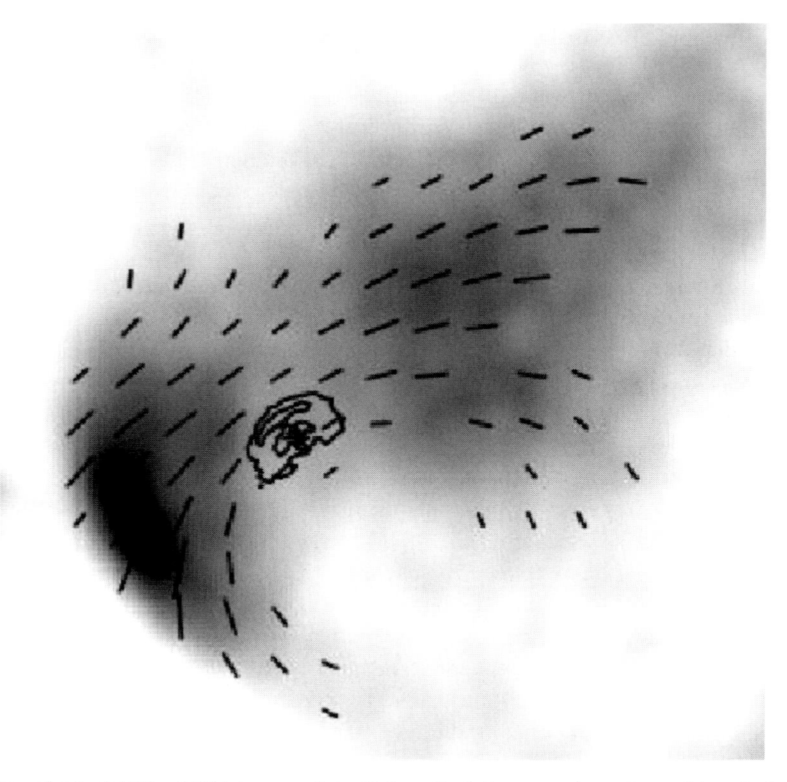

Fig. 7.15. 5 GHz ATCA image of the Vela radio lobes near the pulsar with polarization vectors showing how highly structured the magnetic field is. The contours near the center are of the X-ray nebula as seen with Chandra. Note this is only a small part of the Vela X radio region. (Adapted from Fig. 10 of Dodson *et al.* 2003b)

Dodson *et al.* 2003b), and are themselves just a small part near the edge of the filamentary, flat spectrum Vela X radio complex. It is likely the reverse shock has displaced the larger radio nebula, which is why the proper motion vector of the pulsar is not directed away from the center of Vela X (Blondin *et al.* 2001).

7.6.6 Young composite SNR

The standard theoretical picture of a PWN growing within the expanding shell of the supernova blast wave has been around since the early eighties. Although quite a few young composite SNR were known at that time (Helfand & Becker 1987), in embarrassingly few cases had the exciting pulsar been observed. While the unknown radio beaming could be invoked to account for the lack of observable radio pulsations, this situation made it difficult to confront theoretical expectations with observational realities. Since the mid-nineties, several pulsars have been discovered within young composites by deep radio searches (Camilo *et al.* 2000; Camilo *et al.* 2002b) or with sensitive X-ray searches (Torii *et al.* 1997; Gotthelf *et al.* 2000), representing different stages and scenarios in the early evolution of PWN systems.

Perhaps the system whose observational properties best match the assumptions made in the models is G11.2−0.3. This bright, remarkably circular remnant is often associated with a "Guest Star" observed by the Chinese in AD 386 (Clark & Stephenson 1977) and the age inferred from expansion measurements of the shell support this association (Tam & Roberts 2003). HI absorption measurements place the remnant at a distance of ~5 kpc (Green *et al.* 1988). A 65 ms X-ray pulsar is within a few arcseconds of the geometrical center of the shell (Kaspi *et al.* 2001b), which has a characteristic age of 24 000 yr; over ten times that of the SNR (Torii *et al.* 1997), strongly implying $P_0 \simeq P$ and $\dot{E} \simeq \dot{E}_0$. High-resolution radio and X-ray observations have separated the PWN from the shell and measured its spectrum (Tam *et al.* 2002; Roberts *et al.* 2003).

We therefore have in G11.2−0.3 an example of a composite remnant where the assumptions of near spherical symmetry of the shell and negligible displacement of the pulsar from its birthsite are demonstrably true, the energy output history is well constrained to be nearly constant, the age is known, and the distance is fairly well determined. Given its age and estimates of the swept-up mass, it should be nearing the Sedov phase, so the reverse shock should be nearing the PWN. However, the radio PWN is one of the largest relative to the shell and there is marginal evidence that it is expanding (Tam & Roberts 2003), suggesting the reverse shock has not yet begun to crush the PWN bubble. It therefore should still be expanding supersonically, and may be shock-heating the surrounding ejecta. There is thermal X-ray emission that seems to be morphologically related to the radio PWN, but may also be a region of enhanced shell emission seen in projection (Fig. 7.16, see also color plate section). The velocity of the shell inferred from the thermal emission is consistent with that inferred from the expansion rate and HI distance estimates (Tam & Roberts 2003).

Current estimates of the X-ray and radio spectral indices are consistent with a single cooling break in between. However, given the age of the nebula, the implied magnetic field is unusually high. If it uniformly fills a region of the apparent bubble size, then the required magnetic energy is much larger than could have been supplied by the pulsar. This may indicate that the pulsar wind is highly non-spherical, predominantly coming from within $\sim 10^\circ$ of either the poles or the equator. The narrowness of the X-ray emission may support this interpretation (Roberts *et al.* 2003). Varying X-ray spots are seen in this nebula as well, with apparent motions of $\sim c$, much faster than the expected $c/3$ of a standard post-shock flow. So even with this most "simple" of PWN/SNR systems where our knowledge is most complete, there are anomalies that are not well understood.

Kes 75 is a young ($\tau \lesssim 2000$ yr) distant ($d \sim 19$ kpc; Becker & Helfand 1984) composite SNR in many ways similar to G11.2−0.3. It contains a 324 ms X-ray pulsar with the youngest known characteristic age, ~700 yr (Gotthelf *et al.* 2000). The Chandra images of Kes 75 (Helfand *et al.* 2003b) of both the shell and PWN in both X-rays and radio look very similar to G11.2−0.3, and the pulsars powering both PWN are very X-ray efficient and have unusually broad profiles. However, on closer examination, there are remarkable differences. The X-ray efficiency in the 0.5–10 keV band of the PWN in Kes 75 is ~6.5%, which is the highest known, while that of G11.2−0.3 is only ~0.2%, a much more typical value. The pulsar's spin parameters imply both an unusually large magnetic field ($B \sim 5 \times 10^{13}$ G) and that $P_0 \ll P$, so that the energy input into the PWN has changed dramatically over its short history. The implied size of the shell of Kes 75 requires an enormous expansion velocity ($v \sim 13\,000\,\mathrm{km\,s^{-1}}$) and therefore implies a remarkably energetic explosion or else the neutron star was born spinning very rapidly and a significant fraction of its rotational kinetic energy was somehow transferred to the expanding ejecta (Helfand *et al.* 2003b).

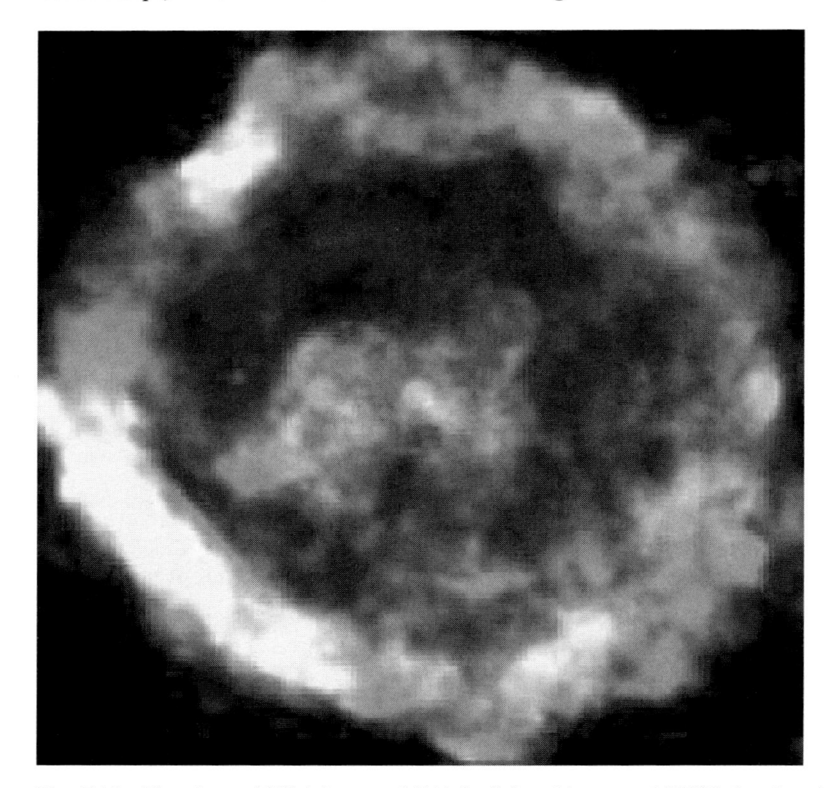

Fig. 7.16. Chandra and VLA image of G11.2−0.3 and its central PWN showing the relationship between the soft thermal X-ray emission (red), the radio synchrotron emission (green) and the hard, non-thermal X-ray emission (blue). The point source at the center is the pulsar, only seen in X-rays. (Adapted from Fig. 1 of Roberts *et al.* 2003) For color version see Plate 7.16.

SNR G292.0+1.8 is an example of a slightly older ($\tau \sim 3000$ yr) composite system. Chandra imaging shows a remnant filled with ejecta material, but there is also a clear X-ray PWN slightly offset from the geometrical center (Hughes *et al.* 2001). In this case there is no clear morphological separation between the radio PWN and shell and it seems likely that the SNR reverse shock has recently begun to interact with the PWN (Gaensler & Wallace 2003).

7.6.7 The Duck and other PWNe with trail morphologies

Several PWNe show a predominantly trail morphology at X-ray, and sometimes radio, wavelengths extending from an X-ray point source, presumably the pulsar, back towards the apparent birth-site. This has been observed in pulsars both within the SNR shell, such as PSR B1853+01 in W44 (Petre *et al.* 2002), and outside, such as coming from PSR B1757−24, the head of the Duck system (Kaspi *et al.* 2001a). X-ray trails have also been seen from sources with Hα bow-shock nebulae, such as the Black Widow binary millisecond pulsar system PSR B1957+20 (Stappers *et al.* 2003, see Fig. 7.17). The apparent radio bow-shock PWNe source associated with the possibly variable gamma-ray source GeV J1809−2328 also shows a clear point source and trail morphology in X-rays (Braje *et al.* 2002). This type of PWNe is most

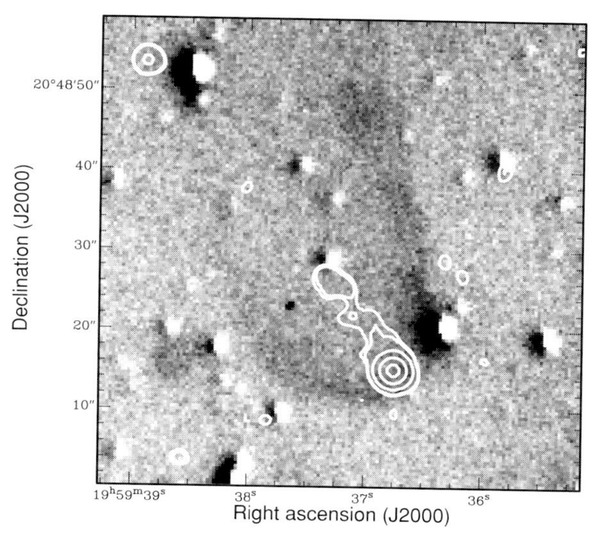

Fig. 7.17. Hα image of Black Widow pulsar PSR B1957+20 with Chandra X-ray contours. (Adapted from Fig. 2 of Stappers *et al.* 2003)

common around pulsars with spin-down energies $\log \dot{E} \lesssim 36.5$ (in erg s^{-1}). These nebulae are sometimes referred to as bow-shock nebulae or ram-pressure confined nebulae, however it is not clear if either of these terms accurately reflect the physical situation in respect to the X-ray and radio nebulae. Therefore, we will simply refer to these nebulae as rapidly moving PWNe (rPWN), where the pulsar's motion is rapid relative to the radial expansion of the nebula and probably relative to the local sound speed.

The Duck is an illustrative case. Upper limits on the proper motion of the pulsar (Gaensler & Frail 2000; Thorsett *et al.* 2002) rule out the X-ray emission being an actual trail, since the synchrotron lifetime of the X-ray emitting particles is too short for the pulsar to have moved the length of the X-ray emitting region in the required time (Kaspi *et al.* 2001a). The X-ray emission is therefore due to a collimated outflow, somehow constrained to move within a cavity created by the pulsar's passage. A spectacular Chandra X-ray image of the Mouse shows the structure of the X-ray nebula to consist of a narrow tail region in a broader, fainter, bow-shaped nebula, which may be too faint to see clearly in the Duck (Gaensler *et al.* 2003c). The extremely long radio tail of the Mouse (Predehl & Kulkarni 1995) is another example where it seems something must continue to confine the flow well beyond where ram-pressure from the pulsar's passage should no longer be relevant. The extraordinary jet/trail coming from PSR B1509−58 (Gaensler *et al.* 2002a) also demonstrates that in many cases pulsar outflows can be narrowly confined over large distances even while still within the parent SNR.

The Black Widow pulsar PSR 1957+20 (Fruchter *et al.* 1988), so called because its wind is believed to be destroying its white dwarf companion, is interesting for several reasons. Unlike the other sources with X-ray trails, it is a millisecond pulsar, demonstrating that old, recycled pulsars also have relativistic winds. It also has an Hα nebula (Kulkarni & Hester 1988), which clearly delineates the forward bow-shock, as well as a narrow trail of non-thermal X-ray emission (Stappers *et al.* 2003), which can serve as a test bed for simulations of rPWNe (Bucciantini 2002).

7.6.8 The gamma-ray connection

The magnetosphere of a young pulsar is one of the few places in the universe where particles can be accelerated such that they will emit high-energy gamma-ray emission through either the synchrotron process, curvature radiation, or inverse Compton scattering. Approximately 1/3 of the known PWNe are coincident with sources of emission at $E >$ 100 MeV observed by the EGRET instrument on the Compton Gamma Ray Observatory (Hartman *et al.* 1999). Some of these are the known gamma-ray pulsars (Thompson *et al.* 1999), and this may be an indication that the potential needed to produce pulsed gamma-ray emission observable at typical Galactic distances is similar to that needed to produce observable PWNe. However, several of the PWNe are associated with unidentified EGRET sources which appear to be variable (Roberts *et al.* 2001; Nolan *et al.* 2003) on timescales of a few months, similar to the synchrotron cooling timescale at these energies, indicating that the PWN may be the source of emission. Practically speaking, gamma-ray sources can serve as tracers of PWNe. Six of the PWNe listed in Tables 7.5 and 7.6 were initially discovered through X-ray imaging of a gamma-ray error box. In addition, the shape of PWNe can be used to infer the viewing angle, and possibly the magnetic inclination angle, of the pulsar (e.g., Ng & Romani 2004), which is critical information for testing models of pulsed gamma-ray emission (Section 7.2). Observations from the upcoming GeV energy telescopes AGILE and GLAST will be able to distinguish between pulsed and unpulsed emission from many of these sources, verify any variability, and locate more PWNe buried deep in the Galactic plane.

Steady, unpulsed emission up to several TeV has been observed towards three pulsars (Crab, Vela, and PSR B1706–44; Weekes 1991; Yoshikoshi *et al.* 1997; Kifune *et al.* 1993) with PWNe, presumably caused by inverse Compton scattering by electrons in the inner nebula. This is direct evidence of very high-energy acceleration and puts constraints on the magnetic field in the inner nebula. As the new generation of ground-based Cerenkov air shower telescopes comes on line, many more PWNe will undoubtedly be observed at energies of $\sim$100 GeV to a few TeV.

7.6.9 Current trends and future directions

The high-resolution X-ray imaging capabilities of the Chandra and XMM-Newton telescopes are allowing detailed study of the structure of many PWNe. The presence of collimated jet-like outflows as a common feature of PWNe is a surprise that most models of PWNe, dominated by a toroidal equatorial flow, did not allow for. New analytical and numerical models are now being developed with varying amounts of power being lost through polar outflows (Khangoulian & Bogovalov 2003; Komissarov & Lyubarsky 2003; Shibata *et al.* 2003). PWNe have also been shown to be dynamical X-ray systems with observable variations on timescales as short as a few weeks (Hester *et al.* 2002). As quantitative measurements of these variations are made in more PWNe the need for the further development of models of acceleration in the inner nebula becomes more acute.

The remarkable diversity of PWNe has also been a surprising result of the new X-ray images. Many of the observations currently archived are too short to distinguish different spectral characteristics of the various features within PWNe. There is a suggestive correlation between the X-ray spectral index of PWNe and the magnetospheric potential (Gotthelf 2003), but whether this will be borne out by further observations and whether there is a different correlation for jet-like and toroidal structures will require deep observations of many more PWNe. The structure of the PWNe also seems to depend on the magnitude and direction of the pulsar's velocity, but so far only a few young pulsars have had their proper motions measured. Radio interferometry campaigns now under way (e.g., Brisken *et al.* 2003a) promise to greatly increase the number of measured pulsar transverse velocities, allowing the determination of how well jets are aligned with velocity vectors (e.g., Dodson *et al.* 2003a), if at all. New 2D and 3D numerical simulations of PWN evolution are beginning to yield intriguing results. For example: relaxing the spherical symmetry outflow constraint but keeping axial symmetry leads to a multiplicity of shock regions that can mimic the Crab morphology (Komissarov & Lyubarsky 2003). Toroidal magnetic outflows can cause an elongation of PWNe with spherical particle outflows, reminiscent of the shape of 3C58 (van der Swaluw 2003), and interactions with the reverse shock can lead to complicated, asymmetric morphologies (Blondin *et al.* 2001; Bucciantini *et al.* 2003).

7.7 X-rays from rotation-powered pulsars in binary systems

For completeness, we mention briefly another related source of X-rays from rotation-powered pulsars – shock emission in binary systems. Although these systems do not obviously satisfy the mandate of this chapter on "isolated neutron stars," the nature of their X-ray emission is very similar to that in a PWN. For high-$\dot{E}$ pulsars in binary systems, the confining medium can be the wind of the companion star. In this case, the confinement, particularly for eccentric binaries, is strongly orbital-phase dependent, leading to a PWN of luminosity (and, presumably, size) that varies systematically with orbital phase. Even circular orbits are of interest since the orientation of the PWN, which, in this case, is not expected to be isotropic, changes with orbital phase. In principle, such systems are an excellent diagnostic of shock acceleration and the pulsar wind as they have regular, repeating and predictable dynamical properties. They also represent a new form of "X-ray binary," one not powered by accretion. However, only two such systems have been detected in X-rays: PSRs B 1259−63 and B1957+20. Their rarity is because of the need to have a high $\dot{E}$ pulsar in a binary system in which the two components come sufficiently close to each other for the shock to be strong, but not too close so that accretion occurs.

The 48 ms radio pulsar PSR B1259−63 is in a 3.4 yr binary orbit having eccentricity 0.87 (Johnston *et al.* 1992). From radio timing, the pulsar is known to have $B = 3 \times 10^{11}$ G and $\tau_c = 3 \times 10^5$ yr. The pulsar's companion, identified by its location within the $< 1''$ pulsar timing error box, is the 10th mag B2Ve star SS 2883. It has mass $\sim 10\,M_\odot$ and radius $\sim 6\,R_\odot$, deduced from its spectral type. The system provides an evolutionary link between the rotation-powered pulsars and the high-mass X-ray binaries. For PSR B1259−63, near periastron, the pulsar approaches its companion to within ~25 Be-star radii. Cominsky *et al.* (1994) detected variable X-rays from the system near apastron. This ruled out standard wind accretion scenarios, magnetospheric emission, and emission from the companion. Cominsky *et al.* (1994) suggested either some form of non-standard accretion or the pulsar wind shocked by the companion wind as possible mechanisms for the X-ray emission. Tavani *et al.* (1994) considered pulsar/Be-star wind interactions in detail, in particular, shock emission at the location of pressure balance between the pulsar and Be-star winds. They suggested this mechanism produced the apastron X-rays, and predicted PSR 1259−63 would be a moderately strong, unpulsed X-ray source near periastron, unable to accrete due to the shock distance from the pulsar being much larger than the accretion radius. Kaspi *et al.* (1995) and Hirayama *et al.* (1996) reported on ASCA observations around the periastron of 1994; Hirayama *et al.* (1999) summarize all the X-ray observations including two more made at the subsequent apastron. The source was clearly seen to increase in intensity by a factor of > 10 near periastron relative to apastron, with the peak luminosity $\sim 10^{34}$ erg s^{-1}. The emission was well described at each epoch by a power law, however the photon index clearly varied from 1.6 to 2.0, with the emission softest at periastron. Tavani and Arons (1997) considered the X-ray radiation mechanisms and interaction geometry in detail, showing that a synchrotron/inverse-Compton scattering model of emission of electron–positron pairs accelerated at the inner shock front of the pulsar cavity and adiabatically expanding in the MHD flow explains well the observed time-variable X-ray flux and spectrum. They conclude that most likely the Be-star spin axis is misaligned with the orbital angular momentum, and its mass outflow rate was constant over the ~ 2 yr period in question. Campana *et al.* (1995) argue that an unusual Galactic X-ray source, LSI+61°303, is a similar system to PSR B1259−63, but with the radio pulsations permanently eclipsed.

PSR B1957+20 is a 1.6 ms recycled radio pulsar in a 9 hr binary system with a low-mass companion (Fruchter *et al.* 1988). The pulsar has $\dot{E} = 10^{35}$ erg s^{-1}. For $\sim 10\%$ of every orbit, the radio pulsations are eclipsed by the wind from the companion, which is being ablated and, eventually, evaporated by the pulsar wind. For this reason, the pulsar is sometimes known as the "Black Widow," as it seems to be destroying the star that gave it new life as a recycled pulsar. The system is surrounded by an Hα bow-shock nebula, oriented with apex in the direction of the pulsar's known proper motion, a result of the interaction of the pulsar wind with the ambient interstellar medium (Kulkarni & Hester 1988). Arons and Tavani (1994) argued that intra-binary shock emission ought to be observable, because of the interaction of the pulsar wind with material being ablated off the companion. They also argued that X-rays might originate from the bow shock. Of course, magnetospheric X-rays might be detectable from this pulsar as well. The first X-ray detection of PSR B1957+20 came from ROSAT, although only a handful of photons were detected (Fruchter *et al.* 1992; Kulkarni *et al.* 1992), and the conclusions that could be drawn were limited. More recently, Chandra observed the PSR B1957+20 system (see Fig. 7.17), and detected both a point source as well as an extended X-ray tail (Stappers *et al.* 2003). The tail emission is likely related to the bow-shock nebula, and represents the first proof that millisecond pulsars have relativistic winds like their much younger counterparts. The origin of the point-source emission is less clear as the observation's time resolution was insufficient to detect pulsations. Planned XMM-Newton observations should be able to decide with certainty whether the point source is due to magnetospheric emission, or if it is from the intra-binary shock, as in PSR B1259−63.

Acknowledgements

The authors thank B. Gaensler and D. Yakovlev for discussions, for careful reading of the manuscript and for many helpful comments. We also thank P. Jaikumar, D. Lai, M. Lyutikov, G. Pavlov, S. Ransom, R. Turolla, M. van Kerkwijk, and S. Zane for helpful conversations, and M. Strickman and J. Dyks for help with figures.

References

Aller, H. D. & Reynolds, S. P. 1985a, in *The Crab Nebula and Related Supernova Remnants* (Cambridge University Press), 75–78

1985b, *ApJ*, **293**, L73

Alpar, M. A., Anderson, P. W., Pines, D., & Shaham, J. 1984, *ApJ*, **276**, 325

Arons, J. 1981, *ApJ*, **248**, 1099

1983, *ApJ*, **266**, 215

Arons, J. & Scharlemann, E. T. 1979, *ApJ*, **231**, 854
Arons, J. & Tavani, M. 1994, *ApJS*, **90**, 797
Arzoumanian, Z., Chernoff, D. F., & Cordes, J. M. 2002, *ApJ*, **568**, 289
Aschenbach, B. 1998, *Nature*, **396**, 141
Atoyan, A. M. 1999, *A&A*, **346**, L49
Baade, W. & Zwicky, F. 1934, *Proc. Nat. Acad. Sci.*, **20**, 254
Bahcall, J. N. & Wolf, R. A. 1965, *ApJ*, **142**, 1254
Bandiera, R., Neri, R., & Cesaroni, R. 2002, *A&A*, **386**, 1044
Baring, M. G. & Harding, A. K. 2001, *ApJ*, **547**, 929
Becker, R. H. & Helfand, D. J. 1984, *ApJ*, **283**, 154
1985, *ApJ*, **297**, L25
Becker, R. H., Helfand, D. J., & Szymkowiak, A. E. 1982, *ApJ*, **255**, 557
Becker, W. 2001, in *X-ray Astronomy: Stellar Endpoints, AGN, and the Diffuse X-ray Background*, AIP Conf. Proc. **599**, eds. N. White, G. Malaguti, & G. G. Palumbo, 13–24
Becker, W. & Aschenbach, B. 2002, in *Neutron Stars, Pulsars, and Supernova Remnants*, eds. W. Becker, H. Lesch, & J. Trümper (Garching bei München: Max-Plank-Institut für extraterrestrische Physik), 64
Becker, W. & Pavlov, G. G. 2002, in *The Century of Space Science*, eds. J. Bleeker, J. Geiss, & M. Huber (Dordrecht: Kluwer)
Becker, W. & Trümper, J. 1997, *A&A*, **326**, 682
1999, *A&A*, **341**, 803
Becker, W., Brazier, K. T. S., & Trümper, J. 1996, *A&A*, **306**, 464
Becker, W., Weisskopf, M. C., Tennant, A. F., *et al.* 2004, *ApJ*, **615**, 908
Bell, J. F., Bailes, M., Manchester, R. N., Weisberg, J. M., & Lyne, A. G. 1995, *ApJ*, **440**, L81
Bietenholz, M. F. & Kronberg, P. P. 1990, *ApJ*, **357**, L13
Bietenholz, M. F., Kassim, N., Frail, D. A., *et al.* 1997, *ApJ*, **490**, 291
Bietenholz, M. F., Frail, D. A., & Hester, J. J. 2001, *ApJ*, **560**, 254
Bignami, G. F., Caraveo, P. A., De Luca, A., & Mereghetti, S. 2003, *Nature*, **423**, 725
Blaes, O. & Madau, P. 1993, *ApJ*, **403**, 690
Blondin, J. M., Chevalier, R. A., & Frierson, D. M. 2001, *ApJ*, **563**, 806
Bocchino, F. & Bandiera, R. 2003, *A&A*, **398**, 195
Braje, T. M. & Romani, R. W. 2002, *ApJ*, **580**, 1043
Braje, T. M., Romani, R. W., Roberts, M. S. E., & Kawai, N. 2002, *ApJ*, **565**, L91
Brinkmann, W. & Ögelman, H. 1987, *A&A*, **182**, 71
Brisken, W. F., Fruchter, A., Goss, W., Herrnstein, R., & Thorsett, S. 2003a, *AJ*, **126**, 3090
Brisken, W. F., Thorsett, S. E., Golden, A., & Goss, W. M. 2003b, *ApJ*, **593**, L89
Bucciantini, N. 2002, *A&A*, **387**, 1066
Bucciantini, N., Blondin, J. M., Del Zanna, L., & Amato, E. 2003, *A&A*, **405**, 617
Burwitz, V., Haberl, F., Neuhäuser, R., *et al.* 2003, *A&A*, **399**, 1109
Camilo, F., Kaspi, V. M., Lyne, A. G., *et al.* 2000, *ApJ*, **541**, 367
Camilo, F. *et al.* 2001, *ApJ*, 557, L51
Camilo, F., Lorimer, D. R., Bhat, N. D. R., *et al.* 2002a, *ApJ*, **574**, L71
Camilo, F., Manchester, R., Gaensler, B., Lorimer, D., & Sarkissian, J. 2002b, *ApJ*, **567**, L71
Camilo, F. *et al.* 2002c, *ApJ*, **571**, L41
Campana, S., Stella, L., Mereghetti, S., & Colpi, M. 1995, *A&A*, **297**, 385
Caraveo, P. A. & Mignani, R. P. 1999, *A&A*, **344**, 366
Caraveo, P. A., Bignami, G. F., DeLuca, A., *et al.* 2003, *Science*, **301**, 1345
Chabrier, G., Potekhin, A. Y., & Yakovlev, D. G. 1997, *ApJ*, **477**, L99
Chakrabarty, D., Pivovaroff, M., Hernquist, L., Heyl, J., & Narayan, R. 2001, *ApJ*, **548**, 800
Chanan, G. A., Helfand, D. J., & Reynolds, S. P. 1984, *ApJ*, **287**, L23
Chen, K. & Ruderman, M. 1993, *ApJ*, **408**, 179
Cheng, K. S. & Zhang, L. 1999, *ApJ*, **515**, 337
Cheng, K. S., Ho, C., & Ruderman, M. 1986, *ApJ*, **300**, 500
Cheng, K. S., Ruderman, M. A., & Zhang, L. 2000, *ApJ*, **537**, 964
Chevalier, R. A. 1982, *ApJ*, **259**, 302
2000, *ApJ*, **539**, L45
Chiu, H. Y. & Salpeter, E. E. 1964, *Phys. Rev. Lett.*, **12**, 413
Clark, D. & Stephenson, F. 1977, *MNRAS*, **179**, 87P

Clark, D. H. & Caswell, J. L. 1976, *MNRAS*, **174**, 267
Comella, J. M., Craft, H., Lovelace, R., Sutton, J., & Tyler, G. 1969, *Nature*, **221**, 453
Cominsky, L., Roberts, M., & Johnston, S. 1994, *ApJ*, **427**, 978
Contopoulos, J., Kazanas, D., & Fendt, C. 1999, *ApJ*, **511**, 351
Crawford, F. & Keim, N. C. 2003, *ApJ*, **590**, 1020
Crawford, F., Kaspi, V. M., Manchester, R. N., *et al.* 1998, in *Proc. Elba Workshop: Neutron Stars and Supernova Remnants*, Vol. 69 (Memorie della Societa Astronomica Italiana), 951–954
Crawford, F., Manchester, R. N., & Kaspi, V. M. 2001, *AJ*, **122**, 2001
Cropper, M., Haberl, F., Zane, S., & Zavlin, V. 2004, *MNRAS*, **351**, 1099
Danner, R. 1998a, *A&AS*, **128**, 331
1998b, *A&AS*, **128**, 349
Daugherty, J. K. & Harding, A. K. 1982, *ApJ*, **252**, 337
1996, *ApJ*, **458**, 278
de Jager, O. C., Harding, A. K., Michelson, P. F., *et al.* 1996a, *ApJ*, **457**, 253
de Jager, O. C., Harding, A. K., Sreekumar, P., & Strickman, M. 1996b, *A&AS*, **120**, C441+
De Luca, A., Mereghetti, S., Caraveo, P. A., *et al.* 2004, *A&A*, **418**, 625
Dedeo, S. & Psaltis, D. 2003, *Phys. Rev. Lett.*, 90, 141101
Deeter, J. E., Nagase, F., & Boynton, P. E. 1999, *ApJ*, **512**, 300
Dickel, J. R., Milne, D. K., & Strom, R. G. 2000, *ApJ*, **543**, 840
Dodson, R., Legge, D., Reynolds, J. E., & McCulloch, P. M. 2003a, *ApJ*, **596**, 1137
Dodson, R., Lewis, D., McConnell, D., & Deshpande, A. A. 2003b, *MNRAS*, **343**, 116
Drake, J. J., *et al.* 2002, *ApJ*, **572**, 996
Dyks, J. & Rudak, B. 2003, *ApJ*, **598**, 1201
Dyks, J., Harding, A. K., & Rudak, B. 2004, *ApJ*, **606**, 1125
Endal, A. S. & Sofia, S. 1978, *ApJ*, **220**, 279
Fesen, R. A. & Kirshner, R. P. 1982, *ApJ*, **258**, 1
Fierro, J. M., Michelson, P. F., Nolan, P. L., & Thompson, D. J. 1998, *ApJ*, **494**, 734
Finley, J. P., Ogelman, H., & Kiziloglu, U. 1992, *ApJ*, **394**, L21
Finzi, A. & Wolf, R. A. 1969, *ApJ*, **155**, 107
Flowers, E. G., Ruderman, M., & Sutherland, P. G. 1976, *ApJ*, **205**, 241
Frail, D. A. & Scharringhausen, B. R. 1997, *ApJ*, **480**, 364
Frail, D. A., Kassim, N. E., Cornwell, T. J., & Goss, W. M. 1995, *ApJ*, **454**, L129
Freire, P. C., Kramer, M., Lyne, A., *et al.* 2001, *ApJ*, **557**, L105
Fruchter, A. S., Stinebring, D. R., & Taylor, J. H. 1988, *Nature*, **333**, 237
Fruchter, A. S., Bookbinder, J., Garcia, M. R., & Bailyn, C. D. 1992, *Nature*, **359**, 303
Gaensler, B. M. & Frail, D. A. 2000, *Nature*, **406**, 158
Gaensler, B. M. & Wallace, B. J. 2003, *ApJ*, **594**, 326
Gaensler, B. M., Bock, D. C.-J., & Stappers, B. W. 2000a, *ApJ*, **537**, L35
Gaensler, B. M., Stappers, B. W., Frail, D. A., & Johnston, S. 1998, *ApJ*, **499**, L69
Gaensler, B. M., Stappers, B. W., Frail, D. A. *et al.* 2000b, *MNRAS*, **318**, 58
Gaensler, B. M., Jones, D. H., & Stappers, B. W. 2002b, *ApJ*, **580**, L137
Gaensler, B. M., Arons, J., Kaspi, V., *et al.* 2002a, *ApJ*, **569**, 878
Gaensler, B. M., Hendrick, S. P., Reynolds, S. P., & Borkowski, K. J. 2003a, *ApJ*, **594**, L111
Gaensler, B. M., Schulz, N., Kaspi, V., Pivovaroff, M., & Becker, W. 2003b, *ApJ*, **588**, 441
Gaensler, B. M., van der Swaluw, E., Camilo, F., *et al.* 2004, *ApJ*, **616**, 383
Gallant, Y. A. & Arons, J. 1994, *ApJ*, **435**, 230
Gamow, G. & Schoenberg, S. 1941, *Phys. Rev.*, **59**, 539
Garmire, G. P., Pavlov, G. G., Garmire, A. B., & Zavlin, V. E. 2000 *IAU circular* 7350
Giacani, E. B., Frail, D. A., Goss, W. M., & Vieytes, M. 2001, *AJ*, **121**, 3133
Goldreich, P. & Julian, W. H. 1969, *ApJ*, **157**, 869
Gonzalez, M. & Safi-Harb, S. 2003, *ApJ*, **591**, L143
Gotthelf, E. V. 2003, *ApJ*, **591**, 361
Gotthelf, E. V. & Wang, Q. D. 2000, *ApJ*, **532**, L117
Gotthelf, E. V., Petre, R., & Vasisht, G. 1999, *ApJ*, **514**, L107
Gotthelf, E. V., Vasisht, G., Boylan-Kolchin, M., & Torii, K. 2000, *ApJ*, **542**, L37
Gotthelf, E. V., Halpern, J. P., & Dodson, R. 2002, *ApJ*, **567**, L125
Green, D. A. & Scheuer, P. A. G. 1992, *MNRAS*, **258**, 833

Green, D. A., Gull, S. F., Tan, S. M., & Simon, A. J. B. 1988, *MNRAS*, **231**, 735
Greiveldinger, C. *et al.* 1996, *ApJ*, **465**, L35
Grindlay, J. E., Camilo, F., Heinke, C., *et al.* 2002, *ApJ*, **581**, 470
Gudmundsson, E. H., Pethick, C. J., & Epstein, R. I. 1983, *ApJ*, **272**, 286
Haberl, F. 2003, in *High Energy Studies of Supernova Remnants and Neutron Stars*, eds. W. Becker & W. Hermsen (astro-ph/0302540)
Haberl, F. 2004, in *Memorie della Societa Astronomica Italiana*
Haberl, F. & Zavlin, V. E. 2002, *A&A*, **391**, 571
Haberl, F., Motch, C., Buckley, D. A. H., Zickgraf, F.-J., & Pietsch, W. 1997, *A&A*, **326**, 662
Haberl, F., Motch, C., & Pietsch, W. 1998, *Astron. Nachr.*, **319**, 97
Haberl, F., Pietsch, W., & Motch, C. 1999, *A&A*, **351**, L53
Haberl, F., Schwope, A. D., Hambaryan, V., Hasinger, G., & Motch, C. 2003, *A&A*, **403**, L19
Haberl, F., Zavlin, V. E., Trüper, J., & Burwitz, V. 2004, *A&A*, **419**, 1077
Hailey, C. J. & Mori, K. 2002, *ApJ*, **578**, L133
Halpern, J. P. & Ruderman, M. 1993, *ApJ*, **415**, 286
Halpern, J. P. & Wang, F. Y.-H. 1997, *ApJ*, **477**, 905
Halpern, J. P., Camilo, F., Gotthelf, E. V., *et al.* 2001, *ApJ*, **552**, L125
Hansen, B. & Phinney, E. S. 1997, *MNRAS*, **291**, 569
Harding, A. K. & Muslimov, A. G. 2001, *ApJ*, **556**, 987
2002, *ApJ*, **568**, 862
Harding, A. K., Muslimov, A. G., & Zhang, B. 2002a, *ApJ*, **576**, 366
Harding, A. K., Strickman, M. S., Gwinn, C., *et al.* 2002b, *ApJ*, **576**, 376
Harnden, F. R., Grant, P. D., Seward, F. D., & Kahn, S. M. 1985, *ApJ*, **299**, 828
Harrus, I., Slane, P., Gaensler, B., *et al.* 2003, in *Young Neutron Stars and their Environment*, IAU Symposium **218**, eds. F. Camilo & B. Gaensler (ASP)
Hartman, R. C. *et al.* 1999, *ApJS*, **123**, 79
Helfand, D. J. & Becker, R. H. 1984, *Nature*, **307**, 215
1987, *ApJ*, **314**, 203
Helfand, D. J., Chanan, G. A., & Novick, R. 1980, *Nature*, **283**, 337
Helfand, D. J., Gotthelf, E. V., & Halpern, J. P. 2001, *ApJ*, **556**, 380
Helfand, D. J., Agüeros, M. A., & Gotthelf, E. V. 2003a, *ApJ*, **592**, 941
Helfand, D. J., Collins, B. F., & Gotthelf, E. V. 2003b, *ApJ*, **582**, 783
Hessels, J. W., Roberts, M. S., Ransom, S. M., *et al.* 2003, in IAU Symposium 218, *Young Neutron Stars and their Environment*, eds. F. Camilo & B. Gaensler (ASP)
Hester, J. J. *et al.* 1996, *ApJ*, **456**, 225
Hester, J. J., Mori, K., Burrows, D., *et al.* 2002, *ApJ*, **577**, L49
Heyl, J. S. & Hernquist, L. 1997, *ApJ*, **491**, L95
2001, *MNRAS*, **324**, 292
Heyl, J. S. & Shaviv, N. J. 2002, *Phys. Rev. D*, **66**, 23002
Heyl, J. S., Shaviv, N. J., & Lloyd, D. 2003, *MNRAS*, **342**, 134
Hibschman, J. A. & Arons, J. 2001, *ApJ*, **546**, 382
Hirayama, M., Nagase, F., Tavani, M., *et al.* 1996, *PASJ*, **48**, 833
Hirayama, M., Cominsky, L. R., Kaspi, V. M., *et al.* 1999, *ApJ*, **521**, 718
Hirotani, K. & Shibata, S. 2001, *MNRAS*, **325**, 1228
Hirotani, K., Harding, A. K., & Shibata, S. 2003, *ApJ*, **591**, 334
Ho, W. C. G. & Lai, D. 2001, *MNRAS*, **327**, 1081
2003, *MNRAS*, **338**, 233
2004, *ApJ*, **607**, 420
Ho, W. C. G., Lai, D., Potekhin, A. Y., & Chabrier, G. 2003, *ApJ*, **599**, 1293
Hughes, J. P., Slane, P. O., & Plucinsky, P. 2000, *ApJ*, **542**, 386
Hughes, J. P., Slane, P. O., Burrows, D. N., *et al.* 2001, *ApJ*, **559**, L153
Johnston, S., Manchester, R. N., Lyne, A. G., *et al.* 1992, *ApJ*, **387**, L37
Jones, D. H., Stappers, B. W., & Gaensler, B. M. 2002, *A&A*, **389**, L1
Kaminker, A. D., Yakovlev, D. G., & Gnedin, O. Y. 2002, *A&A*, **383**, 1076
Kanbach, G. 2002, in *Neutron Stars, Pulsars, and Supernova Remnants*, eds. W. Becker, H. Lesch, & J. Trümper (Garching bei München: Max-Plank-Institut für extraterrestrische Physik), 91
Kanno, S. 1975, *PASJ*, **27**, 287

Kaplan, D. L., Kulkarni, S. R., & Murray, S. S. 2001, *ApJ*, **558**, 270
Kaplan, D. L., van Kerkwijk, M. H., & Anderson, J. 2002a, *ApJ*, **571**, 447
Kaplan, D. L., Kulkarni, S. R., & van Kerkwijk, M. H. 2002b, *ApJ*, **579**, L29
2003a, *ApJ*, **588**, L33
Kaplan, D. L., van Kerkwijk, M. H., Marshall, H. L., *et al.* 2003b, *ApJ*, **590**, 1008
Kargaltsev, O., Pavlov, G. G., Sanwal, D., & Garmire, G. P. 2002, *ApJ*, **580**, 1060
Kaspi, V. M., Manchester, R. N., Johnston, S., Lyne, A. G., & D'Amico, N. 1992, *ApJ*, **399**, L155
Kaspi, V. M., Manchester, R. N., Siegman, B., Johnston, S., & Lyne, A. G. 1994, *ApJ*, **422**, L83
Kaspi, V. M., Tavani, M., Nagase, F., *et al.* 1995, *ApJ*, **453**, 424
Kaspi V. M., Manchester, R. N., Johnston, S., Lyne, A. G., & D'Amico, N. 1996, *AJ*, **111**, 2028
Kaspi, V. M., Bailes, M., Manchester, R. N., *et al.*1997, *ApJ*, **485**, 820
Kaspi, V. M., Gotthelf, E. V., Gaensler, B. M., & Lyutikov, M. 2001a, *ApJ*, **562**, L163
Kaspi, V. M., Roberts, M. S. E., Vasisht, G., *et al.* 2001b, *ApJ*, **560**, 371
Kellner, S. 2002, Ph.D. thesis, Technische Universität München
Kennel, C. F. & Coroniti, F. V. 1984a, *ApJ*, **283**, 694
1984b, *ApJ*, **283**, 710
Kerkwijk van, M. H. & Kulkarni, S. R. 2001, *A&A*, **380**, 221
Kerkwijk van, M. H., Kaplan, D. L., Durant, M., Kulkarni, S. R., & Paerels, F. 2004, *ApJ*, **608**, 432
Khangoulian, D. V. & Bogovalov, S. V. 2003, *Astron. Lett.*, **29**, 495
Kifune, T. *et al.* 1993, *IAU circular* 5905
Kirk, J. G. & Skjæraasen, O. 2003, *ApJ*, **591**, 366
Kirk, J. G., Guthmann, A. W., Gallant, Y. A., & Achterberg, A. 2000, *ApJ*, **542**, 235
Komissarov, S. S. & Lyubarsky, Y. E. 2003, *MNRAS*, **344**, L93
Krause-Polstorff, J. & Michel, F. C. 1985, *MNRAS*, **213**, 43P
Kuiper, L., Hermsen, W., Verbunt, F., & Belloni, T. 1998, *A&A*, **336**, 545
Kuiper, L., Hermsen, W., Verbunt, F., *et al. A&A*, **359**, 615
Kuiper, L., Hermsen, W., Cusumano, G., *et al.* 2001, *A&A*, **378**, 918
Kulkarni, S. R. & Hester, J. J. 1988, *Nature*, **335**, 801
Kulkarni, S. R., Phinney, E. S., Evans, C. R., & Hasinger, G. 1992, *Nature*, **359**, 300
Lai, D. & Ho, W. C. G. 2002, *ApJ*, **566**, 373
2003a, *Phys. Rev. Lett.*, **91**, 071101
2003b, *ApJ*, **588**, 962
Lampland, C. O. 1921, *PASP*, **33**, 79
Larson, M. B. & Link, B. 1999, *ApJ*, **521**, 271
Lattimer, J. M. 1992, in *The Structure and Evolution of Neutron Stars*, eds. D. Pines, R. Tamagaki, & S. Tsuruta (USA: Addison-Wesley)
Lattimer, J. M. & Prakash, M. 2001, *ApJ*, **550**, 426
Lattimer, J. M., Prakash, M., Pethick, C. J., & Haensel, P. 1991, *Phys. Rev. Lett.*, **66**, 2701
Lodenqual, J., Canuto, V., Ruderman, M., & Tsuruta, S. 1974, *ApJ*, **190**, 141
Lu, F. J., Wang, Q. D., Aschenbach, B., Durouchoux, P., & Song, L. M. 2002, *ApJ*, **568**, L49
Lu, F. J., Wang, Q. D., & Lang, C. C. 2003, *AJ*, **126**, 319
Lyne, A. G. & Lorimer, D. R. 1994, *Nature*, **369**, 127
Lyne, A. G. & Manchester, R. N. 1988, *MNRAS*, **234**, 477
Lyne, A. G., Pritchard, R. S., & Smith, F. G. 1988, *MNRAS*, **233**, 667
Lyne, A. G., Pritchard, R. S., Graham-Smith, F., & Camilo, F. 1996, *Nature*, **381**, 497
Lyne, A. G., Manchester, R. N., Lorimer, D. R., *et al.* 1998, *MNRAS*, **295**, 743
Manchester, R. N. 1996, in *Pulsars: Problems and Progress*, IAU Colloquium 160, eds. S. Johnston, M. A. Walker, & M. Bailes (San Francisco: ASP), 193–196
Manchester, R. N., Staveley-Smith, L., & Kesteven, M. J. 1993, *ApJ*, **411**, 756
Marshall, H. L. & Schulz, N. S. 2002, *ApJ*, **574**, 377
Marshall, F. E., Gotthelf, E. V., Zhang, W., Middleditch, J., & Wang, Q. D. 1998, *ApJ*, **499**, L179
Matsui, Y., Long, K. S., & Tuohy, I. R. 1988, *ApJ*, **329**, 838
McGowan, K. E., Kennea, J. A., Zane, S., *et al.* 2003, *ApJ*, **591**, 380
McLaughlin, M. A., Cordes, J. M., Deshpande, A. A., *et al.* 2001, *ApJ*, **547**, L41
Melatos, A. 1998, *Mem. Soc. Astron. Ital.*, **69**, 1009
Melrose, D. B. 2000, in *Pulsar Astronomy – 2000 and Beyond*, IAU Colloquium 177, eds. M. Kramer, N. Wex, & R. Wielebinski (San Francisco: ASP), 721

Mereghetti, S., Bignami, G. F., & Caraveo, P. A. 1996, *ApJ*, **464**, 842
Mereghetti, S., De Luca, A., Caraveo, P. A., *et al.* 2002a, *ApJ*, **581**, 1280
Mereghetti, S., Tiengo, A., & Israel, G. L. 2002b, *ApJ*, **569**, 275
Michel, F. C. 1969, *ApJ*, **158**, 727
Morton, D. C. 1964, *ApJ*, **140**, 460
Motch, C., Haberl, F., Zickgraf, F.-J., Hasinger, G., & Schwope, A. D. 1999, *A&A*, **351**, 177
Motch, C., Zavlin, V. E., & Haberl, F. 2003, *A&A*, **408**, 323
Mukherjee, R., Gotthelf, E. V., Halpern, J., & Tavani, M. 2000, *ApJ*, **542**, 740
Murray, S. S., Slane, P. O., Seward, F. D., Ransom, S. M., & Gaensler, B. M. 2002, *ApJ*, **568**, 226
Muslimov, A. G. & Harding, A. K. 2003, *ApJ*, **588**, 430
2004, *ApJ*, **606**, 1143
Muslimov, A. G. & Tsygan, A. I. 1992, *MNRAS*, **255**, 61
Ng, C. & Romani, R. W. 2004, *ApJ*, **601**, 479
Nicastro, L., Cusumano, G., Löhmer, O., *et al.* 2004, *A&A*, **413**, 1065
Nolan, P. L., Tompkins, W. F., Grenier, I. A., & Michelson, P. F. 2003, *ApJ*, **597**, 615
Ögelman, H., Finley, J. P., & Zimmermann, H. U. 1993, *Nature*, **361**, 136
Olbert, C. M., Clearfield, C., Williams, N., Keohane, J., & Frail, D. 2001, *ApJ*, **554**, L205
Olbert, C. M., Keohane, J. W., Arnaud, K. A., *et al.* 2003, *ApJ*, **592**, L45
Oort, J. & Walraven, T. 1956, *Bull. Astron. Inst. Neth.*, **12**, 285
Ostriker, J. P. & Gunn, J. E. 1969, *ApJ*, **157**, 1395
Ostriker, J. P., Rees, M. J., & Silk, J. 1970, *Astrophys. Lett.*, **6**, 179
Özel, F. 2001, *ApJ*, **563**, 276
2003, *ApJ*, **583**, 402
Pacini, F. & Salvati, M. 1973, *ApJ*, **186**, 249
Page, D. 1995, *ApJ*, **442**, 273
1998, in *The Many Faces of Neutron Stars*, eds. R. Buccheri, J. van Paradijs, & M. A. Alpar (Dordrecht; Boston: Kluwer Academic Publishers), 539
Pavlov, G. G. & Shibanov, I. A. 1978, *Astron. Zh.*, **55**, 373
Pavlov, G. G. & Zavlin, V. E. 2000, in *Pulsar Astronomy – 2000 and Beyond*, IAU Colloquium 177, eds. M. Kramer, N. Wex, & R. Wielebinski (San Francisco: ASP), 613
2003, in *Proc. XXI Texas Symposium on Relativistic Astrophysics*, eds. R. Bandiera, R. Maiolino, & F. Mannucci (World Scientific Publishing Co.), 319
Pavlov, G. G., Zavlin, V. E., Aschenbach, B., Trümper, J., & Sanwal, D. 2000, *ApJ*, **531**, L53
Pavlov, G. G., Zavlin, V. E., Sanwal, D., Burwitz, V., & Garmire, G. P. 2001a, *ApJ*, **552**, L129
Pavlov, G. G., Sanwal, D., Kızıltan, B., & Garmire, G. P. 2001b, *ApJ*, **559**, L131
Pavlov, G. G., Sanwal, D., Garmire, G., & Zavlin, V. 2002a, in *Neutron Stars in Supernova Remnants*, eds. P. Slane & B. Gaensler (San Francisco: ASP), 247
Pavlov, G. G., Zavlin, V. E., & Sanwal, D. 2002b, in *Heraeus Seminar on Neutron Stars, Pulsars, and Supernova Remnants*, MPE Report 278, eds. W. Becker, H. Lesch, & J. Trümper, 273
Pavlov, G. G., Zavlin, V. E., Sanwal, D., & Trümper, J. 2002c, *ApJ*, **569**, L95
Pavlov, G. G., Teter, M. A., Kargaltsev, O., & Sanwal, D. 2003, *ApJ*, **591**, 1157
Perna, R., Narayan, R., Rybicki, G., Stella, L., & Treves, A. 2003, *ApJ*, **594**, 936
Pethick, C. J. 1992, *Rev. Mod. Phys.*, **64**, 1133
Petre, R., Kriss, G. A., Winkler, P. F., & Canizares, C. R. 1982, *ApJ*, **258**, 22
Petre, R., Becker, C. M., & Winkler, P. F. 1996, *ApJ*, **465**, L43
Petre, R., Kuntz, K. D., & Shelton, R. L. 2002, *ApJ*, **579**, 404
Pétri, J., Heyvaerts, J., & Bonazzola, S. 2002, *A&A*, **384**, 414
Porquet, D., Decourchelle, A., & Warwick, R. S. 2003, *A&A*, **401**, 197
Potekhin, A. Y. & Yakovlev, D. G. 2001, *A&A*, **374**, 213
Potekhin, A., Yakovlev, D., Chabrier, G., & Gnedin, O. 2003, *ApJ*, **594**, 404
Prakash, A. Y., Ainsworth, T. L., & Lattimer, J. M. 1988, *Phys. Rev. Lett.*, **61**, 2518
Prakash, M., Prakash, M., Lattimer, J. M., & Pethick, C. J. 1992, *ApJ*, **390**, L77
Predehl, P. & Kulkarni, S. R. 1995, *A&A*, **294**, L29
Psaltis, D., Özel, F., & DeDeo, S. 2000, *ApJ*, **544**, 390
Radhakrishnan, V. & Cooke, D. J. 1969, *Astrophys. Lett.*, **3**, 225
Rajagopal, M. & Romani, R. W. 1996, *ApJ*, **461**, 327
Rajagopal, M., Romani, R. W., & Miller, M. C. 1997, *ApJ*, **479**, 347

Rankin, J. M. 1993, *ApJ*, **405**, 285
Ransom, S. M., Gaensler, B. M., & Slane, P. O. 2002, *ApJ*, **570**, L75
Rees, M. J. & Gunn, J. E. 1974, *MNRAS*, **167**, 1
Reich, W., Fürst, E., & Sofue, Y. 1984, *A&A*, **133**, 4
Reynolds, S. P. & Aller, H. D. 1985, *AJ*, **90**, 2312
Reynolds, S. P. & Chevalier, R. A. 1984, *ApJ*, **278**, 630
Riper van, K. A. & Lamb, D. Q. 1981, *ApJ*, **244**, L13
Riper van, K. A., Link, B., & Epstein, R. I. 1995, *ApJ*, **448**, 294
Roberts, M. S. E., Romani, R. W., Johnston, S., & Green, A. J. 1999, *ApJ*, **515**, 712
Roberts, M. S. E., Romani, R. W., & Johnston, S. 2001, *ApJ*, **561**, L187
Roberts, M. S. E., Romani, R. W., & Kawai, N. 2001, *ApJS*, **133**, 451
Roberts, M. S. E., Tam, C. R., Kaspi, V. M., *et al.* 2003, *ApJ*, **588**, 992
Romani, R. W. 1987, *ApJ*, **313**, 718
1996, *ApJ*, **470**, 469
Romani, R. W. & Yadigaroglu, I.-A. 1995, *ApJ*, **438**, 314
Romani, R. W., Cordes, J. M., & Yadigaroglu, I.-A. 1997, *ApJ*, **484**, L137
Ruderman, M. 2003, in *X-ray and Gamma-ray Astrophysics of Galactic Sources* (astro-ph/0310777)
Ruderman, M. A. & Sutherland, P. G. 1975, *ApJ*, **196**, 51
Rutledge, R. E., Fox, D. W., Bogosavljevic, M., & Mahabal, A. 2003, *ApJ*, **598**, 458
Ryan, E., Wagner, R. M., & Starrfield, S. G. 2001, *ApJ*, **548**, 811
Safi-Harb, S., Ogelman, H., & Finley, J. P. 1995, *ApJ*, **439**, 722
Saito, Y. 1998, Ph.D. thesis, University of Tokyo
Sakurai, I., Kawai, N., Torii, K., *et al.* 2001, *PASJ*, **53**, 535
Sanwal, D., Garmire, G. P., Garmire, A., Pavlov, G. G., & Mignani, R. 2002a, *BAAS*, **200**, 7201
Sanwal, D., Pavlov, G. G., Zavlin, V. E., & Teter, M. A. 2002b, *ApJ*, **574**, L61
Scargle, J. D. 1969, *ApJ*, **156**, 401
Schwope, A. D., Hasinger, G., Schwarz, R., Haberl, F., & Schmidt, M. 1999, *A&A*, **341**, L51
Sedov, L. I. 1959, *Similarity and Dimensional Methods in Mechanics* (New York: Academic Press)
Shibanov, Y. A. & Yakovlev, D. G. 1996, *A&A*, **309**, 171
Shibata, S., Tomatsuri, H., Shimanuki, M., Saito, K., & Mori, K. 2003, *MNRAS*, **346**, 841
Shull, J. M., Fesen, R. A., & Saken, J. M. 1989, *ApJ*, **346**, 860
Slane, P. O. 1994, *ApJ*, **437**, 458
Slane, P. O., Chen, Y., Schulz, N., *et al.* 2000, *ApJ*, **533**, L29
Slane, P., Hughes, J. P., Edgar, R. J., *et al.* 2001, *ApJ*, **548**, 814
Slane, P. O., Helfand, D. J., & Murray, S. S. 2002, *ApJ*, **571**, L45
Slane, P., Zimmerman, E. R., Hughes, J. P., *et al.* 2003, *ApJ*, **601**, 1045
Spitkovsky, A. & Arons, J. 2002, in *Neutron Stars in Supernova Remnants*, eds. P. O. Slane & B. M. Gaensler (San Francisco: ASP), 81
Spruit, H. & Phinney, E. S. 1998, *Nature*, **393**, 139
Stappers, B. W., Gaensler, B. M., Kaspi, V. M., van der Klis, M., & Lewin, W. H. G. 2003, *Science*, **299**, 1372
Staelin, D. H. & Reifenstein, E. C. III, 1968, *Science*, **162**, 1481
Swaluw, E. van der, 2003, *A&A*, **404**, 939
Swaluw, E. van der, Achterberg, A., Gallant, Y. A., & Tóth, G. 2001, *A&A*, **380**, 309
Swaluw, E. van der, Achterberg, A., Gallant, Y., Downes, T., & Keppens, R. 2003a, *A&A*, **397**, 913
Swaluw, E. van der, Downes, T. P., & Keegan, R. 2004, *A&A*, **420**, 937
Tam, C. & Roberts, M. S. E. 2003, *ApJ*, **598**, L27
Tam, C., Roberts, M. S. E., & Kaspi, V. M. 2002, *ApJ*, **572**, 202
Tananbaum, H. 1999, *IAU circular* 7246
Tavani, M. & Arons, J. 1997, *ApJ*, **477**, 439
Tavani, M., Arons, J., & Kaspi, V. M. 1994, *ApJ*, **433**, L37
Tennant, A. F. *et al.* 2001, *ApJ*, **554**, L173
Thompson, D. J. *et al.* 1999, *ApJ*, **516**, 297
Thorsett, S. E., Brisken, W. F., & Goss, W. M. 2002, *ApJ*, **573**, L111
Torii, K., Tsunemi, H., Dotani, T., & Mitsuda, K. 1997, *ApJ*, **489**, 145
Torii, K., Kinugasa, K., Toneri, T., *et al.* 1998, *ApJ*, **494**, L207
Torii, K., Tsunemi, H., Dotani, T., *et al.* 1999, *ApJ*, **523**, L69

Treves, A. & Colpi, M. 1991, *A&A*, **241**, 107
Treves, A., Turolla, R., Zane, S., & Colpi, M. 2000, *PASP*, **112**, 297
Tsuruta, S. 1964, Ph.D. thesis, Columbia University
  1986, *Comments Astrophys.*, **11**, 151
  1998, *Phys. Rep.*, **292**, 1
Tsuruta, S. & Cameron, A. G. W. 1966, *Can. J. Phys.*, **44**, 1863
Tsuruta, S., Teter, M. A., Takatsuka, T., Tatsumi, T., & Tamagaki, R. 2002, *ApJ*, **571**, L143
Tuohy, I. & Garmire, G. 1980, *ApJ*, **239**, 107
Umeda, H., Shibazaki, N., Nomoto, K., & Tsuruta, S. 1993, *ApJ*, **408**, 186
Usov, V. V. & Melrose, D. B. 1995, *Aust. J. Phys.*, **48**, 571
Vasisht, G., Kulkarni, S. R., Anderson, S. B., Hamilton, T. T., & Kawai, N. 1997, *ApJ*, **476**, L43
Velusamy, T., Roshi, D., & Venugopal, V. R. 1992, *MNRAS*, **255**, 210
Wallace, B. J., Landecker, T. L., & Taylor, A. R. 1997, *AJ*, **114**, 2068
Walter, F. M. 2001, *ApJ*, **549**, 433
Walter, F. M. & Lattimer, J. M. 2002, *ApJ*, **576**, L145
Walter, F. M. & Matthews, L. D. 1997, *Nature*, **389**, 358
Walter, F. M., Wolk, S. J., & Neuhauser, R. 1996, *Nature*, **379**, 233
Wang, F. Y.-H., Ruderman, M., Halpern, J. P., & Zhu, T. 1998, *ApJ*, **498**, 373
Wang, Q. D., Gotthelf, E. V., Chu, Y.-H., & Dickel, J. R. 2001, *ApJ*, **559**, 275
Wang, Q. D., Lu, F., & Lang, C. C. 2002, *ApJ*, **581**, 1148
Webb, N. A., Olive, J. F., Barret, D., *et al.* 2004, *A&A*, **419**, 269
Weekes, T. C. 1991, *Space Sci. Rev.*, **59**, 315
Weiler, K. W. 1978, in *Proc. Workshop on Supernovae and Supernova Remnants*, Vol. 49, 545–552
Weiler, K. W. & Panagia, N. 1980, *A&A*, **90**, 269
Weiler, K. W. & Shaver, P. A. 1978, *A&A*, **70**, 389
Weisskopf, M. C. *et al.* 2000, *ApJ*, **536**, L81
Weisskopf, M. C., O'Dell, S. L., Paerels, F., *et al.* 2004, *ApJ*, **601**, 1050
Whiteoak, J. B. Z. & Green, A. J. 1996, *A&AS*, **118**, 329
Woltjer, L., Salvati, M., Pacini, F., & Bandiera, R. 1997, *A&A*, **325**, 295
Yakovlev, D. G., Levenfish, K. P., & Shibanov, Y. A. 1999, *Phys. Usp.*, **42**, 737
Yakovlev, D. G., Kaminker, A. D., & Gnedin, O. Y. 2001, *A&A*, **379**, L5
Yakovlev, D. G., Gnedin, O. Y., Kaminker, A. D., & Potekhin, A. Y. 2002a, in *Proc. 270 Heraeus Seminar on Neutron Stars, Pulsars and Supernova Remnants*, MPE Report 278., eds. W. Becker, H. Lesch, & J. Trümper (Bad Honnef)
Yakovlev, D. G., Kaminker, A. D., Haensel, P., & Gnedin, O. Y. 2002b, *A&A*, **389**, L24
Yakovlev, D. G., Gnedin, O. Y., Kaminker, A. D., Levenfish, K. P., & Potekhin, A. Y. 2004, *Adv. Space Res.*, **33**, 523
Yoshikoshi, T. *et al.* 1997, *ApJ*, **487**, L65
Zane, S., Turolla, R., Stella, L., & Treves, A. 2001, *ApJ*, **560**, 384
Zavlin, V. E. & Pavlov, G. G. 2002, in *Proc. 270 Heraeus Seminar on Neutron Stars, Pulsars and Supernova Remnants*, eds. W. Becker, H. Lesch, & J. Trümper (Bad Honnef), 273
Zavlin, V. E., Pavlov, G. G., & Shibanov, Y. A. 1996, *A&A*, **315**, 141
Zavlin, V. E., Pavlov, G. G., & Trümper, J. 1998, *A&A*, **331**, 821
Zavlin, V. E., Trümper, J., & Pavlov, G. G. 1999, *ApJ*, **525**, 959
Zavlin, V. E., Pavlov, G. G., Sanwal, D., & Trümper, J. 2000, *ApJ*, **540**, L25
Zavlin, V. E., Pavlov, G. G., Sanwal, D., *et al.* 2002, *ApJ*, **569**, 894
Zavlin, V. E., Pavlov, G. G., & Sanwal, D. 2004, *ApJ*, **606**, 444
Zhang, L. & Cheng, K. S. 1997, *ApJ*, **487**, 370
Zwicky, F. 1938, *ApJ*, **88**, 522

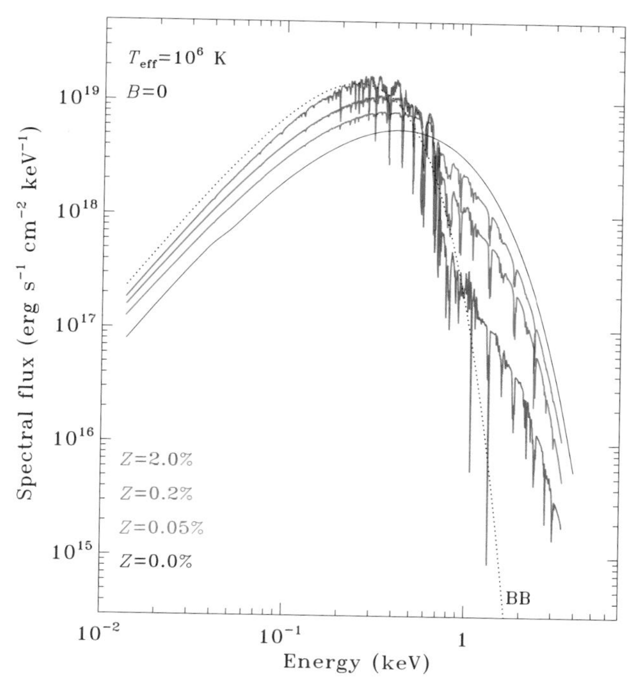

Plate 7.9. Spectra of emergent radiation in non-magnetic neutron-star atmospheres having $T_{\rm eff} = 10^6$ K for different metallicities Z ($Z = 2.0\%$ corresponds to solar metallicity). The corresponding blackbody is shown with a dotted line (From Zavlin & Pavlov 2002).

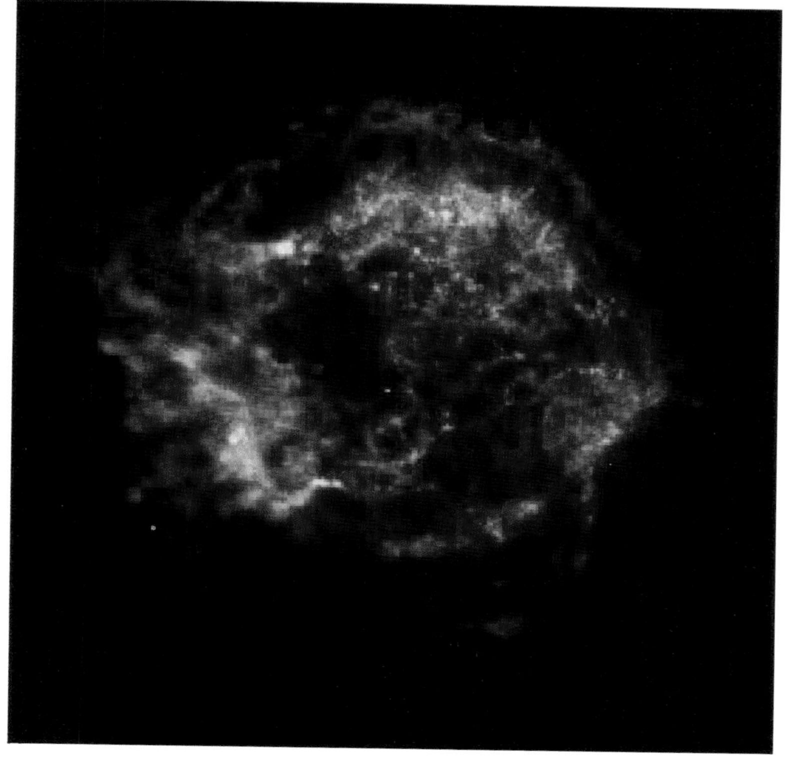

Plate 7.12. "First light" Chandra X-ray image of the supernova remnant Cas A (after Tananbaum 1999). The image is 6′ on a side. The previously unknown point source at the center of the remnant is obvious.

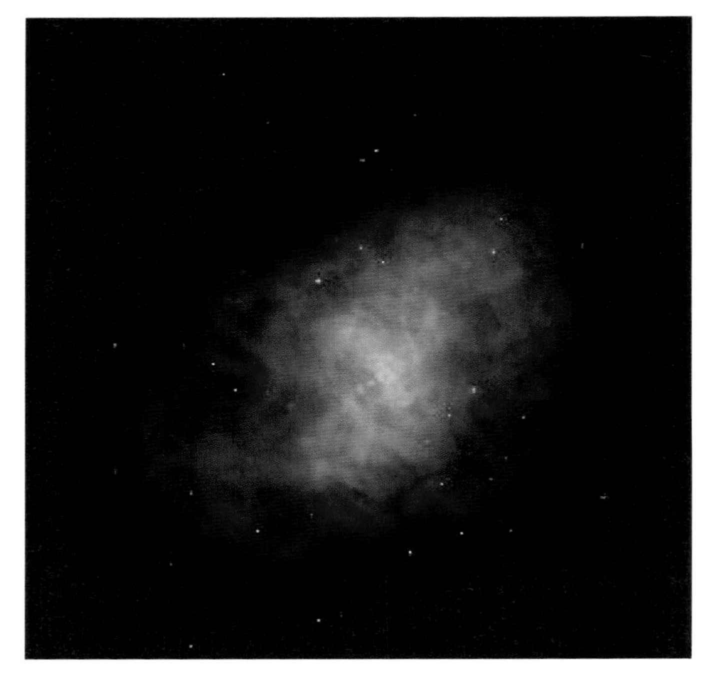

Plate 7.14. A composite image of the Crab Nebula showing X-ray in blue, optical in green, and radio in red. (Credits: X-ray: NASA/CXC/ASU/J. Hester *et al.*; Optical: NASA/HST/ASU/J. Hester *et al.*; Radio: VLA/NRAO; from Chandra website http://chandra.harvard.edu)

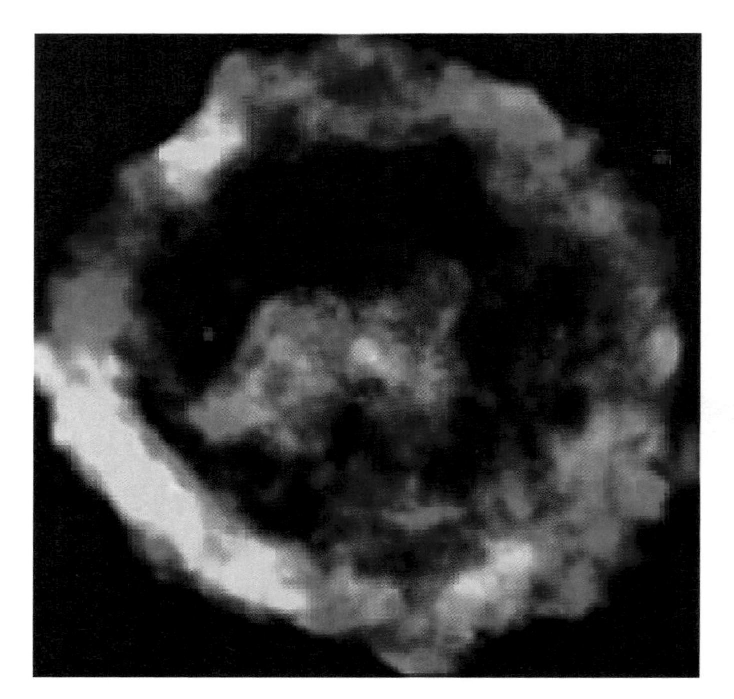

Plate 7.16. Chandra and VLA image of G11.2−0.3 and its central PWN showing the relationship between the soft thermal X-ray emission (red), the radio synchrotron emission (green) and the hard, non-thermal X-ray emission (blue). The point source at the center is the pulsar, only seen in X-rays. (Adapted from Fig. 1 of Roberts *et al.* 2003)

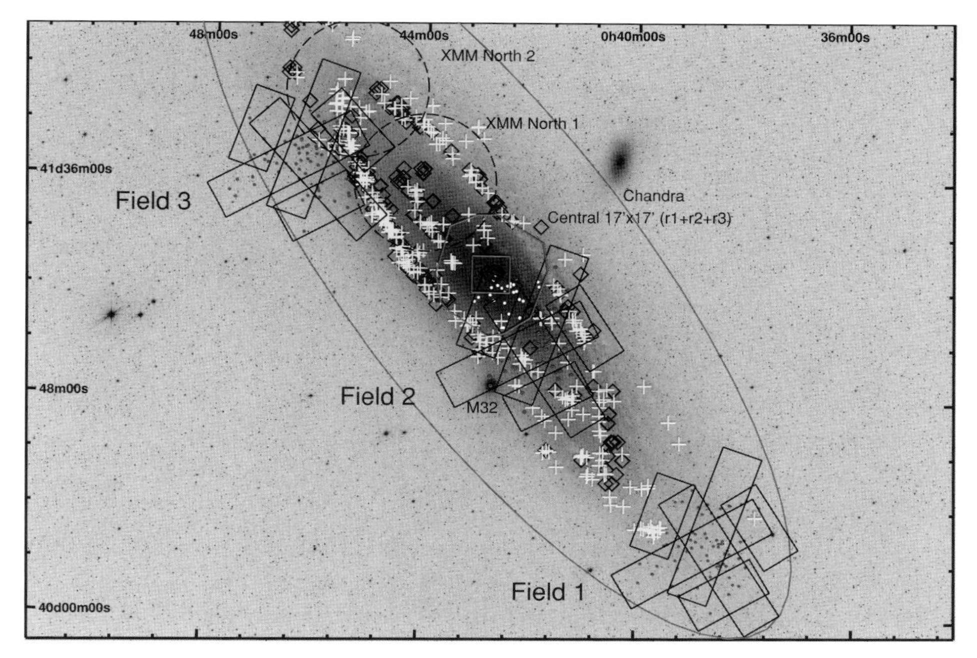

Plate 12.1. Regions of M31 observed with Chandra (rectangles) and XMM-Newton (circles). Dots are detected Chandra sources; yellow crosses and blue diamonds identify supernova remnants and OB associations in the field (not X-ray sources), respectively (from Kong *et al.* 2003).

Plate 12.5. M83 as seen with Chandra. Note the population of point-like sources and the soft diffuse emission (possibly from hot ISM), associated with the spiral arms (from http://chandra.cfa.harvard.edu/photo/2003/1154/index.html).

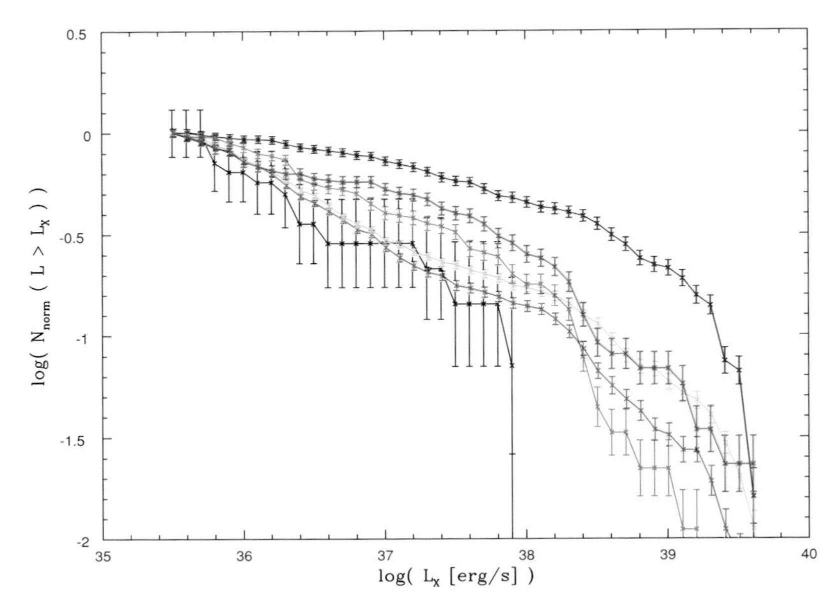

Plate 12.7. Comparison of XRB population models (from Kalogera & Belczynski 2003; also, Belczynski *et al.* 2004) with the observed XLF of NGC 1569 (bottom points; data taken from Martin *et al.* 2002). Models were constructed to match the star-formation history of NGC 1569 (recent starburst duration and metallicity) and model XLFs are shown at different times since the beginning of the starburst. Top to bottom: 10 Myr (blue) 50 Myr (yellow), 110 Myr (red), 150 Myr (cyan), 200 Myr (green). Note that, based on observations in other wavelengths, the age of the starburst is estimated to be 105–110 Myr.

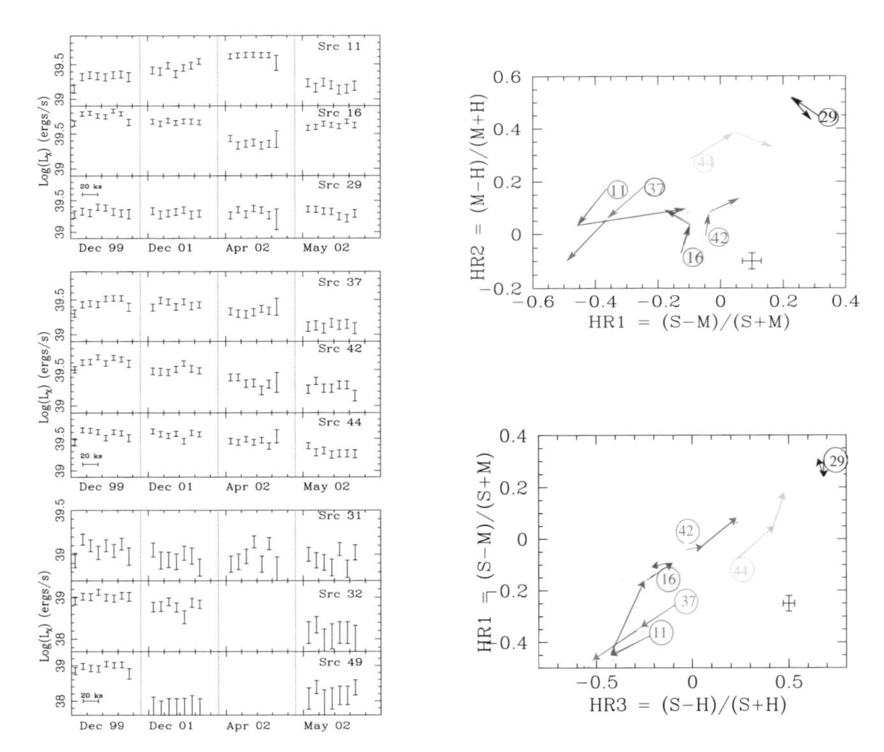

Plate 12.10. *Left*: Chandra lightcurves of the ULXs of The Antennae. *Right*: color–color diagrams of the most luminous sources (Fabbiano *et al.* 2003a).

8

Globular cluster X-ray sources

Frank Verbunt
Universiteit Utrecht

Walter H. G. Lewin
Massachusetts Institute of Technology

8.1 Introduction: some historical remarks

The earliest detections of luminous X-ray sources ($L_X \gtrsim 10^{36}$ erg s^{-1}) in globular clusters were made with the Uhuru and OSO-7 Observatories (Giacconi *et al.* 1972, 1974; Clark, Markert & Li, 1975; Canizares & Neighbours, 1975). About 10% of the luminous X-ray sources in our Galaxy are found in globular clusters. This implies that the probability (per unit mass) of finding a luminous X-ray source in a globular cluster is about two to three orders of magnitude higher than of finding one in the rest of our Galaxy (Gursky 1973; Katz 1975). Clearly, the conditions in globular clusters are very special in that they must be very efficient breeding grounds for X-ray binaries. For reviews that reflect the ideas in the late seventies and early eighties, see Lewin (1980), Lewin & Joss (1983), van den Heuvel (1983) and Verbunt & Hut (1987). At that time there was no evidence for a substantial population of binaries in globular clusters; e.g., Gunn and Griffin (1979) did not find a single binary in a spectroscopic search for radial velocity variations of 111 bright stars in M 3.

Clark (1975) suggested that the luminous cluster sources are binaries formed by capture from the remnants of massive stars. Fabian, Pringle and Rees (1975) specified that they are formed via tidal capture of neutron stars in close encounters with main-sequence stars. Sutantyo (1975) suggested direct collisions between giants and neutron stars as a formation mechanism. Hills (1976) examined the formation of binary systems through star-exchange interactions between neutron stars and primordial binaries of low-mass stars. Hut and Verbunt (1983) compared the relative efficiencies of tidal capture and exchange encounters for neutron stars and for white dwarfs; and showed that the distribution of X-ray sources among globular clusters with different central densities and core sizes is compatible with the formation by close encounters (Verbunt and Hut 1987). The importance of mass segregation, which drives the neutron stars to the core, thereby enhancing the capture rate, was demonstrated by Verbunt and Meylan (1988).

As can be seen from the discovery references in Table 8.1, five luminous globular cluster X-ray sources were known by 1975, eight by 1980, ten by 1982, and thirteen to date. Twelve of these have shown Type I X-ray bursts. Measurements of the black-body radii of the burst sources indicated that they are neutron stars (Swank *et al.* 1977; Hoffman, Lewin & Doty 1977a, b; Van Paradijs 1978). For a review, see Lewin, Van Paradijs & Taam (1995); see also Chapter 3. The absence of luminous accreting black holes in clusters of our Galaxy is presumably a consequence of the small total number of sources, as discussed in Section 8.5.1.2. There is growing evidence that black-hole binaries may exist in globular clusters in several elliptical galaxies (see Section 8.3).

Compact Stellar X-Ray Sources, eds. Walter Lewin and Michiel van der Klis.

Table 8.1. *Some information on the luminous X-ray sources in globular clusters of our Galaxy*

Cluster	Position	Discovery	1st burst	M_λ	P_b	TOXB
NGC 1851	0512−40 [107]	O7[27]	UH[60]	5.6B[41]		UUU
NGC 6440	1745−20 [180]	O7[158]	BS[115]	3.7B[236]		T−N−
NGC 6441	1746−37 [109]	UH[69]	EX[210]	2.4B[42]	5.7 hr[196]	−NN
NGC 6624	1820−30 [133]	UH[69]	ANS[78]	3.0B[2]	11.4 min[206]	UUU
NGC 6652	1836−33 [92]	H2[99]	BS[114]	5.6B[92]	[c]	UUU
NGC 6712	1850−09 [79]	AV[201]	S3[105]	4.5B[106]	20.6 min[d][106]	UUU
NGC 7078-1[a]	2127+12 [239]	Ch[239]		0.7B[6]	17.1 hr[113]	−−−
NGC 7078-2[a]	2127+12 [239]	Ch[239]	Gi[47]	3.1U[239]		−−U
Terzan 1	1732−30 [124]	Ha[156]	Ha[156]			T−−−
Terzan 2	1724−31 [79]	O8[209]	O8[209]			−NU
Terzan 5	1745−25 [93]	Ha[156]	Ha[156]	1.7J[93]		T−U−
Terzan 6	1751−31 [118]	RO[183]	BS[118]		12.36 hr[116]	T−N−
Liller 1	1730−33 [108]	S3[b][149]	S3[104]			T−−−

[a] A luminous X-ray source in NGC 7078 had been found with Uhuru[69], the Chandra observations resolved this source into two sources
[b] X-ray source (the Rapid Burster) discovered before the globular cluster!
[c] 43.6 min period originally assigned to this source is period of fainter X-ray source [92]
[d] or the alias period of 13.2 min
Columns from left to right (1) cluster, (2) rough position (B1950, often used as source name) with reference to the currently most accurate position, (3) the satellite with which the source was discovered as a cluster source, (4) the satellite which detected the first Type I X-ray burst from the source, (5) absolute magnitude with filter of optical counterpart, (6) orbital period, (7) indication (with a "T") whether the source is a transient. The last three columns indicate whether a normal (N) or ultrashort (U) orbital period is suggested by the comparison of optical with X-ray luminosity (column 8, under "O"), the X-ray spectrum (column 9, under "X") and the maximum flux reached during bursts (column 10, under "B"). A "−" in columns 8–10 indicates that no information is available. Satellite names are abbreviated as O(SO-)7, O(SO-)8, UH(URU), H(EAO-)2, A(riel-)V, Ch(andra), Ha(kucho), RO(SAT), S(AS-)3, B(eppo)S(ax), EX(OSAT), Gi(nga). Note that the absolute magnitudes are subject to uncertainties in distance and reddening; also most sources are variable (see Deutsch *et al.* 2000).

Because of the observed correlation between the occurrence of a luminous X-ray source in a globular cluster and a high central density, it was expected early on that these luminous sources would be located close to the cluster centers. These expectations were confirmed by measurements, carried out with the SAS-3 X-ray Observatory, which showed that the positional error circles with radii of 20–30 arcsec (90% confidence) included the optical centers of the clusters (Jernigan & Clark 1979). Later work with the Einstein observatory greatly refined the positional measurements (Grindlay *et al.* 1984). Bahcall and Wolf (1976) have shown that under certain assumptions the average mass of the X-ray sources can be derived from their positions with respect to the cluster center. Even if one accepts the assumptions made, the average mass derived this way for the luminous X-ray sources in globular clusters was not sufficiently accurate to classify these sources, but the result was consistent with the earlier conclusions (see e.g., Lewin 1980; Lewin & Joss 1983) that these are accreting neutron stars (Grindlay *et al.* 1984).

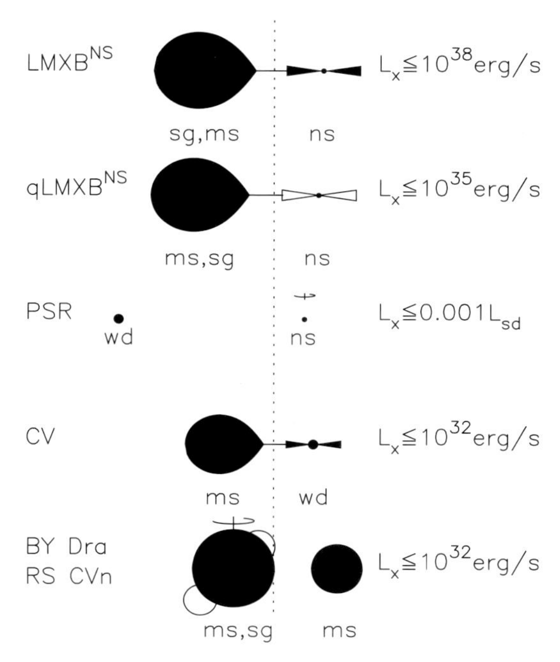

Fig. 8.1. Various types of X-ray sources in globular clusters; sg, ms, wd, and ns stand for subgiant, main-sequence star, white dwarf, and neutron star, respectively. From top to bottom: luminous low-mass X-ray binary, low-luminosity low-mass X-ray binary, recycled radio pulsar (here with a white dwarf companion), cataclysmic variable, and magnetically active binary. L_{sd} stands for spin-down luminosity. Approximate maximum luminosities (in the 0.5–4.5 keV range) are indicated on the right. The low-mass X-ray binaries harboring a neutron star are referred to as $LMXB^{NS}$; when they harbor a back hole, we refer to them as $LMXB^{BH}$, and we refer to both groups together as LMXB.

Sources with $L_X \lesssim 10^{35}$ erg s^{-1} were first found in globular clusters with Einstein (Hertz & Grindlay 1983). More were found with ROSAT, by a variety of authors (Table 8.2); a final, homogeneous analysis of the complete ROSAT data was made by Verbunt (2001). On the basis of these Einstein and ROSAT results, it has gradually become clear that these sources are a mix of various types (see Fig. 8.1). Hertz and Grindlay (1983) suggested that they were mainly cataclysmic variables, and noted that the low-luminosity source in NGC 6440 could be the quiescent counterpart of the luminous transient source in that cluster. Verbunt *et al.* (1984) argued that the more luminous of the low-luminosity sources are all quiescent low-mass X-ray binaries. The first radio pulsar detected as a low-luminosity source in a globular cluster is the pulsar in NGC 6626 (M 28, Saito *et al.* 1997). Finally, Bailyn *et al.* (1990) pointed out that magnetically active binaries also reach X-ray luminosities in the range of the less luminous sources detected with ROSAT.

It was also realized that some of the sources could be unresolved multiple sources; and unresolved emission was found e.g., by Fox *et al.* (1996) in NGC 6341 and NGC 6205. However, it is fair to say that the actual plethora of sources shown by the Chandra observations in virtually every cluster that it observed (Tables 8.2, 8.4) was unpredicted. These observations confirmed that quiescent low-mass X-ray binaries, cataclysmic variables, pulsars, and

Table 8.2. *Observations of low-luminosity sources in globular clusters.*

Cluster	E	R	C/X	Comments
NGC 104/47Tuc	1	5+4	39+66	E=R9=C42 (CV) R5=C58 R7=C46 R6=C56 R10=C27 R11=C25 R13=C2 R19=C30 R4 outside C-frame
NGC 288		1		[199]
NGC 362		2		
Pal 2		0+1		[188]
NGC 1904/M79	1	0+1		E=R
NGC 5139/ωCen	1+4	3+3	3+97	[229] core: EC>(R9a=C6/R9b=C4) both CVs; out-of-core: EB=R7=C3 qLMXB; EA=R3, ED=R4, EE=R5 foreground stars
NGC 5272/M3	1	1		[48] E=R, CV/SSS? also [100], opt.id. [53]
NGC 5824	1	0		R limit just below E detection level
NGC 5904/M5		0+1	10[a]	[88]
NGC 6093/M80		1	9+10	R>(C1/C2/C4/C7· · ·)
NGC 6121/M4		1	12+19	R=C1
NGC 6139		1		
NGC 6205/M13		2+1	2+1	core: RGa=X3 qLMXB; RGb≠X X2≠R out-of-core: RF=X6
NGC 6266/M62		1	45[a]	
NGC 6341/M92		1		[123], [62]
NGC 6352		0+1		[125]
NGC 6366		1	1[a]	[125]
NGC 6388		0+1		
NGC 6397		5+1	9+11	R4a/b/c/d/e=C19/17/23/22/18 R13=C24
NGC 6440	1	2	24	E>(R1>C2/C4/C5· · ·, R2>C1/C3· · ·)
NGC 6541	1	1		
NGC 6626/M28		3+1	12+34	core: (R2a+2b)=C26 R2c=C19 out-of-core: R7=C17
NGC 6656/M22	1+3	1	3+24	core: E=R=X16, opt id.[1] X18≠R out-of-core: E prob. not related to cluster
NGC 6752		4+2	9+8	core: R7a>C4/7/9 R21>C11/12/18 R7b=C1 R22=C6 out-of-core: R6=C3 R14=C2
NGC 6809		1		[125]
NGC 7099		0+1	5[a]	[123]
Total:	8+7	37+18		

[a] number within half-mass radius from [181], detailed analysis not yet published

We list the number of sources found with Einstein (under E), ROSAT (R) and Chandra or XMM (C/X). Numbers following the + sign indicate sources outside the cluster core. Note that the detection limits are very different between clusters. References are Hertz & Grindlay (1983) for Einstein sources, Verbunt (2001) and references therein for ROSAT sources. References for Chandra and XMM-Newton are listed in Table 8.4. Under comments we provide additional references for ROSAT, give occasional source types quiescent (i.e. low-luminosity) low-mass X-ray binary (qLMXB) and cataclysmic variable (CV), and indicate the relation between sources observed by subsequent satellites. = identical; > resolved into multiple sources; A≠B source A not detected by satellite B, due to significant variability. Source numbers under comments are those in the references given.

magnetically active binaries are all X-ray sources in globular clusters, as is discussed in Section 8.4. Whereas some of the Einstein and ROSAT sources are confirmed with Chandra as single sources, others have been resolved into multiple sources; details are given in Tables 8.2.

The positions obtained with ROSAT were sufficiently accurate to find plausible optical counterparts in Hubble Space Telescope (HST) observations in a number of cases. This work was pioneered in NGC 6397 with a search for Hα emitting objects by Cool *et al.* (1993, 1995), and spectroscopic follow-up confirming the classification as cataclysmic variables by Grindlay *et al.* (1995), Cool *et al.* (1998) and Edmonds *et al.* (1999). Plausible candidate counterparts were also found for two X-ray sources in the core of ω Cen (Carson *et al.* 2000). All of these suggested counterparts were confirmed with the more accurate positions obtained with Chandra. In 47 Tuc, of the candidate counterparts suggested by a variety of authors, Verbunt and Hasinger (1998) retain only three, on the basis of more accurate positions of the X-ray sources; these were also confirmed with Chandra. Ferraro *et al.* (1997) suggested ultraviolet stars as counterparts for two sources found by Fox *et al.* (1996) in NGC 6205 (see also Verbunt 2001). An ultraviolet counterpart suggested by Ferraro *et al.* (2000) for a source in NGC 6341 is incompatible with the position of that source (Geffert 1998; Verbunt 2001). Another approach is to look for X-rays from an already known special object. A dwarf nova known since 1941 well outside the central region of NGC 5904 (Oosterhoff 1941) was detected with ROSAT (Hakala *et al.* 1997), and a pulsar in M28 (Lyne *et al.* 1987) was detected with ASCA (Saito *et al.* 1997). Before Chandra, no magnetically active binary was suggested as an optical counterpart for a specific X-ray source.

The luminous X-ray sources in globular clusters are binary systems, and most (if not all) of the low-luminosity X-ray sources are also binary systems or have evolved from them. The presence of binaries is a very important factor in the evolution of a globular cluster (Hut *et al.* 1992). Theoretical considerations and numerical calculations show that a cluster of single stars is unstable against collapse of its core (Hénon 1961). If binaries are present, however, close binary–single star encounters can increase the velocity of the single stars by shrinking the binary orbits. Binaries can therefore become a substantial source of energy for the cluster, sufficient even to reverse the core collapse. Even a handful of very close binaries can significantly modify the evolution of a globular cluster (Goodman & Hut 1989). With a million stars in the cluster as a whole, the number of stars in the core of a collapsed cluster may be only a few thousand. A close binary system, such as an X-ray binary, will have a binding energy that can easily be a few hundred times larger than the kinetic energy of a single star. A dozen such systems, as they were formed, released an amount of energy that is comparable to the kinetic energy of the core as a whole. Encounters between such binaries and other single stars or binaries have the potential to change the state of the core dramatically by increasing or decreasing the core size, and by kicking stars and binaries into the cluster halo or even out of the cluster altogether. The study of the binaries, and X-ray binaries in particular, is therefore of great importance as they play a key role in the cluster's dynamical evolution.

It has been suggested that globular clusters are responsible for the formation of all or some of the low-mass X-ray binaries in our Galaxy, and also those now outside clusters (e.g., Grindlay & Hertz 1985). Specifically, such an origin was suggested by Mirabel *et al.* (2001) for the black-hole X-ray binary XTE J1118+480, and by Mirabel and Rodrigues (2003) for Sco X-1, on the basis of their orbits in the Galaxy. The discovery of very large populations of cluster X-ray sources in other galaxies has rekindled the question of cluster origin for non-cluster sources. In Section 8.3.5 we will give the reasons why we believe that most X-ray binaries in the disk of our Galaxy were formed there, and not in globular clusters.

8.2 The luminous globular cluster X-ray sources in the Galaxy

In Table 8.1, we list some information on the 13 luminous globular cluster X-ray binaries in the Galaxy. A comprehensive study of the X-ray spectra of these luminous sources was made by Sidoli *et al.* (2001), who used BeppoSAX observations in the spectral range between 0.1 and 100 keV. They find that the luminous sources in NGC 1851, NGC 6712 and NGC 6624 have similar spectra. When a two-component model (the sum of a disk-blackbody and a Comptonized spectrum) is used to describe the spectrum, the fitted radii and temperatures are compatible with values expected for radii and temperatures of the inner disk. The spectrum of the luminous source in NGC 6652 is similar, except that some radiation is blocked, possibly by the outer disk (Parmar *et al.* 2001). The spectra of the luminous sources in NGC 6440, NGC 6441, Terzan 2 and Terzan 6 are very different. In the two-component model the inner disk temperature was higher than that of the seed spectrum injected into the Comptonizing plasma, and the inner radius was smaller than those of realistic neutron-star radii. BeppoSAX observed the Rapid Burster in Liller 1 and the luminous source in Terzan 1 when these sources were in a low state; the two luminous sources in NGC 7078 could not be resolved.

Sidoli *et al.* (2001) suggest, on the basis of binary systems whose orbital-periods are known (see Table 8.1), that the two types of spectra correspond to two types of orbital periods: the ultrashort-period systems (observed in NGC 6712 and NGC 6624) and the longer/normal-period systems (observed in NGC 6441 and NGC 7078-1). We classify the sources as ultrashort (orbital period less than 60 min, say) or normal based on this correspondence, in Table 8.1, column (9). It may be noted that this classification does not depend on the physical interpretation of the spectra. The luminous source in Terzan 5 has been added to the suggested ultrashort-period systems, on the basis of its X-ray spectrum as observed with Chandra (Heinke *et al.* 2003a).

It is interesting to compare this tentative classification with two others. The first of these is based on the finding that ultrashort-period systems have a much lower ratio of optical to X-ray flux than systems with longer periods: the optical flux is due to reprocessing of X-rays in the accretion disk; a small accretion disk therefore has a small optical flux (Van Paradijs & McClintock 1994). Thus the absolute visual magnitude, in conjunction with the X-ray luminosity, may be used to estimate whether the orbital period is ultrashort or not. This is done in column (8) of Table 8.1. The other tentative classification scheme is based on the notion that the white dwarf donor stars in ultrashort-period systems do not contain hydrogen. The X-ray bursts of hydrogen-free matter can reach higher luminosities because the Eddington limit is higher in the absence of hydrogen. Kuulkers *et al.* (2003) have carefully investigated the maximum observed luminosities of bursters in globular clusters. On this basis we can also tentatively classify ultrashort-period systems, as we have done in column (10) of Table 8.1.

It is seen that the different classifications are consistent for known ultrashort-period systems in NGC 6624 and NGC 6712 and tentatively classified ultrashort systems in NGC 1851 and NGC 6652 and for the systems with known longer period in NGC 6441 and Terzan 6. Two tentative indicators for the source in Terzan 2 are contradictory.

Five of the thirteen luminous X-ray sources are transients. The source in Terzan 1 was consistently luminous until about 1999, when it switched off (Guainazzi *et al.* 1999). The Rapid Burster in Liller 1 and the luminous source in Terzan 6 are recurrent transients, showing outbursts quite frequently. Intervals of ~6–8 months (Lewin *et al.* 1995) and ~100 days (Masetti 2002) were observed for the Rapid Burster, and ~4.5 months for the luminous source in Terzan 6 (in 't Zand *et al.* 2003). The luminous source in NGC 6440 is a transient whose outbursts have been detected in 1971, 1998 and 2001 (see Section 8.2.1). The transient source in Terzan 5 entered a rare high state in August 2000 (Heinke *et al.* 2003a, and references therein). Interestingly, most (known and suggested) ultrashort-period systems are persistent sources. Note, however, that one of the two periods known for low-luminosity low-mass X-ray binaries is also ultrashort (see Fig. 8.14). (The source in NGC 6652 does occasionally drop below $\sim 10^{36}$ erg s^{-1}, but it is not known by how much.) Whether the above correlations are significant remains to be seen, and will only become evident once more secure orbital periods have been determined.

With Chandra, the positions of the luminous sources have become more accurate. In Fig. 8.2 we show these positions, together with those of the low-luminosity sources that also contain a neutron star. It is seen that some sources, e.g. the luminous source in NGC 6652, are at a large distance from the cluster core.

8.2.1 Notes on individual sources

NGC 1851. The accurate Chandra position for the luminous source in NGC 1851 confirms the previously suggested optical counterpart; this star is very faint, considering the brightness of the X-ray source, which suggests that the binary is an ultrashort-period binary (see Section 8.2 and Table 8.1), i.e., $P_b < 1$ h (Homer *et al.* 2001a).

NGC 6440. The luminous source in NGC 6440 is a transient; outbursts were detected in 1971 with OSO-7 and Uhuru (Markert *et al.* 1975; Forman *et al.* 1976), and again in 1998 and 2001 with BeppoSAX (in 't Zand *et al.* 1999, 2001). The 1998 outburst was followed up with NTT and VLT observations. An optical transient was found at the approximate location of the X-ray transient (Verbunt *et al.* 2000). The 2001 outburst was observed with Chandra (in 't Zand *et al.* 2001), and the source was identified with one of four low-luminosity sources found earlier by Pooley *et al.* (2002b). The 1998 optical and the 2001 X-ray transient are the same source.

NGC 6624. The luminous source in NGC 6624 has an orbital period of 865 s, indicating that the donor is a white dwarf (Verbunt 1987). For such a donor, theory predicts that the orbital period increases with time: $\dot{P}_b/P_b > 8.8 \times 10^{-8}$ yr^{-1}. However, observations made in the period 1967 to 1997 show a decrease in the period, of order $\dot{P}_b/P_b = -5.3 \times 10^{-8}$ yr^{-1} (Van der Klis *et al.* 1993; Chou & Grindlay 2001). This continued decrease cannot be explained by changes in the disk size. However, the X-ray source is located close to the center of the cluster (King *et al.* 1993), and if the central density is high enough, acceleration of the binary in the cluster potential may explain the difference (Chou & Grindlay 2001). Further study is required as discrepancies exist between reported positions for the cluster's center. It is important that the central density of the cluster be determined more accurately. A viable alternative may be that the donor is not a white dwarf, but a stripped core of a slightly evolved main-sequence star (Podsiadlowski, Rappaport & Pfahl 2002).

NGC 6652. Chandra observations of NGC 6652 show three low-luminosity sources in addition to the luminous source. The optical counterpart with a 43.6 min orbital period

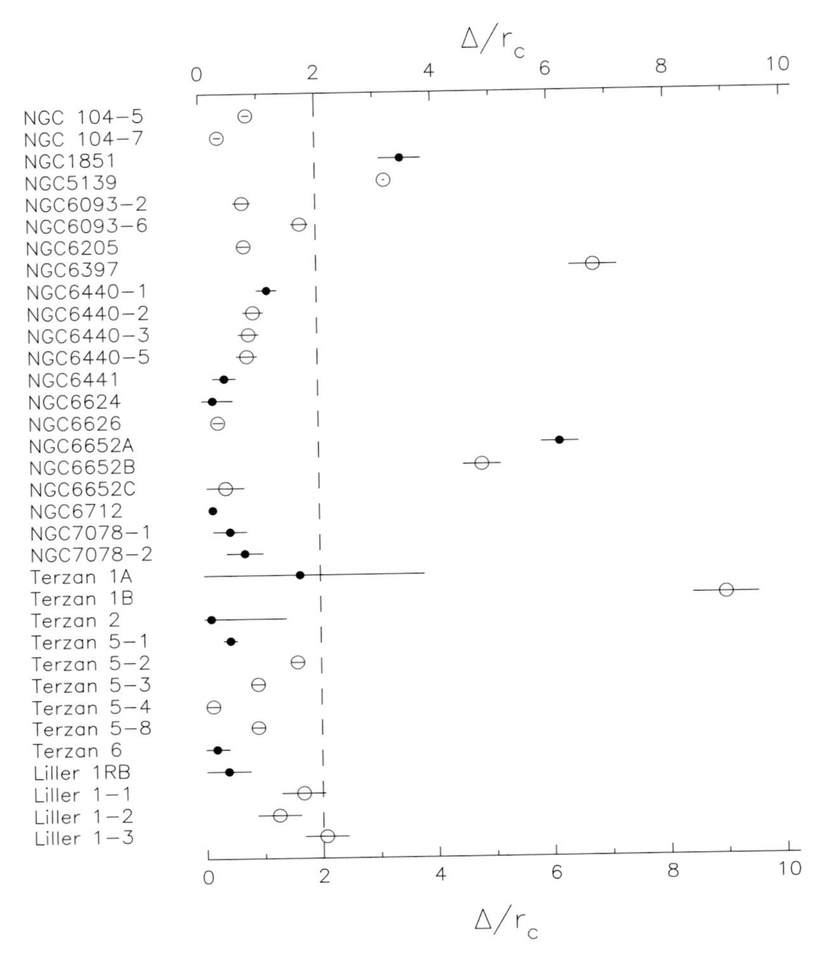

Fig. 8.2. Distance Δ of low-mass X-ray binaries to the center of the globular cluster in which they are located, in units of the core radius r_c. Luminous and low-luminosity sources are indicated with • and ○, respectively. Errors are computed from the uncertainty in the X-ray position and from the uncertainty in the position of the cluster center (assumed to be 1.2″). Core radii and centers are taken from Harris (1996, February 2003 version), except for Terzan 6 (in 't Zand *et al.* 2003). References for the X-ray positions are in Tables 8.1 and 8.2. It is seen that most, but not all, X-ray binaries are within $2r_c$.

previously suggested (Deutsch *et al.* 2000) for the luminous source turns out to be one of the low-luminosity sources instead (Heinke *et al.* 2001). The Chandra data were obtained with the High Resolution Camera and thus do not contain much spectral information. The visual brightness of the new optical counterpart of the luminous X-ray source is still very low; and the suggestion (Deutsch *et al.* 2000) that this source is an ultrashort-period binary stands (see Table 8.1).

NGC 7078. A Chandra observation of NGC 7078 (M15) showed that this cluster contains two luminous sources, at a separation of 3″, seen as a single source in earlier observations with instruments that have less spatial resolution (White & Angelini 2001). The presence of two sources actually had been predicted by Grindlay (1992), as a solution to a puzzle posed by previous observations. The high optical to X-ray flux ratio indicated that the central X-ray source is hidden by the accretion disk, and that only X-rays scattered in our direction by a corona are detected; this implies that the intrinsic X-ray luminosity exceeds the observed luminosity by almost two orders of magnitude (Aurière *et al.* 1984). However, burst observations indicated that the bursts reached the Eddington limit for the distance to M15; this implied that there was no blockage of radiation, and thus that the observed persistent flux was representative for the full luminosity (Dotani *et al.* 1990). The brightest of the two (7078-2, see Table 8.1) is the burster; the optical counterpart is probably a blue star with $U = 18.6$; its position is determined most accurately from its radio counterpart (Kulkarni *et al.* 1990). The less luminous source 7078-1 has the disk corona, and is identified optically with a 17.1 hr partially eclipsing binary (Ilovaisky *et al.* 1993). Its optical brightness and the orbital period – revealed by variable, non-total eclipses – indicate that the donor in this system is a sub-giant. Ultraviolet lines with strong P Cygni profiles indicate extensive mass loss. An analysis of the eclipse timing puts a rough upper limit on the period change of 0.01 d in 22 yr (Naylor *et al.* 1992; Ioannou *et al.* 2003). An extreme ultraviolet flux has been detected from M15. It was believed to come from the X-ray binary AC211, the optical counterpart of 7078-1 (Callanan *et al.* 1999). We suggest that some UV may also come from 7078-2 which allows for a direct view to the center of the accretion disk.

Terzan 1. When Terzan 1 was observed with BeppoSAX in April 1999, the luminosity had dropped to about 2×10^{33} erg s^{-1}, indicating that the luminous source in this cluster had gone into quiescence (Guainazzi *et al.* 1999). Accurate positions for the luminous source had been obtained with EXOSAT ($8''$ accuracy, Parmar *et al.* 1989) and ROSAT ($5''$ accuracy, Johnston *et al.* 1995); remarkably, the source detected with Chandra is not compatible with these positions (Wijnands *et al.* 2002). Probably, all observations of the bright state before 1995 refer to the same source, since the detected luminosities are all similar at, or just below, 10^{36} erg s^{-1} (Skinner *et al.* 1987; Parmar *et al.* 1989; Verbunt *et al.* 1995; Johnston *et al.* 1995). This source was discovered in 1980 during observations with Hakucho; only two bursts were observed in one week. The upper limit to the persistent flux was $\sim 10^{36}$ erg s^{-1} (Makishima *et al.* 1981). It is not clear whether BeppoSAX detected the faint state of the luminous source, or the low-luminosity source found with Chandra.

Terzan 5. Observations of Terzan 5 with Chandra show nine sources in addition to the transient; four of these are probably low-luminosity LMXB^{NS} (Heinke *et al.* 2003a). A possible optical counterpart is a faint blue (in infrared colors) star, at $M_J \simeq 1.7$ when the X-ray source was faint. Heinke *et al.* (2003a) note that the X-ray spectrum when the source is luminous is like those of the luminous sources in NGC 6624 and NGC 6712, and suggest that the source is an ultra-compact binary (see Table 8.1). If that is the case, its high optical flux is surprising. Wijnands *et al.* (2005) find that the spectrum in quiescence (near 10^{33} erg s^{-1}) is dominated by a hard power-law component.

Terzan 6. Extended studies of Terzan 6 with RXTE show that the transient X-ray source in this cluster has fairly frequent outbursts, on average every 140 days (in 't Zand *et al.* 2003). An X-ray position, derived from a Chandra observation, and an improved position for the center of the cluster, found with ESO NTT observations, show that the X-ray source is close to the cluster center. The RXTE observations provide an upper limit to the change in the orbital period: $|\dot{P}/P| < 3 \times 10^{-8}$ yr^{-1}.

Liller 1. The Rapid Burster in Liller 1 is a recurrent transient. It shows a bewildering variety of X-ray behavior. When discovered in 1976 (Lewin *et al.* 1976), it emitted X-rays largely in the form of very frequent bursts (which were later called Type II bursts). The average burst rate was in excess of 10^3 per day; this gave the source its name. There is an approximate linear relation between the burst fluence and the waiting time to the next burst (i.e., the mechanism is like that of a relaxation oscillator). These rapid bursts are the result of spasmodic accretion. Type II bursts have been observed that lasted up to ten minutes with a corresponding waiting time to the next burst of $\sim$1 hr. At times (early in an outburst that typically lasts several weeks), for periods of many days, the Rapid Burster behaved like a normal LMXB (i.e., persistent emission, but no Type II bursts). The Rapid Burster also produces the thermonuclear, Type I, bursts (Hoffman, Marshall & Lewin, 1978). A review of this remarkable source is given by Lewin, Van Paradijs & Taam (1993); see also Section 2.9.5. An accurate Chandra position of the Rapid Burster (Homer *et al.* 2001b) coincides with the radio counterpart (Moore *et al.* 2000). The Einstein position of the Rapid Burster (Hertz & Grindlay 1983) is not compatible with the radio counterpart and with the Chandra position. However, it does coincide with one of three low-luminosity sources also detected with Chandra. Perhaps the low-luminosity source was more luminous at the time of the Einstein observations. On the basis of their luminosities, the low-luminosity sources are probably low-mass X-ray binaries in quiescence (Homer *et al.* 2001b).

8.3 The globular cluster sources outside the Galaxy

In this section, we discuss the very luminous globular cluster X-ray sources observed in galaxies other than our own. The observations we discuss were all done with Chandra, except for the ROSAT observations of M 31. Some of the sources were detected with ROSAT, but the positional accurracy of Chandra allows more secure identifications with globular clusters. Table 8.3 gives an overview of the observations reported so far. The lowest detectable luminosities vary strongly between galaxies. With the exception of M 31 and NGC 5128, however, we are always talking about very luminous sources (the tip of the iceberg). In addition to the sources discussed in this chapter, sources in many other globular clusters associated with other galaxies have been observed but not (yet) recognized as such, e.g., because the required optical cluster studies are not available (see Chapter 12 and Table 12.1).

The number of globular clusters varies widely between galaxies. Precise numbers are difficult to determine: clusters are difficult to detect against the bright background of the central regions of a galaxy, and the cluster distribution may extend beyond the observed area. For example, globular clusters in NGC 4697 have only been identified in an annulus from 1.5 to 2.5 arcmin from the center (Fig. 8.5). And even for nearby M 31 "the size of the globular cluster system is embarrassingly uncertain" (Barmby 2003). Estimates of the total number are often based on an uncertain extrapolation of the measured bright part of the globular cluster luminosity function and depend on the availability of multi-color images that go deep enough to probe a significant portion of the luminosity function (Kundu, private communication). In many galaxies the area in which positions of globular clusters are known with sufficient accuracy for comparison with X-ray positions is limited by the field-of-view of HST-WFPC2 observations: an example is seen in Fig. 8.4.

The number N of globular clusters of a galaxy is sometimes scaled to the total luminosity of the galaxy (derived from absolute magnitude M_V), as a specific frequency S_N, defined as (Harris & van den Bergh 1981)

$$\log S_N = \log N + 0.4(M_V + 15) \qquad (8.1)$$

Table 8.3. *X-ray sources associated with globular clusters in galaxies other than our own*

					HST–FOV					
Galaxy	X	X_g	N	S_N	X	X_g	N	S_N	L_l	L_u
NGC 720[121]	42	12		2.2[134]					38.6	40.0
NGC 1316[130]	81	5		1.7[72]				0.9[70]	37.3	39.3
NGC 1399[3]	214		6450	5.1[46]	45	32	678		37.7	
NGC 1407[241]	160	88		4.0[173]						
NGC 1553[19]	49	2		1.4[5]			1553	0.5[141]	38.3	39.3
NGC 3115[144]	90				36	9				
NGC 4365[144]	149			5.0[134]	44	18[a]	660	2.1[141]		
NGC 4472[143]	135		5900	3.6[191]	72	29	825		37.0	
NGC 4486[126]	174		13450	14[160]	98	60			37.2	39.0
NGC 4649[187]	165			6.9[134]	40	20	497	1.4[141]		
NGC 4697[198]	80	>8	1100	2.5[134]					37.7	39.4
M 31[207]	353	27	500	1.2[11]					35.5	38.3
M 31[44]					90	28			35.3	38.3
NGC 4594[45]	122	32	1900	2.1[192]						
NGC 5128[137]	111	33		2.6[89]		29			36.2	

[a] corrects number given in paper (Kundu, private ommunication).
For each galaxy we list the total number of X-ray sources detected X, the number associated with globular clusters X_g, and the number of globular clusters N with specific frequency S_N (Eq. 8.1); and the same numbers again in a limited field-of-view (FOV; Hubble Space Telescope observations, usually with the WFPC-2 but with the ACS for M 87) where applicable. We also list the logarithm of the X-ray luminosity detection limit, L_l, and the luminosity, L_u, of the most luminous cluster source, in erg s^{-1}. Numbers between [ ] are references.

A "local" specific frequency is often defined for the field-of-view of the HST-WFPC2. The uncertainties in the total number of globular clusters are reflected in large uncertainties of the specific frequencies, and the uncertainty in the distance adds to this. For example, values for NGC 1553 range from 1.22 ± 0.27 to 2.3 ± 0.5 (Bridges & Hanes 1990; Kissler-Patig 1997). Specific frequencies (most are meant to be global) were compiled by Harris (1991), Kissler-Patig (1997), and Ashman & Zepf (1998). Local specific frequencies of globular clusters have been measured in the inner region of 60 galaxies (Kundu & Whitmore 2001a,b).

Many elliptical galaxies, and especially those in the centers of clusters of galaxies, have large numbers of globular clusters (Harris 1991; Ashman & Zepf 1998). Per unit mass, most ellipticals have about twice as many globular clusters as spirals (Zepf & Ashman 1993; Ashman & Zepf 1998). The globular cluster populations in most elliptical galaxies show a bimodal distribution in optical colors (Fig. 8.3). Most of this is due to differences in metallicity, but differences in age may also play a role. Metal-poor clusters are bluer than metal-rich clusters of the same age; at the same metallicity, old clusters are redder than young ones. It has been suggested that the blue metal-poor globular clusters were formed at the proto-galactic epoch, and that the red metal-rich globular clusters resulted from later starbursts, e.g., as a consequence of the mergers that produce the galaxies that we observe today (Ashman & Zepf, 1992; Zepf & Ashman, 1993; for other possibilities see the review

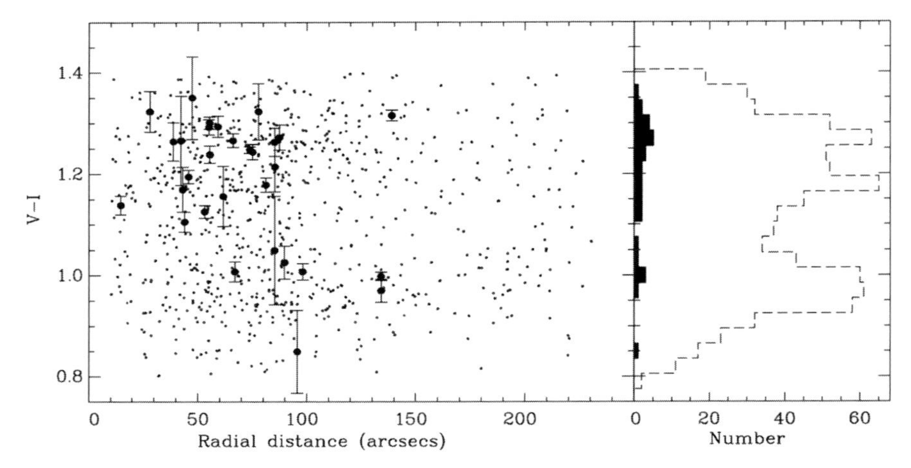

Fig. 8.3. *Left*: *V–I* colors of globular clusters vs. distance from the center of the elliptical galaxy NGC 4472. LMXB–globular-cluster matches are indicated by filled circles. Most of the luminous X-ray sources are located in red globular clusters. *Right*: The optical color distribution is shown with a dashed line; notice the bimodal distribution. The distribution of the globular clusters that house the luminous X-ray sources is also shown. Courtesy of Kundu, Maccarone & Zepf (2002).

by West *et al.* 2004). However, to date there is no convincing evidence for difference in ages of red and blue subsystems (e.g., Puzia *et al.* 2002; Cohen *et al.* 2003; Côté 1999, 2002).

8.3.1 Elliptical galaxies

The most luminous X-ray sources in a galaxy are high-mass X-ray binaries, supernova remnants, and low-mass X-ray binaries. Since elliptical galaxies do not house young stellar populations, virtually all luminous X-ray sources in them will be low-mass X-ray binaries. Table 8.3 provides an overview of the references and results; some additional remarks for individual galaxies follow. We will discuss the X-ray luminosity functions of the globular cluster systems and the reported breaks in some of them in Section 8.3.4.

NGC 1399 is a giant elliptical galaxy in the center of the Fornax Cluster at 20.5 Mpc. A large fraction of the 2–10 keV X-ray emission in an $8' \times 8'$ region is resolved into 214 discrete sources, including many background sources. Thirty-two are in globular clusters (see Fig. 8.4). Many of the globular cluster sources have super-Eddington luminosities (for an accreting neutron star), and their average luminosity is higher than that of the sources not associated with globular clusters. The most luminous source in a globular cluster has an ultra-soft spectrum such as seen in the high state of black hole binaries. This may indicate that some of the most luminous sources are binaries with an accreting black hole, rather than conglomerates of less luminous neutron-star binaries (Angelini *et al.* 2001).

Dirsch *et al.* (2003) find that "within $7'$ the specific frequency of the blue clusters alone is a factor $\sim$3 larger than for the red ones. Outside this radius, both populations have the same high local specific frequency", listed in Table 8.3.

NGC 4697. In this galaxy, most of the X-ray emission is from point sources. The central source, with $L_X = 8 \times 10^{38}$ erg s^{-1}, may be an active nucleus and/or multiple LMXBs (Sarazin *et al.* 2000, 2001).

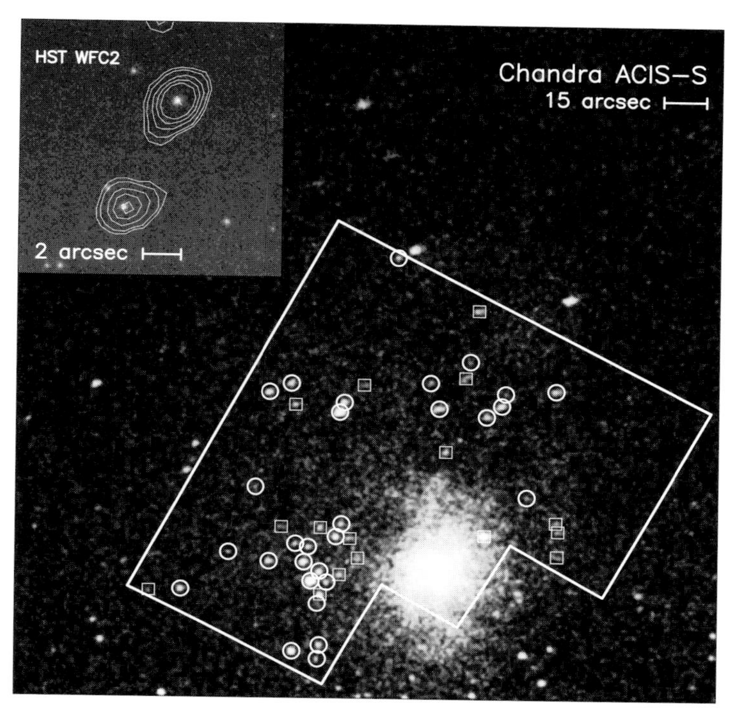

Fig. 8.4. The 0.3–10 keV Chandra image of NGC 1399 centered on an HST pointing, smoothed with a Gaussian of about 0.8″. The white line marks the HST/WFPC2 FOV. The circles show the X-ray source positions that are associated with globular clusters. The squares are the remaining sources. All 45 sources are marked; 38 have a significance in excess of 3σ. The top left image is an example of the Chandra contours overlaid on the HST field. Courtesy Angelini *et al.* (2001).

NGC 4472 is a giant elliptical galaxy. In the inner regions of the galaxy it has been shown that metal-rich red globular clusters are about three times more likely to host a very luminous LMXB than the blue metal-poor ones (Fig. 8.3). The X-ray luminosity does not depend significantly on the properties of the host globular cluster (Kundu *et al.* 2002).

NGC 4365, in the Virgo cluster, is one of a few early-type galaxies whose globular clusters do not have a bi-modal color distribution in $V-I$ (but it does in infrared colors, Puzia *et al.* 2002). Kundu *et al.* (2003) find that the presence of very luminous LMXBs is correlated with metallicity, but not with cluster age. The LMXB fraction per unit mass of the globular clusters is $\sim 10^{-7} M_{\odot}{}^{-1}$. In contrast, Sivakoff *et al.* (2003) find that within the sample of IR-bright globular clusters studied by Puzia *et al.* (2002), the metal-rich, intermediate-age globular clusters are four times more likely to contain LMXBs than the old globular clusters (with an uncertainty of a factor of 2). The luminosity function is a power law with a cutoff at $\sim(0.9$–$2.0) \times 10^{39}$ erg s^{-1}, much higher than the cutoff measured for other ellipticals.

NGC 3115 has a distinct bimodal color distribution of the globular clusters. The metal-poor blue and the metal-rich red globular clusters are both $\sim$12 Gyr old (Puzia *et al.* 2002). There are roughly equal numbers of red and blue globular clusters in the WFPC2 image. Kundu *et al.* (2003) find that the red globular clusters are the preferred sites for LMXB formation, largely as a consequence of their higher metallicity.

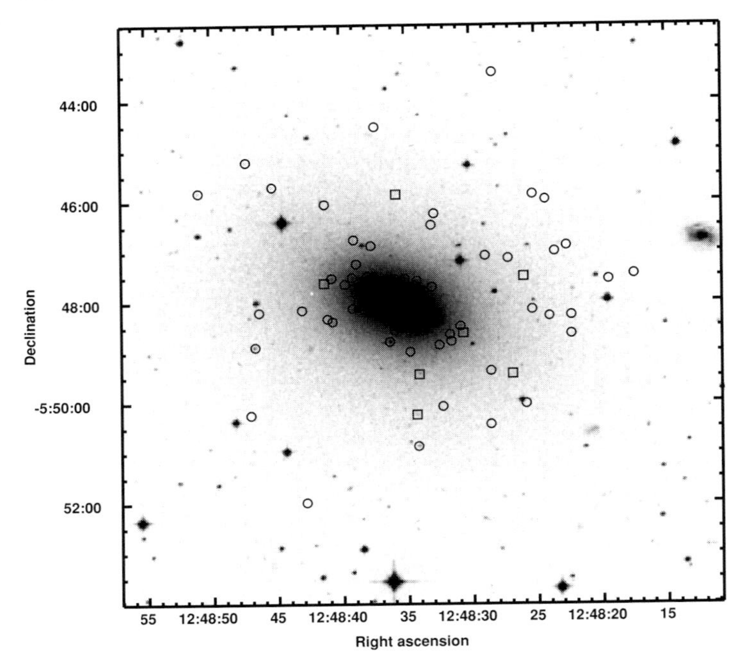

Fig. 8.5. Digital Sky Survey optical image of NGC 4697. The circles show the positions of the X-ray sources detected with Chandra. The squares indicate X-ray sources in known globular clusters. One should note that globular clusters have only been identified in this galaxy in an annulus from 1.5 to 2.5 arcmin from the center. This figure was kindly provided by Craig Sarazin. It is adapted from Figure 3 of Sarazin, Irwin & Bregman (2001).

NGC 1407. White (2002) reported that about 90% of the 160 detected LMXBs have X-ray luminosities that exceed the Eddington limit for neutron stars. He suggests that many may be black-hole binaries (rather than multiple neutron-star binaries within individual globular clusters), since 45% do not reside in globular clusters. To date (September 2004), these results have not yet been published in a refereed journal.

NGC 1553 is an S0 galaxy. 30% of the emission in the 0.3–10 keV band and 60% of the emission in the 2.0–10 keV band is resolved into discrete sources (Blanton, Sarazin & Irwin, 2001). Kissler-Patig (1997) lists a global specific frequency of 2.3 ± 0.5, higher than the value listed in Table 8.3.

NGC 4649 (M60) is a bright elliptical galaxy. It was observed by Randall, Sarazin & Irwin (2003); for details see Table 8.3.

NGC 1316 (Fornax A) is a disturbed elliptical radio galaxy with many tidal tails. Several mergers must have occurred over the past 2 Gyr (see Kim & Fabbiano 2003, and references therein). One of the five globular cluster sources is super-soft. For an adopted distance of 18.6 Mpc, 35% of the sources are above the Eddington limit of a $1.4 M_\odot$ neutron star (Kim & Fabbiano 2003). The luminosity function is well represented by a power law with a slope of -1.3.

NGC 720. Three of the 12 globular cluster sources have X-ray luminosities in excess of 10^{39} erg s^{-1} (at 35 Mpc). It is possible that this galaxy is much closer, and that none of the sources are ultra-luminous (Jeltema *et al.* 2003).

NGC 4486 (M 87) is a giant elliptical galaxy near the center of the Virgo cluster, and has the most populous globular cluster system in the local supercluster. More luminous, redder and denser clusters are more likely to harbor a luminous X-ray source. Metal-rich red globular clusters are about three times more likely to host a luminous LMXB than the blue metal-poor ones; the trend with central density gives strong evidence that encounter rates are important in forming LMXBs in globular clusters. The trend with luminosity can arise as a consequence of the fact that more luminous clusters have higher encounter rates. The X-ray luminosity functions of both globular cluster-LMXBs and non-globular cluster LMXBs are well described by single power laws with an upper cutoff at $\sim 10^{39}$ erg s^{-1} (Jordán *et al.* 2004).

8.3.2 *Spiral galaxies*

Plenty of spirals have been observed with Chandra, but there is very little information on the optical identifications. Globular clusters are hard to find because of the patchy extinction. There is also the difficulty of subtracting the diffuse light of the galaxy. These problems are exacerbated for nearly face-on spirals. In addition to the low-mass X-ray binaries, a spiral galaxy also hosts high-mass X-ray binaries and supernova remnants among the luminous X-ray sources.

M 31 (Andromeda Nebula). The apparent size of M 31 is so big that only ROSAT has studied the whole (Magnier *et al.* 1992; Supper *et al.* 1997). Di Stefano *et al.* (2002) have conducted Chandra observations of $\sim$2560 arcmin^2 in four different areas so as to be representative of the whole. About one-third of the 90 Chandra sources have luminosities (0.5–7 keV) in excess of 10^{37} erg s^{-1}; the most luminous source is probably associated with the globular cluster Bo 375. Its luminosity (0.5–2.4 keV) varied between $\sim 2 \times 10^{38}$ and $\sim 5 \times 10^{38}$ erg s^{-1}. Supper *et al.* (1997) reported regular variations of $\sim$50% on a timescale of $\sim$16 hours. A similar percentage variability was found in the 500 day X-ray lightcurves of two other highly luminous globular clusters in M31, Bo 82 and Bo 86 (Di Stefano *et al.* 2002). Some of the more luminous globular cluster X-ray sources could be multiple sources.

It has been stated on the basis of different data sets that the X-ray luminosity function of globular cluster X-ray sources is different in M 31 than in the Milky Way (Van Speybroeck *et al.* 1979 on the basis of Einstein data; Di Stefano *et al.* 2002), and that it is the same (Supper *et al.* 1997). In Fig. 8.6 we show the normalized cumulative distributions for clusters in the Milky Way and in M 31. The distributions look different, but a Kolmogorov–Smirnov test shows that there is a non-negligible probability, 0.03, that the difference is due to chance. It is therefore possible that the extent to higher luminosities in M 31 is due to the larger number of X-ray sources (and of globular clusters).

M 104 (NGC 4594, Sombrero Galaxy) is an Sa galaxy at a distance of $\sim$8.9 Mpc. Only optically bright globular clusters house the luminous LMXBs detected with Chandra (Di Stefano *et al.* 2003). The majority of the sources with luminosities in excess of 10^{38} erg s^{-1} are located in globular clusters. The luminosity function of X-ray sources in the globular clusters has a cutoff near the Eddington limit for a 1.4 $M_\odot$ neutron star. One globular cluster houses a super-soft source (see Chapter 11). There is a connection between metal-rich, red globular clusters and the X-ray sources. However, the most luminous X-ray sources are equally likely to be located in metal-poor globular clusters with lower optical luminosities. The optically brightest blue globular clusters do not seem to house very luminous X-ray sources.

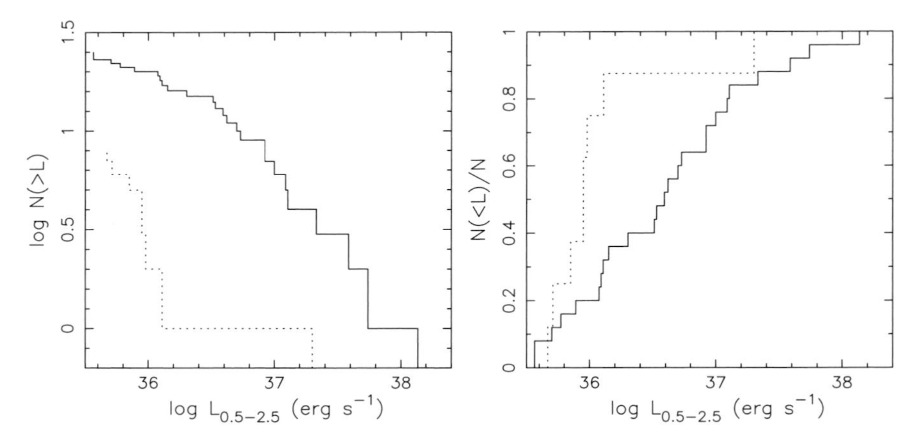

Fig. 8.6. *Left*: Comparison of the cumulative (from high luminosities downward) X-ray luminosity distributions of globular clusters in the Milky Way (dashed curve) and M 31 (solid curve). After Di Stefano *et al.* (2002). *Right*: Normalized cumulative (from low-luminosities upward) X-ray luminosity function for sources with $L_X > 10^{35.5}$erg s^{-1} in the Milky Way (dashed curve) and M 31 (solid curve). The Chandra luminosities given by Di Stefano *et al.* (2002) were multiplied by 0.46 to convert them to the energy range of the ROSAT data from Verbunt *et al.* (1995). The probability that the normalized distributions are the same is 0.03.

8.3.3 NGC 5128: Cen A

This galaxy is probably the result of mergers, and consequently it is somewhat like a mixture between an elliptical and a spiral. Four X-ray sources outside the WFPC2 FOV are coincident with globular clusters (Kraft *et al.* 2001; Minniti *et al.* 2004). 70% of the globular cluster sources have luminosities in excess of 10^{37} erg s^{-1}. There is no indication that any of them are black-hole binaries. The globular cluster X-ray sources are preferentially found in massive globular clusters. Most of the globular clusters that harbor a luminous X-ray source have red colors (metal-rich). NGC 5128 is at a low galactic latitude; there is a lot of foreground extinction. This makes it difficult to get reliable optical data on globular clusters.

8.3.4 Comparison and interpretation

Many galaxies contain a substantially larger number of luminous X-ray sources in globular clusters than our own galaxy (compare Tables 8.1 and 8.3). This can be explained by their larger numbers of globular clusters. The fraction of globular clusters that contains a luminous X-ray source is roughly constant between galaxies, as is the number of X-ray sources in clusters scaled on cluster luminosity or mass (2×10^{-7} $L_{\odot,I}{}^{-1}$ for $L_X > 3 \times 10^{37}$erg s^{-1} Sarazin *et al.* 2003; Kundu *et al.* 2003). Similarly, the larger number of globular cluster X-ray sources in M 31 compared to the Milky Way may be explained by the larger number of clusters (Supper *et al.* 1997; Di Stefano *et al.* 2003). Several authors reported a knee near the Eddington luminosity for an accreting neutron star in the luminosity functions of ellipticals (Sarazin *et al.* 2000, 2001 for NGC 4697; Kundu *et al.* 2002 for NGC 4472; Blanton *et al.* 2001 for NGC 1553, and Randall *et al.* 2003 for NGC 4649). However, Kim and Fabbiano (2004), who corrected the Chandra data for incompleteness, have shown that the luminosity functions for each of the observed elliptical galaxies can be fit with one power law; two power laws do not improve the fit in a significant way. It is interesting to note that, even though no breaks in the individual luminosity functions are significant, if the luminosity functions of all observed ellipticals (containing a total of about 985 point-like sources) are added, a broken power-law fit is a better fit than a single power law; the break is near 5×10^{38} erg s^{-1} (Kim & Fabbiano, 2004; Section 12.4.3).

Clearly, a large number of LMXBs have luminosities substantially above the Eddington luminosity of an accreting neutron star. In analogy with the luminosity distribution in the Milky Way (Grimm *et al.* 2002), this suggests that many of these sources may be accreting black-holes. This suggestion is supported in some cases by the X-ray spectrum, which shows the soft signature of an accreting black-hole (e.g. Angelini *et al.* 2001). The fact that a very luminous accreting black hole is not found in the globular clusters of the Milky Way is probably due to the small number of cluster sources.

An alternative explanation for the LMXBs with luminosities substantially above the Eddington luminosity of an accreting neutron star is provided by Bildsten and Deloye (2004), who note that the Eddington limit for hydrogen-poor gas is higher. They show that the high luminosities can be explained by invoking binaries in which a helium or carbon/oxygen white dwarf of 0.04–0.08 $M_{\odot}$ transfers mass to a neutron star, at orbital periods of 5–10 minutes. From the evolution of such binaries, driven by gravitational radiation, to longer periods and lower mass-transfer rates the expected luminosity function can be computed, and is found to be compatible with the observed luminosity function.

The X-ray sources are found preferably in optically bright clusters (Angelini *et al.* 2001). This could be explained as a scaling with mass (Kundu *et al.* 2002; Sarazin *et al.* 2003). We suggest, however, that the scaling with mass is a proxy for the scaling with the collision number, caused by the strong correlation between mass and collision number. In the Milky Way, the probability of a cluster containing a luminous X-ray source scales better with the collision number than with the mass (Verbunt & Hut 1987; Pooley *et al.* 2003).

In many galaxies, luminous X-ray sources are found preferably in red, metal-rich clusters. Bellazzini *et al.* (1995) demonstrated this for the Milky Way (see Fig. 8.7) and less conclusively for M 31. Di Stefano *et al.* (2003) find in their sample of M 31 clusters that the probability that a cluster contains an X-ray source is not strongly correlated with metallicity. Kundu *et al.* (2002) find that a red cluster in NGC 4472 has a three times higher probability of hosting a luminous X-ray source than a blue cluster. A similar result is found for NGC 4365 by Sarazin *et al.* (2003), and for NGC 3115 by Kundu *et al.* (2003). We consider four suggested explanations. First, if metal-rich clusters are younger, they contain main-sequence stars of higher mass, which are thought to be more efficient in forming an X-ray binary (Davies & Hansen 1998). In NGC 4365, such a young population is indeed present, but it does not show an increased formation rate of X-ray sources (Kundu *et al.* 2003). Also, the preference for metal-rich clusters is observed in the Milky Way and in NGC 3115, where all globular clusters are old. These results show that metallicity, not age, must explain the preference of X-ray sources for red clusters (Kundu *et al.* 2003). Second, a higher X-ray luminosity at higher metallicity would produce a preference for metal-rich clusters in a flux-limited sample. Various models have been suggested to produce higher X-ray luminosities in binaries with a donor of higher metallicity (e.g. Bellazzini *et al.* 1995; Maccarone *et al.* 2004). However, X-ray sources in metal-rich clusters are not observed to be more luminous than those in metal-poor clusters in M 31 (Verbunt *et al.* 1984) or, with less statistical constraint, in NGC 4472 (Maccarone *et al.* 2003). Third, Grindlay (1987) suggests that metal-rich clusters have a flatter initial mass function (and hence more neutron stars). However, such a dependence is not observed

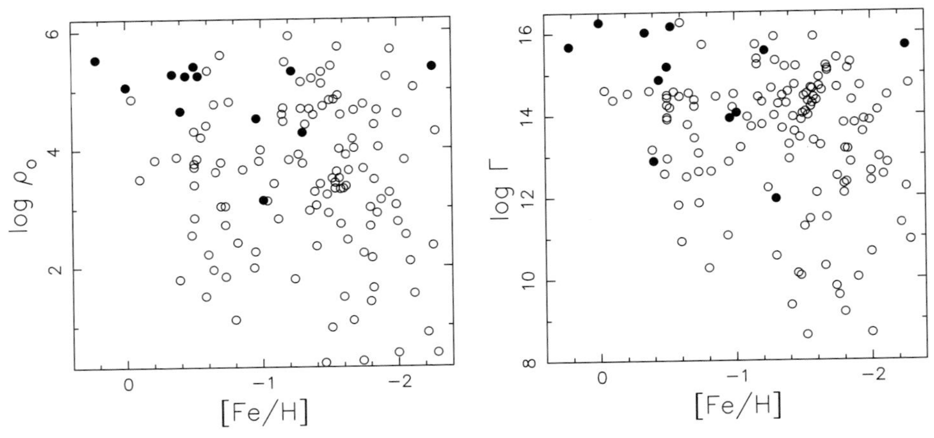

Fig. 8.7. *Left*: central density of globular clusters in the Milky Way as a function of metallicity. Filled circles indicate globular clusters with a luminous X-ray source. Even at the same density there is a preference for high-metallicity clusters. After Bellazzini *et al.* (1995). *Right*: the preference for high-metallicity clusters persists in a plot of collision number (Eqs. 8.5 and 8.6) as a function of metallicity.

in the Milky Way (Piotto & Zoccali 1999). Finally, Bellazzini *et al.* (1995) suggest that the longer lifetimes and larger radii of metal-rich stars enhance their capture rate; the capture probability is proportional to radius (see Eq. (8.5) below), and it must be doubted that the small difference in radii has sufficient effect to explain the observations (Maccarone *et al.* 2004). It is fair to say that the connection between metallicity and the occurrence of LMXBs in globular clusters is not yet well understood.

There is a tendency for X-ray sources in metal-rich globular clusters to have softer X-ray spectra (M 31: Irwin & Bregman 1999; NGC 4472: Maccarone *et al.* 2003).

8.3.5 *Comparison between field and cluster sources*

The X-ray luminosity function of sources in globular clusters is not very different from that of the sources outside globular clusters (Maccarone *et al.* 2003, Sarazin *et al.* 2003). The spatial distribution of X-ray sources outside globular clusters in elliptical galaxies is similar to that of the globular cluster sources. In elliptical galaxies, globular clusters often harbor a very large fraction of all X-ray sources (Table 8.3). This has raised the suggestion that *all* X-ray sources in elliptical galaxies originate in globular clusters (White *et al.* 2002). The field sources could then have been ejected from a cluster, or originate in a cluster that was later destroyed by the galactic tidal field. The demand that a cluster lives long enough to form X-ray binaries, and short enough not to be around now, requires fine tuning. Thus, the ejection hypothesis may be more probable.

This would suggest that a large number of globular clusters translates into a large number of X-ray sources, both in the clusters and (due to ejection) outside them. The fraction of X-ray sources in globular clusters would then be similar for different galaxies. In the Milky Way and in M 31 there are about 10 luminous low-mass X-ray binaries in the disk for each one in a globular cluster. In elliptical galaxies, there is of order 1 low-mass X-ray binary outside clusters for each one in them (see Table 8.3). This indicates that the majority of the disk sources in the Milky Way and M 31, and by extension in spiral galaxies in general, are formed in the disk; although as noted in the Introduction some individual systems may have escaped from globular clusters. (One should note that the HST field of view is much smaller than that of Chandra. Therefore, in comparing the number of X-ray sources associated with globular clusters with those not located in globular clusters (Table 8.3), in all those cases where HST data were needed to identify the clusters, one can only consider the X-ray sources detected in the regions observed with HST.)

For elliptical galaxies the case is less clear. Using optical luminosities of the galaxies and the specific globular cluster frequencies, White, Sarazin and Kulkarni (2002) reported that they found evidence that the sum of the X-ray luminosities of all X-ray sources in ellipticals scales approximately with the number of globular clusters, and they conclude that this indicates that the population outside clusters is formed in the clusters. However, the uncertainties in the specific frequencies may be substantially larger than the values used by these authors, and that makes it difficult to quantify their findings. Kim and Fabbiano (2004) have made a similar study, and caution about the above interpretation.

The fraction of low-mass X-ray binaries in clusters ranges from about 20 to 70% in ellipticals (see Table 8.3). This suggests, in our opinion, that globular clusters alone are not responsible for all low-mass X-ray binaries. In systems with small numbers, the total luminosity can be affected by just a couple of very luminous sources; the number of sources may therefore be a better estimator for the population size than the integrated X-ray luminosity. Clearly, the origin of low-mass X-ray binaries in elliptical galaxies deserves more study.

If the majority of those luminous LMXBs in elliptical galaxies not located in globular clusters are primordial, their luminosities could not have been constant throughout their lifetimes (because the product of age and the required mass-transfer rate would exceed the donor mass). There are two ways out of this lifetime problem: (i) they are not primordial but they were formed in globular clusters, and somehow released into the field, or (ii) the majority of them are transients with a low duty cycle (see Piro & Bildsten 2002). If the latter is the case, follow-up observations with Chandra will be able to reveal the variability if a sufficient number of them have outbursts that last only a few years and not much longer. We may add a third solution, which is that (iii) systems formed from primordial binaries will emerge from their early evolution as neutron stars or black holes with detached main-sequence companions. How long it takes for the binary to turn into an X-ray source then depends on the time required for the orbit to shrink due to loss of angular momentum, or for the donor to expand into a giant after completing its main-sequence evolution (for reviews see Verbunt 1993 and Chapter 16 by Tauris and Van den Heuvel). It may be noted that binaries formed in a globular cluster may also go through a long-lived detached phase (Grindlay 1988).

8.4 Low-luminosity X-ray sources

As already mentioned, a limited number of low-luminosity sources has been detected with Chandra in several clusters that contain a luminous X-ray source. The presence of such a source limits the sensitivity with which low-luminosity sources can be detected, because of the wings of the point spread function. The sharp (<1″) images and high sensitivity of the Chandra observations are best used in clusters that do not contain a luminous source. Such observations show that the central regions of several globular clusters contain dozens of sources. As a typical example, the distribution of the sources in NGC 6440 is concentrated towards the cluster center; while it spreads beyond the core radius, it is fully contained within

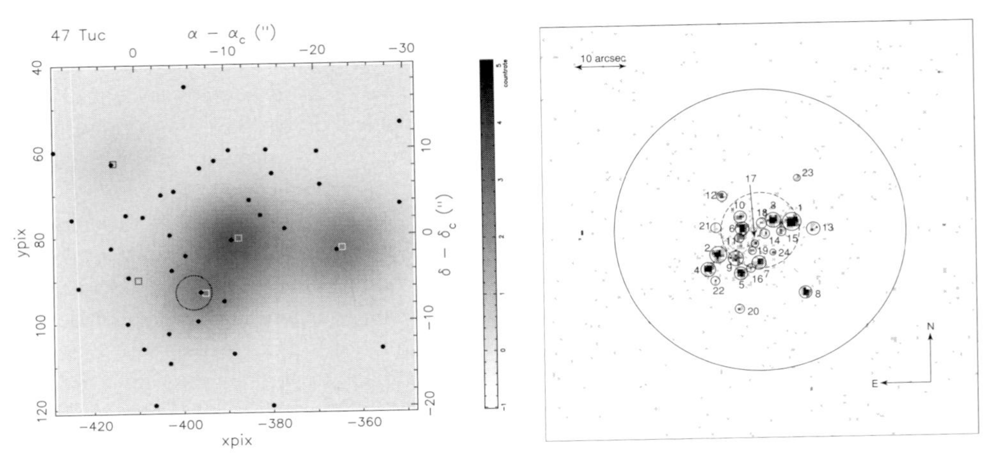

Fig. 8.8. *Left*: The impact of Chandra on the study of low-luminosity X-ray sources in globular clusters is well illustrated by the observations of 47 Tuc. The grey scale of the smoothed ROSAT-HRI countrate indicates the resolution obtained with this instrument (Verbunt & Hasinger 1998). The circle indicates the position (1σ region) of the single Einstein source (Hertz & Grindlay 1983), squares indicate the positions of the five ROSAT sources in this region, filled circles are 39 Chandra positions (Grindlay *et al.* 2001a). *Right*: Projected distribution of X-ray sources in the globular cluster NGC 6440. The dashed and solid lines indicate the core and half-mass radii, respectively. From Pooley *et al.* (2002b). In the case of 47 Tuc, each ROSAT source corresponds to one Chandra source; in the case of NGC 6440, two sources previously found by ROSAT are both resolved into multiple sources.

the half-mass radius (Fig. 8.8). From this spatial distribution alone, it can be safely asserted that almost all sources detected are related to the globular cluster. In clusters with large apparent core radii and/or half-mass radii, a large fraction of the detected sources may be fore- or background sources; an example is ω Centauri.

In trying to determine the nature of all these X-ray sources, we may be guided by our knowledge from previous satellites, in particular ROSAT. Such guidance allows us to make a preliminary classification of a source based on its X-ray flux and spectrum. If a secure optical counterpart is found – which thanks to the accurate source positions of Chandra is often the case whenever sufficiently deep HST observations are available – the classification of a source can be further based on its optical spectrum, and on the ratio of the X-ray and optical fluxes. A secure classification can also be found if the position of a radio pulsar coincides with that of an X-ray source: radio and X-ray positions are so accurate that the probability of a chance coincidence is virtually negligible for these rare objects.

Our discussion of the low-luminosity sources proceeds through the various classes illustrated in Fig. 8.1, namely, low-luminosity low-mass X-ray binaries, recycled radio pulsars, cataclysmic variables, and magnetically active close binaries. An overview of published Chandra observations of low-luminosity sources in globular clusters is given in Table 8.4.

8.4.1 Low-luminosity low-mass X-ray binaries

We consider a low-luminosity low-mass X-ray binary with a neutron star, LMXB$^{\rm NS}$, securely classified when its luminosity is high enough ($L_{\rm X} \gtrsim 10^{32}$ erg s^{-1}) and its X-ray

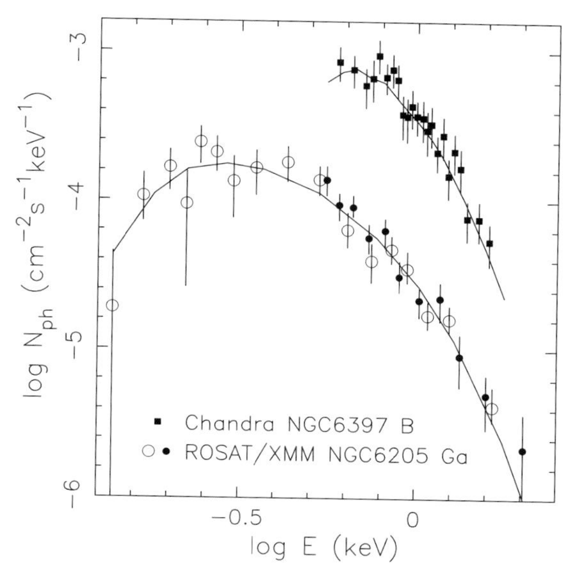

Fig. 8.9. X-ray spectra of low-luminosity X-ray binaries with neutron stars, as observed (i.e., not corrected for interstellar absorption) with ROSAT and XMM for the source Ga in NGC 6205 (M 28) (Gendre *et al.* 2003b; Verbunt 2001); and with Chandra for the source B in NGC 6397 (shifted upwards by 1 decade; in 't Zand, private communication; Grindlay *et al.* 2001b). The solid lines show fits with models for hydrogen atmospheres of neutron stars.

spectrum is soft (blackbody color temperature about 0.1 to 0.3 keV). The reason for this is that most soft X-ray transients in the galactic disk have these properties when they contain a neutron star. Their quiescent X-ray spectra have been roughly described as Planck spectra with a temperature of about 0.3 keV (Verbunt *et al.* 1994), but more correctly should be fitted with model spectra of neutron star atmospheres as have been computed by, e.g., Rajagopal & Romani (1996) and Zavlin *et al.* (1996). For quiescent transients in the disk, such fits give effective temperatures of 0.1–0.2 keV and neutron star radii of roughly 10 km (Rutledge *et al.* 1999). The situation is more problematic if a transient in quiescence has a power-law spectrum and a luminosity in the range $10^{31} - 10^{34}$ erg s^{-1}. In that case, the system could be either a $LMXB^{NS}$ or a low-mass X-ray binary with a black hole, $LMXB^{BH}$ (see Tomsick *et al.* 2003; Wijnands *et al.* 2005). A hard spectrum can also indicate a cataclysmic variable, as may be the case for one or two sources in NGC 6652 and Terzan 1.

Chandra and XMM are sensitive enough to detect luminosities of $L_X \gtrsim 10^{32}$ erg s^{-1} in any cluster that they observe, with sufficient counts to determine whether the spectra are power laws or thermal (i.e., soft). Sources for which fits with neutron star atmosphere models have been shown to give a good description of the X-ray spectrum include X7 in ω Cen (Rutledge *et al.* 2002; see also Gendre *et al.* 2003a), X5 and X7 in 47 Tuc (Heinke *et al.* 2003b), B in NGC 6397 (Grindlay *et al.* 2001b), CX1 in NGC 6440 (in 't Zand *et al.* 2001), and Ga in NGC 6205 (M 13, Gendre *et al.* 2003b). Most of these sources were detected with ROSAT, being (among) the most luminous sources in each cluster (the exception is CX1 in NGC 6440). As noted above, CX1 in NGC 6440 is the transient, detected in the bright state in 1998 and 2001; whether the transient of 1971 was the same source cannot be ascertained. This source supports our premiss that the more luminous ($L_X \gtrsim 10^{32}$ erg s^{-1}) among the low-luminosity soft sources are quiescent accreting neutron stars.

Probable classifications as low-luminosity LMXB^{NS}, based on the ratio of soft to hard counts as detected with Chandra have been suggested for four of the most luminous faint sources in NGC 6440 (Pooley *et al.* 2002b), and in Terzan 5 (Heinke *et al.* 2003a). Further probable identifications are based on the luminosity of the sources: three low-luminosity LMXB^{NS} (in addition to the Rapid Burster) in Liller 1 (Homer *et al.* 2001b), one or two in NGC 6652 (Heinke *et al.* 2001). We want to point out, however, that it cannot be excluded that some of these are black-hole binaries.

A low-mass X-ray binary with a black hole can have a much lower luminosity than a LMXB^{NS}; as an example, for the transient A0620−00 in quiescence $L_X \simeq 10^{30}$ erg s^{-1}, much of which could even be due to the donor in the binary (Verbunt 1996; Bildsten & Rutledge 2000). At such low luminosities, even Chandra or XMM observations cannot provide a secure classification, and consequently we have no information on the number of low-luminosity low-mass X-ray binaries with a black-hole accretor.

So far, only two low-luminosity LMXB^{NS}s in globular clusters have been identified optically, one in 47 Tuc and one in ω Cen (Edmonds *et al.* 2002b; Haggard *et al.* 2004).

8.4.2 Millisecond pulsars

Most identifications of X-ray sources in globular clusters with recycled radio pulsars are based on positional coincidence. The exceptions are the identifications of the pulsar in NGC 6626 (M 28), which is based on the pulse period, and of pulsars in NGC 6397 (XB) and in 47 Tuc (W29/PSR W), which are based on their orbital periods.

The pulsar in M 28 is the only one in a globular cluster that was identified with an X-ray source before the Chandra observations. By comparing the on-pulse X-rays with the off-pulse X-rays, the X-ray spectrum of the pulse could be isolated (Saito *et al.* 1997). Chandra resolves the pulsar from other cluster sources and obtains a phase-averaged power-law spectrum with photon index 1.2 (Becker *et al.* 2003).

Because accurate (timing) positions are not yet available for many of the radio pulsars, it is likely that some of them have been detected in X-rays already but not yet identified as such. In fact, an X-ray source in NGC 6397 was first identified with a possible BY Dra binary (Grindlay *et al.* 2001b); it was then found that this binary houses a radio pulsar (Ferraro *et al.* 2001). Similarly, NGC 6752 CX11 was identified by Pooley *et al.* (2002a) with a possible cataclysmic variable or background galaxy, but now is more probably identified with PSR D in that cluster on the basis of newly determined timing positions (D'Amico *et al.* 2002); positions of X-ray sources are coincident with the timing positions of PSRs C and (marginally) B.

Verbunt *et al.* (1996) showed that for the radio pulsars detected in X-rays with ROSAT, $L_{0.1-2.4\mathrm{keV}} \lesssim 10^{-3} L_{\mathrm{sd}}$, where $L_{\mathrm{sd}} \equiv I\Omega\dot{\Omega}$ is the loss of rotation energy, usually referred to as the spin-down luminosity, with I the moment of inertia and $\Omega \equiv 2\pi/P$. In accordance with this scaling, the radio pulsars detected in X-rays so far are those with the highest L_{sd} of those in the clusters observed with Chandra. Grindlay *et al.* (2002) assume that the electron density in 47 Tuc is homogeneous, and from small differences in dispersion measures determine the position of each pulsar along the line of sight; this is then used to correct the observed

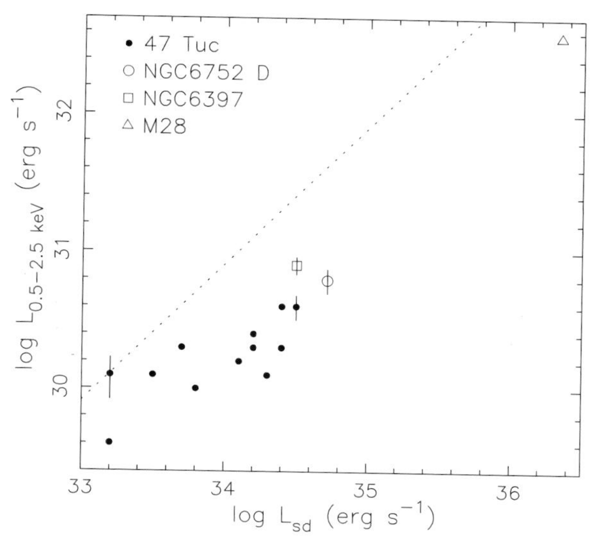

Fig. 8.10. X-ray luminosities as a function of spin-down luminosities, L_{sd}, of radio pulsars in globular clusters. The unresolved pulsar pairs G/I and F/S in 47 Tuc have been omitted. The dotted line indicates $L_{0.1-2.4\,keV} = 10^{-3} L_{sd}$ (Verbunt *et al.* 1996), with a small correction for the different X-ray energy range. The cluster pulsars lie below, but roughly parallel to this relation. Notice M28 in the upper right hand corner. Several 1σ errors are shown; these are computed from Poisson statistics of the detected number of X-ray counts, and do not take into account uncertainties in the spectral energy distribution and/or cluster distance. Data from Grindlay *et al.* (2002), D'Amico *et al.* (2002), Becker *et al.* (2003) and Possenti *et al.* (quoted in Bassa & Stappers 2004).

period derivative for gravitational acceleration in the cluster potential. Comparison of the corrected spin-down luminosities with the X-ray luminosities led Grindlay *et al.* (2002) to suggest that $L_X \propto \sqrt{L_{sd}}$. They further note that the pulsar in NGC 6397 agrees with this (slower) trend, whereas the pulsar in M 28 does not. Grindlay *et al.* argue that the emission of the pulsar in M 28 is mainly magnetospheric in origin, whereas the emission of the other pulsars in globular clusters is mainly thermal emission from the surface of the neutron star.

We reinvestigated the relation between X-ray and the spin-down luminosities for the globular cluster pulsars in Fig. 8.10. We include NGC 6752 D, and the pulsar in M 28. It should be noted that the luminosity of the pulsar in M 28 is only about 20% of the total cluster luminosity as observed with ROSAT (Verbunt 2001; Becker *et al.* 2003), whereas the value used by Grindlay *et al.* (2002) is the total cluster luminosity. Since thermal emission from millisecond pulsars is the result of heating by magnetospheric processes, we prefer not to exclude the magnetospheric X-rays and to retain the pulsar in M 28, and we are inclined to conclude that the general slope of the relation between L_X and L_{sd} is similar to that observed for the pulsars detected in the Galactic disk, with some scatter at the lowest luminosities. The strong downward revision of the spin-down luminosity of the pulsar in NGC 6397 (Possenti *et al.* quoted in Bassa & Stappers 2004) brings this pulsar also in line with the steeper dependence of L_X on L_{sd}.

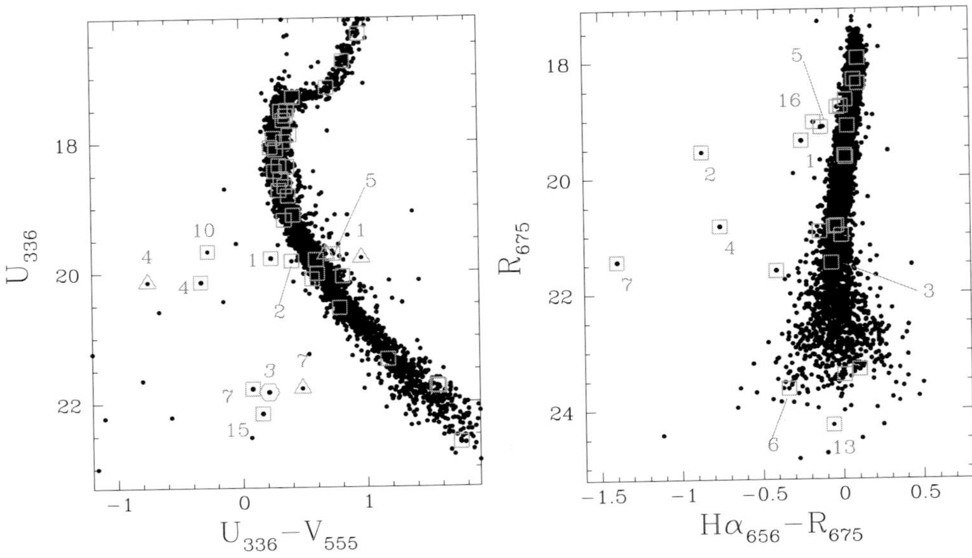

Fig. 8.11. U–V and Hα–R color–magnitude diagrams of the central regions of NGC 6752. Stars within error circles of Chandra X-ray sources are indicated with squares; numbers indicate the corresponding Chandra source. Cataclysmic variables lie to the left of the main sequence in the U–V diagram, i.e., they are blue. When the flux in the narrow Hα filter is higher than in the neighboring continuum (measured in R), the points fall to the left of the main sequence in the Hα–R diagram. Because of variability, the same object may lie in different locations of the color–magnitude diagrams, depending on which data set is used. Updated after Pooley *et al.* (2002a).

8.4.3 Cataclysmic variables

Cataclysmic variables are best identified when an optical counterpart is found. A good indicator is that the optical counterpart is bluer than the main sequence, especially in the ultraviolet; and/or that it has strong $H\alpha$ emission (see Figure 8.11). As an example, such counterparts were identified in NGC 6397, and follow-up spectra show the strong Balmer emission lines prevalent in cataclysmic variables (Cool *et al.* 1995; Grindlay *et al.* 1995; Edmonds *et al.* 1999; note that firm identifications were only possible once Chandra had obtained accurate positions, Grindlay *et al.* 2001b). Quiescent neutron-star low-mass X-ray binaries also have blue spectra with Balmer emission, but can be distinguished from cataclysmic variables through their soft X-ray spectra, and by the fact that they are more luminous than cataclysmic variables (see Section 8.4.1). Optical and ultraviolet color–magnitude diagrams have been used to classify optical counterparts as cataclysmic variables also in NGC 6752 and in 47 Tuc (Pooley *et al.* 2002a; Edmonds *et al.* 2003).

If no optical colors are available, the ratio of X-ray to optical flux provides a good, but not conclusive, indication as to whether a source is a cataclysmic variable, as shown with cataclysmic variables studied in the ROSAT All Sky Survey (Verbunt *et al.* 1997; Verbunt & Johnston 2000). In Fig. 8.12 we show (a measure of) the X-ray luminosity in the 0.5–4.5 keV range as a function of the absolute visual magnitude for X-ray sources in 47 Tuc and in NGC 6752. Only sources that have been classified on the basis of optical/ultraviolet

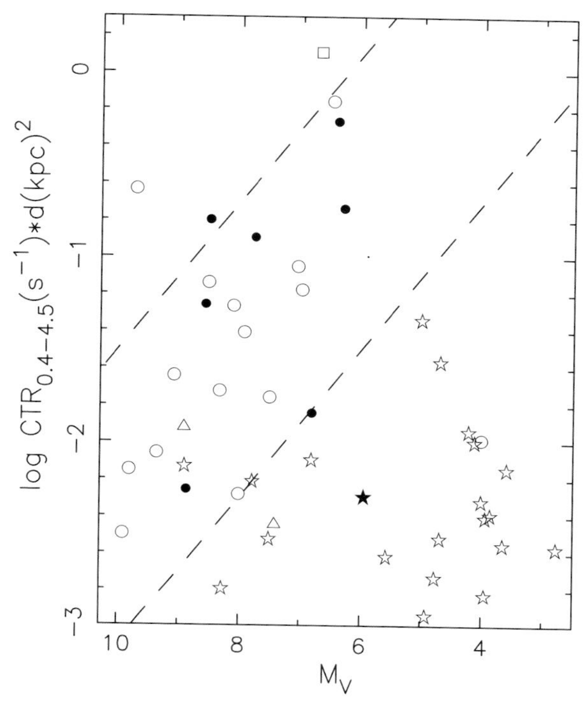

Fig. 8.12. X-ray luminosity as a function of absolute visual magnitude, for optically identified Chandra sources in 47 Tuc (open symbols) and NGC 6752 (filled symbols). Squares, circles, triangles and stars indicate low-luminosity $LMXB^{NS}$, cataclysmic variables, (companions to) recycled pulsars, and magnetically active binaries, respectively. To minimize model dependence, the X-ray luminosity is expressed as the product of Chandra countrate CTR (in the 0.5–4.5 keV band, corrected for interstellar absorption) and the cluster distance d (in kpc) squared. Two dashed lines of constant ratio of X-ray to visual flux roughly separate the low-luminosity low-mass X-ray binaries with neutron stars from the cataclysmic variables; and the latter from the magnetically active binaries (see Verbunt & Hasinger 1998, Pooley *et al.* 2002a). Data from Edmonds *et al.* (2003), Pooley *et al.* (2002a).

color–magnitude diagrams are shown. In the figure we plot the line

$$\log(\mathrm{CTR}_{0.5-4.5\mathrm{keV}} d_{\mathrm{kpc}}^2) = -0.4M_{\mathrm{V}} + 0.9 \tag{8.2}$$

where $\mathrm{CTR}_{0.5-4.5\mathrm{keV}}$ is the number of counts per second in the 0.5–4.5 keV range, and d_{kpc} the distance in kpc. This line roughly separates the cataclysmic variables from magnetically active binaries. A parallel line for an X-ray luminosity which is a factor ~ 40 higher roughly separates the cataclysmic variables from the low-luminosity low-mass X-ray binaries with a neutron star. The figure shows that the ratio of X-ray to optical luminosity is a fairly good classifier of X-ray sources in the absence of more conclusive information.

A further indicator that a source is a cataclysmic variable may be found from optical variability, either orbital or from a (dwarf) nova outburst. Orbital variability may be present in magnetically active binaries too, and thus can be used to classify a source only in combination with other information, such as color–magnitude diagrams, or ratio of X-ray to visual flux.

Two cataclysmic variables were found in NGC 6752 based on periodic variability and Hα emission by Bailyn *et al.* (1996), and were identified with Chandra X-ray sources by Pooley *et al.* (2002a). Variability indicative of dwarf nova outbursts has been detected for several blue objects in 47 Tuc (e.g., Paresce *et al.* 1992; Paresce & De Marchi 1994; Shara *et al.* 1996); these sources have subsequently been identified with Chandra X-ray sources (Grindlay *et al.* 2001a). An optical variable in the core of NGC 6656/M 22 has been identified as a possible dwarf nova, detected in X-rays with Einstein, ROSAT and XMM (Anderson *et al.* 2003; see Table 8.2).

So far, only 47 Tuc, NGC 6397 and NGC 6752 have been studied to such an extent that a large fraction of the X-ray sources in them has been optically identified. Most of them are classified as cataclysmic variables. In ω Cen, several Chandra sources have been identified with (optically detected) cataclysmic variables (Carson *et al.* 2000), but HST observations only cover a small fraction of the cluster. Classifications based only on the X-ray to optical flux ratio must be considered preliminary, as illustrated by the case of NGC 6752 CX11 (see Section 8.4.2).

In general it may be stated that the properties of cataclysmic variables in globular clusters are similar to those of cataclysmic variables in the galactic disk (i.e., in the solar neighborhood; see also Chapter 10). In the galactic disk, distances and interstellar absorption for cataclysmic variables are only inaccurately determined at best. In contrast, for systems in globular clusters these quantities may be set equal to the values for the cluster, which are much better known. Thus comparison between different classes of objects will be more accurate in globular clusters.

As an example, we note that Verbunt and Hasinger (1998) in their analysis of ROSAT observations of 47 Tuc use the ratio of X-ray to visual flux to suggest that 47 Tuc X9, identified with the blue variable V1, is a low-luminosity low-mass X-ray binary with a neutron star. In Fig. 8.12, based on more accurate Chandra data and now secure identifications, the systems with the three highest X-ray to optical flux ratios in 47 Tuc are X10/V3, X7 and X9/V1. X7 is indeed a low-luminosity low-mass X-ray binary with a neutron star, but the hard X-ray spectra of X10 and X9 indicate that they are probably cataclysmic variables. This illustrates the overlap between low-mass X-ray binaries and cataclysmic variables in the X-ray to visual flux ratio.

8.4.4 *Magnetically active binaries*

X-ray sources in globular clusters can be classified as magnetically active binaries when a stellar flare is observed in X-rays; or on the basis of the optical counterpart, when this is a known active binary, or less securely when it lies above the main sequence and/or shows weak Hα emission.

Two OGLE variables in NGC 5139, OGLEGC15 and OGLEGC22, are identified by Cool *et al.* (2002) with Chandra sources (not listed by Rutledge *et al.* 2002, but confirmed by Gendre, private communication). A third OGLE variable in NGC 5139, OGLEGC30, has been detected with XMM (Gendre *et al.* 2003a). Yet another Chandra X-ray source, already detected with ROSAT but not detected with XMM and therefore a variable X-ray source, shows Hα emission, and presumably is also a magnetically active binary (Gendre *et al.* 2003a). Figure 8.12 shows Chandra X-ray sources in 47 Tuc and NGC 6752 that are classified on the basis of color–magnitude diagrams as magnetically active binaries; for many of these binaries in 47 Tuc the orbital lightcurve confirms their identity as coronal X-ray emitters (Edmonds

Table 8.4. *Published Chandra and XMM observations of low-luminosity X-ray sources in globular clusters*

Cluster	Ref.	L_{low}	BX	FX	CV	PSX	(PSR)	BY	N_{tot}
NGC 6440	[180]	2×10^{31}	1	3			(1)		
NGC 6652	[92]	8×10^{32}	1	3	$\rightarrow$1?				
Terzan 1	[243]	3×10^{33}	1	1	$\rightarrow$1?				
Terzan 5	[93]	5×10^{32}	1	4	5		(4)		
Liller 1	[108]	$\sim 10^{34}$	1	3					
47 Tuc	[81]	10^{30}	0	2	>30	15	(22)	26	104
ω Cen	[65]	10^{31}	0	1	>20		(0)	4	~100
NGC 6093	[95]	7×10^{30}	0	2	~15				19
NGC 6121	[14]	1×10^{29}	0	0	3	1	(1)	14	~20
NGC 6205	[66]	2×10^{31}	0	1	4	0	(5)		5
NGC 6397	[82]	3×10^{29}	0	1	9	1	(1)	3	~20
NGC 6626	[15]	2×10^{30}	0	1	~25	1	(1)		
NGC 6656	[237]		0	1?$\leftarrow$	3			~3	
NGC 6752	[179]	2×10^{30}	0	0	10	1	(5)	3	17

For each cluster we give the lowest detectable luminosity (erg s^{-1}, estimated for the range 0.5–2.5 keV), and the estimated numbers of X-ray sources corresponding to luminous low-mass X-ray binaries (BX), low-luminosity low-mass X-ray binaries (FX), cataclysmic variables (CV), recycled pulsars (PSX) (for comparison, we list the number of radio pulsars in the column PSR), and magnetically active binaries (BY). $\rightarrow$1? (1?$\leftarrow$) indicates that one of the sources in the previous (next) column may belong in this column. The final column gives the total number of detected X-ray sources associated with the cluster.

et al. 2003). That care must be taken in classifying sources is shown by the example of NGC 6397 CX12 (see Section 8.4.2).

Interestingly, most magnetically active binaries identified with X-ray sources so far have visual magnitudes higher than or equal to the turnoff stars, implying that they are on the main sequence (BY Dra's). Since the maximum X-ray luminosity of a magnetically active binary scales roughly with the surface area of the stars, this implies that the luminosities of the active binaries in globular clusters are low (typically $L_X < 10^{30}$ erg s^{-1}), compared to systems with giants (RS CVn's), in the Galaxy, which can be up to a hundred times more luminous (Dempsey *et al.* 1993).

8.4.5 *Comparing clusters*

In comparing the different clusters, the limit to which sources can be detected must be taken into account. Low-luminosity low-mass X-ray binaries with a neutron star tend to be more luminous than cataclysmic variables, which in turn tend to be more luminous than magnetically active binaries. This ordering is reflected in the numbers of currently known cataclysmic variables and magnetically active binaries listed in Table 8.4 as a function of the detection limit.

Another number that is important is the estimated number of close encounters between stars in the globular cluster. Pooley *et al.* (2003) show that the number of X-ray sources detected in a globular cluster above an observational threshold of $L_X \simeq 4 \times 10^{30}$ erg s^{-1} (0.5–6 keV)

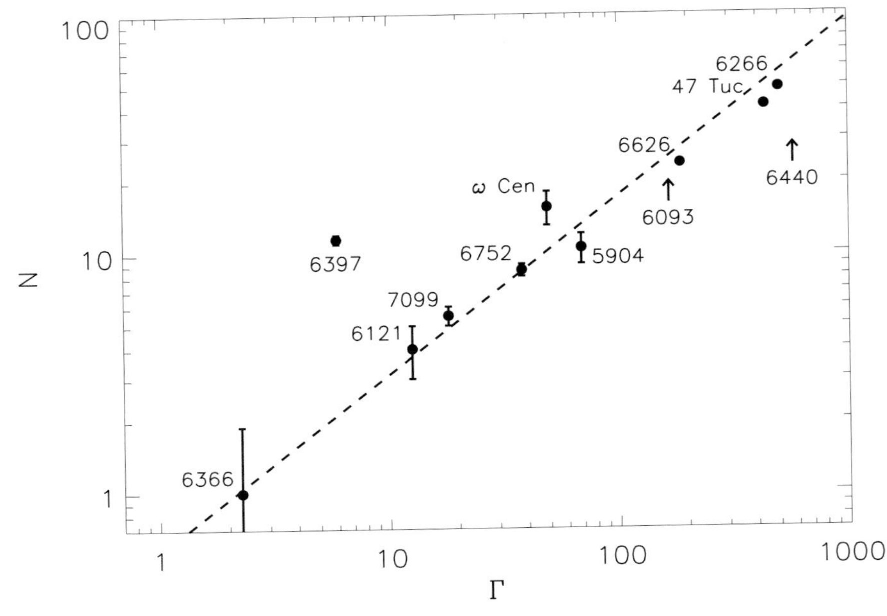

Fig. 8.13. Number N of X-ray sources with $L_X \gtrsim 4 \times 10^{30}$ erg s^{-1} (0.5–6 keV) detected in globular clusters, as a function of the collision number Γ. Γ is a measure of the number of close encounters between stars in a cluster (see Eqs. (8.5) and (8.6)). The luminosity limit implies that most sources are cataclysmic variables. In general N scales quite well with Γ, indicating that cataclysmic variables in globular clusters are formed via close encounters between a white dwarf and another star or a binary. Arrows indicate lower limits. NGC 6397 does not follow the general trend. From Pooley *et al.* (2003).

scales quite well with this number, as shown in Fig. 8.13. Heinke *et al.* (2003d) find that the number of cataclysmic variables alone (at $L_X \gtrsim \times 10^{31}$ erg s^{-1}) possibly increases more slowly with central density than predicted by proportionality to the number of close encounters.

An exception to this scaling is NGC 6397. This cluster has a higher number of neutron star binaries and cataclysmic variables than expected on the basis of its rather low collision number. Remarkably, the number of magnetically active binaries in this cluster is not very high, and this is reflected in a relatively flat X-ray luminosity function (Pooley *et al.* 2002b). If it is true, as argued by Pooley *et al.* (2003), that the high number of neutron star binaries and cataclysmic variables in NGC 6397 is due to its being shocked and stripped in multiple passages through the Galactic disk and/or near the Galactic center, it has to be explained why these mechanisms are more efficient in removing magnetically active binaries than in removing cataclysmic variables and binaries with neutron stars.

8.5 Some remarks on evolution and formation

8.5.1 Evolution

A good first indicator of the evolutionary status of a binary is its orbital period (see Chapter 16 by Tauris & Van den Heuvel and Verbunt 1993 for a more extended discussion of the evolution of X-ray binaries). We show the orbital periods of X-ray emitting binaries

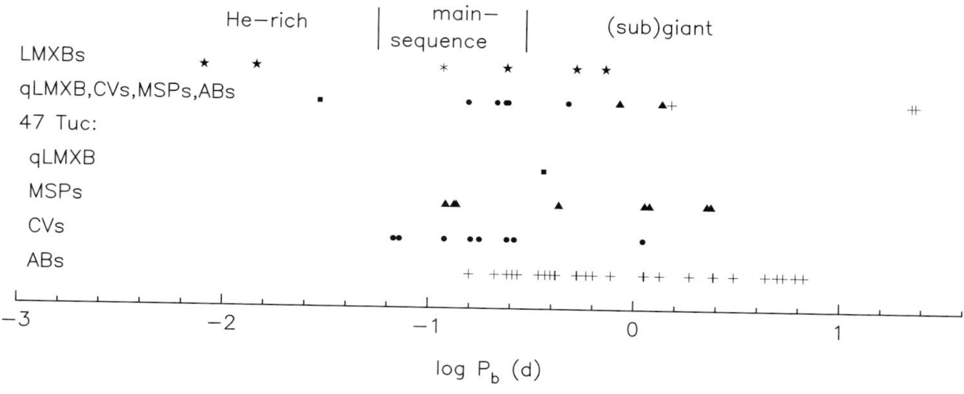

Fig. 8.14. Orbital period distributions of X-ray-detected binaries in globular clusters. Most known orbital periods are for systems in 47 Tuc, and are shown in the lower four rows. The top two rows indicate the luminous X-ray binaries and other binaries in other clusters (with symbols as for 47 Tuc). The period of a cluster source in M 31 is shown with a six-pointed star. The period range in which a main-sequence star can fill its Roche lobe is indicated; systems with shorter periods may contain degenerate stars, with longer periods (sub)giants. Periods from Table 8.1; 47 Tuc: Edmonds *et al.* (2003), Freire *et al.* (2003), Camilo *et al.* (2000); other clusters: Bailyn *et al.* (1996), Deutsch *et al.* (2000), Neill *et al.* (2002), Kaluzny & Thompson (2002), Kaluzny *et al.* (1996), D'Amico *et al.* (2001, 2002); M 31: Trudolyubov *et al.* (2002).

in globular clusters in Fig. 8.14. Most periods known are for binaries in 47 Tuc. It should be noted that there is a selection effect against the discovery of long-period binaries in optical surveys.

The radius R of the Roche lobe of a star with mass M in a binary with a star of mass m is given in units of the distance a between stars as approximately

$$\frac{R}{a} \simeq 0.46 \left(\frac{M}{M+m} \right)^{1/3} \qquad \text{for } M < 0.8m \tag{8.3}$$

Combining this with the third law of Kepler we find

$$P_b \simeq 8.9\,\text{hr} \left(\frac{M_\odot}{M} \right)^{1/2} \left(\frac{R}{R_\odot} \right)^{3/2} \tag{8.4}$$

i.e., the orbital period gives the average density of a Roche-lobe filling star (cf. Section 5.3.1).

The radius of a main-sequence star is roughly given by $R/R_\odot \simeq M/M_\odot$ in the mass range of interest here. With main-sequence stars in old globular clusters limited to masses $M \lesssim 0.8\,M_\odot$, we see that binaries in which mass transfer occurs, i.e., low-mass X-ray binaries and cataclysmic variables, can only have a main-sequence star as the mass donor provided the orbital period is less than about 7 hr. If the orbital period is longer, the donor must be larger than a main-sequence star, i.e., a (sub)giant. It then follows from Fig. 8.14 that, with one exception, all cataclysmic variables in globular clusters can have main-sequence donors. The one exception is AKO9, a cataclysmic variable with a slightly evolved donor in 47 Tuc (e.g. Knigge *et al.* 2003). Of the low-mass X-ray binaries, one may have a main-sequence donor, two binaries must have sub-giant donors; the low-luminosity low-mass X-ray binary in 47 Tuc is probably a sub-giant close to the main sequence. The orbital periods of most active binaries are long enough that even main-sequence stars near the turnoff mass (0.8 $M_\odot$) fit well within the Roche lobes; for those with the shorter periods, both stars must have lower masses to be smaller than their Roche lobes. Two of the low-mass X-ray binaries have ultra-short orbital periods; at such short orbital periods the Roche filling star can be a white dwarf. With $R/R_\odot \simeq 0.01(M/M_\odot)^{-1/3}$, a white dwarf fills its Roche lobe if the orbital period $P_b \simeq 48\,\mathrm{s}(M_\odot/M)$.

The evolution of low-mass X-ray binaries and cataclysmic variables *with main-sequence donors* is driven by the loss of angular momentum $\dot{J}$ from the binary. Writing the angular momentum of the binary as J_b, one finds that the mass-transfer rate $\dot{M}$ is roughly given by $-\dot{M}/M \sim -\dot{J}/J_b$. The loss of angular momentum from gravitational radiation alone is enough to drive mass transfer at a rate of $10^{-10} M_\odot\ \mathrm{yr}^{-1}$; higher mass-transfer rates, as witnessed by luminosities well in excess of $L_X \simeq 10^{36}$ erg s^{-1}, imply other mechanisms. The loss of angular momentum causes the orbit to shrink, and thus the orbital period to become shorter. In binaries *with a (sub-)giant donor*, the mass-transfer rate is very roughly given by the expansion rate of the donor star $-\dot{M}/M \sim \dot{R}/R$. Since the expansion rate of a giant becomes faster as it further ascends the giant branch, this predicts higher mass transfer, i.e., more luminous X-ray emission, for the longest periods. For the two known orbital periods of low-mass X-ray binaries in globular clusters with a sub-giant, expansion of the donor predicts a modest mass transfer of $\sim 10^{-10}\ M_\odot\ \mathrm{yr}^{-1}$. The mass transfer, combined with conservation of angular momentum, causes the orbit to expand, and the orbital period to increase. Enhanced loss of angular momentum from a stellar wind has often been invoked to explain large X-ray luminosities, in binaries with main-sequence or sub-giant donors, but the actual efficiency of this loss mechanism is not known. It is worth noting that many X-ray sources show large variations in their X-ray luminosity on timescales of decades – the transients are an obvious example – indicating that the current mass-transfer rate, even in apparently stable systems, may not be an accurate estimator of mass-transfer rate on an evolutionary timescale.

Something is wrong with the simplest description of binary evolution. This follows, e.g., from the orbital period distribution of the recycled radio pulsars. The expansion of a binary with a sub-giant donor continues until the core of the giant is denuded of its envelope. By then the orbital period has increased by an order of magnitude. The orbital periods of the radio pulsars in 47 Tuc are less than about 2.5 d, suggesting that little if any expansion has occurred during the mass transfer. On the other hand, some pulsar binaries in globular clusters, such as the pulsar binary in M 4, do have periods in excess of a hundred days, with fairly circular orbits, showing that expansion is strong in at least some cases.

What about the ultrashort periods? They may have white-dwarf donors; if so, their orbital period should be increasing. It has been suggested that a collision between a (sub-)giant and a neutron star could lead to expulsion of the giant envelope and leave the neutron star in orbit around the core, which subsequently cools to an under-massive white dwarf. If loss of angular momentum from gravitational radiation pushes the stars closer, mass transfer begins once the white dwarf fills its Roche lobe (Verbunt 1987). Alternatively, it has been suggested that the ultrashort period systems are the outcome of an evolution that starts when a sub-giant starts transferring mass to a neutron star in an orbital period less than ~18 hr (Podsiadlowski *et al.* 2002). Large loss of angular momentum through a stellar wind brings the two stars closer together, and the evolution proceeds to shorter and shorter periods. The minimum period reached through such an evolutionary path is short enough to explain the 11 min period of the $LMXB^{NS}$ in NGC 6624. It is predicted that this binary has a negative period derivative, as observed. There are two problems with this scenario, however. One is that the loss of angular momentum from the giant, required at the start of the mass transfer to convert orbital expansion into orbital shrinking, is rather high; perhaps implausibly high. Second, none of the evolutions along this scenario computed by Pylyser and Savonije (1988) reach the shortest periods within a Hubble time, because it already takes very long for a 1 $M_{\odot}$ star to fill its Roche lobe in a 16 hr period. Van der Sluys *et al.* (2004) investigate this in more detail and find that only binaries in narrow ranges of initial orbital periods and component masses evolve to ultrashort periods within a Hubble time, and that these binaries only spend a small fraction of their life at ultrashort periods; they conclude that no significant population of ultrashort-period binaries in globular clusters can be produced through this evolution channel. The most likely mechanism to produce bright X-ray sources with ultrashort orbital periods is mass transfer from an intermediate-mass donor leading to a common envelope, some time in the past history of the globular cluster (Davies & Hansen 1998; Rasio *et al.* 2000). The result is a binary of the neutron star and the core of the giant, which cools into a white dwarf. In the course of several billion years, gravitational radiation may bring the system into contact.

8.5.1.1 Some specific systems

The orbital period for the low-luminosity low-mass X-ray binary 47 Tuc X5 is too long for a Roche-lobe filling main-sequence donor star with a mass less than the turnoff mass of 0.8 $M_{\odot}$. Edmonds *et al.* (2002b) therefore conclude that the star is smaller than its Roche lobe. We suggest an alternative possibility that the system hosts a 0.8 $M_{\odot}$ sub-giant donor that has recently started to transfer matter to a 1.4 $M_{\odot}$ neutron star. The donor has not yet transferred much of its envelope mass: a low donor mass in an 8.666 hr orbit implies a Roche lobe for the donor that is too small to hold a sub-giant. The system is very sub-luminous for a sub-giant: this is expected for a donor that is losing mass.

PSR 47 Tuc W (Chandra source 29) is a pulsar accompanied by an object whose location in the color–magnitude diagram indicates that it is too big for a white dwarf and too small for a main-sequence star. The orbital lightcurve shows clear heating by the pulsar (Edmonds *et al.* 2002a). If a main-sequence star is heated at constant radius, it moves up and to the left in a color–magnitude diagram, to a location below the main-sequence. If the companion to PSR 47 Tuc W is of this nature, its position about 5 magnitudes below turnoff indicates a very low mass, of an M dwarf. This poses an interesting puzzle for the evolutionary history: if the M dwarf was in the binary from the start, it was too small to transfer mass to the neutron star and spin it up. If on the other hand the main-sequence star was captured by the pulsar tidally or via an exchange encounter, the orbit should be eccentric initially; the question is whether tidal dissipation can circularize the orbit and heat the M dwarf to its current position.

PSR NGC 6397 A is another pulsar accompanied by a low-mass ($\sim$0.25 $M_{\odot}$) companion (Ferraro *et al.* 2003). In this case the companion lies somewhat to the right of the turnoff, at a radius of 1.6(2) $R_{\odot}$ and luminosity 2.0(4)$L_{\odot}$; notwithstanding the proximity of an energetic radio pulsar, the companion shows no sign of heating (Orosz & van Kerkwijk 2003). The position of the companion in the color–magnitude diagram is hard to explain. Orosz and van Kerkwijk invoke a stellar collision, causing a slightly evolved star near the turnoff to lose most of its envelope.

8.5.1.2 *Black holes*

The absence of known very luminous ($L_X \geq 10^{38.5}$ erg s^{-1}, say) low-mass X-ray binaries with a black hole in globular clusters of our Galaxy has led to the suggestion that black holes are efficiently ejected from globular clusters through dynamical processes (Kulkarni *et al.* 1993; Portegies Zwart & McMillan 2000). The discovery of very luminous, soft X-ray sources in globular clusters in other galaxies shows that X-ray binaries with black holes probably exist in globular clusters (see Section 8.3).

There is no evidence that M 15 contains an intermediate mass black hole; an upper limit for the mass of about 10^3 M$_\odot$ can be set both from an analysis of pulsar accelerations in this cluster, and from an analysis of radial velocities of stars close to the center (Phinney 1992; Gerssen *et al.* 2003). A case has been made for a binary in NGC 6752 of two black holes, of which at least one has an intermediate mass (Colpi *et al.* 2002). The argument for this is the presence of a white-dwarf/radio-pulsar binary in the outskirts of the cluster, which most likely was ejected from the cluster core. If the binary was ejected with the white dwarf companion to the pulsar already formed, the very small eccentricity of its orbit implies that the orbit of the other binary involved in the scattering was much larger. To still produce an ejection velocity for the pulsar binary high enough for it to reach the outer cluster region then requires at least one black hole with a mass ~ 100 M$_\odot$ in the scattering binary (Colpi *et al.* 2002). To solidify the case for a binary black hole it would have to be demonstrated that the pulsar indeed belongs to NGC 6752 (as is probable), and that the pulsar binary was ejected before the formation of the white dwarf (which is not obvious). The optical identification of the white dwarf companion to this pulsar shows that the white dwarf is young compared to the age of the globular cluster; this strengthens the case for a scenario in which a binary consisting of a main-sequence star and a neutron star was ejected from the cluster core, and subsequent evolution of the main-sequence star led to circularization of the orbit (Bassa *et al.* 2003).

8.5.2 **Formation**

The rate at which stars with number density n encounter target stars with number density n_c in a cluster with dispersion velocity v is given by (e.g., Hut & Verbunt 1983):

$$\Gamma \propto \int n_c n A v \, \mathrm{d}V \propto \int \frac{n_c n R}{v} \mathrm{d}V \propto \frac{\rho_o^2 r_c^3}{v} R \tag{8.5}$$

where A is the interaction cross section (proportional to R/v^2 because of gravitational focusing), R the radius of the star, ρ_o is the central mass density of the cluster and r_c its core radius. Because the number densities of stars drop rapidly with distance from the cluster center, the integral over volume dV can be approximated by multiplying the central encounter rate with the volume of the cluster core. An analogous equation gives the exchange encounter rate

$$\Gamma_e \propto \int n_c n_b A_b v \, \mathrm{d}V \propto \int \frac{n_c n_b a}{v} \mathrm{d}V \propto \frac{\rho_o^2 r_c^3}{v} a \tag{8.6}$$

where n_b is the number of binaries per unit volume, and a the semi-major axis of the binary. The ratio of tidal capture to exchange encounters is roughly

$$\frac{\Gamma}{\Gamma_e} \sim \frac{R}{a}\frac{n}{n_b} \tag{8.7}$$

The velocity dispersion v is related to the core mass and radius through (a specific version of) the virial theorem (King 1966):

$$v \propto \sqrt{\rho_o}\, r_c \tag{8.8}$$

Therefore (Verbunt 2003)

$$\Gamma \propto {\rho_o}^{1.5} r_c^2 R \qquad \text{and} \qquad \Gamma_e \propto {\rho_o}^{1.5} r_c^2 a \tag{8.9}$$

Because neutron stars are formed with appreciable velocities, a cluster with a high mass is expected to retain a higher fraction of the neutron stars that are formed in it than a cluster with a low mass. In a cluster with strong mass segregation, virtually all the neutron stars will have migrated to the core. Thus a massive cluster with strong mass segregation is expected to have a much higher central number density of neutron stars than a low-mass little-segregated cluster. Thus, the ratio n_c/ρ_o for neutron stars, and through this the proportionality constant for the last members of Eqs. (8.5) and (8.6) will vary widely between clusters (Verbunt & Meylan 1988). On the other hand, white dwarfs are always retained upon formation, and due to their lower masses are less affected by mass segregation. This is probably the reason that the relation between the number of X-ray sources (mainly cataclysmic variables) and Γ is as narrow as shown in Fig. 8.13.

Due to the large number density of stars in a cluster core, an appreciable fraction of neutron stars in that core may be involved in a close encounter with a single star or with a binary. The formation of tidal bulges during passage of a neutron star within above three times the radius of a main-sequence star drains enough energy from the relative motion of the two stars to bind them in a binary. This process is called tidal capture (Fabian *et al.* 1975). Whether it is efficient in the formation of a binary with a neutron star is under debate, because of the large amount of energy residing in the initially very eccentric orbit of the newly formed binary. If the orbit circularizes rapidly because of tides on the main-sequence star, the energy released is enough to (almost) destroy the main-sequence star (Ray *et al.* 1987; Verbunt 1994). Rapid circularization can be avoided if the energy exchange between tides and orbit is chaotic, as is likely in a highly eccentric orbit (Mardling 1995). Mass loss from the main-sequence star due to tidal heating may further limit the damage to the deeper layers of the star.

A neutron star can also be exchanged into a pre-existing binary when it takes the place of one of the binary members in an exchange encounter (Hills 1976). Which of the two mechanisms is more important depends on the number of binaries present in the core and on their orbital period distribution; as well as on the efficiency of the tidal capture process.

If a binary is of a type that very rarely results from the evolution of a primordial binary, then its presence in a globular cluster may be ascribed to formation via a close encounter. Such is the case for binaries with a neutron star. If a binary is very frequently formed from a primordial binary, then it is likely to be primordial also when present in a globular cluster. This is the case for magnetically active close binaries. Cataclysmic variables are somewhere in between, and thus in clusters can be formed both via close encounters and via evolution of a primordial binary. Figure 8.13 shows that the number of binaries with $L_X \gtrsim 4 \times 10^{30}$ erg s^{-1} scales well with the number of encounters in a cluster. Since most binaries with such luminosities are cataclysmic variables, this implies that most cataclysmic variables are in fact formed via close encounters. One reason for this is that evolution from a primordial binary into a cataclysmic variable passes through a stage in which the binary is very wide; such a wide binary is easily unbound in a globular cluster by a passing star and the formation of a cataclysmic variable is prevented (Davies 1997). If the number of cataclysmic variables increases more slowly with central density than as $\rho_o^{1.5}$, as suggested by Heinke *et al.* (2003d), this could suggest that primordial binaries do still contribute to the formation of cataclysmic variables. Remarkably, Jordán *et al.* (2004) find that the probability for a globular cluster associated with NGC 4486 (M 87) to harbor a bright X-ray source also scales with a lower power of ρ_o than the collision number, i.e., roughly as $\Gamma\rho_o^{-0.5}$.

Looking now at the period distribution of the cataclysmic variables and low-mass X-ray binaries in globular clusters, we see that their periods are short, $\lesssim 1$ d. This may indicate that they are formed at short periods, which hints at tidal capture as the main formation process. Some care is necessary before one jumps to conclusions, however. Mass transfer in wide binaries tends to be faster, and thus wide binaries live shorter, and will be less numerous even if their formation rate is the same as that of short binaries. Also, longer periods are more difficult to measure, and some of the many binaries with unknown periods may have long periods. In addition, a wide binary with a neutron star or white dwarf can become closer via encounters with field stars. We doubt that this process is sufficiently efficient, given the observed presence in 47 Tuc of active binaries with periods up to ten days that apparently have avoided further shrinking of their orbits. On the whole we tend to conclude that rumors of the death of the tidal capture model for the formation of binaries with a neutron star and of cataclysmic variables have been much exaggerated.

That exchange encounters do occur in globular clusters is evident from the wide pulsar binaries, such as M 4 PSR A ($P_b = 191$ d, Thorsett *et al.* 1999) and M 53 PSR A (255 d, Kulkarni *et al.* 1991). These are found in clusters with a relatively low central density, which allows long-period binaries to survive (e.g., Verbunt 2003). They must have evolved from binaries with initial periods too long to be formed by tidal capture, in which the neutron star can thus only have entered via an exchange encounter. (M 15 PSR C is an eccentric binary of two neutron stars in the outskirts of M 15, and is another product of an exchange encounter: Phinney & Sigurdsson 1991.)

Acknowledgements

We are very grateful for comments, suggestions and help from Lorella Angelini, Keith Ashman, Pauline Barmby, Cees Bassa, Boris Dirsch, Rosanne Di Stefano, Josh Grindlay, Bill Harris, Piet Hut, Andres Jordán, Arunav Kundu, Erik Kuulkers, Tom Maccarone, Dave Pooley, Katherine Rhode, Craig Sarazin, Rudy Wijnands, and Steve Zepf.

References

1. Anderson, J., Cool, A., & King, I. 2003, *ApJ*, **597**, L137
2. Anderson, S., Margon, B., Deutsch, E., Downes, R., & Allen, R. 1997, *ApJ*, **482**, L69
3. Angelini, L., Loewenstein, M., & Mushotzky, R. 2001, *ApJ*, **557**, L35
4. Ashman, K. & Zepf, S. 1992, *ApJ*, **384**, 50
5. ——— 1998, *Globular Cluster Systems* (Cambridge: Cambridge University Press)

6. Aurière, M., Le Fèvre, O., & Terzan, A. 1984, *A&A*, **138**, 415
7. Bahcall, J. & Wolf, R. 1976, *ApJ*, **209**, 214
8. Bailyn, C., Grindlay, J., & Garcia, M. 1990, *ApJ*, **357**, L35
9. Bailyn, C., Rubenstein, E., Slavin, S., *et al.* 1996, *ApJ*, **473**, L31
10. Barmby, P. 2003, in *Extragalactic Globular Cluster Systems*, ESO Workshop, 143
11. Barmby, P. & Huchra, J. 2001, *AJ*, **122**, 2458
12. Bassa, C. & Stappers, B. 2004, *A&A*, **425**, 1143
13. Bassa, C., Verbunt, F., Van Kerkwijk, M., & Homer, L. 2003, *A&A*, **409**, L31
14. Bassa, C., Pooley, D., Homer, L., *et al.* 2004, *ApJ*, **609**, 755
15. Becker, W., Swartz, D., Pavlov, G., *et al.* 2003, *ApJ*, **594**, 798
16. Bellazzini, M., Pasquali, A., Federici, L., *et al.* 1995, *ApJ*, **439**, 687
17. Bildsten, L. & Rutledge, R. 2000, *ApJ*, **541**, 908
18. Bildsten, L. & Deloye, C. 2004, *ApJ*, **607**, L119
19. Blanton, E., Sarazin, C., & Irwin, J. 2001, *ApJ*, **552**, 106
20. Bridges, T. & Hanes, D. 1990, *AJ*, **99**, 1100
21. Callanan, P., Drake, J., & Fruscione, A. 1999, *ApJ*, **521**, L125
22. Camilo, F., Lorimer, D., Freire, P., Lyne, A., & Manchester, R. 2000, *ApJ*, **535**, 975
23. Canizares, C. & Neighbours, J. 1975, *ApJ*, **199**, L97
24. Carson, J., Cool, A., & Grindlay, J. 2000, *ApJ*, **532**, 461
25. Chou, Y. & Grindlay, J. 2001, *ApJ*, **563**, 934
26. Clark, G. 1975, *ApJ*, **199**, L143
27. Clark, G., Markert, T., & Li, F. 1975, *ApJ*, **199**, L93
28. Cohen, J., Blakeslee, J., & Côté, P. 2003, *ApJ*, **592**, 866
29. Colpi, M., Possenti, A., & Gualandris, A. 2002, *ApJ*, **570**, L85
30. Cool, A., Grindlay, J., Krockenberger, M., & Bailyn, C. 1993, *ApJ*, **410**, L103
31. Cool, A., Grindlay, J., Cohn, H., Lugger, P., & Slavin, S. 1995, *ApJ*, **439**, 695
32. Cool, A., Grindlay, J., Cohn, H., Lugger, P., & Bailyn, C. 1998, *ApJ*, **508**, L75
33. Cool, A., Haggard, D., & Carlin, J. 2002, in *ω Cen, A Unique Window Into Astrophysics*, eds. F. van Leeuwen, J. Hughes, & G. Piotto (ASP Conf. Ser. 265), 277–288
34. Côté, P. 1999, *AJ*, **118**, 406
35. Côté, P., West, M., & Marzke, R. 1999, *ApJ*, **567**, 853
36. D'Amico, N., Possenti, A., Manchester, D., *et al.* 2001, *ApJ*, **561**, L89
37. D'Amico, N., Possenti, A., Fici, L., *et al.* 2002, *ApJ*, **570**, L89
38. Davies, M. 1997, *MNRAS*, **288**, 117
39. Davies, M. & Hansen, B. 1998, *MNRAS*, **301**, 15
40. Dempsey, R., Linsky, J., Fleming, T., & Schmitt, J. 1993, *ApJS*, **86**, 599
41. Deutsch, E., Anderson, S., Margon, B., & Downes, R. 1996, *ApJ*, **472**, L97
42. 1998, *ApJ*, **493**, 765
43. Deutsch, E., Margon, B., & Anderson, S. 2000, *ApJ*, **530**, L21
44. di Stefano, R., Kong, A., Garcia, M., *et al.* 2002, *ApJ*, **570**, 618
45. di Stefano, R., Kong, A., VanDalfsen, M., *et al.* 2003, *ApJ*, **599**, 1067
46. Dirsch, B., Richtler, T., Geisler, D., *et al.* 2003, *AJ*, **125**, 1908
47. Dotani, T., Inoue, H., Murakami, T., *et al.* 1990, *Nature*, **347**, 534
48. Dotani, T., Asai, K., & Greiner, J. 1999, *Publ. Astron. Soc. Japan*, **51**, 519
49. Edmonds, P., Grindlay, J., Cool, A., Cohn, H., Lugger, P., & Bailyn, C. 1999, *ApJ*, **516**, 250
50. Edmonds, P., Gilliland, R., Camilo, F., Heinke, C., & Grindlay, J. 2002a, *ApJ*, **579**, 741
51. Edmonds, P., Heinke, C., Grindlay, J., & Gilliland, R. 2002b, *ApJ*, **564**, L17
52. Edmonds, P., Gilliland, R., Heinke, C., & Grindlay, J. 2003, *ApJ*, **596**, 1177 & 1197
53. Edmonds, P., Kahabka, P., & Heinke, C. 2004, *ApJ*, **611**, 413
54. Fabian, A., Pringle, J., & Rees, M. 1975, *MNRAS*, **172**, 15p
55. Ferraro, F., Paltinieri, B., Fusi Pecci, F., Rood, R., & Dorman, B. 1997, *MNRAS*, **292**, L45
56. Ferraro, F., Paltinieri, B., Rood, R., Fusi Pecci, F., & Buonanno, R. 2000, *ApJ*, **537**, 312
57. Ferraro, F., Possenti, A., D'Amico, N., & Sabbi, E. 2001, *ApJ*, **561**, L93
58. Ferraro, F., Sabbi, E., Gratton, R., *et al.* 2003, *ApJ*, **584**, L13
59. Forbes, D., Grillmair, C., & Smith, R. 1997, *AJ*, **113**, 1648
60. Forman, W. & Jones, C. 1976, *ApJ*, **207**, L177
61. Forman, W., Jones, C., & Tananbaum, H. 1976, *ApJ*, **207**, L25

62. Fox, D., Lewin, W., Margon, B., van Paradijs, J., & Verbunt, F. 1996, *MNRAS*, **282**, 1027
63. Freire, P., Camilo, F., Kramer, M., *et al.* 2003, *MNRAS*, **340**, 1359
64. Geffert, M. 1998, *A&A*, **340**, 305
65. Gendre, B., Barret, D., & Webb, N. 2003a, *A&A*, **400**, 521
66. 2003b, *A&A*, **403**, L11
67. Gerssen, J., van der Marel, R., Gebhardt, K., *et al.* 2003, *AJ*, **125**, 376
68. Giacconi, R., Murray, S., Gursky, H., *et al.* 1972, *ApJ*, **178**, 281
69. 1974, *ApJS*, **27**, 37
70. Gómez, M., Richtler, T., Infante, L., Drenkhahn, G. 2001, *A&A*, **371**, 875
71. Goodman, J. & Hut, P. 1989, *Nature*, **339**, 40
72. Goudfrooij, P., Alonso, M., Maraston, C., & Minniti, D. 2001, *MNRAS*, **328**, 237
73. Grimm, H., Gilfanov, M., & Sunyaev, R. 2002, *A&A*, **391**, 923
74. Grindlay, J. 1987, in *The Origin and Evolution of Neutron Stars*, IAU Symposium No. 125, eds. D. Helfand & J.-H. Huang (Dordrecht: Reidel), 173–185
75. 1988, in *Globular Cluster Systems in Galaxies*, IAU Symposium No. 126, eds. J. Grindlay & G. Davis Philip (Dordrecht: Reidel), 347–366
76. 1992, in *X-ray Binaries and Recycled Pulsars*, NATO ASI C. 377, eds. E. van den Heuvel & S. Rappaport (Dordrecht: Kluwer), 365
77. Grindlay, J. & Hertz, P. 1985, in *Cataclysmic Variables and Low Mass X-ray Binaries*, eds. D. Lamb & J. Patterson (Dordrecht: Reidel), 79–91
78. Grindlay, J., Gursky, H., Schnopper, H., *et al.* 1976, *ApJ*, **205**, L127
79. Grindlay, J., Hertz, P., Steiner, J., Murray, S., & Lightman, A. 1984, *ApJ*, **282**, L13
80. Grindlay, J., Cool, A., Callanan, P., *et al.* 1995, *ApJ*, **455**, L47
81. Grindlay, J., Heinke, C., Edmonds, P., & Murray, S. 2001a, *Science*, **292**, 2290
82. Grindlay, J., Heinke, C., Edmonds, P., Murray, S., & Cool, A. 2001b, *ApJ*, **563**, L53
83. Grindlay, J., Camilo, F., Heinke, C., *et al.* 2002, *ApJ*, **581**, 470
84. Guainazzi, M., Parmar, A., & Oosterbroek, T. 1999, *A&A*, **349**, 819
85. Gunn, J. & Griffin, R. 1979, *AJ*, **84**, 752
86. Gursky, H. July 1973, Lecture presented at the NASA Advanced Study Institute on Physics of Compact Objects, Cambridge, UK
87. Haggard, D., Cool, A., Anderson, J., Edmonds, P., *et al.* 2004, *ApJ*, **613**, 512
88. Hakala, P., Charles, P., Johnston, H., & Verbunt, F. 1997, *MNRAS*, **285**, 693
89. Harris, W. 1991, *ARA&A*, **29**, 543
90. 1996, *AJ*, **112**, 1487
91. Harris, W. & van den Bergh, S. 1981, *AJ*, **86**, 1627
92. Heinke, C., Edmonds, P., & Grindlay, J. 2001, *ApJ*, **562**, 363
93. Heinke, C., Edmonds, P., Grindlay, J., *et al.* 2003a, *ApJ*, **590**, 809
94. Heinke, C., Grindlay, J., Lloyd, D., & Edmonds, P. 2003b, *ApJ*, **588**, 452
95. Heinke, C., Grindlay, J., Edmonds, P., *et al.* 2003c, *ApJ*, **598**, 516
96. Heinke, C., Grindlay, J., Lugger, P., *et al.* 2003d, *ApJ*, **598**, 501
97. Hénon, M., 1961, *Ann. d'Astroph.*, **24**, 369
98. Hertz, P. & Grindlay, J. 1983, *ApJ*, **275**, 105
99. Hertz, P. & Wood, K. 1985, *ApJ*, **290**, 171
100. Hertz, P., Grindlay, J., & Bailyn, C. 1993, *ApJ*, **410**, L87
101. Hills, J. 1976, *MNRAS*, **175**, 1p
102. Hoffman, J., Lewin, W., & Doty, J. 1977a, *MNRAS*, **179**, 57P
103. 1977b, *ApJ*, **217**, L23
104. Hoffman, J., Marshall, H., & Lewin, W. 1978, *Nature*, **271**, 630
105. Hoffman, J., Cominsky, L., & Lewin, W. 1980, *ApJ*, **240**, L27
106. Homer, L., Charles, P., Naylor, T., *et al.* 1996, *MNRAS*, **282**, L37
107. Homer, L., Anderson, S., Margon, B., Deutsch, E., & Downes, R. 2001a, *ApJ*, **550**, L155
108. Homer, L., Deutsch, E., Anderson, S., & Margon, B. 2001b, *AJ*, **122**, 2627
109. Homer, L., Anderson, S., Margon, B., Downes, R., & Deutsch, E. 2002, *AJ*, **123**, 3255
110. Hut, P. & Verbunt, F. 1983, *Nature*, **301**, 587
111. Hut, P., McMillan, S., Goodman, J., *et al.* 1992, *PASP*, **104**, 981
112. Iben, I., Tutukov, A., & Fedorova, A. 1997, *ApJ*, **486**, 955
113. Ilovaisky, S., Aurière, M., Koch-Miramond, L., *et al.* 1993, *A&A*, **270**, 139

114. in 't Zand, J., Verbunt, F., Heise, J., *et al.* 1998, *A&A*, **329**, L37
115. in 't Zand, J., Verbunt, F., Strohmayer, T., *et al.* 1999, *A&A*, **345**, 100
116. in 't Zand, J., Bazzano, A., Cocchi, M., *et al.* 2000, *A&A*, **355**, 145
117. in 't Zand, J., van Kerkwijk, M., Pooley, D., *et al.* 2001, *ApJ*, **563**, L41
118. in 't Zand, J., Hulleman, F., Markwardt, C., *et al.* 2003, *A&A*, **406**, 233
119. Ioannou, Z., van Zyl, L., Naylor, T., *et al.* 2003, *A&A*, **399**, 211
120. Irwin, J. & Bregman, J. 1999, *ApJ*, **510**, L21
121. Jeltema, T., Sarazin, C., Buote, D., & Garmire, G. 2003, *ApJ*, **585**, 756
122. Jernigan, J. & Clark, G. 1979, *ApJ*, **231**, L125
123. Johnston, H., Verbunt, F., & Hasinger, G. 1994, *A&A*, **289**, 763
124. 1995, *A&A*, **298**, L21
125. 1996, *A&A*, **309**, 116
126. Jordán, A., Côté, P., Ferrarese, L., *et al.* 2004 *ApJ*, **613**, 279
127. Kaluzny, J. & Thompson, I. 2002, *AJ*, **125**, 2534
128. Kaluzny, J., Kubiak, M., Szymanski, M., *et al.* 1996, *A&AS*, **120**, 139
129. Katz, J. 1975, *Nature*, **253**, 698
130. Kim, D.-W. & Fabbiano, G. 2003, *ApJ*, **586**, 826
131. 2004, *ApJ*, **611**, 846
132. King, I. 1966, *AJ*, **71**, 64
133. King, I., Stanford, S., Albrecht, R., *et al.* 1993, *ApJ*, **413**, L117
134. Kissler-Patig, M. 1997, *A&A*, **319**, 83
135. Knigge, C., Zurek, D., Shara, M., Long, K., & Gilliland, R. 2003, *ApJ*, **599**, 1320
136. Kraft, R., Forman, W., Jones, C., *et al.* 2000, *ApJ*, **531**, L9
137. Kraft, R., Kregenov, J., Forman, W., Jones, C., & Murray, S. 2001, *ApJ*, **560**, 675
138. Kulkarni, S., Goss, W., Wolszczan, A., & Middleditch, J. 1990, *ApJ*, **363**, L5
139. Kulkarni, S., Anderson, S., Prince, T., & Wolszczan, A. 1991, *Nature*, **349**, 47
140. Kulkarni, S., Hut, P., & McMillan, S. 1993, *Nature*, **364**, 421
141. Kundu, A. & Whitmore, B. 2001a, *AJ*, **121**, 2950
142. 2001b, *AJ*, **122**, 1251
143. Kundu, A., Maccarone, T., & Zepf, S. 2002, *ApJ*, **574**, L5
144. Kundu, A., Maccarone, T., Zepf, S., & Puzia, T. 2003, *ApJ*, **589**, L81
145. Kuulkers, E., den Hartog, P., in 't Zand, J., *et al.* 2003, *A&A*, **399**, 663
146. Larsen, S., Brodie, J., Beasley, M., *et al.* 2003, *ApJ*, **585**, 767
147. Lewin, W. 1980, in *Globular Clusters*, eds. D. Hanes & B. Madore (Cambridge: Cambridge University Press), 315
148. Lewin, W. & Joss, P. 1983, in *Accretion-Driven Stellar X-ray Sources*, eds. W. Lewin & E. van den Heuvel (Cambridge: Cambridge University Press), 41–115
149. Lewin, W., Doty, J., Clark, G., *et al.* 1976, *ApJ*, **207**, L95
150. Lewin, W., van Paradijs, J., & Taam, R. 1993, *Space Sci. Rev.*, **62**, 223
151. 1995, in *X-ray Binaries*, eds. W. Lewin, J. van Paradijs, & E. van den Heuvel (Cambridge: Cambridge University Press), 175–232
152. Lyne, A., Brinklow, A., Middleditch, J., *et al.* 1987, *Nature*, **328**, 399
153. Maccarone, T., Kundu, A., & Zepf, S. 2003, *ApJ*, **586**, 814
154. 2004, *ApJ*, **606**, 430
155. Magnier, E., Lewin, W., van Paradijs, J., *et al.* 1992, *A&AS*, **96**, 379
156. Makishima, K., Ohashi, T., Inoue, H., *et al.* 1981, *ApJ*, **247**, L23
157. Mardling, R. 1995, *ApJ*, **450**, 722, 732
158. Markert, T., Backman, D., Canizares, C., Clark, G., & Levine, A. 1975, *Nature*, **257**, 32
159. Masetti, N. 2002, *A&A*, **381**, L45
160. McLaughlin, D., Harris, W., & Hanes, D. 1994, *ApJ*, **422**, 486
161. Minniti, D., Rejkuba, M., Funes, J., & Akiyama, S. 2004, *ApJ*, **600**, 716
162. Mirabel, I. & Rodrigues, I. 2003, *A&A*, **398**, L25
163. Mirabel, I., Dhawan, V., Mignami, R., *et al.* 2001, *Nature*, **413**, 139
164. Moore, C., Rutledge, R., Fox, D., *et al.* 2000, *ApJ*, **532**, 1181
165. Naylor, T., Charles, P., Hassall, B., Raymond, J., & Nassiopoulos, G. 1992, *MNRAS*, **255**, 1
166. Neill, J., Shara, M., Caulet, A., & Buckley, D. 2002, *AJ*, **123**, 3298
167. Oosterhoff, P. T. 1941, *Ann. Sternwarte Leiden*, **17**, 1

168. Orosz, J. & van Kerkwijk, M. 2003, *A&A*, **397**, 237
169. Paresce, F. & de Marchi, G. 1994, *ApJ*, **427**, L33
170. Paresce, F., de Marchi, G., & Ferraro, F. 1992, *Nature*, **360**, 46
171. Parmar, A., Stella, L., & Giommi, P. 1989, *A&A*, **222**, 96
172. Parmar, A., Oosterbroek, T., Sidoli, L., Stella, L., & Frontera, F. 2001, *A&A*, **380**, 490
173. Perrett, K., Hanes, D., Butterworth, S., *et al.* 1997, *AJ*, **113**, 895
174. Phinney, E. 1992, *Phil. Trans. R. Soc. London A*, **341**, 39
175. Phinney, E. & Sigurdsson, S. 1991, *Nature*, **349**, 220
176. Piotto, G. & Zoccali, M. 1999, *A&A*, **345**, 485
177. Piro, A. & Bildsten, L. 2002, *ApJ*, **571**, L103
178. Podsiadlowski, P., Rappaport, S., & Pfahl, E. 2002, *ApJ*, **565**, 1107
179. Pooley, D., Lewin, W., Homer, L., Verbunt, F., *et al.* 2002a, *ApJ*, **569**, 405
180. Pooley, D., Lewin, W., Verbunt, F., Homer, L., *et al.* 2002b, *ApJ*, **573**, 184
181. Pooley, D., Lewin, W., Anderson, S., *et al.* 2003, *ApJ*, **591**, L131
182. Portegies Zwart, S. & McMillan, S. 2000, *ApJ*, **528**, 17
183. Predehl, P., Hasinger, G., & Verbunt, F. 1991, *A&A*, **246**, L21
184. Puzia, T., Saglia, R., Kissler-Patig, M., *et al.* 2002, *A&A*, **391**, 453
185. Pylyser, E. & Savonije, G. 1988, *A&A*, **191**, 57
186. Rajagopal, M. & Romani, R. 1996, *ApJ*, **461**, 327
187. Randall, S., Sarazin, C., & Irwin, J. 2003, *ApJ*, **600**, 729
188. Rappaport, S., Dewey, D., Levine, A., & Macri, L. 1994, *ApJ*, **423**, 633
189. Rasio, F., Pfahl, E., & Rappaport, S. 2000, *ApJ*, **532**, L47
190. Ray, A., Kembhavi, A., & Antia, H. 1987, *A&A*, **184**, 164
191. Rhode, K. & Zepf, S. 2001, *AJ*, **121**, 210
192. 2004, *AJ*, **127**, 302
193. Rutledge, R., Bildsten, L., Brown, E., Pavlov, G., & Zavlin, V. 1999, *ApJ*, **514**, 945
194. 2002, *ApJ*, **578**, 405
195. Saito, Y., Kawai, N., Kamae, T., *et al.* 1997, *ApJ*, **477**, L37
196. Sansom, A., Dotani, T., Asai, K., & Lehto, H. 1993, *MNRAS*, **262**, 429
197. Sarazin, C., Irwin, J., & Bregman, J. 2000, *ApJ*, **544**, L101
198. 2001, *ApJ*, **556**, 533
199. Sarazin, C., Irwin, J., Rood, R., *et al.* 1999, *ApJ*, **524**, 220
200. Sarazin, C., Kundu, A., Irwin, J., *et al.* 2003, *ApJ*, **595**, 743
201. Seward, F., Page, C., Turner, M., & Pounds, K. 1976, *MNRAS*, **175**, 39P
202. Shara, M., Bergeron, L., Gilliland, R., Saha, A., & Petro, L. 1996, *ApJ*, **471**, 804
203. Sidoli, L., Parmar, A., Oosterbroek, T., *et al.* 2001, *A&A*, **368**, 451
204. Sivakoff, G., Sarazin, C., & Irwin, J. 2003, *ApJ*, **599**, 218
205. Skinner, G., Willmore, A., Eyles, C., *et al.* 1987, *Nature*, **330**, 544
206. Stella, L., Priedhorsky, W., & White, N. 1987, *ApJ*, **312**, L17
207. Supper, R., Hasinger, G., Pietsch, W., *et al.* 1997, *ApJ*, **317**, 328
208. Sutantyo, W. 1975, *A&A*, **44**, 227
209. Swank, J., Becker, R., Boldt, E., *et al.* 1977, *ApJ*, **212**, L73
210. Sztajno, M., Fujimoto, M., van Paradijs, J., *et al.* 1987, *MNRAS*, **226**, 39
211. Thorsett, S., Arzoumanian, Z., Camilo, F., & Lyne, A. 1999, *ApJ*, **523**, 763
212. Tomsick, J., Corbel, S., Fender, R., *et al.* 2003, *ApJ*, **597**, L133
213. Trudolyubov, S., Borozdin, K., Priedhorsky, W., *et al.* 2002, *ApJ*, **581**, L27
214. van den Heuvel, E. 1983, in *Accretion-Driven Stellar X-ray Sources*, eds. W. Lewin & E. van den Heuvel (Cambridge: Cambridge University Press), 303–341
215. van der Klis, M., Hasinger, G., Verbunt, F., *et al.* 1993, *A&A*, **279**, L21
216. van der Sluys, M., Verbunt, F., & Pols, O., 2005, *A&A*, **431**, 647
217. van Paradijs, J. 1978, *Nature*, **274**, 650
218. van Paradijs, J. & McClintock, J. 1994, *A&A*, **290**, 133
219. van Paradijs, J., Verbunt, F., Shafer, R., & Arnaud, K. 1987, *A&A*, **182**, 47
220. van Speybroeck, L., Epstein, A., Forman, W., *et al.* 1979, *ApJ*, **234**, L45
221. Verbunt, F. 1987, *ApJ*, **312**, L23
222. 1993, *ARA&A*, **31**, 93
223. 1994, *A&A*, **285**, L21

224. 1996, in *Compact Stars in Binaries*, IAU Symp. 165, eds. J. van Paradijs, E. van den Heuvel, & E. Kuulkers (Dordrecht: Kluwer Academic Publishers), 333–339
225. 2001, *A&A*, **368**, 137
226. 2003, in *New Horizons in Globular Cluster Astronomy*, eds. G. Piotto, G. Meylan, G. Gjorgovski, & M. Riello (ASP Conf. Ser. 296), 245–254
227. Verbunt, F. & Hasinger, G. 1998, *A&A*, **336**, 895
228. Verbunt, F. & Hut, P. 1987, in *The Origin and Evolution of Neutron Stars*, IAU Symposium No. 125, eds. D. Helfand & J.-H. Huang (Dordrecht: Reidel), 187–197
229. Verbunt, F. & Johnston, H. 2000, *A&A*, **358**, 910
230. Verbunt, F. & Meylan, G. 1988, *A&A*, **203**, 297
231. Verbunt, F., van Paradijs, J., & Elson, R. 1984, *MNRAS*, **210**, 899
232. Verbunt, F., Belloni, T., Johnston, H., van der Klis, M., & Lewin, W. 1994, *A&A*, **285**, 903
233. Verbunt, F., Bunk, W., Hasinger, G., & Johnston, H. 1995, *A&A*, **300**, 732
234. Verbunt, F., Kuiper, L., Belloni, T., *et al.* 1996, *A&A*, **311**, L9
235. Verbunt, F., Bunk, W., Ritter, H., & Pfeffermann, E. 1997, *A&A*, **327**, 602
236. Verbunt, F., van Kerkwijk, M., in 't Zand, J., & Heise, J. 2000, *A&A*, **359**, 960
237. Webb, N., Gendre, B., & Barret, D. 2002, *A&A*, **381**, 481
238. West, M., Côté, P., Marzke, R., & Jordán, A. 2004 *Nature*, **427**, 31
239. White, N. & Angelini, L. 2001, *ApJ*, **561**, L101
240. White, N. & Ghosh, P. 1998, *ApJ*, **504**, L31
241. White, R. 2002, *APS Meeting April 2002* Abstracts 11.010
242. White, R., Sarazin, C. & Kulkarni, S. 2002, *ApJ*, **571**, L23
243. Wijnands, R., Heinke, C., & Grindlay, J. 2002, *ApJ*, **572**, 1002
244. Wijnands, R., Heinke, C., Pooley, D., *et al.* 2005, *ApJ*, **618**, 833
245. Zavlin, V., Pavlov, G., Shibanov, Y. 1996, *A&A*, **315**, 141
246. Zepf, S. & Ashman, K. 1993, *MNRAS*, **264**, 611

9

Jets from X-ray binaries

Rob Fender
University of Amsterdam

9.1 History

Relativistic outflows, or "jets", represent one of the most obvious, important and yet poorly explained phenomena associated with accreting relativistic objects, including X-ray binaries. Originally recognized in images as long, thin structures apparently connected at one end to the nuclei of galaxies, it was soon established that they represent powerful flows of energy and matter away from accreting black holes and back to the Universe at large. From their earliest association with the most luminous sources in the Universe, the active galactic nuclei (AGN), the conclusion could have been drawn that jets were a common consequence of the process of accretion onto relativistic objects. Nevertheless, their association with the analogous accretion processes involving stellar-mass black holes and neutron stars was not systematically explored until the past decade or so.

Although it is now clear that the electromagnetic radiation from X-ray binary jets may extend to at least the X-ray band, historically the key observational aspect of jets is their radio emission. High brightness temperatures (see Section 9.2), "non-thermal" spectra and polarization measurements indicate an origin as synchrotron emission from relativistic electrons. Following the discovery of luminous binary X-ray sources in the 1960s and 1970s, radio counterparts were associated with the brightest of these, e.g., Sco X-1 (Hjellming & Wade 1971a), Cyg X-1 (Hjellming & Wade 1971b) and the outbursting source Cyg X-3 (Gregory *et al.* 1972). However, it was not until radio observations of the strong radio source associated with the unusual binary SS 433 revealed a *resolved* radio source that the field of X-ray binary jets really opened up (Spencer 1979; see also Hjellming & Johnston 1981a,b). Outbursts of "soft X-ray transients" were also often associated with strong, transient radio emission (e.g., A0620–00: Owen *et al.* 1976; GS 1124–583: Ball *et al.* 1995; see also Hjellming & Han 1995; Kuulkers *et al.* 1999; Fender & Kuulkers 2001).

In the 1990s the study of jets from X-ray binaries entered a new phase with the discovery of apparent superluminal motions in the outflow from the bright X-ray transient "microquasar" GRS 1915+105 (Mirabel & Rodríguez 1994; see also Mirabel & Rodríguez 1999; Fender *et al.* 1999a; Rodríguez & Mirabel 1999; Fender *et al.* 2002). For the first time it was clear that the jets from X-ray binaries can also exhibit the kind of significantly relativistic (Lorentz factors $\Gamma \geq 2$, where $\Gamma = (1 - \beta^2)^{-1/2}$ and the velocity parameter $\beta = v/c$) velocities observed in the jets of AGN, and not just the mildly relativistic velocity of $\sim 0.26c$ ($\Gamma = 1.04$) observed in SS 433. Exactly *how* relativistic these jets are will be discussed later. Shortly afterwards a second superluminal galactic source, GRO J1655–40, was discovered (Tingay *et al.* 1995; Hjellming & Rupen 1995).

Compact Stellar X-Ray Sources, eds. Walter Lewin and Michiel van der Klis.
Published by Cambridge University Press.

Since this period detailed investigations of the jets from X-ray binaries, both in the radio band and at shorter wavelengths, have revealed a rich phenomenology and clear patterns of behavior which have provided unique insights into the coupling of accretion and outflow close to relativistic objects. Nevertheless, the deeper we look the more complex the behavior becomes, and this is a rapidly advancing field. In this chapter I shall attempt, subjectively, to describe the state of the research at the beginning of 2003.

In Figs. 9.1 and 9.2 are presented recent sequences of observations of transient relativistic outflows from black hole binaries. Figure 9.1 presents *radio* images of relativistic ejections from three outbursting X-ray binaries on sub-arcsecond angular scales. Figure 9.2 presents *X-ray* images of arcsecond-scale jets moving away from the transient XTE J1550–564 up to four years after the original ejection event, observed with Chandra.

9.2 Physical properties of the jets

In the following I shall briefly outline our understanding of the emission mechanisms in X-ray binary jets, and how we can estimate important physical quantities from the most basic of observations.

9.2.1 Emission mechanism

The radio jets observed from X-ray binaries emit via the synchrotron process. We are drawn to this conclusion by their "non-thermal" spectra, high brightness temperatures and, in some cases, high degree of linear polarization. In the following we will illustrate how some fundamental parameters, e.g., the magnetic field and energy associated with ejection events, can be estimated from basic observations. For a more detailed explanation and exploration of synchrotron emission the reader is directed to, e.g., Longair (1994).

Bright events associated with, for example, X-ray state changes and X-ray transients reveal an optically thin spectrum above some frequency, from which the underlying electron population can be derived. If the underlying electron distribution is a power law of the form $N(E)\,\mathrm{d}E \propto E^{-p}\,\mathrm{d}E$ then observations of the spectral index ($\alpha = \Delta \log S_\nu / \Delta \log \nu$, i.e., $S_\nu \propto \nu^\alpha$; note the lack of a minus sign in this definition contrary to common practice in X-ray spectroscopy) in the optically thin part of the synchrotron spectrum can directly reveal the form of this electron distribution: $p = 1 - 2\alpha$.

Observed optically thin spectral indices, $-0.4 \geq \alpha \geq -0.8$, indicate $1.8 \leq p \leq 2.6$. This is the same range derived for the majority of AGN jets and also for synchrotron emission observed in other astrophysical scenarios, e.g., supernova remnants, and is consistent with an origin for the electron distribution in shock acceleration (e.g., Longair 1994; Gallant 2002).

9.2.2 Minimum energy estimation

Association of a given synchrotron luminosity with a given volume (either by direct radio imaging or by measurement of an associated variability timescale) allows estimation of the minimum energy associated with the synchrotron-emitting plasma (Burbidge 1959), at a corresponding "equipartition" magnetic field.

Longair (1994) gives a clear explanation of the calculation of the minimum energy and corresponding magnetic field, and the interested reader is directed there. Repeating some of his useful formulae, a lower limit to the energy associated with a finite volume V of synchrotron emitting plasma can be obtained from a simple estimate of the monochromatic

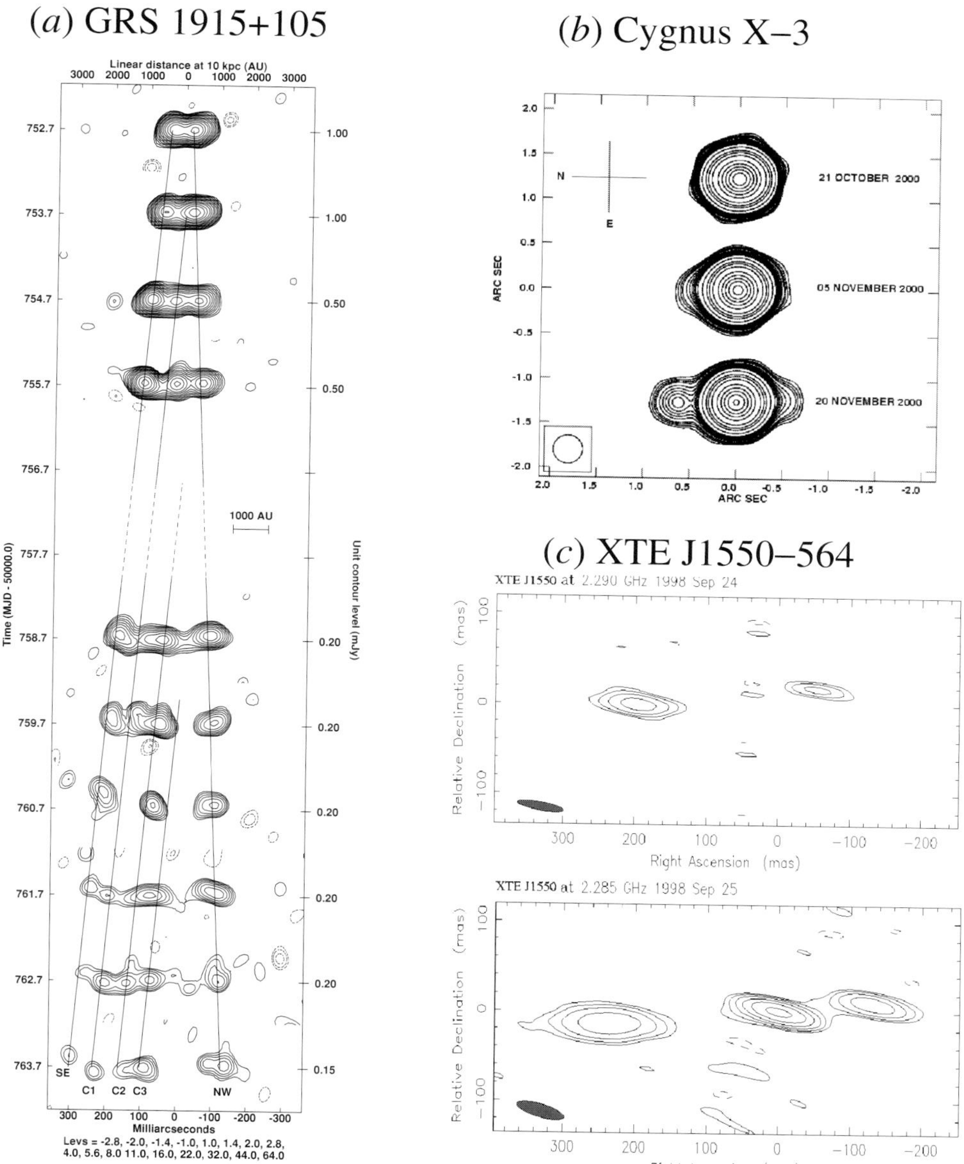

Fig. 9.1. Radio images of relativistic jets from X-ray binaries. Panel (*a*) shows a sequence of images of "superluminal" relativistic ejections from GRS 1915+105 observed with MERLIN (Fender *et al.* 1999a). Panel (*b*) is a sequence of slower, arcsec-scale jets from Cyg X-3 (Martí *et al.* 2002), which may be the jet–ISM interaction zones of the inner, more relativistic jet (Mioduszewski *et al.* 2000). Panel (*c*) presents two VLBI images from Hannikainen *et al.* (2001) of XTE J1550–564 shortly after the major flare in 1998, which was probably responsible for the formation of radio and X-ray lobes (see Fig. 9.2) four years later.

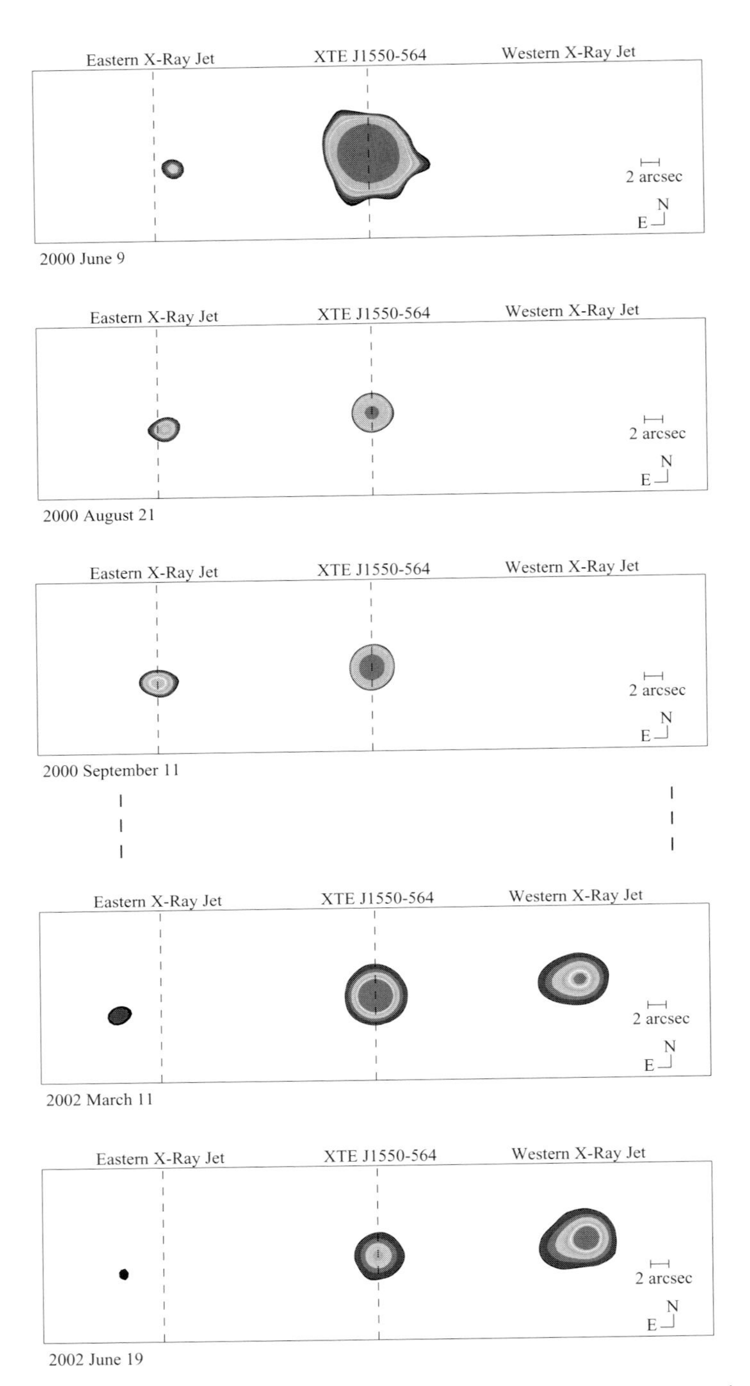

Fig. 9.2. Chandra images of moving X-ray jets from the black hole transient XTE J1550–564 (Corbel *et al.* 2003; see also Kaaret *et al.* 2003 and Tomsick *et al.* 2003). The core is the central component, the "approaching" jet to the left (East) and the "receding" jet to the right (West). These remarkable observations are the first detections of relativistic proper motions in X-rays, and demonstrate unambiguously that X-ray binary jets can accelerate electrons to extremely high energies and as a result are sources of beamed X-rays. Note that in the top panel the apparent fluxes are reduced by the presence of a grating; in fact the eastern jet was brighter at this epoch than at any time subsequently.

luminosity L_ν at a given frequency ν associated with that volume:

$$E_{\rm min} \sim 8 \times 10^6 \eta^{4/7} \left(\frac{V}{\rm cm^3}\right)^{3/7} \left(\frac{\nu}{\rm Hz}\right)^{2/7} \left(\frac{L_\nu}{\rm erg\ s^{-1}Hz^{-1}}\right)^{4/7} \ {\rm erg} \tag{9.1}$$

where $\eta = (1 + \epsilon_{\rm p}/\epsilon_{\rm e})$, and $\epsilon_{\rm p}/\epsilon_{\rm e}$ represents the ratio of energy in protons to that in electrons, and assuming $p = 2$. It is generally assumed that $\epsilon_{\rm p}/\epsilon_{\rm e} \sim 0$ and therefore $\eta \sim 1$, often with little serious justification. In the more common case where we do not image the source but rather infer its size from the rise time Δt of an event (i.e., using $V = (4/3)\pi(c\Delta t)^3$) with a flux density S_ν originating at an estimated distance d, the formula can be rewritten as

$$E_{\rm min} \sim 3 \times 10^{33} \eta^{4/7} \left(\frac{\Delta t}{\rm s}\right)^{9/7} \left(\frac{\nu}{\rm GHz}\right)^{2/7} \left(\frac{S_\nu}{\rm mJy}\right)^{4/7} \left(\frac{d}{\rm kpc}\right)^{8/7} \ {\rm erg} \tag{9.2}$$

The related mean power into the ejection event

$$P_{\rm min} = \frac{E_{\rm min}}{\Delta t} \sim 3 \times 10^{33} \eta^{4/7} \left(\frac{\Delta t}{\rm s}\right)^{2/7} \left(\frac{\nu}{\rm GHz}\right)^{2/7} \left(\frac{S_\nu}{\rm mJy}\right)^{4/7} \left(\frac{d}{\rm kpc}\right)^{8/7} \ {\rm erg\, s^{-1}} \tag{9.3}$$

The minimum energy condition is achieved at so-called "equipartition", when the energy in particles and magnetic field is comparable. This field can be approximated by

$$B_{\rm eq} \sim 30 \eta^{2/7} \left(\frac{S_\nu}{\rm mJy}\right)^{2/7} \left(\frac{d}{\rm kpc}\right)^{4/7} \left(\frac{\Delta t}{\rm s}\right)^{-6/7} \left(\frac{\nu}{\rm GHz}\right)^{1/7} \ {\rm gauss} \tag{9.4}$$

Note that this field is not, as can sometimes be presumed, a *minimum* magnetic field but rather the field corresponding to the minimum energy; i.e., increase or decrease the field and the energy required to produce the observed synchrotron emission increases. The Lorentz factors of electrons (or positrons) emitting synchrotron emission at a given frequency can be estimated by

$$\gamma_{\rm e} \sim 30 \left(\frac{\nu}{\rm GHz}\right)^{1/2} \left(\frac{B}{\rm G}\right)^{-1/2} \tag{9.5}$$

Figure 9.3 shows a radio flare event from the X-ray binary jet source Cyg X-3. The observation is at 15 GHz, has a rise time of $\sim$3500 s, an amplitude of $\sim$200 mJy and Cyg X-3 lies at an estimated distance of $\sim$8 kpc. Using the above approximations we find a minimum energy associated with the event of $E_{\rm min} \sim 5 \times 10^{40}$ erg, and a corresponding mean jet power during the event of $\sim 10^{37}$ erg s^{-1}, many orders of magnitude greater than the observed radiative radio luminosity. The corresponding equipartition field can be estimated as $\sim$0.5 G, in which field electrons radiating at 15 GHz must have Lorentz factors $\gamma \sim 150$.

It should be stressed that the inner regions of jets from X-ray binaries have relativistic Doppler factors (see below) considerably different from unity resulting from relativistic bulk motions, whereas the above estimations are based upon rise times, flux densities and frequencies as measured in the comoving frame. In such cases the observed quantities need to be corrected to the comoving frame before the estimates can be made. In addition, in such cases the kinetic energy associated with the ejection needs to be taken into account. This kinetic energy component is given by

$$E_{\rm kin} = (\Gamma - 1) E_{\rm int} \tag{9.6}$$

i.e., for a bulk Lorentz factor $\Gamma > 2$ (by no means unreasonable – see below) kinetic dominates over internal energy.

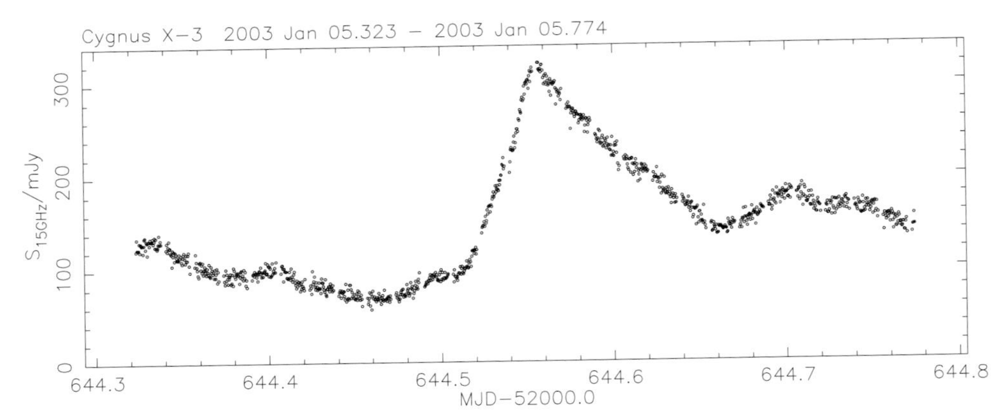

Fig. 9.3. Observation of a radio flare event from the jet source Cyg X-3 at 15 GHz. The rise time of the event ~0.04 d, allows an estimation of the size of the region associated with the event, and thus the minimum energy. Observations from the Ryle Telescope (Guy Pooley, private communication).

9.2.3 Flare events

Flare events such as that presented in Fig. 9.3 are believed to result from the short-term injection of energy and particles into an expanding plasma cloud, presumably in the form of a jet. Such events are characterized by optically thin spectra and are associated with, for example, X-ray transients and persistently flaring sources such as Cyg X-3 and GRS 1915+105. From Fig. 9.3 it is clear that *rise* and *decay* phases can be quite clearly defined. In the "synchrotron bubble" model (van der Laan 1966; Hjellming & Johnston 1988; Hjellming & Han 1995 and references therein) the rise phase corresponds to a decreasing optical depth at frequencies that were initially (synchrotron-)self-absorbed; observational characteristics would be an inverted radio spectrum during the rise phase, and possible Doppler effects on the profile (since the effect takes place in a different frame to the observer). An alternative explanation is that the rise phase represents a finite period of particle injection/acceleration in the outflow; the characteristics of such a phase would be an optically thin spectrum and a duration at least coupled to events more or less in the observer's frame, e.g., the X-ray emission arising from the accretion disk. It seems (to this author) that there are probably observed events of both types. Note that time delays in the propagation of a shock (or other particle acceleration phenomena) through the differing "photospheres" of an outflow may (misleadingly) mimic the "synchrotron bubble" effect (see discussion in Klein-Wolt *et al.* 2002).

The monotonic decay observed after a few days in the radio events from X-ray transients (see below) seems to be primarily due to adiabatic expansion losses, the key signature of which is the same decay rate at all frequencies. Significant loss of energy through the synchrotron emission process itself, or via inverse Compton scattering, results in a more rapid decay at higher frequencies (spectral steepening). The fact that adiabatic losses dominate reveals clearly that the synchrotron radiation observed from such events is only a small fraction of the total energy originally input.

9.2.4 Speed

Mirabel and Rodríguez (1994) first reported apparent superluminal motions from a galactic source, GRS 1915+105. The apparent velocity parameter of the observed motion of the features on the sky, β_{obs}, is related to the observed proper motion by

$$\beta_{obs} \sim \left(\frac{\mu}{170 \text{ mas d}^{-1}}\right)\left(\frac{d}{\text{kpc}}\right) \tag{9.7}$$

This apparent velocity parameter is related to that of the intrinsic velocity β_{int} by

$$\beta_{obs} = \frac{\beta_{int} \sin\theta}{1 \mp \beta_{int}\cos\theta} \tag{9.8}$$

where θ is the angle of the flow to the line of sight ($\mp$ refer to approaching and receding components respectively). Apparent superluminal motion (i.e., $\beta_{obs} > 1$) requires $\beta_{int} \geq 0.7$, indicating that at least mildly relativistic intrinsic velocities are required to achieve the effect (or a badly overestimated distance!). The associated relativistic Doppler shift $1 + \Delta\nu/\nu$ is given by

$$\delta = \Gamma^{-1}(1 \mp \beta_{int}\cos\theta)^{-1} \tag{9.9}$$

where Γ is the bulk Lorentz factor of the flow. This Γ term represents time dilation at relativistic velocities and means that in certain circumstances (probably the case for the superluminal jet sources GRS 1915+105 and GRO J1655–40) *both* jets can be redshifted.

Given observed proper motions of jets, how can we estimate β_{int}? As described in Mirabel & Rodríguez (1994), measurement of μ_{app} and μ_{rec} allows a determination of the following product:

$$\beta_{int}\cos\theta = \frac{(\mu_{app} - \mu_{rec})}{(\mu_{app} + \mu_{rec})} \tag{9.10}$$

where θ is the angle of the ejection to the line of sight and μ_{app}, μ_{rec} are the approaching and receding proper motions respectively (see also Rees 1966; Blandford *et al.* 1977).

Once the proper motions are measured, the angle of ejection, θ, and consequently the intrinsic velocity, β_{int}, are uniquely determined for every distance since

$$\tan\theta = \frac{2d}{c}\left(\frac{\mu_{app}\mu_{rec}}{\mu_{app} - \mu_{rec}}\right) \tag{9.11}$$

and the product $\beta_{int}\cos\theta$ is already known.

The variation of β_{int} and θ as a function of distance for GRS 1915+105 was presented in Fender *et al.* (1999a). There is a maximum distance to the source corresponding to $\beta_{int} = 1$ (i.e., $\Gamma = \infty$):

$$d_{max} = \frac{c}{\sqrt{(\mu_{app}\mu_{rec})}} \tag{9.12}$$

At this upper limit to the distance you also find the maximum angle of the jet to the line of sight,

$$\theta_{max} = \cos^{-1}\frac{(\mu_{app} - \mu_{rec})}{(\mu_{app} + \mu_{rec})} \tag{9.13}$$

In addition to the proper motions and Doppler-shifting of frequencies, there is a boosting effect due to a combination of Doppler and relativistic aberration effects, both contained in the relativistic Doppler factor (Eq. 9.9). An object moving at angle θ to the line of sight with velocity β_{int}(and resultant Lorentz factor Γ) will have an observed surface brightness, δ^k, brighter, where $2 < k < 3$ ($k = 2$ corresponds to the average of multiple events in for example a continuous jet, $k = 3$ corresponds to emission dominated by a singularly evolving event). Therefore the ratio of flux densities from approaching and receding knots – measured at the same angular separation from the core, so as to sample the knots at the same age in their evolution – will be given by

$$\frac{S_{\rm app}}{S_{\rm rec}} = \left(\frac{\delta_{\rm app}}{\delta_{\rm rec}}\right)^{k-\alpha}$$

where α is the spectral index (to compensate for the spectral shape for different Doppler shifts). For a more detailed discussion see, e.g., Blandford *et al.* (1977); Hughes (1991); Mirabel & Rodríguez (1999); Fender (2003).

9.2.4.1 *Observed speeds of steady jets*

There are basically no direct measurements of the speeds associated with the "steady" jets inferred to exist in the low/hard state of black holes (see Section 9.4.1, Chapters 2 and 4), and possibly also in the "plateau" state of GRS 1915+105 (see Fender *et al.* 1999a), and the hard states of some neutron star atoll sources (see Section 2.5). Nevertheless, there are some clues that the jets may be mildly, but not highly, relativistic. Stirling *et al.* (2001), in direct imaging of the milliarcsec-scale jet from Cyg X-1 in the low/hard state, inferred a minimum speed of $\beta \geq 0.6$ based upon the one-sidedness of the jet. Gallo *et al.* (2003) have performed Monte Carlo simulations in order to investigate the effect of significant Doppler boosting on the observed radio:X-ray correlation in the low/hard state (see Fig. 9.9). They found that intrinsic velocities for the radio emitting component of $v > 0.8c$ would probably result in a larger spread in the correlation than is observed – therefore the bulk Lorentz factor Γ of the steady radio-emitting jets is likely to be < 2 (strictly true only for cases in which the X-rays are not significantly beamed).

The observations of the luminous neutron star source Sco X-1 (Fomalont *et al.* 2001a,b) present a fascinating demonstration that the velocity of the flow from the accretion region may be rather different from that observed for the radio-emitting knots. Specifically, an unseen underlying flow with Lorentz factor ≥ 2 is inferred to be powering a particle acceleration zone, which is itself moving away from the binary with a mildly relativistic (and non-constant) speed of $\sim 0.5c$.

9.2.4.2 *Observed speeds of transient jets*

In 1994 VLA observations of apparent superluminal motions from the black hole transient GRS 1915+105 demonstrated unequivocally that X-ray binaries could produce highly relativistic jets (Mirabel & Rodríguez 1994). Since then, a further three or four superluminal sources have been discovered (GRO J1655–40: Tingay *et al.* 1995; Hjellming & Rupen 1995; XTE J1748–288: Rupen *et al.* 1998; XTE J1550–560: Hannikainen *et al.* 2001; Corbel *et al.* 2002; V4641 Sg: Hjellming *et al.* 2000a; Orosz *et al.* 2001), and there is certainly no indication that highly relativistic ejections are unusual for black hole X-ray transients.

But how relativistic are these events? Following Mirabel and Rodríguez (1994) it was widely accepted that X-ray binary jets could be characterized by Lorentz factors ~ 2 (i.e., while significantly relativistic, considerably less so than the most extreme examples of AGN jets). However, in Fender *et al.* (1999) it was shown that a much wider range of bulk Lorentz factors was possible, at least for GRS 1915+105. Fender (2003) has recently shown that direct measurements of proper motions of radio components cannot in practice easily be

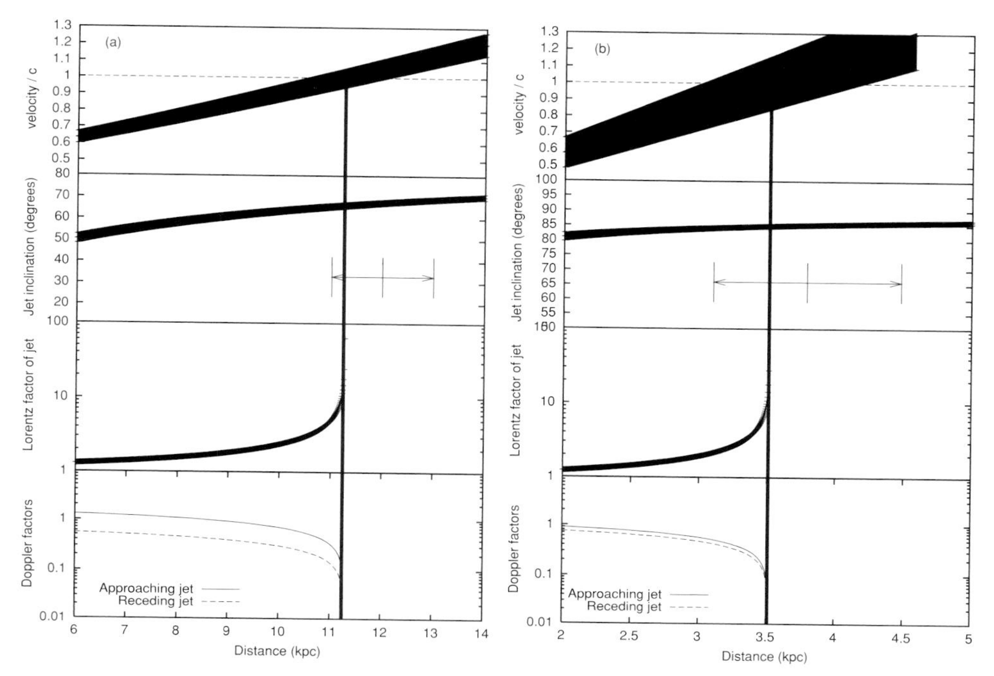

Fig. 9.4. Variation of solutions to velocity, angle to line of sight, Lorentz factor and Doppler factors for (*a*) GRS 1915+105 and (*b*) GRO J6155–40, as a function of distance, based upon observations of proper motions. When compared with the (relatively accurate) distance estimates it is clear that it is very difficult to put an upper limit on the Lorentz or Doppler factors of the flow by such measurements. From Fender (2003).

used to place an upper limit on the Lorentz factor Γ of a highly relativistic flow, due to the steep dependence of this quantity on d near $d_{\max}$ when Γ is large. In Fig. 9.4 the solutions to β, θ, Γ and $\delta_{\mathrm{app,rec}}$ are plotted as a function of distance to the two "superluminal" sources GRS 1915+105 and GRO J1655–40, along with the best distance estimates. It is clear that within uncertainties in the distance estimates (which are already relatively accurate), no upper bound can be set on the Lorentz factor of the jets by observations of proper motions. Nevertheless, Fender and Kuulkers (2001) concluded that the mean bulk Lorentz factor for transients was likely to be ≤ 5 since higher values would probably destroy the observed correlation between radio and X-ray peak fluxes (unless X-rays were also beamed by the same Lorentz factor, implying inclination selection effects in our source lists). There are a couple of caveats to this statement: first, it has been shown at least for XTE J1550–564 that jets decelerate steadily as they propagate away from the binary (Corbel *et al.* 2002; Kaaret *et al.* 2003; see Fig. 9.5); second, the observations of Sco X-1 (Fomalont *et al.* 2001a,b) show us that the Lorentz factor (and hence boosting) of the energizing beam may be very different to that of the actual radio emitting region (consider also V4641 Sgr in this scenario – Orosz *et al.* 2001 and discussion therein).

No proper motions have ever been observed from a confirmed neutron star X-ray transient. The only concrete hint, physical analogies aside, that they may be relativistic, is the lower limit of $\geq 0.1c$ for the arcsec-scale jet of Cir X-1 (Fender *et al.* 1998), which undergoes a transient-like outburst every 16.6 days.

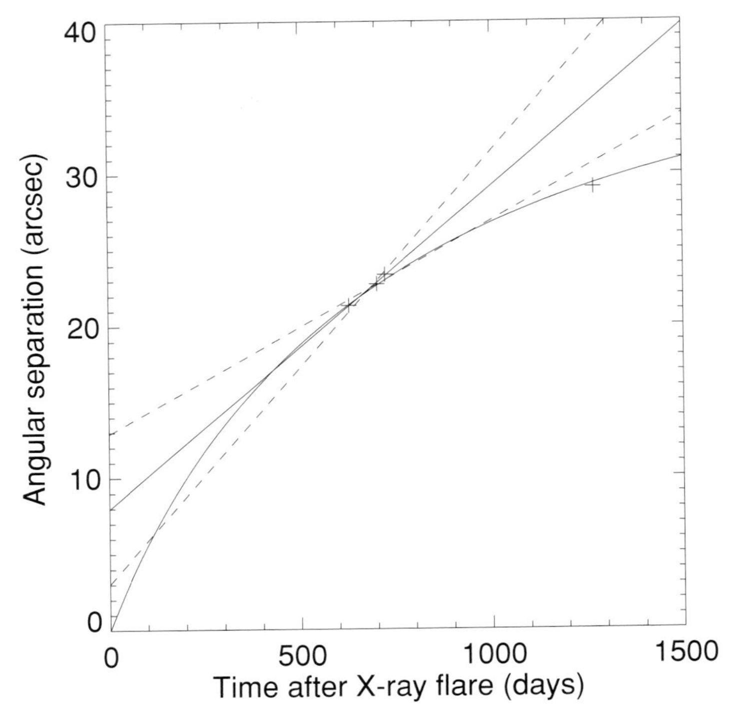

Fig. 9.5. Deceleration of X-ray jets from XTE J1550–564. Comparing a lower limit on the early proper motions on VLBI scales (Hannikainen *et al.* 2001) with subsequent measurements of the X-ray jets (Fig. 9.2) with Chandra, indicates a steady slow-down of the jets. A large fraction of the dissipated kinetic energy seems to be channeled into particle acceleration. From Kaaret *et al.* (2003); see also Corbel *et al.* (2002) and Tomsick *et al.* (2003).

Need the jet velocities be constant? In SS 433 this seems not to be the case – Eikenberry *et al.* (2002) have shown that the velocity of the jet may change by more than 10%. In addition, in XTE J1550–564 (Corbel *et al.* 2002; Fig. 9.2) we clearly observe deceleration of the jet (Fig. 9.5). Since this deceleration probably occurs as a consequence of interactions with the interstellar medium (ISM), it is likely to occur to varying degrees in all X-ray binaries, suggesting that measured velocities may always be a function of time (a relevant point here is that there is nothing to indicate that either the original flare event or the surrounding ISM are particularly unusual in any way).

To summarize, at this stage it seems that the "steady" jets associated with the low/hard state of black holes and, by analogy, possibly with some neutron-star atoll sources are only mildly relativistic. The jets associated with X-ray transients seem almost certain to have considerably higher Lorentz factors which, however, decrease with time as the jet interacts with the ISM (see also Section 9.7). Whether or not there is a smooth continuum of velocities, or a "switch" from mildly to highly relativistic flow speed (e.g., Meier *et al.* 1997) is at present unclear.

9.2.5 Orientation and precession

To date it has been assumed, quite reasonably in the absence of other information, that the jet inclination is perpendicular to the plane of the binary. However, at least two jet sources

(GRO J1655–40 and V4641 Sgr) appear to show significant misalignments (Maccarone 2002 and references therein).

The clearest example of a precessing jet is SS 433. The ∼162.5 day precession of these jets (e.g., Margon 1984; Eikenberry *et al.* 2001) has been assumed to reflect the precession period of the accretion disk (see e.g., Ogilivie & Dubus 2001 for a discussion). Hjellming and Rupen (1995) suggested a precession period for GRO J1655–40 that was very close to the subsequently determined orbital period (Section 5.3.3); similarly there seems to be marginal evidence for precession in the jets of GRS 1915+105 (Fender *et al.* 1999a; see also Rodríguez & Mirabel 1999). Kaufman Bernado *et al.* (2002) and Romero *et al.* (2002) have suggested that precessing jets from X-ray binaries may result in recurrent "microblazar" activity, possibly manifesting itself as high-energy (gamma-ray) flashes as the beam crosses the line of sight. Fender (2003) has discussed the possible signature of precession on the proper motions observed from a jet source.

9.2.6 Composition

Since, with one exception, we have only identified the synchrotron emission from the leptonic (electrons and/or positrons) component in X-ray binary jets (a statement also true for AGN), we have little direct information on their baryonic content (or lack thereof). The one exception is of course SS 433, whose jets are associated with a variety of emission lines in optical, infrared and X-ray spectra (e.g., Margon 1984; Marshall *et al.* 2002).

Why is SS 433 the only jet source with such emission lines? One possible interpretation is that all the other jets (which also seem to have considerably higher bulk velocities than the $\sim 0.26c$ consistently measured for SS 433) have little or no baryonic content and are dominated by electron–positron pairs. This in turn would imply that the majority of the mass in the accretion flow never escapes from the system. It is interesting to note that extended ($\geq$arcsec) X-ray jets have been observed from both SS 433 and XTE J1550–564 (Migliari *et al.* 2002; Corbel *et al.* 2002; Kaaret *et al.* 2003; Tomsick *et al.* 2003; see Fig 9.14). The jets from SS 433 reveal strong emission lines from highly ionized iron and are consistent with thermal emission from a plasma at $\sim 10^7$ K whereas those from XTE J1550–564 reveal a featureless continuum which is consistent with an extrapolation of the synchrotron spectrum from the radio band. Mirabel *et al.* (1997) have discussed effects that would result in atomic emission lines from significantly relativistic jets being very hard to detect, due to extreme Doppler broadening in the jet plasma. In addition, Fender (2003) has shown that the Doppler factors of the jets are very poorly constrained, so that we basically do not know where to look for such lines.

An alternative approach to the composition is to investigate the energetics associated with carrying along a population of "cold" protons in the relativistic flow. Fender and Pooley (2000) did this for the radio–mm–infrared oscillations from GRS 1915+105 (Fig. 9.10) and found the power required to accelerate the proton population to a bulk velocity $\Gamma = 5$ was so large that the ejections were probably at a considerably lower bulk Lorentz factor or did not have a large baryonic component. In a related approach, Celotti and Ghisellini (2003) have concluded that a baryonic component is required for the jets of FRI-type radio galaxies in order to carry most of the power.

Yet another approach to looking for emission lines or balancing energetics is polarization – in particular circular polarization holds the promise of a unique insight into the conditions in the emitting plasma (e.g., Wardle *et al.* 1998; Wardle & Homan 2001). Circular polarization has been detected in the radio emission of three X-ray binaries – SS 433 (Fender *et al.* 2000a), GRS 1915+105 (Fender *et al.* 2002) and GRO J1655–40 (Macquart *et al.* 2002). However, the current state of data and models is not enough to place strong quantitative constraints on the composition of the jets, since the observed circular polarization could arise in both a pair-dominated and baryonic plasma. Right now it seems that we are no closer to convincingly determining the composition of jets from X-ray binaries, and the detection of Doppler-shifted emission (or annihilation) lines from other systems must remain a high priority observation.

9.3 Ubiquity

While clearly an important physical process for some X-ray binaries, in order to establish the broader significance of jets from X-ray binaries it is important to have some idea of their ubiquity. Although it is always preferable to have directly resolved images of jets, in many cases it is enough (or at least the best we can do) to infer the presence of a jet from more circumstantial evidence – in most cases this will be the presence of radio emission with a certain spectrum or type of variability. This approach can be justified by considering the following: the (comoving) brightness temperature T_{B} of an object of physical size R, measured with a flux density S_ν at a frequency ν, and lying at a distance d, is given by the following expression:

$$T_{\mathrm{B}} = 2 \times 10^{13} \left(\frac{S_\nu}{\mathrm{mJy}}\right)\left(\frac{d}{\mathrm{kpc}}\right)^2 \left(\frac{R}{\mathrm{R}_\odot}\right)^{-2} \left(\frac{\nu}{\mathrm{GHz}}\right)^{-2} \mathrm{K} \tag{9.14}$$

Setting a maximum brightness temperature of $T_{\mathrm{B}} \leq 10^{12}$ (above which inverse Compton losses become catastrophic, at least for steady states), this can be rearranged to derive a minimum size for an emitting region, based upon a measured radio flux density and a distance estimate:

$$R \geq 4 \left(\frac{S_\nu}{\mathrm{mJy}}\right)^{1/2} \left(\frac{d}{\mathrm{kpc}}\right)\left(\frac{\nu}{\mathrm{GHz}}\right)^{-1} \mathrm{R}_\odot \tag{9.15}$$

A typical ~5-GHz detection of a "weak" radio counterpart to an X-ray binary is at the ~mJy level, and such sources typically lie at distances of $\geq$5 kpc. Plugging in those numbers produces a minimum size for the emitting region $R \geq 8\ \mathrm{R}_\odot$. Typical binary separations for low-mass X-ray binaries are smaller than this (Section 5.3); even the binary separation of Cyg X-1 – a high-mass X-ray binary in a relatively large 5.6-day orbit – is unlikely to be $\geq 15\ \mathrm{R}_\odot$. Therefore we have a relativistic plasma (since the emission mechanism is synchrotron) with a volume larger than that of the binary system. Such a plasma will be unconfinable by any known component of the binary system, and thus will flow out from the system. Expansion losses will monotonically reduce the flux observed at optically thin frequencies, and this appears to be the case for the "synchrotron bubble" events observed from X-ray transients, repeatedly and clearly resolved by radio interferometers into two-sided outflows. For the steady sources the same expansion losses require that in order to observe persistent radio emission, this plasma must be continually replenished – therefore we are drawn to conclude that an outflow of relativistic plasma is present. In nearly all cases, when this radio-emitting region has been directly resolved, it is in the form of either steady jet-like structures or outflowing "blobs"; by Occam's razor we conclude that this is the most likely scenario for most, if not all, radio emission from X-ray binaries (but see Rupen *et al.* 2002 for the rather different case of CI Cam). Note that it is well known that beamed (i.e., relativistically aberrated) emission can display apparent brightness temperatures $\gg 10^{12}$ K, but invoking relativistic motion to explain away a jet is rather contradictory. Finally, the same simple jet models originally developed for AGN naturally reproduce the spectrum and luminosity of radio emission observed from these systems.

So, allowing ourselves to make the assumption that radio emission is associated with jets, we can draw the following conclusions, which will be discussed in greater detail below:

- *All* black hole systems that are either in the "low/hard" X-ray state (Section 9.4), or are undergoing a major transient outburst (Chapters 2, 4 and 5), are associated with the formation of a jet (albeit possibly of different "types"). Thus the majority of known binary black holes are, or have been in the past, associated with a jet.
- The six brightest low magnetic field *neutron star* systems, the "Z sources", are all associated with jets in some parts of the "Z" track (Section 2.5.2.1). The lower luminosity, low magnetic field systems, which may be crudely lumped together as "atoll" sources (see also Section 2.4.2), *may* be associated with radio emission (although as with black holes there may be bright soft states without jets), implying that the lack of radio detections of the majority is a sensitivity issue.
- The high magnetic field neutron stars, including all but two of the accreting X-ray pulsars, are *not* associated with radio emission

Adding up the numbers, this author concludes that the evidence for a jet is very strong in about 30 X-ray binaries (10–15% of the currently known population), *but* that it is rather likely that jets are present in up to 70% of the systems (basically all except the high magnetic field X-ray pulsars, and a small number of black hole and neutron star systems that are in persistent "soft" states).

9.4 Disk–jet coupling in black hole binaries

One of the richest areas of X-ray binary jets research in the past few years has been the disk–jet coupling, i.e., the relation between inflow and outflow. Some early clues to the phenomenology outlined below were reported earlier in the literature – e.g., some low/hard state transients were known to exhibit flat-spectrum "second stage" radio emission (Hjellming & Han 1995 and references therein) which we would now associate with the compact jet in the core (below). Furthermore, McCollough *et al.* (1999) had reported the bimodal behavior of the radio–X-ray correlations in Cyg X-3, undoubtedly related to the changing disk–jet coupling outlined below.

Black holes exhibit, broadly speaking, several different kinds of X-ray "state" (Chapters 2 and 4). The two most diametrically opposed, which serve to illustrate the relation of jet formation to accretion, can be briefly summarized as:

- **Low/hard (and "off") state:** in this state the X-ray spectrum is dominated by a broadband component which can be fit with a power law of photon index $\sim$1.6, often with a cutoff around 100 keV. Minor additional components to the X-ray spectrum include (sometimes) a weak "blackbody" (accretion disk) component, a "reflection" component and a relatively weak gamma-ray tail. The X-ray power spectra indicate up to 40% r.m.s. variability is present with a "break" at frequencies of around a few Hz.
- **"High/soft" state:** in this state the X-ray spectrum is dominated by a "blackbody" component with a temperature around a few keV, with additional line features and a relatively strong gamma-ray tail. The X-ray power spectrum is much weaker, and can be characterized by a power law with an r.m.s. variability of only a few %.

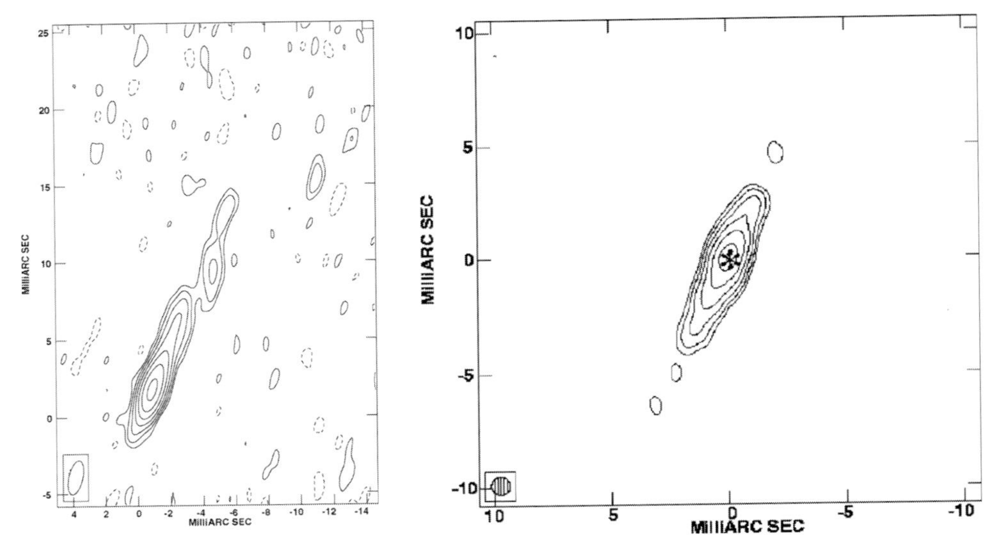

Fig. 9.6. AU-scale jets in persistent hard X-ray states, imaged with the VLBA. The left panel reveals a one-sided jet from Cygnus X-1 in the classical "low/hard" X-ray state (Stirling *et al.* 2001). The right panel shows the quasi-steady jet from GRS 1915+105 in hard "plateau" states (Dhawan *et al.* 2000).

Further details of black hole states may be found in the chapters by van der Klis (Chapter 2), and McClintock and Remillard (Chapter 4).

There are also "intermediate" and "very high" states, which actually both appear to be quasi-steady states that share some of the characteristics of both of the above states, but – crucially for their relation to jet formation – are *much softer* than the regular "low/hard" state. Homan *et al.* (2001) have shown that such states can actually occur at a wide variety of X-ray luminosities.

9.4.1 Steady jets in "low/hard" and "quiescent" states

The radio, and hence jet, properties of the low/hard state black holes can be summarized thus: a "flat" spectrum (spectral index $\alpha \sim 0$) extending throughout the radio band and beyond to higher ν, linear polarization at a level of $\sim$1–3% and variability correlated with the X-ray flux. These broad properties, significantly different from those associated with transient ejection events, are found in every low/hard state source (Fender 2001 and references therein). By analogy with AGN, it has been suggested that these properties could be explained by a compact, self-absorbed jet (Hjellming & Johnston 1988; Falcke & Biermann 1996, 1999; Fender 2001; see also Blandford & Königl 1979). Recently this interpretation has been confirmed by direct imaging of a milliarcsecond-scale jet from Cyg X-1 in the low/hard state (Fig. 9.6 (left), Stirling *et al.* 2001); by analogy it is argued that all low/hard state sources are producing jets.

Furthermore, the hard "plateau" state in GRS 1915+105, which has many similarities to the classical low/hard state, is also associated with a resolved milliarsecond-scale jet (Dhawan *et al.* 2000; Fig. 9.6 (right)), and the two galactic center low/hard state sources 1E 1740.7–2942 and GRS 1758–258 are both associated with large-scale radio lobes, indicating

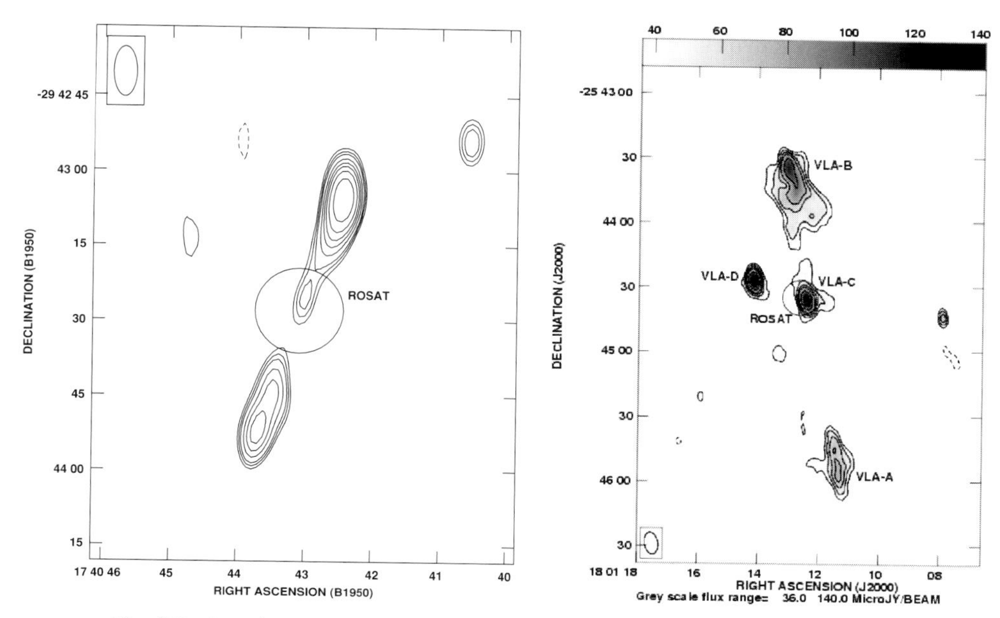

Fig. 9.7. Arcminute-scale radio jets from the galactic center low/hard state sources 1E 1740.7–2942 and GRS 1758–258. Both of these systems spend nearly all their time in the low/hard X-ray state, therefore an interpretation of these lobes is that they result from the long-term action of steady jets on the ISM. From Mirabel *et al.* (1992) and Martí *et al.* (2002).

the long-term action of a jet on the local ISM (Mirabel *et al.* 1992; Martí *et al.* 2002; Fig. 9.7).

9.4.1.1 Spectral extent and jet power

The radio spectrum in the low/hard state is "flat" or "inverted", in the sense that the spectral index $\alpha \geq 0$. This spectral component has been shown to extend to the millimeter regime for two low/hard state sources, Cyg X-1 and XTE J1118+480 (Fender *et al.* 2000b; Fender *et al.* 2001). In Fender (2001) it was suggested that correlated radio–optical (and in fact X-ray) behavior in the low/hard state transient V404 Cyg might suggest an extension of the jet spectral component to the infrared or optical bands. In fact in most, maybe all, low/hard state sources the optical flux densities seem to lie on a rather flat ($\alpha \sim 0$) extension of the radio(–mm) spectrum (e.g., Brocksopp *et al.* 2001; Corbel *et al.* 2001). Jain *et al.* (2001) have observed a secondary maximum in the near-infrared lightcurve of XTE J1550–564 corresponding to a transition to the low/hard state, which they also attribute to synchrotron emission from a jet. Rapid optical variability from XTE J1118+480 in the low/hard X-ray state has also been interpreted as (cyclo-)synchrotron emission (Merloni *et al.* 2000; see also Hynes *et al.* 2003) and may be associated with a sub-relativistic outflow (Kanbach *et al.* 2001; Spruit & Kanbach 2002).

Note that while admittedly not a canonical low/hard state source, there is unambiguous evidence for synchrotron emission from the jet source GRS 1915+105 extending at least to the near-infrared band (Fender *et al.* 1997; Mirabel *et al.* 1998; Eikenberry *et al.* 1998a, 2002; Fender & Pooley 1998, 2000). Not well explained is the correlation in this source between infrared line strength and synchrotron continuum (Eikenberry *et al.* 1998b), indicating a coupling between thermal and non-thermal components. Qualitatively similar infrared flares have been observed from Cyg X-3 (e.g., Mason *et al.* 1986; Fender *et al.* 1996) which with the benefit of hindsight seem likely to be synchrotron in origin. Finally, Sams *et al.* (1996) have observed *extended* infrared emission from GRS 1915+105, which they suggest originates in a jet (while possibly treated with some scepticism at the time, the observation of considerably larger X-ray jets from XTE J1550–564 makes a jet origin seem entirely plausible).

If the flat/inverted radio spectrum is due to self-absorbed synchrotron emission from a conical jet (Blandford & Königl 1979; Hjellming & Johnston 1988) then above some frequency (at which point the whole jet is optically thin) there should be a break to an optically thin spectrum with $-1 \leq \alpha \leq 0$. A compilation of observations of the low/hard state source GX 339–4 appears to have identified just such a cutoff in the near-infrared (Corbel & Fender 2002).

How do we estimate the power associated with this steady, self-absorbed, synchrotron component? Without large amplitude variability, or directly resolved jets, it is not possible to associate a given luminosity with a certain volume, and it is not possible to directly apply standard "minimum energy" arguments (as outlined in Section 9.2). Therefore we must apply other arguments in order to estimate the total jet power. In this case we may estimate the total jet power by (a) carefully measuring the extent of the synchrotron spectrum that it produces, and (b) introducing a radiative efficiency, η, which is the ratio of radiated to total power (in the jet's rest frame). From this we can estimate the jet power as

$$P_{\rm J} \sim L_{\rm J} \eta^{-1} F(\Gamma, i) \tag{9.16}$$

where $L_{\rm J}$ is the radiative luminosity of the jet (i.e., the integral of L_ν over frequency), and $F(\Gamma, i)$ is a correction factor for bulk relativistic motion with Lorentz factor Γ and Doppler factor δ, ($F(\Gamma, i) \sim \Gamma\delta^{-3}$ – see Fender 2001).

Starting from the reasonable assumption that all the emission observed in the radio band is synchrotron in origin, we can try to see how far this spectrum extends to other wavelengths. First, it should be made clear that most systems have not been observed at $\nu < 1$ GHz (although it appears that the flat radio spectrum of Cyg X-1 extends at least as low as 350 MHz – de Bruyn, private communication), and while some low-frequency turnovers may have occasionally been observed, there are no reported cases of a complete cutoff to the synchrotron emission at low radio frequencies. In any case, while a low-frequency cutoff is important for estimating the mass of the ejecta in the (by no means certain) case that there is a proton for each emitting electron, the radiative luminosity is dominated by the high-frequency extent of the synchrotron spectrum.

Possibly the most comprehensive broadband spectrum compiled for a low/hard state source is that for the transient XTE J1118+480, which clearly shows excess emission at near-infrared and probably also optical wavelengths (Hynes *et al.* 2000) and whose radio spectrum smoothly connects to a sub-mm detection at 850 μm (Fender *et al.* 2001). In Fender *et al.* (2001) it is argued that in this case the synchrotron radiative luminosity is already $\geq 1\%$ of the bolometric X-ray luminosity. How important the total jet power is then depends on our estimates for the radiative efficiency, η.

In Fender & Pooley (2000) an estimate of η was made for the radio "oscillation" events from GRS 1915+105, and an upper limit of $\eta \leq 0.15$ obtained. In the original model of Blandford & Königl (1979), it is likely that $\eta \leq 0.15$. In the model of Markoff *et al.* (2001;

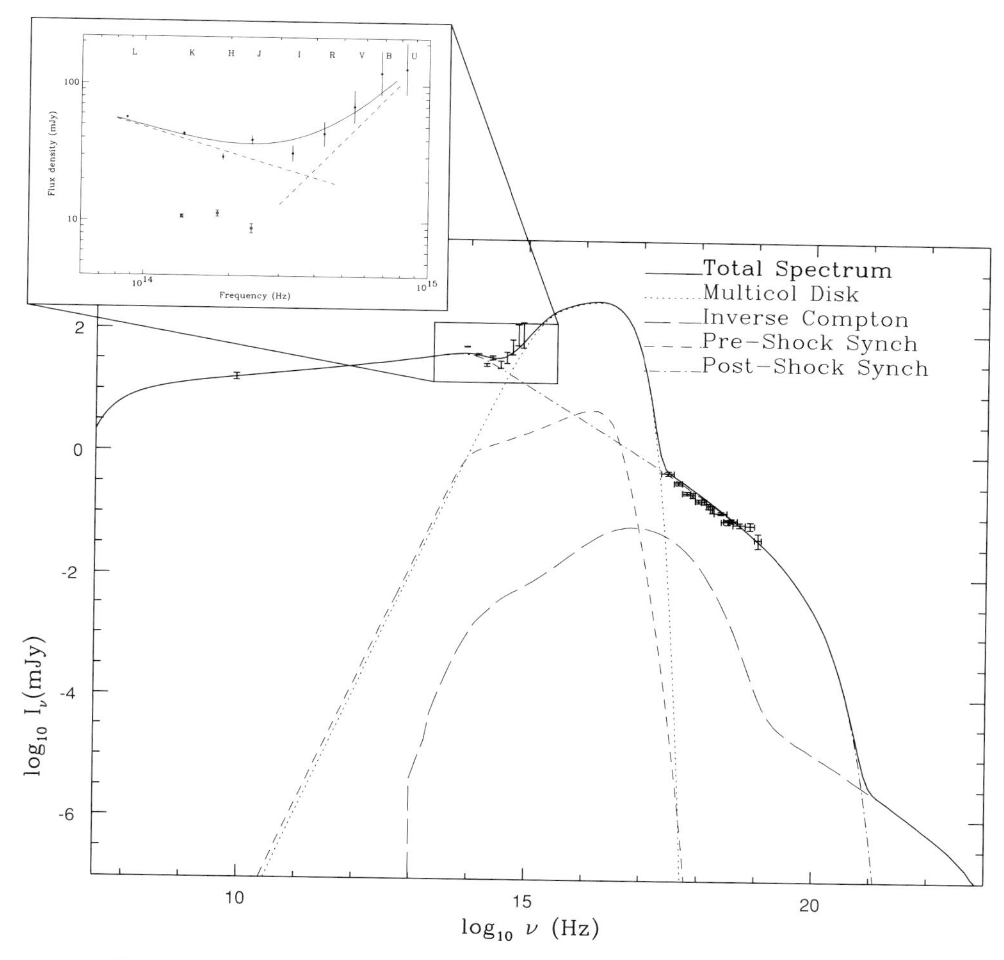

Fig. 9.8. Broadband jet-model fit to the radio–X-ray spectrum of GX 339–4 in the low/hard X-ray state (Markoff *et al.* 2003). The flat spectrum, self-absorbed, synchrotron component extends beyond the radio band and breaks to optically thin emission in the near-infrared (see insert, from Corbel & Fender 2002). An extrapolation of this near-infrared emission connects smoothly to the X-ray power law, suggesting that it may also be optically thin synchrotron emission, contrary to more widely accepted Comptonization models. The broadband spectrum and model fit are comparable to those for XTE J1118+480 while in the same X-ray state (Markoff *et al.* 2001).

specifically for XTE J1118+480) $\eta < 0.1$. Finally it should be noted that Celotti & Ghisellini (2003) estimate $\eta \leq 0.15$ for a sample of AGN. In reality, for the synchrotron process in jets it seems unlikely theoretically that $\eta > 0.2$, and this is backed up by an absence of observational counter-evidence. Therefore, for XTE J1118+480 the power in the jet is likely to be $\geq$10% of the X-ray luminosity. Since all low/hard state sources show a similar broadband spectrum (excluding the influence of different types of mass donor which only affects the near-infrared and optical bands) we are drawn to the conclusion that all low/hard state sources produce powerful jets (Fender 2001).

9.4.1.2 Coupling to X-ray emission

A broad correlation between the radio and X-ray fluxes from a black hole binary in a low/hard state was first noted by Hannikainen *et al.* (1998) for GX 339–4. A similar correlation between radio and X-ray fluxes was found for Cyg X-1 (Brocksopp *et al.* 1999), and Fender (2001) suggested that the radio–X-ray flux ratio was similar for all low/hard state black holes.

In the past couple of years our understanding of this coupling between radio and X-ray emission has advanced significantly. Corbel *et al.* (2000, 2002), in a detailed long-term study of GX 339–4, have found that the radio emission in the low/hard state scales as $L_{\rm radio} \propto L_{\rm X\text{-}ray}^{b}$ where $b \sim 0.7$ for X-rays up to at least 20 keV (possibly steepening towards a linear relationship at the highest X-ray energies). This relation holds over more than three orders of magnitude in soft X-ray flux.

More recently, Gallo *et al.* (2002, 2003) have found almost exactly the same correlation (in both normalization and slope), over a comparable range in X-ray luminosity, for the low/hard state transient V404 Cyg. Furthermore, by compiling data for ten low/hard state sources, it was found that in the luminosity range $10^{-5}L_{\rm Edd} \leq L_{\rm X} \leq 10^{-2}L_{\rm Edd}$ all systems are consistent with the same correlation with a very small scatter (less than one order of magnitude in radio flux), and that above a small percentage of $L_{\rm Edd}$ the radio emission rapidly weakens (Gallo *et al.* 2003). Monte-Carlo simulations of Doppler-boosting effects indicate that such a small spread over such a large range in $L_{\rm X}$ probably restricts the velocity of the jet in the low/hard state to $\beta = v/c \leq 0.8$ (Gallo *et al.* 2003; Fig. 9.9), unless the X-rays are also strongly beamed (in which case strong selection effects are at work).

9.4.1.3 Jets in "quiescence"?

Outside of periods of transient outburst, black hole candidates (BHCs) are typically observed with X-ray luminosities in the range 10^{-6}–10^{-9} Eddington, and are considered to be "quiescent" (see Chapter 4 and, e.g., Garcia *et al.* 2001). Their X-ray spectra are generally not distinguishable from the "low/hard" state however, suggesting that they may also be associated with (relatively) powerful jets. In fact, V404 Cyg – the most luminous quiescent black hole – is clearly associated with a relatively bright and variable radio source (e.g., Hjellming *et al.* 2000b) and GX 339–4 follows the radio–X-ray correlation discussed above down to such X-ray luminosities. Combining the estimates of jet power in the low/hard state with the $L_{\rm radio} \propto L_{\rm X\text{-}ray}^{0.7}$ relation indicates that quiescent BHCs will in fact be "jet-dominated", in the sense that most of the power output will be in the form of an outflow (Fender *et al.* 2003). Combining this result with the greater "radio loudness" (Section 9.5.3) of BHCs compared to neutron star (NS) X-ray binaries can furthermore explain the discrepancy in their quiescent X-ray luminosities (it is observed that NS transients are brighter X-ray sources in quiescence) without any significant advection of accretion power across a black hole event horizon (Fender *et al.* 2003; see also Campana & Stella 2000; Garcia *et al.* 2001; Abramowicz *et al.* 2002 and references therein for a broader discussion of this controversial issue).

9.4.2 *Loss of jet in high/soft states*

The first indication that radio jets are not associated with soft X-ray states can be traced back to Tananbaum *et al.* (1972), in which the appearance of the radio counterpart of Cyg X-1 was associated with a transition from the soft state back to the hard state (see also

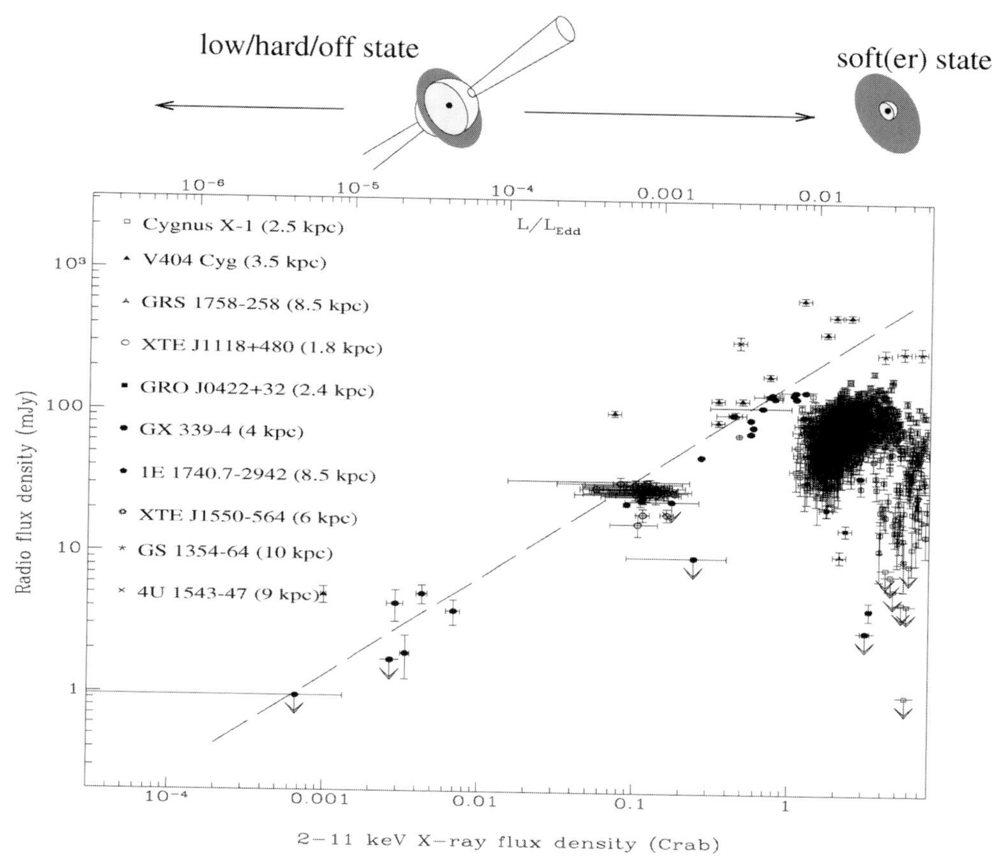

Fig. 9.9. (Quasi-)simultaneous radio and X-ray observations of black hole X-ray binaries, scaled to 1 kpc and corrected for absorption. Below a scaled X-ray flux of a few Crab (corresponding to $\sim$1% of the Eddington luminosity for a 10 $M_{\odot}$ black hole), all black hole binaries follow a correlation of the form $S_{\rm radio} \propto S_{\rm X\text{-}ray}^{0.7}$, in the low/hard and off/quiescent states. The relatively narrow distribution of data around a best-fit relation requires that the bulk Lorentz factor of jets in the low/hard state $\Gamma < 2$. At higher luminosities in the high/soft state the radio emission is strongly suppressed. At still higher luminosities, X-ray transients (including recurrent sources such as Cyg X-3 and GRS 1915+105) produce repeated bright optically thin ejections. The hard plateau state of GRS 1915+105 lies on an extension of the low/hard state coupling. From Gallo *et al.* (2003); see also Corbel *et al.* (2001, 2002).

Hjellming *et al.* 1975). However, while it was surmised that changes in radio emission were associated with changes in the X-ray "state" of X-ray binaries (Hjellming & Han 1995 and references therein), no clear pattern was established (except perhaps in Cyg X-3, where is has been realized for some years that periods of "quenched" radio emission generally preceded large radio outbursts – e.g. Waltman *et al.* 1996).

The situation changed when GX 339–4 spent a year in the high/soft X-ray state in 1998. Radio monitoring of the source in the low/hard state prior to 1998 had already established the existence of a weak, mildly variable radio counterpart (Hannikainen *et al.* 1998), but throughout the soft state no radio counterpart was detected, despite multiple observations

(Fender *et al.* 1999b). The source subsequently returned to the low/hard X-ray state and the weak radio counterpart reappeared (Corbel *et al.* 2000). Here was the strongest evidence that in "soft" disk-dominated states the radio jet was more than an order of magnitude weaker than in the low/hard state, and perhaps non-existent.

Comprehensive radio and X-ray monitoring of Cyg X-1 has revealed that the suppression of the radio emission occurs rather rapidly once at a bolometric luminosity of a small percentage Eddington the transition to the high/soft state occurs (Gallo *et al.* 2003; Maccarone 2003). Given that there are no observed counter-examples, we conclude that the soft X-ray state is never associated with a strong radio jet. This assertion is supported by the detailed studies of GRS 1915+105 reported by Klein-Wolt *et al.* (2002), in which steady "soft" X-ray states are never associated with bright radio emission (Fig. 9.10).

9.4.3 *"Intermediate" and "very high" X-ray states*

While the low/hard and high/soft X-ray states appear to represent both the most diametrically opposed and the most stable of accretion modes associated with black hole XRBs, there are also hybrid states. Both the intermediate and very high (see Chpaters 2 and 4 for more details) states are intermediate in their X-ray hardness between the two aformentioned canonical extremes. It has been suggested that they are the same state, in which case it is an interesting fact that this state can occur over quite a large range in bolometric X-ray luminosity (as can the low/hard and high/soft states, see Homan *et al.* 2001; Sections 2.5.1 and 4.3).

Belloni (1998) suggested that the behavior of GRS 1915+105, oscillating between relatively hard and (two) soft states (see also Belloni *et al.* 2000) was reminiscent of the very high state as observed from other luminous X-ray transients. Since these oscillation events are unambiguously correlated with radio flaring (e.g., Pooley & Fender 1997; Mirabel *et al.* 1998; Klein-Wolt *et al.* 2002 – see Fig. 9.10), a connection was made between this state and episodic jet production.

However, in a very important observation, Corbel *et al.* (2001) have shown that in a transition from the low/hard state to the intermediate state, the radio emission from XTE J1550–564 was reduced by a factor > 50. Furthermore, the state in which the jet from Cyg X-1 is suppressed (see Fig. 9.9) may not be the canonical high/soft state, but the intermediate state (Section 2.10.3 and, e.g., Belloni *et al.* 1996; Miller *et al.* 2002a, but see Gierlinski *et al.* 1999). What remains clear is that when the X-ray spectrum softens the jet weakens or disappears. What needs further investigation is the exact evolution of the X-ray spectral and jet parameters as this occurs, since at present the most comprehensive studies (e.g., Corbel *et al.* 2002; Gallo *et al.* 2003) are based only on flux, not spectral, evolution. In a related work, Pottschmidt *et al.* (2000) report that the magnitude of X-ray time lags in Cyg X-1 is much greater *during* transitions than either before or after, and suggest that this effect may be related to the formation of outflows at these times.

9.4.4 *The highest luminosities and X-ray transients*

X-ray transients typically peak at luminosities greater than those which generally characterize the high/soft state, although often still sub-Eddington (Chen *et al.* 1997). Such high luminosities are, in nearly all cases, very short lived (typically days or less) and the "state" is considerably more difficult to characterize than the canonical low/hard or high/soft states.

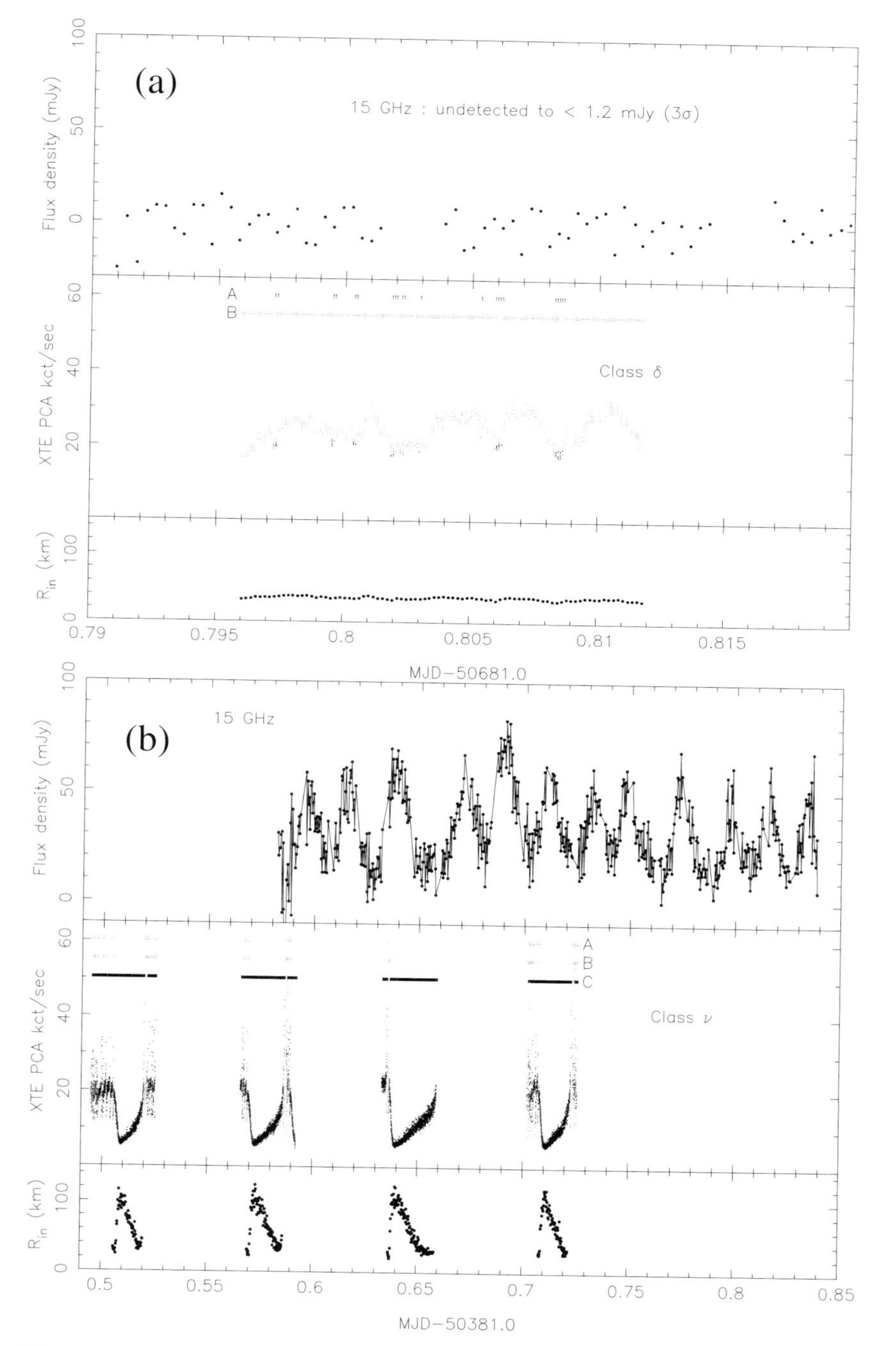

Fig. 9.10. GRS 1915+105 often cycles repeatedly between three X-ray states: A and B are disk-dominated and "soft"; C is much harder (Belloni *et al.* 2000). Panel (a) above shows that when the source only exhibits soft states A and B, the radio emission is very weak; however, when state C is present, Panel (b), the radio emission is much stronger (and in fact there is a one-to-one correspondence between state C "dips" and radio oscillation events). From Klein-Wolt *et al.* (2002).

It was known since the 1970s that bright X-ray transients were associated with transient production of radio emission, whose characteristics could be described at a basic level by "synchrotron bubble" models (Hjellming & Han 1995 and references therein, and also Section 9.2.3). In several, perhaps all, cases, there is evidence for *multiple* ejection events (e.g., Harmon *et al.* 1995; Kuulkers *et al.* 1999; Brocksopp *et al.* 2001). The clearest difference with low/hard state steady jets is the rapid evolution to an optically thin spectrum ($\alpha \sim -0.6$) and monotonic decay (Fender 2001). In addition, linear polarizations of up to a few $\times 10\%$ have been measured (e.g., Fender *et al.* 1999a; Hannikainen *et al.* 2000), and also circular polarization at the $\sim 1\%$ level (see Section 9.2.6). The broad properties of these transient radio events – i.e., the spectral evolution *and* a tendency for multiple ejection events – seem to be similar whether the events are "rare" (e.g., A0620–00, GS 1124–68) or "frequent" (e.g., Cyg X-3, GRS 1915+105).

These ejection events appear to be associated with the change in X-ray state between "off" (which may be analogous to low/hard) and very bright high/soft or very high states (e.g., Harmon *et al.* 1995; Fender & Kuulkers 2001). Some transients actually seem only transit to bright low/hard states and may (e.g., V404 Cyg) or may not (e.g., XTE J1118+480) also display bright optically thin events. One source seems to sit persistently at close to Eddington luminosities, and is a spectacular source of relativistic jets: GRS 1915+105. This source exhibits a wide range of X-ray properties, none of which can be easily classified as normal low/hard or high/soft states (Belloni *et al.* 2000). Its overall X-ray properties may be reminiscent of the very high state (Belloni 1998), but the erratic flips between hard and (two sorts of) soft states is rather unlike any other X-ray binary. However, GRS 1915+105 does fit into the general pattern associating hard X-ray states with jet formation, at least for the "plateau" and "oscillation" events (Dhawan *et al.* 2000; Klein-Wolt *et al.* 2002). Mirabel *et al.* (1998) have suggested that a brief X-ray spike, during which the source X-ray spectrum softens considerably, may indicate the "launch moment" of the jet – this would clearly be an important discovery if true and merits further attention. Cyg X-3 may be displaying similar behavior to GRS 1915+105 – it is certainly accreting at a very high level and almost continuously producing jets – but details of its workings are hidden in the dense wind of its Wolf–Rayet companion.

These radio flares (see Section 9.2.3) have by now been clearly and repeatedly associated with highly relativistic bulk motions (Section 9.2.4). In a comparison of peak radio and X-ray emission from transients, Fender and Kuulkers (2001) found that there appears to be nothing special about the sources in which relativistic jets had been resolved. Therefore it seems reasonable to assume (Occam's razor) that the initial radio emission associated with X-ray transients is always associated with a relativistic outflow. Note also that Garcia *et al.* (2003) have suggested that the largest-scale resolved radio jets may be associated with X-ray transients with relatively long orbital periods.

9.5 Disk–jet coupling in neutron star binaries

As noted in the introduction, radio emission seems to be associated with both Z and atoll type neutron star X-ray binaries, but not with the high-field X-ray pulsars. See Fig. 9.11 for a summary of our current understanding. It is interesting as a historical note that a predictable coupling between X-ray state and radio emission was first suggested for the Z sources (see below), but that in recent years nearly all the attention has switched to the analogous coupling in black hole systems.

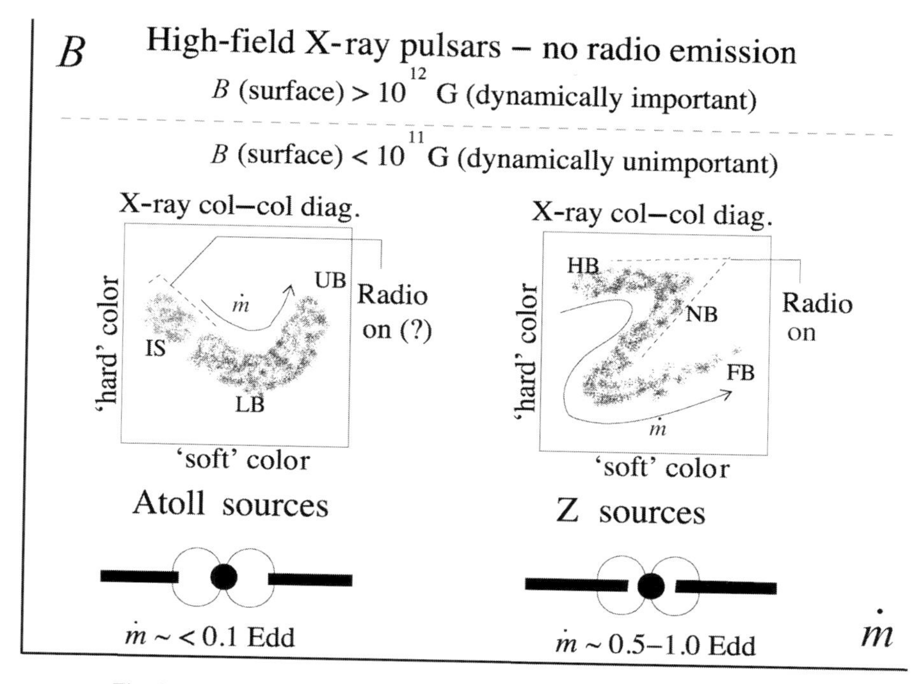

Fig. 9.11. Schematic illustrating our current understanding of the relation between radio emission and X-ray state for the persistent neutron star X-ray binaries.

9.5.1 Z sources

The prototype Z source, Sco X-1, has been known as a variable radio source since the early 1970s (Hjellming & Wade 1971b). This source, together with GX 5-1, GX 17+2, GX 349+2, GX 340+0 and Cyg X-2 form a group of neutron star X-ray binaries accreting at or near to the Eddington limit and exhibiting clear patterns of spectral and timing behavior (Hasinger & van der Klis 1989). It is ironic that an initial association with large-scale radio lobes was disproved by the same proper motion studies (Fomalont & Geldzahler 1991) which subsequently discovered highly relativistic jets on milliarcsecond scales (Bradshaw *et al.* 1999; Fomalont *et al.* 2001a,b). The other five Z sources also have radio counterparts with comparable luminosities (Penninx 1989; Hjellming & Han 1995; Fender & Hendry 2000).

Priedhorsky *et al.* (1986) first suggested that an empirical coupling between X-ray and radio (and optical) emission existed for Sco X-1. Penninx *et al.* (1988) confirmed and refined this pattern of behavior for GX 17+2 and Penninx (1989) suggested that all Z sources would display comparable behavior. The same pattern of behavior has been established for Cyg X-2 (Hjellming *et al.* 1990a) but apparently not in GX 5-1 (Tan *et al.* 1991; but see below for a possible explanation). The radio behavior seems to correlate with position in the Z-shaped track traced out on timescales of hours to days in the X-ray color–color diagram (Section 2.5.2.1; see Fig. 9.11) in the sense that it is strongest on the "horizontal branch" and weakest on the "flaring branch", revealing an apparent anti-correlation with mass accretion rate (or at least, state, cf. Section 2.5) as in the black holes.

As noted above, intensive VLBI campaigns on Sco X-1 have revealed the presence of a relativistic outflow (Bradshaw *et al.* 1999; Fomalont *et al.* 2001a,b – see Fig. 9.12). In

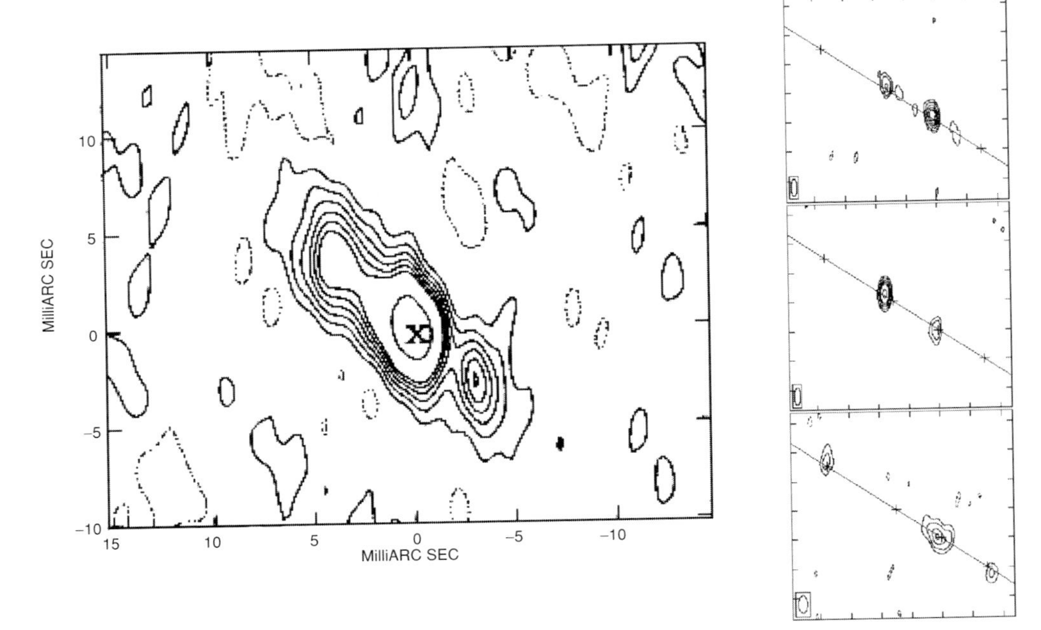

Fig. 9.12. *Left*: A VLBA image of milliarcsecond-scale radio jets from Sco X-1 with the core indicated by X (from Bradshaw *et al.* 1997). *Right*: Multiple sequences of such observations reveal movement of the radio lobes at mildly relativistic velocities ($\sim 0.4c$) while being sporadically energized by a much more relativistic ($\Gamma \geq 2$) beam from the core. Adapted from Fomalont *et al.* (2001a,b).

particular, it seems that following core radio flaring, relativistic ($\Gamma \geq 2$) beams are acting on radio knots, which themselves propagate away from the binary core with mildly relativistic velocities. Given the similarity of the radio properties between the six Z sources, we can fairly confidently conclude that they all have jets; however, since the brightest component is the core, it cannot yet be asserted that the jet–knot interaction is occurring in all of them. Furthermore, caution should be exercised in attempting to associate unresolved radio monitoring (e.g., that performed in the 1990s with the Green Bank Interferometer) with X-ray events (but see Hjellming *et al.* 1990b for a successful experiment) since the delay between core events and subsequent brightening in the knots is comparable to the timescale of motion in the Z – this may be an explanation for the "anomalous" observations of GX 5-1 by Tan *et al.* (1991).

At present there is little study of, and consequently little evidence for, possible extensions of the jet spectrum beyond the radio band in the Z sources (although there are hints of some correlated optical behavior). Estimates of the power in the jets are rather uncertain (Fomalont *et al.* 2001b estimate super-Eddington power in the jets of Sco X-1, but this is based upon the assumption that the major cooling process is synchrotron losses, which is far from clear), and are at present based solely upon radio variability.

9.5.2 *Atoll sources*

It is worth re-stressing here that I am adopting a definition of atoll source to mean all non-Z low magnetic field accreting neutron stars – this is considerably broader than the original definition of Hasinger and van der Klis (1989). Adopting this loose definition, atoll sources are the single largest class of X-ray binary, contributing around 45% of the currently known population (in this classification, atoll includes bursters, dippers, etc.; on this issue see also Sections 2.4.2 and 2.5.2.2). Investigation of their disk–jet coupling, if any, is therefore of paramount interest – not least because they can exhibit hard X-ray states which are very similar to the low/hard states of BHCs, while they of course remain fundamentally different in possessing a solid surface.

Hjellming and Han (1995) list the small number of reported radio detections of atoll sources known at that time. Beyond some weak detections of globular cluster sources, only GX 13+1 was repeatedly detected at a relatively strong level (Garcia *et al.* 1988) – in fact at about the same radio luminosity as the Z sources (Fender & Hendry 2000). However, GX 13+1 seems to be far from a "normal" atoll source (Homan *et al.* 1998; Schnerr *et al.* 2003). It is interesting to note that the three other brightest atoll sources, GX 3+1, GX 9+1 and GX 9+9 have never been detected in the radio band, and spend most of their time in the soft "upper banana" (UB in Fig. 9.11) state.

Most other atoll sources show somewhat harder spectra associated with "island" (IS) X-ray states (similar to the black hole low/hard state). Amongst these, Martí *et al.* (1998) reported repeated detections of the atoll source 4U 1728–34 (GX 354–0) at a level of up to $\sim$0.6 mJy. In recent simultaneous radio and X-ray observations of the same source, Migliari *et al.* (2003) have revealed clear correlations between X-ray luminosity and power spectral properties with the radio flux, establishing for the first time a disk–jet coupling in such systems. Despite their relative faintness in the radio band, it seems that there is a rich phenomenology to be explored in these hard atoll source states.

9.5.2.1 *Neutron star transients*

There are a few detections of radio emission associated with neutron star X-ray transients (see Fender & Kuulkers 2001 for a list). These include an unusual assortment of objects: the recurrent transient Aql X-1 (Hjellming *et al.* 1990c), the first accretion-powered millisecond pulsar SAX J1808.4–3658 (Gaensler *et al.* 1999) and 4U 1730–335 (The Rapid Burster, Moore *et al.* 2000). To hammer home a point made earlier, I consider all these sources to be quite similar in that they are low magnetic field neutron stars accreting, on average, at a considerably sub-Eddington rate, and I call them all atoll sources (it is interesting to note that there has not yet been a NS transient which displayed Z-type properties even at the peak of outburst). The sample for NS transients is considerably poorer than that for BH transients, something which can be at least partially attributed to the fact that they are in general fainter in the radio band (Fender & Kuulkers 2001; see Section 9.5.3).

Cir X-1 can be considered as a recurrent NS transient (perhaps comparable in this respect to the BHC GRS 1915+105 and to Cyg X-3, whose nature is uncertain); it undergoes radio and X-ray flares every 16.6 days, during which periods its X-ray luminosity is super-Eddington. This periodicity is interpreted as heightened accretion during periastron passage of the neutron star in a highly elliptical orbit – essentially this system undergoes repeated, periodic, soft X-ray transient outbursts. The system is associated with an arcsecond-scale one-sided radio jet (Fender *et al.* 1998) embedded within an arcminute-scale radio nebula (Stewart *et al.* 1993). The X-ray classification of Cir X-1 has alternated between Z and atoll types, and at present it is not clear to which category it belongs.

9.5.3 *Black holes versus neutron stars*

There are clearly some broad similarities between black holes and neutron stars in their X-ray–radio coupling. These include:

- An association between states with hard X-ray spectra and strong X-ray variability and the presence of radio emission
- An association between bright X-ray outbursts and radio flare events
- In the brightest cases, the formation of large-scale radio lobes in the ISM

Are there differences between jets from neutron stars and those from black holes? There is at least one. Fender and Kuulkers (2001) have found that defining a quantity "radio loudness" as the peak radio flux of transients, divided by their peak X-ray flux, *black hole transients are more radio loud than neutron stars* (Fig. 9.13). Furthermore, by comparing the data for low/hard state black holes and neutron star Z sources, Fender & Hendry (2000) found a similar difference. In both classes of object, black holes seem to be one or two orders of magnitude more radio loud than neutron star systems. Migliari *et al.* (2003) confirm a difference in radio luminosity by a factor $\sim$30 between the atoll source 4U 1728–34 in a hard state and low/hard state black holes at a comparable Eddington ratio. This difference may be due to greater photon (Compton) cooling of shocked electrons in the neutron star systems, due to the presence of a radiating surface or low-level magnetic field, or perhaps due to some extra source of power (presumably, the black hole spin) in the black hole systems (Fender & Kuulkers 2001). A further possibility (Heinz & Sunyaev 2003; Merloni *et al.* 2003; Falcke *et al.* 2003) is that the radio loudness scales with mass. However in this case, assuming the "stellar mass" black holes are on average five times more massive than the neutron stars, this suggests a rather steeper dependence on mass than considered by these authors.

9.6 High-energy/particle emission from jets

Observations in recent years have revealed unambiguously that jets may be not only *associated* with phases of high-energy emission, but may actually be the *sites of origin* of (some of the) observed emission.

9.6.1 *X-rays*

The possibility of some of the X-ray emission from X-ray binaries arising in jets has already been alluded to in this text, and explicitly suggested in the literature (e.g., Markoff *et al.* 2001; Vadawale *et al.* 2001; Markoff *et al.* 2003; see also Atoyan & Aharonian 1999; Miller *et al.* 2002b). Before discussing this further, it is worth restating the fact that Chandra imaging has unambiguously detected both thermal/emission line (Migliari *et al.* 2002) and hard X-ray spectra (Corbel *et al.* 2002; see also Angelini & White 2003) with jets from X-ray binaries (Fig. 9.14).

In a detailed model, Markoff *et al.* (2001, 2003; see Fig. 9.8 and also Falcke & Biermann 1996, 1999) have suggested that the X-ray power law observed in the low/hard X-ray state may in fact be the optically thin synchrotron emission from the jet which is self-absorbed at lower frequencies. In fact as already noted a break from optically thick to optically thin emission from the jet seems to have been found in the right place for GX 339–4 (Corbel & Fender 2002 – in fact this may have already been noted by Motch *et al.* 1985). This is a radically different interpretation for the origin of X-rays in this state, which are generally ascribed to thermal Comptonization (Chapter 4 and, e.g., Poutanen 1998 and references therein; for more detailed

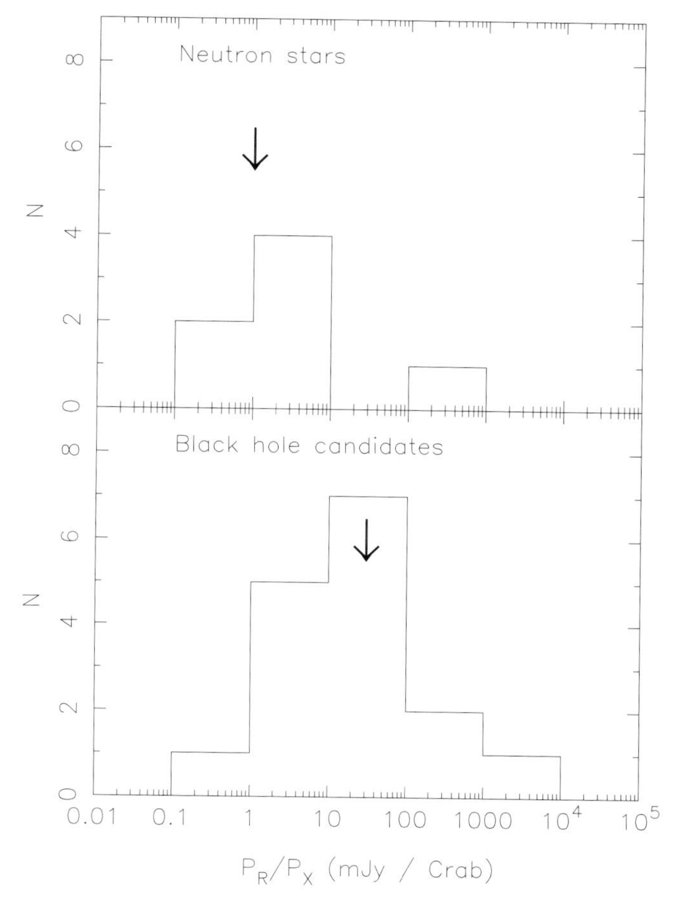

Fig. 9.13. Histograms of the "radio loudness" of neutron star (*top*) and black hole (*bottom*) transients. The black holes are significantly more "radio loud" than the neutron stars, by one to two orders of magnitude. Also indicated by arrows are the mean "radio loudnesses" of the neutron star Z sources and the brighter low/hard state black holes, revealing the same trend. From Fender & Kuulkers (2001).

objections to the model of Markoff *et al.* see Zdziarski *et al.* 2003). The implication of the model, if correct, would be that the majority of power output in the low/hard state is in the form of a jet. Note that in this model the X-ray emitting region would be spatially unresolvable and is not therefore an explanation for the extended X-ray jets observed from XTE J1550–564 and SS 433 (Fig. 9.14). Vadawale *et al.* (2001) have suggested that some component of the X-ray spectrum of GRS 1915+105 may arise in synchrotron emission. Georgonapoulos *et al.* (2002) have suggested that X-ray emission may originate due to Comptonization by jet electrons of photons from the companion star.

Returning to the large-scale X-ray jets, the fact that three have been clearly imaged in the past few years with Chandra (Marshall *et al.* 2002; Migliari *et al.* 2002; Corbel *et al.* 2002; Angelini & White 2003) indicates that they are likely to be rather ubiquitous. The fact that X-ray emission, with a spectrum similar to known "off" state spectra (for XTE J1550–564 and 4U 1755–33) may be associated with beamed, long-lasting jets is of considerable interest.

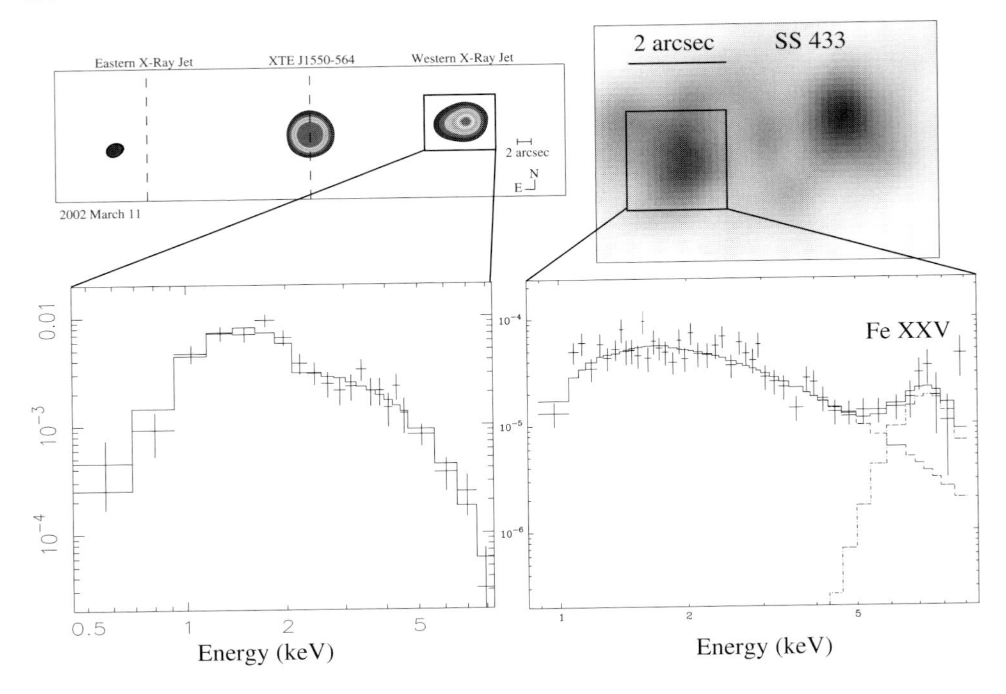

Fig. 9.14. Spatially resolved X-ray spectra of X-ray jets from the black hole transient XTE J1550–564 (*left*) and the persistent powerful jet source SS 433 (*right*). Note the strong emission line (probably Fe XXV) in the SS 433 spectrum, which is clearly not present in the jets of XTE J1550–564. Observations such as these demonstrate unequivocally that jets from X-ray binaries can be sources of both line-rich (thermal) and featureless (non-thermal) X-ray spectra, which may be beamed. Adapted from Corbel *et al.* (2002), Kaaret *et al.* (2003), Migliari *et al.* (2002).

It certainly shows that jets, almost certainly via internal or external shocks, may mimic faint hard states up to several years after the binary source may have completely turned off. These are all extra concerns for interpretations of the quiescent luminosities of transient X-ray binaries: the X-ray jets of XTE J1550–564 (Corbel *et al.* 2002; Kaaret *et al.* 2003; Tomsick *et al.* 2003) are more luminous than most of the quiescent X-ray luminosities for black holes reported in Garcia *et al.* (2001).

9.6.2 High-energy/particle emission

In an important recent work, Paredes *et al.* (2000; see also Ribó *et al.* 2002 and Paredes *et al.* 2002) have reported a convincing association between a massive X-ray binary with persistent radio jets and an unidentified EGRET gamma-ray source. Their favored scenario is that relativistic electrons in the jet Comptonize photons from the binary companion (similar to the model of Georgonapoulous *et al.* 2002). The massive binary and probable jet source LS I +61 303 (Strickman *et al.* 1998; Gregory & Neish 2002 and references therein) may also be associated with a gamma-ray source, with a similar physical origin a possibility.

Heinz and Sunyaev (2002) have discussed the possible contribution of X-ray binary jets to the production of galactic cosmic rays. They conclude that, while in terms of overall energetics such jets are still likely to inject less power into the ISM than supernovae, they may contribute a specific and detectable component to the cosmic ray spectrum. In particular, the shocks in the ISM associated with jets from X-ray binaries will be considerably more relativistic than those associated with the supernovae, and thus may be considerably more efficient at particle acceleration.

Di Stefano *et al.* (2002) have suggested that jets from X-ray binaries could be detectable sources of high-energy neutrinos. Kaiser and Hannikainen (2002) have further suggested that X-ray binary jets may be the origin of a putative redshifted 511 keV annihilation line observed from the direction of the X-ray transient GRS 1124–684 (however, an alternative explanation, which is perhaps more widely accepted, is that the gamma-ray emission was associated with a transition of ^{7}Li – Martin *et al.* 1994; cf. Chapter 5). A possible explanation for the ^{7}Li production is spallation in the companion star atmosphere due to the collision of a misaligned jet (Butt *et al.* 2003).

In the light of the possibility of X-rays directly from jets, several authors have considered the possibility of "micro-blazars" in which jets aligned close to the line of sight could be observed as gamma-ray sources (e.g., Mirabel & Rodriguez 1999; Kaufman Bernado *et al.* 2002; Romero *et al.* 2002).

9.7 Interactions

As has already been alluded to in previous sections, it is becoming clear that interactions between the jet, as launched by the combination of accretion flow plus compact object, and the ambient medium need to be taken into account for a full understanding of both the radiation we observe and the internal physics of the outflows. Of the classes of radio-emitting X-ray binaries only the weakest, the atoll sources, have yet to provide us with a direct example of jet–ISM interactions. These interactions have the potential to act as independent measures of the power associated with jets from X-ray binaries ("calorimeters"), although it has been argued that they may be harder to detect than the corresponding lobes associated with AGN (Heinz 2002; see also Levinson & Blandford 1996). Furthermore, as with AGN, it is possible that some of the presumed shock acceleration may result not from jet–ISM interactions but from internal shocks (Kaiser *et al.* 2000), perhaps resulting from varying flow speeds (see also discussion in Migliari *et al.* 2002 for SS 433).

Considering first the black hole low/hard state sources, as well as the milliarcsecond-scale jet from Cyg X-1 (Stirling *et al.* 2001; Fig 9.6), arcminute-scale ($\equiv$ parsec-scale) jets have been observed from 1E 1740–2942 and GRS 1758–258 (Mirabel *et al.* 1992; Martí *et al.* 2002; see Fig. 9.7). It seems clear that these larger lobes are the result of *in-situ* particle acceleration at the interface between the steady jets and the ISM.

Observations of large-scale radio and X-ray jets from the black hole transient XTE J1550–564 (Corbel *et al.* 2002; Kaaret *et al.* 2003; Tomsick *et al.* 2003) have provided us with unambiguous evidence of broadband particle acceleration at the same time as the jet is decelerating (Figs. 9.2, 9.5). Similarly, a one-sided highly relativistic jet from Cyg X-3 on milliarcsecond-scales (Mioduszewski *et al.* 2001) seems to become a slower-moving, two-sided jet on arcsecond-scales (Martí *et al.* 2001), indicating a deceleration and *in-situ* particle acceleration.

In Sco X-1, the prototype of the Z sources, which are the brightest "persistent" neutron stars, Fomalont *et al.* (2001a,b) have found evidence for the action of an unseen, highly relativistic flow on radio-emitting clouds, which are themselves moving away from the binary

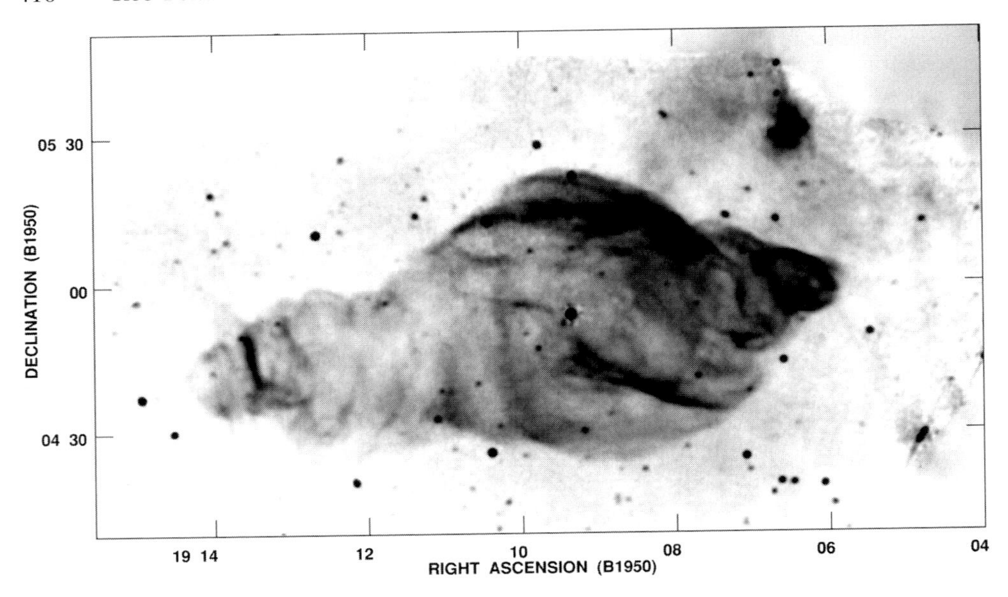

Fig. 9.15. The W50 nebula surrounding the powerful, quasi-continuous jet source SS 433, which seems to have been distorted by the action of the jets over thousands of years. From Dubner *et al.* (1998).

core at mildly relativistic velocities. The recurrent neutron star transient Cir X-1 is associated not only with an asymmetric arcsecond-scale radio jet but with an arcminute-scale radio nebula (Stewart *et al.* 1993; Fender *et al.* 1998) which, given the observed rapid timescale of radio variability from this source, can be unambiguously associated with *in-situ* particle acceleration, almost certainly powered by the jet. The nebula around Cir X-1 provides a good example of the use of such interaction zones as calorimeters – a simple "minimum energy" estimate indicates a total energy in the synchrotron emitting plasma of $>10^{48}$ erg, corresponding to, for example, three thousands years' action at 1% of the Eddington luminosity (see Heinz 2002 for further discussion).

Most spectacularly, the persistent, powerful binary jet source SS 433 powers the degree-scale W50 radio nebula (Fig. 9.15), within which are also located similar-scale X-ray jets (Brinkmann *et al.* 1996). Note that on smaller arcsecond, scales there is already evidence for reheating (Migliari *et al.* 2002), revealing that particle re-acceleration is not only present, but occurs repeatedly at different points in the flow.

In many, perhaps all, of these sources it now seems clear that a picture of a single finite phase of particle acceleration followed by monotonic fading as the source expands and propagates away from its launch site is far too simplistic. Multiple phases of particle accelerations due to shocks – whether internal or external – are perhaps instead the norm. While this necessarily complicates our understanding of the disk–jet coupling (particularly when the various physically distinct sites cannot be spatially resolved), it does, on the other hand, allow us to constrain the power of jets in radio-quiet phases such as the high/soft state. This follows because if these states were producing powerful jets, which for some reason (e.g. extreme Compton cooling) were not radio loud initially, we would still expect the signatures of subsequent shock accelerations to be found.

9.8 Relation to other jet sources

It is a common and useful exercise to compare accretion in X-ray binaries with the analogous processes in related systems, most commonly cataclysmic variables (CVs; see Chapter 10 by Kuulkers *et al.*). In the following I shall briefly compare X-ray binaries to other jet-producing systems. Figure 9.16 indicates schematically possible similarities and differences between jet formation in some of these different classes of object.

9.8.1 Active galactic nuclei

The name "microquasar" (Mirabel *et al.* 1992) clearly reflects the phenomenological similarities between jet-producing X-ray binaries and active galactic nuclei (AGN). Detailed quantitative comparisons are only just beginning to be made, and will no doubt be the subject of many future research papers. At the very roughest level, it is tempting to associate the (disputed) radio loud and radio quiet dichotomy observed in AGN with jet-producing (hard and transient) and non-jet-producing (soft) states in X-ray binaries (see, e.g., Maccarone *et al.* 2003). Furthermore perhaps FRI jet sources can be associated with the low/hard state and FRIIs with transients. Meier (1999; 2001) has considered jet production mechanisms in both classes of object, and drawn interesting parallels. Gallo *et al.* (2003; amongst others!) have made a qualitative comparison between FRIs and low/hard state black hole X-ray binaries and FRIIs and transients.

It is interesting to note that the short timescale disk–jet coupling observed in GRS 1915+105 (Pooley & Fender 1997; Eikenberry *et al.* 1998a; Mirabel *et al.* 1998; Klein-Wolt *et al.* 2001), in its most basic sense – that radio events are preceded by a "dip" and associated spectral hardening in the X-ray lightcurve – may also have an analog in AGN: Marscher *et al.* (2002) have reported qualitatively similar behavior in 3C 120.

Perhaps most exciting is the recent discovery that the power-law relation between radio and X-ray luminosities found for low/hard state BHCs (Corbel *et al.* 2001, 2003; Gallo *et al.* 2003; see Fig 9.9) may be directly relevant for the disk–jet coupling in AGN. Merloni *et al.* (2003) and Falcke *et al.* (2004) have both reported a "fundamental plane" of black hole activity describing the three-way correlation between mass, jet power and accretion power. This plane matches almost perfectly with the Gallo *et al.* (2003) relation between jet and accretion power once the mass term is taken into account, indicating truly similar physics across six to seven orders of magnitude in mass.

9.8.2 Gamma-ray bursts

While current observations allow that X-ray binary jets may on occasion achieve bulk Lorentz factors as large as those of the fastest AGN jets (Fender 2003), gamma-ray bursts (GRBs) appear to belong to another regime, with $\Gamma > 100$ (e.g., Baring & Harding 1997; Lithwick & Sari 2001). While the physics of jet interaction and emission may be similar, being based upon shock acceleration and the synchrotron process, the workings of the jet-producing engine in GRBs are so buried that it is hard to know how to make quantitative comparisons. Nevertheless, such comparisons should be attempted, and the differences between XRB transients, some of which reach super-Eddington rates, and GRBs, may not be as great as currently thought. Since in X-ray binaries we are fairly confident that to some degree the jet activity reflects that in the accretion flow, it may be conceivable that the (highly compressed) patterns of behavior in GRBs (originating in the jet) may reveal similarities with the slower black hole accretion processes observed in XRBs.

9.8.3 Other Galactic jet sources

X-ray binaries aside, there are multiple other sources of jets associated with "stellar"-scale objects within our Galaxy (and presumably others). However, in no other class of sources are there truly relativistic jets associated with accretion.

There are however non-relativistic jets associated with accretion in (at least) young stellar objects (YSOs; e.g., Lada 1985; Reipurth & Bally 2001) and super-soft sources (SSS; see Chapter 11 by Kahabka and van den Heuvel for a full description). The SSS can perhaps be most clearly compared to the X-ray binaries since they seem to be producing highly collimated jets as a result of high accretion rates in a binary (Cowley *et al.* 1998 and references therein), albeit at much lower velocities ($0.01c$ or less). These jets are revealed not by their radio emission but by twin optical/infrared lines originating from the jets (reminiscent of SS 433). The "symbiotic" binary CH Cyg is another interesting source of sub-relativistic jets associated with accretion. These jets *do* emit in the radio band, and may be precessing (Crocker *et al.* 2002); furthermore Sokoloski and Kenyon (2003) have reported a possible disk–jet coupling similar to that found in GRS 1915+105. Finally it is often noted that radio pulsars such as the Crab and Vela seem to be associated with (relativistic) jets and yet are not accreting (e.g., Blandford 2002).

A conclusion that has been drawn from the comparison of X-ray binaries with such diverse galactic objects is that the jet velocity is always comparable to the escape velocity of the accreting object (e.g., Livio 1999; Mirabel & Rodríguez 1999). However, while this seems to hold over the sub-relativistic and mildly relativistic regime, evidence for varying jet speeds from the same black hole, and for $\Gamma > 2$ flows from neutron stars seem to indicate that it may not be a hypothesis which can be extrapolated to the relativistic regime.

9.8.4 Ultraluminous X-ray sources

Ultraluminous X-ray sources are X-ray sources in external galaxies with apparent isotropic luminosities requiring black hole masses of $\sim 100\ M_{\odot}$ or more in order to remain sub-Eddington (i.e., at least a factor of a few more luminous than GRS 1915+105). There are at present three competing explanations for these sources, all involving accretion onto a black hole. If the radiation really is isotropic then "intermediate-mass black holes" are invoked (e.g., Colbert & Mushotzky 1999); alternatively the radiation may be anisotropically emitted from the accreting region (King *et al.* 2001) or relativistically aberrated due to for instance an origin in a jet (Körding *et al.* 2002; see also Georganopoulos *et al.* 2002). At the moment the nature of ULXs remains unclear (see more detailed discussion by King, Chapter 13).

An obvious prediction of the jet model would be radio counterparts to such sources, and there is tantalizing evidence that this may have recently been achieved. Dubus and Rutledge (2002) have suggested that the X-ray source M33 X-8 may be associated with a weak radio source; Kaaret *et al.* (2003) claim to have identified the radio counterpart to an ULX in NGC 5408. While these claims need confirmation, observations of the radio counterparts of such sources will surely provide strong clues to their intrinsic nature.

9.9 On the origin of jets

In this chapter, the observational properties of jets from X-ray binaries have been considered and some broad-ranging empirical relations have been established (most notably the association of jets with hard X-ray states). Such empirical connections require theoretical interpretation and the theory community has in recent years begun to rise to the task (motivated at least in part by a desire to use X-ray binary jets to explain those of AGN). There is certainly no room here to discuss these theoretical developments in detail, but it is worth pointing out some key relevant works.

Blandford and Payne (1982; see also, e.g., Ogilvie & Livio 2001) provided the groundwork for models in which magnetic fields rooted in an accretion flow may produce "radio" jets. The association of the low/hard state with ~steady jet formation has been interpreted by Meier (2001) and Meier *et al.* (2001) as strong evidence for MHD jet formation. In this scenario the strongest jets result from accretion flows with a large scale height, and so the jets are naturally suppressed in high/soft accretion states which are dominated by a geometrically thin accretion disk. Merloni and Fabian (2002) discuss "coronal outflow dominated accretion disks" in which they balance both accretion and outflow powers. Livio *et al.* (2003) have put forward a model in which the hard X-ray states of BHCs represent modes in which the bulk of the accretion energy is deposited into the bulk flow of a relativistic jet. Such "jet dominated" states may be empirically borne out by observations (Fender *et al.* 2003). Lynden-Bell (2003) discusses the formation by magnetized accretion disks of "towers" that can collimate jets. In all these theoretical models a magnetized accretion flow is the basis of a MHD outflow: given the widespread acceptance of the magneto-rotational instability (MRI) as the origin of accretion disk viscosity (e.g., Balbus & Hawley 1991; Turner *et al.* 2002), this highlights the probably key role of magnetic fields in the coupled accretion–outflow system (see, e.g., Kudoh *et al.* 2002 for a discussion of a possible relation between the MRI and jet formation). Figure 9.16 presents four different configurations of accretion with magnetic fields that may result in jet formation. In a rather different but still magnetically oriented approach Tagger and Pellat (1999; see also, e.g., Varniere & Tagger 2002), in the "accretion–ejection instability" model, have suggested that an instability related to the vertical component of magnetic field in the inner regions of accretion disks may result in the transport of energy and angular momentum away from the accretion flow, possibly powering a jet or wind. In this, and the related works of Das *et al.* (2003) and Nobili (2003), the jet should be intimately coupling to the timing properties of the accretor (of course this *is* already empirically observed to a certain extent since the different states of both BHCs and NS X-ray binaries have different timing properties).

As an alternative to magnetic acceleration, radiative acceleration (e.g., the "Compton Rocket" of O'Dell 1981) is unlikely to be able to push jets to the highest observed bulk velocities (Phinney 1982) but may still be operating, via line-locking, in the case of SS 433 (Shapiro *et al.* 1986).

Many variants on radiatively inefficient accretion flows are now beginning to consider outflows as part of their solutions (e.g., Narayan & Yi 1995; Blandford & Begelman 1999; Das 1999; Beckert 2000; Becker *et al.* 2001; Markoff *et al.* 2001). It remains to be seen which, if any, of these models comes closest to reproducing the observational characteristics of accretion onto black holes at a variety of rates, but note that numerical simulations of radiatively inefficient accretion flows also seem to form jets and outflows (Hawley & Balbus 2002). In fact, more than two decades ago Rees *et al.* (1982) discussed a likely connection between "ion-supported tori" (essentially advective flows) and the formation of radio jets. While Rees *et al.* (1982) were motivated by the study of AGN, they noted that "...relativistic jets collimated by tori around stellar-mass black holes may exist within our Galaxy."

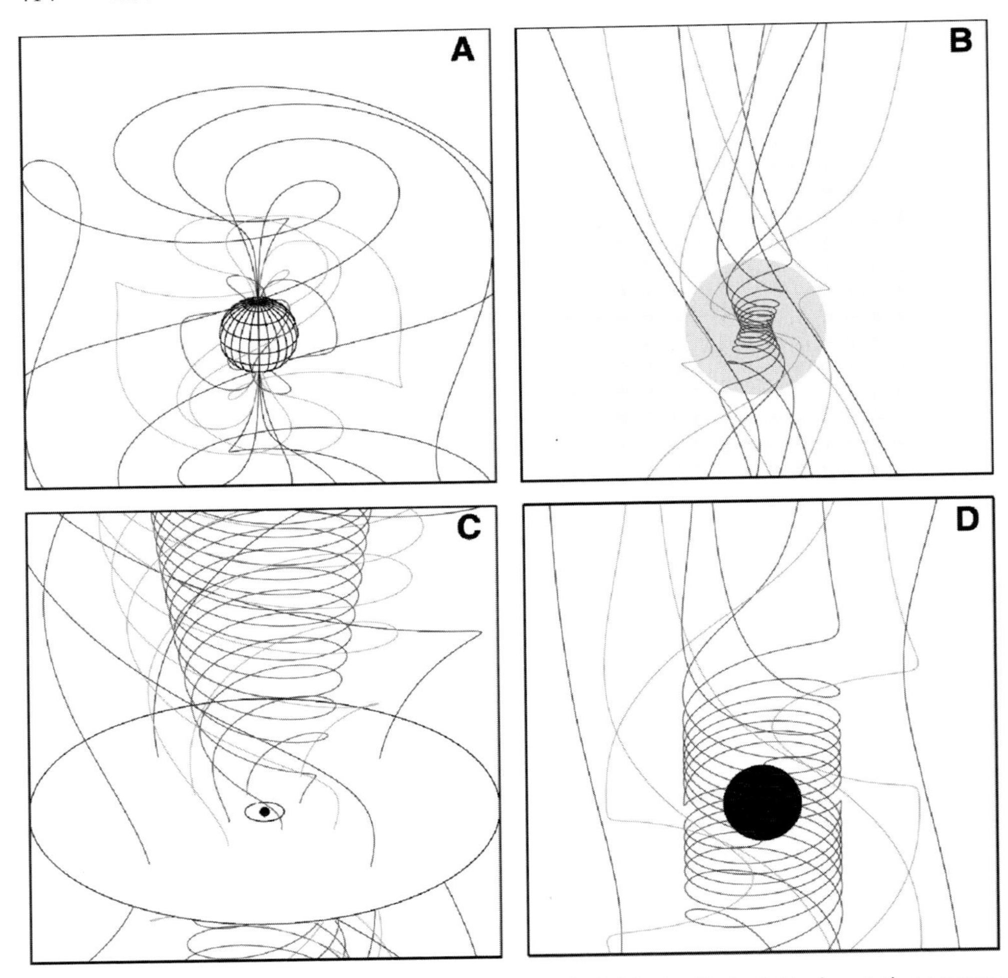

Fig. 9.16. Four ways to make jets with magnetic fields. **A:** dipole field of a rotating neutron star. **B:** A collapsing object drawing and winding up an initially uniform field. **C:** Poloidal magnetic field from a magnetized accretion disk. **D:** Frame-dragging near a rotating black hole resulting in strong coiling of the magnetic field lines. Types **C** and **D**, and possibly also **A**, may be relevant for X-ray binaries; type **A** for isolated pulsars; types **C** and **D** for AGN, and types **B**, **C** or **D** may be relevant for gamma-ray bursts. From Meier *et al.* (2001).

Are we ever going to be able to directly image the jet formation region in X-ray binaries? It seems unlikely – Junor *et al.* (1999) report the direct imaging of jet formation around the (low-luminosity) AGN M87 at a distance of ~ 100 Schwarzschild radii from the black hole. Comparing M87 to X-ray binaries in our own Galaxy, the ratio of distances is so much smaller than the ratio of black hole masses (and therefore Schwarzschild radii), that such imaging will not be possible. For example, a structure of size 100 Schwarschild radii around a 10 $M_{\odot}$ black hole at a distance of 5 kpc would have an angular size of $\sim 10^{-11}$ arcsec! Therefore the key to studying jet *formation* in X-ray binaries will remain in careful multiwavelength studies at the highest time resolution, such as those performed with such success on GRS 1915+105 (e.g., Mirabel *et al.* 1998; Klein-Wolt *et al.* 2002).

9.9.1 *On black hole spin*

It has been suggested both for AGN and for XRBs that the jets may in whole or in part be powered by the spin of the black hole (e.g., Blandford & Znajek 1977; Koide *et al.* 2002), although Livio *et al.* (1999) argue that the energy extracted from the black-hole in this way will never exceed that from the inner regions of the accretion disk. For the black holes in the low/hard state the apparently tight and universal correlation between X-ray and radio fluxes seems to indicate that either:

- Black-hole spin *is not* important for the formation of jets in the low/hard state of black holes. This may be natural if the jets are formed at large distances from the black hole.
- Black-hole spin *is* important, and all the binary black holes have about the same (dimensionless) spin parameter. This may be natural since they all originate in rotating massive stars (cf. radio pulsars).

In this context it may well be the case that even if black hole spin is not important for the low/hard state, it may well still be for the (transient) relativistic ejections that show a much greater scatter (although this may also be attributed to stronger beaming and less comprehensive coverage of lightcurves – Fender & Kuulkers 2001; Gallo *et al.* 2003). Furthermore, it should be noted that these conclusions are rather contrary to those of Cui *et al.* (1998) who conclude that (a) most binary black holes are only slowly spinning, (b) only rapidly spinning black holes produce radio jets.

9.10 Conclusions

In this review I have attempted to summarize our observational understanding of the phenomena of jets from X-ray binaries. In the process I have lightly, but no more, touched on various interpretations currently at large in the literature.

It is interesting to note that, whereas they were poorly investigated or understood one decade ago, these jets are now being considered as possible explanations for many exotic or high-energy phenomena. Not only do they clearly emit from hundreds of MHz to at least several keV, a range of 10^{10} in photon energy, but they may be important sites of particle acceleration in the ISM and even sources of neutrinos. One thing seems clear – they are *powerful* and need to be carefully considered when attempting to describe the physics of accretion onto compact objects. I have no doubt that the next decade will provide yet more excitement and surprises in this field. See Fender and Belloni (2004) and Fender *et al.* (2004) for the most recent (i.e. since this chapter was written) developments in the area of understanding the disk–jet coupling.

Acknowledgements

The author would like to thank Catherine Brocksopp, Stephane Corbel, Elena Gallo, Sebastian Heinz, Thomas Maccarone, Sera Markoff and Simone Migliari for a careful reading of the manuscript and numerous useful suggestions.

References

Abramowicz M. A., Kluzniak W., Lasota J.-P., 2002, *A&A*, **396**, L31
Angelini L., White N. E., 2003, *ApJ*, **586**, L71
Atoyan A. M., Aharonian F. A., 1999, *MNRASm*, **302**, 253
Balbus S. A., Hawley J. F., 1991, *ApJ*, **376**, 214
Ball K., Kesteven M. J., Campbell-Wilson D., Turtle A. J., Hjellming R. M., 1995, *MNRAS*, **273**, 722

Baring M. G., Harding A. K., 1997, *ApJ*, **491**, 663
Becker P. A., Subramanian P., Kazanas D., 2001, *ApJ*, **552**, 209
Beckert T., 2000, *ApJ*, **539**, 223
Belloni T., 1998, *New Astron. Rev.*, **42**, 585
Belloni T., Mèndez M., van der Klis M., *et al.*, 1996, *ApJ*, **472**, L107
Belloni T., Klein-Wolt M., Mèndez M., van der Klis M., van Paradijs J., 2000, *A&A*, **355**, 271
Belloni T., Migliari S., Fender R. P., 2000, *A&A*, **358**, L29
Bildsten L., *et al.*, 1997, *ApJ Supp. Ser.*, **113**, 367
Blandford R., 2002, in *Relativistic Flows in Astrophysics*, Springer Lecture Notes in Physics, vol. 589, p. 227
Blandford R. D., Begelman M. C., 1999, *MNRAS*, **303**, L1
Blandford R. D., Königl A., 1979, *ApJ*, **232**, 34
Blandford R. D., Payne D. G., 1982, *MNRAS*, **199**, 883
Blandford R. D., Znajek R. L., 1977, *MNRAS*, **179**, 433
Blandford R. D., McKee C. F., Rees M. J., 1977, *Nature*, **267**, 211
Bradshaw C. F., Geldzahler B. J., Fomalont E. B., 1997, *ApJ*, **481**, 489
Bradshaw C. F., Fomalont E. B., Geldzahler B. J., 1999, *ApJ*, **512**, L121
Brinkmann W., Aschenbach B., Kawai N., 1996, *A&A*, **312**, 306
Brocksopp C., Fender R. P., Larionov V., *et al.*, 1999, *MNRAS*, **309**, 1063
Brocksopp C., Jonker P. G., Fender R. P., *et al.*, 2001, *MNRAS*, **323**, 517
Burbidge G. R., 1959, *ApJ*, **129**, 849
Butt Y. M., Maccarone T. J., Prantzos N., 2003, *ApJ*, **587**, 748
Campana S., Stella L., 2000, *ApJ*, **541**, 849
Celotti A., Ghisellini G., 2003, *MNRAS*, submitted
Chen W., Shrader C. R., Livio M., 1997, *ApJ*, **491**, 312
Colbert E. J. M., Mushotzky R. F., 1999, *ApJ*, **519**, 89
Corbel S., Fender R. P., 2002, *ApJ*, **573**, L35
Corbel S., Fender R. P., Tzioumis A. K., *et al.*, 2000, *A&A*, **359**, 251
Corbel S. *et al.*, 2001, *ApJ*, **554**, 43
Corbel S., Fender R. P., Tzioumis A. K., *et al.*, 2002, *Science*, **298**, 196
Cowley A. P., Schmidtke P. C., Crampton D., Hutchings J. B., 1998, *ApJ*, **504**, 854
Crocker M. M., Davis R. J., Spencer R. E., *et al.*, 2002, *MNRAS*, **335**, 1100
Cui W., Zhang S. N., Chen W., 1998, *ApJ*, **492**, L53
Das T.K., 1999, *MNRAS*, **308**, 201
Das T. K., Rao A. R., Vadawale S. V., 2003, *MNRAS*, **343**, 443
Di Stefano C., Guetta D., Waxman E., Levinson A., 2002, *ApJ*, **575**, 378
Dhawan V., Mirabel I. F., Rodríguez L. F., 2000, *ApJ*, **543**, 373
Dubner G. M., Holdaway M., Goss W. M., Mirabel I. F., 1998, *AJ*, **116**, 1842
Dubus G., Rutledge R. E., 2002, *MNRAS*, **336**, 901
Eikenberry S. S., Matthews K., Morgan E. H., Remillard R. A., Nelson R. W., 1998a, *ApJ*, **494**, L61
Eikenberry S. S., Matthews K., Murphy T. W. Jr, *et al.*, 1998b, *ApJ*, **506**, L31
Eikenberry S. S., Cameron P. B., Fierce B. W., *et al.*, 2001, *ApJ*, **561**, 1027
Eikenberry S. S., Matthews K., Muno M., *et al.*, 2002, *ApJ*, **523**, L33
Esin A. A., McClintock J. E., Narayan R., 1997, *ApJ*, **489**, 865
Falcke H., Biermann P. L., 1996, *A&A*, **308**, 321
1999, *A&A*, **342**, 49
Falcke H., Körding E., Markoff S., 2004, *A&A*, **414**, 895
Fender R. P., 2001, *MNRAS*, **322**, 31
2003, *MNRAS*, **340**, 1353
Fender, R. P., Belloni, T., 2004, *ARA&A*, **42**, 317
Fender R. P., Hendry M. A., 2000, *MNRAS*, **317**, 1
Fender R. P., Kuulkers E., 2001, *MNRAS*, **324**, 923
Fender R. P., Pooley G. G., 1998, *MNRAS*, **300**, 573
Fender R. P., Pooley G. G., 2000, *MNRAS*, **318**, L1
Fender R. P., Bell Burnell S. J., Williams P. M., Webster A. S., 1996, *MNRAS*, **283**, 798
Fender R. P., Pooley G. G., Brocksopp C., Newell S. J., 1997, *MNRAS*, **290**, L65
Fender, R., Spencer, R., Tzioumis, T., *et al.*, 1998, *ApJ*, **506**, L21

Fender, R. P., Garrington, S. T., McKay, D. J., *et al.*, 1999a, *MNRAS*, **304**, 865
Fender R. *et al.*, 1999b, *ApJ*, **519**, L165
Fender R., Rayner D., Norris R., Sault R. J., Pooley G., 2000a, *ApJ*, **530**, L29
Fender R. P., Pooley G. G., Durouchoux P., Tilanus R. P. J., Brocksopp C., 2000b, *MNRAS*, **312**, 853
Fender R. P., Hjellming R. M., Tilanus R. P. J., *et al.*, 2001, *MNRAS*, **322**, L23
Fender R. P., Rayner D., McCormick G., *et al.*, 2002, *MNRAS*, **336**, 39
Fender R. P., Gallo E., Jonker P. G., 2003, *MNRAS*, **343**, L99
Fender, R. P., Belloni, T., Gallo, E., 2004, *MNRAS*, **355**, 1105
Fomalont E. B., Geldzahler B. J., 1991, *ApJ*, **383**, 289
Fomalont E. B., Geldzahler B. J., Bradshaw C. F., 2001a, *ApJ*, **553**, L27
2001b, *ApJ*, **558**, 283
Gaensler B. M., Stappers B. W., Getts T. J., 1999, *ApJ*, **522**, L117
Gallant Y. A., 2002, in *Relativistic Flows in Astrophysics*, Springer Lecture Notes in Physics, vol. 589, p. 24
Gallo E., Fender R. P., Pooley G. G., 2002, *Proc. 4th Microquasar Workshop: New Views on MICROQUASARS*, Eds. P. Durouchoux, Y. Fuchs, and J. Rodríguez. Center for Space Physics: Kolkata (India), p. 201.
2003, *MNRAS*, **344**, 60
Garcia M. R., Grindlay J. E., Molnar L. A., *et al.*, 1988, *ApJ*, **328**, 552
Garcia M. R., McClintock J. E., Narayan R., *et al.*, 2001, *ApJ*, **553**, L47
Garcia M. R., Miller J. M., McClintock J. E., King A. R., Orosz J., 2003, *ApJ*, **591**, 388
Georganopoulos M., Aharonian F. A., Kirk J. G., 2002, *A&A*, **388**, L25
Gierlinksi M., Zdziarski A. A., Poutanen J., *et al.*, 1999, *MNRAS*, **309**, 496
Gregory P. C., Neish C., 2002, *ApJ*, **580**, 1133
Gregory P. C. *et al.*, 1972, *Nature Phys. Sci.*, **239**, 114
Hannikainen D., Hunstead R. W., Campbell-Wilson D., 1998, *A&A*, **337**, 460
Hannikainen D. C., Hunstead R. W., Campbell-Wilson D., *et al.*, 2000, *ApJ*, **540**, 521
Hannikainen D., Campbell-Wilson D., Hunstead R., *et al.*, 2001, *Ap&SSS*, **276**, 45
Harmon B. A. *et al.*, 1995, *Nature*, **374**, 703
Hasinger G., van der Klis M., 1989, *A&A*, **225**, 79
Hawley J. F., Balbus S. A., 2002, *ApJ*, **573**, 738
Heinz S., 2002, *A&A*, **388**, L40
Heinz S., Sunyaev R., 2002, *A&A*, **390**, 751
2003, *MNRAS*, **343**, L59
Hjellming, R. M., Han, X., 1995, Radio properties of X-ray binaries. In *X-ray Binaries*, Eds. Lewin, W. H. G., van Paradijs, J., van der Heuvel, E. P. J. Cambridge: Cambridge University Press, 308–330
Hjellming R. M., Johnston K. J., 1981a, *Nature*, **290**, 100
1981b, *ApJ*, **246**, L141
1988, *ApJ*, **328**, 600
Hjellming, R. M., Rupen, M. P., 1995, *Nature*, **375**, 464
Hjellming R. M., Wade C. M., 1971a, *ApJ*, **164**, L1
1971b, *ApJ*, **168**, L21
Hjellming R. M., Gibson D. M., Owen F. N., 1975, *Nature*, **256**, 111
Hjellming R. M., Han X., Cordova F. A., Hasinger G., 1990a, *A&A*, **235**, 147
Hjellming R.M. *et al.*, 1990b, *ApJ*, **365**, 681
Hjellming R. M., Han X., Roussel-Dupre D., 1990c, *IAU Circ* 5112
Hjellming R. M. *et al.*, 2000a, *ApJ*, **544**, 977
Hjellming R. M., Rupen M. P., Mioduszewski A. J., Narayan R., 2000b, *ATel* 54
Homan J., van der Klis M., Wijnands R., Vaughan B., Kuulkers E., 1998, *ApJ*, **499**, L41
Homan J., Wijnands R., van der Klis M., *et al.*, 2001, *ApJS*, **132**, 377
Hughes P. A., 1991, (Ed.), *Beams and Jets in Astrophysics*, Cambridge: Cambridge University Press
Hynes R. I., Mauche C. W., Haswell C. A., *et al.*, 2000, *ApJ*, **539**, L37
Hynes R.I. *et al.*, 2003, *MNRAS*, **345**, 292
Jain R. K., Bailyn C. D., Orosz J. A., McClintock J. E., Remillard R. A., 2001, *ApJ*, **554**, L181
Junor W., Biretta J. A., Libio M., 1999, *Nature*, **401**, 891
Kaaret P., Corbel S., Tomsick J. A., *et al.*, 2003, *ApJ*, **582**, 945
Kaaret P., Corbel S., Prestwich A. H., Zezas A., 2003, *Science*, **299**, 365
Kaiser C. R., Hannikainen D. C., 2002, *MNRAS*, **330**, 225

Kaiser C. R., Sunyaev R., Spruit H. C., 2000, *A&A*, **356**, 975
Kanbach G., Straubmeier C., Spruit H. C., Belloni T., 2001, *Nature*, **414**, 180
Kaufman Bernado M. M., Romero G. E., Mirabel I. F., 2002, *A&A*, **385**, L10
King A. R., Davies M. B., Ward M. J., Fabbiano G., Elvis M., 2001, *ApJ*, **522**, L109
Klein-Wolt M., Fender R. P., Pooley G. G., *et al.*, 2002, *MNRAS*, **331**, 745
Koide S., Shibata K., Kudoh T., Meier D. L., 2002, *Science*, **295**, 1688
Körding E., Falcke H., Markoff S., 2002, *A&A*, **382**, L13
Kudoh T., Matsumoto R., Shibata K., 2002, *PASJ*, **54**, 121
Kuulkers E., Fender R. P., Spencer R. E., Davis R. J., Morison I., 1999, *MNRAS*, **36**, 919
Laan, H. van der, 1966, *Nature*, **211**, 1131
Lada C. J., 1985, *ARA&A*, **23**, 267
Levinson A., Blandford R., 1996, *ApJ*, **456**, L29
Lithwick Y., Sari R., 2001, *ApJ*, **555**, 540
Liu Q. Z., van Paradijs J., van den Heuvel E. P. J., 2001, *A&A*, **368**, 1021
Livio M., 1999, *Phys. Rep.*, **311**, 225
Livio M., Ogilvie G. I., Pringle J. E., 1999, *ApJ*, **512**, 100
Livio M., Pringle J. E., King A. R., 2003, *ApJ*, **593**, 184
Longair M. S., 1994, *High Energy Astrophysics, Volume 2 Stars, The Galaxy and the Interstellar Medium*, Cambridge: Cambridge University Press
Lynden-Bell D., 2003, *MNRAS*, **341**, 1360
McCollough M. *et al.*, 1999, *ApJ*, **517**, 951
Maccarone T. J., 2002, *MNRAS*, **336**, 1371
Maccarone T. J., 2003, *A&A*, **409**, 697
Macquart J. P., Wu K., Sault R. J., Hannikainen D. C., 2002, *A&A*, **396**, 615
Margon B., 1984, *ARA&A*, **22**, 507
Markoff S., Falcke H., Fender R. P., 2001, *A&A*, **372**, L25
Markoff S., Nowak M., Corbel S., Fender R., Falcke H., 2003, *A&A*, **397**, 645
Marscher A. P., Jorstad S. G., Gomez J-L., Aller M. F., Terasranta H., Lister M. L., Stirling A. M., 2002, *Nature*, **417**, 625
Marshall H. L., Canizares C. R., Schulz N. S., 2002, *ApJ*, **564**, 941
Martí J., Mirabel I. F., Rodríguez L. F., Chaty S., 1998, *A&A*, **332**, L45
Martí J., Paredes J. M., Peracaula M., 2001, *A&A*, **375**, 476
Martí J., Mirabel I. F., Rodríguez L. F., Smith I. A., 2002, *A&A*, **386**, 571
Martín E. L., Casares J., Molaro P., Rebolo R., Charles P., 1994, *ApJ*, **435**, 791
Mason K. O., Cordova F. A., White N. E., 1986, *ApJ*, **309**, 700
Meier D. L., 1999, *ApJ*, **522**, 753
2001, *ApJ*, **548**, L9
Meier D. L., Edgington S., Godon P., Payne D. G., Lind K. R., 1997, *Nature*, **388**, 350
Meier D. L., Koide S., Uchida Y., 2001, *Science*, **291**, 84
Mèndez M., Belloni T., van der Klis M., 1998, *ApJ*, **499**, L187
Merloni A., Fabian A. C., 2002, *MNRAS*, **322**, 165
Merloni A., Di Matteo T., Fabian A., 2000, *MNRAS*, **318**, L15
Merloni A., Heinz S., di Matteo T., 2003, *MNRAS*, **345**, 1057
Migliari S., Fender R. P., Mèndez M., 2002, *Science*, **297**, 1673
Migliari S., Fender R. P., Rupen M., *et al.*, 2003, *MNRAS*, **342**, L67
Miller J. M. *et al.*, 2002a, *ApJ*, **578**, 348
Miller J. M., Ballantyne D. R., Fabian A. C., Lewin W. H. G., 2002b, *MNRAS*, **335**, 865
Mioduszewski A. J., Rupen M. P., Hjellming R. M., Pooley G. G., Waltman E. B., 2001, *ApJ*, **553**, 766
Mirabel, I. F., Rodríguez, L. F., 1994, *Nature*, **371**, 46
Mirabel, I. F., Rodríguez, L. F., 1999, *ARA&A*, **37**, 409
Mirabel I. F., Rodríguez L. F., Cordier B., Paul J., Lebrun F., 1992, *Nature*, **358**, 215
Mirabel I. F., Bandyopadhyay R., Charles P. A., Shahbaz T., Rodríguez L. F., 1997, *ApJ*, **477**, L45
Mirabel I. F., Dhawan V., Chaty S., *et al.*, 1998, *A&A*, **330**, L9
Moore C. B., Rutledge R. E., For D. W., *et al.*, 2000, *ApJ*, **432**, 1181
Motch C., Ilovaisky S. A., Chevalier C., Angerbault P., 1985, *Space Sci. Rev.*, **40**, 219
Narayan R., Yi I., 1995, *ApJ*, **444**, 231
Nobili L., 2003, *ApJ*, **582**, 954

O'Dell, 1981, *ApJ*, **243**, L147
Ogilivie G. I., Dubus G., 2001, *MNRAS*, **320**, 4850
Ogilvie G. I., Livio M., 2001, *ApJ*, **553**, 1580
Orosz J. A. *et al.*, 2001, *ApJ*, **555**, 480
Owen F. N., Balonek T. J., Dickey J., Terzian Y., Gottesman S. T., 1976, *ApJ*, **203**, L150
Paredes J. M., Martí J., Ribó M., Massi M., 2000, *Science*, **288**, 2340
Paredes J. M., Ribó M., Ros E., Martí J., Massi M., 2002, *A&A*, **393**, L99
Penninx W., 1989, in *Proc. 23rd ESLAB Symposium on Two Topics in X Ray Astronomy. Volume 1: X Ray Binaries*, p. 185
Penninx W., Lewin W. H. G., Zijlstra A. A., Mitsuda K., van Paradijs J., 1988, *Nature*, **336**, 146
Phinney E. S., 1982, *MNRAS*, **198**, 1109
Pooley G. G., Fender R. P., 1997, *MNRAS*, **292**, 925
Pottschmidt K., Wilms J., Nowak M. A., *et al.*, 2000, *A&A*, **357**, 2000
Poutanen J., 1998, in *Theory of Black Hole Accretion Discs*, Eds. Abramowicz, M. A., Björnsson, G., Pringle, J. E., Cambridge: Cambridge University Press, p. 100
Priedhorsky W., Hasinger G., Lewin W. H. G., *et al.*, 1986, *ApJ*, **306**, L91
Rees M. J., 1966, *Nature*, **211**, 468
Rees M. J., Begelman M. C., Blandford R. D., Phinney E. S., 1982, *Nature*, **295**, 17
Reipurth B., Bally J., 2001, *ARA&A*, **39**, 403
Ribó M., Paredes J. M., Romero G. E., *et al.*, 2002, *A&A*, **384**, 954
Rodríguez L. F., Mirabel I. F., 1999, *ApJ*, **511**, 398
Romero G. E., Kaufman Bernado M. M., Mirabel I. F., 2002, *A&A*, **393**, L61
Rupen M., Hjellming R. M., Mioduszewski A. J., 1998, *IAU Circ*. 6938
2002, *New Views on MICROQUASARS, Proc. Fourth Microquasars Workshop*, Eds. P. Durouchoux, Y. Fuchs, and J. Rodriguez. Center for Space Physics: Kolkata (India), p. 213
Sams B., Eckart A., Sunyaev R., 1996, *Nature*, **382**, 47
Schnerr R. S., Reerink T., van der Klis M., *et al.*, 2003, *A&A*, **406**, 221
Shapiro P. R., Milgrom M., Rees M. J., 1986, *ApJS*, **60**, 393
Sokoloski J. L., Kenyon S. J., 2003, *ApJ*, **584**, 1021
Spencer R. E., 1979, *Nature*, **282**, 483
Spruit H. C., Kanbach G., 2002, *A&A*, **391**, 225
Stewart R. T., Caswell J. L., Haynes R. F., Nelson G. J., 1993, *MNRAS*, **261**, 593
Stirling A. M., Spencer R. E., de la Force C. J., *et al.*, 2001, *MNRAS*, **327**, 1273
Strickman M. S., Tavani M., Coe M. J., *et al.*, 1998, *ApJ*, **497**, 419
Tagger M., Pellat R., 1999, *A&A*, **349**, 1003
Tan J., Lewin W. H. G., Hjellming R. M., *et al.*, 1991, *ApJ*, **385**, 314
Tananbaum H., Gursky H., Kellogg E., Giacconi R., Jones C., 1972, *ApJ*, **177**, L5
Tingay, S. J. *et al.*, 1995, *Nature*, **374**, 141
Tomsick J. A., Corbel S., Fender R. P., *et al.*, 2003, *ApJ*, **592**, 933
Turner N. J., Stones J. M., Sano T., 2002, *ApJ*, **566**, 148
Vadawale S. V., Rao A. R., Chakrabarti S. K., 2001, *A&A*, **372**, 793
Varniere P., Tagger M., 2002, *A&A*, **394**, 329
Waltman E. B., Foster R. S., Pooley G. G., Fender R. P., Ghigo F. D., 1996, *AJ*, **112**, 2690
Wardle J. F. C., Homan D. C., 2001, in *Proc. Particles and Fields in Radio Galaxies*, Eds. Robert A. Laing and Katherine M. Blundell, ASP Conf. Ser. **250**, San Francisco: Astronomical Society of the Pacific, p. 152
Wardle J. F. C., Homan D. C., Ojha R., Roberts D. H., 1998, *Nature*, **395**, 457
Zdziarski A. A., Lubinski P., Gilfanov M., Revnivtsev M., 2003, *MNRAS*, **342**, 355

10

X-rays from cataclysmic variables

Erik Kuulkers
ESA/ESTEC, & Aurora Technology BV

Axel Schwope
Astrophysikalisches Institut Potsdam

Andrew Norton
The Open University

Brian Warner
University of Cape Town

10.1 Introduction

Cataclysmic variables (CVs) are a distinct class of interacting binaries, transferring mass from a donor star to a degenerate accretor, a white dwarf (WD). In all observational determinations, and as is required by theory for stable mass transfer, the donor star is of lower mass than the accretor. For comprehensive overviews on the subject of CVs we refer to Hack & La Dous (1993) and Warner (1995).

The majority of CVs have orbital periods, $P_{\rm orb}$, between 75 min and 8 h (see Ritter & Kolb 2003) and consist of Roche lobe-filling main sequence donors and WDs. These are WD analogues of the low-mass X-ray binaries (LMXBs; see Chapter 1). In the period range 8 h–3 d the donors must have larger radii than dwarfs in order to fill their Roche lobes and are therefore evolved subgiants. A few CVs are found with $P_{\rm orb} \sim 200$ d, which require giant donors for them to be lobe-filling. The absence of evolved CVs with periods $\sim$3 to $\sim$200 d is connected with the dynamical instability that results from an initial donor that had a mass larger than about 67% of that of the WD; such binaries will have experienced rapid mass transfer and shortened their periods during a common envelope phase (e.g., Iben & Livio 1993; see also Chapter 16). Beyond $P_{\rm orb} \sim 200$ d, mass-transferring systems also exist. These constitute the *symbiotic binaries* (SBs) and are in general not Roche lobe-filling, but instead consist of a WD orbiting in the wind of a supergiant, and are thus analogues of the high-mass X-ray binaries (see Chapters 1 and 5).

At the short end of the period range a different kind of CV exists, i.e., those in which the mass-losing donors are themselves WDs; they are entirely deficient in hydrogen. These helium-transferring CVs are known, after the type star, as *AM CVn stars* and are observed to have $P_{\rm orb}$ from 60 min down to at least as short as 10 min. They may have evolved by passage through two common-envelope phases, which leaves the cores of both of the component stars exposed (see, e.g, Podsiadlowski *et al.* 2003, and references therein).

The CVs are divided into subtypes. These were originally based entirely on the behavior seen in long-term optical lightcurves, but to this are now added more subtle parameters such as the presence of polarization. In essence, a CV's gross behavior is determined by the rate of mass transfer from the donor, $\dot{M}_{\rm donor}$, and the strength of the magnetic moment, μ, of the WD. Other parameters, such as $P_{\rm orb}$, mass ratio, and chemical abundances, have less effect; but the brightness and spectral variations on an orbital timescale can depend strongly on the inclination.

For $\mu \lesssim 10^{31}$ G cm^{-3} (corresponding to a magnetic field strength $B \lesssim 10^4$ G) the WD is essentially "non-magnetic" (but see Section 10.5.1) and mass is lost from the donor through a relatively narrow stream onto a so-called accretion disk, from where it spirals inwards until

Compact Stellar X-Ray Sources, eds. Walter Lewin and Michiel van der Klis.
Published by Cambridge University Press.

it arrives at the WD (see, e.g., Frank *et al.* 1992) without any significant magnetic influence on the fluid flow. Even for μ up to $\sim 10^{33.5}$ G cm^{-3} such an accretion stream and accretion disk can form, but the inner regions of the disk are removed by the magnetosphere of the WD, within which fluid flow is magnetically channeled. For larger μ no disk can form at all; the accretion stream from the donor couples onto field lines from the WD before the stream can circle around the WD. These latter two configurations, which apply for typical $\dot{M}_{\rm donor} \sim 10^{-10}$–$10^{-8}$ M$_\odot$ yr^{-1}, are known respectively as *intermediate polars* (IPs) for the systems with intermediate field strength, and *polars* (or *AM Her stars*, after the type star) for those with strong fields. The field in polars is so strong that it couples to the field of the donor and forces the WD to corotate with the binary; it also prevents the formation of an accretion disk. In IPs the WD does not corotate. The term *DQ Her star* is often used for IPs of rapid rotation, typically $\lesssim$250 s, but some use the term in place of IP, treating DQ Her as the type star.

Where disks exist there are two principal behaviors, resulting from the different viscosities (and hence ability to transport angular momentum) in cool disks and hot disks (see Lasota 2001; Chapter 13). For low $\dot{M}$ through the disk, the disk is lower in temperature and the viscosity is too low to transport mass through the disk as fast as it arrives from the donor; this state is referred to as *quiescence*. The quiescent disk (which is not stationary) therefore acts as a reservoir of gas, and when a critical density is reached it becomes optically thick, heats up and increases in viscosity, and rapidly transfers gas onto the WD. The resultant release of the gravitational energy of the stored gas gives rise to *dwarf nova* (DN) outbursts, during which the optical luminosity is much enhanced. The DN class is subdivided into *U Gem stars* which have outbursts that last for a few days and recur on timescales of weeks, and *Z Cam stars* where similar outbursts are occasionally interrupted by a standstill at a brightness intermediate between outburst and quiescence. In DN systems with short $P_{\rm orb}$, called *SU UMa stars*, an additional phenomenon occurs – *superoutbursts*, which are typically brighter and last about five times as long as normal outbursts, during which humps in the lightcurves (so-called "superhumps") are present with a period of a few % longer than $P_{\rm orb}$. Superoutbursts are thought to be due to tidal stresses between the outer disk and donor, adding to the higher viscous stress in the outer disk; it may even result in additional mass transfer from the donor.

Outbursts, usually of lower amplitude and short duration compared to the DNe, can arise in the truncated disks of IPs. On the other hand, if $\dot{M}_{\rm donor}$ is high enough it can maintain the disk at high viscosity, producing an equilibrium state that does not undergo normal DN outbursts (though modified outbursts are sometimes seen). These systems are known, from the appearance of their spectra, as *nova-like* systems (NLs). Those showing absorption lines are called *UX UMa stars*, and those showing emission lines are *RW Tri stars*.

The strongly magnetic systems have no disks and therefore lack outbursts. But it is common for polars to show states of low luminosity caused by lowering of $\dot{M}_{\rm donor}$ by as much as two or three orders of magnitude. Other types also show (occasional) low states, especially NLs with $3 < P_{\rm orb}({\rm h}) < 4$, which are known as *VY Scl stars*.

It is supposed that all CVs, with the exception of the helium-transferring AM CVn stars, undergo thermonuclear runaways at the base of the accreted hydrogen-rich layers on the WDs as soon as the layers are massive enough (typically $\sim 10^{-4}$ M$_\odot$). These produce *classical nova* (CN) eruptions and are in some ways the equivalent of X-ray bursts in LMXBs (see Chapter 3). Systems that have shown nova eruptions more than once are referred to as *recurrent novae* (RNe), whereas SBs with nova eruptions are known as *symbiotic novae* (SBNe). Note that some SBNe also recur.

The X-ray behaviors are correlated with the above described optical behavior of the various CV subtypes, but not always positively. During a DN outburst hard X-rays[1] have been seen to increase at the beginning (often with a delay of up to a day after the optical outburst begins), but then are suppressed until near the end of the outburst. Similarly, the high $\dot{M}$ (and, therefore, high accretion luminosity) NLs have relatively low X-ray luminosities, L_X. On the other hand, soft X-ray[1] fluxes are greatly enhanced during DN outbursts, again with a possible delay with respect to the optical, which is caused by the time taken for an outburst to travel from the cooler outer parts of the disk where it started to the inner disk and the WD/disk boundary layer (BL).

The polars have high soft L_X and the IPs have relatively high hard L_X. All of these different X-ray behaviors are simply connected with the optical depth of the BL in the non-magnetic CVs and with the nature of the channeled accretion flow in the magnetic CVs. More details are given in subsequent sections.

X-rays may also be generated in other parts of CV structures: e.g., in the shock waves where gas ejected by nova eruptions meets the interstellar medium; on the hot WD surface after a nova eruption; and in minor contributions from the magnetically active regions on the surface of the donor.

SS Cyg, one of the optically brightest DNe ($m_V \sim 12$–8 mag), was the first CV to be detected in X-rays during a rocket flight when it was in outburst (Rappaport *et al.* 1974). The spectrum was soft with a blackbody temperature, $kT_{bb} < 130$ eV. In its quiescent state it was first detected by ANS, both in soft and hard X-rays. In the soft band, its flux was only a small percentage of that observed during outburst (Heise *et al.* 1978). Since then many more CVs have been detected in X-rays.

We here review X-ray observations of CVs, with some emphasis on what has been achieved in the decade up to 2003. For earlier, more general, reviews we refer the interested reader to, e.g., Córdova & Mason (1983), Hack & La Dous (1993), Córdova (1995) and Warner (1995). We note that many CVs have recently been found in globular clusters; we refer to Chapter 8 for an overview of this subject.

10.2 X-ray emission from non-magnetic CVs

10.2.1 General properties

The collective X-ray properties of CVs have been the subject of several studies using observations with Einstein (Becker 1981; Córdova & Mason 1983, 1984; Patterson & Raymond 1985a; Eracleous *et al.* 1991a,b), ROSAT (Vrtilek *et al.* 1994; van Teeseling & Verbunt 1994; Richman 1996; van Teeseling *et al.* 1996; Verbunt *et al.* 1997), and EXOSAT/ME (Mukai & Shiokawa 1993). For earlier reviews on the X-ray emission from non-magnetic CVs we refer to, e.g., Verbunt (1996) and Mukai (2000). With a few exceptions all non-magnetic CVs radiate at X-ray flux levels $\lesssim 10^{-11}$ erg cm^{-2} s^{-1} (e.g., Patterson & Raymond 1985a (0.2–4 keV); Eracleous *et al.* 1991a (0.1–3.5 keV; 2–10 keV); Mukai & Shiokawa 1993 (2–10 keV); Richman 1996 (0.1–2.4 keV)). This translates to, generally,

[1] With "soft" and "hard" X-rays we refer to X-rays with energies of order 10 eV and of order keV, respectively, unless otherwise noted.

$L_X \simeq 10^{29}$–10^{32} erg s^{-1}. All the CVs below 10^{30} erg s^{-1} are short-period DNe or low-state magnetic CVs (Verbunt *et al.* 1997).

Among the non-magnetic CVs the ratio of the X-ray flux to optical and/or UV flux, F_X/F_{opt}, decreases along the sequence SU UMa stars ($F_X/F_{opt} \sim 0.1$) – U Gem stars – Z Cam stars ($F_X/F_{opt} \sim 0.01$) – UX UMa stars ($F_X/F_{opt} \lesssim 10^{-3}$), due mainly to variations in the optical/UV flux (Verbunt *et al.* 1997; see also van Teeseling & Verbunt 1994; van Teeseling *et al.* 1996). There are a few exceptions, however, such as the double degenerate AM CVn systems (Section 10.2.7). We note that for magnetic CVs F_X/F_{opt} is comparable to that of the SU UMa stars (Verbunt *et al.* 1997). The general pattern is in agreement with Patterson and Raymond (1985a; see also Richman 1996) who find that non-magnetic CVs with high $\dot{M}$ show low F_X/F_{opt}, and in agreement with the fact that F_X/F_{opt} is seen to decrease with increasing P_{orb} (van Teeseling & Verbunt 1994; van Teeseling *et al.* 1996; see also Córdova & Mason 1984).[2] This latter correlation stems from the fact that the UV flux is a strongly increasing function of P_{orb}, which in turn is likely related to $\dot{M}$: a high $\dot{M}$ apparently causes the disk to emit more UV flux, but not more X-ray flux (see, e.g., van Teeseling *et al.* 1996). This general pattern is somewhat perturbed, however, by the anti-correlation between the inclination, i, and the observed X-ray flux (van Teeseling *et al.* 1996; see also Patterson & Raymond 1985a).

An empirical relation between the equivalent width, EW, of the optical Hβ emission line and F_X/F_{opt} exists (Patterson & Raymond 1985a; Richman 1996). This relation predicts F_X/F_{opt} to within a factor of 3. The correlation of F_X/F_{opt} with EW(Hβ) is thought to also reflect an underlying correlation with $\dot{M}$, since EW(Hβ) is known to correlate with the absolute visual magnitude of the disk, which in turn is correlated with $\dot{M}$ (e.g., Patterson 1984; Warner 1995). Thus low $\dot{M}$ systems produce strong Hβ emission lines and a larger F_X/F_{opt}.

The absorption column densities, N_H, as derived from X-ray spectral fits are generally in the range 10^{20}–10^{21} cm^{-2} (e.g., Eracleous *et al.* 1991a; Richman 1996). VW Hyi has one of the lowest values of N_H for any CV (6×10^{-17} cm^{-2}; Polidan *et al.* 1990), which makes it an ideal CV to study in the EUV and soft X-ray range. Comparison of the N_H values with the color excess E_{B-V} derived from the 2200 Å feature (Verbunt 1987), shows that N_H is often higher than predicted on the basis of the average relation derived by Predehl and Schmitt (1995). The excess column may be related to absorbing gas in the CV itself, which can be responsible for some of the orbital variations seen, especially since the CVs displaying highly absorbed spectra are known to have a high inclination (e.g., Eracleous *et al.* 1991a; Verbunt 1996; see also Sections 10.2.3–4).

The X-ray spectral flux distributions within the 0.5–2.5 keV band seem to be fairly similar for most CVs (Verbunt *et al.* 1997), although it appears that SU UMa and UX UMa stars have somewhat softer spectra than other DNe and VY Scl stars in their high state (van Teeseling *et al.* 1996). Individual systems, however, may show significant epoch-to-epoch variability both in luminosity and temperature (e.g., Mukai & Shiokawa 1993). The low signal-to-noise and low resolution X-ray spectra of non-magnetic CVs obtained with early X-ray satellites were generally well described by a bremsstrahlung model (including simple absorption) with typical temperatures in the range 1–5 keV (e.g., Córdova *et al.* 1981; Córdova & Mason

[2] We caution, however, that the measurements for CVs with short orbital periods are biased towards quiescent systems (i.e., CVs with low $\dot{M}$). X-ray observations of high $\dot{M}$ short orbital period CVs could resolve the issue.

1983, 1984; Eracleous *et al.* 1991a; Mukai & Shiokawa 1993). Quiescent DNe are hard X-ray sources with (bremsstrahlung) temperatures of a few keV (e.g., Patterson & Raymond 1985a; Vrtilek *et al.* 1994) up to ~10 keV (e.g., Córdova & Mason 1983). However, systematic residuals in the fitted X-ray spectra observed with, e.g., the Einstein/IPC and ROSAT/PSPC indicated that the X-ray emission is not well described by the single-temperature models, both for low and high $\dot{M}$ CVs, but instead must be described by a range of temperatures (e.g., Eracleous *et al.* 1991a; Richman 1996). For more details, see Sections 10.2.5, 10.3.2 and 10.4.4.

If one assumes that the observed X-ray emission comes from a single-temperature model, the emission measure (i.e., the amount of emitting material), EM,[3] can be determined from the observed photon flux. The EM is not a strong function of $\dot{M}$ (van Teeseling *et al.* 1996). It decreases, however, for CVs with higher i. For CVs with $i < 70°$ this cannot be due to obscuration of the X-ray source by matter in the outer parts of the disk or by the donor. The anti-correlation between EM and i excludes models in which the X-rays are emitted in a relatively large optically thin volume. On the other hand, if the X-rays originate from the inner part of the disk and the scale height of the optically thin X-ray source is not much higher than the disk thickness (as eclipse observations of quiescent DN suggest, see Section 10.2.3), EM could depend on i. Van Teeseling *et al.* (1996) come to the conclusion that in high-inclination systems most of the X-ray flux is absorbed by the disk.

So, what is the origin of the X-ray emission in non-magnetic CVs? In the next section we describe the major source of X-rays in these systems, the boundary layer.

10.2.2 The boundary layer model

In non-magnetic CVs the accretion is governed by the disk. Basic theory predicts that half of the gravitational potential energy of the accreting material is liberated through the viscosity in the disk, while the other half is liberated in a boundary layer (BL) between the disk and the surface layer of the WD (e.g., Shakura & Sunyaev 1973; Lynden-Bell & Pringle 1974; Pringle 1981). Material in the BL moves with Keplerian speeds and collides with the WD which is generally rotating more slowly than break-up velocity. This results in luminosities of the disk and BL of $L_{\rm disk} \simeq L_{\rm BL} \simeq GM_{\rm WD}\dot{M}/2R_{\rm WD}$, where $M_{\rm WD}$ and $R_{\rm WD}$ are the mass and radius of the WD, respectively. For a WD with $M_{\rm WD} = 1\,{\rm M}_\odot$, $R_{\rm WD} = 10^9$ cm and $\dot{M} = 10^{-10}\,{\rm M}_\odot\,{\rm yr}^{-1}$ this amounts to about 4×10^{32} erg s^{-1}. The disk is generally too cool ($kT < 1$ eV) to emit X-rays. It radiates mostly at optical and ultraviolet (UV) wavelengths. The BL, on the other hand, mostly radiates in the extreme ultraviolet (EUV) and in X-rays (e.g., Bath *et al.* 1974b). However, with the modest luminosities the X-rays do not strongly influence the disk, as they do in bright LMXBs.

When $\dot{M}$ is low, such as in DNe in quiescence, the BL is observed to be optically thin. Shocks heat the gas to a temperature of about the virial temperature, $kT_{\rm vir} = GM_{\rm WD}m_{\rm H}/6kR_{\rm WD} \sim 20$ keV. When $\dot{M}$ is high, such as in DNe in outburst or in NLs, the BL is observed to be optically thick. Cooling of the BL is efficient and the X-ray spectrum is thermalized with an approximate blackbody temperature of $kT_{\rm bb} = (GM_{\rm WD}\dot{M}/8\pi\sigma R_{\rm WD}^3)^{1/4} \sim 10$ eV and $L_{\rm X} > 10^{34}$ erg s^{-1} (e.g., Pringle 1977; Pringle & Savonije 1979; Tylenda 1981; Narayan & Popham 1993, Popham & Narayan 1995). The critical $\dot{M}$, $\dot{M}_{\rm crit}$, generally depends on $M_{\rm WD}$ and the viscosity in the disk, and is about $10^{-10}\,{\rm M}_\odot\,{\rm yr}^{-1}$.

[3] Defined as EM $= \int n_e^2\,{\rm d}V$, where n_e is the electron density and V the emitting volume.

Although X-ray observations of CVs are often interpreted within the above described framework of the standard BL model (e.g., many quiescent DNe were found to be modest hard X-ray sources), evidence that BLs really exist in CVs is generally indirect. In the next sections we provide the observational efforts to find this BL, both in quiescence and outburst of DNe, as well as in other high-$\dot{M}$ CVs.

10.2.3 *Quiescent dwarf novae*

For CVs with $\dot{M} < \dot{M}_{\rm crit}$ Patterson and Raymond (1985a) show that the hard X-ray data from Einstein are generally consistent with hot optically thin emission from the BL. The observed temperatures are in the range expected. However, most of these CVs show less BL radiation than predicted (e.g., Pringle *et al.* 1987; Belloni *et al.* 1991; van Teeseling & Verbunt 1994; Vrtilek *et al.* 1994). This also holds for high $\dot{M}$ systems (see Section 10.2.4), and it is referred to as "the mystery of the missing BL" (see Ferland *et al.* 1982). Different explanations for this lack in quiescence have been suggested: disruptions of the inner disk by magnetic fields (Livio & Pringle 1992; Lasota *et al.* 1995; Warner *et al.* 1996); coronal siphon flows (Meyer & Meyer-Hofmeister 1994; Lasota *et al.* 1995); irradiation by the (relatively) hot WD (King 1997); a rapidly rotating WD (e.g., Ponman *et al.* 1995); a rapidly spinning accretion belt (Sion *et al.* 1996); reflection effects and cooling flows (Done & Osborne 1997). Alternatively, the BL largely radiates an additional *very* soft component ($kT \lesssim 10$ eV) which would remain undetectable due to interstellar absorption (e.g., Patterson & Raymond 1985b).

High-inclination CVs provide an opportunity to locate the X-ray emitting regions. When the X-ray source is eclipsed one can constrain its size and location using the orbital phase, $\phi_{\rm orb}$, and the duration of ingress and egress. One factor complicating such observations, however, is that eclipsing CVs tend to be fainter in X-rays than low-inclination CVs, so eclipse studies have been rather count-rate limited (e.g., van Teeseling *et al.* 1996).

X-ray eclipses have been seen during quiescence of the DNe HT Cas (Mukai *et al.* 1997; $i \simeq 81°$), Z Cha (van Teeseling 1997a; $i \simeq 82°$) and OY Car (Pratt *et al.* 1999a; Ramsay *et al.* 2001a; Wheatley & West 2002; $i \simeq 83°$), as well as during a low state in quiescence of HT Cas (Wood *et al.* 1995a). They all occur at the time of the optical WD eclipse. The X-ray ingress and egress are rapid; in OY Car their duration is significantly shorter than in the optical (30 ± 3 s vs. 43 ± 2 s; Wheatley & West 2002). This is consistent with the indication that the total X-ray eclipse in HT Cas has a slightly shorter duration than the optical one (by about 27; Mukai *et al.* 1997). If one assumes that the rapid optical ingress/egress times represent the contact points of the WD with the donor, then the different durations suggest that the X-ray emitting region must be smaller than the WD. The X-rays possibly originate from a broad equatorial belt of which the lower half is absorbed (Mukai *et al.* 1997; van Teeseling 1997a; Ramsay *et al.* 2001a; Wheatley & West 2002). This also explains the observed anti-correlation between i and the EM (van Teeseling *et al.* 1996; see Section 10.2.1). Since the WD is too cool to produce X-rays, a BL must be responsible for the out-of-eclipse X-ray emission (see Mukai *et al.* 1997).

In eclipse there is residual X-ray emission at about 1% of the out-of-eclipse flux, with $L_{\rm X} \sim 3 \times 10^{28}$ erg s^{-1} and a soft spectrum ($kT \sim 1$ keV). The fact that this emission in eclipse is softer than that seen out of eclipse seems to rule out the possibility that the residual flux is BL emission scattered into our line of sight by circumstellar material (Wheatley & West 2002). The luminosity (e.g., Rosner *et al.* 1985; Hempelmann *et al.* 1995), as well as the temperature (e.g., Schmitt *et al.* 1990), are consistent with coronal emission from a cool main-sequence donor. Ramsay *et al.* (2001a), however, argue that the residual emission may come from a weak remnant of a large corona, which is more prominent during outburst (see Section 10.2.4).

Apart from eclipses, dips in the X-ray lightcurves have been observed, up to ~50% deep, during quiescence in U Gem (at $\phi_{orb} \sim 0.3$ and 0.8: Szkody *et al.* 1996, 2000a; $i \sim 65°$), Z Cha ($\phi_{orb} \sim 0.7$–0.8: van Teeseling 1997a), WZ Sge ($\phi_{orb} \sim 0.7$: Patterson *et al.* 1998; $i \sim 75°$), and OY Car ($\phi_{orb} \sim 0.2$–0.5: Ramsay *et al.* 2001a). They are only apparent at low X-ray energies, which indicates absorption effects. Similar kinds of dips have also been found during outburst (see Section 10.2.4).

In order for the dips to be visible at inclinations such as in U Gem the material must be located far from the orbital plane (e.g., Mason *et al.* 1988; Naylor & La Dous 1997). Note that in quiescence the X-ray dips in U Gem were less deep than during outburst. This means that the X-ray emitting region must be only slightly larger than the outburst BL and that the absorbing material that was present at outburst must maintain a similar location in quiescence. The small residual X-ray flux seen may be scattered into the line of sight from high above the plane (by a disk corona or a wind; e.g., Naylor & La Dous 1997; Mason *et al.* 1997; see also Section 10.2.5) or may possibly originate from a hot corona of the donor (Wood *et al.* 1995b). HST observations of OY Car in quiescence show that the UV emission from the WD surroundings is also absorbed by matter above the disk, which is referred to as an "iron curtain" (see Horne *et al.* 1994). However, this curtain does not always seem to exist (Pratt *et al.* 1999a). Note that no dips were found in the quiescent UV lightcurves of OY Car (Ramsay *et al.* 2001a).

Similar kinds of absorption dips have also been seen in IPs (see Section 10.4), as well as in LMXBs (e.g., Mason 1986; Parmar & White 1988; White *et al.* 1995; Kuulkers *et al.* 1998, and references therein; see also Chapter 1). Note that column densities of $\lesssim 10^{19}$ cm^{-2} to $\sim 10^{22}$ cm^{-2} in CVs (e.g., Naylor & La Dous 1997) are sufficient to extinguish soft X-ray emission during the dips, whereas maximum column densities of $> 10^{23}$ cm^{-2} are typically recorded in LMXBs.

A popular model for the dips is the one outlined by Frank *et al.* (1987). They explain the dips as the interaction of the accretion stream with the disk, which splashes material out of the plane to form cool clouds that obscure the radiation produced close to the compact object (see also Armitage & Livio 1996, 1998; Kunze *et al.* 2001). For CVs they predict a single broad dip between phases 0.6 and 0.8, exactly as observed in, e.g., U Gem.

10.2.4 Outbursting dwarf novae and other high-$\dot{M}$ CVs

For most CVs in quiescence $\dot{M}$ onto the WD is of the order of 10^{-12}–10^{-11} M$_\odot$ yr^{-1} (e.g., Patterson 1984; Warner 1995). During an outburst $\dot{M}$ increases by ~2 orders of magnitude, so the disk is likely to cross $\dot{M}_{crit}$. The CV is then expected to change from a hard to a soft X-ray emitter. However, the situation appears to be not that simple, as we will show below.

Soft X-rays have been detected during outbursts of SS Cyg (e.g., Rappaport *et al.* 1974; Mason *et al.* 1978; Córdova *et al.* 1980b; Jones & Watson 1992; Ponman *et al.* 1995), U Gem (e.g., Córdova *et al.* 1984), VW Hyi (van der Woerd *et al.* 1986; Mauche *et al.* 1991; van Teeseling *et al.* 1993; Wheatley *et al.* 1996b), SW UMa (Szkody *et al.* 1988), and Z Cam (Wheatley *et al.* 1996a). Other high-$\dot{M}$ CVs generally do not show the soft component (e.g., Silber *et al.* 1994; van Teeseling *et al.* 1995). When a soft component is present the

X-ray spectra show $kT_{\rm bb} \sim 5$–30 eV; these temperatures are similar to the BL temperatures derived from high resolution EUV and X-ray spectra (Mauche *et al.* 1995: SS Cyg; Long *et al.* 1996: U Gem; Mauche 1996b: VW Hyi; Mauche & Raymond 2000: OY Car; see also Section 10.2.5). Note that not all of the soft component is optically thick (Mauche *et al.* 1995; Long *et al.* 1996). The soft X-ray fluxes increase by a factor of ~100 from quiescence to outburst. However, they are still too low compared to the simple BL models (e.g., Mauche *et al.* 1991; van Teeseling *et al.* 1993; van Teeseling & Verbunt 1994; Ponman *et al.* 1995; Wheatley *et al.* 1996b), similar to the discrepancy seen in quiescence (see Section 10.2.3). A study of the ionization states inferred from the P Cygni lines arising in winds from high $\dot{M}$ CVs led to a similar conclusion (Drew & Verbunt 1985; Hoare & Drew 1991). Various explanations for the discrepancy (or absence) of soft X-ray flux during outburst have been put forward: differences in $N_{\rm H}$ to different systems (e.g., Patterson & Raymond 1985b; Long *et al.* 1996); differences in $M_{\rm WD}$ and the WD rotation (see below); absorption in the disk wind (Jensen 1984; Kallman & Jensen 1985); energy loss in the form of a wind (e.g., Silber *et al.* 1994; Ponman *et al.* 1995). Moreover, changes in the BL temperature can shift most of the flux out of the soft X-ray bandpass (e.g., Córdova *et al.* 1980a; Patterson & Raymond 1985b).

Our knowledge of the evolution of the spectral flux distribution during outbursts of DNe at various wavelengths is mainly based on fragmented (nearly) simultaneous observations. A few dedicated campaigns do exist, however (see, e.g., Pringle *et al.* 1987; Wheatley *et al.* 1996b; Szkody 1999, and references therein). One of the most complete coverages to date of a DN outburst is that of SS Cyg (Mauche & Robinson 2001; Wheatley *et al.* 2000; see Fig. 10.1). We here describe the general behavior seen at EUV and X-ray wavelengths in outbursting DNe.

The soft X-rays lag the optical outburst lightcurve by about 12–36 h during the rise (e.g., Jones & Watson 1992; Mauche & Robinson 2001). This is comparable to that measured in the far-UV ($\lesssim$10 eV; Polidan & Holberg 1984). Wheatley *et al.* (2000) found that the X-ray outburst of SS Cyg started ~18 h *before* the EUV one. The start of the X-ray outburst is marked by a sudden softening of the X-ray spectrum; the rise to soft X-ray maximum is rapid (e.g., Wheatley *et al.* 2000). After reaching maximum early in the outburst, the soft X-ray flux rapidly decreases again (but less fast than the rise). The decrease is more rapid towards shorter wavelengths. The soft X-rays lead the optical lightcurve during the decline, and disappear before the end of the optical outburst (e.g., van der Woerd *et al.* 1986; Mauche & Robinson 2001). The soft X-ray rise and decay times are shorter with respect to the optical (e.g., van der Woerd *et al.* 1986; Jones & Watson 1992).

The initial soft X-ray rise could be the arrival of the heating wave through the disk at the BL, and the sudden spectral softening is as expected in the BL models. The rapid rise time may represent the timescale of the transition between optically thin and thick emission; the less rapid drop at the end may represent the timescale of the inverse process (e.g., Jones & Watson 1992; Wheatley *et al.* 2000).

At the time the soft X-rays appear, the hard X-ray flux is suppressed (e.g., Wheatley *et al.* 2000; Baskill *et al.* 2001). They do not disappear, however. They stay present during the outburst, with somewhat lower temperature and flux than in quiescence. This may be attributed to a density gradient in the optically thick BL, such that there is always a hot optically thin layer which emits hard X-rays (e.g., Patterson & Raymond 1985a; Done & Osborne 1997). The anti-correlation between the soft and hard X-ray flux suggests that we

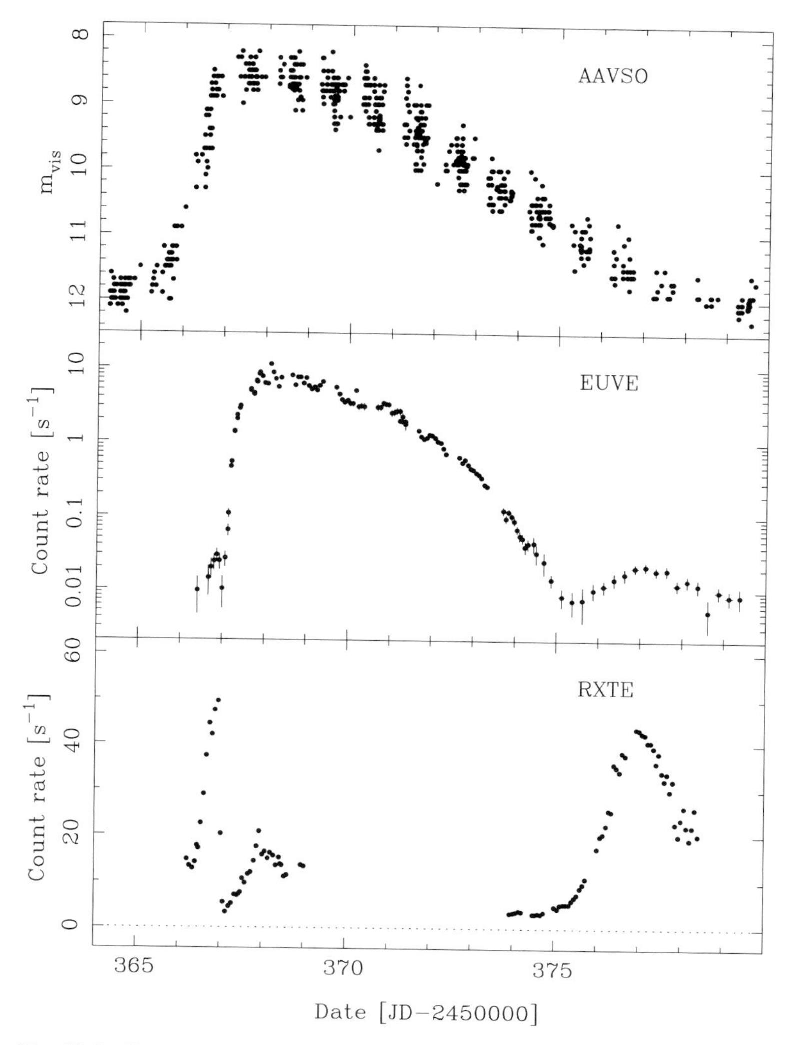

Fig. 10.1. Simultaneous optical (visual and V-band measurements reported to the AAVSO (American Association of Variable Star Observers)), soft X-ray (EUVE) and hard X-ray (RXTE) observations of SS Cyg throughout outburst. Note that the RXTE lightcurve is plotted on a linear scale in order to emphasize the timing of the sharp transitions. From Wheatley *et al.* (2003).

see two physically distinct emission components. The coincidence in the timing shows that they are related, however, and possibly mark the time at which the BL becomes optically thick. The hard X-ray flux during outburst is considerably lower than the soft X-ray flux (e.g., $\sim$0.1% during a superoutburst of VW Hyi; van der Woerd *et al.* 1986) and it generally declines throughout the outburst (e.g., Verbunt *et al.* 1999). The temperature of the hard X-ray component increases from outburst to quiescence (e.g., Hartmann *et al.* 1999). The decline in flux and increase in temperature probably reflects the (slowly) decreasing $\dot{M}$ on the WD (e.g., Jones & Watson 1992; Hartmann *et al.* 1999).

The hard X-ray flux recovers to quiescent levels just at the very end of the optical outburst (Wheatley *et al.* 1996b, 2000; see also Yoshida *et al.* 1992; van Teeseling & Verbunt 1994; Ponman *et al.* 1995). The recovery timescale is slightly longer than that of the optical decline (e.g., Jones & Watson 1992). The hard X-ray flux varies on a timescale of hundreds of seconds with an amplitude of ~100% at the start of the recovery to 50% at the end of that observation, with the hardness ratio staying constant. As the end of the optical outburst is thought to correspond to the cooling of the disk region immediately surrounding the WD, this observation indicates that the hard X-rays originate from an area of the disk very close to the WD (e.g., Wheatley *et al.* 1996b; Verbunt 1996). Note that at the time of the hard X-ray recovery the EUV lightcurve of SS Cyg exhibited a secondary maximum; this extra emission is consistent with the soft tail of the hard X-ray emission (Mauche & Robinson 2001; Wheatley *et al.* 2000, 2002).

The situation is different for the outbursts of U Gem, where both soft and hard X-ray fluxes are higher during outburst than in quiescence (by a factor of ~10–100). The largest increase occurs at EUV wavelengths. This corresponds to the expected increase in an optically thick BL radiating at temperatures near 10 eV at outburst (e.g., Szkody *et al.* 1999). While the optical flux stays constant near maximum, the EUV flux drops (Long *et al.* 1996). Assuming the EUV flux originates from near the WD, this suggests that $\dot{M}$ in the innermost regions of the disk decreases compared to the outer regions (which presumably are still optically thick). During the outburst decline, both the soft and hard X-rays decrease faster than the optical flux (Mason *et al.* 1978; Swank *et al.* 1978; Córdova & Mason 1984; see also Szkody *et al.* 1999). Values derived for $L_{\rm BL}/L_{\rm disk}$ are ~0.5 during quiescence (Szkody *et al.* 1999), which is among the highest for DNe, and ~1 during outburst (Long *et al.* 1996). So, in quiescence, as at outburst, U Gem comes closest to the standard BL model.

There are several reasons why $L_{\rm BL}$ may be larger in U Gem than in systems like VW Hyi (see Long *et al.* 1996, and references therein). Studies (Pringle 1977; Popham & Narayan 1995) have shown that, when $\dot{M}$ is held fixed, the BL temperature and $L_{\rm BL}$ increase substantially with WD mass. On the other hand, rotation of the WD decreases both the amount of energy released and the effective temperature of the BL. The WD in U Gem is more massive (1.0–1.2 $M_\odot$) than WDs in most DNe and VW Hyi in particular ($\sim 0.6\,M_\odot$). It appears to be at most slowly rotating ($v \sin i \lesssim 100\,{\rm km\,s^{-1}}$), while the WD in VW Hyi rotates with a $v \sin i \simeq 400\,{\rm km\,s^{-1}}$ (e.g., Sion *et al.* 2002), which corresponds to 20% of the break-up velocity.

In contrast to quiescence, there are no eclipses in the X-ray and EUV lightcurves of OY Car during outburst (Naylor *et al.* 1988; Pratt *et al.* 1999b; Mauche & Raymond 2000). The NL UX UMa also does not show X-ray and EUV eclipses (Wood *et al.* 1995b). This suggests that the prime X-ray source, probably the BL, is obscured at all orbital phases. From contemporaneous observations at other wavelengths (Naylor *et al.* 1987, 1988), extensive azimuthal structures on the outer disk had been inferred, which may block our view of the BL region. The dips observed in various other CVs (see Section 10.2.3 and below) are also explained with this geometry. The X-rays we see are thought to be emitted or scattered by a more extended source (e.g., Verbunt 1996); e.g., due to a disk corona (Naylor *et al.* 1988) or scattering from a photoionized disk wind (Raymond & Mauche 1991; Mauche & Raymond 2000; see also Section 10.2.5).

Dips in the EUV and X-ray lightcurves during outburst have been seen in U Gem during a normal outburst (at $\phi_{\rm orb} \sim 0.8$; Long *et al.* 1996), as well as during an anomalously long

(∼45 days) outburst (Mason *et al.* 1988). As in quiescence (Section 10.2.3), the dips only occur at low energies, indicating absorption effects. The morphology of the dips changes from cycle to cycle, related to changes in the absorbing material. The dips are deeper at shorter wavelengths, suggesting that the hot central area around the WD is being obscured by cooler material further out. During one dip observed by Mason *et al.* (1988) the X-ray source was completely extinguished in 15 s, putting the absorbing material near the outer edge of the disk.

10.2.5 X-ray spectral features

As mentioned in Section 10.2.1, most of the surveys done so far showed that single and sometimes two-temperature bremsstrahlung models were sufficient to describe the CV X-ray spectra, except for the occasional inclusion of a Gaussian to represent a line near 6.7 keV (see below). This is mainly due either to rather poor energy resolution and/or poor statistics. The use of more realistic models was generally not warranted. With the advent of better resolution, larger collecting area, better photon-counting devices, and broader band passes, it became clear that the spectra are far more complicated. The X-ray spectra of DNe and NLs can probably be best decribed as somewhere between a pure bremsstrahlung model and a pure coronal model (e.g., van Teeseling & Verbunt 1994). Generally, X-ray spectra from non-magnetic CVs are due to hot thermal plasma in the BL, even at high $\dot{M}$ (e.g., Mukai 2000). This is because the shock-heated plasma in the BL must cool from a temperature near 10 keV indicated by the X-ray spectra to the photospheric temperature of the WD (∼2.5 eV). The situation may be further complicated since X-rays from the hottest gas can photoionize cooler gas, altering both the energy balance and the ionization state at intermediate temperatures.

Many CVs, either in quiescence or in outburst, show an emission line near 6.7 keV from the Kα transition of highly ionized Fe, with EW ∼ 0.8–1.0 (e.g., Szkody *et al.* 1990); it is associated with the hard X-ray emitting, optically thin plasma. Line emission near 7.9 keV has been reported just after an outburst of SS Cyg (Jones & Watson 1992) and OY Car (Ramsay *et al.* 2001b). This may be interpreted as thermal Fe-Kβ emission, confirming the origin of line emission from a hot optically thin region. The presence of an absorption edge near 8.3 keV in the Ginga spectrum of SS Cyg implies substantial covering of the hard X-ray emission by the highly ionized gas (possibly a wind). Note that this is hard to reconcile with the picture in which the hard X-rays arise from a hot corona (Yoshida *et al.* 1992). SS Cyg also shows a reflection component both in quiescence and outburst (Done & Osborne 1997). Its contribution is larger in the softer X-ray spectra seen in outburst than in quiescence. This supports models in which the quiescent inner disk is not present or not optically thick, so that the only reflector is the WD surface rather than the WD plus disk. The amount of reflection in outburst is also more consistent with the hard X-rays forming a corona over the WD surface rather than just an equatorial belt as seen in quiescence. Note that a reflection component is absent in OY Car; this possibly is due to the high inclination, so that it may be obscured by the disk (Ramsay *et al.* 2001b).

Although observations with the ASCA satellite showed complex structures in the X-ray spectra, the spectral resolution was still not high enough to resolve individual lines, especially at wavelengths where many lines are expected (e.g., the Fe L complex around 1 keV). First EUVE, and now Chandra and XMM-Newton carry instruments which provide the opportunity to perform detailed temperature diagnostics from individually resolved lines and line ratios.

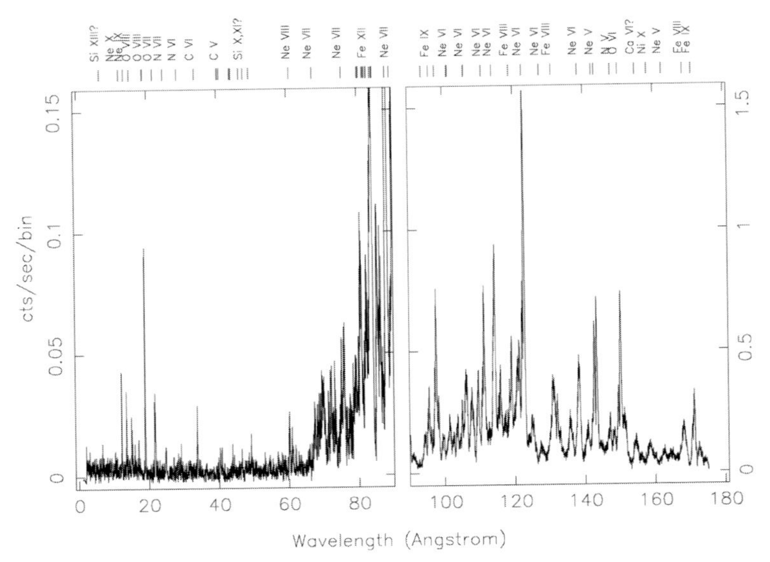

Fig. 10.2. Average Chandra/LETGS spectrum 3.5 d into the superoutburst of WZ Sge. Note the difference in scale between the short- and long-wavelength part of the spectrum. Indicated are many of the line identifications. From Kuulkers *et al.* (2002).

Line ratios can be used to constrain the electron density, electron temperature, and ionization balance (see, e.g., Mauche *et al.* 2001; Szkody *et al.* 2002a, and references therein). The instruments also provide enough velocity resolution to begin to study the effects of velocity broadening, which gives important clues to whether the emitting region is located in a rapidly rotating BL or is close to the more slowly rotating WD.

High resolution X-ray spectra of U Gem in quiescence (Szkody *et al.* 2002a) revealed prominent narrow emission lines of O, Ne, Mg, Si, S, and Fe. The line fluxes, ratios, and widths indicate that the X-ray emission lines arise from a range of temperatures in a high density ($>10^{14}$ cm^{-3}) gas, moving at low (<300 km s^{-1}) velocity, with a small ($<10^7$ cm) scale height compared to the WD radius. This is consistent with the emission coming from an equatorial belt on the WD, as was also inferred from the eclipse lightcurves (see Section 10.2.3).

The EUV/soft X-ray spectra of U Gem in outburst ($\phi_{\rm orb} \sim 0.6$–0.8, Long *et al.* 1996), and OY Car (Mauche & Raymond 2000) and WZ Sge (Kuulkers *et al.* 2002; Fig. 10.2) in superoutburst resemble each other markedly, and are unlike the spectra seen in other DNe, which is attributed to the high inclination of the former sources. Below ~0.2 keV ($\gtrsim$65Å) they show a "forest" of broad (FWHM ~ 800–1200 km s^{-1}) emission lines of intermediate ionization stages of N, O, Ne, Mg and Fe, on top of a continuum (which is weak in OY Car and WZ Sge, and appears to be line-free in U Gem). The phase-resolved spectra of U Gem show that the eclipses affect the continuum more strongly than the lines, implying that the lines are produced in a region of larger extent than that of the continuum, which is presumably formed in the BL. The line identifications alone significantly constrain the physical nature of the emitting plasma. Because all of the above strong lines are resonance lines, good spectral fits are obtained with a model wherein the radiation from the BL and disk is scattered into the line of sight by the system's photoionized disk wind (Mauche & Raymond 2000). Note that the absence of X-ray eclipses in OY Car during superoutburst (Naylor *et al.* 1988) can be understood if much of the X-ray emission we see is light scattered in such an extended wind.

The EUV lines arise from the dominant ionization states of the wind, and their strengths suggest that the wind mass-loss rate in U Gem is a substantial fraction of the WD accretion rate (Long *et al.* 1996). The lines are effectively the emission peaks of P Cygni profiles. This requires the scattering region to be of order 10^{10} cm, similar to the size of the region inferred from studies of the UV P Cygni lines in other DNe and NLs. These P Cygni profiles have been shown to arise in strong winds with terminal velocities of order 3000 km s^{-1} (e.g., Drew & Verbunt 1985; Mauche & Raymond 1987; Woods *et al.* 1990).

WZ Sge's spectrum at higher energies, $\gtrsim$0.2 keV ($\lesssim$65 Å), shows weaker lines of highly ionized ions, with OVIII (0.66 keV; 18.9 Å) standing out. A strong emission line was also seen near 2.4 keV, probably associated with He-like S XV, which cannot be accounted for using an optically thin thermal plasma model (Kuulkers *et al.* 2002).

Because of the recent access to high-quality X-ray spectra, new interpretations are also emerging. This is demonstrated by, e.g., Mukai *et al.* (2003), who show that the Chandra data of the DNe SS Cyg and U Gem and the old nova V603 Aql are well described by a simple cooling-flow model. This in contrast to three IPs which are better described by a photoionization model (see also Section 10.4.4).

10.2.6 VY Scl stars

During optical high states the X-ray spectra of NLs can generally be described by emission with temperatures of $kT_{\rm bb} \sim 0.25$–0.5 keV and $L_{\rm X} \sim 10^{31}$–10^{32} erg s^{-1}. Since VY Scl stars are thought to have a high $\dot{M}$ during the high state, their X-ray spectra are expected to be rather similar to DNe in outburst. However, this seems not to be the case. Distinct differences at other wavelengths exist as well; see Greiner (1999) for an extensive comparison between VY Scl stars and DNe in outburst.

A very soft component ($kT_{\rm bb} \sim 19$ eV) was found in V751 Cyg during an optical low state, at which time the bolometric X-ray luminosity was around 20 times higher ($L_{\rm X} \simeq 5 \times 10^{36}$ erg s^{-1}) than in the high state (Greiner *et al.* 1999; but see Patterson *et al.* 2001). V Sge showed a similar behavior: it is a faint hard X-ray source during optical bright states, while during optical low states it shows X-ray luminosities similar to V751 Cyg (Greiner & van Teeseling 1998). $L_{\rm X}$ is clearly higher than generally observed from CVs (see Section 10.1), and compatible with the lower end of the luminosity distribution of supersoft sources (SSS; see Chapter 11). It was therefore suggested that VY Scl stars in their low states may have a link with SSS (Greiner *et al.* 1999). Not all VY Scl stars in their low state show high values of $L_{\rm X}$, however. Examples are KR Aur with $L_{\rm X} \sim 10^{31}$ erg s^{-1} (Eracleous *et al.* 1991a; Schlegel & Singh 1995) and MV Lyr with $L_{\rm X} \lesssim 5 \times 10^{29}$ erg s^{-1} (Greiner 1999) during a low state.

Moreover, the X-ray spectra during a high (KR Aur) and intermediate (TT Ari) optical state were shown to be poorly described by blackbody radiation; a thermal plasma model described the data better. It was concluded, therefore, that the X-ray spectra of VY Scl stars should be interpreted using the latter model, in both the high and the low state (Mauche & Mukai 2002). This makes the suggested SSS connection less likely (see also Patterson *et al.* 2001).

10.2.7 AM CVn stars

The luminosities of the few AM CVn stars that have been detected in X-rays range from $\sim 10^{28}$ to 5×10^{30} erg s^{-1} (Ulla 1995, and references therein; but see Verbunt *et al.* 1997). The maximum of the overall flux distribution in AM CVn itself peaks around EUV wavelengths; there is no detectable hard X-ray emission (Ulla 1995). The X-ray luminosities of AM CVn stars agree with the coronal luminosities for single stars (e.g., Rosner *et al.* 1985; Hempelmann *et al.* 1995), and possibly with emission from single DB WDs (e.g., Fontaine *et al.* 1982).

Of the few exceptions to the F_X/F_{opt} versus P_{orb} relation (Section 10.2.1), is AM CVn. With $P_{orb} \sim 0.29$ h it has an unexpectedly small ratio of ~0.002. This small ratio might be explained by a high $\dot{M}$, comparable to UX UMa stars (van Teeseling *et al.* 1996). On the other hand, the AM CVn system GP Com has a ratio near unity (van Teeseling & Verbunt 1994). This may be due to the fact that the entire disk in GP Com is in a (low) steady state, in which it will always be optically thin, and will not undergo outbursts similar to that seen in DNe (Marsh 1999).

V407 Vul, a CV related to the AM CVn stars, was recently suggested to be a new type of double-degenerate CV (Marsh & Steeghs 2002). In this CV the mass transfer stream may hit a non-magnetic WD directly due to a very compact orbit of 9.5 min. This results in pulsations in the X-ray flux every 9.5 min, with no X-ray emission in between pulses (suggestive of it being a polar (see Section 10.3), however, neither polarization nor line emission is seen). Its X-ray spectrum is soft ($kT_{bb} \simeq 40$–55 eV; Motch *et al.* 1996; Wu *et al.* 2002). This is explained by the stream breaking into dense blobs which are able to penetrate the photosphere of the WD and therefore become thermalized, giving rise to the soft X-ray emission (Marsh & Steeghs 2002).

10.3 X-ray emission from polars

10.3.1 Introduction

In polars, the originally free-falling matter couples to magnetic field lines somewhere between the two stars and is guided to one or two accretion regions in the vicinity of the magnetic poles. These are the sources of intense X-ray radiation, mainly in the soft X-ray regime, and of cyclotron radiation from IR to UV wavelengths. The observation of pulsed polarized radiation from the cyclotron source led to their nick-names as polars (Krzemiński & Serkowski 1977).

There is no recent review of the X-ray properties of polars in broad generality; the main satellite-related aspects have been reviewed (ROSAT: Beuermann & Thomas 1993; Beuermann & Burwitz 1995; EUVE: Sirk & Howell 1998; Mauche 1999; ASCA: Mukai 1995). Emission from post-shock flows in magnetic CVs is described by Cropper (1990).

Just two polars were known as variable stars before the era of X-ray astronomy began (AM Her, VV Pup) and a very small number were detected in optical spectroscopic surveys before 1999 (AN UMa, CE Gru, MR Ser = PG1550+191). To date about 70 polars are known, the vast majority of them identified as counterparts of serendipitous X-ray sources. Only recently, much deeper optical spectroscopic surveys (Hamburg Schmidt telescope, SDSS) have uncovered new systems in apparently permanent low states of accretion (Reimers *et al.* 1999; Reimers & Hagen 2000; Szkody *et al.* 2002b). They were not or just marginally detected in X-rays, and, due to their low accretion rates, display intriguing cyclotron spectra.

Polars are primarily soft X-ray emitters. Therefore, the all-sky surveys conducted with ROSAT (XRT and WFC) in combination with optical identification programmes permitted the first synoptic view of the magnetic CV sky with high sensitivity. Most polars are found below the 2–3 h CV period gap (Webbink & Wickramasinghe 2002). The gap itself is significantly

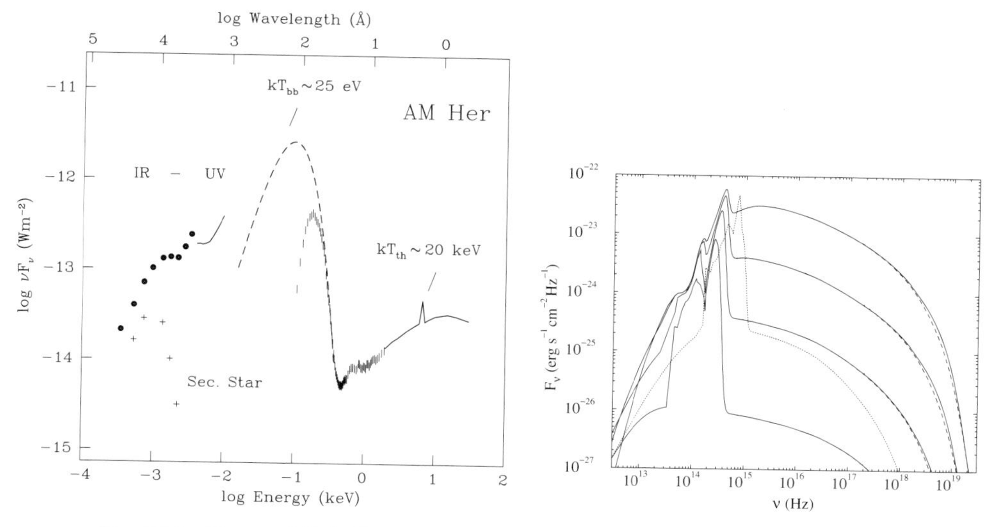

Fig. 10.3. Observed (*left*, AM Her) and theoretical (*right*) spectral energy distributions of polars (Beuermann 1999; Fischer & Beuermann 2001). The models do not take into account the prominent soft X-ray component of reprocessed origin; they account for the primary components of optical cyclotron and hard X-ray bremsstrahlung radiation.

filled in, possibly due to reduced braking by trapping of the wind from the donor within the magnetosphere of the WD (*ibid.*). The space density is of the order of $n_s \simeq 1\text{–}2 \times 10^{-6}\,\mathrm{pc}^{-3}$ for short-period systems and a factor of 10 lower for long-period systems (Beuermann & Schwope 1994).

As shown in Fig. 10.3, polars emit from the IR to the hard X-ray regime. Most of the radiation is accretion-induced. A complete picture therefore requires multi-wavelength observations, preferably obtained contemporaneously because of the inherent high variability on many timescales (from seconds to years). X-ray observations are essential in order to determine the accretion scenarios (which requires one to disentangle the X-ray spectra), the accretion geometries (which requires X-ray observations with full phase coverage), and the accretion history (which requires long-term monitoring in the X-ray domain).

10.3.2 Accretion-induced emission

Matter in the accretion stream is accreted almost vertically on to the magnetic poles of the WD. The accretion process is, therefore, almost always modeled in a one-dimensional quasi-radially symmetric approximation. The presence of, e.g., accretion arcs with corresponding variety of accretion rates and deviations from the radial symmetry due to inclined magnetic field lines is evident from observations but neglected in the modeling for tractability.

Accretion is governed by three parameters, M_{WD}, the accretion rate per unit area $\dot{m}$, and B. The balance between those parameters determines whether the accretion region is heated via a strong hydrodynamic stand-off shock or by particle bombardment, and whether the cooling function is dominated by plasma emission or by cyclotron radiation.

With appropriate boundary conditions, the equations of conservation of mass, momentum and energy can be used to calculate the temperature and density as a function of height, as well as the emerging spectra. Only the one-dimensional non-magnetic case could be solved analytically (Aizu 1973). Present numerical models include cooling by cyclotron emission by solving the fully frequency- and angle-dependent radiative transfer and treat the accretion plasma in a two-fluid approximation (Fischer & Beuermann 2001; see Fig. 10.3); they include gravity, account for its variation within the flow and use up-to-date plasma emission codes (Cropper *et al.* 1998). They take into account pre-shock heating and ionization, which influences the size of the shock jump and the formation of the emerging spectra. However, current models are still one-dimensional and stationary, which limits their direct applicability to observational data.

Multi-temperature plasma emission models were fitted to the hard X-ray spectra of polars with the aim (among others) to estimate $M_{\rm WD}$. A Compton reflection component from the irradiated WD surface, complex absorption (partial covering cold or warm absorbers) in the pre-shock flow or surrounding matter, fluorescent Kα emission and cold interstellar absorption were taken into account (e.g., Done & Magdziarz 1998; Matt *et al.* 2000). Such models give satisfactory fits to the data but tend to predict too high values of $M_{\rm WD}$ compared with dynamical mass estimates (Cropper *et al.* 2000). Van Teeseling *et al.* (1999) question the validity of coronal models for the post-shock emission. High $\dot{M}$ systems may be very optically thick in the resonance lines, and the resulting asymmetric line emission may serve as a diagnostic tool to probe the very inner accretion geometry. X-ray line spectroscopy was used to infer sub-solar abundances of the accreted matter (Done & Magdziarz 1998; Ishida & Ezuka 1999), in unresolved conflict with UV-line spectroscopy (Bonnet-Bidaud & Mouchet 1987).

In the "standard accretion model" (King & Lasota 1979; Lamb & Masters 1979) about half of the X-rays and of the cyclotron radiation are intercepted by the WD surface and are reprocessed as soft X-rays. This simple model predicts about equal luminosities in the bremsstrahlung and cyclotron components on the one hand and the soft emission on the other. Details of this balance depend on the hard X-ray albedo, the irradiation geometry and, observationally, on the viewing geometry. However, since the early days (EXOSAT and Einstein era), a moderate to strong soft X-ray excess over the other components has been observed, creating what was referred to as the "soft X-ray puzzle". The size of the soft excess was difficult to assess exactly, but X-ray flux ratios $F_{\rm soft}/F_{\rm hard}$ up to $\sim$100 were reported. Difficulties in quantifying the soft excess often arise from non-simultaneous observations in the soft and hard spectral bands and from incomplete spectral coverage of the soft component. At temperatures $kT_{\rm bb} \simeq 15$–30 eV, it has its peak emission in the EUV, where most instruments have low sensitivity and interstellar absorption is severe. Apart from a few exceptions which indicate the presence of NeVI absorption edges or NeVII and NeVIII absorption lines, the soft spectra can be well described with a simple blackbody (Mauche 1999). The application of more physical models, e.g., pure-H or solar-abundance stellar atmospheres, does not improve the fits due to the incompleteness of the models and the low signal-to-noise of the data.

Two ways have been proposed to solve the soft X-ray puzzle (Kuijpers & Pringle 1982). The first invokes shredding of the stream into diamagnetic blobs in the magnetospheric interaction region. Subsequent confinement and compression of the blobs leads to highly inhomogeneous accretion of filaments with partly or wholly buried shocks (Frank *et al.* 1988). The primary hard radiation cannot escape freely and the photosphere will be heated from below. Apart from hydrostatic computations of the temperature structure of a one-blob impact (Litchfield & King 1990), this model is not worked out in quantitative detail.

An alternative scenario applies to the low $\dot{M}$ and high B case. In such an environment, cyclotron cooling becomes so efficient that it cools the plasma over a mean free path of the infalling particles, i.e., the shock is resolved and bremsstrahlung is suppressed (bombardment solution, Woelk & Beuermann 1996). ROSAT observations of a large number of polars showed a clear relation between the size of the soft excess and the magnetic field strength in the accretion region (Beuermann & Schwope 1994; Ramsay *et al.* 1994).[4] This correlation was explained either by enhanced fragmentation of the stream in the magnetosphere, i.e., by enhanced blobby accretion, or by enhanced cyclotron cooling, thus supporting either of the two alternatives to the "standard" model. The decomposition of lightcurves in eclipsing systems (e.g., Bailey 1995) and the analysis of soft-to-hard X-ray cross-correlation functions (e.g., Beuermann *et al.* 1991) suggest that regions with low and high $\dot{M}$ co-exist. Consequently, the accretion region cannot be described in terms of just one of the scenarios. Further modification to the "standard" model arises from the fact that the reprocessed component in AM Her is observed with the expected energy content but is detected in the UV instead of the soft X-ray regime, suggesting that the soft X-rays are completely decoupled from the other radiation processes (Gänsicke *et al.* 1995). Similar multi-band investigations including the UV spectral regime are missing for other polars.

10.3.3 X-ray lightcurves

Polars display a rich phenomenology of X-ray lightcurves despite their rather simple accretion geometry. The lightcurves offer large diagnostic potential, since they are modulated by the location and the three-dimensional extent of one or several accretion spots, by stellar eclipses (11 out of 70 systems display stellar eclipses), by non-stationary accretion processes, and by absorption of X-rays within the binary. In the long term the lightcurves are affected by shifts of the accretion regions in longitudinal and/or lateral directions, by changes between one- and two-pole accretion modes (which gives those systems a completely different appearance) and by large-scale variations of $\dot{M}$. Lightcurve changes occur at unpredictable moments and persist for unpredictable durations. It is a common assumption that long-term changes of the accretion rate are related to star spots on the donor at the L_1 (King & Cannizzo 1998) and an attempt has been made to reconstruct a possible pattern of star spots from the accretion history of AM Her (Hessman *et al.* 2000). While irradiation-induced structure is obvious in Doppler tomograms of the donors, these experiments have failed so far to make star spots visible (Schwope 2001).

The main features of the X-ray lightcurve of a polar in a one-pole accretion geometry are exemplified in Fig. 10.4, the ROSAT lightcurve of HU Aqr (Schwope *et al.* 2001). While interpreting these lightcurves one should bear in mind that the spin of the WD is synchronously locked with $P_{\rm orb}$, typically to better than 10^{-6} phase units (i.e., $|P_{\rm spin} - P_{\rm orb}|/P_{\rm orb} < 10^{-6}$). This implies that any phase information can be easily transformed to angular information, either in the frame of the WD or in the frame of the binary star, provided the phase of conjunction of the WD is known. The lightcurve shows a pronounced bright/faint pattern, with the bright phase lasting from phase 0.53 to 1.22 (Fig. 10.4). Further features modulating

[4] Meanwhile, the field strengths of about 45 systems have been measured. Some measurements are based on Zeeman-split Balmer lines from the photosphere or from an accretion halo (Schwope 1996), but most of these measurements are based on the identification of cyclotron harmonic emission lines in low-resolution optical and/or IR spectra, originating from the accretion plasma at one or two accretion regions (for a review see Wickramasinghe & Ferrario 2000).

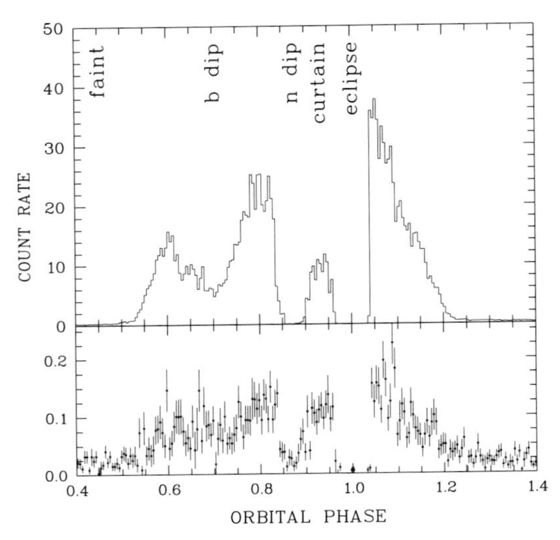

Fig. 10.4. ROSAT X-ray lightcurves of the high-inclination polar HU Aqr. The original photon data comprise 36 ks and were phase-averaged over the 125 min orbital period. Phase zero corresponds to superior conjunction of the WD. The upper panel encompasses the whole spectral band pass (0.1–2.4 keV), the lower panel only the hard X-ray spectral component (0.5–2.0 keV). Several important features around the orbital cycle are indicated. Adapted from Schwope *et al.* (2001).

the bright phase are labeled broad dip (b dip), narrow dip (n dip), curtain, eclipse, and are discussed below. The faint phase is caused by a so-called self-eclipse of the accretion region by the WD. In the faint phase some residual X-ray emission is present, which is probably of scattering origin. The length and phasing of the bright phase allows clues to be drawn on the latitude and longitude of the accretion spot. Most polars have the main accretion spot in the quadrant of the ballistic stream (seen by a hypothetical observer on the WD), i.e., on the leading side of the WD. However, a few systems have their spots on the trailing side of the WD or even on the opposite side from the donor. This implies complex motions of matter in the magnetosphere and it is by no means clear to what extent the simple picture of accretion via Roche-lobe overflow at L_1 is applicable. Recent Doppler tomograms show that even the putative ballistic, freely falling stream in the vicinity of the L_1 might be influenced by the magnetic fields in the binary (Schwarz *et al.* 2002). The spots do not show any preferred latitude, i.e., the spin axis of the WD seems not to be aligned with the rotation axis.

While the hard X-rays are assumed to be formed by (mainly) optically thin plasma radiation, the X-ray brightness should be constant throughout the bright phase. Instead, large fluctuations are seen which are assigned to non-stationary accretion and occultations. The soft X-rays originate from optically thick surfaces and orbital modulations are expected (and observed) to be more marked. The bright phase in general is far from being compatible with a flat accretion spot and heating by accretion blobs rather than irradiation plays an important role. The depth-dependent temperature structure in the vicinity of an accretion filament was computed by Litchfield and King (1990) using similarity to a heat conduction problem. The photosphere is then assumed to be shaped like a mound, but its height proved not to be sufficient to reproduce soft X-ray lightcurves of, e.g., the anomalous state of AM Her (Heise *et al.* 1985), suggesting that hydrodynamic splashes rather than a hydrostatic atmosphere are responsible for the soft X-ray lightcurves.

The bright phase may undergo shifts with respect to binary phase zero. In some cases this reflects the accretion rate dependent penetration of the magnetosphere by the ballistic stream before it gets threaded onto magnetic field lines. In other cases it indicates a small asynchronism of $P_{\rm spin}$ and $P_{\rm orb}$ (Schwope *et al.* 2001, 2002).

There are 11 eclipsing systems; two of them, UZ For and HU Aqr, were found bright enough to resolve the ingress into or the egress from the eclipse in soft X-rays. Detailed modeling shows that the soft X-ray accretion spots have an angular lateral extent of less than 5° (Warren *et al.* 1995; Schwope *et al.* 2001) and a vertical extent of less than $0.05\,R_{\rm WD}$. Hence the soft X-ray accretion region is pillbox-shaped rather than pencil-shaped.

The narrow dip is due to absorption in the accretion stream, which has coupled to the magnetic field and has left the orbital plane. This material may intercept X-rays from the accretion hot spot at certain phases. Those features can be observed only in systems where the accretion stream (or curtain) and the observer are located in the same hemisphere with respect to the orbital plane. Furthermore, the orbital inclination must be sufficiently high and the inclination of the magnetic axis sufficiently low. This ensures that the stream is raised sufficiently high above the plane and the line of sight to the accretion spot crosses the accretion stream. The orbital phase of this feature indicates the azimuth and its width the size of the threading region in the magnetosphere. To a first order, the coupling region is located where the magnetic pressure overcomes the ram pressure of the ballistic stream. A simultaneous hard X-ray and near-IR study of the dips in EF Eri (Watson *et al.* 1989) demonstrates the presence of substantial structure in the dips, implying significant density fluctuations in the stream, either spatial or temporal. The origin of the broad dip centered at an earlier phase remains unclear so far. Its width and X-ray color suggest an origin in warm absorbing matter in close vicinity to the hot accretion spot. Its diagnostic potential needs to be explored.

10.4 X-ray emission from intermediate polars

10.4.1 Introduction

The first of the asynchronous magnetic CVs to be discovered was the remnant of Nova Herculis 1934 – DQ Her – in 1954. Walker (1956) found a highly stable 71.1 s optical modulation with an amplitude of a few hundredths of a magnitude, which disappeared during eclipse. In analogy with the models for X-ray pulsars (Pringle & Rees 1972; see Chapter 7) an accreting oblique dipole rotator was eventually suggested (Bath *et al.* 1974a). The correctness of this model was demonstrated by the discovery of a phase shift in the 71 s signal during eclipse that could be matched to the eclipse of a beam of high energy radiation, emitted by the rotating WD, as it swept over the accretion disk and was reprocessed into optical wavelengths (Warner *et al.* 1972; Patterson *et al.* 1978).

The evident success of this accreting magnetic model led to its adoption when the first X-ray CVs to have two simultaneous periodic modulations were found. For example, AO Psc was observed to have a strong 14.3 min optical periodicity (Warner 1980; Patterson & Price 1980) but a 13.4 min modulation in hard X-rays (White & Marshall 1980). The realization that the frequency difference of these two modulations is equal to the orbital frequency showed that the X-ray period arises from rotation of the WD, but the optical modulation must be caused by the rotating beam being reprocessed from some structure fixed in the rotating

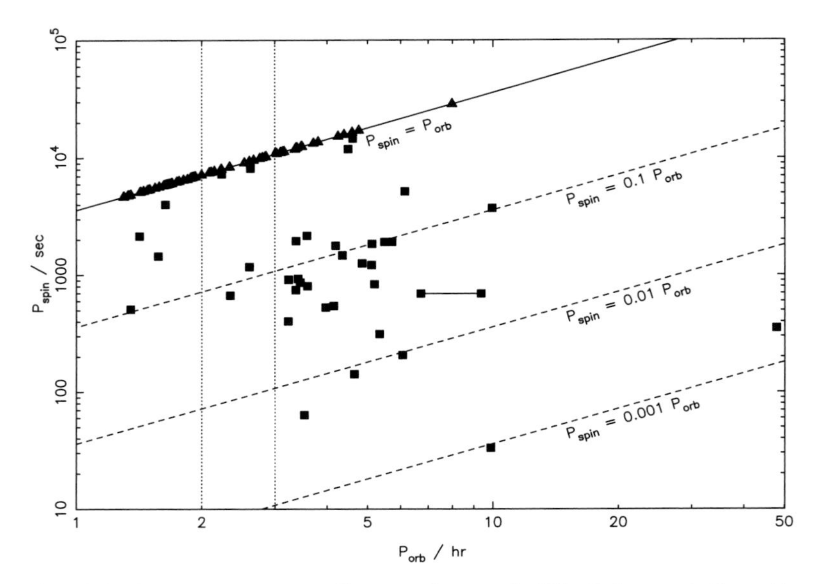

Fig. 10.5. The $P_{\rm spin}$ versus $P_{\rm orb}$ diagram of magnetic CVs. Polars are indicated by triangles; IPs are indicated by squares. From Norton *et al.* (2004).

frame of the binary (e.g., the donor or the thickening of the disk where the stream impacts). It was therefore recognized that only two "clocks" are really present, i.e., $P_{\rm orb}$ and the spin period of the WD, $P_{\rm spin}$.

Because of amplitude modulation, caused largely by geometrical projection effects, a suite of modulations is sometimes seen in optical observations. Denoting $\Omega = 2\pi/P_{\rm orb}$ and $\omega = 2\pi/P_{\rm spin}$, the frequencies $\omega - \Omega$, $\omega - 2\Omega$, $\omega + \Omega$ and $\omega + 2\Omega$ are predicted to occur (Warner 1986). The importance of these sidebands is that they act as proxies for X-rays in those cases where no X-ray modulation has been observed; there are no models other than IPs[5] that explain the presence of orbital sidebands.

Direct observation of an X-ray modulation, usually denoting $P_{\rm spin}$, accompanied by $P_{\rm orb}$, usually obtained from optical photometric or spectroscopic observations, is required to give full conviction to classification as an IP. But an optical periodicity of proven stability (to distinguish from the quasi-periodic oscillations discussed in Section 10.5), and the presence of one or more orbital sidebands, even without any X-ray detection at all, give an irresistible urge for inclusion in the IP lists. The parameter space occupied by the IPs is illustrated in Fig. 10.5.

10.4.2 Modes of accretion

Most IPs are expected to accrete via some form of truncated accretion disk whose inner edge is at the magnetospheric radius. From here, material will attach onto field lines and flow towards the magnetic poles, forming "accretion curtains" above each pole (Rosen *et al.* 1988). Unlike the polars, the accretion flow impacting the WD in IPs will therefore be more extended, occurring over a greater fraction of the WD surface. At some point in the flow, the material will experience a strong shock before settling and cooling mainly

[5] The name "intermediate polar" was introduced by Warner (1983).

by thermal bremsstrahlung. This region is thus the origin of the observed X-ray emission. Modulation at the WD spin period is produced by a combination of self-occultation and varying photoelectric absorption towards the X-ray emission sites.

Some IPs are also believed to accrete (at least in part) directly via a stream, in a similar manner to polars. This stream may overflow a disk (disk-overflow accretion; e.g., Hellier *et al.* 1989; King & Lasota 1991; Armitage & Livio 1996) or replace a disk entirely (stream-fed accretion; Hameury *et al.* 1986). As the stream flips from pole to pole, this will naturally give rise to an X-ray modulation at the sideband frequency $\omega - \Omega$ (Hellier 1991; Wynn & King 1992; Norton 1993). In disk-fed accretion it is likely that the footprints of the field lines onto which the flow attaches are semi-circular X-ray emitting arcs around each magnetic pole, and are fixed on the surface of the WD. By contrast, the field lines to which the stream-fed or disk-overflow accretion attaches are likely to have smaller footprints at each magnetic pole and these will "migrate" around the pole to follow the incoming stream, as a function of the sideband phase (Norton *et al.* 1997).

10.4.3 X-ray lightcurves

In early X-ray observations, the lightcurves of IPs folded at the WD spin period were seen to be roughly sinusoidal and interpreted as largely due to self-occultation of the emission area by the WD (King & Shaviv 1984). First hints that this was not the case came with EXOSAT observations (Mason 1985; Watson 1986; Norton & Watson 1989) which showed that the modulation depth tended to decrease with increasing X-ray energy over the range 1–10 keV, indicating that photoelectric absorption made some contribution to the observed modulation.

The now widely accepted "accretion curtain" model was proposed to explain the data from EX Hya (Rosen *et al.* 1988, 1991). In this model, the emission region is a tall, thin, arc-shaped curtain and the largest X-ray flux is seen when the curtains are viewed from the side (i.e., when a given pole is pointing *away* from the observer). Although EX Hya is an atypical IP, such a model was also successfully applied to the other IPs (e.g., Hellier *et al.* 1991).

Ginga, ROSAT, ASCA and RXTE each observed many IPs, producing lightcurves with extremely high signal-to-noise in many cases. Whilst the pulse profiles of some sources (e.g., EX Hya; Rosen *et al.* 1991) still appear roughly sinusoidal, there are indications that additional structure may be present in others. For example, the combined ROSAT and Ginga pulse profiles of AO Psc and V1223 Sgr show evidence for a small notch superimposed on the peak of the pulse (Taylor *et al.* 1997). Several objects, including GK Per (Ishida *et al.* 1992), XY Ari (Kamata & Koyama 1993), V405 Aur (Allan *et al.* 1996), YY Dra and V709 Cas (Norton *et al.* 1999) show pulse profiles that are double peaked (at least on some occasions). Some of the most complex pulse profiles are those seen from FO Aqr (Norton *et al.* 1992a; see Fig. 10.6), BG CMi (Norton *et al.* 1992b) and PQ Gem (Duck *et al.* 1994) which show narrow notches superimposed on broader modulations and pulse profiles that change significantly with $\phi_{\rm orb}$. Many of these X-ray pulse profiles also vary dramatically on timescales of months or years. For instance, those of GK Per and XY Ari are single peaked in outburst (Watson *et al.* 1985; Hellier *et al.* 1997) but double peaked in quiescence; in other cases, such as V709 Cas, the contributions of various harmonics of the spin frequency are seen to vary (de Martino *et al.* 2001).

It has been noted (e.g., Norton *et al.* 1999) that the IPs exhibiting double-peaked X-ray pulse profiles are mostly those with short $P_{\rm spin}$. The WDs in these objects therefore probably

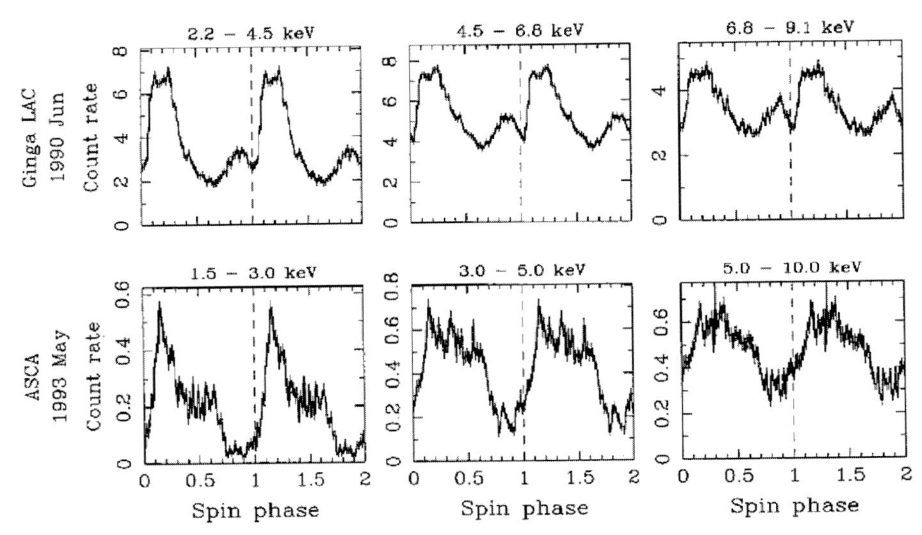

Fig. 10.6. X-ray pulse profiles of FO Aqr. Adapted from Beardmore *et al.* (1998).

have weak magnetic fields, so the magnetospheric radius is relatively small. Consequently the footprints of the disk-fed accretion curtains on the WD surface are relatively large. In contrast to a conventional accretion curtain, the optical depths to X-ray emission are therefore lowest in the direction along the magnetic field lines, and highest in the direction parallel to the WD surface, such that the emission from the two poles conspires to produce double-peaked X-ray pulse profiles (Allan *et al.* 1996; Hellier 1996; Norton *et al.* 1999). Such a pulse profile is therefore *not* a unique indicator of two-pole accretion. Indeed, two-pole accretion onto smaller regions of the WD surface may be considered the "normal" mode of behavior in a disk-fed IP with a longer $P_{\rm spin}$ (and therefore a higher field strength), resulting in a single-peaked pulse profile. Indications of the size of the X-ray emitting region in IPs have come from a study of the deeply eclipsing IP XY Ari. The $\lesssim$2 s egress from eclipse seen in RXTE data limits the accretion region to $\lesssim$0.002 of the WD surface area (Hellier 1997).

Whereas fast rotators with relatively weak fields show double-peaked pulse profiles, several slower rotators with larger fields (and therefore larger magnetospheres) have been seen to exhibit an X-ray sideband modulation (i.e., at a frequency $\omega - \Omega$) at some time. A strong sideband signal is seen in TX Col (Buckley & Tuohy 1989), and FO Aqr (Norton *et al.* 1992a), and weaker signals in AO Psc and V1223 Sgr (Hellier 1992). A dominant sideband period may exist in BG CMi (Norton *et al.* 1992b). Observations of V2400 Oph subsequently confirmed this system as the first truly diskless IP by revealing that its X-ray signal varies *only* at the 1003-s sideband period (Buckley *et al.* 1997; Hellier & Beardmore 2002).

The relative strengths of the X-ray sideband and spin modulation in FO Aqr have been seen to vary on timescales of years (Beardmore *et al.* 1998) and those in TX Col on timescales as short as months (Norton *et al.* 1997). The interpretation is that the relative amounts of accretion occurring via a stream and via a disk vary, possibly due to changes in $\dot{M}$ or other activity on the donor near to the L_1 point.

IPs sometimes show strong orbital modulations in their X-ray lightcurves. The compilation by Hellier *et al.* (1993) showed that FO Aqr, EX Hya, BG CMi and AO Psc all have orbital dips characterized by increased photoelectric absorption around $\phi_{\rm orb} \sim 0.8$. Such orbital modulations have been confirmed by subsequent observations (e.g., Norton *et al.* 1992a,b; Taylor *et al.* 1997; Allan *et al.* 1998). Hellier *et al.* (1993) concluded that the cause was likely to be similar to that in DNe and LMXBs and due to material thrown out of the orbital plane by the stream impact with the disk or the magnetosphere (see Section 10.2.3). Alternatively, or additionally, a spin pulse profile that varies with $\phi_{\rm orb}$ (such as will arise naturally in a stream-fed or disk-overflow model) will naturally give rise to an orbital modulation. It is likely that this effect contributes to the observed orbital modulation, at least in some systems.

10.4.4 X-ray spectra

The relatively low signal-to-noise and low resolution X-ray spectra of IPs obtained with EXOSAT were adequately fitted with single temperature (~tens of keV) bremsstrahlung continua (see also Section 10.2.5) passing through a partial absorber which varied with phase (see the compilation by Norton & Watson 1989). Fluorescent Fe Kα lines were also seen in most IPs (Norton *et al.* 1991). Ginga observations largely confirmed these results but showed that both thermal and fluorescent contributions to the Fe lines were present (Ishida 1991). The higher spectral resolution of ASCA allowed more emission lines to be detected in the X-ray spectra of EX Hya (Ishida & Fujimoto 1995) and AO Psc (Fujimoto & Ishida 1995; Hellier *et al.* 1996), for example, and also showed that V405 Aur, PQ Gem, AO Psc, BG CMi, V2400 Oph, TV Col and V1025 Cen have thermal Fe Kα lines that are broadened by ~200 eV (Hellier *et al.* 1998). In each case, up to three Gaussian lines were required, corresponding to cold, H-like and He-like Fe.

A significant advance in the modeling of IP X-ray spectra came with models that used a multi-temperature emission region, including effects such as reflection from the surface of the WD and partially ionized absorbers (e.g., Cropper *et al.* 1998; Beardmore *et al.* 2000). Using this technique, Cropper *et al.* (1998) fitted the Ginga spectra of nine IPs and determined WD masses for them.

ROSAT discovered a sub-class of IPs, e.g., V405 Aur, PQ Gem and UU Col, characterized by soft X-ray spectra, with blackbody components at $T_{\rm bb}$ of ~tens of eV (Mason *et al.* 1992; Haberl *et al.* 1994; Haberl & Motch 1995; Burwitz *et al.* 1996). The soft X-rays from these objects probably originate, as in polars, from the heated WD surface around the accreting poles. The blackbody fluxes indicate fractional areas of only $\sim 10^{-5}$ of the WD surface for the soft X-ray emission region (Haberl & Motch 1995).

At the time of writing, Chandra and XMM-Newton spectra of IPs are just becoming available and will probably revolutionize our understanding of IP X-ray spectra. For example, Mauche (2002b) shows that line ratios from the Chandra spectrum of EX Hya may be used to determine a plasma temperature that spans the range 0.5 to 10 keV and a plasma density $n \gtrsim 2 \times 10^{14}$ cm^{-3}. Mukai *et al.* (2003) demonstrate that EX Hya's Chandra spectrum is well fit by a simple cooling-flow model, as are those of the DNe U Gem and SS Cyg, and the old nova V603 Aql. In contrast, the Chandra spectra of the IPs V1223 Sgr, AO Psc and GK Per are inconsistent with such a model, but conform with the expectations for line emission from a photoionized plasma (see also Section 10.2.5).

10.5 Rapid oscillations

10.5.1 Dwarf nova oscillations

10.5.1.1 Introduction

The rapid oscillations seen in DQ Her (at 71.1 s, see Section 10.4.1) and AE Aqr are of very high stability ($|\dot{P}|^{-1} > 10^{12}$). In contrast, optical oscillations of low stability

($|\dot{P}|^{-1} \sim 10^{4-6}$) were discovered in some high $\dot{M}$ CVs (namely, DNe in outburst and NLs) by Warner and Robinson (1972). These are known as dwarf nova oscillations (DNOs) and are usually of very low amplitude (typically less than 0.01 mag) and span the range 5–100 s, with a concentration near 25 s. In a given CV they always appear at similar periods. Their short periods indicate a source near to the WD, and there was early expectation that they would be found at short wavelengths. They were indeed later found in the soft X-ray, EUV and UV regions. Only recently, however, have simultaneous EUV and optical observations of DNOs shown that, again as expected, the same phenomenon is being observed in all wavelength regions (Mauche & Robinson 2001) – but frequently the optical modulated flux is merely reprocessed X-ray and EUV radiation.

Before discussing observations of X-ray DNOs we describe a physical model that is gaining acceptance as an explanation of the DNOs. In essence it is an IP model, but the magnetic field lines are connected to the accreted material near the equator of the WD, and not rooted in its interior. Paczyński (1978) pointed out that if the intrinsic field of the WD is low enough ($B \lesssim 10^5$ G: Katz 1975) the accreted material will be able to circulate around the equator of the WD. (The high Q of DQ Her shows that in that CV the field is strong enough to lock the exterior layers to the interior, so the accretion torque is applied to the entire WD.) The shear in the accreting equatorial belt may generate a field strong enough to control the gas flow near the surface of the WD – but $P_{\rm spin}$ of the belt is determined by magnetic coupling to the inner edge of the disk. As $\dot{M}$ waxes and wanes during a DN outburst the inner radius (and Keplerian period) of the disk is first reduced and then increased. The result is a low inertia magnetic accretor (Warner & Woudt 2002), which explains the large range of a DNO period during a DN outburst, and why it is observed to reach a minimum value at the maximum of $\dot{M}$. There is direct spectroscopic evidence for rapidly spinning equatorial belts in DNe during outburst (e.g., Sion *et al.* 1996).

10.5.1.2 Soft X-ray DNOs

Our knowledge of the oscillations in soft X-rays (0.1–0.5 keV) during outbursts of DNe is dominated by observations made of SS Cyg, U Gem and VW Hyi; all relatively nearby and optically bright objects. The first detections were in SS Cyg (Córdova *et al.* 1980b, 1984), U Gem (Córdova *et al.* 1984) and VW Hyi (van der Woerd *et al.* 1987). Modulated soft X-ray emission appears in all of the observed SS Cyg outbursts, but in only one of three observed U Gem outbursts. The hard X-ray emission in SS Cyg is not modulated (Swank 1979). The soft X-ray modulation amplitudes are much greater than in the optical; generally ~25%, but as much as 100% for individual cycles. This shows that much of the accretion luminosity is involved in the modulation process. Other DNOs have been detected, i.e., in the DN HT Cas and the NLs YZ Cnc, RW Sex and AB Dra (Córdova & Mason 1984), but the last three are more probably of the quasi-periodic type discussed in Section 10.5.2. Table 10.1 lists the published studies of the three bright DNe.

The SS Cyg observations by Córdova *et al.* (1980b, 1984) provided the first means of analyzing the short-term temporal variations of the DNOs; in the X-ray region individual cycles can be seen, whereas in optical observations the DNOs are only seen in Fourier transforms (FTs). However, later observations of rare large amplitude optical DNOs, especially those in the DN TY PsA (Warner *et al.* 1989), showed behavior similar to that in X-rays, namely that the DNOs maintain relatively high coherence for a time ΔT and then jump suddenly ($t < 100$ s) to a period typically 0.02 s different. These jumps in period can be in

Table 10.1. *EUV and soft X-ray DNOs in DNe during outburst*

Star	P_{orb} (h)	Period (s)	References
SS Cyg	6.60	9	Córdova *et al.* (1980b)
		10.7	Córdova *et al.* (1984)
		9.6–10.1	Watson *et al.* (1985)
		7.4–10.4	Jones & Watson (1992)
		7.2–9.3	Mauche (1996a)
		2.8[a]	van Teeseling (1997b)
		2.9–8.2	Mauche & Robinson (2001)
		9.1	Mauche (2002a)
U Gem	4.25	25–29	Córdova *et al.* (1984)
		~25	Long *et al.* (1996)
VW Hyi	1.78	14.06–14.4	van der Woerd *et al.* (1987)

[a] A frequency doubling had occurred.

either direction and are superimposed on the steady increase or decrease in period associated with, respectively, decreasing or increasing luminosity. When the luminosity is not changing rapidly, i.e., in NLs or in DNe near maximum light, ΔT can be in excess of an hour; but late in an outburst ΔT decreases to hundreds of seconds and the DNOs become incoherent and difficult or impossible to detect with FT techniques.

In the past five years considerable progress has been made in two areas: the observation of DNOs in the EUV flux (which may be assumed to be a proxy for X-ray modulation) and the extension of studies in the optical. Mauche (1996a, 1997) and Mauche and Robinson (2001) have studied the EUV during outbursts of SS Cyg and discovered several new phenomena, including a frequency doubling of DNOs near maximum of outburst. This shows as a reduction of DNO period from ~6 s to ~3 s; X-ray observations made near maximum of a different outburst of SS Cyg also showed the ~3 s modulation (van Teeseling 1997b). It is possible that the effect is in essence geometrical, with emission from two accretion poles being seen when the inner edge of the disk is very close to the WD surface (Warner & Woudt 2002).

Optical studies of VW Hyi (Woudt & Warner 2002; Warner & Woudt 2002) show the correlation of DNO period with luminosity, detecting for the first time oscillations at ~14 s near maximum, which were previously only seen in X-rays (van der Woerd *et al.* 1987). A rapid slowdown of the DNOs, from a period of ~20 s to ~40 s over about 6 h, coincides with the epoch when the EUV flux plummets almost to zero, and is interpreted as a propeller phase in which accretion is prevented by the magnetic field attached to the rapidly spinning equatorial belt. Following the propeller phase a frequency doubling occurs, which may be a change (at least in visibility) from single-pole to two-pole accretion.

X-ray modulations at 27.87 s in WZ Sge in quiescence (Patterson *et al.* 1998) support the magnetic accretion model for their origin (see, e.g., Warner & Woudt 2002, and references therein). Their behavior in the UV during superoutburst (Knigge *et al.* 2002), for example the lack of coherence and the occurrence of harmonics, also resembles DNO behavior.

10.5.2 Quasi-periodic oscillations

In 1977 a second class of unstable optical oscillations was found during DN outbursts; these have $Q \sim 5$, which means that (as they are spread over a wide range of frequency) they are hard to detect in FTs and were only noticed because of large amplitude in the lightcurve (Patterson *et al.* 1977). Their low coherence gives them the name "quasi-periodic oscillations" (QPOs). They have timescales typically an order of magnitude longer than the DNOs, can be present or absent during outbursts, and are independent of whether DNOs are active. They have been commonly seen in DN outbursts and in NLs, and even occasionally in DNe at quiescence (Warner 1995; Woudt & Warner 2002).

Very few observations of CV QPOs in X-rays have been made. In soft X-rays a very low amplitude signal at 83 s in SS Cyg during one outburst was found, and at 111 s in another (Mauche 1997, 2002a); Córdova and Mason (1984) found a 12% amplitude modulation at 585 s in U Gem during outburst and Ramsay *et al.* (2001a) found a modulation at 2240 s at low energies in OY Car just after the end of an outburst. At higher energies (2–10 keV) large amplitude $\sim$500 s modulations in VW Hyi in the final stages of decline from an outburst were found (Wheatley *et al.* 1996b; see also Section 10.2.4).

A third class of optical oscillations has recently been recognized (Warner *et al.* 2003), which are present during outburst and even very occasionally in quiescence, and change very little in period throughout the outburst of a DN. They are represented by the $\sim$88 s oscillations in VW Hyi (Haefner *et al.* 1979), the $\sim$33 s oscillations in SS Cyg (Patterson 1981) and the $\sim$92 s oscillations in the NL EC 2117$-$54 (Warner *et al.* 2003). They are typically a factor $\sim$4 longer in period than the standard DNOs. They may be represented in X-rays by the $\sim$130 s periodicities seen in U Gem during outburst (Córdova *et al.* 1984; Córdova & Mason 1984).

From the rich phenomenology of optical DNOs and QPOs in DN outbursts (especially VW Hyi) it has been suggested that the QPOs are caused by slow progradely traveling waves in the inner disk; probably close to the inner radius where magnetic channeling begins (Warner & Woudt 2002). Then the optical QPOs are caused by "reflection" and obscuration of radiation from the central regions of the disk and WD.

In the X-ray region quasi-periodic obscuration would account for the modulations listed above. In all of those cases the X-ray QPOs are similar in timescale to the optical QPOs in the same stars; no simultaneous observations in optical and X-rays have yet been made. The newly recognized longer period DNOs probably arise from magnetically controlled accretion onto the body of the WD itself (rather than just its equatorial belt).

10.5.3 The CV two-QPO diagram

The optical DNOs and QPOs in CVs behave in a fashion similar to what is seen in the high and low frequency QPOs in LMXBs (Psaltis *et al.* 1999; Section 2.7), namely that $P_{\rm QPO} \sim 15 P_{\rm DNO}$ (Warner & Woudt 2002; Mauche 2002a; Fig. 10.7). The X-ray DNOs and QPOs in SS Cyg agree with this relationship (Mauche 2002a), and the optical modulations in VW Hyi follow such a relationship over a range of a factor of 2 as the star decreases in brightness at the end of outburst (Woudt & Warner 2002). Furthermore, the CV "two-QPO diagram" (i.e., treating the DNOs as higher frequency QPOs) is an extension of that in X-rays to frequencies three orders of magnitude lower (Warner & Woudt 2002; Warner *et al.* 2003; Fig. 10.7).

10.6 X-ray emission from novae

For earlier X-ray reviews of novae we refer the interested reader to, e.g., Ögelman & Orio (1995), Orio (1999), and Krautter (2002).

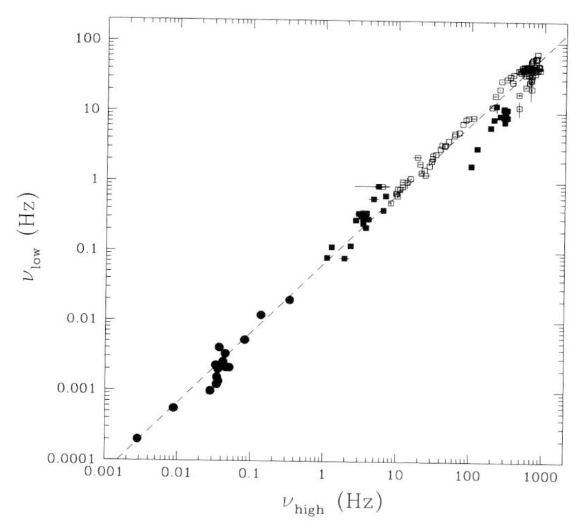

Fig. 10.7. The "two-QPO diagram" for LMXBs (filled squares: black-hole binaries; open squares: neutron-star binaries) and 17 CVs (filled circles). The LMXB data are from Belloni *et al.* (2002). The dashed line marks $P_{QPO}/P_{DNO} = 15$. From Warner *et al.* (2003).

10.6.1 X-ray emission mechanisms

In novae, after the thermo-nuclear runaway in a H-burning shell at the bottom of the accreted layer on the WD, a shock wave may appear and/or a strong radiation driven wind develops (e.g., Webbink *et al.* 1987; Gallagher & Starrfield 1978; Starrfield 1989; Shara 1989; Starrfield *et al.* 1990; Kato & Hachisu 1994). Calculations show that not all the accreted material is ejected, but a substantial fraction (10–90%, depending on the system parameters) can remain on the WD (e.g., Starrfield *et al.* 1972). Residual H burning of the remaining material in a shell on the WD can thus take place, which gives rise to radiation at a constant luminosity, L_{Edd} (referred to as the "constant bolometric luminosity", CBL, phase). If not all of the accreted material is burned and blown away, the WD mass may increase towards the Chandrasekhar mass after a large number of outbursts in RNe, eventually leading to a Type Ia supernova event or to the formation of a neutron star by accretion-induced collapse (e.g., Della-Valle & Livio 1996; Orio & Greiner 1999; but see González-Riestra *et al.* 1998).

A short-lived X-ray emission phase ($\lesssim$1 day) is predicted when the energy of the burning shell first reaches the surface of the WD (Starrfield *et al.* 1990). The energy is then put into the envelope, which leads to a rapid expansion of the WD photosphere up to $\sim$100 $R_\odot$; this leads to a drop in T_{eff}. As the WD photosphere subsequently shrinks back through the ejected material to the equilibrium radius of the WD during the CBL phase, T_{eff} of the photosphere increases again and the peak of the emitted spectrum shifts from visual to UV and finally to the X-ray band (Ögelman *et al.* 1987, 1993; Shore *et al.* 1994; MacDonald 1996; Balman *et al.* 1998, and references therein). At this stage the object is expected to radiate at $L_X \sim 10^{38}$ erg s^{-1} with $kT_{eff} \sim$ 20–85 eV (e.g., Starrfield 1979; Prialnik 1986), very much like a SSS (see Chapter 11). The duration of the CBL phase is expected to last a few years or less, and is proportional to the left-over envelope mass, which in turn is predicted to be inversely proportional to M_{WD} (e.g., Prialnik 1986; Starrfield 1989; Kato & Hachisu 1989, 1994; Kato 1997; see also González-Riestra *et al.* 1998, and references therein). Some novae do not show X-rays during the early stages of outburst. This may be attributed to a large column density of the ejecta and argues in favor of high masses for the expelled gas (e.g., Shore *et al.* 1996). Note that during the first part of the CBL phase the *observed* soft X-ray flux may still increase, due to the clearing of the ejected material as it expands and its density decreases (Krautter *et al.* 1996). Once the H burning ceases, the WD photosphere is expected to cool at a constant radius, and L_X drops gradually. Finally, when novae return to quiescence, accretion may resume again, and they then appear as an ordinary CV. The total energy emitted during all stages is $\sim 10^{44}$–10^{46} erg.

In addition to the soft component due to H burning, shocks in the hot circumstellar material can produce hard X-ray emission[6] (e.g., Brecher *et al.* 1977; Willson *et al.* 1984; Ögelman *et al.* 1987; O'Brien *et al.* 1994; Balman *et al.* 1998, and references therein). Even if there is no shock wave in the outburst, shocks may originate in interacting winds, or they may be due to interaction between ejecta and pre-existing material. For example, in RNe and SBNe the donor is a giant star (often a Mira-type object in SBNe; see also Section 10.7), which has lost a significant amount of material through a wind. Note that CNe typically have main-sequence dwarf donors, which will not have significant stellar winds. Alternatively, shocks may occur in the ejected shells during the different phases of the nova outburst. The expected hard X-ray spectrum is thermal, with $kT_{\rm eff} \sim 0.2$–15 keV, depending on how much time has elapsed since the shock, and how efficient the cooling is, and with $L_X \simeq 10^{33}$–10^{34} erg s^{-1} (e.g., Lloyd *et al.* 1992; O'Brien *et al.* 1994). Additionally a collection of emission lines is expected. Compton scattering of gamma rays produced in the decay of ^{22}Na and ^{26}Al to X-ray energies is expected to also give rise to hard X-ray emission (Starrfield *et al.* 1992; Livio *et al.* 1992; Pistinner *et al.* 1994). However, this may only become important at $kT \gtrsim 6$ keV (Livio *et al.* 1992); moreover, they are probably not the main source of hard X-rays (Starrfield *et al.* 1992).

10.6.2 *Supersoft X-ray emission*

The CN GQ Mus (Nova Mus 1983) was the first nova to show supersoft emission similar to SSS (Ögelman *et al.* 1993; Chapter 11). Subsequently, more novae were found to exhibit such a phase, i.e., the CNe V1974 Cyg (Nova Cyg 1992; Krautter *et al.* 1996; Balman *et al.* 1998), Nova LMC 1995 (Orio & Greiner 1999), V2487 Oph (Nova Oph 1998; Hernanz & Sala 2002), V382 Vel (Nova Vel 1999; Orio *et al.* 2002), and V1494 Aql (Nova Aql 1999 No. 2; e.g., Starrfield *et al.* 2001), and the RN U Sco (Kahabka *et al.* 1999). The soft component has best-fit temperatures of $kT \simeq 30$–80 eV with $L_X \sim 10^{37}$–10^{38} erg s^{-1}. About 20% of the observed novae have shown a supersoft phase (although not all novae have been followed closely enough to exclude such a phase). One of the reasons put forward why not all novae show a supersoft phase is that in most novae all the H might be depleted soon after the outburst, therefore providing no time for a CBL phase (Orio *et al.* 2001a). Simple X-ray fits such as a blackbody do not describe the soft component well (see, e.g., Kahabka *et al.* 1999; Balman & Krautter 2001; Orio *et al.* 2002). This is because the spectral energy distribution of a hot WD differs considerably from a blackbody (e.g., MacDonald & Vennes 1991; Jordan *et al.* 1994; Hartmann & Heise 1997).

[6] Although we here refer to emission typically $\gtrsim 0.1$ keV, we call it *hard* X-ray emission in order to distinguish it from the soft component due to H burning.

The X-ray turn-off times for CNe are generally <3–7 years (e.g., Szkody & Hoard 1994; Orio *et al.* 1996; Hernanz & Sala 2002; see also González-Riestra *et al.* 1998), suggesting that most of the CNe run out of nuclear fuel in the course of a few years after outburst. V1974 Cyg (see Section 10.6.4) turned off only ~18 months after outburst (Krautter *et al.* 1996). The turn-off times are much shorter than the nuclear burning times, and imply that the WD has ejected most of its envelope during or soon after the outburst. Note that GQ Mus turned off ~9–10 years after outburst (Shanley *et al.* 1995); at that time the temperature of the cooling remnant was ~10 eV. This is somewhat longer than most novae, and it is argued that either more mass was left on the WD than generally inferred, or that the short $P_{\rm orb}$, 85.5 min, of GQ Mus and the small donor mass had enhanced the effects of irradiation of the donor (e.g., Diaz & Steiner 1994). This in turn may have induced continued accretion on the WD and prolonged the CBL phase (Ögelman *et al.* 1993; Shanley *et al.* 1995).

The RN CI Aql showed faint ($\sim 10^{33}$ erg s^{-1}) and very soft ($kT \sim 40$–50 eV) X-ray emission, 14 and 16 months after the outburst. However, this is not due to H burning; that phase had already ceased before that time. The observed X-ray emission is suggested to be either due to ionization of the circumstellar material or due to shocks within the wind and/or with the surrounding medium (Greiner & Di Stefano 2002).

In SBNe most of the accreted mass onto the WD is burned quietly for decades, instead of being ejected away quickly as in most CNe. Two SBNe showed supersoft X-ray spectra and luminosities (RR Tel: Jordan *et al.* 1994; Mürset & Nussbaumer 1994; SMC3/RX J0048.4−7332: Kahabka *et al.* 1994; Jordan *et al.* 1996).

10.6.3 Hard X-ray emission

Hard X-ray emission attributed to shocked gas was first detected from the SBN RS Oph in its 1985 outburst. The shocked gas is due to the expanding nova shell colliding with the red-giant wind (e.g., Bode & Kahn 1985). Most CNe and RNe emit hard X-rays in the first months of the outburst with $L_{\rm X}$ peaking at $\sim 10^{33}$–10^{34} erg s^{-1} and bremsstrahlung temperatures in the range 0.5–20 keV (e.g., Orio *et al.* 2001a, and references therein). The eclipsing, fast ($t_3 \simeq 3$ days[7]) CN V838 Her (Nova Her 1991) was detected in hard X-rays 5 days after the outburst, with a $kT \sim 10$ keV bremsstrahlung spectrum (Lloyd *et al.* 1992). The hard emission generally lasts for at least 2 years and even much longer under special circumstances like pre-existing circumstellar material, or a prolonged wind phase (e.g., Orio *et al.* 2001a).

SBNe show generally hard X-ray emission with $L_{\rm X} \lesssim 10^{30}$–$3 \times 10^{33}$ erg s^{-1}. When plotted as time since outburst, there seems to be a general decay law for the X-ray flux in SBNe following outburst (Allen 1981; Kwok & Leahy 1984; Hoard *et al.* 1996; Mürset *et al.* 1997). This is consistent with the decay times of SBNe after the outburst of the order of decades.

10.6.4 V1974 Cyg and V382 Vel

Two CNe, V1974 Cyg and V382 Vel, have been monitored quite frequently in X-rays during the decline from their outburst.

V1974 Cyg was optically a bright nova (Fig. 10.8). It was the first nova to be observed at all wavelengths from gamma rays to radio. It was a moderately fast nova ($t_3 \sim 35$ days) and it

[7] t_3 is the time it takes for the nova to decrease by 3 visual magnitudes.

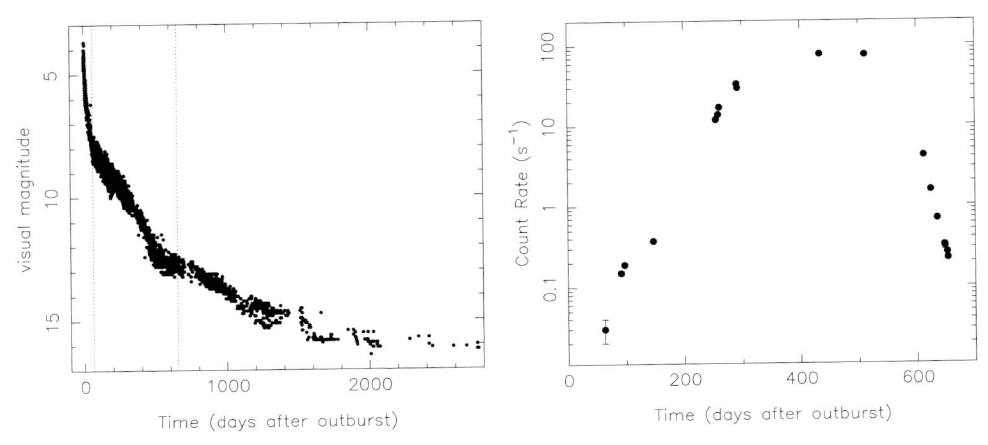

Fig. 10.8. Optical lightcurve (*left*) based on visual and V-band measurements reported to the VSOLJ (Variable Star Observers League in Japan), VSNET (Variable Star NETwork) and AFOEV (Association Française des Observateurs d'Étoiles Variables), and X-ray lightcurve (*right*; from Krautter *et al.* 1996) based on ROSAT/PSPC data (0.1–2.4 keV) of the CN V1974 Cyg (Nova Cyg 1992). The start of the outburst is at February 19, 1992. In the left panel we indicate with dotted lines the time span of the right panel.

was followed from the early rise of the X-ray flux through the time of maximum to the turn-off (Krautter *et al.* 1996; Fig. 10.8). The initial observations showed a hard component with a peak around 1 keV. Subsequently, during the X-ray rise, a much softer component appeared that dominated the spectrum at maximum. This soft component decayed more rapidly (by a factor of $\sim$35) than the hard component. During the early rise $N_{\rm H}$ decreased and then leveled off. The rise in the soft component flux and the decrease in $N_{\rm H}$ is consistent with the clearance of the nova ejecta. The X-ray spectra were best fitted by a WD atmosphere emission model (soft component; $\lesssim$1 keV) and a Raymond–Smith model of thermal plasma (hard component; $\gtrsim$1 keV), see Balman *et al.* (1998). The duration of the CBL is inferred to be $\gtrsim$511 days (Krautter *et al.* 1996).

The hard component spectrum evolved independently of the soft component. The maximum of the hard X-ray emission was reached $\sim$150 days after the outburst with $L_{\rm X} \sim 0.8$–2×10^{34} erg s^{-1}. The time evolution of the hard X-ray flux and the fact that the plasma temperatures decreased from 10 keV to 1 keV suggest emission from shock-heated gas (Balman *et al.* 1998). This component arose from the interaction of the expanding nova wind with density inhomogeneities, as is also indicated by Hubble Space Telescope images within the shell (Paresce *et al.* 1995).

UV observations showed that $\sim$500 days after the outburst the ejecta reached maximum ionization and then started to recombine (Shore *et al.* 1996). This is coincident with the start of the X-ray turn-off observed in X-rays. The decrease in the ionization fraction of the ejecta can be interpreted as a change in the photoionization rate from the central source and is due to a decrease in luminosity and $T_{\rm eff}$ after the cessation of nuclear burning. Two years after the X-ray turn-off a $kT_{\rm eff} \simeq 2$ eV (20 000 K) for the WD was inferred (Shore *et al.* 1997). It is possible that the initial X-ray decline in V1974 Cyg was the result of a cooling WD in which most of the energy was radiated outside the X-ray band pass, rather than representing a return to final quiescence. This implies a longer cooling time, hence a larger amount of mass remaining on the WD (Shore *et al.* 1997).

V382 Vel was a relatively fast nova ($t_3 \sim 10$ days). A short RXTE observation was performed ~3 days after the peak of the optical outburst, but no X-rays were seen. Hard X-rays were detected ~15–17 days after the peak of the optical outburst with $kT \sim 6$–10 keV (Orio *et al.* 2001b; Mukai & Ishida 2001). In the first two months after the outburst it cooled rapidly to $kT \simeq 2.4$ keV (Mukai & Ishida 2001). L_X was $\sim 5 \times 10^{34}$ erg s^{-1} for about 3–5 weeks. As the initially strong intrinsic absorption ($N_H \sim 2 \times 10^{23}$ cm^{-2}; Orio *et al.* 2001b) of the ejected nebula was thinning out, the equivalent N_H decreased to $\sim 2 \times 10^{22}$ cm^{-2} (Mukai & Ishida 2001). Four months later the hard component had cooled to a plasma temperature of $kT < 1$ keV and the absorption column was close to the interstellar value of $N_H \sim 10^{21}$ cm^{-2} (Orio *et al.* 2001b). The hard X-ray emission is interpreted as the result of a shock internal to the nova ejecta. The initial ejecta provide the absorbing column; a layer of later and faster moving ejecta ploughed into the initial ejecta. Note that in the first X-ray observations a weak Fe K line at ≈6.6 keV, with EW ~ 130 eV, was seen. The weakness of the Fe K line is consistent with the shock model, provided that the shocked plasma is not in ionization equilibrium (Mukai & Ishida 2001).

About 6 months after the outburst peak V382 Vel appeared as a bright SSS (Orio *et al.* 2002). The supersoft X-rays were variable by a factor of ~2 on a timescale of minutes. The hard component did not show such variability. Note that variability was also seen in soft X-rays from V1494 Aql, about 10 months after the outburst: it showed a flare which lasted for ~15 min; additionally a periodicity near 42 min was found (Drake *et al.* 2003).

Four months later, the continuum emission from V382 Vel was gone; instead an emission line spectrum, from highly ionized ions, in the supersoft range was inferred. These emission lines presumably have their origin in ionization within the ejected nebula. Note that some of the spectra taken with relatively low resolution may appear softer if there is a strong superimposed nebular emission. The WD then appears to be cooler than obtained from the spectral fits (Orio *et al.* 2002).

V382 Vel was followed for a further period of 8 months, starting 7 months after the outburst. Within a period of less than 6 weeks the flux had dropped by a factor ~200. Thereafter this component continued to decline more slowly, but still somewhat faster than the hard X-rays. The spectra during this period show a wealth of emission lines, which mainly come from H and He-like transitions of Mg, Ne, O, N and C, as well as some Si and Na lines. Most of the lines are broadened with FWHM ~ 2000 km s^{-1}, which is compatible with the velocity of the expanding shell. The He-like triplets of O VII and N VI constrain the plasma temperature to $kT \simeq 39$–43 eV (Burwitz *et al.* 2002b). One possibility is that during the late phase of the nova, when the shell has become transparent, the medium around the system is ionized by the still UV/EUV bright WD. Alternatively, the line-dominated spectrum could result from the interaction of the expanding shell with the circumstellar material, shocks within the expanding shell, or from shocks due to collisions of a fast wind with the interstellar matter (Greiner & Di Stefano 2002).

10.6.5 Quiescent, old novae

Novae in quiescence are often referred to as "old" novae. Sample studies of old novae show X-ray emission with typically $L_X \simeq 10^{30}$–10^{33} erg s^{-1} and $kT \gtrsim 1$ keV, when detected (e.g., Becker 1981, 1989; Becker & Marshall 1981; Córdova & Mason 1983; Orio *et al.* 1993, 2001a; Balman *et al.* 1995; Ögelman & Orio 1995). This is comparable to that found for quiescent DNe (see Section 10.2.1), although in the optical quiescent novae are at least 10 times brighter. Fast novae appear to be brighter than slow novae in quiescence (Becker & Marshall 1981; Orio *et al.* 2001a). Note that V2487 Oph is the first nova to be detected in quiescence both before and after the outburst, with similar X-ray fluxes (Hernanz & Sala 2002). The mass transfer rates derived from L_X are much lower than those inferred from other wavelengths, similar to that found for DNe (see Sections 10.2.3–4). In this case it might be that the BL radiation is emitted entirely in the EUV. The ratio of L_X (0.2–2.4 keV) to L_{opt} varies from 0.005 (DQ Her/Nova Her 1934) to 4.8 (e.g., CP Pup/Nova Pup 1942); for the majority of the systems the value is less than 0.01 (Orio *et al.* 2001a).

GK Per (Nova Per 1901) is the first nova for which a CN shell was detected in X-rays (Balman & Ögelman 1999; Balman 2002). The X-ray nebula is asymmetric and composed of knots/clumps. The temperature and ionization structures do not vary much across the nebula. Its X-ray spectrum is thermal with at least two temperature components, i.e., $kT \sim 0.2$ keV with $L_X \sim 2 \times 10^{31}$ erg s^{-1} and $kT > 30$ keV with $L_X \sim 3 \times 10^{31}$ erg s^{-1}. Distinct Ne IX emission is seen. The knots/clumps are thought to be the result of fragmentation and condensation in the post-shock material. The existence of the X-ray shell in contrast to other CNe is attributed to the high ambient density.

An outlier among the old novae is T Pyx. This RN showed five outbursts between 1890 and 1966. Its outbursts are like those of slow novae, whereas all other RNe show fast nova outbursts. Its luminosity in quiescence is higher than other quiescent novae, $L_X \simeq 1$–3×10^{36} erg s^{-1}. This has been attributed to steady nuclear burning on the WD (e.g., Webbink *et al.* 1987).

10.7 X-ray emission from symbiotic binaries

Symbiotic binaries (SBs; e.g., Kenyon 1986) are divided into two subcategories based on their IR colors. S-type ("stellar") systems have IR colors like those of isolated field red giants, whereas D-type ("dusty") systems have IR colors that are redder, indicative of dust. D-type systems generally contain Mira variables with very high mass-loss rates. The wind is ionized by the WD giving rise to the symbiotic nebula.

SBs show small "outbursts" where the optical increases by ~1–2 mag on timescales of years. The origin of these outbursts is still rather unclear. They may be related to quasi-steady burning of matter on the WD, shell flashes, or unstable accretion. Symbiotic novae (SBNe) are a small subgroup of the SBs. They undergo large amplitude (~7 mag increase) outbursts with durations on the order of decades. These outbursts are thought to be due to thermonuclear runaway events on the surface of the WD. SBNe are discussed in Section 10.6.

Many of the SBs are detected in X-rays. The most X-ray luminous are the D-types, whereas the S-types are typically two orders of magnitude fainter. Also, SBNe are generally brighter in X-rays compared to the other SBs. This is probably because the average luminosities of SBNe are higher than for the other SBs. SBs have been shown to emit either supersoft X-ray emission, and/or hard X-ray emission from an optically thin plasma with $kT \sim 0.25$–1.3 keV and $L_X \sim 10^{30}$–10^{33} erg s^{-1}, or even harder X-ray emission, as from an accreting neutron star (see, e.g., Mürset *et al.* 1997, and references therein).

The hard X-ray emission may be due to the colliding winds of the red giant and the WD (e.g., Willson *et al.* 1984; Kwok & Leahy 1984), jets (Kellogg *et al.* 2001; see also Viotti *et al.* 1987; Leahy & Volk 1995), jets colliding with interstellar material (e.g., Viotti *et al.* 1987; Ezuka *et al.* 1998), X-ray emission from the red giant (e.g., Ezuka *et al.* 1998), or accretion onto the WD (e.g., Leahy & Taylor 1987; Jordan *et al.* 1994). The hard X-ray spectrum of CH Cyg was interpreted as being composed of two components (e.g., Leahy & Volk 1995; Ezuka *et al.* 1998). However, Wheatley (2001) showed that this emission may be solely due to radiation from the WD which is strongly absorbed by a partionally ionized wind from the red giant.

In SBs, Bondi–Hoyle capture of the red-giant wind gives $\dot{M} \sim 10^{-8}\,\mathrm{yr}^{-1}$ onto the WD, which could produce nuclear burning in some systems (e.g., Sion & Starrfield 1994). This might give rise to supersoft X-ray emission, similar to that seen in CNe and RNe (see Section 10.6) and other SSS (see Chapter 11). Depending on how much matter the WD receives and steadily burns, it may be able to increase its mass. SBs have therefore also been put forward as candidates for Type Ia supernovae (e.g., Munari & Renzini 1992). It has been noted, however, that either not enough mass can be accreted from the donor, or a mass outflow from the WD may inhibit such a phase (Mürset & Nussbaumer 1994; Mürset *et al.* 1997).

SBs have indeed shown supersoft X-ray emission (e.g., Mürset *et al.* 1997; see also Section 10.6.2). For example, X-ray spectra during quiescence of AG Dra, which is one of the brightest X-ray sources among the SBs (e.g., Anderson *et al.* 1981), showed a soft component with a temperature of $kT_{\mathrm{bb}} \simeq 14$–$15\,\mathrm{eV}$ (Greiner *et al.* 1997; see also Piro *et al.* 1985; Kenyon 1988). The X-ray luminosity remained constant during quiescence at a level of $L_{\mathrm{X}} \simeq 10^{37}\,\mathrm{erg\,s^{-1}}$ (Greiner *et al.* 1997).

AG Dra was the first SB to be observed during outburst in X-rays (Viotti *et al.* 1995). During the minor ($\Delta V \sim 1$ mag) and major ($\Delta V \sim 2$ mag) outbursts it displays different behavior. During both outbursts the X-ray flux drops; however, kT decreases by a few eV during major outbursts, while it increases by a few eV during minor outbursts. The behavior during major outbursts is explained by expansion (factor 2–6) and cooling of the WD atmosphere, which also explains the anti-correlation between optical/UV and X-ray fluxes (Greiner *et al.* 1997; González-Riestra *et al.* 1999). The cooling could be the result of an increased $\dot{M}$ onto the compact object causing it to expand slowly (Greiner *et al.* 1997). The drop in X-ray flux during minor outbursts might be attributed to the increased absorbing layer between the X-ray source and observer (Friedjung 1988). During a minor outburst Z And also radiated well above $\sim$1 keV, possibly due to the red-giant wind colliding with material ejected during the outburst (Sokoloski *et al.* 2002).

10.8 Concluding remarks

During the last decade there has been a staggering flow of new information, not only in X-rays, but also from other wavelengths. Nowadays, substantial progress in understanding the behavior of CVs is generally made through multi-wavelength efforts. Unfortunately, the space available in this chapter has not allowed us to fully discuss observations across the whole electromagnetic spectrum. Nevertheless, we have shown that our somewhat "restricted" view at EUV and X-ray wavelengths has revealed a great variety of rich phenomena taking place in the environment of an accreting WD.

Substantial improvements in our understanding of CVs at high energies are already coming from the instruments on board Chandra and XMM-Newton and further advances are inevitable over the next few years. These instruments provide the necessary spectral resolution to perform sensible temperature and velocity diagnostics from individually resolved lines and line ratios; something we have previously been used to at optical wavelengths. These exquisite instruments and those which are yet to come will provide us with much more insight into the physical processes that give rise to the rich phenomenology of CVs. This applies to all facets of CV research addressed in this chapter, processes of energy release in magnetic and non-magnetic environments, radiation hydrodynamics, dynamical processes in close binaries and the long-term close binary evolution.

Acknowledgements

We acknowledge discussions with K. Beuermann, J.-P. Lasota, C. Mauche, M. Orio, J. Sokoloski, S. Starrfield and P. Wheatley. AS is supported by the German Bundesministerium für Bildung und Forschung through the Deutsches Zentrum für Luft- und Raumfahrt e.V. (DLR) under grant number 50 OR 9706 8. BW is supported by research funds from the University of Cape Town.

References

Allan, A., *et al.* (1996), *MNRAS* **279**, 1345
Allan, A., Hellier, C., Beardmore, A. (1998), *MNRAS* **295**, 167
Allen, D. A. (1981), *MNRAS* **197**, 739
Anderson, C. M., Cassinelli, J. P., Sanders, W. T. (1981), *ApJ* **247**, L127
Armitage, P. J., Livio, M. (1996), *ApJ* **470**, 1024
Armitage, P. J., Livio, M. (1998), *ApJ* **493**, 898
Aizu, K. (1973), *Prog. Theor. Phys.* **49**, 1184
Bailey, J. (1995), *ASP Conf. Ser.* **85**, p. 10
Balman, S. (2002), *ASP Conf. Ser.* **261**, p. 617
Balman, S., Krautter, J. (2001), *MNRAS* **326**, 1441
Balman, S., Ögelman, H. B. (1999), *ApJ* **518**, L111
Balman, S., Orio, M., Ögelman, H. (1995), *ApJ* **449**, L47
Balman, S., Krautter, J., Ögelman, H. (1998), *ApJ* **499**, 395
Baskill, D. S., Wheatley, P. J., Osborne, J. P. (2001), *MNRAS* **328**, 71
Bath, G. T., Evans, W. D., Pringle, J. E. (1974a), *MNRAS* **166**, 113
Bath, G. T., *et al.* (1974b), *MNRAS* **169**, 447
Beardmore, A. P., *et al.* (1998), *MNRAS* **297**, 337
Beardmore, A. P., Osborne, J. P., Hellier, C. (2000), *MNRAS* **315**, 307
Becker, R. H. (1981), *ApJ* **251**, 626
(1989), in *Classical Novae*, eds. M. F. Bode & A. Evans, New York: Wiley, p. 215
Becker, R. H., Marshall, F. E. (1981), *ApJ* **244**, L93
Belloni, T., *et al.* (1991), *A&A* **246**, L44
Belloni, T., Psaltis, D., van der Klis, M. (2002), *ApJ* **572**, 392
Beuermann, K. (1999), *MPE Report* **272**, p. 410
Beuermann, K., Burwitz, V. (1995), *ASP Conf. Ser.* **85**, p. 99
Beuermann, K., Schwope A. (1994), *ASP Conf. Ser.* **56**, p. 119
Beuermann, K., Thomas, H.-C. (1993), *AdSpR* **13**, 115
Beuermann, K., Thomas, H.-C., Pietsch, W. (1991), *A&A* **246**, L36
Bode, M. F., Kahn, F. D. (1985), *MNRAS* **217**, 205
Bonnet-Bidaud, J. M., Mouchet, M. (1987), *A&A* **188**, 89
Brecher, K., Ingham, W. H., Morrison, P. (1977), *ApJ* **213**, 492
Buckley, D. A. H., Tuohy, I. R. (1989), *ApJ* **344**, 376
Buckley, D. A. H., *et al.* (1997), *MNRAS* **287**, 117
Burwitz, V., *et al.* (1996), *A&A* **310**, 25
(2002a), *ASP Conf. Ser.* **261**, p. 137
(2002b), *AIP Conf. Proc.* **637**, p. 377
Córdova, F. A., 1995, in *X-ray Binaries*, eds. W. H.G. Lewin, *et al.*, Cambridge University Press, p. 331

Córdova, F. A., Mason, K. O. (1983), in *Accretion Driven Stellar X-ray Sources*, eds. W. H. G. Lewin & E. P. J. van den Heuvel, Cambridge University Press, p. 147
(1984), *MNRAS* **206**, 879
Córdova, F. A., *et al.* (1980a), *MNRAS* **190**, 87
(1980b), *ApJ* **235**, 163
Córdova, F. A., Mason, K. O., Nelson, J. E. (1981), *ApJ* **245**, 609
Córdova, F. A., *et al.* (1984), *ApJ* **278**, 739
Cropper, M. (1990), *SSRv* **54**, 195
Cropper, M., Ramsay G., Wu., K. (1998), *MNRAS* **293**, 222
Cropper, M., Wu, K., Ramsay G. (2000), *NewAR* **44**, 57
de Martino, D., *et al.* (2001), *A&A* **377**, 499
Della-Valle, M., Livio, M. (1996), *ApJ* **473**, 240
Diaz, M. P., Steiner, J. E. (1994), *ApJ* **425**, 252
Done C., Magdziarz, P. (1998), *MNRAS* **298**, 737
Done, C., Osborne, J. P. (1997), *MNRAS* **288**, 649
Drake, J. J., *et al.* (2003), *ApJ* **584**, 448
Drew, J., Verbunt, F. (1985), *MNRAS* **213**, 191
Duck, S. R., *et al.* (1994) *MNRAS* **271**, 372
Eracleous, M., Halpern, J., Patterson, J. (1991a), *ApJ* **382**, 290
Eracleous, M., Patterson, J., Halpern, J. (1991b), *ApJ* **370**, 330
Ezuka, H., Ishida, M., Makino, F. (1998), *ApJ* **499**, 388
Ferland, G. J., *et al.* (1982), *ApJ* **262**, L53
Fischer A., Beuermann K. (2001), *A&A* **373**, 211
Fontaine, G., Montmerle, T., Michaud, G. (1982), *ApJ* **257**, 695
Frank, J., King, A. R., Lasota, J.-P. (1987), *A&A* **178**, 137
(1988), *A&A* **193**, 113
Frank, J., King, A. R., Raine, D. (1992), *Accretion Power in Astrophysics*, 2nd edition, Cambridge University Press
Friedjung, M. (1988), *Proc. IAU Coll.* **103**, p. 199
Fujimoto, R., Ishida, M. (1995), *ASP Conf. Ser.* **85**, p. 136
Gallagher, J. S., Starrfield, S. (1978), *ARA&A* **16**, 171
Gänsicke, B. T., Beuermann, K., de Martino, D. (1995), *A&A* **303**, 127
González-Riestra, R., Orio, M., Gallagher, J. (1998), *A&ASS* **129**, 23
González-Riestra, R., *et al.* (1999), *A&A* **347**, 478
Greiner, J. (1999), *A&A* **336**, 626
Greiner, J., Di Stefano, R. (2002), *ApJ* **578**, L59
Greiner, J., van Teeseling, A. (1998), *A&A* **339**, L21
Greiner, J., *et al.* (1997), *A&A* **322**, 576
Greiner, J., *et al.* (1999), *A&A* **343**, 183
Haberl, F., Motch, C. (1995), *A&A* **297**, L37
Haberl, F., *et al.* (1994), *A&A* **291**, 171
Hack, M., La Dous, C. (1993), *Cataclysmic Variables and Related Objects*, NASA SP-507
Haefner, R, Schoembs, R., Vogt, N. (1979), *A&A* **77**, 7
Hameury, J.-M., King, A. R., Lasota, J.-P. (1986), *MNRAS* **218**, 695
Hartmann, H. W., Heise, J. (1997), *A&A* **322**, 591
Hartmann, H. W., *et al.* (1999), *A&A* **349**, 588
Heise, J., *et al.* (1978), *A&A* **63**, L1
(1985), *A&A* **148**, L14
Hellier, C. (1991), *MNRAS* **251**, 693
(1992), *MNRAS* **258**, 578
(1996), *Proc. IAU Coll.* **158**, p. 143
(1997), *MNRAS* **291**, 71
Hellier, C., Beardmore, A. P. (2002), *MNRAS* **331**, 407
Hellier, C., *et al.* (1989), *MNRAS* **238**, 1107
Hellier, C., Cropper, M., Mason, K. O. (1991), *MNRAS* **248**, 233
Hellier, C., Garlick, M. A., Mason, K. O. (1993), *MNRAS* **260**, 299
Hellier, C., *et al.* (1996), *MNRAS* **280**, 877

Hellier, C., Mukai, K., Beardmore, A. P. (1997), *MNRAS* **292**, 397
Hellier, C., Mukai, K., Osborne, J. P. (1998), *MNRAS* **297**, 526
Hempelmann, A., *et al.* (1995), *A&A* **294**, 515
Hernanz, M., Sala, G. (2002), *Science* **298**, 393
Hessman, F. V., Gänsicke, B. T., Mattei, J. A. (2000), *A&A* **361**, 952
Hoard, D. W., Wallerstein, G., Willson, L. A. (1996), *PASP* **108**, 81
Hoare, M. G., Drew, J. E. (1991), *MNRAS* **249**, 452
Horne, K., *et al.* (1994), *ApJ* **426**, 294
Iben, Jr. I., Livio, M. (1993), *PASP* **105**, 1373
Ishida, M. (1991), Ph.D. thesis, University of Tokyo
Ishida, M., Ezuka, H. (1999), *ASP Conf. Ser.* **157**, p. 333
Ishida, M., Fujimoto, R. (1995), *ASP Conf. Ser.* **85**, p. 132
Ishida, M., *et al.* (1992), *MNRAS* **254**, 647
Jensen, K. A. (1984), *ApJ* **278**, 278
Jones, M. H., Watson, M. G. (1992), *MNRAS* **257**, 633
Jordan, S., Mürset, U., Werner, K. (1994), *A&A* **283**, 475
Jordan, S., *et al.* (1996), *A&A* **312**, 897
Kahabka, P., Pietsch, W., Hasinger, G. (1994), *A&A* **288**, 538
Kahabka, P., *et al.* (1999), *A&A* **347**, L43
Kallman, T. R., Jensen, K. A. (1985), *ApJ* **299**, 277
Kamata, Y., Koyama K. (1993), *ApJ* **405**, 307
Kato, M. (1997), *ApJS* **113**, 121
Kato, M., Hachisu, I. (1989), *ApJ* **346**, 424
(1994), *ApJ* **437**, 802
Katz, J. I. (1975), *ApJ* **200**, 298
Kellogg, E., Pedelty, J. A., Lyon, L. G. (2001), *ApJ* **563**, L151
Kenyon, S. J. (1986), *The Symbiotic Stars*, Cambridge University Press
(1988), *Proc. IAU Coll.* **103**, p. 11
King, A. R. (1997), *MNRAS* **288**, L16
King, A. R., Cannizzo, J. K. (1998), *ApJ* **499**, 348
King, A. R., Lasota, J.-P. (1979) *MNRAS* **188**, 653
(1991) *ApJ* **378**, 674
King, A. R., Shaviv, G. (1984), *MNRAS* **211**, 883
Knigge, C., *et al.* (2002), *ApJ* **580**, L151
Krautter, J. (2002), *AIP Conf. Proc.* **637**, p. 345
Krautter, J., *et al.* (1996), *ApJ* **456**, 788
Krzemiński, W., Serkowski, K. (1977), *ApJ* **216**, L45
Kuijpers, J., Pringle, J. E. (1982), *A&A* **114**, L4
Kunze, S., Speith, R., Hessman, F. V. (2001), *MNRAS* **322**, 499
Kuulkers, E., *et al.* (1998), *ApJ* **494**, 753
(2002), *ASP Conf. Proc.* **261**, p. 443
Kwok, S., Leahy, D. A. (1984), *ApJ* **283**, 675
Lamb, D. Q., Masters, A. R. (1979), *ApJ* **234**, L117
Lasota, J.-P. (2001), *NewAR* **45**, 449
Lasota, J. P., Hameury, J. M., Huré, J. M. (1995), *A&A* **302**, L29
Leahy, D. A., Taylor, A. R. (1987), *A&A* **176**, 262
Leahy, D. A., Volk, K. (1995), *ApJ* **440**, 847
Litchfield, S. J., King, A. R. (1990), *MNRAS* **247**, 200
Livio, M., Pringle, J. E. (1992), *MNRAS* **259**, 23P
Livio, M., *et al.* (1992), *ApJ* **394**, 217
Lloyd, H. M., *et al.* (1992), *Nature* **356**, 222
Long, K. S., *et al.* (1996), *ApJ* **469**, 841
Lynden-Bell, D., Pringle, J. E. (1974), *MNRAS* **168**, 603
MacDonald, J. (1996), *Proc. IAU Coll.* **158**, p. 281
MacDonald, J., Vennes, S. (1991), *ApJ* **371**, 719
Marsh, T. R. (1999), *MNRAS* **304**, 443
Marsh, T. R., Steeghs, D. (2002), *MNRAS* **331**, L7

Mason, K. O. (1985), *SSRv* **40**, 99
(1986), *Lecture Notes in Physics* **266**, p. 29
Mason, K. O., *et al.* (1978), *ApJ* **226**, L129
(1988), *MNRAS* **232**, 779
(1992), *MNRAS* **258**, 749
Mason, K. O., Drew, J. E., Knigge, C. (1997), *MNRAS* **290**, L23
Matt, G., *et al.* (2000), *A&A* **358**, 177
Mauche, C. W. (1996a), *ApJ* **463**, L87
(1996b), *Proc. IAU Coll.* **158**, p. 243
(1997), *ApJ* **476**, L85
(1999), *ASP Conf. Ser.* **157**, p. 157
(2002a), *ApJ* **580**, 423
(2002b), *ASP Conf. Ser.* **261**, p. 113
Mauche, C. W., Mukai, K. (2002), *ApJ* **566**, L33
Mauche, C. W., Raymond, J. C. (1987), *ApJ* **323**, 690
(2000), *ApJ* **541**, 924
Mauche, C. W., Robinson, E. L. (2001), *ApJ* **562**, 508
Mauche, C. W., *et al.* (1991), *ApJ* **372**, 659
Mauche, C. W., Raymond, J. C., Mattei, J. A. (1995), *ApJ* **446**, 842
Mauche, C. W., Liedahl, D. A., Fournier, K. B. (2001), *ApJ* **560**, 992
Meyer, F., Meyer-Hofmeister, E. (1994), *A&A* **288**, 175
Motch, C., *et al.* (1996), *A&A* **307**, 459
Mukai, K. (1995), *ASP Conf. Ser.* **85**, p. 119
(2000), *NewAR* **44**, 9
Mukai, K., Ishida, M. (2001), *ApJ* **551**, 1024
Mukai, K., Shiokawa, K. (1993), *ApJ* **418**, 863
Mukai, K., *et al.* (1997), *ApJ* **475**, 812
(2003), *ApJ* **586**, L77
Munari, U., Renzini, A. (1992), *ApJ* **397**, L87
Mürset, U., Nussbaumer, H. (1994), *A&A* **282**, 586
Mürset, U., Wolff, B., Jordan, S. (1997), *A&A* **319**, 201
Narayan, R., Popham, R. (1993), *Nature* **362**, 820
Naylor, T., La Dous, C. (1997), *MNRAS* **290**, 160
Naylor, T., *et al.* (1987), *MNRAS* **229**, 183
(1988), *MNRAS* **231**, 237
Norton, A. J. (1993), *MNRAS* **265**, 316
Norton, A. J., Watson, M. G. (1989), *MNRAS* **237**, 853
Norton, A. J., Watson, M. G., King, A. R. (1991), *Lecture Notes in Physics* **385**, p. 155
Norton, A. J., *et al.* (1992a), *MNRAS* **254**, 705
(1992b), *MNRAS* **258**, 697
(1997), *MNRAS* **289**, 362
(1999), *A&A* **347**, 203
Norton, A. J., Wynn, G. A., Somerscales, R. V. (2004), *ApJ* **614**, 349
O'Brien, T. J., Lloyd, H. M., Bode, M. F. (1994), *MNRAS* **271**, 155
Ögelman, H., Orio, M. (1995), in *Cataclysmic Variables*, eds. A. Bianchini, *et al.*, Dordrecht: Kluwer, p. 11
Ögelman, H., Krautter, J., Beuermann, K. (1987), *A&A* **177**, 110
Ögelman, H., *et al.* (1993), *Nature* **361**, 331
Orio, M. (1999), *Phys. Rep.* **311**, 419
Orio, M., Greiner, J. (1999), *A&A* **344**, L13
Orio, M., *et al.* (1993), *AdSpR* **13**, 351
(1996), *ApJ* **466**, 410
Orio, M., Covington, J., Ögelman, H. (2001a), *A&A* **373**, 542
Orio, M., *et al.* (2001b), *MNRAS* **326**, L13
(2002), *MNRAS* **333**, L11
Paczyński, B. (1978), in *Nonstationary Evolution in Close Binaries*, ed. A. Żytkow, Polish Sci. Publ., Warsaw, p. 89
Paresce, F., *et al.* (1995), *A&A* **299**, 823

Parmar, A. N., White, N. E. (1988), *Mem. Soc. Astr. It.* **59**, 147
Patterson, J. (1981), *ApJS* **45**, 517
(1984), *ApJS* **54**, 443
Patterson, J., Price, C. (1980), *IAU Circ.* 3511
Patterson, J., Raymond, J. C. (1985a), *ApJ* **292**, 535
(1985b), *ApJ* **292**, 550
Patterson, J., Robinson, E. L., Nather, R. E. (1977), *ApJ* **214**, 144
(1978), *ApJ* **224**, 570
Patterson, J., *et al.* (1998), *PASP* **110**, 403
(2001), *PASP* **113**, 72
Piro, L., *et al.* (1985), *IAU Circ.* 4082
Pistinner, S., Shaviv, G., Starrfield, S. (1994), *ApJ* **437**, 794
Podsiadlowski, Ph., Han, Z., Rappaport, S. (2003), *MNRAS* **340**, 1214
Polidan, R. S., Holberg, J. B. (1984), *Nature* **309**, 528
Polidan, R. S., Mauche, C. W., Wade, R. A. (1990), *ApJ* **356**, 211
Ponman, T. J., *et al.* (1995), *MNRAS* **276**, 495
Popham, R., Narayan, R. (1995), *ApJ* **442**, 337
Pratt, G. W., *et al.* (1999a), *MNRAS* **307**, 413
(1999b), *MNRAS* **309**, 847
Predehl, P., Schmitt, J. H. M. M. (1995), *A&A* **293**, 889
Prialnik, D. (1986), *ApJ* **310**, 222
Pringle, J. E. (1977), *MNRAS* **178**, 195
(1981), *ARA&A* **19**, 137
Pringle, J. E., Rees, M. J. (1972), *A&A* **21**, 1
Pringle, J. E., Savonije, G. J. (1979), *MNRAS* **187**, 777
Pringle, J. E., *et al.* (1987), *MNRAS* **225**, 73
Psaltis, D., Belloni, T., van der Klis, M. (1999), *ApJ* **520**, 262
Ramsay, G., *et al.* (1994,) *MNRAS* **270**, 692
(2001a), *A&A* **365**, L288
(2001b), *A&A* **365**, L294
Rappaport, S., *et al.* (1974), *ApJ* **187**, L5
Raymond, J. C., Mauche, C. W. (1991), in *Extreme Ultraviolet Astronomy*, eds. R. F. Malina & S. Bowyer, New York: Pergamon, p. 163
Reimers, D., Hagen, H.-J. (2000), *A&A* **358**, L45
Reimers, D., Hagen, H.-J., Hopp, U. (1999), *A&A* **343**, 157
Richman, H. R. (1996), *ApJ* **462**, 404
Ritter, H., Kolb, U. (2003), *A&A* **404**, 301
Rosen, S. R., Mason, K. O., Córdova, F. A. (1988), *MNRAS* **231**, 549
Rosen, S. R., *et al.* (1991), *MNRAS* **249**, 417
Rosner, R., Golub, L., Vaiana, G. S. (1985), *ARA&A* **23**, 413
Schlegel, E. M., Singh, J. (1995), *MNRAS* **276**, 1365
Schmitt, J. H. M. M., *et al.*, (1990), *ApJ* **365**, 704
Schwarz, R., *et al.* (2002), *ASP Conf. Ser.* **261**, p. 167
Schwope, A. D. (1996), *Proc. IAU Coll.* **158**, p. 189
(2001), *Lecture Notes in Physics* **573**, p. 127
Schwope, A. D., *et al.* (2001), *A&A* **375**, 419
(2002), *A&A* **392**, 541
Shakura, N. I., Sunyaev, R. A. (1973), *A&A* **24**, 337
Shanley, L., *et al.* (1995), *ApJ* **438**, L95
Shara, M. M. (1989), *PASP* **101**, 5
Shore, S. N., *et al.* (1994), *ApJ* **421**, 344
Shore, S. N., Starrfield, S., Sonneborn, G. (1996), *ApJ* **463**, L21
Shore, S. N., *et al.* (1997), *ApJ* **490**, 393
Silber, A., Vrtilek, S. D., Raymond, J. C. (1994), *ApJ* **425**, 829
Sion, E. M., Starrfield, S. G. (1994), *ApJ* **421**, 261
Sion, E. M., *et al.* (1996), *ApJ* **471**, L41
(2002), *ASP Conf. Proc.* **261**, p. 69

Sirk, M. M., Howell, S. B. (1998), *ApJ* **506**, 824
Sokoloski, J. L., *et al.* (2002), *ASP Conf. Proc.* **261**, p. 667
Starrfield, S. (1979), *Proc. IAU Coll.* **53**, p. 274
(1989), in *Classical Novae*, eds. M. F. Bode & A. Evans, New York: Wiley, p. 39
Starrfield, S., *et al.* (1972), *ApJ* **176**, 169
(1990), *Proc. IAU Coll.* **122**, p. 306
(1992), *ApJ* **391**, L71
(2001), *Bull. AAS* **198**, 11.09
Swank, J. H. (1979), *Proc. IAU Coll.* **53**, p. 135
Swank, J. H., *et al.* (1978), *ApJ* **226**, L133
Szkody, P. (1999), *Frontiers Science Ser.* **26**, p. 53
Szkody, P., Hoard, D. W. (1994), *ApJ* **429**, 857
Szkody, P., Osborne, J., Hassall, B. J. M. (1988), *ApJ* **328**, 243
Szkody, P., Kii, T., Osaki, Y. (1990), *AJ* **100**, 546
Szkody, P., *et al.* (1996), *ApJ* **469**, 834
(1999), *ApJ* **521**, 362
(2002a), *ApJ* **574**, 942
(2002b), *AJ* **123**, 413
Taylor, P., *et al.* (1997), *MNRAS* **289**, 349
Teeseling, A. van (1997a), *A&A* **319**, L25
(1997b), *A&A* **324**, L73
Teeseling, A. van, Verbunt, F. (1994), *A&A* **292**, 519
Teeseling, A. van, Verbunt, F., Heise, J. (1993), *A&A* **270**, 159
Teeseling, A. van *et al.* (1995), *A&A* **300**, 808
Teeseling, A. van, Beuermann, K., Verbunt, F. (1996), *A&A* **315**, 467
Teeseling, A. van, Fischer, A., Beuermann, K. (1999), *ASP Conf. Ser.* **157**, p. 309
Tylenda, R. (1981), *Acta Astronomica* **31**, 127
Ulla, A. (1995), *A&A* **301**, 469
Verbunt, F. (1987), *A&AS* **71**, 339
(1996), *MPE Report* **263**, p. 93
Verbunt, F., *et al.* (1997), *A&A* **327**, 602
Verbunt, F., Wheatley, P. J., Mattei, J. A. (1999), *A&A* **346**, 146
Viotti, R., *et al.* (1987), *ApJ* **319**, L7
(1995), in *Cataclysmic Variables*, eds. A. Bianchini, *et al.*, Dordrecht: Kluwer, p. 195
Vrtilek, S. D., *et al.* (1994), *ApJ* **425**, 787
Walker, M. F. (1956), *ApJ* **123**, 68
Warner, B. (1980), *IAU Circ.* 3511
(1983), *Proc. IAU Coll.* **72**, p. 155
(1986), *MNRAS* **219**, 347
(1995), *Cataclysmic Variable Stars*, Cambridge University Press
Warner, B., Robinson, E. L. (1972), *Nature Phys. Sci.* **239**, 2
Warner, B., Woudt, P. A. (2002), *MNRAS* **335**, 84
Warner, B., *et al.* (1972), *MNRAS* **159**, 321
Warner, B., O'Donoghue, D., Wargau, W. (1989), *MNRAS* **238**, 73
Warner, B., Livio, M., Tout, C. A. (1996), *MNRAS* **282**, 735
Warner, B., Woudt, P. A., Pretorius, M. L. (2003), *MNRAS* **344**, 1193
Warren, J. K., Sirk, M. M., Vallerga, J. V. (1995), *ApJ* **445**, 909
Watson, M. G. (1986), *Lecture Notes in Physics* **266**, p. 97
Watson, M. G., King, A. R., Heise, J. (1985), *SSRv* **40**, 127
Watson, M. G., *et al.* (1989), *MNRAS* **237**, 299
Webbink, R., Wickramasinghe, D. T. (2002), *MNRAS* **335**, 1
Webbink, R. F., *et al.* (1987), *ApJ* **314**, 653
Wheatley, P. J. (2001), *AIP Conf. Ser.* **599**, p. 1007
Wheatley, P. J., West, R. G. (2002), *ASP Conf. Proc.* **261**, p. 433
Wheatley, P. J., *et al.* (1996a), *MNRAS* **283**, 101
(1996b), *A&A* **307**, 137
(2000), *NewAR* **44**, P33

Wheatley, P. J., Mauche, C. W., Mattei, J. A. (2003), *MNRAS* **345**, 49
White, N. E., Marshall, F. E. (1980), *IAU Circ.* 3514
White, N. E., Nagase, F., Parmar, A. N. (1995), in *X-ray Binaries*, eds. W. H. G. Lewin, *et al.*, Cambridge University Press, p. 1
Wickramasinghe, D. T., Ferrario, L. (2000), *PASP* **112**, 873
Willson, L. A., *et al.* (1984), *A&A* **133**, 137
Woelk, U., Beuermann, K. (1996), *A&A* **306**, 232
Woerd, H. van der, Heise, J., Bateson, F. (1986), *A&A* **156**, 252
Woerd, H. van der, *et al.* (1987), *A&A* **182**, 219
Wood, J. H., *et al.* (1995a), *MNRAS* **273**, 772
Wood, J. H., Naylor, T., Marsh, T. R. (1995b), *MNRAS* **274**, 31
Woods, A. J., Drew, J. E., Verbunt, F. (1990), *MNRAS* **245**, 323
Woudt, P. A., Warner, B. (2002), *MNRAS* **333**, 411
Wu, K., *et al.* (2002), *MNRAS* **331**, 221
Wynn, G. A., King, A. R. (1992), *MNRAS* **255**, 83
Yoshida, K., Inoue, H., Osaki, Y. (1992), *PASJ* **44**, 537

11

Super-soft sources

P. Kahabka
Max-Planck-Institut für extraterrestrische Physik

E. P. J. van den Heuvel
University of Amsterdam

11.1 Introduction

The luminous super-soft X-ray sources (SSS) were recognized as an important new class of intrinsically bright X-ray sources by Trümper *et al.* (1991) (see also Greiner *et al.* 1991). In fact four of them had already been found in the Magellanic Clouds with the Einstein Observatory around 1980, but they had not been recognized as a separate new class (Long *et al.* 1981; Seward & Mitchell 1981). A careful analysis of the ROSAT data on the first LMC sources showed that while their X-ray luminosities can be as high as the Eddington limit (they range from $\sim 10^{36}$ to 10^{38} erg s^{-1}), their X-ray spectra are extremely soft, typically peaking in the range 20–100 eV, corresponding to blackbody temperatures of $\sim 10^5$ to $\sim 10^6$ K. This is some two orders of magnitude lower than for a classical X-ray binary that contains an accreting neutron star or black hole. Some 40 SSS have been discovered with ROSAT, 16 in the Andromeda Nebula (M31), about a dozen in the Magellanic Clouds, 10 in our own Galaxy and one in NGC 55. Since then, several dozens of SSS have been discovered with BeppoSAX, Chandra and XMM-Newton, mostly in external galaxies. The latter are listed in Section 11.10. A catalog of SSS is given in Greiner (2000a).

In view of the very large interstellar extinction of soft X-rays due to neutral hydrogen, the sources in other galaxies can only be observed when they are near the outer edge of the interstellar hydrogen layer, at the outside of a galaxy. Taking this into account, the total number of sources in M31 and our Galaxy is estimated to be some two orders of magnitude larger than the observed numbers, i.e., of the order of a few thousand (Rappaport *et al.* 1994). The SSS therefore form a major new population of highly luminous X-ray sources in spiral galaxies like our own. The same holds for irregular galaxies like the Magellanic Clouds.

From the observed luminosities and blackbody temperatures of SSS one infers effective stellar radii comparable to those of white dwarfs (WDs). Their observed characteristics are consistent with those of WDs that are steadily or cyclically burning hydrogen-rich matter accreted onto the surface at a rate of the order 10^{-7} $M_\odot$ yr^{-1} (van den Heuvel *et al.* 1992). The required high accretion rates can be supplied by mass transfer on a thermal timescale from a close companion that is more massive than the WD accretor, typically 1.3–3.5 $M_\odot$. These are the so-called close binary super-soft sources (CBSS). Alternatively, the donor star may also be a low-mass red (sub-)giant, these are the symbiotic systems, among which there are also symbiotic (recurrent) novae such as RS Oph and T CrB. Steady burning can also occur in a post-nova phase (CV-type SSS), but only for relatively short timescales, up to decades (Ögelman *et al.* 1993; Krautter *et al.* 1996; Orio & Greiner 1999; Orio *et al.* 2002). All these systems are binaries, but some single star SSS have also been observed, presumably

Compact Stellar X-Ray Sources, eds. Walter Lewin and Michiel van der Klis.
Published by Cambridge University Press.

highly evolved stars on their way to the WD phase, such as PG 1159 stars and nuclei of some planetary nebulae (see Reinsch *et al.* 2002).

Recently it has been suggested, based on similarities in optical variability, that a number of unusual CVs, such as V Sge, T Pyx, WX Cen also belong to the SSS class (Patterson *et al.* 1998; Greiner 2000b).

In this chapter we concentrate on the binary SSS. Optically identified systems are listed in Table 11.1. Additional SSS not included in the table are 1E 1339.8+2837 in the globular cluster M3 (Dotani *et al.* 1999) and RX J0550.9−7151 in the LMC (Reinsch *et al.* 1999). For seven of these systems orbital periods in the range ~9 hr to ~4 d have been determined. It was for these CBSS systems that the "surface nuclear burning" model was first established (van den Heuvel *et al.* 1992). For an earlier review we refer to Kahabka and van den Heuvel (1997).

11.2 Nuclear burning

Nuclear burning is ignited in an envelope of hydrogen-rich matter accreted onto a WD if a critical envelope mass $\Delta M_{\rm crit}$ has been reached which can sustain the high temperature ($\sim 10^8$ K) and pressure ($\gtrsim 10^{18}-10^{20}$ g cm^{-1} s^{-2}) required for nuclear burning, mainly the CNO cycle (Fujimoto 1982a,b). $\Delta M_{\rm crit}$ decreases with increasing WD mass $M_{\rm WD}$ and increasing accretion rate $\dot{M}_{\rm acc}$ and is (for $\dot{M}_{\rm acc} \geq 10^{-10}$ M$_\odot$ yr^{-1} for a WD temperature $T_{\rm WD} = 10^7$ K)

$$\log\left(\frac{\Delta M_{\rm crit}}{\rm M_\odot}\right) \approx A + B\left(\frac{M_{\rm WD}}{\rm M_\odot}\right)^{-1.436} \ln\left(1.429 - \frac{M_{\rm WD}}{\rm M_\odot}\right) + C\left(\log\left(\frac{\dot{M}_{\rm acc}}{\rm M_\odot\ yr^{-1}}\right) + 10\right)^{1.484} \quad (11.1)$$

with $A = -2.862$, $B = 1.542$, and $C = -0.197$ (Prialnik & Kovetz 1995, see also Townsley & Bildsten 2004). The accretion rate onto the WD determines the strength of the outburst. Higher accretion rates lead to less violent outbursts. If the accreted envelope remains on the WD, a steady-state can be achieved if the accretion rate is similar to the nuclear burning rate. The steady-state accretion rate $\dot{M}_{\rm steady}$ has been given by Paczyński and Rudak (1980) and Iben (1982). Hachisu and Kato (2001) give an expression for $\dot{M}_{\rm steady}$, which for a hydrogen content $X = 0.7$ is

$$\dot{M}_{\rm steady} \approx 3.7 \times 10^{-7}\left(\frac{M_{\rm WD}}{\rm M_\odot} - 0.4\right)\ {\rm M_\odot\ yr^{-1}} \quad (11.2)$$

$\dot{M}_{\rm steady} \approx (1-4) \times 10^{-7}$ M$_\odot$ yr^{-1} for $M_{\rm WD} = (0.7-1.4)\,{\rm M_\odot}$. For $\dot{M}_{\rm acc} < \dot{M}_{\rm steady}$ a fraction of the accreted matter is ejected in a nova outburst. Below a threshold $\dot{M}_{\rm low} \approx 0.25\ \dot{M}_{\rm steady}$ all accreted matter will be ejected during a nova outburst. A red giant envelope will form for accretion rates above a critical rate $\dot{M}_{\rm crit} \approx 2M_{\rm steady}$, for which part of the envelope is blown off by a strong wind (see Fig. 11.1).

11.3 Timescales and variability

If a steady-state is sustained, the nuclear burning timescale can be as long as $\sim 10^6$ years (the thermal timescale of the donor, Yungelson *et al.* 1996). A long nuclear burning timescale of $\sim 10^5$ years is required to explain the ionization nebula detected in CAL 83 due to illumination of the local interstellar medium by the super-soft X-ray emission

Table 11.1. *Catalog of optically identified SSS*

Galaxy	Name	Alias	V^a (mag)	T^b (eV)	L^c (10^{37} erg s^{-1})	Type[d]	P_{orb}[e]	i^f (°)	D^g (kpc)	References
Galaxy	RX J0019.8+2156	QR And	12.4	25–37	~0.4	CBSS	15.85 h	16–56	~2	1,2,3,4
	RX J0925.7−4758	MR Vel	17.1	75–94	0.2–10	CBSS	4.03 d	55±10	2–5	5,6,7,8
	V382 Vel	N Vel 1999	2.8–8.0	34–48	40–210	CV-N	3.5 h	–	1.7	9,10
	GQ Mus	N Mus 1983	18	38–43	0.7	CV-N	85.5 m	50–70	4.7	11,12,13
	U Sco	BD−17 4554	8–19	74–76	0.6–24	RN	1.23 d	83±3	6–14	14,15,16,17
	AG Dra	BD+67 922	8.3–9.8	10–15	0.3–1.1	Sy	554 d	30–45	1.6–2.5	18,19,20,21
	V1974 Cyg	N Cyg 1992	4.4–17	34–51	6–37	CV-N	1.95 h	–	2–3	22,23,24
	RR Tel	N Tel 1948	6.7–11	12	1.3	Sy-N	387 d	–	3.6	25,26,27
LMC	CAL 83	LHG 83	16.2–17.1	28–50	0.6–10	CBSS	1.04 d	~20	50	28,29
	CAL 87	LHG 87	19–21	55–76	0.3–0.5	CBSS	10.6 h	~78	50	30,31,4
	N LMC 1995	–	$\gtrsim$10.7	34–39	2.3	CV-N	5.25 h	–	50	32
	RX J0513.9−6951	HV 5682	16.5–17.5	52	9.5	CBSS	18.3 h	~15	50	33,34,35,4
	RX J0527.8−6954	HV 2554	–	18	–	CBSS ?	9.4 h	–	50	36,37
	RX J0537.7−7034	–	~19.7	18–30	>0.6	CV	3.5 h	45–70	50	38,39
SMC	RX J0048.4−7332	SMC 3	15.0	30–35	$\gtrsim$4	Sy-N	~4.4 yr	60–90	60	40,41
	1E 0035.4−7230	SMC 13	20.1–21.5	27–48	0.4–1.1	CV	4.13 h	20–50	60	42,43,44

References: [1]Cowley *et al.* (1998); [2]Beuermann *et al.* (1995); [3]Becker *et al.* (1998); [4]Meyer-Hofmeister *et al.* (1997); [5]Schmidtke *et al.* (2000); [6]Bearda *et al.* (2002); [7]Schmidtke and Cowley (2001); [8]Mennickent *et al.* (2003); [9]Orio *et al.* (2002); [10]Della Valle *et al.* (2002); [11]Diaz *et al.* (1995); [12]Balman and Krautter (2001); [13]Diaz and Steiner (1994); [14]Munari *et al.* (1999); [15]Kahabka *et al.* (1999a); [16]Schaefer and Ringwald (1995); [17]Thoroughgood *et al.* (2001); [18]Tomov *et al.* (2000); [19]Mikolajewska *et al.* (1995); [20]Greiner *et al.* (1997); [21]Gális *et al.* (1999); [22]Rosino *et al.* (1996); [23]Balman *et al.* (1998); [24]Krautter *et al.* (1996); [25]Feast *et al.* (1983); [26]Heck and Manfroid (1985); [27]Jordan *et al.* (1994); [28]Smale *et al.* (1988); [29]Parmar *et al.* (1998); [30]Hutchings *et al.* (1998); [31]Parmar *et al.* (1997); [32]Orio *et al.* (2003); [33]Crampton *et al.* (1996); [34]Reinsch *et al.* (1996); [35]Alcock *et al.* (1996); [36]Greiner *et al.* (1991); [37]Alcock *et al.* (1997b); [38]Greiner *et al.* (2000); [39]Orio and Ögelman (1993); [40]Jordan *et al.* (1996); [41]Kahabka (2004); [42]van Teeseling *et al.* (1998); [43]Kahabka *et al.* (1999b); [44]Crampton *et al.* (1997).

[a]Optical V magnitude; [b]temperature from X-ray spectral fit (for GQ Mus, 9 years after outburst; for RX J0513.9−6951 for the X-ray on-state); [c]bolometric luminosity from X-ray spectral fit (for GQ Mus, 9 years after outburst; for RX J0513.9−6951 for the X-ray on-state); [d]type of the system (CBSS, close-binary SSS; Sy, symbiotic system; N, Nova; RN, recurrent nova; CV, catalclysmic variable); [e]orbital period; [f]binary inclination; [g]source distance.

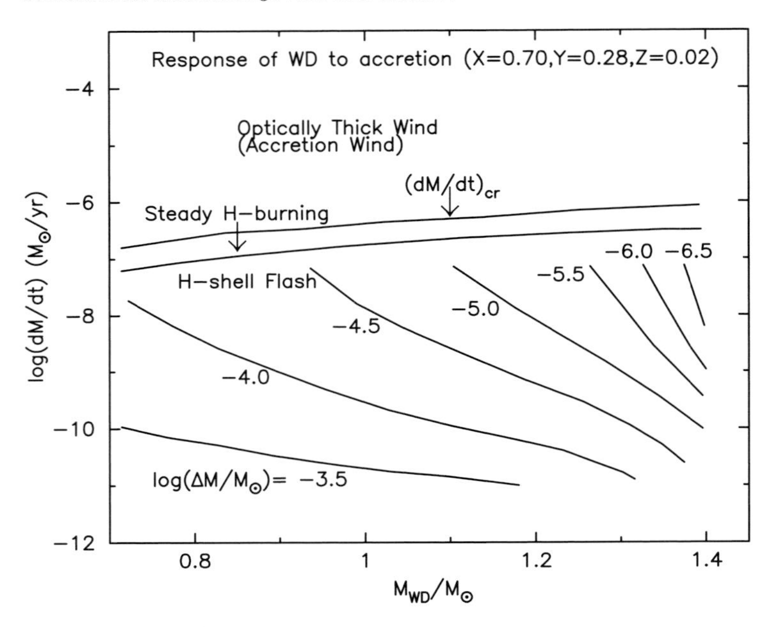

Fig. 11.1. $M_{\rm WD}$–$\dot{M}_{\rm acc}$ plane, regimes of optically thick winds, steady nuclear burning, and flashes (Fig. 2 of Hachisu and Kato 2001; from Fig. 9 of Nomoto 1982). The ΔM values indicate envelope masses (for a given accretion rate) at which burning is ignited.

(Remillard *et al.* 1995). For $\dot{M}_{\rm acc} < \dot{M}_{\rm steady}$ nuclear burning occurs in recurrent shell flashes and lasts for ~1 month to thousands of years (Prialnik & Kovetz 1995; Kahabka 1995; Kato 1997, 1999). For cyclic (time variable) accretion with $\dot{M}_{\rm acc}$ close to $\dot{M}_{\rm crit}$, the timescale of expansion and contraction of the WD envelope is given by the Kelvin–Helmholtz time of the WD envelope

$$\tau_{\rm KH} \approx 3100\ M_{\rm WD} m_{{\rm env},-5}\,(R_9 L_{37})^{-1}\ \text{days} \tag{11.3}$$

with mass $M_{\rm WD}$ (M$_\odot$), envelope mass $m_{{\rm env},-5}$ (10^{-5} M$_\odot$), radius R_9 (10^9 cm), and luminosity L_{37} (10^{37} erg s^{-1}) of the WD. For WD masses of 0.6 to 1.4 M$_\odot$, $\tau_{\rm KH}$ ranges from ≈70 years to ≈10 days. Greiner and Wenzel (1995) detected in the 100-year optical lightcurve of RX J0019.8+2156 non-periodic variations (high and low states) with timescales of ~(20–40) years. There are two SSS for which alternating X-ray "on" and "off" states on shorter timescales have been discovered, RX J0513.9−6951 (Pakull *et al.* 1993; Reinsch *et al.* 1996) and CAL 83 (Alcock *et al.* 1997a; Kahabka 1998; Greiner & Di Stefano 2002). The flux pattern of these sources has been established in the optical due to the multi-year observational Macho campaign. For RX J0513.9−6951 X-ray "on" states of duration ~30 days occur during optical dips of similar duration while X-ray "off" states occur during optical high states. The "on"/"off" pattern has a recurrence time of ~(100−200) days (Southwell *et al.* 1996). Figure 11.2 shows the anti-correlated X-ray and optical behavior of this source.

CAL 83, for which two X-ray off states have been observed (separated by ~4 years), shows an irregular behavior in the optical with switching from high to low and intermediate states, interrupted by dipping with timescales of a few tens of days. A limit-cycle has been proposed for RX J0513.9−6951 to cause transitions between "on" and "off" states in which expansion

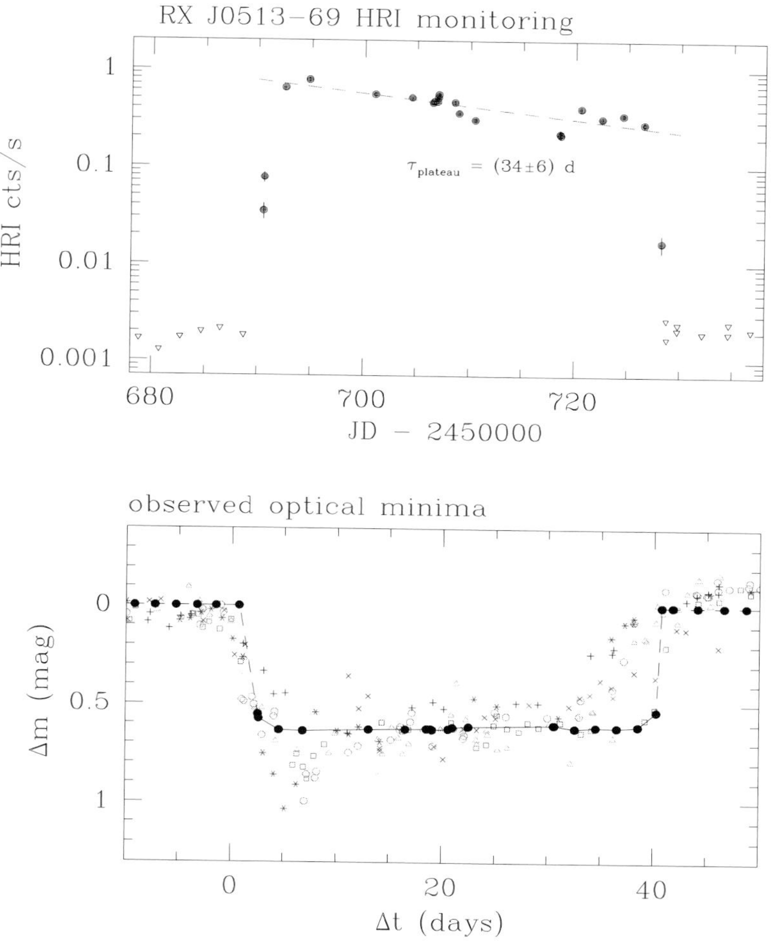

Fig. 11.2. X-ray and optical variability of RX J0513.9−6951. X-ray high states occur during optical low states. Upper panel: X-ray ROSAT HRI count rates during a complete outburst cycle. Lower panel: Optical minima observed with the MACHO project. Filled circles give the lightcurve predicted from the X-ray data (from Reinsch *et al.* 2000).

of the WD atmosphere (due to an increase in $\dot{M}_{acc}$) enhances the irradiation of the accretion disk and the mass flow through the disk (Reinsch *et al.* 2000). The timescale involved is the viscous time of the disk. Hachisu and Kato (2003) explain the fast (a few days) intensity rises and drops in the optical and X-ray lightcurve of RX J0513.9−6951 by the growth and shrinkage of the size of the accretion disk due to the onset and termination of optically thick winds, which are blowing from the WD envelope and occur on a thermal timescale of the WD envelope. It has been proposed that episodic mass loss due to star spot activity of the donor may cause the "on"/"off" pattern of CAL 83 (see Alcock *et al.* 1997a). An explanation proposed by Šimon and Mattei (1999) for V Sge is an irradiation-driven instability of the outer layer of the donor.

Other known stellar cycles are due to red giant and Mira type pulsations which have cycle lengths from ∼ 100 to ∼(300−500) days. SMC 3 is a small amplitude red variable (SARV) with a pulsational period of ~110 days (Kahabka 2004), AG Dra has a bright giant variable with a pulsational period of ~355 days (Gális *et al.* 1999), and the symbiotic nova RR Tel has an AGB Mira variable with a period of ~(350–410) days (Heck & Manfroid 1985).

11.4 Spectra of SSS

The maximum atmospheric temperature $T_{\rm max}$ of steadily nuclear-burning cold ($< 10^7$ K) WDs, which have predicted cooling ages $\gtrsim (1-3) \times 10^8$ years (Yungelson *et al.* 1996), is for $M_{\rm WD} \sim (0.7-1.3)\ {\rm M}_\odot$, according to Iben (1982, see also Suleimanov & Ibragimov 2003)

$$T_{\rm max} \approx 1.4 \times 10^6 \left(0.107 + \left(\frac{M_{\rm WD}}{{\rm M}_\odot} - 0.6\right)^{1.7}\right)\ {\rm K} \tag{11.4}$$

Sion and Starrfield (1994) have calculated evolutionary models for low-mass hot WDs ($M_{\rm WD} < 0.7\ {\rm M}_\odot$) accreting at rates $\sim 10^{-8}\ {\rm M}_\odot\ {\rm yr}^{-1}$ over thousands of years. The WD is heated up during the evolution due to steady nuclear burning and reaches a surface temperature $T_{\rm eff} \sim 3.25 \times 10^5$ K. MacDonald (1996) gives the maximum temperature for a WD after a nova outburst and before turn-off. The WD mass can also be estimated from the bolometric luminosity during the plateau phase of the Hertzsprung–Russell diagram (Iben 1982). The temperature is a better indicator of the WD mass as the luminosity can be reduced due to absorption and scattering.

Effective temperatures in the range ~20 to ~80 eV and bolometric luminosities $\sim 3 \times 10^{36}$ to $\sim 3 \times 10^{38}$ erg s^{-1} (see Table 11.1), have been derived by fitting blackbody, LTE and non-LTE WD model atmospheres (e.g., Hartmann and Heise 1997) to the X-ray spectra of SSS measured with ROSAT (van Teeseling *et al.* 1996; Balman *et al.* 1998), ASCA (Asai *et al.* 1998; Ebisawa *et al.* 2001) and BeppoSAX (Parmar *et al.* 1997, 1998; Hartmann *et al.* 1999; Kahabka *et al.* 1999a,b; Orio *et al.* 2002). Ibragimov *et al.* (2003) applied blanketed LTE WD model atmospheres to ~10 SSS observed with ROSAT and derived spectral parameters for the cold and hot WD approximation. Suleimanov and Ibragimov (2003) showed that most of the investigated SSS are either in the regime of stable nuclear burning or are consistent with cooling WDs. The highly resolved spectrum of CAL 83 measured with XMM-Newton (Paerels *et al.* 2001) is dominated by numerous absorption or emission features (see Fig. 11.3).

The Chandra (Bearda *et al.* 2002) and XMM-Newton (Motch *et al.* 2002) spectra of RX J0925.7−4758 show a wealth of spectral features with emission lines of highly ionized metals (e.g. OVIII and FeXVII), which show P Cygni wind profiles. For a few SSS the neutral hydrogen absorbing column, which is required for the spectral modeling, is available from UV measurements (Gänsicke *et al.* 1998).

11.5 Super-soft novae

A super-soft X-ray phase is observed in ~20% of the classical and recurrent novae and is short lived ($\lesssim$10 years) (Orio *et al.* 2001). GQ Mus has been detected in X-rays and as a SSS with EXOSAT ~(1.3–2.5) years and with ROSAT ~(8.5–9) years after the outburst (Ögelman *et al.* 1987, 1993; Shanley *et al.* 1995). Balman and Krautter (2001) applied CO enhanced WD atmospheric models to the ROSAT observations. The ~1.6 year super-soft X-ray lightcurve could be derived for the classical nova V1974 Cyg during ROSAT observations (Krautter *et al.* 1996). The short duration of the outburst and the ejecta abundances are

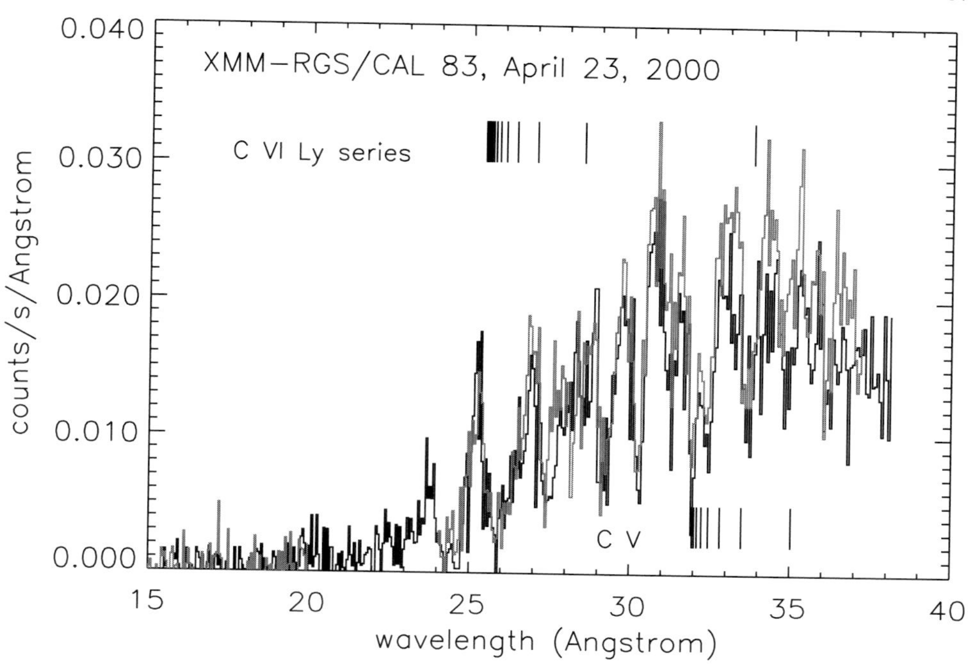

Fig. 11.3. XMM-Newton RGS 1 (black) and 2 (gray) spectrum of CAL 83. The strongest resonance transitions of hydrogen and helium-like C are indicated (from Paerels *et al.* 2001).

consistent with a massive ONe WD. Balman *et al.* (1998) applied ONe (and CO) enhanced model atmospheres and derived the evolution of the nova in the Hertzsprung–Russell diagram. An additional spectrally hard component has been explained as due to interaction of the nova wind with density condensations within the shell (Krautter *et al.* 1996).

Nova LMC 1995 (in the LMC bar) is the third nova for which super-soft X-ray emission has been observed with ROSAT from about half a year until ~3 years after the outburst (Orio and Greiner 1999). With XMM-Newton super-soft emission has been observed ~7 years after the outburst (Orio *et al.* 2003). The bolometric luminosity of the nova at LMC distance was $\sim 2 \times 10^{37}$ erg s^{-1}. Nova LMC 1991 (south-west of the LMC bar) had a super-Eddington luminosity, a large ejecta mass of $\gtrsim 10^{-4}$ $M_{\odot}$, and a low metallicity. A temperature of $\sim(2-4) \times 10^5$ K and a luminosity declining from $\sim 10^{38}$ erg s^{-1} to $\sim(2-4) \times 10^{37}$ erg s^{-1} has been derived during the late phase of the nova from the continuum observed with IUE (Schwarz *et al.* 2001), characteristic for a SSS. Super-soft emission has been discovered with BeppoSAX from the recurrent nova U Sco ~20 days after the 1999 outburst (Kahabka *et al.* 1999a) confirming theoretical predictions of a thermonuclear runaway occurring during a recurrent nova outburst (Kato 1996). Helium abundances of the ejecta He/H of 0.16 ± 0.02 to 4.5 ± 1.7 have been derived for U Sco (e.g., Evans *et al.* 2001; Ijima 2002). In the classical nova V1494 Aql the transition from an emission-line dominated (nebular) spectrum to a spectrum dominated by an optically thick super-soft continuum has been detected with Chandra (Drake *et al.* 2003). Such a transition follows the termination of the optically thick wind phase. The optically thick super-soft spectrum observed for V1494 Aql closely resembled the spectrum of CAL 83 (Paerels *et al.* 2001). In addition a burst lasting for ~1000 s and ~2500 s oscillations have been observed. In the classical nova V4743 Sgr, one month after the optical outburst, a super-soft lightcurve with 22-minute oscillations was observed (Ness *et al.* 2003). In an observation of V382 Vel, half a year after the outburst, a super-soft spectrum was observed, which has been modeled with a non-LTE WD atmosphere, the emission spectrum of a hot diffuse gas, and additional lines of CVI and FeXV (Orio *et al.* 2002).

11.6 The accretion disk, outflows and jets

Super-soft X-ray emission is reprocessed by the accretion disk into UV and optical light (Popham & Di Stefano 1996; Fukue & Matsumoto 2001; Suleimanov *et al.* 1999). Šimon (2003) calculated the dependence of the absolute magnitudes M_V and M_U in the V and U bands of binary sources with super-soft emission on orbital phase and period. He finds that for CBSS the absolute luminosities of the systems increase with the orbital period due to the presence of accretion disks, while for long-period symbiotic systems no pronounced increase in the luminosity is found, arguing against the presence of large accretion disks. Modeling of the optical orbital lightcurve of CAL 87, RX J0019.8+2156, 1E 0035.4−7230, and CAL 83 has been performed by Schandl *et al.* (1997), Meyer-Hofmeister *et al.* (1997), and Kitabatake and Fukue (2002), who find a variation of the disk rim height with orbital phase. This is explained by the intersection of the gas stream with the accretion disk, which causes a bulge (spray) on the disk rim. Quasi-periodic variations in the optical lightcurve (humps) have been observed in RX J0019.8+2156 and RX J0925.7−4758 with a quasi-periodicity of ~1.8 and ~(5.5–7.4) hours respectively (Deufel *et al.* 1999; McGrath *et al.* 2001; Schmidtke *et al.* 2000). These humps have been explained by luminous blobs of material passing through the accretion disk or by changing structures at the disk rim that occult part of the inner disk. The hump periods are consistent with the Keplerian circulation time near the accretion disk rim. In XMM-Newton observations of RX J0925.7−4758 a ~5 hour periodicity has been detected and explained as due to a change in the X-ray emitting area (nuclear-burning WD) by structure on the internal rim of the accretion disk (Motch *et al.* 2002). A very important discovery was that of blue- and redshifted satellites of the HeII λ4686, Hα, and Hβ emission lines, indicating the presence of bipolar jets in RX J0019.8+2156, RX J0513.9−6951, and RX J0925.7−4758. The Doppler velocities of these jets are $v_{\rm bipolar}$ ~885, ~3800, and ~5200 km s^{-1}, respectively (Tomov *et al.* 1998; Becker *et al.* 1998; Southwell *et al.* 1996; Motch 1998). These lines are transient on timescales of months and their velocities give very strong confirmation that we are observing WDs. One can compare these velocities with the terminal velocity of a line-driven wind, $v_\infty^{\rm ld} = 3v_{\rm esc}$, with the escape velocity (Cassinelli 1979)

$$v_{\rm esc} \approx 5160\sqrt{M_{\rm WD}({\rm M}_\odot)/R_{\rm d}(10^9\ {\rm cm})}\ {\rm km\ s}^{-1} \tag{11.5}$$

and of a radiation-driven wind, $v_\infty^{\rm rd} = 0.42v_{\rm esc}$, from an accretion disk surrounding a WD (Hachiya *et al.* 1998; Fukue & Hachiya 1999). In luminous SSS the gaseous envelope is highly ionized and radiation driven winds are expected to be dominating. For $M_{\rm WD} \approx 0.6$ to $1.4\ {\rm M}_\odot$ one derives for radiation driven winds (from the inner disk with radius $R_{\rm d}$) terminal velocities of 1700 to 5700 km s^{-1}, which fit excellently with the line-of-sight velocities of the bipolar outflows. Recently, Bearda *et al.* (2002) found, in the complex Chandra HETGS spectrum of RX J0925.7−4758, P Cygni structure in the line profiles of OVIII (Lyβ) and FeXVII. The highest absorption velocities of ~1500 km s^{-1} are again consistent with radiation driven winds from the inner disk of a massive WD.

11.7 The donor star

In CBSS the donor is assumed to be slightly evolved, between the zero-age main-sequence and the terminal-age main-sequence (TAMS) or slightly beyond (subgiant), and to nearly fill its Roche lobe. For stars at the TAMS with $M_{\rm don} \gtrsim 1.2\,{\rm M}_\odot$,

$$\langle\rho\rangle \approx 20.9(10^{M_{\rm don}})^{-1.54} + 0.0457(10^{M_{\rm don}})^{-0.045}\ {\rm g\ cm^{-3}} \qquad (11.6)$$

assuming a mass–radius relation for solar metallicity stars (Schaller *et al.* 1992). For CAL 83, assuming a Roche-lobe filling donor and $M_{\rm WD} \approx 1.0\ {\rm M}_\odot$, the mean density is $\langle\rho\rangle \approx 0.15\ {\rm g\ cm^{-3}}$, consistent with $M_{\rm don} \gtrsim 1.5\ {\rm M}_\odot$ below the TAMS (van den Heuvel *et al.* 1992). If a significant amount of matter has been transferred during the evolution of the system this mass estimate may not be valid. A commonly used method to constrain $M_{\rm don}$ is by radial velocity emission line diagnostics of the HeII λ4686 line, assuming the line originates predominantly in the accretion disk (Thoroughgood *et al.* 2001; Matsumoto & Mennickent 2000; Becker *et al.* 1998). If the HeII λ4686 line emission is strongly dominated by the mass flow from the donor (Cowley *et al.* 2002), the inferred donor mass may be underestimated. We note that, from the analysis of the radial velocity curves of strong emission lines in the optical spectra of several SSS, low donor masses have been inferred (see Cowley *et al.* 1998). The location of the HeII λ4686 line in the binary system can be inferred from Doppler maps (Deufel *et al.* 1999). The optical spectra of CBSS are dominated by the emission of the bright accretion disk. The donor makes only a minor contribution to the optical spectrum ($\sim$10%), the spectral type can in most cases not be determined. An exception is SSS with optical faint states, such as recurrent novae, where the accretion disk makes only a minor contribution to the overall spectrum, irradiation is strongly reduced, and the donor dominates the optical spectrum. Pritchet and van den Bergh (1977) give for spectral features MgI + Mg H, FeI + CaI, NaI + TiO the equivalent width as a function of spectral type (for giant and main-sequence stars). Anupama and Dewangan (2000) have measured the MgIb and FeI + CaI absorption features in the spectrum of the 1999 outburst of U Sco and have determined the spectral type as K2 IV (sub-giant). Matsumoto and Mennickent (2000) constrained from the non-detection of SiIII λ4525 and MgII λ4481 in the optical spectrum of RX J0925.7−4758 the spectral type of the donor. For an inclination $i \sim 50^\circ$ they find $M_{\rm don} \sim 3.5\ {\rm M}_\odot$, lower than for a nearly Roche-lobe filling solar metallicity star at the TAMS, but consistent with a sub-giant. Drew *et al.* (2003) derived a C/He ratio of $\sim$(0.02–0.06) and a C/O ratio of >1 for QU Car, which would require a carbon star that must have shed most of its envelope to be consistent with the 10.9 hour orbital period.

11.8 Circumbinary material

Kuduz *et al.* (2002) detected in the optical spectrum of RX J0019.8+2156 broad stationary forbidden emission lines of Fe X, OV, and of NV, which show no orbital motion. These lines must originate in material distributed axisymmetrically around the binary center. In addition strong H_2 absorption has been detected in the UV spectrum of RX J0019.8+2156 (Hutchings *et al.* 2001). This indicates the presence of circumbinary material that may be due to winds or jets or to an earlier planetary nebula evolutionary phase of the system.

11.9 Evolution: relation to Type Ia supernovae

The evolution of CBSS undergoing unstable mass transfer ($M_{\rm don} > M_{\rm WD}$) has been calculated by Rappaport *et al.* (1994), Yungelson *et al.* (1996), and Langer *et al.* (2000), see also Bitzaraki *et al.* (2004). An estimate of the mean mass-transfer rate in CBSS is $\dot{M} \approx M_{\rm don}/t_{\rm th}$, with $t_{\rm th} = 3 \times 10^7 (M_{\rm don}/{\rm M}_\odot)^{-2}$ yr, the thermal timescale of the donor (van den Heuvel 1994). For initial donor masses $M_{\rm don} \sim (1.5-2.5)\,{\rm M}_\odot$ this produces mass-transfer rates $\dot{M} \approx (1-5) \times 10^{-7}\,{\rm M}_\odot\,{\rm yr}^{-1}$. An alternative evolutionary scenario for CBSS with very short orbital periods is that of irradiation-driven mass transfer (van Teeseling and King 1998), which could account for systems with periods ($\lesssim$4 hr) (see also Ritter 2000), for which the donor has too small a mass for producing sufficient mass transfer by Roche-lobe overflow.

The WDs in CBSS grow in mass due to their steady nuclear burning of accreted matter. They are therefore candidates for evolving towards a Type Ia supernova (SN Ia), which is expected to occur when a $\sim$1.2 ${\rm M}_\odot$ CO WD grows to the Chandrasekhar limit (see Branch *et al.* 1995; Nomoto *et al.* 2000). The high accretion rates required for steady nuclear burning can be supplied either by slightly evolved main-sequence stars or by low-mass red giants. Population evolution model computations predict occurrence rates in the range $\sim 3.5 \times 10^{-5}\,{\rm yr}^{-1}$ to $\sim(3-6) \times 10^{-3}\,{\rm yr}^{-1}$ (Rappaport *et al.* 1994; Hachisu *et at.* 1996; Li & van den Heuvel 1997; Yungelson & Livio 1998; Fedorova *et al.* 2004; Han & Podsiadlowski 2003). Alternative scenarios proposed for SNe Ia are the off-center (edge-lit) detonation of $\lesssim$0.2 ${\rm M}_\odot$ of hydrogen accreted on top of a CO WD (occurrence rate $\sim(1-5) \times 10^{-3}\,{\rm yr}^{-1}$), the so-called "Type 1.5" scenario, and WD mergers. In view of this range of possibilities for evolving towards a SN Ia, these explosions may form a non-homogeneous class as suggested by recent discoveries of peculiar SNe Ia (see discussion in Fedorova *et al.* 2004). The contributions of the different progenitor channels for SNe Ia to the observationally inferred galactic SN Ia rate of $\sim(1-5) \times 10^{-3}\,{\rm yr}^{-1}$ (Capellaro *et al.* 1997; Capellaro & Turatto 1997) still need more study.

11.10 SSS in nearby galaxies

During the ROSAT survey of M31, 15 SSS and the recurrent super-soft transient RX J0045.4+4154 have been detected (Supper *et al.* 1997; White *et al.* 1995; Section 12.2.1). One transient SSS (RX J0044.0+4118) has been optically identified with a classical nova in M31 (Nedialkov *et al.* 2002). The detected sources have luminosities $\gtrsim 2 \times 10^{37}\,{\rm erg\,s^{-1}}$ and are expected to be connected to massive ($\gtrsim 0.9\,{\rm M}_\odot$) steadily nuclear-burning WDs (Kahabka 1999). During recent surveys performed with Chandra and XMM-Newton five new SSS were detected down to considerably lower luminosities of $\sim(2-8) \times 10^{36}\,{\rm erg\,s^{-1}}$. Two sources, XMM J004308.5+411820 (Shirey 2001) and XMM J004414.0+412204 (Trudolyubov *et al.* 2002), are transients. In addition, the first super-soft pulsator (865 s) has been discovered with XMM-Newton (XMM J004319.4+411759); it is transient and more luminous (Osborne *et al.* 2001; Trudolyubov *et al.* 2001). This period is below the orbital period of accreting main-sequence star binaries. An explanation as a double-degenerate system is unlikely, but a super-soft intermediate polar (M31PSS) appears to be possible (King *et al.* 2002). Three of the five SSS observed after ROSAT are transients. Such a high fraction of transient SSS is consistent with the numbers derived from population synthesis calculations (Yungelson *et al.* 1996).

In M33, seven SSS were detected (Haberl & Pietsch 2001). Eight SSS have been discovered with Chandra in the Sb spiral galaxy M81, at a distance of 3.6 Mpc (Swartz *et al.* 2002). The sources have bolometric luminosities of $\sim 2 \times 10^{36}$ to $\gtrsim 3 \times 10^{38}\,{\rm erg\,s^{-1}}$. Four of the sources are located in the bulge and three sources coincide with spiral arms. The brightest source (N1) is a bulge source with a luminosity $\gtrsim 3 \times 10^{38}$ erg s^{-1} and effective temperature $\sim$(70–90) eV. The very large luminosity would require a massive WD, in a state of luminous outflow. From the age of the bulge population (>8 Gyr), a mass <1 $M_\odot$ is inferred for the stellar companion. The system could be a helium Algol (Iben and Tutukov 1994) or have a red-giant companion. An alternative explanation is an accreting black hole. The source resembles ultra-luminous super-soft X-ray sources found in a few other nearby galaxies. Two further bright sources N2 and N3 with effective temperatures $(7.9-8.5) \times 10^5$ K and $(5.4-7.5) \times 10^5$ K are consistent with massive WDs. Several SSS have been discovered with Chandra in M83 at a distance of ~4 Mpc (Soria 2002). Ten SSS have been discovered with Chandra in the Scd spiral galaxy M101, distance 7.2 Mpc (Pence *et al.* 2001). The three brighter sources have luminosities $\sim 10^{38}$ erg s^{-1} and effective temperatures $\sim(7-11) \times 10^5$ K. Seven fainter sources have effective temperatures $\sim(5.5 \pm 0.2) \times 10^5$ K. Most SSS correlate with spiral arms, one source is located in an interarm region, and another source correlates within 2″ with a globular cluster. In each of the galaxies NGC 55, NGC 300, NGC 1291 and NGC 6822, one SSS has been detected (Schlegel *et al.* 1997; Read & Pietsch 2001; Irwin *et al.* 2002; Eskridge & White 1997). Sarazin *et al.* (2001) discovered with Chandra the first three SSS in an elliptical galaxy, NGC 4697. At a distance of ~16 Mpc these sources must be luminous ($\gtrsim 5 \times 10^{37}$ erg s^{-1}) to be detectable. To evolve towards the TAMS within the age of the elliptical a stellar mass $\sim 1.0\ M_\odot$ is required, below the lower-limit mass for CBSS. These sources may be CV-type SSS or may have a low-mass red-giant companion.

11.11 Summary and conclusions

Super-soft X-ray sources have been discovered in the Magellanic Clouds, the Galaxy, and in about 14 nearby galaxies. A large fraction of these sources is transient. The super-soft transients detected in the Magellanic Clouds are classical and symbiotic novae that have a super-soft phase and systems for which no nova outburst is known but that show in some cases super-soft X-ray "on" and "off" states. Monitoring observations of these sources in X-rays and the optical with MACHO and OGLE have allowed constraints to be put on the binary orbit, the mass transfer through the disk and the mass loss of the donor. Systems such as RX J0513.9–6951, appear to have "too much" mass transfer, as they also eject matter in the form of bipolar jets. This suggests that in their X-ray "off" (optical bright) phase they are bloated. An important puzzle remains the nature of the (few) SSS for which an ultraluminous X-ray phase has been observed, such as the source N1 in M81. The highly resolved X-ray spectra measured for several of the SSS are dominated by numerous spectral features and modeling of these spectra requires sophisticated software for producing synthetic spectra. The X-ray luminosities and spectra set constraints on the masses and radii of the underlying nuclear burning WDs. Several investigations have addressed the question of SSS as progenitors of SNe Ia (or of an accretion induced collapse) and the CBSS are clearly a promising (sub-)class of progenitors.

Acknowledgements

We are indebted to Izumi Hachisu, Frits Paerels and Klaus Reinsch for allowing us to use previously published figures.

References

Alcock, C. *et al.* (1996), *MNRAS* **280**, L49–53
(1997a), *MNRAS* **286**, 483–6
(1997b), *MNRAS* **291**, L13–6
Anupama, G. C. and Dewangan, G. C. (2000), *AJ* **119**, 1359–64
Asai, K., Dotani, T., Nagase, F. *et al.* (1998), *ApJ* **503**, L143–6
Balman, S. and Krautter, J. (2001), *MNRAS* **326**, 1441–7
Balman, S., Krautter, J. and Ögelman, H. (1998), *ApJ* **499**, 395–406
Bearda, H. *et al.* (2002), *A&A* **385**, 511–6
Becker, C. M., Remillard, R.A., Rappaport, S. A. and McClintock, J. E. (1998), *ApJ* **506**, 880–91
Beuermann, K. *et al.* (1995), *A&A* **294**, L1–4
Bitzaraki, O. M., Rovithis-Livaniou, H., Tout, C. A. and van den Heuvel, E. P. J. (2004), *A&A* **416**, 263–80
Branch, D., Livio, M., Yungelson, L. R., Boffi, F. R. and Baron, E. (1995), *PASP* **107**, 1019–29
Capellaro, E. and Turatto, M. (1997), in *Thermonuclear Supernovae*, eds. Ruiz-Lapuente, P. Canal and J. Isern (Dordrecht: Kluwer) p. 77–86
Capellaro, E., Turatto, M., Tsvetkov, D.Yu. *et al.* (1997), *A&A* **322**, 431–41
Cassinelli, J. P. (1979), *ARA&A* **17**, 275–308
Cowley, A. P., Schmidtke, P. C., Crampton, D. and Hutchings, J. B. (1998), *ApJ* **504**, 854–65
(2002), *AJ* **124**, 2233–7
Crampton, D., Hutchings, J. B., Cowley, A. P. *et al.* (1996), *ApJ* **456**, 320–8
Crampton, D., Hutchings, J. B., Cowley, A. P. and Schmidtke, P. C. (1997), *ApJ* **489**, 903–11
Della Valle, M., Pasquini, L., Daou, D. and Williams, R. E. (2002), *A&A* **390**, 155–66
Deufel, B., Barwig, H., Šimić, D., Wolf, S. and Drory, N. (1999), *A&A* **343**, 455–65
Diaz, M. P. and Steiner, J. E. (1994), *ApJ* **425**, 252–63
Diaz, M. P., Williams, R. E., Phillips, M. M. and Hamuy, M. (1995), *MNRAS* **277**, 959–64
Dotani, T., Asai, K. and Greiner, J. (1999), *PASJ* **51**, 519–24
Drake, J. J. *et al.* (2003), *ApJ*, **584**, 448–52
Drew, J. E., Hartley, L. E., Long, K. S. and van der Walt, J. (2003), *MNRAS* **338**, 401–11
Ebisawa, K. *et al.* (2001), *ApJ* **550**, 1007–22
Eskridge, P. B. and White, R. III (1997), *AJ* **114**, 988–96
Evans, A., Krautter, J., Vanzi, L. and Starrfield, S. (2001), *A&A* **378**, 132–41
Feast, M. W., Whitelock, P. A., Catchpole, R. M., Roberts, G. and Carter, B. S. (1983), *MNRAS* **202**, 951–60
Fedorova, A. V., Tutukov, A. V. and Yungelson, L. R. (2004), *Astron. Lett.* **30**, 73–85
Fujimoto, M. Y. (1982a), *ApJ* **257**, 752–66
(1982b), *ApJ* **257**, 767–79
Fukue, J. and Hachiya, M. (1999), *PASJ* **51**, 185–96
Fukue, J. and Matsumoto, K. (2001), *PASJ* **53**, 111–7
Gänsicke, B. T., van Teeseling, A., Beuermann, K. and de Martino, D. (1998), *A&A* **333**, 163–71
Gális, R., Hric, L., Friedjung, M. and Petrík, K. (1999), *A&A* **348**, 533–41
Greiner, J. (2000a), *New Astr.* **5**, 137–41
(2000b), *New Astr. Rev.* **44**, 149–54
Greiner, J. and DiStefano, R. (2002), *A&A* **387**, 944–54
Greiner, J. and Wenzel W. (1995), *A&A* **295**, L5–8
Greiner, J., Hasinger, G. and Kahabka, P. (1991), *A&A* **246**, L17–20
Greiner, J., Bickert, K., Luthardt, R. *et al.* (1997), *A&A* **322**, 576–90
Greiner, J., Orio, M. and Schwarz, R. (2000), *A&A* **355**, 1041–8
Haberl, F. and Pietsch, W. (2001), *A&A* **373**, 438–46
Hachisu, I. and Kato, M. (2001), *ApJ* **558**, 323–50
(2003), *ApJ*, **588**, 1003–8
Hachisu, I., Kato, M. and Nomoto, K. (1996), *ApJ* **470**, L97–100
Hachiya, M., Tajima, Y. and Fukue, J. (1998), *PASJ* **50**, 367–72
Han, Z. and Podsiadlowski, Ph. (2003), *MNRAS* **350**, 1301–9
Hartmann, H. W. and Heise, J. (1997), *A&A* **322**, 591–7
Hartmann, H. W., Heise, J., Kahabka, P., Motch, C. and Parmar, A. N. (1999), *A&A* **346**, 125–33
Heck, A. and Manfroid, J. (1985), *A&A* **142**, 341–5
Heuvel, E. P. J. van den (1994), in *Interacting Binaries*, eds. H. Nussbaumer and A. Orr (Berlin: Springer) p. 263–474

Heuvel, E. P. J. van den, Bhattacharya, D., Nomoto, K. and Rappaport, S. A. (1992), *A&A* **262**, 97–105
Hutchings, J. B., Crampton, D., Cowley, A. P. and Schmidtke, P. C. (1998), *ApJ* **502**, 408–16
Hutchings, J. B., Crampton, D., Cowley, A. P., Schmidtke, P. C. and Fullerton, A. W. (2001), *AJ* **122**, 1572–7
Iben, I. Jr. (1982), *ApJ* **259**, 244–66
Iben, I. Jr. and Tutukov, A. V. (1994), *ApJ* **431**, 264–72
Ibragimov, A. A., Suleimanov, V. F., Vikhlinin, A. and Sakhibullin, N. A. (2003), *Astron. Rep.* **47**, 186–96
Iijima, T. (2002), *A&A* **387**, 1013–21
Irwin, J. A., Sarazin, C. L. and Bregman, J. N. (2002), *ApJ* **570**, 152–64
Jordan, S., Mürset, U. and Werner, K. (1994), *A&A* **283**, 475–82
Jordan, S., Schmutz, W., Wolff, B., Werner, K. and Mürset, U. (1996), *A&A* **312**, 897–904
Kahabka, P. (1995), *A&A* **304**, 227–34
(1998), *A&A* **331**, 328–34
(1999), *A&A* **344**, 459–71
(2004), *A&A* **416**, 57–65
Kahabka, P. and van den Heuvel, E. P. J. (1997), *ARA&A* **35**, 69–100
Kahabka, P., Hartmann, H. W., Parmar, A. N. and Negueruela, I. (1999a), *A&A* **347**, L43–6
Kahabka, P., Parmar, A. N. and Hartmann, H. W. (1999b), *A&A* **346**, 453–8
Kato, M. (1996), in *Supersoft X-Ray Sources*, ed. J. Greiner (Berlin: Springer), p. 15–24
(1997), *ApJS* **113**, 121–9
(1999), *PASJ* **51**, 525–35
King, A., Osborne, J. P. and Schenker, K. (2002), *MNRAS* **329**, L43–6
Kitabatake, E. and Fukue, J. (2002), *PASJ* **54**, 235–40
Krautter, J., Ögelman, H., Starrfield, S., Wichmann, R. and Pfeffermann, E. (1996), *ApJ* **456**, 788–97
Kuduz, M., Reinsch, K., Beuermann, K. and Kube, J. (2002), in *The Physics of Cataclysmic Variables and Related Objects*, eds. B. T. Gänsicke, K. Beuermann and K. Reinsch (San Francisco: ASP), p. 641–2
Langer, N., Deutschmann, A., Wellstein, S. and Höflich, P. (2000), *A&A* **362**, 1046–64
Li, X.-D. and van den Heuvel, E. P. J. (1997), *A&A* **322**, L9–12
Long, K. S., Helfand, D. J. and Grabelsky, D. A. (1981), *ApJ* **248**, 925–44
MacDonald, J. (1996), in *Cataclysmic Variables and Related Objects*, eds. A. Evans and J. H. Wood, (Dordrecht: Kluwer), p. 281–7
Matsumoto, K. and Mennickent, R. E. (2000), *A&A* **356**, 579–84
McGrath, T. K., Schmidtke, P. C., Cowley, A. P., Ponder. A. L. and Wagner, R. M. (2001), *AJ* **122**, 1578–85
Mennickent, R. E., Matsumoto, K. and Diaz, M. (2003), *A&A* **402**, 315–19
Meyer-Hofmeister, E., Schandl, S. and Meyer, F. (1997), *A&A* **321**, 245–53
Mikolajewska, J., Kenyon, S. J., Mikolajewski, M., Garcia, M. R. and Polidan, R. S. (1995), *AJ* **109**, 1289–307
Motch, C. (1998), *A&A* **338**, L13–6
Motch, C., Bearda, H. and Neiner, C. (2002), *A&A* **393**, 913–20
Munari, U. *et al.* (1999), *A&A* **347**, L39–42
Nedialkov, P. *et al.* (2002), *A&A* **389**, 439–45
Ness, J.-U. *et al.* (2003), *ApJ* **594**, L127–30
Nomoto, K. (1982), *ApJ* **253**, 798–810
Nomoto, K., Umeda, H., Kobayashi, C. *et al.* (2000), in *Cosmic Explosions*, eds. S. S. Holt and W. W. Zhang, AIP Conf. Prac.
Ögelman, H., Krautter, J. and Beuermann, K. (1987), *A&A* **177**, 110–6
Ögelman, H., Orio, M., Krautter, J. and Starrfield, S. (1993), *Nature* **361**, 331–3
Orio, M. and Greiner, J. (1999), *A&A* **344**, L13–6
Orio, M. and Ögelman, H. (1993), *A&A* **273**, L56–8
Orio, M., Covington, J. and Ögelman, H. (2001), *A&A* **373**, 542–54
Orio, M., Parmar, A. N., Greiner, J. *et al.* (2002), *MNRAS* **333**, L11–5
Orio, M., Hartmann, W., Still, M. and Greiner, J. (2003), *A&A* **594**, 435–42
Osborne, J. P. *et al.* (2001), *A&A* **378**, 800–5
Paczyński, B. and Rudak, B. (1980), *A&A* **82**, 349–51
Paerels, F., Rasmussen, A. P., Hartmann, H. W. *et al.* (2001), *A&A* **365**, L308–11
Pakull, M. W. *et al.* (1993), *A&A* **278**, L39–42
Parmar, A. N., Kahabka, P., Hartmann, H. W. *et al.* (1997), *A&A* **323**, L33–6
Parmar, A. N., Kahabka, P., Hartmann, H. W., Heise, J. and Taylor, B. G. (1998), *A&A* **332**, 199–203

Patterson, J. *et al.* (1998), *PASP* **110**, 380–95

Pence, W. D., Snowden, S. L., Mukai, K. and Kuntz, K. D. (2001), *ApJ* **561**, 189–202

Popham, R. and DiStefano, R. (1996), in *Supersoft X-Ray Sources*, ed. J. Greiner (Berlin: Springer), p. 65–72

Prialnik, D. and Kovetz, A. (1995), *ApJ* **445**, 789–810

Pritchet, C. and van den Bergh, S. (1977), *ApJS* **34**, 101–14

Rappaport, S., DiStefano, R. and Smith, J. D. (1994), *ApJ* **426**, 692–703

Read, A. M. and Pietsch, W. (2001), *A&A* **373**, 473–84

Reinsch, K., van Teeseling, A., Beuermann, K. and Abbott, T. M. C. (1996), *A&A* **309**, L11–4

Reinsch, K., van Teeseling, A., Beuermann, K. and Thomas, H.-C. (1999), in *Highlights in X-ray Astronomy*, eds. B. Aschenbach and M. J. Freyberg, (MPE report, Garching), p. 70–3

Reinsch, K., van Teeseling, A., King, A. R. and Beuermann, K. (2000), *A&A* **354**, L37–40

Reinsch, K., Beuermann, K. and Gänsicke, B. T. (2002), in *The Physics of Cataclysmic Variables and Related Objects*, eds. B. T. Gänsicke, K. Beuermann and K. Reinsch (San Francisco: ASP), p. 653–4

Remillard, R. A., Rappaport, S. and Macri, L. M. (1995), *ApJ* **439**, 646–51

Ritter, H. (2000), *New Astron. Rev.* **44**, 105–10

Rosino, L., Iijima, T., Rafanelli, P. *et al.* (1996), *A&A* **315**, 463–6

Sarazin, C. L., Irwin, J. A. and Bregman, J. N. (2001), *ApJ* **556**, 533–55

Schaefer, B. E. and Ringwald, F. A. (1995), *ApJ* **447**, L45–8

Schaller, G., Schaerer, D., Meynet, G. and Maeder, A. (1992), *A&AS* **96**, 269–331

Schandl, S., Meyer-Hofmeister, E. and Meyer, F. (1997), *A&A* **318**, 73–80

Schlegel, E. M., Barrett, P. and Singh, K. P. (1997), *AJ* **113**, 1296–309

Schmidtke, P. C. and Cowley, A. P. (2001), *AJ* **122**, 1569–71

Schmidtke, P. C., Cowley, A. P., Taylor, V. A., Crampton, D. and Hutchings, J. B. (2000), *AJ* **120**, 935–42

Schwarz, G. J., Shore, S. N., Starrfield, S. *et al.* (2001), *MNRAS* **320**, 103–23

Seward, F. D. and Mitchell, M. (1981), *ApJ* **243**, 736–43

Shanley, L., Ögelman, H., Gallagher, J. S., Orio, M. and Krautter, J. (1995), *ApJ* **438**, L95–8

Shirey, R. (2001), *IAUC* 7659

Šimon, V. (2003), *A&A* **406**, 613–21

Šimon, V. and Mattei, J. A. (1999), *A&AS* **139**, 75–88

Sion, E. M. and Starrfield, S. G. (1994), *ApJ* **421**, 261–8

Smale, A. P. *et al.* (1988), *MNRAS* **233**, 51–63

Soria, R. (2002), in *High Energy Processes and Phenomena in Astrophysics*, eds. X.-D. Li, Z.-R. Wang and V. Trimble, (San Francisco: ASP)

Southwell, K. A., Livio, M., Charles, P. A., O'Donoghue, D. and Sutherland, W. J. (1996), *ApJ* **470**, 1065–74

Suleimanov, V. F. and Ibragimov, A. A. (2003), *Astron. Rep.* **47**, 197–205

Suleimanov, V. F., Meyer, F. and Meyer-Hofmeister, E. (1999), *A&A* **350**, 63–72

Supper, R. *et al.* (1997), *A&A* **317**, 328–49

Swartz, D. A., Ghosh, K. K., Suleimanov, V., Tennant, A. F. and Wu, K. (2002), *ApJ* **574**, 382–97

Teeseling, A. van and King, A. R. (1998), *A&A* **338**, 957–64

Teeseling, A. van, Heise, J., Kahabka, P. (1996), in *Compact Stars in Binaries*, eds. J. van Paradijs, E. P. J. van den Heuvel, and E. Kuulkers, (Dordrecht: Kluwer), p. 445–50

Teeseling, A. van, Reinsch, K., Pakull, M. W. and Beuermann, K. (1998), *A&A* **338**, 947–56

Thoroughgood, T. D., Dhillon, V. S., Littlefair, S. P., Marsh, T. R. and Smith, D. A. (2001), *MNRAS* **327**, 1323–33

Tomov, T., Munari, U., Kolev, D., Tomasella, L. and Rejkuba, M. (1998), *A&A* **333**, L67–9

Tomov, N. A., Tomova, M. T. and Ivanova, A. (2000), *A&A* **364**, 557–62

Townsley, D. M. and Bildsten, L. (2004), *ApJ* **600**, 390–403

Trudolyubov, S. P., Borozdin, K. N. and Priedhorsky, W. C. (2001), *ApJ* **563**, L119–22

Trudolyubov, S. P., Priedhorsky, W. and Borozdin, K. (2002), *IAUC* 7798

Trümper, J. *et al.* (1991), *Nature* **349**, 579–83

White, N. E., Giommi, P., Heise, J., Angelini, L. and Fantasia, S. (1995), *ApJ* **445**, L125–8

Yungelson, L. and Livio, M. (1998), *ApJ* **497**, 168–77

Yungelson, L., Livio, M., Truran, J. W., Tutukov, A. and Fedorova, A. (1996), *ApJ* **466**, 890–910

12

Compact stellar X-ray sources in normal galaxies

G. Fabbiano
Harvard-Smithsonian Center for Astrophysics

N. E. White
NASA Goddard Space Flight Center

12.1 Introduction

In the 1995 *X-ray Binaries* book edited by Lewin, van Paradijs and van den Heuvel, the chapter on *Normal galaxies and their X-ray binary populations* (Fabbiano 1995) began with the claim that "X-ray binaries are an important component of the X-ray emission of galaxies. Therefore the knowledge gathered from the study of Galactic X-ray sources can be used to interpret X-ray observations of external galaxies. Conversely, observations of external galaxies can provide us with uniform samples of X-ray binaries, in a variety of different environments." This statement was based mostly on the Einstein Observatory survey of normal galaxies (e.g., Fabbiano 1989; Fabbiano, Kim & Trinchieri 1992). Those results have been borne out by later work, yet at the time the claim took a certain leap of faith. Now, nearly a decade later, the sensitive sub-arcsecond spectrally resolved images of galaxies from Chandra (Weisskopf *et al.* 2000), complemented by the XMM-Newton (Jansen *et al.* 2001) data for the nearest galaxies (angular resolution of XMM-Newton is $\sim 15''$), have made strikingly true what was then largely just wishful anticipation.

While a substantial body of ROSAT and ASCA observations exists, which was not included in the 1995 chapter, the revolutionary quality of the Chandra (and to a more limited degree of XMM-Newton) data is such that the present review will be based on these most recent results.

In this chapter we first discuss the emerging awareness of X-ray (0.1–10 keV band, approximately) stellar populations in spiral galaxies: we focus on four well studied galaxies (M31, M81, M83 and M101), and we then discuss the effect of recent widespread star formation on the luminosity functions of the X-ray emitting populations (Section 12.2). We then review the body of observational evidence on the ultra-luminous X-ray sources ($L_X > 10^{39}$ erg s^{-1}) that are associated with active/recent star formation (Section 12.3; see Chapter 13 by King for a review of theoretical work on this subject; see also Chapter 14 by McClintock and Remillard on black hole binaries). We follow with a review of the X-ray population properties of old stellar systems (E and S0 galaxies; Section 12.4). We then discuss the results of correlation analyses of the integrated galaxy emission (Section 12.5), and we conclude with a look at the X-ray evolution of galaxies going back into the deep Universe (Section 12.6).

12.2 X-ray binary populations in spiral galaxies

Because of their proximity, nearby spiral galaxies are where the early work on extragalactic X-ray binary (XRB) populations began (see Fabbiano 1995). For the same reason, these are the galaxies where the deepest samples of sources have been acquired with Chandra and XMM-Newton. Here we will first discuss the recent work done on M31, which, not

Compact Stellar X-Ray Sources, eds. Walter Lewin and Michiel van der Klis.
Published by Cambridge University Press.

surprisingly, is the galaxy that has been studied in most detail. We will then review the results on M81, M83, and M101, to provide examples of the XRB populations in a wider variety of spirals. We conclude this section with a summary of the work on actively star-forming galaxies. We note that this field is evolving rapidly, with an increasing number of galaxies being surveyed and with the sensitivity limit being pushed to fainter fluxes, with ever deeper Chandra observations.

12.2.1 M31

At a distance of only ~700 kpc, M31 (NGC 224, the Andromeda Nebula) is the spiral (Sb) galaxy closest to us. M31 has been observed by virtually all the X-ray observatories since Uhuru, the first X-ray satellite (for a history of the X-ray observations of galaxies, see Fabbiano & Kessler 2001). Starting with the Einstein Observatory and following on with ROSAT, M31 has been the prime target for systematic studies of a population of extra-galactic XRBs, and for comparisons with our own Galactic XRBs (e.g., Long & Van Speybroeck 1983; Trinchieri & Fabbiano 1991; Primini, Forman & Jones 1993; Supper *et al.* 1997, 2001). Chandra and XMM-Newton observations, both by themselves and in combination, are providing new insight on the characteristics of the XRB population of M31. With its sub-arcsecond resolution, Chandra is unique in resolving dense source regions, such as the circum-nuclear region of M31, and detecting faint sources (Garcia *et al.* 2000). Given the proximity of M31 and the relatively low density of luminous XRBs, XMM-Newton provides valuable data on the XRB population of this galaxy, if one excludes the centermost crowded core (Shirey *et al.* 2001).

Source variability and counterparts

Multiple observations of the same fields with these two observatories (and comparison with previous observations) have confirmed the general source variability characteristic of XRBs. XMM-Newton work, following the first statement of source variability (Osborne *et al.* 2001), includes detailed studies of interesting luminous sources. Trudolyubov, Borozdin and Priedhorsky (2001) report the discovery of three transient sources, with maximum X-ray emission in the 10^{37} erg s^{-1} range: a candidate low-mass black hole binary, a source with a long (>1 year) outburst, and a supersoft transient. Trudolyubov *et al.* (2002b) report an 83% modulation with a 2.78 hr period in the X-ray source associated with the globular cluster (GC) Bo 158. Comparison with earlier XMM-Newton observations and with the ROSAT PSPC data, allows these authors to conclude that the modulation is anti-correlated with the source flux, suggesting perhaps a larger less-obscured emission region in the high state. This source resembles Galactic "dip" XRBs, and could be an accreting neutron star. Its period suggests a highly compact system (separation $\sim 10^{11}$ cm).

Widespread source variability is evident from Chandra observations, both from a 47 ks HRC study of the bulge (Kaaret 2002), from a set of eight Chandra ACIS observations of the central $17' \times 17'$ taken between 1999 and 2001 (Kong *et al.* 2002), and from a 2.5 years 17 epochs survey with the Chandra HRC (Williams *et al.* 2004), which also includes the data from Kaaret (2002).

Kong *et al.* find 204 sources, including nine supersoft sources, with a detection limit of $\sim 2 \times 10^{35}$ erg s^{-1}. This detection limit is five times lower than that of the ROSAT HRI catalog (Primini, Forman & Jones 1993), which lists only 77 sources in the surveyed area. They report 22 globular cluster (GC) identifications, 2 supernova remnants, and 9 planetary nebulae associations. By comparing the different individual data sets, they establish that 50% of the sources vary on timescales of months, and 13 are transients. The spectra of the most luminous sources can be fitted with power laws with $\Gamma \sim 1.8$, and, of these, 12 show coordinated flux and spectral variability. Two sources exhibit harder spectra with increasing count rate, reminiscent of Galactic Z sources (e.g., Hasinger & van der Klis 1989). All these characteristics point to an XRB population similar to that of the Milky Way. The HRC survey (Williams *et al.* 2004) reports fluxes and lightcurves for 173 sources, and finds variability in 25% of the sources; 17 of these sources are transients, and two of these are identified with variable HST WFPC2 U band counterparts. One of these two sources is also a transient in the optical and has global properties suggesting a $\sim$10 $M_\odot$ black hole X-ray nova with a period $\geq$9 days. Williams *et al.* (2004) determine that at any given time there are 1.9 ± 1.3 active X-ray transients in M31, and hence they infer that the ratio of neutron star to black hole LMXBs in M31 is $\sim$1, comparable to that in the Galaxy.

Globular cluster sources

The recent X-ray population studies of M31 with Chandra and XMM-Newton demonstrate the importance of large-area surveys of the entire galaxian system (for a review see Chapter 8 by Verbunt and Lewin). A targeted study of GCs with three Chandra fields at large galactocentric radii (Di Stefano *et al.* 2002) revives the old suggestion (Long & Van Speybroeck 1983) that the M31 GC sources are more X-ray luminous than Galactic GC sources. This hypothesis had been dismissed with the ROSAT M31 survey (Supper *et al.* 1997), which however covered only the central 34′ of M31. Di Stefano *et al.* (2002) find that in their fields the most luminous sources are associated with GCs. They detect 28 GC sources, 15 of which are new detections: 1/3 of these sources have L_X(0.5–7 keV) $> 10^{37}$ erg s^{-1}; 1/10 of the sources have L_X(0.5–7 keV) $> 10^{38}$ erg s^{-1}. The X-ray luminosity function (XLF) of the M31 GC sources differs from the Galactic GC XLF, both by having a larger number of sources, and by extending a decade higher in X-ray luminosity (the most luminous M31 GC is Bo 375 with $L_X > 2 \times 10^{38}$ erg s^{-1}; compare with Milky Way GCs that emit less than 10^{37} erg s^{-1}).

Supersoft sources

Supersoft sources (SSS) are very soft X-ray sources, with most of the emission below 1 keV, and spectra that can be fitted with blackbody temperatures of $\leq$100 eV (see Chapter 11 by Kahabka and van den Heuvel). SSS were first discovered in M31 with ROSAT (Supper *et al.* 1997). As noted above, Kong *et al.* (2002) reported nine SSS in their Chandra observations of M31. Recent work by Di Stefano *et al.* (2004) reports 33 SSSs in the same fields surveyed for GCs by Di Stefano *et al.* (2002), of which only two were known since the ROSAT times. Two SSSs are identified with symbiotic stars and two with supernova remnants, but the bulk are likely to be supersoft XRBs. These sources are highly variable, and may be classified in two spectral groups: sources with $kT \leq 100$ eV, and other sources with harder emission, up to $kT \sim 300$ eV. Sixteen of them (on average the most luminous) cluster in the bulge, others are found in both the disk and the halo of M31. Di Stefano *et al.* (2004) point out that some of these sources are detected with luminosities well below 10^{37} erg s^{-1}, the luminosity of a 0.6 $M_\odot$ white dwarf steadily burning hydrogen, and are therefore likely to be lower mass white dwarfs or luminous cataclysmic variables.

The bulge
The XLFs of the global core population (Kaaret 2002 (Chandra HRC); Kong *et al.* 2002 (Chandra ACIS); Trudolyubov *et al.* 2002a (XMM-Newton)) all are in general agreement with each other and with the Einstein (Trinchieri & Fabbiano 1991) and ROSAT studies (Primini, Forman & Jones 1993). However, because of the resolution and sensitivity of Chandra, both Kong *et al.* (2002) and Kaaret (2002) can look at the bulge source population in greater detail than ever before.

Kong *et al.* divide the detected sources into three groups, based on their galactocentric position: inner bulge ($2' \times 2'$), outer bulge ($8' \times 8'$, excluding the inner bulge sources), and disk ($17' \times 17'$, excluding the two bulge regions). When considering the entire bulge population, these authors find a general low-luminosity break of the XLF at $\sim 2 \times 10^{37}$ erg s^{-1}, in agreement with Trudolyubov *et al.* (2002a). However, they also find that the break appears to shift to lower luminosities with decreasing galactocentric radius, going from $0.18 \pm 0.08 \times 10^{37}$ erg s^{-1} in the inner bulge to $2.10 \pm 0.39 \times 10^{37}$ erg s^{-1} in the outermost "disk" region. They note that if the breaks mark episodes of star formation, the more recent of these events must have occurred at larger radii. The slopes of the XLFs also vary (-0.67 ± 0.08 in the center, -1.86 ± 0.40 in the outermost region), but this trend is the opposite of that expected from progressively young populations, where more luminous, short-lived sources, may be found (see e.g., Kilgard *et al.* 2002; Zezas & Fabbiano 2002; Section 12.2.4). Kong *et al.* suggest that the XRB populations of the central regions of M31 may instead all be old (see Trudolyubov *et al.* 2002a), with the shifts of the break resulting from the inclusion of new classes of fainter sources in the inner regions, rather than from a disappearance of the most luminous sources.

Kaaret (2002) contributes to the debate on the nature of the inner bulge sources by investigating their spatial distribution. He shows that the the number of X-ray sources detected in the centermost regions of the bulge ($<100''$) is in excess of what would be expected on the basis of the radial distribution of the optical surface brightness, and suggests that this result may be consistent with a GC origin for the LMXBs.

X-ray source populations in different galaxian fields
With the increased rate of papers on M31, resulting from the XMM-Newton and Chandra surveys of this galaxy, we are now realizing that the X-ray source population of M31 is more varied than previously thought, and that there are correlations between the properties of the X-ray sources and those of the stellar field to which they belong.

In contrast with previous reports (e.g., Trinchieri & Fabbiano 1991; Kong *et al.* 2002), Trudolyubov *et al.* (2002a), by using a larger definition for the radius of the bulge ($15'$), with XMM-Newton observations conclude that, although the XLFs of bulge and disk sources have a similar cumulative slope (-1.3), disk sources are all fainter than $L_X < 2 \times 10^{37}$ erg s^{-1}, while bulge sources can have luminosities as high as $L_X \sim 10^{38}$ erg s^{-1}. They suggest that the most luminous sources are associated with the older stellar population, as in the Milky Way (Grimm, Gilfanov & Sunyaev 2002). However, the fields studied by Trudolyubov *et al.* (2002a) do not include the areas surveyed by Di Stefano *et al.* (2002), where the most luminous GC sources are found (see Fig. 12.1, see also color plate section).

A Chandra ACIS study of XLFs from different regions of M31 (Fig. 12.1, see also color plate section; Kong *et al.* 2003), uses a follow-up of the Di Stefano *et al.* (2002) survey. The results (Fig. 12.2) show that the sources in the central $17' \times 17'$ region are overall more

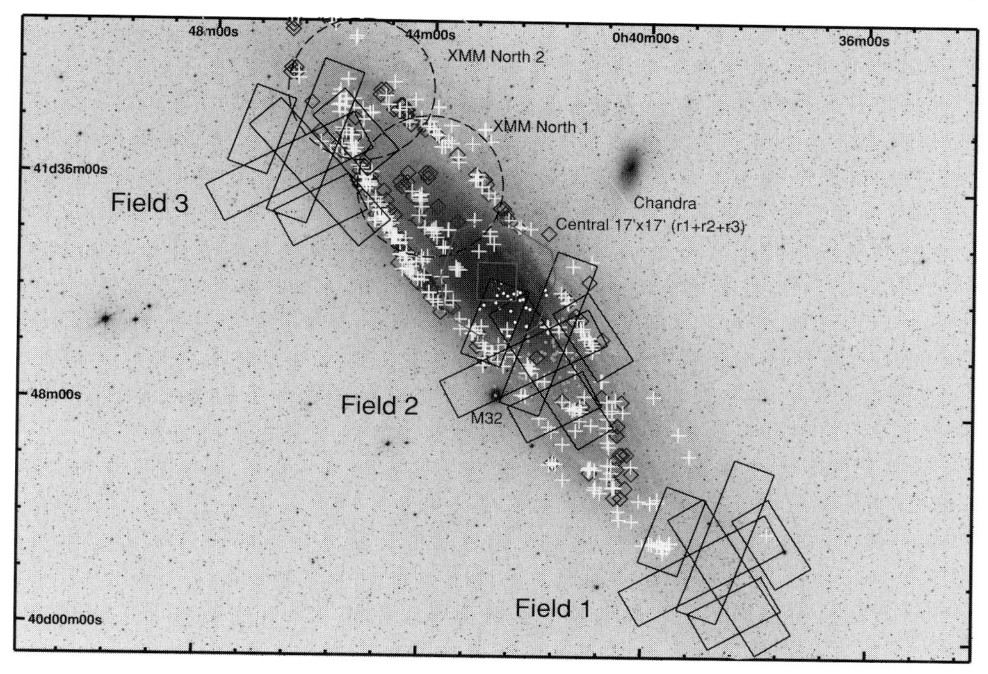

Fig. 12.1. Regions of M31 observed with Chandra (rectangles) and XMM-Newton (circles). Dots are detected Chandra sources; crosses and diamonds identify supernova remnants and OB associations in the field (not X-ray sources), respectively (from Kong *et al.* 2003). For color version see Plate 12.1.

luminous than those from the outer fields (as noticed by Trudolyubov *et al.* 2002a), but only if one removes the GC population, which appears to have a relatively more numerous high luminosity component than the central sources. The slopes of the XLFs of the external fields also vary, and there is an indication that these differences are related to variations in the stellar populations of the different fields: Field 1, which has the steepest slope (cumulative $-1.7^{+0.34}_{-0.15}$) and also the lowest density of X-ray sources, does not appear to have a large young population of stars; Field 2, with the largest X-ray source population and the flattest XLF slope (cumulative -0.9) is in the region with the youngest stellar population. This slope is the closest to that (-0.63 ± 0.13) derived by Grimm, Gilfanov and Sunyaev (2002) for the high-mass X-ray binaries (HMXBs) in the Galaxy; Field 3, with an intermediate XLF slope instead does not appear to cover a large stellar population. The overall integrated slope is instead similar to that found by Grimm *et al.* for the Galactic low-mass X-ray binary (LMXB) population, suggesting that these sources dominate the X-ray emission of M31.

Williams *et al.* (2003), using the Chandra HRC survey of M31, distinguish between a roughly radially symmetric bulge population (within a $7'$ radius) and a field population outside this inner region. They report different XLFs for bulge and disk sources, with a flatter broken power law representing well the disk distribution. Their survey has a wider (although shallower) coverage of the entire M31 galaxy than the Trudolyubov *et al.* (2202a) work, and also covers the southern half of the disk, where the X-ray sources are significantly more luminous than in the northern disk, surveyed with XMM-Newton by Trudolyubov *et al.*

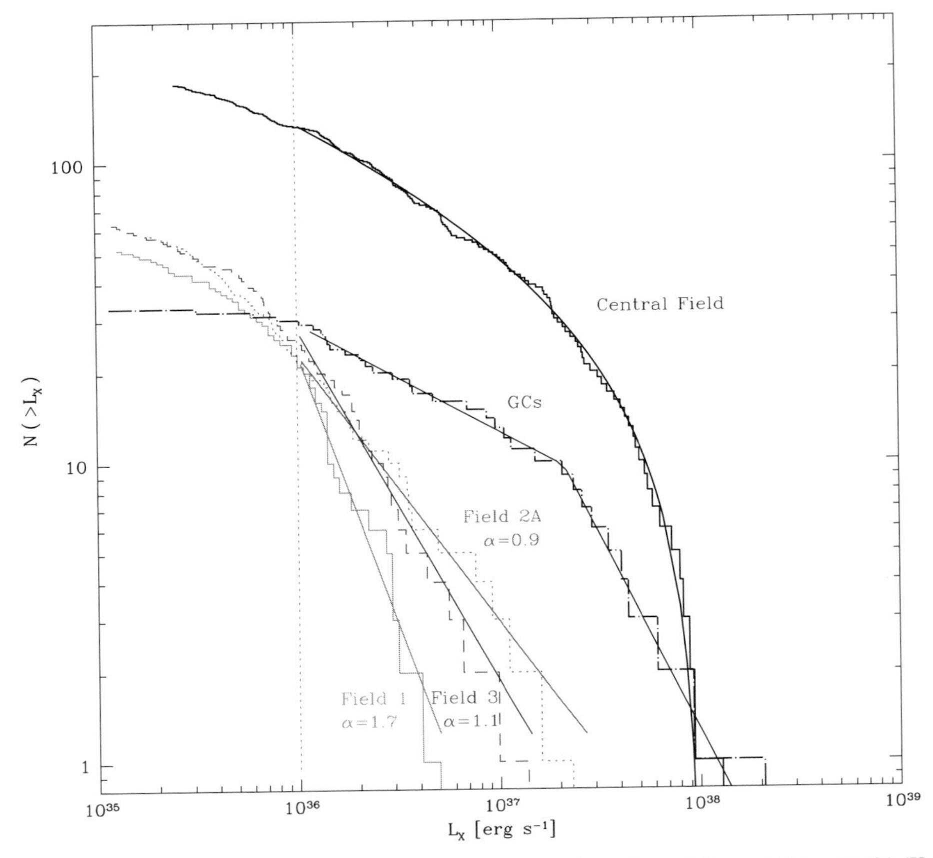

Fig. 12.2. Cumulative XLFs and best-fit power laws from different fields of M31 (Kong *et al.* 2003).

The Trudolyubov *et al.* (2002a), Kong *et al.* (2003), and Williams *et al.* (2004) papers are illuminating in demonstrating the variability of the XLF in different regions, and in pointing out how good spatial sampling and supporting multi-wavelength information are needed to get a complete picture of the XRB population of M31.

12.2.2 M81

As discussed in Fabbiano (1995), M81 (NGC 3031) is a nearby (3.6 Mpc, Freedman *et al.* 1994) Sb galaxy optically similar to M31; however, in X-rays it displays a significantly more luminous population of individual sources (even discounting the nuclear AGN). To get a feel of the progress in sensitivity of X-ray telescopes in the last ~20 years, it is interesting to compare the Einstein observations of M81, where 9 extra-nuclear sources with $L_X \geq 2 \times 10^{38}$ erg s^{-1} were detected (Fabbiano 1988; total ~35 ks exposure time), with the ROSAT results that led to detection of 26 extra-nuclear sources with $L_X > 10^{37}$ erg s^{-1} (Immler & Wang 2001; 177 ks – HRI, 101 ks – PSPC), and finally with the Chandra results: 124 sources detected within the optical D_{25} isophote to a limiting luminosity of $\sim 3 \times 10^{36}$ erg s^{-1} in ~50 ks (Swartz *et al.* 2003).

The Chandra results show that 88% of the non-nuclear emission is resolved into individual sources. The brightest of these sources have luminosities exceeding the Eddington luminosity for a spherically accreting neutron star (see Fabbiano 1995), i.e., they are among the sources dubbed "ultra-luminous X-ray sources" (ULX; see Section 12.3). Of the 66 sources that lie within Hubble Space Telescope (HST) fields, 34 have potential counterparts (but 20 ± 4 chance coincidences are expected). Five sources are coincident with supernova remnants in the spiral arms (including the well studied SN 1993J), but one of them (the ULX X-6) is identified with an XRB, based on its X-ray spectrum. Only four potential GC identifications are found. For one of the M81 sources, Ghosh *et al.* (2001) report a 10-year ROSAT-Chandra X-ray transient lightcurve.

Nine of the sources found in the Chandra observation of M81 are supersoft (SSS; Swartz *et al.* 2002), with L_X(0.2–2.0 keV) in the range of $>2 \times 10^{36} - 3 \times 10^{38}$ erg s^{-1}, and a blackbody emission temperature of 40–80 eV. The fraction of SSS is consistent with the expected values, based on the Galaxy and M31. Four sources are in the bulge and five in the disk; of the latter, four are on the spiral arms. With the exception of the most luminous of these systems, which has a bolometric luminosity $L_{\rm bol} \sim 1.5 \times 10^{39}$ erg s^{-1}, and will be discussed in Section 12.3, all these sources are consistent with the nuclear-burning accreting white dwarf picture of SSS (van den Heuvel *et al.* 1992; see Chapter 11 by Kahabka and van den Heuvel). The SSS associated with the spiral arms tend to have higher emission temperatures, suggesting more massive white dwarf counterparts, which would result from relatively massive stars in a relatively younger stellar population.

The first report of XLF studies in M81 (Tennant *et al.* 2001; Fig. 12.3) showed dramatic differences in the XLFs of bulge and disk sources. While the XLF of the bulge is reminiscent of the bulge of M31, with a relatively steep power-law flattening at $L_X(0.2-8.0 \text{ keV}) < 4 \times 10^{37}$ erg s^{-1}, the XLF of the disk follows an uninterrupted shallow power law (cumulative slope -0.50).

The subsequent more complete study of Swartz *et al.* (2003) confirms the break in the bulge XLF and suggests that it may be due to an aging ~400 Myr old population of LMXBs. The extrapolation of this XLF to lower luminosities can only explain 10% of the unresolved bulge emission, which, however, has the same spatial distribution as the detected bulge sources: besides some gaseous emission, this may suggest an undetected steepening of the XLF due to a yet fainter older population of sources in the central regions. The disk population has different XLFs, depending on the source distance from the spiral arms (Fig. 12.4): in particular, the very luminous ($>10^{38}$ erg s^{-1}) sources responsible for the flat power law are all concentrated on the arms; a break at high luminosities appears when spiral arm sources are excluded. Swartz *et al.* (2003) suggest that these most luminous sources are likely to be very young XRBs resulting from the star formation stimulated by the spiral density waves.

12.2.3 *M83 and M101*

M83 (NGC 5236) and M101 (NGC 5457) are both face-on Sc galaxies. M83 is likely to be a member of the Centaurus group, with a distance of ~4 Mpc (de Vaucouleurs *et al.* 1991); M101 is more distant (~7 Mpc; Stetson *et al.* 1998), but still in the nearby Universe.

M83 is a grand design, barred spiral, with a starburst nucleus. It has been observed extensively in the pre-Chandra era, but here we discuss only the Chandra observations, which are the most relevant for the study of the X-ray source population. M83 was observed with

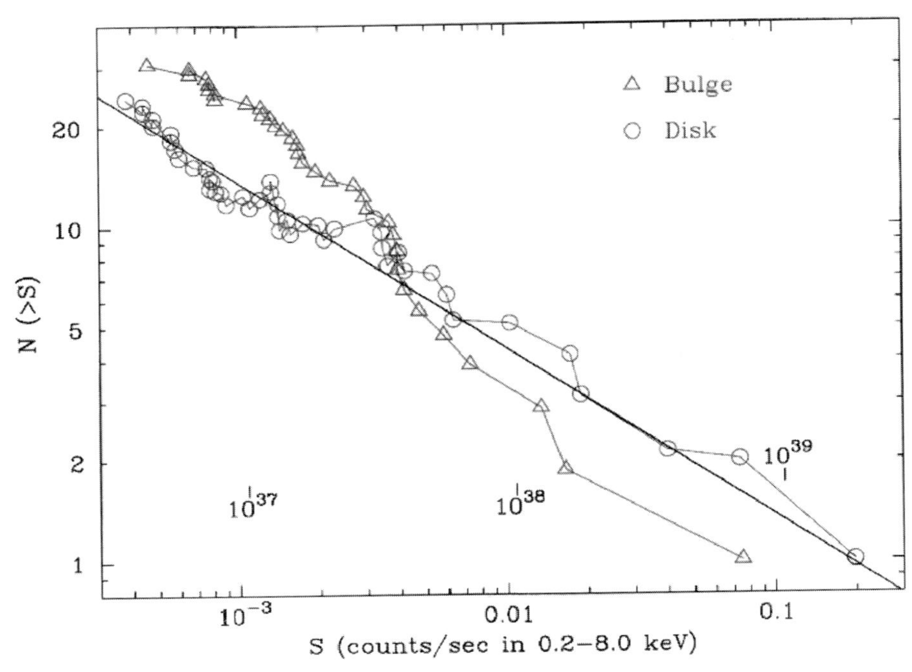

Fig. 12.3. Bulge and disk XLFs for M81. The straight line is the best-fit power law to the disk XLF (Tennant *et al.* 2001).

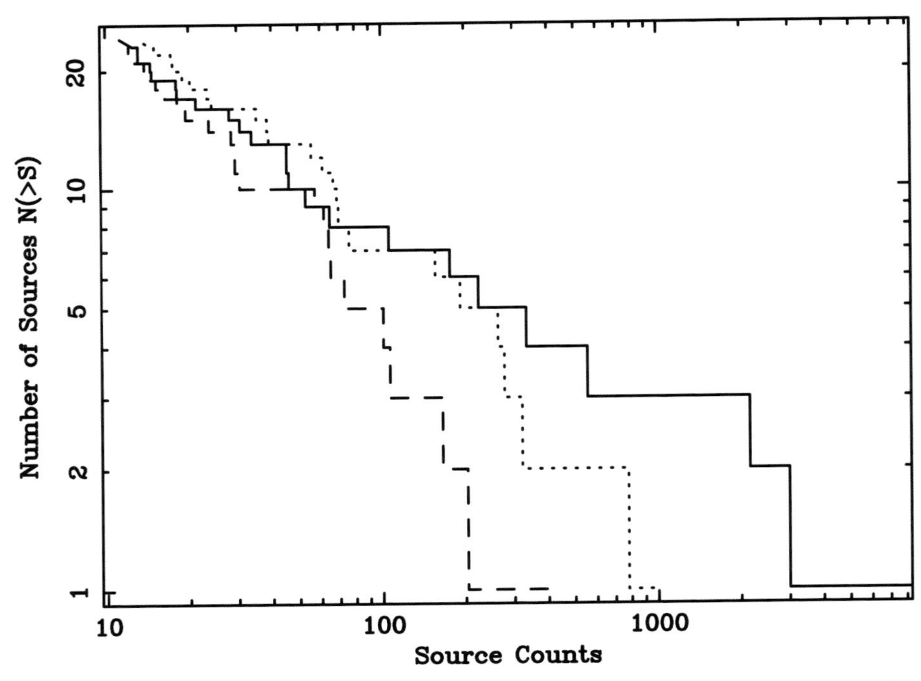

Fig. 12.4. Steepening XLFs of disk sources of M81, at increasing distance from the spiral arms (solid line; Swartz *et al.* 2003).

Fig. 12.5. M83 as seen with Chandra. Note the population of point-like sources and the soft diffuse emission (possibly from hot ISM), associated with the spiral arms (from http://chandra.cfa.harvard.edu/photo/2003/1154/index.html). For color version see Plate 12.5.

Chandra ACIS-S3 for ~50 ks (Fig. 12.5, see also color plate section). Soria and Wu (2002) detect 81 sources in these data, of which 18 had been detected previously with ROSAT; 15 sources are resolved in the previously confused nuclear region, which has the highest source density. The XLF of the sources in the nuclear-bar region, where a young stellar population is likely to prevail, follows a fairly flat unbroken power law (cumulative slope −0.8). The XLF of the disk sources is steeper (slope −1.3), with a break at $\sim 6 \times 10^{37}$ erg s^{-1}, becoming flatter at the lower luminosities. This behavior is reminiscent of the XLFs of the bulges of M31 and M81, and suggests an older XRB population.

In M101, 110 sources (27 of which are expected to be background AGN) were detected in a 98 ks Chandra ACIS-S3 observation, with a limiting luminosity of 10^{36} erg s^{-1} (Pence *et al.* 2001). The sources cluster along the spiral arms, and, interestingly, sources in the inter-arm regions tend to have X-ray colors compatible with AGNs, suggesting contamination by unrelated background objects. Twelve sources are spatially coincident with supernova remnants, but, based on their variability, two of them are identified with XRBs. Eight other luminous sources exhibit variability in the Chandra data, and two more are found variable by comparison with previous ROSAT observations. Ten sources are supersoft, and a correlation

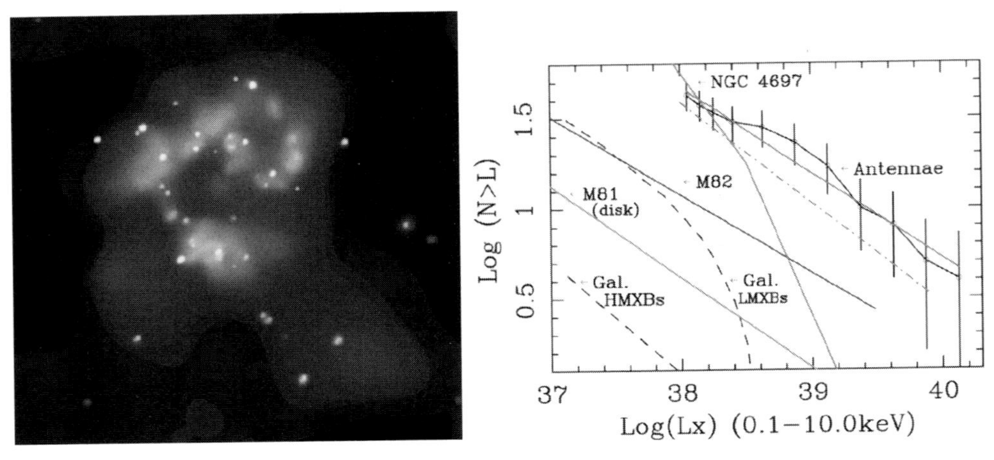

Fig. 12.6. *Left*: Chandra ACIS image of The Antennae (Fabbiano *et al.* 2001). *Right*: the XLF of The Antennae (points with error bars) compared with other galaxies, as labelled. Note the steep XLFs of the Galactic LMXBs (bulge) and of the early-type galaxy NGC 4697 (Zezas & Fabbiano 2002).

between blackbody temperature and total source luminosity is suggested by the data. The XLF of the M101 sources can be modeled with a power law (cumulative slope −0.8) in the 10^{36}–10^{38} erg s^{-1} range.

12.2.4 XRBs in actively starforming galaxies

Observations show flatter XLF slopes (i.e., an increased presence of very luminous sources) in galaxies with more intense star formation. The best example is given by the merger system NGC 4038/39 (The Antennae), where nine ultra-luminous X-ray sources (ULXs; $L_X > 10^{39}$ erg s^{-1}, for a distance of 19 Mpc) were discovered with Chandra (Fabbiano, Zezas & Murray 2001). Other examples of exceptionally luminous sources are found in M82 (Griffiths *et al.* 2000; Kaaret *et al.* 2001; Matsumoto *et al.* 2001), the Circinus galaxy (Smith & Wilson 2001; Bauer *et al.* 2001) and NGC 1365 X-1 (Komossa & Schultz 1998). Consequently, flatter XLFs occur in galaxies with more intense star formation: the cumulative XLF slope is −0.45 in The Antennae (Zezas and Fabbiano 2002; Kilgard *et al.* 2002; Fig. 12.6).

Grimm, Gilfanov and Sunyaev (2003) suggest that the XLFs of starforming galaxies scale with the star formation rate (SFR), thus advocating that HMXBs may be used as a star formation indicator in galaxies. They find that at high SFRs the total X-ray luminosity of a galaxy is linearly correlated to the SFR, and suggest a "universal" XLF of starforming galaxies described by a power law with cumulative slope of $\sim$−0.6 and a cutoff at $L_X \sim$ few × 10^{40} erg s^{-1}. This result of course depends on how well the SFR of a given galaxy is known. This is a subject of considerable interest at this point, since various indicators are differently affected by extinction. The conclusion of a universal slope of the XLF of starforming galaxies may be at odds with the reported correlation between the XLF slope and the 60 μm luminosity from a mini-survey of spiral and starburst galaxies observed with Chandra (Kilgard *et al.* 2002). Also, theoretical models (Kalogera & Belczynski 2003; Belczynski *et al.* 2004) suggest that XLF slopes depend on the age of the starburst, so it is possible that the "universal" XLF slope is not truly universal, but reflects a selection bias, in that the sample used by Grimm, Gilfanov and Sunyaev (2003) may be dominated by starbursts of similar ages.

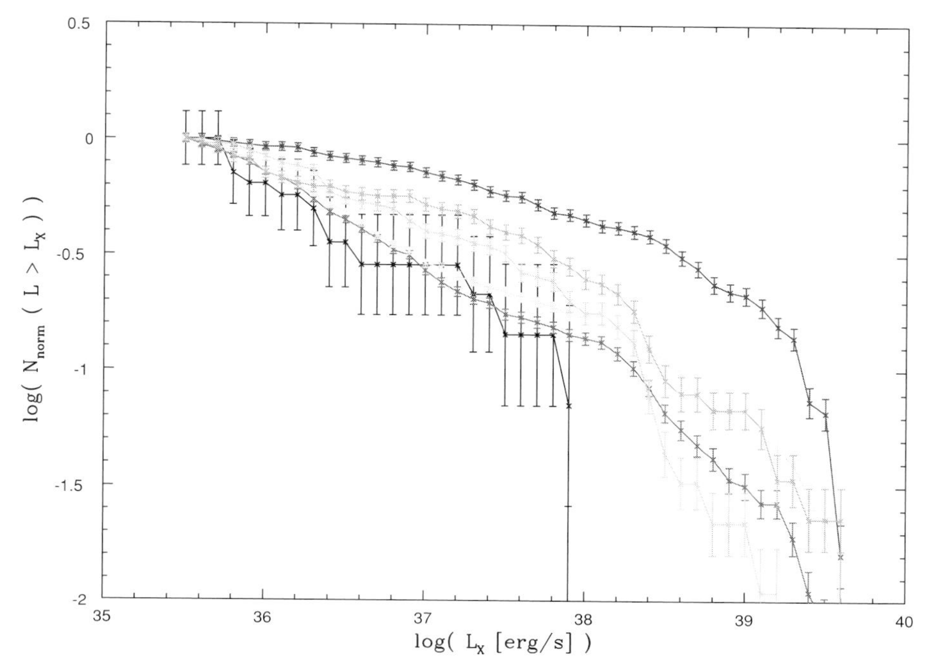

Fig. 12.7. Comparison of XRB population models (from Kalogera & Belczynski 2003; also, Belczynski *et al.* 2004) with the observed XLF of NGC 1569 (bottom points; data taken from Martin *et al.* 2002). Models were constructed to match the star-formation history of NGC 1569 (recent starburst duration and metallicity) and model XLFs are shown at different times since the beginning of the starburst. Top to bottom: 10 Myr, 50 Myr, 110 Myr, 150 Myr, 200 Myr. Note that, based on observations in other wavelengths, the age of the starburst is estimated to be 105–110 Myr. For color version see Plate 12.7.

Comparison of the XLFs for different galaxies, and modeling of the same, provide powerful tools for understanding the nature of the X-ray sources and for relating them to the evolution of the parent galaxy and its stellar population. Early theoretical work has attempted to interpret the XLFs, using ad hoc power-law models, and accounting for aging and impulsive birth of XRB populations (Wu 2001, Kaaret 2002, Kilgard *et al.* 2002). Spurred by the recent observational developments, Kalogera and collaborators have developed the first models of synthetic XLFs, based on XRB evolutionary calculations (Belczynski *et al.* 2004). Such models provide us with a potentially powerful tool for studying the origin and evolution of XRB populations in stellar systems and their connection to galactic environments. A preliminary examination of such models for starburst galaxies (Belczynski *et al.* 2004; see Fig. 12.7, see also color plate section) successfully shows that predictions and consistency checks for the shapes and normalizations of XLFs are possible with theoretical XRB modeling. These new developments demonstrate that the predictions of 1995 are coming true (see Section 12.1).

12.2.5 XLFs and source classification

In summary, with Chandra, X-ray source population studies are finally coming of age. The sub-arcsecond resolution of the Chandra mirrors (Van Speybroeck *et al.* 1997) allows both the separation of discrete sources from surrounding diffuse emission and the detection of much fainter sources than previously possible.

The XLFs of sources in a given system reflect the formation, evolution, and physical properties of the X-ray source population. These differences are evident in different regions of M31, M81 and M83. Comparison of the XLFs of nearby galaxies (and components thereof) with the XLFs of more distant systems provides a general coherent picture, pointing to steeper XLFs in older stellar populations (relative lack of very luminous sources). The XLFs of E and S0 galaxies have cumulative slopes in the range -1.0 to -2.0 (see Section 12.4.3), generally consistent with those of the bulges of M31 and M81. These slopes are significantly steeper than those of sources associated with younger stellar fields in M31, M81, and M83.

Actively starforming galaxies (e.g., M82, The Antennae) are consistent with this picture, having flatter XLFs (an abundance of very luminous sources). A recent study of 32 nearby galaxies extracted from the Chandra archive (Colbert *et al.* 2004) confirms this basic difference between XLFs of old and younger stellar populations, finding cumulative slopes of ~ -1.4 and ~ -0.6 to -0.8 for elliptical and spiral galaxies respectively.

X-ray source classification has also been attempted, on the basis of X-ray color diagrams. While a standard X-ray photometric system does not yet exist, the use of variously defined X-ray colors is becoming a recognized tool (e.g., Zezas *et al.* 2002a; Prestwich *et al.* 2003). The lack of standard photometry, however, makes comparison between different results laborious.

12.3 Ultra-luminous X-ray sources (ULXs)

ULXs are also named super-Eddington sources (see Fabbiano 1989, 1995), super-luminous sources, and intermediate luminosity X-ray objects (IXOs) (Roberts & Warwick 2000; Colbert & Mushotzky 1999; Colbert & Ptak 2003). All these names aim to convey the fact that they are extremely luminous X-ray sources, emitting well in excess of the Eddington luminosity of a spherically accreting and emitting neutron star ($\sim 2 \times 10^{38}$ erg s^{-1}). Usually, sources emitting at $\sim 10^{39}$ erg s^{-1} or above are included in this category. If these sources are emitting isotropically at the Eddington limit, masses in excess of those expected from stellar black holes are implied, up to, in some cases, $\geq 100\ M_\odot$ (e.g., Fabbiano 1989, 1995; Makishima *et al.* 2000). Colbert and Mushotzky (1999) dubbed this type of black holes "intermediate-mass black holes" (IMBH), to distinguish them from the stellar-mass black holes found in Galactic black hole binaries, and also from the supermassive 10^7–$10^9\ M_\odot$ found at the nuclei of galaxies that are responsible for AGNs.

12.3.1 Spectra and spectral variability

Although young supernova remnants may be responsible for ULX emission in some cases (e.g., Fabian & Terlevich 1996), there is now sufficient evidence from spectral and variability data to establish that the majority of ULXs are indeed compact systems, most likely accreting binaries. ASCA X-ray spectra suggested accretion disk emission. These spectra, however, also require temperatures much larger than those expected from black holes of the mass implied by the luminosities of these sources, leading to the suggestion of rotating Kerr black holes (Makishima *et al.* 2000; Mizuno, Kubota & Makishima 2001). In The Antennae ULXs the Chandra spectra (Zezas *et al.* 2002a,b) tend to be hard, and their average co-added spectrum requires both a power-law ($\Gamma \sim 1.2$) and a disk-blackbody component consistent with the ASCA results, with $kT \sim 1.1$ keV. An XMM-Newton survey of 10 galaxies reports ULX spectra consistent with black hole binaries in either high or low state (Foschini *et al.* 2002), but the data quality is too poor for detailed modeling. Similar general spectral results can be found in a Chandra survey of ULXs in different galaxies (Humphrey *et al.* 2003).

However, XMM-Newton high quality spectra of two ULXs in NGC 1313 (X-1 and X-2) led to highly significant detections of soft accretion disk components, with temperatures of $kT \sim 150$ eV, consistent with accretion disks of IMBHs (Miller *et al.* 2003; Fig. 12.8).

The XRB hypothesis is reinforced by observations of correlated luminosity–spectral variability similar to the "high/soft–low/hard" behavior of Cyg X-1 (e.g., in M81 X-9, La Parola *et al.* 2001, with a variety of X-ray telescopes, Fig. 12.9; and in two ULXs in IC 342, Kubota *et al.* 2001 with *ASCA*). However, more recently, Kubota, Done and Makishima (2002) argue that these power-law ULX spectra should not be identified with the low/hard state, but rather may be due to a strongly Comptonized optically thick accretion disk, analogous to the Comptonization-dominated "very high/anomalous state" in Galactic black hole binaries. ASCA observations of one of the IC 342 sources in high state (disk-dominated) revealed a "high/hard–low/soft" low-level variability, with a possible 30–40 hr periodicity, as could be produced by a massive main sequence star orbiting a black hole (Sugiho *et al.* 2001).

With Chandra and XMM-Newton an increasing number of ULXs are being discovered and studied in galaxies. Variability in the Chandra observations of M82 established that the ULXs in this galaxy are likely to be accreting compact objects (Matsumoto *et al.* 2001). The Chandra observations of NGC 3628 (Strickland *et al.* 2001) show the reappearance of the 10^{40} erg s^{-1} variable ULX first discovered with ROSAT (Dahlem, Heckman & Fabbiano 1995). A new transient ULX was discovered in M74 (NGC 628) with XMM-Newton (Soria & Kong 2002). Chandra observations of MF 16 in NGC 6946, formerly identified as an extremely luminous supernova remnant (Schlegel 1994), reveal instead a point-like source with the typical X-ray spectrum of a black hole binary (Holt *et al.* 2003; Roberts & Colbert 2003). Similarly, M81 X-6, which is positionally coincident with a supernova remnant, is identified as an XRB by its X-ray spectrum (Swartz *et al.* 2003). Chandra observations of the nucleus of M33 have revealed a two-component (power-law and disk) spectrum and have established luminosity–spectral variability patterns in this ULX, reminiscent of the black hole binary LMC X-3 (La Parola *et al.* 2003; see also Long, Charles & Dubus 2002); Dubus and Rutledge (2002) compare this source with the Galactic micro-quasar GRS 1915+105.

High/hard–low/soft variability was found in M51 X-7, together with a possible 2.1 hr period (but the time coverage is scant) by Liu *et al.* (2002). Both Cyg X-1 like high/soft–low/hard as well as high/hard–low/soft variability was detected in the population of nine ULXs discovered with Chandra in The Antennae galaxies (Fabbiano *et al.* 2003a; Fig. 12.10, see also color plate section). The latter type of variability can also be found in a few Galactic XRBs (1E 1740.7−2942, GRS 1758−258, GX 339−4, Smith *et al.* 2002; see also the XMM-Newton results on GRS 1758−258, Miller *et al.* 2002). This spectral variability may be indicative of the competition between the relative dominance of the accretion disk versus the innermost hot accretion flow; several scenarios for spectral variability are discussed in Fabbiano *et al.* 2003a and references therein.

12.3.2 Intermediate mass black holes or beamed XRBs?

Although there is clear evidence pointing towards an XRB nature for ULXs, the presence of IMBHs in these systems is by no means universally accepted, and it may be quite possible that ULXs are indeed a heterogeneous population. As discussed above, the ASCA spectra were interpreted by Makishima *et al.* (2000) as evidence for rotating Kerr IMBH, to reconcile the high accretion disk temperature suggested by the model fitting of these spectra with the large black hole masses implied by the bolometric luminosity of the ULXs, which

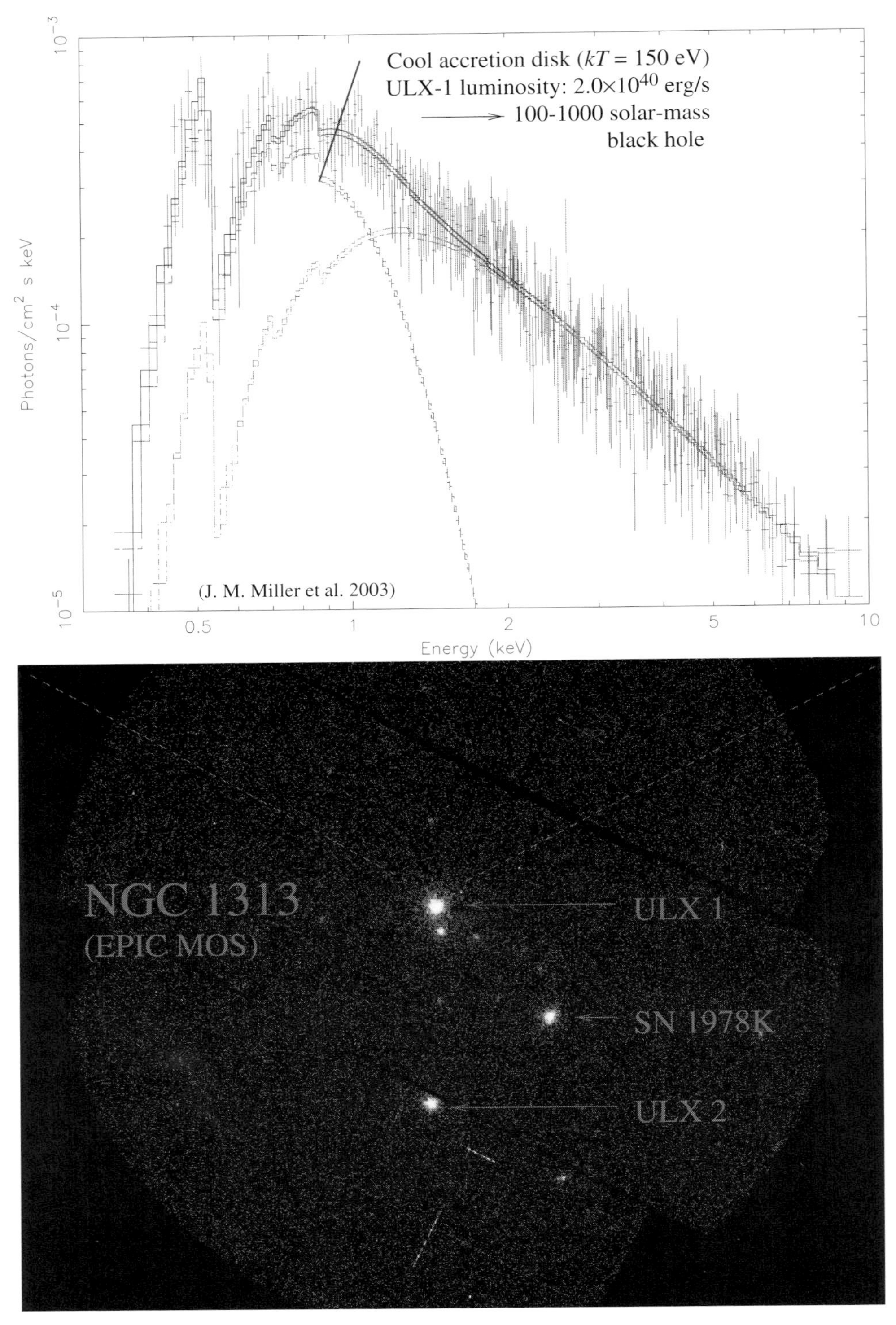

Fig. 12.8. *Top*: X-ray spectrum of ULX-1, compared with best-fit model requiring a cool accretion disk component. *Bottom*: the XMM-Newton image of NGC 1313, showing the position of the two ULXs. (adapted from Miller *et al.* 2003)

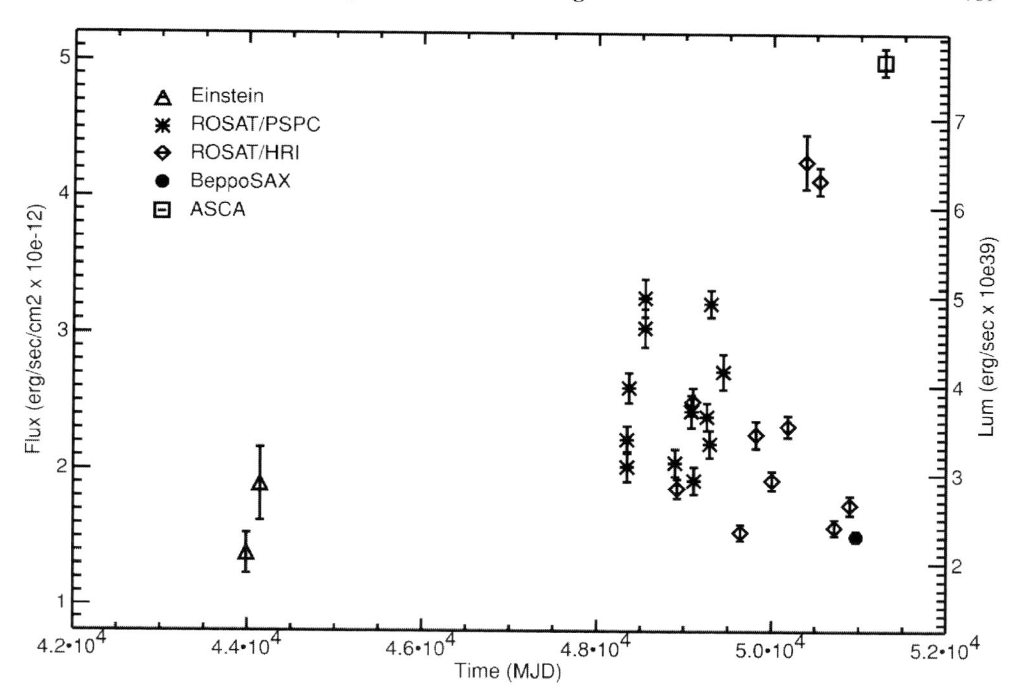

Fig. 12.9. Lightcurve of M81 X-9, covering ~20 yrs of observations (La Parola *et al.* 2001).

would require much cooler disks for a non-rotating IMBH. Colbert and Mushotzky (1999) suggested that these cooler accretion disk components may be present in their ASCA survey of ULXs, but the statistical significance of these early claims is not very high. The Chandra detections of supersoft ULXs (e.g., Swartz *et al.* 2002, in M81; Di Stefano *et al.* 2003b in M104; see also later in this section) could be interpreted as evidence for IMBHs. More important, low-temperature components were discovered in the XMM-Newton spectra of "normal" ULXs: in the NGC 1313 ULXs, which do not require a Kerr black hole, and are entirely consistent with emission from an IMBH accretion disk (Miller *et al.* 2003; Fig. 12.8); and in at least one of the ULXs in The Antennae galaxies ($kT \sim 0.13$ keV) (Miller *et al.* 2004).

Considerable attention has been devoted to an extremely luminous variable 10^{40} erg s^{-1} ULX detected with Chandra near the dynamical center of M82. In the picture of spherical accretion onto an IMBH, the luminosity of this source would imply masses in excess of 100 $M_\odot$ for the accretor. This ULX appears to be at the center of an expanding molecular super-bubble with 200 pc diameter (Matsushita *et al.* 2001). Based on its accurate Chandra position, which is not at the nucleus, Kaaret *et al.* (2001) set an upper limit of $10^5 - 10^6\ M_\odot$ to its mass. Strohmayer and Mushotzky (2003) report quasi-periodic oscillations (QPOs) in the XMM-Newton data of this source. They argue that their discovery suggests emission from an accretion disk, which is incompatible with the radiation being beamed, and therefore implying a less extreme emitted luminosity, as in King *et al.* (2001; see below). On the assumption that the highest QPO frequency is associated with the Kepler frequency at the innermost circular orbit around a Schwarzschild black hole, these authors set an upper limit of $1.87 \times 10^4\ M_\odot$ to the black hole mass: this source could therefore be an IMBH, with mass in the 10^2–$10^4\ M_\odot$ range. The detection of a broad Fe K line is consistent with the IMBH

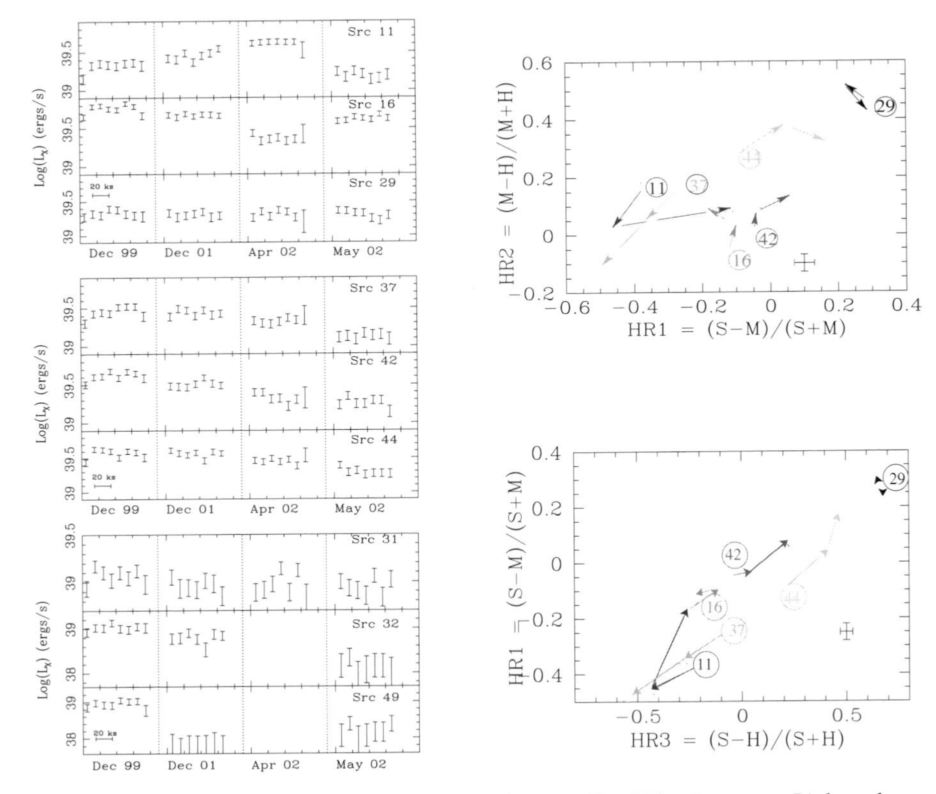

Fig. 12.10. *Left*: Chandra lightcurves of the ULXs of The Antennae. *Right*: color–color diagrams of the most luminous sources (Fabbiano *et al.* 2003a). For color version see Plate 12.10.

scenario. However, as noted by Strohmayer and Mushotzky (2003), the crowded M82 field cannot be spatially resolved with XMM-Newton, making the association of the QPO with the most luminous ULX in the field not entirely proven, and the spectral fit of these data suggests a temperature $kT \sim 3$ keV, much higher than expected from an IMBH accretion disk. Another IMBH candidate is one of two ULXs discovered in the Circinus galaxy (Bauer *et al.* 2001), which displays large and periodic flux variations every 7.5 hr and is well fitted by a multi-color blackbody accretion disk model with $T_{in} = 1.35$ keV, properties consistent with an eclipsing $> 50\ M_\odot$ black hole binary.

As we will discuss below, some results are hard to explain in the IMBH scenario. Two other models have been advanced, which do not require IMBH masses. The large number of ULXs found in The Antennae led to the suggestion that they may represent a normal stage of XRB evolution (King *et al.* 2001). In the King *et al.* (2001) model, the apparent (spherical) accretion luminosity is boosted because of geometrical collimation of the emitting area in thick accretion disks, resulting from the large thermal-timescale mass transfer characterizing the later stages of a massive XRB (see Chapter 13 by King). Exploiting the similarity with Galactic micro-quasars, the jet emission model of Körding *et al.* (2002) produces enhanced luminosity via relativistic beaming. In at least one case, the variable luminous ULX 2E1400.2−4108 in NGC 5408, there is observational evidence pointing to this relativistic jet model: Kaaret *et al.* (2003) find weak radio emission associated with the X-ray source, and argue that both the multi-wavelength spectral energy distribution and the X-ray spectrum are consistent with the Körding *et al.* (2002) scenario.

In some cases at least, the IMBH hypothesis is supported by the association of the ULX with diffuse Hα nebulae, suggesting isotropic illumination of the interstellar medium by the ULX, and therefore absence of beaming (e.g., Pakull & Mirioni 2002 in the case of the NGC 1313 sources, see Miller *et al.* 2003). M81 X-9 is also associated with an optical nebula, which also contains hot gas (La Parola *et al.* 2001; Wang 2002). Wang (2002) considers the possibility that this nebula may be powered by the ULX and also speculates that it may be the remnant of the formation of the ULX. Weaver *et al.* (2002) discuss a heavily absorbed ULX in the nuclear starburst of NGC 253; this source appears to photoionize the surrounding gas. Weaver *et al.* speculate that it may be an IMBH, perhaps connected with either the beginning or the end of AGN activity. However, in at least one case (IC 342 X-1, Roberts *et al.* 2003), there is a suggestion of anisotropic photoionization, which may indicate beamed emission from the ULX.

In The Antennae, comparison with HST data shows that the ULXs are offset from star-forming stellar clusters. While coincidence with a stellar cluster may be due to happenstance because of the crowded fields, the absence of an optical counterpart is a solid result and suggests that the ULXs may have received kicks at their formation (Zezas & Fabbiano 2002), which may be unlikely in the case of a massive IMBH forming in a dense stellar cluster (e.g., Miller & Hamilton 2002). An alternate IMBH scenario, discussed by Zezas and Fabbiano, is that of primordial IMBHs drifting through stellar clusters after capturing a companion (Madau & Rees 2001).

Other optical studies find counterparts to ULXs, and set indirect constraints on the nature of the accretor. A blue optical continuum counterpart to the variable ULX NGC 5204 X-1 was found by Roberts *et al.* (2001), and subsequently resolved by Goad *et al.* (2002) with HST. These authors conclude that the stellar counterpart points to an early-type binary. Similarly, Liu, Bregman and Seizer (2002) find an 08V star counterpart for M81 X-1, a ULX with average $L_X \sim 2 \times 10^{39}$ erg s^{-1}. These counterparts may be consistent with the picture of King *et al.* (2001), of ULXs as XRBs experiencing thermal timescale mass transfer.

Recent results on supersoft variable ULXs suggest that the emitting region may not be associated with the inner regions of IMBH accretion disks in these sources, but may be due to Eddington-driven outflows from a stellar mass black hole. The spectral variability (at constant bolometric luminosity) of the soft ULX P098 in M101 (detected with Chandra; Mukai *et al.* 2003) led to the suggestion of an optically thick outflow from a 15–25 $M_\odot$ black hole, regulated by the Eddington limit. Chandra time monitoring observations of The Antennae have led to the discovery of a variable supersoft source ($kT = 90-100$ eV for a blackbody spectrum), reaching ULX luminosities of 2.4×10^{40} erg s^{-1} (Fabbiano *et al.* 2003b). The assumption of unbeamed emission would suggest a black hole of $\geq 100\ M_\odot$. However, the radiating area would have to vary by a factor $\sim$1000 in this case, inconsistent with gravitational energy release from within a few Schwarzschild radii of a black hole. As discussed in Fabbiano *et al.* (2003b), a surprising possible solution is a white dwarf with $M \sim 1\ M_\odot$, at the Eddington limit, with a variable beaming factor (up to a beaming factor $b \sim 10^{-2}$). A second possible solution involves outflows from a stellar-mass black hole, accreting near the Eddington limit (as in Mukai *et al.* 2003) but with mildly anisotropic radiation patterns ($b \sim 0.1$, as in King *et al.* 2001). Similar sources are reported in M81 (Swartz *et al.* 2002), NGC 300

(Kong & Di Stefano 2003), and other nearby spiral galaxies (Di Stefano & Kong 2003; Kong 2003; see also Di Stefano *et al* 2004 for SSSs in M31).

Transient behavior has been shown to be an important observational diagnostic that could allow us to distinguish between beamed models and IMBH accretion for the origin of ULXs in young, star-forming regions (Kalogera *et al.* 2004). Accretion onto IMBH black holes can lead to unstable disks and hence transient behavior, whereas beamed binary systems have transfer rates that are high enough for the disks to be stable and X-ray emission to be persistent. Therefore, long-term monitoring can prove a valuable and possibly unique tool in unravelling the nature of ULXs.

12.4 XRBs in elliptical and S0 galaxies

As discussed by Fabbiano (1995), XRBs could not be directly detected in E and S0 galaxies with pre-Chandra telescopes, because of the distance of these galaxies and the limited angular resolution of the telescopes. The presence of XRBs in E and S0 galaxies was predicted by Trinchieri and Fabbiano (1985), based on an analogy with the bulge of M31, for which such a population could be detected (Van Speybroeck *et al.* 1979; see also Fabbiano, Trinchieri & Van Speybroeck 1987). This early claim was reinforced by differences in the average spectral properties of E and S0 galaxies with different X-ray-to-optical luminosity ratios, which suggested a baseline X-ray faint XRB emission (Kim, Fabbiano & Trinchieri 1992; Fabbiano, Kim & Trinchieri 1994), and by the ASCA discovery of a hard spectral component in virtually all E and S0 galaxies (Matsushita *et al.* 1994), which, however, could also have been due, at least in part, to accreting massive nuclear black holes (Allen, Di Matteo & Fabian 2000).

The Chandra images (Fig. 12.11) leave no doubt about the presence of rich populations of point-like sources in E and S0 galaxies. Published results, of which the first one is the paper on NGC 4697 by Sarazin, Irwin and Bregman (2000), include point-source detections in a number of galaxies. These source populations have been detected with varying low-luminosity detection thresholds (a function of galaxy distance and observing time). While most of the detected sources have luminosities in the $10^{37} - 10^{39}$ erg s^{-1} range, some were detected at luminosities above 10^{39} erg s^{-1}, in the ultra-luminous-X-ray (ULX) source range (see Section 12.3). A representative summary (limited to papers published or in press as of May 2003) is given in Table 12.1.

The X-ray colors or co-added spectra of these sources are consistent with those of LMXBs (see above references, and Irwin, Athey & Bregman 2003); however, a variety of spectral properties have been reported in some cases, similar to the spectral variety of Galactic and Local Group XRBs, including a few instances of very soft and supersoft (i.e., all photons below ~ 1 keV) sources (e.g., NGC 4697, Sarazin, Irwin & Bregman 2000; M84, Finoguenov & Jones 2002; NGC 1399, Angelini, Loewenstein & Mushotzky 2001; NGC 1316, Kim & Fabbiano 2003). The overall spatial distribution of these sources follows that of the stellar light, but there are exceptions, such as in NGC 720, where the most luminous sources follow arcs (Jeltema *et al.* 2003); in the two galaxies NGC 4261 and NGC 4697, where the X-ray source distributions are highly asymmetric (Zezas *et al.* 2003); and in NGC 4472, where the X-ray source distribution may be more consistent with that of globular clusters (GCs) than of the general field stellar light (Kundu, Maccarone & Zepf 2002; Maccarone, Kundu & Zepf 2003). No firm conclusion on the origin and evolution of these sources exists. Given the old

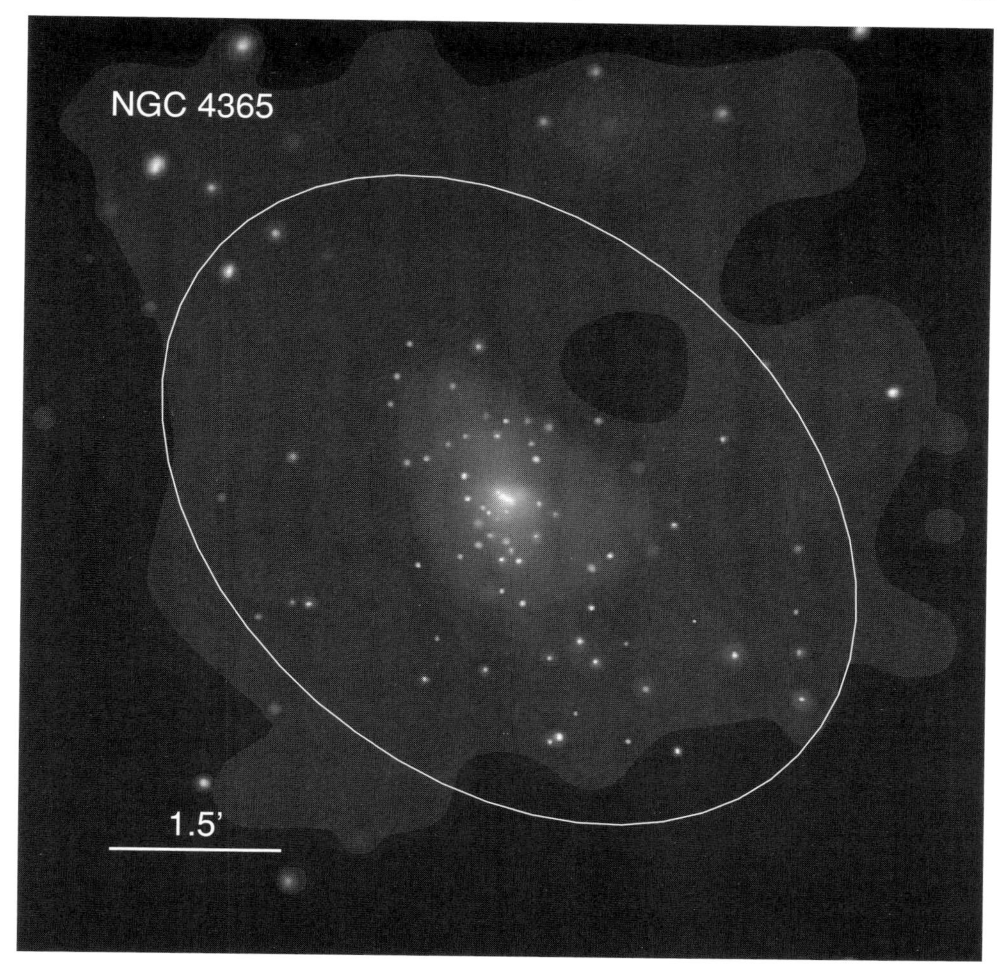

Fig. 12.11. Chandra ACIS image of the Virgo elliptical NGC 4365, using archival data. The white ellipse is the D_{25} isophote (from de Vaucouleurs *et al.* 1991).

stellar population of the parent galaxies, and the lifetimes of LMXBs, it has ben suggested that these sources may be outbursting transients (Piro & Bildsten 2002). Alternatively, more recent formation and evolution in GCs may result in steady sources (Maccarone, Kundu & Zepf 2003). With the exception of NGC 5128, which is near enough to allow detection of sources in the 10^{36} erg s^{-1} luminosity range, and for which multiple observations demonstrate widespread source variability (Kraft *et al.* 2001), the Chandra observations performed so far typically only give a single snapshot of the most luminous part of the XRB population in a given galaxy (Section 8.2). In NGC 5128, a comparison of the two Chandra observations reveals at least five transients (sources that disappear with a dimming factor of at least 10), supporting the Piro and Bildsten scenario.

Chandra observations of highly significant asymmetries in the spatial distribution of X-ray sources in otherwise regular old elliptical galaxies (Zezas *et al.* 2003) may suggest

Table 12.1. *E and S0 galaxies: representative summary of Chandra results*

Name	No. of sources	L_X (erg s^{-1}) band (keV)	Comment
NGC 720	42	$4 \times 10^{38} - 1 \times 10^{40}$ 0.3–7	9 ULX in "arc" pattern 12 associations with GCs (Jeltema *et al.* 2003)
NGC 1291	~50	$<3 \times 10^{38}$ 0.3–10	3 associations with GCs (Irwin *et al.* 2002)
NGC 1316	81	$2 \times 10^{37} - 2 \times 10^{39}$ 0.3–8	$kT \sim 5$ keV average spectrum 5 associations with GCs (Kim & Fabbiano 2003)
NGC 1399	~140	$5 \times 10^{37} - 5 \times 10^{39}$ 0.3–10	70% associated with GCs (Angelini *et al.* 2001)
NGC 1553	49	$1.6 \times 10^{38} - \sim 1 \times 10^{40}$ 0.3–10	X-ray colors consistent with NGC 4697 3 associations with GCs (Blanton *et al.* 2001)
NGC 4374 (M84)	~100	$3 \times 10^{37} - \sim 2 \times 10^{39}$ 0.4–10	spectra consistent with Galactic LMXB (Finoguenov & Jones 2002)
NGC 4472	~120	$1 \times 10^{37} - \sim 1.5 \times 10^{39}$ 0.5–8	40% associated with CGs (Kundu *et al.* 2002)
NGC 4697	~80	$5 \times 10^{37} - 2.5 \times 10^{39}$ 0.3–10	average spectrum $kT \sim 8$ keV 7 (20%) in GCs (Sarazin *et al.* 2001)
NGC 5128 (CenA)	246	$2 \times 10^{36} - 1 \times 10^{39}$ 0.4–10	9 identifications with GCs (Kraft *et al.* 2001)
NGC 5846	~40	$3 \times 10^{38} - 2 \times 10^{39}$ 0.3–10	(Trinchieri & Goudfrooij 2002)

rejuvenation of the stellar population of these galaxies. In NGC 4261, the most significant example, all the detected sources are luminous, above the Eddington limit for a neutron star accretor. If the X-ray sources were standard LMXBs belonging to the dominant old stellar population, we would expect their spatial distribution to be consistent (within statistics) with that of the stellar light. However this is not so, as indicated by Kolmogorov–Smirnov tests and Bayesian block analysis. On the basis of simulations of galaxy interactions (Hernquist & Spergel 1992; Mihos & Hernquist 1996), this result suggests that the luminous XRBs may belong to a younger stellar component, related to the rejuvenating fall-back of material in tidal tails onto a relaxed merger remnants.

12.4.1 ULXs in early-type galaxies

As can be seen from Table 12.1, in early-type galaxies the occurrence of sources with $L_X = 1\text{–}2 \times 10^{39}$ erg s^{-1} is common, although generally limited to a few sources per galaxy. These sources could easily be explained with normal black hole binaries or moderately beamed neutron star binaries (King 2002). In their mini-survey of early-type galaxies observed with Chandra (which include some of the ones listed in Table 12.1), Irwin, Athey and Bregman (2003) find that of the four sources with X-ray luminosities in the 1–2 $\times$ 10^{39} erg s^{-1} range for which they can derive spectra, three have soft spectra, similar to those of black hole binaries in high state (see also Finoguenov & Jones 2002).

Not much can be said about the variability of ULXs in early-type galaxies, because repeated Chandra observations of a given galaxy are not generally available. In the case of NGC 5128, comparison with previous ROSAT images (see Colbert & Ptak 2002) shows considerable flux variability in these very luminous sources: two ULXs were detected in ROSAT observations, both have considerably lower luminosities in the Chandra data (Kraft *et al.* 2001), and one of them may have disappeared.

In general, sources with $L_X > 2 \times 10^{39}$ erg s^{-1} are relatively rare in early-type galaxies as compared to actively starforming galaxies (see Section 12.3). An association of these sources with GCs has been suggested (Angelini, Loewenstein & Mushotzky 2001), but this is not always the case, as exemplified by NGC 720 (Section 8.2.1). This galaxy (Jeltema *et al.* 2003) is peculiar in possessing nine ULXs (this number is of course dependent on the assumed distance, 35 Mpc; see also Section 8.2.1), a population as rich as that of the actively starforming merger galaxies The Antennae (Fabbiano, Zezas & Murray 2001, Zezas & Fabbiano 2002). Only three of these ULXs can be associated with GCs.

Irwin, Athey and Bregman (2003), based on a survey of 15 early-type galaxies (including the bulge of M31), but not NGC 720, conclude that in their sample there is no evidence for the presence of sources with $L_X > 2 \times 10^{39}$ erg s^{-1} in early-type galaxies. The number of these sources detected in their sample is entirely consistent with the expected number of background sources in the area surveyed.

The associations of some ULXs in early-type galaxies with GCs, if real, may support the possibility that a subset of these sources may be associated with IMBH (>10 $M_\odot$) (see Fabbiano 1989 and references therein; Irwin, Athey & Bregman 2003). However, most of the ULXs in early-type galaxies are likely to be lower-mass binaries, given the stellar population of the parent galaxy. King (2002; see also Piro & Bildsten 2002) suggests that they may be a class of ULXs associated with outbursts of soft X-ray transients, resulting in moderately beamed emission from the inner regions of a thick accretion disk. In the case of NGC 720 they may be related to a "hidden" younger stellar population (Jeltema *et al.* 2003).

12.4.2 X-ray sources and globular clusters

The association of X-ray sources in early-type galaxies with GCs has been widely discussed (see Chapter 8 by Verbunt and Lewin). As can be seen from Table 12.1, associations with GCs range from $\leq 10\%$ in most galaxies, $\sim 40\%$ in some Virgo galaxies (NGC 4472, NGC 4649), to 70% in NGC 1399, the dominant galaxy in a group. The statistics are somewhat fraught with uncertainty, since lists of GCs from HST are not available for all the galaxies studied with Chandra, and the detection thresholds differ in different galaxies. However, this association is interesting and has led to the suggestion that all the LMXBs in early-type galaxies may form in GCs, whence they may be expelled if they receive strong enough kicks at their formation, or may be left behind if the GC is tidally disrupted. This suggestion was first advanced by Sarazin, Irwin and Bregman (2000), and was more recently elaborated by White, Sarazin and Kulkarni (2002), on the basis of a correlation of the specific GC frequency with the ratio of the integrated LMXB luminosity to the optical luminosity of 11 galaxies (however,

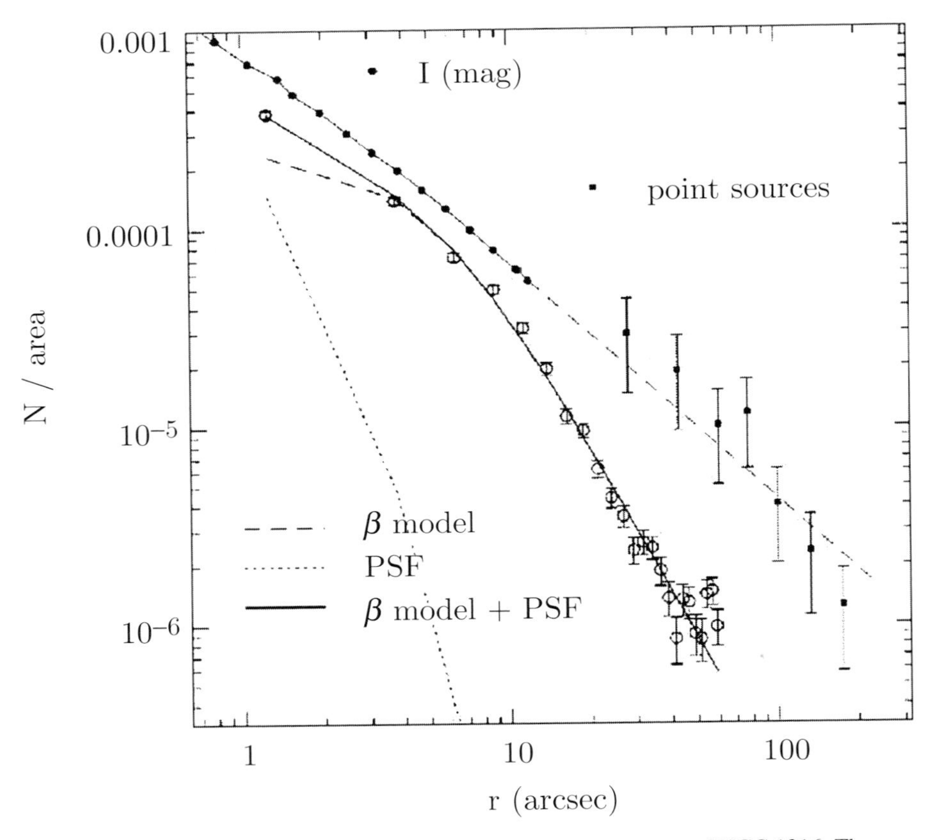

Fig. 12.12. Radial distributions of the emission components of NGC 1316. The gaseous component (hot ISM) is represented by the inner distribution of points. The cumulative XRB contribution is given by the outer set of points; the dashed line through these points is the extrapolation of the stellar (I) surface brightness (Kim & Fabbiano 2003).

see Section 8.3.5). Kundu, Maccarone and Zepf (2002) explore the LMXB–GC connection in NGC 4472, where they find that 40% of the sources detected at $L_X > 1 \times 10^{37}$ erg s^{-1} are associated with GCs. In this galaxy, the fraction of GCs hosting an X-ray source is 4%, the same as in the Galaxy and M31. More luminous, more metal-rich, and more centrally located GCs are more likely to host LMXBs, reflecting both an increased probability of binary formation with the numbers of stars in a GC, and also an effect of metallicity in aiding binary formation (Kundu, Maccarone & Zepf 2002). Sarazin *et al.* (2003) suggest that the fraction of LMXBs associated with GCs increases with the Hubble sequence (from bulges to S0s, Es, cDs). This increase may be related to a constant probability of finding an LMXB in a GC per unit optical luminosity, together with an increasing number of GCs with Hubble type.

While the possibility of LMXB formation in GCs is intriguing, this is still an open question, since evolution of bulge stars may also produce LMXBs (e.g., Kalogera & Webbink 1998; Kalogera 1998; Section 8.3.5). The spatial distribution of the LMXBs, if it follows the optical stellar light (e.g., in NGC 1316, Kim & Fabbiano 2003; Fig. 12.12), would be consistent with this hypothesis. However, in NGC4472 at least, no differences are found in the distributions of X-ray luminosities of the GC sources and the other LMXBs (Maccarone, Kundu & Zepf 2003). In the inner bulge of M31, at radii that even with Chandra cannot be explored in elliptical galaxies because of their distances, the distribution of LMXBs appears more peaked than that of the optical light (Kaaret 2002).

12.4.3 X-ray luminosity functions

The XLFs of the early-type galaxies observed with Chandra are generally steeper than those of star-forming galaxies (see Section 12.2.4), i.e., with a relative lack of luminous HMXBs. These XLFs are generally well fitted with power laws or broken power laws with (cumulative) slopes ranging from -1.0 to -1.8, and breaks have been reported both at $2-3 \times 10^{38}$ erg s^{-1} (however, see Section 8.3.4), the Eddington luminosity of an accreting neutron star (Sarazin, Irwin & Bregman 2000; Blanton, Sarazin & Irwin 2001; Finoguenov & Jones 2002; Kundu, Maccarone & Zepf 2002), and at higher luminosities (10^{39} erg s^{-1}) (Jeltema *et al.* 2003, in NGC 720). While the former break may be related to a transition between neutron star and black hole binaries (Sarazin, Irwin & Bregman 2000), the latter, high-luminosity break, could be produced by a decaying (aging) starburst component from binaries formed in past merging and starbursting episodes (Wu 2001). This possibility was suggested in the case of NGC 720 (Jeltema *et al.* 2003). The XLFs of NGC 5128 (Kraft *et al.* 2001), obtained at different times and reflecting source variability, are well fitted with single power laws in the luminosity range of 10^{37}–10^{39} erg s^{-1}. In NGC 1291 (Irwin, Sarazin & Bregman 2002), no super-Eddington sources are detected.

The effects of detection incompleteness have been considered by Finoguenov and Jones (2002), and have been recently explored extensively by Kim and Fabbiano (2003) in their derivation of the XLF of NGC 1316. Low-luminosity sources may be missed because of higher background/diffuse emission levels in the inner parts of galaxies, and also because of the widening of the Chandra beam at larger radii. Correcting for these effects with an extensive set of simulations, Kim and Fabbiano (2003) found that an apparent $2–3 \times 10^{38}$ erg s^{-1} break in the XLF of NGC 1316 disappeared when incompleteness was taken into account, and the XLF of this galaxy could be represented by an unbroken power law down to luminosities of $\sim 3 \times 10^{37}$ erg s^{-1} (Fig. 12.13). This result shows that caution must be exercised in the derivation of XLFs, and that perhaps some of the previous reports should be reconsidered. If the XLFs extend unbroken to lower luminosities, the amount of X-ray emission from undetected LMXBs in early-type galaxies can be sizeable, as is the case in NGC 1316. This result is important not only for our understanding of the XRB populations, but also for the derivations of the parameters of the hot interstellar medium in these system (see Kim & Fabbiano 2003). Ignoring the contribution to the emission of hidden XRBs results in biases and erroneous results and may give the wrong picture of the overall galaxy dynamics and evolution. Moreover, the dominance at large radii of XRB emission over the hot ISM (see Fig. 12.12) in some (X-ray faint) ellipticals, does also adversely affect mass measurements of these galaxies from low-resolution X-ray data (Kim & Fabbiano 2003).

Kim and Fabbiano (2004) further explored the XLFs of E and S0 galaxies, by deriving bias-corrected XLFs of sources detected in a uniformly selected sample of 14 E and S0 galaxies observed with Chandra ACIS-S3. The entire sample yields 985 point-like X-ray sources, with typical detections of 30–140 sources per galaxy. After correcting for

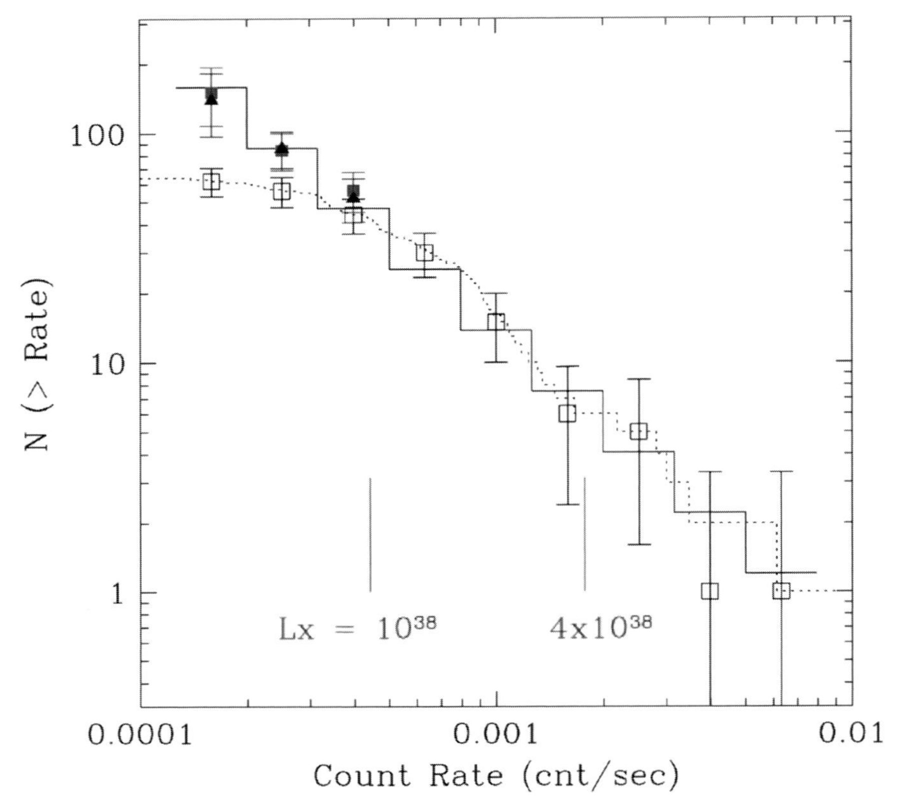

Fig. 12.13. Observed (empty squares) and corrected (filled points) XLFs of NGC 1316 (Kim and Fabbiano 2003).

incompleteness, the individual XLFs are statistically consistent with a single uninterrupted power law, with consistent slope (cumulative, -0.8 to -1.2). A break at or near the neutron star Eddington luminosity, as reported in the literature for some of these galaxies, is not required in any case. Given the uniform XLF shape, Kim and Fabbiano (2004) generated a combined, higher statistics XLF, representative of X-ray sources in elliptical galaxies. Although the combined XLF is marginally consistent with a single power law, a broken power law gives an improved fit. The break luminosity is $5 \pm 1.6 \times 10^{38}$ erg s^{-1} (with an error at 90%), which may be consistent with the Eddington luminosity of neutron stars with the largest possible mass (3 $M_{\odot}$), helium-enriched neutron star binaries, or low-mass stellar-mass black holes. If the steepening of the XLF slope at high luminosities is real and does not mask a step in the XLF, this result would imply a different population of high-luminosity sources, instead of a beaming effect (as in King 2002). This high-luminosity portion of the XLF would then reflect the mass function of black holes in these galaxies. This high-luminosity population does not resemble that of the ultra-luminous X-ray sources detected in star-forming galaxies, where no break in the XLF is present and the XLF is much flatter (see Section 12.2.4).

12.5 Multi-wavelength correlations

Although this chapter is focused on the XRB populations that we can now resolve and study with Chandra in galaxies as distant as ~ 20 Mpc, the study of the integrated emission properties of samples of galaxies (either more distant, or observed at lower resolution) can also give useful information on the average properties of their XRB components. We will summarize here some of these studies, which were pursued mostly by using the samples of galaxies observed with Einstein and ROSAT.

Most of the early work in this area was done by Fabbiano and collaborators, using the first sample of galaxies ever observed in X-rays, the Einstein sample (see reviews in Fabbiano 1989, 1995). Besides suggesting the baseline XRB emission in E and S0 galaxies, which is now confirmed with Chandra (Section 12.4), these results suggested a general scaling of the integrated X-ray emission with the optical luminosity (and therefore stellar population) of the galaxies, and pointed to a strong association of the XRB populations of disk/arm-dominated spirals with the far-IR emission, i.e., the younger component of the stellar population (e.g., Fabbiano, Gioia & Trinchieri 1988; see also David, Jones & Forman 1992). More recent work on the Einstein sample (Shapley, Fabbiano & Eskridge 2001; Fabbiano & Shapley 2002[1]), on ROSAT-observed galaxies (Read & Ponman 2001), and on BeppoSAX and ASCA data (Ranalli, Comastri & Setti 2003) has examined some of these correlations afresh. Given the different pass-bands of these observatories, these studies have a varied sensitivity to the effect of hard XRB emission and soft hot ISM emission in the galaxies.

The Einstein sample is the largest, consisting of 234 S0/a-Irr galaxies observed in the 0.2–4. keV band. The X-ray luminosities are compared with B, H, 12 μm, 60 μm, 100 μm, global FIR, and 6 cm luminosities (Shapley, Fabbiano & Eskridge 2001; Fabbiano & Shapley 2002). Both fluxes and upper limits were used in this work, to avoid obvious selection biases. This work provides baseline distributions of $L_{\rm X}$ and of $L_{\rm X}/L_{\rm B}$ for the entire Hubble sequence (including E and S0 galaxies), and a critical compilation of distances for the sample. Multi-variable correlation analysis shows clear dependencies of the emission properties on the morphological type of the galaxies (and therefore indirectly on the stellar population and star-formation activity). In Sc-Irr galaxies, all the emission properties (including the X-rays) are tightly correlated, suggesting a strong connection to the stellar population. This is not true for S0/a-Sab, where there is a general connection of the X-ray luminosity with the B and H-band emission (stellar population), but not with either radio or FIR. In Sc-Irr galaxies the strongest link of the X-ray emission is a linear correlation with the FIR, suggesting a connection with the star-forming stellar component. This conclusion is reinforced by a correlation between $L_{\rm X}/L_{\rm B}$ and L_{60}/L_{100}, which associates more intense X-ray emission with hotter IR colors.

The X-ray emission/star-formation connection is also discussed as a result of the analysis of a small sample (17 nearby spirals) observed with ROSAT in a softer energy band (0.1–2.0 keV; Read & Ponman 2001), and more recently from the analysis of another small sample (also 17 galaxies) observed in the 2–10 keV band (Ranalli, Comastri & Setti 2003). The advantage of this harder band is that the emission is predominantly due to the XRB population (if the sample does not include AGNs). These authors suggest that the hard X-ray emission can be used as a clean indicator of star formation, because extinction is not a problem at these energies.

[1] Probably the last paper to be published on the Einstein data.

These correlation analyses are now being extended to the XRB populations detected with Chandra. Colbert *et al.* (2004) report good correlations between the total point source X-ray luminosity in a sample of 32 galaxies of different morphological type extracted from the Chandra archive and the stellar luminosity (both B and K bands). While correlations are still present in the spiral and merger/irregular galaxies with FIR and UV luminosities, the ellipticals do not follow this trend and show a clear lack of FIR and UV emission, consistent with their older stellar populations. This result is consistent with the conclusions of Fabbiano and Shapley (2002; see above), which were however based on the analysis of the integrated X-ray luminosity of bulge-dominated and disk/arm-dominated spiral and irregular galaxies.

In summary, there is a correlation between X-ray emission and star-formation rate (SFR) in star-forming galaxies, that may lead to a new indicator of the SFR. However, one has to exercise caution, because this conclusion is only true for star-forming galaxies. In old stellar systems (bulges, gas-poor E and S0s), the X-ray emission is connected with the older stellar population of these systems. This conclusion is also in agreement with the recent studies of XLFs (Sections 12.2 and 12.4.3).

12.6 The X-ray evolution of galaxies

X-ray images of the extra-galactic sky routinely taken with Chandra and XMM-Newton do not typically detect normal galaxies as serendipitous sources in the field. Instead the images reveal a relatively sparse population of point sources, the majority of which are active galactic nuclei (AGN) with a space density of order a thousand per square degree. Normal galaxies are not detected because the X-ray luminosity of normal galaxies is relatively low and the predicted fluxes very faint. However, in the deepest few million second or more exposures made with Chandra (the Chandra Deep Fields – CDFs; Giacconi *et al.* 2002, Alexander *et al.* 2003) faint X-ray emission has been detected from optically bright galaxies at redshifts of 0.1 to 0.5 (Hornschemeier *et al.* 2001). These are amongst the faintest X-ray sources in the CDF, with fluxes of $\sim 10^{-16}$ erg cm^{-2} s^{-1}, corresponding to a luminosity of 10^{39} to 10^{41} erg s^{-1} – the range seen from nearby galaxies (e.g., see Shapley, Fabbiano & Eskridge 2001). Some of these might be galaxies containing a low-luminosity AGN, but most are likely to be part of an emerging population of normal galaxies at faint X-ray fluxes.

The detection sensitivity of Chandra can be increased by "stacking" analysis, i.e., by "stacking" sub-images centered on the positions of galaxies in comparable redshift ranges. This can push the threshold of Chandra to $\sim 10^{-18}$ erg cm^{-2} s^{-1} – equivalent to an effective exposure time of several months or more. Brandt *et al.* (2001) used this technique for 24 Lyman Break galaxies at $z \sim 3$ in the Hubble Deep Field North (Steidel *et al.* 1996) and detected a signal with an average luminosity of 3×10^{41} erg s^{-1} – similar to that of nearby starburst galaxies. Nandra *et al.* (2002) confirmed this result by increasing the number of Lyman Break galaxies to 144 and then extended it to also include 95 Balmer Break galaxies at $z \sim 1$. The Balmer Break galaxies were detected with a lower average luminosity of 7×10^{40} erg s^{-1}, but with a similar X-ray to optical luminosity ratio as the Lyman Break galaxies. Hornschemeier *et al.* (2002) report "stacking" detections of optically luminous spiral galaxies at $0.4 < z < 1.5$.

These Chandra X-ray Observatory detections of normal galaxies at high redshifts have initiated the study of the X-ray evolution of normal galaxies over cosmologically interesting distances. Evolution of the X-ray properties of galaxies is to be expected because the SFR of the Universe was at least a factor of 10 higher at redshifts of 1–3 (Madau *et al.* 1996). Since the X-ray luminosity of galaxies scales with the infrared and optical luminosity (see Section 12.5; Fabbiano, Gioia & Trinchieri 1988; David, Jones & Forman 1992; Shapley, Fabbiano & Eskridge 2001; Fabbiano & Shapley 2002) the increased star formation will have a corresponding impact on the X-ray properties of galaxies at high redshift (White & Ghosh 1998). For spiral galaxies without an AGN, the overall X-ray luminosity in the 1–10 keV band will typically be dominated by the galaxy's X-ray binary population. There is expected to be a corresponding increase in the number of high-mass XRBs associated with the increased SFR. The "detection" of the Balmer and Lyman Break galaxies by Nandra *et al.* (2002) and the factor of 5 increase in the X-ray luminosity from redshift 1 to 3 is consistent with an increasing star-formation rate. Nandra *et al.* (2002) point out that the X-ray luminosity of galaxies provides a new "dust free" method to estimate the star formation rate, as also pointed in the BeppoSAX study of Ranalli, Comastri and Setti (2003), and by Grimm, Gilfanov and Sunyaev (2003).

The low-mass X-ray binary (LMXB) population created by the burst in star formation at $z > 1$ may not emerge as bright X-ray sources until several billion years later (White & Ghosh 1998; Ghosh & White 2001). This is due to the fact that the evolutionary timescales of LMXBs, their progenitors, and their descendants are thought be significant fractions of the time interval between the SFR peak and the present epoch. In addition to an enhancement near the peak ($z \approx 1.5$) of the SFR due to the prompt turn-on of the relatively short-lived massive X-ray binaries, there may be a second enhancement, by up to a factor ~ 10, at a redshift between ~ 0.5 and ~ 1 due to the delayed turn-on of the LMXB population (Ghosh & White 2001). This second enhancement will not be associated with an overall increase in the optical or infrared luminosity of the galaxy, resulting in an increase in the X-ray to optical luminosity ratio. Hornschemeier *et al.* (2002) using the "stacking" technique detected X-ray emission from L_* redshift 0.4 to 1.5 spiral galaxies in the HDF-N. The X-ray to optical luminosity ratios are consistent with those of galaxies in the local Universe (e.g., Shapley, Fabbiano & Eskridge 2001), although the data indicate a possible increase in this ratio by a factor of 2–3.

Ptak *et al.* (2001) discuss the observable consequences of the increased SFR at high redshifts for the X-ray detection of galaxies at redshift >1 in the HDF-N. To do this Ptak *et al.* (2001) used the Ghosh and White (2001) models for the evolution of the underlying X-ray binary populations for several different possible SFR models (the SFR with redshift is not well known). Depending on the SFR model used, the average X-ray luminosity of galaxies in the HDF-N can be an order of magnitude higher than in the local Universe. These model predictions can be translated into a prediction of the number counts versus flux. Figure 12.14 taken from Hornschemeier *et al.* (2003) shows the number counts from the CDF-N (which are dominated by AGN), along with the predictions from Ptak *et al.* (2001) for two different SFR models. The emerging population of optically bright, X-ray faint (OBXF) galaxies detected in the CDF-N is also shown, along with the extension of the source counts to fainter fluxes using a fluctuation analysis of the CDF-N (Miyaji & Griffiths 2002). The predictions are that emission from normal galaxies, largely at redshift of 1–3, will start to dominate the source counts somewhere between fluxes of 10^{-17} and 10^{-18} erg cm^{-2} s^{-1}. The cross on Fig. 12.14 shows the constraint from the stacking analysis of Hornschemeier *et al.* (2002) for relatively nearby spiral galaxies ($z < 1.5$), which is in agreement with the predictions from Ptak *et al.* (2001) for the lower SFR models.

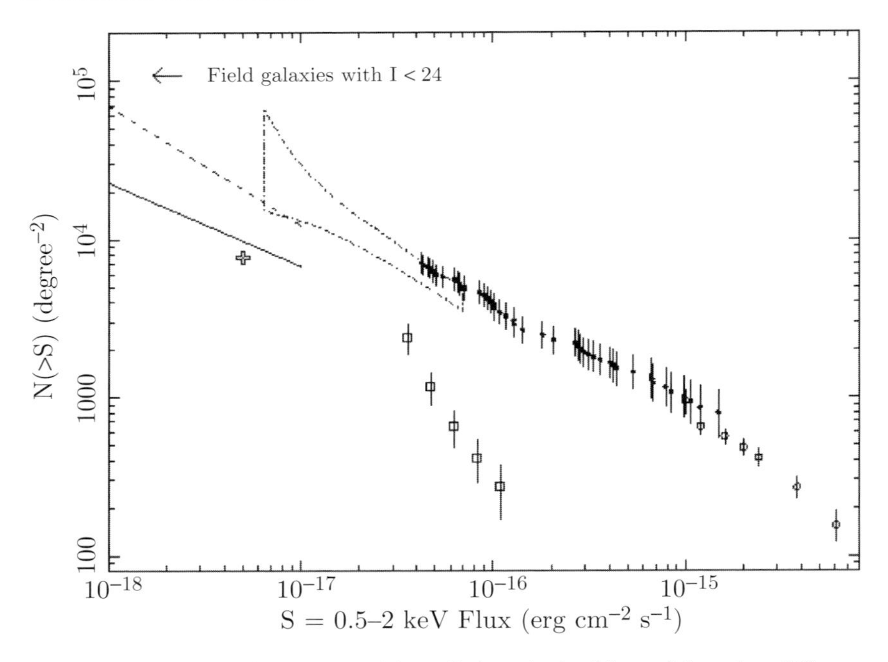

Fig. 12.14. CDF-N number counts, with predictions (at the faint end) based on different SRF at high redshift. The open boxes are the counts from the optically bright, X-ray faint sources – these are mainly normal and starburst galaxies, but some low-luminosity AGN may be present. The cross is the result of the "stacking" analysis using $z \leq 1.4$ galaxies in the CDF-N field. The solid and dashed lines at the lowest fluxes are the predictions of the galaxy number counts from Ptak *et al.* 2001 (from Hornschemeier *et al.* 2003). The leftward pointing arrow indicates the number density of field galaxies at $I = 24$ mag.

Much deeper Chandra exposures of several months or even a year long will be able eventually to reach fluxes of 10^{-18} erg cm^{-2} s^{-1} and directly test the models for the X-ray evolution of galaxies – given the projected long lifetime of Chandra and good luck, these very deep exposures hopefully will eventually happen as the mission matures. To obtain spectra of these galaxies, which are typically at a redshift of 1–3, and to see higher redshift objects at a similar faint flux level, will require 100–1000 times more collecting area with <1 arcsec angular resolution to avoid confusion (e.g., Fabbiano 1990, 1995, 2000; Elvis & Fabbiano 1997; Fabbiano & Kessler 2001). Even more challenging, to resolve an AGN or an offset ULX from more extended emission from the galaxy will require an angular resolution of order 0.1 arcsec. Such mission parameters are technologically extremely challenging, but nonetheless are being pursued by NASA, ESA and ISAS as a long-term goal for X-ray astronomy (Parmar *et al.* 2002, Zhang *et al.* 2002).

12.7 Conclusions

As we have shown in this review, X-ray studies of galaxies are now yielding copious information on the properties of their XRB populations. The classification and study of these different populations is providing a unique tool for understanding the origin and evolution of

XRBs, and for relating these sources to the evolution of the stellar populations of the parent galaxies, both in the nearby and the far-away Universe.

Acknowledgements

This work would not have happened without the vision of Riccardo Giacconi, who pushed forward the high-resolution X-ray telescope concept, and the work of Leon Van Speybroeck, who designed the Chandra optics. We thank the colleagues who have provided figures and comments (Martin Elvis, Ann Hornschemeier, Phil Kaaret, Vicky Kalogera, Andrew King, Albert Kong, Jeff McClintock, Jon Miller, Richard Mushotzky, Doug Swartz, Harvey Tananbaum, Andreas Zezas). This work benefitted by the Aspen Summer Workshop on Compact X-ray Sources (Summer 2002), and would not have been completed without the relentless prodding of Walter Lewin, to whom we are indebted. We acknowledge partial support from the Chandra X-ray Center under NASA contract NAS 8-39073.

References

Alexander, D. M. *et al.* 2003, *AJ*, **126**, 539
Allen, S. W., Di Matteo, T. & Fabian, A. C. 2000, *MNRAS*, **311**, 493
Angelini, L., Loewenstein, M. & Mushotzky, R. F. 2001, *ApJ*, **557**, L35
Bauer, F. E., Brandt, W. N., Sambruna, R. M. *et al.* 2001, *ApJ*, **122**, 182
Bavdaz, M., Peacock, A. J., Parmar, A. N. & Beijersbergen, M. W., Proc. SPIE, Vol. 4497, 31–40 *X-Ray and Gamma-Ray Instrumentation for Astronomy XII*, eds. Kathryn A. Flanagan; Oswald H. Siegmund
Belczynski, K., Kalogera, V., Zezas, A. & Fabbiano, G. 2004, *ApJ*, **601**, L147
Blanton, E. L., Sarazin, C. L. & Irwin, J. A. 2001, *ApJ*, **552**, 106
Brandt, W. N., Hornschemeier, A. E. Schneider, D. P. *et al.* 2001b, *ApJ*, **558**, L5.
Colbert, E. J. M. & Mushotzky, R. F. 1999, *ApJ*, **519**, 89
Colbert, E. J. M. & Ptak, A. F. 2002, *ApJS*, **143**, 25
Colbert, E. J. M., Heckman, T. M., Ptak, A. F. & Strickland, D. K. 2004, *ApJ*, **602**, 231 (astr-ph/0305476)
Dahlem, M., Heckman, T. M. & Fabbiano, G. 1995, *ApJ*, **442**, L49
David, L., Jones, C. & Forman, W. 1992, *ApJ*, **388**, 82
de Vaucouleurs, G., de Vaucouleurs, A., Corwin, H., Jr. *et al.* 1991, *Third Reference Catalogue of Bright Galaxies* (New York: Springer)
Di Stefano, R. & Kong, A. K. H. 2003, *ApJ*, **592**, 884
Di Stefano, R., Kong, A. K. H., Garcia, M. R. *et al.* 2002, *ApJ*, **570**, 618
Di Stefano, R. *et al.* 2004, *ApJ*, **610**, 247
Di Stefano, R., Kong, A. K. H., Van Dalfsen, M. L. *et al.* 2003, *ApJ*, **599**, 1067
Dubus, G. & Rutledge, R. E. 2002, *MNRAS*, **336**, 901
Elvis, M. S. & Fabbiano, G. 1997, in *The Next Generation of X-Ray Observatories*, p. 33 (astro-ph/9611178)
Fabbiano, G. 1988, *ApJ*, **325**, 544
1989, *ARA&A*, **27**, 87
1990, in *High-Energy Astrophysics in the 21st Century*, Taos, NM 1989, ed. P. C. Joss, AIP Conf. Proc. 221, 74
1995, in *X-Ray Binaries*, eds. W. H. G. Lewin, J. van Paradijs and E. P. J. van den Heuvel (Cambridge: Cambridge University Press), 390–416
2000, in *Astrophysical Plasmas: Codes, Models, and Observations*, Proc. Conf. Mexico City, October 25–29, 1999, eds. J. Arthur, N. Brickhouse, and J. Franco, Revista Mexicana de Astronomìa y Astrofìsica (Serie de Conferencias), Volume 9, p. 6–13
Fabbiano, G. & Kessler, M. F. 2001, in *The Century of Space Science*, eds. J. A. M. Bleeker, J. Geiss, M. C. E. Huber (Kluwer: Dordrecht), Vol. 1, p. 561
Fabbiano, G. & Shapley, A., 2002, *ApJ*, **565**, 908
Fabbiano, G., Trinchieri, G. & Van Speybroeck, L. S. 1987, *ApJ*, **316**, 127
Fabbiano, G., Gioia, I. M. & Trinchieri, G., 1988, *ApJ*, **324**, 749
Fabbiano, G., Kim, D.-W. & Trinchieri, G. 1992, *ApJ Suppl.*, **80**, 531
1994, *ApJ*, **429**, 94

Fabbiano, G., Zezas, A. & Murray, S. S. 2001, *ApJ*, **554**, 1035
Fabbiano, G., Zezas, A., King, A. R. *et al.* 2003a, *ApJ Lett.*, **584**, 5
Fabbiano, G., King, A. R., Zezas, A. *et al.* 2003b, *ApJ*, **591**, 843
Fabian, A. C. & Terlevich, R. 1996, *MNRAS*, **280**, L5
Finoguenov, A. & Jones, C. 2001, *ApJ*, **547**, L107
2002, *ApJ*, **574**, 754
Foschini, L., Di Cocco, G., Ho, L. C. *et al.* 2002, *A&A*, **392**, 817
Freedman, W. L. *et al.* 1994, *ApJ*, **427**, 628
Garcia, M. R., Murray, S. S., Primini, F. A. *et al.* 2000, *ApJ*, **537**, L23
Giacconi, R. *et al.* 2002, *ApJS*, **139**, 369
Goad, M. R. Roberts, T. P., Knigge, C. & Lira, P. 2002, *MNRAS*, **335**, L67
Ghosh, P. & White, N. E., 2001, *ApJ*, **559**, L97
Ghosh, K. K., Swartz, D. A., Tennant, A. F. & Wu, K. 2001, *A&A*, **380**, 251
Griffiths, R. E., Ptak, A., Feigelson, E. D. *et al.* 2000, *Science*, **290**, 1325
Grimm, H.-J., Gilfanov, M. & Sunyaev, R. 2002, *A&A*, **391**, 923
2003, *MNRAS*, **339**, 793
Hasinger, G. & van der Klis, M. 1989, *A&A*, **225**, 79
Hernquist, L. & Spergel, D. N. 1992, *ApJ*, **399**, L117
Heuvel, E. P. J. van den, Bhattacharya, D., Nomoto, K. & Rappaport, S. A. 1992, *A&A*, **262**, 97
Holt, S. S., Schlegel, E. M., Hwang, U. & Petre, R. 2003, *ApJ*, **588**, 792
Hornschemeier, A. E. *et al.* 2001, *ApJ* **554**, 742
Hornschemeier, A. E., Brandt, W. N., Alexander, D. M. *et al.* 2002, *ApJ*, **568**, 82
Hornschemeier, A. E. *et al.* 2003, *AJ*, **126**, 575
Humphrey, P. J., Fabbiano, G., Elvis, M., Church, M. J. & Balucinska-Church, M. 2003, *MNRAS*, **344**, 134
Immler, S. & Wang, Q. D. 2001, *ApJ*, **554**, 202
Irwin, J. A., Sarazin, C. L. & Bregman, J. N. 2002, *ApJ*, **570**, 152
Irwin, J. A., Athey, A. E. & Bregman, J. N. 2003, *ApJ*, **587**, 356
Jansen, F. *et al.* 2001, *A&A*, **365**, L1
Jeltema, T. E., Canizares, C. R., Buote, D. A. & Garmire, G. P. 2003, *ApJ*, **585**, 756
Kaaret, P. 2002, *ApJ*, **578**, 114
Kaaret, P., Corbel, S., Prestwich, A. S. & Zezas, A. 2003, *Science*, **299**, 365
Kaaret, P., Prestwich, A. H., Zezas, A. *et al.* 2001, *MNRAS*, **321**, L29
Kalogera, V. 1998, *ApJ*, **493**, 368
Kalogera, V. & Belczynski, K. 2003, ITP talk, http://online.kitp.ucsb.edu/online/clusters_c03/kalogera/
Kalogera, V. & Webbink, R. F. 1998, *Apj*, **493**, 351
Kalogera, V., Henninger, M., Ivanova, N. & King, A. R. 2004, *ApJ*, **603**, L41
Kilgard, R. E., Kaaret, P., Krauss, M. I. *et al.* 2002, *ApJ*, **573**, 138
Kim, D.-W. & Fabbiano, G. 2003, *ApJ*, **586**, 826
2004, *ApJ*, **611**, 846
Kim, D.-W., Fabbiano, G. & Trinchieri, G. 1992, *ApJ*, **393**, 134
King, A. R. 2002, *MNRAS*, **335**, L13
King, A. R., Davies, M. B., Ward, M. J., Fabbiano, G. & Elvis, M. 2001, *ApJ*, **552**, L109
Komossa, S. & Schulz, H. 1998, *A&A*, **339**, 345
Kong, A. 2003, *MNRAS*, **346**, 265
Kong, A. K. H. & Di Stefano, R. 2003, *ApJ Lett.*, **590**, 13
Kong, A. K. H., Garcia, M. R., Primini, F. A. *et al.* 2002, *ApJ*, **577**, 738
Kong, A. K. H., Di Stefano, R., Garcia, M. R. & Greiner, J. 2003, *ApJ*, **585**, 298
Körding, E., Falcke, H. & Markoff, S. 2002, *A&A*, **382**, L13
Kraft, R. P., Kregenow, J. M., Forman, W. R., Jones, C. & Murray, S. S. 2001, *ApJ*, **560**, 675
Kubota, A., Mizuno, T., Makishima, K. *et al.* 2001, *ApJ*, **547**, L119
Kubota, A., Done, C. & Makishima, K. 2002, *MNRAS*, **337**, L11
Kundu, A., Maccarone, T. J. & Zepf, S. E. 2002, *ApJ*, **574**, L5
La Parola, V., Peres, G., Fabbiano, G., Kim, D. W. & Bocchino, F. 2001, *ApJ*, **556**, 47
La Parola, V., Damiani, F., Fabbiano, G. & Peres, G. 2003, *ApJ*, **583**, 758
Liu, J.-F., Bregman, J. N., Irwin, J. & Seitzer, P. 2002, *ApJ*, **581**, L93
Liu, J.-F., Bregman, J. N. & Seitzer, P. 2002, *ApJ*, **580**, L31

Long, K. S. & Van Speybroeck, L. P. 1983, in *Accretion-driven Stellar X-ray Sources*, eds. W. H. G. Lewin & E. P. J. van den Heuvel (Cambridge University Press: Cambridge), 117
Long, K. S., Charles, P. A. & Dubus, G. 2002, *ApJ*, **569**, 204
Maccarone, T. J., Kundu, A. & Zepf, S. E. 2003, *ApJ*, **586**, 814
Madau, P. & Rees, M. J. 2001, *ApJ*, **551**, L27
Madau, P., Ferguson, H. C., Dickinson, M. E. *et al.* 1996, *MNRAS*, **283**, 1388
Makishima, K. *et al.* 2000, *ApJ*, **535**, 632
Martin, C. L. *et al.* 2002, *ApJ*, **574**, 663
Matsumoto, H., Tsuru, T. G., Koyama, K. *et al.* 2001, *ApJ*, **547**, L25
Matsushita, K., Makishima, K., Awaki, H. *et al.* 1994, *ApJ*, **436**, L41
Matsushita, S., Kawabe, R., Matsumoto, H. *et al.* 2001, in ASP Conf. Ser. **249**: *The Central Kiloparsec of Starburst and AGN: The La Palma Connection*, p. 711
Mihos, J. C. & Hernquist, L. 1996, *ApJ*, **464**, 641
Miller, J. M., Wijnands, R., Rodriguez-Pascual, P. M. *et al.* 2002, *ApJ*, **566**, 358
Miller, J. M., Fabbiano, G., Miller, M. C. & Fabian, A. C. 2003a, *ApJ Lett.*, **585**, 37
Miller, J. M., Zezas, A., Fabbiano, G. & Schweizer, F. 2004, *ApJ*, **609**, 728
Miller, M. C. & Hamilton, D. P. 2002, *MNRAS*, **330**, 232
Miyaji, T. & Griffiths, R. E., 2002, *ApJ*, **564**, L5
Mizuno, T., Kubota, A. & Makishima, K. 2001, *ApJ*, **554**, 1282
Mukai, K., Pence, W. D., Snowden, S. L., Kuntz, K. D., 2003, *ApJ*, **582**, 184
Nandra, K., Mushotzky, R. F., Arnaud, K. *et al.* 2002, *ApJ*, **576**, 625
Osborne, J. P. *et al.* 2001, *A&A*, **378**, 800
Pakull, M. W. & Mirioni, L. 2002, Proc. Symp. *New Visions of the Universe in the XMM-Newton and Chandra Era*, 26–30 November 2001, ESTEC, the Netherlands (astro-ph/0202488)
Parmar, A., Hasinger, G. & Turner, M. 2002, presented at COSPAR meeting, E2368P
Pence, W. D., Snowden, S. L., Mukai, K. & Kuntz, K. D. 2001, *ApJ*, **561**, 189
Piro, A. L. & Bildsten, L. 2002, *ApJ*, **571**, L103
Prestwich, A. H., Irwin, J. A., Kilgard, R. E. *et al.* 2003, *ApJ*, **595**, 719
Primini, F. A., Forman, W. & Jones, C. 1993, *ApJ*, **410**, 615
Ptak, A., Griffiths, R., White, N. & Ghosh, P. 2001, *ApJ*, **559**, L91
Ranalli, P., Comastri, A. & Setti, G. 2003, *A&A*, **399**, 39
Read, A. M. & Ponman, T. J. 2001, *MNRAS*, **328**, 127
Roberts, T. P. & Colbert, E. J. M. 2003, *MNRAS*, **341**, L49
Roberts, T. P., Goad, M. R., Ward, M. J. *et al.* 2001, *MNRAS*, **325**, L7
Roberts, T. P., Goad, M. R., Ward, M. J. & Warwick, R. S. 2003, *MNRAS*, **342**, 709
Roberts, T. P. & Warwick, R. S. 2000, *MNRAS*, **315**, 98
Sarazin, C. L., Irwin, J. A. & Bregman, J. N. 2000, *ApJ*, 544, L101
2001, *ApJ*, **556**, 533
Sarazin, C. L., Kundu, A., Irwin, J. A. *et al.* 2003, *ApJ*, **595**, 743
Schlegel, E. M. 1994, *ApJ*, **424**, L99
Shapley, A., Fabbiano, G. & Eskridge, P. B. 2001, *ApJS*, **137**, 139
Shirey, R. *et al.* 2001, *A&A*, **365**, L195
Smith, D. A. & Wilson, A. S. 2001, *ApJ*, **557**, 180
Smith, D. M., Heindl, W. A. & Swank, J. H. 2002, *ApJ*, **569**, 362
Soria, R. & Kong, A. K. H. 2002, *ApJ*, **572**, L33
Soria, R. & Wu, K. 2002, *A&A*, **384**, 99
Speybroeck, L. van, Epstein, A., Forman, W. *et al.* 1979, *ApJ*, **234**, L45
Speybroeck, L. van, Jerius D., Edgar, R. J. *et al.* 1997, *Proc. SPIE*, **3113**, 89
Strohmayer, T. E. & Mushotzky, R. F. 2003, *ApJ*, **586**, L61
Sugiho, M., Kotoku, J., Makishima, K. *et al.* 2001, *ApJ*, **561**, L73
Steidel, C. C., Giavalisco, M., Dickinson, M. & Adelberger, K. L. 1996, *AJ*, **112**, 352
Stetson, P. B. *et al.* 1998, *ApJ*, **508**, 491
Strickland, D. K., Colbert, E. J. M., Heckman, T. M. *et al.* 2001, *ApJ*, **560**, 707
Supper, R., Hasinger, G., Pietsch, W. *et al.* 1997, *A&A*, **317**, 328
Supper, R., Hasinger, G., Lewin, W. H. G. *et al.* 2001, *A&A*, **373**, 63
Swartz, D. A., Ghosh, K. K., Sulemainov, V., Tennant, A. F. & Wu, K. 2002, *ApJ*, **574**, 382
Swartz, D. A., Ghosh, K. K., McCollough, M. L. *et al.* 2003, *ApJ Suppl.*, **144**, 213

Tennant, A. F., Wu, K., Ghosh, K. K., Kolodziejczak, J. J. & Swartz, D. A. 2001, *ApJ*, **549**, L43
Trinchieri, G. & Fabbiano, G. 1985, *ApJ*, **296**, 447
Trinchieri, G. & Fabbiano, G. 1991, *ApJ*, **382**, 82
Trinchieri, G., Goudfrooij, P. 2002, *A&A*, **386**, 472
Trudolyubov, S. P., Borozdin, K. N. & Priedhosky, W. C. 2001, *ApJ*, **563**, L119
Trudolyubov, S. P., Borozdin, K. N., Priedhosky, W. C., Mason, K. O. & Cordova, F. A. 2002a, *ApJ*, **571**, L17
Trudolyubov, S. P., Borozdin, K. N., Priedhosky, W. C. *et al.* 2002b, *ApJ*, **581**, L27
Wang, Q. D. 2002, *MNRAS*, **332**, 764
Weaver, K. A., Heckman, T. M., Strickland, D. K. & Dahlem, M. 2002, *ApJ*, **576**, L19
Weisskopf, M., Tananbaum, H., Van Speybroeck, L. & O'Dell, S. 2000, *Proc. SPIE*, **4012** (astro-ph 0004127)
White, N. E. & Ghosh, P. 1998, *ApJ*, **504**, L31
White, R. E., III, Sarazin, C. L. & Kulkarni, S. R. 2002, *ApJ*, **571**, L23
Williams, B. F., Garcia, M. R., Kong, A. K. H. *et al.* 2004, *ApJ*, **609**, 735
Wu, K. 2001, *Pub. Astron. Soc. Australia*, **18**, 443
Zezas, A. & Fabbiano, G. 2002, *ApJ*, **577**, 726
Zezas, A., Fabbiano, G., Rots, A. H. & Murray, S. S. 2002a, *ApJ Suppl.*, **142**, 239
2002b, *ApJ*, **577**, 710
Zezas, A., Hernquist, L., Fabbiano, G. & Miller, J. 2003, *ApJ*, **599**, L73
Zhang, W., Petre, R. & White, N. E. 2001, *X-ray Astronomy 2000*, ASP Conf. Proc. **234**. Eds. R. Giacconi, S. Serio and L. Stella. San Francisco: Astronomical Society of the Pacific.

13

Accretion in compact binaries

A. R. King
University of Leicester

13.1 Introduction

In a compact binary a black hole, neutron star, or white dwarf accretes from a companion star. These systems have long been a paradigm for accretion theory. Much of our present view of how accretion occurs comes directly from the comparison of theory with observations of these sources. Since theory differs little for other objects such as active galaxies, increasing efforts have recently gone into searching for correspondences in observed behavior. This chapter aims at giving a concise summary of the field, with particular emphasis on new developments since the previous edition of this book.

These developments have been significant. Much of the earlier literature implicitly assumed that accreting binaries were fairly steady sources accreting most of the mass entering their vicinity, often with main-sequence companions, and radiating the resulting accretion luminosity in rough isotropy. We shall see that in reality these assumptions fail for the majority of systems. Most are transient; mass ejection in winds and jets is extremely common; a large (sometimes dominant) fraction of even short-period systems have evolved companions whose structure deviates significantly from the zero-age main sequence; and the radiation pattern of many objects is significantly anisotropic. It is now possible to give a complete characterization of the observed incidence of transient and persistent sources in terms of the disc instability model and formation constraints. X-ray populations in external galaxies, particularly the ultraluminous sources, are revealing important new insights into accretion processes and compact binary evolution.

13.2 Accretion disc theory

Essentially all of the systems discussed here accrete via discs. Accretion disc theory is the subject of many books and reviews (see, e.g., Frank *et al.* 2002 and Pringle 1981). Accordingly this section simply summarizes the main results without giving detailed derivations.

13.2.1 Disc formation

Matter accreting on to a mass M forms a disc if its specific angular momentum J is too large for it to impact the object directly. We define the circularization radius

$$R_{\rm circ} = \frac{J^2}{GM} \qquad (13.1)$$

which is where the matter would orbit if it lost energy but no angular momentum. The condition for disc formation is typically that $R_{\rm circ}$ should exceed the effective size of the

Compact Stellar X-Ray Sources, eds. Walter Lewin and Michiel van der Klis.
Published by Cambridge University Press.

accretor (a parabolic orbit with specific angular momentum J would reach a minimum separation $0.5R_{\rm circ}$). This effective size is identical with the radius of a non-magnetic white dwarf or neutron star, but is of order the magnetospheric radius if there is a dynamically significant magnetic field. For a black hole the effective size is the radius of the last stable circular orbit (Section 2.8.1). If mass transfer occurs via Roche lobe overflow, J is comparable with the specific orbital angular momentum of the binary, $R_{\rm circ}$ is large, and the condition for disc formation is almost always satisfied. The exceptions are some cataclysmic variables where the white dwarf accretor is strongly magnetic (Chapter 10), and double white dwarf systems (Chapter 10) with companion/accretor mass ratios larger than about 0.15 (see, e.g., Nelemans *et al.* 2001).

In the usual case that matter can indeed orbit at $R_{\rm circ}$, disc formation will follow if energy is lost through dissipation faster than angular momentum is redistributed. Since the orbit of lowest energy for a given angular momentum is a circle, matter will follow a sequence of circular orbits about the compact accretor. The agency for both energy and angular momentum loss is called viscosity. For many years the nature of this process was mysterious, but recently a strong candidate has emerged, in the form of the magneto-rotational instability (MRI; Balbus & Hawley 1991). Here a comparatively weak magnetic field threading the disc is wound up by the shear, and transports angular momentum outwards. Reconnection limits the field growth and produces dissipation. Numerical simulations show that this is a highly promising mechanism, and will shortly reach the point of allowing direct comparison with observations.

13.2.2 Thin discs

While viscosity transports angular momentum and thus spreads the initial ring at $R_{\rm circ}$ into a disc, the nature of this accretion disc is determined by the efficiency with which the disc can cool. In many cases this is high enough that the disc is *thin*: that is, its density scale height H obeys

$$H \simeq \frac{c_{\rm s}}{v_{\rm K}} R \ll R \tag{13.2}$$

at disc radius R, where $c_{\rm s}$ is the local sound speed, and

$$v_{\rm K} = \left(\frac{GM}{R}\right)^{1/2} \tag{13.3}$$

is the Kepler velocity, with M the accretor mass. In this state the azimuthal velocity is close to $v_{\rm K}$, and the radial and vertical velocities are much smaller. The properties of being thin, Keplerian and efficiently cooled are all equivalent, and if any one of them breaks down so do the other two.

If the thin disc approximation holds, the vertical structure is almost hydrostatic and decouples from the horizontal structure, which can be described in terms of its surface density Σ. If the disc is axisymmetric, mass and angular momentum conservation imply that the latter obeys a non-linear diffusion equation

$$\frac{\partial \Sigma}{\partial t} = \frac{3}{R}\frac{\partial}{\partial R}\left(R^{1/2}\frac{\partial}{\partial R}[\nu \Sigma R^{1/2}]\right) \tag{13.4}$$

Here ν is the kinematic viscosity, which is usually parametrized as

$$\nu = \alpha c_s H \tag{13.5}$$

where α is a dimensionless number. In a steady state this gives

$$\nu\Sigma = \frac{\dot{M}}{3\pi}\left[1 - \beta\left(\frac{R_{\rm in}}{R}\right)^{1/2}\right] \tag{13.6}$$

where $\dot{M}$ is the accretion rate and the dimensionless quantity β (with $0 \leqslant \beta \leqslant 1$) is specified by the boundary condition at the inner edge $R_{\rm in}$ of the disc. In a steady thin disc, dissipation $D(R)$ per unit surface area is also proportional to $\nu\Sigma$, i.e.,

$$D(R) = \frac{9}{8}\nu\Sigma\frac{GM}{R^3}\left[1 - \beta\left(\frac{R_{\rm in}}{R}\right)^{1/2}\right] \tag{13.7}$$

so that the surface temperature T is independent of the viscosity ν despite being entirely generated by it:

$$T = T_{\rm visc} = \left\{\frac{3GM\dot{M}}{8\pi R^3\sigma}\left[1 - \beta\left(\frac{R_{\rm in}}{R}\right)^{1/2}\right]\right\}^{1/4} \tag{13.8}$$

Integrating Eq. (13.7) over the two disc faces shows that the total power radiated is

$$L = \left(\frac{3}{2} - \beta\right)\frac{GM\dot{M}}{R_{\rm in}} \tag{13.9}$$

Loss of gravitational binding energy contributes $GM\dot{M}/2R_{\rm in}$ here, while the remaining $(1-\beta)GM\dot{M}/R_{\rm in}$ comes from the rotational energy of the star. For example, a disc ending at the radius R_* of a non-rotating star has $R_{\rm in} = R_*$, $\beta = 1$.

13.2.3 Disk timescales

Equation (13.4) shows that Σ changes on a timescale

$$t_{\rm visc} \sim \frac{l^2}{\nu} \tag{13.10}$$

if its spatial gradient is over a length scale l. Hence we would expect a disc to make significant changes in its surface density and thus its luminosity on a timescale $\sim R^2/\nu$, where R is its outer radius. We can use this fact to get an idea of the magnitude of the viscosity in observed discs. In dwarf novae, which are short-period white-dwarf binaries, the disc size is $R \sim 1 - 3 \times 10^{10}$ cm, and surface density changes take a few days. This suggests that $\alpha \sim 0.1$. Encouragingly, numerical simulations of the MRI give comparable answers. There are two other obvious timescales in a disc. The first is the dynamical timescale

$$t_{\rm dyn} \sim \frac{R}{v_{\rm K}} = \left(\frac{R^3}{GM}\right)^{1/2} \tag{13.11}$$

characterizing states in which dynamical equilibrium is disturbed; note that vertical hydrostatic balance is restored on a timescale

$$t_z \sim \frac{H}{c_s} = \frac{R}{v_{\rm K}} = t_{\rm dyn} \tag{13.12}$$

where we have used (13.2). The second is the thermal timescale

$$t_{\rm th} = \frac{\Sigma c_{\rm s}^2}{D(R)} \sim \frac{R^3 c_{\rm s}^2}{GM\nu} = \frac{c_{\rm s}^2}{v_{\rm K}^2}\frac{R^2}{\nu} = \left(\frac{H}{R}\right)^2 t_{\rm visc} \tag{13.13}$$

where we have used (13.7). The alpha-disk parametrization (13.5) can be used to show that

$$t_{\rm visc} \sim \frac{1}{\alpha}\left(\frac{H}{R}\right)^{-2} t_{\rm dyn} \tag{13.14}$$

so we finally have the ordering

$$t_{\rm dyn} \sim t_z \sim \alpha t_{\rm th} \sim \alpha (H/R)^2 t_{\rm visc} \tag{13.15}$$

i.e., dynamical $<$ thermal $<$ viscous.

13.2.4 *Breakdown of the thin disc approximation*

The thin disc approximation discussed above requires the accreting matter to cool efficiently. However, flows with low radiative efficiency on to a black hole can in principle occur, for at least two reasons: the accretion rate $\dot{M}$ may be so low that the inflowing gas has low density and thus a long cooling time, or conversely $\dot{M}$ may be so large that the flow is very optically thick, and radiation is trapped and dragged down the hole. As energy is advected inwards, these flows are called ADAFs (advection-dominated accretion flows). If the accretor is not a black hole, the advected energy must be released near the surface of the accretor. Advection has been invoked to explain observations of quiescent transients (see Section 13.7 below).

Considerable theoretical effort has gone into trying to understand such radiatively inefficient flows. Clearly the flows cannot be geometrically thin, making analytic treatments difficult. For example, such studies often assume a discontinuous change from a thin Keplerian disc to an ADAF at some "transition radius" $R_{\rm tr}$. However, the need to carry off the angular momentum lost in the transition means that any transition region must be extended itself over a size several times the assumed $R_{\rm tr}$. Moreover, the difficulty of applying a predictive theory for the disc viscosity means that $R_{\rm tr}$ is taken as a free parameter in attempts to fit observations, a freedom that would not be present in reality. Recent numerical studies of accretion at low radiative efficiency (e.g., Stone *et al.* 1999; Stone & Pringle 2001) find that very little of the matter flowing in at large radius actually accretes to the black hole. Instead, the time-averaged mass inflow and outflow rates both increase strongly with radius, and almost cancel. The simulations have not so far been run for long enough to reach any kind of steady state: one obvious possibility is that the density may eventually reach values at which radiative cooling does become efficient, leading to a thin disc phase.

13.2.5 *Warping of discs*

An important effect in disc physics (Petterson 1977; Iping & Petterson 1990; Pringle 1996) is that accretion discs tend to warp if exposed to irradiation from a source at their centers, which may be the accretion flow itself. The origin of the warping is that the disc must scatter or re-radiate the incident radiation, which results in a pressure force normal to its surface. If the surface is perturbed from complete axisymmetry the force can increase or decrease in such a way as to cause the perturbation to grow. As the gravitational potential is close to spherical symmetry near the accreting object, it is quite possible for disc material to orbit at angles to the binary plane. A full perturbation analysis (Pringle 1996) corrects an error in the original papers and shows that the condition for warping is

$$L \gtrsim 12\pi^2 \nu_2 \Sigma v_\phi c \tag{13.16}$$

where ν_2 is the vertical kinematic viscosity coefficient.

This inequality shows that we can expect warping in discs in sufficiently luminous systems. We can re-express (13.16) by defining the ratio of the vertical viscosity coefficient ν_2 to the usual radial one ν as $\psi = \nu_2/\nu$. Then (13.6) allows us to write (13.16) as

$$L \gtrsim 12\pi^2 \psi \nu \Sigma v_\phi c = 4\pi \psi \dot{M} v_\phi c \tag{13.17}$$

for a steady disc. Now

$$v_\phi = \left(\frac{GM}{R}\right)^{1/2} = \left(\frac{R_{\rm Schw}}{2R}\right)^{1/2} c \tag{13.18}$$

where $R_{\rm Schw} = 2GM/c^2$ is the Schwarzschild radius of the central star, so combining with (13.17) gives the condition

$$L \gtrsim 4\pi \psi \dot{M} c^2 \left(\frac{R_{\rm Schw}}{2R}\right)^{1/2} \tag{13.19}$$

If we finally assume that the central luminosity L comes entirely from accretion at the steady rate $\dot{M}$ on to a compact object of radius R_*, we can write

$$L \simeq \dot{M} c^2 \frac{R_{\rm Schw}}{R_*} \tag{13.20}$$

and use (13.19) to give

$$\frac{R}{R_*} \gtrsim 8\pi^2 \psi^2 \frac{R_*}{R_{\rm Schw}} \tag{13.21}$$

Equation (13.21) now tells us if warping is likely in various systems. First, it is clearly very unlikely in accretion-powered white dwarf binaries such as CVs, since with $M = 1\,{\rm M}_\odot$, $R_* = 5 \times 10^8$ cm and $\psi \sim 1$ we find the requirement $R \gtrsim 7 \times 10^{13}$ cm, demanding binary periods of several years. (Warping may occur in some supersoft X-ray binaries, where the energy source is nuclear burning rather than the gravitational energy release of the accreted matter.) For neutron stars and black holes by contrast warping is probable if the disc is steady, since with $M = 1\,{\rm M}_\odot$, $R_* = 10^6$ cm, $\psi \sim 1$ we find $R \gtrsim 3 \times 10^8$ cm, while for a black hole with $R_* = R_{\rm Schw}$ warping will occur for $R \gtrsim 8\pi^2 R_{\rm Schw} \sim 2.4 \times 10^7 (M/{\rm M}_\odot)$ cm. It thus seems very likely that discs in LMXBs are unstable to warping, at least for persistent systems. A warped disc shape therefore offers an explanation for the observed facts that persistent LMXB discs are both irradiated and apparently have large vertical extent. The X-ray light curves of persistent LMXBs are strongly structured, implying that the X-rays are scattered by the accretion flow (White & Holt 1982).

Of course, the discussion above only tells us about the possible onset of warping. To see what shape the disc ultimately adopts one must resort to numerical calculations. These have so far only been performed in a highly simplified manner, but do provide suggestive results. For example, one might imagine that warps would be self-limiting, in that a significant warp would geometrically block the very radiation driving the warp. Numerical calculations show that this does not occur: the reason is that warping always starts at large disc radii (cf. 13.21), where the mean disc plane is perturbed away from the original one. Matter flowing inwards from these radii has angular momentum aligned with the perturbed disc plane, and so transfers a "memory" of it to the inner disc. One can show that a warped disc always has a line of nodes following a leading spiral, with the result that the disc can become markedly distorted from its original plane shape without shadowing large areas and arresting the growth of the warp. Of course these calculations are highly simplified, and one might in reality expect the warps to be limited (perhaps by tidal torques), before attaining such distorted shapes. However, it is clear that radiation-induced warping is a promising mechanism for making persistent LMXB discs deviate from the standard picture. An attractive feature of this explanation is that, although the discs deviate *globally* from the standard picture, the thin disc approximation nevertheless continues to apply *locally*, in the sense that the warping is always over length scales much larger than the local scale height H.

13.2.6 Accretion disc stability

Many accreting sources are observed to vary strongly. The clearest examples are dwarf novae (DN), which are binaries in which a white dwarf accretes from a low-mass star, and soft X-ray transients (SXTs), where a black hole or neutron star accretes from a low-mass companion. In both cases the system spends most of its time in quiescence, with occasional outbursts in which it is much brighter. However, beyond this qualitative similarity, there are very clear quantitative differences. In dwarf novae the typical timescales are: quiescence $\sim$ weeks–months, outburst $\sim$ days, and the system luminosity typically rises from $\sim 10^{32}$ erg s^{-1} to $\sim 10^{34}$ erg s^{-1}. In SXTs, the corresponding numbers are quiescence $\sim$1–50 yr or more, outburst $\sim$months, system luminosity rises from $\sim 10^{32}$ erg s^{-1} to $10^{38} - 10^{39}$ erg s^{-1}. Remarkably, it is possible to explain both types of system with a similar model: the SXT version contains only one extra ingredient over that currently accepted for DN.

The basic model at work in both cases is the disc instability picture. There is a huge literature on this subject: Lasota (2001) and Frank *et al.* (2002) give recent reviews. The fundamental idea behind the model is that in a certain range of mass transfer rates, the disc can exist in either of two states: a hot, high viscosity state (outburst) and a cool, low viscosity state (quiescence). In practice these two states correspond to hydrogen existing in ionized or neutral states respectively. The very steep dependence of opacity on ionization fraction and thus temperature makes any intermediate states unstable, and the disc jumps between the hot and cool states on a thermal timescale. In each of these two states it evolves on a viscous timescale. The hierarchy (13.15) shows that this pattern does qualitatively reproduce the observed behavior of long quiescence, short outburst, with rapid transitions between them. Since the basic cause of instability is hydrogen ionization we can immediately deduce the condition for a disc to be stable: it must have no ionization zones. Thus a sufficient condition for suppressing outbursts and making a system persistent is that its surface temperature T should exceed some value $T_{\rm H}$ characteristic of hydrogen ionization (a typical value for $T_{\rm H}$ is 6500 K, depending somewhat on the disc radius). We can assume that this condition is also necessary if the system is to be persistent, i.e., a system is persistent if and only if

$$T_{\rm visc} > T_{\rm H} \tag{13.22}$$

throughout its accretion disc. Since T decreases with disc radius (cf. 13.8) this condition is most stringent at the outer disc edge $R = R_{\rm out}$, so we require

$$T_{\rm visc}(R_{\rm out}) > T_{\rm H} \tag{13.23}$$

for stability. In principle there is another family of persistent sources where the opposite condition

$$T_{\rm visc} < T_{\rm H} \tag{13.24}$$

holds throughout the disc; the condition is tightest close to the inner edge of the disc, where T has a maximum value. There is clearly a strong selection effect against finding such systems, which must be inherently faint, and do not call attention to themselves by having outbursts.

In the following sections we shall see that the occurrence of outbursts implies powerful constraints on the evolution of both CVs and X-ray binaries.

13.3 Dwarf novae: the nature of the outbursts

The disc instability picture described above works quite well when applied to dwarf novae (see the review by Lasota 2001), given the limitations of current treatments of disc viscosity. There is good reason to hope that improvements here will refine the picture further. In particular, two- and three-dimensional disc simulations have given a realistic picture of many disc phenomena that remained obscure in the early one-dimensional calculations. The first example of this was Whitehurst's (1988) investigation of superhumps. Superhumps are a photometric modulation at a period slightly longer than the spectroscopically determined orbital period (Chapter 10). They occur in a subclass of short-period dwarf novae during particularly long outbursts called superoutbursts. They had defied many attempted explanations until Whitehurst's simulations of discs residing in the full Roche potential, rather than simply the field of the accreting star. These revealed that in binaries with sufficiently small secondary-to-primary mass ratios $q = M_2/M_1 \lesssim 0.25$ the disc becomes eccentric and precesses progradely within the binary. Tidal stressing of this disc causes dissipation and thus the superhump modulation.

This picture explains many of the observed superhump properties. Because the modulation results from intrinsic tidal stressing rather than geometrical effects, it is independent of the system inclination. Given that white dwarfs in CVs have masses M_1 lying in a small range $\sim$0.6–1.0 $M_\odot$ the restriction to small mass ratios q means that M_2 must be small ($\lesssim$ 0.15–0.25 $M_\odot$), accounting for the fact that almost all of the superoutbursting systems have periods below the 2–3 hr CV gap. The reason for the restriction to small q was subsequently traced to the fact that the superhump phenomenon is driven by the 3:1 orbital resonance (Whitehurst & King 1991; Lubow 1991); only for small ratios can the disc get large enough to access this resonance.

For some time there were competing explanations for the superoutbursts themselves. One (Vogt 1983) invoked enhanced mass transfer from the secondary triggered by a normal thermal-viscous instability. In contrast Osaki (1989) suggested that in a series of normal outbursts the disc grows in size because accretion on to the white dwarf successively removes matter of low angular momentum. After several such episodes the disc reaches the 3:1 resonant radius, where tides remove angular momentum very effectively and cause more prolonged accretion of a significant fraction of the disc mass. Recent 2-D simulations (Truss *et al.* 2001) show that the latter model does function as suggested, and agrees with observation. Superoutbursts are a direct result of tidal instability. No enhanced mass transfer from the secondary is required to initiate or sustain either the superoutburst or the superhumps, provided that the mass ratio is small enough that the disc can grow to the 3:1 resonant radius.

13.4 Dwarf novae: the occurrence of the outbursts

13.4.1 Short-period dwarf novae

An important question for the disc instability idea is whether it correctly divides observed CV systems into dwarf novae and persistent (novalike) systems (see Chapter 10). In particular all non-magnetic CVs with periods below the well-known gap at 2–3 hr are dwarf novae. To answer this question we have to predict conditions such as mass transfer rate $-\dot{M}_2$ as a function of binary period P, and then check condition (13.22) by setting $\dot{M} = -\dot{M}_2$ in (13.8). Evidently the inequality (13.22) should fail for systems below the gap. In the standard view of CV evolution the secondary stars are assumed to be completely unevolved low-mass main-sequence stars, and CV binaries evolve under angular momentum loss via gravitational radiation and magnetic stellar wind braking (see, e.g., King 1988 for a review; Chapter 16). This picture leads to mass transfer rates $-\dot{M}_2(P)$ which fulfil our expectation above (see Fig. 13.2): the disc instability picture correctly predicts that short-period non-magnetic CVs are dwarf novae.

However things are clearly more difficult for systems above the gap. Here there is a mixture of dwarf novae and novalikes, strongly suggesting that the simple recipe described above for checking (13.22) does not capture the essence of the situation. As the disc instability picture seems to describe outbursts quite well, given a suitable mass transfer rate $-\dot{M}_2$, the most likely resolution of the problem is that the adopted relation for $-\dot{M}_2(P)$ is too simple. This problem is not confined to dwarf novae, but appears to be generic to all CVs above the period gap.

One possibility uses the fact that $-\dot{M}_2(P)$ is the *average* mass transfer rate, taken over timescales $>10^5$ yr. Fluctuations on timescales shorter than this could still be unobservable in any individual system, but lead to an effective spread in instantaneous mass transfer rates. This suggestion has the virtue of leaving intact the existing picture of long-term CV evolution (e.g., the period histogram), which uses $\dot{M}_2(P)$. The fluctuations might themselves result from cycles driven by irradiation of the companion star (King *et al.* 1995). The main problem for this approach is that it tends to predict an almost bimodal distribution of instantaneous mass transfer rates, with the low state too low to give dwarf nova properties in good agreement with observation.

A second quite different idea (King & Schenker 2002; Schenker & King 2002) uses the observed fact that many CV secondaries have spectral types significantly later than would be expected for a ZAMS (zero-age main-sequence) star filling the Roche lobe (Baraffe & Kolb 2000). The idea here is that the spread in $-\dot{M}_2$ reflects real differences in the nature of the secondary star; in many cases this has descended from a star that was originally more massive than the white dwarf, allowing significant nuclear evolution (I shall use the term “evolved” to describe any star whose internal structure deviates significantly from the zero-age main sequence, even if the exterior appearance resembles a ZAMS star.) Moreover, the large mass ratio means that the donor star’s Roche lobe tends to contract as it loses mass, leading to mass transfer on its thermal timescale.

Schenker *et al.* (2002) give an explicit example of this type of evolution, modeling the CV AE Aquarii. This system is not a dwarf nova, but instead has a rapidly spinning magnetic white dwarf which centrifugally expels most of the matter transferred to it. The fact that the spin has not attained some kind of equilibrium in which accretion is allowed strongly suggests that the mass transfer rate has dropped very sharply in the recent past, on a timescale short compared with the observed spindown timescale of $\sim 10^7$ yr. This is highly suggestive of the ending of thermal-timescale mass transfer when the binary mass ratio becomes sufficiently small; the current masses are $M_1 \simeq 0.9\,\mathrm{M}_\odot$, $M_2 \simeq 0.6\,\mathrm{M}_\odot$ (Welsh *et al.* 1995; Casares *et al.* 1996). These masses themselves are suggestive: the white dwarf may have accreted some matter from the companion, implying steady nuclear burning and thus a thermal-timescale mass transfer rate, while the secondary is clearly larger than its main-sequence radius given the current binary period $P = 9.88$ hr (Welsh *et al.* 1993). Finally AE Aqr's ultraviolet spectrum (Jameson *et al.* 1980) shows extremely strong NVλ1238, but the usual corresponding resonance line CIVλ1550 is completely undetectable. As these lines have virtually identical ionization and excitation conditions this is strongly suggestive of an abundance anomaly in which nitrogen is enhanced at the expense of carbon. This in turn is a signature of CNO processing, which clearly requires the star to have been more massive ($\gtrsim 1.5\,\mathrm{M}_\odot$) in the past.

Schenker *et al.* (2002) show that descent from thermal-timescale evolution gives a consistent picture of AE Aqr (see Fig. 13.1). This route therefore offers a way of introducing an intrinsic spread in $-\dot{M}_2(P)$. Generally the mass transfer rates are lower than for ZAMS secondaries, which may explain some of the dwarf novae observed at periods above the gap (Fig. 13.2). In this picture many dwarf novae should show signs of chemical evolution. Evidently a good way of checking for this is to use the ratio of the ultraviolet resonance lines of carbon and nitrogen (CIVλ1550 to NVλ1238) as explained above.

Descendants of thermal-timescale mass transfer are expected on phase-space grounds to make up as much as 50% of short-period CVs (see Schenker *et al.* 2002) and so may account for the incidence of dwarf novae at such periods. The main difficulty for this type of explanation is that as seen from Fig. 13.2 there is a tendency for CVs with evolved secondaries to fill the CV period gap. A possible resolution of this difficulty is that the system moves very rapidly through such periods, but more work is needed to substantiate this.

13.4.2 Long-period dwarf novae

We have so far discussed only short-period ($P \lesssim 12$ hr) CVs. There are a small number of systems at longer periods, the best-studied being GK Per ($P = 48$ hr). All of these systems appear to have outbursts. This is exactly what we would expect from the stability condition (13.23). As the period of a CV increases, so does the disc size $R_{\rm out}$, roughly as the binary separation $a \propto P^{2/3}$. From (13.8) and (13.23) a stable disc then requires

$$-\dot{M}_2 > \dot{M}_{\rm crit} \propto R_{\rm out}^3 \propto P^2 \tag{13.25}$$

At such periods mass transfer is driven by nuclear expansion of the secondary, at a rate of roughly

$$-\dot{M}_2 \propto P \tag{13.26}$$

(see (13.40) below). Hence at sufficiently long periods all systems are likely to be dwarf novae. The coefficients in (13.25) and (13.26) show that this will hold for $P \gtrsim 1$ d, in agreement with observation.

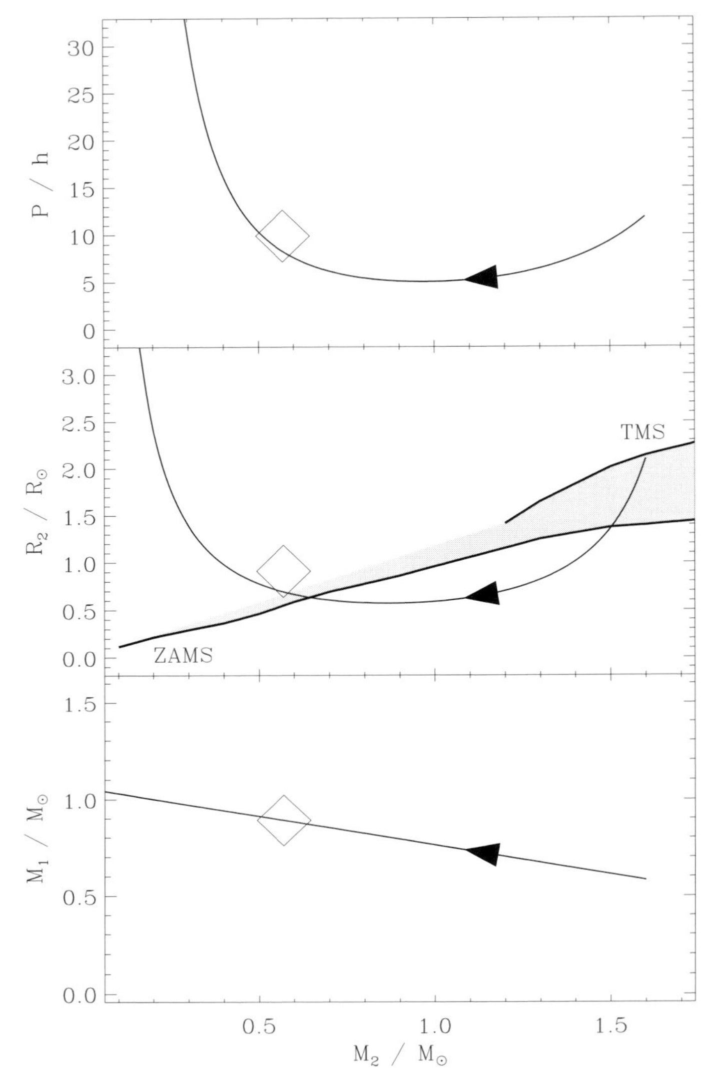

Fig. 13.1. Evolution of a model AE Aqr system. The three panels show the evolution of the orbital period, secondary radius and WD mass. The diamond marks the current position of AE Aqr. The calculations assume that a fraction η of the transferred mass is lost by propeller action (the curve shown has $\eta = 0.3$). The initial condition is a $1.6\,\mathrm{M}_\odot$ secondary star that has almost reached its maximum MS radius. The curves ZAMS, TMS mark the zero-age and turnoff main sequences, and thus the radii of single stars during their main-sequence lives. The evolutionary track shows that the secondary star of AE Aqr is larger than its ZAMS radius at some stages and smaller at others. Without mass loss ($\eta = 1$) the WD grows well beyond the Chandrasekhar limit before reaching the current $P_{\rm orb}$ of AE Aqr, even from the lowest possible WD mass. (From Schenker *et al.* 2002)

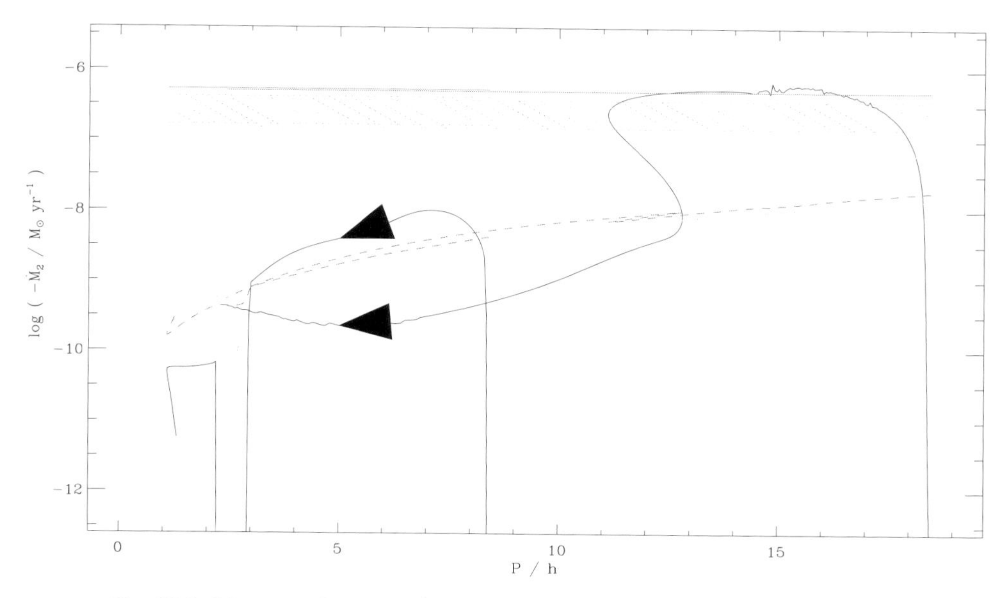

Fig. 13.2. Mass transfer rate $-\dot{M}_2$ versus orbital period P for CVs. The curve starting near $P = 8.5$ h represents a standard CV evolution under magnetic braking and gravitational radiation, beginning with an unevolved secondary star. The mass transfer rate falls to zero in the CV period gap 2 h $\lesssim P \lesssim$ 3 h. The second $-\dot{M}_2(P)$ curve is for a slightly evolved secondary initially more massive than the white dwarf. This undergoes thermal-timescale mass transfer, including an episode of stable nuclear burning of the matter accreting on to the white dwarf (for $-\dot{M}_2 \gtrsim 10^{-7}\,M_\odot\,yr^{-1}$, hatched region) before the mass ratio decreases sufficiently that the system reaches the CV period range with a lower mass transfer rate than the standard evolution. The dashed curves show the disc stability criteria $T_{visc}(R_{out}) > T_H$. These differ because R_{out} is different for the two evolutions. (Figure by Klaus Schenker)

It is interesting to ask why relatively few CVs are seen at these periods. There are several effects selecting against finding them, the strongest probably being that they would only be found in outbursts, which may be rather rare. Nevertheless these systems may be important, as they may offer a channel for making Type Ia supernovae (King *et al.* 2003).

13.5 Soft X-ray transients: the nature of the outbursts

The comparative success of the disc instability idea in explaining dwarf novae, and the qualitative similarities with soft X-ray transients, make it natural to ask if the accretion discs in SXTs are subject to the same instability. The major obstacle here is the vastly different timescales noted earlier. One way of accommodating this mathematically is simply to reduce the disc viscosity in SXTs compared with dwarf novae. However there is no physical motivation for this step, which cannot be regarded as plausible.

A more likely explanation of the long timescales in SXTs uses the observed fact that the accretion discs in SXT outbursts (and indeed in persistent LMXBs) are heavily irradiated. Van Paradijs and McClintock (1994) show that the optical brightness of outbursting SXTs and persistent LMXBs correlates strongly with their X-ray luminosity, and is far greater than would be expected from local viscous dissipation within the disc (cf. 13.8). Irradiation thus raises the surface temperature of the disc, and may potentially stabilize it by removing its ionization zones. The early work on irradiated discs all followed van Paradijs (1996) in using the formula

$$T_{\rm irr}(R)^4 = \frac{\eta \dot{M}_{\rm c} c^2 (1-\beta_a)}{4\pi\sigma R^2}\left(\frac{H}{R}\right)^n \left[\frac{\mathrm{d}\ln H}{\mathrm{d}\ln R} - 1\right] \tag{13.27}$$

where η is the efficiency of rest-mass energy conversion into X-ray heating, $\dot{M}_{\rm c}$ is the central accretion rate, H the disc scale height at disc radius R, β_a is the albedo of the disc faces, and the factor in square brackets lies between 1/8 and 2/7. The index n equals1 or 2 depending on whether there is a central irradiating point source or not; this is discussed further below. The ratio H/R is roughly constant in a disc, so $T_{\rm irr}$ falls off as $R^{-1/2}$. Thus for a large enough disc, $T_{\rm irr}$ dominates the disc's own effective temperature $T_{\rm visc}$, which goes as $R^{-3/4}$. This agrees with our expectations above.

Constructing a self-consistent irradiated disc is a difficult theoretical problem, and attempts to date do not produce results in agreement with observation. In particular, calculations of axisymmetric discs with irradiation (e.g., Kim *et al.* 1999; Dubus *et al.* 1999a; Tuchman *et al.* 1990) tend to show that the inner parts of the disc expand and shadow the outer disc from the irradiation. As most of the disc mass is at large radii, this effect would prevent irradiation having a significant effect on the outbursts. However, since there is abundant observational evidence that LMXB discs *are* strongly irradiated, it is safe to assume this; we shall return later to the question of why axisymmetric disc calculations have difficulty in reproducing this result.

Now let us consider an outburst in a soft X-ray transient. This will be triggered by the ionization instability at some disc radius, and eventually lead to matter accreting strongly on to the black hole or neutron star at the disc center. At this point the X-ray outburst begins: the central X-ray emission irradiates the disc, and heats it. If the disc is small enough the whole of it will now have a surface temperature $T_{\rm irr}$ above $T_{\rm H}$ and be in the hot, high-viscosity state. This reinforces the tendency of mass to accrete inwards, and reduces the local surface density Σ on a viscous timescale. If there were no irradiation, as in a dwarf nova, Σ would eventually at some radius reach the value $\Sigma_{\rm min}$ where the disc must jump back (on a thermal timescale) to the cool, low-viscosity state. This in turn would trigger a cooling wave to move across the disc and return it all to the quiescent state, ending the outburst. However, if the disc is irradiated by the central X-ray source this cannot happen: the local temperature is fixed not by local viscous dissipation, but by irradiation. As this is fueled by the central accretion rate it is a globally rather than locally determined quantity. The disc is trapped in the hot state everywhere until the central accretion rate declines to the point where irradiation can no longer keep the disc in this state. But if there is no cooling wave, the central accretion rate can only decline as a result of the accretion of a significant fraction of the disc mass. It follows that irradiation is likely to prolong disc outbursts and make them use up more of the disc mass. For a given mass transfer rate, the latter effect will lengthen the quiescent intervals in which the disc mass is rebuilt by accretion from the companion star.

This line of reasoning allowed King and Ritter (1998) to give a simple explanation of why SXTs have much longer outburst and quiescent timescales. Their treatment also shows that the X-ray lightcurves are likely to have certain characteristic shapes. An outbursting disc is approximately in a steady state with surface density

$$\Sigma(R) \simeq \frac{\dot{M}_{\rm c}}{3\pi\nu} \tag{13.28}$$

where $\dot{M}_c$ is the central accretion rate. Integrating this gives the total initial mass of the hot zone as

$$M_h = 2\pi \int_0^{R_h} \Sigma R \, dR \simeq \dot{M}_c \frac{R_h^2}{3\nu} \tag{13.29}$$

since the inner disc radius is much smaller than the outer radius R_h reached by the heating front. In (13.29) ν is some suitable average of the kinematic viscosity in the disc, close to its value near R_h. Note that (13.29) is effectively a dimensional relation, and simply asserts the obvious fact that the mass of a steady disc is given by the product of the accretion rate and the viscous time at its outer edge: it does not for example assume that ν is constant through the disc.

As we reasoned above, the only way in which the mass of the hot zone can change is through central accretion, so we have $\dot{M}_c = -\dot{M}_h$, and

$$-\dot{M}_h = \frac{3\nu}{R_h^2} M_h \tag{13.30}$$

or

$$M_h = M_0 e^{-3\nu t/R_h^2} \tag{13.31}$$

where M_0 is the initial mass of the hot zone. This in turn implies that the central accretion rate, and thus the X-ray emission, decays exponentially, i.e.,

$$\dot{M}_c = \frac{R_h^2 M_0}{3\nu} e^{-3\nu t/R_h^2} \tag{13.32}$$

Note that the peak accretion rate at the start of the outburst can be expressed as (disk mass)/(viscous time of the entire hot disc). If the quiescent disc is close to the maximum mass allowed before becoming unstable, this peak rate depends only on the disc size: King & Ritter (1998) find

$$\dot{M}_c(\text{peak}) \simeq 4.8 \times 10^{-8} R_{11}^{7/4} \text{ M}_\odot \text{ yr}^{-1} \tag{13.33}$$

for small (fully irradiated) discs, and

$$\dot{M}_c(\text{peak}) \simeq 4.1 \times 10^{-8} R_{12}^{2} \text{ M}_\odot \text{ yr}^{-1} \tag{13.34}$$

for large discs, where R_{11}, R_{12} are the disc radii in units of 10^{11} cm and 10^{12} cm respectively. These expressions agree with the fact that most SXT outbursts are observed to be close to the Eddington luminosity, even for cases where the accretor is a $\sim$10 $M_\odot$ black hole. The decay constant for outbursts is

$$\tau \simeq 40 R_{11}^{5/4} \text{ d} \simeq 2 R_{12}^{5/4} \text{ yr} \tag{13.35}$$

showing that outbursts can be very prolonged in wide systems.

Eventually $\dot{M}_c$ drops to the point that irradiation cannot keep the outer edge of the disc in the hot state. The outburst proceeds nevertheless, as the central irradiation keeps the inner disc regions ionized. A simple calculation shows that the X-rays then decay linearly rather than exponentially, still on the hot-state viscous timescale. A sufficiently large disc cannot be kept in the hot state by irradiation even at the start of the outburst, and so is always in the linear regime.

A more exact treatment (King 1998) solves the diffusion equation (13.4) for an irradiated disc. The solutions predict steep power-law X-ray decays $L_X \sim (1 + t/t_{visc})^{-4}$, changing to $L_X \sim (1 - t/t'_{visc})^4$ at late times, where t_{visc}, t'_{visc} are viscous timescales. These forms closely resemble the approximate exponential and linear decays inferred above in these two regimes. It is important to realize that the decays are quite different than for unirradiated discs because the viscosity is a function of the central accretion rate rather than of local conditions in the disc.

Since disc size scales with the binary separation, and thus as $P^{2/3}$ by Kepler's law, we arrive at a simple picture in which SXT outbursts in short-period systems begin as exponential, becoming linear near the end of the outburst. In long-period systems the outbursts may be linear throughout. These basic features are generally, but not universally, found in observations. Thus short-period systems such as A0620–00 ($P = 7.8$ hr) have classic "FRED" (fast rise, exponential decay) lightcurves (see Chapter 4), whereas GRO J1744–28 (with $P = 11.8$ d one of the longest-period SXTs) has an entirely linear decay (Giles *et al.* 1996). The exponential–linear dichotomy is examined in detail by Shahbaz *et al.* (1998), and is in good agreement with observation.

While this simple picture is largely correct, there are a number of complications that we should address. We return first to the problem mentioned above, namely that axisymmetric calculations of irradiated discs tend to produce configurations that are strongly self-shadowed, and thus very unlike what is observed. The resolution of this difficulty comes from the realization (Pringle 1996) that, as was discussed in Section 13.2.5 above, strong self-irradiation causes an accretion disc to warp in a non-axisymmetric fashion, with irradiation possible at all radii. Even though this irradiation is relatively patchy (see Fig. 7 of Pringle 1997) it is likely to keep the disc ionized if it is intense enough, i.e., if $T_{irr} > T_H$, because recombination times are long compared with the dynamical time. This suggests a reason why irradiation seems to be unreasonably effective, as discussed above. A second consequence of warping will be very important later on, namely that the radiation field must become quite anisotropic. This follows from considering the disc angular momentum vectors $\mathbf{j}(R)$ measured with respect to the accreting object. An unwarped disc has all the $\mathbf{j}(R)$ vectors parallel to that of the binary, $\mathbf{J}_{orb}$. However, the $\mathbf{j}(R)$ are clearly scrambled in many directions once warping has taken place. Since matter joined the disc with $\mathbf{j}(R)$ parallel to $\mathbf{J}_{orb}$, the radiation field must have changed its own (originally zero) angular momentum to compensate. This must mean that it is anisotropic, and indeed this is what Pringle's calculations reveal (Pringle 1997; Fig. 7). In general the radiation field at infinity is confined to a fairly narrow double cone.

In addition to the exponential or linear X-ray lightcurves discussed above, many SXT outbursts show a secondary maximum (factors of a few) in their lightcurves once the X-rays have declined by a few e-folds. This increase signals the accretion of a new source of mass, which is rather smaller than the original heated disc region giving rise to the basic lightcurve shape. There have been several attempted explanations of this, sometimes invoking extra mass transfer from the secondary star, but a recent 2-D simulation of SXT outbursts (Truss *et al.* 2002) reveals a likely cause. The simulations confirm the simple picture described above as the cause of the main outburst. The secondary maximum results from two effects not included in the arguments above. First, even in a disc irradiated to its outer edge, some of the matter at the outer disc rim will remain in the cool state, because the outer edge of the disc flares and shields it. Second, the mass ratios in all SXTs allow the disc to reach the 3:1

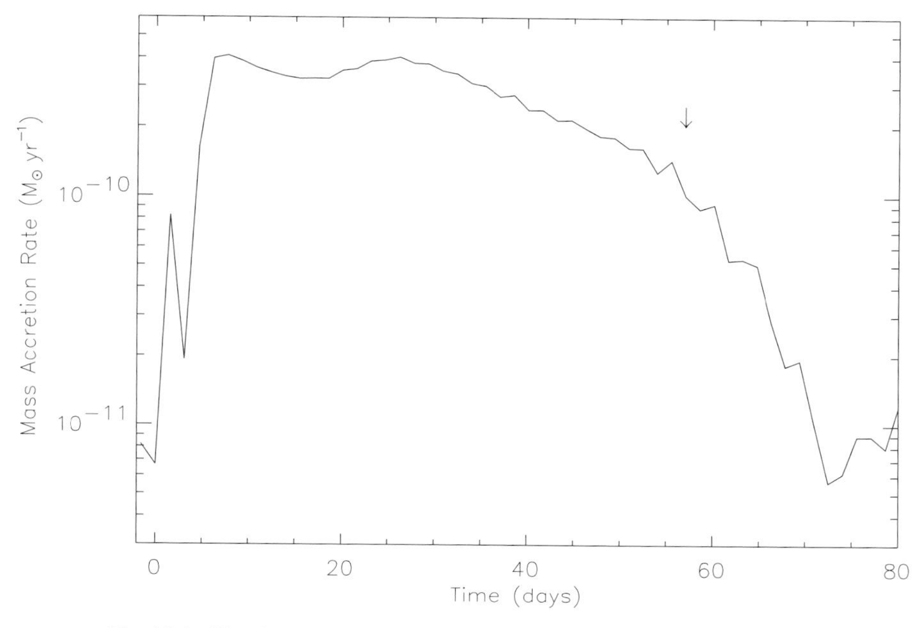

Fig. 13.3. Simulated outburst of a soft X-ray transient showing a secondary maximum, here after ~26 d (Truss *et al.* 2002). Note that the slopes of the lightcurve are similar each side of this maximum.

resonant radius. This is fairly obvious for black-hole systems, where $M_1 \sim 5-10\ \mathrm{M}_\odot$, but we will see shortly that it is true of neutron-star transients also.

The result of including these two effects in the simulations is to produce a second, superoutburst-like increase in the accretion rate through the disc (Fig. 13.3), aided by the strong tidal torque at the resonant radius.

The main effect causing a deviation from linear decays in long-period systems is that, by definition, such systems can have a large fraction of their disc mass permanently in the low-temperature state. This mass reservoir can give rise to hysteresis effects, distorting the simple picture predicting linear decays. The mass reservoir can also allow bursts to recur more frequently than might be expected. For example, GRO J1744–28 was completely undetected in ~30 yr of X-ray astronomy until its first outburst, but then had another outburst only a few months later. For very wide systems the outbursts can also be very long, and indeed some LMXBs usually classified as "persistent" may actually be in outbursts that have lasted for the entire history of X-ray observations. Proof of this comes from the longest-period SXT currently known, GRS 1915+105, which again was not detected until it went into outburst in 1992 (see also Chapter 4). With some variability it has remained bright ever since, showing that outbursts in wide systems can certainly last for at least a decade. The binary period of 33.5 d and mass ~15 $\mathrm{M}_\odot$ (Greiner *et al.* 2001) show that the disc radius here must be $\sim 4 \times 10^{12}$ cm, implying from (13.39) below that $M_{\mathrm{disc,\ max}} \simeq 7 \times 10^{29}$ g. Such a disc could supply the inferred outburst accretion rate $\sim 10^{19}\ \mathrm{g\ s^{-1}}$ for more than 10^3 yr. This is far longer than the decay constant $\tau \sim 10$ yr (13.35). GRS 1915+105 provides an explicit example of the complex behavior of long-period SXTs, as it is observed to vary on the timescale τ but maintains its outburst for longer.

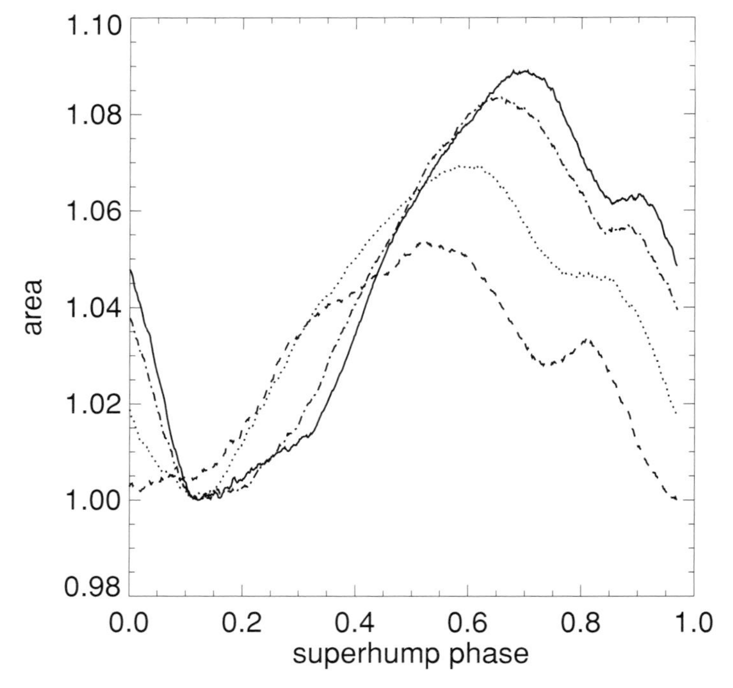

Fig. 13.4. Disk surface area as it varies over the course of a single superhump cycle (Haswell *et al.* 2001). The solid line shows the change in surface area of the entire disc (+ stream), the dot-dash line shows disc region with densities $>1\%$ of the maximum density, dotted line shows regions with density $>5\%$ of the maximum and the dashed line shows regions with density $>10\%$ of the maximum. In this simulation the superhump period is $1.0295 P_{\rm orb}$.

An obvious corollary of these ideas is that there must exist a large unseen population of quiescent transients that have never been observed to have an outburst. In the next section we will see that there is strong evidence for the existence of such objects. Taken together, all these effects show that there is a general tendency for outbursts in long-period systems to show a more complex variety of behaviors than those in short-period systems.

As mentioned above, all SXTs have mass ratios allowing the disc to reach the 3:1 resonant radius, and thus one might expect superhumps to appear. A survey of the optical data on these systems (O'Donoghue & Charles 1996) concluded that indeed superhumps are observed in them. However, at first sight this seems paradoxical, as we have asserted above that superhumps are a modulation of the viscous dissipation in precessing discs, but also that irradiation completely outweighs this dissipation in LMXB (including SXT) discs. The resolution of this problem is that the superhump modulates not only the disc dissipation, but also the disc area (Haswell *et al.* 2001; Section 5.3.6.2). Since a larger area intercepts and reradiates a large fraction of the central X-rays, there should indeed be a superhump modulation of the optical light. In fact, for constant X-ray luminosity the optical lightcurve should simply map the variation of the disc area (Fig. 13.4). The resulting curve agrees well with observation. In general the optical lightcurve is the convolution of the X-ray lightcurve with this area variation.

13.6 Soft X-ray transients: the occurrence of the outbursts

We should next consider the question of when SXT outbursts occur. Here there have been major advances since the last edition of this book. It is now clear that outbursts are extremely prevalent among LMXBs, so that, if anything, persistent systems are rather the exception. This realization has in turn had important consequences for our understanding of compact binary evolution.

These advances stem from van Paradijs's (1996) realization that the correct condition for LMXBs to be stable against outbursts (and thus appear as persistent systems) is

$$T_{\rm irr}(R_{\rm out}) > T_{\rm H} \qquad (13.36)$$

rather than (13.23). In his paper van Paradijs (1996) used the observed X-ray luminosities $L_{\rm X}$ to replace the combination $\eta \dot{M}_{\rm c} c^2$ in the expression (13.27) for $T_{\rm irr}$ (with $n = 1$, appropriate for a strong central irradiating source). He was able to demonstrate that indeed the condition (13.36) correctly divides persistent LMXBs from transient systems.

This success means that we can now use the condition (13.36) with $T_{\rm irr}$ calculated from the evolutionary mean mass transfer rate rather than the observed $L_{\rm X}$, i.e., with $\dot{M}_{\rm c}$ replaced by $-\dot{M}_2$. The results are revealing.

13.6.1 Short-period SXTs

For short orbital periods, $P \lesssim 12$ hr, mass transfer must be driven by angular momentum loss, and we might expect the secondary stars to be unevolved low-mass main-sequence stars. Now Fig. 13.5 shows that if the secondaries in either neutron-star or black-hole LMXBs are unevolved main-sequence stars, all systems are predicted to be persistent. But observation (see Chapter 4) shows that most, if not all, short-period black-hole LMXBs are *transient*, and there are also a number of neutron-star SXTs at such periods.

One possible way of avoiding this conclusion, at least for black-hole systems, is the idea that the irradiation effect might be weaker for a black-hole accretor. Shakura and Sunyaev (1973) pointed out that if there is no strong central source, the irradiation comes only from central disc regions lying in the orbital plane, reducing its effect by a second projection angle $\sim H/R$. Thus the value $n = 2$ would be appropriate in (13.27). King *et al.* (1997b) found that indeed this would make black-hole LMXBs transient even with unevolved secondary stars. However, the X-ray spectra of most black-hole systems have a strong power-law component, which is usually thought to come from a corona. In this case it is unlikely that the weaker irradiation law with $n = 2$ in (13.27) is appropriate for determining disc stability.

If this explanation is abandoned, it seems that the inevitable conclusion (King *et al.* 1996) is that despite appearances, the secondary stars are actually chemically evolved in a large fraction of short-period LMXBs – if not a majority or even a totality in the black-hole case – and hence have lower mass transfer rates. This must mean that they descend from stars with initial masses greater than 0.8 $M_\odot$, and have had time to evolve away from the ZAMS before mass transfer driven by angular momentum loss has pulled them in to short orbital periods. Figures 13.6 and 13.7 show explicitly that this kind of evolution does produce short-period transient systems.

This shift of view parallels the similar shift in our view of CVs (see Section 13.4.1 above), where we now believe that there is a significant admixture ($\lesssim$50%) of evolved secondaries in CVs that have descended via the thermal-timescale mass transfer route. But the shift for short-period LMXBs may be more drastic: it could be that *all* black-hole LMXBs have

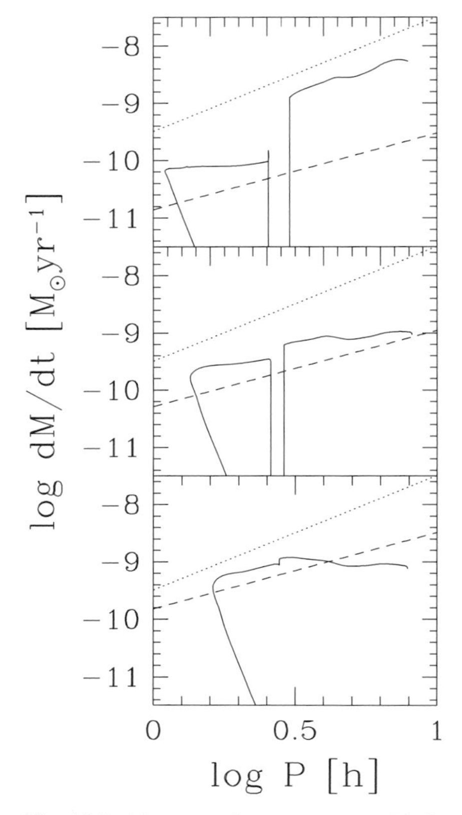

Fig. 13.5. Mass transfer rate versus orbital period for LMXBs with various primary masses and an unevolved secondary of initial mass 1 $M_\odot$ (King *et al.* 1996). The evolution is driven by magnetic braking as long as the secondary has a convective core. The dotted and dashed lines are the critical mass transfer rates given by (13.23) and (13.36) for standard and irradiated discs respectively. Systems with lower mass transfer rates are transient. *Top panel:* 1.4 $M_\odot$ neutron-star primary. *Middle panel:* 10 $M_\odot$ black-hole primary. *Bottom panel*: 50 $M_\odot$ primary. All systems are stable according to the irradiated-disk criterion (13.36). This shows that short-period transients must have evolved secondaries.

evolved secondaries, as so far there is no convincing evidence that *any* black-hole LMXB has an unevolved secondary. Black-hole systems with main-sequence companions would appear among the handful of persistent LMXBs that do not have Type I X-ray bursts (signalling the presence of a neutron star). They are clearly difficult to identify, as in persistent systems one cannot measure a mass function and so get a dynamical mass estimate. However, we will see that even some persistent LMXBs probably have evolved secondaries.

An independent line of argument leading to a similar conclusion comes from the period distribution of LMXBs (Fig. 13.8). This shows no sign of the familiar CV period gap between 2 and 3 hours orbital period. This suggests that in LMXBs accretion either does not cease at 3 hr, or does so in a narrower period range, which may also differ for individual systems. This is just what we see in Figs. 13.6 and 13.7, showing the mass transfer rates for neutron-star and black-hole LMXBs that have reached short periods with secondaries having some degree of nuclear evolution.

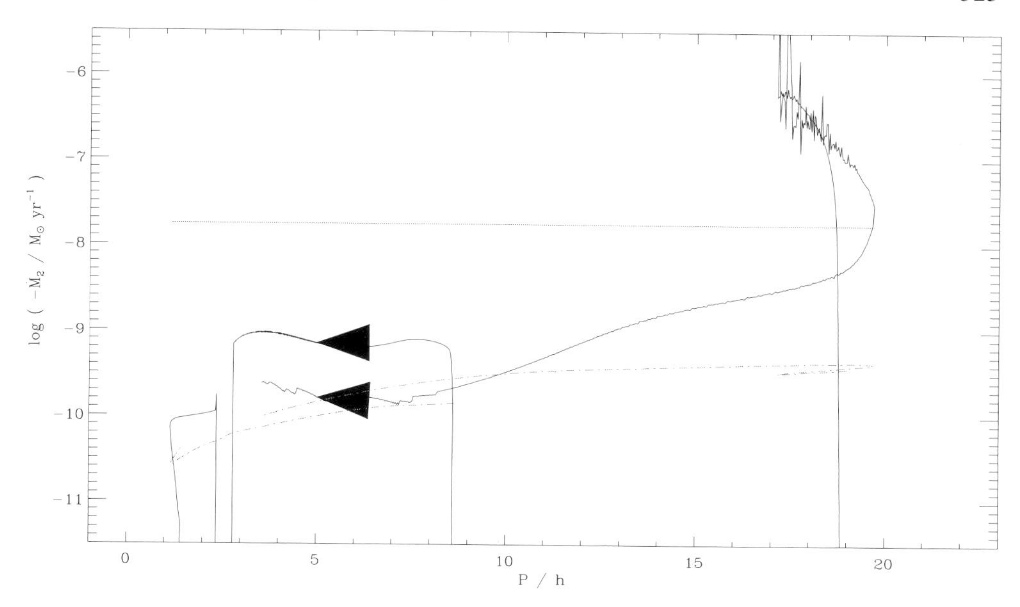

Fig. 13.6. Mass transfer rate $-\dot{M}_2$ versus orbital period P for neutron-star LMXBs ($M_1 = 1.4\,M_\odot$). The curve starting near $P = 8.5$ h represents evolution under magnetic braking and gravitational radiation, beginning with an unevolved secondary star. The mass transfer rate falls to zero in a period gap 2.3 h $\lesssim P \lesssim$ 2.8 h. The second $-\dot{M}_2(P)$ curve begins with a slightly evolved secondary, initially more massive (2.35 $M_\odot$) than the neutron star. This undergoes thermal-timescale mass transfer (the horizontal line shows the Eddington limit) before reaching short periods with a lower mass transfer rate than the standard evolution. The dashed curves show the disc stability criteria $T_{irr}(R_{out}) > T_H$. These differ as the two evolutions have different R_{out}. The unevolved system is always persistent, while the system with the evolved secondary is transient in the period range 5–10 hr. Compare Fig. 13.2. (Figure by Klaus Schenker)

Confirmation of these ideas comes from observations of the short-period black-hole SXT XTE J1118+480 (Haswell *et al.* 2002). The Hubble Space Telescope ultraviolet spectrum of this object shows extremely strong NVλ1238, but CIVλ1550 is completely undetectable. As with AE Aqr (see above) this is a clear sign of CNO processing. The secondary star must have been more massive ($\gtrsim$1.5 $M_\odot$) in the past than the present binary period (4.1 hr) would suggest. It was evidently somewhat chemically evolved when mass transfer began. For much of its life this would not have been obvious from the composition of its outer layers and thus of the mass visible in emission lines: Fig. 13.9 shows that this would have appeared completely normal until the secondary mass got low enough for convection to mix the processed layers into the envelope.

Given the pervasive evidence that short-period SXTs have evolved companions we should now ask why this is so. In the black-hole case the answer seems to go back to evolution of the binary as the black-hole progenitor expands off the main sequence. There are two cases. The binary may be so wide at this stage that the companion never interacts with the primary's envelope as this expansion occurs. Such systems clearly do not reach contact and produce a BHLMXB until the companion has expanded off the main sequence, if at all. These systems thus cannot produce such binaries with unevolved companions. If on the other hand the binary

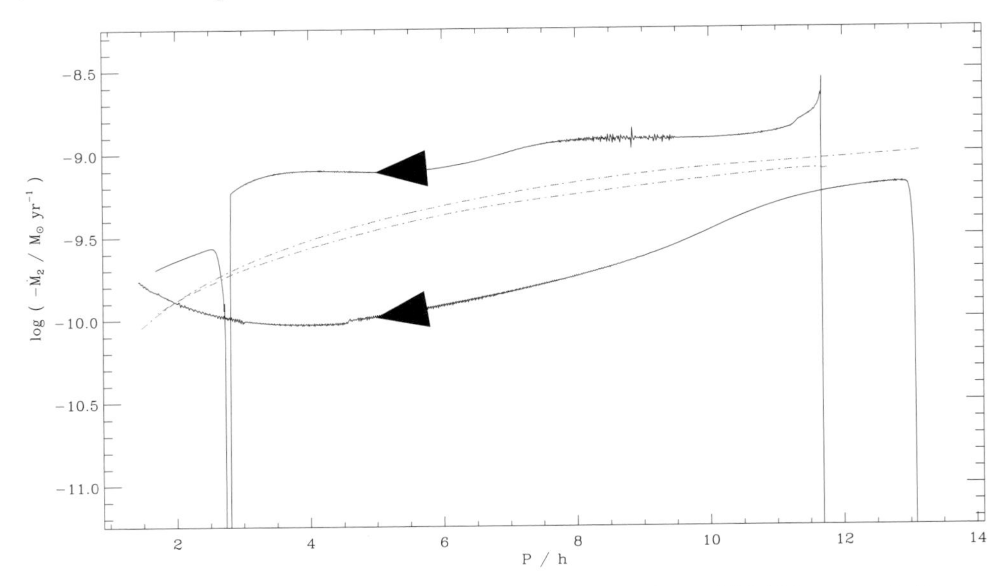

Fig. 13.7. As for Fig. 13.6, but for black-hole LMXBs ($M_1 = 7\,M_\odot$).The curve starting near $P = 11.8$ h represents evolution under magnetic braking and gravitational radiation, beginning with an unevolved secondary star of mass $1.5\,M_\odot$. The curve starting near $P = 13$ h shows the evolution starting with a significantly nuclear-evolved secondary of $1\,M_\odot$. As before the dashed curves are the stability criteria $T_{irr}(R_{out}) > T_H$ in the two cases. Again only the system with the evolved secondary is transient. (Figure by Klaus Schenker)

is relatively narrow, so that the primary's envelope engulfs the companion as it expands, we must ask if the latter can survive this common-envelope (CE) evolution. Friction between the star and the envelope will remove orbital energy and help to unbind the envelope. If the black-hole progenitor, its compact He core and its envelope have masses M_{1p}, M_c, M_e, and R_{1p} is its radius, then equating the loss of orbital energy to the binding energy of the envelope in the standard way (see, e.g., the discussion in Livio 1996) shows that

$$\frac{GM_{1p}M_e}{\lambda R_{1p}} = \alpha\left(\frac{GM_cM_{2i}}{2a_f} - \frac{GM_{1p}M_2}{2a_0}\right) \tag{13.37}$$

where M_{2i} is the companion mass at this point, a_0, a_f are the initial and final orbital separations, and λ, α are the usual weighting and efficiency parameters. For a low-mass companion we have $M_{2i} \ll M_{1p}, M_c, M_e$, and the final separation obeys

$$a_f < \frac{\alpha\lambda}{2}\frac{M_c}{M_{1p}}\frac{M_{2i}}{M_e}R_1 \ll R_{1p}. \tag{13.38}$$

In general the final separation is too small for the companion star to fit inside it, so the stars merge rather than forming a binary. The small mass ratio means that there was insufficient orbital energy available to eject the black-hole progenitor's envelope. Black-hole X-ray binaries must have initial mass ratios $q_i = M_{2i}/M_1$ above some limiting value, and thus post-CE secondary star masses M_{2i} above some minimum value. The uncertainties inherent in any discussion of CE evolution do not allow a precise answer, but it is plausible that this effect limits q_i to values $\gtrsim 0.1$ and thus M_{2i} to values $\gtrsim 1\,M_\odot$. This creates a real possibility that

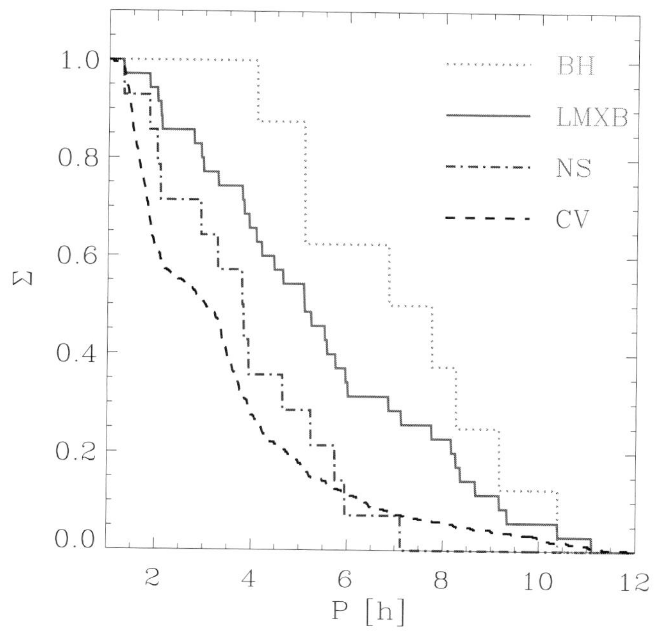

Fig. 13.8. The cumulative period histogram below $P = 12$ hr for LMXBs (solid) compared with that for CVs (dashed). Also shown are the histograms for black-hole LMXBs (dotted) and neutron-star LMXBs (dot-dashed). The CV histogram clearly shows a flatter slope between period $P = 3$ and 2 hours, corresponding to the well-known period gap. There is no significant evidence for such a feature in the LMXB histogram, which is consistent with a uniform distribution. Neutron-star and black-hole systems appear to have significantly different period distributions. (Data from Ritter & Kolb 2003. Figure by Klaus Schenker)

this star has time to undergo significant nuclear evolution before getting into contact and starting to transfer mass (see Fig. 13.11). The first paper to consider this type of evolution was by Pylyser and Savonije (1988).

After the CE phase, the fate of the binary is determined by a competition between various processes which all tend to bring the secondary into contact. The first is nuclear evolution: the star is changing its core composition and eventually increasing its radius on a nuclear timescale $t_{\rm nuc}$. The other timescales $t_{\rm MB}$, $t_{\rm GR}$ describe orbital angular momentum loss via magnetic braking and gravitational radiation respectively, which shrink the binary separation. If $t_{\rm nuc}$ is always shorter than the other timescales, the companion reaches contact as it expands away from the main sequence, and the binary evolves towards longer periods $P \gtrsim 1{-}2$ d with mass transfer driven by nuclear expansion. This is the origin of the long-period LMXBs we shall discuss in the next section. If instead the timescale for orbital angular momentum loss (in practice $t_{\rm MB}$) is shorter than the nuclear timescale $t_{\rm nuc}$, the binary shrinks and reaches contact at a period of a few hours. It becomes a short-period LMXB, with mass transfer driven by angular momentum loss as discussed earlier in this section. However $t_{\rm nuc}$ is only slightly longer than $t_{\rm MB}$ for the masses $M_{2\rm i} \gtrsim 1\,{\rm M}_\odot$ we have inferred for black-hole binaries. (Note that magnetic braking is probably ineffective (i.e., $t_{\rm MB} \to \infty$) for masses $M_{2\rm i} \gtrsim 1.5\,{\rm M}_\odot$.) If the post-CE separation is narrow (i.e., just wide enough to avoid a merger) the binary will reach contact with the secondary still on the main sequence, and produce a persistent

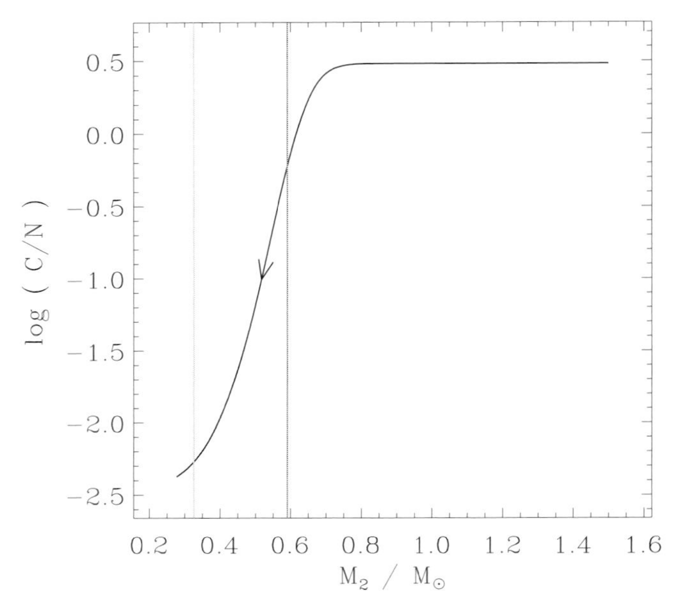

Fig. 13.9. Evolution of the surface abundance ratio C/N for a binary beginning mass transfer from a 1.5 $M_\odot$ main-sequence star on to a 7 $M_\odot$ black hole. The C/N ratio remains at the cosmic value ($\log(C/N) \simeq 0.5$) until the companion mass M_2 is reduced below 0.8 $M_\odot$. Until this point chemical changes are confined to the interior: at the turn-on period of 15 hr the core hydrogen fraction had already been reduced to 28%. The current period of XTE J1118+480 is reached at a mass of 0.33 $M_\odot$, indicated by the left vertical line, while the other near 0.6 $M_\odot$ shows the period of A0620–00. The transferred mass has been accreted by the black hole which has grown beyond 8 $M_\odot$. (Figure from Haswell *et al.* 2002)

BH + MS system (Fryer & Kalogera 2001). Such special initial conditions must make these systems intrinsically rare. In addition they will be very hard to identify, as it is difficult to measure a dynamical mass in a persistent system; they could lurk undetected among LMXBs that do not show Type I X-ray bursts. For the majority of systems, the shrinkage towards contact after the CE phase takes a noticeable fraction of $t_{\rm nuc}$, and the companion is likely to be significantly nuclear-evolved when it comes into contact. At this point further nuclear evolution is frozen, as the star is losing mass on a timescale ($t_{\rm MB}$ or $t_{\rm GR}$) that is shorter than $t_{\rm nuc}$. This argument suggests that we should expect most short-period black-hole LMXBs to have chemically evolved secondaries, accounting for the high fraction of transients among them.

For short-period neutron-star LMXBs the question is obviously more delicate: a majority of them appear to be persistent, and thus could have completely unevolved companions, but there is a non-negligible fraction of transients. It is now well understood (Kalogera & Webbink 1996) that the formation of a neutron-star LMXB is an extremely rare event, requiring highly constrained initial parameters. This results chiefly from the requirement to keep the binary intact when the neutron-star progenitor explodes. King and Kolb (1997) nevertheless found that a significant fraction of short-period neutron-star binaries would have evolved companions and thus be transient if the supernova explosion was assumed fairly symmetrical, and Kalogera *et al.* (1998) extended this result to the case of significantly anisotropic supernovae in which the neutron star receives a kick.

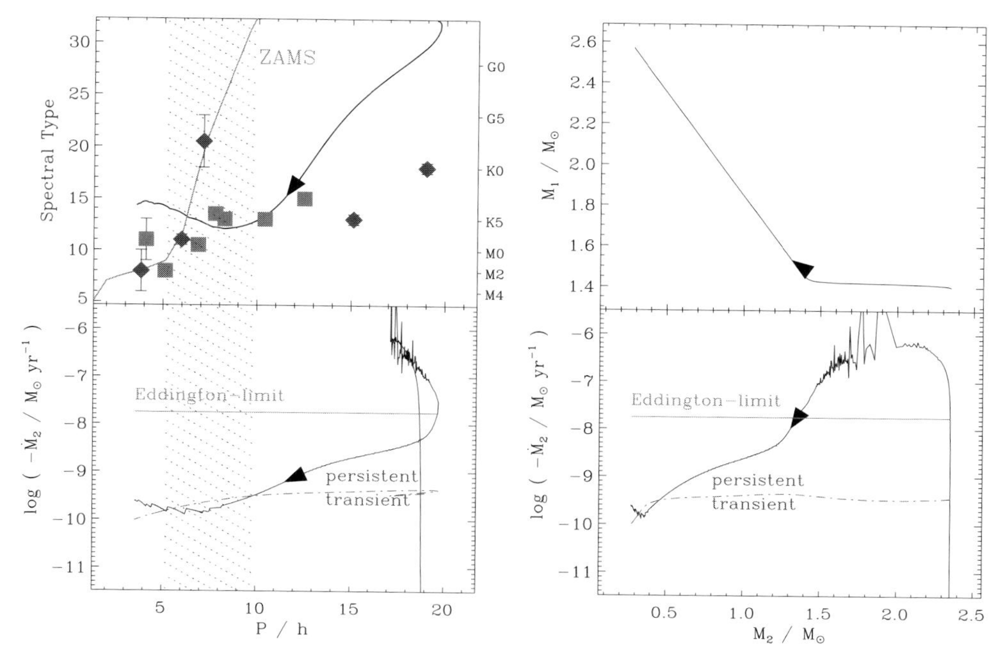

Fig. 13.10. The evolution of short-period neutron-star LMXBs. The initial companion mass is 2.35 $M_\odot$, with core hydrogen fraction 35%. The system first goes through a phase of thermal-timescale mass transfer ($M_2 \gtrsim 1.5\,M_\odot$ in lower right-hand panel, $P \sim 17-20$ hr in lower left-hand panel). After this phase the mass transfer rate drops below the Eddington limit and the neutron star begins to grow significantly in mass (upper right-hand panel). At periods between 10 and 5 hr the system becomes transient (shaded region in left-hand panels). The spectral type of the secondary (solid curve in top left-hand panel) is close to that which a ZAMS secondary would have (lightcurve) at these periods. Varying the mass and initial degree of evolution of the secondary at the onset of mass transfer produces a family of curves displaced horizontally from the solid curve. The squares show persistent neutron-star LMXBs and the diamonds are neutron-star transients.

13.6.2 Post-minimum SXTs

We have seen that a large fraction of neutron-star LMXBs apparently have fairly unevolved companions, and are persistent systems in the $\sim 2-10$ hr period range as expected from Fig. 13.5. However, this figure also shows that such systems are likely to become transient after passing the predicted minimum period ~ 80 min for this type of "CV-like" binary evolution. In the CV case, post-minimum systems are generally regarded as undetectable owing to their faintness. However, straightforward application of the ideas of the last two sections shows that post-minimum NSLMXBs could very well be detectable (King 2000). They are likely to have outbursts reaching $\sim 10^{37}$ erg s^{-1} before declining on an e-folding timescale of a few days. In effect, these systems bring themselves to our notice by saving up their rather feeble mass transfer rates $\sim 10^{-11} M_\odot$ yr^{-1} and using them to produce X-rays for only about 1% of their lifetimes, but with of course 100 times the luminosity. The population of faint transients found by BeppoSAX in a $40^\circ \times 40^\circ$ field around the Galactic center may be drawn from this group. Almost all of these are known to contain neutron stars, signalled by Type I X-ray bursts. Further systematic study of this population may have much to tell us about binary evolution in the Galaxy.

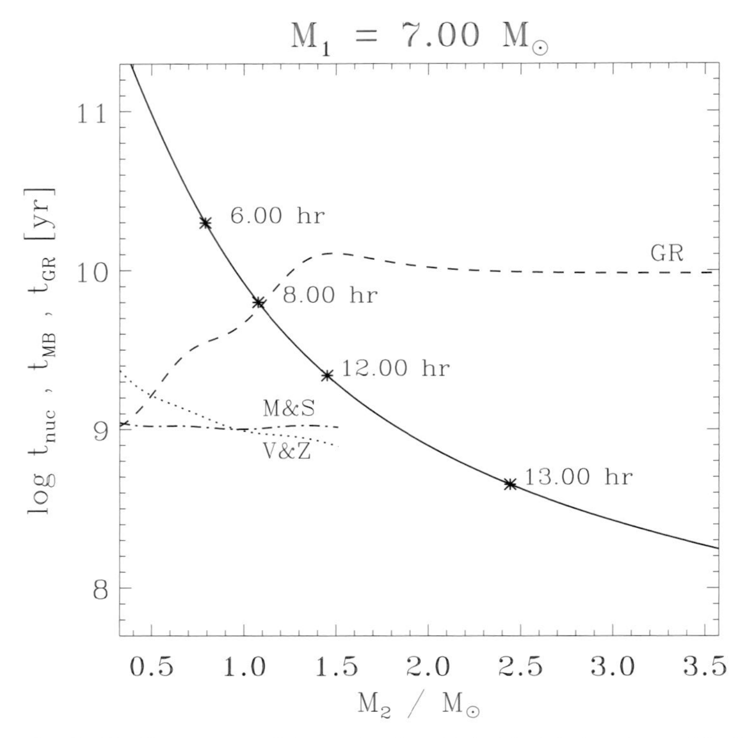

Fig. 13.11. Comparison of timescales for a 7 $M_\odot$ black-hole primary. The full curve gives the nuclear timescale as a function of the secondary mass. At various points the orbital period for a ZAMS star filling its Roche lobe is indicated. The other three curves show the various relevant angular momentum loss timescales ($(-\mathrm{d}\ln J/\mathrm{d}t)^{-1}$): for gravitational radiation (GR, dashed) and magnetic braking according to Mestel and Spruit (M&S, dash-dotted) and Verbunt and Zwaan (V&Z, dash-triple-dotted), all in the version of Kolb (1992). Magnetic braking is assumed to be quenched in stars that have no convective envelopes, i.e., for $M_2 \gtrsim 1.5\,M_\odot$.

13.6.3 On/off transients

Some short-period neutron-star SXTs show variability unlike any other LMXBs. In particular, at least two short-period transients, EXO 0748–68 and GS 1826–24, seem essentially to have simply "turned on", i.e., they were undetected for the first ~30 yr of X-ray astronomy, but have remained "on" ever since their discovery. In compensation, several short-period transients, e.g., X2129+470, X1658−298, have also been observed to turn off during the same time. These on/off transitions cannot result from disc instabilities, as the mass involved is too great. Taking the current bolometric luminosity of X0748−678 as $\sim 10^{37}$ erg s^{-1}, a 10% efficiency of rest-mass energy conversion requires its neutron star to have accreted $\gtrsim 5 \times 10^{25}$ g since 1985 when it was observed to turn on. However, the maximum surface density allowing a disc to remain in the quiescent state leads to a maximum possible quiescent disc mass

$$M_{\mathrm{disc,\,max}} \simeq 10^{-8} R^3 \ \mathrm{g} \qquad (13.39)$$

(e.g., King & Ritter 1998), where R is the outer disc radius in cm. The 3.82 hr period implies a total binary separation of only 9×10^{10} cm, and thus $R \lesssim 5 \times 10^{10}$ cm (assuming a disc filling 90% of the Roche lobe, with a conservative mass ratio $M_2/M_1 \gtrsim 0.1$). From (13.39) this gives $M_{\rm disc,\ max} \lesssim 1.3 \times 10^{24}$ g, far smaller than the mass accreted since 1985. This can only have come from the companion star, implying stable disc accretion during the "on" state.

The fact that these systems are all known (from the presence of Type I X-ray bursts) to contain neutron stars offers a suggestive answer. We see from Fig. 13.10 that even neutron-star systems with quite evolved secondaries only just contrive to lower their mass transfer rates sufficiently to become transients, typically in the period range 5–10 hr. Evidently some of these systems make transitions across this stability curve for intervals that are shortlived in evolutionary terms, but long enough to account for the observed turn-on as a persistent LMXB. This is quite similar to the behavior of a group of CVs known as Z Cam stars, where dwarf nova behavior is from time to time suspended as the system enters a "standstill" (see, e.g., Warner 1995; Chapter 10). This behavior can plausibly be ascribed (e.g., King & Cannizzo 1998) to starspot activity on the secondary star; during outbursting epochs, starspots block enough of the mass transfer region near the inner Lagrange point that the mass transfer rate is reduced to a value slightly below the critical one for disc instability (for the unirradiated discs in CVs). During standstills, enough of these starspots disappear that the mass transfer rate now reaches the regime for stable disc accretion.

The secondaries in short-period LMXBs are probably magnetically active like those in CVs, so it is plausible that a similar effect could cause LMXBs to move between transient and persistent behavior. However, since such a standstill can only occur after an outburst has triggered the transition to the hot disc state, they must be separated by the usual very long quiescent intervals. If the standstills are themselves also very prolonged, and the SXT outbursts during the low mass transfer rate phase are rare or faint, the long-term X-ray behavior of these systems will consist essentially of "off" and "on" states, with only short transitions ("outbursts" and "decays") between them. This on–off behavior is of course just what is observed for several of these systems. Accordingly we identify the on–off transients of Table 13.1 as the LMXB analogues of the Z Cam systems.

13.6.4 Long-period SXTs

For orbital periods $\gtrsim 1-2$ d mass transfer in LMXBs must be driven by the nuclear expansion of the secondary. This star is a low-mass subgiant or giant whose radius and luminosity are determined almost purely by the mass of its helium core, quite independently of its total mass. A simple analytic prescription for this (Webbink *et al.* 1983; King 1988; see also Ritter 1999) shows that the mass transfer rate is

$$-\dot{M}_2 \simeq 4.0 \times 10^{-10} P_{\rm d}^{0.93} m_2^{1.47}\ {\rm M_\odot\ yr^{-1}} \tag{13.40}$$

where $P_{\rm d}$ is the orbital period measured in days and m_2 the total secondary mass in $\rm M_\odot$. Using this together with the stability criterion (13.36) King *et al.* (1997a) showed that most such systems must be transient according to the criterion (13.36): see Fig. 13.12. The only exceptions to this statement are neutron-star systems where the secondary has lost relatively little of its envelope and still has a mass $M_2 \gtrsim 0.8-0.9\,{\rm M_\odot}$. The reason for this propensity of long-period systems to be transient is clear: the accretion disc is so large that central irradiation cannot keep its outer edge ionized.

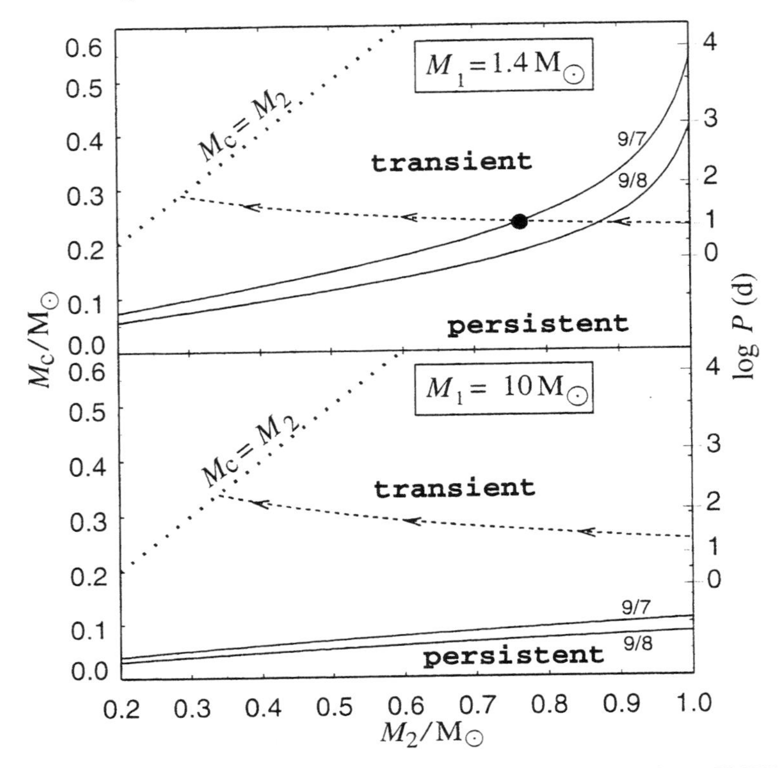

Fig. 13.12. (*Top*) Critical core mass versus total secondary mass for an LMXB with a $1.4\,M_\odot$ neutron-star primary. The two solid curves give the boundaries between persistent and transient behavior for disc scale heights varying as $H \propto R^{9/7}$ and $R^{9/8}$ respectively. The scale on the right-hand axis gives the approximate binary period at which the companion would fill its Roche lobe. Also shown (dashed curve) is the track in the $M_c - M_2$ plane followed by a binary driven by nuclear expansion of the companion. This star has core and total masses $M_c(0) = 0.23\,M_\odot$ and $M_2(0) = 1.5\,M_\odot$ at the start of mass transfer. The positions of known systems Sco X-2, LMC X-2, Cyg X-2 and V395 Car, corresponding to the minimum allowed secondary masses for persistent behavior are all very close to the filled circle. The maximum secondary mass for the neutron-star transient GRO J1744–2844 is again close to $0.75\,M_\odot$ (for slope $9/7$) or close to $0.87\,M_\odot$ (for slope $9/48$). (*Bottom*) Same as at top, but for a $10\,M_\odot$ primary. All these systems are transient. (Figure from King *et al.* 1997a)

This result has a number of important consequences. First, taken together with the results of the previous section, it shows that persistent sources are if anything the exception among LMXBs, being largely confined to a group of short-period neutron-star sources with fairly unevolved companions (see Table 13.1).

Second, we see from (13.40) that essentially all LMXBs with periods longer than a day or so must have mass transfer rates $\gtrsim 10^{-10}\,M_\odot\,yr^{-1}$. In transients, almost all of this mass must attempt to accrete on to the neutron star or black hole during outbursts. With typical duty cycles $\lesssim 10^{-2}$ it follows that the outburst accretion rates must be $\gtrsim 10^{-8}\,M_\odot\,yr^{-1}$, and so at or above the Eddington limit for a neutron star. For longer orbital periods or smaller duty cycles the rates are still higher, and will in general reach those corresponding to the Eddington limit for a black hole. These predictions agree with observations of SXT outbursts: for example

Table 13.1. *Transient and persistent behavior among LMXBs as predicted by the disc instability picture*

Accretor	Companion	$P \lesssim 12$ hr	$P \gtrsim 12$ hr
Neutron star	unevolved	persistent NSLMXB, faint SXT after minimum period	–
Neutron star	evolved	persistent NSLMXB plus some SXT and on/off transients for $P\sim5-10$ hr	SXT unless $M_2 \gtrsim 0.8-0.9\,M_\odot$, progenitors of wide PSR binaries
Black hole	unevolved	formation very rare; persistent BHLMXB	–
Black hole	evolved	SXT	SXT, microquasar

V404 Cygni ($P_d = 6.47$) had a peak outburst luminosity $\sim10^{39}$ erg s^{-1} (see Tanaka & Lewin 1995).

Third, the endpoint of this binary evolution is detectable in many cases, and gives us a test of the theory outlined here. As the companion burns more hydrogen in the shell source surrounding its helium core, the mass of the latter grows, expanding the envelope, increasing the binary period, and driving mass transfer at an increasing rate (cf. 13.40). Eventually all of the envelope mass will be used up, some added to the helium core but most transferred to the compact accretor. We are left with a binary consisting of the low-mass helium white dwarf core of the companion in a wide orbit with the "bare" accretor. The orbital period is given directly by the helium core mass, which specifies the envelope size just before the latter was lost, and thus the period via Roche geometry (Savonije 1987). If the accretor was a neutron star it may have been spun up by accreting angular momentum along with mass. This spin-up can cause the neutron star to turn on again as a radio pulsar, a process known as recycling (Radhakrishnan & Srinivasan 1982). However, if almost all the mass reaches the neutron star in super-Eddington outbursts, the efficiency of both mass and angular momentum gain will be extremely low. This effect may prevent the neutron star spinning up to millisecond periods in systems with a final period $\gtrsim$100 d if the duty cycle is $\lesssim10^{-2}$ (Li & Wang 1998; Ritter & King 2001). This may account for the otherwise suprisingly slow spin rates of some pulsars in long-period binaries. In addition, it appears that there are no millisecond pulsars in wide circular binaries with periods $\gtrsim$200 d (see, e.g., Table 1 in Taam *et al.* 2000). However, the dearth of such binaries may also reflect formation constraints (Willems & Kolb 2002).

Fourth, from the arguments above it appears that radio pulsars in wide circular binaries must descend from transients with shorter orbital periods. Yet although at least 10 pulsar binaries with periods longer than 50 d are known (Taam *et al.* 2000), we do not know of a single neutron-star SXT with an orbital period longer than 11.8 d (GRO J1744−28), even though we would have expected X-ray satellites to see outbursts from such systems anywhere in the Galaxy within the last 30 yr. One obvious reason for this is that the outbursts are very rare. If this is the sole reason for our failure to see outbursts, Ritter and King (2001) estimate that the recurrence times of the outbursts must be at least 300 yr, and probably considerably longer. The idea of such long recurrence times gets strong independent support from observations of the quiescence of the long-lasting transient KS 1731−260 (Wijnands *et al.* 2001; Rutledge *et al.* 2002). Here the neutron star is seen to be so cool that a considerable time, perhaps 10^3 yr, must elapse between outbursts.

13.6.5 *Transient outbursts in high-mass systems*

So far this section has dealt with outbursts in LMXBs. In contrast to those systems, supergiant X-ray binaries do not show disc instability outbursts. The obvious reason for this is that the companion star is itself a potent ionization source, and is able by itself to keep the accretion disc in the hot state. If it has effective temperature T_* and radius R_* then the irradiation temperature on the surface of a disc element at distance $R \gg R_*$ from it is

$$\left(\frac{T_{\rm irr}}{T_*}\right)^4 \simeq \frac{2}{3\pi}\left(\frac{R_*}{R}\right)^3 (1-\beta) \qquad (13.41)$$

where β is the albedo (e.g., Frank *et al.* 2002, Eq. (5.103); note that the star is an extended source of irradiation since its radius R_* is much larger than the local scale height H of the disc). For a disc around a compact star orbiting the massive star in a binary with a circular orbit this gives

$$T_{\rm irr} = 6900 T_{30} R_{10}^{3/4} M_{10}^{-1/4} P_{10}^{-1/2}\ {\rm K} \qquad (13.42)$$

where T_{30}, R_{10}, M_{10}, P_{10} are T_*, R_* and the binary total mass M and period P in units of 3×10^4 K, $10\,{\rm R}_\odot$, $10\,{\rm M}_\odot$ and 10 d respectively, and we have taken $(1-\beta)^{1/4} \simeq 1$.

Comparing $T_{\rm irr}$ with $T_{\rm H} \simeq 6500$ K, this equation suggests that outbursts will be suppressed in HMXBs with O or early B primaries ($T_* \gtrsim 30000$ K, $R_* \sim 20-30\,{\rm R}_\odot$) unless the binary period is longer than ∼10 d. This agrees with the fact that outbursts are not seen in most supergiant X-ray binaries with known orbital periods.

However, disc instabilities may occur in systems with longer orbital periods or high eccentricities e, since at apastron a factor $(1+e)^{-3/4}$ appears on the right-hand side of (13.42). Both possibilities occur in Be X-ray binaries. Here the accretion disc is replenished by a burst of mass transfer as the accretor (apparently always a neutron star) passes close to the Be star's equatorial disc. This burst, or the change in the gravitational potential felt by the disc, is probably the cause of the outbursts usually observed near periastron. Evidently there might in some cases be a second outburst resulting from the thermal-viscous disc instability near apastron, as the disc is allowed to cool there. However, the accretion disc here is clearly not in a steady state, so numerical simulations will be needed to check this idea.

13.7 Quiescent transients and black-hole horizons

One of the main motivations for studying accretion flows is to learn more about the accreting objects. We have seen in the earlier sections that there are very strong indications that many compact binaries, particularly SXTs, do contain black holes. However, all of this evidence is indirect: black holes are a consistent solution, rather than a required one. It would be very interesting to discover *direct* evidence for the defining property of a black hole, namely the lack of a stellar surface. If the systematic difference ($n = 2$ or 1) in the irradiation law (13.27) for black-hole and neutron-star SXTs had held up this would have provided such evidence. However, we saw in Section 13.6.1 that there is little reason to believe this.

A quite separate argument for a systematic BH/NS difference (Narayan *et al.* 1997; Garcia *et al.* 2001) uses the idea (see Section 13.2.4) that an ADAF on to a black hole will be inherently fainter than the same flow on to a neutron star, because the advected energy is released at the stellar surface in the latter case. As we have seen, dynamical mass determinations suggest that some SXTs contain black holes, while others contain neutron stars. There is some observational evidence that the former systems are systematically fainter than the latter in quiescence (but see also Chapter 9), as expected if indeed the two groups have similar ADAFs in this phase.

However, the last requirement is very strong, even granted that ADAFs actually occur in quiescence, which is not entirely settled. We have seen in Section 13.6 that there are systematic differences between black-hole and neutron-star SXTs evolution. Even at similar orbital periods, the two groups probably have different mean mass transfer rates. This in turn may lead to differing outburst/quiescent behavior, undermining the assumption of similar ADAFs in the two cases. It is clear that much more work is needed on these effects if quiescent transients are to provide direct evidence for black-hole horizons.

13.8 Ultraluminous X-ray sources

It has been known for more than 20 years that some external galaxies contain X-ray sources outside their nuclei whose luminosities exceed the Eddington limit for a 1 $M_\odot$ object (Fabbiano 1989). These ultraluminous X-ray sources (ULXs) have attracted considerable interest in recent years (see, e.g., Makishima *et al.* 2000 and references therein; Chapters 4, 5, 9, 12) partly because one simple way of evading the Eddington limit constraint is to assume larger black-hole masses than are generally found as the endpoints of stellar evolution (e.g., Colbert & Mushotzky 1999; Ebisuzaki *et al.* 2001; Miller & Hamilton 2002). Such intermediate-mass ($\sim 10^2 - 10^4$ $M_\odot$) black holes are an ingredient of some pictures of galaxy formation (e.g., Madau & Rees 2001), and thus raised the hope that ULXs might represent such a population.

However, it now appears that although individual ULXs might conceivably harbor intermediate-mass black holes, this cannot be true of the class as a whole. Instead ULXs are probably, in the main, X-ray binaries in rather extreme evolutionary phases. They offer exciting insight into many of the topics discussed in this book.

13.8.1 The nature of the ULX class

There are several lines of argument suggesting that the ULX class involves stellar-mass accretors. These are both negative and positive. The negative arguments concern the difficulties of forming and then feeding intermediate-mass objects. King *et al.* (2001) summarize several of these. A black hole of $10^2 - 10^4\,M_\odot$ cannot result from current stellar evolution, as stars of $\gtrsim 100\,M_\odot$ are subject to huge mass loss if they have any significant metal content, and rapidly reduce their masses to quite modest values before producing black holes. Primordial stellar evolution (i.e., with hydrogen and helium alone) can produce black holes with such masses, but then the question of feeding the hole with accretion becomes critical. As we have seen in Section 13.6.1, black-hole binaries must either be born with separations so wide that reaching contact at all is problematic, or have initial mass ratios above a minimum value $q_i \gtrsim 0.1$. With black-hole masses $M_1 > 100\,M_\odot$ the companion must have $M_{2i} \gtrsim 10\,M_\odot$ and thus long ago have become a compact object itself. Various other routes to making intermediate-mass black holes have been suggested, often invoking mergers within globular clusters. Again the process is rather delicate, as the merged object must not attain the rather low space velocity required to escape the cluster before its mass has built up to the required value. ULXs are not observed to be members of globular clusters, and indeed their incidence is often associated with recent star formation, so the hole must eventually be ejected from the cluster. There is then again the problem of finding a companion to supply the hole with mass: the hole apparently did not achieve this feat in the cluster, even given the high stellar density, but must nevertheless manage it in the field. At the very least these difficulties suggest that the efficiency of finding a companion and thus turning the system on as a ULX must be rather low. The observed numbers of ULXs found in star-forming systems such as the Antennae ($\sim$10) therefore demand rather high formation rates for intermediate-mass black holes if they are to explain the ULX class as a whole.

In addition to these negative arguments, there are some positive arguments favoring a stellar-mass black-hole origin for ULXs. First, in most cases their X-ray spectra are consistent with thermal components at $kT \sim 1-2$ keV. This is a natural temperature for a stellar-mass object (see Fig. 13.16). In addition X-ray spectral transitions typical of such sources are observed in ULXs (e.g., Kubota *et al.* 2001). Second, many ULXs, though not all, are close to regions of star formation (Zezas *et al.* 1999; Roberts & Warwick 2000, Fabbiano *et al.* 2001; Roberts *et al.* 2002). This is consistent with ULXs being the extreme end of an HMXB population formed in such regions. Third, optical identifications (e.g., Goad *et al.* 2002) are consistent with HMXBs.

On the basis of these arguments King *et al.* (2001) suggested that most ULXs were probably mildly (factors $\lesssim 10$) anisotropically emitting X-ray binary systems accreting at close to the Eddington value. This allows apparent luminosities up to $\sim 10^{40}$ erg s^{-1}, compatible with the great majority of claimed ULXs. At first sight this seems to require large numbers of unseen sources, and thus a high birthrate. However, this is not so, as the following analysis shows.

We assume that a compact object of mass M_1 accretes from a mass reservoir (e.g., a companion star) of mass M_2. We denote the mean observed number of ULXs per galaxy as n, the beaming factor as b ($= \Omega/4\pi$, where Ω is the solid angle of emission), the duty cycle (= time that the source is active as a fraction of its lifetime) as d, and define an "acceptance rate" a as the ratio of mass accreted by M_1 to that lost by M_2, i.e., the mean accretion rate $\dot{M}_1 = a(-\dot{M}_2)$. We further define $L_{\rm sph}$ as the apparent X-ray (assumed bolometric) luminosity of a source, given by the assumption of isotropic emission, and let $L_{40} = L_{\rm sph}/10^{40}$ erg s^{-1}. From these definitions it follows that the luminosity

$$L = bL_{\rm sph} = 10^{40} bL_{40} \text{ erg s}^{-1} \tag{13.43}$$

and the minimum accretor mass if the source is not to exceed the Eddington limit is

$$M_1 \gtrsim 10^2 bL_{40}\, \mathrm{M}_\odot \tag{13.44}$$

The total number of such sources per galaxy is

$$N = \frac{n}{bd} \tag{13.45}$$

with a minimum mean accretion rate during active phases of

$$\dot{M}_{\rm active} = \frac{\dot{M}_1}{d} = -\frac{\dot{M}_2 a}{d} > 10^{-6} bL_{40}\, \mathrm{M}_\odot \text{ yr}^{-1} \tag{13.46}$$

The mass loss rate from M_2 is thus

$$-\dot{M}_2 > 10^{-6} \frac{bd}{a} L_{40}\, \mathrm{M}_\odot \text{ yr}^{-1} \tag{13.47}$$

and the lifetime of a source is

$$\tau = -\frac{M_2}{\dot{M}_2} \lesssim 10^6 \frac{m_2 a}{bdL_{40}} \text{ yr} \tag{13.48}$$

with $m_2 = M_2/\mathrm{M}_\odot$, leading to a required birthrate per galaxy

$$B = \frac{N}{\tau} \gtrsim \frac{n}{bd} \cdot \frac{bdL_{40}}{10^6 m_2 a} = 10^{-6} \frac{nL_{40}}{m_2 a} \text{ yr}^{-1} \tag{13.49}$$

The important point to note here is that the required birthrate is independent of beaming (and duty cycle): the greater intrinsic source population N required by $bd < 1$ (cf. 13.45) is compensated by their longer lifetimes (cf. 13.48).

A possible alternative to the idea of mild anisotropy as an explanation for ULXs was proposed by Begelman (2002), who suggested that a magnetized accretion disc might allow luminosities that were genuinely super-Eddington by factors up to ~10 (see also Shaviv 1998, 2000). An observationally motivated objection to this is the existence of neutron stars that have apparently passed through phases of super-Eddington mass transfer without showing signs of significant mass or angular momentum gain, as we might expect if super-Eddington accretion were allowed. The difficulty in spinning up neutron stars in wide circular binaries (Section 13.6.4 above) is an example.

Strong confirmation of the idea that the ULXs represent a population of stellar-mass X-ray binaries comes from work by Grimm *et al.* (2003). They show that the cumulative luminosity functions of nearby starburst galaxies, as well as the Milky Way and Magellanic Clouds, can be fitted by a single form normalized by the star formation rate (SFR), as measured by various conventional indicators. The form

$$N(>L) = 5.4 \times \mathrm{SFR} \times \left(L_{38}^{-0.61} - 210^{-0.61}\right) \tag{13.50}$$

is used in Figs. 13.13 and 13.14, where L_{38} is the X-ray luminosity in units of 10^{38} erg s^{-1}.

These results mean that the ULX population must be some kind of extension of the HMXB/LMXB populations contributing to the luminosity function at lower luminosity. Note that this result is asserted only for the ULX *as a class*. King *et al.* (2001) point out that all the arguments above still allow the possibility that *individual* ULXs could involve intermediate-mass black holes. However, it is probably fair to say that at the time of writing no convincing example is known, with one possible exception: the ULX in M82 is extremely bright ($L_X \sim 10^{41}$ erg s^{-1}; Kaaret *et al.* 2001; Matsumoto *et al.* 2001). It is conceivable that this is the captured nucleus of a dwarf galaxy, which would naturally have had a nuclear black hole of intermediate mass.

13.8.2 Models for ULXs

If ULXs are X-ray binaries, we should ask what causes their unusual appearance, and in particular their defining feature, the apparent super-Eddington luminosity. The immediate cause appears to be a highly super-Eddington mass inflow rate near the accretor, leading to three characteristic features: (i) the total accretion luminosity is of order L_{Edd}; (ii) this is confined to a solid angle $4\pi b \lesssim 4\pi$, making the source apparently super-Eddington when viewed from within this solid angle (even if it is not genuinely super-Eddington); and (iii) the bulk of the super-Eddington mass inflow is either accreted at low radiative efficiency, or more probably, ejected in the form of a dense outflow, probably including relativistic jets.

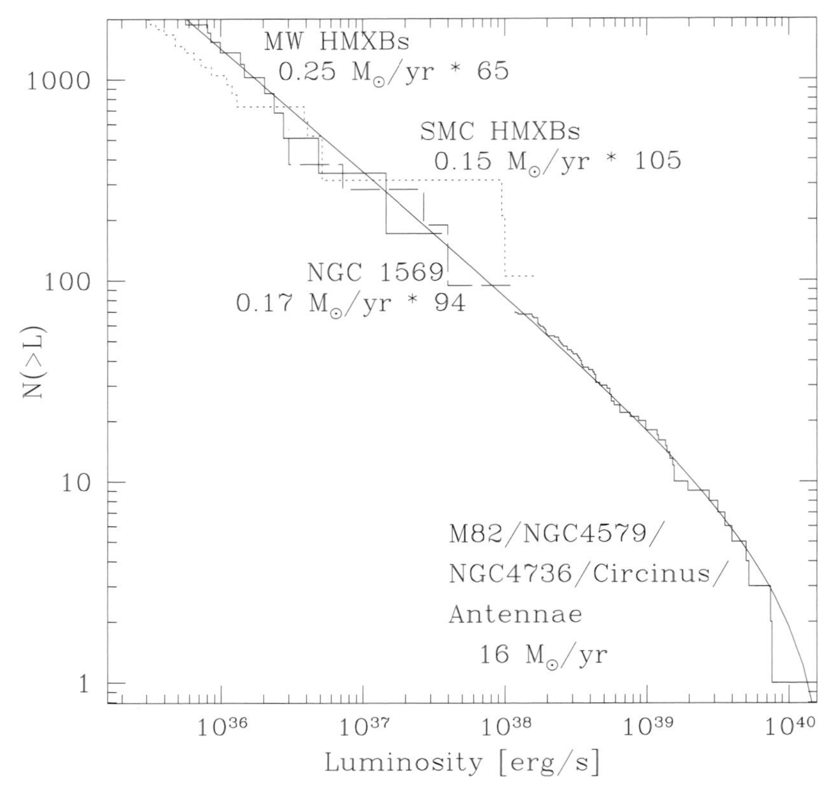

Fig. 13.13. Combined luminosity function of compact X-ray sources in the starburst galaxies M82, NGC 4038/9, NGC 4579, NGC 4736 and Circinus, with a total SFR of $16\,M_\odot\,yr^{-1}$ (above 2×10^{38} erg s^{-1}), and the luminosity functions of NGC 1569, HMXBs in the Milky Way, and in the Small Magellanic Cloud (below 2×10^{38} erg s^{-1}). The thin solid curve is the best fit to the combined luminosity function of the starburst galaxies only, given by (13.50). (Figure from Grimm *et al.* 2003)

A suggested accretion flow with these features (Paczyński & Wiita 1980; Jaroszynski *et al.* 1980; Abramowicz *et al.* 1980) postulates an accretion disc whose inner regions are geometrically thick, and a central pair of scattering funnels through which the accretion radiation emerges. Note that this form of "beaming" does not involve relativistic effects, although Doppler boosting in a relativistic jet has also been suggested as a way of explaining the high luminosities (Koerding *et al.* 2001; Markoff *et al.* 2001). The thick-disk plus funnels anisotropy mechanism explicitly requires a high mass inflow rate near the black-hole or neutron-star accretor, much of which must be ejected, probably some of it in the form of a jet. (In fact, the motivation of the original papers, Jaroszynski *et al.* 1980; Abramowicz *et al.* 1980, was to produce a geometry favoring jet production.)

The identification with super-Eddington mass inflow rates made above allows us to identify the likely ULX parent systems. There are two situations in which X-ray binaries naturally have such rates: phases of thermal-timescale mass transfer, and bright SXT outbursts. The first of these is considered extensively by King *et al.* (2001) and its main features can be summarized briefly here.

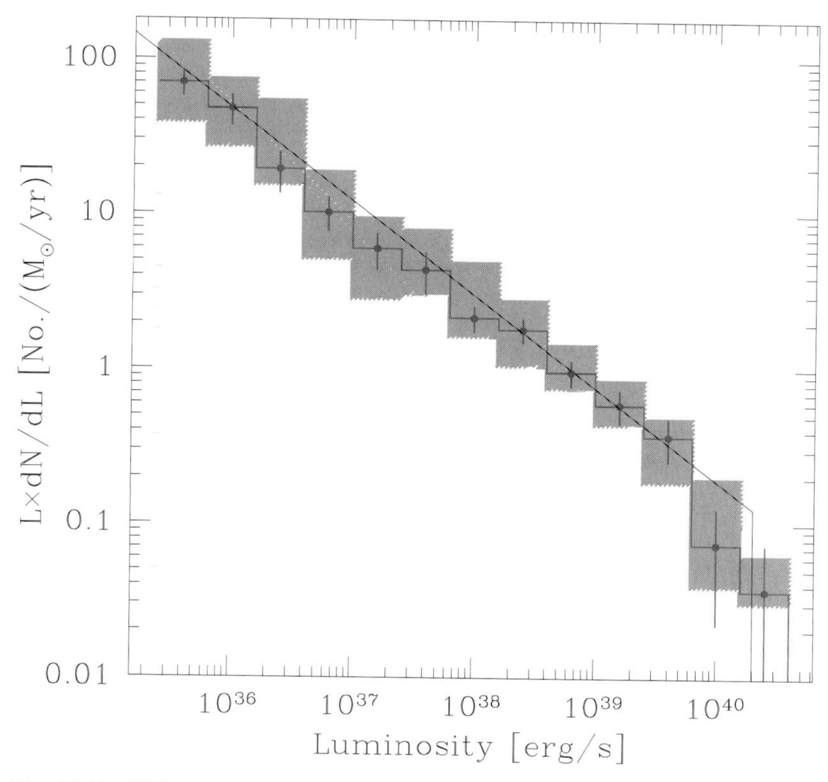

Fig. 13.14. Differential luminosity function of compact X-ray sources obtained from the whole sample of Grimm *et al.* (2003). The straight line is the same as in Fig. 13.13. The shaded area is the 90% confidence level interval obtained from a Monte Carlo simulation allowing for uncertainties in the star formation rate and distances.

Thermal-timescale mass transfer occurs in any Roche-lobe-filling binary where the ratio q of donor mass to accretor mass exceeds a critical value $q_{\rm crit} \sim 1$. Thus all high-mass X-ray binaries will enter this phase once the companion fills its Roche lobe, either by evolutionary expansion, or by orbital shrinkage via angular momentum loss. Depending on the mass and structure of the donor, extremely high mass transfer rates $\dot{M}_{\rm tr} \sim 10^{-7}-10^{-3}\,{\rm M_\odot\,yr^{-1}}$ ensue. SS433 is an example of a system currently in a thermal-timescale mass transfer phase (King *et al.* 2000) which has descended by this route. The idea that SS433 itself might be a ULX viewed "from the side" provides a natural explanation of its otherwise puzzlingly feeble X-ray emission ($L_{\rm X} \sim 10^{36}$ erg s^{-1}, Watson *et al.* 1986).

The binary probably survives the thermal-timescale phase without entering common-envelope (CE) evolution provided that the donor's envelope is largely radiative (King & Begelman 1999). Observational proof of this is provided by Cygnus X-2 (King & Ritter 1999; Podsiadlowski & Rappaport 2000), whose progenitor must have been an intermediate-mass binary (companion mass $\sim 3\,{\rm M_\odot}$, neutron-star mass $\sim 1.4\,{\rm M_\odot}$). CE evolution would instead have engulfed the binary and extinguished it as a high-energy source. The binary would probably have merged, producing a Thorne–Żytkow object.

The birthrates of intermediate and high-mass X-ray binaries are compatible with the observed numbers of ULXs: King *et al.* (2001) show that the birthrates required to explain the

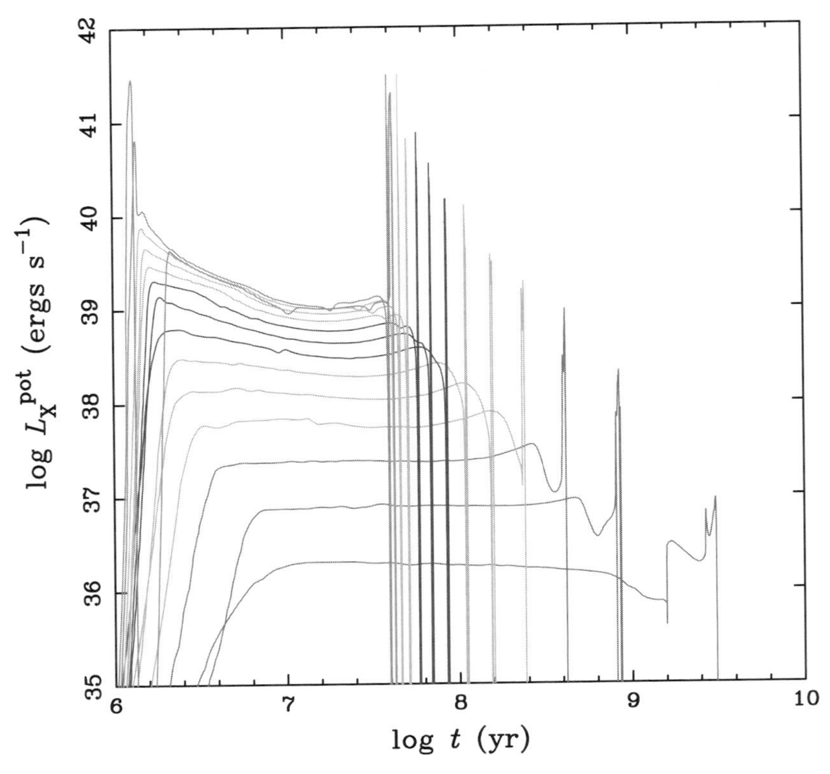

Fig. 13.15. Potential X-ray luminosities as a function of time for 14 black-hole binary evolution sequences. Each curve corresponds to a different initial donor mass ranging from 2 to 17 $M_\odot$. Higher luminosities correspond to higher initial donor mass. In all cases the donor is unevolved at the start of mass transfer, and the black-hole mass is 10 $M_\odot$. The spikes at the start of evolutions with initially massive donors correspond to thermal-timescale mass transfer ($M_2 \gtrsim M_1$), while those at the end show the donor climbing the giant branch. The latter phase lasts about 5% of the total evolution. (Figure from Rappaport *et al.* 2005)

latter are independent of the dimensionless beaming and duty-cycle factors b, d. For massive systems the thermal timescale lasts longer than the preceding wind-fed X-ray binary phase; the fact that there are far fewer observed ULXs than massive X-ray binaries must mean that the beaming and duty-cycle factors obey $bd \ll 1$. This picture also explains the observed association of ULXs with star formation. There is in addition some evidence that the ULXs in the Antennae are on average slightly displaced from star clusters, suggesting that they have acquired significant space velocities as a result of a recent supernova explosion, just like HMXBs (Zezas *et al.* 2002). If this is correct it is a direct demonstration that the masses of ULX systems are not unusually high.

Recently Rappaport *et al.* (2005) have performed extensive population synthesis calculations for black-hole binaries with initial companion masses in the range 2–17 $M_\odot$. Their results imply that super-Eddington mass transfer rates also occur naturally in other situations. Normal main-sequence under core hydrogen burning produces mildly super-Eddington transfer rates for sufficiently massive donors. Much higher transfer rates occur as the donor climbs the giant branch. This phase lasts only about 5% of the entire evolution. Allowing accretion luminosities to exceed L_{Edd} by factors $\lesssim 10$ (as in the suggestion of Begelman 2002) produces ULX populations broadly compatible with observation.

While thermal-timescale and giant-branch mass transfer probably accounts for a significant fraction of observed ULXs, bright SXT outbursts will also produce super-Eddington accretion rates, and are the only possibility for explaining the ULXs observed in elliptical galaxies (King 2002). SXT outbursts in long-period systems are an attractive candidate because they are both bright and long-lasting.

SXT outbursts in systems with such periods are complex because of the large reservoir of unheated mass at the edge of the disc, which can eventually contribute to the outburst (see Section 13.5). Full numerical calculations will be needed to describe this process. However, the trends with increasing P are clear: the outbursts become longer (several decades) and involve more mass, but the quiescent intervals increase more rapidly (several $\gtrsim 10^3$ yr) so that the outburst duty cycle d decreases (Ritter & King 2001). This results in inflow rates that become ever more super-Eddington at large P. Spectacular evidence of super-Eddington accretion is provided by GRS 1915+105, which has been in effectively continuous outburst since 1992. The observed X-ray luminosity $L_X \gtrsim 7 \times 10^{39}$ erg s^{-1} implies that at least $\sim 10^{-6}$ M$_\odot$ has been accreted over this time, requiring a large and massive accretion disc. In line with this, it appears that the binary is wide ($P \simeq 33$ d; Greiner *et al.* 2001). At the reported accretor mass $M_1 = (14 \pm 4)$ M$_\odot$ (Greiner *et al.* 2001) there is little doubt that the current mass inflow near the black hole is highly super-Eddington. Evolutionary expansion of the donor will drive a persistent mass transfer rate $-\dot{M}_2 \sim 10^{-9}(P/\mathrm{d})$ M$_\odot$ yr^{-1} $\sim 3 \times 10^{-8}$ M$_\odot$ yr^{-1} (King *et al.* 1996) which is already close to the Eddington rate $\dot{M}_{Edd} \sim 10^{-7}$ M$_\odot$ yr^{-1}. Given an outburst duty cycle $d \ll 1$, the mean inflow rate $\sim -\dot{M}_2/d$ is $\gg \dot{M}_{Edd}$. Note that we definitely do not look down the jet in GRS 1915+105, which is at about 70° to the line of sight (Mirabel & Rodríguez 1999), so it is quite possible that the apparent luminosity in such directions is much higher than the observed L_X.

The observed ULX population of a given galaxy is a varying mixture of these thermal-timescale and transient types, depending on the star formation history of that galaxy. Thermal-timescale SS433-like systems should predominate in galaxies with vigorous star formation, such as the Antennae, while ULXs in elliptical galaxies must be of the microquasar transient type, as there are no high-mass X-ray binaries. We therefore expect ULXs in ellipticals to be variable. However, the microquasar systems most likely to be identified as ULXs are clearly those with the brightest and longest outbursts, so baselines of decades may be needed to see significant numbers turning on or off. There is some evidence of such variability from the differences between ROSAT and Chandra observations of the same galaxies. The fact that none of the SXTs found in the Galaxy has turned out to be a ULX suggests that the beaming factor b must be $\lesssim 0.1$ for this mode of accretion. This agrees with our conclusion above that $b \ll 1$ for the ULXs in ellipticals.

Evidence that the two suggested classes of ULXs do resemble each other in similarly super-Eddington accretion states comes from Revnivtsen *et al.* (2002), who report RXTE observations of an episode of apparently super-Eddington accretion in the soft X-ray transient V4641 Sgr. Revnivtsen *et al.* remark on the similarity of the object's appearance to SS433 in this phase. One might be discouraged by the apparent suppression of X-rays in this state. However, we are presumably outside the beam of most intense X-ray emission in both cases: neither should actually appear as a ULX. More work is needed on whether the X-ray spectra from these objects are consistent with X-rays leaking sideways from the assumed accretion geometry. Direct evidence that X-ray emission in ULXs is anisotropic is perhaps understandably meagre, but may be suggested by the comparison of optical and X-ray data in NGC 5204 X-1 (Roberts *et al.* 2002), where low-excitation optical spectra are seen from regions close to the ULX.

Both SS433 and the microquasars are distinguished by the presence of jets, at least at some epochs. In SS433 the jets precess with a 164-day period, presumably because of disc warping (Pringle 1996). If looking closely down the jet is required in order to see high luminosities one might expect to see such periods in a class of ULXs. This effect could for example explain the ~106-day modulation seen in the bright source in M33 (M33 X-8, $L_X \sim 10^{39}$erg s^{-1}) by Dubus *et al.* (1999b). However, a beam as narrow as commonly inferred ($\lesssim 1^\circ$) for SS433 would give an unacceptably short duty cycle. If instead it is not necessary to look down the jet to see a high luminosity, this would rule out Doppler boosting as the cause of the latter, and ULXs would not be direct analogues of BL Lac systems.

13.8.3 *Black-hole blackbodies*

The tentative conclusion at the end of the last section suggests another. X-ray binaries and active galactic nuclei share the same basic model, and so far have shown a fairly good correspondence in their modes of behavior. If as suggested above ULXs do not correspond to BL Lac systems, this may mean that we are currently missing a class of each type: there should exist an apparently super-Eddington class of AGN, and a set of X-ray binaries with Doppler-boosted X-ray emission.

A tentative answer to one of these questions has recently emerged. Many black-hole sources emit a substantial fraction of their luminosities in blackbody-like spectral components. It is usual to assume that these are produced in regions at least comparable in size to the hole's Schwarzschild radius, so that a measure of the emitting area provides an estimate of the black-hole mass M. However, there is then no guarantee that the source luminosity (if isotropic) obeys the Eddington limit corresponding to M. King and Puchnarewicz (2002) show that the apparent blackbody luminosity $L_{\rm sph}$ and temperature T must obey the inequality

$$L_{\rm sph} < L_{\rm crit} = 2.3 \times 10^{44} (T/100\ {\rm eV})^{-4}\ {\rm erg\ s}^{-1} \qquad (13.51)$$

(where T_{100} is T in units of 100 eV) for this to hold. This limit is shown in Fig. 13.16. Sources violating it must either be super-Eddington, or radiate anisotropically, or radiate from a region much smaller than their Schwarzschild radii. Not suprisingly, some ULXs appear above the limit. (Note that they are not required to do this to qualify as ULXs: the defining characteristic is simply that their "Eddington masses" (right-hand scale of Fig. 13.16) are $\gg 10\,{\rm M}_\odot$.) The large group of AGN violating the limit are the so-called ultrasoft AGN, which may thus be the AGN analogues of the ULXs. The second question remains: a search for Doppler-boosted X-ray binaries among the ULXs may be rewarding.

13.8.4 *Supersoft ULXs*

Very recently a number of ULXs have been observed with very low spectral temperatures (~50–100 eV) and consequent photospheric sizes (~10^9 cm) much larger than the Schwarzschild radius of a stellar-mass object (e.g., Mukai *et al.* 2003). At first sight these might at last appear as strong evidence for the long-sought intermediate-mass black holes. However, Mukai *et al.* (2003) have pointed out that accretion at rates comparable

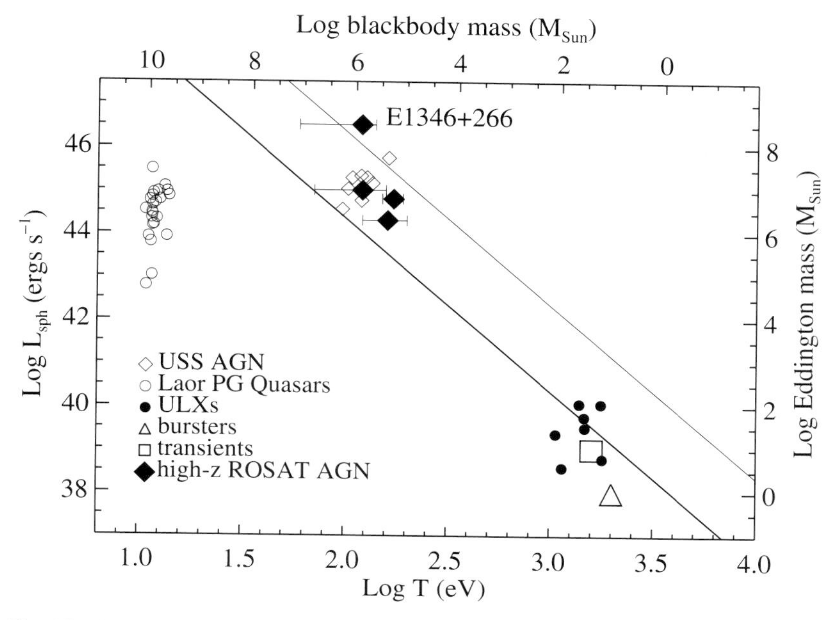

Fig. 13.16. Luminosity and blackbody temperature for bright X-ray blackbody sources: ultrasoft AGN (diamonds), the Laor *et al.* (1997) sample of PG quasars (open circles), ULXs (filled circles). X-ray transients (open square) and bursters (triangle) are shown schematically only, in the interests of clarity. The thick solid line is the limit $L_{\rm sph} = L_{\rm Edd}$. Sources below this line are compatible with the constraints of the Eddington limit, isotropic emission and an emission area no smaller than the Schwarzschild radius, but must have masses M respectively above and below the values given on the right-hand vertical and upper horizontal scales. Sources above the line must either (a) violate the Eddington limit, or (b) have a significantly anisotropic radiation pattern, or (c) emit from a region much smaller than their Schwarzschild radii. The thin line represents a luminosity $L_{\rm sph} = 100 L_{\rm Edd}$. (Figure from King and Puchnarewicz 2002)

to Eddington must lead to outflow, and shown that the opacity of the resulting wind does imply supersoft emission with a photospheric size of this order. The M101 source studied by Mukai *et al.* (2003) has a supersoft luminosity of order 10^{39} erg s^{-1} and so does not require anisotropic emission for a black-hole mass $\gtrsim 10\,{\rm M}_\odot$. However, their analysis is easily extended to the case that an Eddington-limited source blows out a wind confined to a double cone of total solid angle $4\pi b$ about the black-hole axis. Since this wind is the path of lowest optical depth through the accretion flow, the radiation will escape this way also, implying anisotropic emission once again. Mukai *et al.* (2003) assume a constant velocity for the outflowing material as this is likely to achieve escape velocity and coast thereafter. This leads to an equivalent hydrogen column from radius R to infinity of $N_{\rm H} = \dot{M}_{\rm out}/4\pi b v R$ and thus (assuming Compton scattering opacity) a photospheric radius

$$R_{\rm ph} = \frac{3\times 10^8}{b v_9}\dot{M}_{19}\ {\rm cm} \tag{13.52}$$

where v_9 is v in units of 10^9 cm s^{-1} and $\dot{M}_{19}$ is the outflow rate in units of 10^{19} g s^{-1}, the Eddington accretion rate for a $10\,{\rm M}_\odot$ black-hole. Clearly we can again interpret such supersoft ULXs in terms of stellar-mass black-holes.

It is worth noting that the presence of a photosphere of this kind seems inevitable in any source accreting significantly above the Eddington accretion rate $\dot{M}_{\rm Edd}$. A completely general calculation (Pounds *et al.* 2003) shows that

$$\frac{R_{\rm ph}}{R_{\rm s}} = \frac{1}{2\eta b}\frac{c}{v}\frac{\dot{M}_{\rm out}}{\dot{M}_{\rm Edd}} \simeq \frac{5}{b}\frac{c}{v}\frac{\dot{M}_{\rm out}}{\dot{M}_{\rm Edd}} \tag{13.53}$$

where we have taken the accretion efficiency $\eta \simeq 0.1$ at the last step. Since $b \leqslant 1$, $v/c < 1$ we see that $R_{\rm ph} > R_{\rm s}$ for any outflow rate $\dot{M}_{\rm out}$ of order $\dot{M}_{\rm Edd}$. In other words, any black-hole source accreting at above the Eddington rate is likely to have a scattering photosphere at several $R_{\rm s}$.

13.9 Conclusions

Our picture of accretion in compact binary systems has advanced considerably over recent years. In the past it was common to think of these sources as relatively steady systems that accreted most of the mass transferred to them, often from main-sequence companions, and radiated roughly isotropically. It now seems that none of these implicit assumptions is really justified. Transient behavior is extremely widespread, to the point that persistent sources are rather exceptional. Much of the transferred mass is not accreted at all, but blown away from the accretor: jets are only the most spectacular manifestation of a very widespread trait. Even short-period systems often have significantly evolved companions, and there is little evidence for a period gap for short-period LMXBs. Disc warping and other effects can apparently cause many sources to radiate with significant anisotropy.

Despite these complicating effects there are reasons for optimism. One can now give a complete characterization of the observed incidence of transient and persistent sources in terms of the disc instability model and formation constraints. X-ray populations in external galaxies, particularly the ultraluminous sources, are revealing important new insights into accretion processes and compact binary evolution.

Acknowledgements

I thank Ed Colbert, Juhan Frank, Hans-Jochen Grimm, Vicky Kalogera, Uli Kolb, James Murray, Philipp Podsiadlowski, Jim Pringle, Tim Roberts, Martin Ward, Graham Wynn, and particularly Klaus Schenker and Dan Rolfe for much help in the writing of this review. Theoretical astrophysics research at Leicester is supported by a PPARC rolling grant. I gratefully acknowledge a Royal Society Wolfson Research Merit Award.

References

Abramowicz, M. A., Calvani, M., Nobili, L., 1980, *ApJ*, **242**, 772
Balbus, S. A., Hawley, J. F., 1991, *ApJ*, **376**, 214
Baraffe, I., Kolb, U., 2000, *MNRAS*, **318**, 354
Begelman, M. C., 2002, *ApJ*, **568**, L97
Casares, J., Mouchet, M., Martínez-Pais, I. G., Harlaftis, E. T., 1996, *MNRAS*, **282**, 182
Colbert, E. J. M., Mushotzky, R. F., 1999, *ApJ*, **519**, 89
Dubus, G., Lasota, J. P., Hameury, J. M., Charles, P., 1999a, *MNRAS*, **303**, 39
Dubus, G., Long, K., Charles, P. A., 1999b, *ApJ*, **519**, L135
Ebisuzaki, T., Makino, J., Tsuru, T. G., *et al.*, 2001, *ApJ*, **562**, L19
Fabbiano, G., 1989, *ARA&A*, **27**, 87
Fabbiano, G., Zezas, A., Murray, S. S., 2001, *ApJ*, **554**, 1035
Frank, J., King, A. R., Raine, D. J., 2002, *Accretion Power in Astrophysics* 3rd Edn. (Cambridge: Cambridge University Press).

Fryer, C. L., Kalogera, V., 2001, *ApJ*, **554**, 548
Garcia, M. R., McClintock, J. E., Narayan, R., *el al.*, 2001, *ApJ*, **553**, L47
Giles, A. B., Swank, J. H., Jahoda, K., *et al.*, 1996, *ApJ*, **469**, 25
Goad, M. R., Roberts, T. P., Knigge, C., Lira, P., 2002, *MNRAS*, **335**, 67
Greiner, J., Cuby, J. G., McCaughrean, M. J., 2001, *Nature*, **414**, 522
Grimm, H. -J., Gilfanov, G., Sunyaev, R., 2003, *MNRAS*, **339**, 793
Haswell, C. A., King, A. R., Murray, J. R., Charles, P. A., 2001, *MNRAS*, **321**, 475
Haswell, C. A., Hynes, R. I., King, A. R., Schenker, K., 2002, *MNRAS*, **332**, 928
Iping, R. C., Petterson, J. A., 1990, *A&A*, **239**, 221
Jameson, R. F., King, A. R., Sherrington, M. R., 1980, *MNRAS*, **191**, 559
Jaroszynski, M., Abramowicz, M. A., Paczynski, B., 1980, *Acta Astron*, **30**, 1
Kaaret, P., Prestwich, A. H., Zezas, A., *et al.*, 2001, *MNRAS*, **321**, L29
Kalogera, V., Webbink, R. F., 1996, *ApJ*, **458**, 301
1998, *ApJ*, **493**, 351
Kalogera, V., Kolb, U., King, A. R., 1998, *ApJ*, **504**, 967
Kim, S-W.,Wheeler, J. C., Mineshige, S., 1999, *PASJ*, **51**, 393
King, A. R., 1988, *QJRAS*, **29**, 1
1998, *MNRAS*, **296**, L45
2000, *MNRAS*, **317**, 438
2002, *MNRAS*, **335**, 13
King, A. R., Begelman, M. C., 1999, *ApJ*, **519**, L169
King, A. R., Cannizzo, J. K., 1998, *ApJ*, **499**, 348
King, A. R., Kolb, U., 1997, *ApJ*, **481**, 918
King, A. R., Puchnarewicz, E., 2002, *MNRAS*, **336**, 445
King, A. R., Ritter, H., 1998, *MNRAS*, **293**, 42
1999, *MNRAS*, **309**, 253
King, A. R., Schenker, K., 2002, in *The Physics of Cataclysmic Variables and Related Objects*, ASP Conf. Proc., **261**. Eds. B. T. Gänsicke, K. Beuermann, and K. Reinsch (San Francisco: Astronomical Society of the Pacific) p. 233
King, A. R., Frank, J., Kolb, U., Ritter, H., 1995, *ApJ*, **444**, L37
King, A. R., Kolb, U., Burderi, L., 1996, *ApJ*, **464**, L127
King, A. R., Frank, J., Kolb, U., Ritter, H., 1997a, *ApJ*, **484**, 844
King, A. R., Kolb, U., Szuszkiewicz, E., 1997b, *ApJ*, **488**, 89
King, A. R., Taam, R. E. Begelman, M. C., 2000 *ApJ*, **530**, L25
King, A. R., Davies, M. B., Ward, M. J., Fabbiano, G., Elvis, M., 2001, *ApJ*, **552**, L109
King, A. R., Rolfe, D. J., Schenker, K., 2003, *MNRAS*, **341**, 35
Koerding, E., Falcke, H., Markoff, S., Fender, R., 2001 *Astronomische Gesellschaft Abstract Series*, **18**, P176
Kubota, A., Mizuno, T., Makishima, K., *et al.*, 2001, *ApJL*, **547L**, 119
Laor, A., Fiore, F., Elvis, M., Wilkes, B. J., McDowell, J. C., 1997, *ApJ*, **477**, 93
Lasota, J. P., 2001, *NewAR*, **45**, 449
Li, X.-D., Wang, Z.-R., 1998, *ApJ*, **500**, 935
Livio, M., 1996, *Evolutionary Processes in Binary Stars*, NATO ASI Series C, **477**. Eds. R. A. M. J. Wijers, M. B. Davies and C. A. Tout (Dordrecht: Kluwer Academic Publishers) p. 141
Lubow, S. H., 1991, *ApJ*, **381**, 268
Madau, P., Rees, M. J., 2001, *ApJ*, **551**, L27
Makishima, Z., Kubota, A., Mizuno, T., *et al.*, 2000, *ApJ*, **535**, 632
Markoff, S., Falcke, H., Fender, R., 2001, *A&A*, **372**, L25
Matsumoto, H., Tsuru, T. G., Koyama, K., *et al.*, 2001, *ApJ*, **547**, L25
Miller, C., Hamilton, D. P., 2002, *MNRAS*, **330**, 232
Mirabel, I. F., Rodríguez, L. F., 1999, *ARAA*, **37**, 409
Mukai, K., Pence, W. D., Snowden, S. L., Kuntz, K. D., 2003, *ApJ*, **582**, 184
Narayan, R., Garcia, M. R., McClintock, J. E., 1997, *ApJ*, **478**, L79
Nelemans, G., Portegies Zwart, S. F., Verbunt, F., Yungelson, L.R., 2001, *A&A*, **368**, 939
O'Donoghue, D., Charles, P. A., 1996, *MNRAS*, **282**, 191
Osaki, Y., 1989, *PASJ*, **41**, 1005
Paczyński, B., Wiita, P. J., 1980, *A&A*, **88**, 23
Paradijs, J. van, 1996, *ApJ*, **464**, L139

Paradijs, J. van, McClintock, J. E., 1994, *A&A*, **290**, 133
Petterson, J. A., 1977, *ApJ*, **216**, 827
Podsiadlowski, Ph., Rappaport, S., 2000, *ApH*, **529**, 946
Pounds, K. A., Reeves, J. N., King, A. R., *et al.*, 2003, *MNRAS*, **345**, 705 (erratum, *MNRAS*, **356**, 1599)
Pringle, J. E., 1981, *ARAA*, **19**, 137
1996, *MNRAS*, **281**, 357
1997, *MNRAS*, **488**, 47
Pylyser, E. H. P., Savonije, G. J., 1988, *A&A*, **191**, 57
Radhakrishnan, V., Srinivasan, G., 1982, *Curr. Sci.*, **51**, 1096
Rappaport, S. A., Podsiadlowski, Ph., Pfahl, E., 2004, *MNRAS*, **356**, 401
Revnivtsev, M., Gilfanov, M., Churazov, E., Sunyaev, R., 2002, *A&A*, **391**, 1013
Ritter, H., 1999, *MNRAS*, **309**, 360
Ritter, H., King, A. R., 2001, in *Evolution of Binary and Multiple Star Systems*, ASP, Conf. Ser., **229**. Eds. Ph. Podsiadlowski, S. Rappaport, A. R. King, F. D'Antona and L. Burderi (San Francisco: Astronomical Society of the Pacific) p. 423
Ritter, H., Kolb, U., 2003, *A&A*, **404**, 301
Roberts, T., Warwick, R., 2000, *MNRAS*, **315**, 98
Roberts, T. P., Goad, M. R., Ward, M. J., *et al.*, 2001, *MNRAS*, **325**, L7
Roberts, T. P., Warwick, R. S., Ward, M. J., Murray, S. S., 2002, *MNRAS*, **337**, 677
Rutledge, R. E., Bildsten, L., Brown, E. F., *et al.*, 2002, *ApJ*, **580**, 413
Savonije, G. J., 1987, *Nature*, **325**, 416
Schenker, K., King, A. R., 2002, in *The Physics of Cataclysmic Variables and Related Objects*, ASP Conf. Ser., **261**. Eds. B. T. Gänsicke, K. Beuermann, and K. Reinsch (San Francisco: Astronomical Society of the Pacific) p. 242
Schenker, K., King, A. R., Kolb, U., Zhang, Z., Wynn, G. A., 2002, *MNRAS*, **337**, 1105
Shahbaz, T., Charles, P. A., King, A. R., 1998, *MNRAS*, **301**, 382
Shakura, N. I., Sunyaev, R. A., 1973, *A&A*, **24**, 337
Shaviv, N. J., 1998, *ApJ*, **494**, L193
2000, *ApJ*, **532**, L137
Stone, J. M., Pringle, J. E., 2001, *MNRAS*, **322**, 461
Stone, J. M., Pringle, J. E., Begelman, M. C., 1999, *MNRAS*, **310**, 1002
Taam, R. E., King, A. R., Ritter, H., 2000, *ApJ*, **350**, 928
Tanaka, Y., Lewin, W. H. G., 1995, in *X-ray Binaries*, Eds. W. H. G. Lewin, J. van Paradijs and E. P. J. van den Heuvel (Cambridge: Cambridge University Press)
Truss, M. R., Murray, J. R., Wynn, G. A., 2001, *MNRAS*, **324**, L1
Truss, M. R., Wynn, G. A., Murray, J. R., King, A. R., 2002, *MNRAS*, **337**, 1329
Tuchman, Y., Mineshige, S., Wheeler, J. C., 1990, *ApJ*, **359**, 164
Vogt, N., 1983, *A&A*, **118**, 95
Warner, B., 1995, *Cataclysmic Variable Stars* (Cambridge: Cambridge University Press) Chapter 3
Watson, M. G., Stewart, G. C., Brinkmann, W., King, A. R., 1986, *MNRAS*, **222**, 261
Webbink, R. F., Rappaport, S. A., Savonije, G. J., 1983, *ApJ*, **270**, 678
Welsh, W. F., Horne, K., Gomer, R., 1993, *ApJ*, **410**, L39
1995, *MNRAS*, **275**, 649
White, N. E., Holt, S. S., 1982, *ApJ*, **257**, 318
Whitehurst, R., 1988, *MNRAS*, **232**, 35
Whitehurst, R., King, A. R., 1991, *MNRAS*, **249**, 25
Wijnands, R., Miller, J. M., Markwardt, C., Lewin, W. H. G., van der Klis, M., 2001, *ApJ*, **560**, 159
Willems, B., Kolb, U., 2002, *MNRAS*, **337**, 1004
Zezas, A., Fabbiano, G., Rots, A. H., Murray, S. S., 2002, *ApJ*, **577**, 710
Zezas, A., Georgantopoulos, I., Ward, M. J., 1999, *MNRAS*, **308**, 302

14

Soft gamma repeaters and anomalous X-ray pulsars: magnetar candidates

P. M. Woods
Universities Space Research Association

C. Thompson
Canadian Institute for Theoretical Astrophysics

14.1 Introduction

Baade and Zwicky (1934) were the first to envision the formation of neutron stars as the end product of a supernova explosion. Their forward thinking was not vindicated for another three decades, with the discovery of the first radio pulsars by Bell and Hewish (Hewish *et al.* 1968). What Baade and Zwiscky could not have anticipated, however, was the menagerie of astrophysical objects that are now associated with neutron stars. Today, we observe them as magnetically braking pulsars, accreting pulsars in binary systems, isolated cooling blackbodies, sources of astrophysical jets, and emitters of high-luminosity bursts of X-rays. Here, we focus on two of the most extraordinary evolutionary paths of a neutron star, namely soft gamma repeaters (SGRs) and anomalous X-ray pulsars (AXPs).

Soft gamma repeaters were discovered as high-energy transient burst sources; some were later found also to be persistent X-ray pulsars, with periods of several seconds, that are spinning down rapidly. Anomalous X-ray pulsars are identified through their persistent pulsations and rapid spin down; some have also been found to emit SGR-like bursts. In spite of the differing methods of discovery, this convergence in the observed properties of the SGRs and AXPs has made it clear that they are, fundamentally, the same type of object. What distinguishes them from other neutron stars is the likely source of energy for their radiative emissions, magnetism. The cumulative behavior of SGRs and AXPs is now best described by the magnetar model, in which the decay of an ultra-strong magnetic field ($B > 10^{15}$ G) powers the high-luminosity bursts and also a substantial fraction of the persistent X-ray emission.

For many years, the apparent absence of radio pulsars with magnetic fields much exceeding 10^{13} G, and the apparent lack of a good motivation for the existence of much stronger fields in neutron stars, inhibited serious consideration of their astrophysical consequences. It was noted early on that fields as strong as $10^{14}-10^{15}$ G could be present in neutron stars as the result of flux conservation from the progenitor star (Woltjer 1964). Ultra-strong magnetic fields were introduced by hand in simulations of rotating supernova collapse, as a catalyst for energetic outflows (LeBlanc & Wilson 1970; Symbalisty 1984). A related possibility is that ordinary radio pulsars could contain intense toroidal magnetic fields as a residue of strong differential rotation in the nascent neutron star (e.g., Ardelyan *et al.* 1987). Later it was realized that appropriate conditions for true dynamo action could exist in proto-neutron stars (Thompson & Duncan 1993), leading to the formation of a class of ultra-magnetic neutron stars with dissipative properties distinct from those of radio pulsars (Duncan & Thompson 1992). In recent years, pulsar searches have largely closed the observational gap between the dipole fields of radio pulsars and magnetar candidates (Manchester 2004).

Compact Stellar X-Ray Sources, eds. Walter Lewin and Michiel van der Klis.
Published by Cambridge University Press.

We first review the history of this relatively new subfield of high-energy astrophysics. Then we summarize in more detail the burst emission, the persistent X-ray emission of the SGRs and AXPs, their torque behavior, the counterparts observed in other wavebands, and their associations with supernova remnants. Finally, we discuss the magnetar model.

14.1.1 Soft gamma repeaters: a brief history

On 1979 January 7, a burst of soft gamma rays lasting a quarter of a second was detected from SGR 1806−20 by instruments aboard the *Venera* spacecraft – the first observation of a soft gamma repeater. This burst, along with a handful like it recorded over the next few years, were originally classified as a subtype of classical gamma-ray burst (GRB), one with a short duration and a soft spectrum (Mazets & Golenetskii 1981). The locations of three repeaters were obtained from this early data set. It was not until after an intense reactivation of SGR 1806−20 in 1983, however, that the independent nature of these sources was fully appreciated (Hurley 1986; Laros *et al.* 1987). Their propensity to emit multiple bursts (no GRB has yet been shown to repeat); the deficit of high-energy gamma-ray emission; and their similarity to each other merited designation as a new class of astrophysical transient.

The first detection of a SGR burst was soon followed, on 1979 March 5, by an extremely energetic SGR flare (Mazets *et al.* 1979). This extraordinary event began with a bright spike peaking at $\sim 10^{45}$ erg s^{-1} (Golenetskii *et al.* 1984), followed by a 3 min train of coherent 8 s pulsations whose flux decayed in a quasi-exponential manner (Feroci *et al.* 2001). The burst was well localized at the edge of the supernova remnant (SNR) N49 in the Large Magellanic Cloud, and its source is now identified as SGR 0526−66 (Cline *et al.* 1982). The high luminosity, strong pulsations, and apparent association with a SNR strongly suggested that the source was a young, magnetized neutron star with a spin period of 8 s.

Following the announcement of the SGRs, a variety of models were proposed. These included accretion onto magnetized neutron stars (e.g. Livio & Taam 1987; Katz, Toole & Unruh 1994), cometary accretion onto quark stars (Alcock, Farhi & Olinto 1986), as well as thermonuclear energy release on a magnetized neutron star (Woosley & Wallace 1982). Damping of the vibration of a neutron star had been suggested as a mechanism for the March 5 flare (Ramaty *et al.* 1980), but the coupling of the crust to the magnetosphere is much too weak to explain the observed luminosity if $B \sim 10^{12}$ G (Blaes *et al.* 1989). Indeed, the main shortcoming of all these models (e.g., Norris *et al.* 1991) was the lack of an adequate explanation for both the giant flare and the more common recurrent bursts, which last only $\sim$0.1 s and have much lower peak luminosities ($<10^{41}$ erg s^{-1}).

Efforts to understand the nature of the SGRs were constrained by the lack of information about persistent counterparts. This changed with the discovery of persistent X-ray emission from all three known SGRs (Murakami *et al.* 1994; Rothschild, Kulkarni & Lingenfelter 1994; Vasisht *et al.* 1994). Around the same time, the magnetar model was put forth to explain the high-luminosity bursts of the SGRs (Duncan & Thompson 1992; Paczyński 1992; Thompson & Duncan 1995) and the persistent X-ray emission of both the SGRs and the AXPs (Thompson & Duncan 1996). Thompson & Duncan (1996) predicted slow pulsations and rapid spin down from the quiescent X-ray counterparts of the SGRs. A major breakthrough in determining the nature of SGRs was made shortly thereafter (Kouveliotou *et al.* 1998a), with the discovery of 7.5 s pulsations and rapid spin down in the X-ray counterpart to SGR 1806−20. Kouveliotou *et al.* (1998a) interpreted this measurement in terms of the magnetic braking of an isolated neutron star with a $\simeq 10^{15}$ G dipole magnetic field.

Our understanding of SGRs has continued to blossom in recent years in good part due to extensive monitoring campaigns and improved instrumentation. These observations have revealed correlated changes in SGR persistent emission properties during periods of burst activity, dramatic variations in spin down torque, and a much larger collection of bursts of all types. For example, a near carbon-copy of the first giant flare was recorded on 1998 August 27 from SGR 1900+14 (Hurley *et al.* 1999a). In spite of detector advancements, there has been only one additional confirmed SGR discovered after the first three in 1979: SGR 1627−41 emitted more than 100 bursts in 1998 (Kouveliotou *et al.* 1998b; Woods *et al.* 1999a). Two bursts were recorded in 1997 from a fifth candidate source, SGR 1801−23 (Cline *et al.* 2000). Another candidate, SGR 1808−20, was detected once and localized (Lamb *et al.* 2003a) to a position very near, but formally inconsistent with, the direction of SGR 1806−20. It should be cautioned that this burst was recorded during a burst active phase of SGR 1806−20. Note that we have not included in this tally sources first identified as AXPs, and later found to burst like SGRs.

14.1.2 Anomalous X-ray pulsars: a brief history

The first detection of an anomalous X-ray pulsar was made by Fahlman and Gregory (1981), who discovered pulsations from the X-ray source 1E 2259+586 at the center of the SNR CTB 109. This object was first interpreted as a peculiar X-ray binary: its energy spectrum was much softer than is typical of accreting pulsars, and no optical counterpart was detected. Later the source was found to be spinning down in a secular manner (Koyama, Hoshi & Nagase 1987). Its X-ray luminosity was much too high to be powered by the loss of rotational energy from the putative neutron star.

Several other similar sources were discovered in the ensuing 15 years. The objects 1E 2259+586, 1E 1048.1−5937, and 4U 0142+61 were grouped together by Hellier (1994) and Mereghetti and Stella (1995) as possible low-mass X-ray binaries, along with the known short-period binary 4U 1626−67. (The source RX J1838.4−0301 was also included initially, but was later shown not to be an X-ray pulsar.) The salient properties of this class were a narrow range of spin periods (5−9 s), fairly constant X-ray luminosities ($\sim 10^{35}-10^{36}$ erg s^{-1}), no evidence for orbital Doppler shifts and – with the exception of 4U 1626−67 – relatively soft X-ray spectra and steady spin down. However, 4U 1626−67 is also distinguished from the other sources by the detection of optical pulsations of the brightness expected from a compact, accreting binary. In light of these differences, its membership as an AXP has been revoked. Three new AXPs have been discovered since 1996 (1RXS J170849.0−400910, 1E 1841−045, and XTE J1810−197), along with two candidate sources (AX J1845−0258 and CXOU J0110043.1−721134).

The AXPs nonetheless appear to be too young to be low-mass binaries: some are associated with SNR, and they have a small scale height above the Galactic plane (van Paradijs, Taam & van den Heuvel 1995). As noted by Thompson and Duncan (1993, 1996), their "anomalous" property is the mechanism powering their X-ray emission. These authors identified 1E 2259+586, and later the AXP population as a whole, with isolated magnetars powered by the decay of a $\sim 10^{15}$ G magnetic field. The principal competing model postulated that the AXPs are neutron stars surrounded by fossil disks that were acquired during supernova collapse or during a common-envelope interaction (Corbet *et al.* 1995; van Paradijs *et al.* 1995; Chatterjee, Hernquist & Narayan 2000; Chatterjee & Hernquist 2000). Finally, it was also noted that the loss of rotational energy from an isolated, magnetic, high-mass white dwarf is much larger than from a neutron star with the same spin parameters, and could supply the observed X-ray output (Paczyński 1990). However, the apparent youth of the object, and its residence in a SNR, remained puzzling in that interpretation.

The detection of optical and near infrared counterparts to the AXPs, beginning with 4U 0142+61 (Hulleman, van Kerkwijk & Kulkarni 2000), has provided a useful discriminant between the fossil disk and magnetar models. Dim counterparts have now been detected for four AXPs, with an optical/IR luminosity typically one thousandth of that emitted in 2–10 keV X-rays. This constrains any remnant accreting disk to be very compact (e.g. Perna, Hernquist & Narayan 2000). The optical emission of 4U 0142+61 has been found to pulse at the same period as in the X-ray band, with a pulsed fraction that is equal or higher (Kern & Martin 2002). The large pulsed fraction appears problematic in any accretion model, where the optical emission arises from re-processing of the X-rays by a disk. There are no reliable a-priori predictions of optical/infrared emission from magnetars.

The detection of X-ray bursts similar to SGR bursts from at least one, and possibly two, AXPs has confirmed a key prediction of the magnetar model. Two weak bursts were observed from the direction of 1E 1048.1−5937 (Gavriil, Kaspi & Woods 2002); and more than 80 SGR-like bursts were detected from 1E 2259+586 during a single, brief (∼11 ks) observation of the source (Kaspi *et al.* 2003). Overall, at least 10% of the X-ray output of 1E 2259+586 appears to be powered by transient releases of energy, and a much larger fraction in some other AXPs. Although these observations have not yet provided unambiguous proof that the AXPs have ultra-strong magnetic fields, they have confirmed the conjecture that the AXPs and SGRs belong to the same class of neutron stars. In this review, we refer to these sources collectively as magnetar candidates.

14.2 Burst observations

The defining behavior of SGRs is their repetitive emission of bright bursts of low-energy (soft) gamma rays. The most common SGR bursts have short durations (∼0.1 s), thermal spectra, and peak luminosities reaching up to 10^{41} erg s^{-1} – well above the standard Eddington limit of $\sim 2 \times 10^{38}$ erg s^{-1} for a 1.4 $M_{\odot}$ neutron star. In this section, we describe these short bursts. We include the very similar bursts detected from two AXPs, which share several properties with the SGR bursts, including durations, spectra, and energy distribution (Gavriil, Kaspi & Woods 2004). We then review the more extraordinary bursts emitted by the SGRs, including the three giant flares and the intermediate bursts. Unless otherwise stated, the quoted burst luminosities and energies cover photon energies above 20 keV and assume isotropic emission.

14.2.1 Short duration SGR bursts: temporal properties and energy distribution

The properties of the most common SGR bursts do not appear to vary greatly between different periods of activity, or indeed between different sources (e.g., Aptekar *et al.* 2001; Göğüş *et al.* 2001). A burst typically has a faster rise than decay, and lasts ∼100 ms. Four examples from SGR 1806−20, SGR 1900+14, and 1E 2259+586 are shown in Fig. 14.1. A number of bursts are multi-peaked, like the two shown from SGR 1806−20 and SGR 1900+14. Complicated bursts like these can usually be decomposed into burst

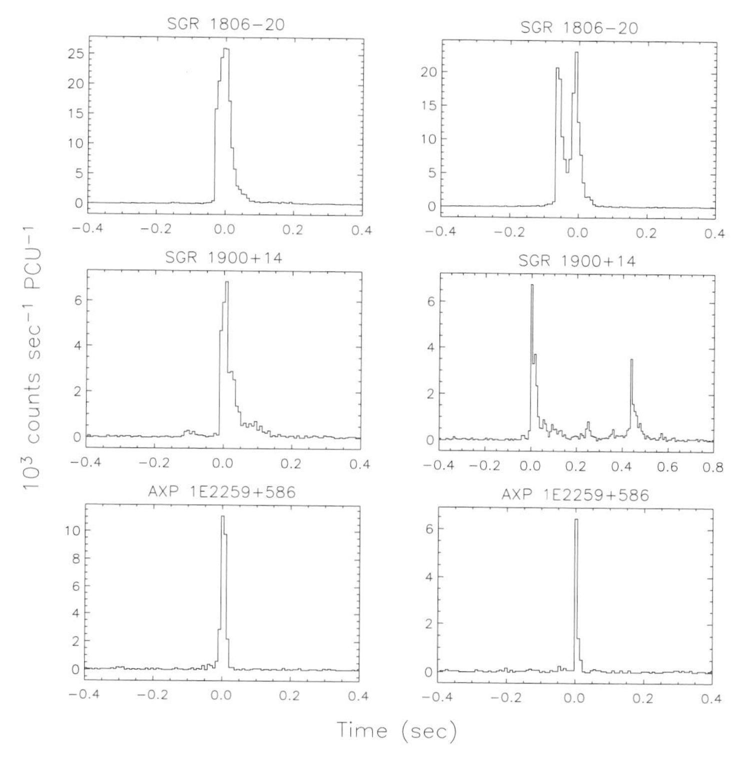

Fig. 14.1. A selection of common burst morphologies recorded from SGR 1806−20, SGR 1900+14 and 1E 2259+586, as observed with the RXTE PCA. All lightcurves display counts in the energy range 2−20 keV, with a time resolution of 7.8 ms. See text for further details.

"units." Göğüş *et al.* (2001) showed that the intervals between sub-peaks have a broad distribution, suggesting that these multi-peaked bursts are a superposition of two (or more) single-peaked burst units close in time.

The morphological uniformity of (the majority of) SGR bursts was noted early on (e.g., Atteia *et al.* 1987; Kouveliotou *et al.* 1987). The burst durations have a narrow distribution: they show a mild positive correlation with burst fluence (e.g., Göğüş *et al.* 2001), but do not vary significantly with photon energy. A sample of 164 bursts recorded from the four known SGRs by the Konus series of gamma-ray detectors is tabulated by Aptekar *et al.* (2001). Durations could be measured for 106 ordinary bursts (Fig. 14.2), with a mean of 224 ms. More recently, the higher flux sensitivity of the Rossi X-ray Timing Explorer PCA has provided larger burst samples for individual sources, at lower photon energies (2−20 keV vs. > 25 keV). In particular, T_{90} burst durations[1] were measured for 190, 455, and 80 bursts from the magnetar candidates SGR 1806−20, SGR 1900+14 (Göğüş *et al.* 2001), and 1E 2259+586 (Gavriil *et al.* 2004), respectively. The mean durations of these samples were 162, 94, and 99 ms. That these values are somewhat lower than in the Konus sample may be due, in part, to the higher mean fluence of the Konus bursts.

[1] The time to accumulate 90 percent of the burst fluence; see Koshut *et al.* (1996).

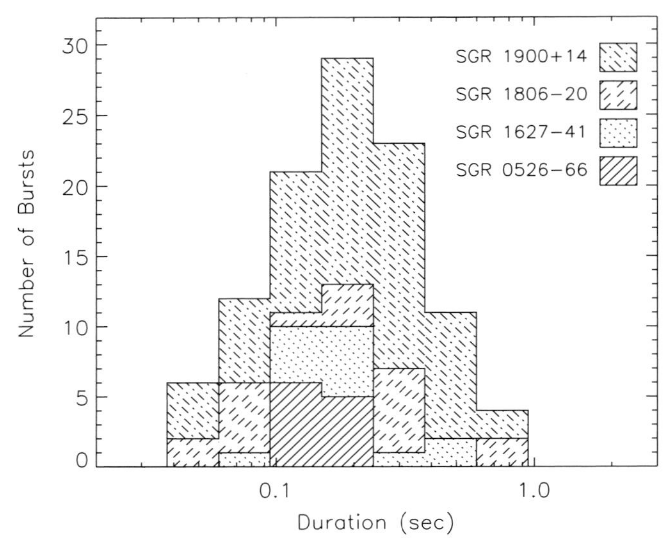

Fig. 14.2. Duration distribution for 106 bursts from the four known SGRs as observed by the Konus detectors (15–100 keV) between 1978 and 2000 (Aptekar *et al.* 2001).

The burst activity in SGRs tends to be concentrated in time. These episodes of enhanced burst activity are referred to as outbursts. They occur at irregular intervals with variable duration and intensity (Fig. 14.3). Within each outburst, the recurrence patterns of individual bursts are just as irregular as those of the outbursts themselves, and differ dramatically from what is observed in X-ray bursts (of either Type I or II) in accreting neutron stars (Chapter 3). There is no correlation between the energy of a given burst and the time to the next burst in either the SGRs (Laros *et al.* 1987; Göğüş *et al.* 1999), or in 1E 2259+586 (Gavriil *et al.* 2004). (Such a correlation is present in the Type II bursts of the Rapid Burster; see, for example, Lewin, van Paradijs & Taam 1993.)

The distribution of waiting times between bursts follows a log-normal function with a mean that depends on the sensitivity of the detector and the strength of the outburst. For example, the waiting times spanned some 7 orders of magnitude during the 1983 activation of SGR 1806−20 with a (logarithmic) mean of $\sim 10^4$ s (Laros *et al.* 1987; Hurley *et al.* 1994). Cheng *et al.* (1996) pointed out that the waiting times between earthquakes show a similar distribution. The waiting times between bursts from SGR 1900+14, 1E 2259+586, and a more recent outburst of SGR 1806−20 are all consistent with a log-normal distribution, although given the lower flux threshold the mean waiting time is only $\sim 10^2$ s in these three samples.

The energies radiated during the common ($\sim$0.1 s) SGR bursts follow a power-law distribution, $dN/dE \propto E^{-5/3}$. Cheng *et al.* (1996) first uncovered this distribution in SGR 1806−20, and pointed out the similarity with the Gutenburg–Richter law for earthquakes. Similar distributions are measured in a variety of other physical systems, including solar flares and avalanches. Subsequently, it has been shown that the other three SGRs and the AXP 1E 2259+586 all possess very similar burst energy distributions. The power-law index is about $-5/3$ in SGR 1900+14, SGR 1627−41, and 1E 2259+586, but is not well constrained

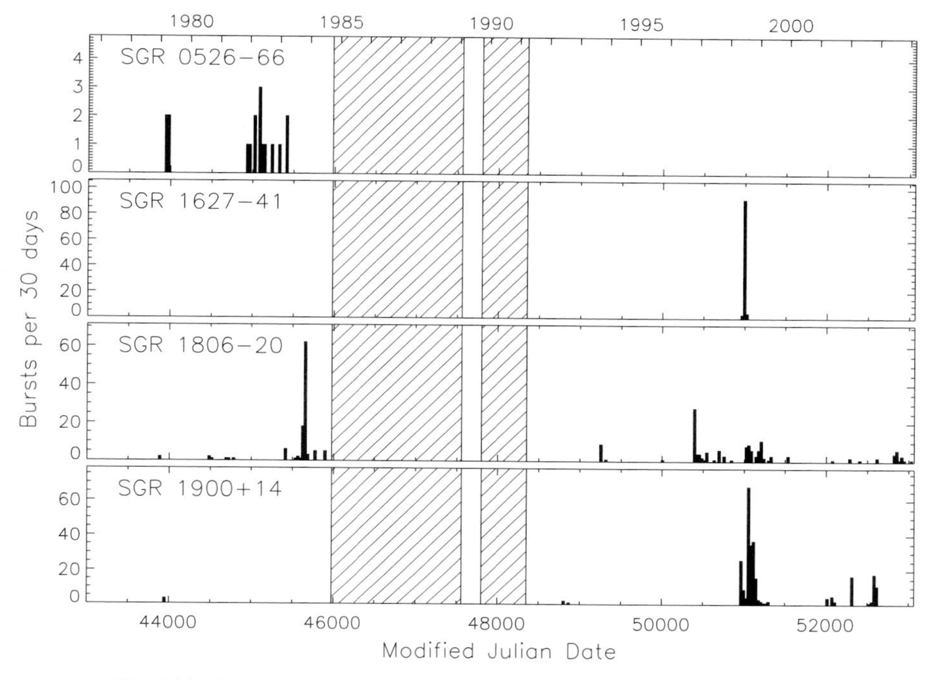

Fig. 14.3. Burst activity history of the four confirmed SGRs. The bursts identified here were detected with a suite of large field-of-view detectors having different sensitivities. The shaded regions indicate epochs where there were no active detectors sensitive to SGR bursts. (IPN data courtesy of Kevin Hurley)

in SGR 0526−66. A possible break from a $-5/3$ index to a somewhat flatter value (-1.4) at low burst energies was measured in a larger sample of bursts from SGR 1806−20 (Göğüş *et al.* 1999).

Series of many short bursts, with extremely small waiting times (multi-episodic bursts) have been observed on rare occasions (Hurley *et al.* 1999b). They involve several tens of bright SGR bursts which are packed into an interval of a few minutes. Intense burst episodes like these are more commonly seen at lower peak flux; but three instances involving high-luminosity SGR bursts have been recorded from SGR 1900+14. The BATSE lightcurve of the 1998 September 1 multi-episodic burst is shown in Fig. 14.4. Note the continuous envelope of emission underlying the most intense portion of the burst episode.

14.2.2 Spectral properties

Bursts from SGRs were discovered using all-sky detectors with little sensitivity below ~30 keV. Above this photon energy, SGR burst spectra are well modeled by optically thin thermal bremsstrahlung (OTTB). The temperatures so obtained fall within the narrow range $kT = 20$–40 keV, indicative of the spectral uniformity of SGR bursts. The spectra of SGR bursts vary weakly with intensity – not only from burst to burst within a given source, but also between sources. This effect was first demonstrated by Fenimore, Laros & Ulmer (1994) for SGR 1806−20, and later by Aptekar *et al.* (2001) in the Konus sample of bursts from four SGR sources.

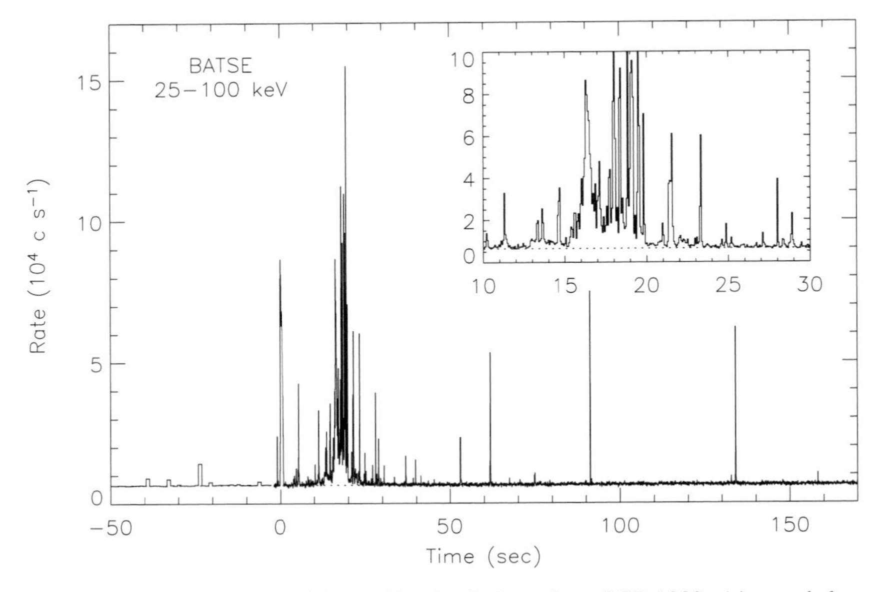

Fig. 14.4. The time history of the multi-episodic burst from SGR 1900+14 recorded on 1998 September 1 with BATSE (25−100 keV). The inset shows a close-up of the most intense part of the lightcurve. The background level is indicated by the dotted line. Note the envelope of emission lasting ∼5−7 s during the most intense phase.

A typical SGR burst spectrum (the solid line in Fig. 14.5) is compared with sample spectra from other extra-solar high-energy burst phenomena, specifically, short gamma-ray bursts, long gamma-ray bursts, X-ray flashes (XRFs), thermonuclear burning or Type I X-ray bursts (XRBs), and spasmodic accretion or Type II X-ray bursts (example shown is a GRO J1744−28 burst spectrum (e.g., Giles *et al.* 1996) – Rapid Burster energy spectra are significantly softer (Lewin *et al.* 1993)). Although the soft end of the spectral distribution of GRBs and XRFs overlaps the SGRs, the durations of the SGR bursts are usually shorter by two orders of magnitude than those of the detected XRFs (whose durations are $\gtrsim$10 s) and otherwise show strong morphological differences with GRBs. Distinguishing the two types of bursts is therefore straightforward in practice. For reviews on XRFs and GRBs, see Chapters 6 and 15. One caveat here is that the initial ∼0.5 s spikes of the giant flares show greater spectral similarities with GRBs, and so extra-galactic SGR flares may be hidden in the BATSE catalog of short-duration GRBs (Duncan 2001).

A shortcoming of the OTTB model is that it over-predicts the flux of photons with energies below ∼15 keV (Fenimore *et al.* 1994). It is doubtful that this spectral rollover is due to a thick column of absorbing material, since the requisite $N_{\rm H}$ is an order of magnitude greater than what is deduced from the persistent X-ray emission. Recently, a 7−150 keV HETE-2 spectrum of a high-fluence burst from SGR 1900+14 was successfully fit by the sum of 4.1 keV and 10.4 keV blackbodies (Olive *et al.* 2003). A similar result was obtained with 1.5−100 keV BeppoSAX spectra of 10 bursts also from SGR 1900+14 (Feroci *et al.* 2004). The temperatures of these lower peak flux bursts are consistent with the HETE-2 burst spectrum – so that the flux ratio of the two blackbody components is approximately constant.

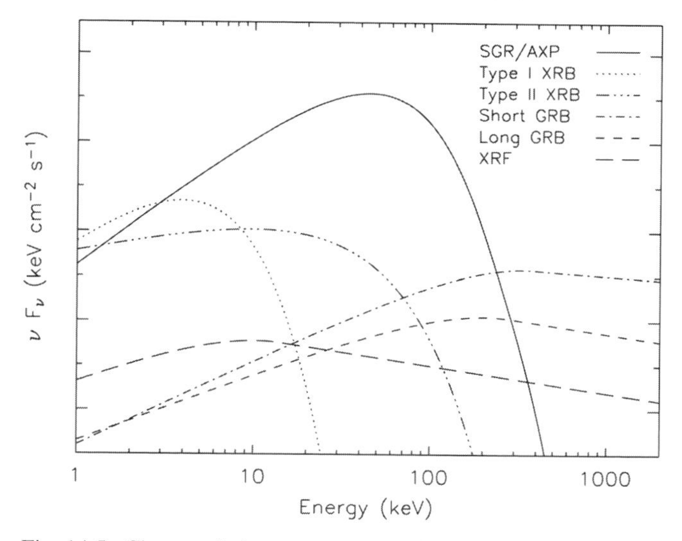

Fig. 14.5. Characteristic energy spectra of high-energy burst phenomena. The SGR/AXP (magnetar candidate) burst spectrum shown is a composite of the two blackbody model that fits burst spectra well below ~50 keV convolved with the OTTB model that better represents the burst spectrum at higher energies. Note that there exists a continuum of peak energies between the softer XRFs and the harder/brighter long GRBs.

Furthermore, the absorbing column measured during the bursts is consistent with the value obtained in quiescence.

The improved sensitivity of RXTE allowed Göğüş *et al.* (2001) to show that the less energetic bursts from SGR 1806−20 and SGR 1900+14 are also slightly harder spectrally. The bursts detected from the AXP 1E 2259+586 have similar spectra to those of the SGRs, although in the AXP it is the brighter bursts which tend to be harder (Gavriil *et al.* 2004).

14.2.3 Giant flares

Giant flares are the most extreme examples of SGR bursts. Their output of high-energy photons is exceeded only by blazars and cosmological gamma-ray bursts, and their luminosities can reach one billion times the Eddington luminosity of a neutron star. The flares begin with a ~1-second spike of spectrally hard emission which decays rapidly into a softer, pulsating tail that persists for hundreds of seconds. These coherent pulsations are at the spin period of the underlying neutron star. The giant flares are rare: only three have been detected from the four known SGRs over 20 years of observation, so the corresponding rate is approximately once per ~15−55 yr (per source). In contrast with GRBs and blazars, there is no evidence for strong beaming in the SGR flares.

The first giant flare was recorded on 1979 March 5 from SGR 0526−66 (Mazets *et al.* 1979) and, indeed, was only the second SGR burst observed. The source is well localized in the LMC (Section 14.6), and so the isotropic energy of the flare was 5×10^{44} erg – some ten thousand times larger than a typical thermonuclear flash. The initial peak of this flare lasted ~0.2 s and had significant structure on timescales shorter than ~2 ms. It was spectrally harder ($kT \sim 250-500$ keV) than the common SGR bursts, and reached a peak luminosity of 4×10^{44} erg s^{-1} (Mazets *et al.* 1979; Fenimore *et al.* 1981). Thereafter, the flux decayed

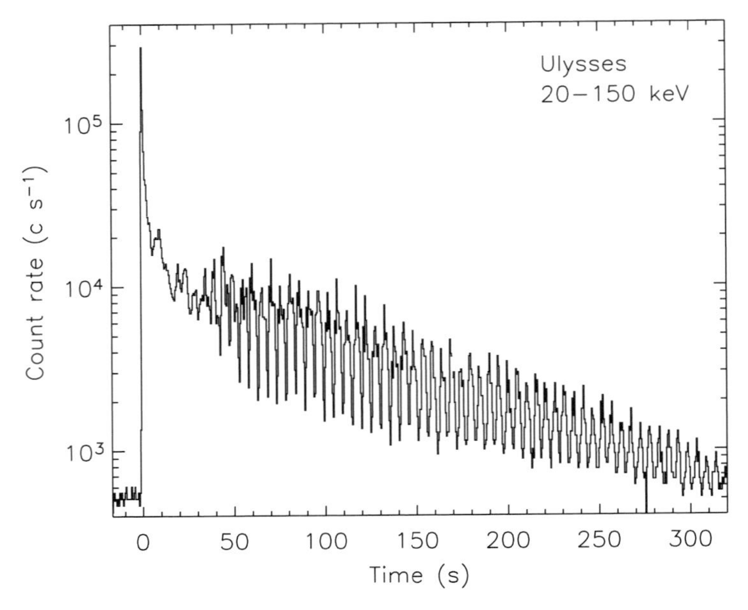

Fig. 14.6. The giant flare from SGR 1900+14 as observed with the gamma-ray detector aboard Ulysses (20−150 keV). Note the strong 5.16 s pulsations clearly visible during the decay. Figure after Hurley *et al.* (1999a).

in a quasi-exponential manner over the next ~2–3 minutes. A reanalysis of the ISEE-3 data using a model of a magnetically confined, cooling fireball (see Section 14.7.2), shows that the data are also consistent with a well-defined termination of the X-ray flux at ~160 s (Feroci *et al.* 2001). The pulsations during this phase of the burst have a period of 8.00 ± 0.05s (Terrell *et al.* 1980). The pulse profile shows two clear peaks per cycle and a change in morphology during the first few cycles. The spectrum of the decaying tail had an OTTB temperature of ~30−38 keV, consistent with the spectra of the recurrent burst emissions from this SGR.

The second giant flare was not recorded until almost 20 years later, on 1998 August 27 from SGR 1900+14 (Hurley *et al.* 1999a; Feroci *et al.* 1999; Mazets *et al.* 1999a; Feroci *et al.* 2001). This event (Fig. 14.6) was, in many respects, a carbon copy of the March 5 flare. It began with a bright spike lasting ~0.35 s, and the X-ray spectrum contained a very hard power-law component $\mathrm{d}N/\mathrm{d}E \propto E^{-1.5}$ in the initial stages. Only a lower bound of 3×10^{44} erg s^{-1} was obtained for its peak luminosity, because the flare saturated nearly all detectors that observed it. In fact, the X-ray flux incident on the night side of the Earth was high enough to force the ionosphere to its day-time level (Inan *et al.* 1999). Its total energy exceeded 10^{44} erg.

The decline in the flux from the August 27 flare was followed to a well-defined termination some 400 s after the initial spike (Feroci *et al.* 2001). The spectrum, after the first 50 s, equilibrated to a (OTTB) temperature of ~30 keV, even while the luminosity continued to decrease by more than an order of magnitude. During this same phase, the lightcurve maintained large-amplitude pulsations with a 5.16 s period, precisely equal to the periodicity that had been previously detected in the persistent X-ray emission of

SGR 1900+14 (Hurley *et al.* 1999c). The pulse maintained a complex four-peaked pattern that gradually simplified into a smooth, single pulse during the final stages of the flare (Mazets *et al.* 1999a).

The third giant flare was detected on 2004 December 27 from SGR 1806–20 (Hurley *et al.* 2005; Mereghetti *et al.* 2005a; Palmer *et al.* 2005; Terasawa *et al.* 2005) following several months of intense burst activity from this source. This flare was the brightest extra-solar gamma-ray transient ever recorded, reaching a peak flux of 5–20 erg cm^{-2} s^{-1} which corresponded to an isotropic peak luminosity of $2-5 \times 10^{47}$ erg s^{-1}, and a total energy of $\sim 4 \times 10^{46}$ erg. As with the other giant flares, the primary spike was spectrally hard, peaking at ~0.5 MeV. The morphology of this flare was very similar to the other two. The main spike was followed by a ~380 s train of high-amplitude pulsations at the spin frequency of the neutron star.

Detailed timing analysis of the 2004 December 27 flare revealed high-frequency quasi-periodic oscillations (QPOs) for the first time in a magnetar candidate (Israel *et al.* 2005). During the middle stages of the tail of the flare, QPOs at frequencies of 18, 30 and 92.5 Hz were detected. These QPOs probably represent internal torsional oscillations of the star, which could be supported either by its crust (Duncan 1998) or by its core magnetic field (Thompson & Duncan 2001).

14.2.4 Intermediate bursts

Intermediate bursts are intermediate in duration, peak luminosity and energy between the common recurrent SGR bursts and the giant flares. They have durations of seconds or longer, and peak luminosities exceeding $\sim 10^{41}$ erg s^{-1}. They tend to have abrupt onsets and, if the duration is less than the rotation period of several seconds, also abrupt end points. The flux generally varies smoothly in between. The short, recurrent bursts (e.g., Göğüş *et al.* 2001) are usually more irregular, which suggests that the emitting particles cool more rapidly. The intermediate bursts are most commonly observed in the days and months following the giant flares, which suggests that they represent some residual energy release by essentially the same mechanism. In a nutshell, these bursts appear to be "aftershocks" of the giant flares.

Time histories of four examples of intermediate bursts are shown in Fig. 14.7. Their isotropic energies range from 10^{41} to 10^{43} ergs. Up until 1998, there were few intermediate bursts recorded from SGRs, most from SGR 0526−66 (Golenetskii *et al.* 1984). Since 1998, several more bursts have been detected from other SGRs which begin to fill in the apparent gap in energy and duration. (The largest was an event recorded on 2001 April 18 from SGR 1900+14; Guidorzi *et al.* 2004.) This suggests that there may be a continuum of burst sizes covering the smallest recurrent bursts all the way up to the giant flares.

The spectra of most intermediate bursts are consistent with the spectra of the short, recurrent bursts and the pulsating tails of the giant flares. The spectra do not vary much, either from burst to burst or within individual bursts. A striking exception to this rule was a very intense ($L_{peak} \sim 10^{43}$ erg s^{-1}) and spectrally hard (kT_{peak} ~120 keV) burst detected from SGR 1627–41 (Mazets *et al.* 1999b; Woods *et al.* 1999a). This burst lasted ~0.5 s and was similar both spectrally and temporally to the initial peaks of the giant flares – but without the extended softer pulsations. Two bursts recorded from SGR 1900+14 during the 1998–9 activation were also spectrally much harder than all other burst emission from this SGR (Woods *et al.* 1999b) with the exception of the initial spike of the August 27 flare. These bursts are, in fact,

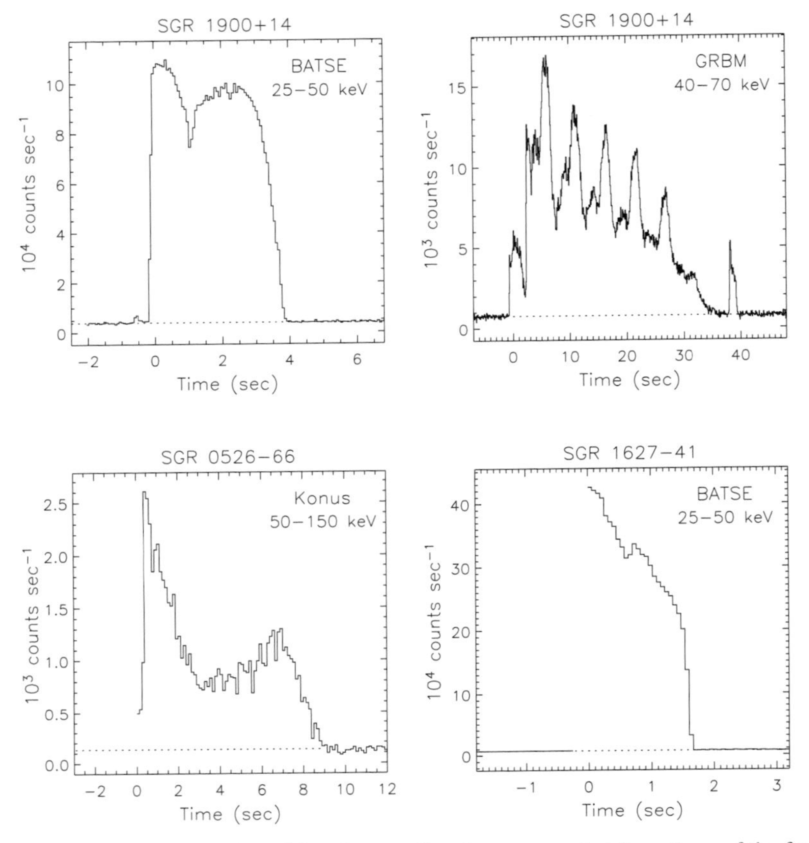

Fig. 14.7. Time histories of four intermediate bursts recorded from three of the four SGRs. Clockwise from upper left: SGR 1900+14 burst recorded with BATSE on 1998 October 28, SGR 1900+14 burst recorded with GRBM on 2001 April 18, SGR 1627−41 burst recorded with BATSE on 1998 June 18, and SGR 0526−66 burst recorded with Konus on 1982 February 27. Energy ranges are shown in each figure panel. The rise of the SGR 1627−41 burst is unresolved due to a gap in the BATSE data. GRBM data courtesy of M. Feroci and F. Frontera. Konus data courtesy of S. Golenetskii.

spectrally and temporally indistinguishable from classical GRBs. They were not exceptionally bright and had durations of ∼1 s with a fast rise and exponential decay. Their spectra were consistent with a power law (photon index ∼–2) whose hardness was anti-correlated with X-ray flux.

14.2.5 Possible spectral features

Discrete features in burst spectra from magnetar candidates have been reported from SGR 0526−66, SGR 1900+14, SGR 1806−20, and 1E 1048.1−5937. It should be emphasized that, as was the case previously with classical gamma-ray bursts, the same spectral feature has not yet been detected in the same burst by independent instruments.

In SGR 0526−66, Mazets *et al.* (1979) reported evidence for a broad peak in the energy spectrum at ∼430 keV during the main peak of the giant flare of March 5. Using RXTE

PCA data, Strohmayer and Ibrahim (2000) discovered a significant emission feature at $\sim$6.7 keV during a precursor to the intermediate burst of 1998 August 29 from SGR 1900+14. An additional feature consistent with twice this energy is seen, but its significance is marginal. Ibrahim, Swank and Parke (2003) presented the analysis of 56 spectra accumulated with the RXTE PCA taken from selected SGR 1806−20 bursts intervals. Of the 56 spectra, a handful showed a statistically significant ($>3\sigma$) absorption feature near 5 keV. These authors have argued that these lines represent proton cyclotron absorption features in a strong magnetic field. In addition, two bursts were recorded with the PCA from the direction of 1E 1048.1−5937 within two weeks of each other late in 2001 (Gavriil *et al.* 2002). In the first of these bursts, a strong emission feature was seen at $\sim$7 keV with less significant features at energies consistent with the first three harmonics.

14.3 Persistent X-ray emission

Historically, one of the defining properties of AXPs was the relative steadiness of their X-ray emission, over a fairly narrow range 10^{35}−10^{36} erg s^{-1}. Over the past several years, however, it has become clear that at least half and possibly most magnetar candidates are variable X-ray sources. Some of the observed variability is clearly driven by burst activity (see Section 14.5), but at least a few sources have shown large changes in luminosity ($\sim$10−100) with little or no detected burst activity. For example, XTE J1810−197 was discovered in 2003 at a luminosity of $\sim 2 \times 10^{36}$ erg s^{-1} (Ibrahim *et al.* 2004), but archival observations from the 1990s found the source in a "low state" with a luminosity two orders of magnitude smaller (Gotthelf *et al.* 2004). One of the AXP candidates, AX J1845−0258, was discovered at a luminosity $\sim 10^{35}$ erg s^{-1} in a 1993 ASCA observation (Torii *et al.* 1998; Gotthelf & Vasisht 1998), yet follow-up observations 3 and 6 years later found the flux 20 times dimmer (Vasisht *et al.* 2000). No high-luminosity SGR-like bursts have ever been seen from either of these sources.

One of the best studied AXPs, 1E 1048.1−5937, has also shown signs of flux variability (e.g., Oosterbroek *et al.* 1998). Weekly monitoring with RXTE has revealed two flares in the pulsed flux lasting several months (Gavriil & Kaspi 2004). Unlike the burst-induced variability, the rises of these flares were resolved and lasted a few weeks. Interestingly, two small bursts were detected near the peak of the first flare (Gavriil *et al.* 2002), but none were seen at any point during the much brighter and longer-lived second flare. Imaging X-ray observations have shown similar varibility in the phase-averaged luminosity, albeit with much sparser sampling (Mereghetti *et al.* 2004). As with XTE J1810−197 and AX J1845−0258, the "baseline" luminosity of 1E 1048.1−5937 is low ($\sim 6 \times 10^{33}$ erg s^{-1}) relative to the average luminosity of the AXP class.

The realization that SGRs and AXPs can enter low states with luminosities of order 10^{33}−10^{34} erg s^{-1} for extended periods of time has important implications on the total number of magnetar candidates in our Galaxy (Section 14.8). Their duty cycle as bright X-ray sources is presently unknown, especially as a function of age. There is still much to be learned about the similarities and differences between magnetar candidates in their dim states, and other low-luminosity X-ray sources such as isolated neutron stars, compact central objects and high-field radio pulsars (Chapter 7). It is possible that some of these other sources occasionally become X-ray bright like the AXPs.

Table 14.1. *X-ray spectral properties of the SGRs and AXPs*

Source[a]	N_H 10^{22} (cm^{-2})	Blackbody temperature (keV)	Photon index	Unabsorbed flux[b] 10^{-11} (erg cm^{-2} s^{-1})	Luminosity[c] 10^{35} (erg s^{-1})
SGR 0526−66	0.55	0.53	3.1	0.087	2.6
SGR 1627−41	9.0	–	2.9	0.027−0.67	0.04−1.0
SGR 1806−20	6.3	–	2.0	1.2−2.0	3.2−5.4
SGR 1900+14	2.6	0.43	1.0−2.5	0.75−1.3	2.0−3.5
CXOU 010043.1−721134	0.14	0.41	–	0.010	0.39
4U 0142+61	0.91	0.46	3.4	8.3	0.72
1E 1048.1−5937	1.0	0.63	2.9	0.41−2.3	0.053−0.25
1RXS J170849−400910	1.4	0.44	2.4	6.4	1.9
XTE J1810−197[d]	1.1	0.67	3.7	0.01−2.2	0.01−2.6
1E 1841−045	2.5	0.44	2.0	1.9	1.1
AX J1845−0258	9	–	4.6	0.04−1.0	0.05−1.2
1E 2259+586	1.1	0.41	3.6−4.2	1.6−5.5	0.17−0.59

[a] Spectral values given for quiescent state only (i.e., periods with no *detected* burst activity)
[b] All fluxes and luminosities integrated over 2.0−10.0 keV
[c] Assumed distances given in Table 14.4
[d] Spectral parameters given were obtained during "high" state of source following its discovery in 2003
References (SGR 0526) Kulkarni *et al.* 2003; (SGR 1627) Kouveliotou *et al.* 2003; (SGR 1806) Mereghetti *et al.* 2000; (SGR 1900) Woods *et al.* 2001; (CXO 0100) Lamb *et al.* 2002; (4U 0142) Patel *et al.* 2003; (1E 1048) Mereghetti *et al.* 2004; (RXS 1708) Rea *et al.* 2003; (XTE 1810) Gotthelf *et al.* 2004; (1E 1841) Morii *et al.* 2003; (AX 1845) Gotthelf & Vasisht 1998, Torii *et al.* 1998, Vasisht *et al.* 2000; (1E 2259) Woods *et al.* 2004

14.3.1 X-ray spectra

The X-ray spectra of SGRs and AXPs (0.5−10 keV) are usually well fit by a two-component model, a blackbody plus a power law, modified by interstellar absorption (Table 14.1). The soft blackbody component is not required in a few sources, but these tend to be dim and/or heavily absorbed (e.g., SGR 1627−41). During quiescence (i.e., outside of bursting activity), the blackbody temperature does not vary greatly between different members of the class (Marsden & White 2001) or with time for individual sources (e.g., Oosterbroek *et al.* 1998). On the other hand, the non-thermal component does show significant variations between different sources (Marsden & White 2001) and with time in a few cases (e.g., Woods *et al.* 2004).

The first systematic study of the X-ray spectra of SGRs and AXPs was performed by Marsden and White (2001), who found that the spectral hardness of the persistent X-ray counterparts of these sources formed a continuum and was positively correlated with the spin-down rate of the pulsar (Fig. 14.8). The varying hardness of the X-ray spectrum with spin-down rate was linked to the non-thermal component of the spectrum.

Until the launch of Integral, very little was known about the persistent X-ray spectra of magnetar candidates above ~10 keV due to the limitations of past instrumentation. Currently, there have been reported Integral detections of persistent hard X-ray emission above ~15 keV

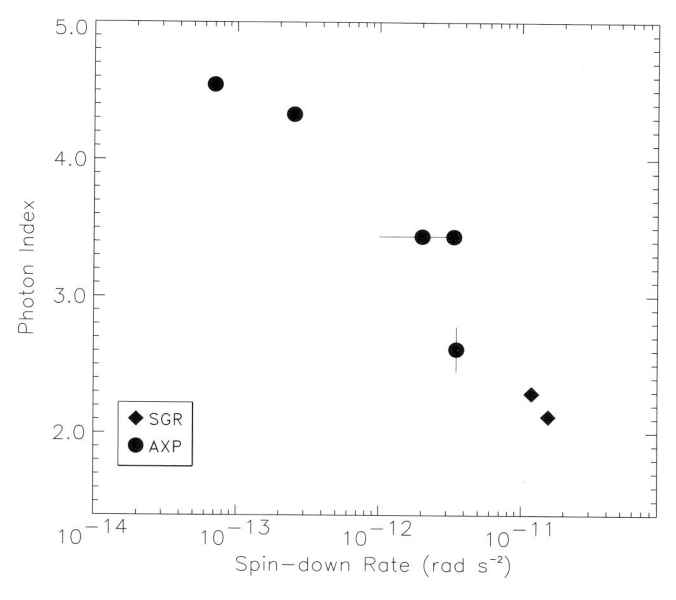

Fig. 14.8. The variation of the single power-law photon index versus spin-down rate $|\dot{\Omega}|$ for each SGR and AXP. The results for objects with more than one observation have been averaged. The photon index decreases (spectral hardness increases) with increasing spin-down rate. Figure from Marsden & White (2001).

from four magnetar candidates (1E 1841−045 (Molkov *et al.* 2004; Bird *et al.* 2004), 1RXS J170849.0−400910 (Revnivtsev *et al.* 2004), SGR 1806−20 (Bird *et al.* 2004), and 4U 0142+61 (den Hartog *et al.* 2004)). Kuiper, Hermsen and Mendez (2004) were the first to show that the hard X-rays detected from 1E 1841−045 do, in fact, originate with the AXP, when they detected pulsed emission with RXTE HEXTE. Interestingly, the pulsed spectrum follows a power law with a photon index −1.0 up to at least 100 keV. Knowledge of the photon distribution with energy above 15 keV is crucial to determining the underlying emission mechanism (see Section 14.7.3).

There are no definite detections of spectral features in the *persistent* X-ray emission of magnetar candidates. Grating spectra from Chandra and XMM-Newton have yielded strong upper limits (<30 eV) on narrow line features for 4U 0142+61 (Juett *et al.* 2002) and 1E 2259+586 (Woods *et al.* 2004) in the energy range 0.5−5 keV. The only reported detection of a spectral feature comes from a BeppoSAX spectrum of 1RXS J170849.0−400910 where Rea *et al.* (2003) find a $\sim 4\sigma$ absorption line at ~8 keV. The absence of spectral features at X-ray energies where proton-cyclotron resonances would occur in magnetar-strength fields places constraints on models of the transmission of heat through the surface, and of surface heating.

14.3.2 *Pulse profiles and pulsed fractions*

The X-ray pulse profiles of magnetar candidates range from simple sinusoids to more complex profiles showing (typically) two maxima per cycle (Fig. 14.9). The observed pulse morphologies of the AXPs are consistent with either one or two hot spots on the surface of a neutron star (Özel 2002), but the spectrally harder SGRs sometimes have more complicated

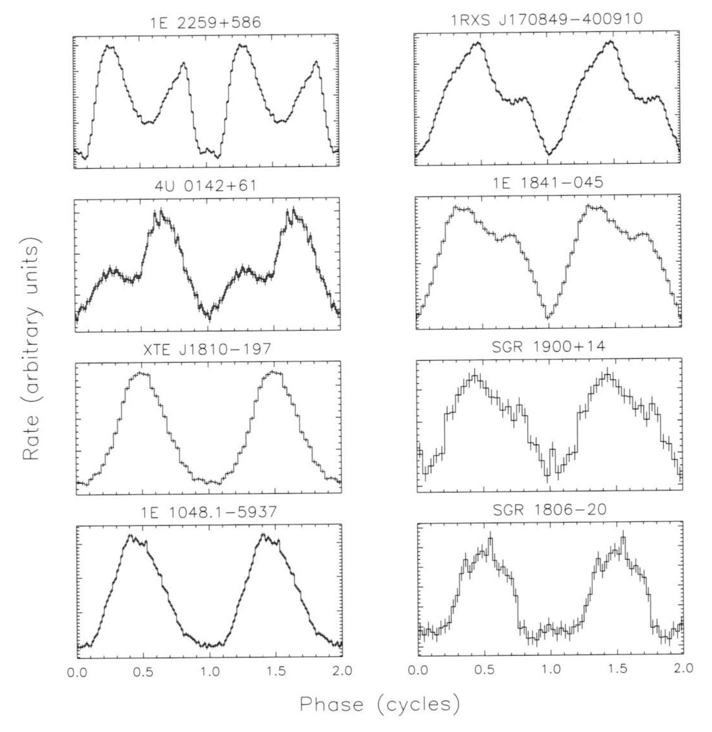

Fig. 14.9. The folded pulse profiles of eight different magnetar candidates. The sources are ranked according to inferred dipole magnetic field strength. Magnetic field increases from top to bottom and left to right. All profiles are of X-rays between 2 and 10 keV as observed with the RXTE PCA. Note that the folded profile of SGR 1900+14 is from after the August 27 flare. AXP pulse profiles courtesy of V. M. Kaspi and F. P. Gavriil.

pulse profiles. In contrast with most accreting X-ray pulsars, the pulse profile of a SGR or AXP often has a weak dependence on photon energy.

The pulse profiles shown in Fig. 14.9 are ranked in order of increasing spin-down rate (from top to bottom and left to right). No strong trend is apparent; but note that the pulse profile of SGR 1900+14 was much more complex before the August 27 flare. The evolution of the pulse morphology in SGR 1900+14 and other magnetar candidates is discussed in Section 14.5.3.

The root-mean-square (rms) X-ray pulsed fractions of magnetar candidates range from 4 to 60% (Table 14.2). Note that in the literature, both peak-to-peak and rms are reported, and the rms values are always less than the peak-to-peak values for any given AXP/SGR pulse profile. Similar to the pulse shape, the pulsed fractions of the AXPs show little or no change with photon energy (0.5−10 keV). Since the relative contribution of the blackbody spectral component to the total photon flux changes from 0% to as much as ~70% over this bandpass, Özel, Psaltis and Kaspi (2001) argued that the two spectral components of the AXPs must by highly correlated or caused by the same physical process. The pulsed fraction also places strong constraints on models in which the 2–10 keV emission of the AXPs is purely due to cooling emission through the surface of the star (Özel *et al.* 2001).

Table 14.2. *Pulse timing properties of the SGRs and AXPs*

Source	Period (s)	Period derivative (10^{-11} s s^{-1})	Magnetic field[a] (10^{14} gauss)	Spin down age[b] (10^3 years)	Pulsed fraction[c] (% rms)
SGR 0526−66	8.0	6.6	7.4	1.9	4.8
SGR 1627−41	6.4?	–	–	–	<10
SGR 1806−20	7.5	8.3−47	7.8	1.4	7.7
SGR 1900+14	5.2	6.1−20	5.7	1.3	10.9
CXOU 010043.1−721134	8.0	–	–	–	10
4U 0142+61	8.7	0.20	1.3	70	3.9
1E 1048.1−5937	6.4	1.3−10	3.9	4.3	62.4
1RXS J170849−400910	11.0	1.9	4.7	9.0	20.5
XTE J1810−197	5.5	1.5	2.9	5.7	42.8
1E 1841−045	11.8	4.2	7.1	4.5	13
AX J1844−0258	7.0	–	–	–	48
1E 2259+586	7.0	0.048	0.60	220	23.4

[a] $B_{\rm dipole} = 3.2 \times 10^{19}\sqrt{P\dot{P}}$ G (the mean surface dipole field)
[b] Characteristic age of pulsar spinning down via magnetic braking ($P/2\dot{P}$)
[c] $f_{\rm rms} = r_{\rm avg}^{-1}[\frac{1}{N}\sum_{i=1}^{N}(r_i - r_{\rm avg})^2 - e_i^2]^{1/2}$, where N = number of phase bins, r_i is the count rate (2−10 keV) in the ith phase bin, e_i is the error in the rate, and $r_{\rm avg}$ is the average count rate
References (SGR 0526) Kulkarni *et al.* 2003; (SGR 1627) Woods *et al.* 1999a; (SGR 1806) Woods *et al.* 2002; (SGR 1900) Woods *et al.* 2002; (CXO 0100) Lamb *et al.* 2003b; (4U 0142) Gavriil & Kaspi 2002; (1E 1048) Kaspi *et al.* 2001; Gavriil & Kaspi 2004; (RXS 170849) Gavriil & Kaspi 2002; (XTE 1810) Ibrahim *et al.* 2004; (1E 1841) Gotthelf *et al.* 2002; (AX 1845) Gotthelf & Vasisht 1998, Torii *et al.* 1998; (1E 2259) Gavriil & Kaspi 2002

It should be kept in mind that emission due to surface heating can be much more strongly beamed (e.g., Basko & Sunyaev 1975); and that cyclotron scattering by persistent electric currents can strongly modify the observed pulse profile in active magnetars (Thompson *et al.* 2002).

14.4 Timing behavior

The spin periods of the SGRs and AXPs are clustered between 5 and 12 seconds, a very narrow range compared with radio pulsars and accreting X-ray pulsars. These sources are all spinning down rapidly and persistently, with fairly short characteristic ages $P/\dot{P} \sim 10^3$–10^5 yrs. The magnitude of the spin-down torque is consistent with magnetic dipole braking of an isolated neutron star with a dipole field of $\sim 10^{14}$–10^{15} G (Fig. 14.10). Although most of the characteristic ages are less than 10^4 yrs, the ages for individual sources should be treated with caution since the spin-down torque has been observed to vary by more than a factor ~4 in some sources (e.g., SGR 1806–20 and SGR 1900+14). The pulse timing properties are summarized in Table 14.2.

Long-term phase-coherent timing of SGRs and AXPs recently became feasible with RXTE. Currently several AXPs have continuous timing solutions, with some dating back to 1998 (e.g., Gavriil & Kaspi 2002). In the case of two AXPs (1E 1048.1−5937 (Kaspi *et al.* 2001); XTE J1810−197 (Ibrahim *et al.* 2004)) and two SGRs (SGR 1806−20 and SGR 1900+14

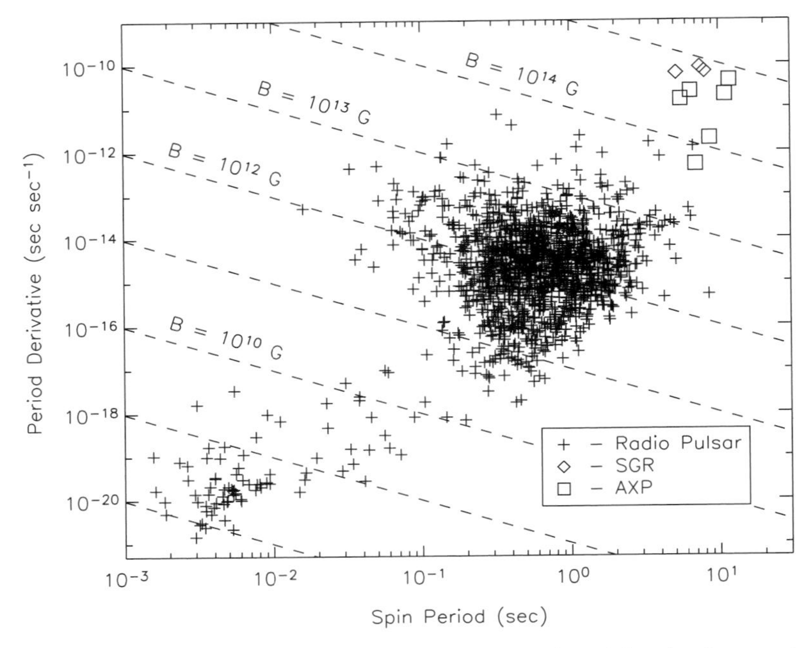

Fig. 14.10. Period versus period derivative for radio pulsars (plus signs), anomalous X-ray pulsars (squares), and soft gamma repeaters (diamonds). Contours of constant inferred magnetic field strength are drawn as diagonal dashed lines. Radio pulsar data courtesy of the ATNF Pulsar Group.[2] AXP and SGR timing data are given in Table 14.2.

(Woods *et al.* 2002)) phase-coherent timing is not always possible but has been obtained over stretches of months to years. Major results of this timing effort have been the discoveries of three glitches from two AXPs and strong timing noise detected in both SGRs and the AXP 1E 1048.1−5937.

The occurrence of large glitches is a natural consequence of the magnetar model (Thompson & Duncan 1996) and was even suspected to be the primary source of timing noise in various models of the AXPs (Usov 1994; Heyl & Hernquist 1999). In 1999, the first glitch from an AXP was observed by RXTE in the source 1RXS J170849.0−400910 (Kaspi, Lackey & Chakrabarty 2000). Since that time, another glitch was detected from the same source (Kaspi & Gavriil 2003; Dall'Osso *et al.* 2003) and one glitch was observed in 1E 2259+586 coincident with a burst active episode (Kaspi *et al.* 2003; Woods *et al.* 2004). No glitches have been directly observed in the SGRs, although SGR 1900+14 has shown evidence for rapid spin-*down* at the time of the August 27 flare (see Section 14.5.4). The magnitude of the angular momentum exchange within the star that one infers for the AXP glitches is more characteristic of Crab-type pulsar glitches than the larger Vela-type glitches, but there are some dissimilarities with radio pulsar glitch behavior (Table 14.3).

The glitch observed from 1E 2259+586 was especially interesting in that it coincided with a SGR-like outburst, and also with changes in the X-ray flux and pulse profile that persisted for months (see Section 14.5.3). There is evidence for a very long-term

[2] http://www.atnf.csiro.au/research/pulsar/psrcat/

Table 14.3. *Properties of the three glitches observed in two AXPs*

Source	1RXS J1708−40	1RXS J1708−40	1E 2259+586
$\Delta\nu/\nu^a$	5.5×10^{-7}	1.4×10^{-7}	3.7×10^{-6}
$\Delta\nu_g/\nu$	–	–	$> 6.1 \times 10^{-6}$
τ_g (days)	–	–	14
$\Delta\nu_d$	–	4.1×10^{-6}	$\sim\Delta\nu_g/\nu$
τ_d (days)	–	50	16
$\Delta\dot{\nu}/\dot{\nu}$	−0.010	$< \lvert 0.001 \rvert$	+0.022
t_{glitch} (MJD TDB)	51444.6	52014.2	52443.1

[a] Frequency denoted by ν and frequency derivative by $\dot{\nu}$. The subscript g indicates the frequency growth terms and d indicates decay terms.
References: (1RXS 1708) Kaspi & Gavriil 2003; Dall'Osso *et al.* 2003; (1E 2259) Woods *et al.* 2004

component of the post-glitch frequency recovery (consistent with a persistent change in torque) in this glitch, as well as in one of the glitches of 1RXS J170849.0−400910. Overall one observes a great diversity in behavior even within this small sample of glitch events, and more extended monitoring is required to unravel the relationship between glitch behavior and burst activity.

In addition to rapid spin down, all SGRs and AXPs have shown significant timing noise: an irregular drift of the spin frequency superposed on the secular spin down trend. Recent timing solutions have shown that most of this noise is not caused by resolved glitches. The existence of timing noise was first noted for the AXPs 1E 2259+586 (e.g., Baykal & Swank 1996) and 1E 1048.1−5937 (e.g., Oosterbroek *et al.* 1998). Its strength was estimated in four AXPs by Heyl & Hernquist (1999), and was found to be marginally consistent with an extrapolation of the correlation between timing noise strength and braking torque observed in radio pulsars (e.g., Arzoumanian *et al.* 1994).

In the RXTE era, the first direct detection of timing noise was made in SGR 1806−20 (Woods *et al.* 2000). This SGR is one of the "noisiest" rotators among SGRs and AXPs and has shown a long-term persistent change in torque, along with large stochastic offsets in the X-ray pulse phase on timescales as short as $\sim 10^4$ s (Woods *et al.* 2002). Although the strength of the timing noise is consistent with some of the "quieter" accreting X-ray pulsars, the shape of the torque power spectrum is more similar to that observed in radio pulsars. Overall, the timing analysis of the AXPs (Kaspi *et al.* 2001; Gavriil & Kaspi 2002) and one other SGR (Woods *et al.* 2002) has revealed a broad range of torque variability, with some evidence for a correlation between the strength of the timing noise and the spin-down rate.

14.5 Burst-induced variability

It has become evident that burst activity in the SGRs can have a persistent effect on the underlying X-ray source. During the 1998 burst activation of SGR 1900+14, the X-ray counterpart became brighter, its energy spectrum was altered, and the pulse shape changed dramatically. Furthermore, the X-ray counterpart to SGR 1627−41 has become progressively dimmer since the one recorded outburst from this SGR in 1998. Finally, the AXP 1E 2259+586 showed a broad array of spectral and temporal changes coincident with its 2002 outburst. We now present some details of the burst-induced variability that is observed in magnetar candidates.

14.5.1 X-ray afterglows and AXP outbursts

Extended X-ray afterglow has been detected following four separate bursts from SGR 1900+14. The first such detection followed the giant flare of 1998 August 27 (Woods *et al.* 2001). One half hour following the flare, the persistent X-ray flux from SGR 1900+14 remained ~700 times brighter than the pre-flare level. The X-ray flux decayed over the next 40 days approximately as a power law in time ($F \propto t^{-\alpha}$ with an exponent $\alpha = 0.71$). The blackbody component of the X-ray spectrum was hotter ($kT = 0.94$ keV) one day into the afterglow phase than it was before the burst ($kT = 0.5$ keV); but 18 days later the power-law component of the spectrum was again dominant.

Afterglows have also been detected from SGR 1900+14 following bursts on 1998 August 29 (Ibrahim *et al.* 2001; Lenters *et al.* 2003 (Fig. 14.11)), 2001 April 18 (Feroci *et al.* 2003), and 2001 April 28 (Lenters *et al.* 2003). A power-law decay is also seen in these cases, with a return to the pre-burst flux level between 10^4 and 10^6 s following the burst. Enhanced thermal emission is typical, with temperatures as high as ~4 keV (corresponding to a hot spot covering ~1% of the neutron star surface). In fact, the afterglow of the 2001 April 28 burst involved only enhanced thermal emission. Within this small sample of afterglows from SGR 1900+14, the 2–10 keV afterglow energy is about 2% of the 25–100 keV burst energy (Lenters *et al.* 2003).

Resolved observations of individual SGR burst afterglows are still rare, because they require pointed X-ray observations coincident with a burst (or very soon thereafter). For example, no soft X-ray afterglow could be detected from SGR 1806–20 following the 2004 December 27 giant flare due to Sun-angle constraints for X-ray telescopes at that time of year. It is much easier to observe the collective effect of SGR burst activity on the persistent X-ray flux, as is seen in the case of SGR 1900+14 (Fig. 14.12).

The detection of X-ray bursts from the AXP 1E 2259+586 (Kaspi *et al.* 2003) was the fortunate result of a long-term monitoring campaign by RXTE. The X-ray flux of this source increased by at least a factor ~20 on 2002 June 18 (Woods *et al.* 2004), during which more than 80 SGR-like bursts were emitted (Gavriil *et al.* 2004). This first component of the flux decay was spectrally hard, contained all of the observed burst activity, and involved only ~1% of the neutron star surface. It decayed within ~1 day, and was followed by a much more gradual flux decay over the following year. This more extended X-ray brightening involved a significant fraction of the warm stellar surface, but only a modest spectral hardening. No bright burst (similar to the intermediate bursts of the SGRs) appears to have preceded this activity.

The longer term flux variability of the SGR and AXP sources is still unclear. Some sources (such as the AXPs 1E 2259+586, 1E 1048.1–5937, and 4U 0142+61) have remained X-ray bright for two–three decades. A previous X-ray brightening of 1E 2259+586 detected 10 years earlier by Ginga (Iwasawa, Koyama & Halpern 1992) allows one to deduce that at least ~10% of the X-ray output of this source is released in transient events. A steady decrease in X-ray flux was also observed in SGR 1627–41 after its outburst in 1998 June/July, but this was followed by a sharper drop a few years later (Kouveliotou *et al.* 2003). The X-ray flux of this source has appeared to level off at a value ($\sim 4 \times 10^{33}$ erg s^{-1}) consistent with the low state levels seen in at least two other sources. It is possible that magnetar candidates become

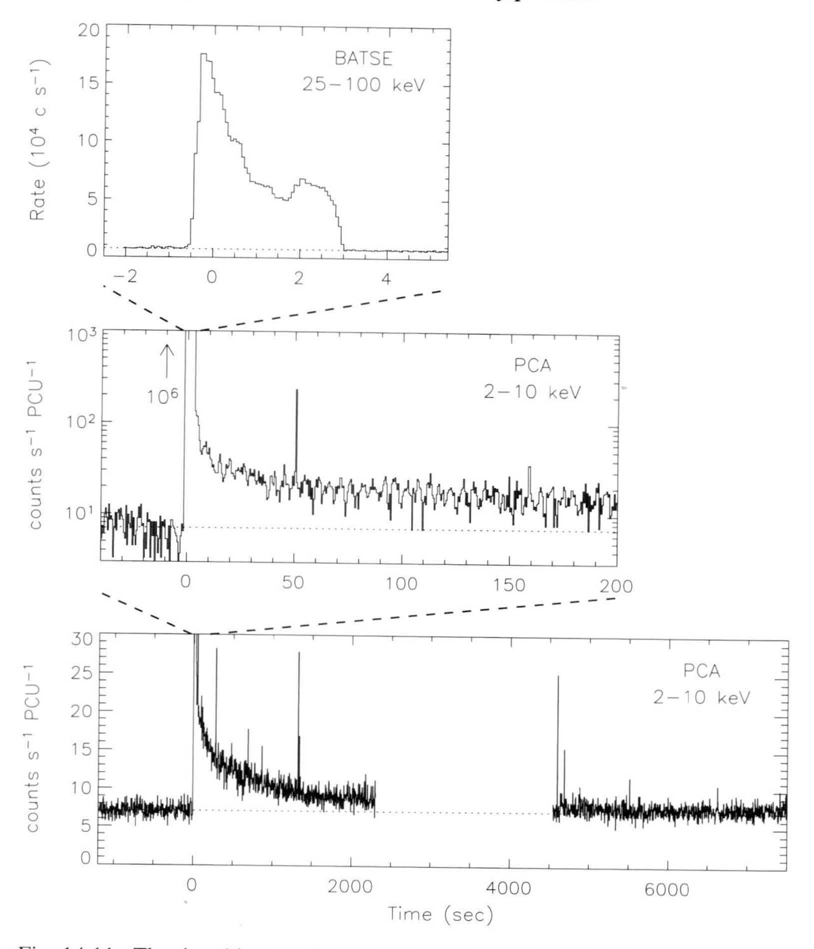

Fig. 14.11. The time history of the August 29 burst and its afterglow. *Top*: The BATSE lightcurve showing the sharp rise and fall of the burst. *Middle*: The PCA lightcurve including the burst (off-scale) and the early portions of the afterglow (T+3 s). There is a sharp discontinuity in the energy spectrum when the high-luminosity burst emission terminates at 3 s (Ibrahim *et al.* 2001) indicating the transition from the burst to the afterglow. Note the clear 5.16 s pulsations in the lightcurve. *Bottom*: The PCA lightcurve over a longer time interval showing the gradual decay of the afterglow. The spectral evolution during the afterglow is presented in Ibrahim *et al.* (2001) and Lenters *et al.* (2003). The horizontal dotted lines in all panels represent the background level.

increasingly intermittent X-ray sources as they age; alternatively, some intermittent sources may have weaker magnetic fields. There is, nonetheless, clear evidence for both short-term and long-term flux variability associated with X-ray outbursts.

14.5.2 Transient counterparts at other wavelengths

The only recorded radio emission from magnetar candidates has been seen following giant flares. A faint radio transient having a spectral index of -0.74 ± 0.15 persisted for 2 weeks following the 1998 August 27 giant flare of SGR 1900+14 (Frail, Kulkarni & Bloom 1999). No radio detection of this SGR was made before or since.

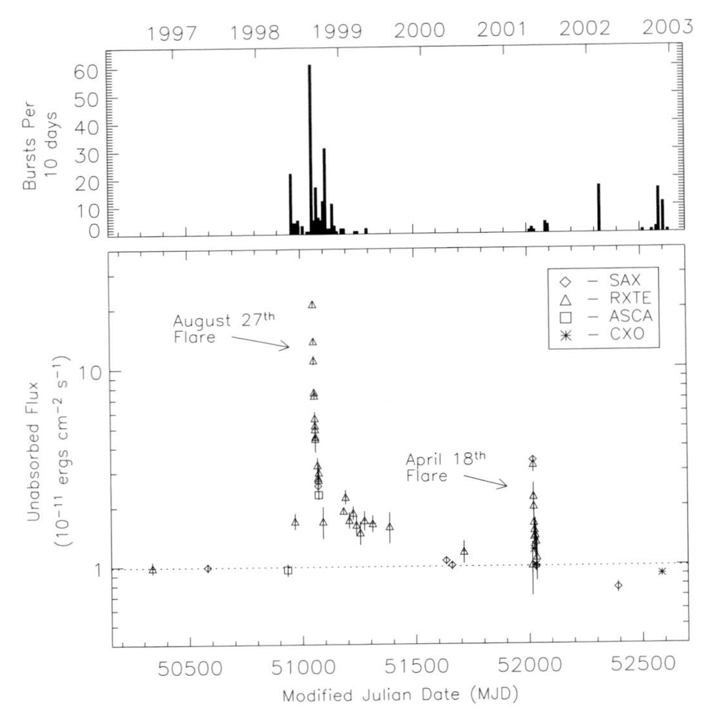

Fig. 14.12. *Top panel*: Burst rate history of SGR 1900+14 as observed with BATSE and the IPN. *Bottom panel*: Persistent/pulsed flux history of SGR 1900+14 covering 5.5 years. The vertical scale is unabsorbed 2–10 keV flux. The pulsed fraction is assumed constant to convert pulsed flux to phase-averaged flux (see Woods *et al.* 2001 for details). The dotted line marks the nominal quiescent flux level of this SGR. Note the strong correlation between the burst activity and the flux enhancements.

A far brighter radio afterglow (factor ~500) was seen following the 2004 December 27 flare from SGR 1806–20 (Gaensler *et al.* 2005; Cameron *et al.* 2005). Up to 20 days past the flare, the transient radio nebula was found to be linearly polarized (1–3%) and expanding at a rate 0.4 d_{15} c (Gaensler *et al.* 2005; Taylor *et al.* 2005). Between 9 and 20 days post-flare, the spectral index was -0.75 ± 0.02 and the flux decayed achromatically in time with a power-law index $\alpha = -2.7$. Beyond 20 days the afterglow re-brightened, possibly due to a transition to the Sedov–Taylor phase of the expansion (Gelfand *et al.* 2005).

Optical and/or infrared (IR) counterparts have been discovered in at least five magnetar candidates (see Section 14.6.4), some of which have shown variability. The IR flux of the AXP 1E 2259+586 increased following its 2002 June outburst (Kaspi *et al.* 2003). A week after the X-ray burst activity, the K-band flux was 3.4 times higher than measured two years earlier. The IR flux decayed in lock-step with the X-ray flux decay and returned to nearly its pre-outburst level within one year (Tam *et al.* 2004; Israel *et al.* 2003a). Similar correlated IR/X-ray flux variability was seen in SGR 1806–20 during the build-up to the giant flare (Kosugi, Ogasawara & Terada 2005).

Automated telescopes such as ROTSE have observed SGRs during burst active periods (Akerlof *et al.* 2000). However, the visual extinction toward the SGRs observed is extremely

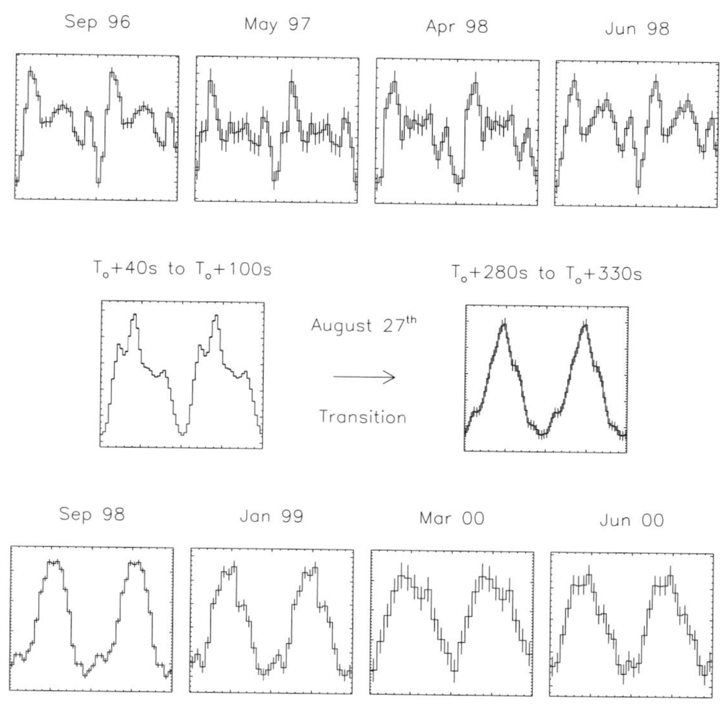

Fig. 14.13. Evolution of the pulse profile of SGR 1900+14 from 1996 September through 2000 April (Woods *et al.* 2001). All panels display two pulse cycles and the vertical axes are count rates with arbitrary units. The two middle panels were selected from Ulysses data (25–150 keV) of the August 27 flare. Times over which the Ulysses data were folded are given relative to the onset of the flare (T_0). The top and bottom rows are integrated over the energy range 2–10 keV. From top-to-bottom, left-to-right, the data were recorded with the RXTE, BeppoSAX, ASCA, RXTE, RXTE, RXTE, BeppoSAX, and RXTE.

high and so the acquired limits on optical burst emission are not constraining. Follow-up observations of SGR bursts with robotic IR cameras with fast photometric capabilities now coming online, such as BLANK, could provide interesting constraints on the burst mechanism.

14.5.3 Changes in X-ray pulse shape and pulsed fraction

Changes in the X-ray pulse profile have been observed in magnetar candidates during periods of intense burst activity. The most profound changes in pulse properties have been observed in SGR 1900+14. In particular, at the time of the giant flare of 1998 August 27, the pulse profile of the persistent emission changed dramatically from a complex, multi-peaked morphology to a simple, nearly sinusoidal morphology (Fig. 14.13). The change in pulse shape has persisted even years after the post-flare afterglow faded away (Göğüş *et al.* 2002). The effectively permanent change in pulse shape observed in SGR 1900+14 argues for a magnetic field reconfiguration at the time of the giant flare (Woods *et al.* 2001). A similar simplification of the pulse shape was observed – at a much higher flux level – over the last few minutes of the August 27 flare (Fig. 14.13). This change in pulse profile was smooth and gradual; indeed, the flux decline was consistent with cooling of a magnetically confined plasma (Feroci *et al.* 2001). Thus, the magnetic field reconfiguration may well have been concentrated in the initial impulsive phase of the flare.

A significant change in pulse shape was also observed in SGR 1806–20 during the 2004 December 27 giant flare (Palmer *et al.* 2005). Unlike the flare of SGR 1900+14, the pulse shape of SGR 1806–20 actually became more complex as the flare progressed. The amplitudes of two small peaks in the pulse profile grew relative to the primary peak during the flare. Similarly, the persistent, soft X-ray pulse profile became more complex following the flare evolving from a single peak to two distinct peaks in 2–10 keV X-rays (Woods *et al.* 2005).

During the 2002 June 18 outburst of 1E 2259+586, the pulse profile evolved rapidly showing large changes in the relative amplitudes of the two peaks (Kaspi *et al.* 2003; Woods *et al.* 2004). Some residual change has persisted for at least one year after the burst activity. In sum, the connection between X-ray outbursts and pulse shape changes can be subtle, and frequent monitoring of magnetar candidates will be required to understand it better.

Large changes in pulsed fraction have been observed in magnetar candidates during periods of burst activity. During burst afterglows of SGR 1900+14, the rms (2–10 keV) pulsed fraction has been observed to rise to $\sim$20–30% from its quiescent value of $\sim$11% (Lenters *et al.* 2003). Intriguingly, the enhanced pulsations remain in phase with the pre-burst pulsations in at least two bursts. This indicates a direct correlation between the source of the pulsed X-ray emission and the active burst region.

A pulsed fraction change was also observed in 1E 2259+586 during its 2002 June outburst (Woods *et al.* 2004). During the RXTE observation where the burst activity was seen, the pulsed fraction (2–10 keV) actually *decreased* to $\sim$15% from the pre-outburst level of 23%. The pulsed fraction recovered to $\sim$23% within $\sim$6 days of the outburst, much more rapidly than the pulse shape recovered.

14.5.4 Connection with timing anomalies

X-ray burst activity appears to have a variety of effects on the spin behavior of magnetar candidates. A comparison of the spin evolution of SGR 1900+14 before and after the August 27 flare showed that the source underwent a transient spin down within an 80 day window that bracketed the flare (Woods *et al.* 1999c). The decrease in spin frequency, by one part in 10^4, was opposite in sign to pulsar glitches. A comparison of the pulse timing during and after the flare indicates that the change in frequency occurred within several hours of the flare (Palmer 2002). The amplitude of the spin down is consistent with an enhanced magnetic torque due to a relativistic outflow of particles (Thompson & Blaes 1998; Frail *et al.* 1999; Thompson *et al.* 2000) if the dipole field of the star is $\sim 10^{15}$ G.

Only the giant flare from SGR 1900+14 has shown direct evidence for burst-induced spin down. This SGR suffered another timing anomaly at the time of the 2001 April 18 flare, but sparse data coverage did not allow for an unambiguous determination of its nature (Woods *et al.* 2003). More than 5 years of timing and burst data for each SGR have revealed longer-term increases in braking torque, of similar magnitudes, that are not however synchronized with burst activity. (Indeed the output in X-ray bursts from SGR 1806–20 was much smaller than from SGR 1900+14 over this time interval.) There is, nonetheless, some tentative evidence for a causal relation between burst activity and torque variability, in that the most burst-active magnetar candidates are also those which show the strongest timing noise.

A timing anomaly of a different type, the glitch of 1E 2259+586 (Section 14.4), coincided with the 2002 June X-ray outburst (Kaspi *et al.* 2003; Woods *et al.* 2004). Since the beginning of the X-ray activity was not observed, it was not possible to determine whether it preceded, overlapped, or followed the onset of the glitch. Nonetheless, the change in rotational energy associated with the glitch was much smaller than the energy released in the X-ray transient, suggesting that the trigger involved some other agent (e.g., the release of magnetic stresses).

In the case of the SGRs, glitches of the magnitude seen in 1E 2259+586 cannot generally be excluded: the timing ephemerides preceding X-ray outbursts are less accurate for these sources. However, X-ray transients associated with glitches of other AXP sources are easier to constrain. No burst emission was seen near the time of either of the two glitches observed in 1RXS J170849.0−400910 (Kaspi & Gavriil 2003). A ~1 day hard-spectrum transient, such as was observed from 1E 2259+586, could easily have been missed. However, the 1E 2259+586 outburst also showed a sustained X-ray afterglow lasting months and a pulse profile change for a somewhat shorter time interval (Woods *et al.* 2004). The regular monitoring of 1RXS J170849.0−400910 would have been sensitive to changes of this magnitude. Continued phase-coherent timing of these objects is needed to determine the extent to which glitches are accompanied by burst activity.

14.6 Locations, SNR associations, and counterparts

A multi-wavelength approach has always proven fruitful for understanding enigmatic astrophysical objects. The SGRs and AXPs are no exception to this rule. Sub-arcsecond determinations of their X-ray locations (Table 14.4) have allowed follow-up observations at radio and optical/IR wavelengths. (In the case of the SGRs, these represent a considerable refinement over previous triangulation of burst emissions using the Interplanetary Network.) We now discuss the results of this collective effort to study the magnetar candidates.

14.6.1 SNR associations

Supernova remnants are the glowing relics of massive explosions produced during the formation of some neutron stars. The surface brightness of a SNR depends upon its age and the density of the local inter-stellar medium. Because of dimming and observational selection (see, e.g., Gaensler & Johnston 1995), not every SNR contains a young pulsar. Likewise, young pulsars are not always found in SNRs. In fact, only one third of young ($<10^5$ yr) pulsars are expected to be found positionally coincident with their associated SNRs, which is entirely consistent with the observed fraction (Chapter 7).

As shown by Gaensler *et al.* (2001), the situation is similar for the AXPs and SGRs as a group. Of the ten confirmed magnetar candidates, only two have solid SNR associations (1E 2259+586 with CTB 109 (Gregory & Fahlman 1980) and 1E 1841−045 with Kes 73 (Vasisht & Gotthelf 1997)). The AXP candidate AX J1845−0258 has a solid association with the SNR G29.6+0.1 (Gaensler *et al.* 1999). Assuming that the SGR/AXP kick velocities incurred at birth are similar to those of radio pulsars, and given the space density of SNRs in the LMC, the association of SGR 0526−66 with the SNR N49 (Cline *et al.* 1982) was shown to be less secure than previously thought: the probability of a chance alignment is about 0.5% (Gaensler *et al.* 2001). These authors argued that other SGR/SNR and AXP/SNR associations reported in the literature were unconvincing, or likely to be chance superpositions.

The ages of the remnants associated with the magnetar candidates are $\sim 10^4$ yr, consistent with other age estimates for most of these objects. Other than containing a magnetar candidate, there is nothing unusual about these two SNRs relative to those of comparable age that are associated with radio pulsars. Note also that the young characteristic age of SGR 1900+14

Table 14.4. *The X-ray positions, reported associations, and the inferred distances of the SGRs and AXPs*

Source	Right ascension[b] (J2000)	Declination (J2000)	Associated SNR/cluster	Distance (kpc)	Galactic scale height (pc)
SGR 0526−66	05^h 26^m 00.89^s	−66° 04′ 36.3″	N49/cluster?	50	n/a
SGR 1627−41	16^h 35^m 51.84^s	−47° 35′ 23.3″	–	11	−21
SGR 1801−23[b]	18^h 00^m 59^s	−22° 56′ 50″	–	∼10	–
SGR 1806−20	18^h 08^m 39.32^s	−20° 24′ 39.5″	cluster	15	−63
SGR 1900+14	19^h 07^m 14.33^s	+09° 19′ 20.1″	cluster	15	+200
CXOU 010043.1−721134	01^h 00^m 43.14^s	−72° 11′ 33.8″	–	57	n/a
4U 0142+61	01^h 46^m 22.42^s	+61° 45′ 02.8″	–	3	−20
1E 1048.1−5937	10^h 50^m 07.14^s	−59° 53′ 21.4″	–	3	−27
1RXS J170849−400910	17^h 08^m 46.87^s	−40° 08′ 52.4″	–	5	+3
XTE J1810−197	18^h 09^m 51.08^s	−19° 43′ 51.7″	–	∼10	–
1E 1841−045	18^h 41^m 19.34^s	−04° 56′ 11.2″	G27.4+0.0	7	−1
AX J1844−0258[c]	18^h 44^m 53^s	−02° 56′ 40″	G29.6+0.1	∼10	+20
1E 2259+586	23^h 01^m 08.30^s	+58° 52′ 44.5″	G109.1−1.0	3	−52

[a] All positions accurate to <1″ unless otherwise noted
[b] Only a very crude IPN location (∼80 arcmin^2 area) exists for this SGR
[c] The positional accuracy for this AXP is a 20″ radius circle
References (SGR 0526) Kulkarni *et al.* 2003; Klose *et al.* 2004; (SGR 1627) Wachter *et al.* 2004; Corbel *et al.* 1999; (SGR 1801) Cline *et al.* 2000; (SGR 1806) Kaplan *et al.* 2001; Fuchs *et al.* 1999; Corbel & Eikenberry 2004; (SGR 1900) Frail *et al.* 1999; Vrba *et al.* 2000; (CXO 0100) Lamb *et al.* 2002; (4U 0142) Patel *et al.* 2003; Hulleman, van Kerkwijk & Kulkarni 2004; (1E 1048) Wang & Chakrabarty 2002; (RXS 170849) Israel *et al.* 2003b; (XTE 1810) Israel *et al.* 2004; (1E 1841) Wachter *et al.* 2004; Vasisht & Gotthelf 1997; (AX 1845) Vasisht *et al.* 2000; Gaensler *et al.* 1999; (1E 2259) Patel *et al.* 2001; Kothes, Uyaniker & Aylin 2002

(∼10^3 yrs) would make the absence of a SNR counterpart surprising, unless its true age were significantly larger (Thompson *et al.* 2000).

14.6.2 Galactic distribution

The distances to the SGRs and AXPs have been estimated in several several different ways, with widely varying degrees of precision (Table 14.4). One SGR (SGR 0526−66) and one AXP candidate (CXOU J0110043.1−721134) are located in the Large and Small Magellanic Clouds, respectively, and have well-determined distances. Two AXPs (1E 2259+586 and 1E 1841−045) are positioned close to the centers of SNR and are, very likely, physically associated (see Section 14.6.2). Both SGR 1806−20 and SGR 1900+14 may be associated with massive star clusters (see Section 14.6.4), each of which has an estimated distance. Most other distances rely on the measurement of interstellar absorption from X-ray spectra, and the intervening distribution of molecular clouds. A more complete discussion of the distance uncertainties is given in Özel *et al.* (2001).

All except two of the magnetar candidates are located within our Galaxy, and are positioned close to the Galactic plane. Their estimated heights above (or below) the Galactic plane for these sources are given in Table 14.4. The small rms scale height ($z_{\rm rms} \simeq 70$ pc) implies a young source population. This is consistent with the young ages inferred from the spin parameters (Table 14.2) and the SNR associations.

14.6.3 Radio limits

Despite deep, sensitive radio observations of most magnetar candidates, no persistent radio emission has been detected from any SGR or AXP. The only recorded radio detections of magnetar candidates were of transient outbursts seen from SGR 1900+14 and SGR 1806−20 in the days and weeks following their respective giant flares (see Section 14.5.2). Pulsed emission from this SGR in 1998 near the X-ray pulse period (5.16 s) was reported using 100 MHz data taken with the BSA (Shitov 1999), but the radio ephemeris disagreed significantly with the X-ray ephemeris (Woods *et al.* 1999c).

The non-detection of magnetar candidates at radio frequencies does not necessarily mean that they are very different from standard pulsars in their radio properties (Gaensler *et al.* 2001). Most pulsars are expected to have fluxes below the current limits for the AXPs. Long-period pulsars tend to have narrower beams (<1 deg), effectively reducing the chances that their beams will cross our line-of-sight. Given the small number of observed SGRs and AXPs, it is not surprising that none are detected.

The Parkes Multi-Beam Survey has revealed a handful of radio pulsars with magnetar strength fields as inferred from their spin parameters (Camilo *et al.* 2000; McLaughlin *et al.* 2003). These pulsars have X-ray luminosities much lower than their AXP/SGR counterparts (Pivovaroff, Kaspi & Camilo 2000), in spite of their similar dipole field strengths. This result has led Pivovaroff *et al.* and others (Camilo *et al.* 2000; McLaughlin *et al.* 2003) to conclude that membership as an AXP (or SGR) requires more than just a strong dipole magnetic field. Nonetheless, it would not be surprising to observe the transition of a high-field radio pulsar to a brighter X-ray state, given the realization that some magnetar candidates can enter X-ray low states for extended periods of time.

14.6.4 Optical and IR counterparts

The first optical detection of a magnetar candidate, the AXP 4U 0142+61, was made by Hulleman *et al.* (2000). They discovered an object with unusual colors spatially coincident with the X-ray position of the AXP. The broadband spectrum of 4U 0142+61 is shown in Fig. 14.14. Since this initial discovery, at least four other optical/IR counterparts with similar characteristics have been discovered for other AXPs and one SGR. The optical/IR properties of the magnetar candidates are given in Table 14.5. See also Hulleman, van Kerkwijk & Kulkarni (2004) for a table of empirically estimated magnitudes of those sources not yet discovered.

The discovery of optical/IR counterparts to magnetar candidates has placed valuable new constraints on their nature. Fast photometry of 4U 0142+61 revealed optical (R band) pulsations with a high pulsed fraction ($\sim$27% peak-to-peak) at the spin frequency of the neutron star (Kern & Martin 2002). The pulsed fraction of the optical pulsations is comparable to or greater than the pulsed fraction at X-ray energies.

Monitoring of the optical/IR counterparts has shown that the IR fluxes of the AXPs vary with time and burst activity (Section 14.5.2). If the IR flux is an indicator of burst emission (e.g., 1E 2259+586 in 2002 June), then IR flux variability seen in 1E 1048.1−5937 (Israel *et al.* 2002) and 4U 0142+61 (Hulleman, van Kerkwijk & Kulkarni 2004) could indicate

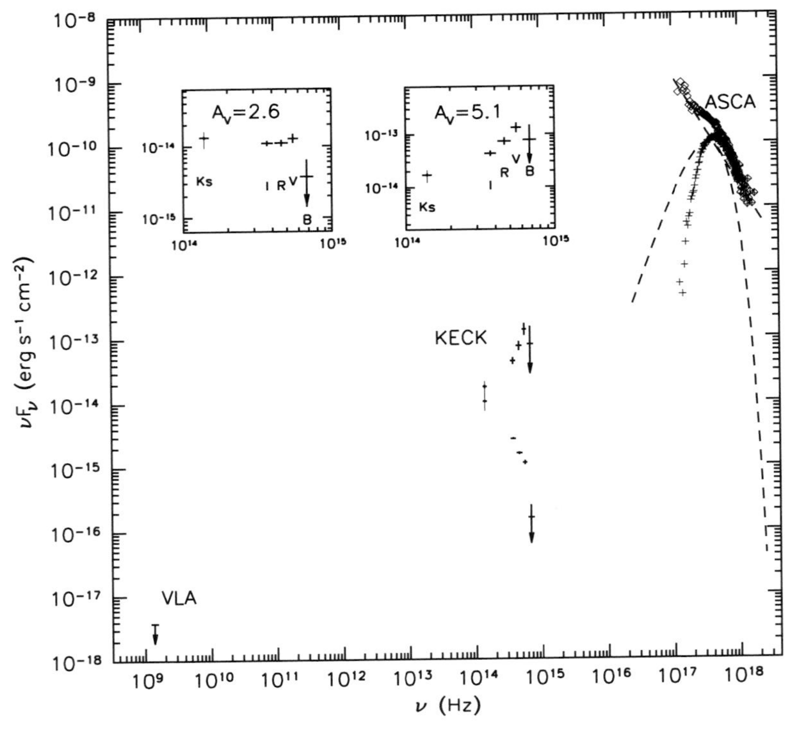

Fig. 14.14. Energy distribution for 4U 0142+61 (Hulleman *et al.* 2004). At low frequencies (10^{14}–10^{15} Hz), the points marked V, R, I, K_s indicate the observed V, R, I, and K_s-band fluxes. The vertical error bars reflect the uncertainties, while the horizontal ones indicate the filter bandwidths. The set of points above the measurements indicate de-reddened fluxes for $A_V = 5.4$, as inferred from the X-ray column density. At high frequencies (10^{17}–10^{18} Hz), the crosses show the incident X-ray spectrum as inferred from ASCA measurements. The diamonds show the spectrum after correction for interstellar absorption, and the two thick dashed curves show the two components used in the fit. Figure courtesy of M. van Kerkwijk and F. Hulleman.

the presence of undetected bursts. Clearly, continued monitoring of magnetar candidates is required to confirm some of these early findings.

Deep IR observations of the fields of two SGRs have revealed the presence of massive star clusters in which the SGRs are possibly embedded. Fuchs *et al.* (1999) found that SGR 1806–20 was positionally coincident with a cluster of massive stars at a distance ~15 kpc. Indeed, a very luminous star is located close to SGR 1806–20 (van Kerkwijk *et al.* 1995), but is not positionally coincident with it (Hurley *et al.* 1999d). Analysis of the other members of this cluster has shown that this particular cluster contains some of the most massive stars in our Galaxy, perhaps even the most luminous star in our Galaxy (Eikenberry *et al.* 2001). Similarly, Vrba *et al.* (2000) found a cluster of massive stars at a distance ~15 kpc surrounding SGR 1900+14, and a young stellar cluster was recently identified in close proximity to SGR 0526–66 (Klose *et al.* 2004). These findings suggest that these SGRs may have very massive progenitors.

Table 14.5. *Optical and IR magnitudes of SGRs and AXPs*

Source	V	R	I	J	H	K	K_s
SGR 0526−66	>27.1	–	>25	–	–	–	–
SGR 1627−41	–	–	–	>21.5	>19.5	–	>20.0
SGR 1806−20[a]	–	–	–				21.0 – 22
SGR 1900+14	–	–	–	>22.8	–	–	>20.8
CXOU 010043.1−721134	–	–	–	–	–	–	–
4U 0142+61	25.6	25.0	23.8	–	–	19.6	20.1
1E 1048.1−5937[a]	–	>24.8	26.2	21.7	20.8	–	19.4−21.3
1RXS J170849−400910[b]	–	–	–	20.9	18.6	–	18.3
XTE J1810−197	–	–	>24.3	–	22.0	–	20.8
1E 1841−045[b]	–	>23	–	–	–	–	19.4
AX J1845−0258	–	–	–	–	–	–	–
1E 2259+586[a]	–	>26.4	>25.6	>23.8	–	–	20.4−21.7

[a] Has shown variability in the infrared
[b] Candidate counterpart
References (SGR 0526) Kaplan *et al.* 2001; (SGR 1627) Wachter *et al.* 2004; (SGR 1806) Kosugi *et al.* 2005; (SGR 1900) Kaplan *et al.* 2002; (CXO 0100) Lamb *et al.* 2002; (4U 0142) Hulleman *et al.* 2000, 2004; (1E 1048) Wang & Chakrabarty 2002; Israel *et al.* 2002; Durant, van Kerkwijk & Hulleman 2003; (RXS 170849) Israel *et al.* 2003b; (XTE 1810) Israel *et al.* 2004; (1E 1841) Wachter *et al.* 2004, Mereghetti *et al.* 2001; (1E 2259) Hulleman *et al.* 2001; Kaspi *et al.* 2003; Israel *et al.* 2003a

Finally, one other avenue open to pursue in optical/IR studies of magnetar candidates is the measurement of proper motion. Hulleman *et al.* (2000) have already placed a 2σ upper limit of 0.03 arcsec yr^{-1} – corresponding to $1400\,(D/10\ \mathrm{kpc})\ \mathrm{km\ s}^{-1}$ – on the proper motion of 4U 0142+61. Some models of magnetar formation (e.g., Duncan & Thompson 1992) suggest high kick velocities incurred at birth. The precise astrometry available with optical/IR observations will allow tighter constraints on the proper motions of the AXPs and SGRs, and shed further light on the physical connection of these sources to nearby SNR.

14.7 Magnetar model

We will organize our discussion of the magnetar model around its predictions, the extent to which they have been verified or falsified, and outline areas in which further advancement of theory is needed to make quantitative comparisons with data. The basic idea of the model is that the variable X-ray emission – the bursts lasting up to $\sim$1000 s and the transient changes in persistent emission observed up to $\sim$1 yr – are powered by the decay of the star's magnetic field. A rms field exceeding $\sim 10^{15}$ G is needed to supply an output of 10^{35} erg s^{-1} extending over 10^4 yr.

We begin by recalling that the AXPs as a group are systematically much brighter thermal X-ray sources than radio pulsars of the same characteristic age. The evidence for repeated bulk heating of the crust, and the relative brightness of the X-ray emission, has bolstered the suggestion that magnetic field decay is the main energy source for that emission. That conclusion is most secure in those magnetar candidates that show large transient swings in X-ray brightness over a period of years. Some SGRs and AXPs have nearly flat 2–10 keV energy spectra, which manifestly cannot be powered primarily by cooling or by spin down. Indeed, it has recently been found that the bolometric output of some AXPs is dominated by a hard, rising energy spectrum up to an energy of (at least) $\sim$100 keV (Kuiper *et al.* 2004).

14.7.1 Magnetic field decay

Several physical effects become important as the magnetic field of a neutron star is raised above $\sim 10^{14}$–10^{15} G (Thompson & Duncan 1996). First, a field stronger than $\sim 10^{15}$ G will, as it decays, significantly raise the temperature of the deep crust and core of the star at an age of $\sim 10^3$–10^4 yr. The rate of drift of the magnetic field and the entrained charged particles can, as a result, be significantly accelerated. Second, elastic stresses in the crust of the star are no longer able to withstand a large departure from magnetostatic equilibrium. The crustal lattice has a finite shear modulus μ, and when the yield strain $\theta_{\rm max}$ is exceeded the lattice will respond in an irreversible manner. The characteristic magnetic field strength is $B_{\rm yield} = (4\pi\theta_{\rm max}\mu)^{1/2} = 2 \times 10^{14}(\theta_{\rm max}/10^{-3})^{1/2}$ G. For example, Hall drift of the magnetic field has qualitatively different consequences in the crust when the field is stronger than $B_{\rm yield}$. The mean electron drift motion that supplies the current also advects the magnetic field, and causes stresses in the crust to slowly build up. When $B > B_{\rm yield}$, irregularities in the magnetic field are damped directly by crustal yielding, rather than by a non-linear coupling to high-frequency modes which suffer ohmic damping. Such a "Hall cascade" plays a key role in facilitating the decay of $\sim 10^{12}$ G magnetic fields in the crust (Goldreich & Reisenegger 1992).

A third, related effect is that when the star contains a strong $\sim 10^{15}$ G toroidal magnetic field, the rate of ejection of magnetic helicity from the interior can be high enough to induce a significant twist on the external poloidal field lines. This effect is especially important following periods of X-ray burst activity, and has been suggested as a source of persistent increases in spin-down rate, and as a source of external heating (Thompson, Lyutikov & Kulkarni 2002).

The heat flux through the stellar crust is also modestly enhanced in strong magnetic fields (van Riper 1988; Heyl & Hernquist 1998; Potekhin & Yakovlev 2001), although the composition plays a more important role in determining the thermal transparency. In particular, the heat flux can be up to several times larger if the surface has a light-element (H or He) composition, than if it is iron (Chabrier, Potekhin & Yakovlev 1997; Heyl & Hernquist 1997).

Key early papers on the microscopic transport of the magnetic field in neutron star interiors are by Haensel, Urpin & Iakovlev (1990), Goldreich & Reisenegger (1992), and Pethick (1992). Sweeping of magnetic fluxoids out of a superconducting core by the interaction with the superfluid vortices has been considered as a mechanism of magnetic field decay in radio pulsars (Ruderman, Zhu & Chen 1998). It may, however, be suppressed if the field is stronger than $\sim 10^{15}$ G and the fluxoids are tightly bunched. Integrated models of magnetar evolution and cooling have been calculated by Thompson & Duncan (1996), Heyl & Kulkarni (1998), Colpi, Geppert & Page (2000), Kouveliotou *et al.* (2003). Heating can have subtle effects on the superfluid properties of the star; in particular, if the critical temperature for the onset of neutron pairing in the core is less than several $\times\ 10^8$ K, then heating will force a significant delay in the transition to core neutron superfluidity (Arras, Cumming & Thompson 2004). After this transition, a neutron star undergoes a significant drop in surface X-ray flux (Yakovlev *et al.* 2001).

It should be emphasized that the behavior of the SGRs and AXPs has not been observed over baselines longer than ~20–30 yr. Some AXPs such as 1E 2259+586 and 4U 0142+61 have sustained bright ($\sim 10^{35}$ erg s^{-1}) thermal X-ray emission over this period of time, which is comparable to or longer than the thermal conduction time across the crust (Gnedin *et al.* 2001). Their duty cycle as bright X-ray sources is presently unknown.

14.7.2 Mechanism for magnetar bursts

Both the short and the long outbursts of magnetars are hypothesized to arise from the direct injection of energy into the magnetosphere, through a rearrangement of the magnetic field and the formation and dissipation of strong localized currents. Magnetar flares are distinguished from solar flares in two key respects (Thompson & Duncan 1995): the magnetic field is anchored in a rigid medium (the lower crust) which has some finite shear strength; and the energy density in the magnetic field is high enough that rapid thermalization of energized charged particles can be expected. (In the short SGR bursts, the rate of release of energy is not rapid enough to effect complete thermalization and drive the photon chemical potential to zero, but it is at the onset of the giant flares; Thompson & Duncan 2001.)

The crust provides a plausible site for the initial loss of equilibrium that triggers an outburst. For example, the relaxation behavior observed over a period of weeks in SGR 1806–20 (Palmer 1999) suggests that the release of energy in successive short SGR bursts is limited by inertial and frictional forces. In addition, bursts similar to short SGR bursts are observed to begin at least two larger events: the August 27 and August 29 flares of SGR 1900+14. The duration of the short bursts is comparable to the time for a torsional deformation to propagate vertically across the crust in a $\sim 10^{14}$ G poloidal magnetic field. The ability of the interior of the star to store much stronger toroidal magnetic fields than the exterior provides a hint that the ensuing burst is driven primarily by a loss of equilibrium in the crust, rather than by reconnection and simplification of non-potential magnetic fields outside the star. Nonetheless, it is likely that both effects will occur in concert, given the magnitude of the energies released.

The initial spikes of the giant flares have been associated with expanding fireballs composed of $e^{\pm}$ pairs and non-thermal gamma rays (Paczyński 1992), and the pulsating tails with thermalized energy which remain confined close to the neutron star by its magnetic field (Thompson & Duncan 1995). In the spikes, the combination of rapid (<0.01 s) variability with a hard non-thermal spectrum points to a low baryon contamination. The argument that most of the flare energy is deposited in the first second comes from (i) the near coincidence between the energy of the initial spike and the energy radiated over the remaining ~300 s of the burst; and (ii) the smooth adiabatic simplification of the pulse profile in the tail of the 1998 August 27 flare, which shows no evidence for secondary impulsive injections of energy that would be associated with a continuing substantial reorientation of the magnetic field. The lower bound on the magnetic moment implied by the confinement of $\sim 10^{44}$ erg is $B R_{\rm NS}^3 \simeq 10^{14}$ G (Thompson & Duncan 2001).

Large-scale deformations of the crust are constrained by its high hydrostatic pressure, but varying implications have been drawn for its elastic response to evolving magnetic stresses. One possibility is that the crust develops a dense network of small-scale (but macroscopic) dislocations, and that the resulting fast ohmic heating of the uppermost layers of the star is what powers the extended afterglow observed following SGR flares (Lyubarsky, Eichler & Thompson 2002). Alternatively Jones (2003) and Lyutikov (2003) raise the possibility that the response of the crust may be more gradual and purely plastic, which would force the main source of energy for an X-ray flare into the magnetosphere. Evidence that the shear deformations of the crust are spatially concentrated comes from the observation of hard thermal X-ray emission – covering $\sim$1% of the surface area of the star – right after the August 29 flare of SGR 1900+14 (Ibrahim *et al.* 2001) and during the transient brightening of 1E 2259+586 (Woods *et al.* 2004).

14.7.3 Burst spectral evolution and afterglow

A trapped thermal fireball (in which the photons have a Planckian distribution at a temperature $\sim$1 MeV) is very optically thick to scattering, given the high density of electron–positron pairs. It releases energy through the contraction of its cool surface – in contrast to the cooling of a material body of fixed surface area. Thus, the X-ray flux is predicted to drop rapidly toward the end of a flare, when the external fireball evaporates (Thompson & Duncan 1995). A simple model of a contracting spherical surface, bounding a fireball with a modest temperature gradient, provides an excellent fit to the 1998 August 27 flare (Feroci *et al.* 2001).

The temperature of the fireball surface is also buffered by a quantum electrodynamic effect: X-ray photons propagating through intense magnetic fields are able to split in two or merge together (Adler 1971). The rate of splitting grows rapidly with photon frequency, but loses its dependence on magnetic field strength when $B \gg B_{\rm QED} = 4.4 \times 10^{13}$ G (Thompson & Duncan 1992). Energy and momentum are both conserved in this process, with the consequence that only one polarization mode can split. As a result, splitting freezes out below a characteristic blackbody temperature of $\sim$12 keV in super-QED magnetic fields (Thompson & Duncan 1995). This is, very nearly, the temperature observed during an extended period of flux decline in the pulsating tail of the 1998 August 27 flare (Feroci *et al.* 2001). In some geometries, double Compton scattering can also be a significant source of photon seeds near the scattering photosphere (Lyubarsky 2002).

The rate of radiative conduction through an electron gas is greatly increased by the presence of a strong magnetic field, which suppresses the opacity of the extraordinary polarization mode (Sil'antev & Iakovlev 1980; Lyubarskii 1987). Thus, the high luminosities of the intermediate flares, and the pulsating tails of the giant flares, also point to the presence of 10^{14}–10^{15} G magnetic fields (Paczyński 1992). This effect can, however, be suppressed by mode exchange near the stellar surface (Miller 1995), and probably requires a confining magnetic field (Thompson & Duncan 1995).

One clear prediction of the trapped fireball model is that $\sim$1% of the trapped energy will be conducted into the surface of the neutron star over the duration of the fireball phase (Thompson & Duncan 1995). This energy can explain the prompt afterglow observed immediately following the intermediate burst on 1998 August 29 (Ibrahim *et al.* 2001), but heat conducted into the crust cannot supply afterglow longer than $\sim 10^4$ s following the burst. The relative importance of such conductive heating for the observed afterglow – as compared with direct bulk heating and continuing relaxation of currents outside the star – is not well understood.

A super-QED magnetic field has other interesting radiative effects. The gyrational energy of a proton or other ion can fall in the keV range: $\hbar eB/m_{\rm p}c = 6.3\,(B/10^{15}\ {\rm G})$ keV, possibly allowing for the formation of absorption features (Zane *et al.* 2001; Özel 2003; but see Ho & Lai 2003 for a discussion of how polarization mode switching can drastically reduce the equivalent width of such a line feature). When the radiative flux out of the star exceeds $\sim 10^{36}$

erg s^{-1}, this means that the radiative force applied at the cyclotron resonance can exceed the force of gravity (Thompson *et al.* 2002). The same large resonant cross-section also allows an ion component of a persistent electric current flowing outside the star to have a measurable influence on the X-ray spectrum through cyclotron scattering.

Searches for X-ray lines during SGR bursts are potentially diagnostic of the burst mechanism and the strength of the magnetic field. Short, low-energy SGR bursts are probably highly localized on the neutron star surface. The magnetosphere is probably at a higher temperature than the surface during a burst, and so proton cyclotron features may be seen in emission in the keV range.

14.7.4 Electrodynamics

Highly non-thermal persistent X-ray emission is observed in the actively bursting SGR sources, and from it one infers the presence of magnetospheric currents much stronger than the rotationally driven Goldreich–Julian current. Although the spin-down power in SGRs 1806−20 and 1900+14 peaks at values approaching the X-ray luminosity during their periods of most extreme spin-down torque, there is no correlation between the two. To power the X-ray emission, the energy that must be dissipated per Goldreich–Julian particle on the open field lines exceeds $\sim 10^8$ MeV. By contrast, if the flux of particles close to the stellar surface is normalized to $cB_{\rm NS}/4\pi e R_{\rm NS}$, then the energy dissipated per particle need not exceed ~ 100 MeV (e.g., the binding energy of an ion to the star).

The persistent changes in X-ray pulse profile observed following SGR bursts could be caused by a change in the emission pattern; or by a change in the distribution of particles that re-scatter the X-rays higher in the magnetosphere (e.g., at ~ 10 neutron star radii where the cyclotron resonance of the electrons is in the keV range). For example, in a magnetosphere threaded by persistent electric currents, the optical depth at the cyclotron resonance of the current-carrying charges is of the order of unity over a continuous range of frequencies (if the poloidal magnetic field is twisted through ~ 1 radian; Thompson *et al.* 2002). In such a situation, an X-ray photon will undergo a significant shift in frequency as it escapes the magnetosphere.

A non-thermal component of the X-ray spectrum is not always needed to fit the persistent emission of the AXPs: given the narrow bandpass being fit, the convolution of two blackbodies sometimes gives an acceptable fit in soft-spectrum sources (Israel *et al.* 2000). Measurements of the AXP emission above several keV are crucial to understanding the physical origin of the high-energy excess.

14.7.5 Torque behavior

In addition to the basic predictions of rapid spin down in the SGRs and bursting activity in the AXPs, theoretical work on magnetars has anticipated some other observed properties of the magnetar candidates. The deduction that the magnetic fields of the SGRs are time-variable suggested that the spin down would also be highly variable (Thompson & Blaes 1998). If the magnetic field is variable on very short timescales, then torque variations will arise from a continuous flux of high-frequency Alfvén waves and particles away from the star (see also Harding, Contopoulos & Kazanas 1999; Thompson *et al.* 2000). This is the most plausible explanation for the transient spin down observed in SGR 1900+14 at the time of the 1998 August 27 flare. But the apparent time lag between bursting activity and large torque variations in SGRs 1900+14 and 1806−20, the observation of long-term (>year) and persistent increases in torque, and the persistence of the changes in X-ray pulse profile following outbursts, are more consistent with the presence of large-scale static currents on the closed magnetospheric field lines (Thompson *et al.* 2002). Torque variations could, in principle, also arise from a change in the fraction of open field lines due to the suspension of a modest amount of material inside the speed-of-light cylinder (Ibrahim *et al.* 2001); but it is difficult to see why such material would not be redistributed or expelled immediately following a bright SGR flare.

The magnetar model is conservative in the sense that the AXPs and SGRs are assumed to be standard neutron stars, distinguished only from radio pulsars (including the high-magnetic field tail of the pulsar population) by the presence of a strong wound-up magnetic field *inside* the star (Thompson & Duncan 2001), and by the active transfer of magnetic helicity across the stellar surface. One way of testing this basic hypothesis is to search for glitches associated with a superfluid component. Large glitches will be triggered in slowly rotating magnetars via the release of magnetic stresses in the crust – either due to sudden unpinning (Thompson & Duncan 1993) or to plastic deformations of the crust during which vortices remain pinned (Thompson *et al.* 2000). In the second case, spin down of the superfluid occurs if the crust is twisted adiabatically about an axis that is tilted with respect to the rotation axis: more superfluid vortices move outward away from the rotation axis than move toward it. A large glitch observed in the AXP 1E 2259+586 (Kaspi *et al.* 2003; Woods *et al.* 2004) provides a nice test of these ideas. The year-long soft X-ray afterglow observed following the glitch suggests that the crust was subject to a smooth, large-scale deformation. Related effects occurring in a superfluid core have also been implicated in the fast timing noise of the SGRs (Arras *et al.* 2004).

14.8 Future directions

We close by outlining some major unsolved problems associated with the SGRs and AXPs.

(1) *What is the birth rate of AXPs and SGRs compared with radio pulsars? What fraction of neutron stars go through a phase of strong magnetic activity?* The selection of magnetar candidates – through their burst activity as SGRs, or through their persistent X-ray pulsations as AXPs – is limited by sensitivity. There are $\sim$10 magnetar candidates in our Galaxy and a conservative estimate of their average age is $\sim 10^4$ years as derived from their spin down. Thus, a *lower limit* to the Galactic birth rate is 1 per 1000 years (Kouveliotou *et al.* 1994; van Paradijs *et al.* 1995), or $\sim$10% of the radio pulsar birth rate (Lyne *et al.* 1998). The birth rate that we infer for AXPs and SGRs depends critically upon the efficiency with which we detect them. The efficiency of detecting low-luminosity bursts such as those from 1E 2259+586 is quite low. The overall efficiency of identifying magnetar candidates has not yet been quantified. The observation of transitions to persistent low-luminosity states suggests that it may be lower than previously thought; but these transient sources could be older on average than the more persistent sources. Given the number of selection effects, we cannot rule out a detection efficiency as low as $\sim$10%, and a birth rate comparable to that of radio pulsars.

(2) *What fraction of short GRBs are extra-Galactic SGR giant flares?* The initial peak of the of 2004 December 27 could have been detected with BATSE out to $\sim 40\,d_{15}$ Mpc (Hurley *et al.* 2005; Palmer *et al.* 205) where d_{15} is the distance to the SGR in units of

15 kpc. The actual contribution of gaint flares to the short GRB rate depends on the maximum energy with which flares are emitted. The observable volume from which flares of (isotropic) energy $E \leq E_{\rm max}$ could be detected scales as $E_{\rm max}^{3/2}$, whereas the upward extrapolation of the SGR burst energy distribution implies that the number of flares emitted *per source* decreases as $E(\mathrm{d}N/\mathrm{d}E) \sim E^{-2/3}$. With these scalings, the fraction of the short GRB population supplied by SGR flares increases as $\sim E_{\rm max}^{0.8}$. The absence of bright galaxies in a handful of short GRBs error boxes (Nakar *et al.* 2005) and the lack of a concentration of short GRBs toward the Virgo cluster (Pamer *et al.* 2005) suggests that the fraction of short GRBs powered by events similar to the December 27 flare may be smaller than $\sim$0.2. However, it is conceivable that yet more energtic flares are released by some distant SGR sources. Swift observations of short GRBs over the coming years will addsess the issue.

(3) *Why are the spin periods of AXPs and SGRs strongly clustered in an interval of 5–12 s?* This clustering suggests a real upper cutoff of $\sim$12 s in the period distribution (Psaltis & Miller 2002). In fact, the observed periods lie close to the upper envelope of the period distribution of radio pulsars, and are consistent with a reduction in torque following the termination of active pair cascades on open magnetic field lines (Thompson *et al.* 2002). Field decay could, in principle, also play a role in determining the observed range of spin periods (Colpi *et al.* 2000). The large $\sim 2 \times 10^5$ yr characteristic age of 1E 2259+586 (which resides in the $\sim 10^4$ yr old SNR CTB 109) provides a strong hint of torque decay in that particular AXP.

(4) *Do the SGRs, AXPs and high B-field radio pulsars form a continuum of magnetic activity, or are they different phases/states of a more uniform class of object?* The heating of a neutron star by a decaying magnetic field is unfortunately sensitive to the configuration of the field. Arras *et al.* (2004) consider a toroidal configuration, and show that as the field strength is reduced from $\sim 10^{15}$ G down to $\sim 10^{14}$ G, the soft X-ray luminosity interpolates between the levels characteristic of AXPs and of radio pulsars. Indeed the minimal magnetic field needed to power N flares of the energy of the 2004 December 27 event over the lifetime of SGR 1806−20 is $2 \times 10^{15}(N/10)^{1/2}$ G. Thus, the observed X-ray emission of middle-aged radio pulsars is consistent with the hypothesis that the AXPs and SGRs have much stronger internal magnetic fields. However, a few AXPs have shown transitions to low-luminosity states where their X-ray output is reduced by a factor $\sim$100 on a timescale of years. Much further exploration of the interplay between magnetic field transport, surface cooling, and superfluidity is required. For example, the relaxation of the crustal magnetic field caused by electron captures on the heavy nuclei in the neutron-drip solid has not yet been explored.

(5) *What can we learn of neutron star matter from observations of SGR and AXP activity? Are SGRs and AXPs fundamentally neutron stars?* The detection of glitches in two AXPs indicates the presence of a superfluid component, whose pinning behavior (as deduced from the post-glitch response) is similar to that observed in radio pulsars. Burst afterglows have the potential to probe the outer layers of magnetars. Lyubarsky *et al.* (2002) argue that the extended afterglow observed following the 1998 August 27 flare is more consistent with the strongly stratified outer crust of a neutron star, than it is with a nearly constant density quark star. It has been suggested that the high luminosities of SGR flares are a result of QCD confinement near the surface of a bare quark star (Usov 2001); but the theoretical motivation for such objects is problematic (e.g., Akmal *et al.* 1998).

(6) *What is the initial spin period of magnetars? Could some magnetars (with millisecond periods) be connected to GRBs?* It is not known whether magnetars and radio pulsars

are distinguished by the initial rotation of the neutron star, or alternatively by the stability properties of the magnetic field. It is possible that all nascent neutron stars develop $\sim 10^{15}$ G magnetic fields through fluid instabilities; but it is also likely that the large-scale order of the magnetic field is correlated with the speed of the rotation. For example, a large-scale helical dynamo is possible when the rotation period is comparable to the timescale of the convective motions, which is $\sim$3 ms for Ledoux convection during the 10 s Kelvin phase of the neutron star (Duncan & Thompson 1992). This led to the prediction of a class of energetic supernovae in which the neutron core deposits $\sim 10^{51} - 10^{52}$ erg of rotational energy by magnetic dipole radiation and later forms a strongly magnetic stellar remnant. The spin energy can be tapped even on the short timescale for the shock to emerge from a compact CO core, when the effects of a neutrino-driven wind are taken into account (Thompson, Chang & Quataert 2004). Whether a proto-magnetar is also a viable source of gamma-ray burst emission (as suggested independently by Usov 1992 and Duncan & Thompson 1992) is more problematic: the net mass released during neutrino cooling is a few orders of magnitude larger than what will quench gamma-ray emission from the expanding relativistic wind.

(7) *What is the evolutionary sequence of magnetars? How do very young ($<10^3$ years) and older ($>10^5$ years) systems manifest themselves? Are there old magnetars in our local neighborhood? in globular clusters?* The limited $\sim 10^4$ yr lifetime of SGR flare activity is a significant constraint on models of magnetic field decay. This lifetime is determined by the microscopic transport processes acting on the magnetic field; *and* by the manner in which the star falls out of magnetostatic equilibrium – about which little is presently understood. Hall drift is not sensitive to temperature and can, for that reason, continue to power a low level of X-ray emission ($\sim 10^{33}$ erg s^{-1}) at an age of $\sim 10^6$ yrs. Rotational energy deposition also becomes more significant at lower flux levels (e.g., Ruderman *et al.* 1998). Evidence has recently been found for broad absorption features in the spectra of a few soft-spectrum dim isolated neutron stars (van Kerkwijk *et al.* 2004; Haberl 2004) perhaps indicative of proton cyclotron absorption in a magnetic field of several $\times 10^{13}$ G (Zane *et al.* 2001; Ho & Lai 2003; Özel 2003). Their evolutionary relation to radio pulsars and magnetars would be elucidated by a detailed cross comparison with the low-luminosity states of magnetars. Detailed timing measurements are likely to be essential to unraveling these interconnections.

(8) *Do magnetars exist in binary systems? If so, how does a magnetar react to accretion? Or, alternatively, do magnetars form only through the sacrifice of a binary companion?* The modest number of binary neutron stars that may have 10^{14} G magnetic fields (e.g., GX 1+4, 2S 0114+650; Li & van den Heuvel 1999) is probably inconsistent with a magnetar birth rate exceeding $\sim$10% of the total neutron star birth rate, unless the magnetars have systematically larger kicks or their dipole fields decay significantly above an age of $\sim 10^5$ yr. There is some evidence for large kicks in SGRs 0525−66 and 1900+14 (Cline *et al.* 1982; Thompson *et al.* 2000), but not in any of the AXPs. The limited $\sim 10^4$ yr lifetime of AXPs as bright X-ray sources makes it unlikely to see bright thermal X-ray emission from magnetars formed in high-mass binaries.

Acknowledgements

We thank Shri Kulkarni for his help and advice in planning this review, and for detailed comments on the manuscript. We also thank Robert Duncan, Fotis Gavriil, Nanda Rea and Marten van Kerkwijk for their comments.

References

Adler, S. L., 1971 *Ann. Phys.* **67**, 599
Akerlof, C., *et al.* 2000, *ApJ*, **542**, 251
Akmal, A., Pandharipande, V. R., & Ravenhall, D. G. 1998, *Phys. Rev. C*, **58**, 1804
Alcock, C., Farhi, E., & Olinto, A. 1986, *Phys. Rev. Lett.* **57**, 2088
Aptekar, R. L., *et al.* 2001, *ApJ*, **137**, 227
Ardelyan, N. V., *et al.* 1987, *Soviet Astr.*, **31**, 398
Arras, P., Cumming, A., & Thompson, C. 2004, *ApJ*, **608**, L49
Arzoumanian, Z., Nice, D. J., Taylor, J. H., & Thorsett, S. E. 1994, *ApJ*, **422**, 671
Atteia, J. L., *et al.* 1987, *ApJ*, **320**, L105
Baade, W. & Zwicky, F. 1934, *Proc. Nat. Acad. Sci.* **20**, 259
Basko M. M. & Sunyaev, R. A. 1975, *A&A*, **42**, 311
Baykal, A. & Swank, J. 1996, *ApJ*, **460**, 470
Bird, A. J., *et al.* 2004, *ApJ*, **607**, L33
Blaes, O., Blandford, R. D., Goldreich, P., & Madau P. 1989, *ApJ*, **343**, 839
Cameron, P. B., *et al.* 2005, *Nature*, **434**, 1112
Camilo, F., *et al.* 2000, *ApJ*, **541**, 367
Chabrier, G., Potekhin, A. Y., & Yakovlev, D. G. 1997, *ApJ*, **477**, L99
Chatterjee, P. & Hernquist, L. 2000, *ApJ*, **543**, 368
Chatterjee, P., Hernquist, L., & Narayan, R. 2000, *ApJ*, **534**, 373
Cheng, B., Epstein, R. I., Guyer, R. A., & Young, C. 1996, *Nature*, **382**, 518
Cline, T. L., *et al.* 1982, *ApJ*, **255**, L45
  2000, *ApJ*, **531**, 407
Colpi, M., Geppert, U., & Page, D. 2000, *ApJ*, **529**, L29
Corbel, S. & Eikenberry, S. S. 2004, *A&A*, **419**, 191
Corbel, S., Chapuis, C., Dame, T. M., & Durouchoux, P. 1999, *ApJ*, **526**, L29
Corbet, R. H. D., *et al.* 1995, *ApJ*, **443**, 786
Dall'Osso, S., *et al.* 2003, *ApJ*, **499**, 485
den Hartog, P., *et al.* 2004, *ATEL* 293
Duncan, R. 1998, *ApJ*, **498**, L45
  2001, in 20th *Texas Symp. on Relativistic Astrophysics*, Eds. J. C. Wheeler & H. Martel, *AIP*, vol. 586, p. 495
Duncan, R. & Thompson, C. 1992, *ApJ*, **392**, L9
Durant, M., van Kerkwijk, M. H., & Hulleman, F. 2003, in *Young Neutron Stars and their Environment*, IAU Symp. **218**, Eds. F. Camillo & B. M. Gaensler, p. 251 (astro-ph/0309801)
Eikenberry, S. S., *et al.* 2001, *ApJ*, **563**, L133
Fahlman G. G. & Gregory, P. C. 1981, *Nature*, **293**, 202
Fenimore, E. E., *et al.* 1981, *Nature*, **289**, 42
Fenimore, E., Laros, J. G., & Ulmer, A. 1994, *ApJ*, **432**, 742
Feroci, M., *et al.* 1999, *ApJ*, **515**, L9
Feroci, M., Hurley, K., Duncan, R. C., & Thompson, C. 2001, *ApJ*, **549**, 1021
Feroci, M., *et al.* 2003, *ApJ*, **596**, 470
  2004, *ApJ*, **612**, 408
Frail, D., Kulkarni, S. & Bloom, J. 1999, *Nature*, **398**, 127
Fuchs, Y., *et al.* 1999, *A&A*, **350**, 891
Gaensler, B. M. & Johnston, S. 1995, *MNRAS*, **277**, 1243
Gaensler, B. M., Gotthelf, E. V., & Vasisht, G. 1999, *ApJ*, **526**, L37
Gaensler, B. M., Slane, P. O., Gotthelf, E. V., & Vasisht, G. 2001, *ApJ*, **559**, 963
Gaenslar, B. M., *et al.* 2005, *Nature*, **434**, 1104
Gavriil, F. & Kaspi, V. M. 2002, *ApJ*, **567**, 1067
  2004, *ApJ*, **607**, 959
Gavriil, F. P., Kaspi, V. M., & Woods, P. M. 2002, *Nature*, **419**, 142
  2004, *ApJ*, **607**, 959
Gelfand, J. D., *et al.* 2005, *ApJ* in press (astro-ph/0503269)
Giles, A. B., *et al.* 1996, *ApJ*, **469**, L25
Gnedin, O. Y., Yakovlev, D. G., & Potekhin, A. Y. 2001, *MNRAS*, **324**, 725

Göğüş, E., *et al.* 1999, *ApJ*, **526**, L93
2001, *ApJ*, **558**, 228
2002, *ApJ*, **577**, 929
Goldreich, P. & Reisenegger, A. 1992, *ApJ*, **395**, 250
Golenetskii, S. V., Ilyinskii, V. N., Mazets, E. P. 1984, *Nature*, **307**, 41
Gotthelf, E. V. & Vasisht, G. 1998, *New Astr.*, **3**, 293
Gotthelf, E. V., *et al.* 2002, *ApJ*, **564**, L31
Gotthelf, E. V., Halpern, J. P., Buxton, M., & Bailyn, C. 2004, *ApJ*, **605**, 368
Gregory, P. C. & Fahlman, G. G. 1980, *Nature*, **287**, 805
Guidorzi, C., *et al.* 2004, *A&A*, **416**, 297
Haberl, F. 2004, *Mem. S. A. It.*, **75**, 454
Haensel, P., Urpin, V. A., & Iakovlev, D. G. 1990, *A&A*, **229**, 133
Harding, A. K., Contopoulos, I., & Kazanas, D. 1999, *ApJ*, **525**, L125
Hellier, C. 1994, *MNRAS*, **271**, L21
Hewish, A., *et al.* 1968, *Nature*, **217**, 709
Heyl, J. S. & Hernquist, L. 1997, *ApJ*, **489**, L67
1998, *MNRAS*, **300**, 599
1999, *MNRAS*, **304**, L37
Heyl, J. S. & Kulkarni, S. R. 1998, *ApJ*, **506**, L61
Ho, W. C. G. & Lai, D. 2003, *MNRAS*, **338**, 233
Hulleman, F., *et al.* 2001, *ApJ*, **563**, L49
Hulleman, F., van Kerkwijk, M. H. & Kulkarni, S. R. 2000, *Nature*, **408**, 689
2004, *A&A*, **416**, 1037
Hurley, K. 1986, Talk presented at the Gamma-Ray Stars Conference in Taos, NM
Hurley, K. J., McBreen, B., Rabbette, M., & Steel, S. 1994, *A&A*, **288**, L49
Hurley, K., *et al.* 1999a, *Nature*, **397**, 41
1999b, *ApJ*, **510**, L107
1999c, *ApJ*, **510**, L111
1999d, *ApJ*, **523**, L37
2005, *Nature*, **434**, 1098
Ibrahim, A., *et al.* 2001, *ApJ*, **558**, 237
Ibrahim, A. I., Swank, J. H., & Parke, W. 2003, *ApJ*, **584**, L17
Ibrahim, A. I., *et al.* 2004, *ApJ*, **609**, L21
Inan, U. S., *et al.* 1999, *GeoRL*, **26**, 3357
Israel, G. L., *et al.* 2000, *ApJ*, **560**, L65
2002, *ApJ*, **580**, L143
2003a, in *Young Neutron Stars and their Environment*, IAU Symp. **218**, Eds. F. Camillo & B. M. Gaensler, p. 247, (astro-ph/0310482)
2003b, *ApJ*, **589**, L93
2004, *ApJ*, **603**, L97
2005, *ApJ*, **628**, L53
Iwasawa, K., Koyama, K. & Halpern, J. P. 1992, *Publ. Astron. Soc. Japan*, **44**, 9
Jones, P. B. 2003, *ApJ*, **595**, 342
Juett, A. M., Marshall, H. L., Chakrabarty, D., & Schulz, N. S. 2002, *ApJ*, **568**, L31
Kaplan, D. L., *et al.* 2001, *ApJ*, **556**, 399
Kaplan, D. L., Kulkarni, S. R., Frail, D. A., & van Kerkwijk, M. H. 2002, *ApJ*, **566**, 378
Kaspi, V. M. & Gavriil, F. P. 2003, *ApJ*, **596**, L71
Kaspi, V. M., Lackey, J. R., & Chakrabarty, D. 2000, *ApJ*, **537**, L31
Kaspi, V. M., *et al.* 2001, *ApJ*, **558**, 253
2003, *ApJ*, **588**, L93
Katz, J. I., Toole, H. A., & Unruh, S. H. 1994, *ApJ*, **437**, 727
Kerkwijk, M. H. van, *et al.*1995, *ApJ*, **444**, L33
2004, ApJ, **608**, 432
Kern, B. & Martin, C. 2002, *Nature*, **417**, 527
Klose, S., *et al.* 2004, *ApJ*, **609**, L13
Koshut, T. M., *et al.* 1996, *ApJ*, **463**, 570
Kosugi, G., Ogaswara, R., and Terada, H. 2005, *ApJ*, **623**, L125

Kothes, R., Uyaniker, B., & Aylin, Y. 2002, *ApJ*, **576**, 169
Kouveliotou, C., *et al.* 1987, *ApJ*, **322**, L21
1994, *Nature*, **368**, 125
1998a, *Nature*, **393**, 235
1998b, *GCN Circ.* 107
2003, *ApJ*, **596**, L79
Koyama, K., Hoshi, R., & Nagase, F. 1987, *PASJ*, **39**, 801
Kuiper, L., Hermsen, W., & Mendez, M. 2004, *ApJ*, **613**, 1173
Kulkarni, S. R., *et al.* 2003, *ApJ*, **585**, 948
Lamb, R. C., Fox, D. W., Macomb, D. J., & Prince, T. A. 2002, *ApJ*, **574**, L29
Lamb, D., *et al.* 2003a, *GCN Circ.* 2351
Lamb, D., Prince, T. A., Macomb, D. J., & Majid, W. A. 2003b, *IAU Circ.* 8220
Laros, J., *et al.* 1987, *ApJ*, **320**, L111
LeBlanc, J. M. & Wilson, J. R. 1970, *ApJ*, **161**, 541
Lenters, G. T., *et al.* 2003, *ApJ*, **587**, 761
Lewin, W. H. G, van Paradijs, J., & Taam, R. E. 1993, *Space Sci. Rev.*, **62**, 223
Li, X.-D. & van den Heuvel, E. P. J. 1999, *ApJ*, **513**, L45
Livio, M. & Taam, R. E. 1987, *Nature*, **327**, 398
Lyne, A. G., *et al.* 1998, *MNRAS*, **295**, 743
Lyubarskii, Y. E. 1987, *Astrophysics*, **25**, 277
Lyubarksy, Y. 2002, *MNRAS*, **332**, 199
Lyubarsky, E., Eichler, D., & Thompson, C. 2002, *ApJ*, **580**, L69
Lyutikov, M. 2003, *MNRAS*, **346**, 540
Manchester, R. N. 2004, *Science*, **304**, 542
Marsden, D. & White, N. E. 2001, *ApJ*, **551**, L155
Mazets, E. P. & Golenetskii, S. V. 1981, *Ap&SS*, **75**, 47
Mazets, E. P., *et al.* 1979, *Nature*, **282**, 587
1999a, *Astron. Lett.*, **25**, 635
1999b, *ApJ*, **519**, L151
McLaughlin, M. A., *et al.* 2003, *ApJ*, **591**, L135
Mereghetti, S. & Stella, L. 1995, *ApJ*, **442**, L17
Mereghetti, S., Cremonesi, D., Feroci, M., & Tavani, M. 2000, *A&A*, **361**, 240
Mereghetti, S., *et al.* 2001, *MNRAS*, **321**, 143
2004, *ApJ*, **608**, 427
2005a, *ApJ*, **624**, L105
2005b, *ApJ*, **628**, 938
Miller, M. C. 1995, *ApJ*, **448**, L29
Molkov, S. V., *et al.* 2004, *Astron. Lett.*, **30**, 534
Morii, M., Sato, R., Kataoka, J., & Kawai, N. 2003, *PASJ*, **55**, L45
Murakami, T., *et al.* 1994, *Nature*, **368**, 127
Nakar, E., Gal-Yam, A., Piran, T., & Fox, D. B. 2005, (astro-ph/0502148)
Norris, J. P., Hertz, P., Wood, K. S., & Kouveliotou, C. 1991, *ApJ*, **366**, 240
Olive, J-F., *et al.* 2003, in *Gamma-ray Burst and Afterglow Astronomy 2001*, Eds. G. R. Ricker & R. K. Vanderspek, AIP, **662**, p. 82
Oosterbroek, T., Parmar, A. N., Mereghetti, S., & Israel, G. L. 1998, *A&A*, **334**, 925
Özel, F. 2002, *ApJ*, **575**, 397
2003, *ApJ*, **583**, 402
Özel, F., Psaltis, D., & Kaspi, V. M. 2001, *ApJ*, **563**, 255
Paczyński, B., 1990, *ApJ*, **365**, L9
1992, *Acta Astron.*, **42**, 145
Palmer, D. M. 1999, *ApJ*, **512**, L113
2002, in *Soft Gamma Repeaters: The Rome 2001 Mini-Workshop*, Eds. M. Feroci & S. Mereghetti, Mem. S. A. It., **73**, no. 2, p. 578
Palmer, D., *et al.* 2005, *Nature*, **434**, 1107
Paradijs, J. van, Taam, R. E., & van den Heuvel, E. P. J. 1995, *A&A*, **299**, L41
Patel, S. K., *et al.* 2001, *ApJ*, **563**, L45
2003, *ApJ*, **587**, 367

Perna, R., Hernquist, L. E., & Narayan, R. 2000, *ApJ*, **541**, 344
Pethick, C. J., 1992, in *Structure and Evolution of Neutron Stars*, Eds. D. Pines, R. Tamagaki, and S. Tsuruta (New York: Addison-Wesley), p. 115
Pivovaroff, M. J., Kaspi, V. M., & Camilo, F. 2000, *ApJ*, **535**, 379
Potekhin, A. Y. & Yakovlev, D. G. 2001, *A&A*, **374**, 213
Psaltis, D. & Miller, M. C. 2002, *ApJ*, **578**, 325
Ramaty, R., *et al.* 1980, *Nature*, **287**, 122
Rea, N., *et al.* 2003, *ApJ*, **586**, L65
Revnivtsev, M. G., *et al.* 2004, *Astr. Lett.*, **30**(6), 382
Riper, K. A. van, 1988, *ApJ*, **329**, 339
Rothschild, R., Kulkarni, S., & Lingenfelter, R. 1994, *Nature*, **368**, 432
Ruderman, M., Zhu, T., & Chen, K. 1998, *ApJ*, **492**, 267
Shitov, Yu.P. 1999, *IAU Circ.* 7110
Sil'antev, N. A. & Iakovlev, D. G. 1980, *Ap&SS*, **71**, 45
Strohmayer, T. E. & Ibrahim, A. I. 2000, *ApJ*, **537**, L111
Symbalisty, E. M. D. 1984, *ApJ*, **285**, 729
Tam, C. R., Kaspi, V. M., van Kerkwijk, M. H. & Durant, M. 2004, *ApJ*, **617**, L53
Taylor, G. B., *et al.* 2005, *ApJ*, in press, (astro-ph/0504363)
Terasawa, T., *et al.* 2005, *Nature*, **434**, 1110
Terrell, J., Evans, W. D., Klebesadel, R. W., & Laros, J. G. 1980, *Nature*, **285**, 383
Thompson, C. & Blaes, O. 1998, *Phys. Rev. D*, **57**, 3219
Thompson C. & Duncan R. C. 1992, in *Compton Gamma-Ray Observatory*, Eds. M. Friedlander, N. Gehrels and D. J. Macomb (AIP: New York), p. 1085
1993, *ApJ*, **408**, 194
1995, *MNRAS*, **275**, 255
1996, *ApJ*, **473**, 322
2001, *ApJ*, **561**, 980
Thompson, C., *et al.* 2000, *ApJ*, **543**, 340
Thompson, C., Lyutikov, M., & Kulkarni, S. R. 2002, *ApJ*, **574**, 332
Thompson, T., Chang, P., & Quataert, E. 2004, *ApJ*, **611**, 380
Torii, K., Kinugasa, K., Katayama, K., & Tsunemi, H. 1998, *ApJ* **503**, 843
Usov, V. V. 1992, *Nature*, **357**, 472
1994, *ApJ*, **427**, 984
2001, *ApJ*, **559**, L135
Vasisht, G. & Gotthelf, E. V. 1997, *ApJ*, **486**, L129
Vasisht, G., Kulkarni, S., Frail, D., & Greiner, J. 1994, *ApJ*, **431**, L35
Vasisht, G., Gotthelf, E. V., Torii, K., & Gaensler, B. M. 2000, *ApJ*, **542**, L49
Vrba, F. J., *et al.* 2000, *ApJ*, **533**, L17
Wachter, S., *et al.* 2004, *ApJ*, **615**, 887
Wang, Z. & Chakrabarty, D. 2002, *ApJ*, **579**, L33
Woltjer, L. 1964, *ApJ*, **140**, 1309
Woods, P. M., *et al.* 1999a, *ApJ*, **519**, L139
1999b, *ApJ*, **527**, L47
1999c, *ApJ*, **524**, L55
2000, *ApJ*, **535**, L55
2001, *ApJ*, **552**, 748
2002, *ApJ*, **576**, 381
2003, *ApJ*, **596**, 464
2004, *ApJ*, **605**, 378
2005, *ATEL*, 407
Woosley, S. E. & Wallace, R. K. 1982, *ApJ*, **258**, 716
Yakovlev, D. G., Kaminker, A. D., Gnedin, O. Y., & Haensel, P. 2001, *Phys. Rep.*, **354**, 1
Zane, S., Turolla, R., Stella, L, & Treves, A. 2001, *ApJ*, **560**, 384

15

Cosmic gamma-ray bursts, their afterglows, and their host galaxies

K. Hurley
University of California, Berkeley

R. Sari and S. G. Djorgovski
California Institute of Technology

15.1 Introduction

Regarded as an astrophysical mystery and a curiosity for decades, cosmic gamma-ray bursts are finally entering the mainstream of astronomy and astrophysics. In the past few years, we have learned that they lie at cosmological distances, and are probably caused, possibly among other things, by the collapses and subsequent explosions of massive stars. Energetically they are roughly analogous to supernovae, to which they may indeed be related in some cases; no new physics needs to be invented to explain their prodigious luminosities. Unlike supernovae, however, they are relatively rare, and their energy output is distributed quite differently over wavelength and time. They can probably be observed out to distances comparable to, or even farther than, those of the most distant quasars, which makes them useful to cosmologists as lighthouses to the early Universe. Finally, too, they hold the promise of revealing properties of early galaxies such as star formation rates and metallicities in ways that are unique. For all of these reasons, in addition to the facts that they signal the formation of black holes and drive ultra-relativistic winds, they have begun to attract the attention of people working in very diverse disciplines. The words "gamma-ray burst" have even begun to enter the vocabulary of the general public, which regards them with a certain morbid fascination.

It was not at all clear a decade ago that the study of gamma-ray bursts (GRBs) had such a promising future. If, as many people then suspected, they were generated by some sort of activity on galactic neutron stars, they would probably not have been observable for more than the tens of seconds of the bursts themselves, and they might well have remained a curiosity. Two popular accounts of how our understanding of GRBs evolved have now appeared (Katz 2002; Schilling 2002), so in this chapter we will forego the intriguing history of the subject, and begin by describing the phenomenology of bursts. The brief gamma-ray emitting phase is followed by a longer duration, long-wavelength "afterglow"; the characteristics and the theory of afterglows are described next. Finally, the observations of afterglows often lead to the identification of the host galaxies of bursts, which are treated in the last section of this chapter.

In this rapidly evolving field, some of the most up-to-date information is found on websites. A non-exhaustive list of them is:

The Interplanetary Network: ssl.berkeley.edu/ipn3/index.html
HETE-II: space.mit.edu/HETE
BATSE: www.batse.msfc.nasa.gov/batse/
BeppoSAX: www.asdc.asi.it/bepposax/

Compact Stellar X-Ray Sources, eds. Walter Lewin and Michiel van der Klis.

Swift: swift.gsfc.nasa.gov
The gamma-ray burst coordinates network: gcn.gsfc.nasa.gov/gcn/
Jochen Greiner's afterglow website: www.mpe.mpg.de/~jcg/grb.html
A radio catalog of gamma-ray burst afterglows: www.aoc.nrao.edu/~frail/grb_public.html

The most recent conference proceedings are *Gamma-Ray Bursts in the Afterglow Era* edited by Costa, Frontera, and Hjorth (2001), *Gamma-Ray Burst and Afterglow Astronomy 2001* edited by Ricker and Vanderspek (2003), *Supernovae and Gamma-Ray Bursts* edited by Weiler (2003), *Gamma-Ray Bursts in the Afterglow Era* edited by Feroci, *et al.* (2004), and *Gamma-Ray Bursts: 30 Years of Discovery* edited by Fenimore and Galassi (2004).

15.2 The big picture

A typical GRB occurs in a star-forming region of a galaxy at a redshift $z \approx 1$. In currently popular models, it is caused by the collapse of a massive star ($\approx$30 solar masses) which has exhausted its nuclear fuel supply. The star collapses to a black hole threaded by a strong magnetic field, and possibly fed by an accretion torus. In this configuration, energy can be extracted through the Blandford–Znajek (1977) mechanism. This energy goes into accelerating shells of matter, once part of the massive star, to ultra-relativistic velocities (Lorentz factors of several hundred). These shells collide with one another as they move outward, producing "internal" shocks in a solar-system sized volume. The shocks accelerate electrons, and the electrons emit synchrotron radiation. In the observer's frame, the radiation appears in gamma rays, and produces a burst with $\approx$20 second duration. If the gamma rays were emitted isotropically, they would account for well over 10^{53} erg of energy in many cases. However, there is evidence that this gamma radiation is strongly beamed, within a cone whose opening angle is only several degrees (Frail *et al.* 2001) and thus that the total energy emitted in this stage is some two orders of magnitude smaller. As the shells continue to move outward, they eventually reach a region of enhanced density. This could be either the interstellar medium or a region which was populated with matter by the massive star in its final stages of evolution. As the shells impinge on this region, they produce "external" shocks, which give rise to a long-lived radio, optical, and X-ray afterglow that may be detectable for years in the radio, weeks to months in the optical, and weeks in X-rays. There is about an order of magnitude less energy in the afterglow than in the burst itself. Initially, this afterglow radiation is beamed, but as the shells decelerate, they spread laterally and the radiation tends towards isotropy. The afterglow tends to fade as a power law with time. However, in many cases, the decline is not completely monotonic; "bumps" can appear in the optical lightcurve, and they have been interpreted either as a supernova-like component or as the result of micro-lensing.

The model described above is known as the "standard fireball model". Such models had been discussed extensively long before the GRB distance scale was known, but the establishment of a cosmological distance scale for bursts brought them into sharp focus (Mészáros 2002). To be sure, there are competing models, as well as variations on this theme, and they cannot be ruled out. Afterglows are only detected for about one-half the bursts. In those cases where they are not detected, the host galaxies cannot be identified, and it is almost impossible to demonstrate that the GRB is due to the collapse of a massive star, as opposed to the merger of two neutron stars, for example.

Because the gamma-rays are beamed, we detect only a small fraction of them. The most sensitive GRB detector flown (BATSE aboard the Compton Gamma-Ray Observatory) detected roughly one burst per day down to its threshold, and missed about one per day due to well-understood effects such as Earth-blocking. Using current estimates of beaming, this implies that the Universe-wide GRB rate is at least 1000/day, and possibly more if there are many weaker bursts that were not detected by BATSE.

15.3 Some technical details

Before the radio, optical, or X-ray afterglow can be identified, the burst must be localized rapidly (i.e., within a day or so) to reasonable accuracy (several tens of arcminutes) during the bursting phase, which typically lasts only several tens of seconds. The two main ways to do this are first, using a coded mask detection system (e.g., BeppoSAX, Costa *et al.* 1997; HETE, Ricker *et al.* 2003) and second, by timing the arrival of the burst at spacecraft separated by interplanetary distances (the Interplanetary Network, Hurley *et al.* 2000). Burst detection rates using these techniques can reach $\approx$0.5 to one per day.

Next, the position of the burst must be communicated rapidly to observers. This is now done almost exclusively through the Gamma-Ray Burst Coordinates Network (GCN, Barthelmy, Cline, & Butterworth 2001), which reaches almost 600 recipients. In the early phases, the afterglow may be bright enough to be detected by amateur astronomers with telescopes of modest size ($m \approx 16$ in the day or so following the burst; in many cases, the burst outshines its host galaxy). This phase is crucial, because the position of an optical afterglow must be determined to arcsecond accuracy or better to allow optical spectroscopy to take place with telescopes of much larger size, as well as to permit deep observations which may reveal the presence of a host galaxy.

In many cases, the X-ray afterglow has been observed within hours of the burst by the same spacecraft as localized the burst, after slewing to the position (BeppoSAX, Costa 2000). In other cases, target-of-opportunity observations have been carried out with a different spacecraft from that which detected the burst, but with much longer delays.

Radio measurements can be carried out at a more leisurely pace, since the radio emission tends to peak days after the event (Frail, Waxman & Kulkarni 2000). But only the largest radio telescopes need apply: typical peak fluxes are at the millijansky level.

Over the years, astronomers have become extremely adept at identifying GRB counterparts, to the point where redshifts have been measured in less than 8 hours from the time of the burst.

15.4 The bursting phase

15.4.1 Gamma-ray burst lightcurves

Gamma-ray bursts are, for a few seconds, the brightest objects in the gamma-ray sky. Figure 15.1 shows an example. Indeed, bursts are so bright that an uncollimated, unshielded detector with a surface area of only 20 cm^2 can detect a burst out to a redshift of $z = 4.5$ (Andersen *et al.* 2000).

Burst durations span about 5 orders of magnitude, from 0.01 to 1000 s. The duration distribution displays a clear bimodality, with short bursts (durations $\approx$0.2 s) comprising around 25% of the total, and long bursts (durations $\approx$20 s) comprising the remainder (Mazets *et al.* 1981b; Dezalay *et al.* 1996; Norris *et al.* 1984; Hurley 1992; Kouveliotou *et al.* 1993). The

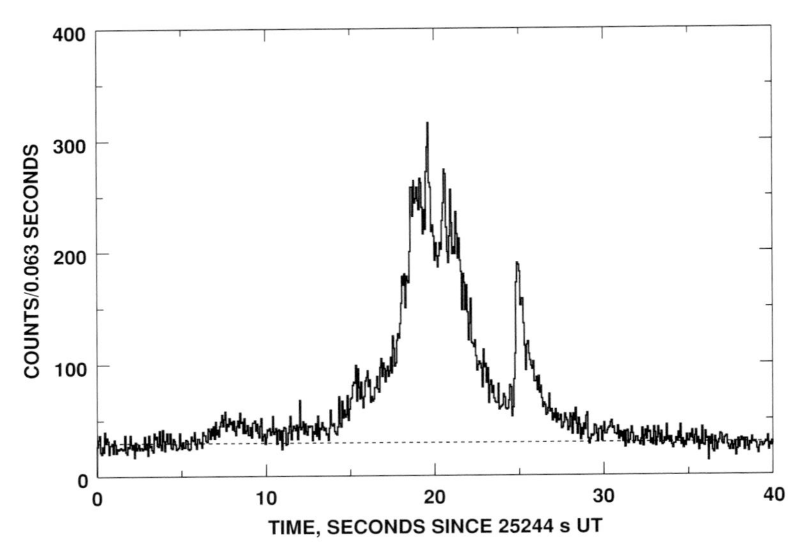

Fig. 15.1. A bright burst observed by the Ulysses GRB experiment on October 8, 2002. The energy range is 25–150 keV. The dashed line indicates the background level. The 25–100 keV fluence of this event was $\approx 8 \times 10^{-5}$ erg/cm^2.

distribution may be described by a lognormal function (McBreen *et al.* 1994). Apart from this, gamma-ray burst lightcurves are generally very different from one burst to another, although certain morphological types have been noted. For example, about 7% of all bursts display a fast-rise, exponential decay morphology (Bhat *et al.* 1994). What determines the shape of the lightcurve in this and other morphologies is unknown.

In the lightcurves of very intense bursts, it is sometimes possible to detect a long, faint tail after the intense emission has ceased (Burenin *et al.* 1999). The count rate in the tail falls as a power law with time. This is a relatively short-lived ($\approx$ 1000 s) gamma-ray afterglow. There is also evidence that the X-ray afterglow starts during the gamma-ray burst in many cases (Frontera *et al.* 2000).

15.4.2 Energy spectra

GRB energy spectra have been measured from $\approx$2 keV (Frontera *et al.* 2000) to 18 GeV (Hurley *et al.* 1994). Even at the highest energies accessible to spark chamber detectors, there is little or no evidence for spectral breaks. Indeed, there is even tantalizing evidence for TeV emission from one burst (Atkins *et al.* 2000). The spectra may be fit over a wide range with various models. One is the so-called "Band model" (Band *et al.* 1993), which has no particular physical derivation. Another is a synchrotron spectrum (Bromm & Schaefer 1999). An example of the latter is shown in Fig. 15.2. The spectrum in this figure is plotted in νF_ν units, which make it clear that the peak of the energy output during the burst, $E_{\rm peak}$, is indeed at gamma-ray energies. The distribution of $E_{\rm peak}$ derived from BATSE data is quite narrow (Mallozzi *et al.* 1995), which is surprising considering the great diversity exhibited by most other GRB characteristics. The extent to which the $E_{\rm peak}$ distribution could be biased due to detector characteristics has been considered. There is presently no compelling evidence that a population of very high $E_{\rm peak}$ GRBs exists (Harris & Share 1998), although not all the

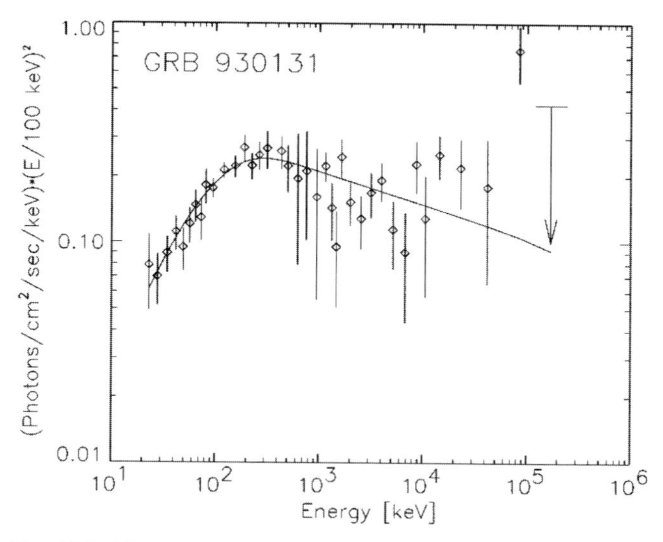

Fig. 15.2. The spectrum of GRB 930131 over four decades, and a synchrotron model fit (Bromm & Schaefer 1999). The spectrum is plotted in νF_ν units, which show the amount of energy per decade. The peak is at an energy $E_{\rm peak} \approx 200$ keV, and the energy output declines only slightly at higher energies.

phase space has been searched for such events. On the other hand, however, there *is* evidence for one or two classes of soft-spectrum bursts, called "X-ray rich GRBs" and "X-ray flashes", which appear to have all the characteristics of gamma-ray bursts, except for the gamma rays above 20 keV or so (Heise *et al.* 2001). These may account for up to 30% of the total bursts. Some of these were in fact detected by BATSE (Kippen *et al.* 2001), but it seems plausible that many were not, leading to a possible bias against these events (see Chapter 6 by Heise and in 't Zand).

An interesting correlation exists between the time histories and the energy spectra of bursts: the short bursts have harder energy spectra than the long bursts (Dezalay *et al.* 1996; Kouveliotou *et al.* 1993). It is presently not understood why this is the case.

15.4.3 Polarization

Polarization in the 0.15–2 MeV gamma-ray emission from GRB 021206 was detected using the data of the RHESSI spacecraft (Coburn & Boggs 2003). This is the only observation of its kind to date, and the level of polarization observed was 80%. One possible explanation for this is that a large-scale, ordered magnetic field is present, with an intensity roughly equal to the equipartition strength.

15.4.4 GRB statistics

It has been known for a long time that the spatial distribution of bursts is isotropic, to varying degrees of statistical uncertainty e.g., Mazets *et al.* 1981a. It has similarly been known that the GRB number–intensity relation, or log N–log S curve, displayed a turnover at low intensities, that is, a paucity of weak events with respect to the $-3/2$ power law expected for a homogeneous distribution in Euclidean space (Mazets *et al.* 1981b). However, it was not until BATSE results became available that the situation was put into perspective and clarified (Pendleton *et al.* 1996; Paciesas *et al.* 1999). With unprecedented statistics and careful attention to instrumental details, the BATSE results confirmed the isotropy of bursts. They also confirmed the turnover in the log N–log S curve, albeit at much smaller intensities than before, indicating that the earlier turnover was due to instrumental effects (mainly the loss of sensitivity to weak events). The present results are consistent with a cosmological population of bursts (e.g., Stern, Atteia & Hurley 2002). By some estimates, GRBs may occur out to redshifts of 10 or more, and current missions such as HETE are capable of detecting them to $z = 8$ (Lamb & Reichart 2000).

It is obviously interesting to sample the GRB population below the BATSE trigger threshold. For example, if there were an epoch of early star formation in the Universe that gave rise to GRB-producing stellar deaths, this might be manifested by an increase in the log N–log S curve at low intensities. Two recent studies have succeeded in exploring the low-intensity population (Kommers *et al.* 2000; Stern *et al.* 2001). By identifying bursts that were too weak to trigger the BATSE detector, they have effectively reduced the threshold by a factor of about 2. However, the conclusions of the studies differ; in one case, there appears to be evidence that the curve continues to rise at low intensities, suggesting that bursts continue to originate from earlier and earlier parts of the Universe, while in the other case, there is evidence for a flattening of the curve. It will require the next generation of GRB detectors to sort this out.

15.4.5 Burst types and classes

There are many *types* of gamma-ray bursts. They are often referred to as *classes*, but it is not known whether they actually originate from different kinds of explosions, or the explosions of different kinds of stars, as opposed to say, originating from different viewing angles or other observing conditions. A brief, non-exhaustive summary of burst types follows.

(a) Long and short bursts. The duration distribution is bimodal, and the energy spectra of the short bursts are harder than those of the long bursts. No radio, optical, or X-ray counterpart has been found for any short event (Hurley *et al.* 2002). It has been speculated that the short bursts might arise from neutron star–neutron star mergers (e.g., Macfadyen & Woosley 1999a), which could take place far from a host galaxy, and lack an interstellar medium (ISM) on which to produce a long-lived afterglow.

(b) Dark bursts. While virtually all long bursts display X-ray afterglows, only about one-half of them have detectable radio or optical afterglows. There are various ways to hide intrinsically bright afterglows and make them undetectable, and it is possible that more than one explanation is required. On the speculative side, possibilities include absorbing the light in the host galaxy, placing the burst at high redshift, or invoking a flat spectral shape (e.g., Hjorth *et al.* 2002). Bursts can also appear to be dark if their afterglows are intrinsically weak and/or rapidly fading. This has actually been observed: GRB 021211 is in this category. It was detected by the HETE spacecraft and its position was circulated to astronomers 22 s after the start of the burst (Crew *et al.* 2002). Observations began with automated telescopes a few minutes later (Park *et al.* 2002). If the first observations had started later, the afterglow might have been undetectable (Fox *et al.* 2003).

(c) Bursts possibly associated with supernovae. The first such event was GRB 980425 (Galama *et al.* 1998c), a burst whose position and time of occurrence were both consistent with those of an optical Type Ic supernova, 1998bw. In this case, however, the burst was quite weak, while paradoxically the redshift of the supernova host galaxy was quite small

($z = 0.008$). In other cases, supernova-like bumps in the afterglow lightcurves have been identified and attributed to underlying supernovae (Bloom *et al.* 1999). The clearest evidence to date comes from GRB 030329, an intense burst at a redshift of $z = 0.17$, which has probably been studied more closely than any other event. Its optical spectrum resembles that of SN 1998bw (Stanek *et al.* 2003), and its afterglow lightcurve displays a supernova-like bump (Hjorth *et al.* 2003); the associated supernova was named 2003dh. There is little doubt now that some GRBs are associated with supernovae, although in many cases, no supporting observational evidence can be found.

(d) X-ray flashes (XRFs). These are bursts that resemble GRBs in almost every respect: durations, spatial distributions, etc. However, they display little or no emission above $\approx$25 keV (Heise *et al.* 2001; Chapter 6). Possibly related to them are the X-ray rich GRBs, which display some gamma-ray emission (there is no widely accepted definition yet for just what ratio of X-ray flux to gamma ray flux constitutes an XRF). One way to eliminate gamma radiation is to redshift the burst. However, one X-ray rich GRB, 021004, has a redshift of only 1.6 (Fox *et al.* 2002), so this cannot be the only explanation.

15.5 GRB theory: the generic picture

The mechanism leading to the phenomenon of GRBs is still a matter of debate. Nevertheless, some basic characteristics are well understood. Below, we show how the observed spectra, energies, and timescales of GRBs have led to a generic model, the so-called fireball shock model that is almost independent of knowledge about the unknown "inner engine".

The extreme characteristics of GRBs, i.e., the observed large energies and short timescales, lead to a paradox, the "compactness problem". An energy of 10^{52} erg is released within a variability time $\delta T \sim 0.1$ s in the form of $\approx$1 MeV photons. This translates into a huge number of photons, $N = 10^{56}$. If we now assume that the energy is released in a small volume of linear dimension $R \leq c\delta T \sim 3 \times 10^9$ cm (which is naively required by the variability timescale), then the optical depth to pair creation would be the number of photons per unit area, multiplied by the Thomson cross section σ_T or

$$\tau \sim \sigma_T \frac{N}{4\pi R^2} \sim 3 \times 10^{11} \gg 1$$

But, if that were true, it would imply that all the photons have created pairs and thermalized. However, the observed spectrum of GRBs, as shown in the previous section is highly non-thermal!

The only known solution to the "compactness" problem is relativistic motion (Goodman 1986; Paczyński 1986). These effects have been considered in detail (Krolik & Pier 1991; Fenimore, Epstein & Ho 1993; Baring & Harding 1997). A critical review of these as well as some new limits are given by Lithwick and Sari (2001). If the emission site is moving relativistically toward the observer with a Lorentz factor γ, then the optical depth is reduced compared to the stationary estimate, due to two effects. First, the size of the source can be larger by a factor of γ^2. This will still produce variability over a short timescale given by $\delta T = R/\gamma^2 c$, since not all of the source is seen because the radiation for a relativistically moving object is beamed (see Fig. 15.3). Second, the photons in the local frame are softer by a factor of γ, and therefore only a small fraction of them, those at the high-energy tail of the GRB spectrum, have enough energy to create pairs. The combination of these two effects reduces the optical depth by a factor of $\sim\gamma^{6.5}$, where the exact power depends on the GRB spectrum (see Lithwick & Sari 2001). Therefore, the optical depth is reduced below

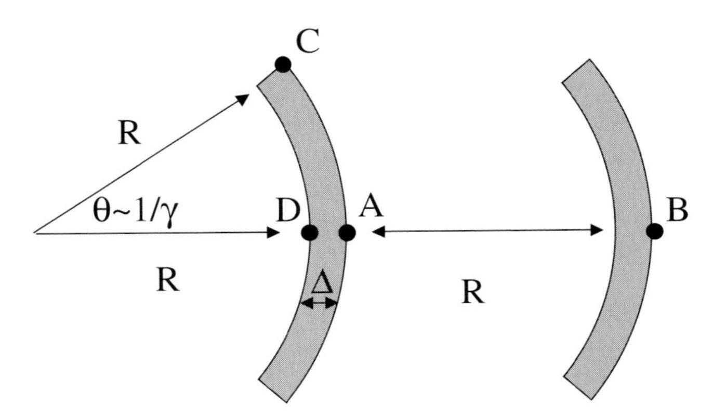

Fig. 15.3. Timescales from an expanding relativistic fireball. The gray area represents the observed section of the fireball that can be seen by an observer located far to the right. The angular opening of that section is $1/\gamma$ due to relativistic beaming. Consider the four photons emitted at points A, B, C, and D. Photons A, C, and D were emitted simultaneously, but photon A will arrive at the observer first, since it is closer to the observer. The arrival-time delay of photons C and D with respect to photon A is simply given by the extra distance they have to travel. Therefore $\delta T_{\mathrm{C-A}} = R(1 - \cos\theta)/c = R/2\gamma^2 c$, and $\delta T_{\mathrm{D-A}} = \Delta/c \sim R/\gamma^2 c$, where we have used the fact that relativistic dynamics of fireballs imply $\Delta \sim R/\gamma^2$. Finally, photon B was emitted long after photon A (about a time R/c later than photon A); however, it is much closer to the observer, resulting in $\delta T_{\mathrm{B-A}} = R/2\gamma^2 c$. All three timescales lead to the expression $R/\gamma^2 c$. A short observed variability timescale can therefore be obtained even for large radius, if the Lorentz factor is sufficiently high. The naive estimate of $R \leq c\delta T$ is, therefore, to be replaced by $R \leq \gamma^2 c\delta T$.

unity, and the "compactness problem" is solved, if the Lorentz factor is larger than about one hundred.

This solution to the compactness problem led to a three stage generic scenario for GRBs. First, a compact source releases about 10^{52} erg, in a small volume of space and on a short timescale. This large concentration of energy expands due to its own pressure (Mészáros & Rees 1993; Piran & Shemi 1993; Piran, Shemi & Narayan 1993). If the rest mass that contaminates the site is not too large, $\lesssim 10^{-5}\ \mathrm{M}_{\odot}$ (the requirement of a small baryonic load), this will result in relativistic expansion with $\gamma > 100$. Finally, at a large enough radius, the kinetic energy (bulk motion) of the expanding material is converted to internal energy and radiated, mainly in gamma-rays. At this stage the system is optically thin and high-energy photons can escape. We now discuss this third stage in some detail.

15.5.1 Internal vs. external shocks

Assume a flow carrying 10^{52} erg as kinetic energy. In order for this to produce photons, the kinetic energy must be converted back into internal energy and radiated away. The flow must therefore, at least partially, slow down. Two scenarios were proposed for this deceleration: external shocks (Mészáros & Rees 1993) and internal shocks (Narayan, Paczyński & Piran 1992; Rees & Mészáros 1994). In the external shocks scenario, the relativistic material is running into some (external) ambient medium, possibly the interstellar medium (ISM) or a stellar wind that was emitted earlier by the progenitor. In the internal shocks scenario the inner engine is assumed to emit an irregular flow consisting of many shells that travel with a variety of Lorentz factors and therefore collide with one another and thermalize part of their kinetic energy.

The property that proved to be very useful in constraining these two possibilities is the variability observed in many of the bursts. In the external shocks scenario, this variability is attributed to irregularities in the surrounding medium, e.g., clouds. Each time the ejecta run into a higher density environment, they produce a peak in the emission. In the internal shocks scenario, the source has to emit many shells, and when two of them collide a peak in the emission is produced. External shocks thus require a complicated surrounding with a relatively simple source that explodes once, while internal shocks require a more complicated source that will explode many times to produce several shells. Due to these very different requirements on the source, the question of internal or external shocks is of fundamental importance in understanding the nature of the phenomenon.

The size of the clouds that the ejecta run into, in the external shocks scenario, has to be very small in order to produce peaks that are narrower than the duration of the burst (Fenimore, Madras & Nayakchin 1996). Sari and Piran (1997a) gave the following argument. The size of the clouds has to be smaller than $R/N\gamma$ to produce peaks that are narrower by a factor of N than the duration of the burst. The number of clouds should be smaller than N otherwise pulses arriving from different clouds will overlap and the amplitude of the variability will be reduced. Finally, due to relativistic beaming, the observable area of the ejecta is $(R/\gamma)^2$. The maximal efficiency of the external shocks scenario is therefore given by

$$\frac{\text{cloud area} \times \text{number of clouds}}{\text{observed shell area}} \leq \frac{1}{N} \sim 1\% \tag{15.1}$$

Since in many bursts $N > 100$, external shocks have a severe efficiency problem in producing highly variable bursts. The problem is even more dramatic if long quiescent periods, which are observed in many bursts (Nakar & Piran 2002), are taken into account. Also, other predictions of external shocks are inconsistent with the observed temporal profile (Ramirez-Ruiz & Fenimore 1999). Moreover, the density ratio between the clouds and their surroundings has to be huge, of the order of $\gamma N^2 \sim 10^6$, in order for the ejecta to be slowed down mainly by the dense clouds rather than by the low-density medium that they are embedded in. Finally, we mention that, despite the above arguments, some still favor other scenarios (Dermer & Mitman 1999; Dar & DeRujula 2000).

Internal shocks do not suffer from these problems. Detailed calculations show that the observed temporal structure from internal shocks closely follows the operation of the inner engine that generated the shells (Kobayashi, Piran & Sari 1997). In this scenario, the source must be variable on timescales shorter than a second and last for as long as 100 seconds, just as the bursts themselves.

The efficiency of internal shocks is largely determined by the ratio of Lorentz factors between different shells that are colliding with each other. The larger the ratio, the larger the efficiency. A simple scenario that demonstrates this is the case of two equal mass shells with Lorentz factors $\gamma_1 \gg \gamma_2 \gg 1$. Conservation of energy and momentum in a collision between the shells leads to a Lorentz factor that is the geometric mean of the initial ones $\sqrt{\gamma_1\gamma_2}$. Therefore, the energy left in the system as non-thermal is a small fraction $\sqrt{\gamma_2/\gamma_1}$ of the initial energy. Beloboradov (2000) has argued that if large Lorentz factor ratios are allowed, the internal shock efficiency is only limited by the fraction of energy in the shock given to the radiating electrons. Kobayashi and Sari (2001) have then shown that multiple collisions

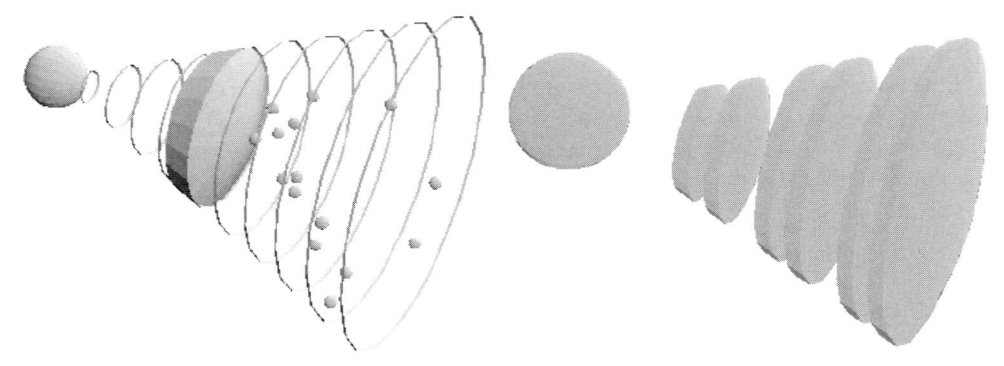

Fig. 15.4. Producing variability by external shocks (*left*) or internal shocks (*right*). In the external shocks scenario, the variability is produced by irregularities in the surroundings. If the surroundings consist of a low-density medium that contains high-density clouds, then whenever the shell hits one of the clouds a peak in the emission is produced. The number of clouds within the observable cone (of angular size $1/\gamma$ due to relativistic beaming) should therefore be roughly the number of observed peaks. The source itself, in this model, needs to produce only a single shell in a single (simple) explosion. However, the external shocks scenario has low efficiency, due to the small total surface area of the clouds when compared to the area of the shell. In the internal shocks case, the temporal structure arises from the source, i.e., the source produces a more complex explosion. There is no efficiency problem, provided that the relative Lorentz factor between shells is large.

between shocks may result in "ultra efficient" internal shocks, in the sense that even more than the fraction of energy given to electrons can be radiated away.

The mechanism by which the thermal energy produced by internal shocks is converted to radiation is almost certainly synchrotron and inverse Compton, since these are the dominant radiation mechanisms at the low densities involved. While both mechanisms probably take place, it is actually not very clear which of the two produces the observed radiation. Synchrotron emission is preferred for several reasons (Sari, Narayan & Piran 1996; Sari & Piran 1997b) and inverse Compton probably produces a higher energy component.

15.6 The afterglow: theory

After the internal shocks produce the GRB, the shell interacts with the surrounding medium and decelerates. Again it emits radiation by synchrotron and inverse Compton. As the flow decelerates, the emission shifts to lower and lower frequencies. This emission, the afterglow, may last on detectable levels for years after the GRB itself!

Afterglow was predicted well before it was observed (Paczyński & Rhoads 1993; Katz, 1994; Vietri, 1997; Mészáros & Rees 1997). The afterglow theory is relatively simple. It deals with the emission on timescales much longer than that of the GRB. The details of the complex initial conditions are therefore forgotten and the condition of the GRB remnant can be described by a self-similar solution with a small number of parameters, such as the total energy and the external density. It is assumed that the electrons are accelerated by the shock into a power-law distribution of electron Lorentz factors $N(\gamma_e) \propto \gamma_e^{-p}$ for $\gamma_e > \gamma_m$. The lower cutoff γ_m of this distribution is set by the assumption that the electrons acquire a fixed fraction, ϵ_e, of the thermal energy. It is also assumed that a considerable magnetic field is built behind the shock, which is again characterized by a certain fraction ϵ_B of equipartition. The energy density behind a relativistic shock is given by $4\gamma^2 n m_p c^2$, where $n = n_1$ cm^{-3} is the proton number density behind the shock, γ is the Lorentz factor of the fluid behind the shock, and m_p is the proton mass. These equipartition assumptions then result in

$$\gamma_m = \frac{p-2}{p-1}\frac{m_p}{m_e}\epsilon_e\gamma \cong 630\epsilon_e\gamma \tag{15.2}$$

$$B = 0.4\sqrt{\epsilon_B n_1}\gamma \text{ gauss} \tag{15.3}$$

where B is the magnetic field, and m_e is the electron mass. The relativistic electrons then emit synchrotron radiation which produces the observed afterglow. The broadband spectrum of such afterglow emission was given by Sari, Piran, and Narayan (1998).

The afterglow synchrotron spectrum can be fully described by the electron energy index p, the peak flux F_m and three characteristic frequencies (ν_m, ν_c, ν_a):

(i) ν_m is the synchrotron frequency of the minimum energy electron, with Lorentz factor γ_m. From synchrotron theory $\nu_m \cong (eB/2\pi m_e c)\gamma_m^2$ in the local frame of the fluid; here e is the electron charge. Transforming this to the observer frame (blue shifted by the Lorentz factor and redshifted by a factor of $(1+z)$) and using Eqs. (15.2) and (15.3) we obtain

$$\nu_m = 1.4\times 10^{13}\,\text{Hz}\,(1+z)^{-1}\left(\frac{\epsilon_e}{0.1}\right)^2\left(\frac{\epsilon_B}{0.1}\right)^{1/2}\left(\frac{\gamma}{10}\right)^4 n_1^{1/2} \tag{15.4}$$

(ii) The cooling time of an electron is inversely proportional to its Lorentz factor γ_e. Therefore, electrons with a Lorentz factor higher than a critical Lorentz factor $\gamma_e > \gamma_c$ cool on the dynamical timescale of the system. This characteristic Lorentz factor is given by the condition $\sigma_T c\gamma^2\gamma_c^2 B^2 t\gamma/6\pi(1+z) = \gamma_c m_e c^2$, and corresponds to the "cooling frequency"

$$\nu_c = 1.2\times 10^{13}\,\text{Hz}\,(1+z)\left(\frac{\epsilon_B}{0.1}\right)^{-3/2}\left(\frac{\gamma}{10}\right)^{-4} n_1^{-3/2} t_{\text{days}}^{-2} \tag{15.5}$$

where t_{days} is the observer time in days. Here we have also taken into account that time is redshifted.

(iii) Below some critical frequency ν_a the flux is self-absorbed and is given by the Rayleigh–Jeans portion of a blackbody spectrum.[1] The self-absorption frequency is given by

$$\nu_{sa} = 93\,\text{GHz}\,(1+z)^{-13/5}\left(\frac{\epsilon_B}{0.1}\right)^{6/5}\left(\frac{\gamma}{10}\right)^{28/5} n_1^{9/5} t_{\text{days}}^{8/5} \tag{15.6}$$

if $\nu_c < \nu_m$, and by

$$\nu_{sa} = 87\,\text{GHz}\,(1+z)^{-8/5}\left(\frac{\epsilon_e}{0.1}\right)^{-1}\left(\frac{\epsilon_B}{0.1}\right)^{1/5}\left(\frac{\gamma}{10}\right)^{8/5} n_1^{4/5} t_{\text{days}}^{3/5} \tag{15.7}$$

if $\nu_c > \nu_m$.

(iv) The normalization of the spectrum is given by the total number of radiating electrons $4\pi R^3 n_1/3$ times the peak flux from a single electron, resulting in

$$F_m = 220\,\text{mJy}\,(1+z)^{-2} d_{L,28}^{-2}\left(\frac{\epsilon_B}{0.1}\right)^{1/2}\left(\frac{\gamma}{10}\right)^8 n_1^{3/2} t_{\text{days}}^3 \tag{15.8}$$

where $d_{L,28}$ is the luminosity distance in units of 10^{28} cm.

[1] Granot, Piran, and Sari (2000b) have found that if $\nu_c < \nu_m$, then the self-absorption frequency actually splits into two: ν_{ac} and ν_{sa}, where an optical depth of unity is produced by non-cooled electrons and all electrons, respectively. In between these two frequencies the spectral slope is $\nu^{11/8}$.

The broadband spectrum of the well-studied GRB 970508 (Galama *et al.* 1998b) is in very good agreement with the theoretical picture. Note that the derivation above is quite general. It does not depend either on the surrounding density profile or on the geometry of the event. Both these effects are hidden in the evolution of the fluid Lorentz factor γ, and the particle density n_1 as a function of time.

The evolution of this spectrum as a function of time depends on the hydrodynamics. The simplest, which describes the observations in some cases quite well, is the adiabatic model with a constant density surrounding medium. The rest mass collected by the shock at radius R is about $R^3\rho$, where ρ is the mass density. On average, the particles move with a Lorentz factor of γ^2 in the observer frame (one factor of γ is the bulk motion and the other is the random thermal motion). Therefore, the total energy is given by $E \propto \gamma^2 R^3 \rho c^2$. Assuming that the radiated energy is negligible compared to the energy of the flow, we obtain $\gamma \propto R^{-3/2}$ or in terms of the observer time, $t = R/\gamma^2 c$, we get $\gamma \propto t^{-3/8}$.

$$\nu_{\rm m} = 6 \times 10^{15}\,\text{Hz}\,(1+z)^{1/2} E_{52}^{1/2} \epsilon_{\rm e}^{2} \epsilon_{B}^{1/2} t_{\rm days}^{-3/2}$$

$$\nu_{\rm c} = 9 \times 10^{12}\,\text{Hz}\,(1+z)^{-1/2} \epsilon_{B}^{-3/2} n_1^{-1} E_{52}^{-1/2} t_{\rm days}^{-1/2}$$

$$\nu_{\rm sa} = 2 \times 10^{9}\,\text{Hz}\,(1+z)^{-1} \epsilon_{\rm e}^{-1} \epsilon_{B}^{1/5} n_1^{3/5} E_{52}^{1/5}$$

$$F_{\rm m} = 20\,\text{mJy}\,(1+z) \epsilon_{B}^{1/2} n_1^{1/2} E_{52} d_{L28}^{-2}$$

If, on the other hand (Chevalier & Li 1999), the density drops as R^{-2} (as is expected if the surroundings are a wind produced earlier by the progenitor of the burst) we get $\gamma \sim t^{-1/4}$. Choosing the parameter A_* to define the normalization of the density as $\rho R^2 = A_* 5 \times 10^{11}$ g/cm results in

$$\nu_{\rm m} = 1.7 \times 10^{14}\,\text{Hz}\,(1+z)^{1/2} E_{52}^{1/2} \epsilon_{\rm e}^{2} \epsilon_{B}^{1/2} t_{\rm days}^{-3/2}$$

$$\nu_{\rm c} = 7 \times 10^{11}\,\text{Hz}\,(1+z)^{-3/2} \epsilon_{B}^{-3/2} A_*^{-2} E_{52}^{1/2} t_{\rm days}^{1/2}$$

$$\nu_{\rm sa} = 1.5 \times 10^{10}\,\text{Hz}\,(1+z)^{-2/5} \epsilon_{\rm e}^{-1} \epsilon_{B}^{1/5} A_*^{6/5} E_{52}^{-2/5} t_{\rm days}^{-3/5}$$

$$F_{\rm m} = 180\,\text{mJy}\,(1+z)^{3/2} \epsilon_{B}^{1/2} A_* E_{52}^{1/2} t_{\rm days}^{-1/2} d_{L28}^{-2}$$

These simple scalings, for the case of a constant density ambient medium, lead to the spectral evolution given in Fig. 15.5. The derivations above use a very simple description of the flow. It represents the fluid as if it had a single magnetic field strength and a single Lorentz factor γ and all of the material is moving directly towards the observer. Also, a very approximate description of the synchrotron emission was used. In reality, of course, the situation is more complicated. There are two effects that must be taken into account. The most dramatic one is the fact that matter slightly off the line of sight does not move directly towards the observer (Waxman 1997b; Panaitescu & Mészáros 1998; Sari 1998). The amount of Lorentz boost from that matter is reduced. Second, fluid elements at different distances from the shock have somewhat different Lorentz factors, magnetic fields and electron energies. These variations can be estimated using the self-similar solution of Blandford and McKee (1976). The outcome of these more detailed calculations is the same scaling laws, but with a more accurate coefficient for the break frequencies as well as an estimate of the shape of the spectrum around each break frequency (Granot, Piran & Sari 2000; Gruzinov & Waxman

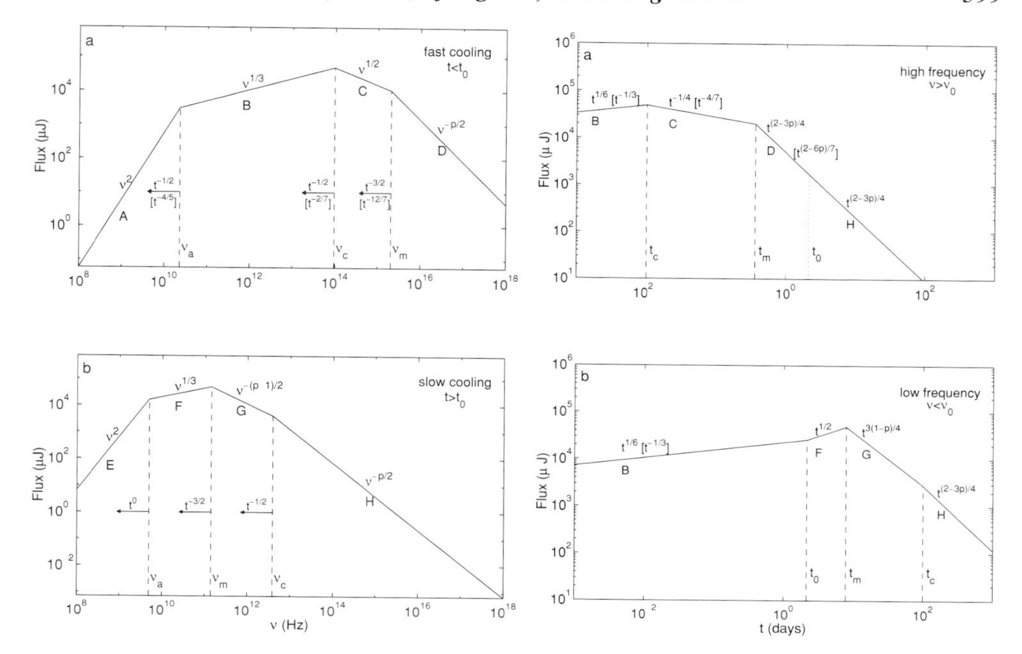

Fig. 15.5. Theoretical spectra (*left*) and lightcurves (*right*) of synchrotron emission from a power-law distribution of electrons for the case of a constant density ambient medium and a spherical explosion. For most cases $p = 2.2$–2.5 fits the observed spectra and lightcurves well.

1999; Granot, Piran & Sari 1999). The equations given above already take these effects into account, and the coefficients given are accurate for $p = 2.2$.

The above scalings assumed adiabatic evolution. At first sight one might think that if the fraction of energy given to the electrons, ϵ_e, is less than unity, then perhaps only a small fraction of the energy can be radiated away. However, the same fireball energy is given again and again to newly shocked electrons. Each time, a fraction ϵ_e can be radiated away, and the overall effect can be large, much above the fraction ϵ_e. Energy losses during the cooling phase can be taken into account (Sari 1998; Cohen, Piran & Sari 1998) using $dE/dR = -(16\pi/3)R^2\epsilon_e\gamma^2 m_p c^2 n$. This results in $E = E_0 \times (t/t_0)^{-17\epsilon_e/12}$ for a constant density environment and $E = E_0 \times (t/t_0)^{-3\epsilon_e/2}$ for a wind environment. These effects are not taken into account in many models but may actually have a significant impact if ϵ_e is not too far below unity. In the case of GRB 000926, energy losses appear to have reduced the energy of the system by a factor of 5 (Harrison *et al.* 2001).

Given the above hydrodynamic evolution, one can construct lightcurves at any given frequency. These will also consist of power laws, changing from one power law to another once the break frequencies pass through the observed band. These predicted power law lightcurves and spectra are in fair agreement with afterglow observations (see Section 15.7).

We have so far considered synchrotron radiation only. Since the optical depth of the system is small, most of the synchrotron photons emitted can be observed. Still, inverse Compton can affect the system in two ways. First, it may add an observable high-energy component. This requires a moderately high density. Second, it may provide an important cooling mechanism, and alter the synchrotron spectrum by its effect on ν_c. The ratio of the inverse Compton (IC) to synchrotron luminosity (a measure of their relative importance for cooling) can be computed very generally (Sari, Narayan & Piran 1996), in a way that does not deal with the details of the spectrum, but depends only on the underlying physical properties of the expanding shock wave. We generalize the derivation given by Sari, Narayan, and Piran (1996) to describe both fast and slow cooling regimes by introducing a parameter η, equal to the fraction of the electron energy that was radiated away (via both synchrotron and IC emission) (Sari & Esin 2001). Then the ratio of luminosities, in the limit of single scattering, is given by

$$x \equiv \frac{L_{\rm IC}}{L_{\rm syn}} = \frac{U_{\rm rad}}{U_B} = \frac{U_{\rm syn}}{U_B} = \frac{\eta U_{\rm e}/(1+x)}{U_B} = \frac{\eta \epsilon_{\rm e}}{\epsilon_B(1+x)} \tag{15.9}$$

where $U_{\rm syn}$, U_B and $U_{\rm e}$ are the energy density of synchrotron radiation, magnetic field and relativistic electrons, respectively. Note that in general $U_{\rm syn} = \eta\beta U_{\rm e}/(1+x)$, where β is the velocity of material behind the shock front (in the frame of the shock); however, for a relativistic shock $\beta \cong 1$. The importance of inverse Compton therefore diminishes quickly when the fireball becomes non-relativistic.

Solving Eq. (15.9) for x we obtain

$$x = \frac{-1+\sqrt{1+4\frac{\eta\epsilon_{\rm e}}{\epsilon_B}}}{2} \cong \begin{cases} \frac{\eta\epsilon_{\rm e}}{\epsilon_B}, & \text{if } \frac{\eta\epsilon_{\rm e}}{\epsilon_B} \ll 1 \\ \left(\frac{\eta\epsilon_{\rm e}}{\epsilon_B}\right)^{1/2} & \text{if } \frac{\eta\epsilon_{\rm e}}{\epsilon_B} \gg 1 \end{cases} \tag{15.10}$$

Modeling afterglow data often suggests that $\epsilon_{\rm e} \gg \epsilon_B$ and therefore inverse Compton may be of importance.

15.7 The afterglow revolution

Motivated by the prediction of a late-time softer radiation (the afterglow), several groups executed rapid radio follow-up observations of GRB error boxes. Detection of a radio afterglow seemed most promising. Not only does the large field of view match well with the large error boxes (several degrees) that were then available on short timescales (within a day), but maximum light was also expected to occur later at longer wavelengths. The best (pre-BeppoSAX era) limits on such afterglow radio emission were obtained for GRB 940301. This GRB triggered an extensive multi-wavelength campaign with ground-based optical and radio observatories from the BATSE/COMPTEL/NMSU Rapid Response Network (McNamara *et al.* 1995). No obvious candidate radio counterparts were found (Frail *et al.* 1994; Koranyi *et al.* 1995; Galama *et al.* 1997a).

15.7.1 The first identifications

The breakthrough came in early 1997, when the Wide-Field Cameras (WFCs; Jager *et al.* 1993) on board the Italian–Dutch satellite BeppoSAX (Piro, Scarsi & Butler 1995) obtained their first quickly available (within hours) accurate positions of GRBs (several arcminutes). This allowed rapid follow-up observations which led to the discoveries of X-ray (Costa *et al.* 1997), optical (van Paradijs *et al.* 1997), millimeter (Bremer *et al.* 1998) and radio (Frail *et al.* 1997) counterparts of GRBs. These observations quickly settled the distance controversy. The first transient optical counterpart, of GRB 970228, is in a faint galaxy with $\sim$0.8″ diameter (Sahu *et al.* 1997). Detection of absorption features in the spectrum of the optical transient (OT) associated with GRB 970508 (Metzger *et al.* 1997) established that this event was at a redshift greater than $z = 0.835$. GRBs come from "cosmological" distances and are thus extremely powerful events. They are by far the most luminous photon sources in the Universe, with (isotropic equivalent) peak luminosities in gamma rays up to 10^{52} erg/s, and total energy budgets up to several 10^{53-54} erg (Kulkarni *et al.* 1998, 1999a) (but see Sections 15.8 and 15.9 for a discussion of collimated outflow, which reduces the inferred total energy). Within the first day, the optical emission is usually brighter than 20th magnitude (some 10 mag brighter (absolute) than the brightest supernovae) and therefore small telescopes can play an important role in measuring the lightcurve. Today, a large worldwide collaboration is observing these events and the data are submitted to the Gamma-Ray Burst Coordinates Network in near-real time, allowing other observatories to react rapidly.

15.7.2 Confirmation of the relativistic blast-wave model

A stringent test of the relativistic blast-wave model came with the discovery of X-ray (Costa *et al.* 1997) and optical afterglow following GRB 970228 (van Paradijs *et al.* 1997; Galama *et al.* 1997b). The X-ray and optical afterglows of GRB 970228 show a power-law temporal decay; this is a trend observed in all subsequent X-ray and optical afterglows, with power-law exponents in the range 1 to 2.

Let us first concentrate on the forward shock and assume slow cooling (the bulk of the electrons do not radiate a significant fraction of their own energy and the evolution is adiabatic); this appears applicable to some observed GRB afterglows at late times ($t > 1$ hr). The simplest assumption is that of spherical symmetry and a constant ambient density. As both the afterglow's spectrum and the temporal evolution of the break frequencies ν_a, ν_m, ν_c are, in the relativistic blast-wave model, power laws (see Section 15.6), the evolution of the flux is also a power law in time. For example, for $\nu_m \leq \nu \leq \nu_c$, the decay of the flux is $F_\nu \propto t_{obs}^{-3(p-1)/4}$, and the power-law spectral slope α relates to the spectral slope β as $\alpha = -3/2\beta$. Several authors (Wijers, Rees & Mészáros 1997; Reichart 1997; Waxman 1997a) showed that to first order this model describes the X-ray and optical afterglow of GRB 970228 very well.

GRB 970508 was the first GRB with a radio counterpart (Frail *et al.* 1997). The radio lightcurves (8.5 and 4.9 GHz) show large variations on timescales of less than a day, but these damp out after one month. This finds a viable explanation in interstellar scintillation (stochastic refraction and diffraction by the fluctuations in the interstellar medium electron density between the source and the observer). The damping of the fluctuations can then be understood as the effect of source expansion on the diffractive interstellar scintillation. Thus a source size of roughly 10^{17} cm was derived (at 3 weeks), corresponding to a mildly relativistic expansion of the shell (Frail *et al.* 1997).

GRB 970508 remains one of the best observed afterglows: the radio afterglow was visible at least 368 days (and with 2.5σ significance on day 408.6 (Frail, Waxman & Kulkarni 2000)), and the optical afterglow up to $\sim$450 days (e.g., Fruchter *et al.* 2000; Galama *et al.* 1998a; Castro-Tirado *et al.* 1998). In addition millimeter (Bremer *et al.* 1998), infrared and X-ray (Piro *et al.* 1998) counterparts were detected. These multi-wavelength observations allowed the reconstruction of the broad radio to X-ray spectrum for this GRB (Galama *et al.* 1998b). Galama *et al.* (1998b) found that the "standard" model provides a successful and consistent description of the afterglow observations over nine decades in frequency, ranging in time from the event until several months later. The synchrotron afterglow spectrum of this GRB allows measurements of the electron energy spectrum p, the three break frequencies (ν_a, ν_m and ν_c), and the flux at the peak, $F_{\rm m}$. For GRB 970508 the redshift z is also known, and all blast-wave parameters could be deduced: the total energy (per unit solid angle) $E = 3.5 \times 10^{52}$ erg, the ambient (nucleon) density $n_1 \approx 5$, the fraction of the energy in electrons $\epsilon_{\rm e} \approx 0.5$ and that of the magnetic field $\epsilon_B = 0.01$ (Wijers & Galama 1999; Granot, Piran & Sari 1999). The numbers themselves are uncertain by an order of magnitude, but the result shows that the "standard" model fits the expectations very well.

Following these first attempts at modeling the broad-band afterglow, more detailed modeling efforts have been made. For example, Panaitescu and Kumar (2001a) have modeled a sample of GRBs with relativistic jets (see Sections 15.8 and 15.9 for a detailed discussion on jets) and find: typical energies of $10^{50} - 10^{51}$ erg, ambient densities ranging from 10^{-3} to 10 cm^{-3}, beaming angles ranging between 1° and 4°, and that a wind-like ambient medium can in some cases be ruled out. GRB 000301C was modeled with a hard electron-energy distribution (Panaitescu 2001); $p = 1.5$ (but see Berger *et al.* 2000) and GRB 010222 also requires a hard electron-energy distribution (Galama *et al.* 2003). Evidence has been presented for an inverse Compton emission component in the afterglow of GRB 000926 (Harrison *et al.* 2001).

The highly relativistic nature of the GRB source (Galama *et al.* 1999) can once more be seen in the extreme brightness temperature of the GRB 990123 optical flash (Akerlof *et al.* 1999) $T_{\rm b} \geq 10^{17}$ K (see Section 15.11), which by far exceeds the Compton limit of 10^{12} K. In this case the optical signal from GRB 990123 was some 18 mag brighter (absolute) than the brightest supernovae. The extreme brightness can be explained by emission from the reverse shock (see Section 15.11).

15.8 Collimated outflow (jets): theory

The hydrodynamic evolution described in Section 15.6 assumed spherical symmetry. However, many astrophysical phenomena, especially those involving extreme energetics, are not spherical but in the form of jets. As we will see, this is most probably the case also for GRBs.

Jets have been discussed extensively in the context of GRBs. First, the similarity between some of the observed features of blazars and AGNs led to the speculation that jets also appear in GRBs (Paczyński 1993). Second, the regions emitting the GRBs as well as the afterglow must be moving relativistically. The emitted radiation is strongly beamed, and we can observe only a region with an opening angle $1/\gamma$ off the line of sight. Emission outside of this very narrow cone is not observed. These considerations have led to numerous speculations on the existence of jets and to attempts to search for the observational signature of jets both during the GRB phase (Mao & Yi 1994) and in the context of the afterglow (Rhoads 1997, 1999; Mészáros, Rees & Wijers 1998).

We begin by clarifying some of the confusing terminology. There are two distinct but related effects. The first, *jets*, describes scenarios in which the relativistic flow emitted from the source is not isotropic but collimated into a finite solid angle. The term jet refers to the geometrical shape of the relativistic flow emitted from the inner engine. The second effect is that of *relativistic beaming*. The radiation from any object that is radiating isotropically in its own rest frame, but moving with a large Lorentz factor γ in the observer frame, is beamed into a small angle $1/\gamma$ around its direction of motion. This is an effect of special relativity. It has nothing to do with the ejecta's geometry (spherical or jet) but only with the fact that the ejecta is moving relativistically. The effect of relativistic beaming allows an observer to see only a small angular extent, of size $1/\gamma$ centered around the line of sight. Since we know the flow is ultra-relativistic (initially $\gamma > 100$), there is no question that the relativistic beaming effect is always relevant for GRBs. The question we are interested in is that of the existence of "jets".

The idealized description of a jet is a flow that occupies only a conical volume with half opening angle θ_0. In fact, the relativistic dynamics is such that the width of the material in the direction of its propagation is much smaller than its distance from the source by a factor of $1/\gamma^2$. The flow, therefore, does not fill the whole cone. Instead it occupies only a thin disk at its base, looking more like a flying pancake (Piran 1999) (see Fig. 15.4). If the "inner engine" emits two such jets in opposite directions then the total solid angle towards which the flow is emitted is $\Omega = 2\pi\theta_0^2$. Whether the relativistic flow is in the form of a jet or a sphere has three important implications.

The total emitted energy Optical observations of afterglows enabled redshift determinations, and therefore reasonably accurate estimates of the distance, D, to these events (the uncertainty is now in the cosmological parameters of the Universe). The so called "isotropic energy" can then be inferred from the fluence F (the total observed energy per unit area at Earth) as $E_{\rm iso} = 4\pi D^2 F$ (taking cosmological corrections into account, $D = d_L/\sqrt{1+z}$ where d_L is the luminosity distance and z is the redshift). The numbers obtained in this way range from 10^{51} erg to 10^{54} erg with the record of 3×10^{54} erg held by the famous GRB 990123. These huge numbers approach the equivalent energy of a solar mass, all emitted in a few tens of seconds!

These calculations assumed that the source emitted the same amount of energy in all directions. If instead the emission is confined to some solid angle Ω then the true energy is $E = \Omega D^2 F$. As we show later Ω is very weakly constrained by the GRB itself and can be as low as 10^{-6}. If so, the true energy in each burst $E \ll E_{\rm iso}$. We will show later that interpretation of the multi-wavelength afterglow lightcurves indeed indicates that some bursts are jets with solid angles considerably less than 4π. The isotropic energy estimates may be fooling us by a few orders of magnitude! Clearly this is of fundamental importance when considering models for the sources of GRBs.

The event rate In its glory days, BATSE detected about one burst per day. With the help of several redshift measurements, or alternatively, with the use of the cumulative brightness distribution (the Log *N*–Log *S* curve), this translates to about 10^{-7} bursts per year per galaxy or 0.5 bursts per Gpc^{-3} per year (Schmidt 1999, 2001). However, if the emission is collimated to $\Omega \ll 4\pi$ then we do not see most of the events. The true event rate is then larger than that measured by BATSE by a factor of $4\pi/\Omega$. Again this is of fundamental importance. Clearly, the corrected GRB event rate must not exceed that of compact binary mergers or the birth rate of massive stars if these are to produce the majority of the observed GRBs.

The physical ejection mechanism Different physical models are needed to explain collimated and isotropic emission. For example, in the collapsar model (e.g., MacFadyen & Woosley 1999b), relativistic ejecta that are believed to create the GRB are produced only around the rotation axis of the collapsing star with a half opening angle of about $\theta_0 \cong 0.1$. Such models would have difficulties explaining isotropic bursts as well as very narrow jets.

Table 15.1. *The spectral index β and the temporal index α as functions of p for a spherical and a jet-like evolution*

	Spectral index β, $F_\nu \propto \nu^{-\beta}$	Lightcurve index α, $F_\nu \propto t^{-\alpha}$	
		Sphere	Jet
$\nu < \nu_a$	$\beta = -2$	$\alpha = -1/2$	$\alpha = 0$
$\nu_a < \nu < \nu_m$	$\beta = -1/3$	$\alpha = -1/2$	$\alpha = 1/3$
$\nu_m < \nu < \nu_c$	$(p-1)/2 \cong 0.7$	$\alpha = 3(p-1)/4 \cong 1.05$	$\alpha = p \cong 2.4$
		$\alpha = 3\beta/2$	$\alpha = 2\beta + 1$
$\nu > \nu_c$	$p/2 \cong 1.2$	$\alpha = (3p-2)/4 \cong 1.3$	$\alpha = p \cong 2.4$
		$\alpha = 3\beta/2 - 1/2$	$\alpha = 2\beta$

Typical values are quoted using $p = 2.4$. The parameter-free relation between α and β is given for each case (eliminating p). The difference in α between a jet and a sphere is always substantial at all frequencies.

15.8.1 The jet break

As the afterglow evolves, γ decreases and it will eventually fall below the initial inverse opening angle of the jet. The observer will notice that some of the sphere is missing from the fact that less radiation is observed. This effect alone will produce a significant break, steepening the lightcurve decay by a factor of $\gamma^2 \propto t^{-3/4}$ even if the dynamics of each fluid element has not changed. The transition should occur at the time $t_{\rm jet}$ when $1/\gamma \cong \theta_0$. Observing this time can therefore provide an estimate of the jet's opening angle according to

$$t_{\rm jet} \approx 6.2\,{\rm hr}\,(1+z)(E_{52}/n_1)^{1/3}(\theta_0/0.1)^{8/3} \qquad (15.11)$$

Additionally, Rhoads (1999) has shown that at about the same time (see however Panaitescu & Mészáros 1999; Mészáros & Rees 1999; Moderski, Sikora & Bulik 2000), the jet will begin to spread laterally so that its opening angle $\theta(t) \sim 1/\gamma$. The ejecta now encounter more surrounding matter and decelerate faster than in the spherical case. The Lorentz factor then decays exponentially with the radius and as $\gamma \propto t^{-1/2}$ with observed time. Taking this into account, the observed break is even more significant. The slow cooling spectrum given in Fig. 15.5 evolves with decreasing peak flux $F_{\rm m} \propto t^{-1}$ and the break frequencies evolve as $\nu_{\rm m} \propto t^{-2}$, $\nu_{\rm c} \propto t^0$ and $\nu_{\rm a} \propto t^{-1/5}$. This translates to a temporal decay at a given frequency given in Table 15.1.

The jet break is a hydrodynamic one. It should therefore appear at the same time at all frequencies – an achromatic break.[2] Though an achromatic break is considered to be a strong signature of a jet, one should keep in mind that any other hydrodynamic transition will also produce an achromatic break. To name a few: the transition from relativistic to non-relativistic dynamics, a jump in the ambient density or the supply of new energy from slower shells that catch up with the decelerated flow. However, the breaks produced by the transition from a

[2] Sari (1997), argued that there may be about a factor of 2 difference in the effective transition time between the four different spectral regimes (e.g., below or above $\nu_{\rm m}$) due to the fact that the emission in these different regimes weighs contributions from various emission radii differently.

spherical-like evolution (when $1/\gamma < \theta_0$) to a spreading jet have a well-defined prediction for the change in the temporal decay indices. The amount of break depends on the spectral regime that is observed. It can be seen from Table 15.1 that the break is substantial ($\Delta\alpha > 0.5$ in all regimes) and should be easily identified.

15.9 Observational evidence for collimated outflow (jets)

The theory of jet evolution and of the resulting lightcurves was worked out before evidence for jets was obtained. In fact, Rhoads (1999), has used this theory to constrain the amount of collimation in GRB 970508, which did not show any significant steepening of the afterglow lightcurve. He concluded that the opening angle of a jet, if it exists, must be more than 30 degrees. We note that if the jet's opening angle is of order unity, the total energy may still be about an order of magnitude lower than the isotropic estimate. However, in this case the break will be "hidden" as it will overlap the transition to non-relativistic dynamics. Based on late time radio data, it was suggested that this is the case for GRB 970508 (Frail, Waxman & Kulkarni 2000).

The first claim for narrow jets in GRBs came from Sari, Piran and Halpern (1999). They noted that the observed decays in GRB afterglows that do not show a break either have a shallow slope $F_\nu \propto t^{-1.2}$ or a very steep slope $F_\nu \propto t^{-2}$. They argued that the rapidly decaying bursts are those in which the ejecta were a narrow jet and the break in the lightcurve were before the first observations. Interestingly, evidence for jets is found when the inferred energy (without taking jets into account) is the largest. This implies that the jets account for a considerable fraction of the wide luminosity distribution seen in GRBs, and the true energy distribution is less wide than it seems to be.

The predicted lightcurve transition (from a regular to a fast decay caused by a jet) has been observed in the optical afterglow of GRB 990123 (Kulkarni *et al.* 1999a; Castro-Tirado *et al.* 1999; Fruchter *et al.* 1999). However, no evidence for such an increase in the decay rate was found in near-infrared K-band observations (Kulkarni *et al.* 1999a). A similar transition was better sampled in afterglow data of GRB 990510 (Fig. 15.6); optical observations of GRB 990510 show a clear steepening of the rate of decay of the light simultaneously in all optical bands between ~3 hours and several days (Harrison *et al.* 1999; Stanek *et al.* 1999) to roughly $F_\nu(t) \propto t^{-2.2}$. Together with radio observations, which also reveal a transition, it is found that the transition is very much frequency-independent; this virtually excludes explanations in terms of the passage of the cooling frequency, but is what is expected in the case of beaming (Harrison *et al.* 1999). Harrison *et al.* (1999) derive a jet opening angle (from the jet break time) of $\theta_0 \cong 0.08$, which for this burst would reduce the total energy in gamma rays to $\sim 10^{51}$ erg.

Frail *et al.* (2001) collected the jet break times for a sample of GRBs with known redshifts. From these, a wide range of jet opening angles is inferred in GRBs: from 3° to more than 25°, with a strong concentration near 4°. This relatively narrow collimation implies that the observed GRB rate has to be corrected for the fact that conical fireballs are visible to only a fraction of observers. Frail *et al.* find that the "true" GRB rate is ~500 times larger than the observed GRB rate. Although the isotropic equivalent energies of GRBs range from about 5×10^{51} to 1.4×10^{54} erg, when one corrects the observed gamma-ray energies for the geometry of the outflow, GRB energies appear narrowly clustered around 5×10^{50} erg (see Fig. 15.7). Similar conclusions were obtained by Piran *et al.* (2001) and Panaitescu and Kumar (2001b).

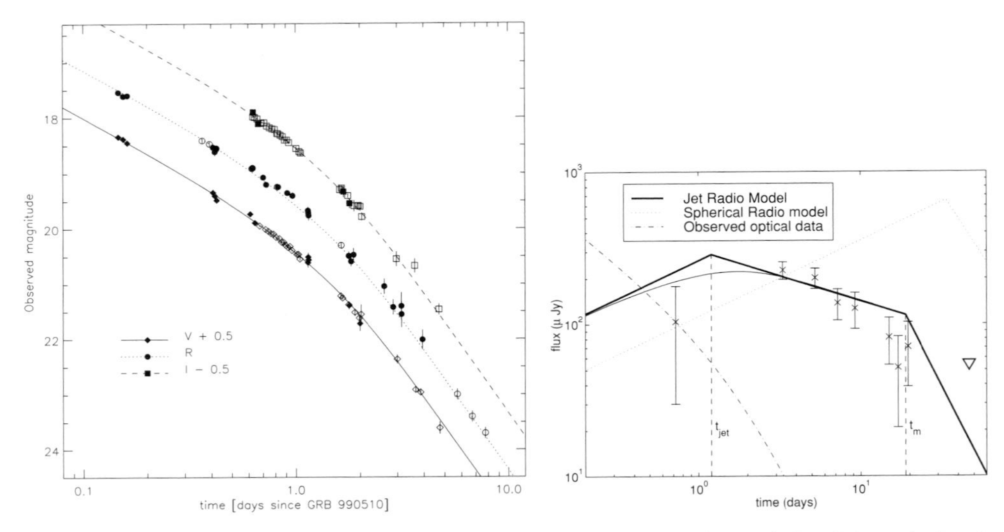

Fig. 15.6. GRB 990510, the "classical" case for a "jet": an achromatic break in optical and radio at $t_{\rm jet} = 1.2$ days implying a jet opening angle $\theta_0 = 0.08$. The temporal slopes before and after the break agree well with the theory if $p = 2.2$. For this burst the isotropic gamma-ray energy $E_{\rm iso} = 2.9 \times 10^{53}$ erg but the "true" total energy is only $E = 10^{51}$ erg. From Harrison *et al.* (1999).

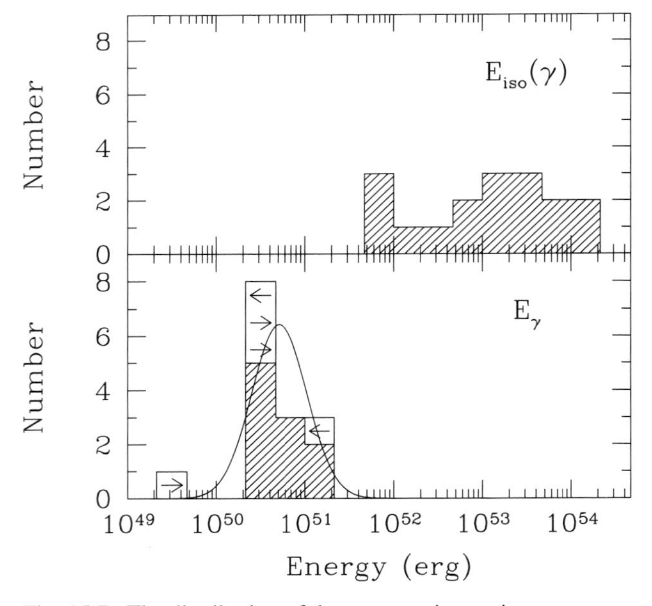

Fig. 15.7. The distribution of the apparent isotropic gamma-ray energy of GRBs with known redshifts (*top*) versus the geometry-corrected energy (*bottom*). While the isotropic energy $E_{\rm iso}$ spans three orders of magnitudes, the geometrically corrected energy, $E_\gamma = E_{\rm iso}\theta^2/2$, is very narrowly distributed. This implies that the sources of GRBs produce roughly the same amount of energy, about 5×10^{50} erg, but that energy is distributed over a variety of angles resulting in a wide distribution of isotropic energies. From Frail *et al.* (2001) and Perna *et al.* (2003).

The central engines of GRBs thus produce approximately a similar amount of energy, and the broad range of fluence and luminosity observed for GRBs appears to be largely the result of a wide variation of opening angles. The reason why this range in angles exists is currently not understood. Our understanding of GRBs has come a long way in the past four years. It is interesting to note that before the redshift era, most models assumed that the events were standard candles with energies of about 10^{51} erg. As more and more redshifts were determined, the energy record increased steadily up to 10^{54} erg. The standard candle hypothesis was abandoned. It is remarkable that now, when more detailed understanding allows us to infer the beaming angles of these explosions, the true energy budget is back at $\sim 10^{51}$ erg, and the explosions are once again standard candles (though not in the same sense as before).

Postnov, Prokhorov and Lipunov (2001), Rossi, Lazzati and Rees (2002), and Zhang and Mészáros (2002) pointed out that another interpretation is possible for the Frail *et al.* result. Instead of a variety of jets with different opening angles, a standard jet can be invoked with energy density per unit solid angle falling away from the axis as θ^{-2}; the differences in the apparent opening angle then come from variations in the orientation of the observer relative to the jet's axis. Perna, Sari & Frail (2003) showed that the distribution of the observed opening angles is consistent with this assumption, adding credence to the universal jet model. If this model is correct, the rate of GRBs is much lower, because it should not be corrected by the factor of 500 of Frail *et al.*; however, the energy is still low, of order 10^{51} erg.

15.10 Polarization: a promising tool

An exciting possibility to further constrain the models and obtain a more direct proof of the geometrical picture of "jets" is to measure linear polarization. Varying polarization at optical wavelengths has been observed in GRB afterglows at the level of a few to ten percent (Covino *et al.* 1999, 2002; Wijers *et al.* 1999; Rol *et al.* 2000; Bersier *et al.* 2003).

High levels of linear polarization are usually the smoking gun of synchrotron radiation. The direction of the polarization is perpendicular to the magnetic field and can be as high as 70%. Gruzinov and Waxman (1999) and Medvedev and Loeb (1999) considered the emission from spherical ejecta, which by symmetry should produce no polarization on the average, except for fluctuations of order a few percent. Polarization is more natural if the ejecta are a "jet" and the line of sight to the observer is within the jet but does not coincide with its axis. In this case, the spherical symmetry is broken (Gruzinov 1999; Ghisellini & Lazzati 1999; Sari 1999), and the polarization produced by synchrotron radiation will not vanish. For simplicity, assume that the magnetic field behind the shock is directed along the shock's plane (the results hold more generally, as long as the magnetic field has a preferred direction). The synchrotron polarization from each part of the shock front, which is perpendicular to the magnetic field, is therefore directed radially.

As long as the relativistic beaming angle $1/\gamma$ is narrower than the physical size of the jet θ_0, one is able to see a full ring and therefore the radial polarization averages out (the first frame, with $\gamma\theta_0 = 4$ of the left plot in Fig. 15.8). As the flow decelerates, the relativistic beaming angle $1/\gamma$ becomes comparable to θ_0 and only a part of the ring is visible; net polarization is then observed. Note that, due to the radial direction of the polarization from each fluid element, the total polarization is maximal when a quarter ($\gamma\theta_0 = 2$ in Fig. 15.8) or when three quarters ($\gamma\theta_0 = 1$ in Fig. 15.8) of the ring are missing (or radiate less efficiently)

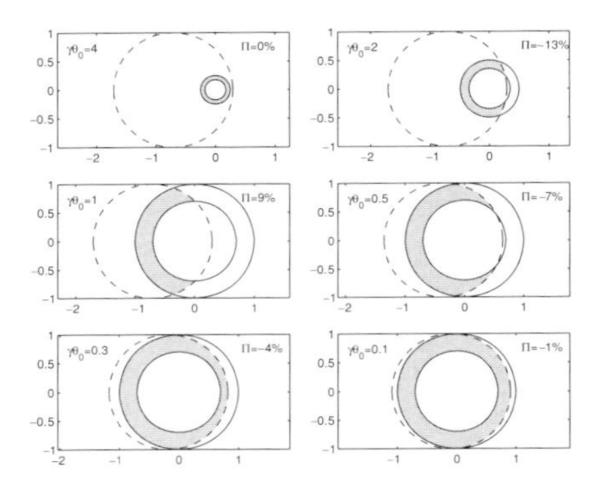

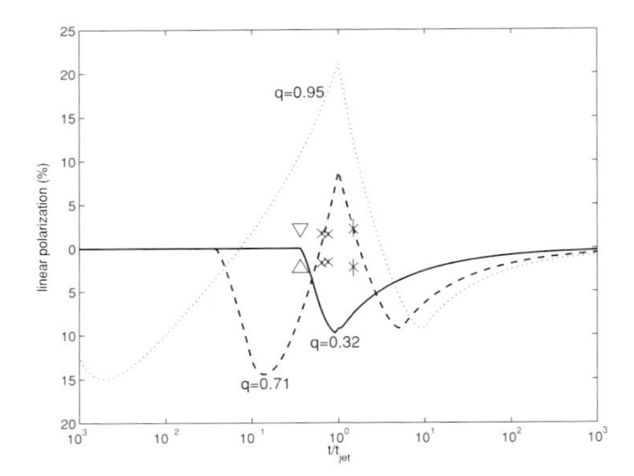

Fig. 15.8. *Left*: Shape of the emitting region. The dashed line marks the physical extent of the jet, and solid lines give the viewable region $1/\gamma$. The observed radiation arises from the gray shaded region. In each frame, the percentage of polarization is given at the top right and the initial size of the jet relative to $1/\gamma$ is given on the left. The frames are scaled so that the size of the jet is unity. *Right*: Observed and theoretical polarization lightcurves for three possible offsets of the observer relative to the jet axis. Observational data for GRB 990510 are marked by crosses, assuming $t_{\rm jet} = 1.2$ days. The upper limit for GRB 990123 is given by a triangle, assuming $t_{\rm jet} = 2.1$ days.

and vanishes for a full and a half ring. The polarization, when more than half of the ring is missing, is perpendicular to the polarization direction when less than half of it is missing.

At late stages the jet expands sideways and since the offset of the observer from the physical center of the jet is constant, spherical symmetry is regained. The vanishing and reoccurrence of significant parts of the ring results in a unique prediction: there should be three peaks of polarization, with the polarization position angle during the central peak rotated by 90° with respect to the other two peaks. If the observer is very close to the center, more than half of the ring is always observed, and therefore only a single direction of polarization is expected. A few possible polarization lightcurves are presented in Fig. 15.8.

15.11 The reverse shock emission: theory and observations

The previous sections discussed the theory and the observations of the "late" afterglow, hours or more after the burst. During that period, most of the energy of the system had already been given to the shocked surroundings, and it is that region that dominates the emission. However, during the first few tens of seconds, the evolution of the Lorentz factor as a function of time is not self-similar. There are two shocks: a forward shock going into the surrounding medium and a reverse shock going into the expanding ejecta. The hydrodynamic details were discussed by Sari and Piran (1995).

During the initial stages, the internal energy stored behind the shocked-surrounding matter and the energy of the shocked ejecta are comparable. However, the temperature of the shocked ejecta is much lower, typically by a factor of $\gamma \sim 10^2$. This results in an additional emission component with a typical frequency lower by a factor of $\gamma^2 \sim 10^4$, which, for typical parameters, is near the optical passband. Contrary to the "standard" late afterglow, this emission is very sensitive to the initial Lorentz factor. Theoretical predictions for such a flash were given in detail by Sari and Piran (1999a, 1999c) and were earlier suggested as a possibility by Mészáros and Rees (1997).

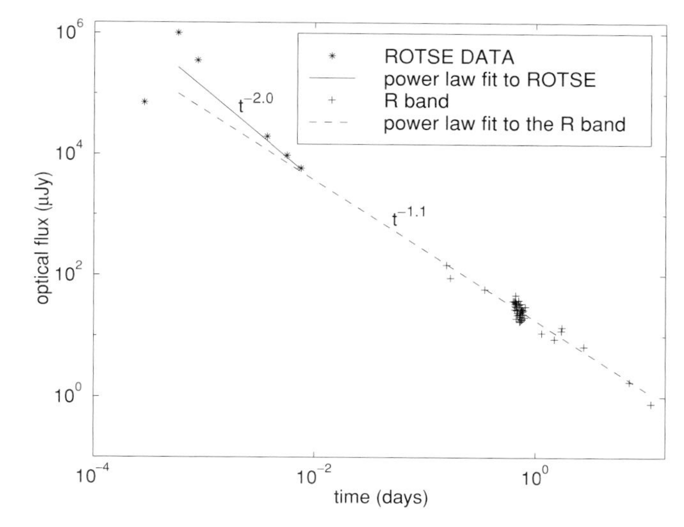

Fig. 15.9. R-band lightcurve of the afterglow of GRB 990123. The ROTSE data show that the optical lightcurve peaked at $m_V \sim 9$ (Akerlof *et al.* 1999). The dashed line indicates a power-law fit to the lightcurve(for $t > 0.1$ days), which has exponent -1.12 ± 0.03. From Galama *et al.* (1999).

One of the most exciting events in the field of afterglow studies was the detection of bright (9th magnitude) optical emission simultaneous with GRB 990123 by the ROTSE team (Akerlof *et al.* 1999). The ROTSE telescope obtained its first images only 22 seconds after the start of GRB 990123 (i.e., during the GRB), following a notification received from BATSE aboard the Compton Gamma-Ray Observatory. The ROTSE observations show that the optical lightcurve peaked at $m_V \sim 9$ magnitudes some 60 seconds after the event (Akerlof *et al.* 1999). After maximum a fast decay followed for at least 15 minutes. The late-time afterglow observations show a more gradual decline (Galama *et al.* 1999; Kulkarni *et al.* 1999a; Castro-Tirado *et al.* 1999; Fruchter *et al.* 1999, Sari & Piran 1999b) (see Fig. 15.9).

The redshift $z = 1.6$, inferred from absorption features in the OT's spectrum, implies that the optical flash would have been as bright as the full Moon had the GRB occurred in the nearby galaxy M31 (Andromeda). A different way to put this in perspective is that the flash was some 18 mag brighter (absolute) than the brightest supernovae. Galama *et al.* (1999) have shown that, if one assumes that the emission detected by ROTSE comes from a non-relativistic source of size ct, then the observed brightness temperature $T_b \geq 10^{17}$ K of the optical flash exceeds the Compton limit of 10^{12} K. This confirms the highly relativistic nature of the GRB source.

The observed optical properties of this event are well described by emission from the reverse shock that initially decelerates the ejecta, provided that the initial Lorentz factor is about 200 (Sari & Piran 1999b; Mészáros & Rees 1999). It takes tens of seconds for the reverse shock to sweep through the ejecta and produce the bright flash. Later, the shocked hot matter expands adiabatically and the emission quickly shifts to lower frequencies and considerably weakens.

The ROTSE observations show that the prompt optical and gamma-ray lightcurves do not track each other (Akerlof *et al.* 1999). In addition, detailed comparison of the prompt optical emission with the BATSE spectra of GRB 990123 (at three epochs for which both optical and gamma-ray information is available) shows that the ROTSE emission is not a simple extrapolation of the GRB spectrum to much lower energies (Galama *et al.* 1999; Briggs *et al.* 1999).

If this interpretation is correct, GRB 990123 would be the first burst in which all three emitting regions have been seen: internal shocks causing the GRB, the reverse shock causing the prompt optical flash, and the forward shock causing the afterglow. The emissions thus arise from three different emitting regions, explaining the lack of correlation between the GRB, the prompt optical and the late-time optical emission (Galama *et al.* 1999) (but see Liang *et al.* 1999).

Another new ingredient that was found in GRB 990123 is a radio flare (Kulkarni *et al.* 1999b). Contrary to all other afterglows, where the radio peaks around a few weeks and then decays slowly, this burst had a fast rising flare, peaking around a day and then decaying quickly. This can be interpreted as emission from the cooling ejecta that were heated earlier on by the reverse shock. Using the Blandford and McKee (1976) self-similar solution to derive the evolution of the ejecta and their emission properties one finds that the typical frequency scales as $\nu_{\rm m}^r \propto t^{-73/48}$ and the flux at that frequency scales as $F_{\rm m}^r \propto t^{-47/48}$ (Sari & Piran 1999a) (see Kobayashi & Sari (2001) for revised scalings when the temperature of the ejecta is non-relativistic). Therefore, within a day the emission from the adiabatically cooling ejecta that produced the 60 s optical flash in GRB 990123 is expected to shift to radio frequencies (Sari & Piran 1999b). Using the observed optical flash and the above scalings, a good fit to the radio data is obtained. The optical flash and the radio flare may therefore be related.

Given the above interpretation of the reverse shock emission, it is important to ask whether GRB 990123 is an exception, or whether the phenomena of radio flares and optical flashes are more common. Radio flares appear to exist in other cases (Frail *et al.* 2001). However, since early radio data are usually sparse, and these events did not have an early optical observation to find the associated optical flash, the interpretation in terms of emission from the reverse shock is less secure than in the case of GRB 990123. In the optical, from robotic optical experiments such as ROTSE and LOTIS, strong upper limits exist for several bursts. The upper limits show that the optical flash does not scale with the fluence of the event (Akerlof *et al.* 2000; Kehoe *et al.* 2001). However, with reasonably small changes in the density or the initial Lorentz factor, those events could have escaped detection (Kobayashi 2000). HETE-II, or new satellites like Swift, provide or will provide accurate positioning on timescales of seconds, and strong constraints on the generality of optical flashes and radio flares will be obtained.

15.12 GRB host galaxies and redshifts

Host galaxies of GRBs serve a dual purpose: they determine the redshifts, which are necessary for a complete physical modeling of the bursts, and they provide some insights about the possible nature of the progenitors, e.g., their relation to massive star formation, etc. The subject has been reviewed previously, e.g., by Djorgovski *et al.* (2001b, 2003).

15.12.1 Overall properties of GRB hosts

As of this writing, plausible or certain host galaxies have been found for all but one or two of the bursts with optical, radio, or X-ray afterglows localized with arcsecond precision.

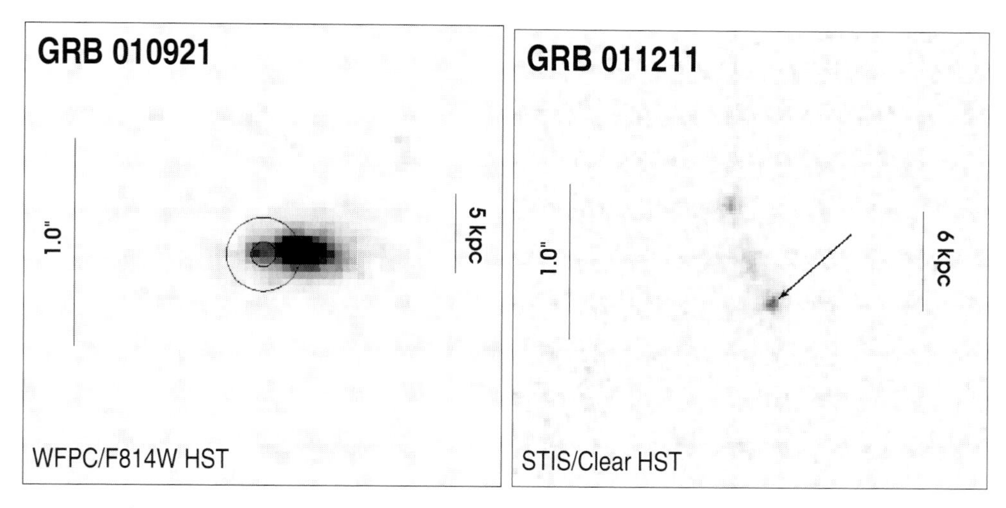

Fig. 15.10. Typical HST images of two GRB host galaxies. The OTs are indicated by the circle (for GRB 010921) and by the arrow (for GRB 011211). Side bars give the projected angular scale, and the physical scale at the source. From Bloom, Kulkarni & Djorgovski (2002), images courtesy of J. S. Bloom.

Two examples are shown in Fig. 15.10. The median apparent magnitude is $R \approx 25$ mag, with tentative detections or upper limits reaching down to $R \approx 29$ mag. The missing cases are at least qualitatively consistent with being in the faint tail of the observed distribution of host galaxy magnitudes.

Down to $R \sim 25$ mag, the observed distribution is consistent with deep field galaxy counts (Brunner, Connolly & Szalay 1999), but fainter than that, complex selection effects may be playing a role. It can also be argued that the observed distribution should correspond roughly to luminosity-weighted field galaxy counts. However, the actual distribution would depend on many observational selection and physical (galaxy evolution) effects, and a full interpretation of the observed distribution of GRB host galaxy magnitudes requires a careful modeling. We note also that the observations in the visible probe the UV in the restframe, and are thus especially susceptible to extinction. However, sub-mm detections of dusty GRB hosts are currently limited by the available technology to only a handful of ultra-luminous sources.

Starting with the first redshift measurement that unambiguously demonstrated the cosmological nature of GRBs (Metzger *et al.* 1997) by late 2002 there were over 30 redshifts measured for GRB hosts and/or afterglows. The median redshift is $\langle z \rangle \approx 1.0$, spanning the range from 0.25 (or 0.0085, if the association of GRB 980425 with SN 1998bw is correct) to 4.5 (for GRB 000131). The majority of redshifts so far are from the spectroscopy of host galaxies, but an increasing number are based on the absorption-line systems seen in the spectra of the afterglows (which are otherwise featureless power-law continua). Figure 15.11 shows two examples. Reassuring overlap exists in several cases; invariably, the highest-z absorption system corresponds to that of the host galaxy, and has the strongest lines. In some cases (a subset of the so-called "dark bursts") no optical transient (OT) is detected, but a combination of the X-ray (XT) and radio transients (RT) unambiguously pinpoints the host galaxy.

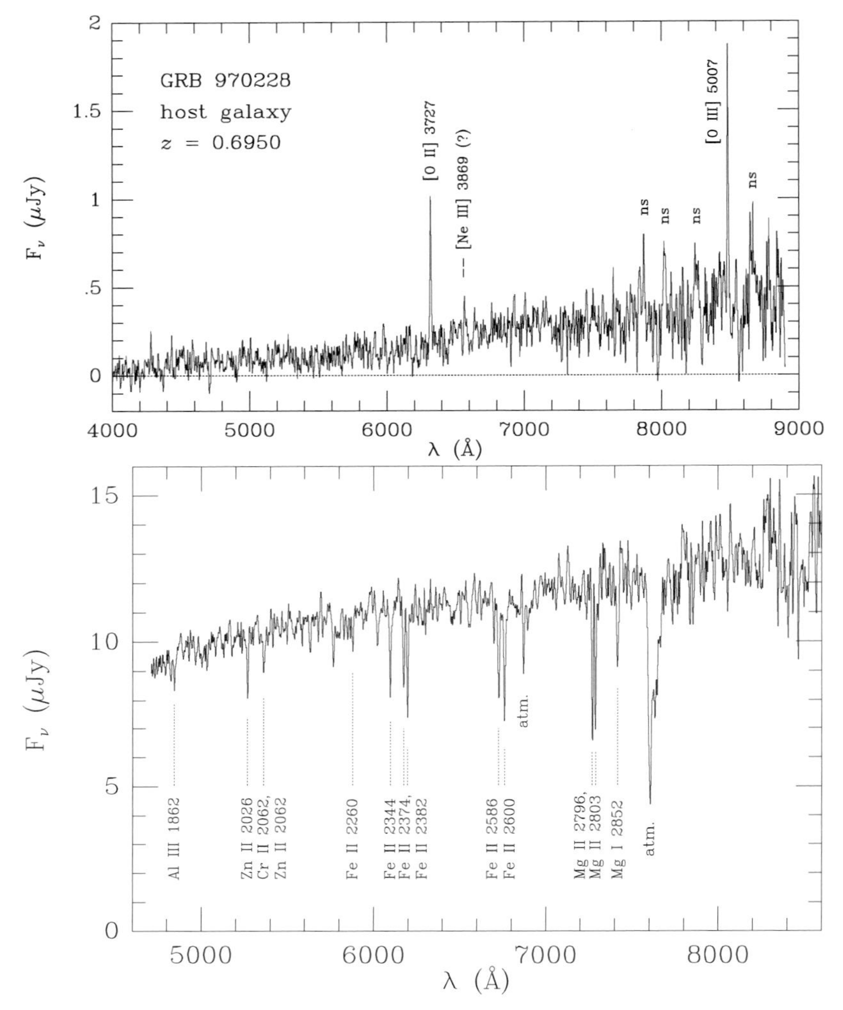

Fig. 15.11. Typical spectra of a GRB host galaxy, showing the standard emission lines indicative of active star formation (for GRB 970228, *top*), and of an OT, showing the strong absorption lines from the ISM in the host galaxy (for GRB 990123, *bottom*). The top figure is from Bloom, Djorgovski & Kulkarni (2001). The bottom figure is reprinted with permission from *Nature* (Kulkarni *et al.* 1999a) copyright 1999 Macmillan Publishers Ltd.

A new method for obtaining redshifts may come from the X-ray spectroscopy of afterglows, using the Fe K line at ~6.55 keV (Piro *et al.* 1999, 2000; Antonelli *et al.* 2000), or the Fe absorption edge at ~9.28 keV (Weth *et al.* 2000; Yohshida *et al.* 1999; Amati *et al.* 2000). Rapid X-ray spectroscopy of GRB afterglows may become a powerful tool for understanding their physics and origins.

Are the GRB host galaxies special in some way? If GRBs are somehow related to massive star formation (e.g., Paczyński 1998; Totani 1997, etc.), it may be worthwhile to examine their absolute luminosities and star formation rates (SFR), or spectroscopic properties in general. This is hard to answer (Krumholz, Thorsett & Harrison 1998; Hogg & Fruchter 1999; Schaefer 2000) from their visible (~ rest-frame UV) luminosities alone: the observed light traces an indeterminate mix of recently formed stars and an older population, and cannot be unambiguously interpreted in terms of either the total baryonic mass, or the instantaneous SFR.

The magnitude and redshift distributions of GRB host galaxies are typical for the normal, faint field galaxies, as are their morphologies (Odewahn *et al.* 1998; Holland 2001; Bloom, Kulkarni & Djorgovski 2002) when observed with the HST: they are often compact, and sometimes suggestive of a merging system (Djorgovski, Bloom & Kulkarni 2002; Hjorth *et al.* 2002), but that is not unusual for galaxies at comparable redshifts.

Within the host galaxies, the distribution of GRB–host offsets follows the light distribution closely (Bloom, Kulkarni & Djorgovski 2002), which is roughly proportional to the density of star formation (especially for the high-z galaxies). It is thus fully consistent with a progenitor population associated with the sites of massive star formation.

Spectroscopic measurements provide direct estimates of recent, massive SFR in GRB hosts. Most of them are based on the luminosity of the [O II] 3727 doublet (Kennicut 1998), the luminosity of the UV continuum at $\lambda_{rest} = 2800$ Å (Madau, Pozzetti & Dickinson 1998), in some cases (e.g., Kulkarni *et al.* 1998) from the luminosity of the Lyα 1216 line, and in others (e.g., Djorgovski *et al.* 1998) from the luminosity of Balmer lines (Kennicut 1998). All of these estimators are susceptible to the internal extinction and its geometry, and have an intrinsic scatter of at least 30%. The observed *unobscured* SFRs range from a few tenths to a few $M_\odot$ yr^{-1}. Applying the reddening corrections derived from the Balmer decrements of the hosts, or from the modeling of the broad-band colors of the OTs (and further assuming that they are representative of the mean extinction for the corresponding host galaxies) increases these numbers typically by a factor of a few. All this is entirely typical for the normal field galaxy population at comparable redshifts. However, such measurements are completely insensitive to any fully obscured SFR components.

Equivalent widths of the [O II] 3727 doublet in GRB hosts, which may provide a crude measure of the SFR per unit luminosity (and a worse measure of the SFR per unit mass), are on average somewhat higher (Djorgovski *et al.* 2001a) than those observed in magnitude-limited field galaxy samples at comparable redshifts (Hogg *et al.* 1998). A larger sample of GRB hosts, and a good comparison sample, matched both in redshift and magnitude range, are necessary before any solid conclusions can be drawn from this apparent difference.

One intriguing hint comes from the flux ratios of [Ne III] 3869 to [O II] 3727 lines: they are on average a factor of 4 to 5 higher in GRB hosts than in star-forming galaxies at low redshifts (Djorgovski *et al.* 2001b). Strong [Ne III] requires photoionization by massive stars in hot H II regions, and may represent indirect evidence linking GRBs with massive star formation.

The interpretation of the luminosities and observed star formation rates is vastly complicated by the unknown amount and geometry of extinction. The observed quantities (in the visible) trace only the unobscured stellar component, or the components seen through optically thin dust. Any stellar and star formation components hidden by optically thick dust cannot be estimated at all from these data, and require radio and sub-mm observations.

Both observational windows, the optical/NIR (rest-frame UV) and the sub-mm (rest-frame FIR) suffer from some biases: the optical band is significantly affected by dust obscuration, while the sub-mm and radio bands lack sensitivity, and therefore uncover only the most prodigiously star-forming galaxies. As of late 2002, radio and/or sub-mm emission powered by obscured star formation has been detected from four GRB hosts (Berger, Kulkarni & Frail 2001; Berger *et al.* 2002b; Frail *et al.* 2002). The surveys to date are sensitive only to the ultra-luminous ($L > 10^{12}\,L_\odot$) hosts, with SFR of several hundred $M_\odot\,yr^{-1}$. Modulo the uncertainties posed by the small number statistics, the surveys indicate that about 20% of GRB hosts are objects of this type, where about 90% of the total star formation takes place in obscured regions.

Given the uncertainties of the geometry of optically thin and optically thick dust, optical colors of GRB hosts cannot be used to make any meaningful statements about their net star formation activity. The broad-band optical colors of GRB hosts are not distinguishable from those of normal field galaxies at comparable magnitudes and redshifts (Bloom, Djorgovski & Kulkarni 2001; Sokolov *et al.* 2001). It is notable that the optical/NIR colors of GRB hosts detected in the sub-mm are much bluer than typical sub-mm selected galaxies, suggesting that the GRB selection may be probing a previously unrecognized population of dusty star-forming galaxies.

On the whole, the GRB hosts seem to be representative of the normal, star-forming field galaxy population at comparable redshifts, and so far there is no evidence for any significant systematic differences between them.

15.12.2 GRB hosts in the context of galaxy evolution

The observed redshift distribution of GRB hosts is about what is expected for an evolving, normal field galaxy population at these magnitude levels. There is an excellent qualitative correspondence between the observations and simple galaxy evolution models (Mao & Mo 1998).

If GRBs follow the luminous mass, then the expected distribution would be approximated by the luminosity-weighted galaxy luminosity function (GLF) for the appropriate redshifts. The hosts span a wide range of luminosities, with a characteristic absolute restframe B-band magnitude $M_{B,*} \approx -20$ mag, approximately half a magnitude fainter than in the GLF at $z \approx 0$, but commensurate with the late-type (i.e., star-forming disk) galaxy population at $z \approx 0$ (Madgwick *et al.* 2002; Norberg *et al.* 2002). This is somewhat surprising, since one expects that the evolutionary effects would make the GRB host galaxies, with a typical $z \sim 1$, brighter than their descendants today. The GRB host GLF also has a somewhat steeper tail than the composite GLF at $z \approx 0$, but again similar to that of the star-forming, late-type galaxies. This is in a broad agreement with the results of deep redshift surveys which probe the evolution of field galaxy populations out to $z \approx 1$ (Lilly *et al.* 1995; Ellis 1997; Fried *et al.* 2001; Lin *et al.* 1999).

The interpretation of these results is complex: the observed light reflects an unknown combination of the unobscured fraction of recent star formation (especially in the high-z galaxies, where we observe the restframe UV continuum) and the stellar populations created up to that point. Our understanding of the field galaxy evolution in the same redshift range as probed by the GRB hosts is still largely incomplete. Different selection effects may be plaguing the field and the GRB host samples. While much remains to be done, it seems that GRB hosts provide a new, independent check on the traditional studies of galaxy evolution at moderate and high redshifts.

15.13 GRBs and cosmology

While interesting on their own, GRBs are now rapidly becoming powerful tools to study the high-redshift Universe and galaxy evolution, thanks to their apparent association with massive star formation, and their brilliant luminosities.

There are three basic ways of learning about the evolution of luminous matter and gas in the Universe. First, a direct detection of sources (i.e., galaxies) in emission, either in the UV/optical/NIR (the unobscured components), or in the FIR/sub-mm/radio (the obscured component). Second, the detection of galaxies selected in absorption along the lines of sight to luminous background sources, traditionally QSOs. Third, diffuse extra-galactic backgrounds, which bypass all of the flux or surface brightness selection effects plaguing all surveys of discrete sources found in emission, but at a price of losing the redshift information, and the ability to discriminate between the luminosity components powered by star formation and powered by AGN. Studies of GRB hosts and afterglows can contribute to all three of these methodological approaches, bringing in new, independent constraints for models of galaxy evolution and of the history of star formation in the Universe.

15.13.1 Dark bursts: probing the obscured star formation history

Within months of the first detections of GRB afterglows, no OTs were found associated with some well-localized bursts despite deep and rapid searches; the prototype "dark burst" was GRB 970828 (Djorgovski *et al.* 2001a). Perhaps the most likely explanation for the non-detections of OTs when sufficiently deep and prompt searches are made is that they are obscured by dust in their host galaxies. This is an obvious culprit if indeed GRBs are associated with massive star formation.

Support for this idea also comes from detections of RTs without OTs, including GRB 970828, 990506, and possibly also 981226 (see Frail *et al.* 2000 and Taylor *et al.* 2000). Dust reddening has been detected directly in some OTs (e.g., Ramaprakash *et al.* 1998; Bloom *et al.* 1998; Djorgovski *et al.* 1998, etc.); however, this only covers OTs seen through optically thin dust, and there must be others, hidden by optically thick dust. An especially dramatic case was the RT (Taylor *et al.* 1998) and IR transient (Larkin *et al.* 1998) associated with GRB 980329 (Yost *et al.* 2002). We thus know that at least some GRB OTs must be obscured by dust.

The census of OT detections for well-localized bursts can thus provide a completely new and independent estimate of the mean obscured star formation fraction in the Universe. Recall that GRBs are now detected out to $z \sim 4.5$ and that there is no correlation of the observed fluence with the redshift (Djorgovski *et al.* 2003), so that they are, at least to a first approximation, good probes of the star formation over the observable Universe.

As of late 2002, there have been ~ 70 adequately deep and rapid searches for OTs from well-localized GRBs. We define "adequate searches" as reaching at least to $R \sim 20$ mag within less than a day from the burst, and/or at least to $R \sim 23-24$ mag within 2 or 3 days; this is a purely heuristic, operational definition, and an intentionally liberal one. In just over a half of such searches, OTs were found. Inevitably, some OTs may have been missed due to an intrinsically low flux, an unusually rapid decline rate (Fynbo *et al.* 2001; Berger *et al.* 2002a), or very high redshifts (so that the brightness in the commonly used BVR bands would be affected by the intergalactic absorption). Thus the *maximum* fraction of all OTs (and therefore massive star formation) hidden by the dust is $\sim 50\%$.

This is a remarkable result. It broadly agrees with the estimates that there is roughly an equal amount of energy in the diffuse optical and FIR backgrounds (see, e.g., Madau 1999). This is contrary to some claims in the literature which suggest that the fraction of the obscured star formation was much higher at high redshifts. Recall also that the fractions of the obscured and unobscured star formation in the local Universe are comparable.

There is one possible loophole in this argument: GRBs may be able to destroy the dust in their immediate vicinity (up to ~10 pc?) (Waxman & Draine 2000; Galama & Wijers 2000), and if the rest of the optical path through their hosts (~kpc scale?) was dust-free, OTs would become visible. Such a geometrical arrangement may be unlikely in most cases, and our argument probably still applies. A more careful treatment of the dust evaporation geometry is needed, but it is probably safe to say that GRBs can provide a valuable new constraint on the history of star formation in the Universe.

15.13.2 GRBs as probes of the ISM in evolving galaxies

Absorption spectroscopy of GRB afterglows is now becoming a powerful new probe of the ISM in evolving galaxies, complementary to the traditional studies of QSO absorption line systems. The key point is that the GRBs almost by definition (that is, if they are closely related to the sites of ongoing or recent massive star formation, as the data seem to indicate) probe the lines of sight to dense, central regions of their host galaxies (~1–10 kpc scale). On the other hand, the QSO absorption systems are selected by the gas cross section, and favor large impact parameters (~10–100 kpc scale), mostly probing the gaseous halos of field galaxies, where the physical conditions are very different.

The growing body of data on GRB absorption systems shows exceptionally high column densities of gas, when compared to the typical QSO absorption systems; only the highest column density DLA systems (themselves ostensibly star-forming disks or dwarfs) come close (Mirabal *et al.* 2002; Savaglio, Fall & Fiore 2002; Castro *et al.* 2003). This is completely consistent with the general picture described above. (We are referring here to the highest redshift absorbers seen in the afterglow spectra, which are presumably associated with the host galaxies themselves; lower redshift, intervening absorbers are also frequently seen, and their properties appear to be no different from those of the QSO absorbers.)

This opens the interesting prospect of using GRB absorbers as a new probe of the chemical enrichment history in galaxies in a more direct fashion than is possible with the QSO absorbers, where there may be a very complex dynamics of gas ejection, infall, and mixing at play.

Properties of the GRB absorbers are presumably, but not necessarily (depending on the unknown geometry of the gas along the line of sight) reflecting the ISM of the circumburst region. Studies of their chemical composition do not yet reveal any clear anomalies, or the degree of depletion of the dust, but the samples in hand are still too small to be really conclusive. Also, there have been few searches for the variability of the column density of the gas on scales of hours to days after the burst, with no clear detections so far. Such an effect may be expected if the burst afterglow modifies the physical state of the gas and dust along the line of sight by the evaporation of the dust grains, additional photoionization of the gas, etc. However, it is possible that all such changes are observable only on very short timescales, seconds to minutes after the burst. In any case, a clear detection of a variable ISM absorption against a GRB afterglow would be a very significant result, providing new insight into the circumstances of GRB origins.

15.13.3 High-redshift GRBs: probing the primordial star formation and re-ionization

Possibly the most interesting use of GRBs in cosmology is as probes of the early phases of star and galaxy formation, and the resulting re-ionization of the Universe at $z \sim 6-20$. If GRBs reflect deaths of massive stars, their very existence and statistics would provide a superb probe of the primordial massive star formation and the initial mass function (IMF). They would be by far the most luminous sources in existence at such redshifts (much brighter than SNe, and most AGN), and they may exist at redshifts where there were *no* luminous AGN. As such, they would provide unique new insights into the physics and evolution of the primordial IGM during the re-ionization era (see, e.g., Lamb & Reichart 2001; Loeb 2002a,b).

There are two lines of argument in support of the existence of copious numbers of GRBs at $z > 5$ or even 10. First, a number of studies using photometric redshift indicators for GRBs suggest that a substantial fraction (ranging from ~10% to ~50%) of all bursts detectable by past, current, or forthcoming missions may be originating at such high redshifts, even after folding in the appropriate spacecraft/instrument selection functions (Fenimore & Ramirez-Ruiz 2000; Reichart *et al.* 2001; Lloyd-Ronning, Fryer & Ramirez-Ruiz 2002).

Second, a number of modern theoretical studies suggest that the very first generation of stars, formed through hydrogen cooling alone, were very massive, with $M \sim 100-1000\ M_\odot$ (Bromm, Coppi & Larson 1999; Abel, Bryan & Norman 2000; Bromm, Kudritzki & Loeb 2001; Bromm, Coppi & Larson 2002; Abel, Bryan Norman 20020). While it is not yet absolutely clear that some as-yet unforeseen effect would lead to a substantial fragmentation of a proto-stellar object of such a mass, a top-heavy primordial IMF is at least plausible. It is also not yet completely clear that the (probably spectacular) end of such an object would generate a GRB, but that too is at least plausible (Fryer, Woosley & Heger 2001). Thus, there is some real hope that significant numbers of GRBs and their afterglows would be detectable in the redshift range $z \sim 5-20$, spanning the era of the first star formation and cosmic re-ionization (Bromm & Loeb 2002).

Spectroscopy of GRB afterglows at such redshifts would provide crucial, unique information about the physical state and evolution of the primordial ISM during the re-ionization era. The end stages of the cosmic re-ionization have been detected by spectroscopy of QSOs at $z \sim 6$ (Becker *et al.* 2001; Djorgovski *et al.* 2001c; Fan *et al.* 2001). GRBs are more useful in this context than the QSOs, for several reasons. First, they may exist at high redshifts where there are no comparably luminous AGN yet. Second, their spectra are highly predictable power laws, without complications caused by the broad Lyα lines of QSOs, and can reliably be extrapolated blueward of the Lyα line. Finally, they would provide a genuine snapshot of the intervening ISM, without an appreciable proximity effect which would inevitably complicate the interpretation of any high-z QSO spectrum (luminous QSOs excavate their Stromgren spheres in the surrounding neutral ISM out to radii of at least a few Mpc, whereas the primordial GRB hosts would have a negligible effect of that type; see, e.g., Lazzati *et al.* (2001).

Detection of high-z GRBs is thus an urgent cosmological task. It requires a rapid search for afterglows, as well as high-resolution follow-up spectroscopy, in both the optical and NIR. Such an effort would be well worth the considerable scientific rewards in the end.

Acknowledgements

SGD and RS wish to thank numerous collaborators, including E. Berger, J. S. Bloom, S. M. Castro, F. Chaffee, D. Fox, D. A. Frail, T. Galama, R. Goodrich, J. Halpern, F. A. Harrison, S. R. Kulkarni, A. Mahabal, P. Price, D. Reichart, A. Soderberg, S. Yost, and many others. Our work was supported by grants from the NSF, NASA, and private donors.

References

Abel, T., Bryan, G. & Norman, M. 2000, *ApJ* **540**, 39
2002, *Science* **295**, 93
Akerlof, C., *et al.* 1999, *Nature* **398**, 400
2000, *ApJL* **532**, L25
Amati, L., *et al.* 2000, *Science* **290**, 953
Andersen, M., *et al.* 2000, *A&A* **364**, L54
Antonelli, L. A., *et al.* 2000, *ApJ* **545**, L39
Atkins, R., *et al.* 2000, *ApJ* **533**, L119
Band, D., *et al.* 1993, *ApJ* **413**, 281
Baring, M. G. & Harding, A. K. 1997, *ApJ* **491**, 663
Barthelmy, S. D. Cline, T. L. & Butterworth, P. 2001, in *Gamma Ray Bursts*, eds. R. M. Kippen, R. S. Mallozzi & G. J. Fishman (New York, AIP Press)
Becker, R., *et al.* (the SDSS collaboration) 2001, *AJ* **122**, 2850
Beloboradov, A. 2000, *ApJ* **539**, 25
Berger, E., *et al.* 2000, *ApJ* **545**, 56
Berger, E., Kulkarni, S. R. & Frail, D. A. 2001, *ApJ*, **560**, 652
Berger, E., *et al.* 2002a, *ApJ* **581**, 981
2002b, *ApJ* **588**, 99
Bersier, D., *et al.* 2003, *ApJ* **583**, L63
Bhat, P., *et al.* 1994, *ApJ* **426**, 604
Blandford, R. & Mc Kee, C. F. 1976, *Phys. Fluids* **19**, 1130
Blandford, R. & Znajek, R. 1977, *MNRAS* **179**, 433
Bloom, J. S., *et al.* 1998, *ApJ* **508**, L21
1999, *Nature* **401**, 453
Bloom, J. S., Djorgovski, S. G. & Kulkarni, S. R. 2001, *ApJ* **554**, 678
Bloom, J. S., Kulkarni, S. R. & Djorgovski, S. G. 2002, *AJ* **123**, 1111
Bremer, M., *et al.* 1998, *A&A* **332**, L13
Briggs, M. S., *et al.* 1999, *ApJ* **524**, 82
Bromm, V. & Loeb, A. 2002, *ApJ* **575**, 111
Bromm, V. & Schaefer, B. 1999, *ApJ* **520**, 661
Bromm, V., Coppi, P. & Larson, R. 1999, *ApJ* **527**, L5
2002, *ApJ* **564**, 23
Bromm, V., Kudritzki, R. & Loeb, A. 2001, *ApJ* **552**, 464
Brunner, R., Connolly, A. & Szalay, A. 1999, *ApJ* **516**, 563
Burenin, R., *et al.* 1999, *A&A* **344**, L53
Castro, S., *et al.* 2003, *ApJ* **586**, 128
Castro-Tirado, A., *et al.* 1998, *Science* **279**, 1011
1999, *Science* **283**, 2069
Chevalier, R. A. & Li, Z. 1999, *ApJL* **520**, L29
Coburn, W. & Boggs, S. 2003, *Nature* **423**, 415
Cohen, E., Piran T. & Sari, R. 1998, *ApJ* **509**, 717
Costa, E. 2000, in *Gamma-Ray Bursts*, eds. R. M. Kippen, R. S. Mallozzi & G. J. Fishman (New York, AIP Press)
Costa, E., *et al.* 1997, *Nature* **387**, 783
Costa, E., Frontera, F. & Hjorth, J. (eds.) 2001, *Gamma-Ray Bursts in the Afterglow Era* (Berlin, Springer Verlag)
Covino, S., *et al.* 1999, *A&A* **348**, L1
2002, *A&A* **392**, 865
Crew, G., *et al.* 2002, *GCN Circ.* 1734
Dado, S., Dar, A. & Rujula, A. 2002, *A&A* **388**, 1079
Dermer, C. & Mitman, K. 1999, *ApJ* **513**, L5
Dezalay, J.-P., *et al.* 1996, *ApJ* **471**, L27
Djorgovski, S. G., *et al.* 1998, *ApJ* **508**, L17
2001a, *ApJ* **562**, 654
2001b, in *Gamma-Ray Bursts in the Afterglow Era*: 2nd Workshop, eds. E. Costa *et al.*, ESO Astrophysics Symposia (Berlin: Springer Verlag) p. 218

2001c, *ApJ* **560**, L5
2002, in *Proc. IX Marcel Grossmann Meeting*, eds. V. Gurzadyan *et al.* Singapore: World Scientific
Djorgovski, S. G., Bloom, J. S. & Kulkarni, S. R. 2003, *ApJ* **591**, L13
Eichler, D., *et al.* 1989, *Nature* **340**, 126
Ellis, R. 1997, *ARAA* **35**, 389
Fan, X., *et al.* (the SDSS collaboration) 2001, *AJ* **122**, 2833
Fenimore, E. & Galassi, M. (eds.) 2004, *Gamma-Ray Bursts: 30 Years of Discovery*, AIP Conf. Proc. **727** (New York: American Institute of Physics)
Fenimore, E. & Ramirez-Ruiz, E. 2000, astro-ph/0004176
Fenimore, E. E., Epstein, R. I. & Ho, C. 1993, *A&AS* **97**, 59
Fenimore, E. E., Madras, C. D. & Nayakchin, S. 1996, *ApJ* **473**, 998
Feroci, M., Frontera, F., Masetti, N., & Piro, L. (eds.) 2004, *Gamma-Ray Bursts in the Afterglow Era*, ASP Conf. Ser. **312** (San Francisco: Astronomical Society of the Pacific)
Fox, D., *et al.* 2002, *GCN GRB Observation Report* 1569
2003, *ApJ* **586**, L5
Frail, D. A., *et al.* 1994, *ApJ* **437**, L43
1997, *Nature* **389**, 261
Frail, D. A., Waxman, E. & Kulkarni, S. R. 2000, *ApJ* **537**, 191
Frail, D. A., *et al.* 2000, *ApJ* **538**, L129
2001, *ApJ* **562**, L55
2002, *ApJ* **565**, 829
Fried, J., *et al.* 2001, *A&A* **367**, 788
Frontera, F., *et al.* 2000, *ApJS* **127**, 59
Fruchter, A., *et al.* 1999, *ApJL* **519**, L13
2000, *ApJ* **545**, 664
Fryer, C., Woosley, S. & Heger, A. 2001, *ApJ* **550**, 372
Fynbo, J., *et al.* 2001, *A&A* **369**, 373
Galama, T. J. & Wijers, R. 2000, *ApJ* **549**, L209
Galama, T. J., *et al.* 1997a, *A&A* **321**, 229
1997b, *Nature* **387**, 479
1998a, *ApJL* **497**, L13
1998b, *ApJL* **501**, L97
1998c, *Nature* **395**, 670
1999, *Nature* **398**, 394
2003, *ApJ* **587**, 135
Ghisellini, G. & Lazzati, D. 1999, *MNRAS* **309**, L7
Goodman, J. 1986, *ApJL* **308**, 46
Granot, J., Piran, T. & Sari, R. 1999, *ApJL* **527**, 236
2000a, *ApJ* **513**, 679
2000b, *ApJ* **534**, L163
Gruzinov A. 1999, *ApJ* **525**, L29
Gruzinov A. & Waxman E., 1999,*ApJ* **511**, 852
Harris, M. & Share, G. 1998, *ApJ* **494**, 724
Harrison, F. A., *et al.* 1999, *ApJL* **523**, L121
2001, *ApJ* **559**, 123
Heise, J., *et al.* 2001, in *Gamma-Ray Bursts in the Afterglow Era*, eds. E. Costa, F. Frontera & J. Hjorth (Berlin, Springer Verlag) p.16
Hjorth, J., *et al.* 2002, *ApJ* **576**, 113
2003, *Nature* **423**, 847
Hogg, D. & Fruchter, A. 1999, *ApJ* **520**, 54
Hogg, D., *et al.* 1998, *ApJ* **504**, 622
Holland, S. 2001, in *Relativistic Astrophysics*, eds. C. Wheeler & H. Martel, AIP Conf. Proc. **586** (New York: AIP) p. 593
Hurley, K. 1992, in *Gamma-Ray Bursts*, eds. W. Paciesas & G. Fishman (New York, AIP Press) p. 3
Hurley, K., *et al.* 1994, *Nature* **372**, 652
2000, *ApJ* **534**, L23
2002, *ApJ* **567**, 447

Jager, R., *et al.* 1993, *Adv. Space Res.* **13**, 12, 315
Katz, J. I. 1994, *ApJ* **422**, 248
2002, *The Biggest Bangs* (New York, Oxford University Press)
Kehoe, R., *et al.* 2001, *ApJ* **554**, L159
Kennicut, R. 1998, *ARAA* **36**, 131
Kippen, R., *et al.* 2001, in *Gamma-Ray Bursts in the Afterglow Era*, eds. E. Costa, F. Frontera & J. Hjorth (Berlin, Springer Verlag) p. 22
Kobayashi, S. 2000, *ApJ* **545**, 807
Kobayashi, S. & Sari, R. 2001, *ApJ* **551**, 934
Kobayashi, S., Piran, T. & Sari, R. 1997, *ApJ* **490**, 92
Kommers, J., *et al.* 2000, *ApJ* **533**, 696
Koranyi, D. M., *et al.* 1995, *MNRAS* **276**, L13
Kouveliotou, C., *et al.* 1993, *ApJ* **413**, L101
Krolik, J. H. & Pier, E. A. 1991, *ApJ* **373**, 277
Krumholtz, M., Thorsett, S. & Harrison, F. 1998, *ApJ* **506**, L81
Kulkarni, S. R., *et al.* 1998, *Nature* **393**, 35
1999a, *Nature* **398**, 389
1999b, *ApJ* **522**, L97
Kumar, P. & Piran, T. 2000, *ApJ* **532**, 286
Lamb, D. & Reichart, D. 2000, *ApJ* **536**, 1
2001, in *Gamma-Ray Bursts in the Afterglow Era*: 2nd Workshop, eds. E. Costa *et al.*, ESO Astrophysics Symposia (Berlin: Springer Verlag) p. 226
Larkin, J., *et al.* 1998, *GCN Circ.* 44
Lazzati, D., *et al.* 2001, in *Gamma-Ray Bursts in the Afterglow Era*: 2nd Workshop, eds. E. Costa *et al.*, ESO Astrophysics Symposia (Berlin: Springer Verlag) p. 236
Li, L. & Paczyński, B. 1998, *ApJL* **507**, L59
Liang, E. P., *et al.* 1999, *ApJL* **519**, L21
Lilly, S., *et al.* 1995, *ApJ* **455**, 108
Lin, H., *et al.* 1999, *ApJ* **518**, 533
Lithwick, Y. & Sari, R. 2001, *ApJ* **555**, 540
Lloyd-Ronning, N., Fryer, C. & Ramirez-Ruiz, E. 2002, *ApJ* **574**, 554
Loeb, A. 2002a, in *Lighthouses of the Universe: The Most Luminous Celestial Objects and Their Use for Cosmology*, eds. M. Gilfanov, R. Sunyaev & E. Churazov (Berlin: Springer Verlag) p. 137
2002b, in *Supernovae and Gamma-Ray Bursters*, ed. K. Weiler (Berlin: Springer Verlag) [astro-ph/0106455] p. 445
MacFadyen, A. & Woosley, S. 1999a, *ApJ* **524**, 262
1999b, *ApJ* **526**, 152
Madau, P. 1999, *ASPCS* **193**, 475
Madau, P., Pozzetti, L. & Dickinson, M. 1998, *ApJ* **498**, 106
Madgwick, D., *et al.* (the 2dF team) 2002, *MNRAS* **333**, 133
Mallozzi, R., *et al.* 1995, *ApJ* **454**, 597
Mao, S. & Mo, H. J. 1998, *A&A* **339**, L1
Mao, S. & Yi, I. 1994, *ApJ* **424**, L131
Mazets, E., *et al.* 1981a, *Ap&SS* **80**, 3
1981b, *Ap&SS* **80**, 119
McBreen, B., *et al.* 1994, *MNRAS* **271**, 662
McNamara, B. E., *et al.* 1995, *Ap&SS* **231**, 251
Medvedev, M. V. & Loeb A. 1999, *ApJ* **526**, 697
Mészáros, P., 2002, *ARAA* **40**, 137
Mészáros, P. & Rees, M. J. 1993, *ApJ* **405**, 278
1997, *ApJ* **476**, 232
1999, *MNRAS* **306**, L39
Mészáros, P. Rees, M. & Wijers, R. A. M. J. 1998, *ApJ* **499**, 301
Metzger, M. R., *et al.* 1997, *Nature* **387**, 879
Mochkovitch, R., *et al.* 1993, *Nature* **361**, 236
Moderski, R., Sikora, M. Bulik, T. 2000, *ApJ* **529**, 151
Mirabal, M., *et al.* 2002, *ApJ* **578**, 818

Nakar, E. & Piran, T. 2002, *MNRAS* **331**, 40
Narayan, R. & Paczyński, B. & Piran, T. 1992, *ApJL* **395**, L83
Norberg, P., *et al.* (the 2dF team) 2002, *MNRAS*, **336**, 907
Norris, J., *et al.* 1984, *Nature* **308**, 434
Odewahn, S. C., *et al.* 1998, *ApJ* **509**, L5
Paciesas, W., *et al.* 1999, *ApJS* **122**, 465
Paczyński, B. 1986, *ApJ* **308**, L43
1993, *Ann. NY Acad. Sci.* **688**, 321
1998, *ApJ* **494**, L45
Paczyński, B. & Rhoads, J. 1993, *ApJ* **418**, L5
Panaitescu, A. 2001, *ApJ* **556**, 1002
Panaitescu, A. & Kumar, P. 2001a, *ApJ* **554**, 667
2001b, *ApJ* **560**, L49
Panaitescu A. & Mészáros, P. 1998, *ApJL* **493**, L31
1999, *ApJ* **503**, 314
Paradijs, J. van, *et al.* 1997, *Nature* **386**, 686
Park, H. -S. *et al.* 2002, *GCN Circ.* 1736
Pendleton, G., *et al.* 1996, *ApJ* **464**, 606
Perna, R., Sari, R. & Frail, D. A. 2003, *ApJ* **594**, 379
Piran, T. 1999, in *Gamma Ray Bursts: The First Three Minutes*, ed. J. Poutanen, ASP Conf. Ser. **190** (San Francisco: ASP) p. 3
Piran, T. & Shemi, A. 1993, *ApJ* **403**, L67
Piran, T., Shemi, A. & Narayan, R. 1993, *MNRAS* **263**, 861
Piran, T., *et al.* 2001, *ApJ* **560**, L167
Piro, L., Scarsi, L. & Butler, R. C. 1995, *Proc. SPIE* **2517**, 169
Piro, L., *et al.* 1998, *A&A* **331**, L41
1999, *A&ASup* **138**, 431
2000, *Science* **290**, 955
Postnov, K. A., Prokhorov, M. E. & Lipunov, V. M. 2001, *Astron. Rep.* **45**, 236
Ramaprakash, A., *et al.* 1998, *Nature* **393**, 43
Ramirez-Ruiz, E. & Fenimore, E. E. 1999, *A&A* **138**, 521
Rees, M. J. & Mészáros, P. 1994, *ApJL* **403**, L93
Reichart, D. E. 1997, *ApJL* **485**, L57
Reichart, D., *et al.* 2001, *ApJ* **552**, 57
Rhoads, J. E. 1997, *ApJL* **478**, L1
1999, *ApJ* **525**, 737
Ricker, G. & Vanderspek, R. (eds.) 2003 *Gamma-Ray Burst and Afterglow Astronomy* (New York: AIP Press)
Ricker, G., *et al.* 2003, in *Gamma-Ray Burst and Afterglow Astronomy*, eds. G. Ricker & R. Vanderspek, AIP Conf. Proc. 662 (New York: AIP Press) p. 3
Rol, E., *et al.* 2000, *ApJ* **544**, 707
Rossi, E., Lazzati, D. & Rees, M. J. 2002, *MNRAS* **332**, 945
Sahu, K. C., *et al.* 1997, *Nature* **387**, 476
Sari, R. 1997, ***ApJL*** 489, L37
1998, *ApJL* **494**, L17
1999, *ApJ* **524**, L43
Sari, R. & Esin A. 2001, *ApJ* **548**, 787
Sari, R. & Mészáros, P. 2000, *ApJ* **535**, L33
Sari, R. & Piran T. 1995, *ApJ* **455**, L143
1997a, *ApJ* **485**, 270
1997b, *MNRAS* **287**, 110
1999a, *ApJ* **520**, 641
1999b, *ApJL* **517**, L109
1999c, *A&A* **138**, 537
Sari, R., Narayan, R. & Piran, T. 1996, *ApJ* **473**, 204
Sari, R., Piran, T. & Halpern, J. 1999, *ApJ* **519**, L17
Sari, R., Piran, T. & Narayan, R. 1998, *ApJL* **497**, L17

Savaglio, S., Fall, S. M. & Fiore, F. 2003, *ApJ* **585**, 638
Schaefer, B. 2000, *ApJ* **532**, L21
Schilling, G. 2002, *Flash! The Hunt for the Biggest Explosions in the Universe* (Cambridge: Cambridge University Press)
Schmidt, M. 1999, *ApJ* **523**, L117
2001, *ApJ* **552**, 36
Sokolov, V. V., *et al.* 2001, *A&A* **372**, 428
Stanek, K. Z., *et al.* 1999, *ApJL* **522**, L39
2003, *ApJ* **591**, L17
Stern, B., *et al.* 2001, *ApJ* **563**, 80
Stern, B., Atteia, J. -L. & Hurley, K. 2002, *ApJ* **304**, 304
Taylor, G. B., *et al.* 1998, *ApJ* **502**, L115
2000, *ApJ* **537**, L17
Totani, T. 1997, *ApJ* **486**, L71
Vietri, M. 1997, *ApJ* **478**, L9
Waxman, E. 1997a, *ApJL* **485**, L5
1997b, *ApJL* **491**, L19
Waxman, E. & Draine, B. 2000, *ApJ* **537**, 796
Weiler, K. (ed.) 2003, *Supernovae and Gamma-Ray Bursts,* Berlin: Springer
Weth, C., *et al.* 2000, *ApJ* **534**, 581
Wijers, R. A. M. J. & Galama, T. J. 1999, *ApJ* **523**, 177
Wijers, R. A. M. J., Rees, M. J. & Mészáros, P. 1997, *MNRAS* **288**, L51
Wijers, R. A. M. J., *et al.* 1999, *ApJL* **523**, L33
Yoshida, A., *et al.* 1999, *A&ASup* **138**, 433
Yost, S., *et al.* 2002, *ApJ* **577**, 155
Zhang, B. & Mészáros, P. 2002, *ApJ* **571**, 876

16

Formation and evolution of compact stellar X-ray sources

T. M. Tauris
Copenhagen University

E. P. J. van den Heuvel
University of Amsterdam

16.1 Introduction and brief historical review

In this chapter we present an overview of the formation and evolution of compact stellar X-ray sources. For earlier reviews on the subject we refer to Bhattacharya & van den Heuvel (1991), van den Heuvel (1994) and Verbunt & van den Heuvel (1995). The observations and populations of high-mass X-ray binaries (HMXBs) and low-mass X-ray binaries (LMXBs) were covered earlier in Chapter 1 by Psaltis.

In our Galaxy there are about 100 bright X-ray sources with fluxes well above 10^{-10} erg cm^{-2} s^{-1} in the energy range 1–10 keV (above the Earth's atmosphere). The distribution of these sources shows a clear concentration towards the Galactic center and also towards the Galactic plane, indicating that the majority do indeed belong to our Galaxy. Furthermore, a dozen strong sources are found in Galactic globular clusters (Section 8.2) and in the Magellanic Clouds. Shortly after the discovery of the first source (Sco X-1, Giacconi *et al.* 1962) Zel'Dovitch and Guseinov (1966), Novikov and Zel'Dovitch (1966) and Shklovskii (1967) suggested that the strong Galactic X-ray sources are accreting neutron stars or black holes in binary systems. (The process of mass accretion onto a supermassive black hole had already been suggested as the energy source for quasars and active galactic nuclei by Salpeter (1964), Zel'Dovitch (1964) and Zel'Dovitch and Novikov (1964).)

The X-ray fluxes measured correspond to typical source luminosities of $10^{34} - 10^{38}$ erg s^{-1} (which is more than 25 000 times the total energy output of our Sun). Table 16.1 lists the rates of accretion required to generate a typical X-ray luminosity of 10^{37} erg s^{-1}. Also listed is the amount of gravitational potential energy released per unit mass ($\Delta U/m = GM/R$) by accretion onto a 1 $M_\odot$ stellar (compact) object, as well as the column density towards the stellar surface (or Schwarzschild radius) in the case of spherical accretion, $\sigma = L_X 4\pi \sqrt{R/(GM)^3}$. The table shows that only for accreting neutron stars and black holes is the column density low enough to allow X-rays to escape, as X-rays are stopped at column densities larger than a few g cm^{-2}. Hence, the strongest Galactic sources cannot be accreting white dwarfs.

The first evidence for an accreting binary came from Webster and Murdin (1972) and Bolton (1972). They discovered that Cyg X-1 is a binary system with an orbital period of 5.6 days. They measured the amplitude of the radial velocity curve (72 km s^{-1}) of the O9.7 supergiant optical counterpart, and independently concluded that the X-ray emission is the result of accretion onto a compact object, which is probably a black hole given the fact that the derived mass of the compact object $> 3\,M_\odot$ when assuming a realistic mass ($\geq 15\,M_\odot$) for the O9.7 supergiant.

Compact Stellar X-Ray Sources, eds. Walter Lewin and Michiel van der Klis.
Published by Cambridge University Press.

Table 16.1. *Energetics of accretion: see text*

Stellar object $1\,M_\odot$	Radius (km)	$\Delta U/mc^2$	$\Delta U/m$ (erg/g)	dM/dt^a ($M_\odot$/yr)	Column density^a (g/cm^2)
Sun	7×10^5	2×10^{-6}	2×10^{15}	1×10^{-4}	140
White dwarf	10 000	2×10^{-4}	1×10^{17}	1×10^{-6}	16
Neutron star	10	0.15	1×10^{20}	1×10^{-9}	0.5
Black hole	3	$0.1 \sim 0.4$	4×10^{20}	4×10^{-10}	0.3

[a] $L_X = 10^{37}$ erg/s

Shortly afterwards, the accreting binary model was nicely confirmed by Schreier *et al.* (1972) who discovered that the regularly pulsing source Cen X-3 is a member of an eclipsing binary system. Its regular X-ray pulsations (indicating a neutron star) have a period of 4.84 s, and the regular X-ray eclipses have a duration of 0.488 days, and repeat every 2.087 days. The pulse period shows a sinusoidal Doppler modulation with the same 2.087 day period and is in phase with the X-ray eclipses, indicating that the X-ray pulsar is moving in a circular orbit with a projected velocity of 415.1 km s^{-1}. Hence, the 4.84 s period of the X-ray pulsations is the rotation period of the neutron star and the binary orbital period is 2.087 days. Using Kepler's third law ($\Omega^2 = GM/a^3$) combined with the eccentricity and measured radial velocity amplitude ($K_x = \Omega\, a_x \sin i/\sqrt{1-e^2}$) of the X-ray pulsar one can determine the so-called mass function of the binary:

$$f(M) = \frac{M_2^3 \sin^3 i}{(M_X + M_2)^2} = \frac{1}{2\pi G}\, K_x^3\, P_{\rm orb}\,(1-e^2)^{3/2} \qquad (16.1)$$

where M_X and M_2 denote the masses of the accreting compact star and its companion star, respectively, and i is the inclination of the orbital angular momentum vector with respect to the line-of-sight to the Earth. For Cen X-3 the mass function is $f = 15.5\,M_\odot$ and thus one can derive a minimum companion mass of about $18\,M_\odot$.

The recognition that neutron stars and black holes can exist in close binary systems came at first as a surprise. It was known that the initially more massive star should evolve first and explode in a supernova (SN). However, as a simple consequence of the virial theorem, the orbit of the post-SN system should be disrupted if more than half of the total mass of the binary is suddenly ejected (Blaauw 1961). For X-ray binaries like Cen X-3 it was soon realized (van den Heuvel & Heise 1972; and independently also by Tutukov & Yungelson 1973) that the survival of this system was due to the effects of large-scale mass transfer that must have occurred prior to the SN.

The formation of low-mass X-ray binaries ($M_{\rm donor} \leq 1.5\,M_\odot$) with observed orbital periods mostly between 11 min and 12 hr, as well as the discovery of the double neutron star system PSR 1913+16 (Hulse & Taylor 1975) with an orbital period of 7.75 hr, was a much tougher nut to crack. How could these stars end up being so close when the progenitor star of the neutron star must have had a radius much larger than the current separation? It was clear that such systems must have lost a large amount of orbital angular momentum. The first models to include large loss of angular momentum were made by van den Heuvel and de Loore (1973) for the later evolution of high-mass X-ray binaries (HMXBs), showing that in this way very close systems like Cyg X-3 can be formed, and Sutantyo (1975) for the origin of LMXBs. The important concept of a "common envelope" (CE) evolution was introduced by Paczyński (1976) and Ostriker (1976). This scenario, as well as that of van den Heuvel and de Loore (1973), could link the massive X-ray binary Cyg X-3 and the binary radio pulsar PSR 1913+16. In this scenario a neutron star is captured by the expansion of a giant companion star and is forced to move through the giant's envelope. The resulting frictional drag will cause its orbit to shrink rapidly while, at the same time, ejecting the envelope before the naked core of the giant star explodes to form another neutron star. It was suggested by Smarr and Blandford (1976) that it is an old "spun-up" neutron star which is observed as a radio pulsar in PSR 1913+16. The magnetic field of this visible pulsar is relatively weak ($\sim 10^{10}$ G, some two orders of magnitude lower than the average pulsar magnetic field) and its spin period is very short (59 ms). Hence, this pulsar is most likely spun-up (or "recycled") in an X-ray binary where mass and angular momentum from an accretion disk is fed to the neutron star (as already suggested by Bisnovatyi-Kogan and Komberg 1974). The other neutron star in the system was then produced by the second supernova explosion and must be a young, strong **B**-field neutron star (it is not observable – either because it has already rapidly spun-down, due to dipole radiation, or because the Earth is not hit by the pulsar beam).

The idea of recycling pulsars was given a boost by the discovery of the first millisecond radio pulsar (Backer *et al.* 1982). As a result of the long accretion phase in LMXBs, millisecond pulsars are believed to be formed in such systems (Alpar *et al.* 1982; Radhakrishnan & Srinivasan 1982). This model was beautifully confirmed by the discovery of the first millisecond X-ray pulsar in the LMXB system SAX 1808.4–3658 (Wijnands & van der Klis 1998). Now four of these accreting millisecond pulsars are known (Galloway *et al.* 2002; Markwardt, Smith & Swank 2003). For a detailed discussion on the evidence for the presence of rapidly spinning weakly magnetized neutron stars in low-mass X-ray binaries we refer to Chapter 2 by van der Klis and Chapter 3 by Strohmayer and Bildsten.

Finally, another ingredient that has important consequences for close binary evolution is the event of a "kick" imparted to newborn neutron stars as a result of an asymmetric SN. They were first applied to binary systems by Flannery and van den Heuvel (1975). There is now ample evidence for the occurrence of such kicks inferred from the space velocities of pulsars and from dynamical effects on surviving binaries (e.g., Dewey & Cordes 1987; Lyne & Lorimer 1994 and Kaspi *et al.* 1996). Furthermore, as argued by Kalogera and Webbink (1998), without kicks the formation of LMXBs with short orbital periods cannot be explained.

In this review we will concentrate on the stellar evolutionary processes that are responsible for the formation of the different types of compact binaries and their evolution. Our focus is mainly on binaries with neutron star or black hole accretors. We refer to other reviews for discussions on CVs and AM CVn systems (Chapter 10), super-soft sources (Chapter 11) and sdB-star binaries (Maxted *et al.* 2001; Han *et al.* 2002). In the next section, however, we will first give a short introduction to the observational properties of the X-ray binaries and binary pulsars which we discuss afterwards (see also Chapter 1 by Psaltis).

16.2 Compact binaries and their observational properties

Over 90% of the strong Galactic X-ray sources appear to fall into two distinct groups: the high-mass X-ray binaries (HMXBs) and the low-mass X-ray binaries (LMXBs). These

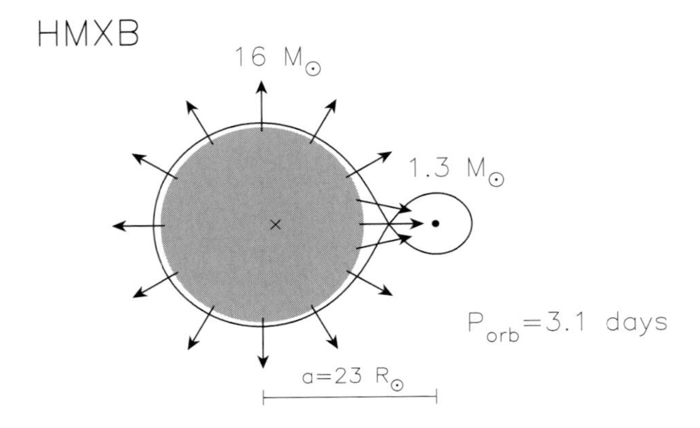

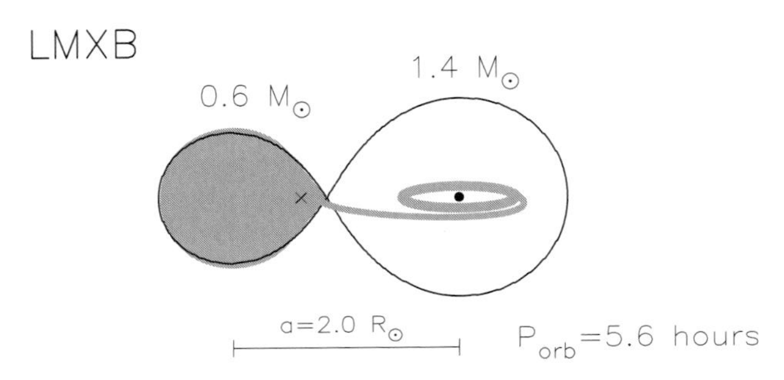

Fig. 16.1. Examples of a typical HMXB (*top*) and LMXB (*bottom*). The neutron star in the HMXB is fed by a strong high-velocity stellar wind and/or by beginning atmospheric Roche-lobe overflow. The neutron star in an LMXB is surrounded by an accretion disk which is fed by Roche-lobe overflow. There is also observational evidence for HMXBs and LMXBs harboring black holes.

two groups differ in a number of physical characteristics (see Fig. 16.1 and Table 16.2 for some examples). As we shall see later on, binary pulsars and single millisecond pulsars are the descendants of the X-ray binaries containing an accreting neutron star.

16.2.1 High-mass X-ray binaries (HMXBs)

There are about 130 known HMXBs (Liu, van Paradijs & van den Heuvel 2000) and 25 have well-measured orbital parameters. There are $\sim$40 pulsating HMXB sources with typical pulse periods between 10 and 300 seconds (the entire observed range spans between 0.069 seconds and 20 minutes). Among the systems with $P_{\rm orb} \leq 10$ days and $e \leq 0.1$ are the strong sources and "standard" systems such as Cen X-3 and SMC X-1. These are characterized by the occurrence of regular X-ray eclipses and double-wave ellipsoidal light variations produced by tidally deformed ("pear-shaped") giant or sub-giant companion stars with masses $>10\,{\rm M}_\odot$. However, the optical luminosities ($L_{\rm opt} > 10^5\,{\rm L}_\odot$) and spectral types of the companions indicate original ZAMS masses $\geq 20\,{\rm M}_\odot$, corresponding to O-type progenitors. The companions have radii $10-30\,{\rm R}_\odot$ and (almost) fill their critical Roche-lobes, see Section 16.4. In a number of pulsating sources, such as X0115+63 (and Her X-1, an intermediate-mass X-ray binary system) there are absorption/emission features in the

Table 16.2. *The two main classes of strong Galactic X-ray sources*

	HMXB	LMXB
X-ray spectra	$kT \geq 15$ keV (hard)	$kT \leq 10$ keV (soft)
Type of time variability	regular X-ray pulsations no X-ray bursts	only a very few pulsars often X-ray bursts
Accretion process	wind (or atmos. RLO)	Roche-lobe overflow
Timescale of accretion	10^5 yr	10^7-10^9 yr
Accreting compact star	high **B**-field NS (or BH)	low **B**-field NS (or BH)
Spatial distribution	Galactic plane	Galactic center and spread around the plane
Stellar population	young, age $<10^7$ yr	old, age $> 10^9$ yr
Companion stars	luminous, $L_{\rm opt}/L_{\rm X} > 1$ early-type O(B)-stars $>10\,M_\odot$ (Pop. I)	faint, $L_{\rm opt}/L_{\rm X} \ll 0.1$ blue optical counterparts $\leq 1\,M_\odot$ (Pop. I and II)

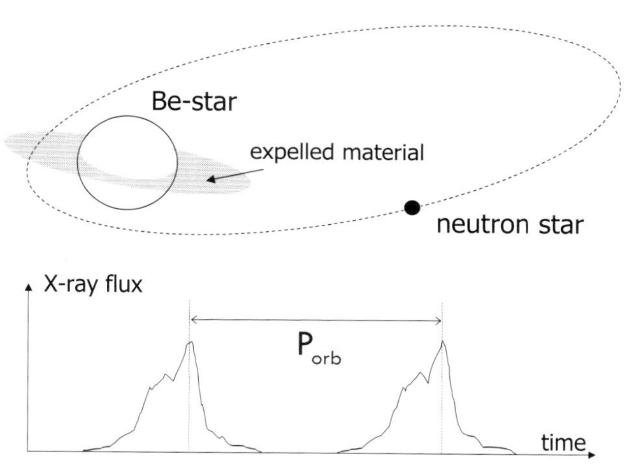

Fig. 16.2. Schematic model of a Be-star X-ray binary system. The neutron star moves in an eccentric orbit around the Be-star which is not filling its Roche-lobe. However, near the periastron passage the neutron star accretes circumstellar matter, ejected from the rotating Be-star, resulting in an X-ray outburst lasting several days.

X-ray spectrum that are most probably cyclotron lines, resulting from magnetic fields with strengths $B \simeq 5 \times 10^{12}$ G (Kirk & Trümper 1983). Among the standard HMXBs, there are at least two systems that are thought to harbor black holes: Cyg X-1 and LMC X-3.

Another group of HMXBs consists of the moderately wide, eccentric binaries with $P_{\rm orb} \simeq 20-100$ days and $e \simeq 0.3-0.5$. A new third (sub-)group has recently been proposed by Pfahl *et al.* (2002). These systems have $P_{\rm orb} \simeq 30-250$ days and small eccentricities $e \lesssim 0.2$. Together these two groups form a separate sub-class of HMXBs: the Be-star X-ray binaries (see Fig. 16.2; first recognized as a class by Maraschi, Treves & van den Heuvel 1976). In the Be-star X-ray binaries the companions are rapidly rotating B-emission stars situated on, or close to, the main sequence (luminosity class III–V). There are more than 50 such systems known making them the most numerous class of HMXBs (see, for example, van den Heuvel & Rappaport 1987 and Chapter 1 by Psaltis for reviews). The Be-stars are deep inside their Roche-lobes, as is indicated by their generally long orbital periods ($\gtrsim$15 days) and by the absence of X-ray eclipses and of ellipsoidal light variations. According to the luminosities and spectral types, the companion stars have masses in the range about 8–20 $M_\odot$ (spectral types O9–B3, III–V). The X-ray emission from the Be-star X-ray systems tends to be extremely variable, ranging from complete absence to giant transient outbursts lasting weeks to months. During such an outburst episode one often observes orbital modulation of the X-ray emission, due to the motion of the neutron star in an eccentric orbit, see Fig. 16.2. The recurrent X-ray outbursts are most probably related to the irregular optical outbursts generally observed in Be-stars, which indicate sudden outbursts of mass ejection, presumably generated by rotation-driven instability in the equatorial regions of these stars (see, for example, Slettebak 1988). While the Be-star X-ray binaries are transient sources (often unobservable for months to years) the "standard" systems are persistent X-ray sources. HMXBs are located along the Galactic plane among their OB-type progenitor stars.

16.2.2 Low-mass X-ray binaries (LMXBs)

Orbital periods have been measured for some 30 of these systems. They range from 11 minutes to 17 days, similar to the orbital periods of cataclysmic variables – see also Chapter 5 by Charles and Coe. Only in the widest few of these systems can one observe the spectrum of the optical companion. In all other systems, the optical spectrum is that of the hot accretion disk. The LMXBs are very seldom X-ray pulsars. The reason is their relatively weak magnetic fields $\sim 10^9 - 10^{11}$ G, which are expected to result from accretion-induced field decay (Taam & van den Heuvel 1986; Geppert & Urpin 1994; Konar & Bhattacharya 1997; Cumming, Zweibel & Bildsten 2001; Bhattacharya 2002). On the other hand, these sources show X-ray bursts (sudden thermonuclear fusion of accreted matter at the surface of the neutron star – see Chapter 3 by Strohmayer and Bildsten), which are suppressed if the magnetic field strength $> 10^{11}$ G (Lewin & Joss 1983). For this reason such bursts are not observed in HMXBs. The discovery of (kilohertz) quasi-periodic oscillations (QPOs) in the X-ray flux of LMXBs has provided a clear timing signature of the accreting neutron stars and black holes in these systems. In the past decade much insight of detailed accretion physics and testing of the general theory of relativity has been revealed by observations and "beat frequency" models (see Chapter 2 by van der Klis for a review). There are more than a dozen LMXBs systems for which there is strong evidence for the presence of a black hole – see Section 16.2.4 and Chapters 4 and 5.

Most of the LMXBs are located in the Galactic bulge and in globular clusters, and thus appear to belong to an old stellar population. They do not show considerable run-away characteristics, as they are confined mostly to within 0.6 kpc of the Galactic plane – however, some LMXBs have transverse velocities in excess of 100 km s^{-1}. It is interesting to notice that the systems in globular clusters must have velocities smaller than the escape velocities from these clusters, which are less than about 30 km s^{-1}.

16.2.3 Intermediate-mass X-ray binaries (IMXBs)

Above we have described the HMXB and LMXB systems with companion stars $> 10\,M_\odot$ and $\lesssim 1\,M_\odot$, respectively. There must also exist a large number of Galactic compact binaries with companion star masses in the interval 1–10 $M_\odot$ as has been argued, e.g., by van den Heuvel (1975). These are the so-called intermediate-mass X-ray binaries (IMXBs). IMXBs have recently been recognized as a class of their own. It is important to be aware that IMXB systems are not easily observed as a result of a simple selection effect against X-ray sources with intermediate-mass companions. The reason is the following (van den Heuvel 1975): "Standard" HMXBs have evolved (sub)giant companions that are massive enough to have a strong stellar wind mass-loss rate (typically $\dot{M}_{\rm wind} \simeq 10^{-6}\,M_\odot\ {\rm yr}^{-1}$) sufficient to power a bright X-ray source, via an accreting neutron star or black hole, for $10^5 - 10^6$ yr. The LMXBs are not wind-fed X-ray sources. These systems experience mass transfer via Roche-lobe overflow (RLO) from their companion star to the compact object. The LMXBs often evolve slowly on a nuclear timescale ($\sim 10^8$–10^9 yr) and the majority of the transferred material is usually funneled onto the compact object, via an accretion disk, yielding accretion rates of 10^{-10}–$10^{-8}\,M_\odot\ {\rm yr}^{-1}$. In IMXBs the companions are not massive enough to produce sufficiently high wind mass-loss rates to power an observable X-ray source. Subsequently when IMXBs evolve through RLO, the relatively large mass ratio between the companion star and a neutron star causes this phase to be short lived – either on a sub-thermal timescale (Tauris, van den Heuvel & Savonije 2000), or the system evolves through a common envelope. In either case, the systems evolve on a timescale of only a few 1000 yr. Furthermore, the very high mass-transfer rates under these circumstances ($\dot{M} > 10^{-4}\,M_\odot\ {\rm yr}^{-1} \gg \dot{M}_{\rm Edd}$) may cause the emitted X-rays to be absorbed in the dense gas surrounding the accreting neutron star. We therefore conclude that HMXBs and LMXBs are naturally selected as persistent X-ray sources fed by a strong stellar wind and RLO, respectively.

Also, from a theoretical point of view we know that IMXBs must exist in order to explain the formation of binary pulsars with heavy CO or ONeMg white dwarf companions (Tauris, van den Heuvel & Savonije 2000). Despite the above mentioned selection effects against IMXBs there are a few such systems with neutron stars detected: Her X-1 and Cyg X-2 are systems of this type. In the latter system the companion presently has a mass of $<1\,M_\odot$, but it is highly over-luminous for this mass, which indicates that it is an evolved star that started out with a mass between 3 and 4 $M_\odot$ at the onset of the mass-transfer phase (Podsiadlowski & Rappaport 2000; King & Ritter 1999). Among the black hole X-ray binaries, IMXBs are more common, for example: GRO J1655–40 ($M_{\rm d} \sim 1.5\,M_\odot$; Beer & Podsiadlowski 2002), 4U 1543–47 ($M_{\rm d} \sim 2.5\,M_\odot$; Orosz *et al.* 1998), LMC X-3 ($M_{\rm d} \sim 5\,M_\odot$; Soria *et al.* 2001) and V 4642 Sgr ($M_{\rm d} \sim 6.5\,M_\odot$; Orosz *et al.* 2001). In these systems the donor star ($M_{\rm d}$) is less massive than the black hole, so mass transfer by RLO is stable.

16.2.4 Soft X-ray transients (SXTs)

A great breakthrough in the discovery of black holes in X-ray binaries came with the discovery by McClintock and Remillard (1986) that the K5V companion of the source A0620–00, in a spectroscopic binary with $P_{\rm orb} = 7.75$ hr, has a velocity amplitude of $> 470\ {\rm km\,s}^{-1}$. This large orbital velocity indicates that even if the K-dwarf had zero mass, the compact object would have mass of $> 3\,M_\odot$ and therefore must be a black hole. Since then over a dozen such systems consisting of a black hole with a low-mass donor star have been discovered. These black hole systems are the so-called soft X-ray transients (SXT) and appear as bright X-ray novae with luminosities $L_X \sim 10^{38}\ {\rm erg\,s}^{-1}$ for several weeks. At the same time, the optical luminosity of these systems brightens by 6 to 10 magnitudes, making them optical novae as well. After the decay of the X-ray and optical emission, the spectrum of a K or G star becomes visible, and the large amplitude variations (≥ 100 km s^{-1}) in the radial velocities of these stars indicate a compact object mass of > 3 M$_\odot$ (exceeding the maximum mass thought possible for neutron stars). Their orbital periods are between 8 hours and 6.5 days (see e.g., Lee, Brown & Wijers 2002 and Chapter 4 by McClintock & Remillard).

16.2.5 Peculiar X-ray binaries

Not all observed X-ray binaries fall into the well-defined classes described above. Among the more intriguing systems are SS433 and Cyg X-3, which both have flaring radio emissions and jets (see, e.g., Chapter 9 by Fender). Cyg X-3 ($P_{\rm orb} = 4.8$ hr) is probably a later evolutionary phase of a wide HMXB (e.g., see Fig. 16.15 in Section 16.8; van den Heuvel & de Loore 1973).

16.2.6 The binary and millisecond radio pulsars

The 50 or so Galactic binary radio pulsars detected so far are generally characterized by short spin periods, P, and small values of the period derivative, $\dot{P}$. This can clearly be seen in a $(P, \dot{P})$-diagram of radio pulsars, see Fig. 16.3. Simple expressions for the surface magnetic dipole field strength, B, and "spin-down" age, τ, are given by the magnetic dipole model of pulsars (see Manchester & Taylor 1977):

$$B = \sqrt{\frac{3c^3 I}{8\pi^2 R^6}\, P\dot{P}} \simeq 3 \times 10^{19} \sqrt{P\dot{P}} \text{ gauss} \tag{16.2}$$

where I and R are the moment of inertia ($\sim 10^{45}$ g cm^2) and radius of the neutron star, respectively, and

$$\tau \equiv P/2\dot{P} \tag{16.3}$$

Figure 16.3 shows that the binary pulsars typically have $B = 10^8 - 10^{10}$ G and $\tau = 1-$10 Gyr, whereas the ordinary isolated pulsars have $B = 10^{12} - 10^{13}$ G and $\tau \lesssim 10$ Myr. As will be discussed in a later section in this chapter, the short spin periods and low values of B are consequences of the recycling process where an old neutron star accretes mass and angular momentum from its companion star via an accretion disk. For theoretical calculations of the dynamical evolution of single pulsars in the $(P, \dot{P})$-diagram we refer to Tauris & Konar (2001).

There are four classes of binary pulsars detected so far (see Table 16.3). The different classes are: (i) high-mass binary pulsars (HMBPs) with a neutron star or ONeMg/CO white dwarf companion, (ii) low-mass binary pulsars (LMBPs) with a helium white dwarf companion, (iii) non-recycled pulsars with a CO white dwarf companion, and finally, (iv) pulsars with an unevolved companion. A few further sub-divisions can be introduced from an evolutionary point of view. The globular cluster pulsars will not be discussed here since these systems are the result of tidal capture or exchange encounters – see Bhattacharya & van den Heuvel (1991) and Chapter 8 by Verbunt & Lewin for a review.

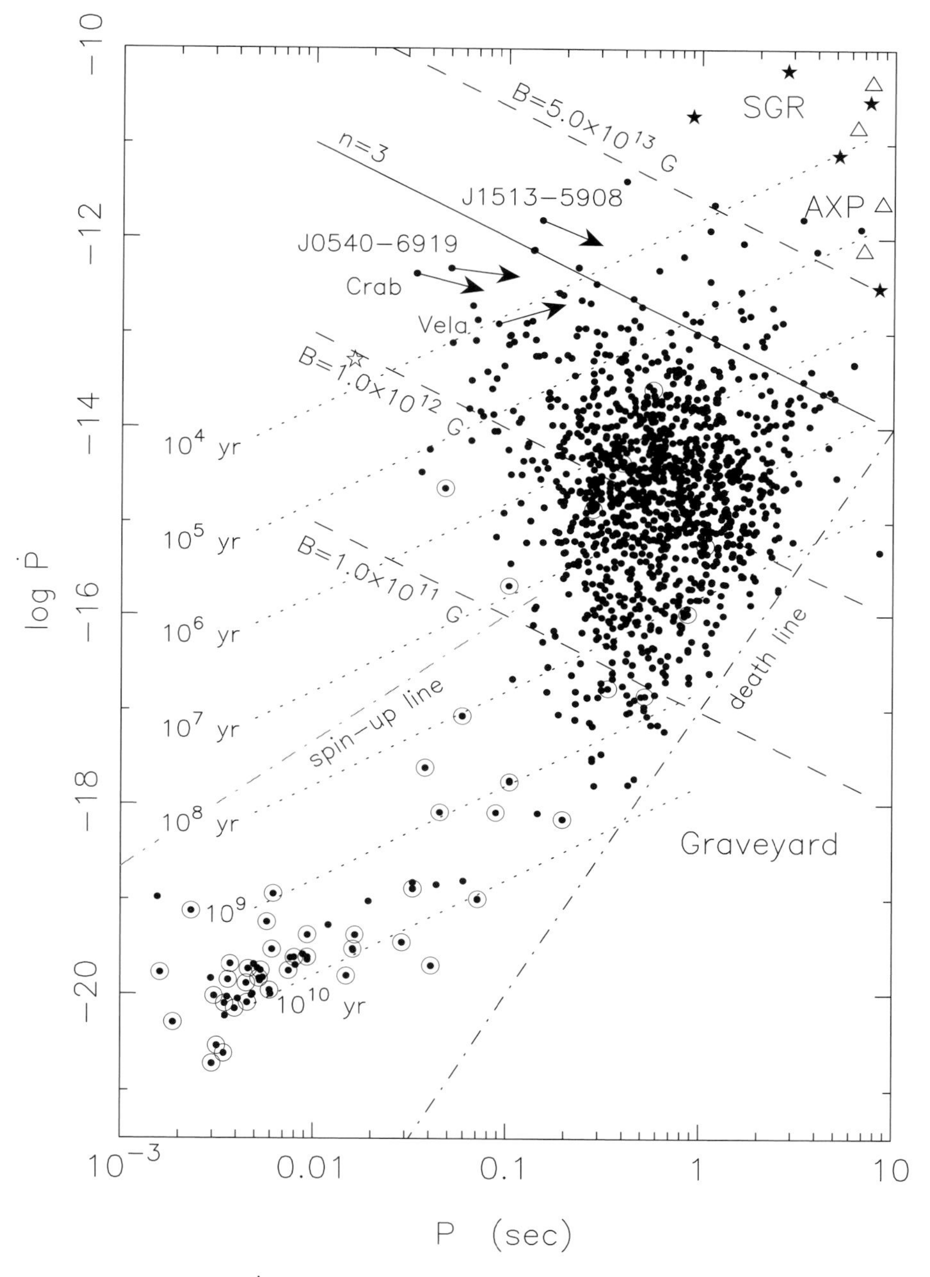

Fig. 16.3. $(P, \dot{P})$-diagram of ~ 1300 observed radio pulsars (ATNF Pulsar Catalogue data). Binary pulsars are marked by a circle. Soft gamma-ray repeaters (SGR) and anomalous X-ray (AXP) pulsars (Chapter 14) are marked by stars and triangles, respectively. Also shown are lines of constant surface dipole magnetic field strength (dashed) and characteristic ages (dotted). The arrows marked on a few young pulsars indicate a measurement of the braking index. The "death line" is the pair-creation limit for generating radio pulses.

Table 16.3. *Main categories and types of binaries with compact objects*

Main type	Sub-type	Observed example	P_{orb}
X-ray binaries			
high-mass donor ($M_{donor} \geq 10\,M_\odot$)	"standard" HMXB	Cen X-3 Cyg X-1	2.087^d (NS) 5.60^d (BH)
	wide-orbit HMXB	X Per	250^d (NS)
	Be-star HMXB	A0535+26	104^d (NS)
low-mass donor ($M_{donor} \leq 1\,M_\odot$)	Galactic disk LMXB	Sco X-1	0.86^d (NS)
	soft X-ray transient	A0620–00	7.75^{hr} (BH)
	globular cluster	X 1820–30	11^{min} (NS)
	millisecond X-ray pulsar	SAX J1808.4–36	2.0^d (NS)
intermediate-mass donor ($1 < M_{donor}/M_\odot < 10$)		Her X-1 Cyg X-2 V 404 Cyg	1.7^d (NS) 9.8^d (NS) 6.5^d (BH)
Binary radio pulsars			
"high-mass" companion ($0.5 \leq M_c/M_\odot \leq 1.4$)	NS + NS (double)	PSR 1913+16	7.75^{hr}
	NS + (ONeMg) WD	PSR 1435–6100	1.35^d
	NS + (CO) WD	PSR 2145–0750	6.84^d
"low-mass" companion ($M_c < 0.45\,M_\odot$)	NS + (He) WD	PSR 0437–4715 PSR 1640+2224	5.74^d 175^d
non-recycled pulsar	(CO) WD + NS	PSR 2303+46	12.3^d
unevolved companion	B-type companion	PSR 1259–63	3.4^{yr}
	low-mass companion	PSR 1820–11	357^d
CV-like binaries			
novae-like systems	($M_{donor} \leq M_{WD}$)	DQ Her SS Cyg	4.7^{hr} 6.6^{hr}
super-soft X-ray sources	($M_{donor} > M_{WD}$)	CAL 83 CAL 87	1.04^d 10.6^{hr}
AM CVn systems (RLO)	(CO) WD + (He) WD	AM CVn	22^{min}
double WD (no RLO)	(CO) WD + (CO) WD	WD1204+450	1.6^d
sdB-star systems	(sdB) He-star + WD	KPD 0422+5421	2.16^{hr}

16.3 Binary stellar evolution and final compact objects

In order to understand how neutron stars, black holes and white dwarfs can be formed in binary systems, a brief overview of the basic elements of the evolution of single stars is necessary. We refer to, e.g., Cox & Giuli (1968) and Kippenhahn & Weigert (1990) for further details.

16.3.1 Summary of the evolution of single stars

The evolution of a star is driven by a rather curious property of a self-gravitating gas in hydrostatic equilibrium, described by the virial theorem, namely that the radiative loss of energy of such a gas causes it to contract and hence, due to release of gravitational potential energy, to increase its temperature. Thus, while the star tries to cool itself by radiating away energy from its surface, it gets hotter instead of cooler (i.e., it has a "negative heat capacity").

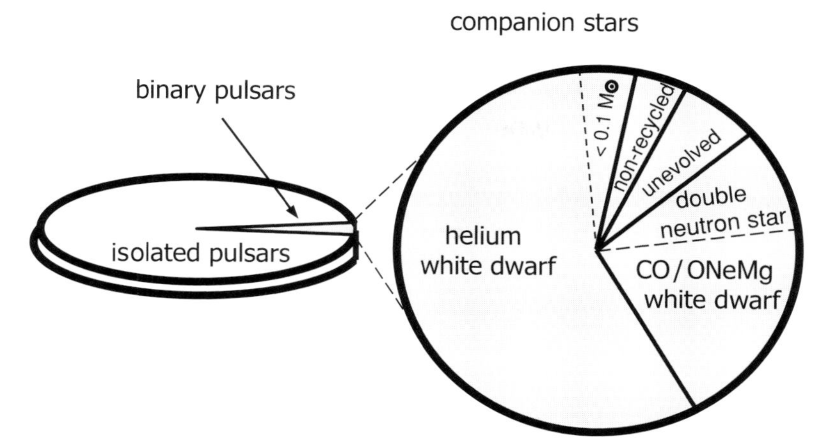

Fig. 16.4. Illustration of the relative distribution of all ~ 1500 radio pulsars observed. About 4% are members of a binary system. The four main classes of binary pulsars are defined according to their formation history – see text.

The more it radiates to cool itself, the more it will contract, the hotter it gets and the more it is forced to go on radiating. Clearly, this "vicious virial cycle" is an unstable situation in the long run and explains why the star, starting out as an interstellar gas globe, must finally end its life as a compact object. In the meantime the star spends a considerable amount of time in the intermediate stages, which are called: "the main sequence", "the giant branch", etc. It is important to realize that stars do *not* shine because they are burning nuclear fuel. They shine because they are hot due to their history of gravitational contraction.

A massive star ($M \gtrsim 10\,\mathrm{M}_\odot$) evolves through cycles of nuclear burning alternating with stages of exhaustion of nuclear fuel in the stellar core until its core is made of iron, at which point further fusion requires, rather than releases, energy. The core mass of such a star becomes larger than the Chandrasekhar limit, the maximum mass possible for an electron-degenerate configuration ($\sim 1.4\,\mathrm{M}_\odot$). Therefore the core implodes to form a neutron star or black hole. The gravitational energy released in this implosion (4×10^{53} erg $\simeq 0.15\, M_{\mathrm{core}}c^2$) is far more than the binding energy of the stellar envelope, causing the collapsing star to explode violently and eject the outer layers of the star, with a speed of $\sim 10^4$ km s^{-1}, in a supernova event. The final stages during and beyond carbon burning are very short-lived (~ 60 yr for a $25\,\mathrm{M}_\odot$ star) because most of the nuclear energy generated in the interior is liberated in the form of neutrinos, which freely escape without interaction with the stellar gas, thereby lowering the outward pressure and accelerating the contraction and nuclear burning.

Less massive stars ($M < 8\,\mathrm{M}_\odot$) suffer from the occurrence of degeneracy in the core at a certain point of evolution. Since for a degenerate gas the pressure only depends on density and not on temperature, there will be no stabilizing expansion and subsequent cooling after the ignition. Hence, the sudden temperature rise (due to the liberation of energy after ignition) causes run-away nuclear energy generation producing a so-called "flash". In stars with $M < 2.3\,\mathrm{M}_\odot$ the helium core becomes degenerate during hydrogen shell burning and, when its core mass M_{He} reaches $0.45\,\mathrm{M}_\odot$, helium ignites with a flash. The helium flash is, however, not violent enough to disrupt the star. Stars with masses in the range $2.3 < M/\mathrm{M}_\odot < 8$ ignite carbon with a flash. Such a carbon flash was believed to perhaps disrupt the entire star in a so-called carbon-deflagration supernova. However, recent observations of white dwarfs in

Table 16.4. *End products of stellar evolution as a function of initial mass*

		Final product	
Initial mass	He-core mass	Single star	Binary star
$<2.3\,M_\odot$	$<0.45\,M_\odot$	CO white dwarf	He white dwarf
$2.3-6\,M_\odot$	$0.5-1.9\,M_\odot$	CO white dwarf	CO white dwarf
$6-8\,M_\odot$	$1.9-2.1\,M_\odot$	O-Ne-Mg white dwarf or C-deflagration SN?	O-Ne-Mg white dwarf
$8-12\,M_\odot$	$2.1-2.8\,M_\odot$	neutron star	O-Ne-Mg white dwarf
$12-25\,M_\odot$	$2.8-8\,M_\odot$	neutron star	neutron star
$>25\,M_\odot$	$>8\,M_\odot$	black hole	black hole

Galactic clusters that still contain stars as massive as $8^{+3}_{-2}\,M_\odot$ (Reimers & Koester 1988; Weidemann 1990) indicate that such massive stars still terminate their life as a white dwarf. They apparently shed their envelopes in the AGB-phase before carbon ignites violently. Furthermore, stars in close binary systems, which are the prime objects in this review, will have lost their envelope as a result of mass transfer via Roche-lobe overflow. This is also the reason why in binary systems the lower ZAMS mass-limit for producing a neutron star is somewhat larger than for an isolated star.

The possible end-products and corresponding initial masses are listed in Table 16.4. It should be noted that the actual values of the different mass ranges are only known approximately due to considerable uncertainty in our knowledge of the evolution of massive stars. Prime causes of this uncertainty include limited understanding of the mass loss undergone by stars in their various evolutionary stages (see Section 16.3.3). To make a black hole, the initial ZAMS stellar mass must exceed at least $20\,M_\odot$ (Fryer 1999), or possibly $25\,M_\odot$. According to MacFadyen, Woosley and Heger (2001), stars $>40\,M_\odot$ form black holes directly (collapsars type I) whereas stars in the interval $25 < M/M_\odot < 40$ produce black holes after a "failed supernova explosion" (collapsars type II). From an analysis of black hole binaries it seems that a mass-fraction of ~ 0.35 must have been ejected in the (symmetric) stellar core collapse leading to the formation of a black hole (Nelemans, Tauris & van den Heuvel 1999). Another fundamental problem is understanding convection, in particular in stars that consist of layers with very different chemical composition. Finally, there is the unsolved question of whether or not the velocity of convective gas cells may carry them beyond the boundary of the region of the star that is convective according to the Schwarzschild criterion. For example, inclusion of this so-called overshooting in evolutionary calculations decreases the lower mass-limit for neutron star progenitors.

16.3.1.1 Three timescales of stellar evolution

There are three fundamental timescales of stellar evolution. When the hydrostatic equilibrium of a star is disturbed (e.g., because of sudden mass loss), the star will restore this equilibrium on a so-called dynamical (or pulsational) timescale:

$$\tau_{\rm dyn} = \sqrt{R^3/GM} \simeq 50\ {\rm min}\ (R/R_\odot)^{3/2}(M/M_\odot)^{-1/2} \tag{16.4}$$

When the thermal equilibrium of a star is disturbed, it will restore this equilibrium on a thermal (or Kelvin–Helmholtz) timescale, which is the time it takes to emit all of its thermal energy content at its present luminosity:

$$\tau_{\rm th} = GM^2/RL \simeq 30\ {\rm Myr}\ (M/{\rm M}_\odot)^{-2} \tag{16.5}$$

The third stellar timescale is the nuclear one, which is the time needed for the star to exhaust its nuclear fuel reserve (which is proportional to M), at its present fuel consumption rate (which is proportional to L), so this timescale is given by

$$\tau_{\rm nuc} \simeq 10\ {\rm Gyr}\ (M/{\rm M}_\odot)^{-2.5} \tag{16.6}$$

In calculating the above mentioned timescales we have assumed a mass–luminosity relation: $L \propto M^{3.5}$ and a mass–radius relation for main-sequence stars: $R \propto M^{0.5}$. Both of these relations are fairly good approximations for $M \geq {\rm M}_\odot$. Hence, it should also be noted that the rough numerical estimates of these timescales only apply to ZAMS stars.

16.3.2 The variation of the outer radius during stellar evolution

Figure 16.5 depicts the evolutionary tracks in the Hertzsprung–Russell diagram of six different stars (50, 20, 12, 5, 2 and 1 ${\rm M}_\odot$). We calculated these tracks using Eggleton's evolutionary code (e.g., Pols *et al.* 1995, 1998). The observable stellar parameters are: luminosity (L), radius (R) and effective surface temperature ($T_{\rm eff}$). Their well-known relationship is given by $L = 4\pi R^2 \sigma T_{\rm eff}^4$. In Fig. 16.6 we have plotted our calculation of stellar radius as a function of age for the 5 ${\rm M}_\odot$ star. Important evolutionary stages are indicated in the figures. Between points 1 and 2 the star is in the long-lasting phase of core hydrogen burning (nuclear timescale). At point 3 hydrogen ignites in a shell around the helium core. For stars more massive than 1.2 ${\rm M}_\odot$ the entire star briefly contracts between points 2 and 3, causing its central temperature to rise. When the central temperature reaches $\sim 10^8$ K, core helium ignites (point 4). At this moment the star has become a red giant, with a dense core and a very large radius. During helium burning it describes a loop in the H-R diagram. Stars with $M \geq 2.3\ {\rm M}_\odot$ move from points 2 to 4 on a thermal timescale and describe the helium-burning loop on a (helium) nuclear timescale following point 4. Finally, during helium shell burning the outer radius expands again and at carbon ignition the star has become a red supergiant on the asymptotic giant branch (AGB).

The evolution of less massive stars ($M < 2.3\ {\rm M}_\odot$) takes a somewhat different course. After hydrogen shell ignition the helium core becomes degenerate and the hydrogen burning shell generates the entire stellar luminosity. While its core mass grows, the star gradually climbs upwards along the red giant branch until it reaches helium ignition with a flash. For all stars less massive than about 2.3 ${\rm M}_\odot$ the helium core has a mass of about 0.45 ${\rm M}_\odot$ at helium flash ignition. The evolution described above depends only slightly on the initial chemical composition and effects of convective overshooting.

16.3.2.1 The core mass–radius relation for low-mass RGB stars

For a low-mass star ($\lesssim 2.3\ {\rm M}_\odot$) on the red giant branch (RGB) the growth in core mass is directly related to its luminosity, as this luminosity is entirely generated by hydrogen shell burning. As such a star, composed of a small dense core surrounded by an extended convective envelope, is forced to move up the Hayashi track its luminosity increases strongly with only a fairly modest decrease in temperature. Hence one also finds a relationship between the giant's radius and the mass of its degenerate helium core – almost entirely independent of the mass present in the hydrogen-rich envelope (Refsdal & Weigert 1971; Webbink, Rappaport & Savonije 1983). This relationship is very important for LMXBs and wide-orbit binary

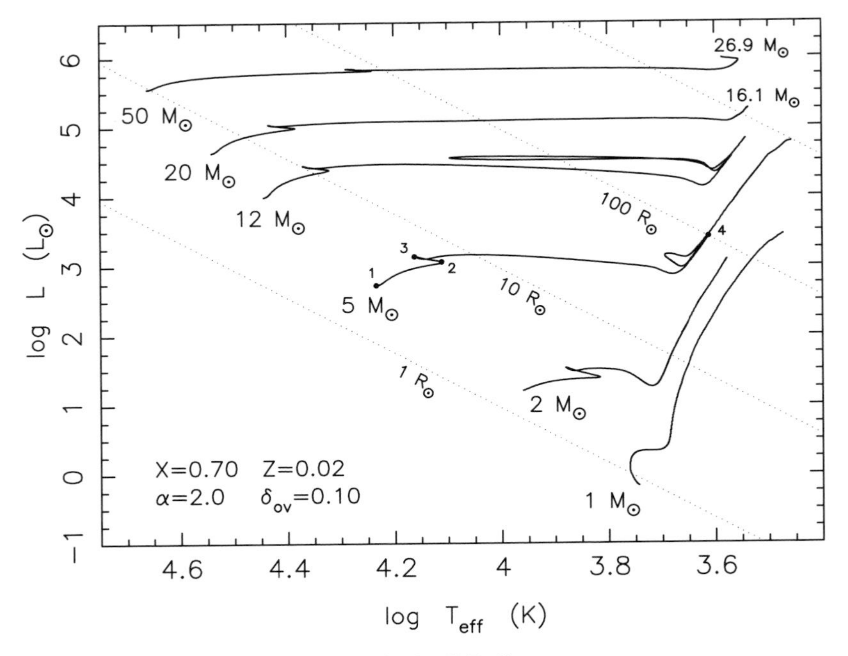

Fig. 16.5. Stellar evolutionary tracks in the H-R diagram.

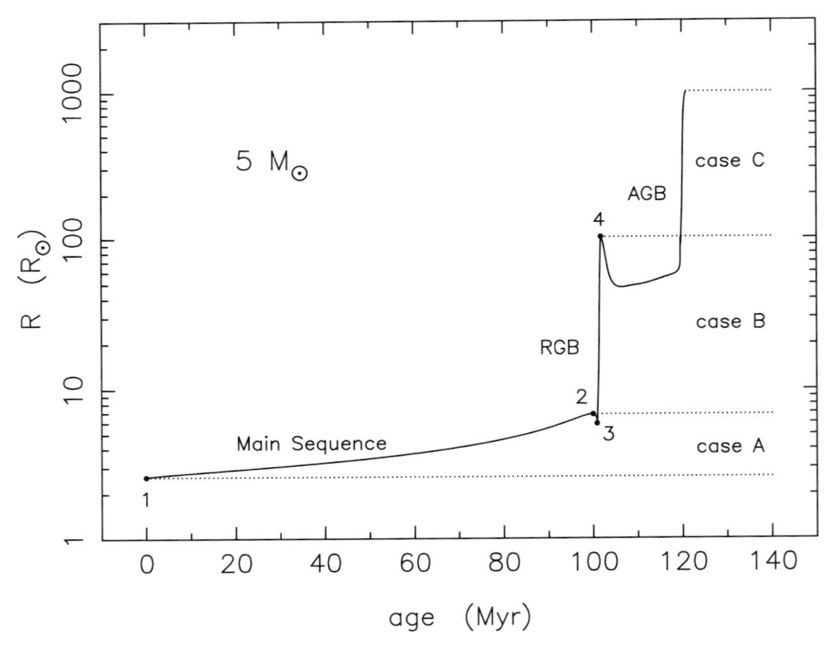

Fig. 16.6. Evolutionary change of the radius of the 5 $M_\odot$ star plotted in Fig. 16.5. The ranges of radii for mass transfer to a companion star in a binary system according to RLO cases A, B and C are indicated – see Section 16.4 for an explanation.

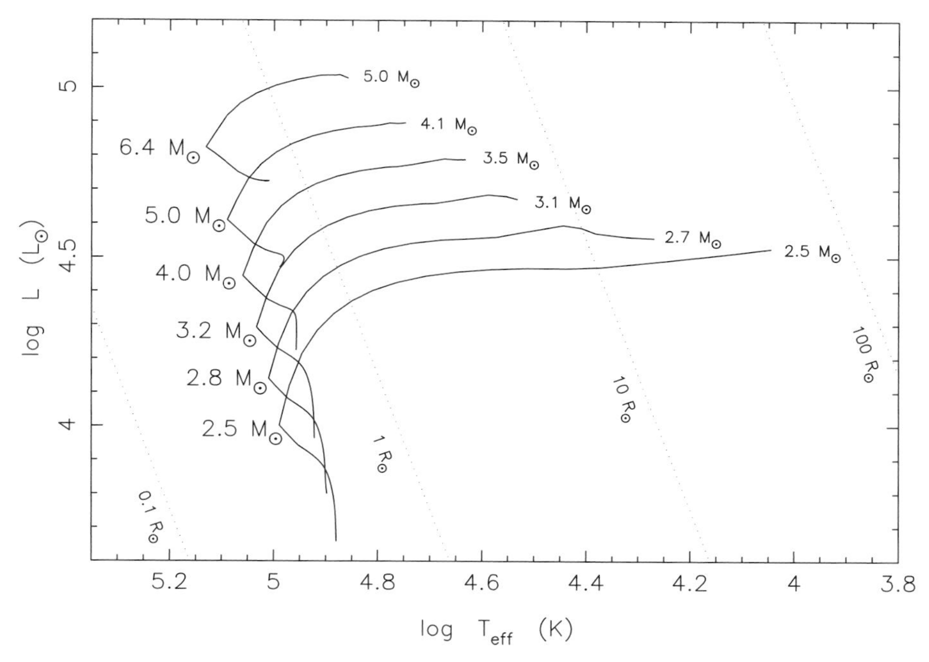

Fig. 16.7. Evolutionary tracks of $2.5-6.4\,M_\odot$ helium stars ($Y = 0.98,\ Z = 0.02$). The final stellar mass (after wind mass loss) is written at the end of the tracks. The expansion of low-mass helium stars in close binaries often results in a second mass-transfer phase (case BB RLO). This plot was made with data provided by O. Pols (2002, private communication).

pulsars since, as we shall see later on, it results in a relationship between orbital period and white dwarf mass.

16.3.3 The evolution of helium stars

For low-mass stars, the evolution of the helium core in post main-sequence stars is practically independent of the presence of an extended hydrogen-rich envelope. However, for more massive stars ($> 2.3\,M_\odot$) the evolution of the core of an isolated star differs from that of a naked helium star (i.e., a star that has lost its hydrogen envelope via mass transfer in a close binary system). Thus, it is very important to study the giant phases of helium star evolution. Pioneering studies in this field are those of Paczyński (1971), Nomoto (1984) and Habets (1986). Of particular interest are the low-mass helium stars ($M_{\rm He} < 3.5\,M_\odot$) since they swell up to large radii during their late evolution – see Fig. 16.7. This may cause an additional phase of mass transfer from the naked helium star to its companion (often referred to as so-called case BB mass transfer). Recent detailed studies of helium stars in binaries have been performed by Dewi *et al.* (2002). Using helium star models ($Z = 0.03,\ Y = 0.97$) calculated by O. Pols (2002, private communication), we fitted the helium star ZAMS radii as a function of mass:

$$R_{\rm He} = 0.212(M_{\rm He}/M_\odot)^{0.654}\,R_\odot \qquad (16.7)$$

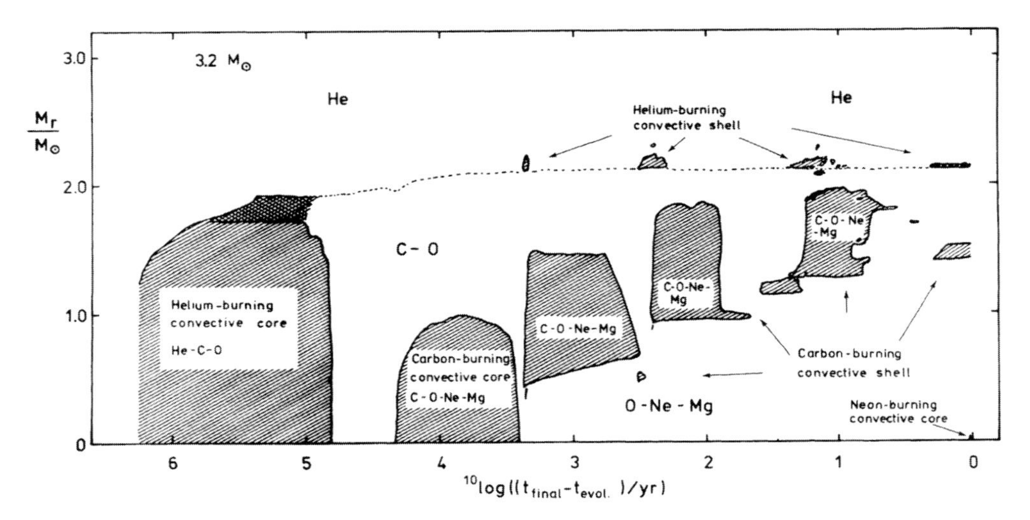

Fig. 16.8. The interior evolution of a 3.2 $M_\odot$ helium star. Hatched regions are convective; double hatched regions are semi-convective. The broken line indicates the region of maximum net energy generation in the helium-burning shell (approx. the boundary between the CO core and its adjacent helium layer). After Habets (1986).

It is important to realize that helium cores in binaries have tiny envelopes of hydrogen ($< 0.01\,M_\odot$) when they detach from RLO. This has important effects on their subsequent radial evolution (e.g., Han *et al.* 2002).

The evolution of more massive helium stars (Wolf–Rayet stars) is also quite important. There is currently no clear agreement on the rate of intense-wind mass loss from Wolf–Rayet stars (e.g., Wellstein & Langer 1999; Nugis & Lamers 2000; Nelemans & van den Heuvel 2001). A best-estimate fit to the wind mass-loss rate of Wolf–Rayet stars is, for example, given by Dewi *et al.* (2002):

$$\dot{M}_{\rm He,\,wind} = \begin{cases} 2.8\times10^{-13}\,(L/{\rm L}_\odot)^{1.5}\ {\rm M}_\odot\,{\rm yr}^{-1}, & \log(L/{\rm L}_\odot)\geq 4.5 \\ 4.0\times10^{-37}\,(L/{\rm L}_\odot)^{6.8}\ {\rm M}_\odot\,{\rm yr}^{-1}, & \log(L/{\rm L}_\odot) < 4.5 \end{cases} \tag{16.8}$$

The uncertainty in determining this rate also affects our knowledge of the threshold mass for core collapse into a black hole (Schaller *et al.* 1992; Woosley, Langer & Weaver 1995; Brown, Lee & Bethe 1999). Very important in this respect is the question of whether the helium star is "naked" or "embedded" – i.e., is the helium core of the massive star surrounded by a thick hydrogen mantle? In the latter case this helium "star" does not lose much mass in the form of a wind and it can go through all burning stages and terminate as a black hole. Single star evolutionary models suggest that this happens above an initial stellar mass of $\geq 19\,M_\odot$, as around this mass a sudden increase in the mass of the collapsing iron core occurs to $\geq 1.9\,M_\odot$ (Woosley & Weaver 1995). In order to form a black hole in a close binary, it is best to keep the helium core embedded in its hydrogen envelope as long as possible, i.e., to start from a wide "case C" binary evolution, as has been convincingly argued by Brown, Lee & Bethe (1999); Wellstein & Langer (1999); Brown *et al.* (2001) and Nelemans & van den Heuvel (2001). In this case, common envelope evolution (Section 16.5) leads to a narrow system consisting of the evolved helium core and the low-mass companion. The helium core collapses to a black hole and produces a supernova in this process, shortly after the spiral-in. When the low-mass

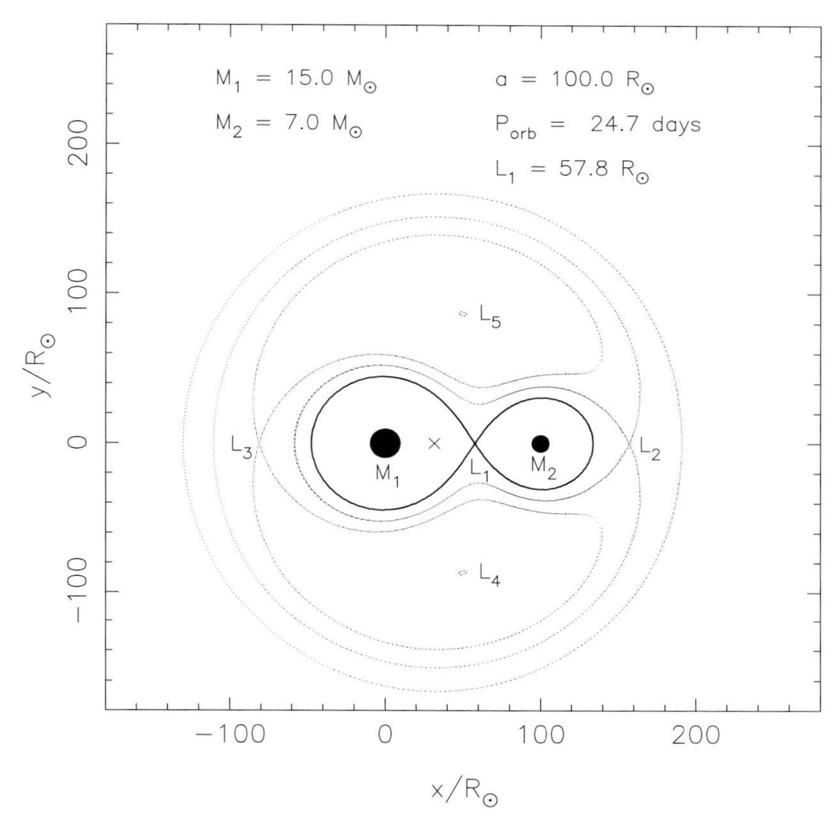

Fig. 16.9. A cross-section in the equatorial plane of the critical equipotential surfaces in a binary. The thick curve crossing through L_1 is the Roche-lobe.

companion evolves to fill its Roche-lobe these systems are observed as soft X-ray transients (SXTs) – see Section 16.2.4.

16.4 Roche-lobe overflow: cases A, B and C

The effective gravitational potential in a binary system is determined by the masses of the stars and the centrifugal force arising from the motion of the two stars around one another. One may write this potential as

$$\Phi = -\frac{GM_1}{r_1} - \frac{GM_2}{r_2} - \frac{\Omega^2 r_3^2}{2} \tag{16.9}$$

where r_1 and r_2 are the distances to the center of the stars with mass M_1 and M_2, respectively; Ω is the orbital angular velocity; and r_3 is the distance to the rotational axis of the binary. It is assumed that the stars are small with respect to the distance between them and that they revolve in circular orbits, i.e., $\Omega = \sqrt{GM/a^3}$. In a binary where tidal forces have circularized the orbit, and brought the two stellar components into synchronized co-rotation, one can define fixed equipotential surfaces in a comoving frame (see e.g., van den Heuvel 1994). The equipotential surface passing through the first Lagrangian point, L_1, defines the "pear-shaped" Roche-lobe – see the cross-section in Fig. 16.9. If the initially more massive star (the donor) evolves to fill its Roche-lobe the unbalanced pressure at L_1 will initiate mass transfer (Roche-lobe overflow, RLO) onto its companion star (the accretor). The radius of the donor's Roche-lobe, R_L, is defined as that of a sphere with the same volume as the lobe. It is a function only of the orbital separation, a, and the mass ratio, $q \equiv M_{\rm donor}/M_{\rm accretor}$, of the binary components. It can be approximated as (Eggleton 1983)

$$\frac{R_L}{a} = \frac{0.49\,q^{2/3}}{0.6\,q^{2/3} + \ln(1+q^{1/3})} \tag{16.10}$$

A star born in a close binary system with a radius smaller than that of its Roche-lobe may, either because of expansion of its envelope at a later evolutionary stage or because the binary shrinks sufficiently as a result of orbital angular momentum losses, begin RLO. The further evolution of the system will now depend on the evolutionary state and structure of the donor star at the onset of the overflow, which is determined by $M_{\rm donor}$ and a, as well as the nature of the accreting star. Kippenhahn and Weigert (1967) define three types of RLO: cases A, B and C. In case A, the system is so close that the donor star begins to fill its Roche-lobe during core-hydrogen burning; in case B the primary star begins to fill its Roche-lobe after the end of core-hydrogen burning but before helium ignition; in case C it overflows its Roche-lobe during helium shell burning or beyond. It is clear from Fig. 16.6 that cases B and C occur over a wide range of radii (orbital periods); case C even up to orbital periods of ~ 10 years. The precise orbital period ranges for cases A, B and C depend on the initial donor star mass and on the mass ratio. Once the RLO has started it continues until the donor has lost its hydrogen-rich envelope (typically $\geq 70\%$ of its total mass) and subsequently no longer fills its Roche-lobe.

16.4.1 The orbital angular momentum balance equation

The orbital angular momentum of a binary system is given by

$$J_{\rm orb} = \frac{M_1 M_2}{M}\,\Omega\, a^2\,\sqrt{1-e^2} \tag{16.11}$$

where a is the separation between the stellar components; M_1 and M_2 are the masses of the accretor and donor star, respectively; $M = M_1 + M_2$ and the orbital angular velocity $\Omega = \sqrt{GM/a^3}$. Here G is the constant of gravity. As mentioned earlier, tidal effects acting on a near-RLO (giant) star will circularize the orbit on a short timescale of $\sim 10^4$ yr (Verbunt & Phinney 1995). In the following we therefore neglect any small eccentricity ($e = 0$). A simple logarithmic differentiation of the above equation yields the rate of change in orbital separation:

$$\frac{\dot a}{a} = 2\,\frac{\dot J_{\rm orb}}{J_{\rm orb}} - 2\,\frac{\dot M_1}{M_1} - 2\,\frac{\dot M_2}{M_2} + \frac{\dot M_1 + \dot M_2}{M} \tag{16.12}$$

where the total change in orbital angular momentum is given by

$$\frac{\dot J_{\rm orb}}{J_{\rm orb}} = \frac{\dot J_{\rm gwr}}{J_{\rm orb}} + \frac{\dot J_{\rm mb}}{J_{\rm orb}} + \frac{\dot J_{\rm ls}}{J_{\rm orb}} + \frac{\dot J_{\rm ml}}{J_{\rm orb}} \tag{16.13}$$

These two equations constitute the orbital angular momentum balance equation and will now be discussed in more detail. The first term on the right-hand side of Eq. (16.13) gives the change in orbital angular momentum due to gravitational wave radiation (Landau &

Lifshitz 1958):

$$\frac{\dot{J}_{\mathrm{gwr}}}{J_{\mathrm{orb}}} = -\frac{32\,G^3}{5\,c^5}\,\frac{M_1 M_2 M}{a^4}\ \mathrm{s}^{-1} \tag{16.14}$$

where c is the speed of light in vacuum. The validity of this mechanism has been beautifully demonstrated in PSR 1913+16, which is an ideal GTR-laboratory (e.g., Taylor & Weisberg 1989). For sufficiently narrow orbits the above equation becomes the dominant term in Eq. (16.13) and will cause a to decrease. Therefore, the orbits of very narrow binaries will tend to shrink continuously, forcing the components into contact. Gravitational radiation losses are a major force driving the mass transfer in very narrow binaries, such as CVs and LMXBs (Faulkner 1971).

The second term in Eq. (16.13) arises due to so-called magnetic braking. The presence of magnetic stellar winds has long been known to decelerate the rotation of low-mass stars (e.g., Kraft 1967; Skumanich 1972; Sonderblom 1983). The loss of spin angular momentum is caused by the magnetic field that exists as a result of chromospheric coronal activity of cool $\lesssim 1.5\,\mathrm{M}_\odot$ stars with sub-photospheric convection zones (Mestel 1984). In tight synchronized binaries, the loss of spin angular momentum occurs at the expense of the orbital angular momentum. As a result the orbital period decreases while the stellar components spin up, due to tidal forces, and approach one another. Based on Skumanich's observations, Verbunt and Zwaan (1981) derived an expression for the effect of the magnetic braking and applied it to LMXBs by extrapolating, the dependence of the magnetic braking on the orbital angular velocity, down to very short orbital periods (of the order ~ hours):

$$\frac{\dot{J}_{\mathrm{mb}}}{J_{\mathrm{orb}}} \simeq -0.5\times 10^{-28}\, f_{\mathrm{mb}}^{-2}\,\frac{k^2 R_2^4}{a^5}\,\frac{GM^3}{M_1 M_2}\ \mathrm{s}^{-1} \tag{16.15}$$

(in cgs units) where R_2 is the radius of the mass-losing star; k^2 is its gyration radius and f_{mb} is a constant of order unity. However, a fundamental law of angular momentum loss is unknown for rapidly rotating stars. Rappaport, Verbunt and Joss (1983) investigated a weaker dependency on the stellar radius. Meanwhile, it now seems that the necessary stellar activity may saturate for rotation periods shorter than ~2–3 days (e.g., Rucinski 1983; Vilhu & Walter 1987), which leads to a much flatter dependence of the angular momentum loss rate on the angular velocity ($\dot{J}_{\mathrm{mb}} \propto \Omega^{1.2}$) than is given by the Skumanich law ($\dot{J}_{\mathrm{mb}} \propto \Omega^3$). Based partly on observational work, Stepien (1995) derived a new magnetic braking law which smoothly matches the Skumanich-law dependence for wide systems to the dependence obtained by Rucinski (1983) for short orbital period ($\lesssim$3 days) systems:

$$\frac{\dot{J}_{\mathrm{mb}}}{J_{\mathrm{orb}}} \simeq -1.90\times 10^{-16}\,\frac{k^2 R_2^2}{a^2}\,\frac{M^2}{M_1 M_2}\,e^{-1.50\times 10^{-5}/\Omega}\ \mathrm{s}^{-1} \tag{16.16}$$

The two formulas above represent a strong and a weak magnetic braking torque, respectively, and their relative strength can be compared in, e.g., Tauris (2001). For a recent discussion see also Eggleton (2001). It should be noted that for many years it was thought that the magnetic field has to be anchored in underlaying radiative layers of a star (Parker 1955). However, recent observations and calculations, e.g., by Dorch and Nordlund (2001), seem to suggest that even fully convective stars still operate a significant magnetic field. This conclusion has important consequences for the explanation of the observed period gap in CVs (Spruit & Ritter 1983). We encourage further investigations on this topic.

The third term ($\dot{J}_{ls}/J_{orb}$) on the right-hand side of Eq. (16.13) describes possible exchange of angular momentum between the orbit and the donor star due to its expansion or contraction. Tauris (2001) calculated the pre-RLO spin–orbit couplings in LMXBs and demonstrated that the sole nuclear expansion of a (sub)giant donor in a tight binary will lead to an orbital period *decrease* by $\sim 10\%$, prior to the onset of the RLO mass transfer, as a result of tidal interactions. This effect is most efficient for binaries with $2 < P_{orb} < 5$ days. In more narrow orbits the donor star does not expand very much and for wide binaries the tidal torque is weak. However, when the effect of magnetic braking is included in the calculations prior to RLO it will dominate the loss of orbital angular momentum if its corresponding torque is relatively strong. The tidal torque in LMXBs can be determined by considering the effect of turbulent viscosity in the convective envelope of the donor on the equilibrium tide (Terquem *et al.* 1998). In very wide orbit LMXBs ($P_{orb} > 100$ days) the orbital separation will always *widen* prior to RLO since the stellar wind mass loss becomes very important for such giant stars. To quantize this effect one can apply the Reimers (1975) wind mass-loss rate:

$$\dot{M}_{wind} = -4 \times 10^{13}\, \eta_{RW} L\, R/M \quad M_{\odot}\, yr^{-1} \tag{16.17}$$

where the luminosity, radius and mass of the mass-losing star are in solar units and $\eta_{RW} \simeq 0.5$ is a mass-loss parameter.

Spin–orbit couplings in X-ray binaries can also help to stabilize the mass-transfer processes in IMXBs with radiative donor stars (Tauris & Savonije 2001). In such systems the effect of pumping spin angular momentum into the orbit is clearly seen in the calculations as a result of a contracting mass-losing star in a tidally locked system. This causes the orbit to widen (or shrink less) and survive the, otherwise dynamically unstable, mass transfer.

The tidal effects in eccentric high-mass binary systems are discussed in, e.g., Witte & Savonije (1999) and Witte (2001). For massive donor stars ($> 8\, M_{\odot}$) one can use the mass-loss rates by, e.g., de Jager, Nieuwenhuijzen & van der Hucht (1988).

Finally, the last term on the right-hand side of Eq. (16.13) represents the change in orbital angular momentum caused by mass loss from the binary system. This is usually the dominant term in the orbital angular momentum balance equation and its total effect is given by

$$\frac{\dot{J}_{ml}}{J_{orb}} = \frac{\alpha + \beta q^2 + \delta\gamma(1+q)^2}{1+q}\,\frac{\dot{M}_2}{M_2} \tag{16.18}$$

where α, β and δ are the fractions of mass lost from the donor in the form of a direct fast wind, the mass ejected from the vicinity of the accretor and from a circumbinary co-planar toroid (with radius, $a_r = \gamma^2 a$), respectively (see van den Heuvel (1994) and Soberman, Phinney & van den Heuvel (1997)). The accretion efficiency of the accreting star is thus given by $\epsilon = 1 - \alpha - \beta - \delta$, or equivalently

$$\partial M_1 = -(1 - \alpha - \beta - \delta)\, \partial M_2 \tag{16.19}$$

where $\partial M_2 < 0$ (M_2 refers to the donor star). These factors will be functions of time as the binary system evolves during the mass-transfer phase.

The general solution for calculating the change in orbital separation during the X-ray phase is found by integration of the orbital angular momentum balance equation (Eq. 16.12). It is often a good approximation to assume $\dot{J}_{gwr}, \dot{J}_{mb} \ll \dot{J}_{ml}$ during short RLO, and if α, β and δ

are constant in time:

$$\frac{a}{a_0} = \Gamma_{\rm ls}\left(\frac{q}{q_0}\right)^{2(\alpha+\gamma\delta-1)} \left(\frac{q+1}{q_0+1}\right)^{\frac{-\alpha-\beta+\delta}{1-\epsilon}} \left(\frac{\epsilon q+1}{\epsilon q_0+1}\right)^{3+2\frac{\alpha\epsilon^2+\beta+\gamma\delta(1-\epsilon)^2}{\epsilon(1-\epsilon)}} \tag{16.20}$$

where the subscript 0 denotes initial values and $\Gamma_{\rm ls}$ is factor of order unity to account for the tidal spin–orbit couplings ($\dot{J}_{\rm ls}$) other than the magnetic braking ($\dot{J}_{\rm mb}$). We remind the reader that $q \equiv M_{\rm donor}/M_{\rm accretor}$.

16.4.2 Stability criteria for mass transfer

The stability and nature of the mass transfer is very important in binary stellar evolution. It depends on the response of the mass-losing donor star and of the Roche-lobe – see Soberman, Phinney & van den Heuvel (1997) for a review. If the mass transfer proceeds on a short timescale (thermal or dynamical) the system is unlikely to be observed during this short phase; whereas if the the mass transfer proceeds on a nuclear timescale it is still able to sustain a high enough accretion rate onto the neutron star or black hole for the system to be observed as an X-ray source for a long time.

When the donor star fills its Roche-lobe, and is perturbed by removal of mass, it falls out of hydrostatic and thermal equilibrium. In the process of re-establishing equilibrium the star will either grow or shrink – first on a dynamical (sound crossing) timescale, and then on a slower thermal (Kelvin–Helmholtz) timescale. But also the Roche-lobe changes in response to the mass transfer/loss. As long as the donor star's Roche-lobe continues to enclose the star the mass transfer is stable. Otherwise it is unstable and proceeds on a dynamical timescale. Hence the question of stability is determined by a comparison of the exponents in power-law fits of radius to mass, $R \sim M^{\zeta}$, for the donor star and the Roche-lobe respectively:

$$\zeta_{\rm donor} \equiv \frac{\partial \ln R_2}{\partial \ln M_2} \quad \wedge \quad \zeta_{\rm L} \equiv \frac{\partial \ln R_{\rm L}}{\partial \ln M_2} \tag{16.21}$$

where R_2 and M_2 refer to the mass-losing donor star. Given $R_2 = R_{\rm L}$ (the condition at the onset of RLO) the initial stability criteria becomes

$$\zeta_{\rm L} \leq \zeta_{\rm donor} \tag{16.22}$$

where $\zeta_{\rm donor}$ is the adiabatic or thermal (or somewhere in between) response of the donor star to mass loss. Note that the stability might change during the mass-transfer phase so that initially stable systems become unstable, or vice versa, later in the evolution (e.g., Kalogera & Webbink 1996). The radius of the donor is a function of time and mass and thus

$$\dot{R}_2 = \left.\frac{\partial R_2}{\partial t}\right|_{M_2} + R_2\,\zeta_{\rm donor}\frac{\dot{M}_2}{M_2} \tag{16.23}$$

$$\dot{R}_{\rm L} = \left.\frac{\partial R_{\rm L}}{\partial t}\right|_{M_2} + R_{\rm L}\,\zeta_{\rm L}\frac{\dot{M}_2}{M_2} \tag{16.24}$$

The second terms on the right-hand sides follow from Eq. (16.21); the first term of Eq. (16.23) is due to expansion of the donor star as a result of nuclear burning (e.g., shell hydrogen burning on the RGB) and the first term in Eq. (16.24) represents changes in $R_{\rm L}$ that are not caused by mass transfer – such as orbital decay due to gravitational wave radiation and tidal spin–orbit couplings. Tidal couplings act to synchronize the orbit whenever the rotation of the donor is perturbed (e.g., as a result of magnetic braking or an increase in the moment of inertia while the donor expands). The mass-loss rate of the donor can be found as a self-consistent solution to Eqs. (16.23) and (16.24) assuming $\dot{R}_2 = \dot{R}_L$ for stable mass transfer.

16.4.3 Response of the Roche-lobe to mass transfer/loss

In order to study the dynamical evolution of an X-ray binary let us consider the cases where tidal interactions and gravitational wave radiation can be neglected. We shall also assume that the amount of mass lost directly from the donor star in the form of a fast wind, or via a circumbinary toroid, is negligible compared to the flow of material transfered via the Roche-lobe. Hence we have $\dot{J}_{\rm gwr} = \dot{J}_{\rm mb} = \dot{J}_{\rm ls} = 0$ and $\dot{J}_{\rm ml}/J_{\rm orb} = \beta q^2/(1+q) \times (\dot{M}_2/M_2)$. This corresponds to the mode of "isotropic re-emission" where matter flows over from the donor star onto the compact accretor (neutron star or black hole) in a conservative way, before a fraction, β, of this material is ejected isotropically from the system with the specific angular momentum of the accretor, i.e., $\dot{M}_1 = -(1-\beta)\dot{M}_2$, $\mathrm{d}J_{\rm orb} = (J_1/M_1)\,\beta\,\mathrm{d}M_2$ and $J_1 = (M_2/M)\,J_{\rm orb}$. In the above formalism this corresponds to $\alpha = \delta = 0$ and $\Gamma_{ls} = 1$. This is actually a good approximation for many real systems (with orbital periods longer than a few days). Following Tauris and Savonije (1999) one can then combine Eqs. (16.10), (16.19) and (16.20) and obtain an analytical expression for $\zeta_{\rm L}$:

$$\begin{aligned}\zeta_{\rm L} &= \frac{\partial \ln R_{\rm L}}{\partial \ln M_2} = \left(\frac{\partial \ln a}{\partial \ln q} + \frac{\partial \ln(R_{\rm L}/a)}{\partial \ln q}\right)\frac{\partial \ln q}{\partial \ln M_2} \\ &= [1+(1-\beta)q]\psi + (5-3\beta)q \end{aligned} \tag{16.25}$$

where

$$\psi = \left[-\frac{4}{3} - \frac{q}{1+q} - \frac{2/5 + 1/3\,q^{-1/3}(1+q^{1/3})^{-1}}{0.6 + q^{-2/3}\ln(1+q^{1/3})}\right] \tag{16.26}$$

In the limiting case where $q \to 0$ (when the accretor is much heavier than the donor star; for example in a soft X-ray transient system hosting a black hole):

$$\lim_{q\to 0} \zeta_{\rm L} = -5/3 \tag{16.27}$$

The behavior of $\zeta_{\rm L}(q,\beta)$ for different X-ray binaries is plotted in Fig. 16.10. This figure is quite useful to get an idea of the stability of the mass transfer when comparing with ζ of the donor star. We see that in general the Roche-lobe, $R_{\rm L}$, increases ($\zeta_L < 0$) when material is transfered from a relatively light donor to a heavy accretor ($q < 0$). In this situation the mass transfer will be stable. Correspondingly $R_{\rm L}$ decreases ($\zeta_{\rm L} > 0$) when material is transfered from a heavy donor to a lighter accretor ($q > 0$). In this case the mass transfer has the potential to be dynamically unstable. This behavior can be understood from the bottom panel of Fig. 16.10 where we plot

$$-\partial\ln(a)/\partial\ln(q) = 2 + \frac{q}{q+1} + q\frac{3\beta-5}{q(1-\beta)+1} \tag{16.28}$$

as a function of q. The sign of this quantity is important since it tells whether the orbit expands or contracts in response to mass transfer (note $\partial q < 0$). It is noticed that the orbit always expands when $q < 1$ and it always decreases when $q > 1.28$ (solving $\partial\ln(a)/\partial\ln(q) = 0$ for $\beta = 1$ yields $q = (1+\sqrt{17})/4 \approx 1.28$). If $\beta > 0$ the orbit can still expand for $1 < q < 1.28$. There is a point at $q = 3/2$ where $\partial\ln(a)/\partial\ln(q) = 2/5$, independent of β. It should also be mentioned for the curious reader that if $\beta > 0$ then, in some cases, it is actually possible for a binary to decrease its separation, a, while increasing $P_{\rm orb}$ at the same time!

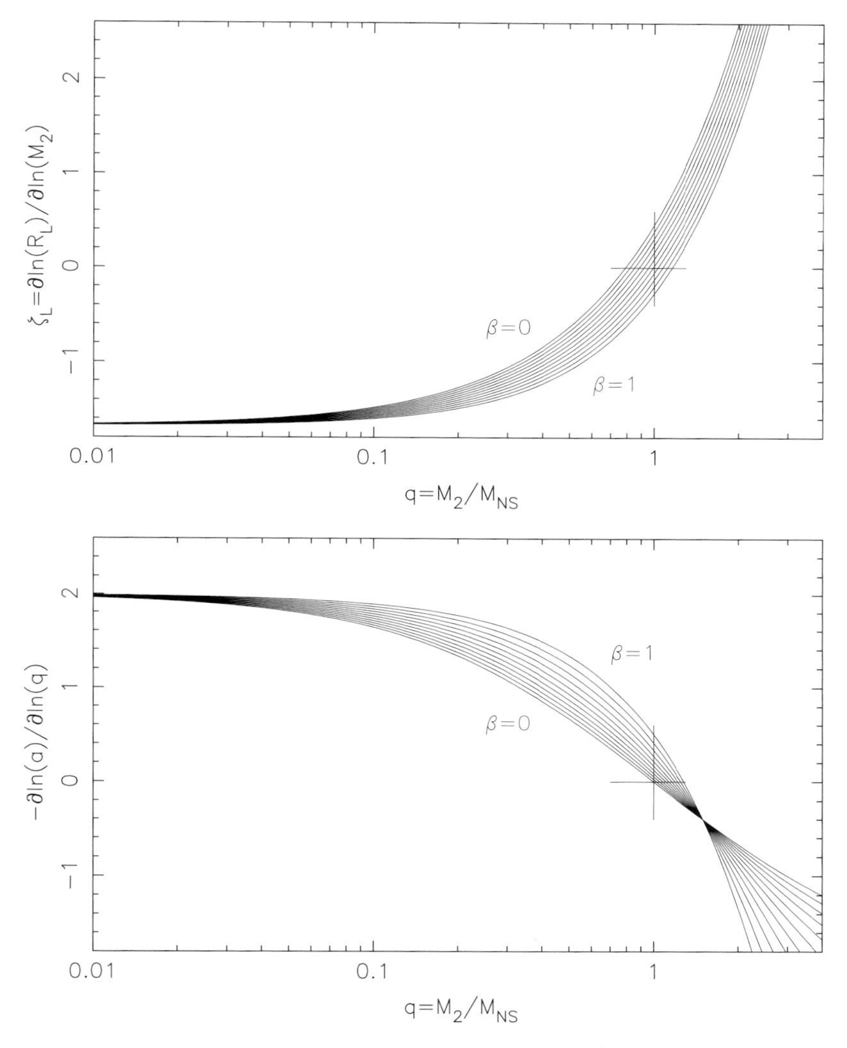

Fig. 16.10. Top panel: the Roche-radius exponent ($R_L \propto M_2^{\zeta_L}$) for LMXBs as a function of q and β. The different curves correspond to different constant values of β in steps of 0.1. Tidal effects are not taken into account and the mass loss is according to the isotropic re-emission model (i.e., $\alpha = 0$, $\delta = 0$). A cross is shown to highlight the cases of $q = 1$ and $\zeta_L = 0$. In the bottom panel we have plotted $-\partial \ln a / \partial \ln q$ as a function of q. When this quantity is positive the orbit widens. This is the case when $q \leq 1$. For more massive donor stars ($q > 1$) the orbit shrinks in response to mass transfer. Since M_2 (and hence q) is decreasing with time the evolution during the mass-transfer phase follows these curves from right to left, although β need not be constant. After Tauris & Savonije (1999).

16.4.4 Response of the mass-losing star: the effect on the binary

The radius of the mass-losing donor star will respond to mass loss. Therefore, in order to obtain a full stability analysis of the mass-transfer process it is important to know whether or not the donor star expands (or contracts) in response to mass loss. This is determined by the stellar structure (i.e., temperature gradient and entropy) of the envelope at the onset of the RLO.

16.4.4.1 Donor stars with a radiative envelope

Donor stars with radiative (or only slightly convective) envelopes will usually shrink (or keep a roughly constant radius) in response to mass loss. Therefore, these stars will give rise to a dynamically stable mass-transfer phase if the mass ratio, q is not too large. Calculations by Tauris and Savonije (1999) show that all LMXBs with donor stars $M_2 \leq 1.8\,\mathrm{M}_\odot$ and a neutron star will have a stable mass-transfer phase. Podsiadlowski, Rappaport and Pfahl (2002) find an upper limit for stability of $2.0\,\mathrm{M}_\odot$. Stars with radiative envelopes are not very evolved and hence they are only found as donor stars in systems with short orbital periods ($P_{\mathrm{orb}} \lesssim 20$ days) at the onset of the RLO. Recently it was demonstrated in detail that even IMXB systems with radiative donor stars $2 < M_2/\mathrm{M}_\odot < 5$ are able to survive extreme mass transfer on a sub-thermal timescale (Tauris, van den Heuvel & Savonije 2000).

16.4.4.2 Donor stars with a convective envelope

The thermal response of a donor star with a deep convective envelope is much more radical. It expands rapidly in response to mass loss due to the super-adiabatic temperature gradient in its giant envelope. This is clearly an unstable situation if the Roche-lobe does not grow accordingly. For systems with $q \gtrsim 1.5$ (heavy donors) the orbital shrinking is so efficient that, in combination with an expanding convective donor, it always leads to the formation of a common envelope and (dynamically unstable) spiral-in evolution. Hence, this is most likely the destiny for all HMXBs.

16.5 Common envelope evolution

A very important stage of the evolution of close binaries is the formation of a common envelope (CE). This phase is accompanied by the creation of a drag force, arising from the motion of the companion star through the envelope of the evolved star, which leads to dissipation of orbital angular momentum (spiral-in process) and deposition of orbital energy in the envelope. Hence, the global outcome of a CE-phase is reduction of the binary separation and often ejection of the envelope. There is strong evidence of orbital shrinkage (as brought about by frictional torques in a CE-phase) in some observed close binary pulsars and white dwarf binaries, e.g., PSR 1913+16, PSR J1756–5322 and L 870–2. In these systems it is clear that the precursor of the last-formed degenerate star must have achieved a radius much larger than the current orbital separation. There are many uncertainties involved in calculations of the spiral-in phase during the CE evolution. The evolution is often tidally unstable and the angular momentum transfer, dissipation of orbital energy and structural changes of the donor star take place on very short timescales ($\sim 10^3$ yr). A complete study of the problem requires very detailed multi-dimensional hydrodynamical calculations. For a general review on common envelopes, see, e.g., Iben & Livio (1993) and Taam & Sandquist (2000).

A simple estimation of the reduction of the orbital separation can be found by equating the binding energy of the envelope of the (sub)giant donor to the required difference in orbital energy (before and after the CE-phase). Following the formalism of Webbink (1984) and de Kool (1990), let $0 < \eta_{\mathrm{CE}} < 1$ describe the efficiency of ejecting the envelope, i.e., of converting orbital energy into the kinetic energy that provides the outward motion of the envelope: $E_{\mathrm{env}} \equiv \eta_{\mathrm{CE}}\ \Delta E_{\mathrm{orb}}$ or,

$$\frac{G M_{\mathrm{donor}} M_{\mathrm{env}}}{\lambda\, a_{\mathrm{i}}\, r_{\mathrm{L}}} \equiv \eta_{\mathrm{CE}} \left[\frac{G M_{\mathrm{core}} M_1}{2\, a_{\mathrm{f}}} - \frac{G M_{\mathrm{donor}} M_1}{2\, a_{\mathrm{i}}} \right] \tag{16.29}$$

yielding the ratio of final (post-CE) to initial (pre-CE) orbital separation:

$$\frac{a_{\rm f}}{a_{\rm i}} = \frac{M_{\rm core}\,M_1}{M_{\rm donor}}\,\frac{1}{M_1 + 2M_{\rm env}/(\eta_{\rm CE}\,\lambda\,r_{\rm L})} \qquad (16.30)$$

where $M_{\rm core} = M_{\rm donor} - M_{\rm env}$; $r_{\rm L} = R_{\rm L}/a_{\rm i}$ is the dimensionless Roche-lobe radius of the donor star so that $a_{\rm i}r_{\rm L} = R_{\rm L} \approx R_{\rm donor}$ and λ is a parameter that depends on the stellar mass–density distribution, and consequently also on the evolutionary stage of the star. The orbital separation of the surviving binaries is quite often reduced by a factor of ~ 100 as a result of the spiral-in. If there is not enough orbital energy available to eject the envelope the stellar components will merge in this process.

16.5.1 The binding energy of the envelope

The total binding energy of the envelope to the core is given by

$$E_{\rm bind} = -\int_{M_{\rm core}}^{M_{\rm donor}} \frac{GM(r)}{r}{\rm d}m + \alpha_{\rm th}\int_{M_{\rm core}}^{M_{\rm donor}} U{\rm d}m \qquad (16.31)$$

where the first term is the gravitational binding energy and U is the internal thermodynamic energy. The latter involves the basic thermal energy for a simple perfect gas $(3\Re T/2\mu)$, the energy of radiation $(aT^4/3\rho)$, as well as terms due to ionization of atoms and dissociation of molecules and the Fermi energy of a degenerate electron gas (Han *et al.* 1994, 1995). The value of $\alpha_{\rm th}$ depends on the details of the ejection process, which are very uncertain. A value of $\alpha_{\rm th}$ equal to 0 or 1 corresponds to maximum and minimum envelope binding energy, respectively. By simply equating Eqs. (16.29) and (16.31) one is able to calculate the parameter λ for different evolutionary stages of a given star.

Dewi and Tauris (2000) and Tauris and Dewi (2001) were the first to publish detailed calculations on the binding energy of the envelope to determine the λ-parameter (however, see also Bisscheroux 1999). Dewi and Tauris (2000) investigated stars with masses $3-10\,{\rm M}_\odot$ and found that while $\lambda < 1$ on the RGB, $\lambda \gg 1$ on the AGB (especially for stars with $M < 6\,{\rm M}_\odot$). Hence, the envelopes of these donor stars on the AGB are easily ejected; with only a relatively modest decrease in orbital separation resulting from the spiral-in. For more massive stars ($M > 10\,{\rm M}_\odot$) $\lambda < 0.1-0.01$ (see Fig. 16.11) and the internal energy is not very dominant (Dewi & Tauris 2001; Podsiadlowski, Rappaport & Han 2003). This result has the important consequence that many HMXBs will not produce double neutron star systems because they coalesce during their subsequent CE-phase (leading to a relatively small merging rate of double neutron star systems, as shown by Voss & Tauris 2003).

It should be noticed that the exact determination of λ depends on how the core boundary is defined (see Tauris & Dewi 2001 for a discussion). For example, if the core boundary (bifurcation point of envelope ejection in a CE) of the $20\,{\rm M}_\odot$ star in Fig. 16.11 is moved out by $0.1\,{\rm M}_\odot$ then λ is typically increased by a factor of ~ 2.

16.5.1.1 The question of hyper-critical accretion onto a NS in a CE

It has been argued that a neutron star spiralling inwards in a common envelope might experience hyper-critical accretion and thereby collapse into a black hole (e.g., Chevalier 1993; Brown 1995; Bethe & Brown 1998; Fryer, Woosley & Hartmann 1999). However, this idea seems difficult to reconcile with the observations of a number of very tight-orbit binary pulsars. For example, the systems PSR J1756–5322 and PSR B0655+64, which have

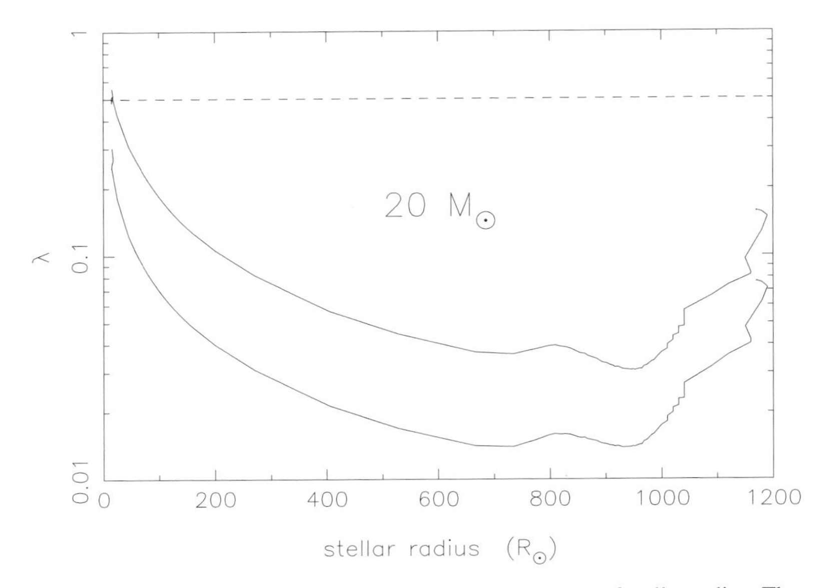

Fig. 16.11. The λ-parameter for a 20 M$_\odot$ star as a function of stellar radius. The upper curve includes internal thermodynamic energy ($\alpha_{\rm th} = 1$) whereas the lower curve is based on the sole gravitational binding energy ($\alpha_{\rm th} = 0$) – see Eq. (16.31). There is a factor ~2 in difference between the λ-curves in accordance with the virial theorem. It is a common misconception to use a constant value of $\lambda = 0.5$ (marked by the dashed line). See text for details.

massive CO white dwarf companions (~0.7 M$_\odot$ and ~1 M$_\odot$) with orbital periods of only 0.45 and 1.06 days, respectively. From an evolutionary point of view there is no other way to produce such systems apart from a CE and spiral-in phase with donor masses between 3 M$_\odot$ and 6 M$_\odot$. From a theoretical point of view it has been argued that the hyper-critical accretion can be inhibited by rotation (Chevalier 1996) and strong outflows from the accretion disk (Armitage & Livio 2000). Finally, it should be noticed that the masses of neutron stars determined in all of the five detected Galactic double neutron star systems are quite similar and close to a value of 1.4 M$_\odot$. Hence, in all these cases it is clear that the first-born neutron star did not accrete any significant amount of material from the common envelope.

16.6 (Asymmetric) supernova explosions in close binaries

After a close binary star has lost its H-envelope, and possibly also its He-envelope, during RLO and/or CE evolution it will collapse and explode in a supernova (SN) if it is massive enough. The critical threshold mass for a helium star to form a neutron star is about 2.8 M$_\odot$ (and somewhat lower for a CO-star, i.e., a helium star that has lost its helium envelope via case BB RLO). This value corresponds roughly to an initial mass of 10 M$_\odot$ (~10 M$_\odot$ for case C, and ~12 M$_\odot$ for case B/A RLO). If the core mass is below this critical threshold mass the star contracts, possibly after a second phase of RLO, and settles peacefully as a cooling white dwarf. If $M_{\rm He} \gtrsim 8$ M$_\odot$ the supernova leaves behind a black hole. All observed neutron stars in binary pulsars seem to have been born with a canonical mass of 1.3 – 1.4 M$_\odot$. Neutron stars in LMXBs may afterwards possibly accrete up to ~1 M$_\odot$ before collapsing further into a black hole (see Section 16.7.3). In very massive X-ray binaries, neutron stars can probably be born with higher masses $\sim 1.8\,M_\odot$ (Barziv *et al.* 2001).

As mentioned in the introduction, there is firm evidence that most newborn neutron stars receive a momentum kick at birth which gives rise to the high velocities observed (typically ~ 400 km s^{-1}), although there also may well be a smaller fraction $\sim$10–20% that receive kicks $\lesssim 50$ km s^{-1} in order to explain the observed neutron stars in globular clusters and the population of very wide-orbit X-ray binaries (Pfahl *et al.* 2002). It is still an open question whether or not black holes also receive a kick at birth. At least in some cases there are observational indications of mass ejection during their formation, as in any successful supernova explosion (Iwamoto *et al.* 1998), and the recent determination of the runaway velocity (112 ± 18 km s^{-1}) of the black hole binary GRO J1655–40 (Mirabel *et al.* 2002) also seems to suggest that the formation of black holes is accompanied by a kick.

In an excellent paper Hills (1983) calculated the dynamical consequences of binaries surviving an asymmetric SN. Tauris and Takens (1998) generalized the problem and derived analytical formulas for the velocities of stellar components ejected from disrupted binaries and also included the effect of shell impact on the companion star. If the collapsing core of a helium star results in a supernova explosion it is a good approximation that the collapse is instantaneous compared with $P_{\rm orb}$. Here we summarize a few important equations. The orbital energy of a binary is given by

$$E_{\rm orb} = -\frac{GM_1M_2}{2a} = -\frac{GM_1M_2}{r} + \tfrac{1}{2}\mu v_{\rm rel}^2 \qquad (16.32)$$

where r is the separation between the stars at the moment of explosion; μ is the reduced mass of the system and $v_{\rm rel} = \sqrt{GM/r}$ is the relative velocity of the two stars in a circular pre-SN binary. The change of the semi-major axis as a result of the SN is then given by (Flannery & van den Heuvel 1975)

$$\frac{a}{a_0} = \left[\frac{1-(\Delta M/M)}{1-2(\Delta M/M)-(w/v_{\rm rel})^2 - 2\cos\theta\,(w/v_{\rm rel})}\right] \qquad (16.33)$$

where $a_0 = r$ and a are the initial and final semi-major axis, respectively; ΔM is the amount of matter lost in the SN; w is the magnitude of the kick velocity and θ is the direction of the kick relative to the orientation of the pre-SN velocity. The orientation of the kick magnitude is probably completely uncorrelated with respect to the orientation of the binary – the escaping neutrinos from deep inside the collapsing core are not aware that they are members of a binary system (see, however, Pfahl *et al.* (2002) and Dewi, Podsiadlowski & Pols (2005) for the hypothesis that the kick *magnitude* may depend on the pre-SN history of the collapsing core). For each binary and a sufficiently high value of w there exists a critical angle, $\theta_{\rm crit}$, for which a SN with $\theta < \theta_{\rm crit}$ will result in disruption of the orbit (i.e., if the denominator of Eq. (16.33) is less than zero).

The sudden mass loss in the SN affects the bound orbit with an eccentricity

$$e = \sqrt{1 + \frac{2E_{\rm orb}L_{\rm orb}^2}{\mu G^2 M_1^2 M_2^2}} \qquad (16.34)$$

where the orbital angular momentum can be derived from (see also Eq. (16.11))

$$L_{\rm orb} = |\mathbf{r}\times\mathbf{p}| = r\mu\sqrt{(v_{\rm rel} + w\cos\theta)^2 + (w\sin\theta\,\sin\phi)^2} \qquad (16.35)$$

Note, in the two equations above $v_{\rm rel}$ is the pre-SN relative velocity of the two stars, whereas the stellar masses and μ now refer to the post-SN values.

Systems surviving the SN will receive a recoil velocity from the combined effect of instant mass loss and a kick. One can easily find this velocity, $v_{\rm sys}$, from conservation of momentum. Let us consider a star with mass $M_{\rm core}$ collapsing to form a neutron star with mass $M_{\rm NS}$ and hence

$$v_{\rm sys} = \sqrt{\Delta P_x^2 + \Delta P_y^2 + \Delta P_z^2}/(M_{\rm NS} + M_2) \tag{16.36}$$

where the change in momentum is

$$\begin{aligned} \Delta P_x &= M_{\rm NS}(v_{\rm core} + w\cos\theta) - M_{\rm core}v_{\rm core} \\ \Delta P_y &= M_{\rm NS}w\sin\theta\cos\phi \\ \Delta P_z &= M_{\rm NS}w\sin\theta\sin\phi \end{aligned} \tag{16.37}$$

and where M_2 is the unchanged mass of the companion star; $v_{\rm core}$ is the pre-SN velocity of the collapsing core, in a center of mass reference frame, and ϕ is the angle between the projection of $\mathbf{w}$ onto a plane $\perp$ to $\mathbf{v}_{\rm core}$ (i.e., $w_y = w\sin\theta\cos\phi$). Beware, if the post-SN periastron distance, $a(1-e)$, is smaller than the radius of the companion star then the binary will merge.

It is important to realize the difference between *gravitational* mass (as measured by an observer) and *baryonic* mass of a neutron star. The latter is $\sim 15\%$ larger for a typical equation-of-state. When considering dynamical effects on binaries surviving a SN this fact is often (almost always) ignored!

16.7 Evolution of LMXBs: formation of millisecond pulsars

Figure 16.12 depicts the formation of an LMXB and millisecond pulsar system. There are now more than 40 binary millisecond pulsars (BMSPs) known in the Galactic disk. They can be roughly divided into three observational classes (Tauris 1996): class A contains the wide-orbit ($P_{\rm orb} > 20$ days) BMSPs with low-mass helium white dwarf companions ($M_{\rm WD} < 0.45\,{\rm M}_\odot$), whereas the close-orbit BMSPs ($P_{\rm orb} \lesssim 15$ days) consist of systems with either low-mass helium white dwarf companions (class B) or systems with relatively heavy CO/O-Ne-Mg white dwarf companions (class C). The last class evolved through a phase with significant loss of angular momentum (either common envelope evolution or extreme mass transfer on a sub-thermal timescale) and descends from IMXBs with donors: $2 < M_2/{\rm M}_\odot < 8$, see Fig. 16.13. The single MSPs are believed to originate from tight class B systems where the companion has been destroyed or evaporated – either from X-ray irradiation when the neutron star was accreting, or in the form of a pulsar radiation/wind of relativistic particles (e.g., van den Heuvel & van Paradijs 1988; Ruderman, Shaham & Tavani 1989; Ergma & Fedorova 1991; Podsiadlowski 1991; Shaham 1992; Tavani 1992). Observational evidence for this scenario is found in eclipsing MSPs with ultra-light companions – e.g., PSR 1957+20 ($P_{\rm orb} = 0.38$ days; $M_2 \simeq 0.02\,{\rm M}_\odot$) and the planetary pulsar PSR 1257+12 (Wolszczan 1994).

For LMXBs it has been shown by Pylyser and Savonije (1988, 1989) that an orbital bifurcation period separates the formation of converging systems (which evolve with decreasing $P_{\rm orb}$ until the mass-losing component becomes degenerate and an ultra-compact binary is formed) from the diverging systems (which finally evolve with increasing $P_{\rm orb}$ until the

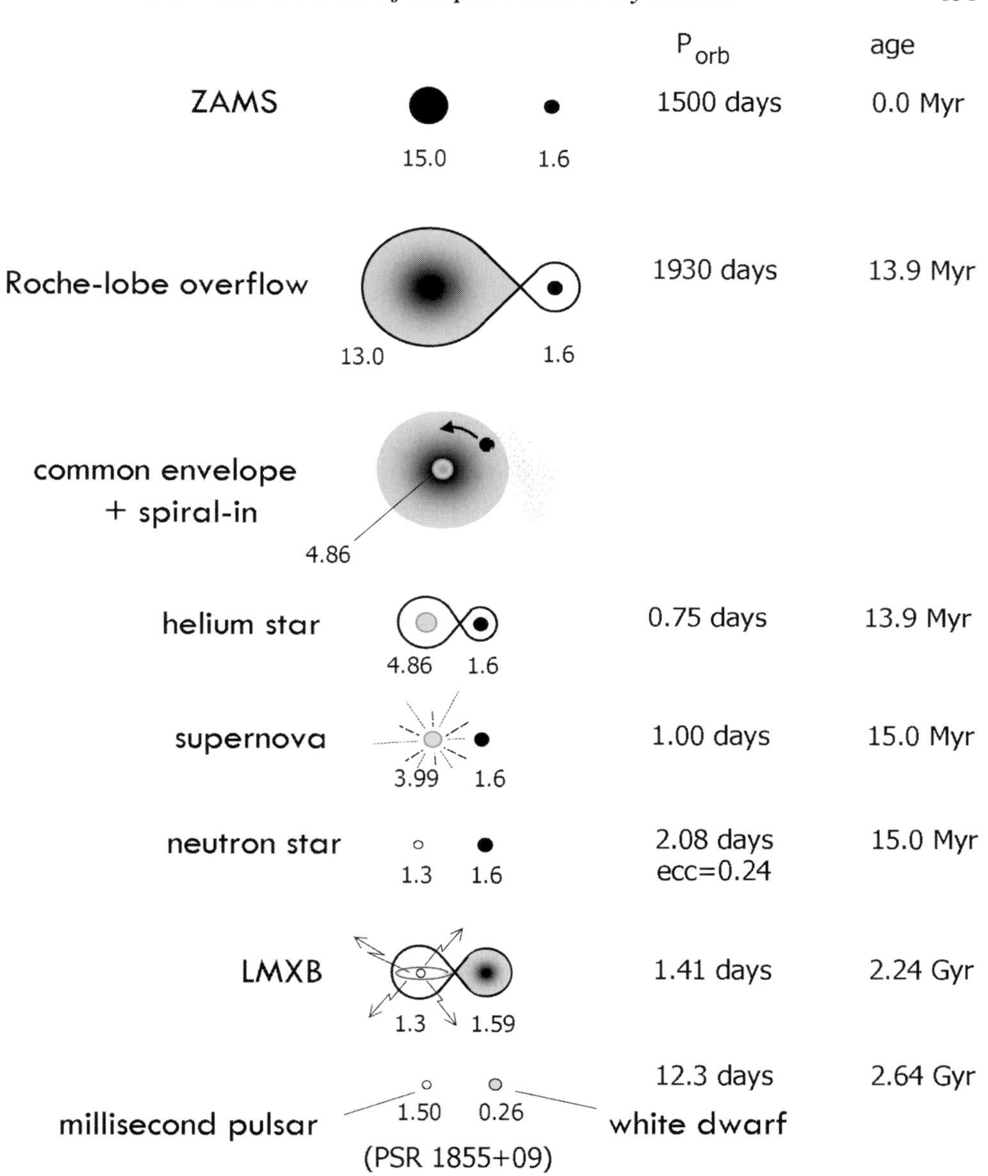

Fig. 16.12. Cartoon depicting the evolution of a binary system eventually leading to an LMXB and finally the formation of a binary millisecond pulsar. Parameters governing the specific orbital angular momentum of ejected matter, the common envelope and spiral-in phase, the asymmetric supernova explosion and the stellar evolution of the naked helium star all have a large impact on the exact evolution. Parameters are given for a scenario leading to the formation of the observed binary millisecond pulsar PSR 1855+09. The stellar masses given are in solar units.

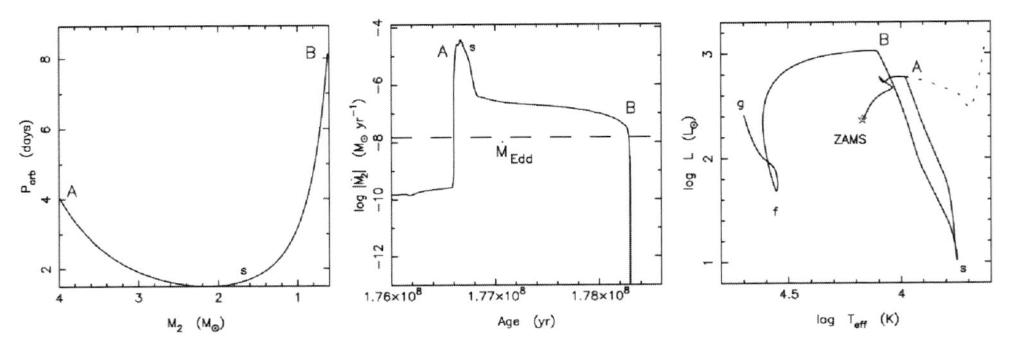

Fig. 16.13. The evolution of an IMXB leading to the formation of a BMSP with a CO WD companion in a close orbit. The initial configuration was a $4\,M_\odot$ donor star and a neutron star with an orbital period of 4 days. The mass-transfer phase is between points A and B. Between points f and g helium is burning in the core of the stripped companion star. After Tauris, van den Heuvel & Savonije (2000).

mass-losing star has lost its envelope and a wide detached binary is formed). This important bifurcation period is about 2–3 days depending on the strength of the magnetic braking torque.

16.7.1 Formation of wide-orbit binary millisecond pulsars

In LMXBs with initial $P_{orb} > 2$ days the mass transfer is driven by internal thermonuclear evolution of the donor star since it evolves into a (sub)giant before loss of orbital angular momentum dominates. These systems have been studied by Webbink, Rappaport and Savonije (1983), Taam (1983), Savonije (1987), Joss, Rappaport and Lewis (1987), Rappaport *et al.* (1995) and more recently Ergma, Sarna and Antipova (1998), Tauris and Savonije (1999) and Podsiadlowski, Rappaport and Pfahl (2002). For a donor star on the red giant branch (RGB) the growth in core mass is directly related to the luminosity, as this luminosity is entirely generated by hydrogen shell burning. As such a star, with a small compact core surrounded by an extended convective envelope, is forced to move up the Hayashi track its luminosity increases strongly with only a fairly modest decrease in temperature. Hence one also finds a relationship between the giant's radius and the mass of its degenerate helium core – almost entirely independent of the mass present in the hydrogen-rich envelope (see Table 16.5). It has also been argued that the core mass determines the rate of mass transfer (Webbink, Rappaport & Savonije 1983). In the scenario under consideration, the extended envelope of the giant is expected to fill its Roche-lobe until termination of the mass transfer. Since the Roche-lobe radius, R_L, only depends on the masses and separation between the two stars, it is clear that the core mass, from the moment the star begins RLO, is uniquely correlated with P_{orb} of the system. Thus the final orbital period ($\sim$2 to 10^3 days) is expected to be a function of the mass of the resulting white dwarf companion (Savonije 1987). Tauris and Savonije (1999) calculated the expected (P_{orb}, M_{WD}) correlation in detail and found an overall best fit:

$$M_{WD} = \left(\frac{P_{orb}}{b}\right)^{1/a} + c \qquad (16.38)$$

Table 16.5. *Stellar parameters for giant stars with* $R = 50.0\,R_\odot$

$M_{\rm initial}/M_\odot$	1.0*	1.6*	1.0**	1.6**
$\log L/L_\odot$	2.644	2.723	2.566	2.624
$\log T_{\rm eff}$	3.573	3.593	3.554	3.569
$M_{\rm core}/M_\odot$	0.342	0.354	0.336	0.345
$M_{\rm env}/M_\odot$	0.615	1.217	0.215	0.514

* Single star ($X = 0.70$, $Z = 0.02$ and $\alpha = 2.0$, $\delta_{\rm ov} = 0.10$).
** Binary star (at onset of RLO: $P_{\rm orb} \simeq 60$ days and $M_{\rm NS} = 1.3\,M_\odot$).
After Tauris & Savonije (1999).

where, depending on the chemical composition of the donor,

$$(a, b, c) = \begin{cases} 4.50 & 1.2 \times 10^5 & 0.120 & \text{Pop. I} \\ 4.75 & 1.1 \times 10^5 & 0.115 & \text{Pop. I+II} \\ 5.00 & 1.0 \times 10^5 & 0.110 & \text{Pop. II} \end{cases} \qquad (16.39)$$

Here $M_{\rm WD}$ is in solar mass units and $P_{\rm orb}$ is measured in days. The fit is valid for BMSPs with helium white dwarf companions and $0.18 \leq M_{\rm WD}/M_\odot \leq 0.45$. The formula depends slightly on the adopted value of the convective mixing-length parameter. It should be noted that the correlation is *independent* of β (the fraction of the transferred material lost from the system – see Section 16.4.1), the mode of the mass loss and the strength of the magnetic braking torque, since the relation between giant radius and core mass of the donor star remains unaffected by the exterior stellar conditions governing the process of mass transfer. However, for the *individual* binary $P_{\rm orb}$ and $M_{\rm WD}$ do depend on these parameters. In Fig. 16.14 we have plotted a theoretical ($P_{\rm orb}$, $M_{\rm WD}$) correlation and also plotted evolutionary tracks calculated for four LMXBs. Although clearly the class of BMSPs with helium white dwarf companions is present, the estimated masses of the BMSP white dwarfs are quite uncertain, since they depend on the unknown orbital inclination angle and the pulsar mass, and no clear observed ($P_{\rm orb}$, $M_{\rm WD}$) correlation has yet been established from the current observations. In particular there may be a discrepancy for the BMSPs with $P_{\rm orb} \gtrsim 100$ days (Tauris 1996).

16.7.2 Formation of close-orbit binary millisecond pulsars

In LMXBs with initial $P_{\rm orb} < 2$ days the mass transfer is driven by loss of angular momentum due to magnetic braking and gravitational wave radiation. The evolution of such systems is very similar to the evolution of CVs – see e.g. Spruit & Ritter (1983); Verbunt & van den Heuvel (1995); Ergma, Sarna & Antipova (1998) and Chapter 10 by Kuulkers *et al.*

16.7.3 Masses of binary neutron stars

In general, the masses of binary pulsars can only be estimated from their observed mass function, which depends on the unknown orbital inclination angle. Only in a few tight systems is it possible to directly measure post-Newtonian parameters (e.g., the general relativistic Shapiro delay) which yield precise values of the stellar masses (see Taylor & Weisberg 1989). For example, in the double neutron star system PSR 1913+16 the (gravitational) masses are known to be 1.441 and $1.387\,M_\odot$. Although the majority of the (rough)

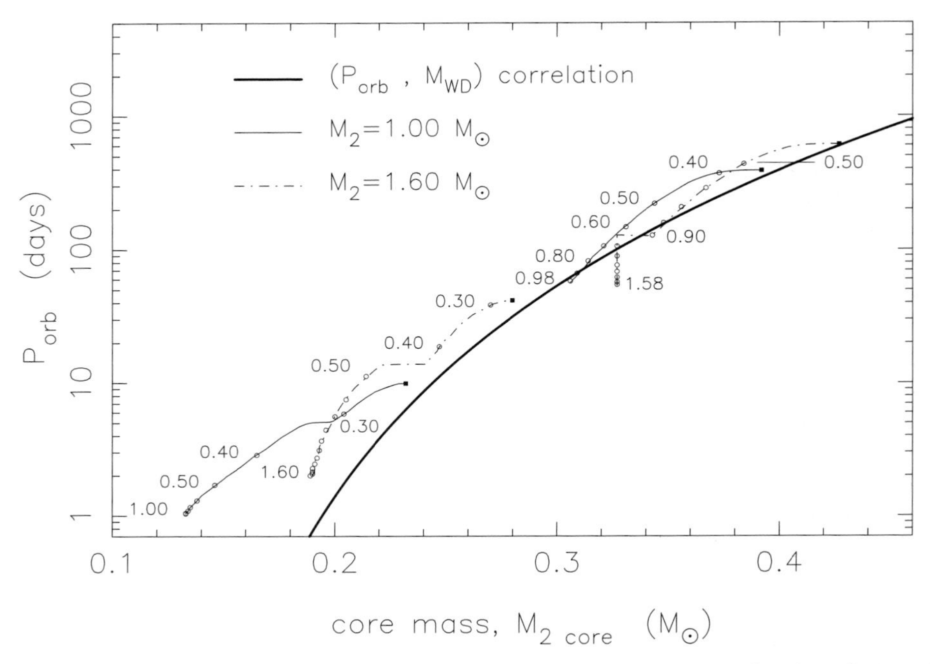

Fig. 16.14. Evolutionary tracks of four LMXBs showing P_{orb} as a function of M_{core} of the donor star. The initial donor masses were 1.0 and 1.6 $M_\odot$ (each calculated at two different initial P_{orb}) and the initial neutron star mass was 1.3 $M_\odot$. The total mass of the donors during the evolution is written along the tracks. At the termination of the mass-transfer process the donor only has a tiny ($\leq$0.01 $M_\odot$) H-envelope and the end-points of the evolutionary tracks are located near the curve representing the (P_{orb}, M_{WD}) correlation for BMSPs. After Tauris & Savonije (1999).

estimated pulsar masses may be consistent with the canonical value of $\sim 1.4\,M_\odot$ (Thorsett & Chakrabarty 1999), one could still expect a spread in neutron star masses from an evolutionary point of view. The recycled pulsars in double neutron star systems, for example, did not have a chance to accrete much material because of the short-lived common envelope and spiral-in phase that these systems evolved through, according to the standard model. Assuming all neutron stars to be born with a mass of $1.3\,M_\odot$, Tauris and Savonije (1999) demonstrated that an (P_{orb}, M_{NS}) anti-correlation would be expected for millisecond pulsars as a simple consequence of the interplay between mass-transfer rate (and thus accretion rate), orbital period and the evolutionary status of the LMXB donor star at the onset of the RLO. However, since this model predicted rather massive ($>2\,M_\odot$) neutron stars in binary millisecond pulsar systems with $P_{orb} \lesssim 30$ days, it failed to explain the mass of PSR B1855+09 ($P_{orb} = 12.3$ days), which is known to be $<1.55\,M_\odot$ from constraints on its Shapiro delay. The authors concluded that this was a proof for the fact that a large amount of matter *must* be lost from the LMXB even for sub-Eddington accretion – probably as a result of either accretion disk instabilities (Pringle 1981; van Paradijs 1996) or the so-called propeller effect (Illarionov & Sunyaev 1985).

The maximum mass of a neutron star depends on the equation of state for dense matter. Barziv *et al.* (2001) reported that the HMXB Vela X-1 has a neutron star mass $\sim 1.86\,M_\odot$. Furthermore, some kHz QPO sources are claimed to host heavy ($>2\,M_\odot$) neutron stars. If this is the case, then all soft EOS, including some kaon condensation EOS (Brown & Bethe 1994; Bethe & Brown 1995), can be ruled out.

16.8 Evolution of HMXBs

16.8.1 Formation of double neutron star/black hole binaries

The formation of a HMXB requires two relatively massive stars ($>12\,M_\odot$). Alternatively the secondary ZAMS star can be less massive initially, as long as it gains sufficient material from the primary star to later end up above the threshold mass for undergoing a supernova explosion (like the primary star). The first mass-transfer phase, from the primary to the secondary star, is usually assumed to be dynamically stable (semi-conservative) if the mass ratio at the onset of the RLO is not too extreme. However, later on *all* HMXBs end up in a common envelope phase, as the neutron star (or low-mass black hole) is engulfed by the extended envelope of its companion, in an orbit that is rapidly shrinking due to heavy loss of orbital angular momentum. As discussed earlier in Section 16.3.3, stellar winds of massive stars, as well as naked helium cores (Wolf–Rayet stars), are some of the most uncertain aspects of the modeling of HMXB evolution. The physical conditions that determine the formation of a neutron star versus a black hole are also still quite unknown. It may well be that core mass is not the only important factor to determine the outcome. Magnetic field and spin of the collapsing core could also play a major role (Ergma & van den Heuvel 1998). Furthermore, it seems clear from observations that there is an overlap in the mass range for making a neutron star versus a black hole.

16.8.2 Gravitational waves and merging NS/BH binaries

The rate of energy loss as a result of gravitational wave radiation (GWR) is given by (in the quadrupole approximation, $a \ll \lambda_{\rm gwr}$)

$$L_{\rm gwr} = \left|\frac{dE}{dt}\right| = \frac{G}{5c^5}\langle \dddot{Q}_{jk}\dddot{Q}_{jk}\rangle\; g(n,e) \;\simeq\; \frac{32G^4}{5c^5}\frac{M^3\mu^2}{a^5}\; f(e) \qquad (16.40)$$

where Q denotes the quadrupole moment of the mass distribution; M is the total mass of the system; μ is the reduced mass and $f(e)$ is a function of the orbital eccentricity (here we have disregarded the dependence on the harmonic number of the wave signal). The energy loss due to GWR can only be subtracted from the orbital energy of the binary and hence the orbital separation will decrease as

$$\dot{a} = \frac{GM\mu}{2E_{\rm orb}^2}\,\dot{E}_{\rm orb} \qquad \left(a = -\frac{GM\mu}{2E_{\rm orb}} \;\wedge\; \dot{E}_{\rm orb} = -L_{\rm gwr}\right) \qquad (16.41)$$

For an eccentric binary:

$$\frac{1}{a}\frac{\mathrm{d}a}{\mathrm{d}t} = -\frac{1}{E}\left.\frac{\mathrm{d}E}{\mathrm{d}t}\right|_{e=0} f(e) \quad \wedge \quad f(e) \simeq \frac{1+\frac{73}{24}e^2+\frac{37}{96}e^4}{(1-e^2)^{7/2}} \qquad (16.42)$$

where the approximate fit for $f(e)$ above is given by Peters (1964).
Now we have an expression for the rate of change in the orbital separation:

$$\dot{a} \simeq -\frac{64G^3}{5c^5}\frac{M^2\mu}{a^3}\,\frac{1+\frac{73}{24}e^2+\frac{37}{96}e^4}{(1-e^2)^{7/2}} \qquad (16.43)$$

which can be transformed into an expression for the merging time, $\tau_{\rm gwr}$ (Peters 1964) as a

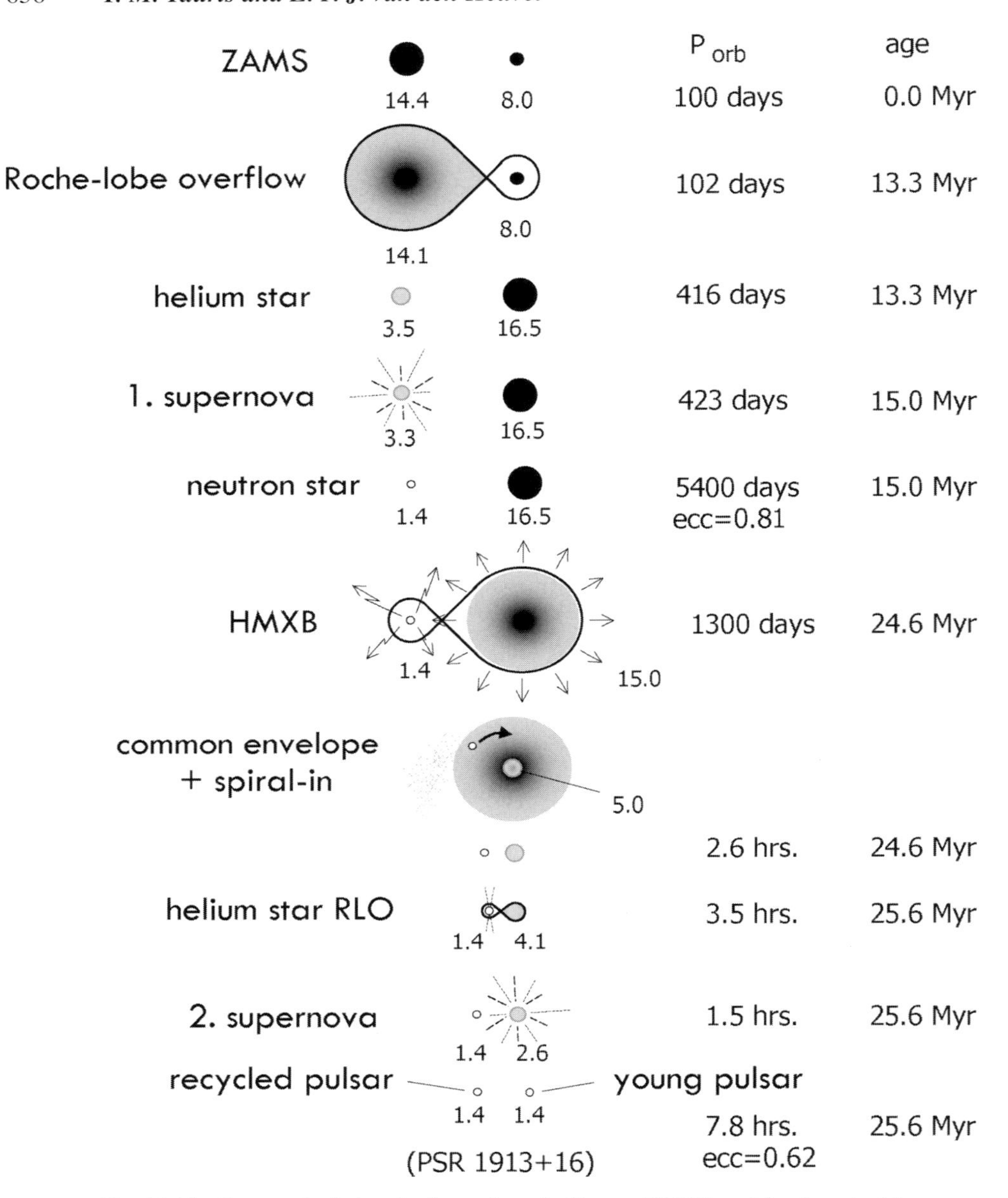

Fig. 16.15. Cartoon depicting the formation of a Be-star/HMXB and finally a double neutron star system. Such a binary will experience two supernova explosions. In a double pulsar system the recycled pulsar is most likely to be observed as a result of its very long spin-down timescale compared to the young pulsar (a factor of $\sim 10^2$). Tight NS–NS systems will coalesce due to gravitational wave radiation. These collisions should be detected by advanced gravitational wave detectors such as LIGO II/VIRGO.

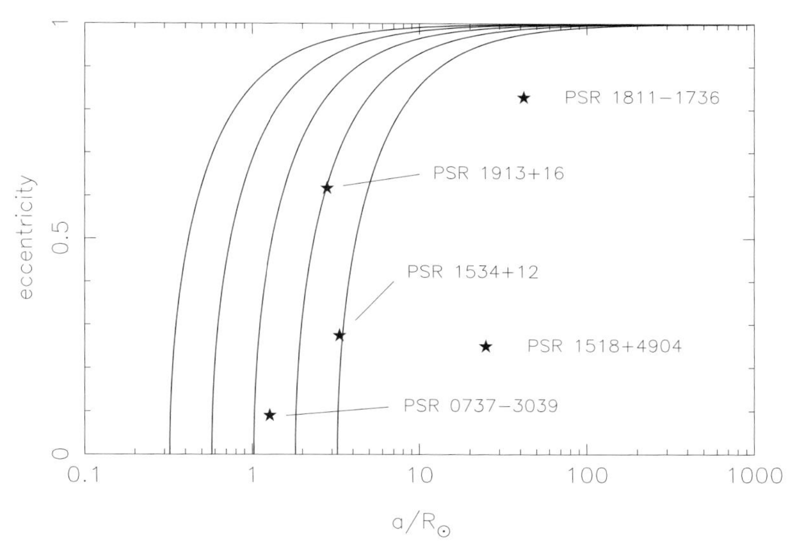

Fig. 16.16. Isochrones for the merging time of double neutron star binaries, as calculated by the authors. The curves correspond to values of (from left to right): 3×10^5 yr, 3 Myr, 30 Myr, 300 Myr and 3 Gyr, respectively. The five detected Galactic double neutron star systems are indicated with ⋆.

function of the initial values (a_0, e_0):

$$\tau_{\rm gwr}(a_0, e_0) = \frac{12}{19}\frac{C_0^4}{\beta} \times \int_0^{e_0} \frac{e^{29/19}[1+(121/304)e^2]^{1181/2299}}{(1-e^2)^{3/2}} \mathrm{d}e \tag{16.44}$$

where

$$C_0 = \frac{a_0\left(1-e_0^2\right)}{e_0^{12/19}}\left[1+(121/304)e_0^2\right]^{-870/2299} \quad \wedge \quad \beta = \frac{64G^3}{5c^5}M^2\mu \tag{16.45}$$

This equation cannot be solved analytically and must be evaluated numerically. The timescale is very dependent on both a and e. Tight and/or eccentric orbits spiral-in much faster than wider and more circular orbits – see Fig. 16.16. For example, we find that the double neutron star system PSR 1913+16 ($P_{\rm orb} = 7.75$ hr, $M_{\rm NS} = 1.441$ and $1.387\ M_\odot$) with an eccentricity of 0.617 will merge in 302 Myr; if its orbit was circular the merger time would be five times longer: 1.65 Gyr! For circular orbits the merging timescale can easily be found analytically: $\tau_{\rm gwr}^{\rm circ} = a_0^4/4\beta$.

16.8.2.1 LISA/LIGO observations of signals from tight NS/BH binaries

The recently discovered double pulsar binary system PSR J0737–3039 (Burgay *et al.* 2003) will play a remarkable role for LISA. It has an orbital period $\sim$2.4 hr and a distance to the Earth of $\sim$0.5 kpc. Continuous emission of gravitational waves should be detected at a strain of $h \approx 5 \times 10^{-21}$ at a frequency of $f \approx 0.23$ mHz. Within one year of integration the signal-to-noise ratio is expected to be $\sim$10. Thus this unique system will become an important LISA calibration source. In comparison, the signals from PSR B1913+16 ($P_{\rm orb} = 7.75$ hr and $d \sim 7$ kpc) are too week, and the wave frequency too small, to be detected by LISA.

However, it is clear from population synthesis studies that many Galactic NS–NS binaries must exist within the detection range of LISA.

The amplitude (or strain) of gravitational waves emitted continuously from a tight binary is given by the sum of the two polarizations of the signal:

$$h = \sqrt{\tfrac{1}{2}\left[h^2_{+,\max} + h^2_{\times,\max}\right]} = \sqrt{\frac{16\pi\, G}{c^3\omega_{\rm gwr}}\frac{L_{\rm gwr}}{4\pi\, d^2}}$$

$$\simeq 1\times 10^{-21}\left(\frac{M_{\rm chirp}}{{\rm M}_\odot}\right)^{5/3}\left(\frac{P_{\rm orb}}{1\ {\rm hr}}\right)^{-2/3}\left(\frac{d}{1\ {\rm kpc}}\right)^{-1} \qquad (16.46)$$

Here the waves are assumed to be sinusoidal with angular frequency, $\omega_{\rm gwr}$, which is ~ 2 times the orbital angular frequency of the binary ($\Omega = 2\pi/P_{\rm orb}$), and $M_{\rm chirp} \equiv \mu^{3/5} M^{2/5}$ is the so-called chirp mass of the system.

As a compact binary continues its inspiral, the gravitational waves will sweep upward in frequency from about 10 Hz to 10^3 Hz, at which point the compact stars will collide and coalesce. It is this last 15 minutes of inspiral, with $\sim 16\,000$ cycles of waveform oscillation, and the final coalescence, that LIGO/VIRGO seeks to monitor. LIGO I and LIGO II are expected to detect NS–NS inspiral events out to a distance of ~ 20 Mpc and ~ 300 Mpc, respectively, according to estimates (Thorne 2001). This corresponds to wave amplitudes of roughly $10^{-20} > h > 10^{-22}$. As a result of the much larger chirp mass for the BH–BH mergers, such binaries will be detected out to a distance luminosity, $d_L \propto M_{\rm chirp}^{5/6}$ (Finn 1998), which is about 4 times larger. Hence, the ratio of detected event rates for BH–BH mergers relative to NS–NS mergers is $4^3 \sim 64$ times larger than the corresponding ratio of such mergers in our Galaxy. As a result BH–BH mergers are expected to be dominant for LIGO detectors as noted by Sipior and Sigurdsson (2002).

The cosmological implications of gravitational wave observations of binary inspiral are also interesting to note (Schutz 1986; Finn 1997). Finally, it should be mentioned that LIGO II is expected to detect burst signals from extra-Galactic supernova explosions as well (Thorne 2001). Investigations of these signals may help to reveal the unknown progenitors of both short and long-duration gamma-ray bursts (GRBs).

16.8.2.2 Galactic merger rates of NS/BH binaries

It is very important to constrain the local merging rate of NS/BH-binaries in order to predict the number of events detected by LIGO. This rate can be determined either from binary population synthesis calculations, or from observations of Galactic NS-NS systems (binary pulsars). Both methods involve a large number of uncertainties (e.g., Kalogera *et al.* 2001 and references therein). Therefore, the current estimates for the Galactic merger rate of NS–NS systems cover a wide range: 10^{-6}–10^{-3} yr^{-1}.

The discovery of the very tight binary pulsar system PSR J0737–3039 indicates that the galactic NS–NS merger rate is almost an order of magnitude larger than previously estimated (Burgay *et al.* 2003; Kalogera *et al.* 2004). In addition, the discovery that in this system both neutron stars are observed as radio pulsars, one recycled and one not-recycled (Lyne *et al.* 2004), has provided a beautiful confirmation of the evolutionary model for the formation of double neutron stars depicted in Fig. 16.15 (Srinivasan & van den Heuvel 1982; Lyne *et al.* 2004; van den Heuvel, 2003, 2004).

In order to extrapolate the Galactic coalescence rate out to the volume of the Universe accessible to LIGO, one can either use a method based on star formation rates or a scaling based on the B-band luminosities of galaxies. Using the latter method Kalogera *et al.* (2001) found a scaling factor of $(1.0-1.5) \times 10^{-2}$ Mpc^{-3}, or equivalently, $\sim$400 for LIGO I (out to 20 Mpc for NS–NS mergers). Since LIGO II is expected to look out to a distance of 300 Mpc (for NS–NS mergers), the volume covered by LIGO II is larger by a factor of $(300/20)^3$ and thus the scaling factor in this case, relative to the coalescence rates in the Milky Way, is about 1.3×10^6. Therefore, the expected rate of detections from NS–NS inspiral events is roughly between 2 and 500 yr^{-1} for LIGO II. However, LIGO I may not detect any NS/BH inspiral event.

Recent studies (Voss & Tauris 2003; Dewi, Pols & van den Heuvel 2005) have included "real λ-values" for the CE-phase and this results in a relatively low Galactic NS–NS merger rate (since fewer systems survive the CE evolution). However, Voss and Tauris (2003) also investigated BH–BH systems and find a relatively high Galactic merger rate of such systems, compared to NS–NS systems and mixed NS/BH systems, and estimate a BH–BH merging detection rate of 840 yr^{-1} for LIGO II.

One should be aware that compact mergers in globular clusters probably also contribute significantly to the total merger rates (Portegies-Zwart & McMillan 2000).

16.9 Spin and B-field evolution of accreting neutron stars

Most X-ray pulsars have spin periods between 10 and 1000 s. In persistent sources, spin periods of the order of seconds are found only in systems where we have clear evidence for the presence of an accretion disk from UV and optical observations. These are systems where (a large part of) the mass transfer is due to RLO: the LMXBs and those few HMXBs in which the supergiant donors are just beginning to overflow their Roche-lobes (Cen X-3, SMC X-1 and LMC X-4). The latter are expected to be powered by a combination of stellar wind and beginning atmospheric RLO (Savonije 1978, 1983). In most of these systems the X-ray pulsars show a secular decrease of their pulse period (spin-up) on a relatively short timescale (10^3-10^5 yr). Although on short timescales episodes of spin-up and spin-down alternate, the average trend is clearly that of spin-up (see, e.g., Bildsten *et al.* 1997). In sources. that are purely wind-fed, the HMXBs with blue supergiant companions that do not yet fill their Roche-lobe, the pulse periods are very long, and they vary erratically in time showing no clear secular trends. This can be explained by the fact that the amount of angular momentum carried by the supersonic winds is negligible, and eddies form in the wind downstream of the neutron star, which alternately may feed co- and counter-rotating angular momentum to it (Taam & Fryxell 1988; Fryxell & Taam 1988; Matsuda *et al.* 1992).

The accretion and spin-evolution of a neutron star in a binary system depend on a number of parameters: the magnetodipole radiation pressure, the ram pressure of the companion star wind, the radius of gravitational capture, the location of the light cylinder, the Alfvén radius, the co-rotation radius, the propeller effect and whether or not an accretion disk is formed – see, e.g., Ghosh & Lamb (1979) and Chapter 13 by King for a detailed description. The old binary neutron stars, which have been spun-up and recycled in LMXBs, reappear as observable millisecond X-ray and radio pulsars. Their so-called equilibrium spin period is given by (see the spin-up line in the $(P, \dot{P})$-diagram in Fig. 16.3)

$$P_{\rm eq} \propto \dot{M}_{\rm NS}^{-3/7} B^{6/7} R_{\rm NS}^{18/7} M_{\rm NS}^{-5/7} \qquad (16.47)$$

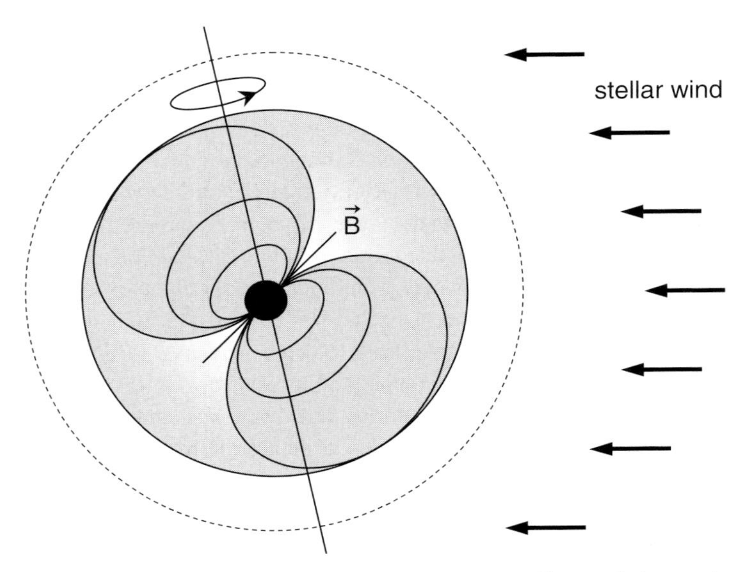

Fig. 16.17. Illustration of the magnetosphere surrounding a wind-accreting neutron star. The rotation period, the magnetic field strength and the ram pressure of the wind determine whether or not accretion onto the neutron star surface is possible. The spin axis and the magnetic field axis are misaligned (thus: a pulsar). X-rays are emitted along the magnetic field axis as the pulsar accretes near its magnetic poles.

As long as $P > P_{\rm eq}$ accretion onto the neutron star is possible. Note that, while ordinary single pulsars have a typical lifetime of $10 - 100$ Myr, the recycled millisecond pulsars have lifetimes of several Gyr and thus continue to light the sky with beamed pulses for a Hubble-time. The reason for this is simply their low values of $\dot{P}$, i.e., relatively weak braking (or radiation-reaction) torques.

16.9.1 The Eddington accretion limit

The typical X-ray luminosities of the LMXBs and HMXBs are in the range $10^{35} - 10^{38}$ erg s^{-1}, corresponding to mass accretion rates onto a neutron star in the range $10^{-11} - 10^{-8}\,M_\odot$ yr^{-1}. When the mass-transfer rate exceeds $\dot{M}_{\rm Edd} \simeq 1.5 \times 10^{-8}\,M_\odot$ yr^{-1} (for spherical accretion of a hydrogen-rich gas), the X-ray luminosity exceeds the so-called Eddington limit at which the radiation pressure force on the accreting matter exceeds the gravitational attraction force of the compact star (Davidson & Ostriker 1973). The concept of an Eddington accretion limit also applies to black holes. At accretion rates $> \dot{M}_{\rm Edd}$ the excess accreting matter will pile up around the compact object and form a cloud optically thick to X-rays, thus quenching the source. Therefore, in the observed "persistent" LMXBs and HMXBs the accretion rates must be in the range $10^{-11} - 10^{-8}\,M_\odot$ yr^{-1}. However, the *mass-transfer rate* from the companion star towards the neutron star may be considerably larger. Calculations show that mass transfer exceeding $10^{-4}\,M_\odot$ yr^{-1} may still be dynamically stable (see Fig. 16.13). In cases where super-Eddington mass transfer occurs, the excess material is ejected in a jet – e.g., as observed in the source SS433 (Chapter 9 and King & Begelman 1999).

16.9.2 Accretion-induced magnetic field decay

In the past decade it has been possible – under certain assumptions – to compute the accretion-induced magnetic field decay of a binary neutron star – see, e.g., Geppert & Urpin (1994), Konar & Bhattacharya (1997) and Cumming, Zweibel & Bildsten (2001). It is often assumed that the magnetic field has been generated in the outer crust by some unspecified mechanism, e.g., by thermo-magnetic effects (Blandford, Applegate & Hernquist 1983), during or shortly after the neutron star formation. The electrical conductivity in the neutron star crust is mainly a function of mass-density, temperature and degree of lattice impurities. By combining the conductive properties with a cooling model, as well as a temperature profile, one can calculate the evolution of the magnetic field by solving the induction equation:

$$\frac{\partial \mathbf{B}}{\partial t} = -\frac{c^2}{4\pi}\nabla \times \left(\frac{1}{\sigma_{\rm el}} \times \nabla \times \mathbf{B}\right) + \nabla \times (\mathbf{v} \times \mathbf{B}) \tag{16.48}$$

where $\mathbf{v}$ is the velocity of accreted material movement and $\sigma_{\rm el}$ is the electrical conductivity of the medium. By choosing a simple vector potential, and introducing the so-called Stokes' stream function, the above equation reduces to a second-order partial differential equation, which can be solved numerically. The current distribution, which is responsible for the B-field, migrates inward as a result of diffusion and enters the highly conducting parts of the neutron star. In that inner region the electrical conductivity is very high (superconducting) and hence the magnetic field will essentially be stable forever – it freezes out at a residual value. The calculated effect of the ohmic dissipation results in final B-fields of $\sim 10^8 - 10^9$ G corresponding to the values estimated from observed millisecond pulsars. A very different model, in which it is assumed that the field is anchored in the superconducting and superfluid neutron star core, was put forward by Ruderman (1998) and collaborators. In this physically very elegant model, the field is driven out of the core by rotational slow-down, and also final fields of $\sim 10^8 - 10^9$ G result. See also the reviews by Bhattacharya & Srinivasan (1995) and Bhattacharya (2002) for further details and alternative models.

In the case of crustal field decay models the next important step would be to solve the evolution self-consistently. That is, to use good stellar evolution models to calculate the mass-transfer rate of the donor star, determine the orbital dynamical response, check whether the material is accreted onto the neutron star or not, determine the crustal temperature from the nuclear burning of accreted material, recalculate the electrical conductivity and its effect on the B-field. The latter then affects the Alfvén radius, and thus the accretion rate and spin period, which again influences the orbital evolution and response of the donor star, etc. However, our state of understanding of the magnetic field decay mechanisms, and of mass and angular momentum loss mechanisms during binary evolution, is still too fragmentary to allow for a detailed quantitative calculation of the evolution of a rotating magnetized neutron star in a binary system.

Acknowledgements

TMT gratefully acknowledges support from the Danish Natural Science Research Council under grant no. 56916.

References

Alpar, M. A., Cheng, A. F., Ruderman, M. A. and Shaham, J. (1982). *Nature* **300**, 728
Armitage, P. J. and Livio, M. (2000). *ApJ* **532**, 540

Backer, D. C., Kulkarni, S. R., Heiles, C., *et al.* (1982). *Nature* **300**, 615
Barziv *et al.* (2001). *A&A* **377**, 925
Beer, M. E. and Podsiadlowski, P. (2002). *MNRAS* **331**, 351
Bethe, H. A. and Brown, G. E. (1995). *ApJ* **445**, L129
(1998). *ApJ* **506**, 780
Bildsten, L., Chakrabarty, D., Chiu, J., *et al.* (1997). *ApJS* **113**, 367
Bisnovatyi-Kogan, G. S. and Komberg, B. V. (1974). *Astron. Zh.* **51**, 373
Bisscheroux, B. (1999). Master's thesis, University of Amsterdam (1999)
Bhattacharya, D. (2002). *JA&A* **23**, 67
Bhattacharya, D. and van den Heuvel, E. P. J. (1991). *Phys. Rep.* **203**, 1
Bhattacharya, D. and Srinivasan, G. (1995), in *X-ray Binaries*, eds. W. H. G. Lewin, J. van Paradijs and E. P. J. van den Heuvel (Cambridge University Press)
Blaauw, A. (1961). *Bull. Astron. Inst. Neth.* **15**, 265
Blandford, R. D., Applegate, J. H. and Hernquist, L. (1983). *MNRAS* **204**, 1025
Bolton, C. T. (1972). *Nature* **235**, 271
Brown, G. E. (1995). *ApJ* **440**, 270
Brown, G. E. and Bethe, H. A. (1994). *ApJ* **423**, 659
Brown, G. E., Lee, C. H. and Bethe, H. A. (1999). *New Astronomy* **4**, 313
Brown, G. E., Heger, A., Langer, N., *et al.* (2001). *New Astronomy* **6**, 457
Burgay, M., D'Amico, N., Possenti, A., *et al.* (2003). *Nature* **426**, 531
Chevalier, R. A. (1993). *ApJ* **411**, L33
(1996). *ApJ* **459**, 322
Cox, J. P. and Giuli, R.T. (1968). *Stellar Structure, vols. I and II* (New York, Gordon and Breach)
Cumming, A., Zweibel, E. and Bildsten, L. (2001). *ApJ* **557**, 958
Davidson, K. and Ostriker, J. P. (1973). *ApJ* **179**, 585
de Jager, C., Nieuwenhuijzen, H. and van der Hucht, K. A. (1988). *A&AS* **72**, 259
de Kool, M. (1990). *ApJ* **358**, 189
Dewey, R. J. and Cordes, J. M. (1987). *ApJ* **321**, 780
Dewi, J. D. M. and Tauris, T. M. (2000). *A&A* **360**, 1043
(2001), in *Evolution of Binary and Multiple Star Systems*, eds. P. Podsiadlowski *et al.* (ASP Conf. Vol. 229) p. 255
Dewi, J. D. M., Pols, O. R., Savonije, G. J. and van den Heuvel, E. P. J. (2002). *MNRAS* **331**, 1027
Dewi, J. D. M., Podsiadlowski, P. and Pols, O. R. (2005). *MNRAS* **363**, L71
Dewi, J. D. M., Pols, O. R. and van den Heuvel, E. P. J. (2005). *MNRAS*, submitted
Dorch, S. B. F. and Nordlund, Å. (2001). *A&A* **365**, 562
Eggleton, P. P. (1983). *ApJ* **268**, 368
(2001), in *Evolution of Binary and Multiple Star Systems*, eds. P. Podsiadlowski *et al.* (ASP Conf. Vol. 229)
Ergma, E. and Fedorova, A. V. (1991). *A&A* **242**, 125
Ergma, E. and van den Heuvel, E. P. J. (1998). *A&A* **331**, L29
Ergma, E., Sarna, M. J. and Antipova, J. (1998). *MNRAS* **300**, 352
Faulkner, J. (1971). *ApJ* **170**, L99
Finn, L. S. (1997), in *Gravitation & Cosmology*, eds. S. Dhurandhar and T. Padmanabhan (Dordrecht, Klüwer) p. 95
(1998). *Phys. Rev.* **D53**, 2878
Flannery, B. P. and van den Heuvel, E. P. J. (1975). *A&A* **39**, 61
Fryer, C. L. (1999). *ApJ* **522**, 413
Fryer, C. L., Woosley, S. E. and Hartmann, D. H. (1999). *ApJ* **526**, 152
Fryxell, B. A. and Taam, R. E. (1988). *ApJ* **335**, 862
Galloway, D. K., Chakrabarty, D., Morgan, E. H. and Remillard, R. A. (2002). *ApJ* **576**, L137
Geppert, U. and Urpin, V. (1994). *MNRAS* **271**, 490
Ghosh, P. and Lamb, F. K. (1979). *ApJ* **234**, 296
Giacconi, R., Gursky, H., Paolini, F. R. and Rossi, B. B. (1962). *Phys. Rev. Lett.* **9**, 439
Habets, G. M. H. J. (1986). *A&A* **167**, 61
Han, Z., Podsiadlowski, P. and Eggleton, P. P. (1994). *MNRAS* **270**, 121
(1995). *MNRAS* **272**, 800
Han, Z., Podsiadlowski, P., Maxted, P. F. L., *et al.* (2002). *MNRAS* **336**, 449

Hills, J. (1983). *ApJ* **267**, 322
Heuvel, E. P. J. van den (1975). *ApJ* **198**, L109
(1994), in *Interacting Binaries*, Saas-Fee course 22 (Heidelberg, Springer) p. 263
(2003). *Nature* **426**, 504
(2004). *Science* **303**, 1143
Heuvel, E. P. J. van den and Heise, J. (1972). *Nature – Physical Science* **239**, 67
Heuvel, E. P. J. van den and de Loore, C. (1973). *A&A* **25**, 387
Heuvel, E. P. J. van den and Rappaport, S. A. (1987), in *Physics of Be-stars*, Proc. IAU Colloq. 92 (Cambridge University Press) p. 291
Heuvel, E. P. J. van den and van Paradijs, J. (1988). *Nature* **334**, 227
Hulse, A. R. and Taylor, J. H. (1975). *ApJ* **195**, L51
Iben, Jr. I. and Livio, M. (1993). *PASP* **105**, 1373
Illarionov, A. F. and Sunyaev, R. A. (1985). *A&A* **39**, 185
Iwamoto, K., Mazzali, P. A., Nomoto, K., *et al.* (1998). *Nature* **395**, 672
Joss, P. C., Rappaport, S. A. and Lewis W. (1987). *ApJ* **319**, 180
Kalogera, V. and Webbink, R. F. (1996). *ApJ* **458**, 301
(1998). *ApJ* **493**, 351
Kalogera, V., Narayan, R., Spergel, D. N. and Taylor, J. H. (2001). *ApJ* **556**, 340
Kalogera, V., Kim, C., Lorimer, D. R., *et al.* (2004). *ApJ* **601**, L179
Kaspi, V. M., Bailes, M., Manchester, R. N., *et al.* (1996). *Nature* **381**, 584
King, A. R. and Begelman, M. C. (1999). *ApJ* **519**, 169
King, A. R. and Ritter, H. (1999). *MNRAS* **309**, 253
Kippenhahn, R. and Weigert, A. (1967). *Z. Astrophys.* **65**, 251
(1990). *Stellar Structure and Evolution* (Heidelberg, Springer)
Kirk, J. G. and Trümper, J. E. (1983), in *Accretion Driven Stellar X-ray Sources*, eds. W. H. G. Lewin and E. P. J. van den Heuvel (Cambridge University Press) p. 216
Konar, S. and Bhattacharya, D. (1997). *MNRAS* **284**, 311
Kraft, R. P. (1967). *ApJ* **150**, 551
Landau, L. D. and Lifshitz, E. (1958). *The Classical Theory of Fields* (Oxford, Pergamon Press)
Lee, C. H., Brown, G. E. and Wijers, R. A. M. J. (2002). *ApJ* **575**, 996
Lewin, W. G. H. and Joss, P. C. (1983), in *Accretion Driven Stellar X-ray Sources*, eds. W. H. G. Lewin and E. P. J. van den Heuvel (Cambridge University Press) p. 41
Liu, Q. Z., van Paradijs, J. and van den Heuvel, E. P. J. (2000). *A&AS* **147**, 25
Lyne, A. G. and Lorimer, D. R. (1994). *Nature* **369**, 127
Lyne, A. G., Burgay, M., Kramer, M., *et al.* (2004). *Science* **303**, 1153
MacFadyen, A. I., Woosley, S. E. and Heger, A. (2001). *ApJ* **550**, 410
Manchester, R. N. and Taylor, J. H. (1977). *Pulsars* (San Francisco, Freeman)
Maraschi, L., Treves, A. and van den Heuvel, E. P. J. (1976). *Nature* **259**, 292
Markwardt, C. B., Smith, E. and Swank, J. H. (2003). *The Astronomer's Telegram* **122**
Matsuda, T., Ishii, T., Sekino, N., *et al.* (1992). *MNRAS* **255**, 183
Maxted, P. F. L., Heber, U., Marsh, T. R. and North, R. C. (2001). *MNRAS* **326**, 1391
McClintock, J. E. and Remillard, R. A. (1986). *ApJ* **308**, 110
Mestel, L. (1984), in *Cool Stars, Stellar Systems, and the Sun*, eds. S. L. Baliunas, L. Hartmann (Berlin, Springer) p. 49
Mirabel, I. F., Mignami, R., Rodrigues, I., *et al.* (2002). *A&A* **395**, 595
Nelemans, G. and van den Heuvel, E. P. J. (2001). *A&A* **376**, 950
Nelemans, G., Tauris, T. M. and van den Heuvel, E. P. J. (1999). *A&A* **352**, L87
Nomoto, K. (1984). *ApJ* **277**, 791
Novikov, I. D. and Zel'Dovitch, Y. B. (1966). *Nuova Cimento Sup.* **4**, 810
Nugis, T. and Larmers, H. J. G. L. M. (2000). *A&A* **360**, 227
Orosz, J. A., Jain, R. K., Bailyn, C. D., *et al.* (1998). *ApJ* **499**, 375
Orosz, J. A., Kuulkers, E., van der Klis, M., *et al.* (2001). *ApJ* **555**, 489
Ostriker J. P. (1976), in *Structure and Evolution of Close Binary Systems*, eds. P. P. Eggleton *et al.* (Dordrecht, Reidel) p. 206
Paczyński B. (1971). *Acta Astron.* **21**, 1
(1976), in *Structure and Evolution of Close Binary Systems*, eds. P. P. Eggleton *et al.* (Dordrecht, Reidel) p. 75

Paradijs, J. van (1996). *ApJ* **464**, L139
Parker, E. N. (1955). *ApJ* **121**, 491
Peters, P. C. (1964). *Phys. Rev* **136**, B1224
Pfahl, E. Podsiadlowski, P., Rappaport, S. A. and Spruit, H. (2002). *ApJ* **574**, 364
Podsiadlowski, P. (1991). *Nature* **350**, 136
Podsiadlowski, P. and Rappaport, S. A. (2000). *ApJ* **529**, 946
Podsiadlowski, P., Rappaport, S. A. and Pfahl, E. (2002). *ApJ* **565**, 1107
Podsiadlowski, P., Rappaport, S. A. and Han, Z. (2003). *MNRAS* **340**, 1214
Pols, O. R., Tout, C. A., Eggleton, P. P. and Han, Z. (1995). *MNRAS* **274**, 964
Pols, O. R., Schröder, K. P., Hurley, J. R., *et al.* (1998). *MNRAS* **298**, 525
Portegies-Zwart, S. F. and McMillan, S. L. W. (2000). *ApJ* **528**, L17
Pringle, J. E. (1981). *ARA&A* **19**, 137
Pylyser, E. and Savonije, G. J. (1988). *A&A* **191**, 57
(1989). *A&A* **208**, 52
Radhakrishnan, V. and Srinivasan, G. (1982). *Current Science* **51**, 1096
Rappaport, S. A., Verbunt, F. and Joss, P. C. (1983). *ApJ* **275**, 713
Rappaport, S. A., Podsiadlowski, P., Joss, P. C., *et al.* (1995). *MNRAS* **273**, 731
Refsdal, S. and Weigert, A. (1971). *A&A* **13**, 367
Reimers, D. (1975), in *Problems in Stellar Atmospheres and Envelopes*, eds. B. Bascheck, W. H. Kegel and G. Traving. (New York, Springer) p. 229
Reimers, D. and Koester, D. (1988). *ESO Messenger* **54**, 47
Rucinski, S. M. (1983). *Observatory* **103**, 280
Ruderman, M. A. (1998), in *The Many Faces of Neutron Stars*, eds. R. Buccheri, J. van Paradijs and A. Alpar. (Dordrecht, Kluwer) p. 77
Ruderman, M. A., Shaham, J. and Tavani, M. (1989). *ApJ* **336**, 507
Salpeter, E. E. (1964). *ApJ* **140**, 796
Savonije, G. J. (1978). *A&A* **62**, 317
(1983), in *Accretion Driven Stellar X-ray Sources*, eds. W. H. G. Lewin and E. P. J. van den Heuvel (Cambridge University Press) p. 343
(1987). *Nature* **325**, 416
Schaller, G., Schaerer, D., Meynet, G. and Maeder, A. (1992). *A&AS* **96**, 269
Schreier, E., Levinson, R., Gursky, H., *et al.* (1972). *ApJ* **172**, L79
Schutz, B. F. (1986). *Nature* **323**, 310
Shaham, J. (1992), in *X-ray Binaries and Recycled Pulsars*, eds. E. P. J. van den Heuvel and S. A. Rappaport. (Dordrecht, Kluwer) p. 375
Shklovskii, I. (1967). *ApJ* **148**, L1
Sipior, M. S. and Sigurdsson, S. (2002). *ApJ* **572**, 962
Skumanich, A. (1972). *ApJ* **171**, 565
Slettebak, A. (1988). *Publ. Astron. Soc. Pac.* **100**, 770
Smarr, L. L. and Blandford, R. D. (1976). *ApJ* **207**, 574
Soberman, G. E., Phinney, E. S. and van den Heuvel, E. P. J. (1997). *A&A* **327**, 620
Sonderblom, D. R. (1983). *ApJS* **53**, 1
Soria, R., Wu, K., Page, M. J. and Sakelliou, I. (2001). *A&A* **365**, L273
Spruit, H. C. and Ritter, H. (1983). *A&A* **124**, 267
Srinivasan, G. and van den Heuvel, E. P. J. (1982). *A&A* **108**, 143
Stepien, K. (1995). *MNRAS* **274**, 1019
Sutantyo, W. (1975). *A&A* **41**, 47
Taam, R. E. (1983). *ApJ* **270**, 694
Taam, R. E. and Fryxell, B. A. (1988). *ApJ* **327**, L73
Taam, R. E. and van den Heuvel, E. P. J. (1986). *ApJ* **305**, 235
Taam, R. E. and Sandquist, E. L. (2000). *ARA&A* **38**, 113
Tauris, T. M. (1996). *A&A* **315**, 453
(2001), in *Evolution of Binary and Multiple Star Systems*, eds. P. Podsiadlowski *et al.* (ASP Conf. Vol. 229)
Tauris, T. M. and Dewi, J. D. M. (2001). *A&A* **369**, 170
Tauris, T. M. and Konar, S. (2001). *A&A* **376**, 543
Tauris, T. M., van den Heuvel, E. P. J. and Savonije, G. J. (2000). *ApJ* **530**, L93

Tauris, T. M. and Savonije, G. J. (1999). *A&A* **350**, 928
Tauris, T. M. and Savonije, G. J. (2001), in *The Neutron Star–Black Hole Connection*, eds. C. Kouveliotou *et al.* (NATO ASI, Dordrecht, Kluwer)
Tauris, T. M. and Takens, R. (1998). *A&A* **330**, 1047
Tavani, M. (1992), in *X-ray Binaries and Recycled Pulsars*, eds. E. P. J. van den Heuvel and S. A. Rappaport (Kluwer, Dordrecht) p. 387
Taylor, J. H. and Weisberg, J. M. (1989). *ApJ* **345**, 434
Terquem, C., Papaloizou, J. C. B., Nelson, R. P. and Lin, D. N. C. (1998). *ApJ* **502**, 588
Thorne, K. S. (2001). LIGO Document Number P-000024-00-D
Thorsett, S. E. and Chakrabarty, D. (1999). *ApJ* **512**, 288
Tutukov, A. V. and Yungelson, L. R. (1973). *Nauchnye Informatsii* **27**, 70
Verbunt, F. and Phinney, E. S. (1995). *A&A* **296**, 709
Verbunt, F. and van den Heuvel, E. P. J. (1995), in *X-ray Binaries*, eds. W. H. G. Lewin, J. van Paradijs and E. P. J. van den Heuvel (Cambridge Univesity Press)
Verbunt, F. and Zwaan, C. (1981). *A&A* **100**, L7
Vilhu, O. and Walter, F. M. (1987). *ApJ* **321**, 958
Voss, R. and Tauris, T. M. (2003). *MNRAS* **342**, 1169
Webbink, R. F. (1984). *ApJ* **277**, 355
Webbink, R. F., Rappaport, S. A. and Savonije, G. J. (1983). *ApJ* **270**, 678
Webster, B. L. and Murdin, P. (1972). *Nature* **235**, 37
Weidemann, V. (1990). *ARA&A* **28**, 103
Wellstein, S. and Langer, N. (1999). *A&A* **350**, 148
Wijnards, R. and van der Klis, M. (1998). *Nature* **394**, 344
Witte, M. G. (2001). Ph.D. thesis, University of Amsterdam
Witte, M. G. and Savonije, G. J. (1999). *A&A* **350**, 129
Wolszczan, A. (1994). *Science* **264**, 538
Woosley, S. E. and Weaver, T. A. (1995). *ApJS* **101**, 181
Woosley, S. E., Langer, N. and Weaver, T. A. (1995). *ApJ* **448**, 315
Zel'Dovitch, Y. B. (1964). *Soviet Physics Doklady* **9**, 195
Zel'Dovitch, Y. B. and Guseinov, O. (1966). *ApJ* **144**, 840
Zel'Dovitch, Y. B. and Novikov, I. D. (1964). *Doklady Academii Nauk SSSR* **158**, 811

Author index

Subject index